HBJ ALGEBRA 2
with TRIGONOMETRY

Arthur F. Coxford

Joseph N. Payne

HBJ

Harcourt Brace Jovanovich, Publishers

New York Chicago San Francisco Atlanta Dallas *and* London

ABOUT THE AUTHORS

ARTHUR F. COXFORD
Professor of Mathematics Education
University of Michigan
Ann Arbor, Michigan

JOSEPH N. PAYNE
Professor of Mathematics Education
University of Michigan
Ann Arbor, Michigan

EDITORIAL ADVISORS

Mrs. Lou Baker
Mathematics Department Chairperson
Arlington High School
Arlington, Texas

Thomas Day
Mathematics Department Chairman
Farmington High School
Farmington, Connecticut

Mrs. Billy Dooley
Mathematics Teacher
Central High School
Columbia, Tennessee

Mrs. Mary Heins
Mathematics Teacher
Coronado High School
El Paso, Texas

Mrs. Alice Mitchell
Mathematics Supervisor
Campbell Union High School District
San Jose, California

Donald Nutter
Chairman, Department of Mathematics
Firestone High School
Akron, Ohio

Mrs. Emily Springer
Formerly Mathematics Department Chairperson
Stephen F. Austin Jr. High School
Amarillo, Texas

August J. Zarcone
Mathematics Instructor
College of DuPage
Glen Ellyn, Illinois

The contributions of **Brother Neal Golden, S.C.,** who wrote the Computer Applications, are gratefully acknowledged.

PICTURE CREDITS

KEY: *t* top; *b* bottom

pages: 25, IBM; 60, UPI; 61, Bijur, Monkmeyer; 171, NASA; 242, Marcia Weinstein; 257 (*t*), Danny Lyon, Magnum; (*b*), HBJ Photo; 457, Hugh Rogers, Monkmeyer; 425, Hal Strong, Shostal; 499, HBJ Photo

Printed in the United States of America

ISBN 0–15–353871–6

Contents

UNIT II POLYNOMIAL RELATIONS AND APPLICATIONS

UNIT IV OTHER RELATIONS AND APPLICATIONS

UNIT V ADVANCED TOPICS AND APPLICATIONS

CHAPTER **1**

Real Numbers

1-1 Variables

Here are some ways of describing a pattern.

Pattern	Using Words	Using a Variable
$0 \times 1 = 0$ $7.6 \times 1 = 7.6$ $\pi \times 1 = \pi$	The product of any real number and 1 is the real number.	$r \times 1 = r$, where r is a real number.

In the expression $r \times 1 = r$, r is a *variable*. A **variable** is a symbol that represents any element of a specified replacement set. By using variables, you can describe many instances of the same pattern.

Recall that an expression such as $3ab$ can be written in several ways.

$$3 \times a \times b \qquad 3 \cdot a \cdot b \qquad 3(a)(b)$$

EXAMPLE 1 Evaluate $12a - 3ab$ for the given replacements for a and b.

a. $a = 5$, $b = 2$ **b.** $a = 10.8$, $b = 1.1$

Solutions:

a. $12a - 3ab = 12(5) - 3(5)(2)$ ← Replace a with 5 and b with 2.

$= 60 - 30$

$= \mathbf{30}$

b. $12a - 3ab = 12(10.8) - 3(10.8)(1.1)$ ← Replace a with 10.8 and b with 1.1.

$= 129.6 - 35.64$

$= \mathbf{93.96}$

An expression that involves variables and constants is an **algebraic expression.** For example, in the algebraic expression

$$7(8x + 5y) - 0.4x^3y^2,$$

x and y are the variables, and 7, 8, 5 and 0.4 are constants.

In the expression $0.4x^3y^2$, x^3 means that x is taken as a factor three times. The number 3 is an **exponent** and x is the **base.**

$$x^3 \text{ means } x \cdot x \cdot x. \qquad y^2 \text{ means } y \cdot y. \qquad z^1 = z.$$

To avoid confusion when evaluating algebraic expressions, you follow the rules agreed upon by mathematicians.

Rules

Order of Operations

1. Evaluate within parentheses or other grouping symbols first.
2. Evaluate expressions with exponents.
3. Perform multiplication and division in order from left to right.
4. Perform addition and subtraction in order from left to right.

EXAMPLE 2 Evaluate $9x^2 + (y^2 - 1)z$ for the given replacements for x, y, and z.

a. $x = 1,\ y = 2,\ z = 3$ **b.** $x = \frac{1}{3},\ y = \frac{7}{4},\ z = 16$

Solutions: **a.**

$$\begin{aligned} 9x^2 + (y^2 - 1)z &= 9(1)^2 + (2^2 - 1)(3) && \leftarrow \text{Apply Rules 1 and 2.} \\ &= 9(1) + (3)(3) && \leftarrow \text{Apply Rule 2.} \\ &= 9 + 9 && \leftarrow \text{Apply Rule 1.} \\ &= \mathbf{18} \end{aligned}$$

b.

$$\begin{aligned} 9x^2 + (y^2 - 1)z &= 9\left(\frac{1}{3}\right)^2 + \left[\left(\frac{7}{4}\right)^2 - 1\right]16 \\ &= 9\left(\frac{1}{9}\right) + \left(\frac{49}{16} - 1\right)16 \\ &= 1 + \left(\frac{33}{16}\right)16 \\ &= 1 + 33 \\ &= \mathbf{34} \end{aligned}$$

CLASSROOM EXERCISES

Evaluate each expression for $x = 2$ and $y = 0.5$.

1. $12x - xy$
2. $\frac{x}{y} - xy$
3. $xy + 100\ y$
4. $\frac{3y}{2} - \frac{1}{x}$

Evaluate each expression for $r = 3$ and $s = \frac{1}{4}$.

5. r^2
6. s^3
7. r^3s^2
8. $4s^3 + \frac{81}{s^3}$
9. $s^4 - \frac{9}{r^2}$

Evaluate. Refer to the rules for order of operations.

10. $(108 \div 9) \div 4 + 3^2$
11. $8 - (15 + 6) \div 3$
12. $[(4 \cdot 12) - (18 \div 6)] \div 15$
13. $[2(24 - 8) \div 4] \div 16$
14. $20 \div (7 - 2) + 5^2 \cdot 3$
15. $[4(5 - 3) - 2(4 - 8)] \div 16$

Evaluate $(x^2 - y^2) + (x + y)(x - y)$ for the given values of x and y.

16. $x = 9,\ y = 5$
17. $x = \frac{1}{3},\ y = \frac{1}{4}$
18. $x = 0.5,\ y = 0.1$
19. $x = 1.1,\ y = 0.3$

WRITTEN EXERCISES

A Evaluate $7a - 2b + \frac{2c}{a}$ for the given values of a, b, and c.

1. $a = 2,\ b = 0,\ c = 1$
2. $a = 3,\ b = 2,\ c = 6$
3. $a = 10,\ b = 4,\ c = 5$

Evaluate $\left(9xy - \frac{1}{y}\right)z$ for the given values of x, y, and z.

4. $x = 3,\ y = 1,\ z = 2$
5. $x = \frac{1}{3},\ y = 4,\ z = 1$
6. $x = 0,\ y = \frac{1}{4},\ z = 2$

Evaluate $\frac{7(2r+t)}{(4t-w)}$ for the given values of r, t, and w.

7. $r = 3, t = 1, w = 2$ **8.** $r = 0.6, t = 1.5, w = 3$ **9.** $r = \frac{1}{4}, t = \frac{1}{8}, w = \frac{1}{16}$

In Exercises 10–14, find x.

10. $7^2 = x$ **11.** $9^3 = x$ **12.** $x^5 = 32$ **13.** $x^3 = 125$ **14.** $x^3 = 64$

Evaluate each expression. Use the rules for order of operations.

15. $16 \div 4 + 7$ **16.** $18 + 24 \cdot 3 - 8$ **17.** $6^2 - 3(3 + 5) \div (3^2 - 3)$

18. $18 \div (6 - 3) + 2^3 \cdot 4$ **19.** $(144 \div 6) \cdot 3^2 - 4^3$ **20.** $(3 \cdot 5 - 128 \div 4)5$

21. $[3(36 \div 9) - (12^2 - 12)]2^2$ **22.** $\frac{1}{3}(48 \div 8) + 5(18 \div 3)2^2$

23. $[12(2 - \frac{1}{3}) - 4] \div (\frac{7}{2} - 3)$ **24.** $5^2[16 \div 4 + 3(8 - 1)] \div 10^2$

Evaluate $3x^2y - yz$ for the given replacements for x, y, and z.

25. $x = 6, y = 9, z = \frac{1}{3}$ **26.** $x = 0.5, y = 12, z = 0.5$ **27.** $x = 8, y = 1, z = 4$

Evaluate $(2a^2 \div b) - (c^2 - a)$ for the given replacements for a, b, and c.

28. $a = 12, b = 8, c = 4$ **29.** $a = 6, b = 3, c = 9$ **30.** $a = \frac{1}{2}, b = \frac{1}{8}, c = 1$

Evaluate $r^2 \div 2st + s^3 - rt$ for the given values of r, s, and t.

31. $r = 10, s = 5, t = 4$ **32.** $r = 1, s = \frac{1}{2}, t = \frac{1}{8}$ **33.** $r = 0, s = \frac{1}{8}, t = 12$

APPLICATIONS: Using Variables

When buying a new car, you can use a formula such as

$$D = 0.80(L - S) + S$$

to approximate the dealer's cost, D, for a new car, where L is the list price (sticker price) of the car and S is the shipping and dealer preparation charge. Use this formula in Exercises 34–39 to find D for the given values of L and S.

34. $L = \$6000$; $S = \$175$ **35.** $L = \$9800$; $S = \$225$ **36.** $L = \$7500$; $S = \$195$

37. $L = \$8200$; $S = \$295$ **38.** $L = \$7150$; $S = \$285$ **39.** $L = \$7200$; $S = \$180$

Some bowling leagues use a "handicap formula" in computing scores for competing bowlers of differing abilities. The formula

$$H = s + 0.8(200 - v)$$

represents a bowler's final handicap score, H, where s is the bowler's actual score (also called the scratch score), and v is the bowler's average score. Use this formula in Exercises 40–45 to find H for the given values of s and v.

40. $s = 163$, $v = 175$ **41.** $s = 165$, $v = 170$ **42.** $s = 136$; $v = 190$

43. $s = 145$; $v = 180$ **44.** $s = 160$; $v = 185$ **45.** $s = 147$; $v = 145$

When a ball is thrown into the air at a speed of 96 feet per second, the distance, d, that it travels in t seconds is given by the formula

$$d = 96t - 16t^2,$$

Use this formula in Exercises 46–52 to find d for each value of t.

46. $t = 0$ **47.** $t = 1$ **48.** $t = 2$ **49.** $t = 3$ **50.** $t = 4$ **51.** $t = 5$ **52.** $t = 6$

B

53. Use your answers to Exercises 46–52 to find how high the ball will go.

54. Use your answers to Exercises 46–52 to find how long the ball will remain in the air.

You can use the following formula

$$A = \frac{72I}{3B(n+1) + I(n+1)}$$

to approximate the annual percentage rate (true rate of interest) on a loan where I = total finance charges, B = amount borrowed, and n = number of monthly payments.

Use this formula in Exercises 55–58 to find A for the given values of I, B, and n. Round your answer to the nearest hundredth.

55. $B = \$1500$; $I = \$300$, $n = 24$

56. $B = \$4000$; $I = \$1000$; $n = 30$

57. $B = \$400$; $I = \$80$, $n = 24$

58. $B = \$900$; $I = \$100$; $n = 12$

C

In Exercises 59–60, use the formula

$$P = N(1 + r)^t$$

to estimate growth at a fixed rate, r, where t is the number of time units and N is the size at time $t = 0$.

59. The sum of \$1000 is invested at 10% per year. Find the value after 2 years.

60. A colony of bacteria grows at the rate of 1% per hour. Suppose 1,000,000 bacteria were present at time $t = 0$. How many would be present in 3 hours?

61. The ratings on a weekly television program showed that the first program had 2,000,000 viewers. The program then lost 20% of its viewers per week. Use the formula $P = N(1 - r)^t$ to find the number of viewers after the fourth program.

Puzzle

Each of six sections of a chain has four links.

The cost of cutting open one link is \$0.50. Joining two links together costs \$1.00. Find the minimum cost of joining the six pieces into one chain.

1-2 Sets of Numbers

Some numbers used as replacements for variables can be described by listing them as sets.

Name	Set Description
Natural Numbers	$N = \{1, 2, 3, \cdots\}$
Whole Numbers	$W = \{0, 1, 2, \cdots\}$
Integers	$I = \{\cdots -2, -1, 0, 1, 2, \cdots\}$

The numbers continue on indefinitely.

The set of integers is the **union,** symbolized by $\cup$, of the set of whole numbers with the **additive inverses,** or **opposites,** of the natural numbers.

$$I = W \cup \{\cdots -3, -2, -1\}$$

A **rational number** is a number that can be expressed in the form $\frac{a}{b}$, where a is an integer and b is a natural number. Shown below are some examples of rational numbers.

$$\frac{2}{3} \qquad -\frac{7}{2} \qquad -3\frac{1}{4} \qquad 6.8 \qquad -9 \qquad 0 \qquad 42$$

You can use *set-builder notation* to describe the set of rational numbers.

$$Q = \{x : x = \frac{a}{b}, a \in I, b \in N\}$$

The set-builder notation above is read, "The set of rational numbers is the set of numbers x, such that $x = \frac{a}{b}$, where a is an element of the set of integers and b is an element of the set of natural numbers." The symbol, $\in$, means "is an element of."

Every rational number can be expressed either as a *terminating decimal* or as a *nonterminating, repeating decimal.*

EXAMPLE 1 Classify each number as a terminating decimal or as a nonterminating, repeating decimal.

a. $\frac{1}{16}$ **b.** $\frac{2}{11}$

Solutions: Divide the numerator by the denominator.

a. $16\overline{)1.0000}$ = .0625, or 0.0625

Thus, $\frac{1}{16}$ is a **terminating decimal.**

b. $11\overline{)2.000000 \cdots}$ = .181818 · · ·, or 0.181818 · · ·

Thus, $\frac{2}{11}$ is a **nonterminating, repeating decimal.**

In **b** of Example 1, the decimal 0.181818 · · · may also be written $0.\overline{18}$. The bar indicates the digits that repeat.

A number that cannot be written as the ratio of an integer and a natural number is an **irrational number.** Further, irrational numbers *cannot* be represented as terminating, or nonterminating, repeating decimals. Here are some examples of irrational numbers.

$$\sqrt{2} \qquad \pi \qquad -\sqrt{7} \qquad \sqrt{50} \qquad -\sqrt{19}$$

The union of the set of rational numbers with the set of irrational numbers is R, the set of **real numbers.**

$$R = Q \cup Ir$$

All rational numbers are real numbers. However, some real numbers are not rational numbers. The figure at the left below shows relationships between the sets of numbers listed at the right.

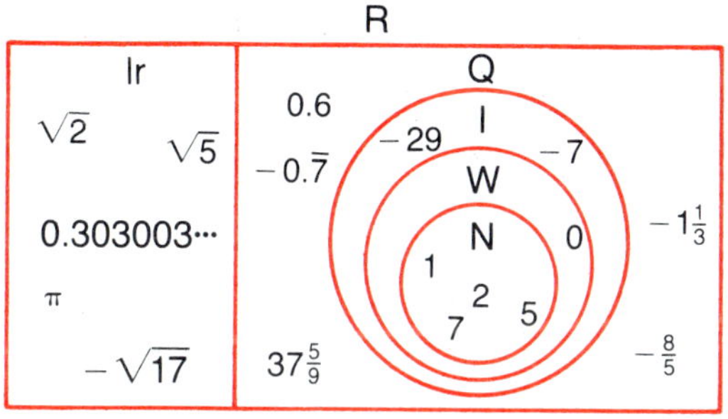

N: Natural numbers
W: Whole numbers
I: Integers
Q: Rational numbers
Ir: Irrational numbers
R: Real numbers

For example, N is a subset of W, or

$N \subset W$ ← **Read: "N is a subset of W."**

because every element of N is also in W.

Also, the *intersection* of N and W, symbolized $N \cap W$, is N because every natural number is also a whole number. The **intersection** of two sets contains all the elements both sets have in common. When two sets have no elements in common, their intersection is $\emptyset$, the **null set** or the **empty set.** The empty set is considered a subset of every set.

EXAMPLE 2 Use the symbol $\in$ to indicate the set or sets to which each number belongs.

a. -5 **b.** $0.\overline{13}$ **c.** 0

Solutions: Use the figure above.

a. $-5 \in$ **I, Q, R** **b.** $0.\overline{13} \in$ **Q, R** **c.** $0 \in$ **W, I, Q, R**

Sets of numbers can be graphed on the number line. On a number line, the point paired with a number is called the **graph** of the number. The number paired with a point on the number line is called the **coordinate** of the point. Example 3 shows the graphs of some subsets of the real numbers.

EXAMPLE 3 Graph each set of numbers.

a. {the integers greater than -2 and less than 6}

b. {the even whole numbers less than 8}

c. {the real numbers greater than -0.5 and less than or equal to 8}

Solutions: **a.**

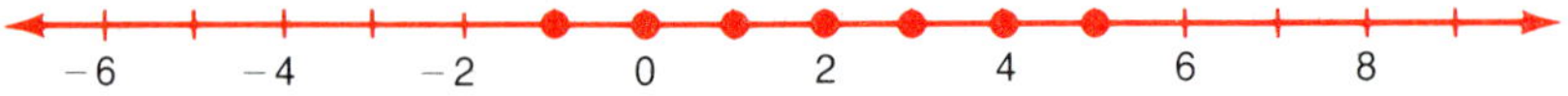

b.

c.

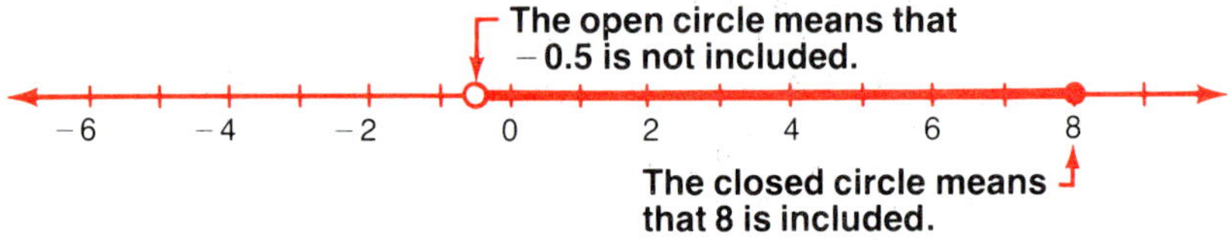

The graph of the real numbers is the entire number line. There is exactly one real number paired with each point on the number line. Also, there is exactly one point on the number line that corresponds to each real number.

CLASSROOM EXERCISES

Classify each number as a terminating decimal, *T*, or as a nonterminating, repeating decimal, *NT*.

1. $\frac{3}{4}$ **2.** $\frac{1}{3}$ **3.** $\frac{9}{10}$ **4.** $\frac{17}{20}$ **5.** $\frac{1}{6}$ **6.** $\frac{7}{8}$

Use the symbol $\in$ to indicate the set(s) to which each number belongs.

7. $-2\frac{2}{3}$ **8.** -100 **9.** 17.08 **10.** $0.\overline{6}$ **11.** π **12.** $-\sqrt{2}$

Name the set of numbers represented by each graph.

13.

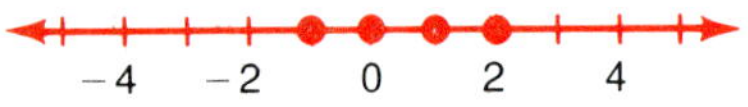

14.

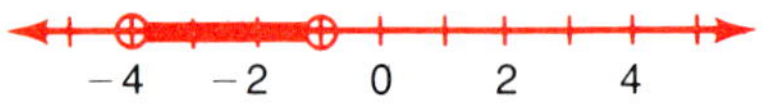

15.

16.

WRITTEN EXERCISES

A For each number in Exercises 1–14:

a. Write a decimal for the number.

b. Classify the number as a terminating decimal, *T*, or a nonterminating, repeating decimal, *NT*.

1. $\frac{9}{20}$ **2.** $\frac{3}{8}$ **3.** $\frac{5}{9}$ **4.** $\frac{6}{11}$ **5.** $-\frac{7}{8}$ **6.** $-\frac{3}{16}$ **7.** $\frac{1}{7}$

8. $\frac{1}{13}$ **9.** $1\frac{1}{2}$ **10.** $2\frac{2}{5}$ **11.** $-5\frac{1}{3}$ **12.** $-4\frac{5}{6}$ **13.** $\frac{3}{11}$ **14.** $\frac{1}{12}$

Use the symbol $\in$ to indicate the set or sets to which each number belongs.

15. 0 **16.** -23 **17.** $\frac{1}{2}$ **18.** $-\frac{2}{3}$ **19.** $\sqrt{5}$ **20.** $\sqrt{25}$ **21.** 1.8

22. $-0.\overline{85}$ **23.** $\sqrt{3}$ **24.** -17 **25.** $0.\overline{5}$ **26.** $-1.0\overline{63}$ **27.** $25\frac{1}{4}$ **28.** $-\pi$

Graph each set of numbers.

29. {whole numbers greater than 3 and less than 7}

30. {odd whole numbers greater than or equal to 5}

31. {real numbers less than 4.5}

32. {integers greater than -4 and less than 2}

33. {even integers greater than or equal to 4 and less than or equal to 3}

34. {real numbers greater than or equal to $-1\frac{3}{4}$}

35. {natural numbers less than 1}

36. {real numbers greater than 0 and less than 1}

37. {integral multiples of 3 greater than 0 and less than 10}

B Complete each statement.

38. $N \cup \{0\} = \underline{?}$ **39.** $Ir \cup Q = \underline{?}$ **40.** $R \cup Ir = \underline{?}$ **41.** $W \cup I = \underline{?}$

42. $Q \cap Ir = \underline{?}$ **43.** $W \cap N = \underline{?}$ **44.** $Ir \cap R = \underline{?}$ **45.** $I \cap W = \underline{?}$

The symbol $\not\subset$ means "is not a subset of." Replace each $\underline{?}$ with $\subset$ or $\not\subset$ to make a true statement.

46. $W \underline{?} N$ **47.** $N \underline{?} W$ **48.** $I \underline{?} R$ **49.** $R \underline{?} I$

50. $Ir \underline{?} Q$ **51.** $Q \underline{?} Ir$ **52.** $W \underline{?} Ir$ **53.** $R \underline{?} W$

Let $A = \{-6, -2.1, -1.75, 0, 2, \pi, 16\}$ and $B = \{-6, -4, -1\frac{3}{4}, 4, 3.\overline{14}, \sqrt{11}\}$. Use sets A and B to list the indicated subsets in Exercises 54–65.

54. {integers in A}

55. {integers in B}

56. {integers in A that are also in B}

57. {natural numbers in A}

58. {irrational numbers in B}

59. {rational numbers in B}

60. {negative numbers in B}

61. {nonnegative integers in A}

62. {positive real numbers in A}

63. {negative real numbers in A}

64. {nonnegative real numbers in A}

65. {nonnegative real numbers in B}

66. {integers in A or in B or in both A and B}

67. {negative even integers in A or in B or in both A and B}

68. {nonnegative real numbers in A or in B or in both A and B}

69. {positive irrational numbers in A or in B or in both A and B}

70. {negative rational numbers in A that are also in B}

71. {irrational numbers in B that are not also in A}

1-3 Operations with Real Numbers

The **replacement sets** for the variables in an algebraic expression or in a formula may contain both positive and negative real numbers.

Recall that **opposites,** such as 2 and -2, are the same distance from 0 on the number line. Opposites have the same **absolute value** (in symbols: $|\ |$).

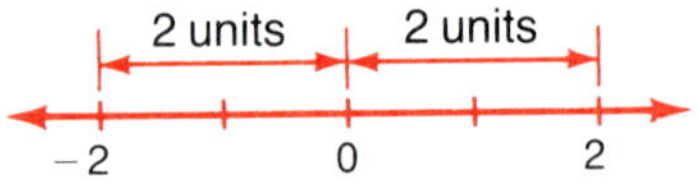

$|2| = 2$ and $|-2| = 2$. $|x| = 2$ means $x = 2$ or $x = -2$.

Definition

Absolute Value

$|x| = x$ for $x \geq 0$ ← **Read "x greater than or equal to 0."**

$|x| = -x$ for $x < 0$ ← **Read "x less than 0."**

EXAMPLE 1 A ship is moving southward at 30 km/hr. Considering north as the positive direction, you can use the formula

$$v_{po} = v_{ps} + v_{so}$$

to find the velocity relative to the ocean of a person walking or running on the deck. In the formula, v_{po} is the velocity of the person relative to the ocean, v_{ps} is the velocity of the person relative to the ship, and v_{so} is the velocity of the ship relative to the ocean.

a. Find v_{po} when a person on deck is walking north at 2 km/hr.

b. Find v_{po} when a person on deck is jogging south at 4.5 km/hr.

Solutions:

a. $v_{po} = v_{ps} + v_{so}$

$v_{po} = 2 + (-30)$ ← **Since $|-30| > |2|$, $-30 + 2 = -(|30| - |2|)$**

$v_{po} = \mathbf{-28}$ **km/hr**

b. $v_{po} = v_{ps} + v_{so}$

$v_{po} = -4.5 + (-30)$ ← $\mathbf{-4.5 + (-30) = -(|-4.5| + |-30|)}$

$v_{po} = \mathbf{-34.5}$ **km/hr**

Subtraction of real numbers is defined in terms of addition. That is, for all real numbers a and b,

$$a - b = a + (-b).$$

EXAMPLE 2 Find the difference between the two temperatures.

a. Aluminum boils at 2450°C and melts at 660°C.

b. Mercury boils at 356.6°C and melts at −38.9°C.

c. Oxygen boils at −183°C and melts at −218°C.

Solutions: **a.** $2450 - 660 = 2450 + (-660) = \mathbf{1790}$ ← **Difference: 1790°.**

b. $356.6 - (-38.9) = 356.6 + [-(-38.9)]$ ← **38.9 is the opposite of −38.9.**

$= 356.6 + 38.9$

$= \mathbf{395.5}$ ← **Difference: 395.5°.**

c. $-183 - (-218) = -183 + [-(-218)]$

$= -183 + 218 = \mathbf{35}$ ← **Difference: 35°.**

Recall that the product of two real numbers having like signs is a positive real number. The product of two real numbers having unlike signs is a negative real number.

Division of real numbers is defined in terms of multiplication. That is, for all real numbers a and b, where $b \neq 0$,

$$a \div b = a \times \frac{1}{b}, \quad \text{or} \quad \frac{a}{b}.$$

← $\frac{1}{b}$ **is the reciprocal of** b.

EXAMPLE 3 Evaluate the expression $\frac{a^3b - 2c}{ac}$ for the given values of a, b, and c.

a. $a = 3$, $b = -1$, $c = -2$ **b.** $a = -1$, $b = -9$, $c = 4$

Solutions: **a.** $\frac{a^3b - 2c}{ac} = \frac{(3)^3(-1) - (2)(-2)}{3(-2)}$ ← $(2)(-2) = -(|2| \cdot |-2|) = -4$

$= \frac{27(-1) - (-4)}{-6}$ ← $-(-4) = (-1)(-4) = 4$

$= \frac{-27 + 4}{-6} = \frac{-23}{-6}$, or $\mathbf{3\frac{5}{6}}$

b. $\frac{a^3b - 2c}{ac} = \frac{(-1)^3(-9) - (2)(4)}{(-1)(4)}$ ← $(-1)^3 = (-1)(-1)(-1) = -1$

$= \frac{(-1)(-9) - 8}{-4}$

$= \frac{9 - 8}{-4} = \frac{1}{-4} = \mathbf{-\frac{1}{4}}$ ← $-\frac{1}{a} = \frac{-1}{a} = \frac{1}{-a}$, $a \neq 0$

CLASSROOM EXERCISES

Evaluate each expression.

1. $|6| + |-6|$ **2.** $|6| - |6|$ **3.** $|-6| + |-6|$ **4.** $|-6| - |-6|$ **5.** $|a| - |-a|$

Add or subtract as indicated.

6. $-19 + 10$ **7.** $-24 + 18$ **8.** $-18 + (-30)$ **9.** $-72 + (-10)$ **10.** $17 + (-20)$

11. $11 + (-12)$ **12.** $-1 + 1$ **13.** $24 - (-8)$ **14.** $-24 - (-8)$ **15.** $-8 - 24$

16. $-8 - (-24)$ **17.** $-3 - 5$ **18.** $-3 + 5$ **19.** $a - (-b)$ **20.** $x - (-9)$

Multiply or divide as indicated.

21. $(-18)(-20)$ **22.** $183(-\frac{1}{3})$ **23.** $(-10)(-1)$ **24.** $\frac{1}{5}(-125)$ **25.** $110 \div (-2)$

26. $-64 \div 4$ **27.** $-96 \div (-16)$ **28.** $\frac{306}{-18}$ **29.** $\frac{-132}{-11}$ **30.** $(0.5)(-3.6)$

WRITTEN EXERCISES

A Evaluate each expression.

1. $|-10|$ **2.** $-|57|$ **3.** $-|-21|$ **4.** $|-8|-|8|$ **5.** $-|13|+|-31|$

6. $|8|-|-6|$ **7.** $3|5-6|$ **8.** $8|9-2|$ **9.** $|-4+1|-2|0|$ **10.** $|-3||4|-7$

Add or subtract as indicated.

11. $16+(-3)$ **12.** $24+(-9)$ **13.** $-12+(-4)$ **14.** $-5+(-3)$ **15.** $14-(-6)$

16. $-19-(-3)$ **17.** $12-(-20)$ **18.** $-13.8+(-4.6)$ **19.** $2\frac{1}{3}+(-1\frac{2}{3})$ **20.** $4.7-5.1$

21. $-3\frac{1}{4}-(-2\frac{1}{8})$ **22.** $\frac{1}{2}-(-\frac{1}{4})$ **23.** $-5.9+3.6$ **24.** $1.3+(-2.4)$ **25.** $-6\frac{3}{8}-4\frac{3}{4}$

26. $-5+6+(-12)+8$ **27.** $10+(-8)+6-(-1)$ **28.** $-14+(-16)-23+(-8)$

29. $-29.6+18.8-(-16.5)$ **30.** $9.1-(-15.6)+(-1.3)$

31. $43.8-0.07-(-16.3)+8.9$ **32.** $|6|-|-7|+(-5)\,|-1|$

33. $-|5|+6\,|7-8|-(-|9|)$ **34.** $-|10|-9+3(|16|)(-|-1|)$

Multiply or divide as indicated.

35. $(-6)8$ **36.** $(-\frac{2}{3})(-\frac{9}{20})$ **37.** $(-1)(-7.6)$ **38.** $(-0.2)(0.8)$ **39.** $-15(1.1)$

40. $28(-\frac{1}{4})$ **41.** $-45 \div 3$ **42.** $-125 \div (-5)$ **43.** $108 \div (-4)$ **44.** $-56 \div \frac{7}{8}$

45. $-0.25 \div 5$ **46.** $2.53 \div (-1.1)$ **47.** $\frac{-135}{3}$ **48.** $\frac{81}{-9}$ **49.** $-\frac{-54}{3}$

Evaluate each expression.

50. $(-2)^2$ **51.** $(-3)^2$ **52.** $(-2)^3$ **53.** $(-4)^4$ **54.** $(-1)^6$ **55.** $(-1)^5$

56. $18 \div (6-3)+(-2)^3(4)$ **57.** $(144 \div 6)(-3)^2-4^3$ **58.** $[(-3 \cdot 5)-(-128 \div 4)]^2$

59. $-[\frac{1}{3}(-48 \div 8)-5\left(\frac{18}{-3}\right)]$ **60.** $[(-8)(-7)(-\frac{1}{2})-8] \div (-2)$ **61.** $[(-3)^2+2 \cdot 5-5^2]^2$

APPLICATIONS: Using Operations with Real Numbers

62. Use the formula in Example 1 to find v_{po} for a person jogging southward at 4 km/hr on the deck of the ship which is moving northward at 32 km/hr.

63. Use the formula in Example 1 to find v_{po} for a person walking northward at 2.5 km/hr on the deck of a ship which is moving northward at 30.5 km/hr.

64. Copper boils at 2300°C and melts at 1083°C. Find the difference between these temperatures.

65. Ammonia boils at −33°C and melts at −75°C. Find the difference between these temperatures.

66. Liquid hydrogen boils at $-252.8°C$ and freezes at $-259.1°C$. Find the difference between these temperatures.

67. Find the difference in age between the first skyscraper built in Chicago in 1884 and the Step Pyramid built in Egypt in 2650 B.C.

68. Find the difference in age between the Parthenon in Athens built in 400 B.C. and the Great Pyramid of Egypt built around 2500 B.C.

The following formula

$$H = 116(i - o)$$

gives the rate of **heat transfer** through a 200 square-foot wall of concrete 6 inches thick, where i and o represent the inside and outside temperatures in Fahrenheit degrees. Heat transfer is measured in British thermal units per hour (abbreviated: Btu/h).

Use this formula in Exercises 69–75, to find the rate of heat transfer for the given values of i and o.

69. $i = 68°F$; $o = 20°F$
70. $i = 70°F$; $o = 43°F$
71. $i = 74°F$; $o = 96°F$
72. $i = 69°F$; $o = 82°F$
73. $i = 66°F$; $o = 8°F$
74. $i = 68°F$; $o = 32°F$

75. In Exercises 69–74, how can you tell whether heat is being gained or lost?

Evaluate each expression for $a = 4$, $b = -3$, $c = -2$, and $d = -10$.

76. $\frac{5b}{2a} - \frac{3a}{7d}$
77. $\frac{3b^2 - cd}{-a}$
78. $\frac{bd}{-a} - \frac{ac}{b}$
79. $\frac{-d^2}{a} + c^3$

Evaluate $\frac{x^2y^3 - 5z^4}{xyz}$ for the given values of x, y, and z.

80. $x = -1$, $y = 3$, $z = -2$
81. $x = 4$, $y = -2$, $z = -1$
82. $x = 6$, $y = \frac{1}{2}$, $z = 1$

B

83. Which of the following expressions are equal? Justify your answer.
   a. $s|-s|$ b. $-s|-s|$ c. $-s|s|$ d. $s|s|$

84. Which of the following expressions are equal? Justify your answer.
   a. $(|-r|)^3$ b. $(-|r|)^3$ c. $(-r)^3$ d. $(-|-r|)^3$

85. Which of the following expressions are equal? Justify your answer.
   a. $\frac{rs}{-t}$ b. $-\frac{rs}{t}$ c. $-\left|\frac{rs}{t}\right|$ d. $-\left|-\frac{rs}{t}\right|$

C

Given the set $T = \{-5, -4, -3, -2, -1, 0, 1, 2, 3, 4, 5\}$, choose the elements of T that will make each statement true.

86. $|x| = 3$
87. $-|x| = 5$
88. $|x - 1| = 5$
89. $|x + 2| = 1$
90. $-2|x| = 6$
91. $\frac{1}{3}|-x| = 12$
92. $2|x| - |x| = |x|$
93. $\frac{|x|}{-3} = 3$
94. $|x| = x$
95. $|x| = -x$
96. $2|x| = 2 + |x|$
97. $|x^2| = |x^3|$

1-4 Postulates for Real Numbers

The following postulates for all real numbers a, b, and c state the fundamental properties for addition and multiplication of real numbers. **Postulates (or axioms)** are statements that are accepted as true without proof.

Postulates for Real Numbers

Postulate	Statement	Example
1. Closure		
for addition	$a + b$ is a unique real number.	$9 + 6 = 15$ and $15 \in R$
for multiplication	$a \cdot b$ is a unique real number.	$-6 \cdot \frac{2}{3} = -4$ and $-4 \in R$
2. Commutative		
for addition	$a + b = b + a$	$3.5 + (-6.2) = -6.2 + 3.5$
for multiplication	$a \cdot b = b \cdot a$	$48 \cdot \frac{1}{3} = \frac{1}{3} \cdot 48$
3. Associative		
for addition	$(a + b) + c = a + (b + c)$	$(3 + 0.4) + 9 = 3 + (0.4 + 9)$
for multiplication	$(ab)c = a(bc)$	$(7 \cdot 8) \cdot 10 = 7 \cdot (8 \cdot 10)$
4. Identity		
for addition	$a + 0 = a$	$-\pi + 0 = -\pi$
for multiplication	$a \cdot 1 = a$	$345 \cdot 1 = 345$
5. Inverse		
for addition	$a + (-a) = 0$	$13.6 + (-13.6) = 0$
for multiplication	$a \cdot \frac{1}{a} = 1, a \neq 0$	$\pi \cdot \frac{1}{\pi} = 1$
6. Distributive		
multiplication over addition	$a(b + c) = ab + ac$, or $(b + c)a = ba + ca$	$9(6 + 5) = 9 \cdot 6 + 9 \cdot 5$, or $(6 + 5)9 = 6 \cdot 9 + 5 \cdot 9$

The word "unique" in the closure postulate means that there is *only one* real number which is the sum or product of two real numbers.

EXAMPLE 1 Name the postulate illustrated by each statement.

Statement	Postulate
a. $(-6 + 9) + 5 = -6 + (9 + 5)$	**a.** Associative, for addition
b. $\frac{4}{3} \cdot \frac{3}{4} = 1$	**b.** Inverse, for multiplication
c. $(-3.1 \cdot 7) \cdot (-2.4) = (-2.4) \cdot (-3.1 \cdot 7)$	**c.** Commutative, for multiplication
d. $-3(8 + 9.3) = -3 \cdot 8 + (-3) \cdot 9.3$	**d.** Distributive
e. $\sqrt{50} + 0 = \sqrt{50}$	**e.** Identity, for addition

The postulates for real numbers can be used to justify operations with algebraic expressions. Unless otherwise stated in this book, the **replacement set for a variable is the set of real numbers.**

EXAMPLE 2 Perform the indicated operations.

a. $8z - 3z$

b. $4(a-3) - (a-8)$

Solutions: a. $8z - 3z = 8z + (-3z)$ ← **Definition of subtraction**

$= [8 + (-3)]z$ ← **By the distributive postulate**

$= \mathbf{5z}$

b. Recall that $-(a-8) = (-1)(a-8)$.

$4(a-3) - (a-8) = 4(a-3) + (-1)(a-8)$

$= 4a - 12 - a + 8$

$= \mathbf{3a - 4}$

In the last step of Example 2**a**, the Substitution Principle allows you to replace the expression $8 + (-3)$ with the *equivalent expression*, 5.

Substitution Principle

For all real numbers a and b where $a = b$, a may be replaced with b.

CLASSROOM EXERCISES

Match each statement in Exercises 1–6 with the postulate for real numbers in **a–g** that it illustrates.

1. $\sqrt{21} + 0 = 21$
2. $-\frac{2}{3}(9 + \frac{57}{2}) = (-\frac{2}{3})(9) + (-\frac{2}{3})(-\frac{57}{2})$
3. $\pi + (-\pi) = 0$
4. $\sqrt{7} \cdot \sqrt{7}$ is a real number
5. $(-\frac{5}{9})(-\frac{9}{5}) = 1$
6. $\sqrt{9} + \sqrt{16}$ is a real number

a. Additive inverse postulate
b. Closure postulate for multiplication
c. Multiplicative inverse postulate
d. Identity postulate for addition
e. Closure postulate for addition
f. Distributive postulate
g. Identity postulate for multiplication

Perform the indicated operations.

7. $17t + 4t$
8. $2(6r + 9)$
9. $27s - 11s + 5$
10. $3(b - 8) + 9b$
11. $6(q - 5) - 8(q - 9)$
12. $23r - (4r + 5)$
13. $-(16c + 9 - 3c)$
14. $5(t - 1) - (t + 1)$
15. $-3(y + 5) - 2(y - 5)$
16. $3[4b - (1 - 2b)]$
17. $-2[-3r + 5(r - 1)]$
18. $-[6q - 5(3 - 8q)]$

WRITTEN EXERCISES

A Name the postulate illustrated by each statement.

1. $6 + 0 = 6$
2. $42 + (-42) = 0$
3. $\frac{8}{9} \cdot \frac{9}{8} = 1$
4. $4(6 + 3) = 4 \cdot 6 + 4 \cdot 3$
5. $3.6 \cdot 1 = 3.6$
6. $4 \cdot \frac{9}{5} = 7\frac{1}{5}$ and $7\frac{1}{5} \in R$
7. $(14 \cdot 12)15 = 14(12 \cdot 15)$
8. $\frac{1}{4} \cdot 64 = 64 \cdot \frac{1}{4}$
9. $32 + 0 = 32$
10. $(1.6 + 4) + 5.1 = 1.6 + (4 + 5.1)$
11. $16 \cdot \frac{1}{16} = 1$
12. $-8.3 + (8.3) = 0$

Perform the indicated operations.

13. $8s + 11s$
14. $9r + 21r$
15. $7x^2 + 3x^2$
16. $6x^2 + (-3x^2)$
17. $-8x - 6x$
18. $-3a - 6a$
19. $-9b + 7b$
20. $-21z + 40z$
21. $42t - 11t + 2$
22. $2.1t - t + 0.3$
23. $-10x + 3x - 8$
24. $-11x - 26x - 8$
25. $10(a + b)$
26. $6(24 - 3)$
27. $-3(x + y)$
28. $-\frac{1}{2}(2x + 4y)$
29. $-3(-5p - 6q)$
30. $-2(y + 6)$
31. $-(8a + b)$
32. $-(-6q + s)$
33. $-2(y - 6) + 4y$
34. $-(p + 3) + 4p - 1$
35. $\frac{1}{5}(10b - 5) - (b + 1)$
36. $-5m - (6 - m)$
37. $3(m - 2) - (2m - 5)$
38. $-(5r + s) - 4r - s$
39. $-10p - (-6p - 3) + 8$
40. $r(7 - 3r) - 2(1 + r)$
41. $7f - 3(2f - 3) + 8$

B

42. $5[4a + 3(a + 2)] - 6$
43. $17 - 4[8x + 2(4x - 3)]$
44. $3(a + 9) - 6[4a - 3(a + 1)]$
45. $-(3a + 4) - 2[3a - 5(a + 3)]$
46. Which real number is its own multiplicative inverse?
47. Which real number has no multiplicative inverse?
48. Is subtraction of real numbers commutative? Given an example to illustrate your answer.
49. Does the associative postulate hold for division of real numbers? Give an example to illustrate your answer.
50. Does the associative postulate hold for subtraction of real numbers? Give an example to illustrate your answer.
51. Is there an identity element for subtraction? If so, what is it?
52. Is there an identity element for division? If so, what is it?

C

53. Is the set of integers closed under multiplication? Give three examples to illustrate your answer.
54. Is the set of whole numbers closed under division? Give 3 examples to illustrate your answer.
55. Is the set of odd integers closed under addition? Give 3 examples to illustrate your answer.
56. Is the set $\{0, 1\}$ closed under multiplication? Illustrate your answer.
57. Is the set $\{-1, 0, 1\}$ closed under addition? Illustrate your answer.
58. Is the set $\{-1, 0, 1\}$ closed under multiplication? Illustrate your answer.

Review

1. Evaluate $3p^3 - (s^2 + 4t)$ when $p = 5$, $s = 6$, and $t = 10$. *(Section 1-1)*

For each number in Exercises 2-7, classify the number as a terminating decimal, *T*, or as a non-terminating, repeating decimal, *NT*. *(Section 1-2)*

2. $\frac{3}{5}$ **3.** $\frac{2}{9}$ **4.** $-\frac{1}{16}$ **5.** $\frac{5}{7}$ **6.** $-\frac{3}{11}$ **7.** $-8\frac{1}{6}$

8. Graph: {odd integers greater than or equal to $-1\frac{1}{4}$}. *(Section 1-2)*

Evaluate each expression. *(Section 1-3)*

9. $27 \div (5 - 14) + (-3)^2$ **10.** $(-4)(3) - (-5 + 8)$ **11.** $(-9)^2 + (-5)(-7) - 8^2$

12. Find the age at which Augustus Caesar, who lived from 63 B.C. to 14 A.D., died. *(Section 1-3)*

Name the postulate illustrated by each statement. *(Section 1-4)*

13. $7\frac{1}{3} + (-7\frac{1}{3}) = 0$ **14.** $(12.6)1 = 12.6$ **15.** $(16 + 2)3 = 3(16 + 2)$

16. $\frac{3}{15} \cdot \frac{15}{3} = 1$ **17.** $-\sqrt{17} + 0 = -\sqrt{17}$ **18.** $a + (-a) = (-a) + a$

CALCULATOR APPLICATIONS

Operations with Real Numbers

To operate with negative numbers on a calculator, first enter the number. Then press the [+/−] key. (On some calculators, this key is marked [sc], for "sign change.") Remember to follow the rules for order of operations given on page 2.

EXAMPLE Use a calculator to evaluate: $15.3 + (-3.2)(4.9)$

SOLUTION First find the product. Then find the sum.

3 [·] 2 [+/−] [×] 4 [·] 9 [=] → -15.68

[+] 1 5 [·] 3 [=] → -0.38

EXERCISES

Use a calculator to perform the indicated operations.

1. $-72 + 35$ **2.** $114 + (-91)$ **3.** $(-46) - (-390)$ **4.** $-112 - 715$

5. $(-3.6)(-7.9)$ **6.** $(5.02)(-0.48)$ **7.** $258 - (18)$ **8.** $(-11,020) - (-76)$

9. $(-9)(-56) - (-3604)$ **10.** $-71.83 + (-9.8)(-9.8)$ **11.** $-9252 + (-724 \div 32)$

REVIEW CAPSULE FOR SECTION 1-5

Solve each equation.

1. $x+9=21$ **2.** $y+8=4$ **3.** $a+1.6=4.3$ **4.** $t+\frac{1}{4}=\frac{1}{2}$

5. $3x=15$ **6.** $\frac{x}{9}=8$ **7.** $0.6r=-12$ **8.** $-x=21$

1-5 Solving Equations

Each of these sentences is an **equation.**

a. $7+3.5=10.5$ **b.** $t+0=21$ **c.** $x-y=5$

Equation **a** is a true statement because the expressions on each side name the same number. Equations **b** and **c** are **open sentences** because they cannot be classified as *true* or *false* until you know which numbers can replace the variables t, x, and y. The **solution set,** or, more simply, **solution** of an open sentence is the set of numbers from the replacement set (or domain) of the variable that makes the sentence true.

EXAMPLE 1 Find the solution set of $4x-5=3$ when the replacement set is $\{-2, -1, 0, 1, 2\}$.

Solution: Evaluate $4x-5=3$ for each element in the replacement set.

$x=-2$: $4(-2)-5\stackrel{?}{=}3$ No

$x=-1$: $4(-1)-5\stackrel{?}{=}3$ No

$x=0$: $4(0)-5\stackrel{?}{=}3$ No

$x=1$: $4(1)-5\stackrel{?}{=}3$ No

$x=2$: $4(2)-5\stackrel{?}{=}3$ Yes ✔

Solution set: $\{2\}$

When the replacement set is finite, you can use the method shown in Example 1 to solve equations. In this book, the replacement set for the variable in an equation is the set of real numbers, unless otherwise stated. For such a replacement set, you often use the Addition and Multiplication Properties for Equations to solve equations.

Properties For Equations

Addition Property for Equations

Adding the same real number to each side of an equation results in an equivalent equation.

Multiplication Property for Equations

Multiplying each side of an equation by the same nonzero real number results in an equivalent equation.

Equivalent equations are equations that have the same solution. To solve an equation, you write equivalent equations until you get the variable alone on one side of the equation.

EXAMPLE 2 Solve and check: $-4x + 7 = -5$

Solution:

$-4x + 7 = -5$ ← **Add −7, the opposite of 7, to each side of the equation.**

$-4x + 7 + (-7) = -5 + (-7)$

$-4x = -12$ ← **Multiply each side by $\left(-\frac{1}{4}\right)$.**

$\left(-\frac{1}{4}\right)(-4x) = \left(-\frac{1}{4}\right)(-12)$

$x = 3$ ← **The variable is alone.**

Check:

$-4x + 7 \stackrel{?}{=} -5$ ← **Replace x with 3.**

$-4(3) + 7 \stackrel{?}{=} -5$

$-12 + 7 \stackrel{?}{=} -5$

$-5 \stackrel{?}{=} -5$ Yes ✔ **Solution set: {3}**

When terms containing the variable occur on both sides of the equation, begin by writing an equivalent equation having the terms with the variable on one side of the equation only.

EXAMPLE 3 Solve and check: $5m - 7 = 2m + 3$

Solution:

$5m - 7 = 2m + 3$ ← **Add (−2*m*) to each side.**

$-2m + 5m - 7 = -2m + 2m + 3$

$3m - 7 = 3$ ← **Add 7 to each side.**

$3m - 7 + 7 = 3 + 7$

$3m = 10$ ← **Multiply each side by $\frac{1}{3}$.**

$m = \frac{10}{3}$ The check is left for you. **Solution set: $\left\{\frac{10}{3}\right\}$**

The first step in solving an equation that contains parentheses is to write an equivalent equation without parentheses.

EXAMPLE 4 Solve and check: $5(3 - 2t) = 28 + 3t$

Solution:

$5(3 - 2t) = 28 + 3t$

$5 \cdot 3 + 5(-2t) = 28 + 3t$ ← **By the distributive postulate**

$15 - 10t = 28 + 3t$ ← **Add −3*t* to each side.**

$15 - 10t + (-3t) = 28 + 3t + (-3t)$

$15 - 13t = 28$ ← **Add −15 to each side.**

$-15 + 15 - 13t = -15 + 28$

$-13t = 13$

$t = -1$ The check is left for you. **Solution set: {−1}**

Always check each element of the solution set in the original equation.

CLASSROOM EXERCISES

1. Find the solution set of $x - 5 = 1$ when the replacement set is $\{1, 2, 3, 4, 5\}$.
2. Find the solution set of $-2x + 3 = 1$ when the replacement set is $\{1, 2, 3, 4, 5\}$.

In Exercises 3–10:

a. State what number you would add to each side as a first step in solving the equation.

b. State the number by which you would multiply each side as a second step in solving the equation.

3. $7c - 3 = 25$
4. $-5 - 3x = 16$
5. $16 + 7h = -33$
6. $\frac{4}{3}n + 10 = 2$
7. $5 = \frac{b}{4} - 3$
8. $\frac{2}{3}x + 5 = 11$
9. $\frac{1}{4}a - 3 = 5$
10. $21 = \frac{4x}{9} + 7$

Solve each equation.

11. $5s - s = 64$
12. $4n - 5 = 30 - 3n$
13. $3t + 12 = 7t$
14. $-21 + c = 8c$

WRITTEN EXERCISES

A

For each equation in Exercises 1–6, find the solution set for the given replacement set.

1. $3x - 10 = 5;\ \{1, 2, 3, 4, 5\}$
2. $4x + 3 = -1;\ \{-2, -1, 0, 1, 2\}$
3. $4x + 3 = 7;\ \{-2, -1, 0, 1, 2\}$
4. $-2x - 5 = -1;\ \{-2, -1, 0, 1, 2\}$
5. $-2x - 5 = -5;\ \{-2, -1, 0, 1, 2\}$
6. $-3x + 2 = -4;\ \{-1, 0, 1, 2, 3\}$

Find the solution set for each equation in Exercises 7–14 when the replacement set is $\{-2, -1, 0, 1, 2, 3, 4\}$.

7. $x^2 - 5 = 1$
8. $x^2 + 3 = 7$
9. $x^2 - 1 = 1$
10. $x^2 + 1 = 1$
11. $2x^2 - 1 = 7$
12. $3x^2 + 5 = 2$
13. $3x^2 - 5 = -2$
14. $-2x^2 + 12 = 12$

Solve and check.

15. $7c + 16 = 2$
16. $-5n - 7 = 28$
17. $2a + 0.7 = 2$
18. $0.2x + 3 = 3.5$
19. $7x - 8 = 13$
20. $-6m - 8 = -32$
21. $18p - 36 = 3$
22. $5 + 15n = 0$
23. $3y = y - 24$
24. $3x = 1 - x$
25. $3x = 48 + 7x$
26. $2m = 18 - m$
27. $4n = n + 4.5$
28. $3.7x - 1.2x = 0.625$
29. $4x - 3 = 3x + 15$
30. $8 + 0.4y = 0.7y - 10$
31. $7q = q + 9.6$
32. $-3n - 2 = 3n - 7$
33. $7(x + 3) = -28$
34. $3(2y - 5) = 9$
35. $2(x - 1) = 7$
36. $5(3 - 2n) = 5n$
37. $10 + 2(3m + 4) = 5m + 14$
38. $4(3c - 2) + 2c = 6$

39. $3n - 4 = 15(\frac{1}{3}n + \frac{2}{5})$
40. $6y - \frac{1}{2}(2y - 8) = 11$
41. $\frac{2}{3}(6m - 9) = 3m + 5$
42. $2.5x = 0.4x - 16.8$
43. $2(3m + 1) + 7 = m - 20$
44. $5(x - 2) - 9 = 1 - 5x$
45. $4(-2m + 1) = 3(m - 5) - 2$
46. $6x + 14 - 3x = 7x - 10 - 2x$

B

47. $3a - 0.2(4a - 1) = 9$
48. $6 - c = 2 - 5(2c - 7)$
49. $5(2y - 3) = -2(3y + y)$
50. $3(5a - 2) = 4(7a + 1) - 1$
51. $10(0.6 - p) = 2(7p + 0.6)$
52. $0.2(3x - 4) = 2(19.7 + 7x)$
53. $3(1 - 2n) + 10 = -4(2n - 3)$
54. $0.4(3 - 2x) = 2.4 - 6(x - 5)$

For each equation in Exercises 55–66, find the solution set when the replacement set is $\{-3, -2, -1, 0, 1, 2, 3\}$.

55. $x^2 - 2x = -1$
56. $x^2 + x = 6$
57. $x^2 + 3x = 0$
58. $4x^2 - 16 = 0$
59. $3x^2 - 15 = -18$
60. $x^2 - 6x = -8$
61. $x^2 + x = 20$
62. $4x^2 + 8x = 12$
63. $x^3 + 4x^2 + x = 16$
64. $x^4 - 9x^2 = 0$
65. $x^3 + 4x^2 - 4x = 4$
66. $3x^3 + 6x^2 = -18$

C Solve and check. The replacement set is {real numbers}.

67. $6m^2 + 9m + 2(m - 3) = 6m^2 + m - 7$
68. $x^2 - 3x - 10 - (x^2 + 4x - 6) = 0$
69. $2n^2 - n - 3 = 2n^2 - 5n - 3$
70. $2(3y - 4) - (2y^2 + 3y) = 10y - 2y^2 + 6$

CALCULATOR APPLICATIONS

Checking Equations

A calculator can be used to check the solutions of equations. When an equation has variables on both sides, evaluate each side separately.

EXAMPLE Equation: $-5(t - 4) = 28 + 3t$ Solution set: $\{-1\}$

SOLUTION Follow the rules for order of operations.

Left side → 1 [+/−] [−] 4 [=] [×] 5 [+/−] [=] 25.

Right side → 3 [×] 1 [+/−] [+] 2 8 [=] 25.

Since the results are the same, the solution is correct.

EXERCISES

Solve each equation. Then use a calculator to check its solution set.

1. $4p = 7p + 15$
2. $8(9z - 11) = 17z$
3. $-4(-2b - 5) = -3(-4b + 12)$
4. $7y - 9 = 3y + 19$
5. $4(n + 2) = n - 10$
6. $5z + 3(z - 2) = 6z - 12$

Problem Solving and Applications

1-6 From Words to Symbols

This table shows some key words that indicate operations used in writing algebraic expressions.

Operation	Key Words
Addition	sum, total, plus, increased by, more than
Subtraction	difference, exceeds, less than, less, minus, decreased by, diminished by, subtracted from
Multiplication	product, multiplied by, times, twice, doubled, tripled
Division	quotient, divided by

To represent a word expression by algebraic symbols, first choose a variable to represent the unknowns. Identify the key word(s) that indicate the operation. Then use algebraic symbols to represent the word expression.

EXAMPLE 1 Write an algebraic expression for each word expression.

a. \$29.50 less than the cost of a coat

b. Five more than 3 times the number of pages

Solutions: **a.** Let $c =$ the cost of the coat.

\$29.50 less than the cost: $c - 29.50$

b. Let $p =$ the number of pages.

Five more than 3 times the number: $3p + 5$

Many word problems involve more than one unknown. In these problems, one condition (Condition 1) tells how the unknowns are related. You can use this condition to represent the unknowns. A second condition (Condition 2) tells what quantities are equal. You can use this condition to write an equation for the problem.

EXAMPLE 2 In a recent year, the cost of electricity used for lighting in a house was \$9.00 less than 6 times the cost of operating a color television set (Condition 1). The total cost for these two items was \$834.50 (Condition 2). Find the cost for lighting.

Solution: [1] Use Condition 1 to represent the unknowns.

Let $c =$ the cost of operating the color television set.

Then $6c - 9 =$ the cost of lighting.

[2] Use Condition 2 to write an equation.

$c + 6c - 9 = 834.50$ ← ***Total* cost**

[3] Solve the equation.

$$c + 6c - 9 = 834.50$$
$$7c - 9 = 834.50$$
$$7c = 843.50$$
$$c = 120.50$$ ← **Don't forget to find 6c − 9.**
$$6c - 9 = 614.00$$

Check: [4] **Condition 1** Is the cost for lighting \$9.00 less than 6 times the cost for the color television? \$614 = 6(\$120.50) − \$9.00? Yes ✓

Condition 2 Is the total cost \$834.50?
\$120.50 + \$614.00 = \$834.50? Yes ✓

The cost for lighting was **\$614.00.**

Always use the original conditions of the problem to check your answer.

Summary

Steps in Solving Word Problems

[1] Identify Condition 1. Use this condition to represent the unknowns.

[2] Identify Condition 2. Use this condition to write an equation for the problem.

[3] Solve the equation.

[4] Check in the original conditions of the problem. Answer the question.

CLASSROOM EXERCISES

Write an algebraic expression for each word expression.

1. Twelve dollars less than the original cost, j, of a pair of jeans
2. The difference between three times the perimeter, p, of a rectangle and 18
3. Twenty more than the number of centimeters of rainfall, n
4. Twice the cost, c, of a lamp increased by \$15
5. The sum of $\frac{2}{3}$ times the number of votes, v, and 1
6. Yesterday's average temperature, t, minus 3°
7. The quotient of 100 points and the number of questions, n
8. The quotient of \$2700 and the number of months, n
9. \$500 less than triple the amount, a, of the investment
10. The quotient of \$250,000 and the number of pages printed, p

WORD PROBLEMS

A Write an algebraic expression for each word expression.

1. Six more than twice the number of seats, *s*
2. The sum of one half the number of plants, *p*, and 4
3. Last years total rainfall, *r*, minus three inches
4. The quotient of the number of books, *b*, and the number of students, *s*
5. The product of $150 and the number of coats, *c*
6. Eight dollars less than the cost of the radio, *r*
7. Twenty more than triple the original number, *v*, of voters
8. Twice the number of customers, *c*, increased by 12
9. The quotient of the cost, *c*, and 21
10. The quotient of $3600 and the number of payments, *p*

Solve each problem. For some problems, Condition 1 is underscored once and Condition 2 is underscored twice.

11. An airline charges $250 more for a first–class ticket than for a coach ticket. Total receipts for 30 coach tickets and 12 first–class tickets were $18,750. Find the cost of each type of ticket.
12. A furniture store sells an armchair for $175 less than its matching sofa. Over a two-day period, total sales for 15 arm chairs and 9 sofas amounted to $19,455. Find the price of an armchair and of a sofa.
13. Gold is selling at $280 per ounce. Claude has $2520 invested in gold. How many ounces does he have?
14. This year there were 2200 more runners in the marathon than last year. A total of 28,100 persons ran in the marathon this year. How many runners were there last year?
15. A family's electric bill for one month was $13.76 less than the gas bill. The total bill for electricity and gas was $105.92. How much was the bill for electricity?
16. In a survey, a group using Toothpaste A had $\frac{1}{5}$ as many perfect checkups as a group using Toothpaste B. The total number of perfect checkups for both groups was 1560. Find the number of perfect checkups for each toothpaste.
17. The population of Fennville tripled in 10 years. Today, Fennville has 93,411 residents. What was the population 10 years ago?
18. The length of a garden is double the width. It takes 68 meters of fencing to enclose the garden. Find its length and its width.
19. The length of each of the equal sides of an isosceles triangle is 42 centimeters less than double the length of the base. The perimeter of the triangle is 96 centimeters. What is the length of each side of the triangle?

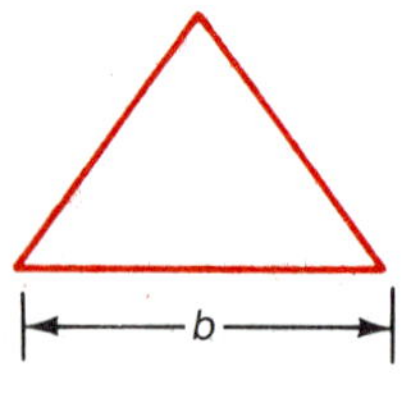

20. An advertisement offers a cookbook for \$7.50 plus two box tops or for \$1.20 plus eight box tops. How much is a box top worth?

B

21. Of the three angles of a triangle, the degree measure of the second is one-half the first and the degree measure of the third is 15 more than the first. Find the measure of each angle of the triangle.

22. A piece of ribbon 60 centimeters long is cut into three different lengths. One piece is 2 centimeters longer than the shortest piece and 2 centimeters shorter than the longest piece. Find the length of each piece.

23. Carla is 5 years older than Nick and Nick is 4 years older than Nancy. The sum of their ages is 46. How old is Carla?

24. Joe bought 3 times as many 20¢ stamps as 10¢ stamps and 6 times as many 10¢ stamps at 5¢ stamps. He spent \$63.75 on stamps. How many of each kind did he buy?

25. Clyde has 68 centimeters of wire with which to make a model of a rectangular solid. The length of the solid will be twice the height. The width will be 1 centimeter more than the height. Into what lengths should he cut the wire?

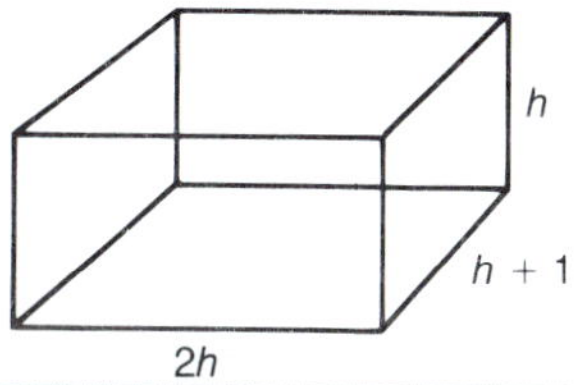

26. Silversole Electronics doubled its order of microcomputers in the second third of the year as compared to the first third of the year. The order for the last third was double that of the second third. If the total order for the year was 1484 microcomputers, how many were ordered in each third?

27. Every six months, the price of fishing equipment has increased by 10% over the previous six months. At the beginning of this year, the price of a certain fishing reel was \$24.20. What was the price of the same reel at the beginning of last year?

28. An automatic coin counter at a bank indicates that \$65 in dimes and quarters was processed. If there were 389 coins, how many were dimes and how many were quarters?

C

29. Two parcels of land have one side in common. One parcel is in the shape of a square and the other is in the shape of an equilateral triangle (three equal sides). If 4200 feet of fencing are required to enclose the land and to separate the two parcels, what are the dimensions of each parcel of land?

30. A mileage test consists of driving at a fixed rate of speed for 2 hours. Then the speed is increased by 10 miles per hour for 1 hour, and finally the latest speed by 5 miles per hour for 3 hours. The distance traveled during the test is 295 miles. What are the rates of speed for each part of the test?

Problem Solving and Applications

1-7 Age Problems/Integer Problems

To represent a person's age a number of years from now, *add* that number of years *to the present age.*

Present age: 17	Age 10 years from now: $17 + 10$, or 27 Age t years from now: $17 + t$

To represent a person's age a number of years ago, *subtract* that number of years *from the present age.*

Present age: 16	Age 7 years ago: $16 - 7$, or 9 Age q years ago: $16 - q$

EXAMPLE 1 Greg is three times as old as Meg (Condition 1). Ten years from now, Greg's age will exceed twice Meg's age at that time by one year (Condition 2). How old are Greg and Meg now?

Solution: [1] Identify Condition 1. Make a table showing the ages now and ten years from now.

	Age Now	Age 10 Years From Now
Meg	r	$r + 10$
Greg	$3r$	$3r + 10$

[2] Use the table and Condition 2 to write an equation.

Greg's age in 10 years	will exceed	twice Meg's age	by	1.
$(3r + 10)$	$-$	$2(r + 10)$	$=$	1

[3] Solve the equation.

$$3r + 10 - 2(r + 10) = 1$$
$$3r + 10 - 2r - 20 = 1$$
$$r - 10 = 1$$
$$r = 11$$
$$3r = 33$$

Don't forget to find 3*r*.

Check: [4] **Condition 1** Is Greg's age three times Meg's age? $33 = 3(11)$? Yes ✓

Condition 2 In 10 years, will Greg's age exceed twice Meg's age by 1? $43 = 42 + 1$? Yes ✓

Thus, Meg is **11** years old and Greg is **33**.

Consecutive integers differ by 1.

$$\begin{array}{cccc} -2, & -1, & 0, & 1, \cdots \\ \downarrow & \downarrow & \downarrow & \downarrow \\ n, & n+1, & n+2, & n+3, \cdots \end{array}$$

Consecutive even integers differ by 2.

$-2, \quad 0, \quad 2, \quad 4, \cdots$

$\downarrow$

$n, n+2, n+4, n+6, \cdots$

Consecutive odd integers also differ by 2.

$-3, \quad -1, \quad 1, \quad 3, \cdots$

$\downarrow$

$n, n+2, n+4, n+6, \cdots$

EXAMPLE 2 The sides of a triangular steel plate can be represented by consecutive odd integers. (Condition 1) The perimeter of the plate is 33 centimeters. (Condition 2) Find the length of each side.

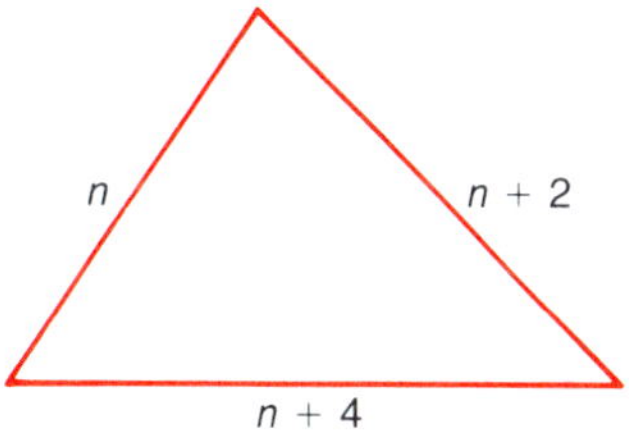

Solution: [1] Let n = the length of the shortest side.
Then $n + 2$ = the length of the second side.
And $n + 4$ = the length of the longest side. ← **Condition 1**

[2] $n + n + 2 + n + 4 = 33$ ← **Condition 2**

[3]
$$3n + 6 = 33$$
$$3n = 27$$
$$n = 9$$ ← **Find the lengths of the other two sides.**
$$n + 2 = 11$$
$$n + 4 = 13$$

Check: [4] **Condition 1** Can the sides be represented as consecutive odd integers? 9, 11, and 13 are consecutive odd integers. ✓

Condition 2 Is the perimeter 33? $9 + 11 + 13 = 33$? Yes ✓

The sides are **9 centimeters, 11 centimeters**, and **13 centimeters** long.

CLASSROOM EXERCISES

Karla's present age is 18.

1. Represent her age t years ago.
2. Represent her age q years from now.
3. Represent the difference between her present age and 5.
4. Represent her age increased by 15.

Tom's present age is 21.

5. How old will Tom be in y years?
6. How old was Tom b years ago?
7. What is 5 times Tom's age decreased by 7?
8. What is twice Tom's age 5 years ago?
9. Represent the 3 consecutive integers following the integer t.
10. Represent the 2 consecutive odd integers following t.
11. Represent the 2 consecutive odd integers *preceding* t.

WORD PROBLEMS

A Solve each problem.

1. Jerry is four times as old as Laura. In 20 years he will be twice as old as Laura. How old are Jerry and Laura?
2. Theresa is three times as old as her sister. In 10 years, she will be twice as old as her sister. How old is Theresa now?
3. Eva Santiago is now four times as old as her son. Four years ago she was ten times as old as her son. How old are Eva and her son now?
4. The sides of a triangle can be represented by three consecutive integers. The perimeter of the triangle is 129 meters. Find the length of each side.
5. Find two consecutive integers whose sum is -45.
6. Find three consecutive integers whose sum is 96.
7. Find three consecutive integers such that the sum of the first and third is 148.
8. The ages of three cousins are consecutive even integers such that the sum of the smallest and greatest integer is 28. How old is each of the cousins?
9. The length and width of a rectangular plot can be represented by consecutive integers. The perimeter of the plot is 38 meters. Find the length and width.
10. When twice the first of three consecutive odd integers is subtracted from three times the third integer, the result is 19. Find the integers.
11. The three sides of a triangular plane route between three cities can be represented by consecutive odd integers. The total distance around the route is 2019 kilometers. Find the distance between the cities.

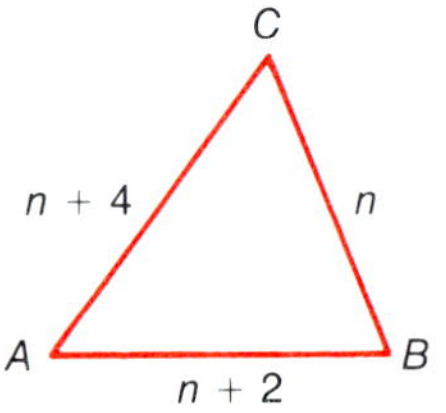

B

12. Tom Jackson is 44 and his son is 11. In t years, Tom will be exactly twice as old as his son. Find t.
13. Eileen who is 20 years old has a grandmother who is 65. How many years ago was the grandmother six times as old as Eileen?
14. The combined ages of two friends is 30 years. Four years ago, the age of one friend exceeded twice the age of the other by one year. Find the present age of each friend.
15. Are there three consecutive odd integers whose sum is 75? If so, find them.
16. Find three consecutive integers whose sum is -75.
17. Find three consecutive odd integers such that four times the smaller is 6 more than the sum of the first and third.
18. When twice the smallest of three consecutive integers is added to the three times the middle integer, the sum is three more than four times the largest. Find the integers.

19. Are there three consecutive odd integers whose sum is 100? If so, find them.

20. The street numbers of the homes of three friends are consecutive even integers. Three times the street number of one friend's home is 82 more than the sum of the street numbers of the homes of the other two. Find the street numbers of the homes for each friend.

21. The number of raffle tickets sold by two homerooms can be represented as consecutive odd integers. If each ticket was sold for \$0.50 and sales from both classes totaled \$752, find the number of tickets sold by each class.

Puzzle

You can use algebraic expressions to perform "mind-reading" tricks.

1. a. Think of a number.

b. Triple the number.

c. Add the original number.

d. Subtract 4.

e. Divide by 4.

f. Add 1.

g. Subtract the original number.

The answer will always be 0. Find how the trick works. (HINT: Use the language of algebra to trace the steps. Thus, **a.** is n, **b.** is $3n$, **c.** is $3n + n$, and so on.

2. a. Write the last two digits of your birth year.

b. Multiply the number by 100.

c. Add 10,000.

d. Add 50.

e. Add the number of the month you were born. (January is 1, February is 2, etc.)

f. Subtract 50.

The answer will indicate the year and the month of birth.
Find how the trick works.

REVIEW CAPSULE FOR SECTION 1-8

Name the postulate illustrated by each equation. *(Pages 14–16)*

1. $a + 8 = 8 + a$

2. $5 \cdot 1 = 5$

3. $4 + (-4) = 0$

4. $(6 \cdot 3) \cdot 4 = 6 \cdot (3 \cdot 4)$

5. $-a + 0 = -a$

6. $-(-a) + (-a) = 0$

7. $-2(a + 3) = -2 \cdot a + (-2) \cdot 3$

8. $6.6 \cdot (14.2 \cdot 9.6) = (6.6 \cdot 14.2) \cdot 9.6$

1–8 Theorems and Proof

The postulates given in Section 1–4 and the definitions found in the chapter state some of the basic properties of numbers. The following postulates, along with the substitution principle (see page 15), relate to the use of "=," the symbol for equality.

Postulates

Postulates of Equality

For all real numbers a, b, and c:

1. $a = a$.	**Reflexive Postulate**
2. If $a = b$, then $b = a$.	**Symmetric Postulate**
3. If $a = b$ and $b = c$, then $a = c$.	**Transitive Postulate**

You can use these postulates and definitions to determine other generalizations about numbers. These generalizations are called **theorems.** A theorem consists of two parts, the if–statement, or **hypothesis,** and the then–statement, or **conclusion.** The hypothesis is what you are given; the conclusion is what you are asked to prove. A **direct proof** uses a logical chain of steps to show that the conclusion follows from the hypothesis.

EXAMPLE 1 Prove the **Cancellation Property for Multiplication.**

If $ac = bc$ for real numbers a, b, c, where $c \neq 0$, then $a = b$.

Given: For real numbers a, b, c, $c \neq 0$, $ac = bc$.

Prove: $a = b$

Proof:

Statements	Reasons
1. $ac = bc$, where $a, b, c \in R$, and $c \neq 0$	**1.** Given (hypothesis)
2. $\frac{1}{c}$ is a real number.	**2.** Multiplicative inverse postulate
3. $ac\left(\frac{1}{c}\right) = bc\left(\frac{1}{c}\right)$	**3.** Multiplication property of equations
4. $a\left(c \cdot \frac{1}{c}\right) = b\left(c \cdot \frac{1}{c}\right)$	**4.** Associative postulate for multiplication
5. $a \cdot 1 = b \cdot 1$	**5.** Inverse postulate for multiplication
6. $\therefore a = b$	**6.** Identity postulate for multiplication

Note that each statement in a proof must be justified by a reason. The reason may be the hypothesis or a definition, a postulate, or a previously proved theorem whose hypothesis is known. Note also that the symbol "$\therefore$" in statement 6 of the proof means "therefore."

The following definitions are used to prove the theorem in Example 2.

Definitions

Even and Odd Numbers

An **even number** is a number of the form $2k$, where k is an integer.

An **odd number** is a number of the form $(2k + 1)$, where k is an integer.

EXAMPLE 2 Prove: If two odd numbers are added, their sum is an even number.

Given: a and b are odd numbers.

Prove: $(a + b)$ is an even number.

Proof:

Statements	Reasons
1. a and b are odd numbers.	1. Given
2. Let $a = 2k + 1$ and $b = 2s + 1$.	2. Definition of odd number
3. Then $a + b = 2k + 1 + 2s + 1$.	3. Addition property of equations
4. $2k + 1 + 2s + 1 = 2k + 2s + 2$	4. Commutative postulate; substitution principle
5. $2k + 2s + 2 = 2(k + s + 1)$	5. Distributive postulate
6. $k + s + 1$ is an integer.	6. Closure postulate for addition
7. $\therefore 2(k + s + 1)$ is an even number.	7. Definition of even number
8. $a + b$ is an even number.	8. Transitive postulate

CLASSROOM EXERCISES

Give the reason for each step of the proof of each theorem.

1. If $a + c = b + c$ for any real numbers a, b, and c, then $a = b$. **(Cancellation Property for Addition)**

Statements
1. $a + c = b + c$ for real numbers a, b, and c.
2. $(a + c) + (-c) = (b + c) + (-c)$
3. $a + [c + (-c)] = b + [c + (-c)]$
4. $a + 0 = b + 0$
5. $a = b$

2. If a, b, c, d, and e arc any real numbers, then $a(b + c + d + e) = ab + ac + ad + ac$.

Statements
1. a, b, c, d and e are real numbers.
2. $a(b + c + d + e) = a[(b + c) + (d + e)]$
3. $a[(b + c) + (d + e)] = a(b + c) + a(d + e)$
4. $a(b + c) + a(d + e) = ab + ac + ad + ae$
5. $a(b + c + d + e) = ab + ac + ad + ae$

WRITTEN EXERCISES

A Give the reason for each step of the proof of each theorem.

1. If a is a real number, $0 - a = -a$.

Statements

1. a is a real number.
2. $-a$ is a real number.
3. $0 - a = 0 + (-a)$
4. $0 + (-a) = (-a) + 0$
5. $(-a) + 0 = -a$
6. $0 - a = -a$

2. If a, b, and c are real numbers, and $a = b$ and $c = b$, then $a = c$.

Statements

1. a, b, and c are real numbers and $a = b$ and $c = b$.
2. $b = c$
3. $a = c$

3. If a, b, and c are real numbers, and $c \neq 0$, then $bc\left(\frac{1}{c}\right) = b$.

Statements

1. b and c are real numbers; $c \neq 0$
2. $\frac{1}{c}$ is a real number.
3. $bc\left(\frac{1}{c}\right) = b\left[c\left(\frac{1}{c}\right)\right]$
4. $b\left[c\left(\frac{1}{c}\right)\right] = b(1)$
5. $b(1) = b$
6. $bc\left(\frac{1}{c}\right) = b$

4. If a, b, and c are real numbers and $a - b = c$, then $a = c + b$.

Statements

1. a, b, c are real numbers; $a - b = c$
2. $a + (-b) = c$
3. $[a + (-b)] + b = c + b$
4. $a + [-b + b] = c + b$
5. $a + 0 = c + b$
6. $a = c + b$

5. If b and c are real numbers, then $(b + c) + (-c) = b$.

Statements

1. b and c are real numbers.
2. $b + c$ is a real number.
3. $-c$ is a real number.
4. $(b + c) + (-c)$ is a real number.
5. $(b + c) + (-c) = b + [c + (-c)]$
6. $(b + c) + (-c) = b + 0$
7. $b + 0 = b$
8. $(b + c) + (-c) = b$

6. If a and b are real numbers, and $b + a = a$, then $b = 0$.

Statements

1. a and b are real numbers; $b + a = a$
2. $a = 0 + a$
3. $b + a = 0 + a$
4. $-a$ is a real number.
5. $b + a + (-a) = 0 + a + (-a)$
6. $b + [a + (-a)] = 0 + [a + (-a)]$
7. $b + 0 = 0 + 0$
8. $b = 0$

7. If a is a real number, then $-(-a) = a$.

Statements

1. a is a real number.
2. $-a$ and $-(-a)$ are real numbers.
3. $-a + a = 0$
4. $-(-a) = -(-a) + 0$
5. $-(-a) = -(-a) + (-a + a)$
6. $-(-a) = [-(-a) + (-a)] + a$
7. $-(-a) + (-a) = 0$
8. $-(-a) = 0 + a$
9. $0 + a = a$
10. $-(-a) = a$

8. If a and b are real numbers and $ab = a$, where $a \neq 0$, then $b = 1$.

Statements

1. a and b are real numbers and $ab = a$, where $a \neq 0$
2. $\frac{1}{a}$ is a real number.
3. $\frac{1}{a}(ab) = \frac{1}{a} \cdot a$
4. $\frac{1}{a}(ab) = 1$
5. $\left(\frac{1}{a} \cdot a\right)b = 1$
6. $1 \cdot b = 1$
7. $b = 1$

9. If a and b are real numbers, then $-[a(-a) + (-b)] = a + b$.

Statements

1. a and b are real numbers.
2. $(-a)$ and $(-b)$ are real numbers.
3. $-[(-a) + (-b)] = -(-a) + [-(-b)]$
4. $-(-a) = a; -(-b) = b$
5. $-[(-a) + (-b)] = a + b$

10. If a and b are even numbers, then $(a + b)$ is an even number.

Statements

1. a and b are real numbers.
2. Let $a = 2k$ and $b = 2s$. Then $a + b = 2k + 2s$.
3. $2k + 2s = 2(k + s)$
4. $k + s$ is an integer.
5. $2(k + s)$ is an even integer.
6. $a + b$ is an even number.

B Prove each statement. All variables represent real numbers.

11. If an even number and an odd number are added, the sum is an odd number.
12. If $a = b$, then $\frac{1}{a} = \frac{1}{b}$, $a \neq 0$, $b \neq 0$.
13. If $a + c = b + d$ and $c = d$, then $a = b$.
14. If $a + b = 0$, then $a = (-b)$. (Uniqueness of additive inverse)
15. If $x + 2 = 9$, then $x = 7$.
16. If $x + (-5) = -2$, then $x = 3$.

C

17. If $x = a + (-b)$, then $x + b = a$.
18. If $a \div b = c$, then $a = c \cdot b$.
19. If $a = c \cdot b$, then $a \div b = c$.
20. If $-a = -b$, then $a = b$.
21. If $a = b$ and $c = d$, then $a + c = b + d$.
22. If $a = b$ and $c = d$, then $ac = bd$.

BASIC: FORMULAS IN PROGRAMS

A BASIC **program** consists of statements, each starting with a number—10, 20, 30, and so on. However, any sequence of increasing positive whole numbers may be used. Statements are usually numbered in multiples of 10 to allow for adding statements later. When all statements have been correctly entered into the computer, typing RUN causes the machine to *execute* (carry out) what the program tells it to do.

In BASIC the following symbols are used for operations.

Operation	BASIC **Symbol**
Addition	+
Subtraction	−
Multiplication	*
Division	/
Raising to a power	↑ or ∧

BASIC follows the usual order of operations: raising to a power first, then multiplication and division, then addition and subtraction, with parentheses used to change the order. The "*" must always be explicitly shown for multiplication, as in `0.80 * (L - S) + S`. Also, a BASIC formula must be typed on one line, with no raised exponents or fractions.

Algebraic Formula	**Formula in** BASIC
$y = 2x^2$	`LET Y = 2 * X↑2`
$k = (a + 7)(b - 3)$	`LET K = (A + 7) * (B - 3)`
$m = \dfrac{a - b}{c - d}$	`LET M = (A - B)/(C - D)`
$z = x^2 + 2x - 1$	`LET Z = X↑2 + 2 * X - 1`

Problem: *Given the list price of a car, L, and the shipping and dealer preparation charge, S, write a program which computes and prints the approximate dealer's cost, D. The formula for computing the dealer's cost is*

$$D = 0.80(L - S) + S.$$

Program:

```
10 REM L = LIST PRICE OF THE CAR
20 REM S = SHIPPING & DEALER PREPARATION CHARGE
30 REM D = DEALER'S COST
40 REM
50 PRINT "WHAT IS THE LIST PRICE OF THE CAR";
60 INPUT L
```

```
70 PRINT "WHAT IS THE SHIPPING & DEALER PREPARATION CHARGE";
80 INPUT S
90 LET D = 0.80 * (L - S) + S
100 PRINT "THE DEALER'S COST IS $"; D
110 PRINT "ANY MORE CARS (1 = YES, 0 = NO)";
120 INPUT X
130 IF X = 1 THEN 50
140 END
```

The formula for computing the dealer's cost (line 90)

This line tells the computer that there are no more lines in the program. (line 140)

Analysis: Here is an explanation of the statements in the program.

Statements 10–40: REM is short for "remark." REM statements are used to add explanatory comments to a program, not for the computer (which ignores REM statements when executing a program), but for people reading the program. Statement 40 is a "blank" remark separating the comments from the actual program which follows. REM statements may appear anywhere in a program.

Statements 50–60: During execution of a program, when the computer comes to a PRINT statement, it displays the message in quotation marks on the screen or on a printer. The purpose of statement 50 is to "prompt" the person running the program to enter the appropriate number. Statement 60 causes the computer to print a question mark (?) and then wait for a number to be entered from the keyboard. When the user has typed a number and pressed the ENTER or RETURN key (depending on the machine) the computer stores the value entered as variable L.

Statements 70–80: These statements repeat the work of statements 50 and 60, but for the variable S (shipping and dealer preparation charge).

Statement 100: Now that all values needed for the calculation have been entered, the LET statement gives the formula for computing the dealer's cost. (See page 4.) In a LET statement the left side of the "=" sign must be only a variable. The right side, however, may contain any legitimate algebraic expression.

Statement 110: The computer will print the value of D it computed in statement 90. Notice that the last D in statement 100 is outside the quotation marks, which causes the *value* of the variable D to be printed rather than the letter "D" itself.

Statements 110–140: Statement 110 asks the user if there is more data to be entered and gives a code for answering the question. Statement 120 accepts the 1 or 0 the person types and calls this value "X." Then, in statement 130, the computer checks the value of X. If X = 1, there is more data to be entered, and the computer is sent back to statement 50 to repeat the steps of the program. If X is not 1, the computer ignores "THEN 50" in statement 130 and instead moves to statement 140, END. At this point the computer would stop executing the program, print "READY" or a similar code word or symbol, and wait for further instructions.

EXERCISES

A Write each formula as a BASIC LET statement.

1. $y = 2x - 3$

2. $p = 2(l + w)$

3. $k = \dfrac{mn}{c + 2}$

4. $z = -6x^2$

5. $r = (4s - t)^3$

6. $t = (r + s)(x - 8)$

7. $s = \dfrac{x - a^3}{3y}$

8. $r = -7 + \dfrac{b}{c - d}$

9. $y = 2x^2 - 3x + 6$

Run the program on pages 34–35 for the following values of L and S.

10. $L = \$10{,}000$; $S = \$250$

11. $L = \$8500$; $S = \$200$

12. $L = \$11{,}200$; $S = \$335$

Write a BASIC program for each problem.

13. Given a bowler's average score, v, and actual score, s, use the formula

$$H = s + 0.8(200 - v)$$

to compute and print the bowler's handicap. (See page 4.)

14. A ball is thrown into the air at a speed of 96 feet per second. Given the time, t, that the ball has traveled, use the formula

$$d = 96t - 16t^2$$

to compute and print the distance the ball has traveled. (See page 5.)

15. Given the length and width of a rectangle, compute and print the perimeter of the rectangle. Use the following formula.

$$p = 2l + 2w$$

16. Given the number of wins and losses, compute and print a team's winning percentage. Assume that there are no ties. Use the following formula.

$$P = w \div (w + l) \cdot 100$$

17. Compute and print the average of three given real numbers. Use the following formula.

$$A = \frac{N1 + N2 + N3}{3}$$

18. Calculate and print a worker's gross weekly pay, given the rate per hour, r, overtime rate per hour, o, number of regular hours, h, and number of overtime hours worked in a week, n. Use the following formula.

$$P = (r \cdot h) + (o \cdot n)$$

19. Given the length of a side of a square, compute and print the perimeter and the area of the square. Use the following formulas.

$$p = 4s \qquad A = s^2$$

20. Input the radius of a circle. Then compute and print the circumference and the area of the circle. Use the following formulas. (NOTE: Usually there is no key for π on a computer keyboard. Use 3.14159 for π.)

$$C = 2\pi r \qquad A = \pi r^2$$

Review

Solve and check. *(Section 1–5)*

1. $0.3y - 2 = 1.6$
2. $-3p + 2 = -16$
3. $12.4x - 2 = 18.2x + 9.6$
4. $8(t - 4) = -56$
5. $4(y - 1) + 2 = 8$
6. $3p - 7 + 10p = 8p + 3$
7. The price of a bottle with a cork is \$1.10. The bottle costs one dollar more than the cork! Find the price of the cork. *(Section 1–6)*
8. The ages of Jake and Jeff are consecutive odd integers. The sum of their ages is 32. How old are Jake and Jeff? *(Section 1–7)*
9. Give the reason for each step of the proof of this theorem. *(Section 1–8)*

Theorem: If a is an even number and b is a odd number, then $(a + b)$ is an odd number.

Statements	Reasons
1. a is an even number and b is an odd number.	1. Given
2. Let $a = 2k$ and $b = 2s + 1$. Then $a + b = 2k + (2s + 1)$.	2. ?
3. $2k + (2s + 1) = (2k + 2s) + 1$	3. ?
4. $(2k + 2s) + 1 = 2(k + s) + 1$	4. ?
5. $k + s$ is an integer.	5. ?
6. $2(k + s) + 1$ is an odd integer.	6. ?
7. $a + b$ is an odd number.	7. ?

Chapter Summary

IMPORTANT TERMS

Absolute value (p. 10)
Additive inverse (p. 6)
Algebraic expression (p. 2)
Associative postulate (p. 14)
Axiom (p. 14)
Base (p. 2)
Closure postulate (p. 14)
Commutative postulate (p. 14)
Conclusion (p. 30)
Direct proof (p. 30)
Distributive postulate (p. 14)
Empty (null) set (p. 7)
Equivalent equations (p. 18)
Exponent (p. 2)
Hypothesis (p. 30)
Identity postulate (p. 14)
Integer (p. 6)
Intersection of sets (p. 7)
Inverse postulate (p. 14)
Irrational number (p. 7)
Natural number (p. 6)
Nonterminating repeating decimal (p. 6)
Open sentence (p. 18)
Opposite (p. 6)
Postulate (p. 14)
Rational number (p. 6)
Real number (p. 7)
Replacement set (p. 10)
Set-builder notation (p. 6)
Solution set (p. 18)
Terminating decimal (p. 6)
Theorem (p. 30)
Union of sets (p. 6)
Variable (p. 2)
Whole number (p. 6)

IMPORTANT IDEAS

1. **Order of Operations**
   [1] Evaluate within parentheses or other grouping symbols first.
   [2] Evaluate expressions with exponents.
   [3] Perform multiplication and division in order from left to right.
   [4] Perform addition and subtraction in order from left to right.

2. There is exactly one real number paired with each point on the number line. There is exactly one point on the number line that corresponds to each real number.

3. For all real numbers a and b, $a - b = a + (-b)$.

4. For all real numbers a and b, where $b \neq 0$, $a \div b = a \times \frac{1}{b}$.

5. **Substitution Principle:** For all real numbers a and b, where $a = b$, a may be replaced with b.

6. **Addition Property for Equations:** Adding the same real number to each side of an equation results in an equivalent equation.

7. **Multiplication Property of Equations:** Multiplying each side of an equation by the same nonzero real number results in an equivalent equation.

8. **Steps in Solving Word Problems**
   [1] Identify Condition 1. Use this condition to represent the unknowns.
   [2] Identify Condition 2. Use this condition to write an equation for the problem.
   [3] Solve. Check in the original conditions of the problem.
   [4] Answer the question.

9. **Postulates of Equality:** For all real numbers a, b, and c:

a. $a = a$	Reflexive Postulate
b. If $a = b$, then $b = a$.	Symmetric Postulate
c. If $a = b$ and $b = c$, then $a = c$.	Transitive Postulate

Chapter Objectives and Review

Objective: *To evaluate algebraic expressions using the rules for order of operations (Section 1–1)*

Evaluate $p^3 \div 3qr^2 - (p - r^2)$ for the given replacements for p, q, and r.

1. $p = 2,\ q = 20,\ r = 4$ **2.** $p = \frac{1}{3},\ q = \frac{1}{4},\ r = \frac{1}{2}$ **3.** $p = 0,\ q = 4,\ r = 5$

When a ball is thrown into the air at a speed of 96 feet per second, the velocity, r, at which it travels after t seconds is given by the formula

$$r = 96 - 32t.$$

Use this formula in Exercises 4–9 to find r for each value of t.

4. $t = 0$ **5.** $t = \frac{1}{2}$ **6.** $t = 1$ **7.** $t = 1\frac{1}{2}$ **8.** $t = 2$ **9.** $t = 3$

Objective: *To identify the subsets of real numbers and the relationships between them (Section 1–2)*

For each number in Exercises 10–16,

a. write a decimal for the number.

b. classify the number as a terminating decimal (T) or as a nonterminating, repeating decimal (NT).

10. $\frac{3}{13}$ **11.** $\frac{5}{6}$ **12.** $\frac{8}{9}$ **13.** $\frac{3}{7}$ **14.** $-4\frac{1}{5}$ **15.** $-\frac{5}{11}$ **16.** $-\frac{7}{12}$

Use the symbol ∈ to indicate the set or sets to which each number belongs.

17. -17 **18.** $\frac{1}{8}$ **19.** $\sqrt{3}$ **20.** $-\frac{\pi}{2}$ **21.** $0.\overline{51}$ **22.** -18 **23.** $\frac{0}{\pi}$

Objective: *To add, subtract, divide, and multiply using positive and negative numbers (Section 1–3)*

Add or subtract as indicated.

24. $-7 - 4$ **25.** $12 + (-14)$ **26.** $-6 - (-9)$ **27.** $-2 + 8$ **28.** $-1 + (-9)$

Multiply or divide as indicated.

29. $36(-\frac{1}{9})$ **30.** $-96 \div 4$ **31.** $-256 \div (-16)$ **32.** $(-7)(-\frac{1}{14})$ **33.** $-56(0.02)$

Evaluate each expression.

34. $(-3)^4(0.2) - (-42 \div 6)^2$ **35.** $[(-9)(4)(-\frac{1}{8}) + \frac{27}{-9}]^2$ **36.** $99 \div (15 - 4) + (-1)^9(7)$

Objective: *To use the postulates for real numbers (Section 1–4)*

Name the postulate illustrated by each statement.

37. $\frac{6}{7} \cdot \frac{7}{6} = 1$ **38.** $(8 \cdot 7)(-4) = 8[7 \cdot (-4)]$ **39.** $-6.2 + 6.2 = 6.2 + (-6.2)$

Perform the indicated operations.

40. $6y + 14y$ **41.** $-8t + 3t$ **42.** $-7(a + b)$ **43.** $-(x + 2) + 3x + 2$

Objective: *To use the properties for equations to solve equations (Section 1–5)*

Solve and check.

44. $-6x + 4 = 34$ **45.** $12m + 1 = 49$ **46.** $7p - 4 = 20 + 10p$

47. $2y + 1.7 = 3.5 + 4y$ **48.** $16c - (9 - 4c) = -49$ **49.** $7(-5x + 4) = 18x + 66 - 3x$

Objective: *To use equations to solve word problems (Section 1–6)*

50. Last year a watch company sold 12,000 more model A watches than model B watches. It sold 66,000 model A and B watches in all. How many model A watches and model B watches were sold?

51. Thirty-six yards of fencing are used to enclose two small gardens that are in the shape of an equilateral triangle and a rectangle respectively (see diagram). Find the length of one side of the triangular garden.

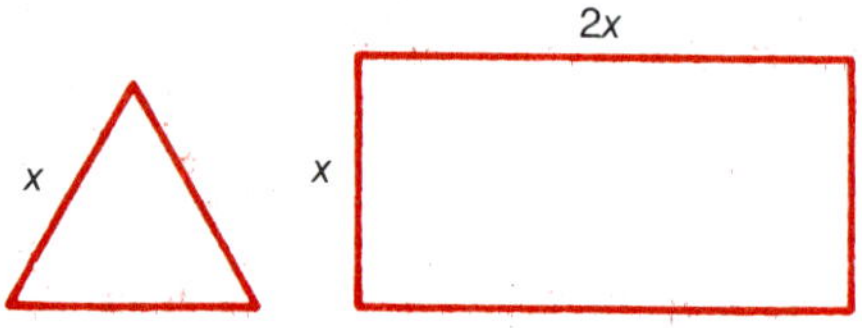

Objective: *To solve age problems and integer problems (Section 1–7)*

52. Alexandro is 6 years older than his wife, Carla. Sixteen years ago Alexandro was one and one-half times as old as Carla. How old are they now?

53. The sum of three consecutive even integers is -138. Find the integers.

Objective: *To prove theorems about real numbers (Section 1–8)*

54. Give a reason for each step of the proof of the following theorem. (HINT: As one of the reasons, use Classroom Exercise 1 on page 31.)

Theorem: If a is a real number, then $a \cdot 0 = 0$.

Statements	**Reasons**
1. $a \cdot 0 + 0 = a \cdot 0$	1. Identity postulate, addition
2. $a \cdot 0 = a \cdot (0 + 0)$	2. ?
3. $a \cdot (0 + 0) = a \cdot 0 + a \cdot 0$	3. ?
4. $a \cdot 0 + 0 = a \cdot 0 + a \cdot 0$	4. ?
5. $0 + a \cdot 0 = a \cdot 0 + a \cdot 0$	5. ?
6. $0 = a \cdot 0$	6. ?
7. $a \cdot 0 = 0$	7. ?

Chapter Test

Evaluate $a^2 - b(3 - c) + b^3$ for the given replacements for a, b, and c.

1. $a = 3,\ b = 6,\ c = 2$ **2.** $a = 0.2,\ b = 0.4,\ c = 1.2$ **3.** $a = 12,\ b = 0,\ c = 100$

Use the symbol $\in$ to indicate the set or sets to which each number belongs.

4. $\sqrt{615}$ **5.** $0.292929 \cdots$ **6.** -17.0000 **7.** $\pi(\sqrt{2} + 3)$

Evaluate each expression.

8. $(-1)^5 - (-8 - 4) - \frac{49}{-7}$ **9.** $56 \div (-8) + (-5)^3$ **10.** $(0.4)^2 - (0.1 - 0.7)^2$

Name the postulate illustrated by each statement.

11. $6(4+8) = 6 \cdot 4 + 6 \cdot 8$ **12.** $(2 \cdot 5)(-4) = -4(2 \cdot 5)$ **13.** $1 + 0 = 1$

14. Perform the indicated operations: $4y - (y - 8) + 7$

Solve and check.

15. $-2 + 5y = 38$ **16.** $9x - 7 = 13 - 4x$ **17.** $x + 14 - 5x = (2 - x)3$

18. A dining table with four chairs sells for $1670. The table alone sells for $230 more than the combined price of the chairs. Find the price of the table and of each chair.

19. Find three consecutive odd integers such that the sum of the smallest and largest is 162.

20. Give a reason for each step of the proof of the following theorem.

Theorem: $0 = -0$ ← **In words: Zero is its own opposite.**

Statements	Reasons
1. $0 + 0 = 0$	1. Identity postulate, addition
2. $0 = 0 + (-0)$	2. ?
3. $0 = -0 + 0$	3. ?
4. $0 + 0 = -0 + 0$	4. ?
5. $(0 + 0) + (-0) = (-0 + 0) + (-0)$	5. ?
6. $0 + [0 + (-0)] = -0 + [0 + (-0)]$	6. ?
7. $0 + 0 = -0 + 0$	7. ?
8. $0 = -0$	8. ?

Preparing for College Entrance Tests

Accuracy and **speed** are important in College Entrance tests. If you can recognize situations in which the basic postulates for the real numbers can be applied, it is often possible to reduce the amount of time needed to determine a correct answer.

REMEMBER: The fewer the computations, the less possibility there is for error!

EXAMPLE $\frac{9(17-4) - 5(17-4)}{4} = \underline{\ ?\ }$ **(a)** 4 **(b)** 13 **(c)** 17 **(d)** 68

Think: $9(17-4) - 5(17-4)$ is like $9x - 5x$, where $x = (17 - 4)$.

Solution: By the distributive postulate, $9x - 5x = (9 - 5)x = 4x$.

Thus, $\frac{9(17-4) - 5(17-4)}{4} = \frac{\overset{1}{\cancel{4}}(17-4)}{\underset{1}{\cancel{4}}} = 17 - 4$, or 13 **Answer: b**

Choose the best answer. Choose *a*, *b*, *c*, or *d*.

1. $\dfrac{12(16-11)-3(16-11)}{9} = \underline{\ ?\ }$

(a) 5 (b) 9 (c) 10 (d) 45

2. $\dfrac{6(8+9)+18(9+8)}{12} = \underline{\ ?\ }$

(a) 2 (b) 17 (c) 32 (d) 34

3. $\dfrac{9(\frac{1}{2}-\frac{1}{4})-6(\frac{1}{2}-\frac{1}{4})}{3} = \underline{\ ?\ }$

(a) $\frac{1}{4}$ (b) $\frac{1}{2}$ (c) $\frac{3}{4}$ (d) 3

4. $15(4-5)-16(4-5)+(4-5) = \underline{\ ?\ }$

(a) -1 (b) 0 (c) 1 (d) 2

5. $4(13+11)-6(13+11)+11+13 = \underline{\ ?\ }$

(a) -48 (b) -24 (c) -2 (d) 24

6. $-7(34-8)+8(34-8)-34+8 = \underline{\ ?\ }$

(a) -26 (b) -16 (c) 0 (d) 26

7. $\dfrac{35+35+35+35+35}{5} = \underline{\ ?\ }$

(a) 7 (b) 28 (c) 35 (d) 147

8. $\dfrac{\frac{2}{3}y+\frac{2}{3}y+\frac{2}{3}y}{3y} = \underline{\ ?\ }$

(a) $\frac{2}{3}$ (b) $\frac{2}{3}y$ (c) 1 (d) $\frac{3}{2}y$

9. $(1\frac{1}{2}-1\frac{3}{4})-(2\frac{1}{2}-1\frac{3}{4}) = \underline{\ ?\ }$

(a) -1 (b) $-\frac{1}{2}$ (c) 0 (d) $\frac{1}{2}$

10. $(3\frac{2}{3}n+1\frac{1}{2}n)-(1\frac{2}{3}n+\frac{1}{2}n) = \underline{\ ?\ }$

(a) $2\frac{5}{6}n$ (b) $3n$ (c) $3\frac{1}{6}n$ (d) $4n$

11. If $9 \cdot 9 \cdot 9 = 3 \cdot 3 \cdot t$, then $t = \underline{\ ?\ }$

(a) 9 (b) 27 (c) 81 (d) 243

12. If $16 \cdot 16 \cdot 16 = 8 \cdot 8 \cdot q$, then $q = \underline{\ ?\ }$

(a) 32 (b) 64 (c) 128 (d) 256

13. If $1 \cdot 2 \cdot 3 \cdot 4 \cdot 5 \cdot 6 \cdot 7 \cdot 8 \cdot 9 \cdot 10 \cdot 11 = 39{,}916{,}800$, then $12 \cdot 11 \cdot 10 \cdot 9 \cdot 8 \cdot 7 \cdot 6 \cdot 5 = \underline{\ ?\ }$

(a) 9,979,200 (b) 19,958,400 (c) 39,916,800 (d) 79,833,600

CHAPTER

2 Equations and Inequalities

2–1 Solving Equations for Variables

Some equations contain more than one variable. When solving such an equation for one variable in terms of the others, be careful to exclude values which could make a denominator equal 0. REMEMBER: *Division by zero is not defined.*

EXAMPLE 1 Solve $x(a-b)+d=2c$ for x. Indicate any restrictions on the values of the variables.

Solution:

$x(a-b)+d=2c$ ← **Add $(-d)$ to each side.**

$x(a-b)=2c-d$ ← **Multiply each side by $\frac{1}{a-b}$.**

$$x=\frac{2c-d}{a-b}$$

Since $a-b=0$ when $a=b$, $\mathbf{x=\frac{2c-d}{a-b},\ a\neq b}$.

Formulas are equations which show how certain quantities are related. It is sometimes useful to solve a formula for an unknown quantity in terms of the known quantities.

EXAMPLE 2 The formula $W=ft+r$ relates W, the weight of fuel carried by a jet, with f, the weight of fuel burned per number of hours, t, and r, the weight of reserve fuel.

a. Solve the formula for f.

b. Find f when $W=50{,}913$ pounds, $r=19{,}404$ pounds, and $t=2\frac{1}{4}$ hours.

Solutions:

a. $W=ft+r$ ← **Add $(-r)$ to each side.**

$W-r=ft$

$\frac{W-r}{t}=f$ or $f=\frac{W-r}{t}$, $t\neq 0$

b. $f=\frac{W-r}{t}$ ← **Replace W with 50,913, r with 19,404, and t with 2.25.**

$$f=\frac{50{,}913-19{,}404}{2.25}$$

$f=\frac{31{,}509}{2.25}$ $f=$ **14,004** The jet burns **14,004 pounds** of fuel per hour.

CLASSROOM EXERCISES

Solve each equation for y. Indicate any restrictions on the values of the variables.

1. $ey+f=g$

2. $\frac{y}{c}=\frac{9}{b-a}$

3. $1+(a+b)y=c-d$

Solve each formula for the indicated variable.

4. $V = lwh$, for h
5. $I = \frac{E}{R}$, for E
6. $A = p(1 + rt)$, for p
7. $S = 180(n - 2)$, for n
8. $C = \frac{E}{R_1 + R_2}$, for E
9. $A = \frac{1}{2}h(a + b)$, for h

WRITTEN EXERCISES

A Solve each equation for x in terms of the other variables. Indicate any restrictions on the values of the variables.

1. $x(a - b) = 2c$
2. $b + c = a - x$
3. $rtx = c$
4. $p = \frac{1}{2}x$
5. $rx + s = t$
6. $3x + 2a = x + 4a$
7. $mx + 9 = m^2$
8. $ax - 8 = a^2$
9. $(a - b)x = c$
10. $\frac{x}{a} - 1 = b$
11. $\frac{x}{b} + c = d - f$
12. $\frac{x}{a} = \frac{b}{c}$
13. $\frac{a}{x} = \frac{5}{e}$
14. $\frac{sx}{r} + a = b$
15. $\frac{f}{h} = \frac{9}{x}$
16. $\frac{cx}{a - b} = 2$
17. $\frac{r + s}{x} = b$
18. $(m - r) = \frac{t + 1}{x}$

Solve each formula for the indicated variable.

19. $s = \frac{1}{2}gt^2$, for g
20. $E = mc^2$, for m
21. $A = p(1 + rt)$, for r
22. $C = \frac{ka - b}{a}$, for k
23. $S = \frac{a}{1 - r}$, for a
24. $F = \frac{mu^2}{gr}$, for r
25. $V = \frac{1}{3}\pi r^2 h$, for h
26. $S = 2\pi r^2 + 2\pi rh$, for h
27. $S = vt - \frac{1}{2}gt^2$, for v
28. $\frac{PV}{n} = c$, for V
29. $S = \frac{n}{2}(a + l)$, for l
30. $V = \frac{4}{3}\pi r^3$, for r^3
31. $E^2 = \frac{JWhr}{t}$, for r
32. $R = \frac{Wh - x}{h}$, for W
33. $r = \frac{v^2 ph}{a}$, for p
34. $T = mg - mf$, for f
35. $R = \frac{EI}{M}$, for E
36. $C = \frac{E}{R_1 + R_2}$, for R_2

APPLICATIONS: Solving Equations for a Variable

The formula $\boldsymbol{W = I^2R}$ relates the number of watts, W, the current, I, in amperes, and the resistance, R, in ohms.

37. Solve the formula for R.
38. Find R when $W = 120$ watts and $I = \frac{1}{2}$ ampere.

The formula $\mathbf{c = 12.5a + 2000}$ relates the estimated cooling capacity, c, of an air conditioner and the area, a, of the space to be cooled.

39. Solve the formula for a.
40. Find a when $c = 9000$ Btu (British thermal units).

The formula $s = -16t^2 + vt$ relates the vertical height, s, of an object launched upward, the time, t, the object is in flight, and the velocity, v, of the object at launch.

41. Solve the formula for v.

42. Find v when $s = 300$ meters and $t = 8$ seconds.

The formula $S = \frac{1780Ad}{r}$ relates the greatest safe load, S, that a rectangular steel beam can bear to the cross–sectional area, A, of the beam, the depth, d, of the beam, and the distance, r, between the beam's supports.

43. Solve the formula for A.

44. Find A when $S = 12{,}816$ pounds, $d = 6$ inches, and $r = 30$ feet.

The formula $b = \frac{r^2}{30F}$ relates the estimated car's speed, r, and the coefficient of friction of the driving surface, F.

45. Solve the formula for F.

46. Find F when $b = 76$ feet and $r = 57$ miles per hour.

The formula $C = \frac{5}{9}(F - 32)$ relates the temperature in Celsius degrees, C, to the temperature in Fahrenheit degrees, F.

47. Solve the formula for F.

48. Find F when $C = 68°$.

The formula $A = \frac{1}{2}h(b_1 + b_2)$ relates the area of a trapezoid, A, the height h, and the two parallel bases, b_1 and b_2.

49. Solve the formula for b_2.

50. Find b_2 when $A = 46.8$, $h = 5.2$, and $b_1 = 8.1$.

The formula $s = \frac{1}{2}gt^2$ relates the distance, s, that a falling object travels, the acceleration due to gravity, g, and the length of time, t, that the object falls.

51. Solve the formula for g.

52. Find g when $s = 396.9$ meters and $t = 9$ seconds.

The formula $s = 180°(n - 2)$ relates the sum of the measures of the angles of a convex polygon, s, and the number of sides, n, of the polygon.

53. Solve the formula for n.

54. Find n when $s = 5580°$.

B

The formula $A = p(1 + rt)$ relates the total amount (principal plus interest), A, in a savings account to the principal, p, the rate of interest, r, and the number of years, t, that the principal has been earning interest.

55. Solve the formula for t.

56. Find t when $A = \$4392$, $p = \$3600$, and $r = 5\frac{1}{2}\%$.

The formula $\boldsymbol{H = 1.13A(i - o)}$ relates the rate of heat transfer, H, through a glass window, the area of the window, A, and the temperature inside, i, and outside, o.

57. Solve the formula for i.

58. Find i when $H = -153.68$ Btu/h, $A = 8$ square feet, and $o = 87°$F.

The formula $\boldsymbol{T = \frac{24I}{B(n+1)}}$ relates the approximate true rate of interest, T, the total finance charges (interest paid on borrowed money), I, the amount borrowed, B, and the number of monthly payments, n.

59. Solve the formula for I.

60. Find I when $T = 19.2\%$, $B = \$400$, and $n = 24$ months.

C The formula $\boldsymbol{D = (\frac{b}{2})^2(s)(n)}$ relates the displacement of an automobile engine, D, the diameter of the engine's bore, b, the engine's stroke, s, and the number of cylinders in the engine, n.

61. Solve the formula for s.

62. Find s when $D = 2562.24$ cubic centimeters, $b = 8$ centimeters, and $n = 6$ cylinders. (HINT: Use $\pi = 3.14$).

The formula $\boldsymbol{R = \frac{S + F + P}{S + P}}$ relates the mass ratio of a spacecraft, R, the vehicle's mass, S, the mass of the fuel, F, and the mass of the cargo (or payload), P.

63. Solve the formula for S.

64. Find S when $R = 18$, $F = 1{,}700{,}000$ kilograms, and $P = 25{,}000$ kilograms.

65. Prove: For all real numbers a, b, and c, where $a \neq 0$, if $ax + b = c$, then $x = \frac{c - b}{a}$.

66. Prove: For all real numbers a and b where $a \neq 0$, if $ax + b = ax$, then $b = 0$.

67. Prove: For all real numbers a and b where $a \neq 0$, if $ax + b = 0$, then $ax = -b$.

REVIEW CAPSULE FOR SECTION 2–2

Evaluate. *(Pages 10–13)*

1. $|3.8|$ **2.** $|-4|$ **3.** $-|-3|$ **4.** $-|18 - 25|$

Solve each equation. *(Pages 10–13)*

5. $|p| = 2$ **6.** $|b| = 12$ **7.** $|-k| = \frac{1}{2}$ **8.** $|q| = 0$

9. $|-z| = 3$ **10.** $|\frac{n}{2}| = 12$ **11.** $|2l| = 6$ **12.** $|3a| = 9$

2–2 Solving Absolute Value Equations

The solution of the equation

$$|x| = 2$$

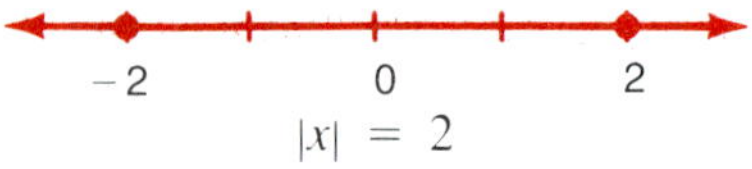

can be read from the graph at the right. The solutions are −2 and 2. That is, the equation

$$|x| = 2 \text{ is equivalent to } x = -2 \text{ \underline{or} } x = 2.$$

A sentence such as "$x = -2$ <u>or</u> $x = 2$" is a **compound sentence.** A compound sentence formed by joining two sentences with the word "or" is a **disjunction.** For a disjunction to be true, *at least* one of the sentences joined by "or" must be true.

To solve an equation involving absolute value, write the equation as an equivalent compound sentence. Then solve each part.

EXAMPLE 1 Solve and check: $|2x + 5| = 11$.

Solution: Write the equivalent compound sentence.

$2x + 5 = 11$	or	$2x + 5 = -11$
$2x = 6$		$2x = -16$
$x = 3$		$x = -8$

Check:

$|2x + 5| = 11$	← **Replace x with 3.**	$|2x + 5| = 11$	← **Replace x with −8.**
$|2(3) + 5| \stackrel{?}{=} 11$		$|2(-8) + 5| \stackrel{?}{=} 11$	
$|6 + 5| \stackrel{?}{=} 11$		$|-16 + 5| \stackrel{?}{=} 11$	
$11 = 11$ Yes ✓		$11 = 11$ Yes ✓	

Solution set: $\{-8, 3\}$

For an equation such as $3|6m| - 8 = 1$, be sure to get the absolute value expression alone on one side of the equation before writing the equivalent compound sentence.

EXAMPLE 2 Solve and check: $3|6m| - 8 = 1$.

Solution:

$3|6m| - 8 = 1$ ← **Add 8 to each side.**

$3|6m| = 9$ ← **Multiply each side by $\frac{1}{3}$.**

$|6m| = 3$ ← **Write an equivalent compound sentence.**

$6m = 3$	or	$6m = -3$
$m = \frac{1}{2}$		$m = -\frac{1}{2}$

The check is left for you. **Solution set: $\left\{-\frac{1}{2}, \frac{1}{2}\right\}$**

Sometimes one or more of the answers obtained as solutions to an equation do not check in the original equation. Such answers are only **apparent solutions** (not solutions at all). Thus, it is important to check all answers in the *original* equation.

EXAMPLE 3 Solve and check: $|3a + 14| = 5a + 2$.

Solution:

$$3a + 14 = 5a + 2 \qquad \text{or} \qquad 3a + 14 = -(5a + 2)$$
$$14 = 2a + 2 \qquad\qquad 3a + 14 = -5a - 2$$
$$12 = 2a \qquad\qquad 16 = -8a$$
$$6 = a \qquad\qquad -2 = a$$

Check:

$$|3a + 14| = 5a + 2 \qquad\qquad |3a + 14| = 5a + 2$$
$$|3(6) + 14| \stackrel{?}{=} 5(6) + 2 \qquad\qquad |3(-2) + 14| \stackrel{?}{=} 5(-2) + 2$$
$$|18 + 14| \stackrel{?}{=} 30 + 2 \qquad\qquad |-6 + 14| \stackrel{?}{=} -10 + 2$$
$$|32| \stackrel{?}{=} 32 \quad \text{Yes} \checkmark \qquad\qquad |8| \stackrel{?}{=} -8 \quad \text{No}$$

Solution set: {6}

CLASSROOM EXERCISES

Write an equivalent compound sentence for each absolute value equation. Do *not* solve the equation.

1. $|x| = 9$ **2.** $|3y| = 12$ **3.** $|n + 7| = 3$ **4.** $|n - 4| = 2$

5. $2|7t| = 70$ **6.** $|4p - 3| = 13$ **7.** $8|6d| + 4 = 16$ **8.** $|2k - 1| = 4k + 6$

Solve and check.

9. $|y - 1| = 5$ **10.** $4|t| = 20$ **11.** $2|r + 4| = 16$ **12.** $|2m - 1| = 5$

13. $|\frac{1}{2}k + 1| = 8$ **14.** $7|8p| - 6 = 15$ **15.** $4|2s| + 8 = 12$ **16.** $12|3y| - 3 = 6$

17. $|2s - 1| = 3s$ **18.** $|5c + 3| = 4c$ **19.** $|3p - 3| = 7p + 10$ **20.** $|6q + 5| = 2q - 8$

WRITTEN EXERCISES

A Solve and check.

1. $|-4k| = 8$ **2.** $|6p| = -24$ **3.** $-|3w| = 5$

4. $|x + 6| = 2$ **5.** $|3y - 5| = 4$ **6.** $|8 - 2t| = 6$

7. $|\frac{1}{3}(6k - 2)| = 6$ **8.** $|3(4m - 5)| = 30$ **9.** $|2w| + 7 = 4$

10. $|4x| - 3 = \frac{2}{3}$ **11.** $2|4s| + 6 = 14$ **12.** $-5|3d| - 2 = 10$

13. $|\frac{1}{2}(k + 2)| - 8 = 4$ **14.** $|3(p - 1)| - 6 = 8$ **15.** $\frac{1}{2}|3r + 2| - 4 = 11$

16. $4|4p - 2| + 1 = 13$ **17.** $|2w + 7| = 3w$ **18.** $|4k + 9| = 7k$

19. $|3x + 10| = 5x + 6$ **20.** $|5c - 2| = 2c + 7$ **21.** $|3(2n + 3)| = \frac{1}{2}n$

22. $|2(3y - 2)| = 3y - 8$ **23.** $|2x - 15 + x| = 2x$ **24.** $|8f + 4 + 3f| = 5f - 3$

B List the elements in each set. Then graph the solution set. (See page 6 for a reminder on how to read set-builder notation.)

25. $\{n : |\frac{1}{3}n + 4| = 8\}$
26. $\{y : |2y - 3| = 9\}$
27. $\{k : |8k + 6| = -5\}$
28. $\{r : |\frac{1}{5}r - 7| = 1\frac{1}{2}\}$
29. $\{b : 6|4b| + 3 = 6\}$
30. $\{f : \frac{1}{2}|9f| - 7 = 11\}$
31. $\{q : 4|5q + 6| = 12\}$
32. $\{z : \frac{1}{8}|3z + 6| - 2 = 10\}$
33. $\{y : |7y - 3| = 4y + 6\}$
34. $\{t : |4t - \frac{1}{2}| = 7t + \frac{3}{4}\}$
35. $\{a : |2a - 12 + 5a| = 6a\}$
36. $\{p : |5(3p + 6)| = p - 1\}$
37. What is the value of $\frac{x}{|x|}$ for any positive real number?
38. What is the value of $\frac{x}{|x|}$ for any negative real number?

C Use the definition of absolute value and the separate cases $x \geq 0$ and $x < 0$ to prove these theorems for all real numbers, x.

39. Theorem: $|x| = |-x|$
40. Theorem: $|x|^2 = x^2$

Solve and check.

41. $|y - 1| = |2y - 3|$
42. $|g - 1| = |2g - 8|$
43. $|m - 1| = -|3m + 1|$

Review

Solve each equation for x in terms of the other variables. Indicate any restrictions on the values of the variables. *(Section 2–1)*

1. $t = 3x$
2. $-cx = b$
3. $(m + n)x = 4p$
4. $\frac{3}{y} = \frac{4}{x}$
5. $\frac{t}{x} = 7$
6. $\frac{4x}{m} - p = c$

Solve each formula for the indicated variable. *(Section 2–1)*

7. $pV = nRT$, for R
8. $A = 4\pi r^2$, for r^2
9. $\frac{R_1 + R_2}{R_1R_2} = \frac{1}{R}$, for R

Solve and check. *(Section 2–2)*

10. $|5 - 3y| = 25$
11. $4|3x| - 2 = 10$
12. $|6a - 4| = 3a + 2$

REVIEW CAPSULE FOR SECTION 2–3

Graph each inequality on the real-number line.

1. $p > 4$
2. $p < 0$
3. $q > \frac{1}{2}$
4. $n < -2.5$
5. $r \leq -1$
6. $k \geq 6\frac{1}{3}$
7. $z \leq -0.5$
8. $s \geq -7\frac{2}{3}$

2–3 Solving Inequalities

An **inequality** is a mathematical sentence that uses $>$ (greater than), $<$ (less than), or $\neq$ (is not equal to). The sentences below are inequalities.

$$\mathbf{13 + t > 48} \qquad \mathbf{12q - 15 < 9} \qquad \mathbf{7t + 8 \neq 20}$$

Properties similar to those used to solve equations can be used to solve inequalities. Adding the same positive or negative number to each side of an inequality results in an equivalent inequality. **Equivalent inequalities** have the same solution(s).

Addition Property

> For all real numbers a, b, and c,
>
> **if $a < b$, then $a + c < b + c$.**

Multiplying each side of an inequality by the same positive number results in an equivalent inequality.

Multiplication Property ($c > 0$)

> For all real numbers a, b, and c,
>
> **if $a < b$ and $c > 0$, then $ac < bc$.**

Example 1 shows how to apply both these properties.

EXAMPLE 1 Solve $2x + 3 < 7$. Graph the solution set.

Solution:

$2x + 3 < 7$ ← **Add (−3) to each side.**

$2x + 3 + (-3) < 7 + (-3)$

$2x < 4$ ← **Multiply each side by $\frac{1}{2}$.**

$\frac{1}{2}(2x) < \frac{1}{2}(4)$

$x < 2$

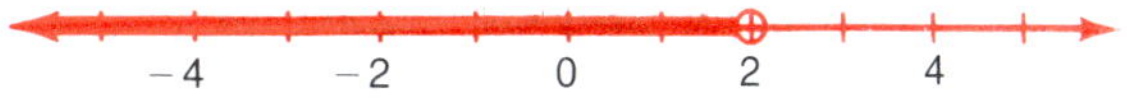

Solution set: {real numbers less than 2}

You can use set-builder notation to write the solution to Example 1 as

$$\{x : x < 2, x \in R\}.$$

The following table shows that multiplying each side of an inequality by a negative number *reverses the order of the inequality.*

Inequality	Multiply by 2.	Multiply by −2.	Multiply by −3.
$5 < 7$	$10 < 14$	$-10 > -14$	$-15 > -21$
$-1 < 0$	$-2 < 0$	$2 > 0$	$3 > 0$
$-9 < -8$	$-18 < -16$	$18 > 16$	$27 > 24$

The table suggests this property.

Multiplication Property ($c < 0$)

> For all real numbers a, b, and c,
> **if $a < b$ and $c < 0$, then $ac > bc$.**

Example 2 shows how to apply this property.

EXAMPLE 2 Solve $5 + 4(x - 1) > 7x + 13$. Graph the solution set.

Solution:

$$5 + 4(x - 1) > 7x + 13$$
$$5 + 4x - 4 > 7x + 13$$
$$4x + 1 > 7x + 13 \quad \longleftarrow \text{Add } (-7x) \text{ to each side.}$$
$$-3x + 1 > 13 \quad \longleftarrow \text{Add } (-1) \text{ to each side.}$$
$$-3x > 12 \quad \longleftarrow \text{Multiply each side by } \left(-\tfrac{1}{3}\right). \text{ Reverse the inequality.}$$
$$\left(-\frac{1}{3}\right)(-3x) < \left(-\frac{1}{3}\right)(12)$$
$$x < -4$$

Solution set: {real numbers less than -4}

The symbol, $\leq$, means "less than or equal to." The symbol, $\geq$, means "greater than or equal to."

EXAMPLE 3 Solve $4x - 10.3 \leq 21x - 1.8$. Graph the solution set.

Solution:

$$4x - 10.3 \leq 21x - 1.8$$
$$-21x + 4x - 10.3 \leq -21x + 21x - 1.8$$
$$-17x - 10.3 \leq -1.8$$
$$-17x - 10.3 + 10.3 \leq -1.8 + 10.3$$
$$-17x \leq 8.5$$
$$-\frac{1}{17}(-17x) \geq -\frac{1}{17}(8.5) \quad \longleftarrow \text{Multiply each side by } \left(-\tfrac{1}{17}\right). \text{ Reverse the inequality.}$$
$$x \geq -0.5$$

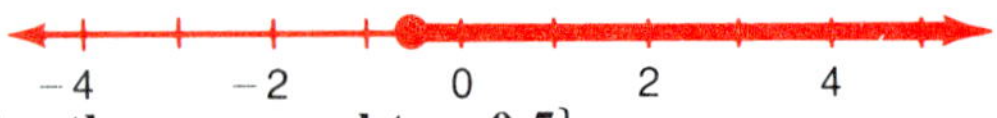

Solution set: {real numbers greater than or equal to -0.5}.

CLASSROOM EXERCISES

Write an equivalent inequality as the first step in solving each inequality.

1. $n + 4 > 16$	**2.** $p - 3 < 12$	**3.** $r - \frac{1}{2} < -2\frac{1}{2}$	**4.** $c + 2.3 > 1.6$
5. $3t + 5 > 11$	**6.** $4n - 1 > -4$	**7.** $7a + \frac{2}{3} < 10$	**8.** $4h - 7 < -2.2$

Write the words *Same* or *Reverse* to describe the direction of an inequality *after* each given operation.

9. Add -6. **10.** Multiply by 2. **11.** Multiply by $-\frac{1}{3}$. **12.** Multiply by -1.

State which procedure was used to obtain the second inequality from the first.

13. $x - 10 < 12;\ x < 22$
14. $y + 8 > 0;\ y > -8$
15. $-5t > 9;\ t < -\frac{9}{5}$
16. $\frac{2}{3}p > 18;\ p > 27$
17. $-t < -6;\ t > 6$
18. $\frac{s}{-2} < 0;\ s > 0$

WRITTEN EXERCISES

A Solve each inequality. Graph its solution set.

1. $s - 2 > 10$
2. $y + 5 < -3$
3. $3m < -9$
4. $\frac{1}{2}d > -16$
5. $3t - 16 > 4$
6. $5a + 7 < -2$
7. $2m + 4 < 6$
8. $9w - 7 > -25$
9. $-4p + 1 > 17$
10. $5 - 3k > 4\frac{1}{2}$
11. $-2(q + 4) < -8$
12. $5(6 - n) < 12$
13. $4s + 3 < 6s - 9$
14. $2p - 4 > 7p + 16$
15. $-3(w + 4) < -w - 9$
16. $5(k - \frac{1}{2}) > 8k + 2\frac{1}{4}$
17. $3q + 4 \geq 10$
18. $8a - 9 \leq -11$
19. $-6(p + 4) \leq 12$
20. $\frac{1}{3}(8s - 4) \leq 5$
21. $3c - 2.5 \geq 5c + 0.9$
22. $9k + 7 \geq 3k - 4$
23. $-4c - 5 \geq 7(4c - 3)$
24. $2(y - 1) - 4 \leq 8y$

B

25. $(4 - 2t) - (4 - t) < (2t - 1) + (t - 3)$
26. $7(2n + 5) - 6(n + 8) > 7$
27. $w - 2(5 - 2w) \geq 6w - 3\frac{1}{2}$
28. $6x^2 - 8 > 2x(3x - 4)$
29. $7 - [4 - (3y - 2)] < 13 + 3y$
30. $2(\frac{1}{2}p - 4) - 5p \leq -3p - \frac{1}{4}(6p - 10)$
31. $2\frac{2}{3}n + \frac{2}{3}(5n - 4) > -\frac{1}{3}(8n + 7)$
32. $q(q + 6.3) + 0.5 \leq 1.8q - 8.5 + q^2$

Solve each inequality for the given replacement set.

33. $\frac{p}{2} - 1 < 3 - p,\ p \in N$
34. $\frac{y - 3}{y} < 1,\ y \in R$

C

35. $1 \leq \frac{2 - 3t}{3} < 4,\ t \in Q$
36. $3 - a < \frac{a}{2} - 1 \leq 10,\ a \in R$

Graph each solution set. REMEMBER: *The **union** of sets A and B (A ∪ B) consists of all the elements in set A or set B or in both. The **intersection** of sets A and B (A ∩ B) consists of all those elements common to sets A and B.*

37. $\{u : u \leq 5\} \cap \{u : u > 5\}$
38. $\{q : q \leq -1\} \cap \{q : q \leq 4\}$
39. $\{x : x < -2\} \cup \{x : x > 2\}$
40. $\{s : s > 0\} \cup \{s : s \geq 2\}$
41. $\{x : 2x + 1 < 5\} \cup \{x : 2x - 1 > 9\}$
42. $\{x : 5x - 2 < 3\} \cup \{x : 2x - 6 < 4\}$
43. $\{x : 3x - 5 \leq 4\} \cap \{x : -2x + 2 \leq -4\}$
44. $\{x : x - 1 \geq 0\} \cap \{x : 2x + 5 \leq 7\}$

Problem Solving and Applications

2–4 Problems Involving Inequalities

The following table shows some key words that indicate which inequality symbol to use in writing certain algebraic expressions.

Key Words	Meaning	Algebraic Expression
Fred's taxable income, i, is *at least* \$27,500.	$i = 27{,}500$ <u>or</u> $i > 27{,}500$	$i \geq 27{,}500$
The *minimum* fee, f, for a mailing list correction is \$1.00.	$f = 1.00$ <u>or</u> $f > 1.00$	$f \geq 1.00$
The population, p, of Lee City, is *at most* 2,500,000.	$p = 2{,}500{,}000$ <u>or</u> $p < 2{,}500{,}000$	$p \leq 2{,}500{,}000$
The *maximum* speed, s, is 55 miles per hour.	$s = 55$ or $s < 55$	$s \leq 55$
Jane's earnings, e, are *at least* \$28,500 and *at most*, \$30,000.	$28{,}500 \leq e$ <u>and</u> $e \leq 30{,}000$	$28{,}500 \leq e \leq 30{,}000$
Temperatures, t, today will be *between* 20°C and 25°C.	$20 < t$ and $t < 25$	$20 < t < 25$

To solve word problems involving inequalities, follow the steps of the summary on page 23.

EXAMPLE 1 A hospital aide society wishes to raise *at least* \$75,000 by selling raffle tickets (Condition 2). Each ticket will cost \$3.25 (Condition 1) and total expenses will amount to \$3000 (Condition 2). How many tickets must be sold in order to raise *at least* \$75,000?

Solution: [1] Use Condition 1 to represent the unknowns.

Let $t =$ the number of tickets sold.

Then $3.25t =$ the value of tickets sold (total sales).

[2] Use Condition 2 to write an inequality.

$3.25t - 3000 \geq 75{,}000$ ← **Total Sales − Expenses ≥ 75,000**

[3] Solve for t.

$$3.25t - 3000 \geq 75{,}000$$
$$3.25t \geq 78{,}000$$
$$t \geq 24{,}000$$

At least **24,000 tickets** must be sold.

Recall the definition of **average** or **mean.**

$$\text{Average} = \frac{\text{Sum of scores}}{\text{Number of scores}}$$

EXAMPLE 2 A bowler's scores in five successive games were 145, 133, 152, 160, and 148 (Condition 2). What must be the score on the next game (Condition 1) to obtain an average score between 150 and 155 (Condition 2)?

Solution:

[1] Let t = the bowler's score on the sixth game.

[2] **Minimum Average Score** < **Bowler's Average Score** < **Maximum Average Score**

$$150 < \frac{t + 145 + 133 + 152 + 160 + 148}{6} < 155$$

[3] Express the inequality with "and." Then solve.

$150 < \frac{t + 738}{6}$	and	$\frac{t + 738}{6} < 155$
$900 < t + 738$		$t + 738 < 930$
$162 < t$		$t < 192$

The bowler's score must be **between 162 and 192**; that is, $\mathbf{162 < t < 192}$.

To check the inequality in Example 2, choose several numbers between 162 and 192. Check whether each number satisfies the inequality by replacing t in the inequality with that number.

CLASSROOM EXERCISES

Write an algebraic expression for each word expression.

1. Wind speeds, w, will be *at least* 29 kilometers per hour.
2. The length, l, of an adult fin whale, is *at most* 24 meters.
3. Projected sales, s, for the month of October, will be *between* \$2,500,000 and \$3,125,000.
4. The population, p, of Scarwick is estimated to be *at most* 675,000.
5. The body length, l, of an adult rock wren is *between* 13 and 15 centimeters.
6. The *maximum* amount, a, for a COD collection charge is \$400.
7. The *minimum* speed, s, is 30 miles per hour.
8. For a specified fee, the amount, a, of a money order must be *at least* \$50.01 and *at most* \$500.
9. The amount, m, of the deposit is *at least* \$500 and *at most* \$10,000.

WORD PROBLEMS

A Solve each problem. For some problems, Condition 1 is underscored once and Condition 2 is underscored twice.

1. A Theater Club wishes to raise at least $85,000 by giving a benefit performance. Tickets for the performance will cost $4.50 each. Expenses will amount to $5000. How many tickets must be sold?
2. A company plans to spend at most $40,000 in an advertising campaign. The plan calls for $28,000 to be spent on TV commercials and the remainder on T-shirts which will be given away. The shirts will cost the company $2.00 each. What is the greatest number of T-shirts that can be given away?
3. A bicycle shop makes a profit of $75 on each bicycle sold. Overhead expenses for one week amount to $3450. How many bicycles must be sold in one week to make a profit of at least $3000?
4. Attendance at the first two games of a season was 1200 and 1500. What must be the attendance at the next game so that the average attendance for the three games will be between 1400 and 1500?
5. A batter's average for one year was .225. His average for the second year was .231. What must be his average for the following year if he wishes to have a minimum three-year average of .230?
6. A student scored 78, 81, and 83 on three tests. What score must the student achieve on the next test in order to have an average of at least 82?
7. The length of a pool is 2 meters more than twice the width. The perimeter of the pool is at most 70 meters. Find the maximum length of the pool.

B

8. Alma Jones has exactly 168 feet of fencing with which to enclose a vegetable garden and to separate it into three equal parts. The maximum length of the garden is 32 feet. Find the maximum width of each section.
9. Karl Snyder left an estate with an estimated worth of at most $225,000. One-third of the estate was left to a favorite charity and the rest was to be divided equally among 6 grandchildren. What are the maximum amounts to be paid to the charity and to each of the grandchildren?
10. An assembly line worker at a sneaker factory can box an average of 126 to 148 pairs of sneakers per hour. Carol estimates that if she boxed 50 less than twice her present rate she would be working at the average rate. What is Carol's rate now?

C

11. The formula $M = \frac{(n-2)180}{n}$ relates M, the degree measure of each interior angle of a regular polygon, to n, the number of sides of the polygon. Each interior angle of a regular polygon is at least 165°. Find the minimum number of sides of the polygon.

REVIEW CAPSULE FOR SECTION 2–5

Solve and check. *(Pages 48–50)*

1. $|6a| = 30$ **2.** $2|y| = 14$ **3.** $|2x - 5| = 19$ **4.** $|4r + 5| = 25$

5. $|6x + 5| = 5x$ **6.** $\frac{1}{2}|6y| - 5 = 7$ **7.** $|2(3m - 6)| = 30$ **8.** $-4|2b| - 8 = 12$

2–5 Solving Absolute Value Inequalities

Inequalities sometimes include the absolute value of an expression involving the variable. To solve such inequalities, recall that solving equations with absolute value involves writing an equivalent compound sentence. For example,

$|x| = 4$ means $x = -4$ or $x = 4$.

$|x| > 4$ means $x < -4$ or $x > 4$.

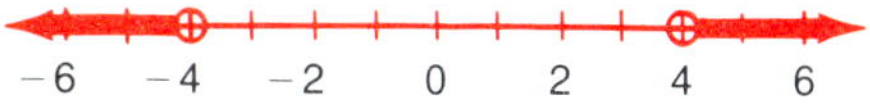

EXAMPLE 1 Solve and graph: $|3y - 5| > 1$.

Solution: $|3y - 5| > 1$ means $(3y - 5) < -1$ or $(3y - 5) > 1$.

$3y - 5 < -1$	or	$3y - 5 > 1$
$3y < 4$		$3y > 6$
$y < \frac{4}{3}$		$y > 2$

Solution set: $\left\{y : y < \frac{4}{3} \text{ or } y > 2\right\}$

The graph at the right shows that the distance of x from 0 is less than 3. That is, $|x| < 3$ means

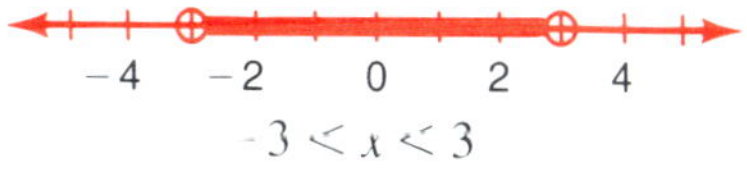

$x > -3$ and $x < 3$. ← **Compound sentence**

This is also written as $-3 < x < 3$.

EXAMPLE 2 Solve and graph: $|2n - 3| < 7$.

Solution: $|2n - 3| < 7$ means that $(2n - 3) > -7$ and $(2n - 3) < 7$.

$2n - 3 > -7$	and	$2n - 3 < 7$
$2n > -4$		$2n < 10$
$n > -2$		$n < 5$

Solution set: $\{n : -2 < n < 5\}$

Recall from Section 2–2 that a compound sentence formed by joining two sentences with the word "or" is a *disjunction*. A compound sentence such as $x > -3$ and $x < 3$ which is formed by joining two sentences with the word "and" is a **conjunction.** For a conjunction to be true, both of the sentences joined by "and" must be true.

EXAMPLE 3 Solve and graph: $\frac{1}{2}|t+9| \le 3$.

Solution: $\frac{1}{2}|t+9| \le 3$ ⟵ **Multiply each side by 2.**

$|t+9| \le 6$ ⟵ **Write the equivalent compound sentence.**

$|t+9| \le 6$ means $(t+9) \ge -6$ and $(t+9) \le 6$

$t+9 \ge -6$ and $t+9 \le 6$

$t \ge -15$ | $t \le -3$

Solution set: $\{t : -15 \le t \le -3\}$

CLASSROOM EXERCISES

Write an equivalent compound sentence for each absolute value inequality.

1. $|y| > 6$
2. $|z| < 7$
3. $|2x-1| > 5$
4. $|5x+2| < 17$
5. $\left|\frac{z}{3}\right| < 9$
6. $|n-6| > 2$
7. $|t-3| \le 5$
8. $|2y+3| \ge 5$

Match each open sentence with its corresponding graph.

9. $x < -1$ or $x > 3$
10. $x > -3$ and $x \le 1$
11. $0 < x < 3$
12. $x \le -2$ or $x > 2$

a.

b.

c.

d.

e.

WRITTEN EXERCISES

A Write an equivalent compound sentence for each inequality. Then label each sentence as a disjunction, *D*, or as a conjunction, *C*.

1. $|x| < 4$
2. $|y| > 2$
3. $|c-3| \le 5$
4. $|n-6| \ge 8$
5. $|2t-3| \ge 7$
6. $|q+7| \le 3$
7. $|2a+3| \le 1$
8. $|3c-7| \ge 2$

Solve and graph. The replacement set is the set of real numbers.

9. $|n| > 6$ **10.** $|y| < 8$ **11.** $|x-3| < 1$ **12.** $|t+4| > 2$

13. $|6-m| < 4$ **14.** $|4x+3| > 6$ **15.** $|2a+3| \leq 1$ **16.** $|3c-7| \geq 2$

17. $|5-2t| \leq 3$ **18.** $|3-2y| > 10$ **19.** $2|x+4| > 10$ **20.** $3|2x-4| < 20$

B

21. $|\frac{2}{3}d-1| > 5$ **22.** $|1-(3-t)| < 2$ **23.** $|4-(m-1)| \leq 7$ **24.** $|-2-5w| \geq 8$

25. $|3n+5| < 4$ **26.** $|6-a| < 2a$ **27.** $|2c-1| \leq 3-c$ **28.** $|3w+4| \geq w+7$

29. $\frac{3}{5}|y-6| \leq 9-\frac{2}{5}y$ **30.** $2|5b|-8 < 22$ **31.** $\left|\frac{3a-9}{2}\right| \leq 6$ **32.** $\left|\frac{p+8}{4}\right| > 2$

33. $|-3x-1| < 3$ **34.** $|-2x+1| > 3$ **35.** $|-\frac{1}{3}x+1| \leq 2$ **36.** $|-\frac{2}{5}x-3| \geq 1$

Use absolute value notation to describe each set.

37. $\{t : -3 < t < 3\}$ **38.** $\{n : n \leq -6\}$ and $\{n : n \leq 6\}$

39. $\{a : a < -4\}$ or $\{a : a > 4\}$ **40.** $\{c : (3c+4) \leq -1\}$ or $\{c : (3c+4) \geq 1$

C Solve and graph. The replacement set is the set of real numbers.

41. $1 \leq |x| < 4$ **42.** $5 < |w| \leq 8$ **43.** $3 < |x-2| < 5$

44. $1 \leq |x-4| \leq 6$ **45.** $5 < |2y-3| < 7$ **46.** $6 \leq |18-3y| \leq 10$

Puzzle

In the figure at the right, find six dots such that no two of these six lie on the same line segment.

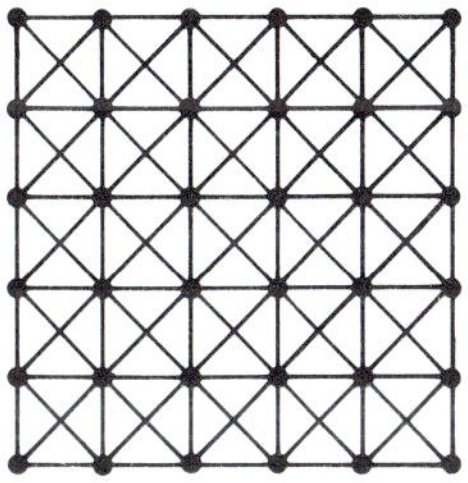

REVIEW CAPSULE FOR SECTION 2–6

State the postulate or property illustrated by each statement. *(Pages 14–16, 18–21, 51–53)*

1. $3+(-3)=0$ **2.** If $-1 < a$, then $-1+4 < a+4$.

3. If $-6 < -1$, $(-2)(-6) > (-2)(-1)$. **4.** If $0 < 1$, then $5(0) < 5(1)$.

5. If $1+8=9$ and $9=6+3$, then $1+8=6+3$.

6. $6[5(-4)] = (6 \cdot 5)(-4)$

7. If x is a real number, then $-x$ is a real number.

8. If $x-9=-2$, then $x-9+9=-2+9$.

Using Statistics: Tables and Formulas

People in many careers gather and analyze statistics for use in decision making or for reading tables to find new information. Two of the statistics that **meteorologists** gather and analyze are the seasonal temperatures and wind speeds of different regions of the country. A combination of low temperatures and high winds can make you feel colder than the actual temperature. The temperature you feel is called the **wind chill temperature.** Meteorologists use the table below to obtain the wind chill temperatures from the data they have gathered.

Winds in MPH	Temperatures in °F									
	35	30	25	20	15	10	5	0	−5	−10
5	33	27	21	19	12	7	0	−5	−10	−15
10	22	16	10	3	−3	−9	−15	−22	−27	−34
15	16	9	2	−5	−11	−18	−25	−31	−38	−45
20	12	4	3	−10	−17	−24	−31	−39	−46	−53
25	8	1	7	−15	−22	−29	−36	−44	−51	−59
30	6	−2	−10	−18	−25	−33	−41	−49	−56	−64
35	4	−4	−12	−20	−27	−35	−43	−52	−58	−67
40	3	−5	−13	−21	−29	−37	−45	−53	−60	−69
45	2	−6	−14	−22	−30	−38	−46	−54	−62	−70

EXAMPLE 1 **a.** Find the wind chill temperature when winds are blowing at 25 miles per hour and the actual temperature is 20°F.

b. Use the following formula to convert the answer to the nearest degree.

$$C=\frac{5}{9}(F-32) \longleftarrow F=-15$$

Solution: **a.** From the table, the wind chill temperature is −15°F. This means that the 25 mile-per-hour wind makes the temperature of 20°F feel like −15°F with no wind.

b. $C=\frac{5}{9}(-15-32)$

$=\frac{5}{9}(-47)=-26$°C (to the nearest degree)

Certain meteorologists, called **climatologists,** study past records of temperatures, wind speeds, sunlight, rainfall, and humidity for patterns that may emerge for a given geographical area, and relationships that may exist between any of these measurements. In this way, formulas are derived to express the relationships that are found. The following formula is one of these.

$$C=33-\frac{(10\sqrt{r}+10.45-r)(33-t)}{22.1} \longleftarrow$$

C = wind chill temperature in °C
t = actual temperature in °C
r = wind speed in meters/second

EXAMPLE 2 Use a scientific calculator to find the wind chill temperature when the actual temperature is 4.4°C and the wind speed is 15.6 meters per second.

Solution: $C = 33 - \frac{(10\sqrt{15.6} + 10.45 - 15.6)(28.6)}{22.1}$ ← **(33 − 4.4)**

33 [−] [(] [(] 10 [×] 15.6 [√] [+] 10.45 [−] 15.6 [)] [×]

[(] 28.6 [)] [÷] 22.1 [)] [=] -11.448846

Thus, the wind chill temperature is **−11.4°C**. Subtracting the actual temperature from the wind chill temperature,

$$-11.4 - 4.4 = -15.8$$

it feels 15.8°C colder than the actual temperature.

EXERCISES

For Exercises 1–8,

a. Use the table to find the wind chill temperature.

b. Convert it to the nearest Celsius degree.

	Winds in MPH	Temperature	Wind Chill
1.	10	0°F	?
2.	20	20°F	?
3.	40	25°F	?
4.	20	10°F	?
5.	15	15°F	?
6.	30	−5°F	?
7.	25	−10°F	?
8.	35	5°F	?

For Exercises 9–12,

a. Use the formula in Example 2 to find the wind chill temperature.

b. State how much colder the wind chill temperature makes it feel.

	Winds in Meters per Second	Temperature	Wind Chill
9.	4.5	−1.1°C	?
10.	8.9	−9.4°C	?
11.	10.6	5.0°C	?
12.	6.7	−15.0°C	?

2–6 Properties of Order/Proof

The following postulate, based on the one-to-one correspondence between real numbers and points on the number line, allows you to compare real numbers.

Postulate

Comparison Postulate

For any real numbers a and b, exactly one of the following is true.

$$a < b \qquad a = b \qquad a > b$$

For $a < b$, the fact that you must move from left to right along the number line to go from a to b suggests the following definition.

REMEMBER: $a < b$ and $b > a$ **are equivalent statements.**

Definition

For real numbers a and b,

$$a < b \quad (\text{or } b > a)$$

if and only if **there is a positive real number** c **such that**

$$a + c = b.$$

Note also that the definition uses the words **"if and only if."** This means that the definition includes two statements.

For real numbers a and b:

1. If there is a positive real number c such that $a + c = b$, then $a < b$.
2. If $a < b$, then there is a positive real number c such that $a + c = b$.

When asked to prove a theorem written in *if and only if*–form, you must remember to prove *both* statements.

EXAMPLE 1 Write as two theorems: A real number a is positive *if and only if* $a > 0$.

Solution: Write one theorem in if–then form. Then interchange the hypothesis and conclusion.

1. If a is a real number and $a > 0$, then a is positive.
2. If a real number, a, is positive, then $a > 0$.

You can use the definition of "less than" and the closure postulate for addition of real numbers to prove the following theorem.

Theorem 2-1

Transitive Theorem for Order

For real numbers a, b, and c,

if $a < b$ **and** $b < c$**, then** $a < c$.

EXAMPLE 2 Prove the **Transitive Theorem for Order.**

Given: For real numbers a, b, and c, $a < b$ and $b < c$.

Prove: $a < c$

Proof:

Statements	Reasons
1. a, b, and c are real numbers; $a < b$; $b < c$	1. Given
2. There are positive real numbers k and m such that $a + k = b$ and $b + m = c$.	2. Definition of "less than"
3. $(a + k) + m = b + m$	3. Addition property of equality
4. $(a + k) + m = c$	4. Substitution principle
5. $a + (k + m) = c$	5. Associative postulate for addition
6. $(k + m)$ is a positive real number.	6. Closure postulate for addition
7. $a < c$	7. Definition of "less than."

The definitions, postulates, and theorems listed in Chapters 1 and 2 can be used to prove other theorems involving inequalities.

EXAMPLE 3 Prove: If $a < b$ for real numbers a and b, then $a - b < 0$.

Given: For real numbers a and b, $a < b$.

Prove: $a - b < 0$

Proof:

Statements	Reasons
1. For real numbers a and b, $a < b$.	1. Given
2. $a + (-b) < b + (-b)$	2. Addition property of inequalities
3. $a + (-b) < 0$	3. Inverse postulate for addition
4. $a - b < 0$	4. Definition of subtraction

You can *disprove* a statement or theorem if you can find *one instance* when the statement is false. This is called a **counterexample.**

EXAMPLE 4 Prove or disprove: If $a > b$, then $a^2 > b^2$.

Solution: Let $a = -5$ and $b = -12$. ← $a > b$

But $(-5)^2 < (-12)^2$. ← $a^2 \ngtr b^2$

Thus, the statement is **disproved.**

CLASSROOM EXERCISES

In Exercises 1–10:
a. Complete each statement.
b. State the definition, postulate, or theorem that justifies each answer.

1. If $-4 < -1$, then $-4 + 3$ $\underline{?}$ -1.
2. If $13 \neq 6 + 5$ and $13 \not< 6 + 5$ (is not less than) $6 + 5$, then 13 $\underline{?}$.
3. If $18 > 7$, then $7 + 11$ $\underline{?}$ 18.
4. If $-2 < 0$ and $0 < 5$, then -2 $\underline{?}$.
5. If $p \not< q$ and $p \not> $ (is not greater than) q, then p $\underline{?}$.
6. If $-5 > -7$ and $-7 > -11$, then -11 $\underline{?}$.
7. If $-19 < 0$, then there exists a positive number c such that $\underline{?}$.
8. If $p < t + 5$ and $t + 5 < -7$, then p $\underline{?}$.
9. If $q + r = t$, then q $\underline{?}$ and r $\underline{?}$.
10. If $6 < h + 1$, then there exists a positive number a such that $\underline{?}$.

WRITTEN EXERCISES

A In Exercises 1–5, write each theorem as two statements.

1. For real numbers a and b, $a < b$ if and only if $(b - a)$ is positive.
2. For real numbers a and b, $a = b$ if and only if $b - a = 0$.
3. For real numbers a and b, $b < a$ if and only if $-(b - a)$ is positive.
4. For real numbers a and b, $a \leq b$ if and only if $a < b$ or $a = b$.
5. For real numbers a and b, $a \geq b$ if and only if $a > b$ or $a = b$.

In Exercises 6–13:
a. Write the Given and the Prove.
b. Give a reason for each step of the proof.

6. If a, b, and c are real numbers and $a < b$, then $a + c < b + c$.

Statements

1. For real numbers a, b, c, $a < b$.
2. There is a positive real number t, such that $a + t = b$.
3. $a + t + c = b + c$
4. $(a + c) + t = b + c$
5. $a + c < b + c$

7. If a and b are real numbers and $a - b < 0$, then $a < b$.

Statements

1. For real numbers a and b, $a - b < 0$.
2. $a + (-b) < 0$
3. $a + (-b) + b < 0 + b$
4. $a + [-b + b] < 0 + b$
5. $a + 0 < + b$
6. $a < b$

8. If a is a real number and $-a > 0$, then $a < 0$.

Statements

1. For any real number a, $-a > 0$.
2. $-a + a > 0 + a$
3. $0 > 0 + a$
4. $0 > a$, or $a < 0$

9. If a is a real number and $a < 0$, then $-a > 0$.

Statements

1. For any real number a, $a < 0$.
2. $a + (-a) < 0 + (-a)$
3. $0 < 0 + (-a)$
4. $0 < -a$, or $-a > 0$

10. If a and b are real numbers and $a < 0$ and $b < 0$, then $ab > 0$.

Statements

1. For real numbers a and b, $a < 0$ and $b < 0$.
2. $a \cdot b > b \cdot 0$
3. $a \cdot b > 0$

11. If a is a real number and $a \neq 0$, then $a^2 > 0$.

Statements

1. For any real number a, $a \neq 0$.
2. $a > 0$ or $a < 0$
3. If $a > 0$, then $a \cdot a > 0$.
4. If $a < 0$, then $a \cdot a > 0$.
5. $\therefore a^2 > 0$

12. If a, b, and c are real numbers and $a < b$ and $c > 0$, then $ac < bc$.

Statements

1. For real numbers a, b, and c, $a < b$ and $c > 0$.
2. There is a positive real number q, such that $a + q = b$.
3. $(a + q) \cdot c = b \cdot c$
4. $ac + qc = bc$
5. qc is a positive real number.
6. $ac < bc$

13. If a, b, c, d are positive real numbers and $a > b$ and $c > d$, then $ac > bd$.

Statements

1. a, b, c, d are positive real numbers; $a > b$, and $c > d$
2. $ac > bc$
3. $bc > bd$
4. $\therefore ac > bd$

14. Prove: If $a > b$ for real numbers a and b, then $a - b > 0$. (HINT: See Example 3.)

15. Prove: If $a - b > 0$ for real numbers a and b, then $a > b$. (HINT: See Exercise 7.)

16. Write the theorems in Example 3 and in Exercise 7 as one theorem in *if and only if*-form.

17. Prove or disprove:
If a, b, c, d are real numbers and $a > b$, and $c > d$, then $a - c > b - d$.

18. Prove or disprove: If a, b, c are real numbers and $c \neq 0$, then $ac > bc$.

19. Prove or disprove: If a, b, c are real numbers and $c \neq 0$, then $\frac{a}{c} < \frac{b}{c}$.

B In Exercises 20–28, prove the given theorem.

20. If a is a positive real number, then $a^3 > 0$.

21. If a is a real number and $a < 0$, then $|a| = -a$.

22. If a and b are positive real numbers and $a < b$, then $a^2 < b^2$.

23. If a and b are real numbers and $a > 0$ and $b > 0$, then $ab > 0$.

24. If a and b are real numbers and $a > 0$ and $b < 0$, then $ab < 0$.

C

25. If a and b are negative real numbers and $a < b$, then $a^2 > b^2$.

26. For $a \neq 0$, $a > 0$ if and only if $\frac{1}{a} > 0$.

27. If a, b, c are real numbers and $a < b$ and $c < 0$, then $ac > bc$.

28. If a and b are real numbers and $a \leq b$ and $b \leq a$, then $a = b$.

Review

Graph each inequality. *(Section 2–3)*

1. $5x - 2 < 13$ **2.** $6w - 4 \geq 10 - 8w$ **3.** $-4(y + 3) \leq 17 - y$

Solve each problem. *(Section 2–4)*

4. A string orchestra is to have at least 36 players. The number of violists is to be twice the number of double bass players and the number of violinists is to to be three times the number of violists. Find the smallest number of double bass players, violists, and violinists.

5. A gardener has 32 meters of fencing material to enclose a rectangular garden. The width of the garden is to be $\frac{3}{5}$ the length. Find the maximum width and length of the garden.

Solve and graph. The replacement set for the variables is the set of real numbers. *(Section 2–5)*

6. $|x| > 1$ **7.** $|t - 1| \leq 3$ **8.** $|4y - 1| < 8$ **9.** $2|2 - p| \geq 13$

10. Prove: If $a + c < b + c$ for real numbers a, b, and c, then $a < b$. *(Section 2–6)* (HINT: See Written Exercise 6 on page 64.)

Puzzle

In a stack of 12 of the same type of coins, one is counterfeit. It is not the same weight as any of the other 11 coins. Using a double pan balance three times only, find the coin that is lighter or heavier than the other 11 coins.

HINT: Start by balancing two groups of 4 coins.

COMPUTER APPLICATIONS

BASIC: SOLVING INEQUALITIES

Problem: *Write a program to solve any inequality of the following form. Include the possibilities that the solution set may be all real numbers or the empty set.*

$$ax + b < c$$

```
100 PRINT "ENTER VALUES FOR A, B, C (IN THAT ORDER,
SEPARATED BY COMMAS)";
110 INPUT A, B, C
120 IF A = 0 THEN 190
130 LET X = (C - B)/A
140 IF A < 0 THEN 170
150 PRINT "X < ";X
160 GOTO 230
170 PRINT "X > ";X
180 GOTO 230
190 IF B < C THEN 220
200 PRINT "NO SOLUTION"
210 GOTO 230
220 PRINT "ALL REAL NUMBERS ARE SOLUTIONS."
230 PRINT "ANY MORE INEQUALITIES TO SOLVE (1 = YES, 0 = NO)";
240 INPUT Z
250 IF Z = 1 THEN 100
260 END
```

The programmer must solve $ax + b < c$ for x for the computer. (refers to statement 130)

Analysis

Statement 120: If A = 0, the solution set will be either the null set or all real numbers. So when A = 0, the computer is told to jump to statement 190, where it will compare B and C to decide between these two possibilities.

Statement 130: The computer reaches this statement when A ≠ 0.

Statements 140–180: The inequality sign must be reversed if A < 0. In this case, statement 140 sends the computer to statement 170, which prints the answer with a ">" sign. If A > 0, the computer moves to statement 150, which prints the answer with a "<" sign. In either case, the computer is told (in statements 160 and 180) to jump to statement 230, which asks the user if there are more inequalities to be solved.

Statements 190–220: The computer can get to statement 190 only from statement 120, when A = 0. In this case, the inequality becomes B < C, which will be true either all the time or never. So in statement 190, if B < C, the computer goes to statement 220 and prints `ALL REAL NUMBERS ARE SOLUTIONS.` But if B is not less than C, statement 200 is executed, printing `NO SOLUTION`. In either event, the computer moves to statement 230 to see if the user has any more data to enter.

In `IF . . . THEN` statements in `BASIC`, the `IF` part contains an equation or inequality in which the following symbols are used.

Algebraic Symbol	$=$	$<$	$>$	$\leq$	$\geq$	$\neq$
`BASIC` Symbol	`=`	`<`	`>`	`<=`	`>=`	`<>`

Either or both sides of the equation or inequality in the `IF` part may include operation symbols. For example, `40 IF A*X < B + C THEN 100` is a valid statement.

EXERCISES

A Run the program on page 60 for the following values of a, b, and c.

1. $a = 2;\ b = 5;\ c = 9$ **2.** $a = -2;\ b = 5;\ c = 9$ **3.** $a = 0;\ b = 7;\ c = 8$

4. Write `REM` statements that can be added to the program on page 60 to explain what the program is doing. For example,

```
135 REM LINES 140-180 DECIDE WHETHER TO REVERSE THE < SIGN.
```

5. Change the problem on page 60 to solve $ax + b \leq c$. What statements in the program must be changed and how?

In Exercises 6–7, write a `BASIC` program for each problem.

6. Solve equations of the form $Ax + B = Cx + D$. Include the possibilities that the solution set may be all real numbers or the null set.

7. Given a real number, print its absolute value. Do not use any "built-in" absolute value function that your computer may have.

The formula $\boldsymbol{W = ft + r}$ relates W, the weight of fuel carried by a jet, with f, the weight of fuel burned per number of hours, t, and r, the weight of reserve fuel. (See page 44.)

8. Given f, t, and r, calculate and print W. To show the proper units in the output, use a statement such as `PRINT "THE WEIGHT OF FUEL IS"; W;" POUNDS".`

9. Given W, t, and r, find and print f. Label the output as pounds per hour.

10. Given W, f, and r, find and print t. Label the output as hours.

The formula $\boldsymbol{W = I^2R}$ relates the number of watts, W, the current, I, in amperes, and the resistance, R, in ohms. (See page 45.)

11. Given I and R, calculate and print W. Label the output as watts.

12. Given W and I, calculate and print R. Label the output as ohms.

B In Exercises 13–15, write a program to solve any inequalities of the given form. Include the possibilities that the solution set may be all real numbers or it may be empty.

13. $|Ax + B| = C$ **14.** $|Ax + B| < C$ **15.** $|Ax + B| > C$

Chapter Summary

IMPORTANT TERMS

Apparent solution (p. 49)
Average (p. 55)
Compound sentence (p. 48)
Conjunction (p. 58)
Counterexample (p. 63)
Disjunction (p. 48)
Equivalent inequalities (p. 51)
Formula (p. 44)
Inequality (p. 51)
Mean (p. 55)

IMPORTANT IDEAS

1. **Addition Property for Inequalities**
   For real numbers a, b, and c, if $a < b$, then $a + c < b + c$.
2. **Multiplication Properties for Inequalities**
   a. For real numbers a, b, and c, if $a < b$ and $c > 0$, then $ac < bc$.
   b. For real numbers a, b, and c, if $a < b$ and $c < 0$, then $ac > bc$.
3. **Comparison Postulate**
   For all real numbers a and b, exactly one of the following is true.
   $a < b \qquad a = b \qquad a > b$
4. **Definition of "$a < b$"** For real numbers a and b, $a < b$ if and only if there is a positive real number c such that $a + c = b$.
5. **Transitive Theorem for Order** For real numbers a, b, and c, if $a < b$ and $b < c$, then $a < c$.
6. A theorem can be disproved by finding one instance for which the theorem is false. This is called **proof by counterexample.**

Chapter Objectives and Review

Objective: *To solve an equation having more than one variable for one of the variables (Section 2–1)*

Solve each equation for x in terms of the other variables. Indicate any restrictions on the values of the variables.

1. $-p + q = r - x$ **2.** $u + \frac{x}{v} = s - t$ **3.** $\frac{u - v}{x} = a$ **4.** $\frac{ax}{c} - d = e$

The formula $v^2 = 2gh$ relates the velocity, v, of an object that has fallen a distance h. The acceleration due to gravity is represented by g, which has a constant value of 9.8 m/sec^2.

5. Solve the formula for h. **6.** Find the value of h when $v = 98$ m/sec.

Objective: *To solve equations that involve absolute value (Section 2–2)*

Solve and check.

7. $|-3p| = 81$ **8.** $|x - 4| = 3$ **9.** $3|5t| - 7 = 45$ **10.** $-3|4m| + 2 = 17$
11. $|8 - 3x| = 31$ **12.** $|2(x - 2)| = x$ **13.** $|8x + 1| = 3x + 16$ **14.** $|5(3 - x)| = 3 - 2x$

Objective: *To solve inequalities and graph their solution sets (Section 2–3)*

Solve each inequality. Graph its solution set.

15. $x - 3 > 4$ **16.** $-(t - 1) > 5$ **17.** $5p + 6 \le 2 - 3p$ **18.** $-4b - 3 \le 4(3b - 2)$

Objective: *To solve problems that involve inequalities (Section 2–4)*

19. An automobile manufacturer can provide no more than 95 cubic feet of passenger space and luggage space for one line of automobiles. The amount of passenger space must be 75 cubic feet more than the luggage space. Find the maximum amount of passenger space and of luggage space.

20. Pat has scores of 80, 86, and 99 on three tests. What score must Pat achieve on the next test in order to have an average of at least 90?

21. For an official United States flag, the ratio of length to width must be exactly 19:10. A manufacturer plans to produce a United States flag that has a perimeter of no more than 70 centimeters. Find, to the nearest tenth of a centimeter, the maximum width and length of the flag.

Objective: *To solve and graph inequalities that involve absolute value (Section 2–5)*

Solve and graph. The replacement set for the variables is the set of real numbers.

22. $|x| > 2$ **23.** $|7 - t| < 4$ **24.** $|3w + 2| \le 5$ **25.** $|4 - 3x| \ge 10$

Objective: *To prove or disprove statements that involve inequalities (Section 2–6)*

26. For the theorem below:

a. Write the Given and the Prove.
b. Give a reason for each step of the proof.

If $ac < bc$ and $\frac{1}{c} > 0$, then $a < b$.

Statements	**Reasons**
1. For real numbers a, b, c, $ac < bc$ and $\frac{1}{c} > 0$.	1. ?
2. $(ac)\frac{1}{c} < bc\left(\frac{1}{c}\right)$	2. ?
3. $a\left(c \cdot \frac{1}{c}\right) < b\left(c \cdot \frac{1}{c}\right)$	3. ?
4. $a \cdot 1 < b \cdot 1$	4. ?
5. $a < b$	5. ?

27. Prove or disprove:

If a, b, and c are real numbers with $\frac{1}{c} < 0$ and $ac > bc$, then $a < b$.

Chapter Test

1. Solve for x in terms of π, p, q, and t: $q - \frac{px}{\pi} = t$. Indicate any restrictions on the values of the variables.

In Exercises 2–3, the formula $Ft = mv - mv_0$ relates the average force, F, with which an object is struck, the time interval, t, during which the force acts, the mass, m, of the object, the initial velocity, v_0, and the terminal velocity, v, of the object.

2. Solve the formula for v.
3. An object with a mass of 2 kilograms is at rest $(v_0 = 0)$. It is then struck with an average force of 2000 newtons during a time interval of 0.0004 seconds. Find the terminal velocity, v, of the struck object.

Solve and check.

4. $|13 - x| = 10$
5. $|7x + 2| = x + 14$
6. $|5x + 6 + 2x| = 4x + 39$
7. Solve and graph: $5x + 2 \geq 20 - 4x$.
8. A computer store has overhead expenses of \$1000 each week. It makes a profit of \$120 on each computer sold. How many computers must be sold in one week to make a profit of at least \$1400?
9. Solve and graph: $|2t - 1| \leq 6$. The replacement set for the variable is the set of real numbers.
10. Prove or disprove: If a and b are real numbers and $a^2 < b^2$, then $a < b$.

Preparing for College Entrance Tests

It is not always necessary to solve an equation for the variable in order to answer a question about the variable.

REMEMBER: It is important to watch for ways to save yourself time and work on College Entrance tests!

EXAMPLE If $6x + 12 = 5$, what is the value of $x + 2$?

(a) $-\frac{19}{6}$ (b) $-1\frac{1}{6}$ (c) $\frac{5}{6}$ (d) $3\frac{1}{6}$

Think: $6x + 12 = 6(x + 2)$. Solve $6(x + 2) = 5$ for $x + 2$.

Solution: $6(x + 2) = 5$

$x + 2 = \frac{5}{6}$ **Answer: c**

Choose the best answer. Choose *a*, *b*, *c*, or *d*.

1. If $2x - 8 = 1$, what is the value of $x - 4$?
   (a) $-\frac{1}{2}$ (b) $\frac{1}{2}$ (c) $4\frac{1}{2}$ (d) $8\frac{1}{2}$

2. If $10y - 5 = -2$, what is the value of $2y - 1$?

(a) $-\frac{5}{2}$ (b) $-\frac{4}{5}$ (c) $-\frac{2}{5}$ (d) $\frac{3}{10}$

3. If $2n - 1 = \frac{2}{3}$, what is the value of $n - \frac{1}{2}$? (HINT: $2n - 1 = 2(n - \frac{1}{2})$)

(a) $1\frac{1}{3}$ (b) 1 (c) $\frac{5}{6}$ (d) $\frac{1}{3}$

4. If $\frac{n}{6} = 8$, what is the value of $\frac{n}{8}$?

(a) 6 (b) $1\frac{1}{3}$ (c) $\frac{3}{4}$ (d) $\frac{1}{6}$

5. If $\frac{4y}{3} = -8$, what is the value of $\frac{1}{2}y$?

(a) -6 (b) -3 (c) $-\frac{1}{3}$ (d) 3

6. If $2a + 2b = -1$ and $6a - 2b = 5$, which of the following is true?

(a) $3a - b = 5$ (b) $a + b > 3a - b$ (c) $a + b < 3a - b$ (d) $a + b = -2$

7. If $2x + 4y = 5$ and $6x - 12y = 15$, which of the following is true?

(a) $2y = 5$ (b) $x + 2y < 5$ (c) $x + 2y > 5$ (d) $2x = 5$

8. If $y = x + 4$, what is the value of $|y - x| + |x - y|$?

(a) 0 (b) 4 (c) 8 (d) 16

9. If $y = -(x + 2)$, what is the value of $\frac{1}{2}|x + y|$?

(a) $\frac{1}{2}$ (b) 1 (c) 2 (d) 4

10. Two less than four times a certain number is 0.8. What is one less than twice the number?

(a) 0.4 (b) 1.2 (c) 1.6 (d) 4.0

11. If $\frac{4n}{5} = 6$, what is the value of $\frac{2n}{3}$?

(a) $11\frac{1}{4}$ (b) $7\frac{1}{2}$ (c) 5 (d) $\frac{5}{6}$

12. Three less than six times a certain number is 0.9. What is five less than ten times the number?

(a) 0.3 (b) 1.5 (c) 0.5 (d) 15

13. Four more than eight times a certain number is 112. What is five more than ten times the number?

(a) 24 (b) 135 (c) 140 (d) 32

14. If $y = -(x - 1)$, what is the value of $\frac{1}{3}|y + x|$?

(a) -1 (b) $-\frac{1}{3}$ (c) $\frac{1}{3}$ (d) 0

15. If $\frac{7n}{9} = 10$, what is the value of $\frac{14n}{3}$?

(a) 20 (b) 60 (c) 28 (d) 180

16. If $r = -(t + 1) + 7$, what is the value of $-\frac{1}{3}|r + t|$?

(a) -6 (b) 6 (c) -2 (d) 2

CHAPTER 3 Graphing

3–1 The Coordinate Plane

The two perpendicular lines which are the **axes** of the coordinate plane intersect at a point called the **origin.** The points of the plane can be placed in one-to-one correspondence with *ordered pairs* of real numbers (x, y). The first number, x, is the **abscissa** and the second number, y, is the **ordinate.** The two numbers are the **coordinates** of the point.

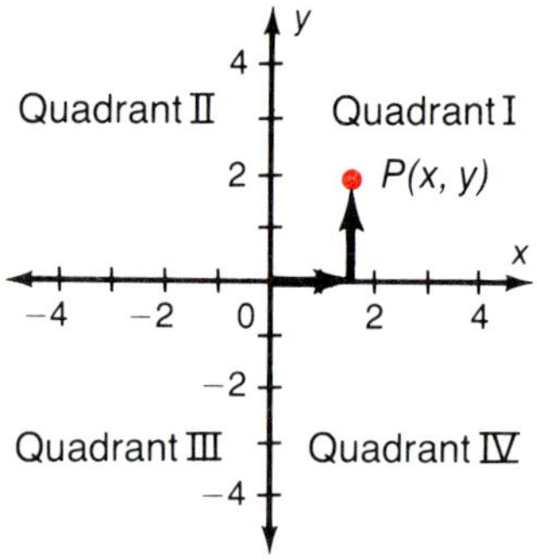

You locate a point by identifying x, its horizontal distance from the origin, and y, its vertical distance from the x axis.

EXAMPLE Graph the points associated with each ordered pair. State the quadrant in which each point lies.

a. $A(-2, 4)$ **b.** $B(4, -2)$ **c.** $C(0, -3)$

Solutions: **a.** $A(-2, 4)$: 2 units to the left of 0 and 4 units upward from the x axis.

Thus, $A(-2, 4)$ lies in **Quadrant II.**

b. $B(4, -2)$: 4 units to the right of 0 and 2 units downward from the x axis.

Thus, $B(4, -2)$ lies in **Quadrant IV.**

c. $C(0, -3)$: On the y axis, and 3 units below the x axis.

Since $C(0, -3)$ lies on an axis of the coordinate plane, it is **not in any quadrant.**

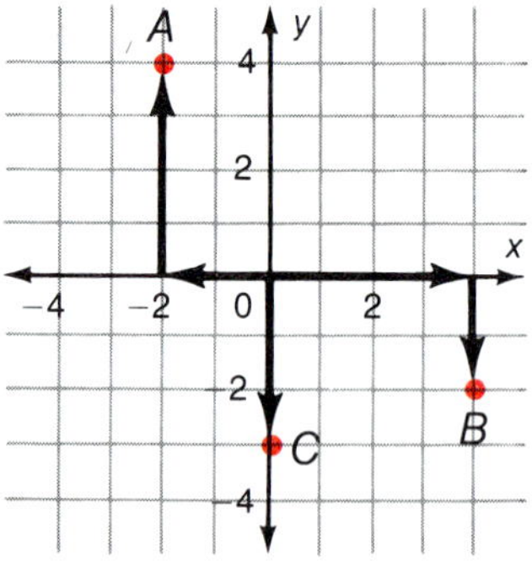

CLASSROOM EXERCISES

State the abscissa and the ordinate of each point.

1. $(-1, 5)$ **2.** $(6, -3)$ **3.** $(0, -8)$ **4.** $(0, 0)$ **5.** $(9, -2)$

State the quadrant in which each point lies.

6. $(3, -5)$ **7.** $(-4, -2)$ **8.** $(3, 4)$ **9.** $(-7, 3)$ **10.** $(-3, 1)$

Name the axis on which each point lies.

11. $(0, 3)$ **12.** $(3, 0)$ **13.** $(-3, 0)$ **14.** $(0, -3)$ **15.** $(0, 0)$

State the coordinates of each point.

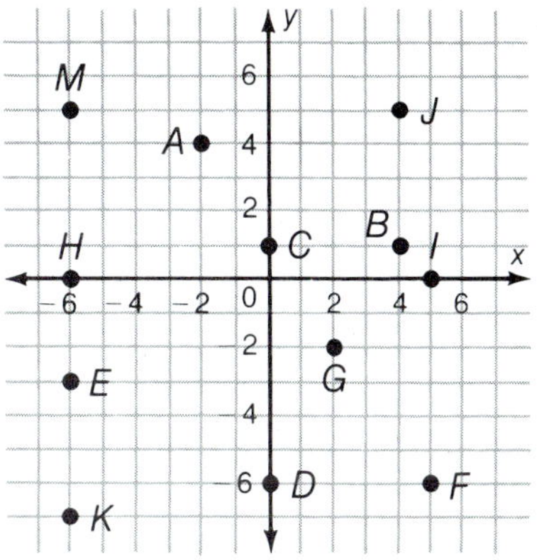

16. *A*	**17.** *B*	**18.** *C*	**19.** *D*
20. *E*	**21.** *F*	**22.** *G*	**23.** *H*
24. *I*	**25.** *J*	**26.** *K*	**27.** *M*

WRITTEN EXERCISES

A In Exercises 1–12, graph the point associated with each ordered pair. State the quadrant in which each point lies.

1. $A(3, 5)$	**2.** $B(-2, 1)$	**3.** $C(3, -2)$	**4.** $D(-1, -4)$
5. $E(6, 0)$	**6.** $F(0, -5)$	**7.** $G(0, 0)$	**8.** $H(-3, 0)$
9. $K(0, 4)$	**10.** $L(2, 3)$	**11.** $J(-4, \frac{3}{2})$	**12.** $P(4, -4)$

In Exercises 13–24, write the coordinates of each point. Refer to the graph.

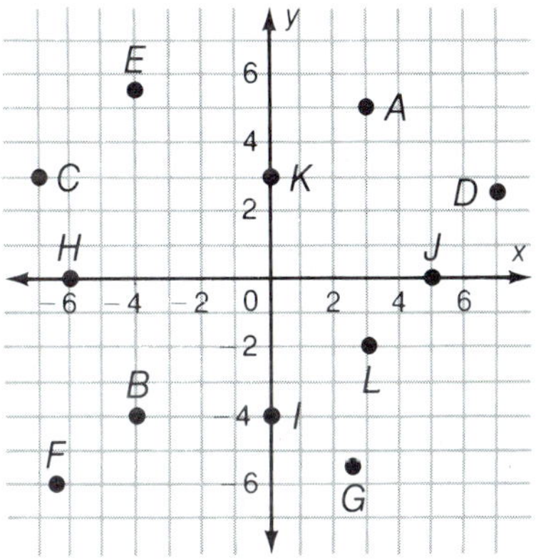

13. *A*	**14.** *B*	**15.** *C*
16. *D*	**17.** *E*	**18.** *F*
19. *G*	**20.** *H*	**21.** *I*
22. *J*	**23.** *K*	**24.** *L*

25. Locate several points on the x axis. What is the ordinate, or y value, of these points? What is the ordinate of every point on the x axis?

26. Locate several points on the y axis. What is the abscissa, or x value, of these points? What is the abscissa of every point on the y axis?

B

27. Describe the graph of the set of all ordered pairs of real numbers whose ordinates are 8; whose ordinates are -2; whose ordinates are 0.

28. Describe the graph of the set of ordered pairs of real numbers whose abscissas are 3; whose abscissas are -1; whose abscissas are 0.

29. Describe the graph of the set of ordered pairs of real numbers whose abscissas and ordinates are equal.

30. Describe the graph of the set of ordered pairs of real numbers whose abscissas and ordinates are additive inverses of each other.

In each of Exercises 31–34, the given coordinates represent the three vertices of a quadrilateral (figure with four sides). Find the coordinates of the fourth vertex, D, so that $ABCD$ is the required quadrilateral. In Exercise 33, there are three possible locations for D. In Exercise 34, there are two possible locations for D.

31. $A(2, 2)$, $B(5, 2)$, $C(5, 4)$; rectangle

32. $A(-2, -1)$, $B(-4, -1)$, $C(-4, -3)$; square

33. $A(-2, 4)$, $B(-4, 2)$, $C(-6, 2)$; parallelogram

34. $A(3, 3)$, $B(5, 1)$, $C(8, 1)$; isosceles trapezoid

C

35. Find the coordinates of the point of intersection of the line passing through $P(-2, -3)$ and $Q(4, 3)$ and the line passing through $A(3, 0)$ and $B(5, 2)$.

36. Graph several ordered pairs of real numbers in Quadrants I and II such that the ordinate of each point equals the square of its abscissa. Connect the points with a smooth curve. Describe the graph.

37. Graph several ordered pairs of real numbers in Quadrants I and II such that the ordinate of each point equals the absolute value of the abscissa. Describe the graph.

38. One line is parallel to the x axis and 5 units above it. A second line is parallel to the y axis and 6 units to its left. What are the coordinates of the point of intersection of the two lines?

Puzzle

Look at the object shown below. Name two different things that you see.

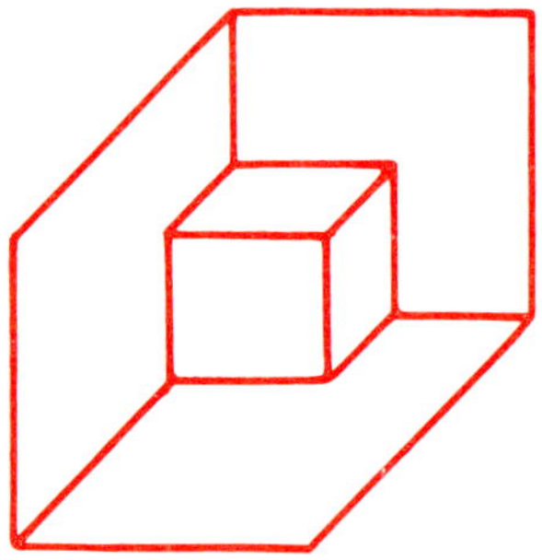

REVIEW CAPSULE FOR SECTION 3–2

Solve each formula for the indicated variable. Indicate any restrictions on the variables. *(Pages 44–47)*

1. $p = a + b + c$, for a **2.** $V = lwh$ for h **3.** $E = mc^2$, for m

4. $A = P + Prt$, for r **5.** $E = am(T - t)$, for T **6.** $V = \pi h(R^2 - r^2)$, for h

3–2 Graphing Equations with Two Variables

Equations such as

$$3x + 2y = 9 \qquad \text{and} \qquad y = 7x - 2$$

x and y are the variables.

are equations in *two variables*. The solutions of such equations are all ordered pairs (x, y) that satisfy the equations (make the equations true). The **graph** of such an equation is the set of points whose coordinates are solutions of the equation.

EXAMPLE 1 Graph $3x - 4y = 2$.

Solution:

[1] Solve the equation for y.

$$3x - 4y = 2$$
$$-4y = 2 - 3x$$
$$y = -\frac{1}{2} + \frac{3}{4}x, \text{ or } y = \frac{3}{4}x - \frac{1}{2}$$

[2] Make a table of ordered pairs. Choose at least 3 convenient values for x; that is, numbers such as -4, 0, 2, and so on.

x	-4	-2	0	2	4
y	$-3\frac{1}{2}$	-2	$-\frac{1}{2}$	1	$2\frac{1}{2}$

[3] Graph the points.
Draw a line to connect the points.

The graph is a **straight line.**

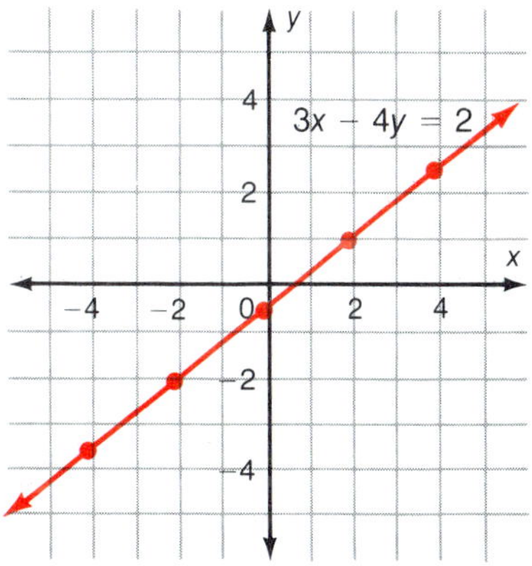

Any point that lies on the graph of $3x - 4y = 2$ is a solution of this equation. Any point that does *not* lie on the graph of $3x - 4y = 2$ is *not* a solution of the equation.

Definition

> Any equation of the form
> $$Ax + By = C$$
> where A, B, and C are real numbers and A and B do not *both* equal zero, is a **linear equation** in two variables.

The following lists two important facts relating to linear equations.

> The graph of every linear equation is a straight line.
> Every straight line in the coordinate plane is the graph of a linear equation.

EXAMPLE 2 Graph: **a.** $y = 4$ **b.** $x = -3$

Solutions: **a.** $y = 4$ is the same as $0 \cdot x + y = 4$

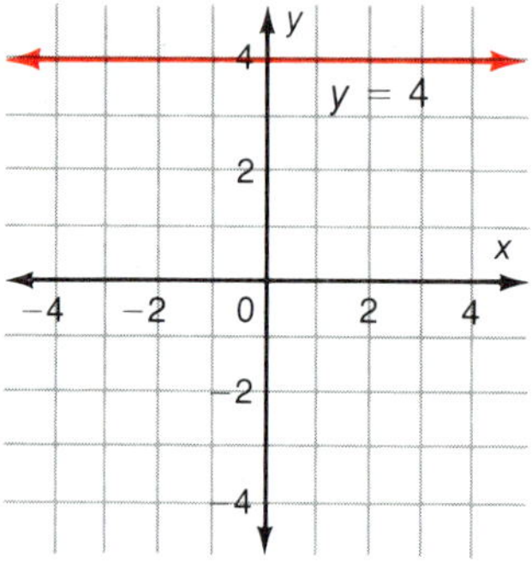

The graph is a **horizontal line 4 units above the *x* axis.**

b. $x = 3$ is the same as $x + 0 \cdot y = -3$

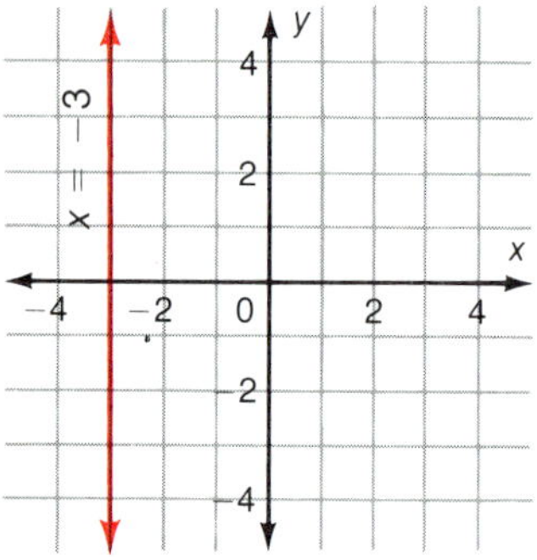

The graph is a **vertical line 3 units to the left of the *y* axis.**

CLASSROOM EXERCISES

In Exercises 1–4, solve each equation for *y*.

1. $3x - y = 9$ **2.** $2x + 5y = 10$ **3.** $y + 2x = 0$ **4.** $y - 7x = 0$

In Exercises 5–10, state whether the given point lies on the graph of the equation.

5. $x + y = 8$; $(0, 0)$ **6.** $x - y = 4$; $(6, 2)$ **7.** $2x = 8 - 3y$; $(4, 0)$
8. $4y = 10 - x$; $(2, 2)$ **9.** $y = -1$; $(2, -1)$ **10.** $x = 1$; $(-5, 0)$

In Exercises 11–15, describe the graph of each equation as a *vertical line* or as a *horizontal line.*

11. $x = 2$ **12.** $y = -5$ **13.** $y = -1$ **14.** $x = -8$ **15.** $y = 0$

WRITTEN EXERCISES

A Graph each equation.

1. $3y = 2x - 1$ **2.** $3y = -2x + 5$ **3.** $\frac{2}{3}y = 5x + 1$
4. $\frac{5}{2}y = 18 - \frac{3}{2}x$ **5.** $2x + 3y = 7$ **6.** $-3x - 2y = 5$
7. $1.5x + 3 - 2y = y$ **8.** $2x - 3y + 5 = x$ **9.** $2x - 1 = 3y + 2$
10. $2x + 1 = 6 - y$ **11.** $4(y - 3) = -x - 2$ **12.** $5(x - 1) - 4(y + 1) = -1$
13. $y = |2x + 1|$ **14.** $y = -|3x + 1|$ **15.** $y = |5x|$ **16.** $y = -|-2x|$

In Exercises 17–22, find k such that the graph of the resulting equation will pass through the point with the given coordinates.

17. $x+y=k;\ P(1,5)$ **18.** $x-y=k;\ Q(0,-3)$ **19.** $2x+y=k;\ R(2,-1)$

20. $3x+4y=k;\ T(-2,-\frac{1}{2})$ **21.** $2x=-5y+k;\ R(6,-2)$ **22.** $2x+3y=k;\ S(-1,-1)$

B Graph each pair of equations on the same pair of coordinate axes. Estimate the coordinates of the points of intersection. If the graphs do not intersect, write *N*.

23. $3x-y=3$, $-x+y=1$ **24.** $2x-3y=5$, $4x-6y=5$ **25.** $2y-x=3$, $y-5=0$ **26.** $2y+x=3$, $x+4=0$

CALCULATOR APPLICATIONS

Evaluating Formulas

A scientific calculator with memory storage, [M+], memory recall, [MR], and square, [x^2], keys will help you to evaluate formulas.

EXAMPLE Evaluate $y=\dfrac{m^2-4w}{t_1+t_2}$ when $m=98$, $w=5$, $t_1=0.4$, and $t_2=512$.

SOLUTION Evaluate the denominator first. Store this value in the memory.

[.] 4 [+] 5 1 2 [=] [M+] → 512.4

9 8 [x^2] [−] 4 [×] 5 [=] [÷] [MR] [=] → 18.704137

EXERCISES

1. Evaluate $F=\dfrac{mu^2}{gr}$, when $m=9.8$, $u=495$, $g=0.045$, and $r=52$.
2. Evaluate $S=\dfrac{rl-a}{r-1}$, when $r=1.5$, $l=100{,}000$, and $a=50$.
3. Evaluate $p=kb+\dfrac{k}{q^2}$, when $k=22.6$, $b=0.042$, and $q=19.141$.

REVIEW CAPSULE FOR SECTION 3–3

Evaluate $|a-b|$ for the given values of a and b. *(Pages 10–13)*

1. $a=9,\ b=12$ **2.** $a=-3,\ b=4$ **3.** $a=6,\ b=-1$ **4.** $a=-1,\ b=-1$

Evaluate $\dfrac{a-b}{2}$ for the given values a and b. *(Pages 10–13)*

5. $a=(x-1),\ b=(x+3)$ **6.** $a=(x+4),\ b=(x-2)$ **7.** $a=(x-8);\ b=(3x-4)$

Evaluate $\dfrac{a-b}{a+b}$ for the given values of a and b. *(Pages 10–13)*

8. $a=-3,\ b=7$ **9.** $a=(x-1);\ b=(x+2)$ **10.** $a=(x+3);\ b=(x-2)$

3–3 Slope of a Line

The steepness or grade of a road is measured by a ratio that compares the *vertical rise* of the road to every 100 feet of *horizontal distance*.

Grade: $\frac{12}{100}$, or 12%

For a highway, the maximum grade or slope is

$\frac{12}{100}$. ← **Vertical change** (12) ← **Horizontal change** (100)

In mathematics, the measure of the steepness or slant of a line is called **slope.** To find the slope of a line, you choose any two points on the line and compute this ratio.

$$\text{Slope} = \frac{\textbf{vertical change}}{\textbf{horizontal change}} = \frac{\textbf{difference of ordinates}}{\textbf{difference of abscissas}}$$

Definition

The **slope, *m*,** of a nonvertical line that contains the points $P_1(x_1, y_1)$ and $P_2(x_2, y_2)$ is

$$m = \frac{y_2 - y_1}{x_2 - x_1}, \text{ or } \frac{y_1 - y_2}{x_1 - x_2}.$$

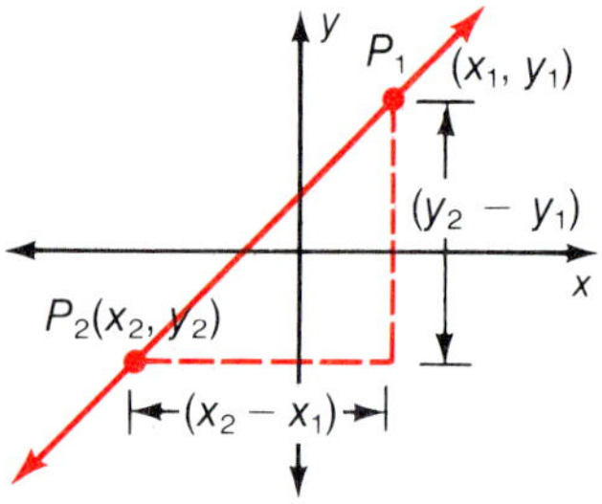

Given two points on a line, you can use either form of the definition to find the slope of the line.

EXAMPLE 1 Find the slope of the line that contains the points $P_1(-2, 9)$ and $P_2(-5, -6)$.

Solution:

Method 1

$$m = \frac{y_2 - y_1}{x_2 - x_1}$$

$$m = \frac{-6 - 9}{-5 - (-2)}$$

$$m = \frac{-15}{-3}$$

$m = 5$ ← **Same slope** →

Method 2

$$m = \frac{y_1 - y_2}{x_1 - x_2}$$

$$m = \frac{9 - (-6)}{-2 - (-5)}$$

$$m = \frac{15}{3}$$

$m = 5$

The **ratio of the vertical change to the horizontal change is everywhere the same,** no matter which two points are chosen to compute the ratio.

EXAMPLE 2 Graph the line that has a slope of $-\frac{2}{3}$ and contains the point $R(-3, -1)$.

Solution: [1] Graph $R(-3, -1)$.

[2] Use the given slope to graph two other points on the line.

Start at R. Since $-\frac{2}{3} = \frac{-2}{3}$, move 2 units down and 3 units to the right. Label this point $Q(0, -3)$.

[3] Start at Q. Move 2 units down, and 3 units to the right. Label this point $P(3, -5)$.

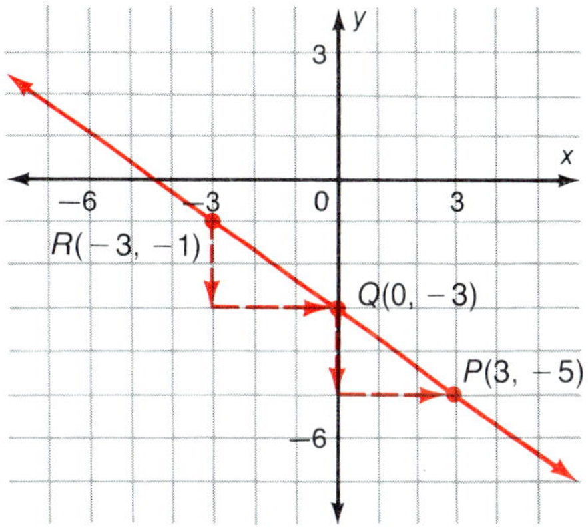

[4] Draw a straight line connecting points R, Q, and P.

In Example 2, you could have drawn line RQ after completing Step 2. However, finding a third point P provides a check on the accuracy of your work.

The slope of a nonvertical line can be positive, negative, or zero. For vertical lines, the slope is not defined.

Positive Slope

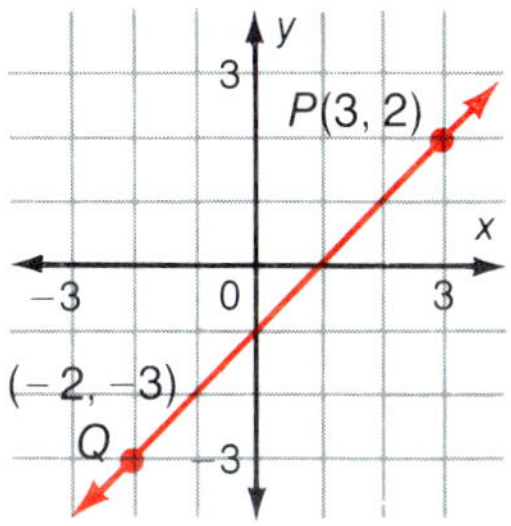

The line *rises* from left to right.

$m = \frac{2-(-3)}{3-(-2)}$

$m = \frac{5}{5} = \mathbf{1}$

Negative Slope

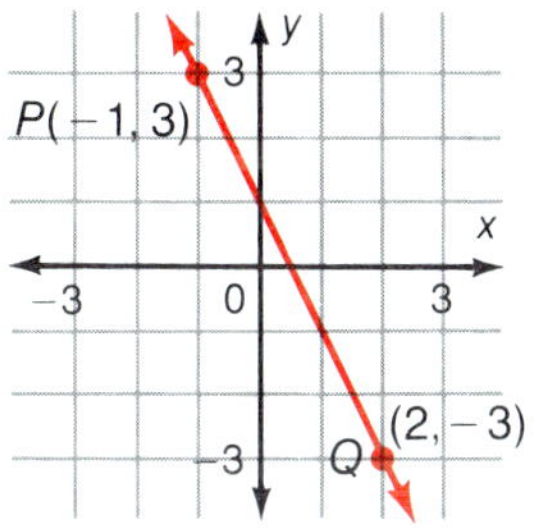

The line *falls* from left to right.

$m = \frac{3-(-3)}{-1-2}$

$m = \frac{6}{-3} = \mathbf{-2}$

Zero Slope

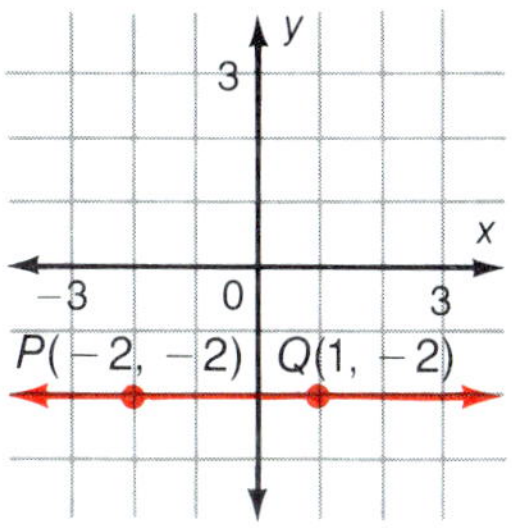

The line is *horizontal*.

$m = \frac{-2-(-2)}{1-(-2)}$

$m = \mathbf{0}$

Not Defined

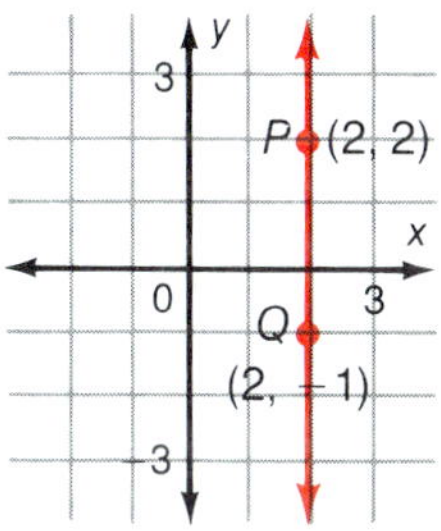

The line is *vertical*.

$m = \frac{2-(-1)}{2-2} = \frac{3}{0}$

No denominator can equal 0. Thus, the slope is **not defined.**

CLASSROOM EXERCISES

Find the slope of the line containing the given points.

1. (0, 0); (3, 5) **2.** (−6, −4); (0, 0) **3.** (2, 3); (3, 3) **4.** (6, −9); (−2, −2)

Find the slope of each line from its graph.

5. 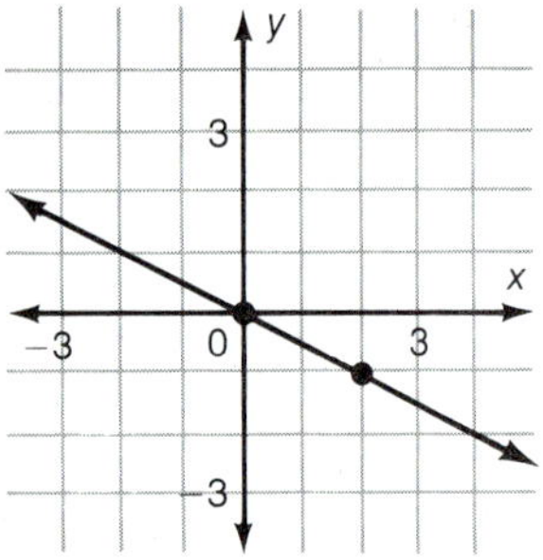

6. 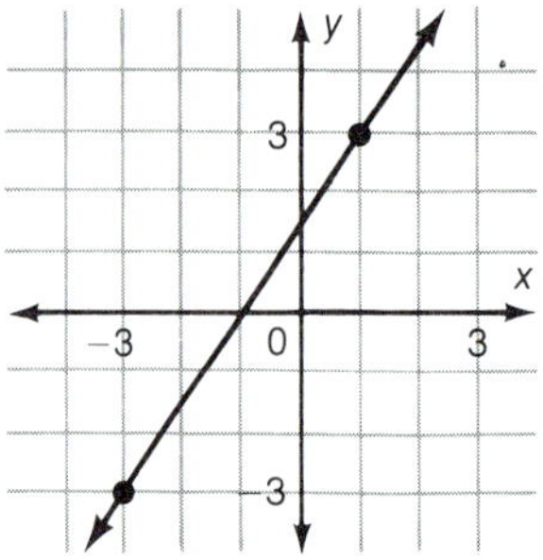

7. 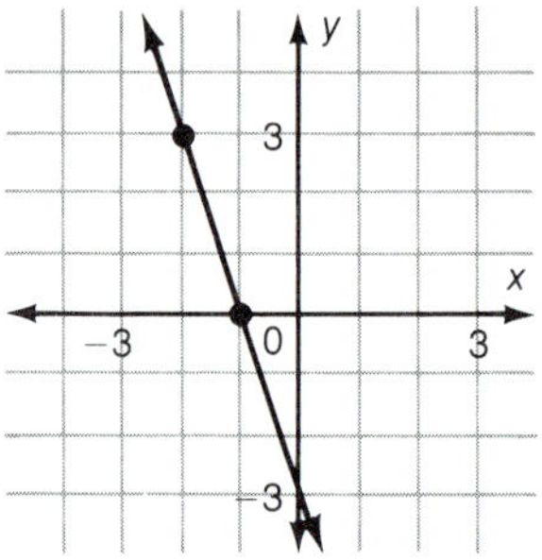

Describe the slope of each graph as positive, negative, zero, or not defined.

8. The graph of the line falls from left to right.
9. The graph is a vertical line.
10. The graph is a horizontal line.
11. The graph of the line rises from left to right.

WRITTEN EXERCISES

A Find the slope of the line containing the two given points.

1. (6, 3); (4, 1)
2. (3, 4); (9, 12)
3. (−2, 5); (−4, 11)
4. (8, −3); (−2, 7)
5. (−4, −6); (2, −3)
6. (5, 8); (−3, −4)
7. (−3, 5); (4, 7)
8. (2, −6); (−1, 9)

Use the figure at the right for Exercises 9–13.

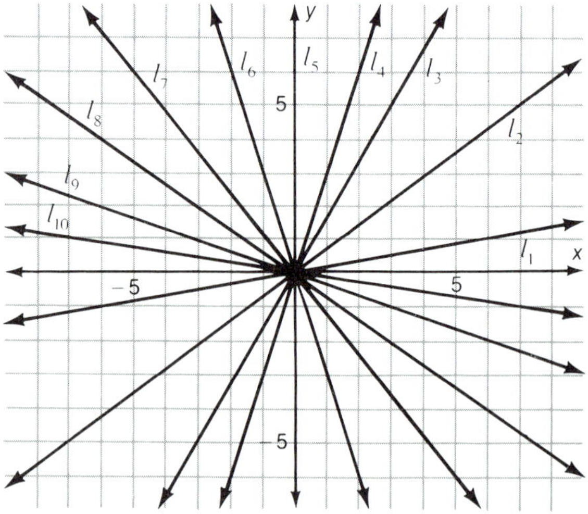

9. Find the slope of each of the lines l_1, l_2, l_3, l_4, l_5, l_6, l_7, l_8, l_9, and l_{10} given in the figure. Note that l_5 coincides with the y axis.
10. Consider each of the lines as a different position of the same line that rotates counterclockwise around the origin. Beginning at a position that coincides with the x axis as the line passes from position l_1 through l_4, is the slope positive or negative? Is the slope increasing or decreasing?
11. As the line passes from position l_6 through l_{10}, is the slope positive or negative? Increasing or decreasing?
12. For which position of the line is the slope equal to 0?
13. For which position of the line is the slope undefined?

In each of Exercises 14–22, graph the line having the given slope and passing through the given point.

14. $m = \frac{1}{2}$; $P(3, 4)$ **15.** $m = \frac{3}{4}$; $T(2, 1)$ **16.** $m = \frac{2}{3}$; $R(-4, 3)$

17. $m - -\frac{2}{3}$; $V(5, -2)$ **18.** $m = -\frac{3}{4}$; $Q(-3, -2)$ **19.** $m = -\frac{3}{7}$; $Y(0, 0)$

20. $m = -1$; $N(0, -2)$ **21.** $m = \frac{4}{3}$; $S(-2, 4)$ **22.** $m = -\frac{2}{3}$; $A(-1, 1)$

23. Graph the line passing through the point $B(5, 2)$. The slope of the line is not defined.

24. Graph the line passing through the point $B(5, 2)$ and having a slope of zero.

B

25. Determine the value of x such that the line passing through the points $P(4, 3)$ and $Q(x, 8)$ has a slope of -4.

26. Determine the value of y such that the line passing through the points $C(-2, -7)$ and $D(2, y)$ has a slope of $\frac{3}{2}$.

27. Show that the points $A(-4, -2)$, $B(1, 1)$ and $C(6, 4)$ are collinear (lie on the same line) by showing that $m_{AB} = m_{BC} = m_{AC}$.

In each of Exercises 28–31, find the slope of the line containing the given points.

28. $A(a, b + c)$; $C(a + b, b + c)$ **29.** $D(a, b)$; $F(a, b + c)$

30. $E(a + b, c + d)$; $K(a - b, -c - d)$ **31.** $R(2a, 3b)$; $T(a, -b)$

32. Find the slopes of the sides of the triangle formed by joining the points $A(2, 4)$, $B(6, -4)$, and $C(-4, -2)$.

C

33. Find x such that points $A(-3, -4)$, $B(1, -2)$ and $C(x, 3)$ are collinear.

34. A line contains the points $A(0, 0)$ and $B(2000, -3000)$. Name another point on the line.

Puzzle

Five friends, Luis, Bill, Jan, Omar, and Judy, are sitting in one row of seats. Neither Luis nor Bill is sitting next to Omar. Neither Luis nor Bill is sitting next to Jan. Neither Omar nor Bill is sitting next to Judy. Judy is sitting just to the right of Jan.

Find the seating arrangement from left to right.

REVIEW CAPSULE FOR SECTION 3–4

Graph each equation. *(Pages 77–79)*

1. $2y - x = 0$ **2.** $4x - 2y = 6$ **3.** $-x + 2y = 10$ **4.** $3y - 2x = 18$

Problem Solving and Applications

3–4 Using Linear Graphs

Property that decreases in value over a period of time is said to **depreciate** in value. For example, suppose that a car purchased for $7200 is resold one year later for $5900. Then the depreciation is

$$\$7200 - \$5900 = \$1300.$$

When the loss in value over a specified period of time is a *fixed per cent* of the original value the depreciation is called **linear depreciation.**

EXAMPLE Karl Robins purchased a small apartment house for $90,000. For tax purposes, Karl assumes a depreciation of $6\frac{2}{3}\%$ per year on the property.

a. Write an equation for the value, V, of the property after t years.

b. Graph the equation.

c. Use the graph to determine how long it will take for the value of the property to depreciate to $30,000.

d. Use the graph to estimate how long it will take for the value of the property to depreciate to $0.00. Round your answer to the nearest whole number of years.

Solutions:

a. Present value of the property: $90,000.

Depreciation in t years: $\$90{,}000 \cdot 0.06\frac{2}{3} \cdot t = 6000t$ ← Present Value − Depreciation in t Years

Value after t years: $V = \mathbf{90{,}000 - 6000}t$

b. Make a table of three ordered pairs. Graph the equation.

t	0	3	5
V	90,000	72,000	60,000

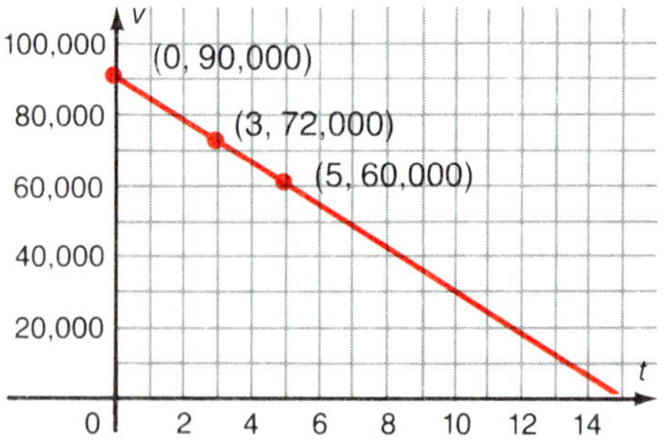

c. From the graph $V = 30{,}000$ when t is about 10. Thus, it will take about **10 years.**

d. From the graph, $V = 0$ when t is about 15. Thus, it will take about **15 years.**

Note that the slope of the equation $V = 90{,}000 - 6000t$ is -6000. This represents a constant yearly loss of $6000 in the value of the property.

CLASSROOM EXERCISES

1. A piece of property worth $16,000 depreciates linearly by 8% per year. Write an equation for the value of P of the property after t years.
2. An item purchased for $5000 depreciates linearly at the rate of 6% per year. Write an equation for the value I after t years.
3. A piece of antique jewelry is purchased for $500 and increases in value at a constant rate of 15% per year. Write an equation for the value J after t years.
4. In a certain year, a department store has sales of $5,570,000 per year. Sales are expected to increase at a constant rate of 6% per year. Write an equation for the amount of sales, S, after t years.

WORD PROBLEMS

A

1. A salesperson paid $12,000 for a car. Suppose that the car depreciates linearly at a rate of 15% per year.
   a. Write an equation for the value V of the car after t years.
   b. Graph the equation.
   c. Use the graph to determine how long it will take for the value of the car to depreciate to $3000.
   d. Use the graph to estimate how long it will take for the car to lose all of its value. Round your answer to the nearest whole number of years.
2. A farmer purchases a piece of machinery for $40,000. The value of the machinery depreciates linearly at a rate of 8% per year.
   a. Write an equation for the value M of the machinery after t years.
   b. Graph the equation.
   c. Use the graph to estimate how long it will take for the machinery to depreciate to one-half its purchase value. Round your answer to the nearest whole number of years.
   d. Use the graph to estimate how long it will take for the machinery to lose all of its value. Round your answer to the nearest whole number of years.
3. Suppose that a person borrows $20,000 at a simple interest rate of 15% per year.
   a. Write an equation for that amount, A, that will be owed after t years. (HINT: Amount owed = $20,000 + Interest for t years.)
   b. Graph the equation. Use $t = 0$, $t = 2$, and $t = 5$.
   c. Use the graph to determine how much will be owed after 10 years.
   d. Use the graph to estimate how much will be owed after 15 years.
   e. Use the graph to estimate how much will be owed after 18 years.

4. Suppose that a person borrows \$30,000 at a simple interest rate of 20% per year.

   a. Write an equation for the amount, A, that will be owed after t years.

   b. Graph the equation.

   c. Use the graph to determine how much will be owed after 15 years.

   d. Use the graph to estimate how much will be owed after 25 years?

5. A communications satellite transmits 1.8 million messages per year. Assume that the number of messages will increase at a constant rate of 10% per year.

   a. Write an equation for the number of messages, N, that will be transmitted after t years.

   b. Graph the equation.

   c. Use the graph to estimate how many messages will be transmitted 2 years from now. Round your answer to the nearest million.

   d. Use the graph to estimate how many messages will be transmitted 5 years from now. Round your answer to the nearest million.

6. Suppose that a person's starting salary at a new job is \$20,000 per year. Assume that salary increases will be at a constant rate of 12% per year.

   a. Write an equation for the amount of salary, S, after t years.

   b. Graph the equation.

   c. Use the graph to estimate the person's salary after 5 years. Round your answer to the nearest thousand dollars.

   d. Use the graph to estimate the person's salary after 15 years. Round your answer to the nearest thousand dollars.

7. Suppose that a taxi fare is \$1.00 plus 25¢ for each $\frac{1}{5}$ mile traveled.

   a. Write an equation for the cost in dollars, C, for traveling m miles.

   b. Graph the equation.

   c. Use the graph to estimate to the nearest mile how far a person could travel for \$5.00.

   d. Use the graph to estimate to the nearest dollar the cost of a taxi ride of $8\frac{1}{2}$ miles.

B

8. Suppose that an executive's yearly salary is \$250,000. Assume that the salary will not increase for 5 years and that the inflation rate is 8% per year.

   a. Write an equation for the salary's real buying power, P, after t years.

   b. From the equation, what is the decrease in buying power per year due to inflation?

   c. Graph the equation.

   d. Use the graph to estimate when the real buying power of the salary will be less than \$200,000.

9. Suppose that a worker's yearly salary is \$30,000 and that salary increases will be at the rate of 12% per year. Assume that the inflation rate is 8% per year.
   a. Write an equation for the real buying power of the salary after t years.
   b. Graph the equation.
   c. From the equation, what is the real increase in the buying power of the salary each year?
   d. Use the graph to estimate when the real buying power of the salary will be greater than \$35,000.

Review

Graph the point associated with each ordered pair. State the quadrant in which each point lies. *(Section 3–1)*

1. $A(6, 2)$ 2. $M(-2, 4)$ 3. $P(3, 1)$ 4. $B(-4, -3)$

Graph each equation. *(Section 3–2)*

5. $2y = 3x + 4$ 6. $4y = -x + 3$ 7. $y = |3x + 2|$

Determine the slope of the line passing through the given points. *(Section 3–3)*

8. $A(2, 4);\ B(-1, 8)$ 9. $D(-3, 7);\ E(5, -2)$ 10. $G(\frac{1}{2}, 4);\ H(-\frac{3}{2}, -5)$

Graph the line having the given slope and passing through the given point. *(Section 3–3)*

11. $m = 4;\ P(2, 3)$ 12. $m = -3;\ Q(-1, 2)$ 13. $m = -\frac{2}{3};\ R(2, -4)$

14. Determine the value of x such that the line passing through $A(x, 12)$ and $B(1, -6)$ has a slope of $-\frac{2}{5}$. *(Section 3–3)*
15. Suppose that a person borrows \$40,000 at a simple interest rate of 18% per year. *(Section 3–4)*
   a. Write an equation for the amount, A, that will be owed after t years.
   b. Graph the equation.
   c. Use the graph to determine how much will be owed after 5 years.

REVIEW CAPSULE FOR SECTION 3-5

Complete. (Pages 10–13)

1. $\frac{1}{5} \cdot (-5) = \underline{\ ?\ }$ 2. $\frac{2}{3} \cdot \left(-\frac{3}{2}\right) = \underline{\ ?\ }$ 3. $\frac{a}{b} \cdot \left(-\frac{b}{a}\right) = \underline{\ ?\ }$ 4. $-\frac{2q}{7} \cdot \frac{7}{2q} = \underline{\ ?\ }$

5. $8 \cdot \underline{\ ?\ } = -1$ 6. $-\frac{5}{9} \cdot \underline{\ ?\ } = -1$ 7. $\frac{4h}{5} \cdot \underline{\ ?\ } = -1$ 8. $-\frac{p}{q} \cdot \underline{\ ?\ } = -1$

Using Statistics:
Measures of Central Tendency

Market researchers collect and analyze data from a **sample** or subset of a larger **population** in order to discover characteristics or trends that will help them sell their product. The *mean* and the *median* are two **measures of central tendency** often used by market researchers to describe a characteristic that is of concern to them.

Definitions

> The **mean** is the arithmetic average.
>
> The **median** is the middle measure in a set of data that contains an odd number of elements. When the number of elements is even, the **median** is the mean of the two middle values.

One method for collecting data to analyze is **systematic sampling** in which every *r*th member of the population is chosen for the sample.

Problem: Julio Ruiz, a market researcher for Olympic Films, is gathering information about the ages of people who go to see science fiction movies. He is interested in the average age of these people so that he can gear the advertising campaign for Olympic's upcoming science fiction movie to that age group. He takes a sample at Marlowe's Cinema by stopping every tenth person and asking their age. The data below represent the researcher's results.

17	42	21	78	16	21	31
29	29	16	49	19	81	16
69	69	18	31	22	14	21
75	42	78	18	41	22	16
18	80	42	42	42	16	16
21	19	18	44	18	22	14
49	17	16	18	18	18	17

Analysis: Julio then computed the mean and median to the nearest integer.

$$\text{Mean} = \frac{\textbf{sum of the measures}}{\textbf{number of measures}} = \frac{\Sigma x}{n} = \frac{1566}{49} = 32$$

The median is the 25th number when the data is ordered from least to greatest (there are 49 elements). Therefore, the median is **21.** To determine which average to use, Julio structured a table using intervals for the ages and the number of people, or **frequency**, in each interval. With this he structured a **histogram** in order to get a "picture" of the distribution.

NOTE: Since each interval has a common boundary, a boundary age, such as 20, 30, and so on, is in the interval to the right.

Age Interval	Frequency
10–20	22
20–30	9
30–40	2
40–50	9
50–60	0
60–70	2
70–80	3
80–90	2

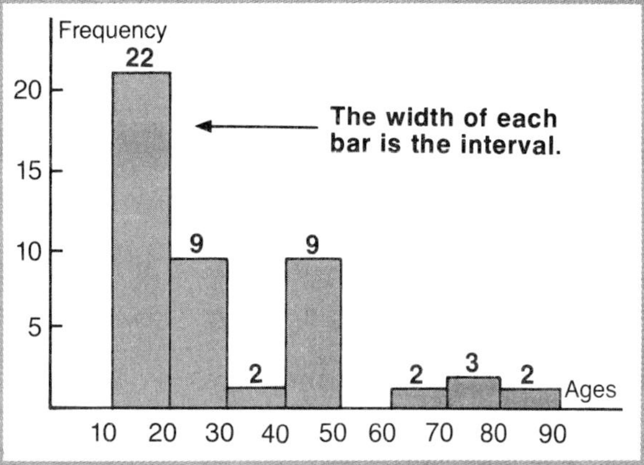

The histogram shows that the ages are not distributed symmetrically, but are **skewed.** When this happens, the median is the best measure of central tendency.

Conclusion: Julio will recommend that the advertising campaign be aimed primarily at people around the age of 21.

EXERCISES

Calculate the mean and median to the nearest integer. State which measure of central tendency gives a better description of the sample. Explain your answer.

1. The following numbers were drawn from a table of random numbers.

28	53	58	66
37	36	78	41
87	35	13	38
61	46	1	83
47	15	96	3
57	84	66	47
37	81	46	4
86	78	55	12
93	76	91	83

2. These scores represent time (in hours) between failures in aircraft radios.

108	168	152	74	85
136	52	34	62	137
150	175	121	136	126
120	143	174	164	48
42	127	148	81	137
243	74	122	86	128
139	123	86	85	125
103	115	132	49	72
110	82	111	137	77

3. The coach of a high school football team measured the weights (in pounds) of every 25th boy on the enrollment list.

120	210	194	167	205
127	185	176	155	157
148	169	145	176	118
162	175	161	189	171
148	184	176	171	187
183	138	136	201	152

4. A used-car dealer wanted an estimate of the price being paid for used cars in his community. He collected this data.

4500	3500	6700	2800	6500
4700	4800	2800	4700	4600
4500	4500	4500	2700	4500
4800	2800	3700	4600	4600
4000	3700	2800	3700	2800
2800	5800	5700	4500	5800

3–5 Parallel and Perpendicular Lines

In a plane, lines *having the same slope are parallel.* Conversely, parallel lines have the same slope.

Vertical lines (lines with undefined slope) are also parallel.

EXAMPLE 1 Show that line l_1, which contains the points $A(-3, -4)$ and $B(5, -2)$ is parallel to line l_2 which contains the points $C(-4, 2)$ and $D(4, 4)$.

Solution: [1] Find the slope of each line.

For line l_1:

$$m_1 = \frac{y_2 - y_1}{x_2 - x_1}$$

$$m_1 = \frac{-2-(-4)}{5-(-3)}$$

$$m_1 = \frac{2}{8} = \frac{1}{4}$$

For line l_2:

$$m_2 = \frac{y_2 - y_1}{x_2 - x_1}$$

$$m_2 = \frac{4-2}{4-(-4)}$$

$$m_2 = \frac{2}{8} = \frac{1}{4}$$

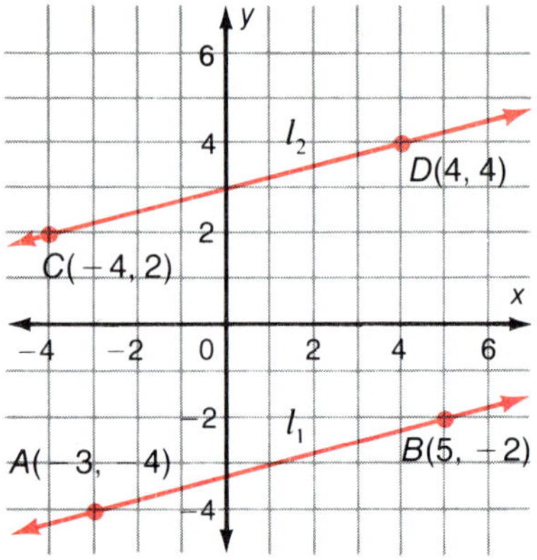

[2] Compare the slopes.

Since $m_1 = m_2$, the lines are **parallel.**

Two numbers are **negative reciprocals** of each other if their product is −1. For example, since

$$\frac{2}{3} \times -\frac{3}{2} = -1,$$

$\frac{2}{3}$ and $-\frac{3}{2}$ are negative reciprocals.

Recall that two lines in a plane are **perpendicular** if they intersect at right angles. Thus, any vertical line such as $x = 3$ is perpendicular to any horizontal line such as $y = -2$.

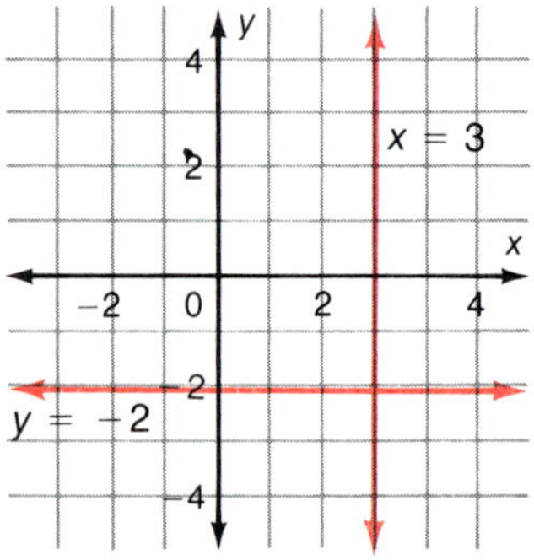

Two nonvertical lines can also intersect at right angles.

Theorem 3–1 **If two nonvertical lines are perpendicular, their slopes are negative reciprocals of each other.**

Theorem 3–2 **If the slopes of two lines are negative reciprocals of each other, the lines are perpendicular.**

Theorem 3–1 states that if two lines l_1 and l_2 are perpendicular to each other, then

$$\boldsymbol{m_1 \cdot m_2 = -1}, \quad \text{or} \quad \boldsymbol{m_1 = -\frac{1}{m_2}}.$$

Theorem 3–2 states that if two lines have slopes m_1 and m_2 such that $m_1 \cdot m_2 = -1$ or $m_1 = -\frac{1}{m_2}$, then the lines are perpendicular.

EXAMPLE 2 Show that line l_1 which contains the points $A(0, 3)$ and $B(4, 0)$ is perpendicular to line l_2 which contains points $C(0, -2)$ and $D(3, 2)$.

Solution: [1] Find the slope of each line.

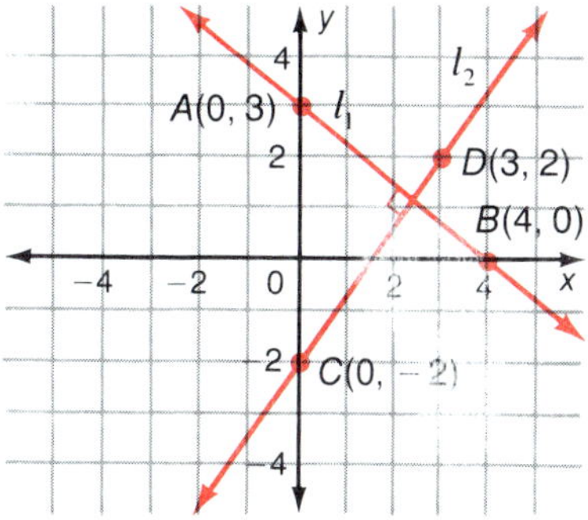

For line l_1:

$$m_1 = \frac{0-3}{4-0}$$

$$m_1 = -\frac{3}{4}$$

For line l_2:

$$m_2 = \frac{2-(-2)}{3-0}$$

$$m_2 = \frac{4}{3}$$

[2] Find $m_1 \cdot m_2$.

$$m_1 \cdot m_2 = -\frac{3}{4} \cdot \frac{4}{3} = \mathbf{-1}$$

Since $\boldsymbol{m_1 \cdot m_2 = -1}$, the lines are **perpendicular.**

CLASSROOM EXERCISES

In Exercises 1–4, find the slope of all lines parallel to the line containing the two given points.

1. (2, 0); (−4, 1) **2.** (7, −3); (4, 2) **3.** (−4, 3); (1, 7) **4.** (0, 0); (−2, −1)

In Exercises 5–8, find the slope of all lines perpendicular to the line containing the two given points.

5. (−6, 5); (0, 0) **6.** (−1, −3); (−5, −4) **7.** (2, 4); (6, 7) **8.** (1, −4); (1, 7)

WRITTEN EXERCISES

A Classify the lines determined by the two pairs of points as *parallel, perpendicular,* or *neither* parallel nor perpendicular.

1. (−2, 7), (3, 6) and (4, 2), (9, 1)

2. (0, 0), (−5, 3) and (5, 2), (0, 5)

3. (2, 5), (8, 7) and (−3, 1), (−2, −2)

4. (5, 3), (−5, −2) and (6, −2), (4, 5)

5. Show that the line which contains the points $P(2, 4)$ and $Q(6, 7)$ is parallel to the line containing the points $A(6, 2)$ and $B(10, 5)$.

6. Show that the line which contains the points $Q(-3, -6)$ and $R(6, -3)$ is parallel to the line containing the points $D(0, \frac{2}{3})$ and $E(-4, -\frac{2}{3})$.

7. Show that the line which contains the points $D(0, \frac{2}{3})$ and $E(-6, -\frac{2}{3})$ is perpendicular to the line containing the points $Q(-1, 4)$ and $T(1, -5)$.
8. Show that the line which contains the points $T(-1, -7)$ and $S(2, 2)$ is perpendicular to the line containing points $A(1, -\frac{4}{3})$ and $B(3, -2)$.
9. Find the value of x such that the slope of the segment with endpoints at $M(x, 5)$ and $Q(3, -1)$ is 4.
10. The line containing $E(a, 3)$ and $F(2, 0)$ is parallel to the line that contains $D(2, 8)$ and $C(-3, -4)$. Find a.
11. Find the value of k if the line through $R(5, -3)$ and $S(0, k)$ is perpendicular to the line through $T(-3, 2)$ and $V(-2, -5)$.
12. Show that the triangle whose vertices have the coordinates $A(3, 2)$, $B(8, 16)$, and $C(11, 4)$ is a right triangle.

B

13. The vertices of a quadrilateral are $A(3, 1)$, $B(5, 6)$, $C(7, 6)$ and $D(10, 2)$. Show that the diagonals of the quadrilateral are perpendicular to each other.
14. Determine the values of y such that $Q(0, y)$ is the vertex of the right angle of a right triangle whose other vertices are $S(-3, 1)$ and $R(-4, 8)$.
15. Show that $P(-9, 3)$, $Q(2, 1)$, $R(8, 9)$ and $S(-3, 11)$ are the vertices of a parallelogram by showing that opposites sides are parallel.
16. Show that $P(-3, -1)$, $Q(-1, -3)$, $R(4, 2)$, and $S(2, 4)$ are the vertices of a rectangle by showing that opposite sides are parallel and that one pair of adjacent sides is perpendicular.

The vertices of a parallelogram have coordinates $A(-2, 4)$ $B(2, 6)$, $C(7, 2)$ and $D(a, 0)$. Use this information in Exercises 17–19.

17. Find the slope of $\overline{AB}$.
18. Express the slope of $\overline{DC}$ in terms of a.
19. Use your answers to Exercises 17 and 18 to find a.
20. Show that $L(-3, 5)$, $Q(0, 2)$, $T(-2, -5)$ and $V(-10, 3)$ are the vertices of a trapezoid.

C

21. Complete the following proof by supplying the missing reasons.

Given: $l_1 \| l_2$, and m_1 the slope of l_1, m_2 the slope of l_2. **Prove:** $m_1 = m_2$

Statements

1. $l_1 \| l_2$
2. Construct $\overline{FE}$ perpendicular to the x axis at E.
3. $\angle ODB \cong \angle ECF$
4. $\triangle BOD$ and $\triangle FEC$ are right triangles.
5. $\triangle BOD \sim \triangle FEC$
6. $\frac{OB}{OD} = \frac{EF}{EC}$
7. $m_1 = m_2$

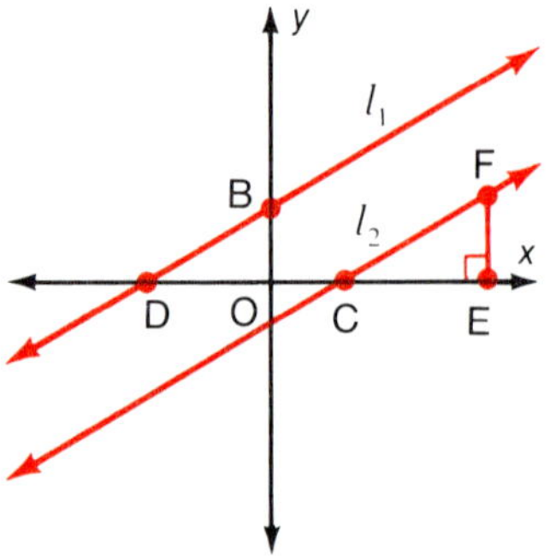

3–6 The Equations of a Line

When you know the slope of a line and the coordinates of a point on the line, you can write an equation for the line.

Let $P_1(x_1, y_1)$ be any point on a line with slope m. Choose any other point $P(x, y)$ on the line. Then, by the definition of slope,

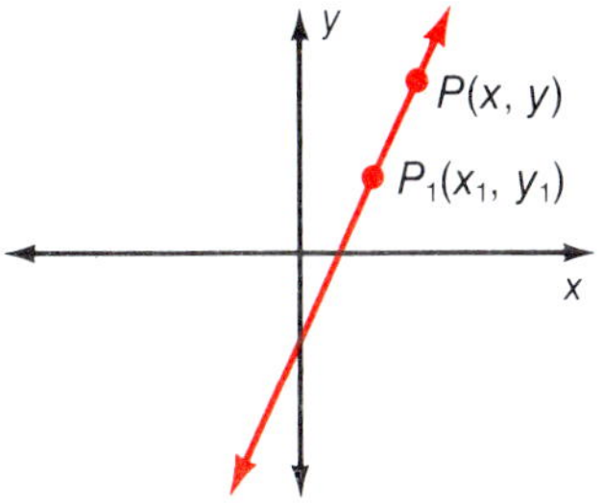

$$m = \frac{y - y_1}{x - x_1}$$ ← **Multiply each side by $(x - x_1)$, $x \neq x_1$.**

$m(x - x_1) = y - y_1$ or $y - y_1 = m(x - x_1)$.

This equation is called the **Point–Slope Form** of a linear equation.

Theorem 3–3

Point–Slope Form of a Linear Equation

If the slope of a nonvertical line is a real number m and a point $P_1(x_1, y_1)$ lies on the line, the equation of the line is

$$y - y_1 = m(x - x_1) \text{ for all } x, y \in R.$$

Also, $y - y_1 = m(x - x_1)$ where $x, y \in R$ is the equation of a nonvertical line having slope m and passing through the point (x_1, y_1).

EXAMPLE 1 **a.** Write the equation of the line having slope $-\frac{1}{2}$ and passing through the point $P(2, -3)$.

b. Graph the line.

Solutions: **a.** Use the point–slope form of the equation.

$$y - y_1 = m(x - x_1)$$ ← $y_1 = -3$, $m = -\frac{1}{2}$, $x_1 = 2$

$$y - (-3) = -\frac{1}{2}(x - 2)$$

$$y + 3 = -\frac{1}{2}(x - 2)$$

b. Graph $P(2, -3)$.

Since $-\frac{1}{2} = \frac{-1}{2}$, start at P.

Move 1 unit down, then 2 units to the right.

Label this point Q. Draw line PQ.

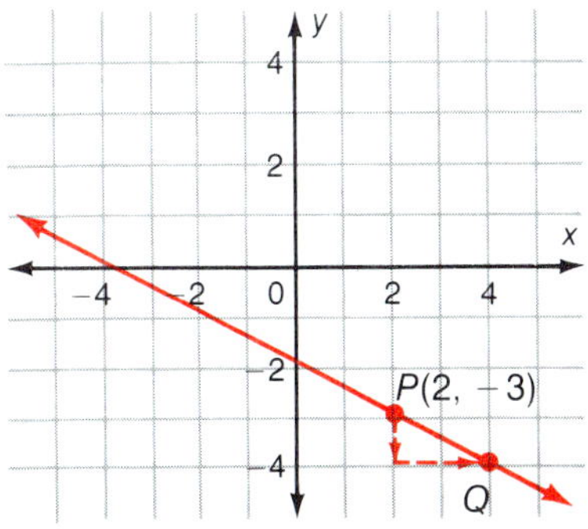

When graphing a line in the coordinate plane, it is sometimes convenient to determine where it crosses the x axis and the y axis. The *abscissa* of the coordinates of the point where a line crosses the x axis is called the **x intercept.** The *ordinate* of the coordinates of the point where the line crosses the y axis is the **y intercept.**

EXAMPLE 2 Write the equation of the line having slope -3 and y intercept 5.

Solution: Since the y intercept is 5, the point $(0, 5)$ is on the graph.

Use the point–slope formula to write the equation.

$y - y_1 = m(x - x_1)$ ← **$x_1 = 0, y_1 = 5, m = -3$**

$y - 5 = -3(x - 0)$

$y - 5 = -3x$, or $\mathbf{y = -3x + 5}$

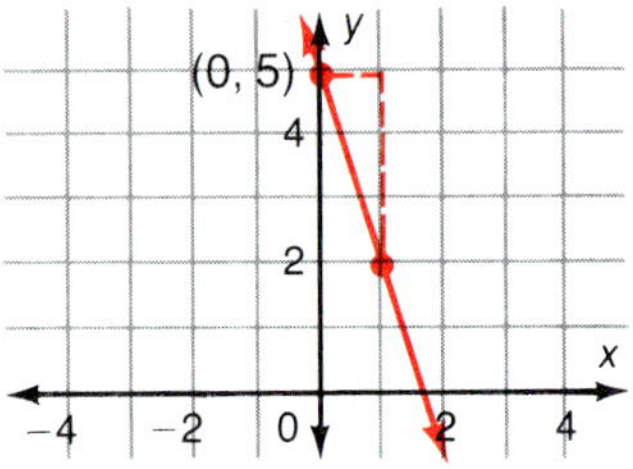

The equation $y = -3x + 5$ is written in **slope–intercept form.** Note that the slope of the line, -3, and the y intercept, 5, can be read directly from the equation.

Theorem 3–4

Slope–Intercept Form of a Linear Equation

If the slope of a nonvertical line is m and its y intercept is b, the equation of the line is

$$y = mx + b, \text{ for } x, y, b \in R.$$

Also, $y = mx + b$ where $x, y, b \in R$ is the equation of a line with slope m and y intercept b.

EXAMPLE 3 **a.** Find the slope and y intercept of the graph of $2x + 3y = 6$.

b. Graph the line.

Solutions: **a.** Write the equation in slope–intercept form.

$2x + 3y = 6$

$3y = 6 - 2x$ ← **$m = -\frac{2}{3}, b = 2$**

$y = 2 - \frac{2}{3}x$ or $y = -\frac{2}{3}x + 2$ ← **Solve for y.**

Slope: $-\dfrac{2}{3}$ $\qquad$ y intercept: **2**

b. Graph $P(0, 2)$.

Since $m = -\dfrac{2}{3} = \dfrac{-2}{3}$, start at P.

Move 2 units down, then 3 units to the right. Label this point Q. Draw line PQ.

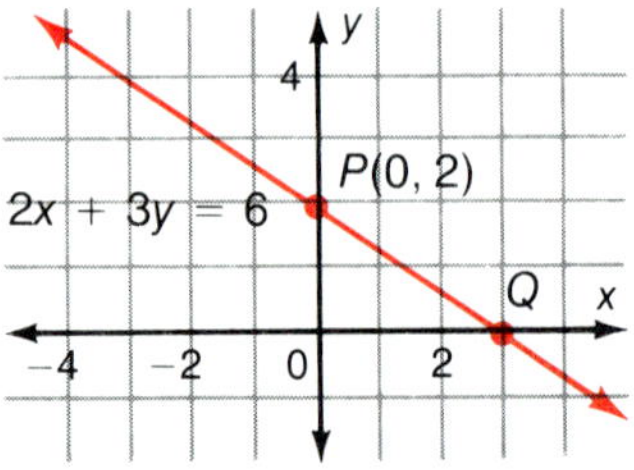

The equation of any line in the coordinate plane can be written in a more general form.

General Form of a Linear Equation

Any line in the coordinate plane can be written in the form $Ax + By = C$, where $A, B, C \in R$ and A and B are not both zero.

Also, $Ax + By = C$, where $A, B, C \in R$ and A and B are not both zero, is the equation of a line in the coordinate plane.

EXAMPLE 4 Write the equation of the line containing the points $R(1, 1)$ and $Q(5, 9)$. Write the equation in the form $Ax + By = C$.

Solution: [1] Use the given points to find the slope of the line.

$$m = \frac{9-1}{5-1} = \frac{8}{4} = 2$$

[2] Let $P(x, y)$ be any other point on the line. Use the slope and one of the given points to write the equation in point–slope form.

$y - y_1 = m(x - x_1)$ ← **$m = 2, x_1 = 5, y_1 = 9$**

$y - 9 = 2(x - 5)$

$y - 9 = 2x - 10$

$\mathbf{-2x + y = -1}$ ← **$Ax + By = C$**

In Example 4, the same equation would have been obtained if point $R(1, 1)$ had been used.

CLASSROOM EXERCISES

Use the given point and slope to write an equation of the line in point–slope form.

1. $R(2, 11); m=3$ **2.** $Q(6, -1); m=5$ **3.** $S(-2, 3); m=-\frac{1}{2}$ **4.** $T(-1, -3); m=\frac{3}{4}$

Write the slope and y intercept of each line.

5. $y = -3x + 1$ **6.** $y = \frac{2}{5}x - 3$ **7.** $y = -x$ **8.** $y - 5x = 1$

Use the given slope and y intercept to write an equation of the line in slope–intercept form.

9. $m=-5, b=1$ **10.** $m=1; b=3$ **11.** $m=-\frac{1}{2}, b=0$ **12.** $m=\frac{2}{5}, b=-\frac{1}{5}$

Write each equation in the form $Ax + By = C$.

13. $y=-2x+7$ **14.** $y-8=-2(x+5)$ **15.** $2y-3x+1=0$ **16.** $3x=7y$

WRITTEN EXERCISES

A In Exercises 1–12, use the given point and slope to write an equation of the line in point–slope form.

1. $P(5, 2); m=3$ **2.** $Q(3, -5); m=2$ **3.** $Q(-2, 7); m=-3$ **4.** $R(0, 3); m=\frac{2}{3}$

5. $P(-2, 3); m=-\frac{3}{2}$ **6.** $T(-3, 0); m=-5$ **7.** $A(3, -4); m=0$ **8.** $S(-1, -5); m=\frac{3}{4}$

9. $T(-2, 4); m=-\frac{4}{3}$ **10.** $S(1, -1); m=\frac{1}{2}$ **11.** $T(-2, 3); m=1$ **12.** $Q(4, -4); m=0$

In Exercises 13–20, write the equation of the line having the given slope and *y* intercept. Write the equation in slope intercept form.

13. $m=-2; b=1$ **14.** $m=\frac{2}{3}; b=-5$ **15.** $m=-\frac{1}{2}, b=-1$ **16.** $m=0, b=4$

17. $m=3, b=\frac{1}{2}$ **18.** $m=-\frac{1}{4}; b=8$ **19.** $m=-\frac{1}{3}; b=1$ **20.** $m=1, b=0$

In Exercises 21–28, write each equation in slope–intercept form. Give the slope and *y* intercept of each line. Then draw the graph.

21. $2x + y = 4$ **22.** $3x - y = 8$ **23.** $x - 2y = 0$ **24.** $2x + 3y = -6$

25. $3x + 4y = 12$ **26.** $2x - 6y = 2$ **27.** $x + 3y = 0$ **28.** $4x - 5y = 10$

In Exercises 29–36, write the equation of the line containing each of the following pairs of points. Write your answer in the form $Ax + By = C$.

29. $(3, 1); B(5, 6)$ **30.** $P(-5, 3); Q(8, 1)$ **31.** $R(-5, -2); S(7, 5)$ **32.** $S(0, 0); R(6, 8)$

33. $C(0, 0); F(-5, 1)$ **34.** $F(0, 8); C(8, 0)$ **35.** $D(-5, -3); G(0, 0)$ **36.** $B(5, 3); D(5, 9)$

In Exercises 37–40, classify the graphs of each pair of equations as parallel, *P*, perpendicular, *D*, or neither parallel nor perpendicular, *N*.

37. $2x - 3y = 6$ and $10x - 15y = 6$ **38.** $5x + 4y = 0$ and $8x - 10y = 7$

39. $2x + 3y = 0$ and $x = \frac{2}{3}y$ **40.** $2x + 4y - 1 = 0$ and $x + 4y = 17$

Show that the graphs of the following pairs of equations are parallel.

41. $y = 2x + 8$ and $y - 2x = 5$ **42.** $3x + y = 0$ and $6x + 2y = 9$

Show that the graphs of the following pairs of equations are perpendicular.

43. $y = 2x + 8$ and $2y + x = 1$ **44.** $3x + y = 4$ and $x - 3y = 4$.

B

45. Write the equation of the line that contains the origin and is parallel to the graph of $2x + 3y = 5$.

46. Write the equation of the line that contains the point $E(3, -5)$ and is parallel to the graph of $3x + 4y = 12$.

47. Write the equation of the line that contains the point $Q(-4, -7)$ and is perpendicular to the graph of $2x - 3y = 6$.

48. Write the equation of the line that contains the point $T(3, 2)$ and is parallel to the x axis.

49. Write the equation of the line that is parallel to the y axis and contains the point $P(3, 2)$.

50. Write the equation of the line that is parallel to the x axis and 3 units below it.

51. Write the equation of the line parallel to the y axis and 7 units to the left of it.

Let $A(3, -1)$, $B(-5, 5)$, and $C(2, 6)$ be the vertices of triangle ABC.

52. Find the equations of lines AB, BC, and AC.

53. What is the equation of the line parallel to line BC through point A?

54. Find the equation of the line perpendicular to line AC through point B.

55. Find the equation of the line perpendicular to line AC through point C.

C

56. Prove that the line in the xy plane whose x intercept is $(a, 0)$ and whose y intercept is $(0, b)$ is defined by

$$\frac{x}{a} + \frac{y}{b} = 1,$$

where $x, y, a, b \in R$ and $a \neq 0$. This is the **intercept form** of a linear equation.

57. The vertices of a triangle are $A(3, 3)$, $B(6, 2)$ and $C(8, -2)$. Find the equations of the lines that contain the altitudes of triangle ABC.

In Exercises 58–60, write each equation in the form $Ax + By = C$. In each case, identify A, B, and C.

58. $y = mx + b$

59. $\dfrac{y - y_1}{x - x_1} = m$

60. $y - c = 0$

61. $\dfrac{x}{a} + \dfrac{y}{b} = 1$

REVIEW CAPSULE FOR SECTION 3-7

1. Write the equation of the line with slope 3 and passing through the origin. *(Pages 93–97)*

2. As the values of x increase in the equation for Exercise 1, do the values of y increase or decrease? *(Pages 77–79)*

3. As the values of x decrease in the equation for Exercise 1, do the values of y increase or decrease? *(Pages 77–79)*

4. Write the equation of the line with slope -2 and passing through the origin. *(Pages 93–97)*

5. As the values of x increase in the equation for Exercise 3, do the values of y increase or decrease? *(Pages 77–79)*

6. As the values of x decrease in the equation for Exercise 3, do the values of y increase or decrease? *(Pages 77–79)*

Problem Solving and Applications

3–7 Direct Variation and Proportion

The table and the graph below show that the amount of sales tax, t, varies directly as the amount of purchase, p, for a constant tax rate, $r = 6\% = 0.06$.

p	t	$r = \frac{t}{p}$
\$100	\$ 6	$\frac{6}{100} = 0.06$
\$200	\$12	$\frac{12}{200} = 0.06$
\$350	\$21	$\frac{21}{350} = 0.06$
\$900	\$54	$\frac{54}{900} = 0.06$

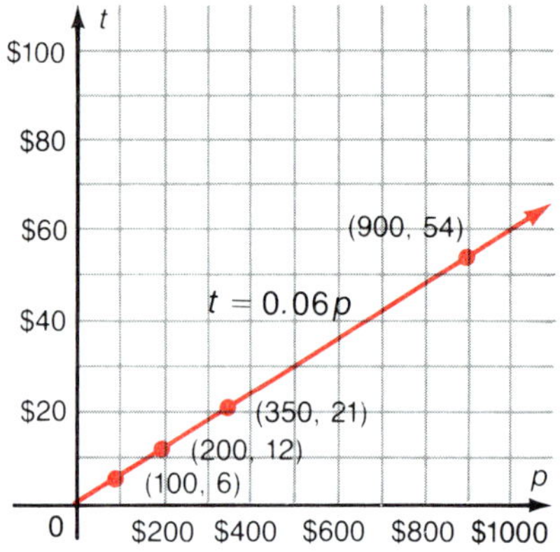

The ratio $\frac{t}{p}$ is the same for each ordered pair (p, t). That is,

$$\frac{t}{p} = 0.06, \text{ or } t = 0.06p.$$

We say that t **varies directly** as p or t **varies with** p. The constant, 0.06, is called the **constant of variation.**

Definition

> A **linear variation,** or a **direct variation,** is defined by an equation of the form
>
> $$y = mx \quad \text{or} \quad y = kx$$
>
> where m (or k) is a nonzero constant.

EXAMPLE 1 If y varies directly as x and $y = -9$ when $x = 3$, find y when $x = -4$.

Solution: [1] Find the constant of variation; that is, find $\frac{y}{x}$.

$\frac{y}{x} = k$ ← **Replace x with 3 and y with −9.**

$\frac{-9}{3} = -3$ ← **Constant of variation.**

[2] Write the equation for direct variation with $k = -3$.

$y = -3x$ ← **Replace x with −4.**

$y = -3(-4)$

$y = \mathbf{12}$

Note that the ordered pairs (0, 0) and (1, k) satisfy the equation $y = kx$. Thus, the graph of a direct variation equation passes through the origin and has a slope of k. For example, the graph of the equation

$$y = 4x$$

passes through the origin, through the point (1, 4), and has a slope of 4.

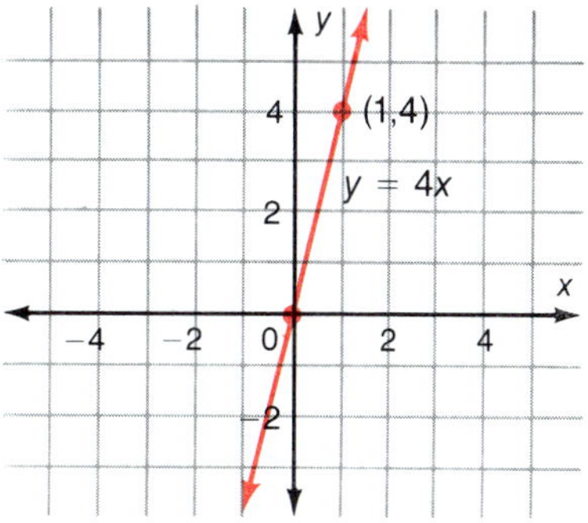

Suppose that (x_1, y_1) and (x_2, y_2) are any two points on this graph. Then, for $x_1 \neq 0$, $x_2 \neq 0$,

$$\frac{y_1}{x_1} = 4 \quad \text{and} \quad \frac{y_2}{x_2} = 4.$$

Thus, $\dfrac{y_1}{x_1} = \dfrac{y_2}{x_2}$.

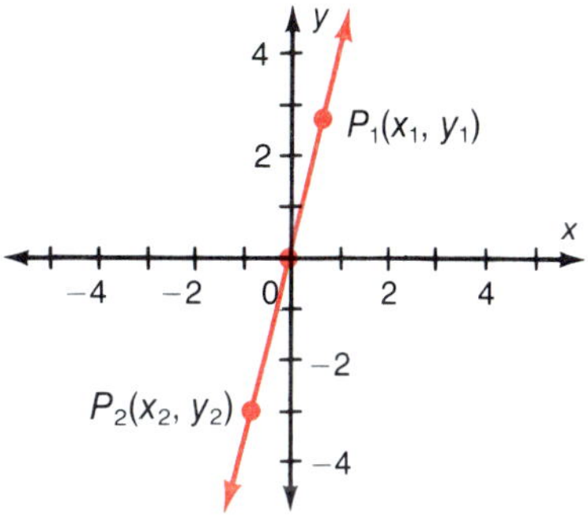

These two equal ratios are called a **proportion.** The proportion can also be written:

Means

$$y_1 : x_1 = y_2 : x_2.$$

Extremes

Read: "y_1 is to x_1 as y_2 is to x_2."

y_1 and x_2 are called the **extremes** of the proportion and x_1 and y_2 are called the **means.**

Since $\dfrac{y_1}{x_1} = \dfrac{y_2}{x_2}$ can be written as $y_1x_2 = x_1y_2$,
the product of the means of a proportion equals the product of the extremes.

You can use proportions to solve problems involving direct variation, especially if you are not required to find the constant of variation.

EXAMPLE 2 The distance sound travels in air varies directly with the time in seconds. If sound takes 18.5 seconds to travel 5550 meters, how far will it travel in one minute?

Solution: Let d = the distance in meters that sound travels in t seconds.

$$\frac{d_1}{t_1} = \frac{d_2}{t_2}$$ ← **$d_1 = 5550$ m; $t_1 = 18.5$ sec; $d_2 = ?$; $t_2 = 1$ min $= 60$ sec**

$$\frac{5550}{18.5} = \frac{d_2}{60}$$ ← **Solve for d_2.**

$$5550(60) = 18.5d_2$$ ← **Product of means = product of extremes**

$$18{,}000 = d_2$$ Sound will travel **18,000 meters** in one minute.

CLASSROOM EXERCISES

In Exercises 1–5, determine whether each equation represents a direct variation. Answer *Yes* or *No*.

1. $\frac{y}{x} = -3$ **2.** $y = \frac{1}{3}x$ **3.** $x - y = 0$ **4.** $xy = 1$ **5.** $x = -\frac{1}{2}y$

For each ordered pair in Exercises 6–10, y varies directly as x. Find k, the constant of variation.

6. $(16, 4)$ **7.** $(8, -12)$ **8.** $(1, -\frac{1}{5})$ **9.** $(b, 3b)$ **10.** $(-q, q)$

Write an equation for each direct variation. Use k as the constant of variation.

11. The total cost T, in dollars for electricity varies directly with the number, n, of kilowatt–hours used.

12. The circumference, C, of a circle varies directly as the length, d, of its diameter.

13. The cost, B, of building a house varies directly as the floor area, A, of the house.

14. The amount of commission, c, a real–estate agent receives varies directly as the amount of sales, s.

15. The volume, v, of a gas kept at constant pressure varies directly as the absolute temperature, T.

WRITTEN EXERCISES

A In Exercises 1–6, y varies directly as x.

1. If $y = -6$ when $x = 3$, find y when $x = 8$.

2. If $y = 72$ when $x = -9$, find y when $x = 36$.

3. If $y = 2$ when $x = 10$, find y when $x = 27$.

4. If $y = -1$ when $x = 3$, find y when $x = -8$.

5. If $y = 3.5$ when $x = 14$, find y when $x = 8.4$.

6. If $y = 7\frac{1}{2}$ when $x = 2\frac{1}{4}$, find y when $x = 2\frac{2}{5}$.

Write an equation for each direct variation. Use k as the constant of variation.

7. The total cost, C, of gasoline varies directly as the number of liters, l.

8. The distance, d, traveled in a given number of hours varies directly as the average rate, r.

9. For rectangles having the same base, the area, A, varies directly as the height, h.

10. The resistance, R, of a copper wire varies directly as the length, l, of the wire.

11. The perimeter, P, of an equilateral triangle varies directly as the length of a side, s.

In Exercises 12–15:

a. Determine the ratio $\frac{y}{x}$ for each ordered pair in the set of ordered pairs.

b. Tell whether each set of ordered pairs is in direct variation.

c. Write a function for each direct variation.

12. $\{(4, -2), (5, -2\frac{1}{2}), (6, -3), (7, -3\frac{1}{2})\}$

13. $\{(4, -6), (2, -3), (-8, 12), (12, -8)\}$

14. $\{(-8, -4), (-4, -1), (2, \frac{1}{2}), (4, 1)\}$

15. $\{(-2, -10), (2\frac{1}{2}, \frac{1}{2}), (1, 5), (3, 15)\}$

APPLICATIONS: Using Direct Variation

Use a proportion to solve each problem.

16. The voltage, E, in an electrical current varies directly as the current, i. If $i = 40$ when $E = -0.06$, find E when $i = 6$.

17. The pressure, p, on a submarine varies directly as its depth, d, below the surface of the ocean. If $p = 64$ when $d = 100$, find p when $d = 2640$.

18. A family decided to keep a record of the number of phone calls made. In the first week in October, 38 calls were made. At this rate, how many will be made in one year?

19. A recipe for 3 servings calls for $\frac{1}{2}$ liter of water. How many liters should be used for 5 servings?

20. A pitcher in baseball allows 10 runs in 15 innings. At this rate, how many runs does the pitcher allow per 9-inning game?

21. The distance a spring is stretched is directly proportional to the mass of the object attached to the spring. If an object with mass 50 kilograms stretches a spring a distance of 19 centimeters, find the mass of an object that will stretch the spring a distance of 4.75 centimeters.

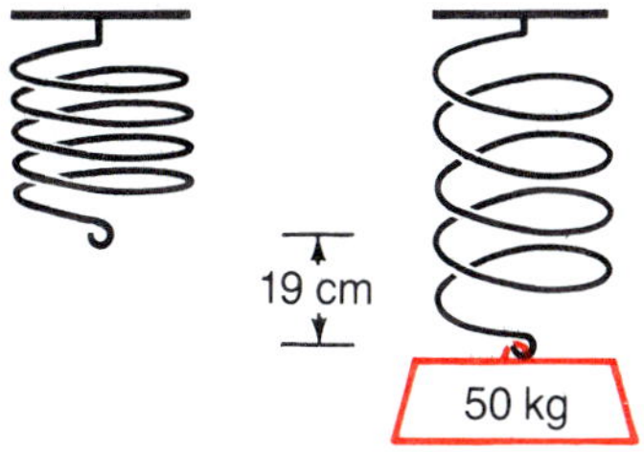

22. A car travels 245 kilometers on 35 liters of gasoline. How far will the car travel on a full tank of 100 liters?

23. In a certain area, 0.9 centimeters of snow fell in 3 hours. At this rate, how much snow would fall in 7 hours?

24. A broker receives a commission of \$4800 from the sale of real estate worth \$120,000. At the same rate, what would be the commission on the sale of \$1,000,000 worth of real estate?

25. Out of a group of 128 people, 17 are joggers. At this rate, how big a group could be expected to have 75 joggers?

Solve each proportion for x.

26. $\frac{x}{5} = \frac{3}{20}$

27. $\frac{5}{4} = \frac{x}{12}$

28. $\frac{30}{4x} = \frac{10}{2x}$

29. $\frac{1}{3} = \frac{x}{x + 8}$

30. $\frac{21 - x}{x} = \frac{2}{1}$

31. $\frac{3}{12 - x} = \frac{1}{3}$

32. $\frac{4}{x} = \frac{5}{x + 2}$

33. $\frac{x + 1}{3} = \frac{7x - 1}{5}$

B Solve each proportion for x.

34. $x : 7 = 3 : 8$ **35.** $4 : 9 = x : 2$ **36.** $(3x + 2) : 2 = (x - 1) : 1$

37. $(2x + 3) : 4 = (x - 5) : 3$ **38.** $6 : (x + 2) = 4 : (2x - 3)$ **39.** $8 : (x - 2) = 6 : (x - 3)$

40. If $y = kx$ and $k > 0$, does y increase or decrease as x increases?

41. If $y = kx$ and $k > 0$, does y increase or decrease as x decreases?

42. If $y = kx$ and $k < 0$, does y increase or decrease as x increases?

43. If $y = kx$ and $k < 0$, does y increase or decrease as x decreases?

C

44. If $\frac{x}{y} = \frac{a}{b}$, show that $\frac{x}{y} = \frac{x + a}{y + b}$. **45.** If $\frac{x}{y} = \frac{a}{b}$, show that $\frac{x - a}{y - b} = \frac{a}{b}$.

46. If $\frac{x}{y} = \frac{a}{b}$, show that $\frac{x}{a} = \frac{y}{b}$. **47.** If $\frac{x}{y} = \frac{a}{b}$, show that $\frac{y}{x} = \frac{b}{a}$.

CALCULATOR APPLICATIONS

Direct Variation

The constant key, [K], on a scientific calculator will help you to solve problems that involve direct variation.

EXAMPLE The equation $\mathbf{y = 2.1911x}$ describes the growth of x dollars invested at 16% interest compounded quarterly for 5 years. Find the growth of the following investments.

a. \$250 **b.** \$775

SOLUTION Store the constant. 2 [·] 1 9 1 1 [×] [K]

To evaluate, enter the x value and press [=].

a. 2 5 0 [=] → 547.775

b. 7 7 5 [=] → 1698.1025

EXERCISES

In Exercises 1–6, find the growth of each investment.
Equation: $\mathbf{y = 1.8061x}$ Meaning: 12% compounded quarterly for 5 years

1. \$450 **2.** \$700 **3.** \$995 **4.** \$1275 **5.** \$2040 **6.** \$5780

REVIEW CAPSULE FOR SECTION 3–8

Solve. Graph the solution set on the number line. *(Pages 51–53)*

1. $2x + 5 < 11$ **2.** $5x \le 4x - 9$ **3.** $-5x < 4x - 9$

4. $-3x + 1 \ge -8$ **5.** $4(x - 5) \le 6x - 5$ **6.** $25 \le -5(4 - x)$

3–8 Graphing Linear Inequalities

The graph of a linear equation identifies three sets of points in the coordinate plane.

1. The set of points in the half–plane **above** the line.
2. The set of points in the half–plane **below** the line.
3. The set of points **on** the line. The line is the **boundary** or **edge** of each half–plane.

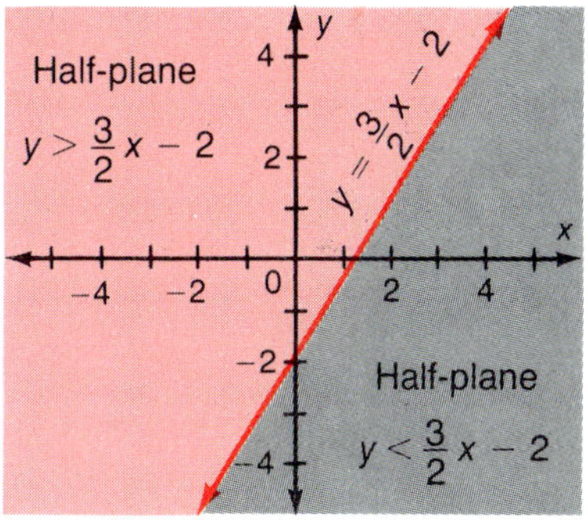

The equation of the line in the graph is $y = \frac{3}{2}x - 2$. The coordinates of every point in the plane makes one of these statements true.

$$y = \tfrac{3}{2}x - 2 \qquad y > \tfrac{3}{2}x - 2 \qquad y < \tfrac{3}{2}x - 2$$

EXAMPLE 1 Graph $2y - x > -6$.

Solution: [1] Solve for y.

$2y - x > -6$

$2y > x - 6$

$y > \frac{1}{2}x - 3$

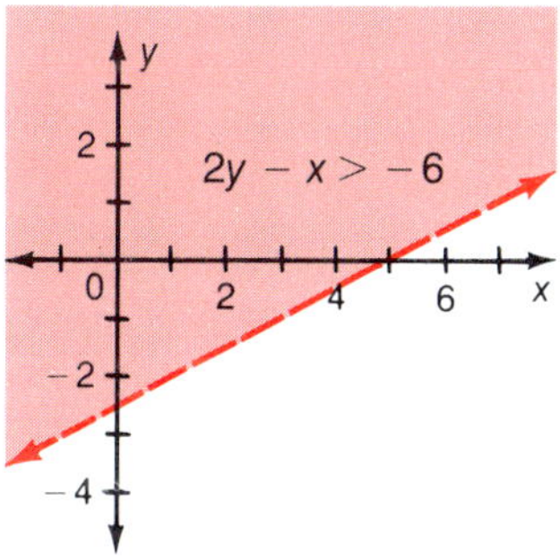

[2] Graph $y = \frac{1}{2}x - 3$.

[3] Choose a point in each half–plane.

Test to determine which half-plane contains the points which satisfy $2y - x > -6$.

Try (0, −6)	Try (0, 0).
$2(-6) - 0 > -6$?	$2(0) - 0 > -6$?
$-12 > -6$? No	$0 > -6$? Yes ✔

All points in the half-plane **above** the line make $2y - x > -6$ true. The graph is shown by the **shaded half–plane.**

In Example 1, the graph of $2y - x = -6$ is drawn as a dashed line to show that points on the line are not in the solution set of $2y - x > -6$. The graph is an **open half-plane.** For the graph of $2y - x \geq -6$ the $2y - x = -6$ is drawn as a solid line. The The graph is a **closed half–plane.**

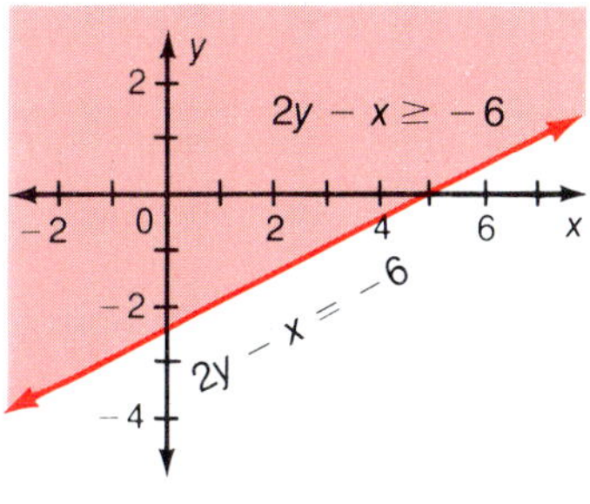

EXAMPLE 2 Graph: **a.** $-3y \geq 6$ **b.** $2x > 7$

Solutions: **a.** $-3y \geq 6$

$y \leq -2$

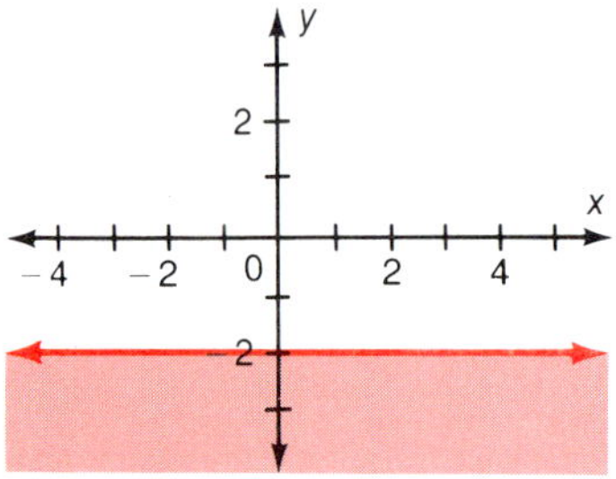

b. $2x > 7$

$x > \frac{7}{2}$

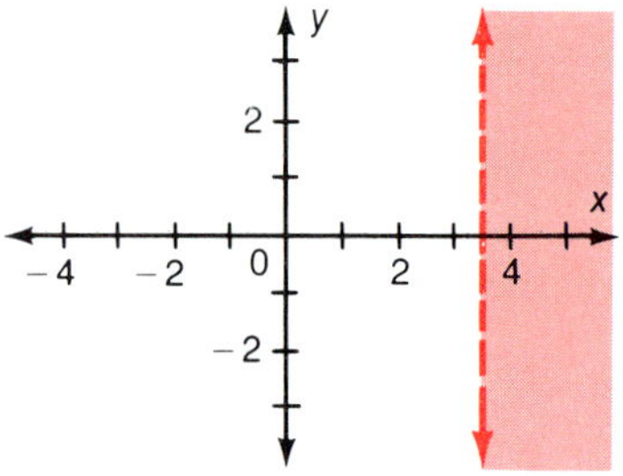

CLASSROOM EXERCISES

Tell which of the given points lie on the graph of each inequality.

$$\{(-3, -6), (0, 0), (\tfrac{1}{5}, -5), (3, 8), (4, 2)\}$$

1. $y > 0$ **2.** $x \geq 4$ **3.** $2x + y < 3$ **4.** $5x - y \geq 6$

Describe the graph of each inequality as an open half-plane, *O*, or a closed half-plane, *C*.

5. $y > x - 3$ **6.** $y \leq 3x - 9$ **7.** $y \geq 2$ **8.** $4x - y < 5$

Write the inequality that defines the shaded portion of each plane.

9.

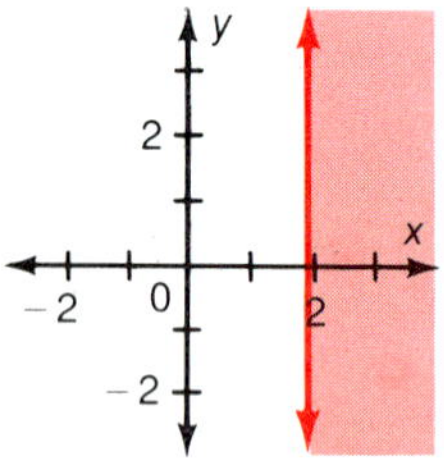

10.

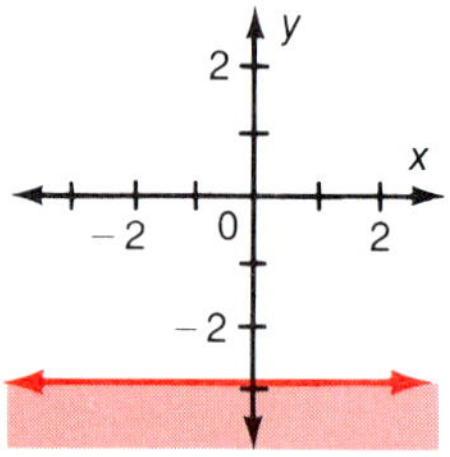

11.

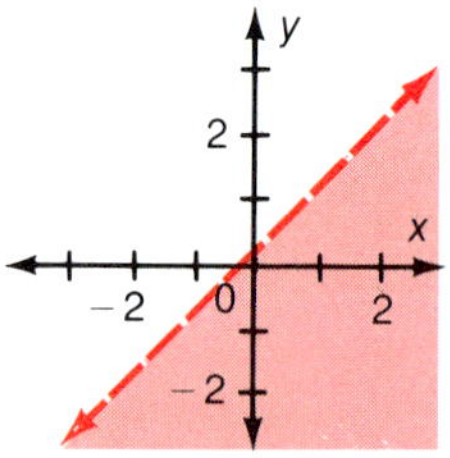

12.

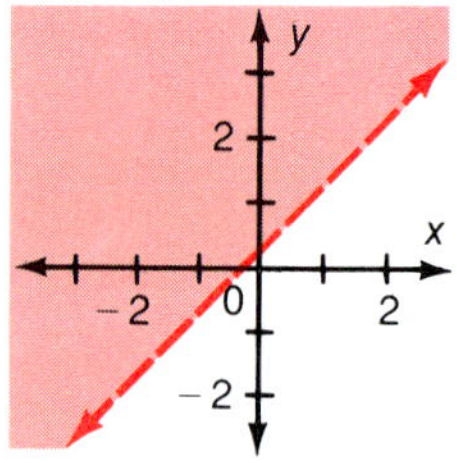

WRITTEN EXERCISES

A Graph each inequality.

1. $x \geq -2$ **2.** $y \leq -4$ **3.** $y \geq 0$ **4.** $x \leq 0$

5. $y > 3x - 2$ **6.** $y < -2x + 5$ **7.** $x + 2y > 4$ **8.** $2x + y < 6$

9. $x + 3y < 6$ **10.** $-2x + y > 8$ **11.** $y \geq 2x - 3$ **12.** $y \leq -4x + 7$

13. $x - 2y \leq 3$ **14.** $3x + y \geq 4$ **15.** $2x + 3y - 6 \leq 0$ **16.** $3x + 2y - 6 \leq 0$

B Graph each inequality.

17. $y \geq |x|$ **18.** $y < |2x|$ **19.** $y < |x| + 2$ **20.** $|x| + y \geq 3$

In Exercises 21–26, graph both inequalities in the same plane. Shade the region containing the points whose coordinates satisfy both inequalities.

21. $x < 3$ and $y > -2$

22. $x > -1$ and $y \leq 4$

23. $x + 3y > 6$ and $y \leq 4$

24. $2x - y < 8$ and $x - 2y < -2$

25. $x < 0$ and $y < 0$

26. $x - 2y > 6$ and $x - 2y < 6$

C Graph each set of points.

27. The set of all points in the coordinate plane that are 4 units from the origin.

28. The set of all points in the coordinate plane in the interior of the circle whose center is the origin and whose radius is 5 units.

29. The set of all points in the coordinate plane that are 6 units from the point $(3, -4)$.

30. The set of all points in the coordinate plane with x coordinates between 1 and 4 and with y coordinates between 1 and 6.

31. The set of all points in the coordinate plane that lie below the line on which the coordinates of every point are additive inverses.

32. The set of all points on a line that contains $Q(2, -3)$ and has a slope of $\frac{2}{5}$.

Review

1. Show that the line that contains the points $S(4, 2)$ and $T(6, -4)$ is parallel to the line containing the points $A(-1, 1)$ and $B(-2, 4)$. *(Section 3–5)*

2. Find the value of x such that the line through $M(x, 7)$ and $N(2, -3)$ is perpendicular to the line through $G(7, 2)$ and $H(2, 3)$. *(Section 3–5)*

Write the equation of the line having the given slope and y intercept. Write each equation in the slope–intercept form, and in the form $Ax + By = C$. *(Section 3–6)*

3. $m = 3$; $b = 1$ **4.** $m = \frac{1}{2}$; $b = 4$ **5.** $m = -6$; $b = -2$ **6.** $m = -\frac{1}{3}$; $b = 3$

In Exercises 7–8, y varies directly as x. *(Section 3–7)*

7. If $y = -4$ when $x = 10$, find y when $x = 25$.

8. If $y = \frac{1}{2}$ when $x = 3$, find y when $x = 12$.

9. An investment of \$12,500 earns \$906.25 in one year. At the same rate, how much will an investment of \$18,000 earn in one year? *(Section 3–7)*

Graph each inequality. *(Section 3–8)*

10. $x \geq -3$ **11.** $y \geq 4$ **12.** $y \leq -2x + 3$ **13.** $x \leq -3y - 6$

BASIC: GENERATING ORDERED PAIRS FOR GRAPHING

A computer can produce a table of ordered pairs that can be used to graph the solution set of an equation.

Problem: *Given values of m and b in the equation $y = mx + b$, write a program which prints a table of ordered pairs that satisfy the equation.*

```
100 PRINT "FOR THE EQUATION Y = MX + B, WHAT ARE M AND B";
110 INPUT M, B
120 PRINT
130 PRINT " X";TAB(10);" Y"
140 PRINT
150 FOR X = -5 TO 5
160 LET Y = M * X + B
170 PRINT X; TAB(10); Y
180 NEXT X
190 PRINT
200 PRINT "ANY MORE EQUATIONS (1 = YES, 0 = NO)";
210 INPUT Z
220 IF Z = 1 THEN 100
230 END
```

Analysis

Statements 130 and 170: The TAB function acts like the tab (tabulator) key on a typewriter. TAB(10) tells the computer, when printing a line of output, to move right to position 10 (from the left margin) of the screen or printer. In this way the headings X and Y are positioned correctly over the two columns of values.

Statements 120 and 140: These commands cause the computer to skip lines before and after the headings X and Y. 120 PRINT is equivalent to 120 PRINT"   " (that is, print nothing on that line).

Statements 150–180: The FOR and NEXT statements set up a loop that the computer executes a prescribed number of times. X is first set to −5, Y is computed (statement 160), and the ordered pair printed (statement 170). The statement NEXT X tells the machine to add one to X, changing its value to −4. The loop is now executed again. The program continues in this way for X equal to −3, −2, 1, 0, 1, 2, 3, and 4. When X = 5, the loop is executed one last time. Then the computer leaves the loop and moves to statement 190, which skips a line before the computer asks if there is another equation (statements 200–220).

To obtain more ordered pairs from the program, change statement 150 to the following.

```
150 FOR X = -5 TO 5 STEP .5
```

"STEP .5" means that instead of adding one to obtain the next value X, the computer adds 0.5. Thus X will then take on the values −5, −4.5, −4, −3.5, · · · , 4, 4.5, and 5. At the end of a FOR statement, if no STEP is listed, it is understood to be 1.

More complicated FOR statements are possible, such as the following.

```
110 FOR J = I + 1 TO N
235 FOR K = A - B TO C * C STEP 2
340 FOR D = 100 TO 1 STEP -1
432 FOR R = 2 * X TO X + 10 STEP J/2
```

EXERCISES

A

In Exercises 1–6, solve for *y*. Then use the program on page 106 to generate ordered pairs that satisfy the equation.

1. $2x + y = 10$ **2.** $x - y = 17$ **3.** $5y = 15x - 12$

4. $3x + 1.5y - 15 = 0$ **5.** $4y - 7x = 12$ **6.** $x = 10 - 6y$

7. Modify the program on page 106 so that the user enters A, B, and C of the equation $Ax + By = C$. The program then generates ordered pairs that satisfy the equation (for x from −5 to 5).

In Exercises 8–13, write a BASIC program for each problem.

8. Given (x_1, y_1) and (x_2, y_2), the coordinates of two points, print the slope of the line through the two points. Include the possibility that the slope may be undefined.

9. Given the slope of a line, print the slope of any line perpendicular to the given line. Include the possibility that the slope of the original line or of the perpendicular line may be undefined.

10. Suppose that y varies directly as x. Given an ordered pair and another value of x, compute and print the value of y that corresponds to the second x value.

B

11. Given the slope of a line and the coordinates of a point on the line, print the equation of the line in the form $y = mx + b$. Print the equation "neatly." For example, print `Y = 2 X - 5` and not `Y = 2 X + -5`; print `Y = -5 X` and not `Y = -5 X + 0`.

12. Given two points on a line, print the equation of the line in the form $y = mx + b$. If the line is vertical, print the equation in the form $x = c$. Follow the instructions in Exercise 11 about printing the equation neatly.

13. Given the coordinates of a point, print the quadrant in which the point lies or the axis on which it lies. For (0, 0) print THE ORIGIN.

Chapter Summary

IMPORTANT TERMS

Abscissa (p. 74)
Closed half-plane (p. 103)
Constant of variation (p. 98)
Coordinates (p. 74)
Direct variation (p. 98)
Extremes of a proportion (p. 99)
Horizontal line (p. 78)
Linear equation (p. 77)
Means of a proportion (p. 99)
Open half-plane (p. 103)
Ordinate (p. 74)
Origin (p. 74)
Proportion (p. 99)
Quadrant (p. 74)
Slope (p. 80)
Vertical line (p. 78)
x axis (p. 74)
x intercept (p. 94)
y axis (p. 74)
y intercept (p. 94)

IMPORTANT IDEAS

1. The points of the plane can be placed in one-to-one correspondence with ordered pairs of real numbers.
2. To graph a linear equation in two variables:
   [1] Solve the equation for y.
   [2] Make a table of ordered pairs.
   [3] Graph the points. Connect them with a straight line.
3. The graph of every linear equation is a straight line.
   Every straight line in the coordinate plane is the graph of a linear equation.
4. For any straight line, the ratio of the vertical change to the horizontal change is always the same.
5. A line with positive slope *rises* from left to right. A line with negative slope *falls* from left to right.
6. The slope of a horizontal line is 0. The slope of a vertical line is undefined.
7. If two lines have the same slope, the lines are parallel. If two lines have slopes that are negative reciprocals, the lines are perpendicular.
8. The equation of a straight line can be written in several forms:

   Point–slope form: $y - y_1 = m(x - x_1)$

   Slope–intercept form: $y = mx + b$

   General form: $Ax + By = C$
9. In a proportion, the product of the means equals the product of the extremes.
10. The graph of a linear equation divides the coordinate plane into two half-planes.

Chapter Objectives and Review

Objective: *To graph a point given its coordinates, and to determine the coordinates of a point from its graph (Section 3–1)*

Graph each point. State the quadrant in which each point lies.

1. $A(2, 6)$ **2.** $B(3, -1)$ **3.** $C(-\frac{1}{2}, 0)$ **4.** $D(0, -8)$ **5.** $E(-6, -4)$ **6.** $F(-5, 2)$

Objective: *To graph a linear equation (Section 3–2)*

Graph each equation.

7. $3x - y = 5$ **8.** $3(y - 2) = 4 - x$ **9.** $7x - 1 = 2y$

10. $-2 - x = 2(x - y)$ **11.** $\frac{2}{3}(6 - 3x) = 4(1 - y)$ **12.** $y - 1 = |3x|$

Objective: *To determine the slope of a line (Section 3–3)*

Find the slope of the line containing the two given points.

13. $A(-3, 0)$ and $B(1, 1)$ **14.** $C(-2, 4)$ and $D(3, -1)$ **15.** $E(-4, -3)$ and $G(9, -2)$

Objective: *To use linear graphs to solve problems (Section 3–4)*

A stamp collection is valued at $12,000. Each year it increases in value at a constant rate of 6% of its original value. Use this information to answer Exercises 16–18.

16. Write an equation for the value V of the collection after t years.

17. Graph the equation.

18. Use the graph to estimate the value of the collection after 8 years.

Objective: *To determine whether two lines are parallel or whether they are perpendicular (Section 3–5)*

State whether the two given lines are parallel or perpendicular. Give a reason for each answer.

19. The lines $2x - y = 3$ and $2(1 - y) = 7 - 4x$

20. The line $y - 3x = 7$ and the line containing the points $V(1, 5)$ and $N(4, 4)$

Objective: *To determine the equation of a line given* **(a)** *the slope and a point on the line, or* **(b)** *the slope and y intercept of the line, or* **(c)** *two points on the line (Section 3–6)*

21. Write the equation of the line having slope -6 and passing through the point $Q(0, 8)$.

22. Write the equation of the line having slope $\frac{7}{2}$ and y intercept 4.

23. Write the equation of the line containing the points $R(-1, 11)$ and $S(6, -3)$.

Objective: *To use a proportion to solve direct variation problems (Section 3–7)*

24. If p varies directly with q and $p = 1500$ when $q = -20$, find q when $p = 375$.

25. If d varies directly as r and $d = 35$ when $r = 14$, find d when $r = 6$.

26. A baseball player has hit 26 home runs in 182 times at bat. At this rate, how many home runs will he hit if he bats 357 times this season?

Objective: *To graph linear inequalities (Section 3–8)*

Graph each inequality.

27. $x > 3y + 4$ **28.** $y \geq -4x - 3$ **29.** $2(y - 4) \leq 4(x - 1)$

30. $2(x - 3y) < 4 - y$ **31.** $|x| < y - 3$ **32.** $x \geq |2y| - 3$

Chapter Test

1. State the ordinate of the point $P(-\frac{1}{2}, \frac{3}{4})$.

2. In which quadrant does the point $R(\frac{3}{8}, -1\frac{1}{3})$ lie?

Graph each equation.

3. $7 - 2y = 5x$ **4.** $5x - y = 3(1 + x)$ **5.** $2y + 1 = |x|$

6. What is the slope of the line segment whose endpoints are $P(6, -3)$ and $Q(-2, 1)$?

7. Find the equation of the line passing through the point $S(2, -1)$ and having a slope of $\frac{3}{2}$.

8. A truck purchased for \$21,000 depreciates linearly in value at the rate of 7% per year.

a. Write an equation for the value M of the truck after n years.

b. Graph the equation.

c. Use the graph to estimate the value of the truck after 4 years.

9. Show that the line containing the points $A(4, 5)$ and $B(10, 9)$ is perpendicular to the line containing the points $C(0, -1)$ and $D(-2, 2)$.

10. Find the value of k such that the line through the points $P(7, -1)$ and $Q(k, 1)$ is parallel to the line through the points $S(-3, 1)$ and $T(2, 2)$.

Use the given point and slope to write an equation of the line in point-slope form.

11. $P(4, -4)$, $m = -\frac{3}{2}$ **12.** $Q(-2, 5)$, $m = \frac{1}{2}$

13. Write an equation of the line through the points $R(-7, -3)$ and $S(0, -2)$. Write your answer in the form $Ax + By = C$.

14. The cost C of renting office space varies directly with the floor area rented. If it costs \$160,000 to rent 5000 square meters of space, how much office space can be rented for \$256,000?

15. Graph the inequality $x + 2y < -6$.

More Challenging Problems

1. Solve: $|x^2 + 3x + 4| = |x^2 - 2x - 8|$
2. What is the smallest possible value of xy if $2x - 4 \geq 6$ and $8 - y \leq 5$?
3. If $3 < x < 8$, $-10 < y < -7$, and $F = \frac{3x+2}{y+1}$, what is the largest possible integral value for F?
4. The sum of five numbers is 130. If the numbers are arranged in order, the difference between any two consecutive numbers is d. Find the numbers.
5. A farmer divided his sheep among his three sons. Alfred got 20% more than John, and Alfred got 25% more than Charles. John's share was 3600. How many sheep did Charles get?
6. The sum of the ages of Paul, Bill, and John is 53. Paul's age now is twice Bill's age 5 years ago, and Paul is 2 years older than John. What is the age of the oldest of the three?

Preparing for College Entrance Tests

On College Entrance tests, it is important to recognize problems that can be solved by using proportions.

REMEMBER: Compare the *same units* in the *same order*.

EXAMPLE If x tires cost y dollars, how many tires can be bought with z dollars?

(a) $\frac{y}{xz}$ **(b)** $\frac{xy}{z}$ **(c)** $\frac{xz}{y}$ **(d)** $\frac{x+y}{z}$

Think: $\frac{x \text{ tires}}{y \text{ dollars}} = \frac{?}{z \text{ dollars}}$ ← **Tires** ← **Dollars**

Solution: Let n = the number of tires.

$$\frac{x}{y} = \frac{n}{z}$$

$$yn = xz$$

$$n = \frac{xz}{y}$$ **Answer: c**

Choose the best answer. Choose *a*, *b*, *c*, or *d*.

1. If it takes s sacks of grain to feed c chickens, how many sacks of grain are needed to feed k chickens?

(a) $\frac{ck}{s}$ **(b)** $\frac{sk}{c}$ **(c)** $\frac{cs}{k}$ **(d)** $\frac{c}{sk}$

2. If a machine fills 150 bottles in m minutes, how many minutes will it take to fill b bottles?

(a) $\frac{mb}{150}$ (b) $\frac{150m}{b}$ (c) $\frac{150b}{m}$ (d) $\frac{150}{mb}$

3. If Jim earns d dollars in h hours, how many dollars will he earn in $h + 20$ hours?

(a) $\frac{20d}{h}$ (b) $d + \frac{20d}{h}$ (c) $21d$ (d) $\frac{dh}{h + 20}$

4. If $6n$ cans fill $\frac{n}{2}$ cartons, how many cans does it take to fill 2 cartons?

(a) $\frac{3}{2}n^2$ (b) $6n^2$ (c) 24 (d) $24n$

5. A car travels d kilometers in 30 minutes. At that rate, how many hours will it take the car to travel $10d$ kilometers?

(a) 5 (b) $50d$ (c) $20d^2$ (d) 300

6. For nonzero numbers a, b, and c, $2a = 3b$ and $6b = 7c$. Then $\frac{a}{c} =$ __?__

(HINT: Since $2a = 3b$, $4a = 6b$. But $6b = 7c$. Then $4a = 7c$.)

(a) $\frac{9}{7}$ (b) $\frac{4}{7}$ (c) $\frac{7}{9}$ (d) $\frac{7}{4}$

7. For nonzero numbers x, y, and z, $3x = 4y$ and $12y = 5z$. Then $\frac{z}{x} =$ __?__

(a) $\frac{16}{5}$ (b) $\frac{9}{5}$ (c) $\frac{6}{5}$ (d) $\frac{5}{16}$

8. For nonzero numbers p, q, and r, $2p = \frac{q}{3}$ and $3p = \frac{r}{4}$. Then $\frac{q}{r} =$ __?__

(a) $\frac{2}{1}$ (b) $\frac{1}{4}$ (c) $\frac{1}{2}$ (d) $\frac{3}{4}$

9. For nonzero numbers a, b, c, and d, $2a = b$, $3b = 2c$, and $c = 3d$. Then $\frac{a}{d} =$ __?__

(a) $\frac{1}{2}$ (b) $\frac{3}{2}$ (c) 2 (d) 1

10. 56 : 70 as 12 : __?__

(a) 14 (b) 15 (c) 18 (d) 20

11. For nonzero numbers p, q, r, and s, $\frac{p}{q} = \frac{r}{s}$. Which of the following *must* be true?

(a) $\frac{p}{r} = \frac{s}{q}$ (b) $\frac{r}{s} = \frac{q}{p}$ (c) $\frac{p - q}{q} = \frac{s - r}{r}$ (d) $\frac{p + q}{p} = \frac{r + s}{r}$

12. The length of each side of a square is s meters. What is the ratio of the perimeter of the square (in meters) to its area (in square meters)?

(a) $\frac{s}{4}$ (b) $\frac{4}{1}$ (c) $\frac{4}{s}$ (d) $\frac{2}{1}$

CHAPTER **4**

Systems of Sentences

4-1 Solving Linear Systems: Graphing

Shown below are two systems of linear equations and a system of linear inequalities.

Systems of Equations

$$\begin{cases} x + y + 1 = 0 \\ 2x + y + 2 = 0 \end{cases} \qquad \begin{cases} x + 2y + z = 8 \\ 2x - 2y + z = 4 \\ -x + y + z = -1 \end{cases}$$

System of Inequalities

$$\begin{cases} x - y < 8 \\ x + 2y > 6 \end{cases}$$

Each system is actually a *compound sentence.*

$$\begin{cases} x + y + 1 = 0 \\ 2x + y + 2 = 0 \end{cases} \text{ means } x + y + 1 = 0 \text{ \underline{and} } 2x + y + 2 = 0.$$

For an ordered pair to be a solution of the system, the ordered pair must be a solution of each sentence in the system.

Definition

> The **solution set of a system of sentences** is the set of all ordered pairs that makes **all** the sentences true.

Geometrically, the solution set of a system of sentences is the set of points **common** to the graphs of the sentences of the system.

EXAMPLE 1 Solve by graphing: $\begin{cases} x - 2y = 7 \\ x + y = -2 \end{cases}$

Solution:

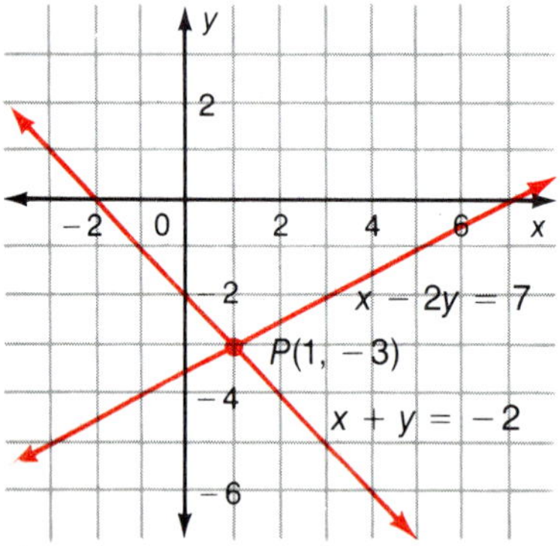

Graph each equation.

The graphs intersect at *P*(1, −3).

Check:

$$\begin{aligned} x - 2y &= 7 \\ 1 - 2(-3) &\stackrel{?}{=} 7 \\ 1 + 6 &\stackrel{?}{=} 7 \\ 7 &\stackrel{?}{=} 7 \quad \text{Yes } \checkmark \end{aligned} \qquad \begin{aligned} x + y &= -2 \\ 1 + (-3) &\stackrel{?}{=} -2 \\ 1 - 3 &\stackrel{?}{=} -2 \\ -2 &\stackrel{?}{=} -2 \quad \text{Yes } \checkmark \end{aligned}$$

Replace *x* with 1 and *y* with −3 in each equation.

Solution set: $\{(1, -3)\}$

You can also use a system of sentences to solve word problems.

EXAMPLE 2 The sum of two numbers, x and y, is 9 and their difference is 3. Write a system of equations to describe the relationship. Use a graph to solve the system.

Solution: $\begin{cases} x + y = 9 \\ x - y = 3 \end{cases}$ ← **The sum of x and y is 9.** ← **The difference is 3.**

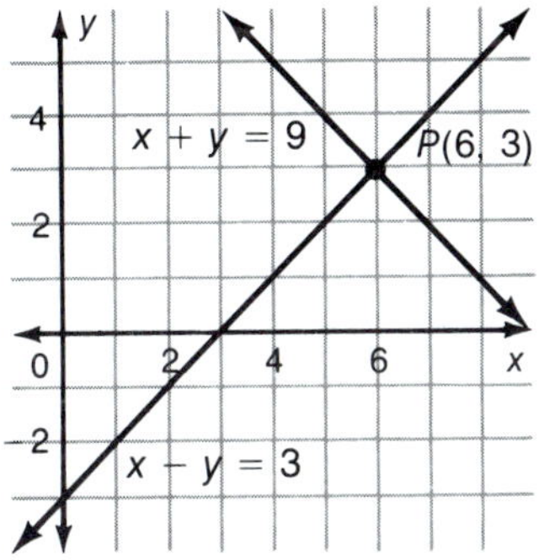

The graphs intersect at $P(6, 3)$.

Check: $x + y = 9$ | $x - y = 3$

$6 + 3 \stackrel{?}{=} 9$ | $6 - 3 \stackrel{?}{=} 3$

$9 \stackrel{?}{=} 9$ Yes ✓ | $3 \stackrel{?}{=} 3$ Yes ✓

Solution set: $\{(6, 3)\}$

It is not always possible to find exact solutions when solving a system of equations graphically. More precise techniques will be developed in Sections 4–2, 4–3, and 4–4.

CLASSROOM EXERCISES

Determine whether the given ordered pair is a solution of the system.

1. $\begin{cases} x + 3y = 3 \\ 2x + y = -4 \end{cases}$; $(-2, 2)$
2. $\begin{cases} x + 2y - 2 = 0 \\ x = -2 \end{cases}$; $(-2, 2)$
3. $\begin{cases} x + 6y = -6 \\ y + x = 5 \end{cases}$; $(-4, 1)$

Write a system of equations in two variables to describe each relationship.

4. The sum of two numbers, x and y, is 14; their difference is 4.
5. The product of two numbers, x and y, is 6. Twice the second minus the first equals 1.

WRITTEN EXERCISES

A In Exercises 1–9, solve each system by graphing.

1. $\begin{cases} x + y = 0 \\ x - y = 3 \end{cases}$
2. $\begin{cases} y = x + 4 \\ y = 2x + 4 \end{cases}$
3. $\begin{cases} y = 6 \\ 2x + y = -2 \end{cases}$
4. $\begin{cases} y = \frac{1}{2}x - 3 \\ x = 0 \end{cases}$
5. $\begin{cases} x + 3 = 0 \\ y - 7 = 0 \end{cases}$
6. $\begin{cases} x = y \\ 2y + 3x = -10 \end{cases}$
7. $\begin{cases} 4x + 3y = 0 \\ 2x - 6y = 5 \end{cases}$
8. $\begin{cases} 4x + 2y = 6 \\ 3x - 4y = 10 \end{cases}$
9. $\begin{cases} 2x + y = -2 \\ 4x + 2y = -1 \end{cases}$

In Exercises 10–16, write an equation to describe each relationship. Solve the systems in Exercises 10, 11, 13, and 14.

10. Two numbers, x and y; their sum is 5 and their difference is 1.

11. Two numbers, n_1 and n_2; their sum is 12 and their difference is 3.

12. Two numbers, a and b; the sum of their reciprocals is 24 and the difference of their reciprocals is 4.

13. Two numbers, p and q; the sum of the first and twice the second is 34, while the difference of the first and four times the second is 4.

14. Two acute angles, x and y; their sum is 87° and their difference is 19°.

15. Two numbers, r and s; the difference of their squares is 0 and the sum of the numbers is 0.

16. Two numbers, a and b; their product is 12, while the difference of the first and twice the second is 2.

APPLICATIONS: Using Linear Systems

B Write a system of equations to describe each relationship. Solve the system by graphing.

17. The dimensions of a rectangular yard with area A are l and w. The area increases by 12 square meters when the length is increased by 2 meters. The area decreases by 8 square meters when the width is decreased by 1 meter. Find the dimensions of the yard.

18. Rowing downstream, a person travels 6 kilometers in 1 hour. The return trip takes 2 hours. The person's rowing rate is r; the rate of the stream is s. Find the person's rowing rate and the rate of the stream.

19. A coin bank contains x dimes and y quarters. There are 22 coins in the bank. The value of the coins is \$3.25. Find the number of dimes and quarters in the bank.

20. Sam buys n bananas at \$0.20 each and q tomatoes at \$0.15 each, and spends \$2.20. He buys 12 items in all. Find the number of bananas and tomatoes he buys.

21. The sum of two numbers c and d is 17. Twice c is 2 less than d. Find the two numbers.

REVIEW CAPSULE FOR SECTION 4–2

Write each equation in the form $Ax + By = C$. *(Pages 77–79)*

1. $2y = 7x - 3$ **2.** $3x = 4 - 5y$ **3.** $y + 2x - 1 = 0$ **4.** $4x - 7y + 3 = 0$

Write the additive inverse and the multiplicative inverse for each number. *(Pages 18–21)*

5. -2 **6.** $\frac{2}{3}$ **7.** $-\frac{3}{8}$ **8.** $1\frac{3}{4}$ **9.** $\frac{1}{7}$ **10.** $-\frac{1}{5}$

4–2 Solving Linear Systems: Addition Method

The solution set of the following system is $\{(1, -3)\}$.

$$\begin{cases} x - 2y = 7 & \mathbf{1} \\ x + y = -2 & \mathbf{2} \end{cases}$$

If you add Equations **1** and **2**, right side to right side and left side to left side, you get Equation **3**.

$$(x - 2y) + (x + y) = 7 + (-2)$$
$$2x - y = 5 \qquad \mathbf{3}$$

Verify that $(1, -3)$ is also a solution of Equation **3**.

If you multiply Equations **1** and **2** by any nonzero real number and add the results, the ordered pair $(1, -3)$ will also be a solution of the new equation.

$$\begin{array}{rcrl} 5(x - 2y = 7) & \longrightarrow & 5x - 10y = 35 & \longleftarrow \textbf{Add.} \\ -3(x + y = -2) & \longrightarrow & -3x - 3y = 6 & \\ \hline & & 2x - 13y = 41 & \quad \mathbf{4} \end{array}$$

Verify that $(1, -3)$ is also a solution of Equation **4**.

Equations **1**, **2**, **3**, and **4** can be combined in pairs to form **equivalent systems.** They are equivalent systems because the solution set, $\{(1, -3)\}$, is the same for both systems.

For example, the following are equivalent systems.

$$\begin{cases} x - 2y = 7 & \mathbf{1} \\ x + y = -2 & \mathbf{2} \end{cases}$$

and

$$\begin{cases} 2x - y = 5 & \mathbf{3} \\ 2x - 13y = 41 & \mathbf{4} \end{cases}$$

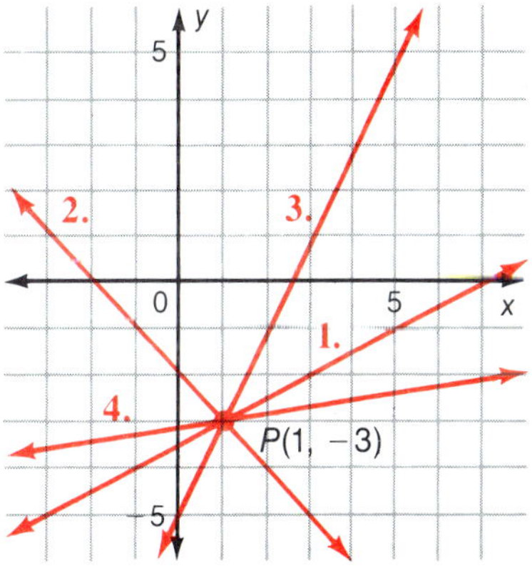

Geometrically, these equations can be represented by four lines passing through the same point,

$$P(1, -3)$$

This suggests the following theorem.

Theorem 4–1

> If an ordered pair (a, b) is in the solution set of the two linear equations $A_1x + B_1y = C_1$ and $A_2x + B_2y = C_2$, then the sum of any multiples of these equations has (a, b) in its solution set.

You use Theorem 4–1 when you solve a system of equations by the *Addition Method*.

Summary

> **Addition Method for Solving Linear Systems**
>
> Choose multiples of the equations that eliminate one of the variables when the equations are added.

To eliminate one of the variables, the coefficients of that variable in both equations must be additive inverses.

EXAMPLE 1 Solve the following system by eliminating the y terms.

$$\begin{cases} 2x + y = 1 & \textbf{1} \\ x + 3y = 13 & \textbf{2} \end{cases}$$

Solution:

$-6x - 3y = -3$ ← **Multiply 1 by −3.**

$x + 3y = 13$ ← **Multiply 2 by 1.**

$-5x = 10$ ← **Solve for x.**

$x = -2$

$2x + y = 1$ ← **Solve for *y* by replacing x with −2 in one of the original equations.**

$2(-2) + y = 1$

$-4 + y = 1$

$y = 5$

Check:

$x + 3y = 13$ ← **Replace x with −2 and *y* with 5 in equation 2.**

$-2 + 3(5) \stackrel{?}{=} 13$

$13 \stackrel{?}{=} 13$ Yes ✓

Solution set: $\{(-2, 5)\}$

In Example 2, the system is solved by eliminating the x terms.

EXAMPLE 2 Solve: $\begin{cases} 3x + 2y = 5 & \textbf{1} \\ 4x + 4y = 10 & \textbf{2} \end{cases}$

Solution: Multiply Equation 1 by −4 and Equation 2 by 3.

$-4(3x + 2y = 5) \rightarrow -12x - 8y = -20$

$3(4x + 4y = 10) \rightarrow 12x + 12y = 30$

$4y = 10$

$y = 2\frac{1}{2}$ ← **Solve for x by replacing *y* with $2\frac{1}{2}$ in one of the original equations.**

$3x + 2y = 5$

$3x + 2(2\frac{1}{2}) = 5$

$3x + 5 = 5$

$3x = 0$

$x = 0$ The check is left for you.

Solution set: $\{(0, 2\frac{1}{2})\}$

In Example 2 you could have eliminated y rather than x. The solution set would be identical.

CLASSROOM EXERCISES

In Exercises 1–4, determine the number by which you would multiply each side of each equation in order to eliminate the *y* terms.

1. $\begin{cases} 3x - 5y = 7 \\ x + 10y = 0 \end{cases}$
2. $\begin{cases} x - 4y = 2 \\ 2x + 5y = 1 \end{cases}$
3. $\begin{cases} 7y - 2x = 6 \\ x + y = -3 \end{cases}$
4. $\begin{cases} 9y - 3x = -6 \\ 3y - 2x = -1 \end{cases}$

In Exercises 5–8, determine the number by which you would multiply each side of each equation in order to eliminate the *x* terms.

5. $\begin{cases} 2x + y = 6 \\ x - 3y = 4 \end{cases}$
6. $\begin{cases} 3x - y = 1 \\ 2x + 3y = 2 \end{cases}$
7. $\begin{cases} 2y = 6x - 4 \\ 3x - y = 2 \end{cases}$
8. $\begin{cases} 5y = 3 + 3x \\ 10y = 18 - 6x \end{cases}$

Solve by the addition method.

9. $\begin{cases} x + y = 6 \\ x - y = 2 \end{cases}$
10. $\begin{cases} x + y = -6 \\ 2x - y = 2 \end{cases}$
11. $\begin{cases} 4x + 2y = -8 \\ x - 2y = -7 \end{cases}$
12. $\begin{cases} -2x + y = 6 \\ 2x + y = -2 \end{cases}$

WRITTEN EXERCISES

A Solve each system by the addition method.

1. $\begin{cases} 4x + 3y = 17 \\ 2x + 3y = 13 \end{cases}$
2. $\begin{cases} 5x + y = 14 \\ 2x + y = 5 \end{cases}$
3. $\begin{cases} x - 3y = 7 \\ x - 5y = 13 \end{cases}$
4. $\begin{cases} 3x + y = 15 \\ 3x + 7y = 15 \end{cases}$
5. $\begin{cases} x + y = 9 \\ 5x - y = 3 \end{cases}$
6. $\begin{cases} 4a + b = 14 \\ 6a + b = 16 \end{cases}$
7. $\begin{cases} a + 3b = 7 \\ a + b = 5 \end{cases}$
8. $\begin{cases} m + 5n = 11 \\ m - n = 5 \end{cases}$
9. $\begin{cases} 2x - 3y = -7 \\ 4x - 5y = -9 \end{cases}$
10. $\begin{cases} 2x + 3y = -6 \\ 3x + 2y = 1 \end{cases}$
11. $\begin{cases} 2x + 5y = 0 \\ 3x - 2y = -19 \end{cases}$
12. $\begin{cases} 8x + 5y = -13 \\ 3x - 2y = -1 \end{cases}$
13. $\begin{cases} 2x + y = 1 \\ 3x - 2y = 8 \end{cases}$
14. $\begin{cases} 4x + 3y = 0 \\ 2x - 6y = 5 \end{cases}$
15. $\begin{cases} 3x + 2y = 9 \\ 2x - 6y = 6 \end{cases}$
16. $\begin{cases} 4w - z = 22 \\ 6w - 7z = 44 \end{cases}$
17. $\begin{cases} b = 3a - 5 \\ 3a - b = 8 \end{cases}$
18. $\begin{cases} 2p + 5q = 11 \\ 3p - 2q = 4 \end{cases}$
19. $\begin{cases} 0.1c - 0.2d = 0.1 \\ 0.4c + 0.6d = 8 \end{cases}$
20. $\begin{cases} 0.04a + 0.05b = 44 \\ a + b = 1000 \end{cases}$
21. $\begin{cases} g = 2(8 + h) \\ 3h = 2(1 + 2g) \end{cases}$

B Solve the following compound sentences.

22. $2x + 3y = -1$ <u>and</u> $5x + 6y = -3$
23. $9x + 2y = -12$ <u>and</u> $3x - 4y = -11$
24. $\{(x, y) : 3x + 2y = 5$ <u>and</u> $6x - 4y = -2\}$
25. $\{(x, y) : 6x - 3y = -14$ <u>and</u> $3x + 12y = 38\}$
26. $\{(x, y) : 7x + 4y = 8$ <u>and</u> $4x - 11y = 9\}$
27. $\{(x, y) : 9x + 8y - 6 = 0$ <u>and</u> $5x - 2y + 16 = 0$

In Exercises 28–33, solve each system for x and y.

28. $\begin{cases} x+y=a+b \\ x-y=a-b \end{cases}$

29. $\begin{cases} x+y=a \\ x-y=b \end{cases}$

30. $\begin{cases} 3x+2y=a \\ x-y=b \end{cases}$

31. $\begin{cases} 2x+3y=13a \\ 3x-4y=-6a \end{cases}$

32. $\begin{cases} x+y=2w \\ 3x-y=2(w+h) \end{cases}$

33. $\begin{cases} 4x+y=-2c \\ 2x-3y=3c \end{cases}$

C

34. Solve the general linear system at the right for x and y in terms of a_1, b_1, c_1, and c_2.

$$\begin{cases} a_1x+b_1y=c_1 \\ a_2x+b_2y=c_2 \end{cases}$$

35. Find the coordinates of the vertices of the triangle formed by the graphs of the following lines.

$x-y=-3 \qquad 3x+4y=5 \qquad 6x+y=17$

36. Show that the graphs of the following lines form a parallelogram. Give the coordinates of the vertices of the parallelogram.

$2x-y=4 \qquad x+3y=6 \qquad 2x-y=8 \qquad x+3y=1$

37. Find an equation of the line which passes through the origin and through the intersection of the lines defined by $4x+y=2$ and $2x-3y=8$.

In Exercises 38–41, consider the system: $\begin{cases} 4x+2y-8=0. \\ -3x-4y+1=0 \end{cases}$

The set of all sums of multiples of two linear equations has a family of lines with a common point as graphs. The family of lines related to this system may be defined by the equation $s(4x+2y-8)+r(-3x-4y+1)=0$, where s and r are any real numbers not both zero.

38. Graph the original pair of equations.

39. On the same coordinate axes, graph the equations you get by substituting the given values of r and s in $s(4x+2y-8)+r(-3x-4y+1)=0$.

a. $r=1$ and $s=1$

b. $r=2$ and $s=1$

c. $r=2$ and $s=3$

40. Do the lines in Exercise 35 have a point in common? If so, name the point.

41. Let $P(x_1, y_1)$ be the point common to both $4x+2y-8=0$ and $-3x-4y+1=0$. Show that no matter what r and s are (not both zero), the resulting line contains the point $P(x_1, y_1)$.

REVIEW CAPSULE FOR SECTION 4–3

In Exercises 1–4, replace y with $(2-3x)$. Then solve for x. *(Pages 10–13, 44–47)*

1. $x+y=2$
2. $2x+3y=12$
3. $5x-7y=14$
4. $2x-y+8=0$

In Exercises 5–8, replace x with $(1-4y)$. Then solve for y. *(Pages 10–13, 44–47)*

5. $2x+y=-5$
6. $y-x=9$
7. $y-x+3=0$
8. $y-\frac{1}{2}x=0$

4–3 Solving Linear Systems: Substitution Method

To solve a linear system by the **substitution method,** follow these steps.

Summary

Substitution Method for Solving Linear Systems

1. Solve one equation of the system for one variable in terms of the other variable.
2. Substitute this value in the other equation and solve.
3. Substitute the value from step 2 in one of the original equations. Solve for the value of the second variable.
4. Check in the original equations of the system.

EXAMPLE Solve and check: $\begin{cases} 2x - 3y = 13 & \quad 1 \\ 3x + y = 3 & \quad 2 \end{cases}$

Solution:

1. $3x + y = 3$ ← **Solve Equation 2 for *y*.**

$y = 3 - 3x$

2. $2x - 3y = 13$ ← **In Equation 1, replace *y* with 3 − 3*x*.**

$2x - 3(3 - 3x) = 13$

$2x - 9 + 9x = 13$

$11x = 22$

$x = 2$

3. $3x + y = 3$ ← **In Equation 2, find *y* when *x* = 2.**

$3(2) + y = 3$

$y = -3$ ← **Check (2, −3) in both equations.**

Check:

4. $2x - 3y = 13$

$2(2) - 3(-3) \stackrel{?}{=} 13$

$13 \stackrel{?}{=} 13$ Yes ✓

$3x + y = 3$

$3(2) - 3 \stackrel{?}{=} 3$

$3 \stackrel{?}{=} 3$ Yes ✓

Solution set: {(2, −3)}

CLASSROOM EXERCISES

Solve each equation for *y* in terms of *x*.

1. $y + x = 6$ **2.** $2x - y = -4$ **3.** $x + \frac{y}{6} = 7$ **4.** $5y - 8 = 3x$

Solve each equation for *x* in terms of *y*.

5. $x - y = 8$ **6.** $2x + y = 6$ **7.** $\frac{x}{2} - \frac{y}{3} = 1$ **8.** $3y + 2x = 17$

WRITTEN EXERCISES

A Solve by the substitution method.

1. $\begin{cases} 2x + y = 7 \\ y = 3 \end{cases}$
2. $\begin{cases} y = 2x + 4 \\ x = -2 \end{cases}$
3. $\begin{cases} x - 2y = 8 \\ x = 5 - y \end{cases}$
4. $\begin{cases} x = 3y \\ x - y = 8 \end{cases}$
5. $\begin{cases} 2x + 7y = 1 \\ 2x = 10 + 2y \end{cases}$
6. $\begin{cases} 5x + y = 7 \\ 3x + 2y = 0 \end{cases}$
7. $\begin{cases} x + y = 6 \\ 2x - 3y = 2 \end{cases}$
8. $\begin{cases} 2x + y = -4 \\ 3x + 2y = -5 \end{cases}$
9. $\begin{cases} 3x - 2y = -16 \\ 2x + y = -9 \end{cases}$

B Rewrite each equation so that it has integral coefficients. Then solve the system. Use the method you prefer.

10. $\begin{cases} \frac{x}{2} + \frac{y}{4} = 4 \\ \frac{2x}{3} + \frac{3y}{2} = 11\frac{2}{3} \end{cases}$
11. $\begin{cases} \frac{3x}{2} + \frac{y}{4} = 7 \\ \frac{x}{5} - \frac{2y}{3} = 2\frac{1}{3} \end{cases}$
12. $\begin{cases} \frac{x}{3} + \frac{y}{5} = -\frac{1}{5} \\ \frac{2x}{3} - \frac{3y}{4} = -5 \end{cases}$
13. $\begin{cases} \frac{a}{2} - \frac{2b}{3} = 2\frac{1}{3} \\ \frac{3a}{2} + 2b = -25 \end{cases}$
14. $\begin{cases} 2x - 1.5 = y \\ 2y + 0.5 = 3x \end{cases}$
15. $\begin{cases} a - \frac{7b}{3} = -21 \\ a + \frac{b}{5} = 17 \end{cases}$
16. $\begin{cases} x - 2.3 = -5y \\ 3y = 2x - 2 \end{cases}$
17. $\begin{cases} 10x + 5y = 3.5 \\ 3x + 0.6 = 4y \end{cases}$
18. $\begin{cases} 2x - 3y = 0.1 \\ 5y - 0.3 = 3x \end{cases}$

C

19. Find an equation of the line which passes through $P(3, 2)$ and through the point of intersection of $y = -\frac{1}{4}x + \frac{1}{2}$ and $x + \frac{3}{2}y = -\frac{1}{2}$.
20. Find an equation of the line which passes through $Q(1, 3)$ and through the point of intersection of $y = 2 - 4x$ and $2x - 3y = 8$.
21. Find an equation of the line which passes through $Q(0, 0)$ and through the point of intersection of $\frac{8}{x} + \frac{3}{y} = 11$ and $-\frac{7}{x} - \frac{2}{y} = -4$.

REVIEW CAPSULE FOR SECTION 4-4

Write an equation to represent each statement. *(Pages 22–29)*

1. The sum of the tens digit, t, and the units digit, u, of a number is 16.
2. The cost of x crates of oranges at \$10.50 per crate and y crates of peaches at \$12.90 per crate is \$775.
3. A coin bank contains x nickels and y dimes. There are 42 coins in all.
4. The total value of x dimes and y quarters is \$18.90.
5. The cost of r crates of peas at \$10.50 per crate and t crates of apples at \$11.00 per crate is \$1615.00.

Problem Solving and Applications

4–4 Using Linear Systems to Solve Word Problems

EXAMPLE 1 A trucker buys crates of apples and pears to sell at a Farmer's Market. The apples cost \$6.00 per crate and the pears cost \$5.50 per crate. Each crate weighs 25 pounds. If the truck can carry 5000 pounds and the trucker has \$1180 to spend, how many crates of each kind of fruit can he buy?

Solution: [1] Represent the variables.

Let $a=$ the number of crates of apples.

Let $b=$ the number of crates of pears.

[2] Organize the given information in a table.

Item	Number of Crates	Total Cost	Total Weight
Apples	a	$6.00a$	$25a$
Pears	p	$5.50p$	$25p$

[3] Identify the conditions. Write an equation for each condition.

$$\begin{cases} 6a + 5.5p = 1180 \\ 25a + 25p = 5000 \end{cases}$$

$6a + 5.5p = 1180$ ← **Condition 1: Cost of apples + Cost of pears = Total cost**

$25a + 25p = 5000$ ← **Condition 2: Weight of apples + Weight of pears = Total weight**

[4] Solve the system.

$a =$ **160 crates of apples** $p =$ **40 crates of pears**

Check: [5] Check in the original conditions of the problem.

Condition 1: $(160 \times \$6.00) + (40 \times 5.50) = \$1180$? Yes ✓

Condition 2: $(160 \times 25) + (40 \times 25) = 5000$? Yes ✓

These terms are used in some motion problems.

Tail wind: A wind blowing in the same direction as a plane is heading

Head wind: A wind blowing in the opposite direction in which a plane is heading.

Speed upstream: The speed of a boat headed in a direction opposite to that of the current

Speed downstream: The speed of a boat headed in the same direction as the current

EXAMPLE 2 With a given tailwind, a plane traveled 3840 kilometers in 4 hours. On the return trip, the same wind increased the plane's travel time by 1 hour. Find the speed of the plane going and returning.

Solution: [1] Let $x =$ the speed of the plane (in still air) in km/h.
Let $y =$ the speed of the wind in km/h.

[2]

Trip	Rate (r)	Time (t)	Distance $d = rt$
Going (with tailwind)	$x + y$	4	$4(x + y)$
Returning (with headwind)	$x - y$	5	$5(x - y)$

[3] $\begin{cases} 4(x + y) = 3840 \\ 5(x - y) = 3840 \end{cases}$ ← **Condition 1**, ← **Condition 2**

[4] $x = 864$ km/h $\quad y = 96$ km/h

[5] The check is left for you.

Speed of plane going: $864 + 96 =$ **960 km/h**

Speed of plane returning: $864 - 96 =$ **768 km/h**

Summary

To use linear systems to solve word problems involving two variables:

[1] Use two different variables to represent the unknowns.

[2] Organize the given information in a table, if possible.

[3] Identify the conditions of the problem. Write an equation to represent each condition.

[4] Solve the system.

[5] Check in the original conditions of the problem.

CLASSROOM EXERCISES

Complete each table. Then write a system of equations that could be used to solve each problem.

1. A fruit store manager plans to make baskets of oranges and apples to sell as gifts. Each basket will contain 24 pieces of fruit and will sell for \$7.50. The manager wants to get 25¢ for each orange, 20¢ for each apple, and \$1.95 for the basket. How many oranges will there be in each basket?

Item	Quantity	Value
Apples	x	?
Oranges	y	?

2. A boat can travel 24 miles downstream in 2 hours. It can travel the same distance upstream in 3 hours. Find the rate, r, of the boat in still water and the rate, c, of the current.

Direction	Rate	Time	Distance
Downstream	$r + c$	?	?
Upstream	?	?	?

3. A coin bank contains 30 coins in dimes and nickels. Their combined value is \$2.10. How many coins of each kind are there?

Item	Number	Value
Dimes	x	?
Nickels	y	?

4. The sum of the digits of a two-digit number is 11. When the digits are reversed, the new number formed is 20 less than twice the original number. Find the original number.

Tens Digit	Units Digit	Original Number	New Number
t	u	$10t + u$	?

WORD PROBLEMS

A Use a system of two linear equations to solve each problem.

1. A seed company sells 1-ounce packages containing a mixture of lettuce seeds. Green Stem seeds sell for 50¢ per ounce and White Lime seeds sell for 35¢ per ounce. The 1-ounce mixed package sells for 40¢. How many ounces of each kind of seed are in the package?
2. An advertiser plans to spend \$127,000 on 20 one-minute TV commercials. A one-minute spot before 9 P.M. costs \$8000; after 9 P.M., the same spot costs \$5000. How many one-minute spots before and after 9 P.M. can be bought?
3. A piano tuner can tune an upright in 3 hours and a spinet in 2 hours. She charges \$40 for tuning an upright and \$30 for tuning a spinet. The tuner wishes to work 40 hours per week and to earn \$560 per week. How many of each kind of piano must she tune each week?
4. An apartment building contains 80 units. Some of these are one-bedroom units renting for \$425 per month and some are two-bedroom units renting for \$550 per month. When the building is entirely rented, the total monthly rental income is \$38,375. How many apartments of each type are there?
5. An 18% alcohol solution is mixed with a 45% alcohol solution to produce 12 ounces of a 36% alcohol solution. How many ounces of the 18% solution and the 45% solution must be used?
6. An alloy containing 60% copper is made by melting together two alloys that are 25% copper and 75% copper, respectively. How many kilograms of each alloy must be used to produce 200 kilograms of the 60% alloy?
7. The sum of two numbers is 52. Twice the first number is 17 more than the second. Find the numbers.

8. Twice one number is 28 more than a second number. When 23 is added to the second number, the result is 12 less than three times the first number. What are the numbers?

9. The tens digit of a two-digit number is three less than twice the units digit. The sum of the digits of the number is 12. Find the number.

10. The sum of the digits of a two-digit number is 7. When the digits are reversed, the new number is two more than twice the original number. Find the original number.

11. The units digit of a two-digit number is 2 more than four times the tens digit. When the digits are reversed, the new number is 13 more than three times the original number. What is the original number?

12. Two angles are complementary (their sum is 90°). One angle is 36° more than twice the other. Find the measure of each angle.

13. Two angles are supplementary (their sum is 180°). One angle is 28° less than three times the other. Find the measure of each angle.

14. The perimeter of a rectangular swimming pool is 84 meters. The width of the pool is $\frac{3}{4}$ the length. What are the length and width of the pool?

15. A collection of coins consisting of dimes and quarters amounts to $5.35. Twice the number of dimes exceeds the number of quarters by 5. How many of each kind of coin were there?

16. Thelma cashed a check for $880 at her bank. She received $20-bills and $10-bills. There were 8 fewer $10-bills than $20-bills. How many bills of each type did she receive?

17. A coin bank contains $37 in quarters and dimes. The number of dimes is 10 more than twice the number of quarters. How many dimes and how many quarters are there?

18. A real estate broker wishes to invest $81,000 in buying land. He buys a total of 120 acres, paying $500 an acre for one lot, and $800 an acre for an adjacent lot. How many acres were there in each lot?

19. Jim Youngblood invested part of $20,000 at 12% annually and the rest at 14%. The total income from these investments is $2560. How much did Jim invest at each rate?

20. Linda Velez invested a sum of money, part at 14% and the rest at 15%. Her total interest income per year from the investment is $11,900. The amount invested at 15% is $2000 more than the amount invested at 14%. How much did she invest at each rate?

21. Horace Trask invested a certain amount in bonds yielding 12% per year and twice the first amount in bonds yielding 16% per year. His total annual income from these investments is $3520. How much did he invest in each type of bond?

22. Jaime has invested $15,000, part at 6.5% interest per year and the rest at 9% per year. Total yearly income from both investments is $1200. How much does Jaime have invested at each rate?

23. Traveling downstream, a boat covers a distance of 27 kilometers in 3 hours. Traveling upstream, it takes the boat twice as much time to cover two-thirds the distance. Find the rate of the boat in still water and the rate of the current.

24. Flying with a tailwind, a plane travels 3600 kilometers in 6 hours. Flying at the same speed and with the same wind blowing, the plane can make the return trip in 7.2 hours. Find the speed of the plane going and returning.

25. A ferry can travel 9.6 miles downstream in 2 hours. Going upstream, the ferry can travel $\frac{3}{4}$ of this distance in the same amount of time. What is the rate of the ferry in still water and what is the rate of the current?

CALCULATOR APPLICATIONS

Checking Systems of Equations

To use a scientific calculator to check the solutions to a system of equations, you may need to use the *parentheses keys,* (and).

EXAMPLE Check the solution to the following system.

$$\begin{cases} 4(x + y) = 3840 \\ 5(x - y) = 3840 \end{cases}$$

Solutions: $x =$ **864;** $y =$ **96**

SOLUTION Replace x and y in the equations.

$4(864 + 96) = 3840$ $\qquad$ $5(864 - 96) = 3840$

4 [×] [(] 8 6 4 [+] 9 6 [)] [=] $\quad$ 3840.

5 [×] [(] 8 6 4 [−] 9 6 [)] [=] $\quad$ 3840.

EXERCISES

Use a scientific calculator to check the solutions to Exercises 23–25 above.

REVIEW CAPSULE FOR SECTION 4-5

Give the slope and y intercept of the graph of each line. *(Pages 80–83, 93–97)*

1. $3x + y = 6$
2. $x - 5y = 0$
3. $4x + 3y = 12$
4. $2y - 3x = 6$

Classify the graphs of each pair of equations as parallel *P*, or not parallel, *NP*. *(Pages 90–92)*

5. $2x - 3y = 6$ and $8x - 12y = 24$
6. $2x + 3y - 1 = 0$ and $4x = 6y + 2$
7. $y - 5x = 3$ and $2y - 6 = 10x$
8. $3x - y = 8$ and $12x - 48 = 4y$

4–5 Consistent and Inconsistent Systems

When the graphs of two linear equations are drawn in the coordinate plane, they may be related to each other as shown below.

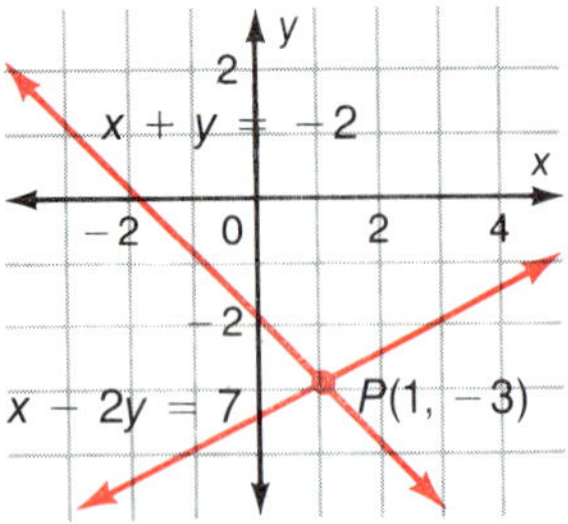

Figure 1

$$\begin{cases} x + y = -2 & \quad 1 \\ x - 2y = 7 & \quad 2 \end{cases}$$

The graphs intersect in *exactly* one point.

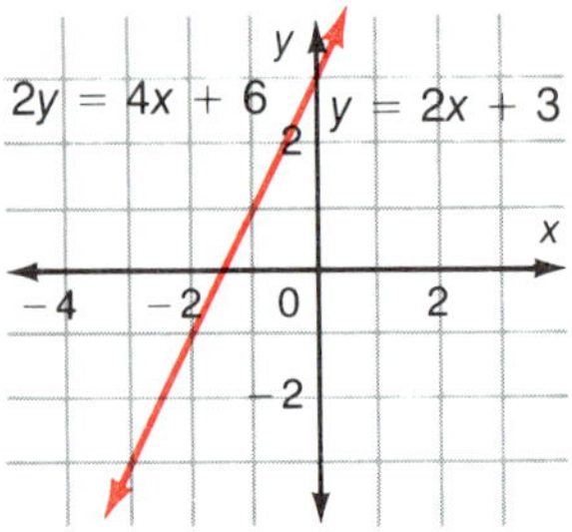

Figure 2

$$\begin{cases} y = 2x + 3 & \quad 1 \\ 2y = 4x + 6 & \quad 2 \end{cases}$$

The graphs coincide; that is, they have an *infinite number* of points in common.

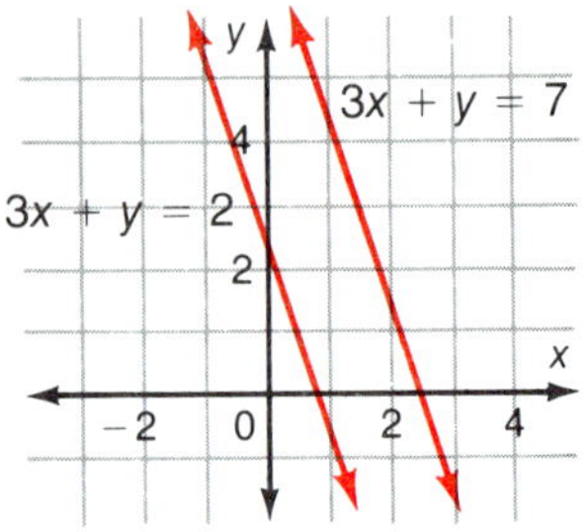

Figure 3

$$\begin{cases} 3x + y = 2 & \quad 1 \\ 3x + y = 7 & \quad 2 \end{cases}$$

The graphs are parallel; that is, they have *no* points in common.

The graphs in Figures 1 and 2 have *at least one point* in common. The systems are said to be **consistent.** The graphs in Figure 3 have *no point* in common. This system is said to be **inconsistent.**

Definitions

> A **consistent system** of equations or inequalities is one whose solution set contains at least one ordered pair.
>
> An **inconsistent system** of equations or inequalities is one whose solution set is the empty set.

Write the equations of the systems above in slope-intercept form.

System 1

$$\begin{cases} y = -x - 2 & \quad 1 \\ y = \frac{1}{2}x - \frac{7}{2} & \quad 2 \end{cases}$$

System 2

$$\begin{cases} y = 2x + 3 & \quad 1 \\ y = \frac{4}{2}x + \frac{6}{2} & \quad 2 \end{cases}$$

System 3

$$\begin{cases} y = -3x + 2 & \quad 1 \\ y = -3x + 7 & \quad 2 \end{cases}$$

For System 1, exactly one ordered pair satisfies both equations. For this system,

$$m_1 = -1 \quad \text{and} \quad m_2 = \frac{1}{2}. \qquad \text{Thus, } \boldsymbol{m_1 \neq m_2}.$$

For System 2, every ordered pair that satisfies Equation 1 also satisfies Equation 2. The system is **dependent.** For this system,

$$m_1 = 2 \quad \text{and} \quad m_2 = 2. \qquad \text{Also, } b_1 = 3 \quad \text{and} \quad b_2 = 3.$$

Thus, $\boldsymbol{m_1 = m_2}$ **and** $\boldsymbol{b_1 = b_2}$.

For System 3, no ordered pair satisfies *both* equations. For this system,

$m_1 = -3$ and $m_2 = -3$. Also, $b_1 = 2$ and $b_2 = 7$.

Thus, $\boldsymbol{m_1 = m_2}$ and $\boldsymbol{b_1 \neq b_2}$.

EXAMPLE Give the number of solutions for each system.

a. $\begin{cases} x + y = 3 \\ x = 5 - y \end{cases}$ b. $\begin{cases} 2x = 5y + 4 \\ 4x - 10y = 8 \end{cases}$ c. $\begin{cases} 3x + 7y = -4 \\ 2x + 5y = 3 \end{cases}$

Solutions: **a.** Both lines have the same slope, -1. The y intercepts, 3 and 5, are different. Therefore, the graphs are parallel lines, and the system has **no solution.**

b. Both lines have the same slope, $\frac{2}{5}$, and the same y intercept, $-\frac{4}{5}$. Therefore, the graphs coincide, and the system has an **infinite number of solutions.**

c. The lines have slopes of $-\frac{3}{7}$ and $-\frac{2}{5}$. Therefore, their graphs intersect in exactly one point and the system has **one solution.**

Summary

Properties of a Linear System of Two Equations

$\boldsymbol{y = m_1x + b_1}$ and $\boldsymbol{y = m_2x + b_2}$

Description	Slopes and y intercepts	Graphs	Solutions
Consistent	$m_1 \neq m_2$	Intersect in one point	One
Dependent	$m_1 = m_2$; $b_1 = b_2$	Coincide	Infinite number
Inconsistent	$m_1 = m_2$; $b_1 \neq b_2$	Parallel	None

CLASSROOM EXERCISES

Use the graph of each system to classify it as consistent, *C*, or inconsistent, *I*. When a system is also dependent, write *D*.

1.
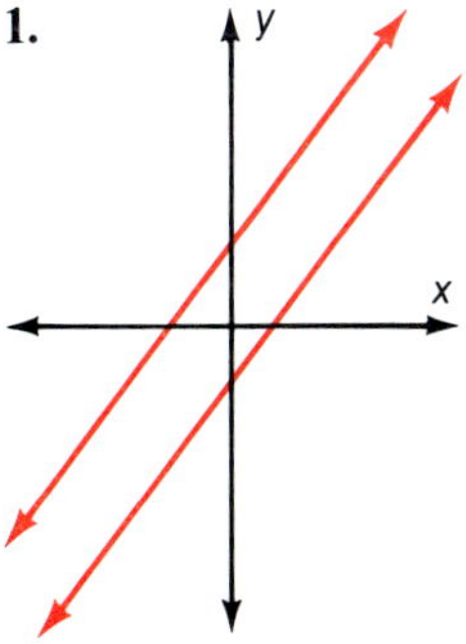

2.
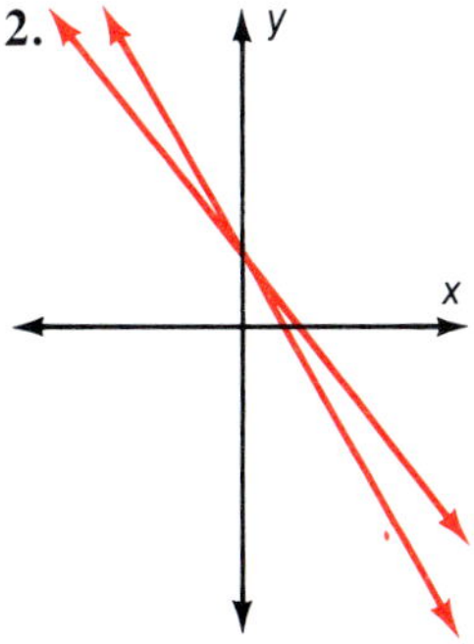

3.
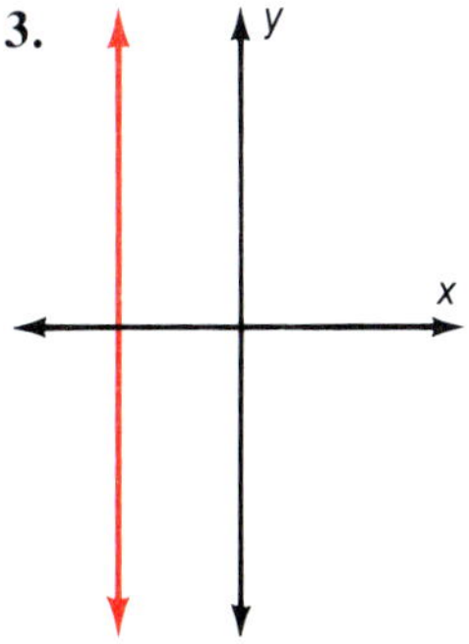

4.
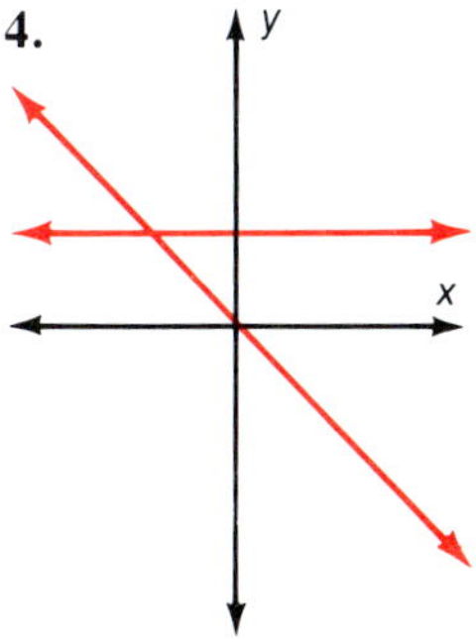

WRITTEN EXERCISES

A In Exercises 1–8, graph each system. Use the graph to classify each system as consistent, *C*, or inconsistent, *I*. When a system is also dependent, write *D*.

1. $\begin{cases} x - y = 8 \\ 3x - 3y = 12 \end{cases}$ **2.** $\begin{cases} x + y = 3 \\ 4x + 4y = 12 \end{cases}$ **3.** $\begin{cases} x + y = 12 \\ 3y = x \end{cases}$ **4.** $\begin{cases} 6x + 5y = 7 \\ 3x - 7y = 13 \end{cases}$

5. $\begin{cases} x - y = 4 \\ 3x = 4 + 2y \end{cases}$ **6.** $\begin{cases} x - y = 4 \\ 3x + 3y = 12 \end{cases}$ **7.** $\begin{cases} 2x + 3y = 2 \\ y = -\frac{3}{4}x \end{cases}$ **8.** $\begin{cases} x = 4y \\ \frac{x}{4} + y = 20 \end{cases}$

In Exercises 9–16, find the slope and *y* intercept of each line. Give the number of solutions for each system.

9. $\begin{cases} x + y = 6 \\ x - y = 2 \end{cases}$ **10.** $\begin{cases} 2x + y = 5 \\ y = 8 - 2x \end{cases}$ **11.** $\begin{cases} 3x + y = 5 \\ 2y = 10 - 6x \end{cases}$ **12.** $\begin{cases} 4x - 3y = 7 \\ 12x - 9y = 7 \end{cases}$

13. $\begin{cases} y = x \\ 4y - 4x = 9 \end{cases}$ **14.** $\begin{cases} 2x + 3y = 8 \\ 3x + 2y = 10 \end{cases}$ **15.** $\begin{cases} y = 12 - x \\ x - y = 2 \end{cases}$ **16.** $\begin{cases} 2x - 5y = 4 \\ 4x = 8 + 10y \end{cases}$

B Solve and check. Use any method you choose.

17. $\begin{cases} \frac{x}{2} - \frac{y}{9} = 3 \\ y = \frac{9}{2}x - 27 \end{cases}$ **18.** $\begin{cases} \frac{5x}{2} + \frac{2y}{3} = -21 \\ \frac{x}{3} - \frac{y}{9} = -1 \end{cases}$ **19.** $\begin{cases} \frac{2x}{5} + \frac{3y}{4} = -4\frac{1}{5} \\ \frac{3x}{5} = \frac{3y + 2}{4} + \frac{7}{10} \end{cases}$

20. $\begin{cases} \frac{x}{3} + 3y = 15 \\ \frac{x}{4} + 4y = 13 \end{cases}$ **21.** $\begin{cases} \frac{2x}{3} + \frac{3y}{4} = -10 \\ y + 34 = \frac{5x}{6} \end{cases}$ **22.** $\begin{cases} \frac{3}{2}x - \frac{4}{3}y = \frac{11}{3} \\ \frac{1}{4}x - \frac{2}{3}y = -\frac{7}{6} \end{cases}$

Indicate whether each system is consistent or inconsistent.

23. $\{(x, y)\colon 2x + y = 10\} \cap \{(x, y)\colon 2x + y = 5\}$

24. $\{(x, y)\colon x + 3y = 7\} \cap \{(x, y)\colon 2x + 6y = 14\}$

C

25. For the linear system,

$$\begin{cases} y = m_1x + b_1 \\ y = m_2x + b_2 \end{cases}$$

show that the system has a unique solution when $m_1 \neq m_2$.

26. For the linear system in Exercise 25, show that when $m_1 = m_2$ and $b_1 \neq b_2$, the solution set is the empty set.

Use the following linear system for Exercises 27 and 28.

$$\begin{cases} 6x - 4y = 8 \\ -9x + 6y = a \end{cases}$$

27. Find the values of *a* for which the system has an infinite number of solutions.

28. Find the value(s) of *a* for which the system has no solution.

Review

In Exercises 1–4, solve each system by graphing. *(Section 4–1)*

1. $\begin{cases} x + y = 5 \\ x - y = 3 \end{cases}$
2. $\begin{cases} x = -2 \\ 3x + y = -4 \end{cases}$
3. $\begin{cases} y = -x \\ 3x - y = 12 \end{cases}$
4. $\begin{cases} 3x - 4y = 13 \\ 2x + 3y = -14 \end{cases}$

In Exercises 5–8, solve each system by the addition method. *(Section 4–2)*

5. $\begin{cases} 5x + 2y = 4 \\ 3x + 2y = 8 \end{cases}$
6. $\begin{cases} x + y = 3 \\ 3x - y = 11 \end{cases}$
7. $\begin{cases} 3x - 6y = -78 \\ 5x + 8y = 14 \end{cases}$
8. $\begin{cases} 7x + 2y = 13 \\ 3x - 5y = -6 \end{cases}$

In Exercises 9–12, solve each system by the substitution method. *(Section 4–3)*

9. $\begin{cases} 3x - y = 13 \\ x = 4 \end{cases}$
10. $\begin{cases} 2x + 4y = 14 \\ y = 2 - x \end{cases}$
11. $\begin{cases} x + y = -8 \\ 2x - 8y = 4 \end{cases}$
12. $\begin{cases} 3x + y = 8 \\ 4x + 3y = 0 \end{cases}$

In Exercises 13–15, use a system of two linear equations in two variables to solve each problem. *(Section 4–4)*

13. A 25% salt solution is mixed with a 10% salt solution to produce 100 grams of a 16% salt solution. How many grams were there of the 25% salt solution and of the 10% salt solution?
14. The units digit of a two–digit number is two more than twice the tens digit. When the digits are reversed, the new number is seven more than twice the original number.
15. A family invested \$20,000 in two securities. One pays 15% per year and the other pays 12% per year. The interest for a certain year amounted to \$2580. How much was invested at each rate?

In Exercises 16–19, find the slope and *y* intercept of each line. Give the number of solutions for each system. *(Section 4–5)*

16. $\begin{cases} y = -x \\ 3x + 3y = 0 \end{cases}$
17. $\begin{cases} y - x = 0 \\ y = x + 7 \end{cases}$
18. $\begin{cases} 2y - 3x = 4 \\ y = \frac{2}{3}x + 2 \end{cases}$
19. $\begin{cases} y + 5x = 6 \\ y + 6 = -5x \end{cases}$

REVIEW CAPSULE FOR SECTION 4–6

Write each equation in the form $Ax + By = C$. *(Pages 93–97)*

1. $6x = 9 - 2y$
2. $y = 12 - 5x$
3. $4x + y - 3 = 0$
4. $3y - 8x + 12 = 0$

Simplify. *(Pages 14–16)*

5. $g(bf - ec)$
6. $-h(af - cd)$
7. $-b(de - gf) - c(-de + gf)$

Using Statistics:

Frequency Distribution

Agricultural engineers are concerned with conservation and management of soil and water resources. They must gather and analyze statistics to plan for the future use of these resources. Such is the case below.

PROBLEM: The farmers of New Town are concerned about the trend they see in the mean (average) annual temperatures in their town. They claim that the years from 1979 through 1983 were extremely warm. An agricultural engineer was consulted to help them find out if this is true or not.

Analysis: The information in the following table was gathered.

Mean Annual Temperatures for New Town

Year	Mean Temp.	5-year Mean	Year	Mean Temp.	5-year Mean	Year	Mean Temp.	5-year Mean	Year	Mean Temp.	5-year Mean
1879	48.0			49.7	49.0	**1933**	51.9	50.2		51.3	51.4
	47.0			48.9	48.9		50.8	50.7		51.0	51.5
	47.0		**1908**	48.9	48.7		49.6	50.7		54.0	52.0
	49.0			48.7	48.8		49.3	50.4		51.4	52.0
1883	48.0	47.8		50.3	49.3		50.6	50.6	**1963**	52.7	52.1
	48.0	47.8		49.2	49.2	**1938**	48.4	49.9		53.1	52.4
	47.9	48.0		50.7	49.6		50.7	49.9		54.6	53.2
	49.3	48.4	**1913**	48.7	49.5		50.9	50.2		52.0	52.8
	48.3	48.3		49.4	49.7		50.6	50.4		52.0	52.9
1888	50.8	48.9		49.2	49.4		51.5	50.6	**1968**	50.9	52.5
	52.5	49.8		46.7	48.9	**1943**	52.8	51.3		52.6	52.4
	52.7	50.1		48.8	48.6		51.8	51.6		50.2	51.5
	50.8	51.0	**1918**	50.4	48.9		51.1	51.6		52.6	51.7
	51.7	51.7		48.3	48.7		49.8	51.4		51.6	51.5
1893	49.9	51.5		50.9	49.0		50.2	51.2	**1973**	51.9	51.7
	48.8	50.8		50.1	49.7	**1948**	50.4	50.7		50.5	51.3
	47.6	49.8		50.3	50.0		51.6	50.7		50.0	51.5
	48.9	49.4	**1923**	50.5	50.0		51.8	50.8		51.7	51.4
	47.4	48.5		49.9	50.3		50.9	51.0		51.4	51.4
1898	48.3	48.2		52.3	50.6		48.8	50.7	**1978**	51.7	51.2
	48.7	48.2		49.4	50.5	**1953**	51.7	51.0		50.8	51.5
	47.2	48.1		51.1	50.6		51.0	50.8		51.9	51.6
	50.2	48.4	**1928**	49.4	50.4		50.6	50.6		51.9	51.6
	49.3	48.7		47.9	50.0		51.7	50.8		51.9	51.6
1903	50.0	49.0		49.8	49.5		51.5	51.3	**1983**	53.0	51.9
	48.6	49.0		50.9	49.8	**1958**	52.1	51.4			
	47.6	49.1		49.3	49.7						

The mean temperatures of each year from 1879 through 1983 are in the second column of the table. The third column contains the mean temperatures for each group of 5 years ending with the measurement. For example, the mean temperature of the years 1909 through 1913 was 49.5. Using the same method, it can be seen that the mean temperature for the years 1979 through 1983 was 51.9.

A **frequency distribution** of the 5-year means was constructed and is shown in the dot diagram below.

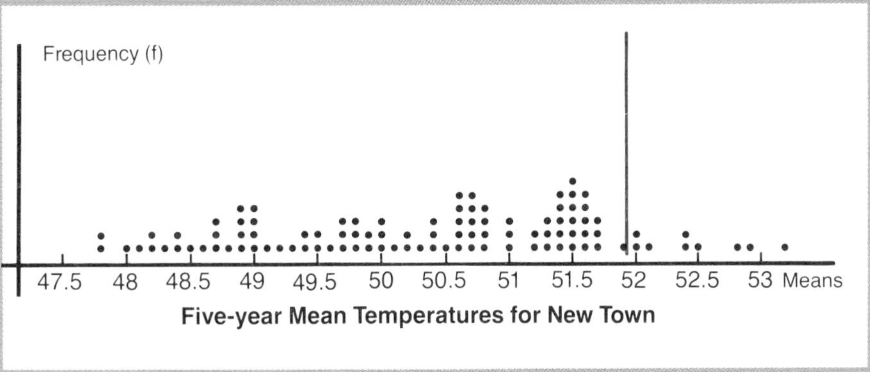

A line has been drawn at 51.9. There are 9 entries to the right of that line and 91 entries to the left of the line. The value of 51.9 is in the top $\frac{10}{100}$, or 10% of the 5-year means. This means that over the period being considered, 90% of the 5-year mean temperatures were below 51.9.

Conclusion: Thus, the farmers are correct. The 5-year mean of 51.9 is unusually high and there is a definite warming trend. They should ask the agricultural engineer to help them plan for future water conservation and crop management.

EXERCISES

1. Draw the dot diagram of the frequency distribution of the 5-year means for the years from 1960 through 1983. (Hint: Start with 5-year mean at 1964.)
2. What per cent of the observations in this time period are below 51.9?
3. Based on the answer to Exercise 2, would you say that 51.9 is an unusually high 5-year mean for this time period? (In the upper 20%?)
4. What does the answer to Exercise 3 tell you about using statistics?

The data for the total seasonal rainfall in Groverton for the 70 years from 1904 through 1983 is given below (read horizontally).

19	12	18	11	18	5	20	12	17	18
11	15	18	9	24	12	12	20	27	21
12	24	15	22	21	18	9	13	13	17
15	16	18	22	24	23	13	22	11	23
9	8	28	20	20	14	9	19	8	17
16	17	9	19	18	19	14	10	13	8
15	9	11	18	17	17	25	17	20	21

5. Construct a frequency distribution and its dot diagram for this data.
6. What per cent of the observations lie below 20?
7. Would you consider 20 a high, average, or low amount of rainfall over the time period being considered?

4-6 Solving Linear Systems: Determinants

General formulas for the solution of a system of two linear equations in two variables can be determined. Consider a system written in general form where a_1, b_1, c_1, a_2, b_2, and c_2, are real numbers.

$$\begin{cases} a_1x + b_1y = c_1 & \quad \mathbf{1} \\ a_2x + b_2y = c_2 & \quad \mathbf{2} \end{cases}$$

← **Multiply each side by b_2.**
← **Multiply each side by $-b_1$.**

$$\begin{array}{r} a_1b_2x + b_1b_2y = c_1b_2 \\ -a_2b_1x - b_1b_2y = -c_1b_1 \\ \hline \end{array}$$

← **Add.**

$$a_1b_2x - a_2b_1x = c_1b_2 - c_2b_1$$

← **Solve for x.**

$$(a_1b_2 - a_2b_1)x = c_1b_2 - c_2b_1$$

← **By the distributive postulate**

$$x = \frac{c_1b_2 - c_2b_1}{a_1b_2 - a_2b_1},\ a_1b_2 - a_2b_1 \neq 0$$

Similarly, by multiplying Equation 1 by $-a_2$ and Equation 2 by a_1, you can find the general solution for y.

$$y = \frac{a_1c_2 - a_2c_1}{a_1b_2 - a_2b_1},\ a_1b_2 - a_2b_1 \neq 0$$

These values can be checked in Equations 1 and 2.

There is a convenient way to represent the numerators and denominators for x and y.

Definition

> For all real numbers a_1, a_2, b_1, b_2,
>
> $$\begin{vmatrix} a_1 & b_1 \\ a_2 & b_2 \end{vmatrix} = a_1b_2 - a_2b_1.$$

In the definition, the square array of numbers set off with bars is called a **determinant.** Since the determinant has two rows and two columns, it is a 2×2 (read: two by two) determinant.

$$\begin{vmatrix} a_1 & b_1 \\ a_2 & b_2 \end{vmatrix}$$

← **Row 1**
← **Row 2**

Column 1 **Column 2**

EXAMPLE 1 Evaluate each determinant.

Solutions:

a. $\begin{vmatrix} 2 & 1 \\ 3 & 2 \end{vmatrix} = 2(2) - (3)(1)$

$= 4 - 3 = \mathbf{1}$

b. $\begin{vmatrix} 0 & 3 \\ -1 & -4 \end{vmatrix} = 0(-4) - (-1)(3)$

$= 0 - (-3) = \mathbf{3}$

You can use 2×2 determinants to solve systems of linear equations in two variables.

Compare the numerators and denominators of the general solution for x and y with the definition of a 2×2 determinant.

Numerator for x	**Numerator for y**	**Denominator for x and y**
$c_1b_2 - c_2b_1 = \begin{vmatrix} c_1 & b_1 \\ c_2 & b_2 \end{vmatrix}$	$a_1c_2 - a_2c_1 = \begin{vmatrix} a_1 & c_1 \\ a_2 & c_2 \end{vmatrix}$	$a_1b_2 - a_2b_1 = \begin{vmatrix} a_1 & b_1 \\ a_2 & b_2 \end{vmatrix}$

Replaces the coefficients of the x terms

Replaces the coefficients of the y terms

Thus, $x = \dfrac{\begin{vmatrix} c_1 & b_1 \\ c_2 & b_2 \end{vmatrix}}{\begin{vmatrix} a_1 & b_1 \\ a_2 & b_2 \end{vmatrix}}$ and $y = \dfrac{\begin{vmatrix} a_1 & c_1 \\ a_2 & c_2 \end{vmatrix}}{\begin{vmatrix} a_1 & b_1 \\ a_2 & b_2 \end{vmatrix}}$

The denominators are the same.

Notice that the numerator for the general solution of x consists of the constant terms and of the coefficients of the y terms. Similarly, the numerator for the general solution of y consists of the coefficients of the x terms and of the constant terms.

EXAMPLE 2 Use determinants to solve: $\begin{cases} 2x = 5 + 3y \\ 3y = 8 - x \end{cases}$

Solution: Write the equations in general form: $\begin{cases} 2x - 3y = 5 \\ x + 3y = 8 \end{cases}$

$$x = \frac{\begin{vmatrix} 5 & -3 \\ 8 & 3 \end{vmatrix}}{\begin{vmatrix} 2 & -3 \\ 1 & 3 \end{vmatrix}} \qquad y = \frac{\begin{vmatrix} 2 & 5 \\ 1 & 8 \end{vmatrix}}{\begin{vmatrix} 2 & -3 \\ 1 & 3 \end{vmatrix}}$$

$$x = \frac{(5)(3) - (8)(-3)}{(2)(3) - (1)(-3)} \qquad y = \frac{(2)(8) - (1)(5)}{(2)(3) - (1)(-3)}$$

$$x = \frac{15 + 24}{6 + 3} = \mathbf{\frac{39}{9}} \qquad y = \frac{16 - 5}{6 + 3} = \mathbf{\frac{11}{9}}$$

The solution of a system of linear equations in two variables can be generalized in **Cramer's Rule.**

Theorem 4–2

Cramer's Rule

For all real numbers a_1, a_2, b_1, c_1, c_2, the solution to

$$\begin{cases} a_1x + b_1y = c_1 \\ a_2x + b_2y = c_2 \end{cases} \text{ is } (x, y) \text{ where}$$

$$x = \frac{\begin{vmatrix} c_1 & b_1 \\ c_2 & b_2 \end{vmatrix}}{\begin{vmatrix} a_1 & b_1 \\ a_2 & b_2 \end{vmatrix}} \quad \text{and} \quad y = \frac{\begin{vmatrix} a_1 & c_1 \\ a_2 & c_2 \end{vmatrix}}{\begin{vmatrix} a_1 & b_1 \\ a_2 & b_2 \end{vmatrix}}, \quad \begin{vmatrix} a_1 & b_1 \\ a_2 & b_2 \end{vmatrix} \neq 0$$

CLASSROOM EXERCISES

Evaluate each determinant.

1. $\begin{vmatrix} 3 & -7 \\ 5 & 2 \end{vmatrix}$
2. $\begin{vmatrix} 9 & 0 \\ 3 & -4 \end{vmatrix}$
3. $\begin{vmatrix} \frac{1}{2} & -\frac{1}{3} \\ 6 & 10 \end{vmatrix}$
4. $\begin{vmatrix} p & t \\ -v & m \end{vmatrix}$

In Exercises 5–8, represent the solution, (x, y), of each system of equations by a determinant. Do *not* evaluate the determinant.

5. $\begin{cases} 2x - y = 8 \\ x + 2y = 9 \end{cases}$
6. $\begin{cases} 9x - 4y = -11 \\ 6x - 5y = 8 \end{cases}$
7. $\begin{cases} 3x + 4y - 10 = 0 \\ 2y - x = 0 \end{cases}$
8. $\begin{cases} 2x = 14 - 4y \\ 3y = 10 - x \end{cases}$

WRITTEN EXERCISES

A Evaluate each determinant.

1. $\begin{vmatrix} 5 & 2 \\ 6 & 3 \end{vmatrix}$
2. $\begin{vmatrix} 3 & 7 \\ 4 & 1 \end{vmatrix}$
3. $\begin{vmatrix} -4 & 8 \\ 2 & 3 \end{vmatrix}$
4. $\begin{vmatrix} -1 & -2 \\ 6 & 7 \end{vmatrix}$
5. $\begin{vmatrix} 3 & -2 \\ -9 & 6 \end{vmatrix}$
6. $\begin{vmatrix} 1 & -1 \\ -1 & 1 \end{vmatrix}$
7. $\begin{vmatrix} p & 3 \\ -1 & p \end{vmatrix}$
8. $\begin{vmatrix} \frac{1}{4} & -\frac{1}{3} \\ \frac{1}{2} & 0 \end{vmatrix}$

Solve for *t*.

9. $\begin{vmatrix} t & 1 \\ -1 & 3 \end{vmatrix} = 7$
10. $\begin{vmatrix} 2 & 5 \\ 1 & -t \end{vmatrix} = 9$
11. $\begin{vmatrix} 6 & 3 \\ -t & 2 \end{vmatrix} = -1$
12. $\begin{vmatrix} 5 & t \\ 8 & 0 \end{vmatrix} = 8$

Use Cramer's Rule to solve each system.

13. $\begin{cases} 3x - 2y = 1 \\ x - 5y = 4 \end{cases}$
14. $\begin{cases} 2x + 3y = 2 \\ -2x + 6y = 1 \end{cases}$
15. $\begin{cases} 2x + y = 6 \\ x - y = 4 \end{cases}$
16. $\begin{cases} 3x + 2y = 1 \\ 4x - y = 3 \end{cases}$
17. $\begin{cases} x = 1 + 2y \\ 3x = 5 - y \end{cases}$
18. $\begin{cases} 3x + 4y = 10 \\ 2y = x \end{cases}$
19. $\begin{cases} 3y - 2x = 10 \\ 3x + 6y = 25 \end{cases}$
20. $\begin{cases} 2x + 5y = 6 \\ x = 2y - 5 \end{cases}$
21. Use Cramer's Rule to solve the systems in Exercises 5–8 of the Classroom Exercises.

B Use Cramer's rule to solve each system. When a system is inconsistent, write *Inconsistent.* When a consistent system is dependent, write *Dependent.*

22. $\begin{cases} 2x + y = 8 \\ 2x + y = 5 \end{cases}$
23. $\begin{cases} 3x + y = 5 \\ 2y = 10 - 6x \end{cases}$
24. $\begin{cases} x + y = 12 \\ x - y = 2 \end{cases}$
25. $\begin{cases} 2x = 5y + 4 \\ 4x - 10y = 8 \end{cases}$

APPLICATIONS: Using Determinants to Solve Word Problems

Write a system of equations in two variables to represent each problem. Then use determinants to solve the system.

26. Twice the shorter side of a parallelogram exceeds the longer side by 5 centimeters. One-third the sum of the shorter side and 9 exceeds one-fifth the longer side by 3. Find the lengths of the sides.

27. Alma Rodriguez has \$6.50 in quarters and dimes. After spending 6 quarters and 8 dimes for bus fare, the number of quarters was the same as the number of dimes. How many of each coin did Alma have originally?

28. When 9 is added to both numerator and denominator of a fraction, it equals $\frac{6}{7}$. When 7 is subtracted from both, the fraction equals $\frac{2}{3}$. Find the fraction.

29. A fraction is equivalent to $\frac{3}{4}$. When the numerator is increased by 2 and the denominator is decreased by 6, the result is $\frac{5}{6}$ of the reciprocal of the original fraction. Find the original fraction.

30. Two families who live 200 miles apart are to meet for a picnic lunch. One family travels at a rate of 35 miles per hour and the other at a rate of 45 miles per hour. If both families leave at the same time and travel toward each other from opposite directions, how long will it take them to meet?

C Use Cramer's Rule to solve each system.

31. $\begin{cases} \frac{1}{x} + \frac{1}{y} = 7 \\ \frac{1}{x} - \frac{1}{y} = 1 \end{cases}$ **32.** $\begin{cases} \frac{4}{x} + \frac{9}{y} = -1 \\ \frac{6}{x} - \frac{12}{y} = 7 \end{cases}$ **33.** $\begin{cases} \frac{2}{x} + \frac{6}{y} = 5 \\ \frac{7}{x} - \frac{4}{y} = -20 \end{cases}$ **34.** $\begin{cases} ax + by = c \\ dx + cy = f \end{cases}$

35. Prove: If the lines defined by $a_1x + b_1y = c_1$ and $a_2x + b_2y = c_2$ are parallel, then $\begin{vmatrix} a_1 & b_1 \\ a_2 & b_2 \end{vmatrix} = 0 \quad (a_1, a_2, b_1, b_2, c_1, c_2, \in \text{R})$.

36. Prove: If $\begin{vmatrix} a_1 & b_1 \\ a_2 & b_2 \end{vmatrix} = 0$, then the lines defined by $a_1x + b_1y = c_1$ and $a_2x + b_2y = c_2$ (where $a_1, a_2, b_1, b_2, c_1, c_2 \in \text{R}$) are parallel or coincident.

CALCULATOR APPLICATIONS

Cramer's Rule

To use a scientific calculator to solve a system by Cramer's Rule, you use the memory storage, [M+], memory recall, [MR], and the sign-change, [+/−] keys.

EXAMPLE Use Cramer's Rule to solve this system for x: $\begin{cases} 6x + 5y = 7 \\ 3x - 7y = 13 \end{cases}$

SOLUTION By Cramer's Rule, $x = \dfrac{7(-7) - (5)(13)}{6(-7) - (5)(3)}$. Evaluate the denominator *first*.

6 [×] 7 [+/−] [−] 5 [×] 3 [=] [M+] 7 [×]

7 [+/−] [−] 5 [×] 1 3 [=] [÷] [MR] [=]   **2.**

EXERCISES

Use a calculator to solve Exercises 13–20 on page 136.

4-7 Third-Order Determinants

A first-degree equation in three variables can be written as follows.

$$Ax + By + Cz = D \longleftarrow \textbf{Standard form}$$

The solution set of a system of such equations is the set of **ordered triples,** (x, y, z), that satisfies all the equations of the system.

You can use third-order determinants to find these solutions. Other methods of solving a system of linear equations in three variables will be described in Chapter 13.

Definition For all real numbers $a_1, b_1, c_1, a_2, b_2, c_2, a_3, b_3, c_3$, the **determinant,**

$$\begin{vmatrix} a_1 & b_1 & c_1 \\ a_2 & b_2 & c_2 \\ a_3 & b_3 & c_3 \end{vmatrix} \quad \text{is equal to}$$

$$a_1b_2c_3 + a_2b_3c_1 + a_3b_1c_2 - a_1b_3c_2 - a_2b_1c_3 - a_3b_2c_1.$$

A convenient way to evaluate a third-order determinant is shown at the right. The first two columns are repeated at the right of the determinant.

$$\left|\begin{matrix} a_1 & b_1 & c_1 \\ a_2 & b_2 & c_2 \\ a_3 & b_3 & c_3 \end{matrix}\right| \begin{matrix} a_1 & b_1 \\ a_2 & b_2 \\ a_3 & b_3 \end{matrix}$$

Add the products found from descending arrows.

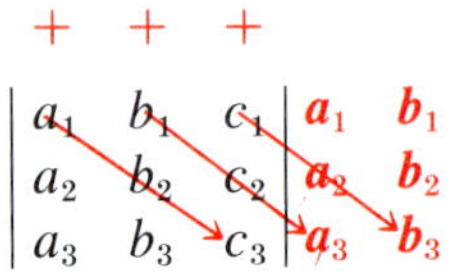

Subtract the products found from ascending arrows.

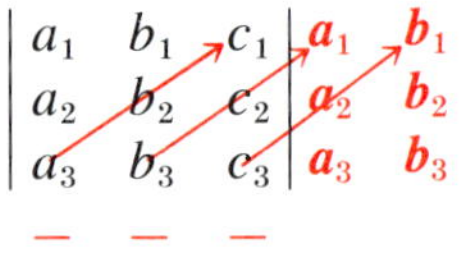

$$a_1b_2c_3 + b_1c_2a_3 + c_1a_2b_3 - a_3b_2c_1 - b_3c_2a_1 - c_3a_2b_1 =$$
$$a_1b_2c_3 + a_2b_3c_1 + a_3b_1c_2 - a_1c_2b_3 - a_2b_1c_3 - a_3b_2c_1$$

EXAMPLE 1 Evaluate: $\begin{vmatrix} 3 & -2 & 5 \\ -1 & 4 & -6 \\ 1 & 7 & -3 \end{vmatrix}$

Solution:

$$\begin{matrix} + & + & + \\ \left|\begin{matrix} 3 & -2 & 5 \\ -1 & 4 & -6 \\ 1 & 7 & -3 \end{matrix}\right| & \begin{matrix} 3 & -2 \\ -1 & 4 \\ 1 & 7 \end{matrix} \\ - & - & - \end{matrix} = -36 + 12 - 35 - 20 + 126 + 6 = \mathbf{53}$$

The solution of the general system of three linear equations in three variables

$$\begin{cases} a_1x + b_1y + c_1z = d_1 \\ a_2x + b_2y + c_2z = d^2 \\ a_3x + b_3y = c_3z = d_3 \end{cases}$$

can be calculated from these determinants.

$$x = \frac{\begin{vmatrix} d_1 & b_1 & c_1 \\ d_2 & b_2 & c_2 \\ d_3 & b_3 & c_3 \end{vmatrix}}{\begin{vmatrix} a_1 & b_1 & c_1 \\ a_2 & b_2 & c_2 \\ a_3 & b_3 & c_3 \end{vmatrix}} \qquad y = \frac{\begin{vmatrix} a_1 & d_1 & c_1 \\ a_2 & d_2 & c_2 \\ a_3 & d_3 & c_3 \end{vmatrix}}{\begin{vmatrix} a_1 & b_1 & c_1 \\ a_2 & b_2 & c_2 \\ a_3 & b_3 & c_3 \end{vmatrix}} \qquad z = \frac{\begin{vmatrix} a_1 & b_1 & d_1 \\ a_2 & b_2 & d_2 \\ a_3 & b_3 & d_3 \end{vmatrix}}{\begin{vmatrix} a_1 & b_1 & c_1 \\ a_2 & b_2 & c_2 \\ a_3 & b_3 & c_3 \end{vmatrix}}$$

These equations are called **Cramer's Rule** for systems of three linear equations in three variables.

EXAMPLE 2 Use Cramer's Rule to solve this system.

$$\begin{cases} 2x + y + 3z = -2 \\ 5x + 2y = 5 \\ 2y + 3z = -13 \end{cases}$$

Solution:

$$x = \frac{\left|\begin{array}{rrr|rr} -2 & 1 & 3 & -2 & 1 \\ 5 & 2 & 0 & 5 & 2 \\ -13 & 2 & 3 & -13 & 2 \end{array}\right|}{\left|\begin{array}{rrr|rr} 2 & 1 & 3 & 2 & 1 \\ 5 & 2 & 0 & 5 & 2 \\ 0 & 2 & 3 & 0 & 2 \end{array}\right|} = \frac{(-12 + 0 + 30) - (-78 + 0 + 15)}{(12 + 0 + 30) - (0 + 0 + 15)}$$

$$x = \frac{18 + 63}{27} = \mathbf{3}$$

$$y = \frac{\left|\begin{array}{rrr|rr} 2 & -2 & 3 & 2 & -2 \\ 5 & 5 & 0 & 5 & 5 \\ 0 & -13 & 3 & 0 & -13 \end{array}\right|}{\text{Denominator}} = \frac{(30 + 0 - 195) - (0 + 0 - 30)}{27}$$

$$y = \frac{-135}{27} = \mathbf{-5}$$

$$z = \frac{\left|\begin{array}{rrr|rr} 2 & 1 & -2 & 2 & 1 \\ 5 & 2 & 5 & 5 & 2 \\ 0 & 2 & -13 & 0 & 2 \end{array}\right|}{\text{Denominator}} = \frac{(-52 + 0 - 20) - (0 + 20 - 65)}{27}$$

$$z = \frac{-27}{27} = \mathbf{-1}$$

Thus, the solution set is $\{(\mathbf{3, -5, -1})\}$. The check is left for you.

CLASSROOM EXERCISES

Write the determinants you would use to solve each system for *x*, *y*, and *z*. Do *not* evaluate the determinants.

1. $\begin{cases} 2x + y - z = 3 \\ 4x - y + 4z = 0 \\ -3y + 2z = 6 \end{cases}$

2. $\begin{cases} x + y - z = 2 \\ 6x + y + z = 4 \\ 4x - y + 3z = 0 \end{cases}$

3. $\begin{cases} x - y + 2z = 6 \\ 2x + y + z = 3 \\ x - y + 3z = -9 \end{cases}$

WRITTEN EXERCISES

A Evaluate each determinant.

1. $\begin{vmatrix} 3 & 4 & -1 \\ 1 & 1 & 5 \\ 2 & 5 & 3 \end{vmatrix}$

2. $\begin{vmatrix} 3 & 0 & 5 \\ -1 & -5 & -3 \\ 4 & -2 & 6 \end{vmatrix}$

3. $\begin{vmatrix} 3 & -2 & 5 \\ -1 & 4 & -6 \\ 0 & 7 & -3 \end{vmatrix}$

4. $\begin{vmatrix} -8 & 6 & 4 \\ 3 & -5 & 3 \\ 2 & 1 & 0 \end{vmatrix}$

Solve each equation.

5. $\begin{vmatrix} z & -2 & 4 \\ 1 & 3 & -1 \\ 3z & 5 & 5 \end{vmatrix} = 0$

6. $\begin{vmatrix} 3 & 2t & 3t \\ 1 & -2 & -1 \\ 7 & 4 & 5 \end{vmatrix} = 12$

7. $\begin{vmatrix} 3 & 5 & -2 \\ 1 & 2 & -1 \\ q & -3 & 4q \end{vmatrix} = 0$

Use Cramer's Rule to solve each system.

8. $\begin{cases} 2x + y - z = 0 \\ x - y + z = 6 \\ x + 2y + z = 3 \end{cases}$

9. $\begin{cases} 2x + y - z = 2 \\ x + y + z = 7 \\ x + 2y + z = 4 \end{cases}$

10. $\begin{cases} x + y + z = 2 \\ 2x - y + z = 0 \\ x + 2y - z = 4 \end{cases}$

11. $\begin{cases} 2x + y + z = 4 \\ x + 3y + 2z = 12 \\ 3x + 2y + 3z = 16 \end{cases}$

12. $\begin{cases} x + 2y - z = 5 \\ 2x + z = -1 \\ 3x - 4y - 2z = 7 \end{cases}$

13. $\begin{cases} 2x + y = 43 \\ x + 3z = 47 \\ y - z = 14 \end{cases}$

B

14. $\begin{cases} 0.4x + 0.3y + 0.6z = 17 \\ 0.6x + 0.5y + 0.8z = 24 \\ x + 0.2y + 0.3z = 13 \end{cases}$

15. $\begin{cases} 0.4x - 0.5y - 0.2z = 8 \\ 0.3x + 0.2y - 0.3z = 2.9 \\ 0.5x - 0.5y = 6.5 \end{cases}$

16. $\begin{cases} 4x + y = 5b \\ 3x + 2z = 0 \\ 2y - 3z = 3b \end{cases}$

APPLICATIONS: Using Cramer's Rule

Write a linear system in three variables to represent each problem. Then use Cramer's Rule to solve the system.

17. Twice the height of a rectangular box is 1 centimeter less than the width. The length of the box is two centimeters less than the sum of the width and height. The sum of the length, width, and height is 18 centimeters. Find the length, width, and height of the box.

18. The sum of three numbers is 49. The second number is 3 times the first and the third number is 1 more than twice the difference between the second and the first. Find the numbers.

19. Ed Carter paid \$5.85 for some 20¢-stamps, 5¢-stamps, and 3¢-stamps. He bought 93 stamps in all. The number of 20¢-stamps was 7 less than $\frac{1}{2}$ the number of 3¢-stamps. How many stamps of each kind did Ed buy?

20. One shopper bought 3 pounds of butter, 4 pounds of cheese and 2 loaves of bread for \$12.99. A second shopper bought 2 pounds of the same brand of butter, 3 pounds of the same brand of cheese, and 3 loaves of the same brand of bread for \$10.21. A third shopper bought 4 pounds of butter, 2 pounds of cheese, and 4 loaves of bread for \$11.90. Find the cost of one pound of butter, of one pound of cheese, and of one loaf of bread.

21. A local newspaper has 40 pages available for advertising in each issue. Full-page ads cost \$200, half-page ads cost \$100, and quarter-page ads cost \$50. For a certain issue, the paper sold \$3250 worth of these ads. The number of quarter-page ads was 5 times the number of full-page ads. How many ads of each kind were sold?

C The denominator, *D*, of the third-order determinants in Cramer's Rule is called the **determinant of the coefficients.** When $D = 0$, either the system is inconsistent or the system has an infinite number of solutions.

In Exercises 22–24, evaluate *D*. When $D = 0$, state whether the system is *inconsistent* or has an *infinite number* of ordered triples in its solution set.

22. $\begin{cases} x + y = -1 \\ 3x + 2y = 1 \\ 2x - y = 3 \end{cases}$

23. $\begin{cases} x - y - z = 2 \\ 2x - 2y - 2z = 5 \\ x + 2y + 3z = 4 \end{cases}$

24. $\begin{cases} 5x - 3y - 2z = 5 \\ x + y - z = 3 \\ 2x - 6y + z = -5 \end{cases}$

Puzzle

On a shopping trip, Amanda spent half the money in her purse. When she got home, she discovered that she had as many pennies as she had dollars and half as many dollars as she had cents when she left home. How much money did Amanda have at the start of her shopping trip if she had 3 pennies when she got home?

REVIEW CAPSULE FOR SECTION 4-8

Write an equation or a compound sentence to represent each statement. *(Pages 74–76)*

1. The set of all points in the plane that are 3 units to the right of the *y* axis.

2. The set of all points in the plane that are 2 units to the left of the *y* axis.

3. The set of all points in Quadrant I of the coordinate plane.

4. The set of all points in Quadrant IV of the coordinate plane.

4–8 Solving Systems of Linear Inequalities

The solution set of the system of inequalities

$$\begin{cases} x - y < 0 \\ 2x + y > 4 \end{cases}$$

is the set of ordered pairs which are the coordinates of all points in the *intersection* (overlapping) of two plane regions.

EXAMPLE 1 Solve the system above by graphing.

Solution: [1] To graph $x - y < 0$, first graph the related equation $x - y = 0$. This line is dashed to show that it is *not* in the solution set of $x - y < 0$.

Substitute the coordinates of at least two points to find the half-plane where $x - y < 0$ is true.

Test	Result	
(2, 0)	$x - y < 0;\ 2 - 0 \ < 0$ is false. ←	**The half-plane below**
(−2, 0)	$x - y < 0;\ -2 - 0 < 0$ is true. ←	**The half-plane above**

Shade the half-plane *above* the line for $x - y = 0$. (Figure 1)

The graph for $2x + y < 4$ is found in the same way. (Figure 2)

[2] The solution set of the system is the intersection of these two graphs. (Figure 3)

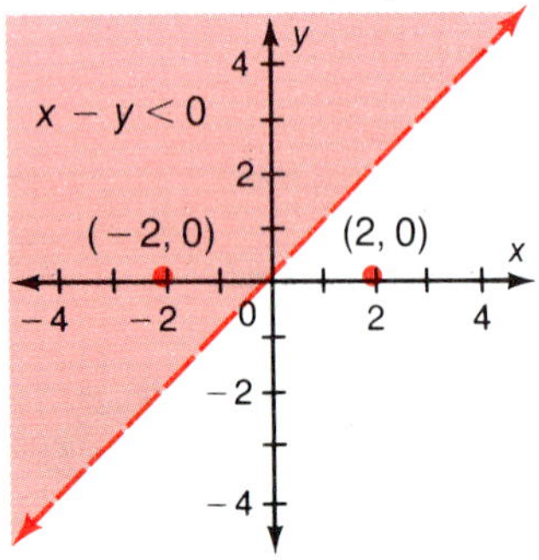

Figure 1

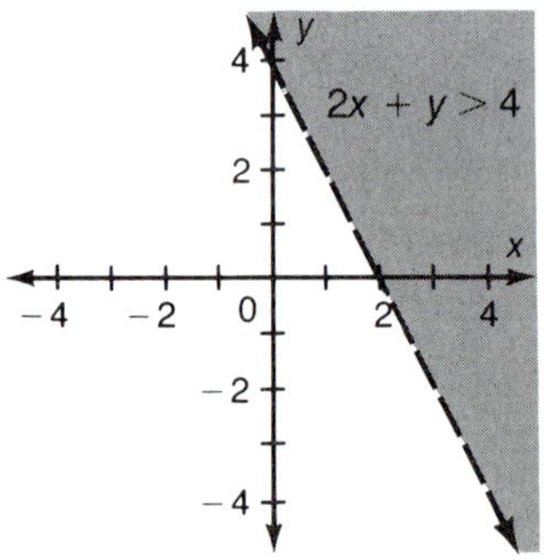

Figure 2

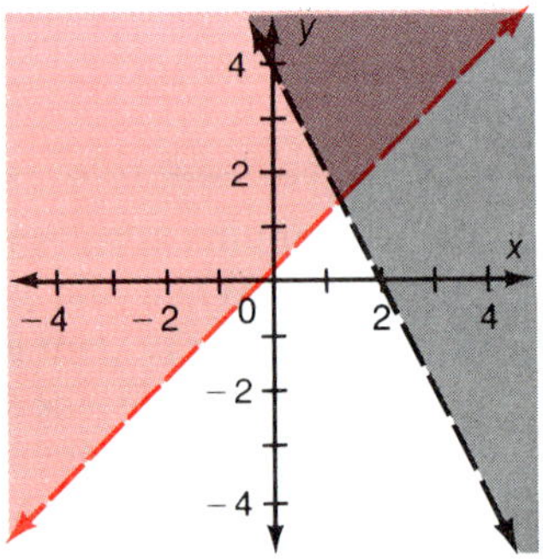

Figure 3

The following summarizes how to find the solution of a system of linear inequalities by graphing.

Summary

To solve a system of linear inequalities by graphing:

[1] Graph each inequality in the system.

[2] Use heavier shading to show the intersection (overlapping) of the two graphs. The coordinates of all points in this region are the solution of the system.

EXAMPLE 2 Solve by graphing: $\begin{cases} x + y \leq -2 \\ x - 2y \geq 7 \end{cases}$.

Solution: The solution set is the intersection of the half-planes satisfying $x + y \leq -2$ and $x - 2y \geq 7$.

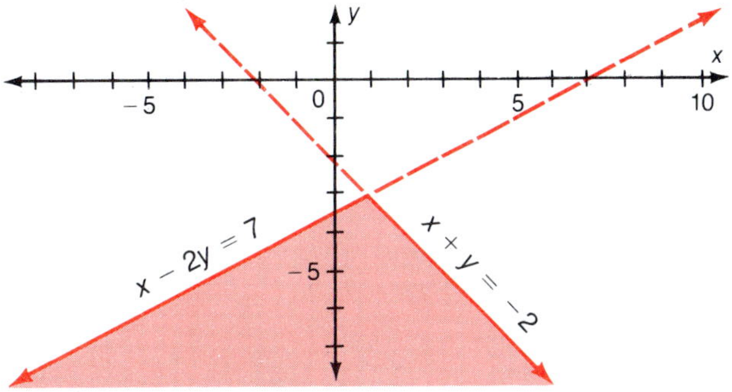

All points on the line $x + y = -2$ are part of the solution set of $x + y \leq -2$. Likewise, all points on the line $x - 2y = 7$ are part of the solution set of $x - 2y \leq 7$. Therefore, the solution set is the **shaded region of the graph and the rays** (solid portions of the lines) **that border this region.**

The union of a line and one of its half-planes, such as the graph of $x + y \leq 2$, is a **closed half-plane.**

Follow a similar procedure to graph a system of three inequalities in two variables.

EXAMPLE 3 Solve by graphing: $\begin{cases} x \geq 0 \\ y \geq 1 \\ x + y \leq 4 \end{cases}$.

Solution: Graph each of the inequalities of the system. The solution set of each inequality is a closed half-plane.

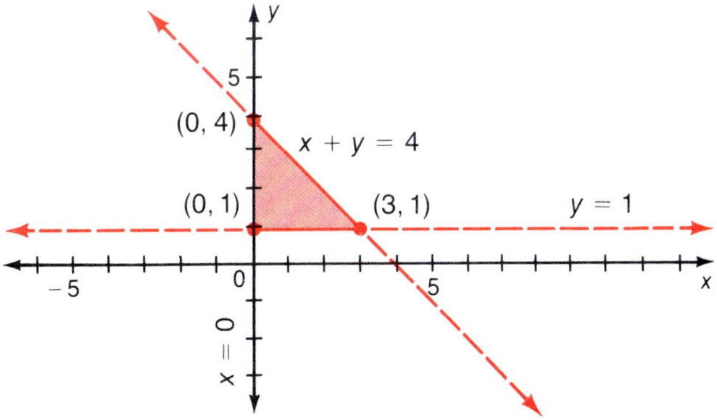

The solution set can be described as the **triangle whose vertices are (0, 1), (3, 1), and (0, 4), together with the interior of the triangle.**

CLASSROOM EXERCISES

The graphs of $y = 18 - 5x$, $y = x$, and $y = -2$ form the 7 regions shown at the right when graphed on the same coordinate axes.

In Exercises 1–8, name the region or regions that form the solution set of the given system.

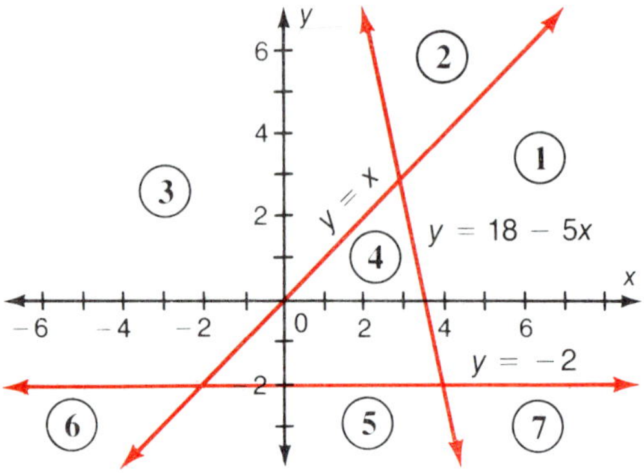

1. $\begin{cases} y > -2 \\ y < x \end{cases}$

2. $\begin{cases} y < x \\ y < 18 - 5x \end{cases}$

3. $\begin{cases} y < -2 \\ y < x \end{cases}$

4. $\begin{cases} y > x \\ y > -2 \end{cases}$

5. $\begin{cases} y < x \\ y > 18 - 5x \end{cases}$

6. $\begin{cases} y > 18 - 5x \\ y > -2 \end{cases}$

7. $\begin{cases} y < -2 \\ y < x \\ y > 18 - 5x \end{cases}$

8. $\begin{cases} y < x \\ y < 18 - 5x \\ y > -2 \end{cases}$

WRITTEN EXERCISES

A Solve each system by graphing.

1. $\begin{cases} x + 2y \le 5 \\ x - 3y \le 2 \end{cases}$

2. $\begin{cases} 3x - 2y \ge 4 \\ x + y \ge 4 \end{cases}$

3. $\begin{cases} 2x - y < 2 \\ x + 3y > 3 \end{cases}$

4. $\begin{cases} 5x - y < 5 \\ x - y > 2 \end{cases}$

5. $\begin{cases} x + 3y < 1 \\ 2x + 6y < 4 \end{cases}$

6. $\begin{cases} x + 3y < 1 \\ 2x + 6y > 4 \end{cases}$

7. $\begin{cases} x + y > 4 \\ x - y > 6 \end{cases}$

8. $\begin{cases} x + y > 2 \\ x - y < 5 \end{cases}$

9. $\begin{cases} x \ge 2 \\ y - x \ge 0 \\ x + y \le 8 \end{cases}$

10. $\begin{cases} x \ge 1 \\ y \ge 1 \\ x + y \le 5 \end{cases}$

11. $\begin{cases} x \ge 1 \\ y \le 1 \\ x - y \le 7 \end{cases}$

12. $\begin{cases} x \le -1 \\ y \ge 1 \\ 2y - x \le 6 \end{cases}$

13. $\begin{cases} x \le 0 \\ y \le 0 \\ x + 2y \ge -6 \end{cases}$

14. $\begin{cases} x \le 0 \\ 3y + 2x \ge 0 \\ 3y - 2x \le 12 \end{cases}$

15. $\begin{cases} y + 4x \le 25 \\ 2y - x \le 5 \\ 5y + 2x \ge 17 \end{cases}$

16. $\begin{cases} 5y - x \le 8 \\ y + 5x \le 12 \\ 3y + 2x \ge -3 \end{cases}$

17. $\begin{cases} x \ge 0 \\ 2y + 3x \le 8 \\ 4y - 5x \ge 24 \end{cases}$

18. $\begin{cases} x + y \ge 4 \\ x - y \le 1 \\ x + 2y \le 8 \end{cases}$

19. $\begin{cases} y > 1 \\ y < 2x \\ y < x + 3 \end{cases}$

20. $\begin{cases} y < x + 4 \\ y < 2x \\ y < 1 \end{cases}$

B

21. $\begin{cases} 2x + 3y \le 8 \\ 5x - 4y \ge 10 \end{cases}$

22. $\begin{cases} 4y + 3x \ge 0 \\ 3y - 2x \le 4 \end{cases}$

23. $\begin{cases} 2y + 3x \ge -4 \\ 5y - x \le 6 \end{cases}$

24. $\begin{cases} y - 2x \ge 6 \\ y + x \ge -3 \end{cases}$

25. $\begin{cases} x + y \ge 7 \\ x + y \ge 2 \\ x \ge 0 \\ y \ge 0 \end{cases}$

26. $\begin{cases} x \ge 0 \\ y \ge 0 \\ y + x \le 6 \\ y + 3x \le 12 \end{cases}$

27. $\begin{cases} x \ge 0 \\ 2y - x \le 8 \\ 4y + 3x \le 36 \\ 8y - 3x \ge 0 \end{cases}$

28. $\begin{cases} x \ge 0 \\ y \le 0 \\ y < 2x \\ x + y > 1 \end{cases}$

C

29. $\begin{cases} |y| \ge 2 \\ |x| \le 1 \end{cases}$

30. $\begin{cases} |x| < 3 \\ |x + y| < 1 \end{cases}$

31. $\begin{cases} |x + y| \ge -1 \\ |x + y| < -1 \end{cases}$

32. $\begin{cases} |x| + |y| = 4 \\ |y| \ge 4 \end{cases}$

Problem Solving and Applications

4–9 Linear Programming

Managers and production engineers try to minimize costs and maximize profit. In such situations, there are usually a large number of conditions called **constraints** that must all be satisfied simultaneously.

Finding a maximum or minimum value that satisfies all of the given conditions is called **linear programming.** These problems are called *linear* because the conditions involve straight lines. It has been proved that **maximum and minimum values occur at one of the corners or vertices of the polygon formed by the graph of the region that contains the points satisfying the constraints.**

EXAMPLE Reynold's Hauling Company has 7 six–ton trucks, 4 ten–ton trucks and 9 drivers. The company has been contracted to haul 360 tons of asphalt per day for a road construction job. The six–ton trucks can make 8 trips a day while the ten–ton trucks can make 6 trips per day. A six–ton truck costs $15 per day and a ten–ton truck costs $24 per day, not including drivers' salaries.

a. Graph the region that satisfies the constraints.

b. If all 9 drivers are to work on the job, how many trucks of each type should be used to minimize the cost?

Solution:

a. [1] List the constraints.

1. No more than 7 six-ton trucks are available.
2. No more than 4 ten–ton trucks are available.
3. No more than 9 trucks can be used because there are only 9 drivers.
4. A six–ton truck can make no more than 8 trips and a ten–ton truck can make no more than 6 trips to haul 360 tons per day.

[2] Write a system of inequalities that describes the constraints.

Let $x =$ the number of six–ton trucks.
Let $y =$ the number of ten–ton trucks.

$$\begin{cases} \textbf{1. } x \le 7 \\ \textbf{2. } y \le 4 \\ \textbf{3. } x + y \le 9 \\ \textbf{4. } 48x + 60y \ge 360 \end{cases}$$

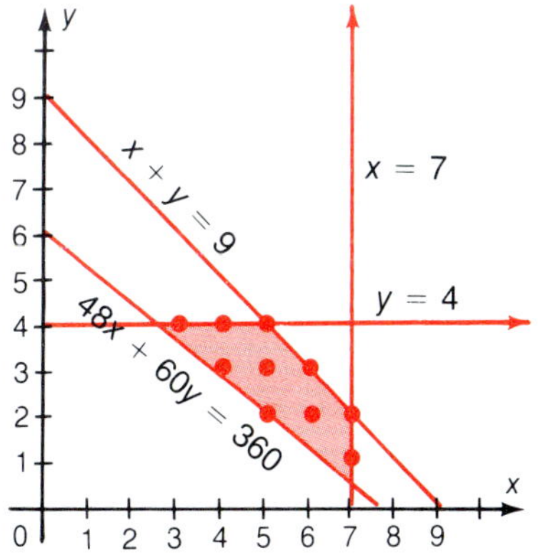

[3] Graph the system. Any point in the shaded region will satisfy the four constraints.

Since there must be an integral number of trucks, the only points of concern are those whose coordinates are integers.

These points are:

(3, 4) (4, 4) (4, 3) (5, 2) (5, 3) (5, 4) (6, 2) (6, 3) (7, 1) (7, 2)

b. Since all 9 drivers are to work on the job, the only points to be considered are (5, 4) and (7, 2). Note that these points occur at corners or vertices of the polygon graphed on page 145.

[4] To determine which point or points will minimize the cost, write an equation that describes the cost, C.

$$C = 15x + 24y$$

[5] Determine which point (or points) will minimize the cost, C.

Possible Solutions	(5, 4)	(7, 2)
Cost	171	153

The table shows that the minimum cost to do the job with 9 drivers is **$153**, using **7 six-ton trucks and 2 ten-ton trucks.**

CLASSROOM EXERCISES

Solve by the linear programming method.

1. Maximize $P(x, y) = 5x + 3y$

Constraints

$$\begin{cases} x + y \le 6 \\ x - y \le 4 \\ x \ge 0 \\ y \ge 0 \end{cases}$$

2. Minimize $C(x, y) = 2x + y$

Constraints

$$\begin{cases} x + y \ge 6 \\ x - y \ge 4 \\ x \ge 0 \\ y \ge 0 \end{cases}$$

3. Maximize $P(x, y) = 3x + y$

Constraints

$$\begin{cases} 2x + y \le 3 \\ 3x + 4y \le 12 \\ x \ge 0 \\ y \ge 0 \end{cases}$$

4. Minimize $C(x, y) = x + y$

Constraints

$$\begin{cases} x + 2y \ge 3 \\ 3x + 4y \ge 8 \\ x \ge 0 \\ y \ge 0 \end{cases}$$

WORD PROBLEMS

A

1. Suppose the constraints in a manufacturing problem are $x \ge 0$, $y \ge 0$, $y + \frac{1}{4}x \ge 2$, $y + x \ge 5$ and $y + 2x \ge 7$ and cost $= 2y + 3x$. Find x and y to minimize the cost.

2. A carpenter makes bookcases in two sizes, large and small. It takes 6 hours to make a large bookcase and 2 hours to make a small one. The profit on a large bookcase is $50 and the profit on a small bookcase is $20. The carpenter can spend only 24 hours per week making bookcases and must make at least two of each size per week. How many of each size must be made per week in order to provide maximum profit?

3. The annual cost of heating a building is $3 for every square foot of window and $1 for every square foot of wall or roof. A local building code requires that the area of the windows must be at least $\frac{1}{8}$ the area of the walls and roof. To the nearest square foot, what is the largest surface area a building can have if its annual heating cost is not to exceed $1000?

4. A tire manufacturer has 1000 units of raw rubber to use in producing radial tires for passenger cars and tractor tires. Each radial tire requires 5 units of rubber; each tractor tire requires 20 units. Labor costs are \$8 for a radial tire and \$12 for a tractor tire. Suppose that the manufacturer does not wish to pay more than \$1500 in labor costs and wishes to make a profit of \$10 per radial tire and \$25 per tractor tire. How many of each kind of tire should be made in order to maximize profits?

B

5. A furniture company makes two kinds of wooden table legs, one plain and one fancy. Each plain leg takes 2 hours on a lathe and 1 hour of sanding. Each fancy leg takes 1 hour on the lathe and 4 hours of sanding. The 4 lathes and 6 sanding machines are used 12 hours a day. Each plain leg nets \$3 in profits and each fancy leg nets \$5 in profit. If the company can sell all the table legs it makes, how many of each kind should be produced each day in order to maximize profit?

Review

Evaluate each determinant. *(Section 4–6)*

1. $\begin{vmatrix} 4 & 1 \\ 2 & 0 \end{vmatrix}$ 2. $\begin{vmatrix} -3 & 5 \\ 6 & 7 \end{vmatrix}$ 3. $\begin{vmatrix} \frac{1}{2} & -\frac{1}{2} \\ -\frac{1}{2} & \frac{1}{2} \end{vmatrix}$ 4. $\begin{vmatrix} 0.2 & 0.1 \\ -0.3 & 0.4 \end{vmatrix}$

5. Use Cramer's Rule to solve this system: $\begin{cases} 2x - y = 5 \\ x + 2y = 25 \end{cases}$ *(Section 4–6)*

Evaluate each determinant. *(Section 4–7)*

6. $\begin{vmatrix} 1 & 1 & 1 \\ 0 & -1 & 1 \\ 1 & 1 & -1 \end{vmatrix}$ 7. $\begin{vmatrix} 2 & 5 & 0 \\ -1 & 6 & 8 \\ 3 & 0 & 0 \end{vmatrix}$ 8. $\begin{vmatrix} 4 & -2 & 6 \\ 5 & -1 & 3 \\ 6 & 5 & 5 \end{vmatrix}$ 9. $\begin{vmatrix} 6 & 1 & 6 \\ 5 & 2 & -4 \\ 1 & -9 & 1 \end{vmatrix}$

10. Use Cramer's Rule to solve this system: $\begin{cases} x + y - 2 = 0 \\ x + 2y + z = -3 \\ 2x - y - 8z = 9 \end{cases}$ *(Section 4–7)*

Solve each system by graphing. *(Section 4–8)*

11. $\begin{cases} x - y \le 6 \\ x - 2y \le 3 \end{cases}$ 12. $\begin{cases} x - 3y < -5 \\ 2x + 5y > 1 \end{cases}$ 13. $\begin{cases} x \ge -1 \\ 3x + 7y \le -4 \\ 2x + 5y \ge -3 \end{cases}$ 14. $\begin{cases} x + y \ge 2 \\ x - y \ge -1 \\ 2x - y \le 1 \end{cases}$

15. A baker wants to maximize the protein in a 2-cup batch of cookies. The recipe specifies that at least half of the ingredients must be oatmeal and the rest wheat germ. There are 10 grams of protein in a cup of oatmeal and 16 grams of protein in a cup of wheat germ. How much of each ingredient should be used? *(Section 4–9)*

BASIC: SOLVING SYSTEMS OF EQUATIONS

You can write a BASIC program to solve a system of two linear equations by using **Cramer's Rule** (page 135).

Problem: *Write a program which uses Cramer's Rule to solve systems of two linear equations in two variables.*

```
100 REM   SOLVE   A1 * X + B1 * Y = C1
110 REM           A2 * X + B2 * Y = C2   USING CRAMER'S RULE
120 REM
130 PRINT "ENTER THE COEFFICIENTS OF THE FIRST EQUATION";
140 INPUT A1, B1, C1
150 PRINT "ENTER THE COEFFICIENTS OF THE SECOND EQUATION";
160 INPUT A2, B2, C2
170 LET D = A1 * B2 - A2 * B1
180 IF D = 0 THEN 230
190 LET X = (B2 * C1 - B1 * C2)/D
200 LET Y = (A1 * C2 - A2 * C1)/D
210 PRINT "SOLUTION IS (";X;",";Y;")"
220 GOTO 240
230 PRINT "NO UNIQUE SOLUTION"
240 PRINT "ANY MORE SYSTEMS TO SOLVE (1 = YES, 0 = NO)";
250 INPUT Z
260 IF Z = 1 THEN 130
270 END
```

D is the determinant of the system. (line 170)

Output: The following is the output from a sample run of the program above.

```
RUN
ENTER THE COEFFICIENTS OF THE FIRST EQUATION? 1,1,1
ENTER THE COEFFICIENTS OF THE SECOND EQUATION? 2,2,2
NO UNIQUE SOLUTION

ANY MORE SYSTEMS TO SOLVE (1 = YES, 0 = NO)? 1
ENTER THE COEFFICIENTS OF THE FIRST EQUATION? 1,1,1
ENTER THE COEFFICIENTS OF THE SECOND EQUATION? 2,2,-7
NO UNIQUE SOLUTION

ANY MORE SYSTEMS TO SOLVE (1 = YES, 0 = NO)? 1
ENTER THE COEFFICIENTS OF THE FIRST EQUATION? 1,1,1
ENTER THE COEFFICIENTS OF THE SECOND EQUATION? 2,-1,2
SOLUTION IS ( 1 , 0 )

ANY MORE SYSTEMS TO SOLVE (1 = YES, 0 = NO)? 0
READY
```

Analysis

Statements 180 and 230: If D = 0, the computer prints `NO UNIQUE SOLUTION`. The program does not decide whether the system has no solution or infinitely many solutions. (See Exercise 18 below.)

Statements 190–210: If $D \neq 0$, then X and Y are computed using Cramer's Rule. In statement 210, because the parentheses and the comma are inside quotation marks, the solution is printed as an ordered pair. (Notice the output for the last set of data in the sample run.)

EXERCISES

A In Exercises 1–15, write each system in the following form.

$$\begin{cases} A1x + B1y = C1 \\ A2x + B2y = C2 \end{cases}$$

Then use the program on page 148 to solve each system.

1. $\begin{cases} 3x = 2y + 8 \\ 5y = 4x - 6 \end{cases}$
2. $\begin{cases} y = 5x + 1 \\ x - 2y + 6 = 0 \end{cases}$
3. $\begin{cases} x = y + 6 \\ 4y + 4x = 15 \end{cases}$
4. $\begin{cases} 2x - y = 5 \\ x + 2y = 25 \end{cases}$
5. $\begin{cases} 2x - y = -3 \\ 4x - y = -2 \end{cases}$
6. $\begin{cases} 2x - 3y = 13 \\ 3x + y = 3 \end{cases}$
7. $\begin{cases} y = -3x + 4 \\ 3x = -y + 6 \end{cases}$
8. $\begin{cases} x = -3y + 1 \\ 3y + 2x = 5 \end{cases}$
9. $\begin{cases} y = 2x - 3 \\ x - 2y = 9 \end{cases}$
10. $\begin{cases} 3x - 14 = y \\ 2x + 4y = 14 \end{cases}$
11. $\begin{cases} 3x - 2y = 8 \\ -2x + 3y + 12 = 0 \end{cases}$
12. $\begin{cases} y + 3x = 7 \\ y = x - 13 \end{cases}$
13. $\begin{cases} 5x - 2 = -3y \\ y - 3x = -2 \end{cases}$
14. $\begin{cases} -3y + 10 = 5x \\ 2x - 8 = -3y \end{cases}$
15. $\begin{cases} 5y - 7 = -4x \\ 5x + 6y = 8 \end{cases}$

B In Exercises 16–17, write a `BASIC` program for each problem.

16. Use Cramer's Rule to solve systems of three linear equations in three variables.
17. Given an ordered pair (x, y), determine whether it is in the common solution set of the following system of inequalities.

$$\begin{cases} x > 0 \\ y > 0 \\ x + y < 9 \\ 48x + 60y \geq 360 \end{cases}$$

C

18. Expand the program on page 148 so that, when D = 0, it prints whether the system has no solution or infinitely many solutions. For the first set of data in the sample run of the program on page 148, the revised program should print `DEPENDENT SYSTEM--INFINITELY MANY SOLUTIONS`. For the second set of data in the sample run, the revised program should print `INCONSISTENT SYSTEM--NO SOLUTION`.

Chapter Summary

IMPORTANT TERMS

Addition method (p. 118)
Closed half-plane (p. 143)
Consistent systems (p. 128)
Constraints (p. 145)
Dependent systems (p. 128)
Determinant (p. 134)
Inconsistent systems (p. 128)
Linear programming (p. 145)
Substitution method (p. 121)
System of equations (p. 114)
System of inequalities (p. 142)

IMPORTANT IDEAS

1. The **graph of the solution set of a system of sentences** is the set of points common to the graphs of the sentences of the system.

2. **Addition Method for Solving Linear Systems.** Choose multiples of the equation that eliminate one of the variables when the equations are added.

3. **Substitution Method for Solving Linear Systems**

   [1] Solve one equation of the system for one variable in terms of the other variable.

   [2] Substitute this value in the other equation and solve.

   [3] Substitute the value from step [2] in one of the original equations. Solve for the value of the second variable.

   [4] Check in the original equations of the system.

4. To use linear systems to solve **word problems involving two variables:**

   [1] Use two different variables to represent the unknowns.

   [2] Organize the given information in a table, when possible.

   [3] Identify the conditions of the problem. Write an equation to represent each condition.

   [4] Solve the system.

   [5] Check the original conditions of the problem.

5. A linear system is either **consistent** or **inconsistent.** If the graphs of the two equations of a consistent system coincide, then the system is also **dependent.**

6. For all real numbers, a_1, b_1, a_2, b_2, $\begin{vmatrix} a_1 & b_1 \\ a_2 & b_2 \end{vmatrix} = a_1b_2 - a_2b_1$.

   For all real numbers a_1, b_1, c_1, a_2, b_2, c_2, a_3, b_3, c_3,

$$\begin{vmatrix} a_1 & b_1 & c_1 \\ a_2 & b_2 & c_2 \\ a_3 & b_3 & h_3 \end{vmatrix} = a_1b_2c_3 + a_2b_3c_1 + a_3b_1c_2 - a_1b_3c_2 - a_2b_1c_3 - a_3b_2c_1.$$

7. **Cramer's Rule** for systems of two linear equations in two variables

   The solution of the system $\begin{cases} a_1x + b_1y = c_1 \\ a_2x + b_2y = c_2 \end{cases}$ is

$$x = \frac{\begin{vmatrix} c_1 & b_1 \\ c_2 & b_2 \end{vmatrix}}{\begin{vmatrix} a_1 & b_1 \\ a_2 & b_2 \end{vmatrix}}, \quad y = \frac{\begin{vmatrix} a_1 & c_1 \\ a_2 & c_2 \end{vmatrix}}{\begin{vmatrix} a_1 & b_1 \\ a_2 & b_2 \end{vmatrix}}.$$

8. **Cramer's Rule** for systems of three linear equations in three variables

The solution of the system $\begin{cases} a_1x + b_1y + c_1z = d_1 \\ a_2x + b_2y + c_2z = d_2 \\ a_3x + b_3y + c_3z = d_3 \end{cases}$ is

$$x = \frac{\begin{vmatrix} d_1 & b_1 & c_1 \\ d_2 & b_2 & c_2 \\ d_3 & b_3 & c_3 \end{vmatrix}}{\begin{vmatrix} a_1 & b_1 & c_1 \\ a_2 & b_2 & c_2 \\ a_3 & b_3 & c_3 \end{vmatrix}}, \quad y = \frac{\begin{vmatrix} a_1 & d_1 & c_1 \\ a_2 & d_2 & c_2 \\ a_3 & d_3 & c_3 \end{vmatrix}}{\begin{vmatrix} a_1 & b_1 & c_1 \\ a_2 & b_2 & c_2 \\ a_3 & b_3 & c_3 \end{vmatrix}}, \quad z = \frac{\begin{vmatrix} a_1 & b_1 & d_1 \\ a_2 & b_2 & d_2 \\ a_3 & b_3 & d_3 \end{vmatrix}}{\begin{vmatrix} a_1 & b_1 & c_1 \\ a_2 & b_2 & c_2 \\ a_3 & b_3 & c_3 \end{vmatrix}}.$$

9. To solve a **system of linear inequalities** by graphing:

   [1] Graph each inequality in the system.

   [2] Use heavier shading to show the intersection of the two graphs. The coordinates of all points in this region are the solutions of the system.

10. **Linear programming** furnishes a technique of finding maximum and minimum values of a linear relation subject to certain constraints expressed as linear inequalities.

Chapter Objectives and Review

Objective: *To solve a system of linear equations graphically (Section 4–1)*

In Exercises 1–4, solve each system by graphing.

1. $\begin{cases} x + y = -2 \\ x - y = 0 \end{cases}$
2. $\begin{cases} y = 4 \\ x - 2y = -3 \end{cases}$
3. $\begin{cases} x - 2 = 0 \\ y + 5 = 0 \end{cases}$
4. $\begin{cases} 3x - y = -4 \\ 2x - 3y = 2 \end{cases}$

Objective: *To solve a system of linear equations by the addition method (Section 4–2)*

In Exercises 5–8, solve each system by the addition method.

5. $\begin{cases} x + 2y = 12 \\ x + 3y = 18 \end{cases}$
6. $\begin{cases} 3x - 4y = -14 \\ 3x + 2y = 16 \end{cases}$
7. $\begin{cases} 7x + 10y = 18 \\ 5x - 16y = -4 \end{cases}$
8. $\begin{cases} 6x - 3y = 2 \\ 5x + 2y = 1 \end{cases}$

Objective: *To solve a system of linear equations by the substitution method (Section 4–3)*

9. $\begin{cases} y = -4 \\ 2x - 7y = 5 \end{cases}$
10. $\begin{cases} y = -2x \\ x + 4y = 21 \end{cases}$
11. $\begin{cases} x + y = 18 \\ 4x - 5y = 5 \end{cases}$
12. $\begin{cases} 6x - y = 7 \\ x + 2y = 8 \end{cases}$

Objective: *To use a system of two linear equations in two variables to solve a problem (Section 4–4)*

13. An automobile dealer makes \$400 on each large size automobile that he sells and \$250 on each compact. One year he made \$34,750 by selling 100 large size and compact cars. How many of each kind did he sell?

14. The units digit of a certain number is 5 more than twice the tens digit. If the digits of the number are reversed, then the new number is 54 more than the original number. Find the original number.

15. The perimeter of a rectangular garden is 80 meters. The width of the garden is one-fourth the length. Find the width and length.

16. Sarah Putnam has 18 coins made up entirely of quarters and half dollars. The total value is \$7.50. Find the number of each kind of coin.

17. Mrs. Randolph invested part of her earnings at 12% and another part at 9%. Her yearly income from the investments was \$375. If she had reversed the investments, her income would have been \$360. How much did she invest at each rate?

18. Flying with a tailwind, a plane traveled 6300 kilometers in 7 hours. On the return trip, the same wind decreased the plane's travel time by 1 hour. Find the speed of the plane coming and going.

Objective: *To identify linear systems as consistent or inconsistent and to tell whether a consistent system is also dependent (Section 4–5)*

In Exercises 19–22, graph each system. Use the graph to classify each system as consistent, *C*, or inconsistent, *I*. When a system is also dependent, write *D*.

19. $\begin{cases} x = y \\ 5x + 5y = 0 \end{cases}$

20. $\begin{cases} y = x - 3 \\ x - y = 1 \end{cases}$

21. $\begin{cases} 3y = 4x - 1 \\ x = \frac{4}{3}y \end{cases}$

22. $\begin{cases} x - 2y = 4 \\ y = \frac{1}{2}x - 2 \end{cases}$

Objective: *To use Cramer's Rule to solve a linear system of two equations in two unknowns (Section 4–6)*

Use Cramer's Rule to solve each system.

23. $\begin{cases} 3x + y = 10 \\ 2x - 3y = 25 \end{cases}$

24. $\begin{cases} 3x - 5y = 0 \\ 2x + y = 0 \end{cases}$

25. $\begin{cases} 3x + 2y = 2 \\ 2x + 3y = -2 \end{cases}$

26. $\begin{cases} 4x + 3y = -6 \\ 5x = 2y + 3 \end{cases}$

Objective: *To use Cramer's Rule to solve a linear system of three equations in three unknowns (Section 4–7)*

Use Cramer's Rule to solve each system.

27. $\begin{cases} x - y + z = 4 \\ x + 2y - z = 5 \\ 2x - y + z = 9 \end{cases}$

28. $\begin{cases} x + y - z = -9 \\ 2x - y + 2z = 16 \\ 3x + 2y + z = 3 \end{cases}$

29. $\begin{cases} 2x - y + z = 0 \\ x - 2y - z = 15 \\ 3y + z = 0 \end{cases}$

30. $\begin{cases} x + 2z = 2 \\ 2y + 3z = 4 \\ 2x - y = -8 \end{cases}$

Objective: *To solve a system of linear inequalities by graphing (Section 4–8)*

Solve each system by graphing.

31. $\begin{cases} x + y \le 4 \\ x - y \le 2 \end{cases}$

32. $\begin{cases} 2x - y > -2 \\ x + 4y > 1 \end{cases}$

33. $\begin{cases} y \ge -2 \\ 2x - 3y \le 3 \\ x + 2y \le 4 \end{cases}$

34. $\begin{cases} y \le \frac{1}{2}x \\ x + y \ge 1 \\ x + 5y \ge -11 \end{cases}$

Objective: *To solve problems by using linear programming (Section 4-9)*

35. Randi devotes 10 hours per week to making belts and handbags for sale. One hour is required to make a belt and two hours are required for a handbag. The material for a belt costs \$6 and the material for a handbag costs \$9. Randi can afford to spend \$50 per week for materials. The profit on one belt is \$20 and the profit on one handbag is \$45. How many belts and how many handbags should Randi sell in order to obtain the largest possible profit?

Chapter Test

1. Solve by graphing: $\begin{cases} x + 2y = 1 \\ 7x - 4y = -20 \end{cases}$

2. Solve by the addition method: $\begin{cases} 5x + 11y = -7 \\ -3x + 2y = 30 \end{cases}$

3. Solve by the substitution method: $\begin{cases} y = 6x \\ 8x - 3y = 25 \end{cases}$

4. A lighting engineer plans to install 35 light fixtures in a new supermarket. The total wattage for the fixtures is 4000 watts. Only 100-watt fixtures and 150-watt fixtures are available. How many of each type of fixture will be used?

5. Dan Simm has part of his \$4500 savings invested in a certificate of deposit and the rest invested in a money market mutual fund. Last year his money earned 8% return from the certificate of deposit and an average of 11% from the money market fund to give a total return of \$414 on both investments. How much money was there in each type of investment?

6. Graph the following system: $\begin{cases} y = 2x + 1 \\ 2x - y = 2 \end{cases}$. Then use the graph to classify the system as either *consistent* or *inconsistent*.

7. Use Cramer's Rule to solve: $\begin{cases} 4x - 3y = 9 \\ -5x + 2y = 1 \end{cases}$

8. Use Cramer's Rule to solve: $\begin{cases} 3x + 3y - z = 9 \\ x - 3y - 3z = 7 \\ 3x - y + 3z = -7 \end{cases}$

9. Solve by graphing: $\begin{cases} x \geq -1 \\ x + y \geq 4 \\ 3x + y \leq 8 \end{cases}$

10. Janet Swifteagle wants to purchase up to 20 minutes of computer time from a time-sharing computer company. She needs at least 7 minutes of time from Computer A. The time-sharing company will rent Computer A only if the user agrees also to rent Computer B for at least 5 minutes. The rent for Computer A is \$500 per minute and the rent for Computer B is \$200 per minute. In order to minimize the total cost, how much computer time should Janet buy using each machine?

Review of Word Problems: Chapters 1–4

1. A rectangular pool has a perimeter of 44 meters. The length of the pool is 12 meters greater than the width. How wide is the pool? *(Section 1–6)*
2. Karen's present age exceeds her daughter's age by 25 years. In 15 years Karen will be twice as old as her daughter. How old is each person? *(Section 1–7)*
3. Find three consecutive integers whose sum is -57. *(Section 1–7)*
4. John scored 76, 91, 82 and 95 on four tests. What score must John achieve on the next test in order to have an average of at least 85? *(Section 2–4)*
5. James Grant purchased a condominium for \$85,000. The value of the condominium increases at a constant rate of 12% per year.
   a. Write an equation for the value, V, of the property after t years. Graph the equation.
   b. Use the graph to estimate how long it will take for the value of the condominium to increase to \$150,000. Round your answer to the nearest whole number of years. *(Section 3–4)*
6. A baseball player hits 3 home runs in every 40 times at bat. At this rate, how many times must he bat before hitting 60 home runs? *(Section 3–7)*
7. The Champion Sporting Good Company manufactures two grades of golf balls, the Gofar, which sells for 80 cents each, and the Sure–Par, which sells for \$1.10 each. The company received \$3715 for an order of 4250 balls. How many of each kind were ordered? *(Section 4–4)*
8. The tens digit of a two–digit number is one less than twice the units digit. The sum of the digits of the number is 14. Find the number. *(Section 4–4)*
9. Two angles are supplementary (their sum is 180°). One angle is 18° more than five times the other. Find the measure of each angle. *(Section 4–4)*
10. A merchant makes a bank deposit of \$350 with 51 five– and ten–dollar bills. How many bills of each type were deposited? *(Section 4–4)*
11. Betty invested a sum of money, part at 12% and the rest at 16%. Her total interest income per year from the investment is \$2480. The amount invested at 12% is \$1500 less than the amount invested at 16%. How much did she invest at each rate? *(Section 4–4)*
12. Traveling downstream a boat can cover 32 kilometers in 4 hours. Going upstream, it travels only half of this distance in 4 hours. What is the rate of the boat in still water and what is the rate of the current? *(Section 4–4)*
13. A manufacturer of appliances can produce at most 20 washing machines and at most 30 dryers per day. The maximum total hours of labor that the manufacturer can assign daily is 96. Each washing machine requires 3 hours of labor; each dryer requires 2 hours of labor. The profit on each washing machine is \$30, and the profit on each dryer is \$40. Find the number of washing machines and dryers the company should produce for maximum profit each day. *(Section 4–9)*

Cumulative Review: Chapters 1–4

Choose the best answer. Choose *a*, *b*, *c*, or *d*.

1. Find the value of $3x^2 - 2x(x - y)$ when $x = 1$ and $y = 3$.

a. -13 **b.** 19 **c.** -125 **d.** 7

2. Which statement is false?

a. $N \subset I$ **b.** $W \subset Q$ **c.** $W \subset N$ **d.** $\{0\} \not\subset N$

3. Find the value of $|-3| \cdot (-\frac{1}{5})$.

a. -0.6 **b.** 0.6 **c.** -15 **d.** 15

4. When are $|x - 4|$ and $x - 4$ equal?

a. When $x \geq 4$ **b.** When $x \leq 4$ **c.** When $x = 0$ **d.** When $x = -4$

5. Name the real number postulate illustrated by $\frac{3}{5}(x - 5) = (x - 5)\frac{3}{5}$.

a. Additive Inverse Postulate **b.** Commutative Postulate for Multiplication

c. Distributive Postulate **d.** Multiplicative Inverse Postulate

6. Find the solution set for $\frac{2}{3}x - 5 = \frac{8}{3}x - 4$, when the replacement set is I.

a. $\{-2\}$ **b.** $\{2\}$ **c.** $\{-\frac{1}{2}\}$ **d.** $\{\ \}$ or ϕ

7. The average germination time, g, for fuchsia is two more than three times the germination time, c, for cabbage. The difference in germination time is 10 days. Which equation best represents these conditions?

a. $3g - 2 + g = 10$ **b.** $3c - 2 - c = 10$

c. $3c + 2 - c = 10$ **d.** $3g + 2 - g = 10$

8. Solve: $|3k + 8| = 7k$

a. $-\frac{4}{5}$ **b.** 2 **c.** 2 or $-\frac{4}{5}$ **d.** -2 or $\frac{4}{5}$

9. Solve: $-5z - 3 < -2z - 8$

a. $z < \frac{5}{3}$ **b.** $z < -\frac{5}{3}$ **c.** $z > \frac{5}{3}$ **d.** $z > -\frac{5}{3}$

10. The RPT wishes to raise at least \$4,000 to purchase new band uniforms. A raffle is planned with tickets selling at \$2.50 each. Expenses amount to \$300. Which inequality best represents these conditions?

a. $2.5t + 300 \leq 4000$ **b.** $2.5t - 300 \leq 4000$

c. $2.5t + 300 \geq 4000$ **d.** $2.5t - 300 \geq 4000$

11. Mary had scores of 78 and 88 on two quizzes. The final quiz will count for two quizzes. What is the lowest score she can get and still have an average of 85?

a. 89 **b.** 88 **c.** 87 **d.** 86

12. Solve: $|x - 2| < 41$

a. $x > -39$ <u>and</u> $x < 43$ **b.** $x > -43$ <u>and</u> $x < 39$

c. $x < -39$ <u>or</u> $x < 43$ **d.** $x > -43$ <u>or</u> $x < 39$

13. What is the slope of the graph of $5x - 2y = 3$?

a. $\frac{5}{2}$ **b.** $-\frac{5}{2}$ **c.** $\frac{2}{5}$ **d.** $-\frac{2}{5}$

14. What is the location of the point $P(-8, 0)$?

a. In quadrant II
b. In quadrant III
c. On the y axis
d. On the x axis

15. Which point is *not* on the graph of $5x - 2y = 3$?

a. $A(1, -1)$
b. $B(1, 1)$
c. $C(-1, -4)$
d. $P(0, -\frac{3}{2})$

16. A car depreciates 25% the first year and linearly at a rate of 10% per year every year after the first. What is the value of a \$10,000 automobile at the end of 5 years?

a. \$5000
b. \$4500
c. \$7500
d. \$3750

17. How are the graphs of $2x - 3y = 5$ and $3x + 2y = 5$ related?

a. They are parallel.
b. They are perpendicular.
c. They each contain $P(\frac{25}{13}, 1)$.
d. They have the same y intercept.

18. Write the equation of the line with slope $\frac{2}{3}$ and with y intercept $\frac{4}{3}$.

a. $3y - 2x = 4$
b. $3y + 2x = 4$
c. $2y - 3x = 4$
d. $2y + 3x = 4$

19. If y varies directly as x, which statement is false?

a. The graph contains $P(0, 0)$.
b. $x \div y$ is a constant.
c. $x \cdot y$ is a constant.
d. $y = k \cdot x$.

20. Which of the following best describes the solution set for this system?

$$\begin{cases} x + y = 5 \\ 2x + y = 3 \end{cases}$$

a. Infinite
b. Empty
c. A single point
d. Exactly two points

21. Solve: $\begin{cases} 3x - 2y = 5 \\ x - y = 2 \end{cases}$

a. $\{(1, 1)\}$
b. $\{(1, -1)\}$
c. $\{(-1, 1)\}$
d. $\{(2, -1)\}$

22. Art and Diane paddled their canoe a distance of 2 miles up the Manistee River in an hour. They rode the current back to their starting point in $\frac{1}{3}$ hour. What was their paddling rate in miles per hour?

a. 4
b. 6
c. 8
d. 10

23. If y varies directly as x and $x = 5$ when $y = -\frac{5}{2}$, find y when $x = 2$.

a. 2
b. 1
c. 0
d. -1

24. Evaluate: $\begin{vmatrix} 2 & 3 \\ -1 & 4 \end{vmatrix}$

a. 5
b. 11
c. -11
d. -5

25. For the system $\begin{cases} 2x - 3y = 5 \\ x + 3y = 8 \end{cases}$, which determinant can be used to find x?

a. $\dfrac{\begin{vmatrix} 5 & 2 \\ 8 & 1 \end{vmatrix}}{\begin{vmatrix} 2 & -3 \\ 1 & 3 \end{vmatrix}}$
b. $\dfrac{\begin{vmatrix} 2 & 5 \\ 1 & 8 \end{vmatrix}}{\begin{vmatrix} 2 & 5 \\ 1 & 8 \end{vmatrix}}$
c. $\dfrac{\begin{vmatrix} 5 & -3 \\ 8 & 3 \end{vmatrix}}{\begin{vmatrix} 2 & -3 \\ 1 & 3 \end{vmatrix}}$
d. $\dfrac{\begin{vmatrix} 5 & -3 \\ 8 & 3 \end{vmatrix}}{\begin{vmatrix} -3 & 2 \\ 3 & 1 \end{vmatrix}}$

Preparing for College Entrance Tests

Choose the best answer. Choose *a*, *b*, *c*, or *d*.

1. $\dfrac{17(4-7)+7(4-7)}{6} = \underline{\ \ ?\ \ }$

(a) -15 **(b)** -12 **(c)** 4 **(d)** 12

2. If p pounds of steak serves q adults, how many adults can be served with r pounds of steak?

(a) $\dfrac{pr}{q}$ **(b)** $\dfrac{qr}{p}$ **(c)** $\dfrac{q}{pr}$ **(d)** $\dfrac{r}{qp}$

3. $\dfrac{26(19-3)-7(19-3)}{16} = \underline{\ \ ?\ \ }$

(a) 16 **(b)** 18 **(c)** 19 **(d)** 48

4. $\dfrac{\frac{3}{4}n+\frac{3}{4}n+\frac{3}{4}n+\frac{3}{4}n}{6} = \underline{\ \ ?\ \ }$

(a) $\frac{1}{2}n$ **(b)** $\frac{2}{3}n$ **(c)** $\frac{3}{4}n$ **(d)** $2n$

5. If $\frac{x}{2}$ bottles fill $\frac{x}{12}$ cartons, how many cartons are needed for 6 bottles?

(a) 1 **(b)** $\dfrac{x}{2}$ **(c)** $\dfrac{x^2}{4}$ **(d)** 36

6. $19-36+2(36-19)-3(36-19) = \underline{\ \ ?\ \ }$

(a) 34 **(b)** 17 **(c)** 0 **(d)** -34

7. For nonzero real numbers a, b, and c, $4a = 5b$ and $10b = 15c$.

Then $\dfrac{a}{c} = \underline{\ \ ?\ \ }$

(a) $\frac{15}{8}$ **(b)** $\frac{5}{6}$ **(c)** $\frac{8}{15}$ **(d)** $\frac{6}{5}$

8. If $a = \frac{1}{3}$ and $b = (\frac{1}{2}-a)(\frac{1}{3}-a)(\frac{1}{4}-a)(\frac{1}{5}-a)$, which of the following must be true?

(a) $b < 0$ **(b)** $b > a$ **(c)** $a + b < 0$ **(d)** $b = 0$

9. $\dfrac{8 \cdot 8 \cdot 8 \cdot 8}{8+8+8+8} = \underline{\ \ ?\ \ }$

(a) 1 **(b)** 64 **(c)** 128 **(d)** 256

10. $(8\frac{1}{2} - 5\frac{5}{8}) - (4 - 5\frac{5}{8}) = \underline{\ \ ?\ \ }$

(a) $1\frac{1}{4}$ **(b)** $2\frac{1}{8}$ **(c)** 4 **(d)** $4\frac{1}{2}$

11. For nonzero real numbers x, y, z, and w, $w = 2x$, $x = 2y$, and $y = 2z$.

Then $\dfrac{w}{z} = \underline{\ \ ?\ \ }$

(a) $\frac{1}{4}$ **(b)** $\frac{4}{1}$ **(c)** $\frac{8}{1}$ **(d)** $\frac{1}{16}$

12. If $25 \cdot 25 \cdot 25 = 5 \cdot 5 \cdot 5 \cdot k$, then $k =$ __?__

(a) 5 (b) 25 (c) 125 (d) 625

13. 168:252 as 6:__?__

(a) 4 (b) 9 (c) 18 (d) 36

14. The ratio of the length of a rectangle to its perimeter is 1:3. What is the ratio of the length of the rectangle to its width? (HINT: Perimeter of rectangle $= 2l + 2w$)

(a) 2:1 (b) 3:1 (c) 3:2 (d) 1:1

15. The ratio of the width of a rectangle to its perimeter is 1:5. What is the ratio of the width of the rectangle to its length?

(a) 3:2 (b) 5:2 (c) 2:5 (d) 2:3

16. The ratio of the length of a rectangle to its perimeter is 2:7. What is the ratio of the width of the rectangle to its length?

(a) 3:7 (b) 4:7 (c) 3:4 (d) 4:3

17. If $1 + 2 + 3 + 4 + 5 + 6 + 7 + 8 + 9 = 45$, then $10 + 9 + 8 + 7 + 6 + 5 =$ __?__

(a) 35 (b) 45 (c) 46 (d) 55

18. What is the total cost of 1.5 liters of apple cider at \$0.88 per liter and 1.5 liters of grape juice at \$1.12 per liter?

(a) \$1.50 (b) \$3.00 (c) \$4.50 (d) \$14.88

19. For nonzero real numbers p, q, r, and s, $pr = qs$. Which of the following must be true?

(a) $\frac{p}{r} = \frac{s}{q}$ (b) $\frac{p+s}{s} = \frac{r+q}{q}$ (c) $\frac{p}{q} = \frac{r}{s}$ (d) $\frac{p+q}{q} = \frac{r+s}{r}$

20. If n dogs eat q cans of food each day, how many cans of food will be needed to feed w dogs each day?

(a) $\frac{n}{qw}$ (b) $\frac{qn}{w}$ (c) $\frac{qw}{n}$ (d) $\frac{q}{nw}$

21. For nonzero real numbers a, b, c, and d, $\frac{b-a}{b} = \frac{d-c}{d}$. Which of the following must be true?

(a) $\frac{a}{c} = \frac{d}{b}$ (b) $ac = bd$ (c) $\frac{b}{a+b} = \frac{d}{c+d}$ (d) $\frac{a-1}{b} = \frac{c-1}{d}$

22. What is the total cost of $3\frac{1}{2}$ pounds of apples at 44¢ per pound and $3\frac{1}{2}$ pounds of bananas at 56¢ per pound?

(a) \$1.75 (b) \$3.00 (c) \$3.50 (d) \$7.00

23. If $\frac{4n}{5} = 6$, what is the value of $\frac{2n}{3}$?

(a) $11\frac{1}{4}$ (b) $7\frac{1}{2}$ (c) 5 (d) $\frac{5}{6}$

CHAPTER **5**

Relations and Functions

5–1 Relations and Functions

The table at the right shows how the average number of miles per gallon for *all vehicles* in the United States has increased in four recent years. You can write this information as a set of ordered pairs. The first coordinate of each ordered pair represents the year. The second coordinate represents the average mileage.

Year	Average Mileage
1979	12.58
1980	12.85
1981	13.00
1982	13.05

$$\{(1979, 12.58), (1980, 12.85), (1981, 13.00), (1982, 13.05)\}$$

A set of ordered pairs is a *relation.* The set of first coordinates is the *domain* of the relation. The set second coordinates is the *range.*

Domain	Range
{1979, 1980, 1981, 1982}	{12.58, 12.85, 13.00, 13.05}

Definitions

A **relation** is a set of ordered pairs.

For a relation, the set of its first elements is the **domain.** The variable associated with the domain is the **independent variable.**

For a relation, the set of its second elements is the **range.** The variable associated with the range is the **dependent variable.**

EXAMPLE 1 Find the range of the relation $2x = y - 4$ for the domain $\{-2, -1, 0, 1\}$.

Solution: Solve $2x = y - 4$ for the dependent variable y.

$y = 2x + 4$

x	y
−2	?
−1	?
0	?
1	?

$y = 2(-2) + 4 = 0$

$y = 2(-1) + 4 = 2$

$y = 2(0) + 4 = 4$

$y = 2(1) + 4 = 6$

x	y
−2	0
−1	2
0	4
1	6

The range is {**0, 2, 4, 6**}.

When the domain of a relation is not specified, assume it to be the set of real numbers.

For the relation in Example 1, each element in the domain is paired with *exactly one* element of the range. Thus, the relation is a **function.** The function is shown at the right as a mapping.

Mapping

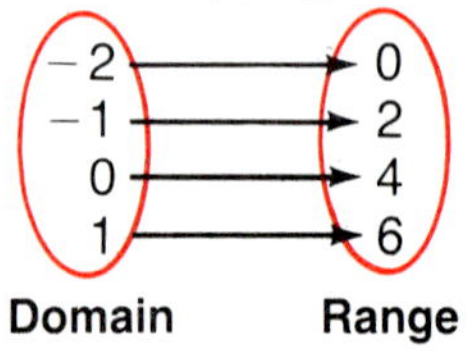

Definition

> A **function** is a relation such that for every element in the domain, there is *exactly one* element of the range.

EXAMPLE 2 Tell whether each mapping represents a function. Give a reason for each answer.

a.

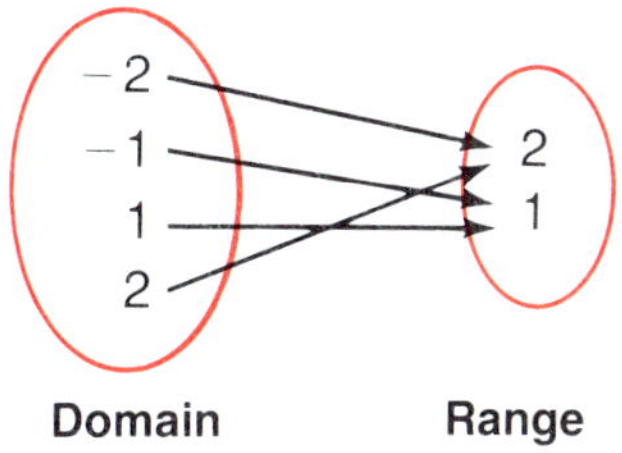

b.

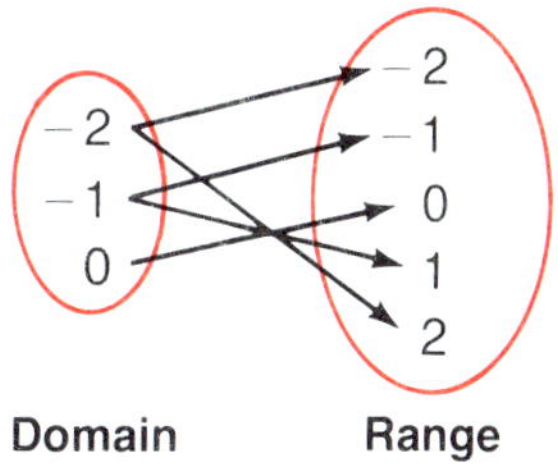

Solutions: **a.** The mapping represents a **function** because each element in the domain is paired with one and only one element of the range.

b. The mapping does **not** represent **a function,** because the element −2 in the domain is paired with two elements, −2 and 2, in the range.

Also, −1 in the domain is paired with two elements, −1 and 1, in the range.

You can also use a graph to tell whether a relation is a function. This is called the *Vertical Line Test.*

Vertical Line Test

> Imagine a vertical line moving across the graph of a relation.
>
> If the line intersects the graph anywhere in **more than one point,** the relation is **not** a function.

EXAMPLE 3 Determine which of the relations graphed below represent a function. Give a reason for each answer.

a.

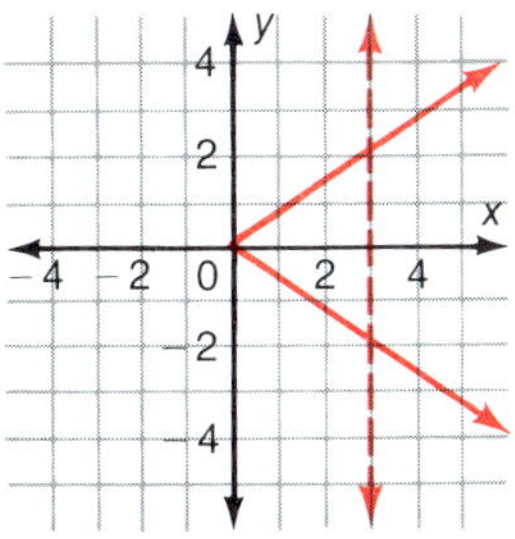

b.

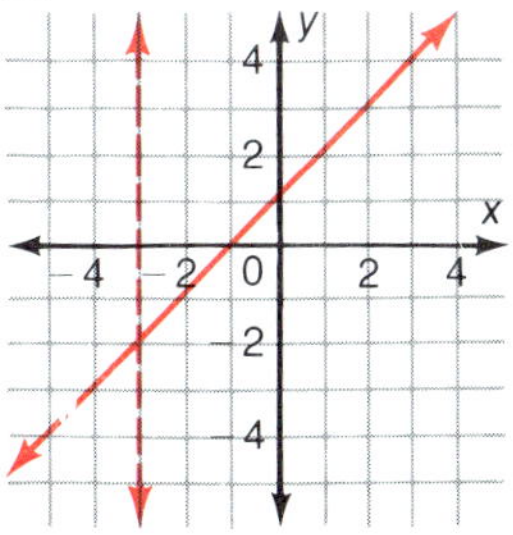

c.

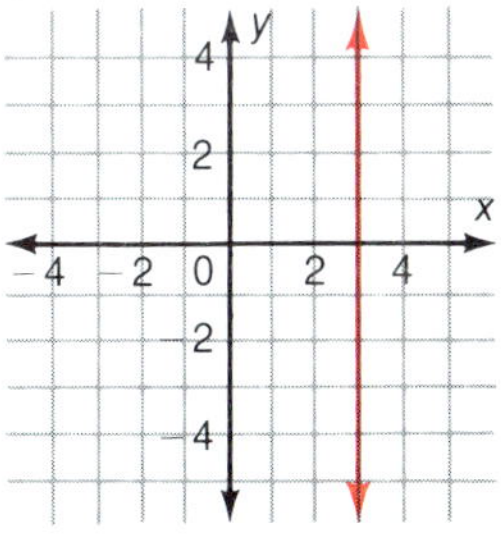

Solutions: **a.** **No.** A vertical line will intersect the graph in more than one point

b. **Yes.** Any vertical line intersects the graph in one point only.

c. **No.** A vertical line intersects the graph in *every* point on the graph.

CLASSROOM EXERCISES

State the domain and range of each relation.

1. $\{(1, 4), (2, 3), (5, 6), (8, 13)\}$

2. $\{(4, 3), (6, -3), (2, -1), (-2, -1)\}$

3. $\{(-1, 1), (2, 4), (-1, 7), (2, 9)\}$

4. $\{(0, 1), (0, 0), (5, 3), (-2, 6)\}$

Determine whether each mapping or graph represents a function. Give a reason for each answer.

5.

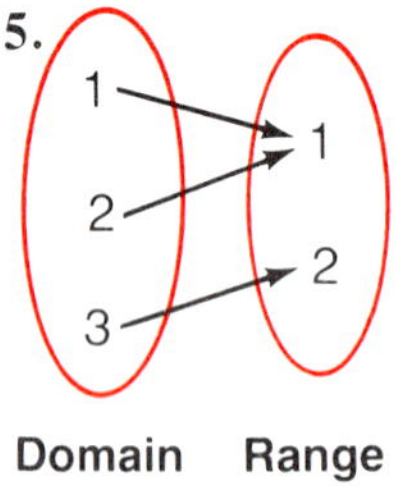

6.

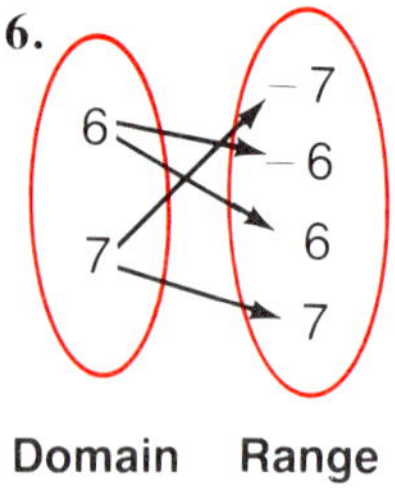

7.

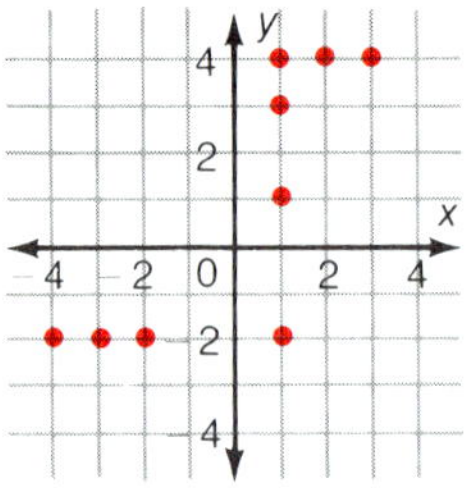

8.

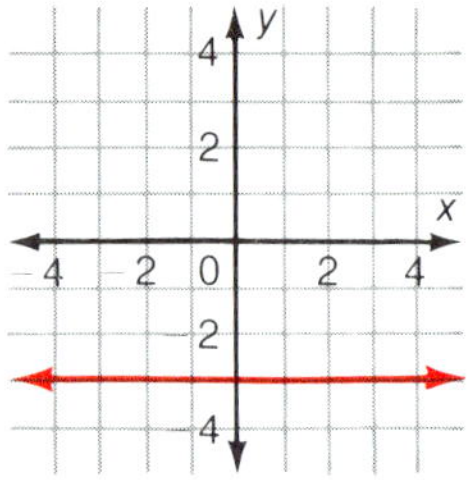

WRITTEN EXERCISES

A State the domain and range of each relation.

1. $\{(1, 4), (3, 8), (5, 12), (8, 16)\}$

2. $\{(4, 5), (5, 4), (4, -5), (5, -4)\}$

3. $\{(-2, 2), (-1, 1), (0, 0), (1, 1), (2, 2)\}$

4. $\{(9, 3), (9, -3), (4, 2), (4, -2)\}$

5. $\{(-2, 2), (-1, 1), (0, 0), (1, -1)\}$

6. $\{(2, 3), (2, -1), (2, 6)\ (2, 9)\}$

Find the range of each relation for the given domain.

Relation	Domain	Relation	Domain
7. $y = 7x$	$\{1, 2, 3\}$	**8.** $y = 12 - 5x$	$\{-6, -5, -4, -3\}$
9. $y = 8 - 3x$	$\{0, 1, 2, 3\}$	**10.** $y = 2x + 4$	$\{-6, -5, -4, -3\}$
11. $y - 3x = 0$	$\{1, 2, 3\}$	**12.** $y + 4x = 0$	$\{1, 2, 3\}$
13. $y = 0 \cdot x - 9$	$\{-1, 0, 1\}$	**14.** $y = \|x\|$	$\{-4, -2, -1, 0, 1\}$
15. $y = -\|x\|$	$\{-4, -2, -1, 0, 1\}$	**16.** $y = 2\|x\| - 1$	$\{-2, 0, 2, 4\}$

Determine whether each mapping represents a function. Give a reason for each answer.

17.

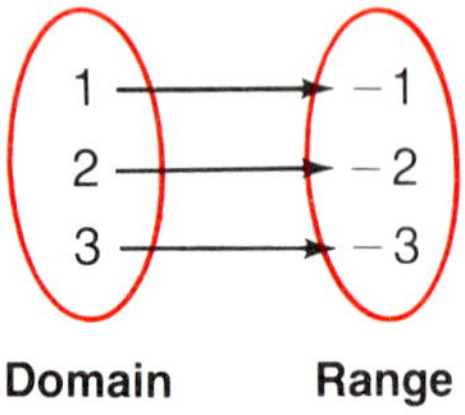

18.

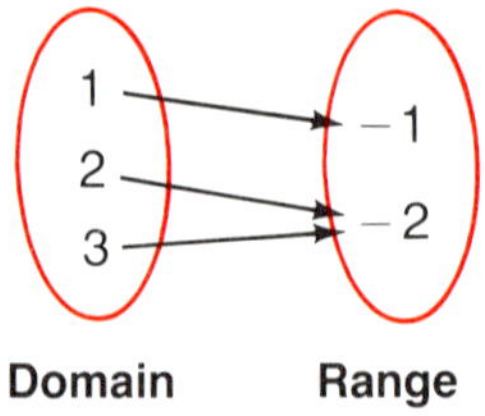

19.

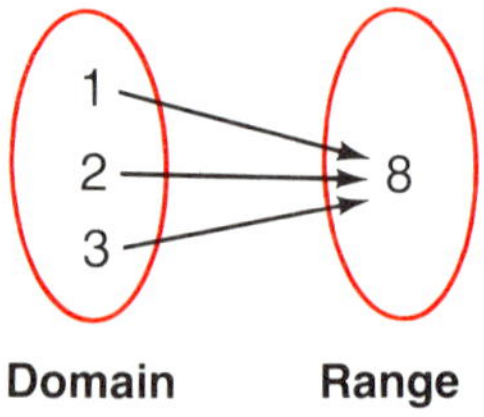

20.

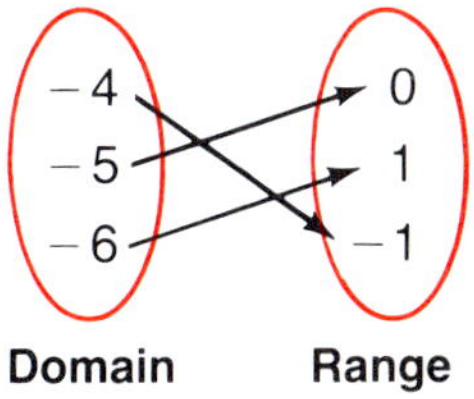

21.

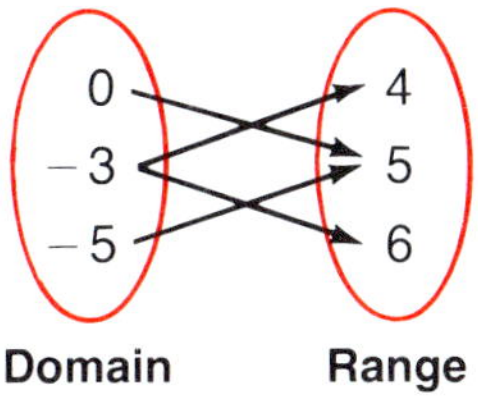

22.

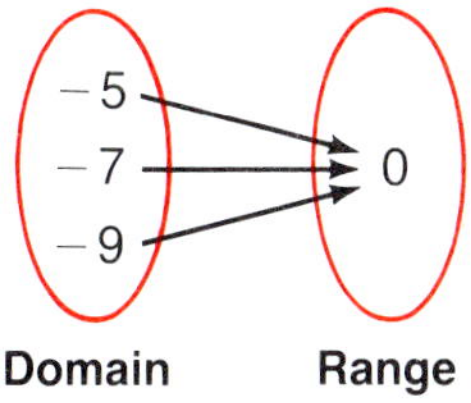

Use the Vertical Line Test to determine whether each graph represents a function.

23.

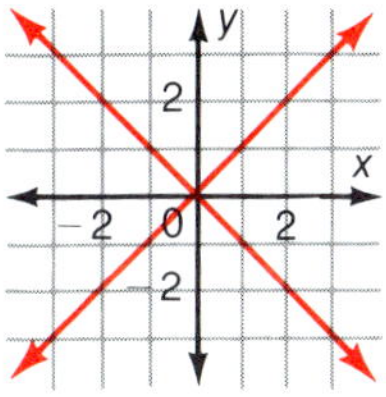

24.

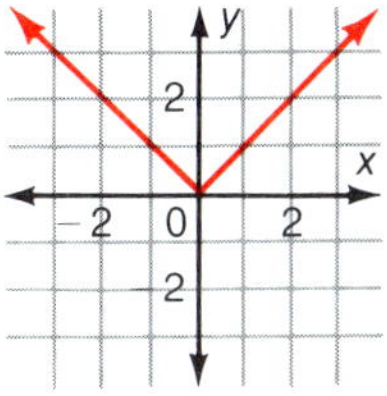

25.

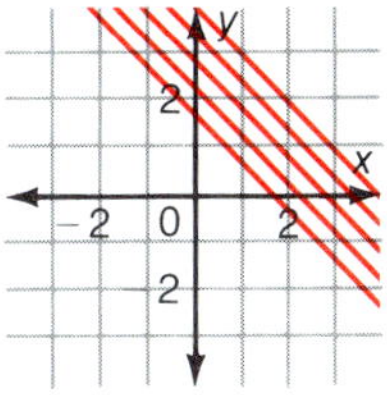

26.

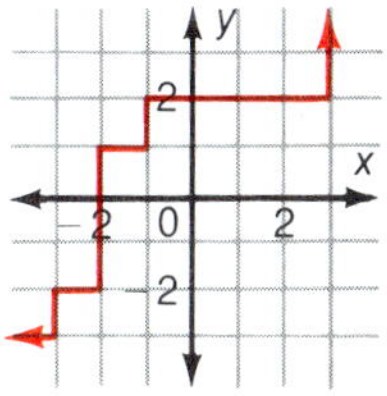

27.

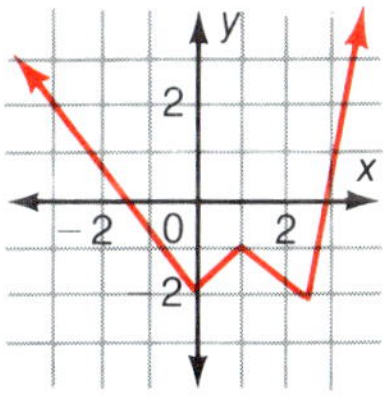

28.

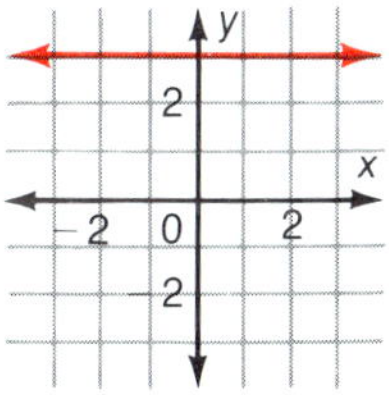

29.

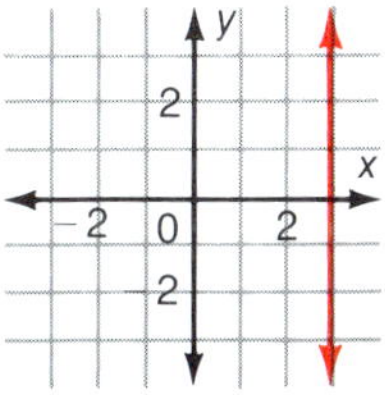

30.

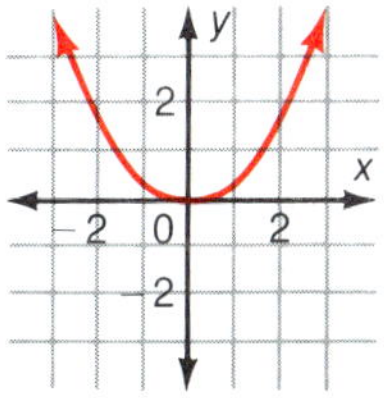

B

In Exercises 31–36:

a. Draw a mapping diagram for each relation.
b. Use the diagram to determine whether each relation is a function.

31. $\{(1, 3), (2, 3), (-1, 4), (-6, 4)\}$

32. $\{(-4, -4), (-1, -1), (0, 0), (1, 1)\}$

33. $\{(-4, 4), (-1, 1), (0, 0), (1, -1)\}$

34. $\{(-1, -4), (-1, 0), (-1, 2), (-1, 4)\}$

35. $\{(-3, -2), (-1, -2), (0, -2), (3, -2)\}$

36. $\{(4, -2), (1, -1), (1, 1), (4, 2)\}$

In Exercises 37–44:
a. Graph each relation for the domain $\{-3, -2, -1, 0, 1, 2, 3\}$.
b. Tell whether each relation is a function.

37. $y - x = 0$

38. $y + x = 0$

39. $y - x = 1$

40. $y + 3x = 0$

41. $y = 2x - 3$

42. $x + y = 10$

43. $y = x^2$

44. $y = -x^2$

In Exercises 45–56:
a. State the domain and range of each relation.
b. Tell whether each relation is a function.

45.

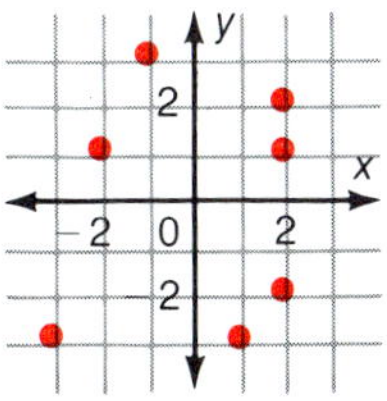

46.

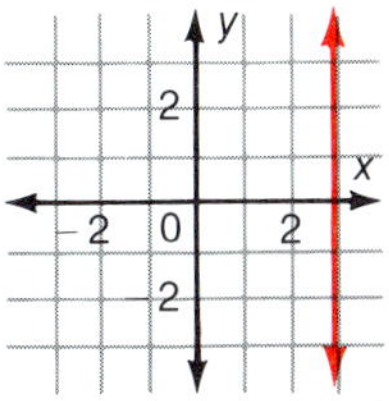

47.

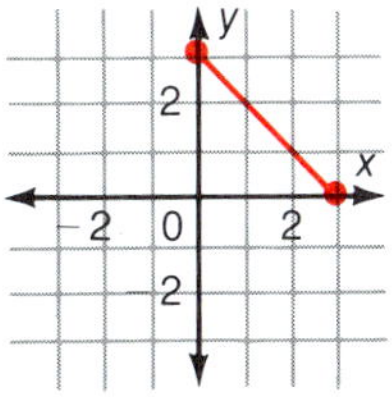

48.

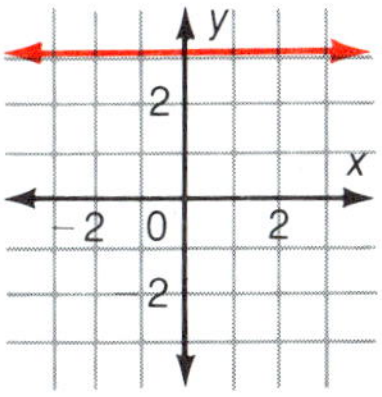

49. $y = x$ **50.** $y = -x$ **51.** $y - x = 1$ **52.** $y + 3x = 0$

53. $y = 0 \cdot x - 4$ **54.** $x = 0 \cdot y + 1$ **55.** $y = |x|$ **56.** $y = -|x|$

C In Exercises 57–65:

a. Graph each relation. The domain is the set of real numbers.

b. Give the range of each relation.

57. $y = |x| + 3$ **58.** $y = -3 - |x|$ **59.** $y < x$

60. $y \geq x + 5$ **61.** $y \leq 2x - 3$ **62.** $|y| \leq 3|x|$

63. $|y| < 3$ **64.** $|x| > -1$ **65.** $|y| = -|x| + 1$

CALCULATOR APPLICATIONS

Domain and Range

A calculator with a sign-change key will help you to compute the range for a relation.

EXAMPLE Compute the range for the relation defined by $y = -\frac{7}{8}x - 2$ for the domain $\{-2\frac{3}{8}, -2, \frac{1}{25}, 4\}$.

SOLUTION Write $-2\frac{3}{8}$ as a fraction: $-\frac{19}{8}$.

7 [÷] 8 [+/−] [×] 1 9 [÷] 8 [+/−] [−] 2 [=]   0.078125

Verify that −0.25, −2.035, and −5.5 are also in the range.

EXERCISES

Compute the range of each relation for the given domain.

1. $y = -4x + \frac{1}{5}$; D: $\{-3, \frac{1}{4}, 1, 7\frac{1}{16}\}$

2. $y = -x + 8$; D: $\{-9, -2\frac{3}{4}, -\frac{17}{25}, 3\}$

3. $y = 5x - 10$; D: $\{-6, -4, \frac{5}{8}, 3\}$

4. $y = -\frac{5}{2}x + 3$; D: $\{-7, -\frac{3}{50}, 1\frac{1}{16}, 2\}$

5. $y = \frac{3}{40}x - 6$; D: $\{-5\frac{3}{10}, -\frac{11}{16}, 4, 5\frac{1}{20}\}$

6. $y = -\frac{7}{16}x - 2\frac{3}{4}$; D: $\{-1\frac{5}{8}, -\frac{9}{25}, \frac{1}{16}, 3\}$

REVIEW CAPSULE FOR SECTION 5-2

Find the value of $9 - 5x^2$ for each value of x. *(Pages 10–13)*

1. 0 **2.** 7 **3.** −1 **4.** c **5.** $-d$

Find the value of $3x^2 - 2x + 1$ for each value of x. *(Pages 10–13)*

6. 4 **7.** −3 **8.** 0 **9.** a **10.** $-b$

5–2 Representing Relations and Functions

You can think of a relation as a set of ordered pairs. When you wish to emphasize the *rule* of a relation, it is sometimes useful to think of it as "mapping" its domain onto its range.

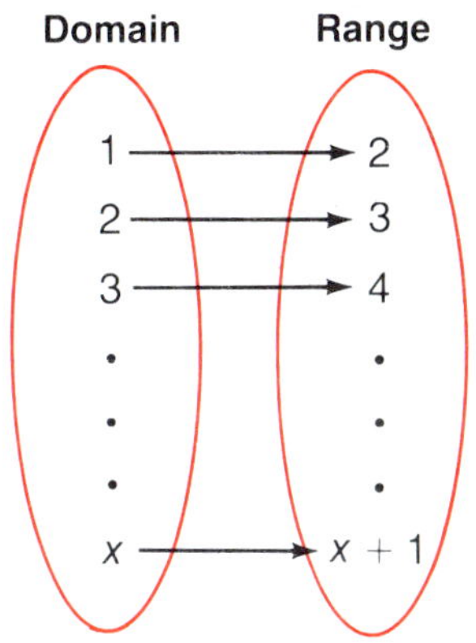

x maps onto $x + 1$, or

$$x \longrightarrow x + 1$$

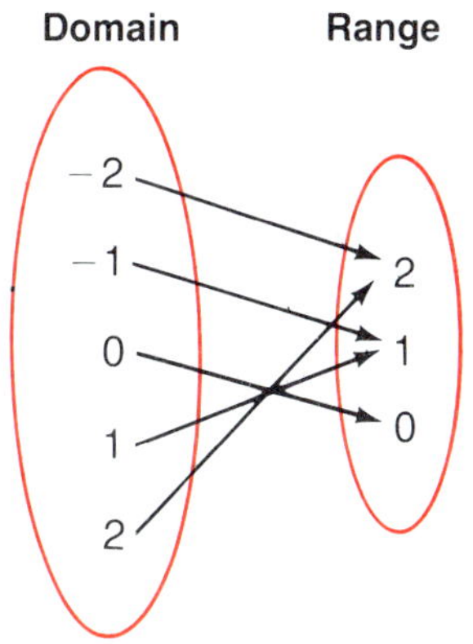

x maps onto $|x|$, or

$$x \longrightarrow |x|$$

Function notation is used to represent relations which are functions. Some letters commonly used to represent functions are f, g, h, F, G, and H. In function notation, the mapping on the left above can be written as

$$f{:}x \longrightarrow x + 1 \qquad \text{or} \qquad f(x) = x + 1.$$

Thus, if x represents an element in the domain of a function, then

$f(x)$ ⟵ **Read "f of x" or "f at x."**

is the corresponding element of the range. The ordered pairs of the function can be written as

$$(x, f(x)) \qquad \text{or as} \qquad (x, y).$$

EXAMPLE For $f(x) = 3x - 1$, find:

a. $f(-4)$ **b.** $f(8)$ **c.** $f(a + b)$

Solutions: **a.** Since $f(x) = 3x - 1$, $f(-4) = 3(-4) - 1$. ⟵ **Replace x with (−4).**

$f(-4) = -12 - 1$

$f(-4) = \mathbf{-13}$

b. $f(x) = 3x - 1$

$f(8) = 3(8) - 1$

$f(8) = \mathbf{23}$

c. $f(x) = 3x - 1$

$f(a + b) = 3(a + b) - 1$

$f(a + b) = \mathbf{3a + 3b - 1}$

Shown below are five functions that occur often in mathematics. For these functions, the letter D represents the domain; the letter R represents the range.

SPECIAL FUNCTIONS

Constant Function

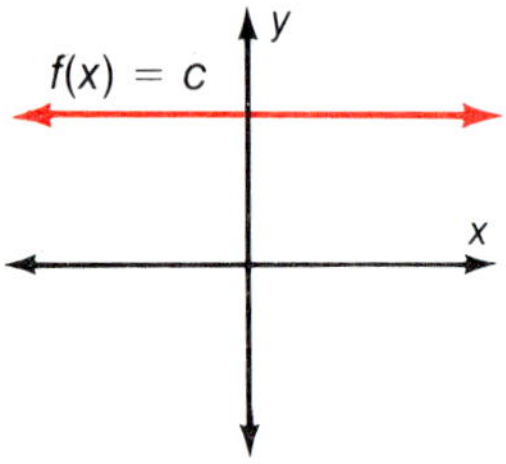

Rule: $f{:}x \longrightarrow c$ or $f(x) = c$

D:{Real numbers}

R:{c}

Identity Function

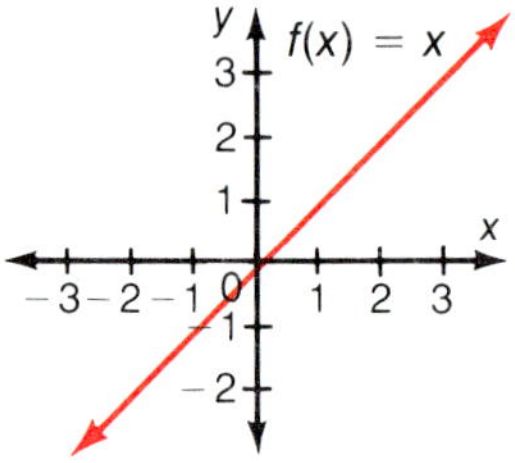

Rule: $f{:}x \longrightarrow x$ or $f(x) = x$

D:{Real numbers}

R:{Real numbers}

Absolute Value Function

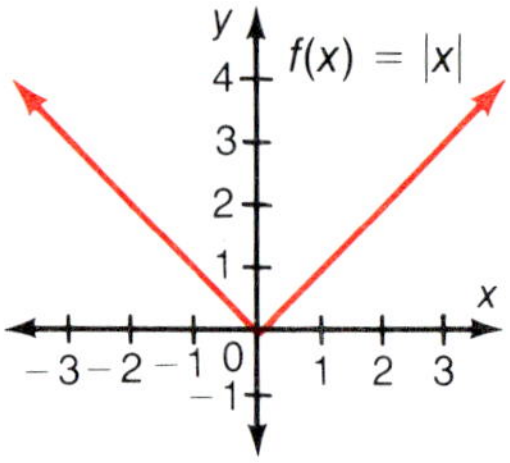

Rule: $f{:}x \longrightarrow |x|$ or $f(x) = |x|$

D:{Real numbers}

R:{Nonnegative real numbers}

Greatest Integer Function

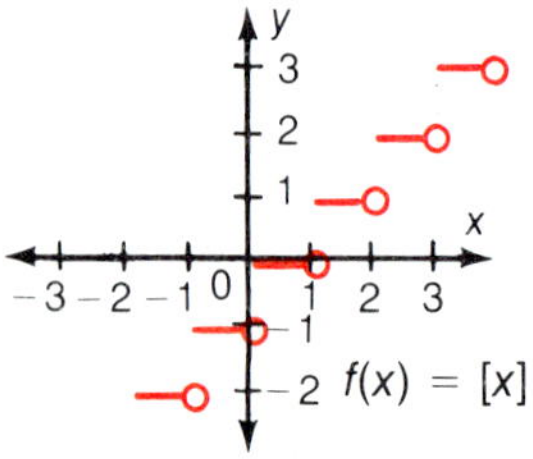

Rule: $f{:}x \longrightarrow [x]$ or $f(x) = [x]$

D:{Real numbers}

R:{Integers}

Signum Function

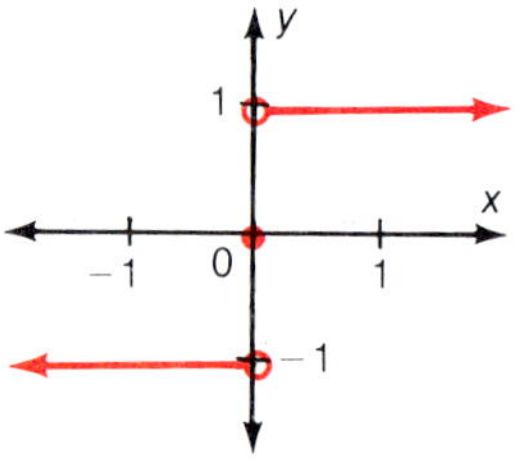

Rule:

$$f(x) = \begin{cases} -1, \text{ for } x < 0 \\ 0, \text{ for } x = 0 \\ 1, \text{ for } x > 0 \end{cases}$$

D:{Real numbers}

R:{−1, 0, 1}

The symbol "[ ]" means "the greatest integer less than or equal to x."

The constant function, $f(x) = c$, and the identity function, $f(x) = x$, are examples of *linear functions*. Every **linear function** can be expressed in the form

$$y = mx + b \quad \text{or} \quad f(x) = mx + b.$$

For the constant function, $m = 0$ and $b = f(x)$.

For the identity function, $m = 1$ and $b = 0$.

The graph of a linear function whose domain is the set of real numbers is a straight line.

CLASSROOM EXERCISES

Use function notation to give the rule for each mapping.

1.
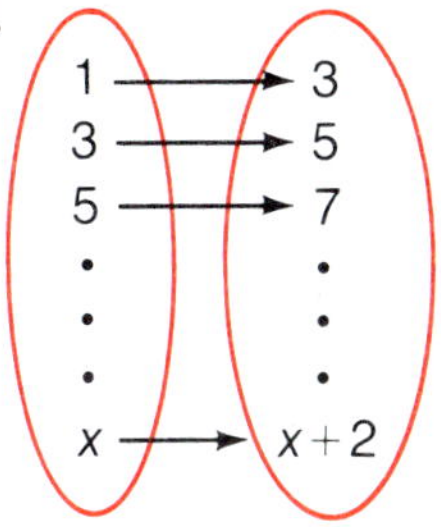

2.
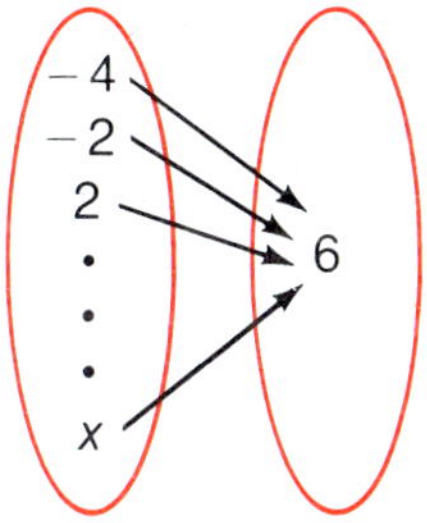

3.
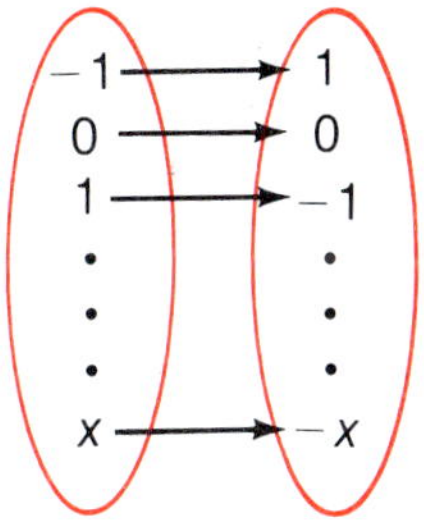

4.
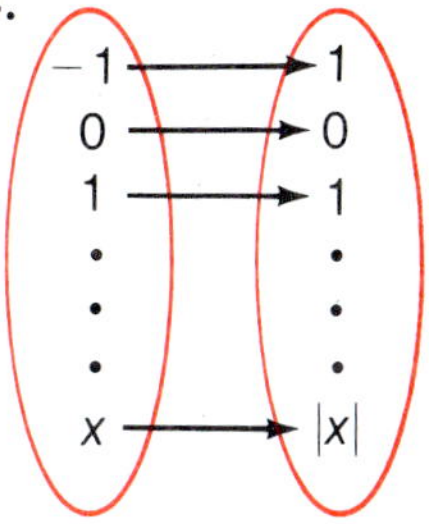

For $f(x) = 6 - 4x$, find each of the following.

5. $f(1)$ **6.** $f(-1)$ **7.** $f(0)$ **8.** $f(\frac{1}{2})$ **9.** $f(-0.25)$ **10.** $f(a)$

For $g(x) = 8 - 3x^2$, find each of the following.

11. $g(0)$ **12.** $g(2)$ **13.** $g(-3)$ **14.** $g(-\frac{1}{3})$ **15.** $g(t)$ **16.** $g(h)$

WRITTEN EXERCISES

A Express the rule for each mapping in function notation in two ways.

1. x maps onto $2x + 1$.
2. x maps onto $1 + |x|$.
3. x maps onto $x^2 - 1$.
4. x maps onto $-x$.
5. x maps onto $\frac{1}{x}$.
6. x maps onto itself.

For $h(x) = 3x - 5$, find each of the following.

7. $h(-5)$ **8.** $h(0)$ **9.** $h(1)$ **10.** $h(\frac{1}{3})$ **11.** $h(2)$ **12.** $h(5)$

For $g(x) = 2x^2 - 3x$, find each of the following.

13. $g(0)$ **14.** $g(1)$ **15.** $g(-1)$ **16.** $g(\frac{1}{2})$ **17.** $g(\frac{1}{3})$ **18.** $g(c)$

For $g(x) = \dfrac{2x - 1}{3}$, find each of the following.

19. $g(0)$ **20.** $g(\frac{1}{2})$ **21.** $g(-\frac{1}{2})$ **22.** $g(t)$ **23.** $g(-t)$ **24.** $g(a + b)$

For $f(x) = |x|$, find each of the following.

25. $f(2)$ **26.** $f(-1)$ **27.** $f(3.5)$ **28.** $f(0)$ **29.** $f(-\pi)$ **30.** $f(t)$

For $f(x) = [x]$, find each of the following.

31. $f(0)$ **32.** $f(-2.1)$ **33.** $f(3.7)$ **34.** $f(\pi)$ **35.** $f(-\pi)$ **36.** $f(2.6)$

In Exercises 37–42, match the graph of each relation with the equation below that defines the relation.

A: $y = x$ B: $y = -x$ C: $y = |x|$

D: $y = -|x|$ E: $x = |y|$ F: $x = -|y|$

37.

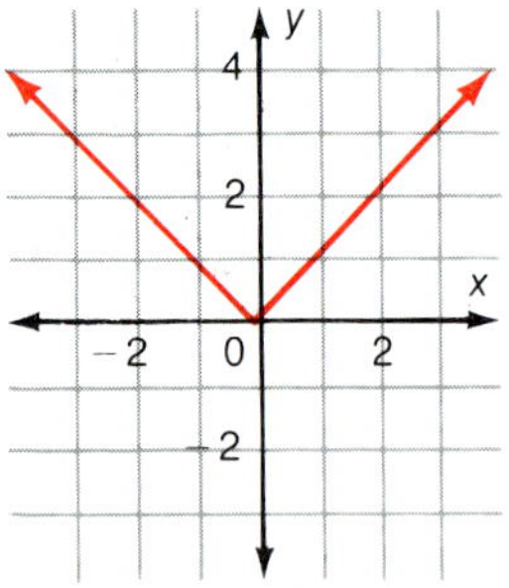

38.

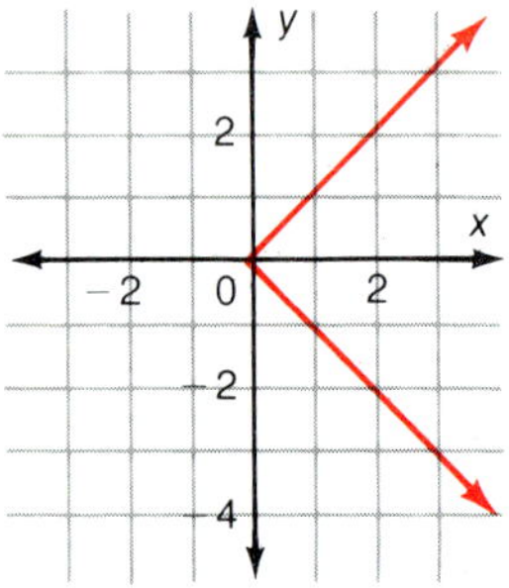

39.

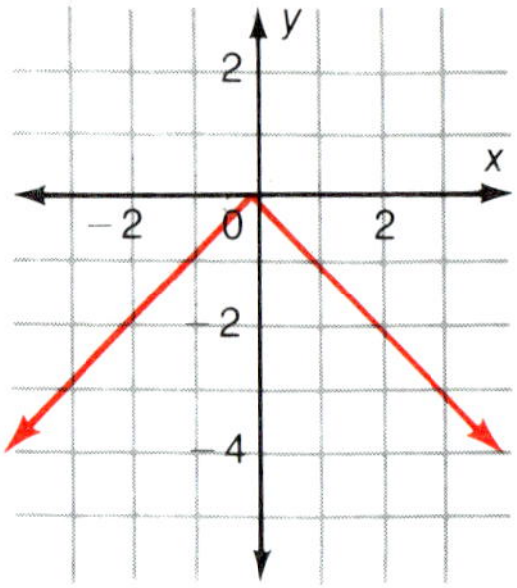

40.

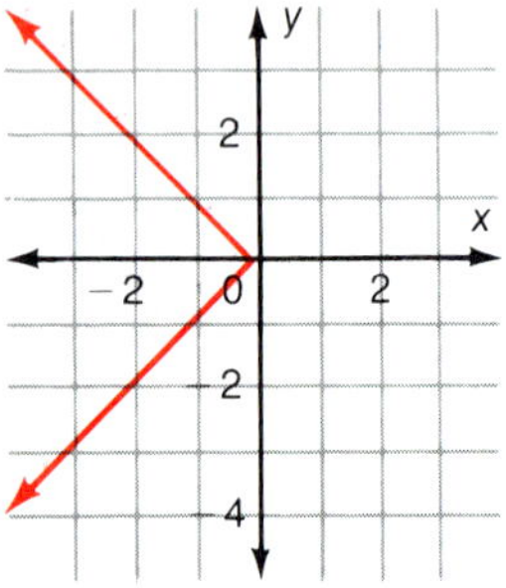

41.

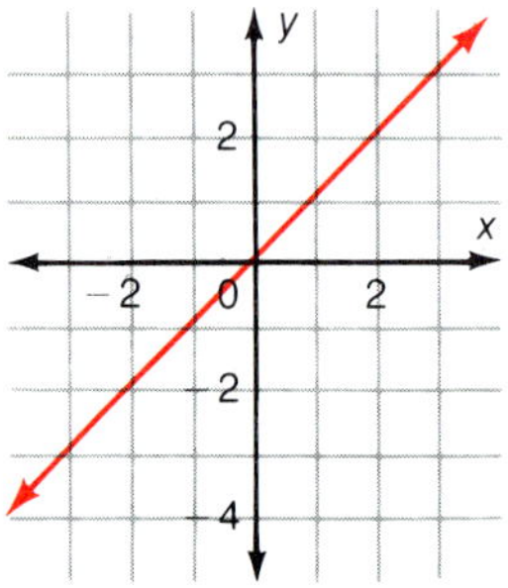

42.

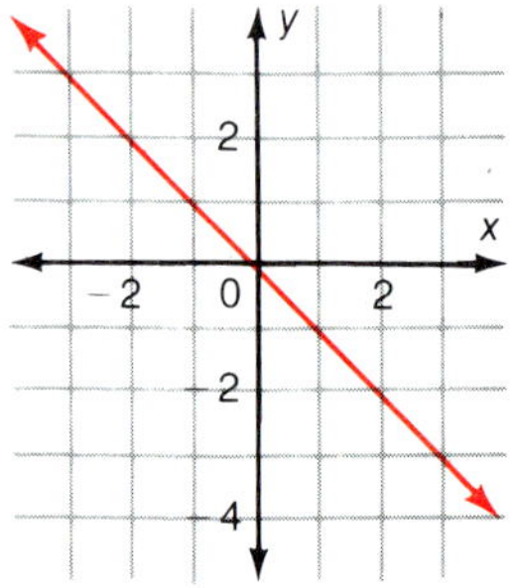

Graph each function.

43. $g(x) = |x| + 1$

44. $h(x) = |x| - 1$

45. $f(x) = -[x]$

46. $g(x) = \begin{cases} 1, \text{ for } x < 0 \\ 0, \text{ for } x = 0 \\ -1, \text{ for } x > 0 \end{cases}$

47. $h(x) = \begin{cases} -2, \text{ for } x < 0 \\ -1, \text{ for } x = 0 \\ 3, \text{ for } x > 0 \end{cases}$

48. $f(x) = \begin{cases} 0, \text{ for } x > 0 \\ -1, \text{ for } x = 0 \\ -3, \text{ for } x < 0 \end{cases}$

B Draw a mapping diagram to represent each function. The domain for each function is $\{-1, 0, 2, 3\}$.

49. $f(x) = 1 - 2x$

50. $h(x) = |x|$

51. $g(x) = x^2$

52. $f(x) = |x - 4|$

53. $g(x) = -3$

54. $h(x) = 4 - |x|$

In Exercises 55–58, use the function $h(x) = x$ over the domain $\{1, 2, 3, 4, 5, 6\}$.

55. What is the range of h?

56. Does $h(2) + h(3) = h(2 + 3)$?

57. Does $h(2) \cdot h(3) = h(2 \cdot 3)$?

58. Does $h(5x) = 5h(x)$?

In Exercises 59–62, use the function $g(x) = x$ over the domain $\{-4, -3, -2, -1, 0, 1, 2, 3, 4, 5, 6\}$.

59. What is the range of g?

60. Does $g(-2) + g(2) = g(0)$?

61. Does $g(-3) \cdot g(-2) = g(6)$?

62. Does $-1 \cdot g(x) = g(-x)$?

For each of the following functions, use function notation to give a rule for the pairing over the given domain.

63. $\{(-1, 1), (-2, 2), (1, -1), (2, -2)\}$ **64.** $\{(1, 3), (3, 9), (-2, -6), (-4, 12)\}$

65. $\{(2, 0), (0, 0), (-1, 0), (-3, 0)\}$ **66.** $\{(5, -\frac{1}{5}), (3, -\frac{1}{3}), (\frac{1}{2}, -2), (\frac{1}{4}, -4)\}$

67. $\{(-1, -3), (0, -1), (1, 1), (2, 3)\}$ **68.** $\{(-3, 5), (0, 5), (1, 5), (4, 5)\}$

69. $\{(-2, 4), (-1, 1), (0, 0), (1, 1)\}$ **70.** $\{(-2.5, -3), (-1.9, -2), (0.5, 0), (1.9, 1)\}$

For $g(x) = |x| - x + [x]$, find each of the following.

71. $g(0)$ **72.** $g(1)$ **73.** $g(-1)$ **74.** $g(\pi)$ **75.** $g(-2.3)$

C If $f(x) = x^2$, $g(x) = -x$ and $h(x) = 3x - 1$, find each of the following.

76. $f[g(x)]$ **77.** $h[f(x)]$ **78.** $g[h(x)]$ **79.** $h[g(x)]$

Use these functions in Exercises 80–81.

$f(x) = c$ $f(x) = x$ $f(x) = |x|$

$f(x) = [x]$ $f(x) = hx$ $f(x) = mx + b$

80. For which of these functions does $f(x_1) + f(x_2) = f(x_1 + x_2)$?

81. For which of these functions does $f(x_1) \cdot f(x_2) = f(x_1 \cdot x_2)$?

82. If f is a function such that $f(x + h) = f(x) + f(h)$ for all $x, h \in R$, show that $f(0) = 0$.

83. For the function in Exercise 82, show that $f(-h) = -f(h)$.

Puzzle

Given: $g(x) = x_1 + x_2 + x_3 + x_4$

$h(x) = x_1x_2 + x_2x_3 + x_3x_4 + x_4x_1$

$t(x) = x_1x_2x_3 + x_2x_3x_4 + x_3x_4x_1 + x_4x_1x_2$

$q(x) = x_1x_2x_3x_4$

Problem: Express $T = \frac{1}{x_1} + \frac{1}{x_2} + \frac{1}{x_3} + \frac{1}{x_4}$ in terms of $g(x)$, $h(x)$, $t(x)$, and $q(x)$.

REVIEW CAPSULE FOR SECTION 5–3

Find the slope of the line containing the given points. *(Pages 80–83)*

1. $P(1, 2)$; $Q(5, 6)$ **2.** $A(-3, -4)$; $B(-1, 5)$ **3.** $R(1, a)$; $T(-1, b)$

Write the equation of the line passing through the given points. Write the equation in slope-intercept form. *(Pages 93–97)*

4. $T(1, 4$; $S(3, 5)$ **5.** $R(0, -1)$; $T(-1, -2)$ **6.** $A(-4, -4)$; $B(9, 9)$

Problem Solving and Applications

5–3 Relations and Functions

Many real-world applications can be expressed as linear functions in two variables. The equation or graph of the function is a "model" for the application. The model can be used to predict other values of the function.

EXAMPLE A team of home economists found that after 6 minutes of cooking time, the moisture content of a hamburger was 48% by weight. After ten minutes of cooking time, the moisture content was 40%.

a. Assume that the relation is linear. Graph the relation.

b. Write an equation for the relation. Use t for the number of minutes of cooking time and c for the moisture content (in per cent).

c. Use the equation to predict the moisture content of the hamburger after 20 minutes of cooking time.

Solutions: **a.**

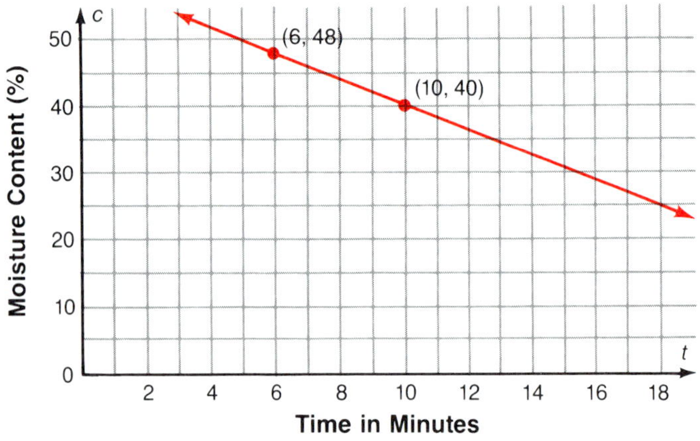

b. Use the point-slope form of a linear equation.

$m = \frac{48 - 40}{6 - 10}$

$m = \frac{8}{-4}$, or $\mathbf{-2}$

$y_2 - y_1 = m(x_2 - x_1)$ ← **Point-slope form**

$40 - c = -2(10 - t)$

$40 - c = -20 + 2t$

$-c = 2t - 60$

$\mathbf{c = 60 - 2t}$

c. Find c when $t = 20$.

$c = 60 - 2t$

$c = 60 - 2(20)$

$c = \mathbf{20}$

After 20 minutes, the moisture content will be **20%**.

WRITTEN EXERCISES

A A group of scientists tested the effect of weightlessness on an astronaut's reaction time. After 5 hours of weightlessness, it took the astronaut 0.7 second to respond to a flashing light by pressing a button. After 13 hours of weightlessness, the astronaut's reaction time was 2 seconds. Use this information in Exercises 1–4.

1. Assume that the relation is linear. Graph the relation.
2. Write an equation for the relation. Use w for the number of hours of weightlessness and t for the reaction time in seconds.
3. Use the equation to predict the astronaut's reaction time when $w = 24$.
4. Use the equation to predict when the astronaut's reaction time will be 1 minute.

A set of experimental data shows that at a temperature of 21°C, a working person uses 3000 calories per day. At a temperature of 15°C, the same person would use 3180 calories per day. Use this information in Exercises 5–9.

5. Assume that the function is linear. Graph the function.
6. Write an equation for the function. Use T for temperature and c for the number of calories.
7. Use the equation to predict how many calories per day would be used by a working person when the temperature is 50° (tropical climate).
8. Use the equation to predict how many calories per day would be used by a working person when the temperature is −50°C (polar climate).
9. Use the equation to predict the temperature range when a working person would be using between 4000 and 5000 calories per day.

Regular air fares in the United States are a linear function of the distance traveled. In a recent year, the cost of a ticket from Miami to Cleveland (1200 miles) was $480. The cost of a ticket from Denver to Los Angeles (1000 miles) was $400. Use this information in Exercises 10–13.

10. Write an equation for the function. Let m represent the number of miles and let a represent the cost of the one-way air fare between the two cities.
11. Use the equation in Exercise 10 to approximate the air fare from Chicago to Dallas (800 miles).
12. Use the equation in Exercise 10 to approximate the air fare from San Diego to Washington, D.C (2300 miles).
13. Use the equation in Exercise 10 to approximate the air fare from Cincinnati to St. Louis (300 miles).

In 1982, U.S. postage rates for first-class letters were 20¢ for the first ounce or fraction of an ounce, and 17¢ for each additional ounce or fraction of an ounce. The total weight could not exceed 12 ounces.

14. Graph the "postage function" described above.

15. What is the maximum value of the function?

16. What is the minimum value of the function?

17. Use the graph to determine the cost of mailing a set of documents weighing $8\frac{1}{2}$ ounces.

18. Use the graph to determine the cost of mailing a manuscript weighing $11\frac{3}{4}$ ounces.

B The graph at the right shows the velocity (speed) of a model rocket from launch to landing. After launch, the propellant burns for several seconds, boosting the rocket upward. After burnout, the rocket coasts upward a short time before beginning to fall back to earth. Shortly after the rocket starts downward (negative velocity), a small explosive charge opens a parachute. The parachute slows the rocket's fall and softens the impact on landing. Use this information and the graph in Exercises 19–25.

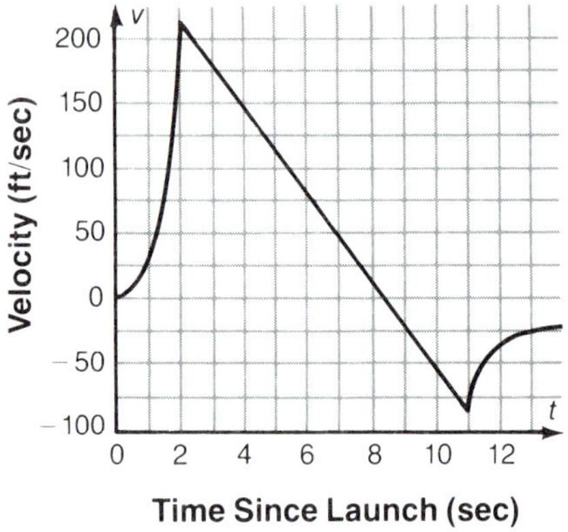

19. What was the rocket's maximum upward velocity when the propellant burned out?

20. How many seconds did it take the rocket to reach this maximum velocity?

21. The rocket begins to fall toward earth when $v = 0$. How many seconds after launch is this?

22. When did the rocket reach its greatest distance above the earth?

23. How many seconds after launch did the parachute open?

24. What was the velocity of the rocket when the parachute opened?

25. For how many seconds did the rocket fall before the parachute opened?

REVIEW CAPSULE FOR SECTION 5–4

Give the domain and range of each function. *(Pages 160–164)*

1. $y = 2x + 1$ **2.** $y = -|x|$ **3.** $y = 2x^2$ **4.** $y = 1$

Find the value of $2x - 1$ for each value of x. *(Pages 10–13)*

5. t **6.** $t + 1$ **7.** $t - 3$ **8.** t^2 **9.** $-t$ **10.** $-t^2$

Review

In Exercises 1–4, find the range of each relation for the given domain. *(Section 5–1)*

	Relation	Domain		Relation	Domain
1.	$y = -x$	$\{-1, 0, 4\}$	**2.**	$y = 2x - 1$	$\{-3, -1, 0, 2\}$
3.	$y = 0 \cdot x - 4$	$\{-6, -4, 0, 3, 9\}$	**4.**	$y = -\lvert x\rvert + 1$	$\{-2, -1, 0, 1, 2, 3\}$

In Exercises 5–10, tell whether each mapping or graph represents a function. Give a reason for your answer. *(Section 5–1)*

5.

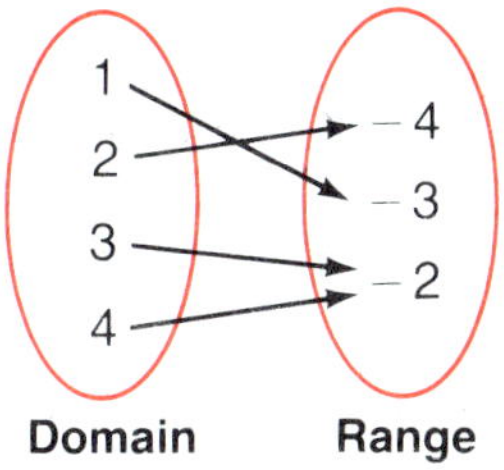

6.

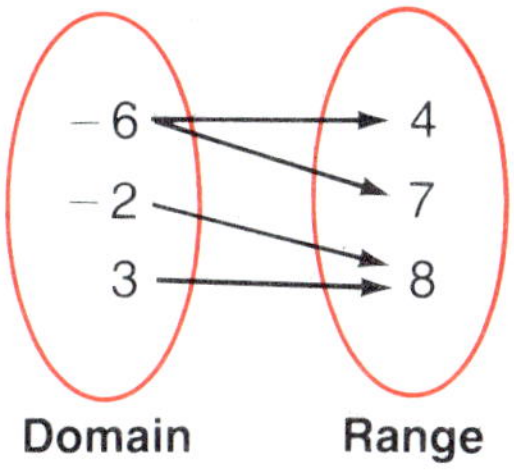

7.

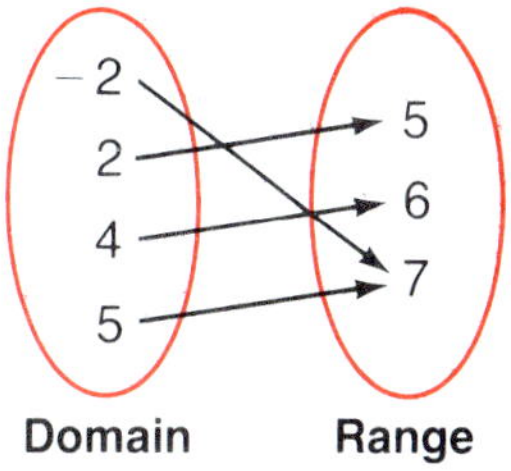

8.

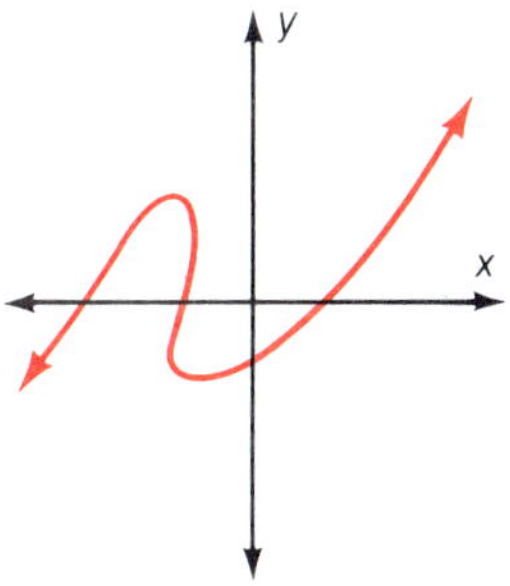

9.

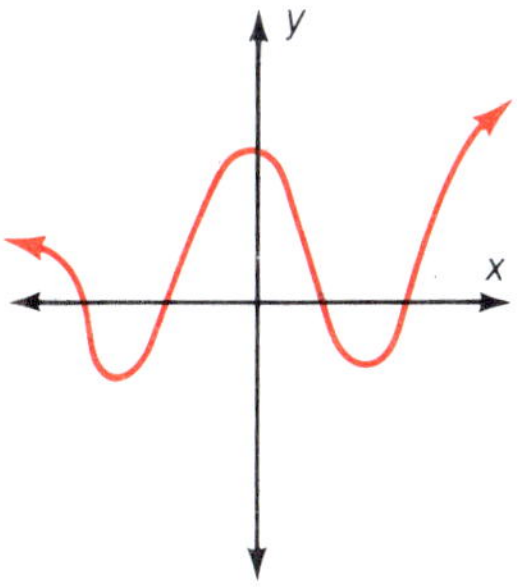

10.

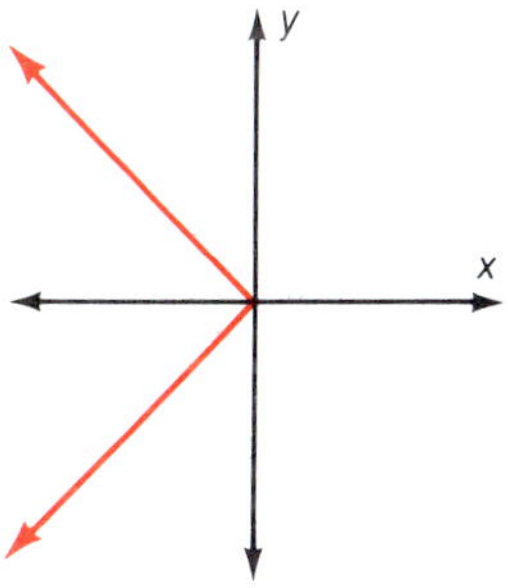

For $f(x) = 2x + |x|$, find each of the following. *(Section 5–2)*

11. $f(0)$ **12.** $f(3)$ **13.** $f(-1)$ **14.** $f(-8)$ **15.** $f(4.2)$ **16.** $f(-x)$

A sealed container of air is connected to a pressure gauge. When the air's temperature is lowered to 0° Celsius, the pressure gauge reads 0.95 atmosphere. (*One atmosphere* is the pressure of air under certain standard physical conditions.) When the air is cooled further to −100°, the pressure gauge reads 0.60 atmosphere.

Use this information in Exercises 17–20. *(Section 5–3)*

17. Assume that the function is linear and that 0.60 and 0.95 are in the function's domain. Graph the function.

18. Write an equation for the function. Use p for the pressure (atmospheres) and T for temperature (degrees Celsius).

19. Use the graph to predict the temperature of the air when the air pressure is 0.35 atmospheres.

20. Use the graph to predict the temperature of the air when the air pressure is 0.

5–4 Composite Functions

In the figure below, the function f maps $\{3, 2, 1\}$ onto $\{0, 4, 5\}$. Then the function g maps $\{0, 4, 5\}$ onto $\{-1, 7, 9\}$. In these mappings, the set $\{0, 4, 5\}$ plays two roles. It is *both* the range of f and the domain of g.

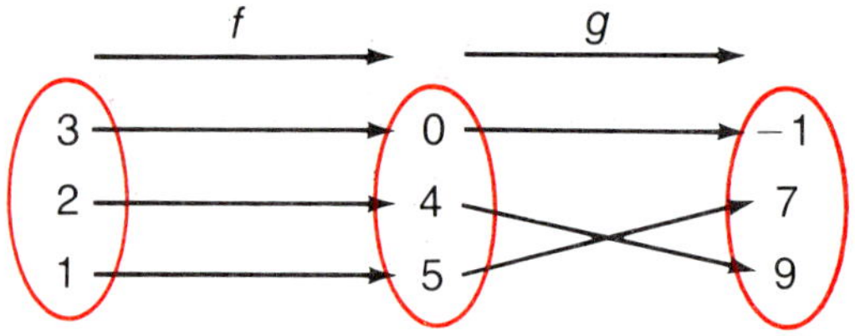

When a set of elements is both the range of one function *and* the domain of a second function, a third function is defined. In the figure.

3 maps onto 0 and 0 maps onto −1. So 3 maps onto −1.
2 maps onto 4 and 4 maps onto 9. So 2 maps onto −9.
1 maps onto 5 and 5 maps on to 7. So 1 maps onto 7.

The new function maps $\{3, 2, 1\}$ onto $\{-1, 7, 9\}$ where $\{0, 4, 5\}$ is the link between the two sets. The new function whose ordered pairs are $\{(3, -1), (2, 9), (1, 7)\}$ is called the *composite function*,

$g \circ f$. ← **Read: "*g* composition *f*" or "the composite of *g* and *f*."**

Definition

> If g and f are functions such that the range of f is in the domain of g, then the **composite function** $g \circ f$ can be described as
> $$[g \circ f](x) = g(f(x)).$$

EXAMPLE 1 For $f(x) = x + 3$ and $g(t) = 2t + 5$ find each of the following.

a. The equation that defines $g(f(x))$ **b.** $g(f(2))$

Solutions: **a.** To find $g = (f(x))$, first replace $f(x)$ with $x + 3$.

$g(f(x)) = g(x + 3)$ ← **Evaluate *g* for $t = x + 3$.**

$= 2(x + 3) + 5$ ← **Simplify.**

$= 2x + 6 + 5$

$= \mathbf{2x + 11}$

b. Since $g(f(x)) = 2x + 11$,

$g(f(2)) = 2(2) + 11$

$g(f(2)) = 4 + 11 = \mathbf{15}$

It is important to note that $g \circ f$ need not equal $f \circ g$. Thus, composition of functions is *not commutative*.

EXAMPLE 2 For $f(x) = x + 1$ and $g(x) = \frac{1}{x-1}$, $x \neq 1$, show that $f(g(x)) \neq g(f(x))$.

Solution: $g(f(x)) = g(x+1)$ ← **Evaluate *g* for x = x + 1.**

$$= \frac{1}{(x+1)-1}$$

$$= \mathbf{\frac{1}{x}}$$

$f(g(x)) = f\left(\frac{1}{x-1}\right)$ ← **Evaluate *f* for $x = \frac{1}{x-1}$**

$= \frac{1}{x-1} + 1$ ← $1 = \frac{x-1}{x-1}$; $\frac{1}{x-1} + \frac{x-1}{x-1} = \frac{1+x-1}{x-1}$

$$= \frac{1+x-1}{x-1}$$

$= \mathbf{\frac{x}{x-1}}$ Since $\frac{1}{x} \neq \frac{x}{x-1}$, $\mathbf{g(f(x)) \neq f(g(x))}$.

In certain instances, however, $g(f(x))$ can equal $f(g(x))$.

EXAMPLE 3 For $h(x) = 4x - 3$ and $p(x) = \frac{x+3}{4}$, show that $h(p(x)) = p(h(x))$.

Solution: $h(p(x)) = h\left(\frac{x+3}{4}\right)$ ← **Evaluate *h* when x is replaced with $\frac{x+3}{4}$.**

$$= 4\left(\frac{x+3}{4}\right) - 3$$

$$= x + 3 - 3 = \mathbf{x}$$

$p(h(x)) = p(4x-3)$ ← **Evaluate *p* when x is replaced with 4x − 3.**

$$= \frac{(4x-3)+3}{4}$$

$= \frac{4x}{4} = \mathbf{x}$ Thus, $\mathbf{h(p(x)) = p(h(x))}$.

CLASSROOM EXERCISES

In Exercises 1–4,:

a. Use sets *A*, *B*, and *C* to draw mapping diagrams where *f* maps *A* onto *B* and *g* maps *B* onto *C*.

b. Use the mapping diagram to write the ordered pairs of the composite function $g \circ f$.

1. $A = \{1, 2, 3\}$; $B = \{3, 5, 6\}$; $C = \{3, 5, 7\}$;
$f(1) = 3$; $f(2) = 5$; $f(3) = 6$; $g(3) = 7$; $g(5) = 5$; $g(6) = 3$

2. $A = \{1, 2, 3\}$; $B = \{4, 5\}$; $C = \{6, 7\}$;
$f(1) = 4; f(2) = 5; f(3) = 4; g(4) = 6; g(5) = 7$

3. $A = \{1, -2, 3\}$; $B = \{4, 6, -8\}$; $C = \{10, -3, 7\}$;
$f(1) = 4; f(-2) = 6; f(3) = -8; g(4) = 7; g(6) = -3; g(-8) = 10$

4. $A = \{-5, 1, 2\}$; $B = \{2, -4, -7\}$; $C = \{1, -5, 2\}$;
$f(-5) = 2; f(1) = -4; f(2) = -7; g(2) = -5; g(-4) = 1; g(-7) = 2$

5. State the domain and range of each composite function in Exercises 1–4.

WRITTEN EXERCISES

A In Exercises 1–8:

a. Write the equation that defines $g(f(x))$.

b. Find $g(f(3))$. **c.** Find $g(f(-2))$. **d.** Find $g(f(a))$.

1. $f(x) = x$, $g(t) = t - 1$
2. $f(x) = x$, $g(t) = \frac{1}{t}$
3. $f(x) = -x$, $g(t) = 3t - 2$
4. $f(x) = 2x - 1$, $g(t) = 3t + 4$
5. $f(x) = x$, $g(t) = -t$
6. $f(x) = \frac{1}{2}x + 3$, $g(t) = 2t - 5$
7. $f(x) = 5x + 2$, $g(t) = -\frac{1}{t}$
8. $f(x) = |x|$, $g(t) = t + 1$

In Exercises 9–16:

a. Write the equation that defines $f(g(x))$.

b. Find $f(g(-1))$. **c.** Find $f(g(5))$. **d.** Find $f(g(a + b))$.

9. $g(x) = x - 1$, $f(t) = t + 2$
10. $g(x) = x$, $f(t) = t - 2$
11. $g(x) = x$, $f(t) = 2 - t$
12. $g(x) = -\frac{1}{4}$, $f(t) = t$
13. $g(x) = x - 1$, $f(t) = \frac{1}{t - 1}$
14. $g(x) = \frac{1}{2}x - 1$, $f(t) = -4t$
15. $g(x) = |x|$, $f(t) = -2t - 3$
16. $g(x) = -|x|$, $f(t) = 1 - t$

In Exercises 17–24:

a. Find $g(f(x))$. **b.** Find $f(g(x))$.

17. $f(x) = 3x$, $g(x) = -x + 1$
18. $f(x) = |x|$, $g(x) = 2x$
19. $f(x) = \frac{1}{x}$, $g(x) = 2x - 3$
20. $f(x) = \frac{1}{1 - x}$, $g(x) = x + 1$
21. $f(x) = x + 5$, $g(x) = x + 2$
22. $f(x) = x^2$, $g(x) = -\frac{1}{x}$
23. $f(x) = 2 - |x|$, $g(x) = \frac{1}{x}$
24. $f(x) = 4x - 3$, $g(x) = \frac{|x|}{4}$

In Exercises 25–32:

a. Find $g(f(x))$ and $f(g(x))$.

b. Tell whether $g(f(x)) = f(g(x))$ or $f(g(x)) \neq g(f(x))$.

25. $f(x) = x$, $g(x) = -x$
26. $f(x) = x$, $g(x) = \frac{1}{x}$
27. $f(x) = x$, $g(x) = -\frac{1}{x}$
28. $f(x) = 3x$, $g(x) = \frac{x}{3}$
29. $f(x) = x$, $g(x) = -\frac{x}{2}$
30. $f(x) = x + 3$, $g(x) = x - 3$
31. $f(x) = x - 5$, $g(x) = 5 - x$
32. $f(x) = x - 7$, $g(x) = x + 7$

APPLICATIONS: Using Composite Functions

B In order to remain unnoticed as she paced off distances, the detective Miss Sycamore was fitted with a skirt that restricted her stride to 75 centimeters.

33. Write an algebraic rule for the function that expresses the distance, d, in centimeters, as a function of the number of paces, p.

34. Write an algebraic rule for the function that expresses the distance, d, in meters, as a function of the number of paces, p.

Folklore suggests that if you count the number of cricket chirps in 15 seconds and add that number to 38, you will have a good estimate of the Fahrenheit temperature.

35. Express F, the Fahrenheit temperature, in terms of n, the number of cricket chirps per minute.

36. Express n as a function of F.

37. Express n as a function of C, the Celsius temperature. (HINT: $C = \frac{5}{9}(F - 32)$)

In Exercises 38–41, graph the indicated composite function.

38. $g(h(x))$, where $g(x) = -x + 4$ and $h(x) = 2x - 5$

39. $h(g(x))$, where $h(x) = x$ and $g(x) = x^2$

40. $f(h(x))$, where $f(x) = -|x|$ and $h(x) = \frac{1}{x}$

41. $h(f(x))$, where $h(x) = -x$ and $f(x) = 2x - 1$

C For $f(x) = x$, $g(x) = \frac{1}{x}$, and $h(x) = 1 - \frac{1}{x}$, find each of the following.

42. $g \circ f$ **43.** $f \circ g$ **44.** $h \circ f$ **45.** $h \circ g$ **46.** $h \circ g \circ f$ **47.** $h \circ f \circ g$

48. Find two functions f and g such that $f \circ g = g \circ f$.

49. **a.** Graph the functions in Exercise 48 in the same coordinate plane.

b. On the same coordinate plane, graph the line $y = x$.

c. Compare the graphs of $f \circ g$ and $g \circ f$ with respect to the line $y = x$.

50. For $f(x) = ax + b$ and $g(x) = \frac{1}{a}x - \frac{b}{a}$, show that $f(g(x)) = g(f(x)) = x$.

51. For $f(x) = 2x$, find $g(x)$ such that $f(g(x)) = -x$.

REVIEW CAPSULE FOR SECTION 5–5

Solve each equation for y. *(Pages 44–47)*

1. $y + 3x = -1$ **2.** $2y = 3x$ **3.** $5x - y = 6$ **4.** $5 = 3x + y$

5. $5x - y = 0$ **6.** $2x + y - 1 = 0$ **7.** $2y - 6 = 3x$ **8.** $x + y - 3 = 0$

5–5 Inverse Relations

Recall from Section 5–1 that a relation, Q, is a set of ordered pairs. If the elements of the ordered pairs of Q are interchanged, the result is another set of ordered pairs, called an *inverse relation*. The new relation is written as Q^{-1} (Q inverse).

Relation	Domain	Range
Q: $\{(0, 3), (1, 4), (2, 7), (3, 12)\}$	$\{0, 1, 2, 3\}$	$\{3, 4, 7, 12\}$
Q^{-1}: $\{(3, 0), (4, 1), (7, 2), (12, 3)\}$	$\{3, 4, 7, 12\}$	$\{0, 1, 2, 3\}$

Note that the domain of Q is the range of Q^{-1}, and the range of Q is the domain of Q^{-1}.

Definition

Relations Q and Q^{-1} are **inverse relations** if and only if for every ordered pair (x, y) in Q, there is an ordered pair (y, x) in Q^{-1}.

Example 1 illustrates this fact.

The inverse of a function is not always a function.

EXAMPLE 1 Let $Q = \{(2, 1), (3, 2), (4, 2), (5, 3)\}$

a. Find Q^{-1}.

b. Determine whether Q and Q^{-1} are functions.

Solutions: a. Interchange the elements of the ordered pairs of Q.

Q^{-1}: $\{(1, 2), (2, 3), (2, 4), (3, 5)\}$.

b. Each element in the domain of Q is paired with exactly one element of the range, $\therefore$ **Q is a function.**

In Q^{-1}, the ordered pairs $(2, 3)$ and $(2, 4)$ have the same first element, $\therefore$ **Q^{-1} is *not* a function.**

To find the inverse of a function defined by an equation, first interchange x and y in the original equation. Then solve the new equation for y.

EXAMPLE 2 For the function f defined by $y = 2x + 1$, find f^{-1}.

Solution:

$y = 2x + 1$ ← **Interchange x and y.**

$x = 2y + 1$ ← **Solve for y.**

$2y = x - 1$

$y = \frac{1}{2}x - \frac{1}{2}$ ← **Equation of f^{-1}.**

In the figure at the right, P is the midpoint of $\overline{AB}$, and line q is perpendicular to $\overline{AB}$ at P. Then line q is the *perpendicular bisector* of $\overline{AB}$ and points A and B are said to be *symmetric* with respect to line q.

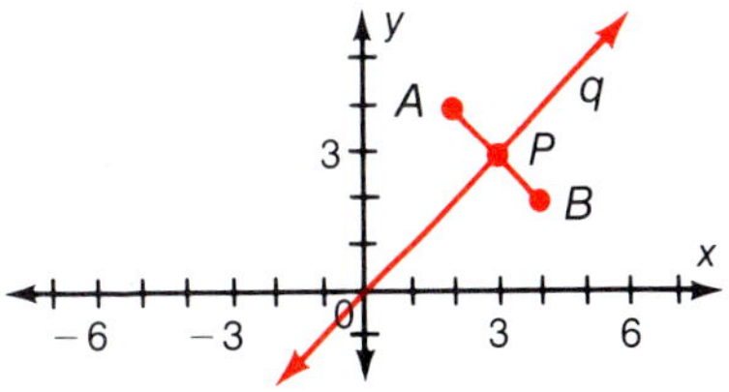

Definition

> Two points A and B are said to be **symmetric with respect to** a line q if and only if q is the **perpendicular bisector** of line segment AB.

Theorem 5–1 relates this definition to points on the graphs of a relation and its inverse.

Theorem 5–1

> **Symmetry of a Relation and its Inverse Relation**
>
> A point on the graph of a relation and the corresponding point on the graph of its inverse are symmetric with respect to the line $y = x$.

You can use this theorem as an aid in graphing.

EXAMPLE 3 Graph $f(x) = 2x + 1$ and its inverse $f^{-1}(x)$ on the same coordinate axes.

Solution:

[1] Graph $f(x) = 2x + 1$.

[2] Graph $y = x$. Use a dashed line.

[3] Choose points on f. Locate points on f^{-1} symmetric to the points on f with respect to $y = x$.

f		f^{-1}
$(-1, -1)$	⟶	$(-1, -1)$
$(0, 1)$	⟶	$(1, 0)$
$(2, 5)$	⟶	$(5, 2)$

[4] Join the points on f^{-1}.

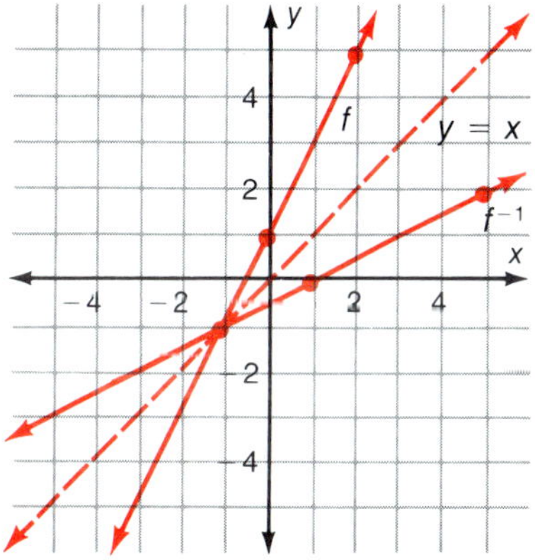

CLASSROOM EXERCISES

Write the inverse of each function. Indicate whether the inverse is also a function.

1. $\{(1, 1), (2, 3), (3, 4), (4, 5)\}$

2. $\{(0, 2), (2, 2), (3, 4), (1, 5)\}$

3. $\{(1, 2), (2, 1), (3, 2), (4, 3)\}$

4. $\{(-1, -2), (-3, -2), (-1, -4), (0, 6)\}$

Write the equation that defines the inverse of each function.

5. $y = -3x - 1$ **6.** $y = \frac{1}{2}x + 4$ **7.** $y = 9$ **8.** $y = 5x$

WRITTEN EXERCISES

A Write the inverse of each function. Indicate whether the inverse is a function.

1. $\{(-1, 3), (-2, 3), (-4, 3), (-5, 0)\}$ **2.** $\{(1, 1), (2, 0), (3, -1), (4, -2), (5, -3)\}$

3. $\{(\frac{1}{2}, \frac{2}{3}), (\frac{2}{3}, \frac{3}{4}), (\frac{3}{4}, \frac{4}{5}), (\frac{4}{5}, \frac{5}{6})\}$ **4.** $\{(0.1, -1), (0.2, -3), (0.3, -2), (0.4, 1)\}$

Write the equation that defines the inverse of each function.

5. $y = x + 2$ **6.** $y = -x$ **7.** $y = 2x$ **8.** $y = -x + 1$ **9.** $y = -3x - 4$

10. $y = 5x + 6$ **11.** $y = \frac{1}{2}x$ **12.** $y = \frac{1}{2}x + 3$ **13.** $y = 3$ **14.** $y = 0$

Determine whether the following pairs of functions are inverses of each other.

15. $y = x + 2;\ y = x - 2$ **16.** $y = 2x + 3;\ y = \frac{1}{2}x + 3$

17. $y = 3x + 1;\ y = 3x - \frac{1}{3}$ **18.** $y = 5x + 4;\ y = \frac{1}{5}x - \frac{4}{5}$

19. $y - x = 0;\ y + x = 0$ **20.** $y - 2x = 1;\ y - \frac{1}{2}x - \frac{1}{2} = 0$

21. $y - 2x = 3;\ y + 2x - 3 = 0$ **22.** $y = \frac{1}{2}x;\ y - 2x = 0$

23. $y - x = \frac{1}{2};\ y - x = \frac{1}{2}$ **24.** $y - cx = b;\ y - \frac{1}{c}x = \frac{b}{c}$

In Exercises 25–32, graph each function, f, and its inverse, f^{-1}, on the same coordinate axes.

25. $f = \{(2, 1), (1, 3), (5, -2), (-3, -4)\}$ **26.** $f(x) = -x$

27. $f(x) = 2x - 3$ **28.** $f(x) = -3x + 5$

29. $f(x) = x$ **30.** $f(x) = |x|$

31. $f(x) = -|x|$ **32.** $f(x) = \pm\sqrt{x},\ x \geq 0$

B

33. If a relation is a function, then no vertical line intersects the graph of the relation in more than one point. How can you tell from the graph of a relation whether the inverse is a function?

In Exercises 34–41, determine whether the inverse of each relation is a function.

34.

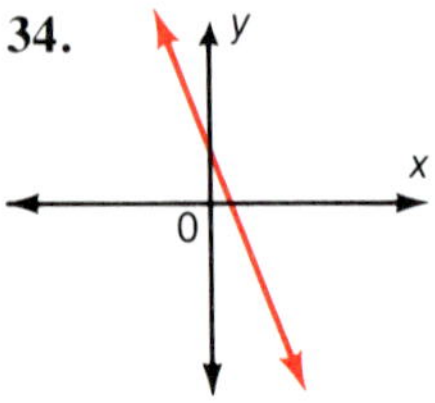

35.

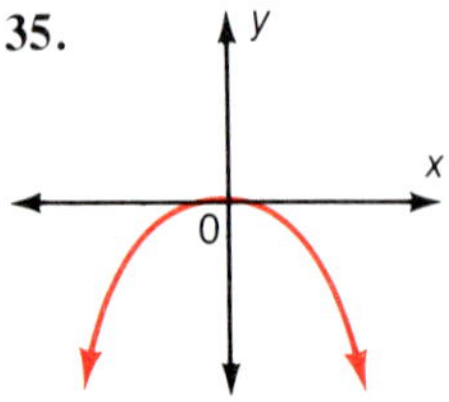

36.

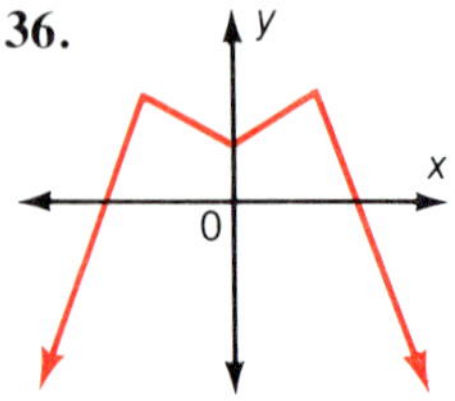

37.

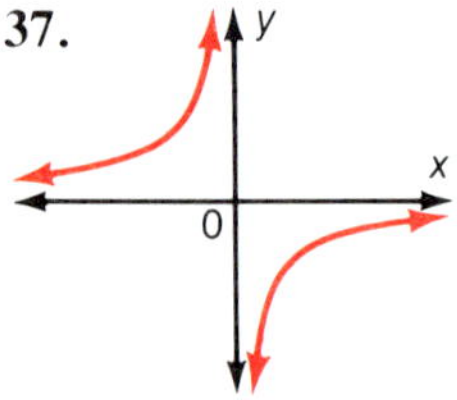

38. 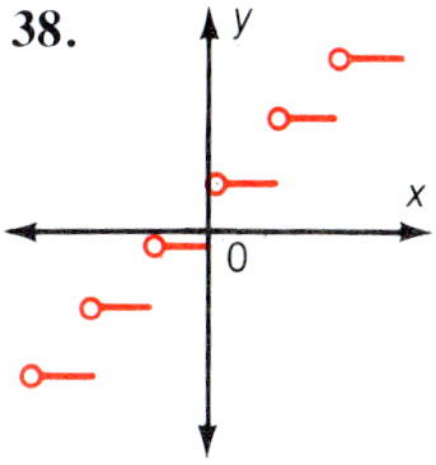

39. 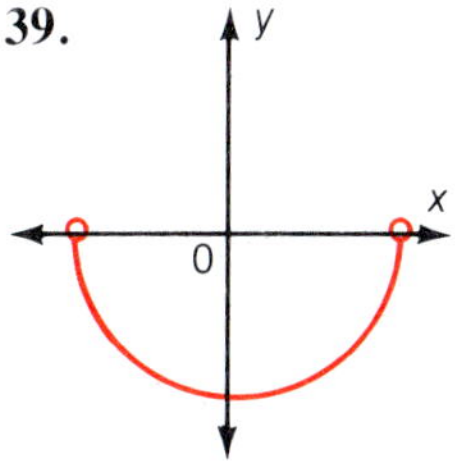

40. 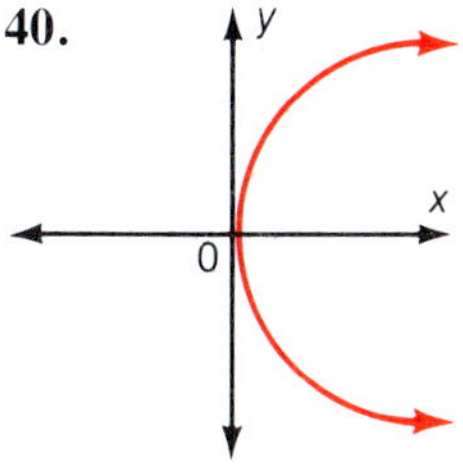

41. 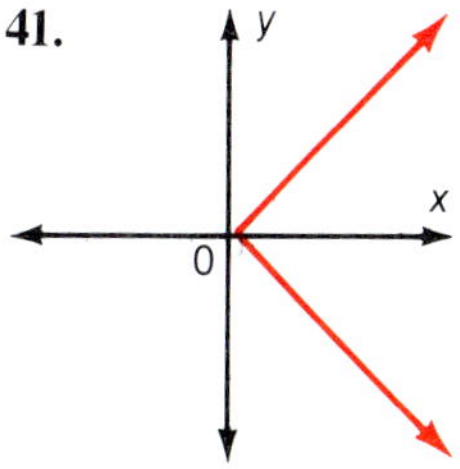

C In Exercises 42–44, determine whether each pair of graphs defines inverse functions. Justify your answer.

42. 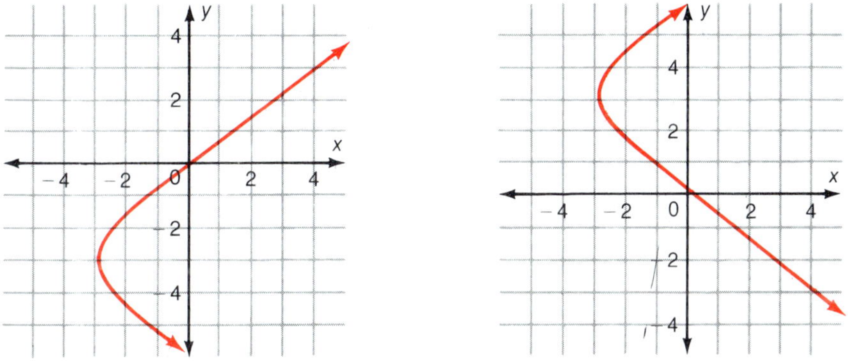

43. 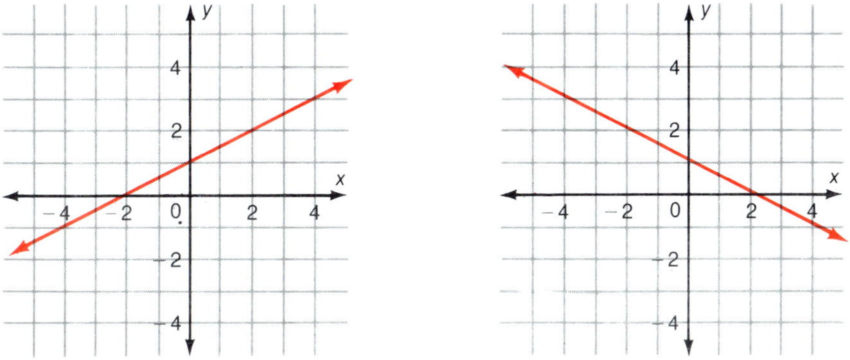

44. 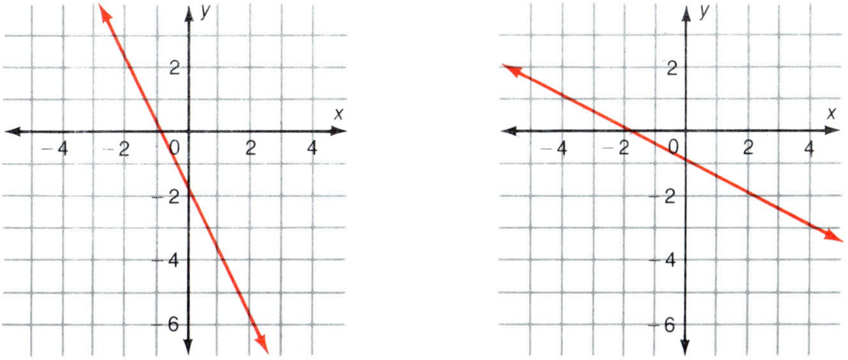

45. Give an example to show that if two functions are inverses of each other, then $f(f^{-1}(x)) = x$ and $f^{-1}(f(x)) = x$.

Review

In Exercises 1–4, find each of the following.

a. The equation that defines $g(f(x))$

b. $g(f(3))$ **c.** $g(f(-2))$ **d.** $g(f(a))$ **e.** $g(f(-b))$

(Section 5–4)

1. $f(x) = 2x$
$g(t) = t + 3$

2. $f(x) = \frac{1}{3}(x + 4)$
$g(t) = t - 1\frac{1}{3}$

3. $f(x) = -x$
$g(t) = \frac{1}{t}$

4. $f(x) = 5x - 2$
$g(t) = \frac{1}{2}t$

In Exercises 5–6, write the inverse of each function. Indicate whether the inverse is a function. *(Section 5–5)*

5. $\{(-2, -2), (-1, 5), (0, 9), (3, 5)\}$

6. $\{(-3, -1), (1, -3), (4, 7), (9, 2)\}$

In Exercises 7–12, graph each function, f, and its inverse, f^{-1}, in the same coordinate plane. *(Section 5–5)*

7. $\{(-4, 4), (-2, 2), (0, 0)\}$

8. $f(x) = -2x$

9. $f(x) = \frac{1}{3}x + 1$

10. $f(x) = -(x + 2)$

11. $f(x) = |x - 2|$

12. $f(x) = 2$

Chapter Summary

IMPORTANT TERMS

Absolute value function (p. 166)
Composite function (p. 174)
Constant function (p. 166)
Dependent variable (p. 160)
Domain (p. 160)
Function (p. 161)
Greatest integer function (p. 166)
Identity function (p. 166)
Independent variable (p. 160)
Inverse relation (p. 178)
Linear function (p. 166)
Range (p. 160)
Relation (p. 160)
Signum function (p. 166)

IMPORTANT IDEAS

1. **Vertical Line Test:** If a vertical line moving across the graph of a relation intersects the graph anywhere in more than one point, then the relation is *not* a function.
2. **A relation** can be represented as a mapping of its domain onto its range.
3. The graph of a **linear function** whose domain is the set of real numbers is a straight line.
4. **Linear equations** can be used to represent many scientific and everyday applications.
5. The **composite function** $g \circ f$ is not the same as the composite function $f \circ g$. Thus, composition of functions is *not commutative*.
6. **Symmetry of a Relation and its Inverse:** A point on the graph of a relation and the corresponding point on the graph of its inverse are symmetric with respect to the line $y = x$.

Chapter Objectives and Review

Objective: *To determine whether a given relation is a function (Section 5–1)*

Tell whether each mapping represents a function. Give a reason for each answer.

1.

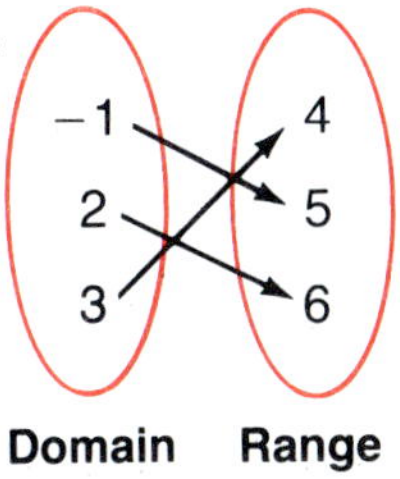

2.

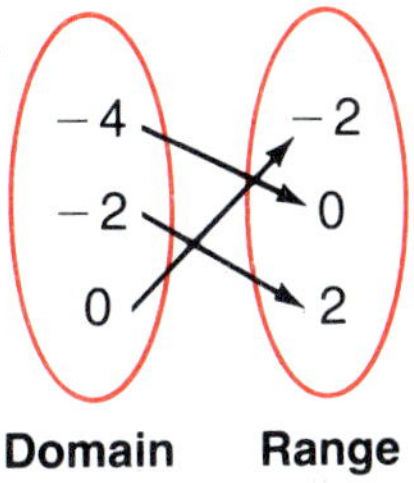

3.

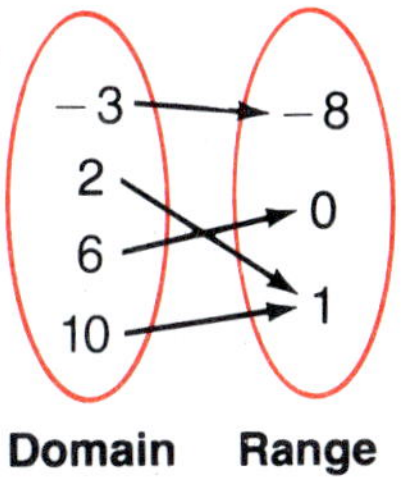

4.

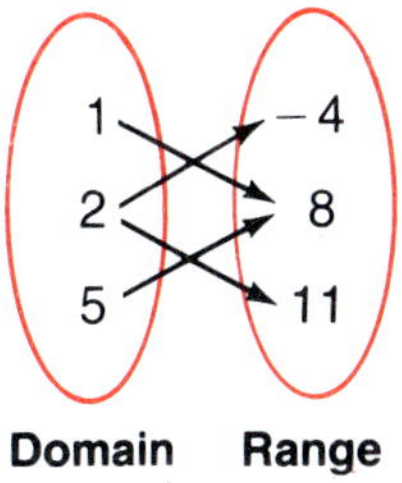

Use the Vertical Line Test to tell whether each graph represents a function.

5.

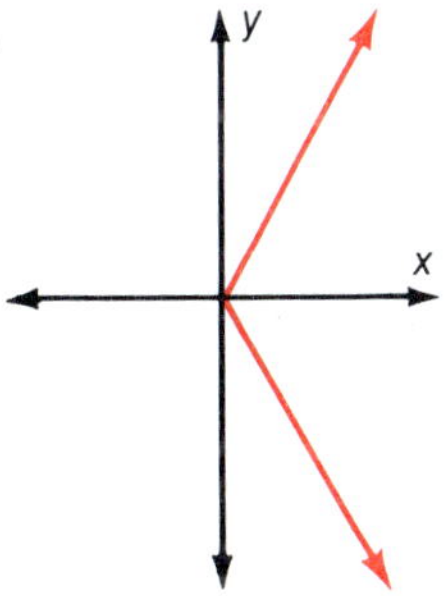

6.

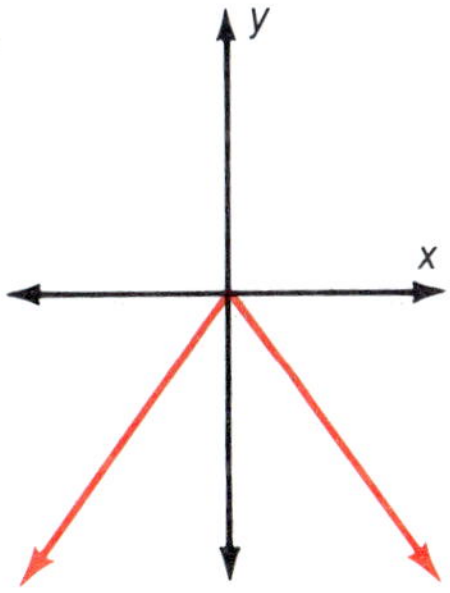

7.

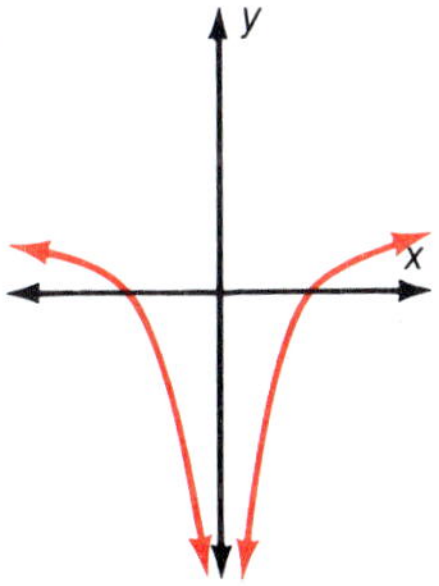

8.

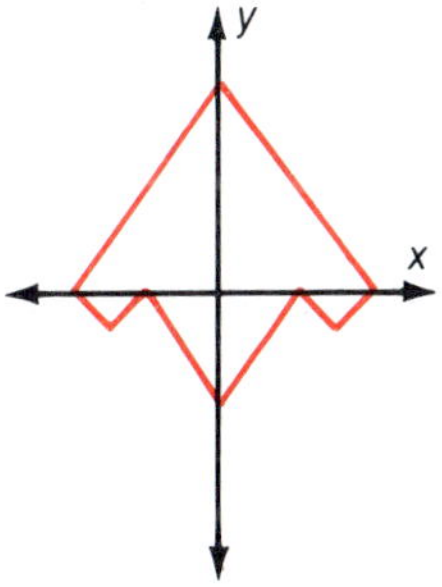

Objective: *To use function notation to represent and graph functions (Section 5–2)*

For $f(x) = \frac{x-2}{4}$, find each of the following.

9. $f(2)$ **10.** $f(-2)$ **11.** $f(\frac{1}{4})$ **12.** $f(0.4)$ **13.** $f(-t)$ **14.** $f(s+t)$

Graph each function.

15. $f(x) = 2 - x$ **16.** $g(x) = |x + 2|$ **17.** $h(x) = 7$ **18.** $f(x) = 1 + [x]$

Objective: *To use the linear function as a model for applications (Section 5–3)*

A 6-year old car has a value of \$5000. Three years earlier, the same car had a value of \$7500. Use this information in Exercises 19–22.

19. Assume that the function is linear. Graph the function.

20. Write an equation for the function. Use t for the age of the car in years and v for the value of the car in dollars.

21. Use the equation to predict how much the car will be worth when the car is 10 years old.

22. Use the equation to predict how old the car will be when $v = 0$.

Objective: *To determine a composite function (Section 5–4)*

In Exercises 23–26, find each of the following.

a. The equation that defines $g(f(x))$

b. $g(f(-4))$ **c.** $g(f(8))$ **d.** $g(f(t))$ **e.** $g(f(-t))$

23. $f(x) = x + 2$
$g(x) = -x$

24. $f(x) = \frac{1}{x}$
$g(x) = |x|$

25. $f(x) = 3x + 6$
$g(x) = \frac{1}{3}x - 2$

26. $f(x) = |x - 5|$
$g(x) = -\frac{1}{x}$

Objective: *To determine the inverse of a function (Section 5–5)*

In Exercises 27–28, write the inverse of each function. Indicate whether the inverse is a function.

27. $\{(-6, 4), (3, 4), (0, 4), (5, 4)\}$

28. $\{(2, 3), (3, 4), (4, 5), (5, 5)\}$

Write the equation that defines the inverse of each function.

29. $y = -x + 2$ **30.** $y = \frac{1}{5}x - 2$ **31.** $y = -8$ **32.** $y = -5x$

Chapter Test

1. Determine whether the mapping below represents a function.

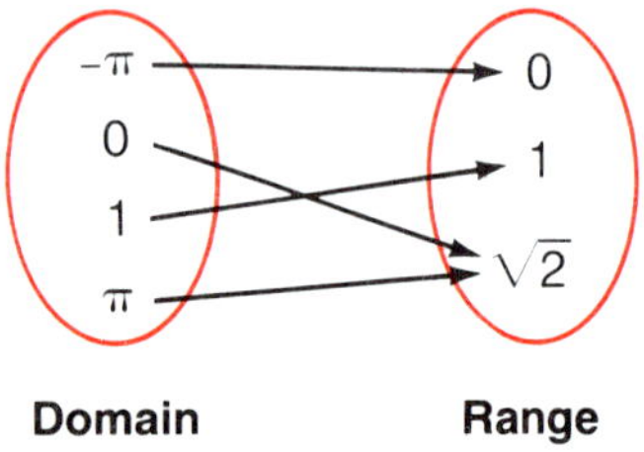

2. Use the Vertical Line Test to determine whether the graph below represents a function.

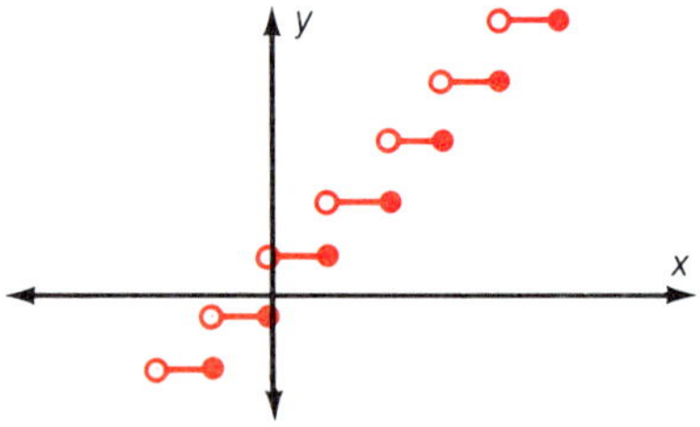

For $h(x) = \frac{1}{3}(x + 2)$, find each of the following.

3. $h(-2)$ **4.** $h(-3)$ **5.** $h(\frac{1}{2})$ **6.** $h(0.1)$ **7.** $h(-x)$ **8.** $h(x - 2)$

9. Graph: $f(x) = |3 - x|$.

A small bag of wet sand is thrown downward from the top of the Leaning Tower of Pisa. The initial speed of the sand is 3 meters per second. At the end of one-half second, the speed of the sand is 7.9 meters per second.

10. Assume that the function is linear. Graph the function.

11. Write an equation for the function. Use t for the time (seconds) and r for the speed (meters per second).

12. The sand lands at the base of the tower 2.9 seconds after it is thrown. Find the speed of the bag of sand when it hits the ground.

In Exercises 13–16, find each of the following.

a. The equation that defines $g(f(x))$

b. $g(f(2))$ **c.** $g(f(-4))$ **d.** $g(f(-t))$ **e.** $g(f(s + t))$

13. $f(x) = -4x$
$g(x) = -x$

14. $f(x) = |x + 4|$
$g(x) = \frac{1}{x}$

15. $f(x) = 6$
$g(x) = 3x - 1$

16. $f(x) = x + 1$
$g(x) = 4$

Write the equation that defines the inverse of each function.

17. $y = \frac{1}{4}x$ **18.** $y = 6x - 4$ **19.** $y = -5$ **20.** $y = \frac{1}{3}x + 9$

More Challenging Problems

1. Find the solution set of $2 \leq |x - 1| \leq 5$.
2. Let $n = x - y^{x-y}$. Find n when $x = 2$ and $y = -2$.
3. $f(x) = \dfrac{x + 1}{x - 1}$. Write $f(f(x))$ in simplest form.
4. If $x + z = y$, what is the value of $|x - y| + |y - x|$?
5. If $f(x) = 2x + 1$ and $g(x) = 3x - 1$, find $f \circ g - g \circ f$.
6. If f and g are functions such that $f(x) = 2x - 3$ and $f(g(x)) = x$, find $g(x)$.
7. Which quadrants entirely contain the set of points satisfying the pair of inequalities $y > 2x$ and $y > 4 - x$?
8. Let $f(x) = x^2 + 3x + 2$ and let S be the set of integers $\{0, 1, 2, \ldots 25\}$. Find the number of members s of S such that $f(s)$ has remainder zero when divided by 6.
9. Let n be the number of pairs of values of b and c such that $3x + by + c = 0$ and $cx - 2y + 12 = 0$ have the same graph. Find n and the values of b and c of each pair.
10. If x men working x hours a day for each of x days produce x articles, find the number of articles (not necessarily an integer) produced by y men working y hours a day for each of y days.
11. In racing over a given distance d at uniform speed, A can finish 20 yards ahead of B, B can finish 10 yards ahead of C, and A can finish 28 yards ahead of C. Find d in yards.
12. Jill and Joan bought identical boxes of stationery. Jill used hers to write one-sheet letters and Joan used hers to write three-sheet letters. Jill used all her envelopes and had 50 sheets of paper left. Joan used all her paper and had 50 envelopes left. Find the number of sheets of paper in each box.

Preparing for College Entrance Tests

When comparing two quantities or expressions, it is sometimes helpful to choose real numbers that satisfy the conditions of the problem and substitute them in the given expressions.

REMEMBER: If you can find at least one number that makes a statement false, then you have shown that the statement is *not always true.*

EXAMPLE If n is a real number and $n < 0$, which of the following statements must be true?

(a) $n < \frac{1}{n}$ **(b)** $5 + n > -(5 + n)$ **(c)** $n^2 > \frac{1}{n^2}$ **(d)** $n < \frac{1}{n^2}$

Think: Two choices involve n^2, and n^2 is always positive. Thus, check the choices involving n^2 first.

Solution: **(c)** $n^2 > \frac{1}{n^2}$ ⟵ **Not true when $n = -\frac{1}{2}$.**

(d) $n < \frac{1}{n^2}$ ⟵ **Always true since $n < 0$ and $\frac{1}{n^2} > 0$**

(a) $n < \frac{1}{n}$ ⟵ **Not true when $n = -\frac{1}{2}$.**

(b) $5 + n > -(5 + n)$ ⟵ **Not true when $n < -5$.** **Answer: d**

Choose the best answer. Choose *a*, *b*, *c*, or *d*. (Assume that all variables represent real numbers.)

1. If $x < 0$, which of the following must be true?

(a) $x - 3 < 3x$ **(b)** $x - 3 < 3 - x$ **(c)** $-3x < x^2$ **(d)** $x^3 > x + 3$

2. If $y > 0$, which of the following must be true?

(a) $2y > \frac{1}{2}y$ **(b)** $-(y - 2) < -y + 2$ **(c)** $-y^2 > (-y)^2$ **(d)** $y^2 > 2y$

3. If $m < 0$, which of the following must be true?

(a) $m - 2 > 2m$ **(b)** $-3m > m - 3$ **(c)** $m^4 > m^2$ **(d)** $(m - 1)^2 > (1 - m)^2$

4. If $r > 0$, which of the following must be true?

(a) $2r > \frac{2}{r}$ **(b)** $r^2 - 1 > 1 - r^2$ **(c)** $-r < r - \frac{1}{2}$ **(d)** $\frac{r - 1}{r} < 1 + \frac{1}{r}$

5. If $x \neq 0$, which of the following must be true?

(a) $\frac{x - (x + x)}{x} > 0$ **(b)** $\frac{x + (x - x)}{x} < 0$ **(c)** $\frac{x - (x - x)}{x} < 0$ **(d)** $\frac{x + (x + x)}{x} > 0$

6. If $mp < 0$, which of the following must be true?

(a) $\frac{m}{p} - 1 < 0$ **(b)** $m + p > 0$ **(c)** $m - p < 0$ **(d)** $p^2 - m^2 > 0$

CHAPTER 6 Polynomials

Sections

Features

Review and Testing

6–1 Polynomial Expressions

The symbol 5^3 means that 5 is taken as a factor three times.

Exponent

Base → $5^3 = 5 \cdot 5 \cdot 5 = 125$ ← **3rd power of 5**

An expression such as 5^3 is a monomial. A **monomial** in the variable x is an expression of the form

$$ax^n$$

where a is a real number and n is a whole number. The number a is a **coefficient** (or **numerical coefficient**) of the monomial. When $n = 0$, $x^0 = 1$.

A **polynomial** is a monomial or the sum of monomials.

EXAMPLE Identify each expression as a polynomial, P, or not a polynomial, NP. Give a reason for each answer.

	Expression	P or NP	Reason
a.	$-7x^2$	P	Definition of polynomial
b.	$3y^3 - \frac{1}{2}y$	P	Definition of polynomial
c.	$z^{\sqrt{5}}$	NP	The exponent is not a whole number.
d.	125	P	Definition of polynomial $(a = 125, n = 0)$

Polynomials can be classified by the number of terms.

Polynomial	Number of terms	Name
$\frac{1}{2}x$	1	Monomial
$x^2 + y^2$	2	Binomial
$\sqrt{2}xy - \sqrt{7}x + 1$	3	Trinomial

The **degree of a monomial** is the *sum* of the exponents of its variables. The degree of a monomial such as 125 is 0. The **degree of a polynomial** is the *greatest degree* of its terms.

Polynomial	Term of greatest degree	Degree of polynomial
$10x^2y^5$	$10x^2y^5$	$2 + 5 = 7$
$x^4 + 10x^2y^5$	$10x^2y^5$	$2 + 5 = 7$
$-x^4 + 10x^2y^5 + y^9$	y^9	9
$x + 2xy + y$	$2xy$	$1 + 1 = 2$
5	5	0
$x^3 - y^3$	x^3 or y^3	3
$17x$	$17x$	1

Polynomials can also be classified by their coefficients.

Polynomials	Coefficients	Polynomial over the set of
$-5x^2 + 3y^2$	$-5, 3$	integers
$\frac{2}{3}x^2 - 6xy + \frac{1}{2}y^2$	$\frac{2}{3}, -6, \frac{1}{2}$	rational numbers
$\sqrt{3}z^2 - 5$	$\sqrt{3}, -5$	real numbers

The terms of a polynomial in simplest form are usually written in order of decreasing degree of one of the variables from left to right.

$$x^3 + 3x^2y + 3xy^2 + y^3 \longleftarrow$$ **Decreasing order of *x***

CLASSROOM EXERCISES

Identify each expression as a polynomial *P*, or not a polynomial, *NP*. Give a reason for each answer.

1. -16 **2.** $-7\sqrt{x}$ **3.** $(-\frac{1}{3})^2z^5$ **4.** $3 + 4x^{\sqrt{6}}$
5. $6 - 5y + 2z$ **6.** $\sqrt{5}t^4 + t^2$ **7.** $2 + 5x^3$ **8.** $\frac{1}{4}x - y^2$
9. $9y^5 - 6y^3 + 9$ **10.** $2y^4 - 2y^{\sqrt{2}}$ **11.** $3x^{\sqrt{3}}y^2$ **12.** $4x^2 + 3x - 7$

WRITTEN EXERCISES

A Identify each expression as a polynomial, *P*, or as not a polynomial, *NP*. Give a reason for each answer.

1. 3 **2.** $\sqrt{3}$ **3.** $\sqrt{5}x$ **4.** $\sqrt{7}x$
5. $2x^2$ **6.** $-3x^2$ **7.** $-\frac{1}{2}x^2 + 5xy$ **8.** $2.7x + 6y$
9. $2x^5 + y$ **10.** $\sqrt{3}x^2 + \frac{1}{5}$ **11.** $5x^{\sqrt{2}}$ **12.** $\sqrt{5}z^5 - 1$
13. $\sqrt{x} + 3$ **14.** $-26xyz^3$ **15.** $\frac{x}{2} + 6x$ **16.** $3\frac{1}{3}x^2 - 4x + \sqrt{2}$
17. $2y^5 + x^{\sqrt{7}}$ **18.** $14x^2 + 3y^3$ **19.** $x + 2xy + z$ **20.** $2x^{\sqrt{5}} + \sqrt{5}$

Identify each polynomial as a monomial, a binomial, or a trinomial. Then state the degree of the polynomial.

21. k^3 **22.** $-7y^4$ **23.** $3 + 16y^2 + x$ **24.** $2p^2 + 3q^2$
25. $11mn + 16m$ **26.** $3xyz - 2x^2y^2z^2$ **27.** $-3t + 6t^2 - 2t^3$ **28.** $\sqrt{6}x + y^5$

In Exercises 29–34, state the numerical coefficients of each term of the given polynomial. Then classify each polynomial by its coefficients.

29. $3x^3 + 2y^2 + 7$ **30.** $-\frac{1}{2}x^4 + 10xy$ **31.** $\sqrt{2}z^7 - 3$
32. $1.3rs^5 - 7.1r^2s + 0.1$ **33.** $\frac{1}{2}a - \frac{1}{3}b - \frac{1}{4}c$ **34.** $x^3 + y^3 - z^3$

In Exercises 35–37, write each polynomial in order of decreasing powers of *y*.

35. $x^3 + 3x^2y + 3xy^2 + y^3$ **36.** $10x^5 + 3x^2y^4$ **37.** $x + 2xy + z$

6–2 Addition and Subtraction of Polynomials

Terms such as $3ab^2$, $7ab^2$ and $-\sqrt{5}ab^2$ are *like terms* because their variable factors, ab^2, are exactly alike. **Like terms** are exactly alike or differ only in their *numerical* coefficients. To add polynomials with like terms, use the distributive postulate.

EXAMPLE 1 Add: $2x^2y + 5x^2y$.

Solution: $2x^2y + 5x^2y = (2+5)x^2y$ ← **Distributive postulate**

$= \mathbf{7x^2y}$

Sometimes it is more convenient to add polynomials by arranging like terms in the same vertical column, as in Method 2 in Example 2.

EXAMPLE 2 Add: $(7z^2 - 2z^3 - 1) + (10z^4 - 8 + z^3 - 5z^2)$.

Solutions: **Method 1: Horizontal Format**

$(7z^2 - 2z^3 - 1) + (10z^4 - 8 + z^3 - 5z^2)$

$= (-2z^3 + 7z^2 - 1) + (10z^4 + z^3 - 5z^2 - 8)$ ← **Commutative postulate for +**

$= 10z^4 + (-2z^3 + z^3) + (7z^2 - 5z^2) + (-1 - 8)$ ← **Distributive post.**

$= 10z^4 + (-2+1)z^3 + (7-5)z^2 + (-1-8)$

$= \mathbf{10z^4 - z^3 + 2z^2 - 9}$

Method 2: Vertical Format

$$\begin{array}{r} -2z^3 + 7z^2 - 1 \\ +\ 10z^4 + z^3 - 5z^2 - 8 \\ \hline \mathbf{10z^4 - z^3 + 2z^2 - 9} \end{array}$$

← **Start with the term of highest degree.**
← **Write like terms in the same column.**
← **Add each column.**

To subtract a polynomial, add its additive inverse. The **additive inverse of a polynomial** can be found *by multiplying the polynomial by* -1. Thus, the additive inverse of $x^3 - 6x$ is

$$-(x^3 - 6x) = (-1)(x^3 - 6x) = -x^3 + 6x.$$

EXAMPLE 3 Subtract: $9a^4 - 3a^3 - 6 - (4a^4 - 2a^3 + 9)$.

Solutions: **Method 1: Horizontal Format**

$9a^4 - 3a^3 - 6 - (4a^4 - 2a^3 + 9)$ ← **Add the additive inverse of $(4a^4 - 2a^3 + 9)$.**

$= 9a^4 - 3a^3 - 6 + (-1)(4a^4 - 2a^3 + 9)$

$= 9a^4 - 3a^3 - 6 - 4a^4 + 2a^3 - 9$

$= 9a^4 - 4a^4 - 3a^3 + 2a^3 - 6 - 9$

$= (9-4)a^4 + (-3+2)a^3 + (-6-9)$

$= \mathbf{5a^4 - a^3 - 15}$

Method 2: Vertical Format

$$\begin{array}{r} 9a^4 - 3a^3 - \ \ 6 \\ -4a^4 + 2a^3 - \ \ 9 \\ \hline 5a^4 - \ \ a^3 - 15 \end{array}$$

$9a^4 - 3a^3 - 6$ ← **Arrange the terms in descending order.**

$-4a^4 + 2a^3 - 9$ ← $-1(4a^4 - 2a^3 + 9) = -4a^4 + 2a^3 - 9$

$5a^4 - a^3 - 15$ ← **Add each column.**

CLASSROOM EXERCISES

For each polynomial in Exercises 1–6, identify the like terms.

1. $4r^2s - 6r^2 + 2r^2s + 5$

2. $xy + 2x + 4xy$

3. $-2u + 11w - 6u$

4. $-4a^2 + 9a + 2a^2 - 6$

5. $7x^2yz - 2xy^2z + x^2yz$

6. $5abc + 6ab + abc - ac$

Add.

7. $4zt^2 + 11zt^2$

8. $3x^2 + 4x^2$

9. $(11n + 3) + (n - 8)$

10. $(3x + 5) + (-5x + 2)$

11. $(3y + 4) + (7y - 3)$

12. $(13r^2 + 2rs - 7) + (5rs + 16)$

Subtract.

13. $(4x - 5y + 18) - (6x - 5y - 2)$

14. $(5z^2 + z) - (6z^2 + z)$

15. $(2t^2 + 8q^2 - 3) - (-2 + 4t^2 - 7q^2)$

16. $(11b^3r - r^2 + 2) - (3b^2 - r^2 + 2)$

17. $(6n^2 - 5nw + 7w^2) - (4nw - w^2 + 3n^2)$

18. $(4z - 9z^4 - z^2) - (-z^2 + 4z - 9z^4)$

WRITTEN EXERCISES

A Add.

1. $$\begin{array}{r} y^2 + 9y + 6 \\ +5y^2 - 3y + 5 \\ \hline \end{array}$$

2. $$\begin{array}{r} 4z^3 + 5z^2 - 2z + 3 \\ +(-2z^3) - z^2 + 7z - 8 \\ \hline \end{array}$$

3. $$\begin{array}{r} a^3 - a^2 + 1 \\ + \quad 5a^2 - 7 \\ \hline \end{array}$$

4. $$\begin{array}{r} 3b^4 - 2b^3 + 19b^2 \quad 3b - 13 \\ + \quad - b^3 - 12b^2 \quad + 9 \\ \hline \end{array}$$

5. $2x^2 + 5x^2$

6. $3yz + 7yz$

7. $4.5rt^2 + rt^2$

8. $2\frac{1}{3}m + \frac{1}{3}m$

9. $6m^2n + 10m^2n$

10. $8p + (-14p)$

11. $3q^2r + (-q^2r)$

12. $\frac{1}{5}xyz + (-\frac{1}{5}xyz)$

13. $(2x + 3) + (x + 5)$

14. $(3x + 2) + (-4x - 3)$

15. $(7x + 4) + (-4x - 2)$

16. $(2x - 1) + (x + 4)$

17. $(5x^2 - 7x + 3) + (4x^2 + 2x)$

18. $(y^2 + 2y - 5) + (8y^2 - 5y + 9)$

19. $(3y^3 + 7y^2 + 5y) + (9y^3 - 18y^2)$

20. $(c^3 - 3c^2 + 8c + 1) + (-2c^3 + 5c^2 - 2c - 3)$

21. $(18a^3 - 5a^2 - 6a + 2) + (7a^3 - 8a + 9)$

22. $(2x^4 - 3x^2 + 2 - 5x) + (4x^2 + 2x - 7x^4 + 6)$

Subtract.

23. $\begin{array}{l} a^3 \qquad - 1 \\ \underline{a^3 - a^2 \qquad} \end{array}$

24. $\begin{array}{r} x^2 + 7x - 13 \\ \underline{(-9x^2) + 3x + 15} \end{array}$

25. $\begin{array}{l} x^3 - 2x^2 + 5x - 7 \\ \underline{x^3 + 3x^2 - 3x + 11} \end{array}$

26. $\begin{array}{r} y^3 - 2y + 7 \\ \underline{5y - 5} \end{array}$

27. $(7r + 8) - (3r + 2)$

28. $(14m + 3n) - (n - 6m)$

29. $(9p + 7) - (15p + 13)$

30. $15 - (20d + 7)$

31. $(-5x - 7y + 9) - (2x + 4y - 3)$

32. $(7x^3 - 5y^2 - 2) - (5x^3 - 2y^2 + 4)$

33. $(y^4 + 9y^2 - 11y) - (13y^3 - 5y^4 + 19y - 1)$

34. $(k^3 - 3k^2 + 6 + 5k) - (3k - 4k^2 + 2k^3 - 8)$

35. $(3a^5 - 2a^3 + 4a^2 - 7) - (-2a^3 + 3a^5 - 7 + 4a^2)$

36. $(9 + 3t^2 - t + 5t^4) - (2t - 3 + 8t^2 - 3t^4)$

37. $(-x^2 - y^2) - (-2x^2 + 3xy - 2y^2)$

38. $(9x^2y^2 - 5xy + 25y^2) - (5x^2y^2 + 10xy - 9y^2)$

B

39. From $x - 2y$, take $-x + y - z$.

40. Subtract $6x^3 - 4x^2 + 2x - 5$ from 0.

41. Find the sum of $3x^2y^5$, $5x^2y^5$, and $-6x^2y^5$.

42. Add $(0.5a^2 - 0.7b^2 - 0.3)$, $(1.7a^2 - 4.6b^2 - 0.9)$, and $(3a^2 + 4b^2 - 2)$.

43. From the sum of $(2x^2 + 3xy - 5y^2)$ and $(6y^2 - 4x^2)$, subtract $(8y^2 - 7xy)$.

44. By how much does $(\frac{3}{4}x^2 - 0.5x + 7)$ exceed $(\frac{1}{2}x^2 + \frac{1}{4}x - 5)$?

45. What polynomial must be subtracted from $2k^2 - k - 1$ to give $k - k^2$?

C The set of polynomials is a **commutative group under addition** because it satisfies the five postulates listed in Column II. Match each statement in Column I with a postulate from Column II. (*P*, *Q*, and *R* are polynomials and 0 is the zero polynomial.)

Column I: Statements	Column II: Postulates
46. $P + Q = Q + P$	a. Closure
47. $P + 0 = P$	b. Commutativity
48. $P + (-P) = 0$	c. Associativity
49. $P + Q$ is a polynomial.	d. Existence of an identity element
50. $(P + Q) + R = P + (Q + R)$	e. Existence of an additive inverse for each polynomial

51. Verify each statement in Exercises 46–50, by using the polynomial $(3x + 5)$ for P, $(2x^2 - 4x + 4)$ for Q, and $(x^2 + 5x + 6)$ for R.

52. Give an example to show that the set of polynomials is not commutative under subtraction. (HINT: Use the polynomials P and Q from Exercise 51. Show that $P - Q \neq Q - P$.)

53. Give an example to show that the set of polynomials is not associative under subtraction.

6–3 Multiplication of Polynomials

Since x^2 means $x \cdot x$ and x^3 means $x \cdot x \cdot x$,

$$x^2 \cdot x^3 = \underbrace{x \cdot x}_{\substack{2 \\ \text{factors}}} \cdot \underbrace{x \cdot x \cdot x}_{\substack{3 \\ \text{factors}}}$$

← **2 + 3, or 5 factors**

Thus, $x^2 \cdot x^3 = x^{2+3} = \mathbf{x^5}$.

This suggests the following theorem.

Theorem 6–1

Exponent Theorem for Multiplication

If m and n are positive integers and a is a real number, then

$$a^m \cdot a^n = a^{m+n}.$$

You can use Theorem 6–1 together with the associative and commutative postulates for multiplication to multiply monomials.

EXAMPLE 1 Multiply: $(4a^2b)(-3a^3b^2c)$.

Solution:

$(4a^2b)(-3a^3b^2c) = 4 \cdot -3 \cdot a^2 \cdot a^3 \cdot b \cdot b^2 \cdot c$ ← **By the commutative and associative postulates for ×**

$= (4 \cdot -3)(a^2 \cdot a^3)(b \cdot b^2)(c)$ ← **By the associative postulate for ×**

$= -12a^{2+3}b^{1+2}c$ ← **By Theorem 6–1**

$= \mathbf{-12a^5b^3c}$

Since a^2 means $a \cdot a$, $(a^2)^3$ means $(a \cdot a)(a \cdot a)(a \cdot a)$, or a^6. This suggests the following theorem.

Power Theorems

If m and n are positive integers and a and b are real numbers,

Theorem 6–2 then $(a^m)^n = a^{mn}$,

Theorem 6–3 and $(ab)^m = a^m b^m$.

EXAMPLE 2 Multiply: $(2x^3y^2)^3(-3x^2y)^2$.

Solution: Simplify the given power of each monomial. Then multiply.

$(2x^3y^2)^3(-3x^2y)^2 = [2^3(x^3)^3(y^2)^3][(-3)^2(x^2)^2(y)^2]$ ← **By Theorem 6–3**

$= (8x^9y^6)(9x^4y^2)$ ← **By Theorem 6–2**

$= \mathbf{72x^{13}y^8}$ ← **By Theorem 6–1**

You can use the distributive postulate to multiply a monomial and a polynomial.

EXAMPLE 3 Multiply: $-cd^2(c^2 + cd + d^2)$

Solution: Multiply each term of the trinomial by $(-cd^2)$.

$$-cd^2(c^2 + cd + d^2) = -cd^2(c^2) + (-cd^2)(cd) + (-cd^2)(d^2)$$ ← **By the distributive postulate**

$$= (-c \cdot c^2 \cdot d^2) + (-c \cdot c \cdot d^2 \cdot d) + (-c \cdot d^2 \cdot d^2)$$

$$= \mathbf{-c^3d^2 - c^2d^3 - cd^4}$$

To find the product of two binomials, use the distributive postulate twice. It is sometimes convenient to use a vertical format.

EXAMPLE 4 Multiply: $(5x - 4)(2x + 7)$

Solutions: Multiply each term of $(2x + 7)$ by $5x$. Then multiply each term by (-4).

Method 1: Horizontal Format

$$(5x - 4)(2x + 7) = 5x(2x + 7) + (-4)(2x + 7)$$ ← **By the distributive postulate**

$$= 5x(2x) + 5x(7) + (-4)(2x) + (-4)(7)$$ ← **By the distributive postulate**

$$= 10x^2 + 35x - 8x - 28$$

$$= \mathbf{10x^2 + 27x - 28}$$

Method 2: Vertical Format

Write the binomials in descending powers of the same variable. Write like terms in the same column.

$$\begin{array}{r} 5x - 4 \\ \underline{2x + 7} \\ 35x - 28 \\ \underline{10x^2 - 8x } \\ \mathbf{10x^2 + 27x - 28} \end{array}$$

$35x - 28$ ← **7(5x − 4)**

$10x^2 - 8x$ ← **2x(5x − 4)**

Study the pattern in finding the product of two binomials. This is sometimes called the **FOIL method.**

	F	O I	L
F, O, I, L	Product of first terms +	Sum of product of outer and inner terms +	Product of last terms
$(5x - 4)(2x + 7) =$	$(5x \cdot 2x)$ +	$(5x \cdot 7 + (-4) \cdot 2x)$ +	$(-4 \cdot 7)$
	$\mathbf{10x^2}$ +	$\mathbf{27x}$ +	$\mathbf{-28}$

You can use the Foil method to multiply two binomials by inspection.

EXAMPLE 5 Multiply by inspection.

a. $(4x+5)(x-9)$ **b.** $(2c-9)(3c-4)$

Solutions:

F O I L

a. $(4x+5)(x-9) = 4x^2 - 36x + 5x - 45 = \mathbf{4x^2 - 31x - 45}$

F O I L

b. $(2c-9)(3c-4) = 6c^2 - 8c - 27c + 36 = \mathbf{6c^2 - 35c + 36}$

You will find it useful to memorize these special binomial products.

Binomial Products

Square of a Binomial Sum: $(a+b)^2 = a^2 + 2ab + b^2$

Square of a Binomial Difference: $(a-b)^2 = a^2 - 2ab + b^2$

Product of a Sum and Difference: $(a+b)(a-b) = a^2 - b^2$

CLASSROOM EXERCISES

Multiply.

1. $(-2xy)(3x^2y^3)$
2. $7c(9-6d)$
3. $(5rt^2)^2(2r^3)$
4. $(-7xy^3)^4$
5. $(x+5)(x+7)$
6. $(5q+3)(qy-1)$
7. $(2d-3)(d-8)$
8. $(2r-3)(5r-9)$
9. $(2t+1)^2$
10. $(a-9)^2$
11. $(d+6)(d-6)$
12. $(z+\frac{3}{4})(z-\frac{3}{4})$

WRITTEN EXERCISES

A Find each product.

1. $2x^3(-3x^7)$
2. $a^3 \cdot 4a^2$
3. $7s^2t^5(-8st^2)$
4. $5a^7 \cdot \sqrt{2}a^5$
5. $(-3mn^2)(2m^2n)$
6. $(-5c^4d^3)(-2c^2d)$
7. $(-2c^2)(3cd)^2$
8. $(2mn)(-3n)^3$
9. $(5t^3)^2(-2s^2)^3$
10. $(2x^2yz^3)^2(-3xy^3z^2)^3$
11. $mn(5m-3n)$
12. $cd(7c+4d)$
13. $-r^2t^2(r^3t - rt^5)$
14. $5xy^2(2x-3y)$
15. $x(6x-7y^2)$
16. $-2(a^2+6a+10)$
17. $4xy^2(3xy - x^2y + 5)$
18. $2ab(9a^3b^4 + 3a^2b - 2b^4)$
19. $(ef)(3e^2)(8f)$
20. $(3x^2y)(-4xy)(5xy^2)$
21. $(-3a^2b)(-2ab^2)(-a^2b^2)$
22. $(2x+5)(3x+4)$
23. $(2x+5)(3x-4)$
24. $(2x-5)(3x-4)$
25. $(5x-2)(2x-3)$
26. $(3z+1)(z^2-7)$
27. $(a^2+b^2)(a^2-2b^2)$
28. $(n+6)^2$
29. $(2m-7)^2$
30. $(a+b)^2$
31. $(x-y)^2$
32. $(2x+5y)^2$
33. $(7t-3r)^2$
34. $-(2a-3b)^2$
35. $(x-y)(x+y)$
36. $(3m+4n)(3m-4n)$
37. $(3c-7d)(2c+d)$
38. $(x-3)(7x-9)$
39. $(2x-2)(x^2+6x-9)$

B Find each product.

40. $(a+b)^3$ **41.** $(r-1)^3$ **42.** $(2m+n)^3$

43. $(2a-3b)(2a+3b)(4a^2+9b^2)$ **44.** $(\frac{2}{3}x-\frac{3}{5})(\frac{2}{3}x+\frac{3}{5})(4x^2+9)$

C The set of polynomials is a commutative group under addition (see page 192). It is not a commutative group under multiplication.

45. Which of the following postulates for a commutative group does not hold for polynomials?

a. Closure under multiplication

b. Commutativity under multiplication

c. Associativity under multiplication

d. Existence of a multiplicative identity

e. Existence of a multiplicative inverse for each nonzero polynomial

Review

Identify each expression as a polynomial, *P*, or not a polynomial, *NP*. *(Section 6–1)*

1. 4 **2.** $\sqrt{7}$ **3.** $\sqrt{5}x^2$ **4.** $9+2x^2$

In each polynomial, identify the term of greatest degree. Then write the degree of the polynomial. *(Section 6–1)*

5. $2x^2+3x$ **6.** $4x^3y^2+9x^2y$ **7.** $7x^5y^2+3x^2y^6$ **8.** $4+2x^2+3y^3$

Add or subtract as indicated. *(Section 6–2)*

9. $(3x^2+2x+4)+(4x^3+3x-2)$ **10.** $(2t^2-3t+7)+(2t-4t^2+9-8t^3)$

11. $(16m-7)-(-2m^2+9m-8)$ **12.** $(3a^2+4a-11)-(6-10a^4+6a-3a^2)$

Find each product. *(Section 6–3)*

13. $4x^2(-2x^8)$ **14.** $(6t^3)^4(3v^2)^6$ **15.** $(2c-3d)(3c+7d)(4ab)$

REVIEW CAPSULE FOR SECTION 6–4

Complete. *(Pages 10–13)*

1. $5a^2=5a(\underline{\ ?\ })$ **2.** $4ab=4a(\underline{\ ?\ })$ **3.** $4ab=a(\underline{\ ?\ })$

4. $9x^2y=9xy(\underline{\ ?\ })$ **5.** $9x^3y=9x^2y(\underline{\ ?\ })$ **6.** $14r^2s^2=2rs(\underline{\ ?\ })$

7. $3uvw=v(\underline{\ ?\ })$ **8.** $21s^3=7s(\underline{\ ?\ })$ **9.** $20ab=4a(\underline{\ ?\ })$

10. $20ab=2b(\underline{\ ?\ })$ **11.** $15m^3n=3n(\underline{\ ?\ })$ **12.** $2xy^2=y(\underline{\ ?\ })$

13. $\frac{1}{3}t^2v=2v(\underline{\ ?\ })$ **14.** $130q^2t=13t(\underline{\ ?\ })$ **15.** $1.1c^2d^3=cd(\underline{\ ?\ })$

6–4 Factoring Polynomials Over the Integers

Factoring reverses the steps you use to find products. To factor a polynomial over the integers, you express it as a product of two or more factors having integral coefficients. The first step in factoring a polynomial is to identify the **greatest common factor** (abbreviated: **GCF**) other than $+1$ and -1.

EXAMPLE 1 Find the GCF of $6x^4y^3$, $18x^3y^3$, and $30x^2y^4$.

Solution: [1] Find the GCF of the numerical coefficients.

GCF: $2 \cdot 3$, or 6

$$2 = ② \cdot ③$$
$$6 = ② \cdot ③ \cdot 3$$
$$10 = ② \cdot ③ \cdot 5$$

} **Prime Factorization**

[2] Compare the powers of any variable that is a factor of the three monomials.

Choose the power which has the smallest exponent.

Powers of Variable			Choice
x^4	x^3	x^2	x^2
y^3	y^3	y^4	y^3

[3] GCF: $6x^2y^3$ ← **Greatest monomial which is a factor of $6x^4y^3$, $18x^3y^3$, and $30x^2y^4$**

EXAMPLE 2 Factor $2x^4y^3 - 6x^3y^3 - 10x^2y^4$.

Solution: GCF: $2x^2y^3$

$$2x^4y^3 - 6x^3y^3 - 10x^2y^4 = 2x^2y^3 \cdot x^2 - 2x^2y^3 \cdot 3x - 2x^2y^3 \cdot 5y$$
$$= 2x^2y^3(x^2 - 3x - 5y)$$

The polynomial $x^2 - 3x - 5y$ has no polynomial factors other than itself, its opposite, and ± 1. Hence, it is a *prime polynomial.*

A polynomial is **prime over the integers** or **factored completely over the integers** when it cannot be factored further into monomials or polynomials with terms that have integral coefficients.

In this text, to factor over a given set *always means to factor completely over that set.*

Sometimes polynomials have a common binomial factor.

EXAMPLE 3 Factor over the integers: $a^2b + a^2 + 2b + 2$.

Solution:

$$a^2b + a^2 + 2b + 2 = a^2(b + 1) + 2(b + 1)$$ ← **$b + 1$ is a common binomial factor.**

$$= (b + 1)(a^2 + 2)$$ ← **Prime polynomial**

A **perfect square trinomial** is a trinomial which is the square of a binomial. Thus, a perfect square trinomial has two equal binomial factors.

Factoring Perfect Squares

$$a^2 + 2ab + b^2 = (a + b)^2$$
$$a^2 - 2ab + b^2 = (a - b)^2$$

EXAMPLE 4 Factor over the integers.

a. $81t^2 + 18t + 1$ **b.** $9y^2 - 24y + 16$

Solutions: Follow the patterns for factoring perfect squares.

a. $81t^2 + 18t + 1 = (9t)^2 + \underline{2(9t)(1)} + (1)^2$ ← **Twice the product of the square roots of the first and last terms**

$= (9t + 1)(9t + 1)$, or $\mathbf{(9t + 1)^2}$

b. $9y^2 - 24y + 16 = (3y)^2 - 2(3y)(4) + (4)^2$

$= (3y - 4)(3y - 4)$, or $\mathbf{(3y - 4)^2}$

In Section 6-3, you saw that $(a + b)(a - b) = a^2 - b^2$. Hence, to factor the difference of two squares, take the square root of each square. Then write the product of their sum and difference.

EXAMPLE 5 Factor over the integers: $25r^2 - 49s^2$

Solution: $25r^2 - 49s^2 = (5r)^2 - (7s)^2$ ← **Difference of squares**

$= \mathbf{(5r + 7s)(5r - 7s)}$ ← **Prime polynomials**

CLASSROOM EXERCISES

Find the GCF of the given monomials.

1. $35a^2$ and $7a$ **2.** $25x^2y$ and $35xy^2$ **3.** $12r^2t^3$ and $18rt^4$

4. $a^2b^3c^2$, ab^2c, and $a^3b^2c^3$ **5.** $48x^4y^2$, $24x^2y^3$, and $75x^5y^2$

Factor over the integers. One factor is written for you.

6. $4ab + b = \underline{\ ?\ }(4a + 1)$ **7.** $x^4y^2 - x^2y^3 = x^2y^2(\underline{\ ?\ } - \underline{\ ?\ })$

8. $t^2 - r^2 = (t + r)(\underline{\ ?\ } - \underline{\ ?\ })$ **9.** $a^3 + 6a^2 + 9a = a(a + \underline{\ ?\ })^2$

10. $x^3 - 2x^2 + x = \underline{\ ?\ }(x - 1)^2$ **11.** $s^2 - 64 = (\underline{\ ?\ } + \underline{\ ?\ })(s - 8)$

12. $2a - 2b - ac + bc = 2(\underline{\ ?\ } - \underline{\ ?\ }) - c(\underline{\ ?\ } - \underline{\ ?\ })$
$= (a - b)(\underline{\ ?\ } - \underline{\ ?\ })$

13. $3a^3 + 3a^2 - 4a - 4 = 3a^2(\underline{\ ?\ } + \underline{\ ?\ }) - 4(\underline{\ ?\ } + \underline{\ ?\ })$
$= (a + 1)(\underline{\ ?\ } - \underline{\ ?\ })$

WRITTEN EXERCISES

A Find the GCF of the monomials.

1. $6x$ and 12
2. $18a^2$ and $9a$
3. $4c$ and $2bc$
4. $3ab^2y$ and $3ay$
5. $2xy^2z$ and z^2y
6. $36xy^2z$ and $27xy^2z^2$
7. $10x^2y^2$, $5xy$, and $-5y^3$
8. $2a^3b^2$, $-2a^3c$, and $2a^3c^2$
9. $10r^2t^2$, $5rt$, and $20r^3$
10. $2x^2y$, $4a^2x$, and $6a^2y^2$

Factor. One factor is written for you.

11. $2c^3 - 2c^2 = \underline{\ ?\ }(c - 1)$
12. $3ab - 6ab^2 + 9b = \underline{\ ?\ }(a - 2ab + 3)$
13. $3x^2 - x = x(\underline{\ ?\ } - \underline{\ ?\ })$
14. $a^3 + 3a^2b + ab^2 = a(\underline{\ ?\ } + \underline{\ ?\ } + \underline{\ ?\ })$
15. $5x^3 - 10x^2 + 35x = \underline{\ ?\ }(x^2 - 2x + 7)$
16. $15r^2t^3 + 24rt^4 = \underline{\ ?\ }(5r + 8t)$

Factor each polynomial over the integers.

17. $8y + 16b$
18. $7x - 7$
19. $9r^2 + 18r$
20. $3a^2 + 3a$
21. $2x^2 + 8$
22. $2x^2 + 4$
23. $5ax - 5a^2x$
24. $14m^2 + 20mn$
25. $3a^3 - 3a^2b - 3ab^3$
26. $3a^2b + 3ab^2 + 3b^4$
27. $2x^2 + 3x^3 + 4x^4y$
28. $24a - 16b + 24c$

Factor by finding the common binomial factor.

29. $a(x + y) - b(x + y)$
30. $x(a - b) + y(a - b)$
31. $3(2x - 5) - a(2x - 5)$
32. $3a(b + c) - 5(b + c)$
33. $cx + cy + bx + by$
34. $y^3 + 2y^2 + 3y + 6$
35. $1 - y + y^2 - y^3$
36. $b^3 + b^2 + b + 1$

Factor over the integers.

37. $x^2 - 18x + 81 = (x - \underline{\ ?\ })^2$
38. $4a^2 + 4a + 1 = (\underline{\ ?\ } + 1)^2$
39. $r^2 - 10rs + 25r^2 = (r - \underline{\ ?\ })^2$
40. $9a^2 - 12ay + 4y^2 = (3a - \underline{\ ?\ })^2$
41. $25x^2 - 60x + 36$
42. $a^2 + 4ab + 4b^2$
43. $s^2 - 8st + 16t^2$
44. $4x^2 + 20x + 25$
45. $4n^2 + 12n + 9$
46. $25x^2 - 40xz + 16z^2$
47. $36m^2 + 12m + 1$
48. $49x^2 + 112xy + 64y^2$

Factor as the product of the sum and difference of squares. In Exercises 49 and 50, one factor is written for you.

49. $x^2 - 4 = (x + 2)(\underline{\ ?\ } - \underline{\ ?\ })$
50. $a^2 - 36b^2 = (\underline{\ ?\ } + \underline{\ ?\ })(a - 6b)$
51. $9x^2 - b^2$
52. $16b^2 - 9$
53. $9x^2 - 4$
54. $25 - x^2y^2$
55. $x^4 - y^2$
56. $z^2 - 16x^2y^2$
57. $25a^4 - 4$
58. $a^2 - 25b^2$
59. $169 - 49x^2$
60. $16 - y^2$
61. $9 - 16b^4$
62. $x^2y^2z^2 - 36$

Factor over the integers.

63. $16x^2 - 1$

64. $14r^4 + 28r^2 - 35r$

65. $x^2y^2 - a^2$

66. $9c^2 + 12c + 4$

67. $144 - t^2$

68. $36q^2 + 12q + 1$

69. $2c(3c + 1) - (3c + 1)$

70. $p^2 + 2p + 7p + 14$

71. $49 - 36z^2$

72. $t^2 - 18t + 81$

73. $3s + r - 6s^2 - 2rs$

74. $15a^3b^3 - 18a^5b^2 + 24ab^4$

75. $30r^2t - 24rt^2 + 36r^3t$

76. $r^2(d - c) + r(d - c)$

77. $25c^2 - 10c + 1$

78. $4r^2 - 625$

79. $64e^2 + 48e + 9$

80. $6x + xy + 6y + y^2$

APPLICATIONS: Using Factoring

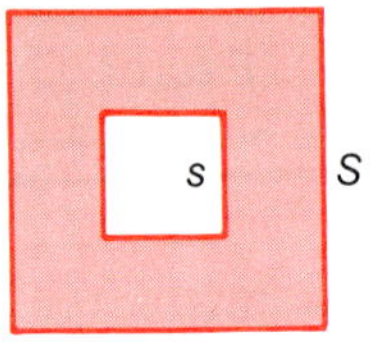

81. The area of the shaded part of the figure at the right is $S^2 - s^2$. Write this in factored form and then find the area of the shaded part when S is 1.5 meters and s is 0.6 meter.

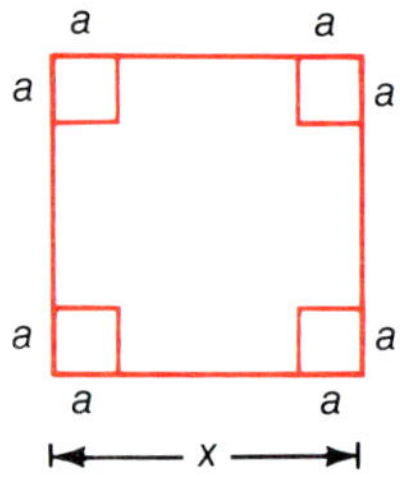

82. The diagram at the right is the plan for a square metal box without a cover. The small squares at the corners are to be cut from the original square piece of metal before the sides are bent up. Find the number of square inches of metal that is used for the box if the original piece is 10.5 inches on a side and each small square is 2.5 inches on a side. Use $A = x^2 - 4a^2$.

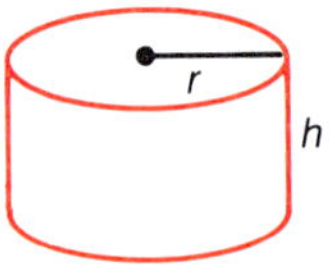

83. The formula for the total surface of a cylinder is $S = 2\pi r^2 + 2\pi rh$. Write S in factored form. How much tin would be needed to make a can with $r = 4$ cm and $h = 6$ cm? (Use $\pi = 3.14$)

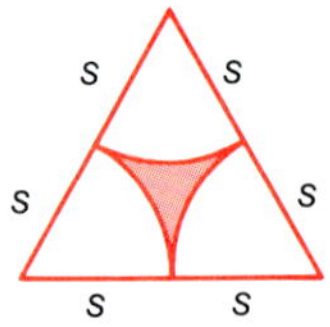

84. The area of the triangle is $s^2\sqrt{3}$. The area of each circular sector is $\frac{\pi s^2}{6}$. Write an expression for the area of the shaded portion. Factor it and find its value for $s = 3$ cm. (Use $\pi = 3.14$ and $\sqrt{3} = 1.73$.)

B Factor over the integers.

85. $b^2(b + 1) + (b + 1)$

86. $(a - b)^2 - (a + b)(a - b)$

87. $x^3 + x^2y - xy^2 - y^3$

88. $3ax - 6ay - 8by + 4bx$

89. $a(x - y) - b(y - x)$

90. $3(4x - 2) - b(2 - 4x)$

91. $a(x + y) - b(y + x)$

92. $5a(x - 3y) - 2b(3y - x)$

C

93. $(a + b)^2 - c^2$

94. $a^2 - (b + c)^2$

95. $(2x + 3y)^2 - 9z^2$

96. $(2a - b)^2 - c^2$

97. $(m^2 + 2mn + n^2) - a^2$

98. $(a^2 + 6a + 9) - 9b^2$

99. $1 - x^2 - 2xy - y^2$

100. $36x^2 - 9y^2 + 6y - 1$

REVIEW CAPSULE FOR SECTION 6–5

In Exercises 1–10, find the integers a and b which satisfy the given conditions. *(Pages 26–29)*

1. $a + b = 12$; $a \cdot b = 35$; $a > b$
2. $a + b = -12$; $a \cdot b = 35$; $a < b$
3. $a + b = -3$; $a \cdot b = -40$; $a < b$
4. $a + b = -18$; $a \cdot b = 80$; $a > b$
5. $a + b = 8$; $a \cdot b = 15$; $a < b$
6. $a + b = -15$; $a \cdot b = 54$; $a > b$
7. $a + b = -1$; $a \cdot b = -42$; $a > b$
8. $a + b = 4$; $a \cdot b = -21$; $a < b$
9. $a + b = -2$; $a \cdot b = -63$; $a < b$
10. $a + b = 29$; $a \cdot b = 54$; $a < b$

6–5 Factoring Quadratic Trinomials

A trinomial such as $x^2 + 5x + 6$ is a **quadratic trinomial** because the degree of the trinomial is 2. In general, quadratic trinomials are factored *by finding two binomial factors whose product is the trinomial.*

To factor a quadratic trinomial, you reverse this pattern.

$$(x + r)(x + s) = x \cdot x + x \cdot s + r \cdot x + r \cdot s$$
$$= x^2 \quad + (s + r)x \quad + r \cdot s$$

Thus, to factor $x^2 + 5x + 6$ over the integers, you find two integers r and s such that $s + r = 5$ and $r \cdot s = 6$.

$$x^2 + 5x + 6 = (x + 3)(x + 2)$$ ← **$3+2=5$; $3 \cdot 2=6$**

Example 1 shows that when the last term of the trinomial is negative, the integers r and s will have opposite signs.

EXAMPLE 1 Factor over the integers: $x^2 + 3x - 10$.

Solution:

[1] $x^2 + 3x - 10 = (x \quad)(x \quad)$ ← **Factor x^2.**

[2] Identify two integers r and s such that $r + s = 3$ and $r \cdot s = -10$.

Trial factors $(r \cdot s)$:	$1(-10)$	$-1(10)$	$2(-5)$	$-2(5)$
Corresponding sum $(r + s)$:	-9 No	9 No	-3 No	**3 Yes** ✓

[3] $x^2 + 3x - 10 = \mathbf{(x - 2)(x + 5)}$

Check: [4] $(x - 2)(x + 5) = x^2 + 5x - 2x - 10$
$$= x^2 + 3x - 10$$

The coefficient of x^2 in the quadratic trinomial $3x^2 - 10x + 8$ is not 1. To factor this trinomial, you must find the factors of both $3x^2$ and 8 such that the sum of the outer and inner products is the middle term, $-10x$. The procedures for factoring quadratic trinomials of this type are shown in Examples 2 and 3.

EXAMPLE 2 Factor over the integers: $3x^2 - 10x + 8$.

Solution: [1] $3x^2 - 10x + 8 = (3x \quad)(x \quad)$ ← **Factor $3x^2$.**

[2] Since -10 is negative and 8 is positive, both factors of 8 must be negative. Write possible pairs of factors.

Trial Factors	Outer Product	+	Inner Product	=	Middle Term
$(3x - 8)(x - 1)$	$-3x$	+	$-8x$	=	$-11x$ No
$(3x - 1)(x - 8)$	$-24x$	+	$-x$	=	$-25x$ No
$(3x - 2)(x - 4)$	$-12x$	+	$-2x$	=	$-14x$ No
$(3x - 4)(x - 2)$	$-6x$	+	$-4x$	=	**$-10x$ Yes** ✓

[3] $3x^2 - 10x + 8 =$ **$(3x - 4)(x - 2)$** The check is left for you.

In Example 3, the coefficients of the second and third terms are both negative.

EXAMPLE 3 Factor over the integers: $6y^2 - 5y - 25$.

Solution: [1] The factors of $6y^2$ could be $6y \cdot y$ or $3y \cdot 2y$.

$6y^2 - 5y - 25 = (6y \quad)(y \quad)$ or $(3y \quad)(2y \quad)$

[2] The factors of -25 will have opposite signs.

Trial Factors	Outer Product	+	Inner Product	=	Middle Term
$(6y + 5)(y - 5)$	$-30y$	+	$5y$	=	$-25y$ No
$(6y - 5)(y + 5)$	$30y$	+	$-5y$	=	$25y$ No
$(3y + 5)(2y - 5)$	$-15y$	+	$10y$	=	**$-5y$ Yes** ✓

[3] $6y^2 - 5y - 25 =$ **$(3y + 5)(2y - 5)$** The check is left for you.

Some quadratic trinomials cannot be factored over the integers.

EXAMPLE 4 Factor over the integers: $2x^2 + 4x + 3$.

Solution: [1] $2x^2 + 4x + 3 = (2x \quad)(x \quad)$

[2]

Trial Factors	Outer Product	+	Inner Product	=	Middle Term
$(2x + 3)(x + 1)$	$2x$	+	$3x$	=	$5x$ No
$(2x + 1)(x + 3)$	$6x$	+	x	=	$7x$ No

Since there are no other possible factors, $2x^2 + 4x + 3$ is a **prime polynomial over the integers.**

CLASSROOM EXERCISES

Factor over the integers.

1. $x^2 + 8x + 15 = (x + 3)(x + \underline{?})$
2. $t^2 + 3t - 4 = (t + 4)(t - \underline{?})$
3. $10c^2 - 7c - 12 = (5c + \underline{?})(2c - \underline{?})$
4. $3w^2 - 16w - 12 = (3w + \underline{?})(w - \underline{?})$
5. $3a^2 + 7a + 2 = (3a + \underline{?})(\underline{?} + \underline{?})$
6. $12y^2 - 7y + 1 = (4y - \underline{?})(\underline{?} - \underline{?})$

WRITTEN EXERCISES

A Factor over the integers.

1. $x^2 - x - 42$
2. $c^2 + 4c - 21$
3. $r^2 + 4r - 12$
4. $-8x + x^2 - 20$
5. $e^2 + e - 90$
6. $-2a - 63 + a^2$
7. $m^2 - 9m + 8$
8. $y^2 + 10y + 24$
9. $s^2 - 11s + 18$
10. $t^2 - 2t - 8$
11. $x^2 + 2x - 35$
12. $a^2 - 4a - 21$
13. $2y^2 - 3y - 5$
14. $2b^2 + 9b - 18$
15. $3a^2 + 10a - 8$
16. $6c^2 - c - 15$
17. $2c^2 + 7c - 30$
18. $e^2 - 18e + 72$
19. $3b^2 + 5b + 2$
20. $2b^2 + 3 - 7b$
21. $n^2 - 7n + 6$
22. $2y^2 - y - 15$
23. $2y^2 - 7y - 15$
24. $a^2 + 18 - 9a$
25. $x^4 - x^2y - 6y^2$
26. $a^6 + 3a^3 - 28$
27. $a^4 - 2a^2y^2 - 3y^4$
28. $4y^6 - 17y^3 - 15$
29. $x^8 + 4x^4 - 21$
30. $6x^6 + x^3 - 12$

B Factor over the rational numbers.

31. $x^2 + \frac{3}{4}x + \frac{1}{8}$
32. $x^2 - \frac{2}{3}x + \frac{1}{9}$
33. $x^2 + 0.5x + 0.06$
34. $a^{2m} + 6a^m - 16$
35. $a^2 + 0.5a - 0.24$
36. $2y^{2x} - 3y^x - 5$
37. $x^2 + (a + b)x + ab$
38. $x^2 - rx + sx - rs$
39. $z^2 + mz + kz + km$

C Factor over the integers.

40. $(x + y)^2 + 6(x + y) - 72$
41. $(c + d)^2 - 7(c + d) - 18$
42. $a^{2m} + 6a^{m-16}$
43. $2y^{4x} - 3y^{2x} - 5$
44. $18x^{4t} - 15x^{2t} - 12$
45. $57 - 22c^{3n} + c^{6n}$
46. The trinomials $2x^2 - 5x + c$ and $4x^2 - 4x + 1$ both have $ax + b$ as a factor. Find the integral values of a, b, and c.

Puzzle

Each letter represents a digit. More than one solution is possible. Find one solution.

(I)(AM)(NOT) = SURE

Using Statistics: Median Line of Fit

In **Business Administration**, managers are concerned with efficiency. Jeanine Curran is the manager of an automobile parts supply house. She must be able to predict an arrival time for a shipment of parts to the customer. From records of past deliveries she constructs the table below in which x is the distance in miles and y is the transportation time in days.

x	210	290	350	480	490	730	780	850	920	1,010
y	5	7	6	11	8	11	12	8	15	12

One way to make predictions from this data is to do the following:

Procedure

1. Graph the points in the coordinate plane. This is a **scatter diagram.**
2. Partition the data into three nearly equal sets such that there are an odd number of points in the first and third sets. For the above data Jeanine uses sets of 3, 4, and 3 points.
3. In both the first and last sets, locate the *median* x and y values. In sets with an odd number of elements, the **median** is the middle value.
4. Graph the two points arising from the median values and draw the line containing them. This is the **median line of fit.**
5. Obtain the equation of the line by using the methods in Section 3–6.
6. Make predictions by substituting x values in the equation and finding y values.

EXAMPLE Use Jeanine Curran's information to plot a scatter diagram to find the median line of fit and predict the time it will take to transport parts a rail distance of 600 miles.

Solution:

1. Construct the scatter diagram.
2. Partition the data:

x	y
210	5
290	7
350	6

x	y
480	11
490	8
730	11
780	12

x	y
850	8
920	15
1010	12

3. Find the median values.

median x: 290 median x: 920
median y: 6 median y: 12

4. Graph the points (290, 6) and (920, 12) and draw the line.

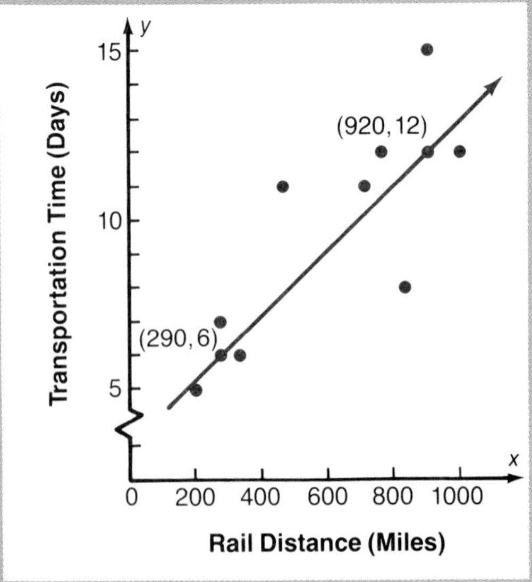

5 Find the equation.

$$m = \frac{12-6}{920-290} = \frac{6}{630} \approx 0.01$$

$$y - 6 = 0.01(x - 290)$$

$$y - 6 = 0.01x - 2.9$$

$$y = 0.01x + 3.1$$

6 Substitute $x = 600$

$$y = 0.01(600) + 3.1$$

$$y = 6 + 3.1 = 9.1 \text{ days}$$

Thus, it will take at least 9 days to make a delivery a distance of 600 miles.

EXERCISES

For Exercises 1–4,

a. Construct a scatter diagram.

b. Find and graph the median line of fit.

c. Determine the equation of the line.

d. Make a prediction for the given value of *x* to the nearest whole number.

1.

Pages	x	10	20	20	30	40	50	60	100	110	120
Hours to Type	y	6	8	12	16	18	21	32	46	60	58

About how many hours are needed to type 70 pages?

2.

Minutes Between Rest Periods	x	0.5	2.0	2.4	3.3	4.2	4.4	5.5	7.9	9.6	13.1
Pounds Lifted per Minute	y	910	680	540	590	520	600	350	400	230	90

About how many pounds can be lifted when $x = 5.0$ minutes?

3.

Tons of Crunchola	x	100	150	170	180	220	240	300	350	400	500
Total Cost (thousands of dollars)	y	65	70	65	85	115	125	140	190	200	200

Find the predicted total cost for 200 tons of crunchola.

4. A statistician for the Civil Aeronautics Board wants to find an equation that relates destination distance to freight charge for a standard sized crate. She takes a sample of 10 freight invoices from different companies and constructs the following table.

Distance (miles)	x	500	600	900	1000	1200	1400	1600	1700	2200	2300
Charge (nearest dollar)	y	31	45	40	60	50	68	65	79	95	105

Find the predicted charge (to the nearest dollar) if the crate's destination is 1500 miles away.

6–6 More on Factoring

The polynomial $x^3 + 8$ is the sum of two cubes; the polynomial $x^3 - 8$ is the difference of two cubes. The pattern for factoring the sum and difference of two cubes is given below.

Sum and Difference of Two Cubes

$$a^3 + b^3 = (a + b)(a^2 - ab + b^2)$$

$$a^3 - b^3 = (a - b)(a^2 + ab + b^2)$$

EXAMPLE 1 Factor over the integers.

a. $8n^3 + 27$ **b.** $27x^3 - 64y^3$

Solutions: **a.** Follow the pattern for $a^3 + b^2$.

$8n^3 + 27 = (2n)^3 + (3)^3$ ← **a = 2n; b = 3**

$= [(2n + 3)][(2n)^2 - (2n)(3) + 3^2]$

$= \mathbf{(2n + 3)(4n^2 - 6n + 9)}$

b. Follow the pattern for $a^3 - b^3$.

$27x^3 - 64y^3 = (3x)^3 - (4y)^3$ ← **a = 3x; b = 4y**

$= [3x - 4y][(3x)^2 + (3x)(4y) + (4y)^2]$

$= \mathbf{(3x - 4y)(9x^2 + 12xy + 16y^2)}$

To find the *prime factors* of a polynomial over a specified set, it is sometimes necessary to apply *more than one factoring technique*.

EXAMPLE 2 Factor over the integers.

a. $2x^2y - 10xy - 28y$ **b.** $-9p^3 + 30p^2 - 25p$

Solutions: First identify any common factors.

a. $2x^2y - 10xy - 28y = 2y(x^2 - 5x - 14)$ ← **Factor the trinomial.**

$= \mathbf{2y(x + 2)(x - 7)}$

b. $-9p^2 + 30p - 25 = -1(9p^2 - 30p + 25)$

$= (-1)(3p - 5)(3p - 5)$

$= \mathbf{-(3p - 5)^2}$

In Example 2b, note that it may be helpful to factor -1 from each term.

In factoring a polynomial completely, you may have to use *the same factoring technique more than once.*

EXAMPLE 3 Factor $4a^4 - 4b^4$ over the integers.

Solution:

$$4a^4 - 4b^4 = 4(a^4 - b^4)$$
$$= 4((a^2)^2 - (b^2)^2) \quad \leftarrow \text{Difference of squares}$$
$$= 4(a^2 + b^2)(a^2 - b^2) \quad \leftarrow (a^2 - b^2) \text{ is the difference of squares.}$$
$$= \mathbf{4(a^2 + b^2)(a + b)(a - b)}$$

These steps can be useful in factoring polynomials. The number of terms in the polynomial may indicate a possible factoring procedure.

Steps in Factoring

1. Identify any common monomial factors.
2. Check for special products.
3. **a.** For two terms: Check for the difference of squares and check for the sum and difference of cubes.

   b. For three terms: Check for perfect squares. Also check for quadratic trinomials that factor as two binomials.

   c. For four terms: Try grouping to find a common binomial factor.

Some polynomials that are prime over the integers can be factored further over the real numbers. For example $x^2 - 3$ is a prime polynomial over the integers. However, it can be factored over the real numbers.

$$x^2 - 3 = (x + \sqrt{3})(x - \sqrt{3})$$

CLASSROOM EXERCISES

Factor over the integers.

1. $x^3 - 8$
2. $y^3 + 27$
3. $a^3 + 1$
4. $5p^2 + 5$
5. $11a^2 + 44b^2$
6. $7x^2 - 7q^2$
7. $90y^2 - 1000$
8. $15r^4 - 60r^2$
9. $48z^3 - 12z$
10. $-45r^2 + 5s^2$
11. $-7a^2 - 42a - 63$
12. $6a^2 - 24a - 72$
13. $6b^3 + b^2 - 15b$
14. $-a^4 + 4a^3 - 4a^2$
15. $-3x^5 + 27x^3$

WRITTEN EXERCISES

A Factor over the integers.

1. $m^3 - n^3$
2. $c^3 - 1$
3. $a^3 - 27b^3$
4. $27x^3 + 8y^3$
5. $a^3b^3 + 27c^3$
6. $64 - x^3$
7. $1 - 64y^3$
8. $q^6 - 27$
9. $z^6 + 125$
10. $\frac{1}{8}a^3 - \frac{1}{27}b^3$
11. $y^3 + 64z^3$
12. $x^3y^3 - z^3$

13. $3g^2 - 21g - 24$
14. $n^3 + n^2 - 12n$
15. $18k^3 - 60k^2 + 50k$
16. $98m^2 - 28m + 2$
17. $2r^2 - 200$
18. $5a^2 - 5b^2$
19. $2x^2 - 32y^2$
20. $75f^2 - 147$
21. $6 - s - s^2$
22. $-3d^2 - 15d + 18$
23. $x^4 - 81$
24. $144x^4 - 81y^4$
25. $4ax^2 + 4ax + a$
26. $-q^4 + 21q^2 - 4q^3$
27. $4a^2 + 16b^2$
28. $2z^2 + 11z + 15$
29. $ab^2 + 14a^2b + 45a^3$
30. $7rs - s^2 + 18r^2$

B

31. $-27cd - 42c - 3cd^2$
32. $6r^2s^2 - 6rs - 36$
33. $3y^4 - 8y^2 - 35$
34. $k^6 - 10k^3p^2 + 25p^4$
35. $(m^2 + 2m + 1) - n^2$
36. $(x - 1)^2 - (q + 1)^2$
37. $(a^2 - b^2) - 3(a + b)$
38. $(x - y)(x^2 + y^2) + 2xy(x - y)$
39. $(2a + 1)(a^2 + 2) - 3a(2a + 1)$
40. $(a^2 - b^2) + (a - b)(b - a)$
41. $(a^2 - b^2) - (a^2 - 6ab + 5b^2)$
42. $x^2 - y^2 - (x^2 - 2xy + y^2)$
43. $a^6 + b^6$
44. $n^6 + 8$
45. $16c^4 + 2cd^3$
46. $a^6 - b^6$
47. $16x^5 + 54x^2y^3$
48. $(x^2 + 4x + 4) - 1$
49. $16a^4 - 54^a$
50. $y^6 - 1$

C

51. $49x^{2n} + 14x^n + 1$
52. $r^{2n}s^{6n} - 25$
53. $3z^{4n} - 10z^{2n} + 3$
54. $k^{4n+1} - k^{2n+1} - 6k$

Factor over the real numbers.

55. $x^2 - 3$
56. $2y^2 - 9$
57. $5b^2 - 7$
58. $t^4 - 4$
59. $x^4 - 3x^2 - 40$
60. $4x^2 - 11$
61. $7y^2 - 11b^2$
62. $9 - a^4$

Review

Factor over the integers. *(Section 6–4)*

1. $16c^2 - 9d^2$
2. $25a^2 + 40a + 16$
3. $d^2 - (e + f)^2$
4. $ab^2 + b^2c + 2a + 2c$
5. $12bc + 3b - 4c + 1$
6. $b^2(c + 1) - 2b(c + 1)$

(Section 6–5)

7. $x^2 - 3x - 28$
8. $2d^4 - 5d^2f - 12f^2$
9. $6r^2 + 21s^2 - 27rs$
10. $x^4 + 2x^2 - 8$
11. $2a^2 - 7ab - 15b^2$
12. $6x^6 + 17x^3y - 14y^2$

(Section 6–6)

13. $8t^3 - 27v^3$
14. $6c^4 - 96d^4$
15. $(a^2 + 2ab + b^2) - d^2$

REVIEW CAPSULE FOR SECTION 6–7

Evaluate each expression for $r = 12$, $s = 5$, and $t = -2$. *(Pages 10–13)*

1. $\frac{r - t}{r + t}$
2. $\frac{3r + 2s}{3t}$
3. $\frac{3r - 2s}{3s + r}$
4. $\frac{r^2 + 3s^2}{-t^2}$

6–7 Division of Polynomials

The following examples illustrate the basic exponent theorems for division.

$\frac{2^6}{2^2} = \frac{64}{4} = 16$, or $\mathbf{2^4}$ So $\frac{2^6}{2^2} = \mathbf{2^4}$ ← $\mathbf{2^{6-2}}$

$\frac{3}{3^3} = \frac{3}{27} = \frac{\mathbf{1}}{\mathbf{9}}$, or $\frac{\mathbf{1}}{\mathbf{3^2}}$ So $\frac{3}{3^3} = \frac{1}{3^2}$ ← $\frac{\mathbf{1}}{\mathbf{3^{3-2}}}$

$\frac{5^2}{5^2} = \frac{25}{25} = \mathbf{1}$ So $\frac{5^2}{5^2} = 1$ ← $\mathbf{5^{2-2} = 5^0}$

The basic exponent theorems for division are generalized below.

Exponent Theorems for Division

For nonnegative integers m and n, and for any real number a, $a \neq 0$:

Theorem 6–4 $\frac{a^m}{a^n} = a^{m-n}$, where $m > n$.

Theorem 6–5 $\frac{a^m}{a^n} = \frac{1}{a^{n-m}}$, where $m < n$.

Theorem 6–6 $\frac{a^m}{a^m} = a^0 = 1$.

EXAMPLE 1 Divide: **a.** $\frac{-8x^2}{x^7}$ **b.** $12a^7 \div -2a^4$.

Solutions:

a. $\frac{-8x^2}{x^7} = \frac{-8}{1} \cdot \frac{x^2}{x^7}$

$= -8 \cdot \frac{1}{x^{7-2}}$ ← **By Theorem 6–5**

$= \frac{-8}{x^5}$, or $-\frac{\mathbf{8}}{\mathbf{x^5}}$

b. $12a^7 \div -2a^4 = \frac{12a^7}{-2a^4}$

$= \frac{12}{-2} \cdot \frac{a^7}{a^4}$

$= -6 \cdot a^{7-4}$ ← **By Theorem 6–4**

$= \mathbf{-6a^3}$

In general, **you divide a polynomial by a monomial by dividing each term of the polynomial by the monomial.**

EXAMPLE 2 Divide $x^3 + x^2 - 2x + 8$ by x.

Solution:

$$\frac{x^3 + x^2 - 2x + 8}{x} = \frac{x^3}{x} + \frac{x^2}{x} - \frac{2x}{x} + \frac{8}{x}$$

$$= x^{3-1} + x^{2-1} - 2x^{1-1} + \frac{8}{x}$$ ← **By Theorems 6–4 and 6–6**

$$= \mathbf{x^2 + x - 2 + \frac{8}{x}}$$

You also use the Exponent Theorems to divide a polynomial by a polynomial. Remember that you can add and subtract like terms only.

EXAMPLE 3 Divide $x^2 + 5x + 4$ by $x + 1$.

Solution:

[1]
$$\begin{array}{r|l} & x \\ x+1 & \overline{x^2+5x+4} \\ & \underline{x^2+\ \ x} \\ & \quad\ \ 4x+4 \end{array}$$

- $x^2 \div x = x$. Write x in the quotient.
- $x(x + 1) = x^2 + x$
- $(x^2 + 5x + 4) - (x^2 + x) = 4x + 4$

[2]
$$\begin{array}{r|l} & x\ +4 \\ x+1 & \overline{x^2+5x+4} \\ & \underline{x^2+\ \ x} \\ & \quad\ \ 4x+4 \\ & \quad\ \ \underline{4x+4} \\ & \qquad\quad 0 \end{array}$$

- $4x \div x = 4$. Write 4 in the quotient.
- $4(x + 1) = 4x + 4$
- $(4x + 4) - (4x + 4) = 0$

Thus, $(x^2 + 5x + 4) \div (x + 1) = \mathbf{x + 4}$.

You can check the answer to Example 3 by using this fact from arithmetic.

(Divisor)(Quotient)	+ Remainder =	Dividend
$(x + 1)(x + 4)$	$+ \quad 0$	$= x^2 + 5x + 4$

Before dividing, always check whether the terms of the dividend and the divisor are arranged in **descending powers of the same variable.**

EXAMPLE 4 Divide $2a - 5a^2 + 4a^4 - 10$ by $2a - 3$.

Solution: $2a - 5a^2 + 4a^4 - 10 = 4a^4 + 0a^3 - 5a^2 + 2a - 10$

$$\begin{array}{r|l} & 2a^3+3a^2+2a\ +4 \\ 2a-3 & \overline{4a^4+0a^3-5a^2+2a-10} \\ & \underline{4a^4-6a^3} \\ & \qquad\ 6a^3-5a^2+2a-10 \\ & \qquad\ \underline{6a^3-9a^2} \\ & \qquad\qquad\quad 4a^2+2a-10 \\ & \qquad\qquad\quad \underline{4a^2-6a} \\ & \qquad\qquad\qquad\quad 8a-10 \\ & \qquad\qquad\qquad\quad \underline{8a-12} \\ & \qquad\qquad\qquad\qquad\quad 2 \end{array}$$

- $4a^4 \div 2a = 2a^3$. Write $2a^3$ in the quotient.
- $2a^3(2a - 3)$
- $6a^3 \div 2a = 3a^2$. Write $3a^2$ in the quotient.
- $3a^2(2a - 3)$
- $4a^2 \div 2a = 2a$. Write $2a$ in the quotient.
- $2a(2a - 3)$
- $8a \div 2a = 4$. Write 4 in the quotient.
- $4(2a - 3)$
- Remainder

Thus, $(4a^4 - 5a^2 + 2a - 10) \div (2a - 3) = \mathbf{2a^3 + 3a^2 + 2a + 4 + \frac{2}{2a - 3}}$, or

$4a^4 - 5a^2 + 2a - 10 = \mathbf{(2a - 3)(2a^3 + 3a^2 + 2a + 4) + 2}$.

The check in Example 4 is left for you.

CLASSROOM EXERCISES

Divide. No denominator equals zero.

1. $\frac{x^7}{x^4}$ **2.** $\frac{-a^{10}}{a^7}$ **3.** $\frac{m^{15}}{m^{20}}$ **4.** $\frac{15a^5}{-5a^2}$ **5.** $\frac{3x^2 + 9x}{3x}$

6. $(36r^4t^4 - 18r^3t^3 + 27r^2t) \div 9r^2t$ **7.** $(6a^2b^2 + 3ab^2 - 9a^2b) \div 3ab$

WRITTEN EXERCISES

A Divide. No denominator equals zero.

1. $\frac{a^3}{a}$ **2.** $\frac{c^6}{c^9}$ **3.** $\frac{6y^5}{8y^2}$ **4.** $\frac{42x^5y^2}{-6x^4y^2}$

5. $\frac{11g - 5}{g}$ **6.** $\frac{6x^2 - 3x}{x}$ **7.** $\frac{m^2 - 2m}{-m}$ **8.** $\frac{5x^2y^2 - 15xy^3}{5xy}$

9. $\frac{36a^5 + 16a^2b^3 - 12a^2b^2}{-4a^2}$ **10.** $\frac{35z^5 + 40z^3 - 45z}{-5z}$

11. $\frac{20r^5 - 28r^4 + 35r^3}{-4r^3}$ **12.** $\frac{y^4 - 16y^3 + 32y^2}{-8y^2}$

13. $(b^2 - 5b + 6) \div (b - 2)$ **14.** $(y^2 - 5y - 24) \div (y + 3)$

15. $(x^2 + 6x - 55) \div (x - 5)$ **16.** $(3a^2 + 14a + 15) \div (a + 3)$

17. $(9x^2 - 16) \div (3x - 4)$ **18.** $(2b^2 + 13b - 24) \div (2b - 5)$

19. $(6m^2 - 11m + 7) \div (3m - 4)$ **20.** $(8b^2 - 35 - 6b) \div (4b + 7)$

21. $(12a^2 - 11a - 10) \div (3a - 5)$ **22.** $(2y + 24y^2 - 15) \div (4y - 3)$

23. $(x^3 + 8) \div (x + 2)$ **24.** $(3x^3 - 10x^2 + 5x - 6) \div (x - 3)$

25. $(15y^3 - y^2 - 18y + 8) \div (3y - 2)$ **26.** $(2n^3 - 13n + 15) \div (n + 3)$

27. $(a^3 - 3a^2 + 5a + 1) \div (a - 2)$ **28.** $(3x^3 - 13x^2 + 6x - 8) \div (x - 4)$

B

29. $(6x^3 + 5x^2 - x + 4) \div x + 4$ **30.** $(n^4 - n^2 - 42) \div (n^2 + 6)$

31. $(8y^3 + 125) \div (5 + 2y)$ **32.** $[3(x + y)^2 - 5(x + y)] \div (x + y)$

33. $(-2n^4 - 5n^3 + 10n^2 + 16) \div (n + 4)$ **34.** $[(6r^4s^3t^2)^3 - 8r^3s^4t^2 + 16r^3s^5] \div (2rs)^3$

C

35. $(x^{2n} - x^n + x^{n-1}) \div x^n$ **36.** $(2^{2n} - 2^n + 2^{n-1} - 2^{n-2}) \div (-2^n)$

37. $(a^3 + b^3 + c^3) \div (a + b + c)$ **38.** $(2x^{3a} - 6x^{2a}y^a + 6x^ay^{2a} - 2y^{3a}) \div (2x^a - 2y^a)$

39. $(2x^5 - x^4) \div (x^2 - 2x + 1)$ **40.** $(x^{4a} + x^{2a}y^{2b} + y^{4b}) \div (x^{2a} + x^ay^b + y^{2b})$

41. Prove Theorem 6–4. **42.** Prove Theorem 6–5. **43.** Prove Theorem 6–6.

44. Is division of polynomials closed? Illustrate your answer.

45. Is division of polynomials commutative? Illustrate your answer.

Find the value of k that will make the second polynomial a factor of the first.

46. $2s^3 - 10s^2 + ks + 66$; $s - 3$ **47.** $-6t^4 + kt^3 + 3t^2 - t + 2$; $t + 2$

48. $kn^5 - 3n^4 + n^3 - 2n + 5$; $n - 1$ **49.** $32r^5 + k$; $2r + 1$

6–8 The Remainder and Factor Theorems

Compare the remainder obtained in the division problem on the left with the value of the polynomial when $x = 3$.

$$\begin{array}{r} 3x - 1 \\ x - 3 \overline{\smash{)}\, 3x^2 - 10x + 8} \\ \underline{3x^2 - 9x} \\ -x + 8 \\ \underline{-x + 3} \\ 5 \end{array} \qquad \begin{array}{l} 3x^2 - 10x + 8 \\ 3(3)^2 - 10(3) + 8 \\ 27 - 30 + 8 \\ 5 \end{array}$$

Same number

The remainder in the division problem and the value of the polynomial are the same. To see why, recall that saying the quotient of 21 divided by 5 is 4 with a remainder of 1 means

$$4 \cdot 5 + 1 = 21.$$

Similarly,

$$(3x - 1)(x - 3) \quad + \quad 5 \quad = 3x^2 - 10x + 8.$$

(Quotient)(Divisor) + Remainder = Dividend

Thus, the two expressions above have the same value for *all* replacements for x. In particular, when $x = 3$,

$$(3x - 1)(x - 3) + 5 = 5 \quad \text{and} \quad 3x^2 - 10x + 8 = 5.$$

This is expressed as a theorem for any polynomial in x of positive degree and any divisor $(x - a)$.

Theorem 6–7

Remainder Theorem

When a polynomial in x is divided by $(x - a)$, the remainder equals the value of the polynomial when $x = a$.

EXAMPLE 1 Use the Remainder Theorem to find the remainder when $x^3 + 2x^2 - 6x + 7$ is divided by $x + 3$.

Solution: [1] Write $x + 3$ in the form $x - a$. Then identify a.

$x + 3 = x - (-3)$ Thus, $a = -3$.

[2] Find the value of the polynomial when $x = -3$.

$$\begin{aligned} x^3 + 2x^2 - 6x + 7 &= (-3)^3 + 2(-3)^2 - 6(-3) + 7 \\ &= -27 + 18 + 18 + 7 \\ &= \mathbf{16} \end{aligned}$$

Replace x with (−3).

The remainder is **16**.

Example 1 shows you how to find the remainder in a division problem without dividing. You can use a procedure called **synthetic division** to find both the quotient and the remainder in a division problem.

EXAMPLE 2 Use synthetic division to find the quotient and the remainder when $4x^4 + 3x^2 - 2x + 1$ is divided by $x - 2$.

Solution: [1] Write the dividend in descending powers of x.

$4x^4 + 0x^3 + 3x^2 - 2x + 1$ ← **Write the missing x^3– term. with a coefficient of 0.**

[2] Write the coefficients of the dividend and the value of a on line 1 as shown.

$a = 2$ →						
2	4	0	3	−2	1	**Line 1**
		8 (4×2)	16 (8×2)	38 (19×2)	72 (36×2)	**Line 2**
	4	8	19	36	73	**Line 3**

$4 + 0 = 4$ $\quad 0 + 8 = 8 \quad$ $3 + 16 = 19$

$4 \times 2 = 8$ $\quad 8 \times 2 = 16 \quad$ $19 \times 2 = 38$, and so on.

[3] Read the coefficients of the quotient and the remainder on line 3.

Quotient: $4x^3 + 8x^2 + 19x + 36$ Remainder: 73

In Example 2, the quotient $4x^3 + 8x^2 + 19x + 36$ is sometimes called a **depressed polynomial.**

When the remainder in a division problem is 0, the divisor is a factor of the dividend (polynomial).

Theorem 6–8

Factor Theorem

The binomial $(x - a)$ is a factor of a polynomial in x if and only if the value of the polynomial is 0 when $x = a$.

The Factor Theorem can help you to identify the factors of a polynomial.

EXAMPLE 3 Determine whether $x + 4$ is a factor of $x^3 + 10x^2 + 29x + 20$.

Solution: Use synthetic division to divide.

$a = -4$ →				
−4	1	10	29	20
		−4	−24	−20
	1	6	5	**0** ← **The remainder is 0.**

Thus, **$x + 4$ is a factor** of $x^3 + 10x^2 + 29x + 20$.

When all the terms of a polynomial have integral coefficients and the coefficient of the term of highest degree is 1, then for $x - a$ to be a factor of the polynomial, a must divide evenly into the constant term.

Example 4 shows you how you can use this fact as well as the Factor Theorem and synthetic division to factor some polynomials.

EXAMPLE 4 Factor completely over the integers: $x^3 - 9x^2 + 24x - 20$.

Solution: [1] Write the possible factors of the constant term, 20.

$$\pm1, \pm2, \pm4, \pm5, \pm10, \pm20$$

[2] Use synthetic division to test each factor.

1	1	−9	24	−20	← **Test $a = 1$.**
		1	−8	16	
	1	−8	16	−4	← **Since the remainder is −4, $a - 1$ is *not* a factor.**
2	1	−9	24	−20	← **Test $a = 2$.**
		2	−14	20	
	1	−7	10	0	← **Since the remainder is 0, $a - 2$ is a factor.**

Thus, $x^3 - 9x^2 + 24x - 20 = (x^2 - 7x + 10)(x - 2)$ ← **Factor $x^2 - 7x + 10$.**

$= (x - 5)(x - 2)(x - 2)$

CLASSROOM EXERCISES

In Exercises 1–3, use the Remainder Theorem to find the remainder when each polynomial is divided by $(x - 3)$.

1. $2x^3 - 7x^2 - 4x + 5$ **2.** $2x^3 + 9x^2 + 8x - 3$ **3.** $x^4 + x^3 - 6x + 9$

Determine whether the second polynomial is a factor of the first. Answer *Yes* or *No*.

4. $y^3 - 8y^2 + 19y - 12;\ y - 4$

5. $z^3 - 27z + 10;\ z + 5$

6. $x^3 - 8;\ x - 2$

7. $a^4 - 27a^2 + 14a + 120;\ a - 12$

8. $m^4 - 8m^2 + 16;\ m + 2$

9. $y^3 + 81;\ y + 3$

WRITTEN EXERCISES

A In Exercises 1–6, use the Remainder Theorem to find the remainder.

1. $(x^3 + 2x^2 - 3x + 1) \div (x - 1)$

2. $(5n^3 - 2n^2 + n + 4) \div (n + 1)$

3. $(4t^3 - 3t^2 + 2t + 1) \div (t - 2)$

4. $(2a^3 - 4a^2 - 3a + 2) \div (a + 4)$

5. $(m^4 - 3m^3 - 2m^2 + m + 5) \div (m + 2)$

6. $(2x^4 - 4x^3 + x^2 + x + 1) \div (x - 3)$

In Exercises 7–14, use synthetic division to find the quotient and the remainder.

7. $(y^3 + y^2 - 20) \div (y + 2)$

8. $(a^4 + 6a^2 - 7a + 12) \div (a - 2)$

9. $(2b^3 - 7b^2 - 4b + 5) \div (b - 3)$

10. $(2z^3 + 10z^2 - 2z + 8) \div (z - 3)$

11. $(2x^4 - 3x^3 - 6x^2 - 8x - 3) \div (x - 3)$

12. $(2x^3 + 3x^2 + 4x - 8) \div (x - \frac{1}{2})$

13. $(6y^3 + 11y^2 - y - 3) \div (y - \frac{1}{2})$

14. $(2a^4 + 5a^3 - a^2 + 4a - 2) \div (a + 3)$

In Exercises 15–20, determine whether the second polynomial is a factor of the first.

15. $y^3 - 8y^2 + 19y - 12,\ y - 3$

16. $a^3 + 5a^2 + 8a + 4,\ a + 2$

17. $3x^4 - 2x^3 + 5x^2 - 2x + 1,\ x - 1$

18. $n^4 - 27n^2 + 14n + 120,\ n - 4$

19. $2m^3 + 3m^2 - 7m - 3,\ m - 2$

20. $6x^4 - 7x^3 + 3x^2 - 7x - 9,\ x + 1$

Determine whether the second polynomial is a factor of the first. If it is, divide to find the quotient. Then factor the quotient.

21. $x^3 - 10x^2 + 27x - 18,\ x - 1$

22. $s^3 - 2s^2 - s + 2,\ s - 2$

23. $z^3 - 8z^2 + 21z - 18,\ z - 3$

24. $y^3 + 10y^2 + 29y + 20,\ y + 4$

Factor over the integers.

25. $y^3 - 8y^2 + 19y - 12$

26. $a^3 + 5a^2 + 8a + 4$

27. $y^3 - 11y^2 + 31y - 21$

28. $s^3 - 6s^2 + 11s - 6$

29. $x^4 - 8x^2 + 16$

30. $x^3 - 27x + 10$

31. $x^4 - 27x^2 + 14x + 120$

32. $x^4 - 10x^3 + 35x^2 - 50x + 24$

33. $a^3 + 1$

34. $m^3 - n^3$

35. $c^3 - 1$

36. $b^3 + 8c$

37. $a^3 - 27b^3$

38. $27x^3 + 8y^3$

39. $a^3b^3 + 27c^3$

40. $64 - x^3$

B Find the value of k that makes the second polynomial a factor of the first.

41. $x^3 + 3x^2 - x + k;\ x - 2$

42. $kx^3 - 2x^2 + x - 6;\ x + 3$

43. $n^4 - kn^2 + 5n - 1;\ n - 1$

44. $a^3 - ka^2 + 2a + k - 3;\ a + 2$

45. $c^3k^2 + 4ck - 5;\ c - 1$

46. $n^4 - 2n^3 + kn^2 - k^2n + k - 6;\ n + 1$

REVIEW CAPSULE FOR SECTION 6–9

Solve. *(Pages 26–29)*

1. The sum of two numbers is 12. Let x represent one number. Represent the second number in terms of x.
2. The larger of two integers exceeds the smaller by 3. Represent the integers in terms of y.
3. The smaller of two integers is 5 less than twice the larger. Represent the integers in terms of t.
4. The sum of three consecutive odd integers is 39. Represent the integers in terms of g.
5. The sum of two consecutive even integers is 132. Represent the integers in terms of z.

Square each binomial as indicated. Simplify when possible. *(Pages 193–196)*

6. $(x - 5)^2$ **7.** $(8 - x)^2$ **8.** $b^2 + (b + 1)^2$ **9.** $y^2 + (y + 3)^2$ **10.** $(r + 6)^2 - r^2$

6–9 Solving Polynomial Equations by Factoring

Some polynomial equations can be solved by factoring. The method applies the following theorem.

Theorem 6–9

> **Zero Product Theorem**
>
> If a and b are real numbers and $a \cdot b = 0$, then either $a = 0$ or $b = 0$ or both a and b equal 0.

EXAMPLE 1 Solve and check: $x^2 + 5x + 6 = 0$.

Solution:

$x^2 + 5x + 6 = 0$ ← **Factor.**

$(x + 3)(x + 2) = 0$ ← **If $a \cdot b = 0$, then $a = 0$ or $b = 0$.**

$x + 3 = 0$ or $x + 2 = 0$

$x = -3$ | $x = -2$

Check: Replace x with (-3).

$$x^2 + 5x + 6 = 0$$
$$(-3)^2 + 5(-3) + 6 \stackrel{?}{=} 0$$
$$9 - 15 + 6 \stackrel{?}{=} 0$$
$$0 \stackrel{?}{=} 0 \quad \text{Yes} \checkmark$$

Replace x with (-2).

$$x^2 + 5x + 6 = 0$$
$$(-2)^2 + 5(-2) + 6 \stackrel{?}{=} 0$$
$$4 - 10 + 6 \stackrel{?}{=} 0$$
$$0 \stackrel{?}{=} 0 \quad \text{Yes} \checkmark$$

Solution set: $\{-3, 2\}$

Unless otherwise stated, the replacement set for the variable is the set of real numbers.

EXAMPLE 2 Solve and check: $x^3 - 4x^2 + x + 6 = 0$.

Solution: [1] Use synthetic division to find the factors of $x^3 - 4x^2 + x + 6$. Try $a = -1$.

$$\begin{array}{r|rrrr} -1 & 1 & -4 & 1 & 6 \\ & & -1 & 5 & -6 \\ \hline & 1 & -5 & 6 & 0 \end{array}$$

← **Since the remainder is 0, $x + 1$ is a factor.**

Thus, $x^3 - 4x^2 + x + 6 = (x + 1)(x^2 - 5x + 6)$ ← **Factor $x^2 - 5x + 6$.**

$= (x + 1)(x - 3)(x - 2)$

[2] Solve the equation.

$$x^3 - 4x^2 + x + 6 = 0$$
$$(x + 1)(x - 3)(x - 2) = 0$$

$x + 1 = 0$ or $x - 3 = 0$ or $x - 2 = 0$

$x = -1$ | $x = 3$ | $x = 2$

The check is left for you. **Solution set:** $\{-1, 2, 3\}$

Sometimes you can use factoring to solve number problems.

EXAMPLE 3 Find two numbers whose sum is 8 and the sum of whose squares is 34.

Solution: Let x = one number.
Then $8 - x$ = the other number. ← **Use the unknowns to write an equation.**

$$x^2 + (8 - x)^2 = 34$$

$$x^2 + 64 - 16x + x^2 = 34 \quad \leftarrow \textbf{Combine like terms.}$$

$$2x^2 - 16x + 64 = 34 \quad \leftarrow \textbf{Add } -34 \textbf{ to each side.}$$

$$2x^2 - 16x + 30 = 0 \quad \leftarrow \textbf{Factor.}$$

$$2(x^2 - 8x + 15) = 0$$

$$2(x - 5)(x - 3) = 0$$

$$x - 5 = 0 \quad \text{or} \quad x - 3 = 0$$

$$x = 5 \quad \quad x = 3 \quad \leftarrow \mathbf{8 - 5 = 3 \text{ and } 8 - 3 = 5.}$$

The numbers are **5** and **3**.
REMEMBER: to check in the original conditions of the problem.

CLASSROOM EXERCISES

Solve and check.

1. $x^2 - 2x - 63 = 0$
2. $y^2 - 9 = 0$
3. $25x^2 - 1 = 0$
4. $t^2 - 5t + 4 = 0$
5. $6r^2 - 7r - 5 = 0$
6. $144q^2 - 1 = 0$
7. $\frac{1}{4}s^2 - \frac{25}{169} = 0$
8. $3x^2 - 10x + 8 = 0$
9. $2x^2 - 5x - 3 = 0$

WRITTEN EXERCISES

A Solve and check.

1. $x^2 - x - 6 = 0$
2. $x^2 - 4 = 0$
3. $x^2 - 9x + 20 = 0$
4. $x^2 - 5x - 14 = 0$
5. $4x^2 - 1 = 0$
6. $9x^2 - 1 = 0$
7. $2x^2 + 9x - 5 = 0$
8. $x^2 + 6x - 27 = 0$
9. $x^2 - 121 = 0$
10. $3x^2 - 4x - 15 = 0$
11. $6x^2 + 7x + 2 = 0$
12. $x^3 - 6x^2 + 11x - 6 = 0$
13. $x^3 - 7x - 6 = 0$
14. $x^3 + 6x^2 + 11x + 6 = 0$
15. $x^3 - 2x^2 - x + 2 = 0$
16. $x^3 + 7x^2 + 7x - 15 = 0$
17. $x^3 - x^2 - 4x + 4 = 0$
18. $a^3 - 9a^2 + 26a - 24 = 0$
19. $2y^3 + y^2 - 13y + 6 = 0$
20. $b^3 + b^2 - 9b - 9 = 0$
21. $x^4 - 10x^2 + 9 = 0$

APPLICATIONS: Using Factoring

22. Find two numbers whose sum is 9 and the sum of whose squares is 41.
23. Find two numbers whose sum is 23 and the sum of whose squares is 269.

24. Find two numbers whose sum is 8 and whose product is 15.

25. Find two numbers whose product is 480 and whose difference is 4.

B

26. The larger of two integers is 3 more than the smaller. The sum of the squares of the two integers is 89. Find the integers.

27. The square of the first of three consecutive odd integers is 9 more than 6 times the sum of the second and third. Find the integers.

28. Two positive integers are in the ratio 1:2. The product of the integers added to their sum equals 5. Find the integers.

C

29. Prove Theorem 6–9.

CALCULATOR APPLICATIONS

Factoring Polynomials

To use a calculator to factor polynomials, you first rewrite the polynomial in a special way.

EXAMPLE Factor $x^3 - 9x^2 + 24x - 20$.

SOLUTION Write the polynomial by repeatedly factoring out x.

$$x^3 - 9x^2 + 24x - 20 = [x^2 - 9x + 24]x - 20$$
$$= [(\mathbf{x} - \mathbf{9})\mathbf{x} + \mathbf{24}]\mathbf{x} - \mathbf{20}$$

Replace x in the factored expression with a trial factor, such as 5.

5 [−] 9 [=] [×] 5 [+] 2 4 [=]

[×] 5 [−] 2 0 [=]　　　0.

Since there is no remainder, **(x − 5) is a factor.** Now try 4 as a factor.

4 [−] 9 [=] [×] 4 [+] 2 4 [=]

[×] 4 [−] 2 0 [=]　　　-4.

Thus, **(x − 4) is not a factor.**

EXERCISES

1. Find one other factor for $x^3 - 9x^2 + 24x - 20$.

Use a calculator to factor each polynomial.

2. $x^3 - 3x^2 - 6x + 8$
3. $x^3 - x^2 - 25x + 25$
4. $x^3 - x^2 - 17x - 15$
5. $x^3 + 4x^2 - 17x - 60$
6. $x^3 + 6x^2 + 11x + 6$
7. $x^3 - 3x^2 - 10x + 24$

Problem Solving and Applications

6–10 Area

Many problems involving area can be represented by a polynomial equation. The domain of the variable in these problems is often restricted by practical reasons.

EXAMPLE 1 Floor tiles are made with a square mold that can be adjusted to form other rectangles.

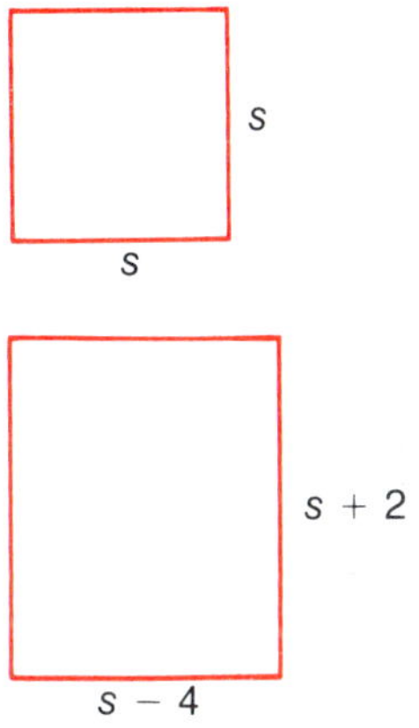

A manufacturer decreases the width of a mold by 4 centimeters and increases the length by 2 centimeters (Condition 1). The resulting tile has an area of 832 square centimeters (Condition 2). Find the length of the original side of the mold.

Solution: [1] Use Condition 1 to represent the variables.

Let s = the length of a side of the mold.
Then the length of the tile is $s + 2$.
The width of the tile is $s - 4$.

[2] Use Condition 2 to write an equation for the problem.

$(s + 2)(s - 4) = 832$ ⟵ **Area = length × width.**

$s^2 - 2s - 8 = 832$

$s^2 - 2s - 840 = 0$ ⟵ **Factor.**

$(s - 30)(s + 28) = 0$

$s - 30 = 0$ or $s + 28 = 0$

$s - 30$ | $s = -28$ ⟵ **The length cannot be negative.**

Thus, the length of the original side of the mold was **30 centimeters.**

Example 2 shows how to solve another type of area problem.

EXAMPLE 2 A swimming pool having the shape of a rectangle is 80 feet long and 50 feet wide (Condition 2). A cement walk of uniform width (Condition 1) and having an area of 1104 square feet (Condition 2) surrounds the pool. Find the width of the walk.

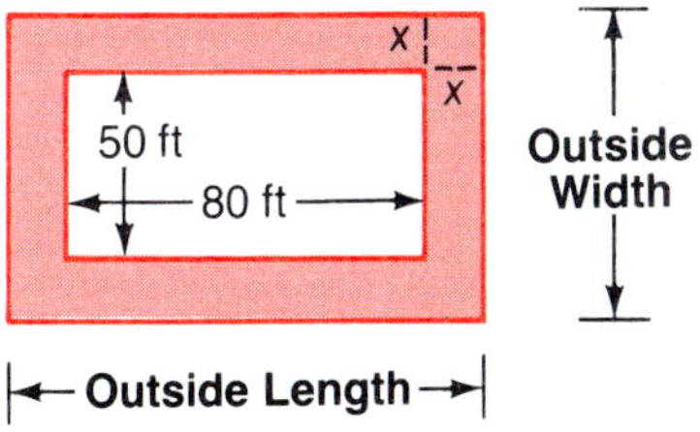

Solution: [1] Let x = the width of the walk.
Then the outside length is $80 + 2x$.
The outside width is $50 + 2x$.

[2] $(80 + 2x)(50 + 2x) - 1104 = (50)(80)$ ← Total Area − Area of Walk = Area of Pool.

$4000 + 160x + 100x + 4x^2 - 1104 = 4000$ ← Combine like terms.

$4x^2 + 260x - 1104 = 0$

$4(x^2 + 65x - 276) = 0$

$4(x - 4)(x + 69) = 0$

$x - 4 = 0$ or $x + 69 = 0$

$x = 4$ $\qquad$ $x = -69$ ← The width cannot be negative.

The width of the walk is **4 feet.**

WORD PROBLEMS

A Solve each problem.

1. A painter decreases the size of a square piece of canvas by cutting 3 centimeters from the length and 5 centimeters from the width. The resulting piece has an area of 120 square centimeters. Find the length of the original piece.
2. A rectangular oil painting is 15 inches wide and 36 inches long. The painting is enclosed by a frame of uniform width. The area of the frame is 472 square inches. Find the width of the frame.

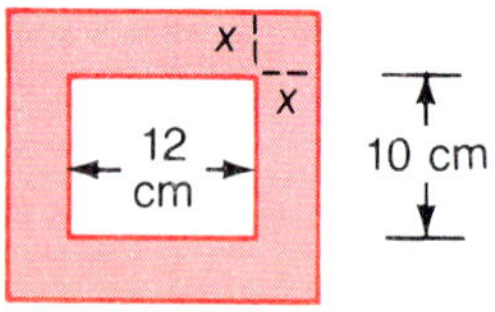

3. Matting for framing is to be cut to include a picture at its center as shown. The total area is three times the area of the picture. Find the width of the matting.

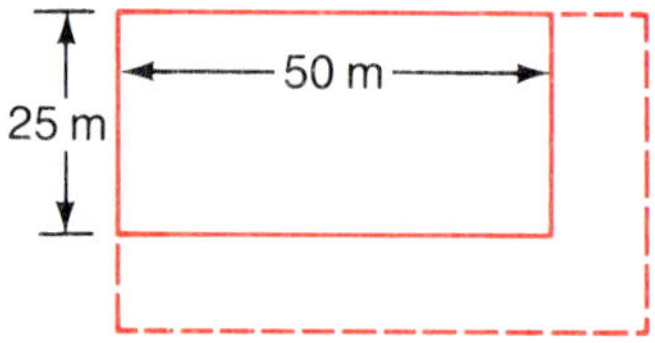

4. Art's vegetable garden is in the shape of a rectangle that is 50 meters long and 25 meters wide. Art adds strips of uniform width to two sides as shown, so that the total area is 2100 square meters. Find the width of the strips.
5. A rectangular ice skating rink is 30 meters wide and 60 meters long. Strips of equal width are cut from two adjacent sides of the rink so that the resulting area is 800 square meters less than the original area. Find the width of the strips.
6. A rectangular picture frame has an area of 240 square centimeters. A photograph 10 centimeters long and 8 centimeters wide is to be placed in the frame so that the distance from the top of the photograph to the top of the frame is twice the distance from one side of the photograph to the edge of the frame. Find these distances.
7. A rectangular urban development project is 3 kilometers wide and 4 kilometers long. A wooded strip of uniform width borders the development. The area of the wooded strip is 10 times that of the development. Find the width of the wooded strip.

BASIC: FACTORING TRINOMIALS

In BASIC, deciding whether one number is a factor of another number involves the INT ("integer") function. INT(X) gives the largest integer less than or equal to X. For example, INT(6.5) = 6; INT(-3.2) = −4. To determine whether an integer A is a factor of an integer B, test whether B/A = INT(B/A). For instance, if B = 36 and A = 4, then B/A = 9 and INT(B/A) = 9. Since B/A = INT(B/A), 4 is a factor of 36.

Problem: *Write a* BASIC *program to find the factors over the integers of trinomials of the form* $Ax^2 + Bx + C$ $(A \neq 0)$. *Allow for the possibility that a trinomial may not be factorable.*

```
100 REM   FACTOR A*X*X + B*X + C, IF POSSIBLE, INTO
110 REM   (D*X + E)(F*X + G)
120 REM
130 PRINT "ENTER THE COEFFICIENTS OF THE TRINOMIAL";
140 INPUT A, B, C
145 IF A = 0 THEN 130
150 FOR Q = 1 TO ABS(A)
160 IF A/Q <> INT(A/Q) THEN 340
170 LET D = Q
180 LET F = A/D
190 FOR R = 1 TO ABS(C)
200 IF C/R <> INT(C/R) THEN 330
210 LET E = R
220 LET G = C/R
230 IF D * G + E * F = B THEN 370
240 IF E = G THEN 300
250 LET E = G
260 LET G = R
270 IF D * G + E * F = B THEN 370
280 IF C < 0 THEN 330
290 IF B > 0 THEN 330
300 LET E = -E
310 LET G = -G
320 IF D * G + E * F = B THEN 370
330 NEXT R
340 NEXT Q
350 PRINT "CANNOT BE FACTORED OVER THE INTEGERS"
360 GOTO 380
370 PRINT "D =";D;" E =";E;" F =";F;" G =";G
380 PRINT "ANY MORE TRINOMIALS TO FACTOR (1 = YES, 0 = NO)";
390 INPUT Z
400 IF Z = 1 THEN 130
410 END
```

Analysis

Statements 150–340: This FOR-NEXT loop searches for a value of Q that is a factor of A. When such a value is found, D is set equal to Q (statement 170) and F to A/D (statement 180). Then another loop is started (statement 190) to search for a factor R of C. If a factor R is found, E is set equal to R (statement 210) and G to C/R (statement 220). In statement 230, the "cross product" DG + EF is tested to see whether it equals B, the coefficient of the middle term of the trinomial. If it does, the computer jumps to statement 370 and prints D, E, F, and G. If it does not, the values of E and G are interchanged (statements 250–260) and the cross product tested again in statement 270. Another possibility is that $C < 0$, and E and G are opposite in sign. In this case, the computer jumps to statement 330 and tries a new value of R. On the other hand, if $C > 0$ and $B < 0$ (statement 290), the negatives of E and G might produce the correct factors (statements 300–320).

Statement 350: If all possible values of Q in combination with all possible values of R have been tried and failed, then the trinomial cannot be factored over the integers.

EXERCISES

A Run the program on page 221 for the following trinomials.

1. $2x^2 + 13x + 15$ **2.** $2x^2 - 13x$ **3.** $x^2 - 16$

4. $6x^2 + 13x - 28$ **5.** $6x^2 - 13x - 28$ **6.** $30x^2 + 27x - 6$

Write a BASIC program for each problem.

7. Given integers x and y, determine whether x is a factor of y.

8. Expand the program in Exercise 7 so that it also determines whether y is a factor of x.

9. Given two positive integers, print their **greatest common factor** (GCF).

B

10. Two positive integers are **relatively prime** if their greatest common factor is 1. Given two positive integers, determine whether they are relatively prime.

11. Reverse the problem on page 221. That is, given D, E, F, and G, compute the product $(Dx + E)(Fx + G)$. Print the trinomial "neatly." For example, print `X↑2 + 5 X - 7` and not `1X↑2 + 5 X + -7`; print `6X↑2 - 4 X` and not `6X↑2 + -4 X + 0`, and so on.

12. Given three positive integers, print their greatest common factor (GCF).

C

13. Expand the program on page 221 so that it prints the factors in the form $(Dx + E)(Fx + G)$. For example, if 2, −13, 15 are the input values, the output will be `(X - 5)(2 X - 3)`. Printing this output as `(1 X + -5) (2 X + -3)` is not acceptable.

Review

Divide. No denominator equals zero. *(Section 6–7)*

1. $\frac{b^5}{b^2}$ **2.** $\frac{33a^6}{11a^8}$ **3.** $\frac{42x^{12} - 18x^4}{3x^2}$ **4.** $\frac{21y^6 - 14y^4 + 28y^2}{-7y}$

5. $(x^2 - 12x + 20) \div (x - 2)$ **6.** $(6b^3 - 8b^2 - 35b - 18) \div (3b + 2)$

Determine whether the second polynomial is a factor of the first. *(Section 6–8)*

7. $y^3 + 2y^2 - 14y + 12,\ y - 2$ **8.** $t^4 + 9t^3 + 16t^2 + t + 21,\ t + 3$

Factor over the integers. *(Section 6–8)*

9. $a^3 - 8a^2 + 15a - 18$ **10.** $s^3 - 6s^2 - 4s + 24$

Solve and check. *(Section 6–9)*

11. $x^2 - 16 = 0$ **12.** $x^2 + 3x - 10 = 0$ **13.** $b^3 - 6b^2 - b + 30 = 0$

14. Find two numbers whose sum is 10 and the sum of whose squares is 52. *(Section 6–9)*

15. A walk of uniform width is being built around a rectangular garden with dimensions 20 meters by 25 meters. The area of the walk will be 196 square meters. Find the width of the walk. *(Section 6–10)*

Chapter Summary

IMPORTANT TERMS

Binomial (p. 188)
Coefficient (p. 188)
Degree of a polynomial (p. 188)
Depressed polynomial (p. 213)
Exponent (p. 188)
Like terms (p. 190)
Monomial (p. 188)
Perfect square trinomial (p. 197)
Polynomial (p. 188)
Prime polynomial (p. 197)
Quadratic trinomial (p. 201)
Trinomial (p. 188)

IMPORTANT IDEAS

1. **Exponent Theorem for Multiplication:** If m and n are positive integers and a is a real number, then

$$a^m \cdot a^n = a^{m+n}.$$

2. **Power Theorems:** If m and n are positive integers and a and b are real numbers, then

$$(a^m)^n = a^{mn}$$

and

$$(ab)^m = a^m b^m.$$

3. Special binomial products

Square of a Binomial Sum: $(a + b)^2 = a^2 + 2ab + b^2$

Square of a Binomial Difference: $(a - b)^2 = a^2 - 2ab + b^2$

Product of a Sum and a Difference: $(a + b)(a - b) = a^2 - b^2$

4. Factoring perfect square trinomials

$$a^2 + 2ab + b^2 = (a + b)^2$$
$$a^2 - 2ab + b^2 = (a - b)^2$$

5. Factoring the sum or difference of two cubes

$$a^3 + b^3 = (a + b)(a^2 - ab + b^2)$$
$$a^3 - b^3 = (a - b)(a^2 + ab + b^2)$$

6. **Exponent Theorems for Division:** For nonnegative integers m and n, and for any real number a, $a \neq 0$:

$$\frac{a^m}{a^n} = a^{m-n} \quad \text{where } m > n$$

$$\frac{a^m}{a^n} = \frac{1}{a^{n-m}} \quad \text{where } m < n$$

$$\frac{a^m}{a^m} = a^0 = 1$$

7. **Remainder Theorem:** When a polynomial in x is divided by $(x - a)$, the remainder equals the value of the polynomial when $x = a$.

8. **Factor Theorem:** The binomial $(x - a)$ is a factor of a polynomial in x if and only if the value of the polynomial is 0 when $x = a$.

9. You can use synthetic division to find both the quotient and the remainder in a division problem.

10. **Zero Product Theorem:** If a and b are real numbers and $a \cdot b = 0$, then either $a = 0$ or $b = 0$ or both a and b equal 0.

Chapter Objectives and Review

Objective: *To identify the degree of a polynomial (Section 6–1)*

In each polynomial, identify the term of greatest degree. Then write the degree of the polynomial.

1. $3y^2 - 5x^3$
2. $3yz - x^5 + \frac{1}{2}z^2$
3. $x^5y^2 + x^8 - y^3z^6$

Objective: *To add and subtract polynomials (Section 6–2)*

Add or subtract as indicated.

4. $(y^2 + x) + (3x - 7)$
5. $(t^2 - 4t^4 + 5) + (t^3 - 3t^2 + 7t)$
6. $(4s - r^2s) - (2r + 3r^2s - s)$
7. $(u - w^5 + u^2w) - (3u - 5u^2w - 7w^5)$

Objective: *To multiply polynomials (Section 6–3)*

Find each product.

8. $(-s^2)(2r^2s)$ **9.** $(3a^3b^2)(-ab^4)$ **10.** $bc(4b^2 - c)$

11. $(2x - 5)(x + 4)$ **12.** $(x + 3)(x^2 - 2x + 7)$ **13.** $(y - 3)^3$

Objective: *To factor a given polynomial over the integers (Section 6–4)*

Factor over the integers.

14. $7z^2t - 21z^2 + zt^2$ **15.** $9x - 3x^2y$ **16.** $b^3c - 2b^2c^4 + 3b^4c^2$

17. $r^3s^2 + 3s^3 - 6rs^2$ **18.** $4(t + 3s) - b(t + 3s)$ **19.** $z^3 - 2z^2 + 5z - 10$

20. $x^2 - 4a^2$ **21.** $81 - 16t^2$ **22.** $c^2d^2 - 121$

Objective: *To factor a quadratic trinomial over the integers (Section 6–5)*

Factor over the integers.

23. $b^2 + 6b - 7$ **24.** $2x^2 - 5x - 3$ **25.** $14x^2 - 3x - 5$

Objective: *To factor a polynomial as the sum or difference of two cubes (Section 6–6)*

Factor over the integers.

26. $27 - y^3$ **27.** $64 + y^6$ **28.** $125a^3 + 1$ **29.** $2r^4 - 54r$

Objective: *To find the quotient of two polynomials (Section 6–7)*

Divide. No denominator equals zero.

30. $\frac{y^3}{y^5}$ **31.** $\frac{9a^2}{3a^6}$ **32.** $\frac{p^3 - 2p}{p}$ **33.** $\frac{3zt + 21z^2t}{3t}$

34. $(d^2 - 4d - 21) \div (d + 3)$ **35.** $(2x^2 - 27x - 45) \div (x - 15)$

Objective: *To use the Remainder Theorem (Section 6–8)*

Use the Remainder Theorem to find the remainder.

36. $(x^3 + 4x^2 - 7x + 3) \div (x - 1)$ **37.** $(n^4 + 2n^3 + n^2 + n - 6) \div (n + 3)$

Objective: *To use synthetic division (Section 6–8)*

Use synthetic division to find the quoteint and remainder.

38. $(x^3 - 7x^2 + 4x - 3) \div (x - 2)$ **39.** $(b^4 - 3b^2 + 5b - 1) \div (b + 3)$

Objective: *To use the Factor Theorem (Section 6–8)*

Determine whether the second polynomial is a factor of the first. If it is, divide to find the quotient. Then factor the quotient.

40. $s^3 - 4s^2 - s + 4;\ s - 4$ **41.** $y^3 + 6y^2 - 13y - 40;\ y - 3$

Objective: *To solve polynomial equations by factoring (Section 6–9)*

Solve and check.

42. $x^2 - 2x - 63 = 0$

43. $x^3 - 2x^2 - 5x + 6 = 0$

Objective: *To use polynomial equations to solve area problems (Section 6–10)*

44. A rectangular flower bed is 7 meters long and 3 meters wide. It is surrounded by a grass border of uniform width. The area of the grass border is 24 square meters. How wide is the grass border?

45. A vacant lot is 35 feet wide and 50 feet long. Strips of equal width are cut from the north and east sides of the lot, to make a concrete sidewalk. The area left for building is 246 square meters less than the original area. Find the width of the strip.

Chapter Test

In each polynomial, identify the term of greatest degree. Then write the degree of the polynomial.

1. $1.3x^5 - 9y^3$

2. $3z - t^5 + z^2t^4 + z^4t$

3. Add: $(x^3 - 5x^2 - 6x + 7) + (2x^2 - 14 + 3x - x^3)$

4. Subtract: $(3m^4 - 2m^3 + 6m + 1) - (2m^4 - 3m^2 + 8m - 5)$

Find the product.

5. $2tv^2(t^3 - 3v + t^2v - \frac{1}{2}t)$

6. $(3x + 7)(x - 6)$

Factor over the integers.

7. $3ab^2 + 6b^2$

8. $x^2 - x - 20$

9. $6x^2 + 41x - 7$

10. $3a + 3b - xa - xb$

11. $x^3 + 2x^2 + 3x + 6$

12. $4a^2 - 25a^2b^2$

13. Divide $6x^2 - x + 2$ by $2x + 1$. The denominator is not equal to zero.

Use the Remainder Theorem to find the remainder when the first polynomial is divided by the second.

14. $x^4 - 3x^3 + 7$; $x + 2$

15. $-2x^3 + x^2 - 4$; $x - 3$

Use the Factor Theorem to determine whether $x + 2$ is a factor of each polynomial.

16. $2x^2 - 5x - 3$

17. $2x^3 - 9x^2 - 11x + 30$

18. Use synthetic division to find the quotient and remainder when $-x^4 + 7x^2 - 4 + 3x$ is divided by $x + 2$.

19. Solve and check: $6t^2 + 8t - 8 = 0$.

20. A rectangular picture frame is 40 inches long and 32 inches wide. All sides of the frame are of uniform width, and it holds a picture of area 1140 square inches. How wide are the sides of the frame?

More Challenging Problems

1. If $a \neq b$, show that $a^2 + b^2 > 2ab$.

2. $x^2 + 2x + 5$ is a factor of $ax^4 + bx^2 + c$. What are the values of a, b, and c?

3. The expression $n^3 - n$ is divisible by some largest integer b for all possible values of n. Find b.

4. If $x = 2$, what is the value of $x + x(x^x)$?

5. If $(x - y)^2 = 30$ and $xy = 17$, find $x^2 + y^2$.

6. When $5x^{13} + 3x^{10} - k$ is divided by $x + 1$, the remainder is 20. Find k.

7. When simplified, how many terms are there in the expansion of $[(x + 3y)^2(x - 3y)^2]^2$?

8. Show that if $a + b + c = 1$, then $(a + b)(a + c) = a + bc$.

9. Solve for x: $(3x - 2)(2x - 3) = 2(x - 3)(3x + 5) + 1$.

10. Show that $9x + 5y$ is a multiple of 17 whenever $2x + 3y$ is a multiple of 17.

11. Knowing that a man was born in the first half of the nineteenth century and that he was x years old in the year x^2, can you determine the year of his birth?

12. When 25, a perfect square, is divided by 8, the least positive remainder is 1. Find, if possible, an integer that is a perfect square and whose least positive remainder is 3 when it is divided by 8.

13. Dan can do a task in 10 hours, Pinky in 12 hours, and José in 13 hours. How long would it take to do the job if Dan and Pinky work for 3 hours and then Pinky and José finish the job?

Preparing for College Entrance Tests

Given a system of equations (or a compound sentence with "and"), you can sometimes answer questions on College Entrance tests without solving the system. The key is in recognizing a binomial product or the factors of such a product.

EXAMPLE If $x + y = 2$ and $x - y = 1$, find $x^2 - y^2$.

(a) 1 (b) $1\frac{3}{4}$ (c) 2 (d) 3

Think: $x^2 - y^2 = (x + y)(x - y)$

Solution: $x^2 - y^2 = (x + y)(x - y)$ ← **From the given equations: $x + y = 2$; $x - y = 1$**

$= (2)(1) = 2$ **Answer: c**

Choose the best answer. Choose *a*, *b*, *c*, or *d*.

1. If $m + n = 9$ and $m - n = 4$, find $m^2 - n^2$.
   (a) 4 (b) 5 (c) 36 (d) 65
2. If $a + 2b = 6$ and $a - 2b = 3$, find $a^2 - 4b^2$.
   (a) 0 (b) 18 (c) 25 (d) 36
3. If $x + y = \frac{1}{2}$ and $x - y = -\frac{3}{4}$, find $x^2 - y^2$.
   (a) $-\frac{5}{16}$ (b) $-\frac{3}{8}$ (c) $\frac{1}{2}$ (d) $\frac{5}{4}$
4. If $x - (y + 1) = 8$ and $x + (y + 1) = 6$, find $x^2 - (y + 1)^2$.
   (a) 28 (b) 45 (c) 48 (d) 50
5. If $-t + s = 5$ and $t + s = 5$, find $t^2 - s^2$.
   (a) -25 (b) 0 (c) 25 (d) 50
6. If $p^2 - 9q^2 = 40$ and $p - 3q = 4$, find $p + 3q$.
   (a) $3\frac{1}{3}$ (b) 10 (c) 30 (d) 36
7. If $p + q = 5$ and $2pq = 8$, find $p^2 + q^2$. (HINT: $(p + q)^2 = p^2 + 2pq + q^2$)
   (a) 1 (b) 9 (c) 13 (d) 17
8. If $m - n = 8$ and $2mn = -24$, find $m^2 + n^2$.
   (a) 40 (b) 52 (c) 88 (d) 192
9. If $xy = 24$ and $x^2 + y^2 = 73$, find $(x - y)^2$.
   (a) 5 (b) 25 (c) 49 (d) 61
10. If $x + y = 4$ and $x - y = 2$, find $x^2 + y^2$. (HINT: Solve the system.)
    (a) 4 (b) 6 (c) 8 (d) 10
11. If $x + y = 9$ and $y - x = 7$, find $x^2 + y^2$.
    (a) 8 (b) 1 (c) 65 (d) 64

CHAPTER **7**

Rational Expressions

7–1 Simplifying Rational Expressions

Rational numbers and rational expressions are similar. *Rational expressions* are usually written as the ratio of two polynomials.

Rational Numbers	Rational Expressions
$\frac{314}{100}$, or 3.14	$\frac{2a^2+5a+1}{a+6}$, $a \neq -6$
$\frac{-17}{1}$, or -17	$\frac{2m-7}{1}$, or $2m-7$

Definition

A **rational expression** is an expression of the form $\frac{P}{Q}$, where P and Q are polynomials, and $Q \neq 0$.

A rational expression is **simplified** or **reduced to lowest terms** when its numerator and denominator have no common factors except 1 and -1. Theorem 7–1 is used to simplify rational expressions.

Theorem 7–1

If the numerator and denominator of a rational expression are multiplied or divided by the same nonzero polynomial, then the original rational expression and the resulting expression are equivalent.

To simplify a rational expression, first *identify the greatest common factor* (GCF) of the numerator and denominator. Then divide *both* the numerator and denominator by the GCF.

EXAMPLE 1 Simplify: $\frac{10x^4yz^2}{16xy^2z^2}$

Solution:

$$\frac{10x^4yz^2}{16xy^2z^2} = \frac{(2xyz^2)(5x^3)}{(2xyz^2)(8y)} \quad \longleftarrow \textbf{GCF: } 2xyz^2$$

$$= \frac{2xyz^2}{2xyz^2} \cdot \frac{5x^3}{8y} = 1 \cdot \frac{5x^3}{8y}, \text{ or } \mathbf{\frac{5x^3}{8y}}$$

Sometimes the greatest common factor is a polynomial.

EXAMPLE 2 Simplify: $\frac{x^2-9}{x^2+x-6}$

Solution:

$$\frac{x^2-9}{x^2+x-6} = \frac{\overset{1}{\cancel{(x+3)}}(x-3)}{\underset{1}{\cancel{(x+3)}}(x-2)} \quad \longleftarrow \textbf{GCF: } (x+3)$$

$$= \mathbf{\frac{x-3}{x-2}} \quad \longleftarrow \frac{1(x-3)}{1(x-2)}$$

The key in Example 3 is in recognizing that

$$1 - a = (-1)(a-1) = -(a-1).$$

EXAMPLE 3 Simplify: $\frac{a^2-1}{1-a}$

Solution: $\frac{a^2-1}{1-a} = \frac{(a+1)(a-1)}{1-a}$ ← $(1-a) = (-1)(a-1)$

$= \frac{(a+1)\overset{1}{\cancel{(a-1)}}}{(-1)\underset{1}{\cancel{(a-1)}}}$ ← GCF: $(a-1)$

$= \frac{a+1}{-1}$, or $-(a+1)$

Assume throughout the rest of this book that no denominator in a rational expression equals zero.

CLASSROOM EXERCISES

Simplify each rational expression.

1. $\frac{105xy}{60xy}$
2. $\frac{35xy^2r^2s^2}{140y^2rs}$
3. $\frac{ab^5}{a^3b}$
4. $\frac{a^2(a-5)}{a(a^2-25)}$
5. $\frac{b(4-b^2)}{b^2(2+b)}$
6. $\frac{(12+6a)(25-a^2)}{(15-3a)(4-a^2)}$
7. $\frac{-3(x-2)}{9(2-x)}$
8. $\frac{3(x^2-1)}{(x^2+1)(1-x)}$

WRITTEN EXERCISES

A Simplify each rational expression.

1. $\frac{6y}{y^2}$
2. $\frac{ax}{ay}$
3. $\frac{ax^2y}{ax^2z}$
4. $\frac{c(x+y)}{d(x+y)}$
5. $\frac{r(x-y)^2}{t(x-y)}$
6. $\frac{-3x^2y^3}{6x^3y^2}$
7. $\frac{12(a+b)}{3(2a+b)}$
8. $\frac{6abx}{9a^2x^2}$
9. $\frac{2a^3b^2x^3}{6a^2b^2x}$
10. $\frac{10x^4yz^2}{16xy^2z^2}$
11. $\frac{60x^2yz^4}{24x^2z^4}$
12. $\frac{ax+ay}{bx+by}$
13. $\frac{x^2-y^2}{x^2+2xy+y^2}$
14. $\frac{x^2-16}{x^2+6x+8}$
15. $\frac{x^2-5x}{x^2-7x+10}$
16. $\frac{-(x-2)(x+3)(x+4)}{(x-2)(x+3)(x+4)}$
17. $\frac{2x-8}{4-x}$
18. $\frac{-x-y}{-(x+y)}$
19. $\frac{ax-ay}{y-x}$
20. $\frac{a^2-1}{a^2+3a+2}$
21. $\frac{a^2-b^2}{b-a}$
22. $\frac{-3a^2+12}{a^2-4a+4}$
23. $\frac{(x+y)(2x-2y)}{(y-x)(3x+3y)}$
24. $\frac{a(b^2-a^2)}{(a-b)^2}$

Simplify each rational expression.

25. $\frac{2y^2 - 18}{18 - 6y}$

26. $\frac{2a^2 - ab - 6b^2}{3a^2 - 7ab + 2b^2}$

27. $\frac{15x^2 + 2xy - y^2}{12x^2 + xy - y^2}$

28. $\frac{a^2 - 7a + 12}{a^2 - 2a - 3}$

29. $\frac{3t^4 - 21t^3 + 36t^2}{t^3 - 8t^2 + 15t}$

30. $\frac{r^2 - 3rs - 10s^2}{r^2 - 25s^2}$

31. $\frac{a^2 - b^2}{(b - a)^2}$

32. $\frac{2x^2 - 8}{x^3 - 8}$

33. $\frac{4t^2 + 17rt - 15r^2}{16t^2 - 9r^2}$

34. $\frac{3x + 6}{x^4 - 16}$

35. $\frac{a^2 - b^2}{a^3 + b^3}$

36. $\frac{ad + bd}{a^2 + 2ab + b^2}$

B

37. $\frac{2x + 3(x - 2)}{4x + 5(x - 2)}$

38. $\frac{16x + 3x^2 - 35}{33x + 5x^2 - 14}$

39. $\frac{3x^3 + 3y^3}{2x^3 - 2x^2y + 2xy^2}$

40. $\frac{(s^2 + 1)(s - 1)^2}{s^4 - 1}$

41. $\frac{x^5 + y^5}{x^3 + y^3}$

42. $\frac{x^3 - 4x^2 + 8x - 8}{x^2 - 4x + 4}$

43. $\frac{6a^3 - 4a^2 - 3a + 2}{3a^2 + a - 2}$

44. $\frac{5q - 45u^2}{5q(1 + 27u^2)}$

45. $\frac{a^2 + ab}{a(a + b) - 2(a + b)}$

The following exercises contain errors made by students in simplifying rational expressions. For each exercise, find the error. Then simplify correctly when possible.

46. $\frac{3(a + b)}{z + 2(a + b)} = \frac{3}{z + 2}$

47. $\frac{x + y}{x^2 + y^2} = \frac{1}{x + y}$

48. $\frac{2a + b}{a + b} = 2$

49. $\frac{a + b + c}{a + b} = c$

50. $\frac{x}{x + y} = \frac{1}{x + y}$

51. $\frac{x^2 + a^2}{z^2 + a^2} = \frac{x^2}{z^2}$

52. $\frac{(x + y)}{(x + y)(x + y)} = x + y$

53. $\frac{a + b}{x + y} \cdot \frac{x + y}{a + b} = 0$

54. $\frac{a - b}{b - a} = 1$

Give the value of the variable for which the denominator of the rational expression will be zero. The replacement set is the set of real numbers.

55. $\frac{x}{x^2 - 4}$

56. $\frac{t}{t^2 + 9}$

57. $\frac{y - 3}{6y^2 - 7y - 20}$

58. $\frac{b^2 - 2b}{b^2 + 4}$

C

59. Define equality for the two rational expressions $\frac{P}{Q}$ and $\frac{R}{S}$. Give an example to illustrate your definition.

60. Prove that for the rational expression, $\frac{P}{Q}$, $\frac{P}{Q} = \frac{KP}{KQ}$, $K \neq 0$, $Q \neq 0$.
(HINT: Use your definition from Exercise 59.)

Puzzle

Find x. (The three dots mean that the pattern continues without end.)

$$x + \cfrac{1}{x + \cfrac{1}{x + \cfrac{1}{x + \cdots}}} = 2$$

7–2 Multiplication of Rational Expressions

Multiplication of rational expressions is similar to multiplication of rational numbers.

$$\frac{2}{3} \cdot \frac{5}{7} = \frac{2 \cdot 5}{3 \cdot 7} = \mathbf{\frac{10}{21}}$$

Definition

For rational expressions, $\frac{P}{Q}$ and $\frac{R}{S}$,

$$\frac{P}{Q} \cdot \frac{R}{S} = \frac{P \cdot R}{Q \cdot S}.$$

To multiply rational expressions, first apply the definition. Then simplify when necessary.

EXAMPLE 1 Multiply: **a.** $\frac{3a}{b} \cdot \frac{c}{2d}$ **b.** $\frac{3x}{4y} \cdot \frac{9y^2}{12x^3}$

Solution: **a.** $\frac{3a}{b} \cdot \frac{c}{2d} = \frac{3a \cdot c}{b \cdot 2d}$ ← **By definition**

$$= \mathbf{\frac{3ac}{2bd}}$$

b. $\frac{3x}{4y} \cdot \frac{9y^2}{12x^3} = \frac{3x \cdot 9y^2}{4y \cdot 12x^3}$ ← **By definition**

$$= \frac{\overset{1}{\cancel{(3xy)}}(9y)}{\underset{1}{\cancel{(3xy)}}(16x^2)}$$ ← **GCF: 3xy**

$$= \mathbf{\frac{9y}{16x^2}}$$

Example 2 shows multiplication of rational expressions having polynomials in their numerators and denominators.

EXAMPLE 2 Multiply: $\frac{ab^2 - b^3}{a^3 + a^2b} \cdot \frac{a^2 - ab - 2b^2}{a^2 - 2ab + b^2}$

Solution: $\frac{ab^2 - b^3}{a^3 + a^2b} \cdot \frac{a^2 - ab - 2b^2}{a^2 - 2ab + b^2} = \frac{(ab^2 - b^3)(a^2 - ab - 2b^2)}{(a^3 + a^2b)(a^2 - 2ab + b^2)}$ ← **Factor.**

$$= \frac{b^2(a - b)(a - 2b)(a + b)}{a^2(a + b)(a - b)(a - b)}$$ ← **GCF: (a − b)(a + b)**

$$= \frac{\overset{1}{\cancel{(a - b)}}\overset{1}{\cancel{(a + b)}}b^2(a - 2b)}{\underset{1}{\cancel{(a - b)}}\underset{1}{\cancel{(a + b)}}a^2(a - b)}$$

$$= \mathbf{\frac{b^2(a - 2b)}{a^2(a - b)}}$$

CLASSROOM EXERCISES

Multiply.

1. $\frac{1}{3} \cdot \frac{x-3}{2}$
2. $\frac{5}{b^2} \cdot \frac{6}{10}$
3. $\frac{9}{2} \cdot \frac{3x-5}{3}$
4. $\frac{2a}{3b} \cdot \frac{27b^3}{16a}$
5. $\frac{a+b}{9} \cdot \frac{3}{(a+b)^2}$
6. $\frac{(r+t)^3}{18} \cdot \frac{2}{(r+t)^2}$
7. $\frac{r^4 s}{t} \cdot \frac{t}{r^2 s}$
8. $\frac{n^2-9}{n+3} \cdot \frac{(n+3)^2}{(n-3)^2}$

WRITTEN EXERCISES

A Multiply.

1. $\frac{4}{a^2} \cdot \frac{a}{8}$
2. $\frac{4}{5} \cdot \frac{-5n}{2}$
3. $3 \cdot \frac{x+3}{5}$
4. $\frac{2}{5} \cdot \frac{x+2}{4}$
5. $\frac{3x^2}{2} \cdot \frac{3x-5}{3x}$
6. $12a^3x \cdot \frac{2a+3}{4a}$
7. $\frac{6}{8} \cdot \left(\frac{2a+3}{3}\right)$
8. $12 \cdot \left(\frac{x}{2}+5\right)$
9. $\frac{6x}{5} \cdot \left(-\frac{10-x}{3}\right)$
10. $\frac{2ab}{3c^2} \cdot \frac{a^2}{c}$
11. $\frac{a^3y^2}{2x^3} \cdot 4x^3$
12. $\frac{7a}{3b} \cdot \frac{4b}{14a^2}$
13. $\frac{4a^2}{3bc} \cdot \frac{12b^2c}{6a^3}$
14. $\frac{3x^4}{4a^2y^2} \cdot \frac{8a^6y^3}{6x^6}$
15. $\frac{7x^2y}{3a^2b} \cdot \frac{6a}{14x^2}$
16. $(x^2-16) \cdot \frac{3(x+4)^3}{x-4}$
17. $\frac{9-d^2}{d+3} \cdot \frac{d}{d-3}$
18. $\frac{c^2-3c-10}{(c-2)^2} \cdot \frac{c-2}{c-5}$
19. $\frac{x^4-y^4}{(x-y)^2} \cdot \frac{y-x}{x^2-y^2}$
20. $\frac{e^2-24-2e}{e^2-30-e} \cdot \frac{(e+5)^2}{e^2-16}$
21. $\frac{x^3+y^3}{(x+y)^3} \cdot \frac{x^2-y^2}{x^3-y^3}$
22. $\frac{a^2-b^2}{(a+b)^2} \cdot \frac{3a+3b}{6a(b-a)}$
23. $\frac{a^2-1}{a^2+2a+1} \cdot \frac{1+a}{1-a}$
24. $\frac{9-a^2}{12+a-a^2} \cdot \frac{a-4}{a-3}$
25. $\frac{a^2-b^2}{2a+2b} \cdot \frac{a^2-ab}{(b-a)^2}$
26. $\frac{c^2+4c}{c^2+3c-4} \cdot \frac{2c-2}{c^3}$
27. $\frac{3r+2}{5r^2-r} \cdot \frac{10r^2+3r-1}{6r^2+r-2}$

B

28. $\frac{ab^2-b^3}{a^3+a^2b} \cdot \frac{a^2-ab-2b^2}{a^2-2ab+b^2}$
29. $\frac{x^2+7xy+10y^2}{x^2+6xy+5y^2} \cdot \frac{x+y}{x^2+4xy+4y^2} \cdot \frac{x+2y}{1}$
30. $\frac{x^3-1}{5x-5} \cdot \frac{10}{3x^2+3x+3}$
31. $\frac{x^2-y^2}{x^3-3x^2y+2xy^2} \cdot \frac{xy-2y^2}{y^2+xy} \cdot \frac{x(x-y)}{(x-y)^2}$
32. $\frac{a^2-b^2}{(a-b)^2} \cdot \frac{a^2+b^2}{(a+b)^2}$
33. $\frac{30-11t+t^2}{9t-6t^2+t^3} \cdot \frac{t^2+2t-15}{t^2-9}$

C

34. $\frac{r^2+rt}{r^2-rt+t^2} \cdot \frac{r^3-t^3}{r^6-t^6}$
35. $\frac{x^6+y^6}{x^3-y^3} \cdot \frac{x-y}{x^2+y^2}$
36. $\frac{1+8x^3}{(2-x)^2} \cdot \frac{4x-x^2}{1-4x^2} \cdot \frac{2-5x+2x^2}{(1-2x)^2+2x}$
37. What is the multiplicative inverse of $\frac{P}{Q}$?
38. Is the set of rational expressions closed with respect to multiplication? Give several examples to justify your answer.

7-3 Division of Rational Expressions

Division of rational expressions is similar to division of rational numbers.

Theorem 7-2

Division of Rational Expressions

For rational expressions $\frac{P}{Q}$ and $\frac{R}{S}$,

$$\frac{P}{Q} \div \frac{R}{S} = \frac{P}{Q} \cdot \frac{S}{R} = \frac{PS}{QR}.$$

To divide by a rational expression, multiply by its multiplicative inverse (reciprocal).

EXAMPLE 1 Divide: $\frac{7x}{12yz^2} \div \frac{49x^3}{18y^2z^3}$.

Solution:

$$\frac{7x}{12yz^2} \div \frac{49x^3}{18y^2z^3} = \frac{7x}{12yz^2} \cdot \frac{18y^2z^3}{49x^3}$$ ← **By Theorem 7-2**

$$= \frac{7x \cdot 18y^2z^3}{12yz^2 \cdot 49x^3}$$ ← **By definition**

$$= \frac{\overset{1}{\cancel{(42xyz^2)}}(3yz)}{\underset{1}{\cancel{(42xyz^2)}}(14x^2)}$$ ← **GCF: $42xyz^2$**

$$= \mathbf{\frac{3yz}{14x^2}}$$ ← **Lowest terms**

Example 2 shows division by a rational expression having binomials in its numerator and denominator.

EXAMPLE 2 Divide: $\frac{x^2 + 4xy + 3y^2}{x^2 - y^2} \div \frac{x + 3y}{x + y}$

Solution:

$$\frac{x^2 + 4xy + 3y^2}{x^2 - y^2} \div \frac{x + 3y}{x + y} = \frac{x^2 + 4xy + 3y^2}{x^2 - y^2} \cdot \frac{x + y}{x + 3y}$$ ← **By Theorem 7-2**

$$= \frac{(x^2 + 4xy + 3y^2)(x + y)}{(x^2 - y^2)(x + 3y)}$$ ← **Definition of multiplication**

$$= \frac{(x + 3y)(x + y)(x + y)}{(x + y)(x - y)(x + 3y)}$$ ← **GCF: $(x + 3y)(x + y)$**

$$= \frac{\overset{1}{\cancel{(x + 3y)}}\overset{1}{\cancel{(x + y)}}(x + y)}{\underset{1}{\cancel{(x + 3y)}}\underset{1}{\cancel{(x + y)}}(x - y)} = \mathbf{\frac{x + y}{x - y}}$$

CLASSROOM EXERCISES

Divide. No denominator equals zero.

1. $\frac{2 \cdot 4}{5} \div \frac{3 \cdot 6}{15}$
2. $\frac{7b}{6c} \div \frac{9bc}{4}$
3. $\frac{5(x-3)}{2x} \div \frac{x^2-9}{x}$
4. $\frac{y^2+3y+2}{y^5} \div \frac{y+2}{y^3}$
5. $\frac{25a^2}{(a-b)^2} \div \frac{5a}{a-b}$
6. $\frac{x-7}{3x^2-8x+4} \div \frac{3x-21}{6x^2-24}$

WRITTEN EXERCISES

A Divide. No denominator equals zero.

1. $\frac{7}{x} \div \frac{81}{x^2}$
2. $\frac{1}{z} \div \left(-\frac{5}{z^2}\right)$
3. $-\frac{t^{3r}}{s} \div \frac{t^{3r}}{s}$
4. $\frac{x^2y^4}{6a^2} \div \frac{x^4y}{10a^2}$
5. $\frac{14x^2}{10b^2} \div \frac{21x^2}{15b^2}$
6. $\frac{5a}{12yz^2} \div \frac{15a^2}{18y^2z^2}$
7. $\frac{-4(a-2)}{-2(x+5)} \div \frac{3(a-2)}{x(x+5)}$
8. $4(n-3) \div \frac{12(n-3)}{100}$
9. $\frac{3(x+5)}{7x} \div \frac{3x(x+5)}{49x}$
10. $\frac{x^2+2x}{y^2-9} \div \frac{x+2}{y+3}$
11. $\frac{x^2y^3-x^2y^2}{a+b} \div \frac{3y-3}{(a+b)^2}$
12. $\frac{c^2-b^2}{x^2-y^2} \div \frac{c+b}{y-x}$
13. $\frac{a^2-16}{a+7} \div \frac{3a-12}{a^2-49}$
14. $\frac{10x^2}{(x+y)^2} \div \frac{5x}{x+y}$
15. $\frac{5}{2a+3b} \div \frac{10}{4a^2-9b^2}$
16. $\frac{2c}{d-5} \div \frac{16c^5}{5-d}$
17. $\frac{8a^2}{a^2-16} \div \frac{4a}{4-a}$
18. $\frac{7y^2-21y}{x+5} \div \frac{9-y^2}{(x+5)^2}$
19. $\frac{x^2-9}{x^2-x-2} \div \frac{x-3}{x^2+x-6}$
20. $\frac{x^2+5x+6}{3x+12} \div \frac{x+3}{x^2+4x}$
21. $\frac{3a^2+30a+75}{a-5} \div \frac{a}{a^2-25}$
22. $\frac{x^2+2x+1}{x^2-4} \div \frac{x^2-x-2}{x+2}$
23. $\frac{d^2+6d}{d^2+2d-8} \div \frac{d^2-36}{d+4}$
24. $\frac{h^2+4h}{h^2+3h-4} \div \frac{h^3}{2h-2}$

B

25. $\frac{x^2-x-2}{x^2-x-6} \div \frac{2x-x^2}{2x+x^2}$
26. $\frac{y^3+1}{x^2-4y^2} \div \frac{y^2-y+1}{x-2y}$
27. $\frac{x^2-1}{x^2+2x+1} \div \frac{1-x}{5+x}$
28. $\frac{(a-b)(b-c)}{c-a} \div \frac{c-b}{a-c}$
29. $\frac{x^2+6xy+5y^2}{x^2+4xy+4y^2} \div \frac{x+y}{x+2y}$
30. $\frac{x^3-6x^2+8x}{x^2-8x+16} \div \frac{2x-4}{10x^2-40x}$
31. $\frac{8x^3-27y^3}{2x+3y} \div \frac{4x^2+6xy+9y^2}{4x^2-9y^2}$
32. $\frac{y^3-8}{y^3-4y} \div \frac{y^2+2y+4}{y^3+2y^2}$

C

33. The following sentence defines division of real numbers.

    For real numbers a, b, and c where $b \neq 0$, $a \div b = c$ means $c \cdot b = a$.

    Write a similar definition for division of rational expressions.

34. Prove Theorem 7–2. (HINT: Use your definition from Exercise 33.)

REVIEW CAPSULE FOR SECTION 7-4

Write the additive inverse of each expression. *(Pages 190–192)*

1. $17m$ **2.** $9n - 9$ **3.** $y + 1$ **4.** $x - 9y$

Simplify each expression. *(Pages 14–16, 193–196)*

5. $7(r - s) - 5(r - s)$

6. $(t - 4)(t - 5) - (t - 7)(t - 2)$

7. $ab - a(a + b)$

8. $(3y + 1)(3y + 1) - (5y^2 + 2y)$

7-4 Addition and Subtraction

The definitions of addition and subtraction of rational expressions are similar to those for addition and subtraction of rational numbers.

Definitions

Addition of Rational Expressions

If $\frac{P}{Q}$ and $\frac{R}{Q}$ are rational expressions, then

$$\frac{P}{Q} + \frac{R}{Q} = \frac{P + R}{Q}.$$

Subtraction of Rational Expressions

If $\frac{P}{Q}$ and $\frac{R}{Q}$ are rational expressions, then

$$\frac{P}{Q} - \frac{R}{Q} = \frac{P - R}{Q}.$$

EXAMPLE 1 Add or subtract as indicated.

a. $\frac{5n}{n+1} + \frac{3n-2}{n+1}$ **b.** $\frac{5n}{n+1} - \frac{3n-2}{n+1}$

Solutions:

a. $\frac{5n}{n+1} + \frac{3n-2}{n+1} = \frac{5n + (3n-2)}{n+1}$

$= \mathbf{\frac{8n - 2}{n + 1}}$

b. $\frac{5n}{n+1} - \frac{3n-2}{n+1} = \frac{5n - (3n-2)}{n+1}$

$= \frac{5n - 3n + 2}{n+1}$

$= \frac{2n + 2}{n+1}$

$= \frac{2\cancel{(n+1)}^{1}}{\cancel{(n+1)}_{1}} = \mathbf{2}$

When the denominators differ, you find the **least common denominator *(LCD)*** before adding or subtracting.

EXAMPLE 2 Find the LCD.

a. 36 and 54 **b.** $3(x^2 - 4)$ and $9(x^2 - 4x + 4)$

Solutions: [1] Find the greatest common factor (GCF).

a. $36 = 2 \cdot \mathbf{2 \cdot 3 \cdot 3}$
$54 = \mathbf{2 \cdot 3 \cdot 3} \cdot 3$
GCF: $\mathbf{2 \cdot 3 \cdot 3}$, or **18**

b. $3(x^2 - 4) = \mathbf{3}(x + 2)\mathbf{(x - 2)}$
$9(x^2 - 4x + 4) = \mathbf{3} \cdot 3 \cdot \mathbf{(x - 2)}(x - 2)$
GCF: $\mathbf{3(x - 2)}$

[2] Find the LCD. The LCD is the product of the GCF and the other factors.

a. LCD: $\underline{18}(\underline{2 \cdot 3})$, or **108**
GCF ↑ ↑ Other factors

b. LCD: $[3(x - 2)][3(x + 2)(x - 2)]$
$\mathbf{9(x - 2)(x + 2)(x - 2)}$

To add or subtract rational expressions with unlike denominators, you write equivalent rational expressions having the LCD as denominator. Then you add or subtract as indicated.

EXAMPLE 3 Add: $\dfrac{7m}{6a^2b^4} + \dfrac{5}{9a^3b^2}$

Solutions: LCD: $18a^3b^4$ Multiply $\dfrac{7m}{6a^2b^4}$ by $\dfrac{3a}{3a}$ and $\dfrac{5}{9a^3b^2}$ by $\dfrac{2b^2}{2b^2}$.

$$\frac{7m}{6a^2b^4} + \frac{5}{9a^3b^2} = \frac{7m}{6a^2b^4} \cdot \frac{3a}{3a} + \frac{5}{9a^3b^2} \cdot \frac{2b^2}{2b^2}$$

$$= \frac{21am}{18a^3b^4} + \frac{10b^2}{18a^3b^4}$$ ← **The denominators are the same.**

$$= \mathbf{\frac{21am + 10b^2}{18a^3b^4}}$$ ← **By definition**

When you are working with three or more denominators, you find the LCD by working with two denominators at a time.

EXAMPLE 4 Find the LCD of $(a - 3)$, $(a^2 - 9)$, and $(a^2 - a - 6)$.

Solutions: [1] Find the LCD of $(a - 3)$ and $a^2 - 9$.

$(a - 3)$ $a^2 - 9 = (a - 3)(a + 3)$
GCF: $(a - 3)$ LCD: $(a - 3)(a + 3)$

[2] Find the LCD of $(a - 3)(a + 3)$ and $a^2 - a - 6$.

$(a - 3)(a + 3)$ $a^2 - a - 6 = (a - 3)(a + 2)$
GCF: $(a - 3)$ LCD: $\mathbf{(a - 3)(a + 3)(a + 2)}$

EXAMPLE 5 Simplify: $\frac{1}{a-3}+\frac{3a-4}{a^2-9}-\frac{2a-3}{a^2-a-6}$

Solution: LCD: $(a-3)(a+3)(a+2)$ ← **From Example 4**

$$\frac{1}{a-3}+\frac{3a-4}{a^2-9}-\frac{2a-3}{a^2-a-6}=\frac{1}{(a-3)}\cdot\frac{(a+3)(a+2)}{(a+3)(a+2)}+\frac{(3a-4)}{(a+3)(a-3)}\cdot\frac{(a+2)}{(a+2)}-\frac{(2a-3)}{(a-3)(a+2)}\cdot\frac{(a+3)}{(a+3)}$$

$$=\frac{(a+3)(a+2)}{(a-3)(a+3)(a+2)}+\frac{(3a-4)(a+2)}{(a+3)(a-3)(a+2)}-\frac{(2a-3)(a+3)}{(a+3)(a-3)(a+2)}$$

$$=\frac{(a^2+5a+6)+(3a^2+2a-8)-(2a^2+3a-9)}{(a-3)(a+3)(a+2)}$$

$$=\frac{a^2+5a+6+3a^2+2a-8-2a^2-3a+9}{(a-3)(a+3)(a+2)}$$

$$=\mathbf{\frac{2a^2+4a+7}{(a-3)(a+3)(a+2)}}$$ ← **Lowest terms**

Summary

Steps for Addition and Subtraction of Rational Expressions

1. Find the LCD.
2. Write equivalent rational expressions having the LCD as denominator.
3. Add or subtract as indicated.

CLASSROOM EXERCISES

Add or subtract as indicated.

1. $\frac{9}{x+y}+\frac{21}{x+y}$
2. $\frac{2}{a+b}-\frac{1}{a+b}$
3. $\frac{3x-11}{2}+\frac{3x+11}{2}$
4. $\frac{3x-11}{2}-\frac{3x+11}{2}$
5. $\frac{x+2}{x-4}+\frac{3x-4}{x-4}$
6. $\frac{x+2}{x-4}-\frac{3x-4}{x-4}$

For each pair of rational expressions in Exercises 7–12:

a. Find the LCD.

b. Rewrite the expressions so that they have the LCD as denominator.

7. $\frac{2}{3a};\ \frac{3}{5a^2}$
8. $\frac{a+b}{3a^3b^2};\ \frac{5}{4a^4b}$
9. $\frac{x+2}{4};\ \frac{x-5}{5}$
10. $\frac{x-2}{(x+2)^2};\ \frac{5}{3(x+2)}$
11. $\frac{2x-3}{2x-5};\ \frac{25x^2}{4x^2-25}$
12. $\frac{2}{1-x};\ \frac{x^2}{1-x^2}$

WRITTEN EXERCISES

A Add or subtract as indicated.

1. $\frac{8m}{m^2 - n^2} + \frac{17m}{m^2 - n^2}$

2. $\frac{8m}{m^2 - n^2} - \frac{17m}{m^2 - n^2}$

3. $\frac{x}{y} - \frac{10}{y}$

4. $\frac{4a - 2}{a^2} + \frac{3a - 7}{a^2}$

5. $\frac{3n - 7}{5} - \frac{9n - 9}{5}$

6. $\frac{6p - 8}{9} - \frac{7p - 9}{9}$

7. $\frac{3x - 4}{a + b} + \frac{7 - 9x}{a + b}$

8. $\frac{x - 7y}{12} - \frac{x - 9y}{12}$

9. $\frac{2x - 15}{26} - \frac{2x - 19}{26}$

Find the LCM of each pair of expressions.

10. $ab^2;\ a^3b$

11. $3xy^2;\ 6x^3y$

12. $x^2 - xy;\ xy - y^2$

13. $a^3;\ a^2 - ab$

14. $a^3 + b^3;\ a + b$

15. $(x + 2)^2;\ 3(x + 2)$

16. $x - 1;\ x^2 - 2x + 2$

17. $2x - 5;\ 4x^2 - 25$

18. $x^2 + 7x + 12;\ x^2 + 6x + 8$

Add or subtract as indicated.

19. $\frac{a}{x^2y} + \frac{b}{xy^2}$

20. $\frac{5a + 2}{a} + \frac{3b - 1}{b}$

21. $\frac{3}{8mn^2} + \frac{2}{3mn}$

22. $\frac{t}{p + q} + \frac{t}{p - q}$

23. $\frac{m}{r - s} + \frac{3m}{r + s}$

24. $\frac{3}{y + 2} + \frac{5}{y - 4}$

25. $\frac{3a - b}{6} - \frac{3a - 2b}{4}$

26. $\frac{2s - t}{s} - \frac{5s + 2t}{t}$

27. $\frac{3}{m - 2} - \frac{2}{m + 5}$

28. $\frac{9}{a + b} - \frac{6}{b + a}$

29. $\frac{5}{3c - 12} - \frac{3}{7c - 28}$

30. $\frac{4}{3a - 6} + \frac{7}{5a - 10}$

31. $\frac{3}{3z - 10} + \frac{2}{z - 2}$

32. $\frac{ab}{a^2 - b^2} + \frac{a}{a - b}$

33. $\frac{17}{x + y} - \frac{20}{y - x}$

34. $\frac{10}{2t - 1} - \frac{6}{1 - 2t}$

35. $\frac{2}{c^2 - 4} - \frac{3}{2 - c}$

36. $\frac{t - 4}{t - 2} - \frac{t - 7}{t - 5}$

37. $3n + 1 + \frac{1}{3n - 1}$

38. $\frac{3b + 1}{b - 2} - \frac{4b + 1}{b - 3}$

39. $\frac{3}{3x - 4} - \frac{5}{5x + 6}$

B

40. $\frac{2r}{r + 3} + \frac{9 - r}{r + 3} - \frac{r + 6}{r + 3}$

41. $\frac{x}{x + y} - \frac{2y^2}{x + y} + \frac{y}{x + y}$

42. $\frac{2x + 8}{x - 3} + \frac{6x - 4}{x - 3} - \frac{4x - 8}{x - 3}$

43. $\frac{x^2 - 3}{x + 2} - \frac{x^2 + 5}{x + 2} - \frac{-2x - 4}{x + 2}$

44. $\frac{x}{x - y} - \frac{2x}{x + y} - \frac{2xy}{x^2 - y^2}$

45. $\frac{3}{t^2 - 4} - \frac{4}{t^2 - 3t + 2}$

46. $\frac{3a}{a + b} - \frac{3b}{a - b} + \frac{a^2 + b^2}{a^2 - b^2}$

47. $\frac{n + 5}{n^2 - 8n + 16} - \frac{n - 4}{n^2 - 2n - 8}$

48. $\frac{2}{x^2 - y^2} + \frac{2}{(x + y)^2} + \frac{2}{(x - y)^2}$

49. $\frac{n + 2}{6n^2 - n + 12} + \frac{3n + 4}{2n^2 + n - 6}$

50. $\frac{1}{x^2 - x - 2} + \frac{1}{x^2 + 2x + 1}$

51. $\frac{t - 7}{t^2 + 4t - 5} - \frac{t - 9}{t^2 + 3t - 10}$

C Simplify each complex fraction.

Example: $\dfrac{a+\dfrac{ab}{c}}{\dfrac{a^2}{d}-a}=\dfrac{\dfrac{ac+ab}{c}}{\dfrac{a^2-ad}{d}}$ ← **Find the LCD of the numerator and the LCD of the denominator.**

$=\dfrac{\left(\dfrac{ac+ab}{c}\right)(cd)}{\left(\dfrac{a^2-ad}{d}\right)(cd)}$ ← **Multiply both numerator and denominator by the LCD of c and d.**

$=\dfrac{(ac+ab)d}{(a^2-ad)c}$ ← **Simplify.**

$=\dfrac{\overset{1}{\cancel{a}d}(b+c)}{\underset{1}{\cancel{a}c}(a-d)}=\mathbf{\dfrac{d(b+c)}{c(a-d)}}$

52. $\dfrac{\frac{1}{2}+\frac{1}{3}}{\frac{1}{2}-\frac{1}{3}}$

53. $\dfrac{2-\frac{1}{6}}{3+\frac{1}{4}}$

54. $\dfrac{1+\frac{x}{y}}{1-\frac{x}{y}}$

55. $\dfrac{2-\frac{m}{3}}{5+\frac{m}{8}}$

56. $\dfrac{\frac{1}{x}-1}{1+\frac{1}{y}}$

57. $\dfrac{a+\frac{a}{b}}{1-\frac{1}{b^2}}$

58. $\dfrac{\frac{m}{n}-1}{\frac{1}{m}+\frac{1}{n}}$

59. $\dfrac{c-d}{c-\frac{d^2}{c}}$

60. $\dfrac{\frac{1}{x-y}-1}{2-\frac{x}{x-y}}$

61. $\dfrac{1}{\frac{1}{a}+\frac{1}{b}+\frac{1}{c}}$

62. $\dfrac{1-\frac{r^2}{s^2}}{1-\frac{2r}{s}+\frac{r^2}{s^2}}$

63. $\dfrac{1+\frac{1}{t+1}}{1+\frac{3}{t-1}}$

Add or subtract as indicated.

64. $\dfrac{3x+y}{6x^2-5xy-6y^2}-\dfrac{x-3y}{8x^2-14xy+3y^2}$

65. $\dfrac{m+1}{m^2+9m+14}-\dfrac{m-2}{m^2-3m-10}+\dfrac{m+2}{m^2+2m-35}$

66. $\dfrac{3a-b}{4a^2-11ab-3b^2}+\dfrac{a+b}{a^2-4ab+3b^2}+\dfrac{5a+b}{4a^2-3ab-b^2}$

67. For real numbers a, b, and c, $a-b=c$ if and only if $a=c+b$. Use this definition to show that for rational expressions, $\dfrac{P}{Q}$ and $\dfrac{R}{Q}$, $\dfrac{P}{Q}-\dfrac{R}{Q}=\dfrac{P}{Q}+\dfrac{-R}{Q}$.

68. Prove that if $\dfrac{P}{Q}$ and $\dfrac{R}{S}$ are rational expressions, then $\dfrac{P}{Q}+\dfrac{R}{S}=\dfrac{PS+QR}{QS}$.

69. Prove: For integers A, B, r, and t, $\dfrac{Ax+B}{(x-1)(x-2)}=\dfrac{r}{(x-1)}+\dfrac{t}{(x-2)}$. (HINT: Solve for r and t in terms of A and B.)

Using Statistics: Standard Deviation

Psychologists collect and analyze data relating to test scores. The psychologist at the right is using the technique of observation through a one-way mirror to collect data. One of the measures of concern to psychologists is the standard deviation. This is a measure of how the data disperses, or spreads out, from the mean.

Definition For a set of data with n elements, in which x_i represents each element, $1 \leq i \leq n$, and $\bar{x}$ represents the mean of the data, the standard deviation is found by the following formula. The Greek letter σ (sigma) represents the standard deviation.

$$\sigma = \sqrt{\frac{\text{sum of the squares of the deviations from the mean}}{\text{number of elements in the set}}} = \sqrt{\frac{\sum_{i=1}^{n}(x_i - \bar{x})^2}{n}}$$

The graph at the right illustrates the ideal distribution for test results. In this model, the mean, median, and mode have the same value. The graph is bell-shaped and symmetrical. Such a graph is called the **normal curve.** This graph is that of the **standard normal distribution** with a mean of 0, and a standard deviation of 1. In a normal distribution, 68% of the data falls within 1 standard deviation of the mean, about 95% within 2 standard deviations, and about 99% within 3 standard deviations.

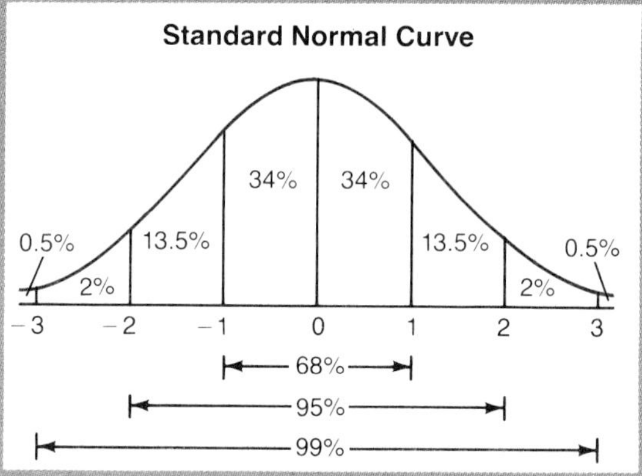

Thus, the more dispersed the data, the larger σ will be.

EXAMPLE 1 Find σ to the nearest tenth for this data: 46, 52, 60, 49, 53, 70.

Solution: [1] Find $\bar{x}$: $\bar{x} = \frac{330}{6} = 55$

[2] $$\sigma = \sqrt{\frac{\sum_{i=1}^{6}(x_i - \bar{x})^2}{6}}$$

$$= \sqrt{\frac{(-9)^2 + (-3)^2 + (5)^2 + (-6)^2 + (-2)^2 + (15)^2}{6}}$$

$$= \sqrt{\frac{380}{6}} = \sqrt{63.3} = 7.956 \approx \mathbf{8.0}$$

EXAMPLE 2 The staff psychologist at Miller, Inc. administered a mechanical aptitude test to 200 applicants for training positions in the company. The results were distributed normally with a mean of 54 and a standard deviation of 8. Anyone with a score above 70 will be accepted as a trainee. How many people will probably be accepted?

Solution: [1] How many standard deviations is 70 from 54?

$70 - 54 = 16$ and $16 \div 8 = 2$.

Thus 70 is 2 standard deviations from 54.

[2] Using the graph on page 242, you can see that 2.5% of the observations are above 2 standard deviations from the mean.

2.5% of 200 = $0.025 \times 200 =$ **5.000**

Thus, **5** people will probably be accepted into the training program.

EXERCISES

1. The scores of two classes on the same test are given below. Find σ for each class, and explain why they differ.

Class A		Class B	
82	65	82	87
85	59	81	81
41	84	85	85
93	87	85	86
87	92	84	88
88	95	87	87
91	98	88	84
99	99	89	91

2. The heights in inches of the players on two teams are shown below. Find σ for each team and explain why they differ.

Team A		Team B	
76	72	80	76
74	74	68	75
70	70	72	76
72	74	76	76
71	75	76	76
76	74	75	77
75	74	73	78
75	74	76	76

The lifetimes of 20,000 light bulbs are normally distributed. The mean lifetime is 400 days and the standard deviation is 50 days. Use this information for Exercises 3–10.

3. What percentage of light bulbs can be expected to last less than 350 days?
4. What percentage of bulbs will probably last between 350 and 450 days?
5. How many light bulbs can be expected to last less than 300 days?
6. How many light bulbs can be expected to last longer than 450 days?
7. How many light bulbs can be expected to last more than 300 days and less than 450 days?
8. How many light bulbs can be expected to last more than 350 days and less than 500 days?
9. How many light bulbs can be expected to last not more than 500 days?
10. How many light bulbs can be expected to last not more than 350 days?

Review

Simplify each rational expression. *(Section 7–1)*

1. $\frac{8x}{x^3}$ **2.** $\frac{4b^2d^3}{8bd^6}$ **3.** $\frac{3x^2 - 27}{6x^2 - 6x - 72}$ **4.** $\frac{5x^3 - 5y^3}{10x^2 + 20xy + 10y^2}$

Multiply. *(Section 7–2)*

5. $\frac{9t^5r^7}{5s^2v^3} \cdot \frac{40s^6v^2}{27t^4r^8}$ **6.** $\frac{b^2 - 16}{b + 2} \cdot \frac{b^3 + 6b^2 + 8b}{b^2 - 8b + 16}$ **7.** $\frac{x^3 - 49x}{(x - 2)^2} \cdot \frac{x^2 - 4}{x^2 - 5x - 14}$

Divide. No denominator equals zero. *(Section 7–3)*

8. $\frac{4t - 8}{t} \div \frac{t^2 - 4}{3t^2 + 6t}$ **9.** $\frac{a^3 - 1}{a^2 - 9b^2} \div \frac{2a^2 + 2a + 2}{a^2 - ab - 6b^2}$ **10.** $\frac{(b^2 - 4)}{(b + 2)^2} \div \frac{2 - b}{3b^2 + b - 10}$

Add or subtract as indicated. *(Section 7–4)*

11. $\frac{6 + 3z}{2z - 5} + \frac{9 - 2z}{2z - 5}$ **12.** $\frac{4}{3x - 2} - \frac{5}{2x + 5}$ **13.** $\frac{x - 3}{x^2 + 5x - 6} - \frac{x - 4}{x^2 + 3x - 4}$

CALCULATOR APPLICATIONS

Evaluating Rational Expressions

To use a calculator to evaluate a rational expression, rewrite the numerator and denominator as you did on page 218.

EXAMPLE Evaluate $\frac{2a^3 + 4a^2 + 3a + 7}{a^3 + 2a^2 - 9a - 18}$ when $a = 4$.

SOLUTION Rewrite the expression. $\frac{2a^3 + 4a^2 + 3a + 7}{a^3 + 2a^2 - 9a - 18} = \frac{[(2a + 4)a + 3]a + 7}{[(a + 2)a - 9]a - 18}$

Evaluate the denominator first. Store its value.

4 [+] 2 [=] [×] 4 [−] 9 [=] [×] 4 [−] 1 8 [=] [M+] 2 [×] 4 [+] 4 [=]

[×] 4 [+] 3 [=] [×] 4 [+] 7 [=] [÷] [MR] [=] 5.0238095

Evaluate Exercises 37, 38, 40, 42, and 43 on page 232 when the variable equals −3.

REVIEW CAPSULE FOR SECTION 7–5

Simplify. *(Pages 193–196)*

1. $(x + 1)(2x + 3) - (x - 1)(2x - 3)$ **2.** $(n - 5)(n + 5) + (n + 15)(n + 5)$

3. $(x + 3)(x - 2) - 3x(x - 5)$ **4.** $2x(x + 3) - (x + 7)(x + 3)$

7–5 Solving Equations with Rational Expressions

To write an equation with rational coefficients as one with integral coefficients, multiply each side of the equation by the LCD.

EXAMPLE 1 Solve: $\frac{1}{4}x - \frac{2}{3} = \frac{1}{8}x$

Solution:

$$\frac{1}{4}x - \frac{2}{3} = \frac{1}{8}x$$ ← **Multiply each side by 24, the LCD.**

$$24\left(\frac{1}{4}x - \frac{2}{3}\right) = 24\left(\frac{1}{8}x\right)$$

$$24\left(\frac{1}{4}x\right) - 24\left(\frac{2}{3}\right) = 24\left(\frac{1}{8}x\right)$$ ← **By the distributive postulate.**

$$6x - 16 = 3x$$

$$3x = 16$$

$$x = \frac{16}{3}$$

Check:

$$\frac{1}{4}x - \frac{2}{3} = \frac{1}{8}x$$ ← **Replace x with $\frac{16}{3}$.**

$$\frac{1}{4}\left(\frac{16}{3}\right) - \frac{2}{3} \stackrel{?}{=} \frac{1}{8}\left(\frac{16}{3}\right)$$

$$\frac{4}{3} - \frac{2}{3} \stackrel{?}{=} \frac{2}{3}$$

$$\frac{2}{3} \stackrel{?}{=} \frac{2}{3}$$ Yes ✓ **Solution set:** $\left\{\frac{16}{3}\right\}$

Follow the same procedure when variables occur in the denominator. REMEMBER: No denominator can equal zero.

EXAMPLE 2 Solve: $\frac{1}{3z} + \frac{1}{8} = \frac{4}{3z}$

Solution:

$$\frac{1}{3z} + \frac{1}{8} = \frac{4}{3z}$$ ← **Multiply each side by 24z, the LCD.**

$$24z\left(\frac{1}{3z} + \frac{1}{8}\right) = 24z\left(\frac{4}{3z}\right)$$

$$24z\left(\frac{1}{3z}\right) + 24z\left(\frac{1}{8}\right) = 24z\left(\frac{4}{3z}\right)$$ ← **By the distributive postulate**

$$8 + 3z = 32$$

$$3z = 24$$

$$z = 8$$ The check is left for you. **Solution set:** $\{8\}$

When you multiply an equation having fractional coefficients by a polynomial, the solution set of the resulting equation may include numbers which are *not* solutions of the original equation. These numbers are called **apparent solutions.** This is why it is important to check all solutions of the resulting equation in the *original equation.*

EXAMPLE 3 Solve: $\frac{2x+3}{x-1} - \frac{2x-3}{x+1} = \frac{10}{x^2-1}$

Solution: LCD: $(x+1)(x-1)$

$$\frac{2x+3}{x-1} - \frac{2x-3}{x+1} = \frac{10}{x^2-1}$$ ← **Multiply each side by (x + 1)(x − 1).**

$$\left[(x+1)(x-1)\right]\left[\frac{2x+3}{x-1} - \frac{2x-3}{x+1}\right] = \left[(x+1)(x-1)\right]\left[\frac{10}{x^2-1}\right]$$

$$\left[(x+1)\overset{1}{\cancel{(x-1)}}\right]\left[\frac{2x+3}{\underset{1}{\cancel{x-1}}}\right] - \left[\overset{1}{\cancel{(x+1)}}(x-1)\right]\left[\frac{2x-3}{\underset{1}{\cancel{x+1}}}\right] = \left[\overset{1}{\cancel{(x+1)}}\overset{1}{\cancel{(x-1)}}\right]\left[\frac{10}{\underset{1}{\cancel{(x+1)}}\underset{1}{\cancel{(x-1)}}}\right]$$

$$(x+1)(2x+3) - (x-1)(2x-3) = 10$$ ← **Multiply.**

$$2x^2 + 5x + 3 - (2x^2 - 5x + 3) = 10$$ ← **Simplify.**

$$2x^2 + 5x + 3 - 2x^2 + 5x - 3 = 10$$

$$10x = 10$$

$$x = 1$$

Check: $\frac{2x+3}{x-1} - \frac{2x-3}{x+1} = \frac{10}{x^2-1}$ ← **Replace x with 1.**

$\frac{2+3}{0} - \frac{2-3}{2} \stackrel{?}{=} \frac{10}{0}$ Since no denominator can be zero, the number 1 is an apparent solution. The solution set is the **empty set,** or **ϕ.**

CLASSROOM EXERCISES

In each of Exercises 1–6, find the LCD of the denominators.

1. $\frac{1}{3}x + \frac{1}{2}x = 12$
2. $\frac{1}{6}n - \frac{1}{4}n = 9$
3. $\frac{3}{a} + \frac{5}{7} = 2$
4. $\frac{5(m+14)}{9m} = 5$
5. $\frac{2}{3y} - 6 = \frac{5}{y}$
6. $\frac{3}{c} = \frac{5}{4c} + \frac{1}{c^2}$

WRITTEN EXERCISES

A Solve and check each equation.

1. $\frac{n}{3} + \frac{n}{4} = 21$
2. $\frac{2n}{3} + \frac{3n}{4} = 51$
3. $\frac{x}{4} - \frac{3}{2} = \frac{x}{6}$
4. $\frac{1}{2}x - \frac{1}{6}x = x - 16$
5. $\frac{7n}{6} - \frac{2n}{3} = n - 4$
6. $\frac{n}{3} - \frac{2n}{3} + \frac{7}{6} = \frac{13}{6}$

7. $\frac{2n}{3}+\frac{5n}{12}-\frac{n}{6}-\frac{5n}{9}=\frac{13}{2}$

8. $\frac{10}{x}-2=\frac{5-x}{4x}$

9. $\frac{3n}{8}-\frac{1}{4}=\frac{n}{3}-\frac{11}{24}$

10. $\frac{3x}{4}-12=\frac{3(x-12)}{5}$

11. $\frac{3}{10}x+8=\frac{5}{12}(x+8)$

12. $4m-\frac{6m+3}{2}=8$

13. $\frac{3x+1}{2}-\frac{3x-4}{3}=\frac{3x+1}{4}$

14. $4x-\frac{5(3-2x)}{3}=17$

15. $\frac{x}{x+5}=\frac{1}{2}$

16. $\frac{5}{2x}=\frac{9-2x}{8x}+3$

17. $\frac{z}{z-3}=2$

18. $\frac{a-2}{a+3}=\frac{3}{8}$

19. $\frac{a-2}{a+3}=\frac{3}{5}$

20. $\frac{11-x}{12+2x}=\frac{4}{7}$

21. $\frac{x-1}{x-2}=1.5$

22. $\frac{9}{x-3}=\frac{7}{x-5}$

23. $\frac{6}{x-2}=\frac{5}{x-3}$

24. $\frac{x-1}{2}-\frac{3x-4}{2}=\frac{5x-3}{8}$

25. $\frac{8x-5}{3}+7=\frac{3x+6}{4}-\frac{3x-7}{6}+\frac{1}{4}$

26. $40-x+\frac{5}{6}(40-x)=2x-11$

B

27. $\frac{2y-1}{2}-\frac{y+2}{2y+5}=\frac{6y-5}{6}$

28. $\frac{3}{x^2-4}=\frac{-2}{5x+10}$

29. $\frac{3}{x+2}+\frac{12}{x^2-4}=\frac{-1}{x-2}$

30. $\frac{3n-1}{3}-\frac{2n+3}{2}=\frac{n-3}{15}$

31. $3\left(\frac{15x}{4}-2\right)=5x-\frac{7}{2}$

32. $\frac{x+3}{x-5}+\frac{6+2x^2}{x^2-7x+10}=\frac{3x}{x-2}$

33. $\frac{n-5}{n+5}+\frac{n+15}{n-5}=\frac{25}{25-n^2}+2$

34. $\frac{x}{12}=\frac{1}{3}\left(17-\frac{2x}{3}\right)-\frac{1}{3}\left(15-\frac{3x}{4}\right)$

C Solve for x.

35. $\frac{2}{3}\left(\frac{x}{a}-1\right)=\frac{3}{4}\left(\frac{x}{a}+1\right)$

36. $\frac{x}{a-x}-\frac{a-x}{x}=\frac{a}{x}$

37. Write an argument to support the following statement.

If r is in the solution set of an equation with rational expressions, then r is in the solution set of the equation formed when both sides of the original equation are multiplied by a nonzero polynomial.

REVIEW CAPSULE FOR SECTION 7-6

In Exercises 1–6, first identify the unknowns. Then write the indicated fraction. *(Pages 26–29)*

1. The denominator of a fraction is 2 less than the numerator.
2. The sum of the numerator and denominator of a fraction is 71.
3. The numerator of a fraction is 1 less than three times the denominator.
4. The denominator of a fraction exceeds twice the numerator by 5.
5. The numerator and denominator of a fraction are in the ratio 2 : 5.

Problem Solving and Applications

7–6 Equations with Rational Expressions

Rational equations are useful in solving some number problems.

EXAMPLE 1 The denominator of a fraction is 1 less than 4 times the numerator. When the numerator is doubled and the denominator is increased by 6, the resulting fraction equals $\frac{2}{5}$. Find the original fraction.

Solution: [1] Use Condition 1 to represent the unknowns.
Let $n =$ the numerator.
Then $4n - 1 =$ the denominator.

[2] Make a table to organize the information.

Original Fraction	New Fraction
$\frac{n}{4n-1}$	$\frac{2n}{(4n-1)+6}$, or $\frac{2n}{4n+5}$

[3] Use Condition 2 and the table to write an equation.
$$\frac{2n}{4n+5} = \frac{2}{5}$$

[4] Solve the equation.
$$[5(4n+5)]\left[\frac{2n}{4n+5}\right] = [5(4n+5)]\frac{2}{5}$$
$$5(2n) = (4n+5)2$$
$$10n = 8n + 10$$
$$2n = 10$$
$$n = \mathbf{5}$$
$$4n - 1 = 4 \cdot 5 - 1 = \mathbf{19}$$

The original fraction was $\mathbf{\frac{5}{19}}$. The check is left for you.

Some number problems involve ratios.

EXAMPLE 2 The dimensions of an oil painting are 17.5 centimeters by 31 centimeters. The painting is to be framed so that the outside dimensions of the framed painting will be in the ratio 5:8 (Condition 2). Find the width of the frame (Condition 1). (Assume uniform width.)

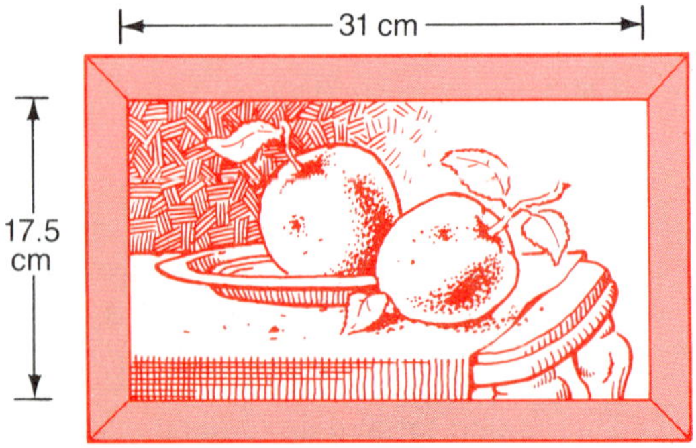

Solution: [1] Let t = the width of the frame.

[2]

	Width	Length
Painting	17.5	31
Painting + Frame	$17.5 + 2t$	$31 + 2t$

[3] $\frac{17.5 + 2t}{31 + 2t} = \frac{5}{8}$ ← Ratio of outside dimensions is $\frac{5}{8}$.

[4] Multiply each side by $8(31 + 2t)$, the LCD.

$$8(17.5 + 2t) = 5(31 + 2t)$$
$$140 + 16t = 155 + 10t$$
$$6t = 15$$
$$t = \frac{15}{6}, \text{ or } 2.5$$

The width is **2.5** centimeters.

CLASSROOM EXERCISES

Complete the table.

1. The denominator of a fraction is 5 more than the numerator. When the numerator is increased by 3 and the denominator is increased by 4, the resulting fraction equals $\frac{1}{2}$.

Original Fraction	New Fraction
?	?
Equation	?

2. The numerator of a fraction is 4 less than the denominator. When both numerator and denominator are increased by 6, the resulting fraction equals $\frac{2}{3}$.

Original Fraction	New Fraction
?	?
Equation	?

3. The denominator of a fraction exceeds the numerator by 1. When the same integer is added to both numerator and denominator, the resulting fraction equals $\frac{8}{9}$.

Original Fraction	New Fraction
?	?
Equation	?

WORD PROBLEMS

A Solve each problem.

1. The numerator of a fraction exceeds the denominator by 8. The fraction which results from subtracting 3 from both the numerator and denominator equals 3. Find the original fraction.

2. The denominator of a fraction is one less than twice the numerator. Adding 7 to both the numerator and denominator results in a fraction which equals $\frac{7}{10}$. Find the original fraction.

3. The sum of the numerator and denominator of a fraction is 95. Reduced to lowest terms, the fraction equals $\frac{5}{14}$. Find the fraction.

4. The denominator of a fraction is one more than twice the numerator. When the numerator is increased by 8 and the denominator is increased by 10, the resulting fraction equals $\frac{2}{5}$. Find the original fraction.

5. The dimensions of a photograph are 12.5 centimeters by 17 centimeters. The photograph is to be framed so that the outside dimensions of the framed photograph will be in the ratio $\frac{39}{48}$. Find the width of the frame. (Assume uniform width.)

6. Clara invested $200,000 more in a business than her partner, Faye. If each had invested $100,000 more, the ratio of their investment would be 2:3. Find the amount each invested.

7. Clyde Burt has $\frac{2}{3}$ as much money invested at 11% as the amount invested at 12%. The total yearly return from these investments was $11,600. How much was invested at each rate?

8. The numerator of a fraction is 6 less than the denominator. When the numerator is increased by 3 and the denominator is increased by 2, the resulting fraction equals $\frac{3}{4}$. Find the numerator of the original fraction.

B

9. The sum of a number and its reciprocal is 2. Find the number(s).

10. The sum of a number and 6 times its reciprocal is 5. Find the number(s).

11. The sum of the reciprocal and twice the square of the reciprocal of a number is 3. Find the number(s).

12. A Senior Citizen Club plans an outing for which total costs amount to $480. When 6 members cancelled, each member had to pay an additional $4. How many members went on the outing?

Cost	Number of Members	Cost Per Member
480	m	?
480	$m - 6$	?

13. An art class plans a trip for which the total costs are $160. When two members cancel out, the cost increases by $4.00 per person. How many persons went on the trip?

14. Total tour costs for a group of people amounted to $800. When 4 additional people decided to go on the tour, the cost per person decreased by $10. How many went on the tour?

15. A manufacturer wishes to receive $4800 for a batch of items. When 1200 items were found to be defective, the selling cost per item increased by $0.20. How many items were manufactured?

16. A store owner plans to sell a number of toys for $48. When the shipment arrived, 16 toys were found to be damaged and the selling price was increased by $0.10 per toy. How many toys were in the shipment?

REVIEW CAPSULE FOR SECTION 7–7

Represent the given variable as indicated. *(Pages 123–127)*

1. A bus travels for t hours at an average rate of 80 kilometers per hour. Represent the distance traveled in terms of t.
2. A cyclist travels for 3 hours at an average rate of r miles per hour. Represent the distance traveled in terms of r.
3. A plane travels 1000 kilometers in t hours. Represent the speed of the plane in terms of t.
4. A car travels 500 miles at an average rate of r miles per hour. Represent the time for the trip in terms of r.

Problem Solving and Applications

7–7 Distance/Rate/Time

Recall the formula which relates distance, rate, and time.

$$d = rt$$

This formula is useful in solving motion problems.

EXAMPLE 1 A salesperson travels between two cities by car in 1 hour less than it takes to make the trip by bus. The average speed when traveling by car is 64 kilometers per hour. When traveling by bus, the average speed is 48 kilometers per hour. Find the distance between the two cities.

Solution: [1] Represent the unknown.

Let $y =$ the distance in kilometers.

[2] Make a table to organize the information.

	Distance	Rate	Time
Car	y	64	$\frac{y}{64}$
Bus	y	48	$\frac{y}{48}$

[3] Use the table to write an equation.

$$\frac{y}{64} + 1 = \frac{y}{48}$$ ← **Time for Car + 1 = Time for Bus**

[4] Solve the equation.

$$192\left(\frac{y}{64}\right) + 192 \cdot 1 = 192\left(\frac{y}{48}\right)$$ ← **LCD: 192**

$$3y + 192 = 4y$$

$$192 = y$$ The distance is **192 kilometers.**

When a plane travels in air, its actual speed is *increased by a tailwind* and *decreased by a headwind.* Similarly, the speed of a boat is increased when it is traveling in the same direction as the current and decreased when the boat is traveling against the current.

EXAMPLE 2 Flora Babcock, an airplane pilot, finds that it takes $\frac{2}{3}$ as much time to travel 450 kilometers with a tailwind as it takes to travel the same distance against the wind. The speed of her plane in still air is 200 kilometers per hour. Find the speed of the wind.

Solution: [1] Let $x =$ the speed of the wind in kilometers per hour.

[2]

	Distance	Rate	Time
With a Tailwind	450	$200 + x$	$\frac{450}{200 + x}$
With a Headwind	450	$200 - x$	$\frac{450}{200 - x}$

[3] $\frac{450}{200 + x} = \frac{2}{3}\left(\frac{450}{200 - x}\right)$ ⟵ $\begin{pmatrix}\text{Time with}\\ \text{a tailwind}\end{pmatrix} = \frac{2}{3}\begin{pmatrix}\text{Time with}\\ \text{a headwind}\end{pmatrix}$

[4] $\frac{450}{200 + x} = \frac{300}{200 - x}$ ⟵ LCD: $(200 + x)(200 - x)$

$$(200 - x)(450) = (200 + x)300$$

$$90{,}000 - 450x = 60{,}000 + 300x$$

$$30{,}000 = 750x$$

$40 = x$ The speed of the wind is **40 kilometers per hour.**

CLASSROOM EXERCISES

In Exercises 1–2, complete the table.

1. Mary jogs the same distance each day. Her average jogging speed is 8 kilometers per hour (abbreviated: km/h). When she decreases her speed by 2 km/h, she adds $\frac{1}{2}$ hour to her jogging time. How far does she jog each day?

Distance	Rate	Time
y	8	$\frac{y}{8}$
y	?	?
Equation	?	

2. A plane that has a speed of 450 km/h in still air takes off and heads west at a time when the wind is blowing from the west at a rate of 50 km/h. How far west can the plane go and return in 3 hours?

Distance	Rate	Time
x	$450 - 50$	$\frac{x}{400}$
x	?	?
Equation	?	

WORD PROBLEMS

A Solve each problem.

1. A business consultant travels between two cities by plane in 10 hours less than it takes to make the trip by train. The average speed of the plane is 480 km/h and the average speed of the train is 80 km/h. Find the distance between the cities.
2. It takes 24 minutes less to travel to an airport by taxi than it does to make the trip by taking a bus. The average speed when traveling by taxi is 80 km/h and the average speed when traveling by bus is 60 km/h. Find the distance to the airport.
3. A bus trip of 300 kilometers would have taken $\frac{4}{5}$ as long if the average speed had been increased by 15 km/h. What was the speed of the bus?
4. The speed of a boat in still water is 9 km/h. It takes the boat twice as long to travel 36 kilometers upstream on the Hurm River as it does to travel the same distance downstream. Find the rate of the current.
5. An airplane pilot found that it took $\frac{2}{3}$ as much time to travel 1800 kilometers with a 50 km/h tailwind as it took to travel the same distance with a 100 km/h headwind. Find the speed of the plane in still air.
6. A plane takes off and flies east at a rate of 300 km/h. On the return trip flying west, the plane's speed is 250 km/h. The plane has enough fuel for a safe flying time of $3\frac{1}{2}$ hours. How far east can the plane fly? Round your answer to the nearest whole kilometer.
7. A man has $3\frac{1}{4}$ hours in which to give some friends a tour of the surrounding countryside. How far from the house can the tour extend if the speed on the trip out is 25 km/h and the speed on the return trip is 40 km/h?
8. Gloria jogged from her home to Rock Creek, a distance of 16 miles. Then she walked home at a rate that was $\frac{2}{5}$ of her jogging rate. The round trip took 7 hours. Find Gloria's jogging and walking rates for the trip.
9. A light plane has a speed of 120 mph in still air. On a certain day, the pilot flies 700 miles with a tailwind. On the return trip, the plane covers $\frac{5}{7}$ of this distance in the same time. What is the speed of the wind?
10. The average speeds of two cars are in the ratio 4:5. Each car travels 100 miles. The faster car covers this distance in 30 minutes less than the slower car. Find the average speed of each car.

B

11. Julio intends to drive his motorbike to a friend's house 200 kilometers away. If he increases his planned rate of speed by 10 km/h, he can decrease his travel time by 40 minutes. What is his planned rate of speed?
12. The faster of two runners on a circular race course can complete a lap in 30 seconds. The slower runner can complete a lap in 40 seconds. The runners start from the same spot at the same time. How long will it take the faster runner to gain a lap on the slower runner?

13. The current in Tomahawk Lake has a speed of 2 km/h. Jeff Eagle can paddle his canoe 24 kilometers upstream and 24 kilometers back to his starting point in the same number of hours as he can paddle 50 kilometers in still water. What is Jeff's speed in still water?

C

14. In how many minutes after 3 o'clock will the hands of a clock first be together? (HINT: Let x represent the number of units that the minute hand travels. Then $\frac{x}{12}$ represents the number of units that the hour hand travels.)

15. In how many minutes after 5 o'clock will the hands of a clock first be together?

16. In how many minutes after 12 o'clock will the hands of a clock first be 20 minute spaces apart?

17. In how many minutes after 8 o'clock will the hands of a clock first be opposite each other?

CALCULATOR APPLICATIONS

Evaluating Continued Fractions

The reciprocal key, [1/x], on a scientific calculator will help you to evaluate a **terminating continued fraction.**

EXAMPLE Evaluate: $1+\cfrac{1}{2+\cfrac{1}{3+\cfrac{1}{4+\cfrac{1}{5}}}}$

SOLUTION Start at the bottom and work upward.

5 [1/x] [+] 4 [=] [1/x] [+] 3 [=] [1/x] [+] 2 [=] [1/x] [+] 1 [=]

1.433121

EXERCISES

Evaluate each continued fraction.

1. $1+\cfrac{1}{2+\cfrac{1}{1+\cfrac{1}{5}}}$

2. $1+\cfrac{1}{1+\cfrac{1}{1+\cfrac{1}{1+\cfrac{1}{7}}}}$

3. $2+\cfrac{1}{1+\cfrac{1}{2+\cfrac{1}{1+\cfrac{1}{3}}}}$

4. $2+\cfrac{1}{2+\cfrac{1}{2+\cfrac{1}{2+\cfrac{1}{9}}}}$

5. $4+\cfrac{1}{3+\cfrac{1}{2+\cfrac{1}{1+\cfrac{1}{5}}}}$

6. $2+\cfrac{1}{3+\cfrac{1}{4+\cfrac{1}{5+\cfrac{1}{11}}}}$

Problem Solving and Applications

7–8 Work/Rate Problems

Equations with rational coefficients occur frequently in work problems. Work problems deal with persons or machines that work at different rates of speed.

Unit Rate	×	Number of Units	=	Amount
60 km/hour	×	5 hours	=	300 kilometers
50 pages/day	×	3 days	=	150 pages
$\frac{1}{4}$ field/week	×	4 weeks	=	1 field

In many work problems, the amount of work done is the **complete job**, or **1.**

EXAMPLE 1 Rosita can type the articles for the school newspaper in 5 hours. It takes Janine 8 hours to type the same number of articles (Condition 2). How long will it take Rosita and Janine to type the articles if they work together (Condition 1)?

Solution: [1] Let $h =$ the number of hours it takes working together.

[2]

	Rate per Hour (Unit Rate)	Number of Hours (Number of Units)	Work Done (Amount)
Rosita	$\frac{1}{5}$ per hour	h	$\frac{1}{5} \cdot h$, or $\frac{h}{5}$
Janine	$\frac{1}{8}$ per hour	h	$\frac{1}{8} \cdot h$, or $\frac{h}{8}$

[3] Use Condition 2 and the table to write an equation.

$$\frac{h}{5} + \frac{h}{8} = 1 \quad \longleftarrow \quad \text{Work Done by Rosita} + \text{Work Done by Janine} = \text{Complete Job}$$

[4] Solve the equation.

$$\frac{h}{5} + \frac{h}{8} = 1 \quad \longleftarrow \quad \textbf{LCD: 40}$$

$$40\left(\frac{h}{5}\right) + 40\left(\frac{h}{8}\right) = 40(1)$$

$$8h + 5h = 40$$

$$13h = 40$$

$$h = \frac{40}{13}, \text{ or } 3\tfrac{1}{13}$$

It would take them $3\frac{1}{13}$ hours working together.

The check is left for you.

In work problems, it is useful to determine the unit rate.

Information	Unit Rate
A printing press completes a job in 15 hours.	$\frac{1}{15}$ per hour
A pump drains a lake in 24 hours.	$\frac{1}{24}$ per hour
A painter paints a room in h hours.	$\frac{1}{h}$ per hour

EXAMPLE 2 Two pumps working together can lower the water level in a dam by 1 meter in $7\frac{1}{5}$ hours. One pump working alone can lower the water level by 1 meter in 12 hours. How long will it take the second pump working alone to lower the water level by the same amount?

Solution: [1] Let h = the number of hours for Pump 2 working alone.

Then $\frac{1}{h}$ = work done per hour (rate) by Pump 2.

[2]

	Rate Per Hour	Number of Hours	Work Done
Pump 1	$\frac{1}{12}$ meter per hour	$7\frac{1}{5}$ hours	$\frac{1}{12} \cdot \frac{36}{5}$, or $\frac{3}{5}$
Pump 2	$\frac{1}{h}$ meter per hour	$7\frac{1}{5}$ hours	$\frac{1}{h} \cdot \frac{36}{5}$, or $\frac{36}{5h}$

[3] $\frac{3}{5} + \frac{36}{5h} = 1$ ← **Work Done by Pump 1** + **Work Done by Pump 2** = **Complete Job**

[4] $5h\left(\frac{3}{5}\right) + 5h\left(\frac{36}{5h}\right) = 5h(1)$

$$3h + 36 = 5h$$

$$36 = 2h$$

$$18 = h$$

Working alone, Pump 2 would take **18 hours** to lower the water level by 1 meter. The check is left for you.

CLASSROOM EXERCISES

Write an equation for each problem. Let t = time, r = rate, and A = work done.

1. Steve mows $\frac{1}{4}$ of a golf course in one day. How much of the course can he mow in t days?
2. Martina can address envelopes at the rate of $\frac{1}{5}$ box per hour. How many boxes can she address in t hours?

3. Sue can address envelopes at the rate of $\frac{1}{4}$ box per hour. How many boxes can Martina and Sue address in t hours by working together?

4. The rates for filling a tank with oil from 3 different pipes are $\frac{1}{4}$ tank/hour, $\frac{1}{5}$ tank/hour, and $\frac{1}{6}$ tank/hour. If all three pipes are open for t hours, how full is the tank after t hours?

WORD PROBLEMS

A

1. Fred can sweep the snow from a sidewalk in 4 minutes while Sarah can do it in 3 minutes. Working together, how long will it take them to sweep the sidewalk? (HINT: Fred's rate is $\frac{1}{4}$ sidewalk/minute. Sarah's rate is $\frac{1}{3}$ sidewalk/minute.)

2. If Ann can sweep the leaves from a yard in 6 minutes and Grace can do it in 8 minutes, how much can they get done working together for 3 minutes?

3. Joan can paint a house in 6 days. Tom can paint it in 9 days. If Joan works alone for 2 days and then is joined by Tom, how long would it take to paint the house? (HINT: The equation is $\frac{1}{6}t_1 + (\frac{1}{6}t_2 + \frac{1}{9}t_2) = 1$ where t_1 is the time Joan works alone, and t_2 is the time they work together.)

4. Tom can process a batch of bills in 8 hours. It takes Bill 10 hours to do the same job. How long would it take them together to complete the work?

5. Sue can assemble a bicycle in 10 minutes. Her sister takes 15 minutes to do the same job. How long would it take if they worked together?

6. Sam Chu can plow one of his fields with a tractor in 4 days. It takes his neighbor 12 days to plow the same field. How long will it take Sam Chu if his neighbor helps him?

7. Ruth can mow a certain strip of lawn with her mower in 20 minutes, and Jane can mow it with hers in 30 minutes. How long would it take if they did this work together?

8. If one pipe can fill a tank in 3 hours, a second pipe in 4 hours, and a third in 5 hours, how long will it take to fill the tank if all three pipes are being used?

9. Sue Ellers, working alone, can paint a house in 12 days. If Clyde Gallup helps, it takes only 4 days. How long will it take Clyde Gallup alone?

10. Art can do a job alone in 8 days. After he has been working alone for 2 days, he is joined by Pete. They finish the job together in 2 more days. How long would it take Pete to do the work alone?

B

11. Ellen can drive her car over a route in 4 hours, and Lila can drive her car over the same route in $3\frac{1}{2}$ hours. How long will it take them to meet if they start at at opposite ends at the same time?

12. A large pipe can fill a tank in 5 hours and a smaller pipe can fill it in 8 hours. A third pipe can empty the tank in 10 hours. How long would it take to fill the tank if all three pipes are open?

13. Steve, Jose, and Clyde can do a piece of work in 4 hours. If Steve can do the same piece of work in 10 hours, and Jose in 12 hours, how many hours will it take Clyde to do the work?

14. Sue's, Lucy's and Helen's rates of work are $\frac{1}{10}$ job/hour, $\frac{1}{12}$ job/hour, and $\frac{1}{13}$ job/hour. Set up the equation for Sue and Lucy if they work together 3 hours, and then are joined by Helen. How long will they take to complete the job?

15. One gardner can plant a flower bed in 4 hours. Another gardner can do the same job in $4\frac{1}{2}$ hours. How long would it take them working together?

C

16. Joe does $\frac{2}{3}$ of a job in 4 hours. Kara can do $\frac{3}{4}$ of what remains to be done in 1 hour. Gloria can do all of what remains to be done in 20 minutes. How long would it take to do the job if the three worked together?

Review

Solve and check each equation. *(Section 7–5)*

1. $3b - \frac{6b+7}{4} = 9$

2. $\frac{c-3}{c+7} = \frac{4}{15}$

3. $\frac{a}{a-5} = \frac{1}{a+5} - \frac{17}{25-a^2}$

Solve each problem.

4. The numerator of a fraction is 4 less than the denominator. If the numerator is increased by 20 and the denominator is decreased by 8, the result equals 9. Find the original fraction. *(Section 7–6)*

5. A science class planned a field trip for which the total costs were $80. When four additional students decided to go on the trip, the cost per student was reduced by $1. How many students went on the trip? *(Section 7–6)*

6. A trip of 500 miles would have taken $\frac{4}{5}$ as long if the average speed of the car had been increased by 10 miles per hour. What was the speed of the car? *(Section 7–7)*

7. Working alone, Sam can paint a room in 6 hours. Tim can paint the room in 4 hours. How long would it take Sam and Tim to complete the job if they worked together? *(Section 7–8)*

8. Juanita can make a patchwork quilt in 7 days. After working on the quilt for two days, Juanita is joined by Linda. They finish in 3 more days. How long would it take Linda to make the quilt alone? *(Section 7–8)*

BASIC: DISTANCE/RATE/TIME

Problem: *Write a program which computes distance, rate, or time. The user will enter three values, with a 0 indicating the unknown quantity.*

```
100 REM   D = DISTANCE, R = RATE, T = TIME
110 REM   GIVEN ANY TWO OF THE THREE, COMPUTE THE THIRD.
120 REM
130 PRINT "ENTER DISTANCE (IN MILES), RATE (IN MPH), TIME
(IN HOURS)"
140 PRINT "IN THAT ORDER WITH 0 FOR THE UNKNOWN QUANTITY";
150 INPUT D, R, T
160 IF D = 0 THEN 230
170 IF R = 0 THEN 250
180 IF T <> 0 THEN 210
190 LET T = D/R
200 GOTO 260
210 PRINT "YOU DID NOT ENTER 0 FOR ANY QUANTITY. TRY AGAIN."
220 GOTO 150
230 LET D = R * T
240 GOTO 260
250 LET R = D/T
260 PRINT "DISTANCE =";D;" MILES; RATE =";R;" MPH; TIME
=";T;" HOURS"
270 PRINT "ANY MORE PROBLEMS (1 = YES, 0 = NO)";
280 INPUT Z
290 IF Z = 1 THEN 130
300 END
```

The following is the output from a sample run of the program above.

```
RUN
ENTER DISTANCE (IN MILES), RATE (IN MPH), TIME (IN HOURS)
IN THAT ORDER WITH 0 FOR THE UNKNOWN QUANTITY? 300,60,0
DISTANCE = 300  MILES; RATE = 60  MPH; TIME = 5  HOURS

ANY MORE PROBLEMS (1 = YES, 0 = NO)? 1
ENTER DISTANCE (IN MILES), RATE (IN MPH), TIME (IN HOURS)
IN THAT ORDER WITH 0 FOR THE UNKNOWN QUANTITY? 435,55,2.6
YOU DID NOT ENTER 0 FOR ANY QUANTITY. TRY AGAIN.
? 1320,0,16
DISTANCE = 1320 MILES; RATE = 82.5 MPH; TIME = 16 HOURS

ANY MORE PROBLEMS (1 = YES, 0 = NO)? 0
READY
```

Analysis

Statements 160–250: If D = 0, the computer goes to statement 230 and computes D as `R * T`. If R = 0, statement 250 is used to compute `R = D/T`. If the computer reaches statement 180, then $D \neq 0$ and $R \neq 0$. Thus, if $T \neq 0$ also, the user failed to enter 0 for any of the three quantities. Statement 210 prints a message to this effect and asks the user to try again. Control then returns to statement 150. On the other hand, returning to statement 180, if T = 0, the computer moves to statement 190 and computes `T = D/R`.

Statements 260–290: Regardless of which formula is used in statements 190, 230, or 250, statement 260 prints the three quantities. Then the standard steps follow for determining whether the user wishes to continue.

EXERCISES

A Run the program on page 259 for the following values.

1. $r = 36$; $t = 17.5$; $d = ?$ **2.** $d = 957$; $r = 55$; $t = ?$ **3.** $d = 3987.5$; $t = 29$; $r = ?$

Write a BASIC program for each problem.

4. An important formula in electronics is $E = IR$, which relates E, the voltage in volts, I, the amount of current in amps, and R, the resistance in ohms. Given any two of these three quantities, find the third. The user will enter a 0 for the unknown quantity.

5. The formula $d = \frac{m}{v}$ relates density, d, mass, m, and unit volume, v. Given values for any two of these variables, find the third. The user will enter 0 for the unknown quantity.

6. Given two positive integers, compute and print their **least common multiple** (LCM).

7. Given the numerator and denominator of a fraction, print the fraction in lowest terms. For example, for an input of 6, 8, print `6/8 = 3/4`.

In Exercises 8–11, the input consists of integers A, B, C, and D, with B, C, and $D \neq 0$. Compute and print in lowest terms the following.

8. The sum `A/B + C/D`.

9. The difference `A/B - C/D`.

10. The product `A/B * C/D`.

11. The quotient of `A/B` divided by `C/D`.

B

12. Expand the program on page 259 so that it checks whether two or more of the three inputs are 0. If they are, print a suitable message and have the user try again to enter correct values.

13. Given three positive integers, compute and print their least common multiple (LCM).

Chapter Summary

IMPORTANT TERMS
Apparent solution (p. 246)
Least common denominator (p. 238)
Lowest terms (p. 230)
Rational expression (p. 230)

IMPORTANT IDEAS

1. If the numerator and denominator of a rational expression are multiplied or divided by the same nonzero polynomial, then the original rational expression and the resulting expression are equivalent.
2. To simplify a rational expression, first *identify the greatest common factor* (**GCF**) of the numerator and denominator. Then divide *both* the numerator and denominator by the GCF.
3. If $\frac{P}{Q}$ and $\frac{R}{S}$ are rational expressions, then
$$\frac{P}{Q} \cdot \frac{R}{S} = \frac{P \cdot R}{Q \cdot S}$$
and
$$\frac{P}{Q} \div \frac{R}{S} = \frac{P}{Q} \cdot \frac{S}{R} = \frac{P \cdot S}{Q \cdot R}$$
4. If $\frac{P}{Q}$ and $\frac{R}{Q}$ are rational expressions, then
$$\frac{P}{Q} + \frac{R}{Q} = \frac{P + R}{Q}$$
and
$$\frac{P}{Q} - \frac{R}{Q} = \frac{P - R}{Q}$$
5. Steps for addition and subtraction of rational expressions:
   [1] Find the LCD.
   [2] Write equivalent rational expressions having the LCD as denominator.
   [3] Add or subtract as indicated.
6. To write an equation with rational coefficients as one with integral coefficients, multiply each side of the equation by the least common denominator (LCD).
7. Distance/rate/time formula: $d = rt$

Chapter Objectives and Review

Objective: *To simplify rational expressions (Section 7–1)*

Simplify each rational expression.

1. $\frac{x^2y}{3xy}$
2. $\frac{6 - 4x}{2x - 3}$
3. $\frac{a^3 + 8}{4 - a^2}$
4. $\frac{3x^2 + x - 14}{9x + 21}$

Objective: *To multiply rational expressions (Section 7–2)*

Multiply.

5. $\frac{2a}{3} \cdot \frac{9}{ab^2}$

6. $\frac{4}{b^2 - 25} \cdot \frac{7b - 35}{8}$

7. $\frac{1 + t}{t^2 + 3t - 4} \cdot \frac{t^2 + 4t}{t^3}$

Objective: *To divide rational expressions (Section 7–3)*

Divide. No denominator equals zero.

8. $\frac{81}{xy} \div \frac{21}{x^3y}$

9. $\frac{3b}{9a^2 - 25b^2} \div \frac{6}{6a + 10b}$

10. $\frac{6t^2 - 7t - 5}{t^2} \div \frac{2t + 1}{3t^3}$

Objective: *To add and subtract rational expressions (Section 7–4)*

Perform the indicated operations.

11. $\frac{5x}{3z^2} - \frac{2x}{3z^2}$

12. $\frac{b}{(a - b)^2} + \frac{ab}{3a - 3b}$

13. $\frac{3s}{3r^2 - 5rs + 2s^2} + \frac{s - r}{r^2 - rs - 2s^2}$

14. $\frac{3}{5 + p} - \frac{2p + 1}{25 - p^2} + \frac{2p}{5 - p}$

Objective: *To solve equations involving rational expressions (Section 7–5)*

Solve and check each equation.

15. $\frac{x}{7} + \frac{x}{8} = \frac{15}{4}$

16. $\frac{8}{3x} + \frac{2}{9} = \frac{4}{x}$

17. $\frac{3}{t} - \frac{3}{5t} = \frac{6}{5}$

18. $\frac{s - 4}{s + 2} = \frac{5}{8}$

19. $\frac{8}{x + 2} - \frac{2}{x - 1} = \frac{5}{9}$

20. $\frac{3x}{8} - \frac{x}{10} = \frac{x - 1}{2} + \frac{1}{20}$

Objective: *To use equations with rational expressions to solve problems (Sections 7–6, 7–7, 7–8)*

21. The denominator of a fraction exceeds the numerator by 5. When 2 is added to both numerator and denominator, the resulting fraction equals $\frac{2}{3}$. Find the original fraction.

22. The sum of the numerator and denominator of a fraction is 66. When reduced to lowest terms the fraction is equal to $\frac{2}{9}$. Find the fraction.

23. Lars jogs uphill to work, averaging 8 miles per hour. Jogging home by the same route, downhill, he averages 11 miles per hour. The trip home takes him 15 minutes ($\frac{1}{4}$ hour) less. How far does he live from work?

24. Kate started out riding her bike at 15 miles per hour. After a while she had a flat tire, and she continued walking at 4 miles per hour. If she covered a total of 53 miles in 5 hours, how much of the distance did she cover on foot?

25. Robin can paint a room in 9 hours. Pat can paint the same room in 11 hours. How long would it take them working together to paint the room?

26. One gardener can plant a flower bed in 5 hours. Another gardener can plant it in 7 hours. How long would it take both of them to plant it together?

Chapter Test

Simplify each rational expression.

1. $\frac{5x^2z^3t}{10x^2z}$

2. $\frac{3y - 12}{4 - y}$

3. $\frac{3a + 6b}{8b + 4a}$

4. $\frac{r^2 + 2r - 8}{(2 - r)(r^2 + 8r + 16)}$

Perform the indicated operations.

5. $\frac{5}{2t + 1} + \frac{1}{6t + 3}$

6. $\frac{2m}{m^2 + 3m + 2} - \frac{2}{m + 2}$

7. $\frac{5q + 5t}{t^2 - 4q^2} \cdot \frac{2t - 4q}{t + q}$

8. $\frac{3a + 2}{10a^2 - 7a + 1} \div \frac{3a^2 - 13a - 10}{4a - 2}$

9. $\frac{2y}{y^2 + 4} + \frac{3y - 1}{y^4 - 16} - \frac{5}{y - 2}$

10. $\frac{c + d}{c - d} \cdot \frac{c^2 - d^2}{10c + 5d} \cdot \frac{5}{c + d}$

Solve and check each equation.

11. $\frac{1}{x - 1} - \frac{7}{15} = \frac{2}{x + 1}$

12. $\frac{20 - x}{5x + 4} = \frac{3}{11}$

13. The denominator of a fraction is one more than twice the numerator. When 10 is added to both the numerator and the denominator, the resulting fraction is equal to $\frac{3}{5}$. Find the original fraction.

14. Tim swam 800 meters upstream. Returning downstream the same distance took him $\frac{4}{5}$ as long. Tim's swimming speed in still water is 18 meters per minute. Find the rate of the current.

15. Janice can paint a room in 12 hours if she uses a brush, and in 8 hours if she uses a roller. She begins to paint with a brush and then changes to a roller. If she finishes painting the room in 9 hours, how long did she work with the brush?

Review of Word Problems: Chapters 1–7

1. Three years from now Alex will be twice as old as Gwen is then. Alex is 23 years old now. How old is Gwen? *(Section 1–7)*
2. Barbara invested \$16,000, part at 9% and part at 12%. Her total interest income per year from the investment is \$3360. How much did Barbara invest at each rate? *(Section 4–4)*
3. With a given tailwind, a plane traveled 400 miles in 2 hours. On the return trip, the same wind decreased the travel time by 1 hour. Find the speed of the plane going and returning. *(Section 4–4)*

After 10 miles of a marathon, 40 entrants had dropped out. After 15 miles, the number of dropouts totaled 90. Use this information in Exercises 4–7. *(Section 5–3)*

4. Assume that the relation is linear. Graph the relation.
5. Write an equation for the relation. Use d for the distance in miles and a for the number of dropouts from the race.
6. Use the equation to predict how many entrants will drop out after 22 miles.
7. Suppose the race began with 2000 entrants. About how many entrants will remain in the race after 26 miles?
8. The area of the shaded part of the figure at the right is $\pi a^2 - \pi b^2$. Write this expression in factored form. Then find the area of the shaded part when a is 9 meters and b is 5 meters. *(Section 6–4)*

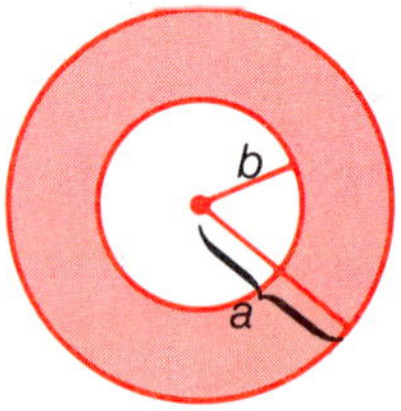

9. The smaller of two integers is 6 less than the larger. The sum of the squares of the two integers is 260. Find the integers. *(Section 6–9)*
10. The dimensions of a rectangular garden are 12 feet by 8 feet. How wide a border of uniform width should be dug around the sides of the garden in order to double the area of the garden? *(Section 6–10)*

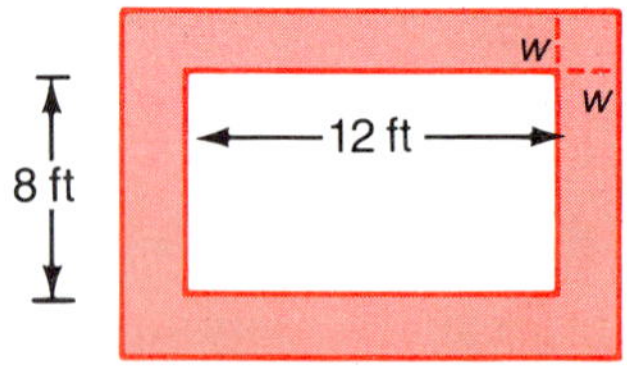

11. The sum of a positive number and 10 times its reciprocal is 7. Find the numbers. *(Section 7–6)*
12. A train travels 300 kilometers in the same time that a bus travels 200 kilometers. The speed of the train is 20 kilometers per hour faster than the speed of the bus. What is the speed of the train? *(Section 7–7)*
13. Working alone a carpenter can build a shed in $1\frac{1}{2}$ days. Working together with a helper, the carpenter can complete the same job in 1 day. How long would it take the helper, working alone, to complete the job? *(Section 7–8)*

Cumulative Review: Chapters 5–7

Choose the best answer. Choose *a*, *b*, *c*, or *d*.

1. If the replacement set for x in the relation $y = 7 - 2x$ is the set of whole numbers, what is the range of the relation?
   - **a.** $\{1, 2, 3, 4, \cdots\}$
   - **b.** $\{7, 6, 5, 4, \cdots\}$
   - **c.** $\{0, 1, 2, 3, 4, 5, 6, 7\}$
   - **d.** $\{7, 5, 3, 1, -1, \cdots\}$

2. Which is the graph of a function?

a.

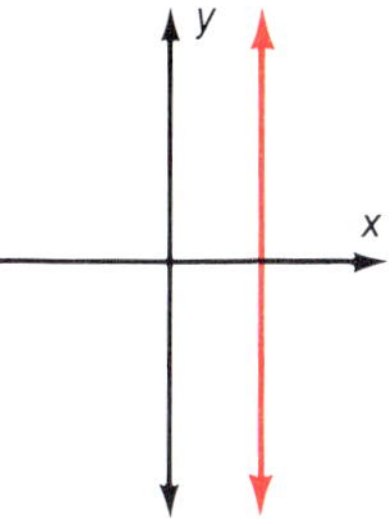

b.

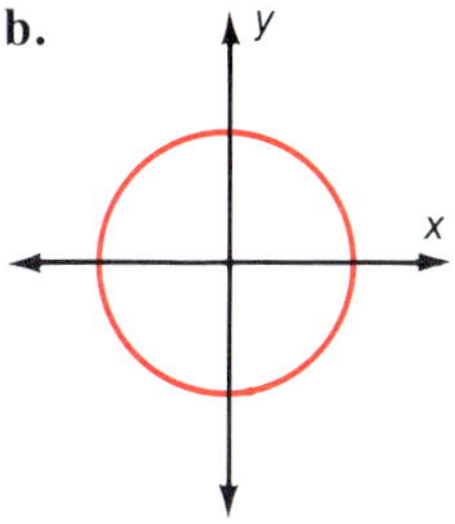

c.

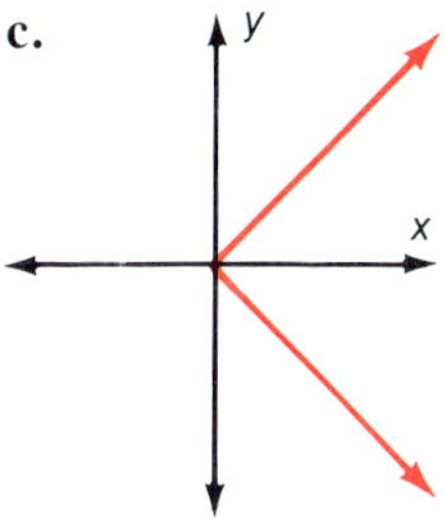

d.

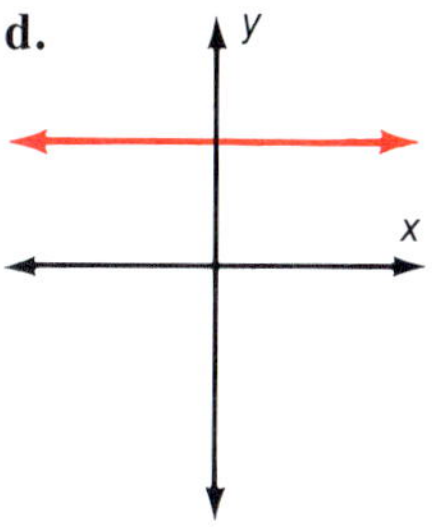

3. If $h(x) = -2x^2 + 2x - 2$, find $h(a - 1)$.
   - **a.** $2a^2 - 6a + 6$
   - **b.** $-2(a^2 - 3a + 3)$
   - **c.** -2
   - **d.** $-2a^2 + 2a - 2$

4. The cost, c, of printing 200 copies of s signatures of 16 pages each is $c = 100s + 350$ dollars. How much would it cost to print 200 copies of a book with 96 pages?
   - **a.** \$1050
   - **b.** \$10,350
   - **c.** \$950
   - **d.** Insufficient data

5. Find the inverse relation of $y = \frac{3}{2}x - 9$.
   - **a.** $y = \frac{2}{3}x + 18$
   - **b.** $x = \frac{2}{3}y + 6$
   - **c.** $y = \frac{2}{3}x + 6$
   - **d.** $2y = 3x - 18$

6. Add: $(3x^3 - 2x^2 + 3) + (5x^2 - 2x + 4)$
   - **a.** $8x^3 - 4x + 7$
   - **b.** $3x^3 + 5x^2 - 4x + 7$
   - **c.** $3x^3 + 3x^2 - 2x + 7$
   - **d.** $8x^2 + 7$

7. Subtract: $(2x^2 + 3x - 5) - (3x^3 + x - 2)$
   - **a.** $-3x^3 + 2x^2 + 2x - 3$
   - **b.** $-x^2 + 2x - 7$
   - **c.** $-3x^3 + 2x^2 + 2x - 7$
   - **d.** $x^3 + 2x - 3$

8. Multiply: $(-2x + 3)(5x + 4)$
   - **a.** $10x^2 - 7x + 12$
   - **b.** $-10x^2 + 23x - 12$
   - **c.** $-10x^2 + 7x + 12$
   - **d.** $-10x^2 + 7x - 12$

9. List all the common factors of $3x(x - 3)$ and $-9x^2(x - 3)$.
   - **a.** 3 and x only
   - **b.** 3 and $(x - 3)$ only
   - **c.** 3 and x^2 only
   - **d.** 3, x and $(x - 3)$ only.

10. Choose the binomial that is a factor of $b^3 + b^2 + b + 1$.
    - **a.** $b^2 - 1$
    - **b.** $b + 1$
    - **c.** $b - 1$
    - **d.** $b^2 + 2$

11. Which of the following is a perfect square trinomial?

a. $16x^2 - 32xy + 64y^2$ b. $16x^2 + 32xy + 64y^2$
c. $16x^2 - 64xy + 64y$ d. $16x^2 + 64xy + 64y^2$

12. Find a binomial factor of $x^2 + \frac{1}{4}x - \frac{1}{8}$.

a. $x - \frac{1}{4}$ b. $x + \frac{1}{4}$ c. $x - \frac{1}{8}$ d. $x + \frac{1}{8}$

13. Find a trinomial factor of $8n^3 - 27$.

a. $4n^2 + 12n^2 + 9$ b. $4n^2 - 12n + 9$
c. $4n^2 + 6n + 9$ d. $4n^2 + 6n - 9$

14. Find the first term of the quotient when $6x^3y - 19x^2y + 10$ is divided by $-2x + 5$.

a. $-3x^2y$ b. $-3x^3y^2$ c. $3x^2y$ d. $3x^3y^2$

15. The remainder when a polynomial, $P(x)$, is divided by $x - 3$ is -17. Which of the following is true?

a. $(x - 3)$ is a factor of $P(x)$. b. $P(-3) = -17$
c. $P(3) = 17$ d. $P(3) = -17$

16. Simplify: $\dfrac{60x^2yz^4}{24x^3y^2z^2}$

a. $\dfrac{60z}{24xy}$ b. $\dfrac{5z^2}{2xy}$ c. $\dfrac{10z^2}{4xy}$ d. $\dfrac{5z^2x}{2x^2y}$

17. Multiply: $\left(\dfrac{9xy}{2x^2y^2}\right)\left(\dfrac{5xy^4}{3x}\right)$

a. $\dfrac{15y^3}{2x}$ b. $\dfrac{45x^2y^4}{6x^3y^2}$ c. $\dfrac{15y^3x}{2}$ d. $\dfrac{15x}{2y^3}$

18. Divide: $\dfrac{x^2 + 5x + 6}{3x + 12} \div \dfrac{x + 2}{x^2 + 4x}$

a. $\dfrac{x(x + 4)}{3(x + 3)}$ b. $\dfrac{x^2 + 4x}{3}$ c. $\dfrac{x^2 + 3x}{3}$ d. $\dfrac{x^2 + 2x}{3}$

19. Add: $\dfrac{2r}{r + 3} + \dfrac{5 - r}{r - 3} - 1$

a. $\dfrac{-4(r + 6)}{r^2 - 9}$ b. $\dfrac{4(r + 6)}{r^2 - 9}$ c. $\dfrac{4(r - 6)}{r^2 - 9}$ d. $\dfrac{-4(r - 6)}{r^2 - 9}$

20. Subtract: $\dfrac{t - 7}{(t + 5)(t - 1)} - \dfrac{t - 9}{(t + 5)(t - 2)}$

a. $\dfrac{t + 6}{(t + 5)(t - 1)(t - 2)}$ b. $\dfrac{t - 5}{(t + 5)(t - 1)(t - 2)}$
c. $\dfrac{1}{(t - 1)(t - 2)}$ d. $\dfrac{t}{(t - 1)(t - 2)}$

21. Fred jogged 20 miles and walked slowly back. If the walking rate was $\frac{1}{5}$ the jogging rate and the round trip took 12 hours, what was his jogging rate?

a. 6 b. 10 c. 14 d. 18

Preparing for College Entrance Tests

Choose the best answer. Choose *a*, *b*, *c*, or *d*.

1. If $4x - 12 = 10$, what is the value of $x - 3$?

(a) $2\frac{1}{2}$ **(b)** $3\frac{1}{3}$ **(c)** $4\frac{3}{4}$ **(d)** $5\frac{1}{2}$

2. If $p > 0$ and $q < 0$, which of the following must be true?

(a) $pq < p^2$ **(b)** $p^2 < q^2$ **(c)** $p + q < p^2$ **(d)** $p - q > q^2$

3. If $x + y = 12$ and $x - y = -6$, find $x^2 - y^2$.

(a) -72 **(b)** -60 **(c)** 36 **(d)** 108

4. $\dfrac{54 + 54 + 54 + 54 + 54 + 54}{54 \cdot 54} = \underline{\ ?\ }$

(a) 216 **(b)** 108 **(c)** 1 **(d)** $\frac{1}{9}$

5. If m and p are real numbers and $mp = 0$, which of the following must be true?

(a) $m = 0$ **(b)** $p = 0$ **(c)** $m = 0$ <u>or</u> $p = 0$ **(d)** $m = 0$ <u>and</u> $p = 0$

6. An automobile was driven d kilometers using g liters of gasoline. At that rate, how many liters of gasoline will be used for $d + 100$ kilometers?

(a) $\dfrac{d(d + 100)}{g}$ **(b)** $g + \dfrac{100g}{d}$ **(c)** $g(d + 100)$ **(d)** $\dfrac{dg}{d + 100}$

7. If $\frac{3x}{7} = 9$, what is the value of $\frac{x}{3}$?

(a) $\frac{9}{7}$ **(b)** 7 **(c)** 21 **(d)** 63

8. If $-a + b = -2$ and $a + b = 8$, find $a^2 - b^2$.

(a) -112 **(b)** -60 **(c)** -16 **(d)** 16

9. $(3.1x - 12.4x) - (3.1x + 12.4x) = \underline{\ ?\ }$

(a) $-24.8x$ **(b)** $-6.2x$ **(c)** 0 **(d)** $6.2x$

10. If x and y are real numbers, with $x < y$ and $x + y < 0$, which of the following must be true?

(a) $y < 0$ **(b)** $x < 0$ **(c)** $|x| < |y|$ **(d)** $xy > 0$

11. If a, b, and c are nonzero numbers and $6a = 8b$ and $4b = 9c$, then $\frac{a}{c} = \underline{\ ?\ }$.

(a) $\frac{1}{3}$ **(b)** $\frac{3}{4}$ **(c)** $\frac{3}{2}$ **(d)** $\frac{3}{1}$

12. If $3m + 3p = 3$ and $4m + 2p = 6$, which of the following is true?

(a) $m + p < 2m + p$ **(b)** $2m + p < 3$

(c) $m + p > 2m + p$ **(d)** $m + p > 1$

13. If $m^2 - 4n^2 = 32$ and $m + 2n = 8$, find $m - 2n$.

(a) -4 **(b)** $\frac{1}{4}$ **(c)** 4 **(d)** 16

14. If $a = \frac{1}{2}$ and $b = (1 + a)(1 - 2a)(1 + 3a)(1 - 4a)$, which of the following is true?

(a) $b < 0$ (b) $b > a$ (c) $b = 0$ (d) $a = b$

15. If $a = b - 1$, what is the value of $|a - b| - |b - a|$?

(a) -2 (b) -1 (c) 0 (d) 1

16. $\dfrac{12(\frac{2}{3} - \frac{1}{6}) - 6(\frac{2}{3} - \frac{1}{6})}{3} =$ __?__

(a) $\frac{1}{3}$ (b) $\frac{2}{3}$ (c) 1 (d) $1\frac{1}{3}$

17. If $cd = -8$ and $c + d = 2$, find $c^2 + d^2$.

(a) -16 (b) 20 (c) 65 (d) 68

18. Six less than nine times a certain number is 12. What is two less than three times the number?

(a) 1 (b) 4 (c) $8\frac{1}{2}$ (d) 20

19. If n is a positive integer, then $n : n^2$ as $n + 1 :$ __?__

(a) $(n + 1)^2$ (b) $n(n + 2)$ (c) $n + 2$ (d) $n(n + 1)$

20. If $n \neq 0$ and $n^2 - n = n$, what is the value of $n - 1$?

(a) -1 (b) 0 (c) 1 (d) 2

21. If $x = -\frac{3}{4}$ and $y = (3x - 4)(4x - 3)(3x + 4)(4x + 3)$, which of the following is true?

(a) $y < x$ (b) $y = 0$ (c) $x - y = 0$ (d) $y > 0$

22. For nonzero real numbers, a, b, c, and d, $a = 2b$, $6b = 9c$, and $3c = 2d$. Then $\frac{d}{a} =$ __?__

(a) $\frac{1}{12}$ (b) $\frac{2}{9}$ (c) $\frac{1}{3}$ (d) $\frac{1}{2}$

23. If $a + b = Q$ and $a^2 - b^2 = N$, where neither Q nor N equals zero, find $a - b$ in terms of Q and N.

(a) QN (b) $\dfrac{Q}{N}$ (c) $Q^2 - N^2$ (d) $\dfrac{N}{Q}$

24. If $a + b = R$ and $a - b = N$, where neither R nor N equals zero, find $a^2 + b^2$ in terms of R and N.

(a) $R^2 + N^2$ (b) $2(R^2 + N^2)$ (c) $\dfrac{R^2 + N^2}{2}$ (d) RN

25. If $y - x = S$ and $x + y = T$, where neither S nor T equals 0, find $x^2 + y^2$ in terms of S and T.

(a) $\dfrac{S^2 - T^2}{2}$ (b) $\dfrac{S^2 + T^2}{2}$ (c) $S^2 + T^2$ (d) $\dfrac{S^2 + T^2}{4}$

26. If $pr = -32$ and $p + r = 4$, find $p^2 + r^2$.

(a) 16 (b) 48 (c) 80 (d) -128

CHAPTER **8**

Radicals/ Quadratic Functions

8–1 Radicals

Recall that every positive real number has two square roots.

Definition

> If n is a real number and $a^2 = n$, then
> $$a = \sqrt{n} \quad \text{or} \quad a = -\sqrt{n}.$$

The symbol, $\sqrt{\ }$, is used to indicate square roots. The symbol, $\sqrt[n]{\ }$, indicates an nth root.

Index → $\sqrt[n]{625}$ ← **Radical**

Radicand (points to 625)

Some numbers have more than one real nth root. For example, 64 has two square roots, 8 and -8. To avoid confusion, mathematicians have agreed that $\sqrt[n]{a}$ should have just one meaning for each integer n, and any real number a.

> **The principal nth root of a is a positive number unless n is odd and $a < 0$.**

$\sqrt{64} = 8$ ← **Principal square root of 64**

$-\sqrt{64} = -8$ $\qquad \pm\sqrt{64} = \pm 8$

$\sqrt[3]{-125} = -5$ ← **Principal cube root of -125**

EXAMPLE Simplify: **a.** $\sqrt{25x^2}$ **b.** $\sqrt[3]{-729x^6}$ **c.** $\sqrt[4]{16x^8}$

Solutions:

a. $\sqrt{25x^2} = |5x|$, or $5|x|$ ← **Since the principal nth root is positive, use $|x|$.**

b. $\sqrt[3]{-729x^6} = \sqrt[3]{(-9)^3(x^2)^3} = -9x^2$ ← **x^2 is always positive.**

c. $\sqrt[4]{16x^8} = \sqrt[4]{2^4(x^2)^4} = 2x^2$

The following definition will help you to find $\sqrt[n]{a}$.

Definition

> **Principal nth Root**
>
> For any real number a and any integer n, $n > 1$,
>
> **a.** $\sqrt[n]{a}$ is a positive real number or 0 when a is positive or equal to 0.
>
> **b.** $\sqrt[n]{a}$ is a negative real number when a is negative and n is odd.

When a is negative and n is even, $\sqrt[n]{a}$ does not represent a real number. In this case, $\sqrt[n]{a}$ represents a *complex number*. Complex numbers will be discussed in Chapter 9.

CLASSROOM EXERCISES

Simplify.

1. $\sqrt{121}$ 2. $\sqrt{36x^2}$ 3. $-\sqrt{100b^2}$ 4. $\sqrt{25x^{16}}$ 5. $\sqrt[3]{-8}$
6. $\sqrt[3]{125a^6}$ 7. $-\sqrt[4]{z^4}$ 8. $\sqrt{(x-1)^2}$ 9. $-\sqrt[5]{32}$ 10. $\sqrt[5]{-32}$
11. $\sqrt{144x^2y^4}$ 12. $\sqrt[3]{27a^3b^{12}}$ 13. $-6\sqrt{81}$ 14. $\pm\sqrt[4]{81a^4b^8}$ 15. $\sqrt{z^2+10z+25}$

WRITTEN EXERCISES

A Simplify.

1. $\sqrt{16}$ 2. $-\sqrt{25}$ 3. $\sqrt{36}$ 4. $\pm\sqrt{169}$ 5. $-\sqrt{100}$
6. $\sqrt{\frac{1}{4}}$ 7. $\sqrt{\frac{9}{25}}$ 8. $\pm\sqrt{196}$ 9. $2\cdot\sqrt{16}$ 10. $-4\cdot\sqrt{36}$
11. $\sqrt[3]{-216}$ 12. $\sqrt[3]{-1}$ 13. $-\sqrt[3]{-64}$ 14. $-\sqrt[3]{1000}$ 15. $\sqrt[3]{343}$
16. $\sqrt[4]{81}$ 17. $-\sqrt[4]{625}$ 18. $\pm\sqrt[4]{1296}$ 19. $\sqrt[5]{32}$ 20. $\sqrt[5]{-243}$
21. $\sqrt{x^2}$ 22. $\sqrt{y^6}$ 23. $\sqrt{x^4}$ 24. $-\sqrt{y^{16}}$ 25. $\sqrt{4a^2}$
26. $\sqrt{4a^4}$ 27. $\pm\sqrt{25a^4}$ 28. $\sqrt{9b^8}$ 29. $\sqrt[3]{8x^3}$ 30. $\sqrt[3]{-27x^6}$
31. $\sqrt[3]{-125b^{15}}$ 32. $\sqrt[4]{a^{12}b^8}$ 33. $\sqrt[4]{c^{20}d^{24}}$ 34. $\sqrt[3]{0.008z^3}$
35. $-\sqrt[3]{0.027b^6}$ 36. $\sqrt{(a+b)^2}$ 37. $-\sqrt{(y-9)^2}$ 38. $\pm\sqrt{(5-t)^4}$
39. $\sqrt{a^2+2a+1}$ 40. $\sqrt{y^2+12y+36}$ 41. $\sqrt{(t-1)^2}$ 42. $-\sqrt{c^2-8c+16}$

B

43. $\sqrt{36b^{36}}$ 44. $\sqrt{\frac{1}{m^2}}$ 45. $\sqrt{\frac{36a^2b^4}{49m^6}}$ 46. $\sqrt{\frac{a^4}{(2x)^6}}$
47. $\sqrt{\frac{(a+b)^4}{(a-b)^4}}$ 48. $\sqrt[3]{-8(x+y)^3}$ 49. $\sqrt[5]{32(x-y)^{10}}$ 50. $\sqrt[4]{81(1-a)^8}$
51. $\sqrt[3]{(x-y)^3}$ 52. $\sqrt[3]{-(x-y)^6}$ 53. $-\sqrt[4]{(r^3+s^3)^4}$ 54. $\sqrt[5]{-32(a-b)^{20}}$

C

55. $\sqrt{4x^{2a}}$ 56. $\sqrt{25a^{2x-4}}$ 57. $\sqrt{x^{8b-16}}$ 58. $\sqrt{x^n}$ 59. $\sqrt[n]{(x+y)^n}$

REVIEW CAPSULE FOR SECTION 8-2

Write each expression as a product of two factors, one of which is a perfect square monomial. *(Pages 197–200)*

Example: $125a^2b^3 = \mathbf{25a^2b^2 \cdot 5b}$

1. 20 2. 150 3. $9s^2t^5$ 4. $162x^3y$ 5. $12a^5b^6c^9$
6. $250a^3b^6$ 7. $81a^4b^3$ 8. $375x^6y^7$ 9. $216p^4q^5$ 10. $54c^2d^5$
11. $320x^{10}$ 12. $32x^3y^9$ 13. $875r^9t^8$ 14. $5a^3b^5$ 15. $49a^{2n}$

8–2 Simplifying Radicals

The following illustrates an important property of radicals.

$$\sqrt{100} = \mathbf{10} = \sqrt{25 \cdot 4} \quad \text{and} \quad \sqrt{25} \cdot \sqrt{4} = 5 \cdot 2 = \mathbf{10}$$

$$\text{Thus, } \sqrt{\mathbf{25 \cdot 4}} = \sqrt{\mathbf{25}} \cdot \sqrt{\mathbf{4}}.$$

Theorem 8–1

Multiplication Theorem for Radicals

For all positive real numbers a and b and for any integer n, $n > 1$,

$$\sqrt[n]{a} \cdot \sqrt[n]{b} = \sqrt[n]{ab}.$$

To simplify square roots, find any factors of the radicand that are perfect squares. REMEMBER: 121 is a perfect square because $11 \cdot 11 = 121$.

EXAMPLE 1 Multiply. Simplify the product when possible.

a. $\sqrt{12} \cdot \sqrt{24}$ **b.** $5\sqrt{2} \cdot 4\sqrt{3}$

Solutions: Apply the Multiplication Theorem for Radicals.

a. $\sqrt{12} \cdot \sqrt{24} = \sqrt{12 \cdot 24}$ ← **By Theorem 8–1**

$= \sqrt{288}$

$= \sqrt{144 \cdot 2}$ ← **144 is a perfect square.**

$= \mathbf{12\sqrt{2}}$

b. $5\sqrt{2} \cdot 4\sqrt{3} = (5 \cdot 4)(\sqrt{2} \cdot \sqrt{3})$

$= 20\sqrt{2 \cdot 3}$

$= \mathbf{20\sqrt{6}}$

In Example 1b, $\sqrt{6}$ cannot be simplified because 6 is not a perfect square and none of its factors, other than 1, is a perfect square.

To simplify nth roots, find any factors that are nth powers.

EXAMPLE 2 Multiply: $\sqrt[4]{5a^3b^5} \cdot \sqrt[4]{125a^2b^3}$

Solution: $\sqrt[4]{5a^3b^5} \cdot \sqrt[4]{125a^2b^3} = \sqrt[4]{5 \cdot 125 \cdot a^3 \cdot a^2 \cdot b^5 \cdot b^3}$ ← **By Theorem 8–1**

$= \sqrt[4]{625 \cdot a^5 \cdot b^8}$ ← **Find factors that are powers of 4.**

$= \sqrt[4]{625a^4b^8} \cdot \sqrt[4]{a}$

$= \mathbf{5a^2b^2\sqrt[4]{a}}$

The following illustrates a similar property for division of radicals.

$$\text{Since } \frac{\sqrt{25}}{\sqrt{16}} = \frac{5}{4} \quad \text{and} \quad \sqrt{\frac{25}{16}} = \frac{5}{4}, \quad \frac{\sqrt{\mathbf{25}}}{\sqrt{\mathbf{16}}} = \sqrt{\frac{\mathbf{25}}{\mathbf{16}}}.$$

Theorem 8–2

Division Theorem for Radicals

For all real numbers a and b where $a \geq 0$ and $b > 0$, and for any integer n, $n > 1$,

$$\frac{\sqrt[n]{a}}{\sqrt[n]{b}} = \sqrt[n]{\frac{a}{b}}.$$

EXAMPLE 3 Divide and simplify: **a.** $\sqrt{96} \div \sqrt{8}$ **b.** $\frac{\sqrt{135x^5}}{\sqrt{5x}}$

Solutions:

a. $\sqrt{96} \div \sqrt{8} = \sqrt{\frac{96}{8}}$ ← **By Theorem 8–2**

$= \sqrt{12}$ ← **Simplify.**

$= \sqrt{4} \cdot \sqrt{3}$

$= 2\sqrt{3}$

b. $\frac{\sqrt{135x^5}}{\sqrt{5x}} = \sqrt{\frac{135x^5}{5x}}$

$= \sqrt{27x^4}$

$= \sqrt{9x^4} \cdot \sqrt{3}$

$= 3x^2\sqrt{3}$

When the denominator of a rational expression contains a radical, a procedure called **rationalizing the denominator** can be used to express the denominator as a rational number.

EXAMPLE 4 Divide $\sqrt{6}$ by $\sqrt{12}$. Rationalize the denominator.

Solution: $\sqrt{6} \div \sqrt{12} = \sqrt{\frac{6}{12}}$ ← **By Theorem 8–2**

$= \sqrt{\frac{1}{2}}$, or $\frac{1}{\sqrt{2}}$

$= \frac{1}{\sqrt{2}} \cdot \frac{\sqrt{2}}{\sqrt{2}}$ ← **Multiplying by $\frac{\sqrt{2}}{\sqrt{2}}$ will make the denominator a rational number.**

$= \frac{1 \cdot \sqrt{2}}{\sqrt{2 \cdot 2}} = \frac{\sqrt{2}}{2}$

CLASSROOM EXERCISES

Multiply. Simplify the product when possible.

1. $\sqrt{3} \cdot \sqrt{5}$
2. $5\sqrt{2} \cdot 6\sqrt{3}$
3. $\sqrt{3} \cdot \sqrt{3}$
4. $3\sqrt{5} \cdot 4\sqrt{5}$
5. $\sqrt{a} \cdot \sqrt{b^3}$
6. $\sqrt{9x^3} \cdot \sqrt{16xy^4}$
7. $\sqrt{2a} \cdot \sqrt{3b}$
8. $\sqrt[3]{8} \cdot \sqrt[3]{16p^6q^3}$

Divide and simplify.

9. $\sqrt{6} \div \sqrt{3}$
10. $\sqrt{8} \div \sqrt{2}$
11. $\sqrt{18b^5} \div \sqrt{9b^3}$
12. $\sqrt{98r^5} \div \sqrt{7r}$

Rationalize the denominator.

13. $\sqrt{\frac{7}{8}}$
14. $\sqrt{\frac{35}{6}}$
15. $\sqrt{\frac{1}{3}}$
16. $-\frac{1}{\sqrt{5}}$
17. $\sqrt{1\frac{1}{5}}$

WRITTEN EXERCISES

A Multiply. Simplify the product when possible.

1. $\sqrt{2} \cdot \sqrt{5}$
2. $\sqrt{8} \cdot \sqrt{2}$
3. $\sqrt{5} \cdot \sqrt{5}$
4. $2\sqrt{2} \cdot 2\sqrt{6}$
5. $3\sqrt{2} \cdot 5\sqrt{2}$
6. $5\sqrt{3} \cdot 3\sqrt{2}$
7. $6\sqrt{2} \cdot 2\sqrt{6}$
8. $(\sqrt{5})(-\sqrt{5})$

9. $4\sqrt{18} \cdot 5\sqrt{3}$
10. $6\sqrt{12} \cdot 10\sqrt{2}$
11. $3\sqrt{4} \cdot 3\sqrt{4}$
12. $(-5\sqrt{2})(5\sqrt{2})$
13. $\sqrt{y^2} \cdot \sqrt{y^3}$
14. $\sqrt{3y} \cdot \sqrt{2y}$
15. $\sqrt{5z^2} \cdot \sqrt{4z^2}$
16. $\sqrt{3b} \cdot \sqrt{12b^4}$
17. $\sqrt{4a} \cdot \sqrt{ab}$
18. $\sqrt{4e^3g} \cdot \sqrt{6eg^2}$
19. $\sqrt{8} \cdot \sqrt{6x^3}$
20. $\sqrt{98x} \cdot \sqrt{2x}$
21. $\sqrt{90x} \cdot \sqrt{5x^2y}$
22. $\sqrt{12ab} \cdot \sqrt{3a^3b}$
23. $\sqrt{3x^3} \cdot \sqrt{18y^2}$
24. $\sqrt{cd} \cdot \sqrt{dg}$
25. $\sqrt[3]{2} \cdot \sqrt[3]{4}$
26. $\sqrt[4]{16} \cdot \sqrt[4]{16}$
27. $7\sqrt[3]{8} \cdot 5\sqrt[3]{27}$
28. $\sqrt[5]{27} \cdot 2\sqrt[5]{9}$
29. $\sqrt[4]{a^3b} \cdot \sqrt[4]{ab^3}$
30. $\sqrt[4]{2y} \cdot \sqrt[4]{5y^3}$
31. $\sqrt[3]{25a^6} \cdot \sqrt[3]{25b^9}$
32. $\sqrt[3]{8a} \cdot \sqrt[3]{8b}$

Divide and simplify.

33. $\frac{\sqrt{136}}{\sqrt{8}}$
34. $\frac{\sqrt{125}}{\sqrt{5}}$
35. $\frac{\sqrt{40}}{\sqrt{2}}$
36. $\frac{\sqrt{128}}{\sqrt{8}}$
37. $\frac{\sqrt{6}}{\sqrt{3}}$
38. $\frac{\sqrt{8}}{\sqrt{2}}$
39. $\frac{\sqrt{18}}{\sqrt{2}}$
40. $\frac{\sqrt{21}}{\sqrt{7}}$
41. $\frac{\sqrt{27}}{\sqrt{3}}$
42. $\frac{\sqrt{18}}{\sqrt{3}}$
43. $\frac{\sqrt{60}}{\sqrt{20}}$
44. $\frac{\sqrt{20}}{\sqrt{5}}$
45. $\frac{\sqrt{40}}{\sqrt{5}}$
46. $\frac{8\sqrt{6}}{4\sqrt{2}}$
47. $\frac{10\sqrt{8}}{2\sqrt{2}}$
48. $\frac{\sqrt{50}}{\sqrt{25}}$
49. $\frac{\sqrt{91}}{\sqrt{13}}$
50. $\frac{\sqrt{98}}{\sqrt{7}}$

Rationalize the denominator.

51. $\sqrt{\frac{1}{5}}$
52. $\sqrt{\frac{1}{6}}$
53. $\sqrt{\frac{1}{10}}$
54. $\sqrt{\frac{5}{6}}$
55. $\sqrt{\frac{3}{11}}$
56. $\sqrt{\frac{2}{7}}$
57. $\sqrt{\frac{1}{a}}$
58. $\sqrt{\frac{3}{b}}$
59. $\sqrt{\frac{4}{q}}$
60. $\sqrt{\frac{2}{ab}}$
61. $\sqrt{\frac{3}{xy}}$
62. $\sqrt{\frac{1}{a^2b}}$

Divide. Express all denominators as rational numbers.

63. $\frac{\sqrt{78}}{\sqrt{18}}$
64. $\frac{\sqrt{44}}{\sqrt{99}}$
65. $\frac{8\sqrt{32}}{2\sqrt{50}}$
66. $\frac{\sqrt{1000}}{\sqrt{8}}$
67. $\frac{\sqrt{121}}{\sqrt{72}}$
68. $\sqrt{\frac{36}{27}}$
69. $\frac{\sqrt{15y}}{\sqrt{5}}$
70. $\frac{\sqrt{3x}}{\sqrt{x}}$
71. $\frac{\sqrt{35a^5}}{\sqrt{7a^2}}$
72. $\frac{\sqrt{12x^2}}{\sqrt{5x}}$
73. $\frac{\sqrt{32x^3}}{\sqrt{8x}}$
74. $\frac{\sqrt{21a^5}}{\sqrt{2a}}$

B Multiply. Simplify the product when possible.

75. $(\sqrt{3})^2$
76. $(4\sqrt{5})^2$
77. $(-4\sqrt{2})^2$
78. $-(4\sqrt{2})^2$
79. $3(\sqrt{3} - \sqrt{2})$
80. $3(\sqrt{2} + \sqrt{3})$
81. $\sqrt{7}(2 - \sqrt{7})$
82. $5\sqrt{5}(5\sqrt{5} - 3\sqrt{2})$

Divide. Express all denominators as rational numbers.

83. $\sqrt{3} \div \sqrt{\frac{1}{3}}$
84. $\sqrt{3} \div \sqrt{\frac{3}{4}}$
85. $\sqrt{\frac{2}{3}} \div \frac{\sqrt{3}}{\sqrt{2}}$
86. $\sqrt{1\frac{1}{4}} \div \sqrt{\frac{4}{5}}$

C

87. Prove: *For all real numbers a and b,* $\sqrt{a} \cdot \sqrt{b} = \sqrt{ab}$.

88. Prove: *For all real numbers a and b where* $a \geq 0$ *and* $b > 0$, $\frac{\sqrt{a}}{\sqrt{b}} = \sqrt{\frac{a}{b}}$.

89. Prove: *For all nonnegative real numbers a and b,* $\sqrt{a^2 + b^2} \leq a + b$.

(HINT: Assume $\sqrt{a^2 + b^2} > a + b$. Then show that this leads to a contradiction.)

REVIEW CAPSULE FOR SECTION 8–3

Add or subtract as indicated. *(Pages 190–192)*

1. $a^2 - 3ab + b^2 - 2a^3 - a^2 + 3b^2 + 5ab$ **2.** $3 - 5r^3 + 6r - (r^3 - 7)$

3. $2x^3 - 5x^2 - 3x + 1 + 3x^2 + 3x - 1$ **4.** $3q^2 - (q + 9) - 7q^2 + 4q$

Multiply. *(Pages 193–196)*

5. $(x + y)(x - y)$ **6.** $(9q - 7r)(9q + 7r)$ **7.** $(2x + 1)(x - 8)$

8–3 Radicals: Addition/Subtraction

The distributive postulate can be used to add or subtract like radicals. Like radicals have the *same index* and the *same radicand.*

Same index

$$3\sqrt[3]{7} + 2\sqrt[3]{7} = (3 + 2)\sqrt[3]{7}$$ ← **By the distributive postulate**

Same radicand

$$= 5\sqrt[3]{7}$$

Sometimes radicands must be simplified in order to obtain like radicals.

EXAMPLE 1 Add or subtract as indicated: $\sqrt{8} - \sqrt{12} + 5\sqrt{2}$

Solution: Simplify $\sqrt{8}$ and $\sqrt{12}$ in order to obtain like radicands.

$$\sqrt{8} - \sqrt{12} + 5\sqrt{2} = \sqrt{4}\cdot\sqrt{2} - \sqrt{4}\cdot\sqrt{3} + 5\sqrt{2}$$

$$= 2\sqrt{2} - 2\sqrt{3} + 5\sqrt{2}$$

$$= (2\sqrt{2} + 5\sqrt{2}) - 2\sqrt{3}$$ ← **By the commutative and associative postulates**

$$= (2 + 5)\sqrt{2} - 2\sqrt{3}$$ ← **By the distributive postulate**

$$= \mathbf{7\sqrt{2} - 2\sqrt{3}}$$

As Example 1 shows, $7\sqrt{2}$ and $2\sqrt{3}$ cannot be combined into one term because their radicands are not alike.

You can also simplify products involving sums or differences of radicals.

EXAMPLE 2 Multiply: $(\sqrt{3} + \sqrt{2})(2\sqrt{3} + 3\sqrt{2})$

Solution:

$$(\sqrt{3} + \sqrt{2})(2\sqrt{3} + 3\sqrt{2}) = \sqrt{3}(2\sqrt{3} + 3\sqrt{2}) + \sqrt{2}(2\sqrt{3} + 3\sqrt{2})$$

$$= \sqrt{3}(2\sqrt{3}) + \sqrt{3}(3\sqrt{2}) + \sqrt{2}(2\sqrt{3}) + \sqrt{2}(3\sqrt{2})$$

$$= 6 + 3\sqrt{6} + 2\sqrt{6} + 6$$ ← **Combine like terms.**

$$= \mathbf{12 + 5\sqrt{6}}$$

Rationalizing a denominator sometimes involves sums or differences of radicals. In this case, you use the fact that

$$(a+b)(a-b)=a^2-b^2.$$

That is, $(\sqrt{2}-\sqrt{5})(\sqrt{2}+\sqrt{5})=(2)^2-(5)^2=2-5$, or $\mathbf{-3}$.

Expressions such as $(\sqrt{2}+\sqrt{5})$ and $(\sqrt{2}-\sqrt{5})$ are called **conjugates**.

Thus, to rationalize the denominator of an expression such as $\frac{3+\sqrt{5}}{1-\sqrt{5}}$, multiply the numerator and denominator by the conjugate of $1-\sqrt{5}$.

EXAMPLE 3 Rationalize the denominator: $\frac{3+\sqrt{5}}{1-\sqrt{5}}$

Solution: Multiply both numerator and denominator by $1+\sqrt{5}$.

$$\frac{3+\sqrt{5}}{1-\sqrt{5}}=\frac{3+\sqrt{5}}{1-\sqrt{5}}\cdot\frac{1+\sqrt{5}}{1+\sqrt{5}}$$

$$=\frac{3+3\sqrt{5}+\sqrt{5}+5}{(1)^2-(\sqrt{5})^2}$$

$$=\frac{8+4\sqrt{5}}{-4}$$

$$=\frac{-4(-2-\sqrt{5})}{-4}=\mathbf{-2-\sqrt{5}}$$

CLASSROOM EXERCISES

Add or subtract as indicated.

1. $5\sqrt{7}+2\sqrt{7}$ **2.** $17\sqrt{3}-\sqrt{3}$ **3.** $\sqrt{8}+\sqrt{32}$ **4.** $5\sqrt{2}-\sqrt{63}$

Multiply.

5. $(5+\sqrt{2})(5-\sqrt{2})$ **6.** $(\sqrt{7}-9)(\sqrt{7}+9)$

7. $(1+\sqrt{5})(3-\sqrt{5})$ **8.** $(4+2\sqrt{7})(3+19)$

Rationalize each denominator.

9. $\frac{5}{1+\sqrt{7}}$ **10.** $\frac{1}{\sqrt{8}-5}$ **11.** $\frac{a}{\sqrt{3}-\sqrt{2}}$ **12.** $\frac{y}{\sqrt{5}+\sqrt{7}}$

WRITTEN EXERCISES

A Add or subtract as indicated.

1. $4\sqrt{3}-2\sqrt{3}$ **2.** $5\sqrt{2}-3\sqrt{2}$ **3.** $6\sqrt{a}-5\sqrt{a}$ **4.** $8\sqrt{2}+\sqrt{2}$

5. $\sqrt{3}+\sqrt{27}$ **6.** $\sqrt{20}+\sqrt{12}$ **7.** $\sqrt{2}+\sqrt{8}$ **8.** $\sqrt{98}-\sqrt{50}$

9. $2\sqrt{12}-3\sqrt{48}$ **10.** $7\sqrt{18}-\sqrt{50}$ **11.** $5\sqrt{12}+3\sqrt{27}$ **12.** $2\sqrt{99}-\sqrt{176}$

13. $5\sqrt{18}+6\sqrt{2}$ **14.** $3\sqrt{45}-2\sqrt{50}$ **15.** $32\sqrt{b^3}-5\sqrt{b^5}$ **16.** $\sqrt{a^2b}+\sqrt{bc^2}$

17. $6\sqrt{20} - \sqrt{45} + 2\sqrt{5}$

18. $2\sqrt{72} - 5\sqrt{20} - \sqrt{98}$

19. $6\sqrt{2} + 7\sqrt{3} + 5\sqrt{2}$

20. $6\sqrt{5} + 8\sqrt{20} - \sqrt{80}$

21. $3 + 4\sqrt{2} - 5\sqrt{50}$

22. $\frac{\sqrt{3}}{5} + \frac{\sqrt{5}}{3}$

23. $\sqrt{\frac{3}{4}} + \sqrt{\frac{1}{3}} - \sqrt{\frac{1}{5}}$

24. $2\sqrt{45} - \frac{2}{3}\sqrt{\frac{1}{5}}$

25. $\frac{-3 + \sqrt{2}}{2} + \frac{-3 - \sqrt{2}}{2}$

26. $\frac{5 + \sqrt{7}}{6} + \frac{5 - \sqrt{7}}{6}$

27. $\sqrt{\frac{y}{7}} + \sqrt{\frac{y}{28}}$

28. $\sqrt{\frac{t}{3}} - \sqrt{\frac{t}{12}}$

29. $2\sqrt[3]{3} + \sqrt[3]{4}$

30. $2\sqrt[3]{3} + \sqrt[3]{24}$

31. $4\sqrt[3]{16} + 3\sqrt[3]{54}$

32. $2\sqrt[3]{32} - \sqrt[3]{108} - \sqrt[3]{\frac{1}{2}}$

Multiply.

33. $(3\sqrt{2} + 2\sqrt{3})(5\sqrt{2} - \sqrt{3})$

34. $(\sqrt{6} + 8)(\sqrt{6} - 3)$

35. $(5\sqrt{2} - 2\sqrt{3})^2$

36. $(a + b\sqrt{2})^2$

37. $(\sqrt{5} + 4)(\sqrt{5} - 4)$

38. $(\sqrt{6} + \sqrt{3})(\sqrt{6} - \sqrt{3})$

39. $(2\sqrt{3} - 3\sqrt{2})(2\sqrt{3} + 3\sqrt{2})$

40. $(2\sqrt{7} - 5)(2\sqrt{7} + 5)$

41. $(-\sqrt{3} - \sqrt{5})(-\sqrt{3} + \sqrt{5})$

42. $(-3\sqrt{5} - 2\sqrt{3})(-3\sqrt{5} + 2\sqrt{3})$

43. $\left(\frac{3 + 2\sqrt{5}}{2}\right)\left(\frac{3 - 2\sqrt{5}}{2}\right)$

44. $\left(\frac{6 + 3\sqrt{2}}{4}\right)\left(\frac{6 - 3\sqrt{2}}{4}\right)$

Rationalize the denominator.

45. $\frac{2}{3 - \sqrt{5}}$

46. $\frac{12}{\sqrt{5} - 2}$

47. $\frac{4}{\sqrt{5} + \sqrt{7}}$

48. $\frac{3}{\sqrt{2} - \sqrt{5}}$

49. $\frac{2 + \sqrt{3}}{1 - \sqrt{3}}$

50. $\frac{2 - \sqrt{3}}{2 + 3\sqrt{3}}$

51. $\frac{1 + 5\sqrt{2}}{4\sqrt{2} - 3}$

52. $\frac{2 + \sqrt{11}}{3\sqrt{11} - 7}$

53. $\frac{7 - \sqrt{5}}{2\sqrt{7} + 3\sqrt{5}}$

54. $\frac{\sqrt{5} + 2}{2\sqrt{5} - 4}$

55. $\frac{\sqrt{3} - 1}{\sqrt{3} + 1}$

56. $\frac{\sqrt{6} - \sqrt{2}}{\sqrt{6} + \sqrt{2}}$

B Add or subtract as indicated. Express denominators as rational numbers.

57. $\frac{\sqrt{a} + \sqrt{b}}{\sqrt{a} - \sqrt{b}} - \frac{\sqrt{a} - \sqrt{b}}{\sqrt{a} + \sqrt{b}}$

58. $\frac{\sqrt{x} + \sqrt{y}}{\sqrt{x} - \sqrt{y}} + \frac{\sqrt{x} - \sqrt{y}}{\sqrt{x} + \sqrt{y}}$

59. $\frac{r}{\sqrt{r + 1}} - \sqrt{r + 1}$

60. $\frac{r}{\sqrt{r - 1}} + \sqrt{r - 1}$

C Factor completely over the set of real numbers.

61. $y^2 + 8y\sqrt{2} + 30$

62. $m^2 + 2m\sqrt{3} - 24$

63. $t^2 + 2t\sqrt{5} + 5$

64. $18r^2 + 5r\sqrt{3} - 6$

65. Given that $2x = -1 + \sqrt{5}$ and that $x^2 + x$ is an integer, find x.

66. A set of numbers forms a group under multiplication if it satisfies these four postulates: closure, associativity, existence of an identity element, and existence of a multiplicative inverse for each number. Show that the set of nonzero numbers of the form $a + b\sqrt{2}$ where a and b are rational numbers, forms a group under multiplication.

8–4 Solving Equations with Radicals

Equations such as $\sqrt{x} = 7$ and $\sqrt{4x+4} = 12$ are **radical equations.** To solve an equation in which one term contains a variable in a radicand, you get that term alone on one side of the equation.

EXAMPLE 1 Solve and check: $\sqrt{4x+4} = 12$

Solution:

$$\sqrt{4x+4} = 12$$

← **Square both sides to remove the $\sqrt{\ }$.**

$$(\sqrt{4x+4})^2 = (12)^2$$
$$4x + 4 = 144$$
$$4x = 140$$
$$x = 35$$

Check:

$$\sqrt{4x+4} \stackrel{?}{=} 12$$
$$\sqrt{4(35)+4} \stackrel{?}{=} 12$$
$$\sqrt{144} \stackrel{?}{=} 12$$
$$12 \stackrel{?}{=} 12 \quad \text{Yes } \checkmark$$

Solution set: $\{35\}$

Raising both sides of an equation to the same exponent does not always produce an equivalent equation. Thus, numbers which appear to be solutions may not satisfy the original equation. These are called **apparent solutions.** Always check carefully to eliminate any apparent solutions.

EXAMPLE 2 Solve and check: $\sqrt{x} = -2$

Solution:

$$(\sqrt{x})^2 = (-2)^2$$
$$x = 4$$

Check:

$$\sqrt{4} \stackrel{?}{=} -2$$
$$2 \neq -2$$

The number 4 is an *apparent* solution. The solution set is ϕ.

When two terms of a radical equation contain variables in the radicand, express the equation with one of these terms on each side.

EXAMPLE 3 Solve and check: $\sqrt{2x-3} - \sqrt{5x-21} = 0$

Solution: Add $\sqrt{5x-21}$ to each side.

$$\sqrt{2x-3} - \sqrt{5x-21} = 0$$
$$\sqrt{2x-3} = \sqrt{5x-21}$$
$$(\sqrt{2x-3})^2 = (\sqrt{5x-21})^2$$
$$2x - 3 = 5x - 21$$
$$-3x = -18$$
$$x = 6$$

Solution set: $\{6\}$

Check:

$$\sqrt{2x-3} - \sqrt{5x-21} \stackrel{?}{=} 0$$
$$\sqrt{12-3} - \sqrt{30-21} \stackrel{?}{=} 0$$
$$\sqrt{9} - \sqrt{9} \stackrel{?}{=} 0$$
$$0 \stackrel{?}{=} 0 \quad \text{Yes } \checkmark$$

CLASSROOM EXERCISES

Solve and check.

1. $\sqrt{x} = 3$
2. $\sqrt{x} = -3$
3. $\sqrt{y+3} = 7$
4. $\sqrt{x-2} = 4$
5. $\sqrt{x-5} + \sqrt{x} = 5$
6. $\sqrt{x-1} + 4 = 0$
7. $\sqrt{3q-1} - 2\sqrt{8-2q} = 0$

WRITTEN EXERCISES

A Solve and check.

1. $\sqrt{x} = 6$
2. $\sqrt{5a} = 10$
3. $\sqrt{3y} + 4 = 5$
4. $5 - \sqrt{2m} = 3$
5. $15 = 3\sqrt{x}$
6. $\sqrt{x-1} = 4$
7. $\sqrt{2n-5} = 3$
8. $\sqrt{2x+5} - 7 = -4$
9. $-7 + \sqrt{2a-3} = -4$
10. $2\sqrt{3x-15} = 16$
11. $3\sqrt{2y+1} + 5 = 0$
12. $2\sqrt{3x-15} = 6$
13. $2\sqrt{y} - 7\sqrt{2} = 0$
14. $\sqrt{8+2x} = -6$
15. $\sqrt{5x-7} - \sqrt{x+10} = 0$
16. $\sqrt{12y-3} - \sqrt{5y+2} = 0$
17. $\sqrt{2a+6} - \sqrt{2a-5} = 0$
18. $\sqrt[3]{2a} = 8$
19. $\sqrt[3]{y+4} = 3$
20. $\sqrt[3]{a+1} = 2$
21. $\sqrt[4]{x+1} = 2$
22. $\sqrt[4]{3y} + 8 = 11$
23. $5 + \sqrt[3]{a} = 1$
24. $2\sqrt{2n-3} + 4 = 1$
25. $\sqrt{x^2-9} = 4$
26. $\sqrt{1+x^2} = 5$
27. $3 = \dfrac{6}{\sqrt{x}}$
28. $2 = \dfrac{8}{\sqrt{5a-4}}$
29. $\sqrt{8+2x} = -6$

B

30. $\sqrt{y-2} = \dfrac{5}{\sqrt{y-2}}$
31. $\sqrt{x^2+3} = x+1$
32. $\sqrt{4y^2-1} = 2y+3$
33. $\sqrt{4x-5} = \sqrt{6x-5}$
34. $\sqrt{2n-5} + \sqrt{2n+3} = 4$
35. $\sqrt{z+4} + \sqrt{z-4} = 4$
36. $\sqrt{r-3} = \sqrt{2} \cdot \sqrt{r}$
37. $\sqrt{q-6} = 3 + \sqrt{q}$
38. $\sqrt{p-8} - \sqrt{p} = -4$

The formula $t = \sqrt{\dfrac{2s}{g}}$ gives the time, t, in seconds that it takes a body starting from rest to fall s meters, where g is the acceleration due to gravity.

39. Solve the formula for s.
40. How far will a body fall in 3 seconds? Use $g = 9.81$ m/sec².

C

41. $\sqrt{z+4} + \sqrt{z-4} = 2\sqrt{z-1}$
42. $\sqrt{4t+5} - \sqrt{t+4} = \sqrt{t-1}$

REVIEW CAPSULE FOR SECTION 8–5

Evaluate each expression for the given values of the variables. *(Pages 2–5)*

1. $|x_1 - x_2|$ for $x_1 = -9$ and $x_2 = -2$
2. $|x_2 - x_1|$ for $x_1 = -9$ and $x_2 = -2$
3. $a^2 + b^2$, for $a = \sqrt{5}$ and $b = \sqrt{3}$
4. $c^2 - b^2$, for $c = 13$ and $b = 12$
5. $\sqrt{(x_1 - x_2)^2 + (y_1 - y_2)^2}$, for $x_1 = 4$, $x_2 = -1$, $y_1 = -3$, and $y_2 = 7$
6. $\dfrac{y_1 + y_2}{2}$, for $y_1 = 3$ and $y_2 = -11$

8–5 Distance Formula/Midpoint Formula

Recall that the distance between points A and B on a number line is the length of line segment AB.

$\overline{AB}$ ← **Line segment AB or segment AB** AB ← **Distance between A and B or length of $\overline{AB}$.**

Distance does not involve direction; it is always a nonnegative number.

Definition

> The **distance between two points** P_1 and P_2 with coordinates x_1 and x_2 on a real number line, is $|x_1 - x_2|$ or $|x_2 - x_1|$. That is,
>
> $$P_1P_2 = |x_1 - x_2| \text{ or } |x_2 - x_1|.$$

EXAMPLE 1 Find each distance.

a. AB **b.** BA **c.** AC

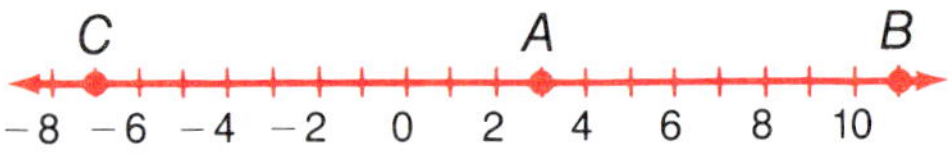

Solutions: **a.** $AB = |3 - 11| = |-8| = \mathbf{8}$ **b.** $BA = |11 - 3| = \mathbf{8}$

c. $AC = |3 - (-7)| = |3 + 7| = \mathbf{10}$

To determine the distance between any two points in the coordinate plane, you use the Pythagorean Theorem.

Theorem 8–3

> **Pythagorean Theorem**
>
> In a right triangle, if the length of the hypotenuse is c and the lengths of the legs are a and b, then,
>
> $$c^2 = a^2 + b^2.$$

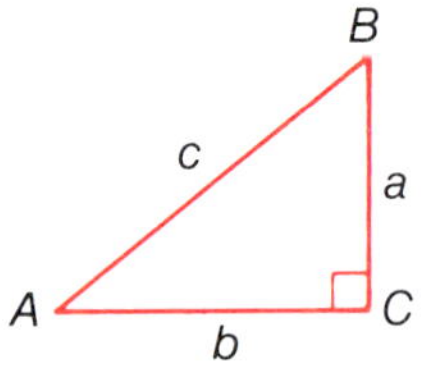

Use this theorem and the definition above in Example 2.

EXAMPLE 2 For the points $A(3, 4)$ and $B(-4, -2)$, find AB.

Solution: Graph points A and B. Draw $\overline{AB}$.

Through B, draw a line parallel to the x axis. Through A, draw a line parallel to the y axis. The intersection of these two lines is $C(3, -2)$.

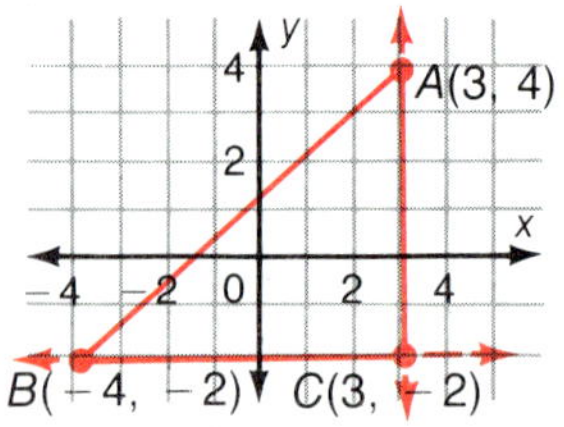

ABC is a right triangle. ← **$\overline{AC}$ is perpendicular to $\overline{BC}$.**

$AB^2 = AC^2 + BC^2$ ← **By the Pythagorean Theorem**

$AB^2 = |4 - (-2)|^2 + |-4 - 3|^2$ ← ***AC:* Abscissas are the same; subtract the ordinates. *BC:* Ordinates are the same; subtract the abscissas.**

$AB^2 = 36 + 49 = 85$

$AB = \sqrt{85}$

The method of Example 2 is generalized in the following theorem.

Theorem 8–4

Distance Formula for Two-Space

The distance between two points $P_1(x_1, y_1)$ and $P_2(x_2, y_2)$ is given by the formula

$$P_1P_2 = \sqrt{(x_1 - x_2)^2 + (y_1 - y_2)^2}.$$

You can use the distance formula to derive the midpoint formula.

Theorem 8–5

Midpoint Formula

The coordinates of the midpoint $M(x, y)$ of the line segment with endpoints $P_1(x_1, y_1)$ and $P_2(x_2, y_2)$ are

$$x = \frac{x_1 + x_2}{2} \text{ and } y = \frac{y_1 + y_2}{2}.$$

EXAMPLE 3 Find the coordinates of the midpoint of the segment with endpoints at $A(3, -5)$ and $B(-5, 11)$.

Solution:

$$x = \frac{x_1 + x_2}{2} \qquad y = \frac{y_1 + y_2}{2}$$

$$x = \frac{3 + (-5)}{2} = \frac{-2}{2} = -1 \qquad y = \frac{-5 + 11}{2} = \frac{6}{2} = 3$$

Coordinates of midpoint: **(−1, 3)**

CLASSROOM EXERCISES

Find the distance between the points with the given coordinates.

1. $A(-2, 5)$; $B(-2, 0)$
2. $A(7, -1)$; $B(-3, -1)$
3. $A(7, -1)$; $B(7, 3)$
4. $A(8, -1)$; $B(-3, -1)$
5. $A(-1, -5)$; $B(6, 2)$
6. $A(a, b)$; $B(3, 3)$

Give the coordinates of the midpoint of the line segment with the given endpoints.

7. $A(2, 0)$; $B(4, 6)$
8. $A(3, 9)$; $B(-5, -1)$
9. $A(2, -2)$; $B(5, 3)$

WRITTEN EXERCISES

A Find the distance between the points with the given coordinates.

1. $A(1, 2)$; $B(4, 3)$
2. $A(-4, 1)$; $B(-5, 4)$
3. $P(2, -6)$; $Q(7, -5)$
4. $P(3, 6)$; $Q(-1, -2)$
5. $C(4, -3)$; $D(-4, 3)$
6. $E(-3, 3)$; $F(0, -4)$
7. $A(7, 1)$; $B(6, 2)$
8. $H(5, 2)$; $I(2, 2)$
9. $P(-2, -5)$; $Q(-2, 4)$
10. $M(4, -3)$; $N(-1, -3)$
11. $R(a, b)$; $S(c, d)$
12. $T(a, b)$; $U(c, b)$

In Exercises 13–14, find the perimeter of each triangle whose vertices have the given coordinates.

13. $A(0, 8)$, $B(-6, 0)$, $C(15, 0)$

14. $A(-3, -1)$, $B(1, 2)$, $C(1, -1)$

15. Use the distance formula to show that $P(-4, 1)$, $Q(5, 4)$ and $R(2, -2)$ are the vertices of an isosceles triangle.

16. Use the distance formula and the Pythagorean Theorem to show that $A(-4, 1)$, $B(5, 4)$ and $C(2, -2)$ are the vertices of a right triangle.

17. Determine whether the triangle whose vertices are $C(-2, 3)$, $D(1, -1)$ and $E(3, 3)$ is isosceles, equilateral, or both.

18. Determine whether the triangle whose vertices are $F(7, 3)$, $H(6, 9)$ and $L(2, 3)$ is isosceles or scalene. (HINT: No two sides of a scalene triangle are equal in length.)

19. Find the length of the diameter of a circle whose end points have the coordinates $A(3, 5)$ and $B(7, -4)$.

20. Find the length of the radius of a circle that has a center $O(3, 4)$ and passes through $P(2, -5)$.

21. Find the lengths of the diagonals of a rectangle with vertices $P(-6, -5)$, $Q(5, -5)$, $R(5, -8)$ and $S(-6, -8)$.

22. Find the lengths of the diagonals of an isosceles trapezoid whose vertices are $A(-1, 5)$, $B(2, 5)$, $C(5, 2)$, and $D(-4, 2)$.

Find the coordinates of the midpoint of the segment with the given endpoints.

23. $A(2, 1)$; $B(-9, 3)$

24. $A(10, 9)$; $B(-5, 6)$

25. $A(-1, -3)$; $B(-5, -7)$

26. $A(-4, 8)$; $B(8, -2)$

27. $A(2, 2)$; $B(-8, -7)$

28. $A(14, -2)$; $B(7, -9)$

29. $A(\frac{1}{2}, -\frac{7}{2})$; $B(4, -9)$

30. $A(a, 0)$; $B(b, 0)$

31. $A(0, a)$; $B(0, b)$

32. One endpoint of a segment is $(6, 0)$ and the midpoint is $(8, 2)$. Find the coordinates of the other endpoint.

33. One endpoint of a segment is $(-8, -12)$ and the midpoint is $(-4, -2)$. Find the coordinates of the other endpoint.

34. Find the coordinates of the center of a circle, when the endpoints of one of its diameters are $(-8, 2)$ and $(10, -4)$.

35. If the center of a circle is $(3, 2)$ and one endpoint of a diameter is $(6, 6)$, what are the coordinates of the other endpoint?

36. Show that the diagonals of the parallelogram whose vertices are $P(1, 2)$, $Q(4, 2)$, $R(-1, -3)$, and $S(-4, -3)$ bisect each other. (HINT: Show that the midpoints of the diagonals coincide.)

37. The vertices of a triangle are $A(4, 6)$, $B(-2, -4)$; and $C(-8, 2)$. Show that the length of the line segment that connects the midpoints of $\overline{AC}$ and $\overline{AB}$ is $\frac{1}{2}BC$.

B

38. Use the distance formula to show that $P(-1, -3)$, $Q(2, 1)$ and $R(5, 5)$ lie on a straight line. (HINT: Show that the longest distance is the sum of the two shorter distances.)

39. Use the distance formula to show that the points $P(-2, 1)$, $Q(2, 2)$ and $R(-1, -1)$ do not lie on the same straight line.

40. Use the distance formula to show that $M(-1, 4)$, $N(-5, 3)$, $R(1, -3)$, and $Q(5, -2)$ are the vertices of a parallelogram.

41. Write an equation to describe all points $P(x, y)$ in a plane that are r units from the origin.

42. Find the perimeter of the triangle formed by joining the midpoints of the sides of triangle ABC given $A(-3, 0)$, $B(5, 0)$ and $C(1, 8)$.

43. The vertices of a trapezoid are $A(2, 3)$, $B(-1, 3)$, $C(-4, -1)$ and $D(5, -1)$. Show that the length of the segment that joins the midpoints of the non-parallel sides is equal to one-half the sum of the lengths of the parallel sides.

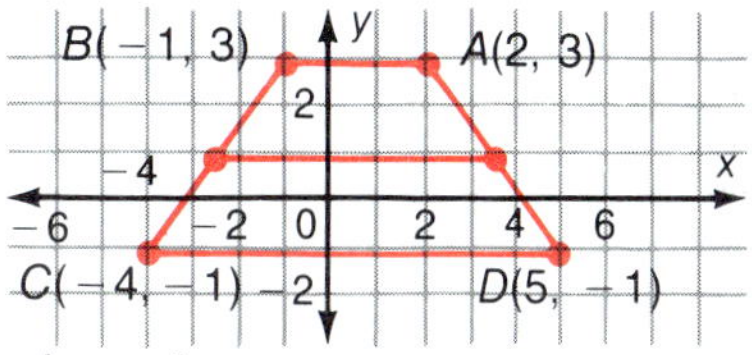

44. Find the value of a and b if the coordinates of the endpoints of a segment are $(a - 3, b + 1)$ and $(a + 7, b + 17)$ and its midpoint is $(3, 3)$.

45. The vertices of a triangle are $A(2, 3)$, $B(6, 9)$ and $C(10, -5)$. Find the length of the median from A to side $\overline{BC}$.

46. Find the coordinates of a point P that divides the segment from $A(2, 3)$ to $B(12, 8)$ so that the ratio $\frac{AP}{PB} = \frac{2}{3}$. Check by using the Distance Formula. (HINT: Consider $\overline{AB}$ the hypotenuse of right triangle BAC whose legs are parallel to the axes. The line that contains P and is parallel to the x axis intersects $\overline{BC}$ at some point $E(12, y)$. The line through P parallel to the y axis intersects $\overline{AC}$ at $D(x, 3)$. Then $\frac{AD}{DC} = \frac{2}{3}$ and $\frac{CE}{EB} = \frac{2}{3}$.

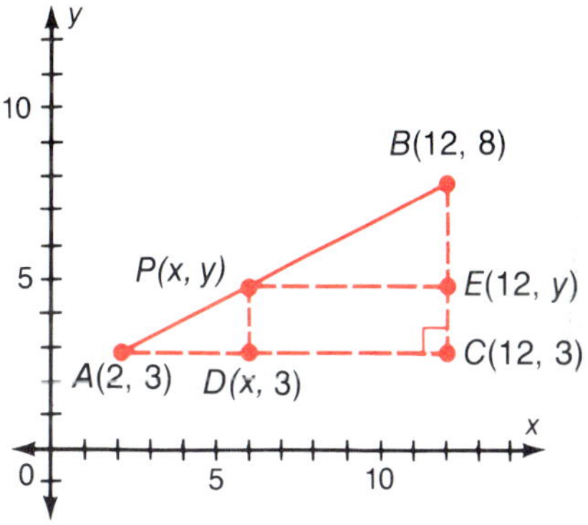

47. What are the coordinates of $P(x, y)$ that divides the segment from $C(5, 2)$ to $D(12, 16)$ so that the ratio $\frac{PC}{PD} = \frac{3}{4}$? Use the Distance Formula to check.

48. Find the coordinates of $P(x, y)$ that divides the segment from $E(2, -4)$ to $F(9, 3)$ so that $\frac{PE}{PF} = \frac{2}{5}$. Check your answer by using the distance formula.

C

49. Find the coordinates of $P(x, y)$ that divides the segment from $E(x_1, y_1)$ to $F(x_2, y_2)$ so that $\frac{PE}{PF} = \frac{m}{n}$.

50. Prove Theorem 8–4.

51. Prove Theorem 8–5.

REVIEW CAPSULE FOR SECTION 8–6

Find the slope of the line containing the given points. *(Pages 80–83)*

1. $P(0, 0)$; $Q(4, 6)$ 2. $P(-5, -3)$; $Q(0, 0)$ 3. $P(-1, -8)$; $Q(4, -1)$

In Exercises 4–6, find the slope of all lines parallel to the line containing the given points. *(Pages 90–92)*

4. $A(3, 0)$; $B(-5, 1)$ 5. $A(-1, -4)$; $B(2, 9)$ 6. $A(0, 0)$; $B(1, 3)$

In Exercises 7–9, find the slope of all lines perpendicular to the line containing the given points. *(Pages 90–92)*

7. $R(5, 8)$; $T(4, 7)$ 8. $R(-3, -1)$; $T(-4, -5)$ 9. $R(1, -5)$; $T(-1, 7)$

8–6 Coordinate Proofs

The distance formula and the midpoint formula are used in many coordinate proofs. When planning a coordinate proof, try to place one vertex of the geometric figure at the origin and at least one side along one of the coordinate axes.

EXAMPLE Prove that the midpoint of the hypotenuse of a right triangle is equidistant from all three vertices.

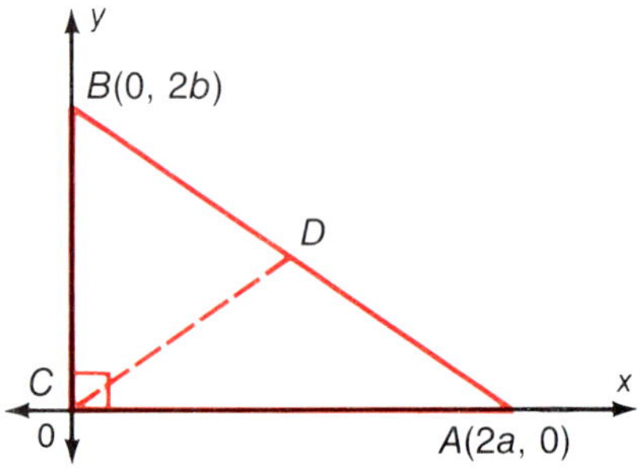

Proof: Place right triangle ABC with C, the right angle, at the origin and let the legs of the triangle lie on the coordinate axes.

Let the coordinates of A be $(2a, 0)$ and $B(0, 2b)$. (Using $2a$ and $2b$ will avoid fractions.)

Let D be the midpoint of $\overline{AB}$. Then by the Midpoint Formula,

$$x \text{ coordinate: } \frac{2a+0}{2} = a \qquad y \text{ coordinate: } \frac{0+2b}{2} = b.$$

Thus, the coordinates of D are (a, b). By the Distance Formula,

$$AD = \sqrt{(a-2a)^2 + (b-0)^2} = \sqrt{a^2+b^2}, \text{ and}$$

$$CD = \sqrt{(a-0)^2 + (b-0)^2} = \sqrt{a^2+b^2}.$$

Thus, since $AD = BD$ by the definition of midpoint, **D is equidistant from the vertices of the triangle.**

CLASSROOM EXERCISES

In Exercises 1–6, place each figure in the coordinate plane. Give the coordinates of the vertices.

1. Square
2. Rectangle
3. Parallelogram
4. Isosceles right triangle
5. Isosceles triangle
6. Equilateral triangle

WRITTEN EXERCISES

A

The vertices of isosceles trapezoid $ABCD$ are $A(0, 0)$, $B(a, 0)$, $C(a-c, b)$ and $D(c, b)$.

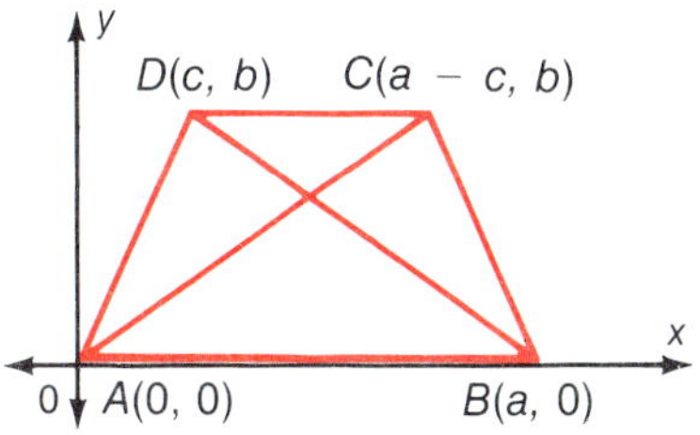

1. Prove that $\overline{AB}$ is parallel to $\overline{DC}$ and that $\overline{AD}$ is not parallel to $\overline{BC}$.
2. Prove that the lengths of $\overline{AC}$ and $\overline{BD}$ are equal.
3. Prove that sides AD and BC are equal in length.

The coordinates of the vertices of triangle ABC are $A(0, 0)$, $B(a, 0)$ and $C(b, c)$.

4. Prove that the line segment joining the midpoints of sides AC and BC is parallel to side AB.
5. Prove that the length of the line segment joining the midpoints of sides AC and BC is one-half the length of side AB.
6. Prove that point $P_3\left(\frac{a+c}{2}, \frac{b+d}{2}\right)$ is equidistant from points $P_1(a, b)$ and $P_2(c, d)$.
7. Prove that the lengths of the diagonals of a rectangle are equal.
8. Prove that the diagonals of a parallelogram bisect each other.
9. Prove that the lengths of the opposite sides of a parallelogram are equal.
10. Prove that the line segments joining the midpoints of the three sides of an isosceles triangle form an isosceles triangle.
11. Use the midpoint formula to prove that the line $y = x$ is the perpendicular bisector of the line segment that joins the points $P(a, b)$ and $Q(b, a)$.

B

12. Prove that the segments joining the midpoints of the sides of any quadrilateral, taken in order, form a parallelogram. (HINT: Let the vertices be at $A(0, 0)$, $B(2c, 0)$, $C(2d, 2e)$, and $D(2b, 2c)$.
13. Prove that the median to the base of an isosceles triangle is perpendicular to the base.
14. Prove the length of the median of a trapezoid is one-half the sum of the lengths of the bases.

C

15. Prove that the medians of a triangle meet at a point that is two-thirds the distance from each vertex to the midpoint of the opposite side.

16. Show that an equation of the perpendicular bisector of the segment with endpoints $P(a, b)$ and $Q(c, d)$ is given by

$$2(c-a)x + 2(d-b)y + a^2 + b^2 - c^2 - d^2 = 0$$

Review

Simplify. *(Section 8–1)*

1. $-\sqrt{144}$
2. $\sqrt[3]{216x^6}$
3. $\sqrt[4]{16c^{24}d^8}$
4. $\sqrt{\frac{81a^8b^{10}}{25c^4d^2}}$
5. $\sqrt[3]{-(a+b)^9}$

Perform the indicated operations. *(Section 8–2)*

6. $3\sqrt{24} \cdot 5\sqrt{3}$
7. $\sqrt{5c^3} \cdot \sqrt{20c^5}$
8. $\sqrt[4]{2y^6z} \cdot \sqrt[4]{8y^2z^3}$
9. $\sqrt{5}(3-2\sqrt{5})$
10. $\frac{\sqrt{54}}{\sqrt{8}}$
11. $\frac{6\sqrt{24}}{4\sqrt{10}}$
12. $\frac{\sqrt{20y}}{\sqrt{5}}$
13. $\frac{\sqrt{40x^3}}{\sqrt{8x}}$
14. $\frac{\sqrt{25ab}}{\sqrt{ab}}$
15. $\frac{\sqrt{27b^3}}{\sqrt{3b}}$

(Section 8–3)

16. $4\sqrt{20} - 6\sqrt{45} + 8\sqrt{80}$
17. $4\sqrt[3]{81} + \sqrt[3]{375} - 9\sqrt[3]{3}$
18. $\sqrt{\frac{a}{5}} + \sqrt{\frac{a}{20}}$
19. $\frac{6+\sqrt[3]{9}}{5} - \frac{6-\sqrt[3]{9}}{5}$
20. $(4\sqrt{3}+2\sqrt{2})(4\sqrt{3}-2\sqrt{2})$
21. $(6\sqrt{a}+5\sqrt{b})(6\sqrt{a}-5\sqrt{b})$

Rationalize the denominator. *(Section 8–3)*

22. $\frac{6}{1+\sqrt{2}}$
23. $\frac{8+\sqrt{2}}{\sqrt{3}-5}$
24. $\frac{2+3\sqrt{5}}{\sqrt{2}+6}$
25. $\frac{\sqrt{5}-\sqrt{3}}{\sqrt{5}+\sqrt{3}}$

Solve and check. *(Section 8–4)*

26. $\sqrt{6b} = 12$
27. $3\sqrt{4x+5} = 15$
28. $\sqrt{a^2+3} = a-2$
29. Find the length of the diameter of the circle that has its center at $Q(-3, 6)$ and passes through $G(1, 3)$. *(Section 8–5)*
30. Prove that the line segments joining the midpoints of opposite sides of a quadrilateral bisect each other. *(Section 8–6)*

REVIEW CAPSULE FOR SECTION 8–7

For each function, determine y for the given value of x. *(Pages 2–5)*

1. $y = x^2;\ x = -1$
2. $y = -x^2;\ x = -5$
3. $y = -x^2 + 2;\ x = \frac{1}{2}$
4. $y = x^2 + 2x + 7;\ x = -3$
5. $y = 2x^2 - 8x + 10;\ x = 4$
6. $y = x^2 - 8x + 12;\ x = \frac{3}{4}$

8–7 Quadratic Functions

In general, a **quadratic function** is a function that may be defined by

$$y = ax^2 + bx + c,$$

where a, b, and c are real number constants, and $a \neq 0$. The restriction $a \neq 0$ assures that $y = ax^2 + bx + c$ defines a *nonlinear function*. The graph of a quadratic function is a **parabola.**

EXAMPLE 1 Graph the quadratic function $y = x^2$.

Solution: [1] Make a table of ordered pairs.

x	−3	−2	−1	0	1	2	3
y	9	4	1	0	1	4	9

[2] Graph the ordered pairs.

[3] Join the points with a smooth curve.

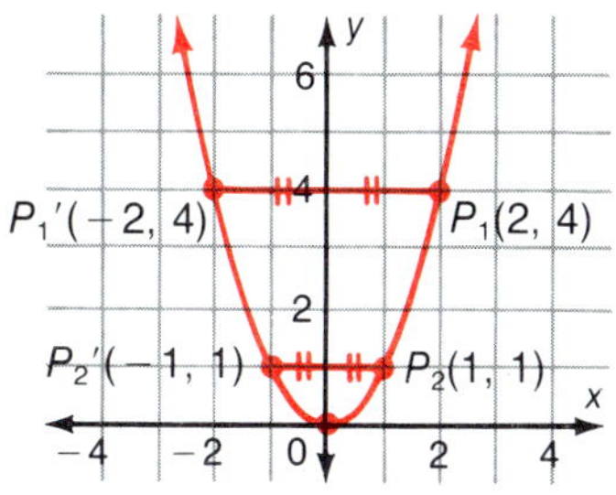

The graph of $y = x^2$ is a curve which is **symmetric** with respect to the y–axis. For each point P on the graph, there is a corresponding point P' such that the y axis is the perpendicular bisector of segment PP'. When this is so, the points P and P' are said to be **mirror images** of each other. Every parabola is symmetric with respect to some line. This line is called the **axis of symmetry.** The axis of symmetry intersects the parabola in a single point called the **turning point,** or **vertex,** of the parabola. The vertex is its own mirror image with respect to the axis of symmetry. The vertex is the lowest, or **minimum,** point of the parabola if the parabola opens upward. It is the highest, or **maximum,** point if the parabola opens downward.

EXAMPLE 2 The equation of a parabola is $y = x^2 + 2x + c$. Find the value of c if the graph of the parabola contains $(1, -3)$.

Solution:

$y = x^2 + 2x + c$ ← **Function**

$-3 = (1)^2 + 2(1) + c$ ← **Replace x with 1 and y with −3.**

$-3 = 3 + c$ ← **Solve for c.**

$\mathbf{-6} = c$ Thus, $\mathbf{c = -6}$; the equation of the parabola is $\mathbf{y = x^2 + 2x - 6}$.

CLASSROOM EXERCISES

Classify each function as linear, L, or quadratic, Q.

1. $f(x) = 2x + 7$
2. $f(x) = x^2 + 2x + 7$
3. $f(x) = 2x$
4. $f(x) = 2x^2 + x$
5. $f(x) = 6x^2 + 1$
6. $f(x) = (a - x)^2$

WRITTEN EXERCISES

A For each function, make a table of values for integral values of x, $-5 \leq x \leq 5$. Use the table to graph each function.

1. $y = x^2 + 1$ **2.** $y = x^2 + 2x + 1$ **3.** $y = x^2 - 5x + 6$

In Exercises 4–7, the equation of a parabola and a point through which the graph of the parabola passes are given. For each equation, find the value of the specified coefficient.

4. $y = x^2 + bx - 9$; $A(-2, -5)$; $b =$ _?_

5. $y = 2x^2 + 3x + c$; $H(1, 0)$; $c =$ _?_

6. $y = ax^2 + 3x - 2$; $J(2, 24)$; $a =$ _?_

7. $y = x^2 + bx + 3$; $N(-1, 8)$; $b =$ _?_

Each figure below is the graph of a quadratic function.

A

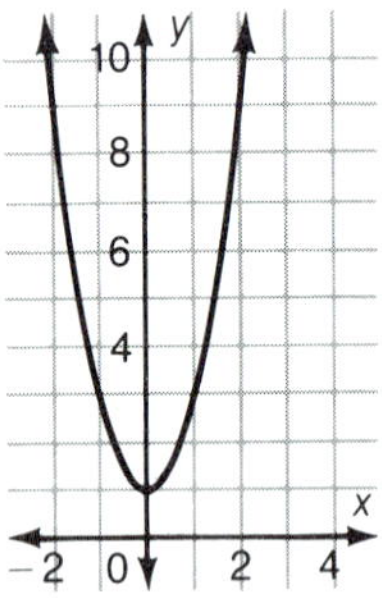

B

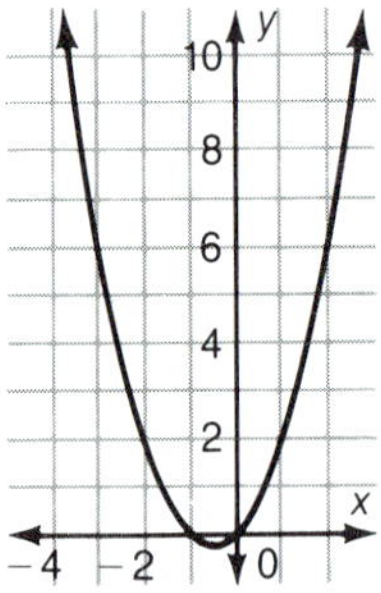

C

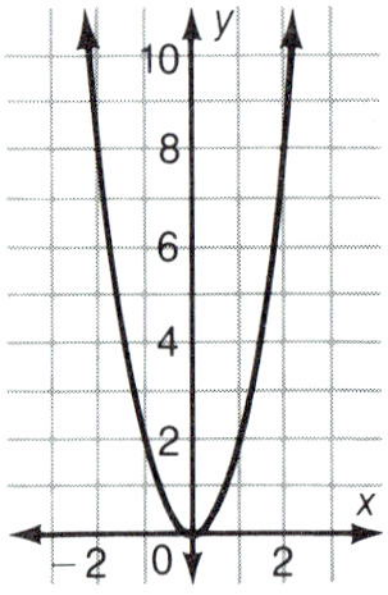

D

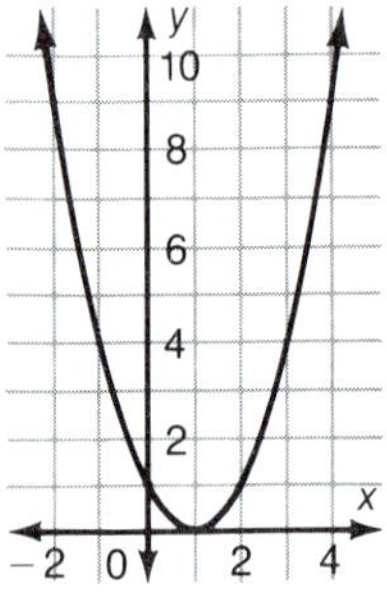

Identify the function graphed above that contains each of the following sets of ordered pairs.

8. $\{(-2, 8), (-1, 2), (0, 0), (1, 2), (2, 8)\}$

9. $\{(-2, 9), (-1, 3), (0, 1), (1, 3), (2, 9)\}$

10. $\{(-2, 9), (-1, 4), (0, 1), (1, 0), (2, 1)\}$

11. $\{(-2, 2), (-1, 0), (0, 0), (1, 2), (2, 6)\}$

B In Exercises 12–13, graph the given quadratic functions on the same set of coordinate axes. Compare the graphs by indicating the axis of symmetry and the vertex of each parabola. Tell whether the parabolas are the same size and shape.

12. $y = x^2$; $y = x^2 - 4$; $y = x^2 - 8x + 12$

13. $y = 2x^2$; $y = 2x^2 - 8x + 8$; $y = 2x^2 - 8x + 10$

14. For $f(x) = ax^2 + bx + c_1$ and $g(x) = ax^2 + bx + c_2$, a and b have the same value in both functions. How do the graphs of the functions compare?

15. For $f(x) = a_1x^2 + c$ and $g(x) = a_2x^2 + c$, c has the same value for both functions and $b = 0$. How do the graphs of the functions compare?

APPLICATIONS: Using Quadratic Functions

C In Exercises 17–19, round answers to the nearest hundredth.

16. An arrow is shot upward with a velocity of 10 meters per second. The height, h, of the arrow above the ground after t seconds is given by the equation, $h = 10t - 4.9t^2$. Graph this function.

17. In Exercise 16, when is the arrow at its highest point above the ground? When does it hit the ground?

18. The height of a rocket t seconds after being given an upward velocity of 80 meters per second is $s = 80t - 4.9t^2$. Graph this function.

19. In Exercise 18, how many seconds after firing does the rocket hit the ground? What is its highest altitude above the ground?

CALCULATOR APPLICATIONS

Evaluating Radicals

The square root key on a calculator can be used to evaluate roots other than square roots.

EXAMPLE Evaluate: **a.** $\sqrt[4]{x}$ when $x = 44$ **b.** $\sqrt{x + \sqrt{x + \sqrt{x + \sqrt{x}}}}$ when $x = 2$

SOLUTIONS **a.** $\sqrt[4]{44}$ means $\sqrt{\sqrt{44}}$.

4 4 [√] [√] → 2.5755096

b. Start with 2 at the extreme right. Then work to the left.

2 [√] [+] 2 [=] [√] [+] 2 [=] [√] [+] 2 [=] [√] → 1.9903695

EXERCISES

Evaluate each of the following.

1. $\sqrt[4]{2073}$ **2.** $\sqrt[4]{719}$ **3.** $\sqrt[4]{3663}$ **4.** $\sqrt[4]{234{,}256}$ **5.** $\sqrt[4]{2{,}220{,}222}$ **6.** $\sqrt[4]{7.2904}$

Evaluate the expression in Example **b** for each value of x.

7. $x = 3$ **8.** $x = 4$ **9.** $x = 6$ **10.** $x = 15$ **11.** $x = 22$ **12.** $x = 513$

REVIEW CAPSULE FOR SECTION 8-8

Tell whether the given points lie on the graph of each given function. *(Pages 160–164)*

1. $y = x^2$; $(-2, 4)$, $(2, 4)$

2. $y = -\frac{1}{2}x^2$, $(-6, -18)$, $(6, -18)$

3. $y = -\frac{3}{2}x^2$; $(-8, -96)$, $(8, -96)$

4. $y = -4x^2 + 3$; $(-1, 7)$, $(1, -7)$

8–8 The Role of a in $y = ax^2$

The following compares the graphs of $y = x^2$ and $y = -x^2$.

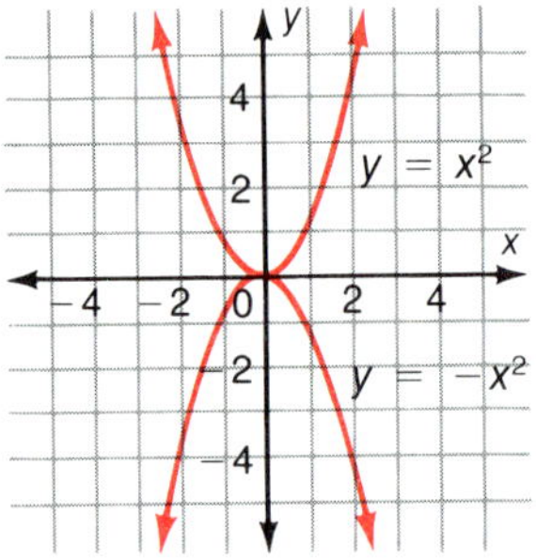

Graph of $y = x^2$	Graph of $y = -x^2$
Opens upward.	Opens downward.
Vertex: (0, 0)	Vertex: (0, 0)
Vertex is a minimum point.	Vertex is a maximum point.

The graphs have the same size and shape; that is, the parabolas are **congruent.**

Now compare the graphs of $y = ax^2$ and $y = -ax^2$ when $|a|$ equals $\frac{1}{2}$, 1, 2, and 3. Each graph is symmetric with respect to the y axis and each has its vertex at the origin. Some parabolas are "steeper" than others, but, for a given value for a, the graphs of $y = ax^2$ and $y = -ax^2$ are congruent.

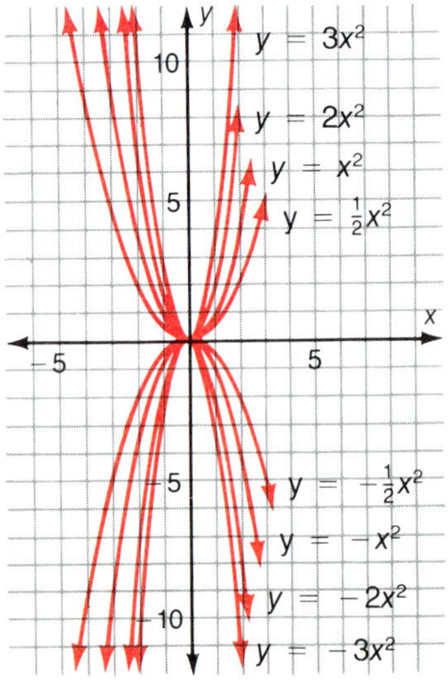

The following table summarizes the effects of the changes in a.

Value of a	Graph of $y = ax^2$
$a > 0$	Opens upward. Vertex: (0, 0) Vertex is a minimum.
$a < 0$	Opens downward Vertex: (0, 0) Vertex is a maximum.
$\|a\| > 1$	Steeper than $y = x^2$.
$\|a\| < 1$	Less steep than $y = x^2$.

EXAMPLE 1 **a.** Without graphing, classify the graphs of $y = 3x^2$, $y = -\frac{1}{4}x^2$, $y = -2x^2$, $y = -7x^2$, and $y = \frac{1}{8}x^2$ as opening upward or opening downward.

b. List the functions in order of the steepness of their graphs, beginning with the least steep.

Solutions: **a.** Opens upward: $y = 3x^2$, $y = \frac{1}{8}x^2$ ← $a > 0$

Opens downward: $y = -\frac{1}{4}x^2$, $y = -2x^2$, $y = -7x^2$ ← $a < 0$

b. Steepness

$y = \frac{1}{8}x^2$, $y = -\frac{1}{4}x^2$, $y = -2x^2$, $y = 3x^2$, $y = -7x^2$ ← $|\frac{1}{8}| < |\frac{1}{4}| < |-2| < |3| < |-7|$

EXAMPLE 2 a. Sketch the graph of $y = \frac{1}{8}x^2$.

b. Write the equation of the axis of symmetry and the coordinates of the vertex.

c. Tell whether the vertex is a maximum point or a minimum point.

Solutions: a. Make a table of ordered pairs.

x	−4	−2	−1	0	1	2	4
y	2	$\frac{1}{2}$	$\frac{1}{8}$	0	$\frac{1}{8}$	$\frac{1}{2}$	2

Graph the ordered pairs. Join the points with a smooth curve.

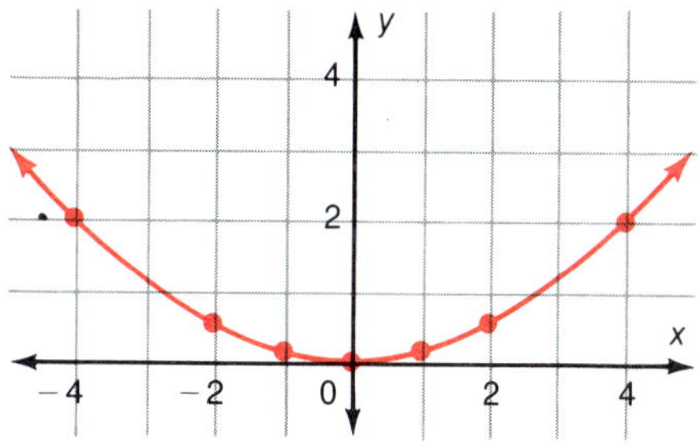

b. Axis of symmetry: $x = 0$

Vertex: **(0, 0)**

c. The vertex is a **minimum point.**

CLASSROOM EXERCISES

Classify the graph of each function as opening upward, *U*, or opening downward, *D*.

1. $y = 5x^2$
2. $y = \frac{1}{3}x^2$
3. $y = -\frac{1}{2}x^2$
4. $y + 2x^2 = 0$
5. $f(x) = 10x^2$
6. $f(x) = -1\frac{1}{2}x^2$

Make a table of values for each of the following, where $-4 \leq x \leq 4$ and x is an integer.

7. $y = 5x^2$
8. $y = -\frac{1}{3}x^2$
9. $y - 3x^2 = 0$
10. $x^2 + 2y = 0$

WRITTEN EXERCISES

A

1. Which graph opens downward, that of $y = 3x^2$ or that of $y = -3x^2$?
2. Which function has the steeper graph, $y = x^2$ or $y = 2x^2$?
3. Which function has the steeper graph, $y = 2x^2$ or $y = \frac{1}{2}x^2$?
4. State in your own words how to determine whether a parabola is "wide" or "narrow" and whether it opens upward or downward.
5. Arrange the following functions in order of the steepness of their graphs, beginning with the least steep.

   a. $f_1(x) = 3x^2$ b. $f_2(x) = 24x^2$ c. $f_3(x) = -\frac{1}{2}x^2$

   d. $f_4(x) = -2x^2$ e. $f_5(x) = \frac{1}{4}x^2$ f. $f_6(x) = -x^2$
6. For what values of a will the graph of $y = ax^2$ have a maximum point?

7. What is the minimum point of the graph of $y = 4x^2$? of $y = 4x^2 + 6$?
8. What is the maximum point of the graph of $y = -4x^2$? of $y = -4x^2 + 3$?
9. Compare the graphs of $y = 2x^2$ and $y = -2x^2$.
10. Do the graphs of $y = -3x^2$ and $y = 3x^2$ have the same size and shape? Explain.

Graph the functions defined by the following equations. Write the equation of the axis of symmetry of each parabola, give the coordinates of its vertex, and indicate whether the vertex is a maximum point or a minimum point.

11. $y = 4x^2$
12. $y = \frac{1}{4}x^2$
13. $y = -\frac{1}{4}x^2$
14. $y = -4x^2$
15. $y = 2x^2$
16. $y = -2x^2$
17. $y = -\frac{1}{2}x^2$
18. $y = \frac{1}{2}x^2$

B

19. Sketch a parabola that contains the points $P(2, 6)$ and $Q(-3, 13\frac{1}{2})$ and has its vertex at the origin.
20. Sketch a parabola that has its minimum point at the origin and contains the point $P(1, 5)$.
21. Sketch a parabola that contains the points $A(1, 5)$ and $B(2, 8)$ and has its vertex at the point $V(0, 4)$.
22. Sketch a parabola that has a maximum point at $R(0, -3)$ and also contains the point $T(-3, -7)$.

C

23. Using y, x, and a, write a general equation for the parabola whose vertex is at the origin and whose axis of symmetry is the x axis.
24. For which values of a will the parabolas in Exercise 23 open to the right?
25. For which values of a will the parabolas in Exercise 23 open to the left?
26. Are the parabolas defined by the general equation in Exercise 23 functions? Give a reason for your answer.
27. Sketch a parabola symmetric to the x axis which contains the point $P(0, 0)$ and for which $a = -2$.
28. Sketch a parabola symmetric to the x axis which passes through the points $C(0, 0)$ and $D(5, 3)$. What is the value of a?

REVIEW CAPSULE FOR SECTION 8-9

Solve each problem. *(Pages 98–102)*

1. If y varies directly as x and $y = 50$ when $x = 5$, find y when $x = 10$.
2. If x varies directly as y and $y = 27$ when $x = 6$, find y when $x = 8$.
3. If D varies directly as t and $D = 30$ when $t = \frac{1}{2}$, find t when $D = 20$.
4. The perimeter, P_1, of an equilateral triangle varies directly as the length, s_1, of a side of the triangle. Write this variation in proportion form. Use P_1, P_2, s_1, and s_2 as the variables.

Problem Solving and Applications

8–9 Variation as the Square

The equation $A = \pi r^2$ indicates that the area of a circular region varies **directly** as the square of its radius, r. The general equation for a direct variation as the square of x is

$$y = kx^2.$$

The coefficient, k, is the **constant of variation.**

EXAMPLE 1 If y varies directly as x^2 and $y = 12$ when $x = 2$, find y when $x = 5$.

Solution:

$y = kx^2$	← **Model**
$12 = k(2)^2$	← **When $x = 2$, $y = 12$.**
$3 = k$	← **Replace k in $y = kx^2$.**
$y = 3x^2$	← **Find y when $x = 5$.**
$y = 3(5)^2 = 75$	

Variation as the square can also be expressed as a *proportion.*

EXAMPLE 2 If y varies directly as x^2 and $y = 12$ when $x = 2$, find y when $x = 5$.

Solution:

$y = kx^2$ or $k = \dfrac{y}{x^2}$	← **Model**
$\dfrac{y_1}{x_1^2} = \dfrac{y_2}{x_2^2}$	← **Ratio of y to x^2 is always the same.**
$\dfrac{12}{2^2} = \dfrac{y_2}{5^2}$	← **$x_1 = 2$; $y_1 = 12$** **$x_2 = 5$; $y_2 =$ __?__**
$4y_2 = 25 \times 12$	← **Solve for y_2.**
$y_2 = 75$	

CLASSROOM EXERCISES

In Exercises 1–7, y varies directly as x^2.

1. Find k if $y = 8$ when $x = 2$.
2. Find k if $y = 27$ when $x = 3$.
3. Find k if $y = -9$ when $x = -3$.
4. Find k if $y = 16$ when $x = -4$.

Use a proportion to solve Exercises 5–7.

5. If $y = 10$ when $x = 4$, find y when $x = 3$.
6. If $y = 10$ when $x = 3$, find y when $x = 4$.
7. If $y = -20$ when $x = -3$, find y when $x = -4$.

WRITTEN EXERCISES

A In Exercises 1–4, y varies directly as the square of x.

1. If $y = 12$ when $x = 5$, find y when $x = 10$.
2. If $y = -5$ when $x = 4$, find y when $x = 8$.
3. If $y = 27$ when $x = 3$, find y when $x = -5$.
4. If $y = 3.5$ when $x = 2.3$ find y when $x = 7$.
5. If y varies directly as the square of x and $y = 4$ when $x = 6$, find y when $x = 9$.
6. If a varies directly as the square of b and $a = 1$ when $b = 8$, find b when $a = 4$.
7. If r varies directly as the square of t and $t = 6$ when $r = 9$, find t when $r = 4$.
8. If P varies directly as the square of q and if $P = 5$ when $q = 4$, find P when $q = 8$.
9. The formula for the kinetic energy of a moving object is $E = KV^2$. Write this formula as a proportion using (E_1, V_1) and (E_2, V_2).
10. The area of a square varies directly as the side of the square, $A = ks^2$. Write this relation as a proportion.

B Use the formula $A = \pi r^2$ for Exercises 11–13.

11. How does the value of A change when r increases?
12. How does the value of A change when r decreases?
13. How does the value of A change when r is multiplied by 3?
14. Make a table of ordered pairs that satisfy $y = kx^2$. Use integral values from -5 to 5 for x.
15. Use the table of Exercise 14 to find the difference between successive values for kx^2. Are the differences the same?
16. Find the difference between the successive values obtained in Exercise 15. Are these "second differences" the same?
17. As x increases by 1 in $y = kx^2$, what generalization can you make about how the second differences change for y?

APPLICATIONS: Using Direct Variation

In Exercises 18–20, write an equation for each statement. Use k as the constant of variation.

18. The distance, d, that an object falls varies directly as the square of the time, t.
19. The kinetic energy, E, of a moving object varies directly as the square of its velocity, v.
20. The surface area S, of a sphere, varies directly as the square of the radius, r.
21. If an object is dropped from a height, the approximate distance, d, in feet it will fall in t seconds is expressed by the formula, $d = 16t^2$. How does d change if t is doubled? tripled? halved?

22. The formula for the surface area of a sphere is $S = 4\pi r^2$. If the value of r is tripled, how does S change? If the value of r is divided by 4, by what is the value of S divided?

23. The area of one circular region is 20 square inches. Find the area ($A = \pi r^2$) of a second circular region if the ratio of the radius of the first circle to that of the second is 2 to 3.

24. The distance in meters that a body falls from rest varies as the square of the number of seconds it has fallen. If a body falls 78.08 meters in 4 seconds, how far will it fall in 7 seconds?

25. Stopping distance after brakes are applied varies directly as the square of the velocity. How many times as great is the stopping distance when the velocity is 60 miles per hour than when it is 30 miles per hour? If the velocity is doubled, how is the stopping distance changed?

26. Is a circular pizza with a 16–centimeter diameter worth exactly twice the price of a pizza with a 10–centimeter diameter?

27. The distance, d, in meters an object will fall in t seconds is given approximately by $d = 4.9t^2$. A ball is dropped from a 200–meter cliff. How long will it take to hit the water below?

Puzzle

A customer in a shoe store bought a pair of sneakers worth \$35 with a \$50 bill. Donna Perado, the salesperson, couldn't change the bill. She went next door to Fulton's Car Wash to get change. She returned and gave the customer the sneakers and the change. Later, Sam Fulton discovered that the \$50 bill was counterfeit, and asked for \$50 from Donna. Of course, she gave him the money. How much did Donna's store lose in all?

REVIEW CAPSULE FOR SECTION 8–10

Solve. *(Pages 165–169)*

1. For $f(x) = 3x^2 - 2x$, find $f(2)$, $f(3)$, and $f(1)$.
2. For $g(x) = -2x^2 + 5x - 3$, find $g(-2)$, $g(4)$, and $g(t-2)$.

Determine the y intercept of each function. *(Pages 93–97)*

3. $h(x) = x$ 4. $g(x) = -\frac{3}{5}x$ 5. $r(x) = -|x| + 3$ 6. $s(x) = 2x - 7$

8–10 The Role of c in $y = ax^2 + bx + c$

The following table compares the graphs of these functions.

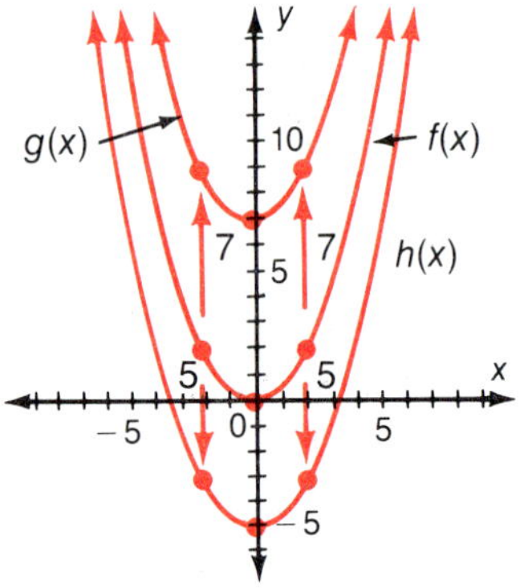

$$f(x) = \frac{1}{2}x^2$$

$$g(x) = \frac{1}{2}x^2 + 7$$

$$h(x) = \frac{1}{2}x^2 - 5$$

$f(x) = \frac{1}{2}x^2$	$g(x) = \frac{1}{2}x^2 + 7$	$h(x) = \frac{1}{2}x^2 - 5$
Opens upward	Opens upward	Opens upward
Vertex: (0, 0)	Vertex: (0, 7)	Vertex: (0, −5)
Vertex is a minimum point.	Vertex is a minimum point.	Vertex is a minimum point.
Axis of symmetry: $x = 0$	Axis of symmetry: $x = 0$	Axis of symmetry: $x = 0$

The graphs of the three functions are congruent, but they have different positions in the plane. Each graph can be made to coincide with the others simply by "sliding" it up or down the y axis. Thus, the graphs differ in the y coordinate of the vertex of each parabola.

EXAMPLE 1 **a.** Use the graph of $f(x) = x^2 - 2x$ at the right to sketch the graph of $g(x) = f(x) + 3$.

b. Write an equation that defines $g(x)$.

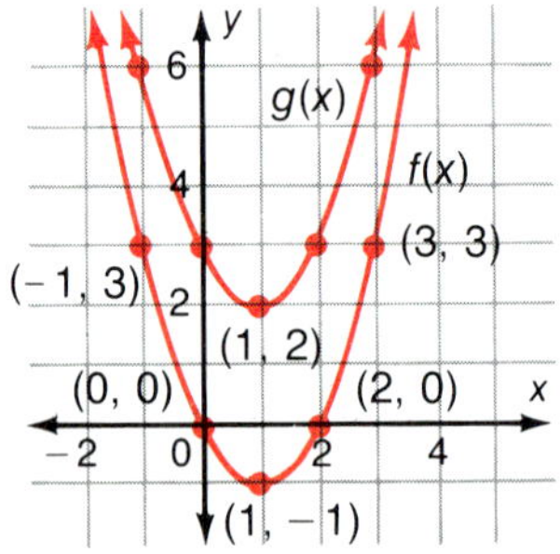

Solutions: **a.**

x	−1	0	1	2	3
f(x)	3	0	−1	0	3
f(x) + 3	6	3	2	3	6

b. $g(x) = f(x) + 3$ ⟵ **Equation of $g(x)$.**

$g(x) = x^2 - 2x + 3$ ⟵ **Replace $f(x)$ with $x^2 - 2x$.**

Example 1 illustrates the relationship between $f(x) = ax^2 + bx$ and $g(x) = ax^2 + bx + c$. The graph of $g(x)$ is congruent to the graph of $f(x)$, and may be graphed from $f(x)$ by adding c to each y coordinate of $f(x)$.

$$g(x) = \underline{x^2 - 2x} + 3 = \underline{f(x)} + 3 \qquad g(x) = \underline{ax^2 + bx} + c = \underline{f(x)} + c$$

This amounts to "sliding" or **translating** the graph of $f(x)$ a distance of c units up or down the y axis.

EXAMPLE 2 If $f(x) = ax^2 + bx + c$, find its y intercept.

Solution: $f(x) = ax^2 + bx + c$

$f(0) = 0 + 0 + c$ ← **For the y intercept, $x = 0$.**

$f(0) = c$ y intercept: c

EXAMPLE 3 The vertex of the graph of $f(x) = -3x^2 + 6x$ is at $V(1, 3)$. Determine the coordinates of the vertex, V', of the graph of $g(x) = -3x^2 + 6x - 3$.

Solution: The functions are congruent. ← **For each function, $a = -3$, $b = 6$.**

$V'(1, 0)$ ← **V' is 3 units below V.**

CLASSROOM EXERCISES

The graphs of each of the following functions are parabolas.

$f_1(x) = 2x^2$	$f_2(x) = -3x^2 + 1$	$f_3(x) = \frac{1}{2}x^2$
$f_4(x) = -2x^2 - 1$	$f_5(x) = 2x^2 + 6$	$f_6(x) = 3x^2 + 4$

1. Which parabolas are congruent?
2. Which parabolas open in the same direction?
3. Determine the coordinates of the vertex of each parabola.

WRITTEN EXERCISES

A

The function $f_2(x)$ is defined by the equation $y = \frac{1}{2}x^2$. The graphs of the functions $f_1(x)$ and $f_3(x)$ are congruent to the graph of $f_2(x)$.

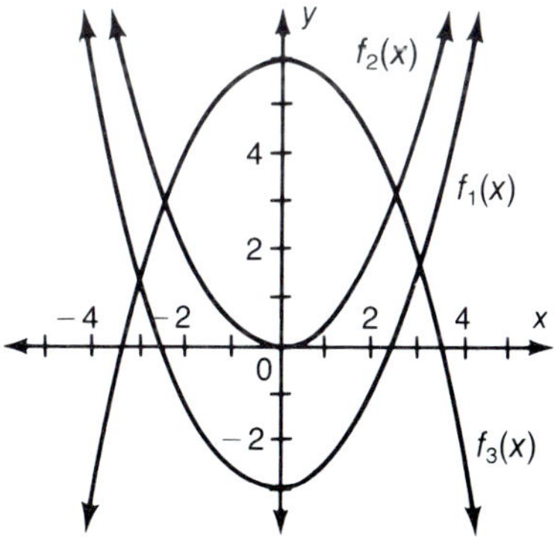

1. List the coordinates of the vertex of each parabola.
2. Write the equation that defines $f_1(x)$ and $f_3(x)$.

In Exercises 3–6, determine the y intercept of the graph of each function.

3. $y = 2x^2 + 7x - 4$ 4. $y = 5x^2$ 5. $y = -3x^2 + 2x + 8$ 6. $y = x^2 + 2x$
7. Are the graphs of the parabolas $y = 3x^2$, $y = -3x^2 + 2$ and $y = 3x^2 + 5$ congruent? Explain.

Sketch the graphs of $y = x^2$ and $y = -x^2$ on the same set of coordinate axes. Then sketch each of the following functions on the same set of axes. Give the coordinates of the vertex of each parabola.

8. $y = x^2 + 2$ **9.** $y = x^2 - 3$ **10.** $y = -x^2 - 4$

11. $y = x^2 + 6$ **12.** $y = -x^2 + 1$ **13.** $y = -x^2 - \frac{1}{2}$

Sketch the graph of $g(x)$ in exercises 14–16 by moving the graph of the given function f the required number of units up or down on its axis of symmetry. Write the equation that defines $g(x)$.

14. Up 2 units **15.** Down 4 units **16.** Up $1\frac{1}{2}$ units

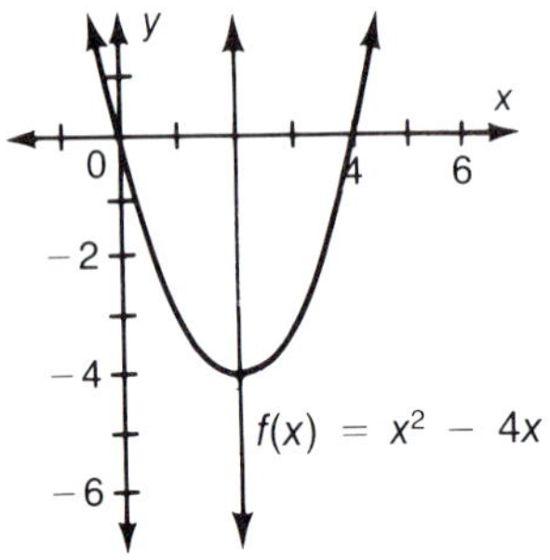

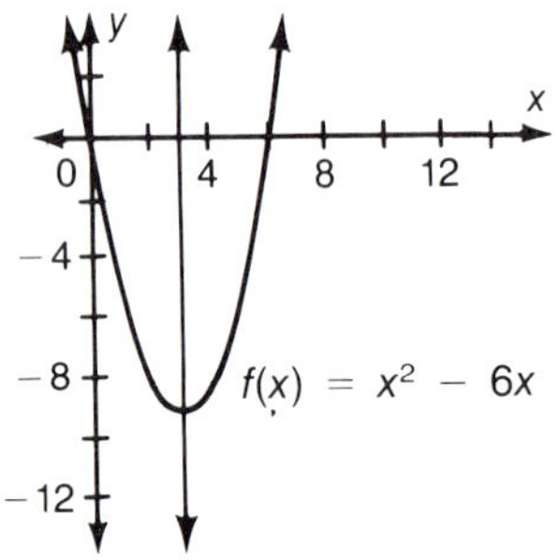

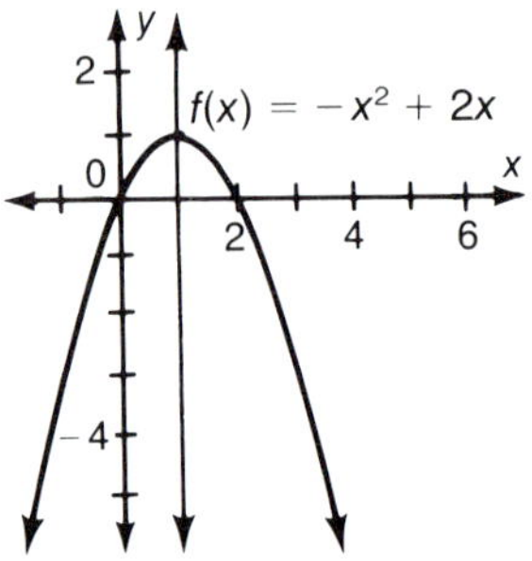

17. How are the graphs of $y = 2x^2$ and $y = -2x^2 + 3$ alike? How are they different?

18. In what direction does the graph of $y = 2x^2 - \frac{1}{2}$ open? What are the coordinates of its vertex?

B

19. What effect does changing $y = 3x^2 - 4$ to $y = -3x^2$ have on the shape of the graph? on the vertex of the graph? on the direction in which the graph opens? Verify by drawing the graph of each function on the same coordinate axes.

20. Find the coordinates of the vertex of the graphs of $y = \frac{1}{2}x^2 + 4$ and $y = 4x^2 + \frac{1}{2}$. Is the graph of the first parabola more or less steep than the graph of the second parabola?

21. If the parabola defined by $y = 5x^2 + c$ passes through the point $(-2, 7)$, determine c.

22. Determine the values of a, b, and c, such that the graph of $y = ax^2 + bx + c$ passes through the origin.

C

23. Are the graphs of $x = ay^2 + by + c_1$ and $x = ay^2 + by + c_2$ congruent? Are they functions? Explain.

24. Sketch the graphs of $x = 2y^2$, $x = 2y^2 - 3$, $x = 2y^2 + 5$ and $x = -2y^2 - 3$ on the same coordinate axes. Describe the effect of changing c from 0 to -3 to 5. How are the graphs of $x = -2y^2 - 3$ and $x = 2y^2 - 3$ related?

25. Describe how to obtain the graphs of $x = ay^2 + by + c$ and $x = ay^2 + by - c$ from the graph of $x = ay^2 + by$. Sketch graphs to illustrate your answer.

REVIEW CAPSULE FOR SECTION 8–11

Evaluate. *(Pages 165–169, 272–274)*

1. $f(x) = x^2 + 6x$ for $x = -3$
2. $g(x) = x^2 + 6x + 12$ for $x = 0$
3. $h(x) = -5x^2 + 3x + 1$ for $x = (t - 1)$
4. $s(x) = ax^2 + bx + c$ for $x = (x - h)$
5. $t(x) = 2x^2 + 4x - 5$ for $x = 2\sqrt{5}$
6. $r(x) = 4 - x^2$ for $x = (1 - \sqrt{3})$

8–11 The Role of b in $y = ax^2 + bx + c$

When the term bx is included in $y = ax^2 + bx + c$, the axis of symmetry of the graph is no longer the y axis, but a *line parallel to the y axis*. The vertex of the parabola lies on the axis of symmetry.

The table below compares the graphs of

$$y = x^2, \quad y = x^2 + 6x, \quad \text{and} \quad y = x^2 + 6x + 12.$$

$y = x^2$

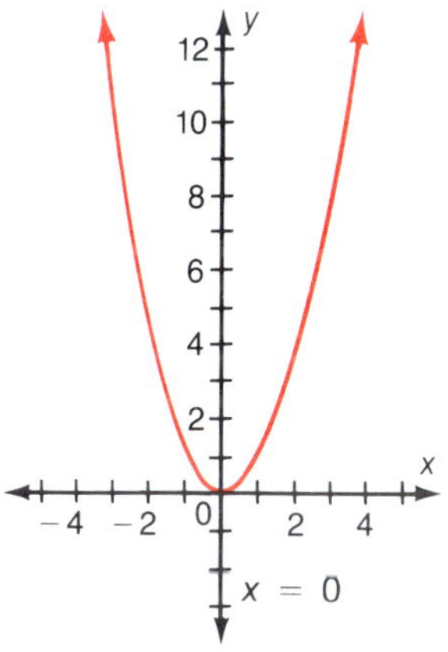

$y = x^2 + 6x$

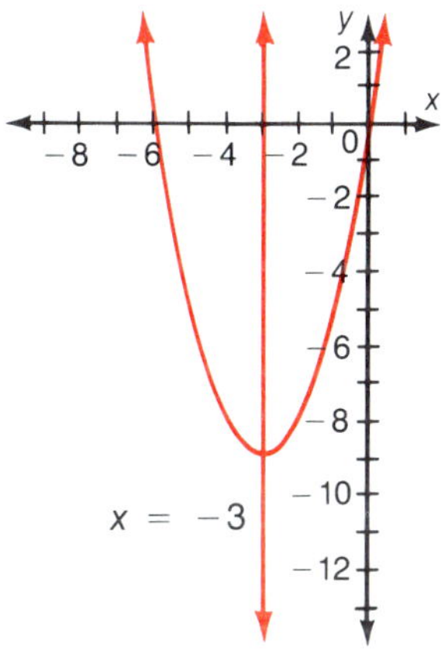

$y = x^2 + 6x + 12$

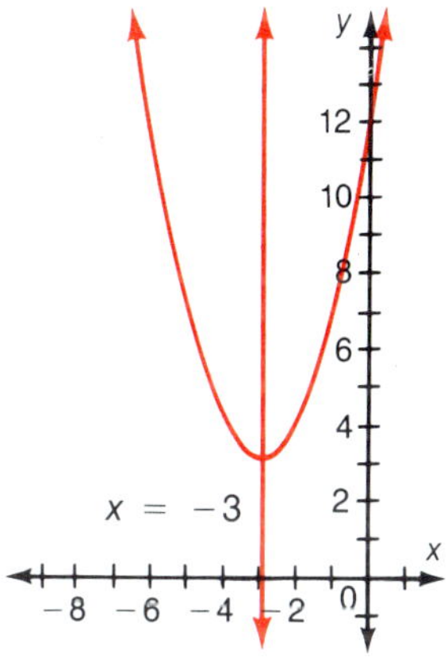

	$y = x^2$	$y = x^2 + 6x$	$y = x^2 + 6x + 12$
Direction	Opens upward	Opens upward	Opens upward
Axis of symmetry	$x = 0$	$x = -3$	$x = -3$
Vertex	$(0, 0)$	$(-3, -9)$	$(-3, 3)$
Maximum or minimum point	Minimum	Minimum	Minimum

Including the term bx in the quadratic function has the effect of sliding the axis of symmetry (and the curve) to the right or left.

Example 1 illustrates how to read the axis of symmetry from the equation. Note that points P_1 and P_2 on the x axis are equidistant from the axis of symmetry.

EXAMPLE 1 Find the equation of the axis of symmetry for $y = ax^2 + bx$.

Solution:

$ax^2 + bx = 0$ ← **Let $y = 0$.**

$x(ax + b) = 0$ ← **Factor.**

$x = 0 \quad \text{or} \quad ax + b = 0$ ← **Solve for x.**

$x = 0 \quad \text{or} \quad x = -\frac{b}{a}$

$P_1(0, 0);\ P_2\left(-\frac{b}{a}, 0\right)$ ← **Two points on the graph**

$x = \frac{0 + \left(-\frac{b}{a}\right)}{2};\ y = \frac{0 + 0}{2}$ ← **By Theorem 8–5**

$M\left(-\frac{b}{2a}, 0\right)$ Equation of axis of symmetry: $x = -\frac{b}{2a}$

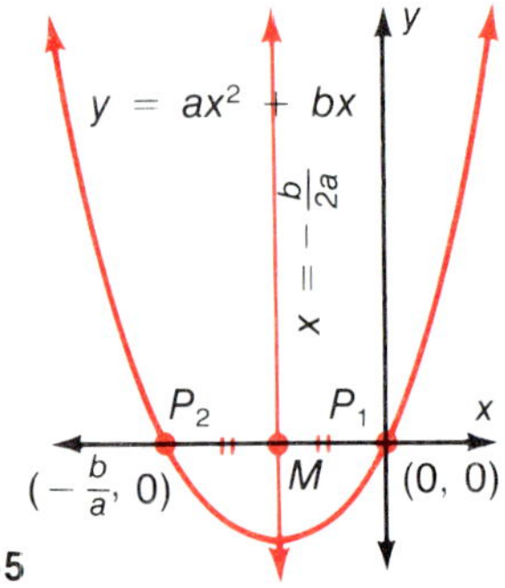

Since changes in the value of c have the effect of raising or lowering the parabola, the equation of the axis of symmetry is the same for $y = ax^2 + bx + c$. This fact and Example 1 lead to the following theorem.

Theorem 8–6

Axis of Symmetry

The equation of the axis of symmetry of a parabola defined by $y = ax^2 + bx + c$, $a \neq 0$, is

$$x = -\frac{b}{2a}.$$

EXAMPLE 2

a. Find the equation of the axis of symmetry and the coordinates of the vertex of the graph of $y = 4x^2 - 8x - 5$.

b. Give the maximum or minimum value of the function.

c. Find two other points on the parabola.

d. Sketch the parabola.

Solutions:

a. $x = -\frac{b}{2a} = -\frac{-8}{2 \cdot 4} = 1$ ← **By Theorem 8–6**

$\mathbf{x = 1}$ ← **Axis of symmetry**

b. $y = 4(1)^2 - 8(1) - 5$ ← **Find y when x = 1.**

$y = 4 - 8 - 5 = -9$

Vertex: **(1, −9)**

Minimum value: −9 ← **a > 0; opens upward.**

c. $\mathbf{P(0, -5)}$ ← **When x = 0, y = −5.**

$\mathbf{P'(2, -5)}$ ← **Image point of P**

d. Join V, P, and P' with a smooth curve.

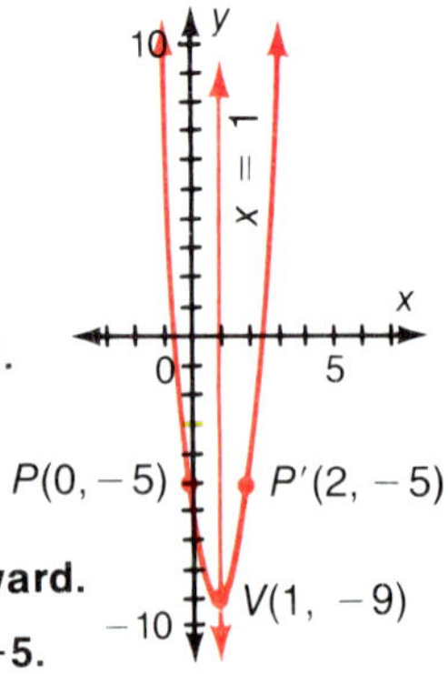

The following summarizes the roles of a, b, and c for the quadratic function $y = ax^2 + bx + c$.

Summary

For the quadratic function $y = ax^2 + bx + c$, $a \neq 0$:

1. Any two quadratic functions having the same value of a are congruent.
   - **a.** When $a > 0$, the curve opens upward.
   - **b.** When $a < 0$, the curve opens downward.
   - **c.** When $|a| > 1$, the graph is steeper than the graph of $y = x^2$.
   - **d.** When $|a| < 1$, the graph is less steep than the graph of $y = x^2$.
2. Changes in the value of c move the curve up and down on its axis symmetry. The y intercept of the graph is c.
3. Changes in the value of b move the curve to the right or left along the x axis or along a line parallel to the x axis.
4. The equation of the axis of symmetry is $x = -\frac{b}{2a}$.
5. The coordinates of the vertex are $\left(-\frac{b}{2a}, f\left(-\frac{b}{2a}\right)\right)$.
6. The maximum or minimum value of the function is $f\left(-\frac{b}{2a}\right)$.

CLASSROOM EXERCISES

For each parabola, find the equation of the axis of symmetry.

1. $y = x^2$ **2.** $y = -x^2 + 7x$ **3.** $y = x^2 - 4x - 5$ **4.** $y = 2x^2 - 4x + 5$

For each parabola, determine the coordinates of the vertex. Indicate whether the point is a maximum or minimum.

5. $y = 3x^2 + 1$ **6.** $y = -5x^2 + 4$ **7.** $y = 5x^2 - 4$ **8.** $y = -2x^2 + x - 3$

9. On the same coordinate axes, sketch the graphs of $y = -2x^2$, $y = -2x^2 + x - 3$, and $y = 2x^2 + x - 3$.

WRITTEN EXERCISES

A For each function in Exercises 1–11:

a. Determine the equation of the axis of symmetry.
b. Determine the coordinates of the vertex of the graph of the parabola.
c. Tell whether the function has a maximum or minimum.

1. $y = x^2 - x - 2$ **2.** $y = x^2 + 5x + 6$ **3.** $f(x) = x^2 + 4$ **4.** $y = 2x^2 + x - 3$

5. $y = x^2 + x - 2$ **6.** $g(x) = 4 - x^2$ **7.** $y = 2 - x - 3x^2$ **8.** $f(x) = x^2$

9. $g(x) = 3x^2 + 6x + 1$ **10.** $y = 2x^2 + 4x - 5$ **11.** $y = -5x^2 + 3x + 1$

Sketch each parabola by graphing **(a)** the axis of symmetry, **(b)** the vertex, and **(c)** one other point on the parabola and the point symmetric to it with respect to the axis of the parabola (image point).

12. $y = x^2 + x - 6$ **13.** $y = x^2 - 4x$ **14.** $y = x^2 - 7x + 6$

15. $y = 2x^2 - 2x - 12$ **16.** $y = 6x^2 - 9x - 6$ **17.** $y = 4x^2 - 4x - 3$

18. How is the graph of $y = x^2 - 4x + 1$ like the graph of $y = 4x - x^2 - 1$? How is it different?

19. If b is changed from -8 to 8 in $y = 2x^2 - 8x + 1$ while a and c remain the same, how does this affect the position of the axis of symmetry and of the vertex?

20. The graph of a parabola is congruent to the graph of $y = 2x^2 - 8x + 3$, opens in the same direction, has the same y intercept, and is 2 units to the left of $y = 2x^2 - 8x + 3$. Write the equation of the parabola.

21. The graph of a parabola is congruent to the graph of $y = 3x^2 - 6x + 4$, has the same y intercept, opens downward, and is 3 units to the right of $y = 3x^2 - 6x + 4$. Write the equation of the parabola.

C

22. Show that if the abscissa of the vertex of the parabola defined by $y = ax^2 + bx + c$ is $-\frac{b}{2a}$, then the ordinate of the vertex is $-\frac{b^2}{4a} + c$.

23. Show that $y = a(x - h)^2 + k$ defines a parabola whose vertex is at (h, k).

24. The equation $x = ay^2 + by + c$ defines a parabola with an axis of symmetry defined by $y = -\frac{b}{2a}$. Illustrate this by sketching the graphs of $x = y^2 - 4y + 2$ and $x = -y^2 - 4y + 2$.

CALCULATOR APPLICATIONS

Minimum Value of a Quadratic Function

The expression $\mathbf{c - \frac{b^2}{4a}}$ represents the minimum value of the quadratic function $y = ax^2 + bx + c$, when $a > 0$.

EXAMPLE Find the approximate minimum value of $y = 21x^2 + 9x + 52$.

SOLUTION $c - \frac{b^2}{4a} = 52 - \frac{9^2}{(4)(21)}$

First evaluate $\frac{9^2}{(4)(21)}$ and store the value.

9 [x²] [÷] 4 [÷] 2 1 [=] [M+] 5 2 [−] [MR] [=] 51.035714

EXERCISES

Find the approximate minimum values of the functions in Exercises 12–17 above.

Problem Solving and Applications

8–12 Maximum and Minimum

Quadratic functions can be useful models in situations where it is desirable to maximize or minimize the use of materials.

EXAMPLE A rain gutter is made by bending up the edges of a piece of aluminum 25 centimeters wide so that the cross-sectional area is a maximum. Find how much the edges of the piece should be bent to maximize the area and find the maximum area.

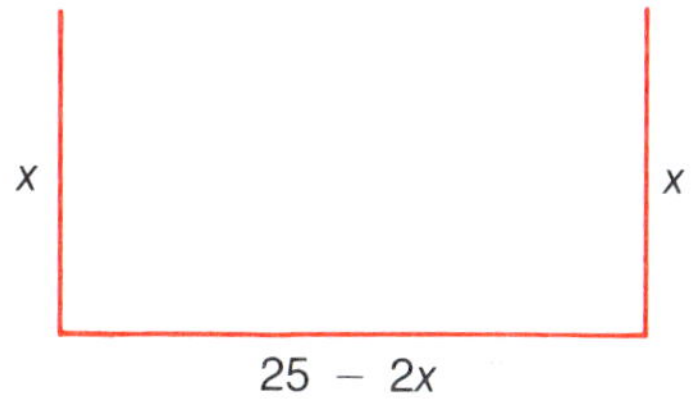

Solution: [1] $A(x) = y = x(25 - 2x)$ ← **Model for cross-sectional area**

$y = 25x - 2x^2$

[2] Since $a < 0$, the graph opens downward. Thus, the vertex is a maximum point.

[3] $x = -\frac{25}{-4} = 6.25$ ← **Axis of symmetry:** $x = -\frac{b}{2a}$

[4] $V(6.25, f(6.25))$ ← **Coordinates of vertex**

V**(6.25, 78.125)** ← **When** $x = 6.25$, $y = 78.125$.

Thus, the maximum cross-sectional area of **78.125 square centimeters** occurs when the edge of the 25-centimeter strip is bent up **6.25 cm.**

CLASSROOM EXERCISES

State whether the graph of each function contains a maximum point or a minimum point.

1. $f(x) = x^2 + x$ **2.** $g(x) = -x^2 + 3x$ **3.** $-x^2 + 3x = h(x)$ **4.** $5x - x^2 = p(x)$

Write a quadratic function to represent each problem.

5. Find two numbers whose sum is 16 and whose product is a maximum.

6. Find the length and width of a rectangle with a perimeter of 20 centimeters such that its area is a maximum.

7. Find two numbers such that their sum is 20 and the sum of their squares is a minimum.

8. Find two numbers such that their sum is 10 and the sum of their squares is a minimum.

WORD PROBLEMS

A

1. A rectangular garden is to be enclosed on three sides by fencing; the fourth side is a side of the house. What is the largest garden that can be enclosed by 30 meters of fencing? (HINT: Write the area as a function of x and find the maximum value of the function.)

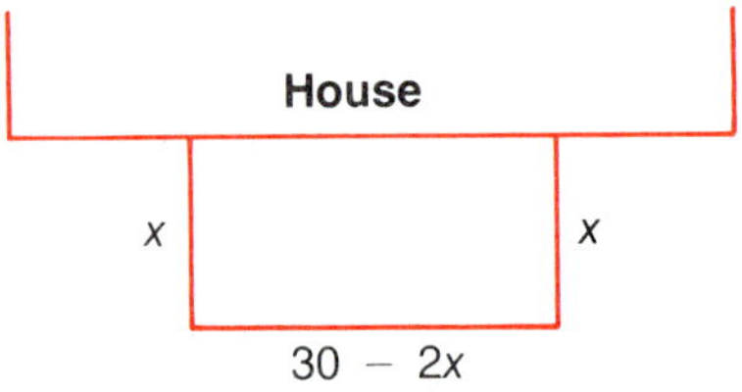

2. If you throw a ball upward with a velocity of v meters per second, its distance in meters, s, above the starting point in t seconds is shown by experiment to be $s = vt - 4.9t^2$. If the ball is thrown with an initial velocity of 30 meters per second, this formula becomes $s = 30t - 4.9t^2$.
   a. Is the graph of this quadratic function a parabola that opens downward or upward?
   b. How many seconds will it take the ball to reach its highest point?
   c. How high will it go?
3. How high will a projectile rise if shot vertically with an initial velocity of 300 meters per second? (See the formula in Exercise 2.) How long will it take the projectile to reach its maximum height?
4. A farmer's daughter wishes to fence in a rectangular poultry yard. She has 200 meters of fencing and wants to enclose the maximum area. What should be the dimensions of the yard so that the enclosed region has a maximum area?

B

5. Find two numbers such that their sum is 20 and the sum of their squares is a minimum.
6. Find two numbers whose sum is 10 and whose product is a maximum.
7. Find two numbers such that their sum is 30 and the sum of their squares is a minimum.

C

8. Jon has two dogs. He wants to make two pens so that each dog will have the same size pen and as much ground space as possible. If Jon has 200 meters of fencing and the pens are to be rectangular, what should the dimensions be? (HINT: Part of the 200 meters of fencing will be used to divide the pens.)

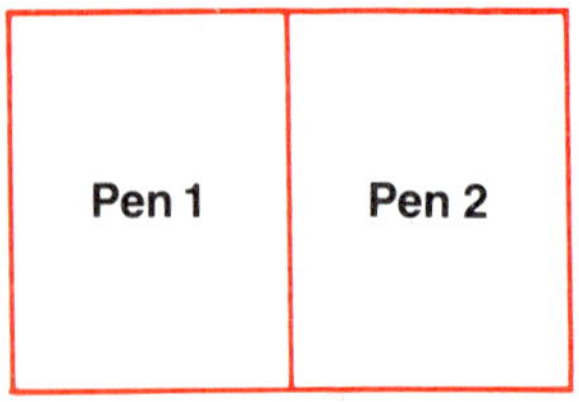

9. A tinsmith wishes to make a double gutter for two liquids from a strip of tin 26 centimeters wide by folding up the edges and folding in the middle. The cross sections of the parts are to be rectangular and have the same area. Find the dimensions that will give maximum carrying capacity.

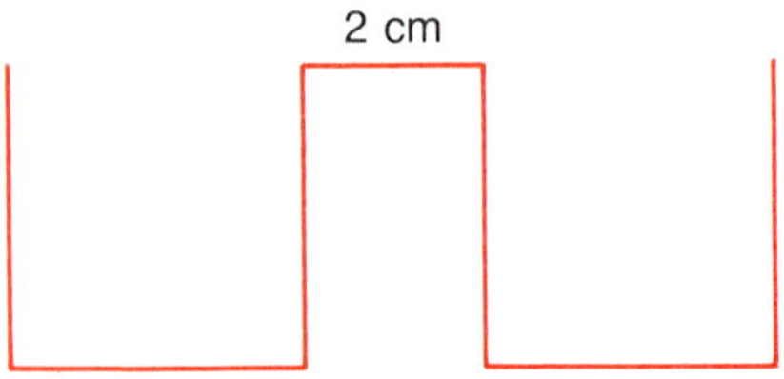

10. If, in Exercise 9, the center wall is to be one centimeter higher than the outer edges, what dimensions will produce the maximum carrying capacity?

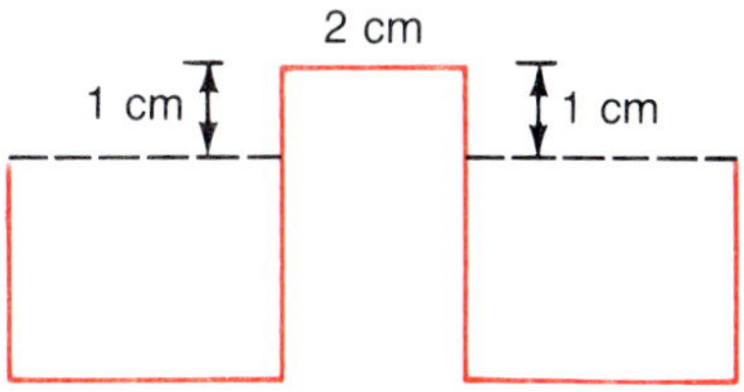

11. Given a fixed perimeter, P, of a rectangle, what are the dimensions of the sides that will give the greatest possible area?

12. An open box is to be made from a rectangular sheet of metal by cutting 2-centimeter squares from each other. If the perimeter of the base of the box must be 40 centimeters, what is the maximum possible volume of the box?

Review

For each function, make a table of values for integral values of x where $-6 \le x \le 6$. Use the table to graph each function. *(Section 8–7)*

1. $y = x^2 + 2$
2. $y = x^2 + 4x + 4$
3. $y = x^2 + 3x - 4$

In Exercises 4–5, the equation of a parabola and a point on the parabola are given. Find the value of the specified coefficient. *(Section 8–7)*

4. $y = x^2 - bx + 3$; $A(2, -1)$; $b =$ ___?___
5. $y = x^2 + 7x + c$; $B(-1, 6)$; $c =$ ___?___
6. Which function has the steeper graph, $y = 3x^2$ or $y = \frac{1}{3}x^2$? *(Section 8–8)*
7. Sketch a parabola that contains the points $A(1, 7)$ and $B(2, 12)$ and has its vertex at $V(0, 5)$. *(Section 8–8)*

In Exercises 8–9, y varies directly as x^2. *(Section 8–9)*

8. If $y = 45$ when $x = 3$, find y when $x = 7$.
9. If $y = 2.6$ when $x = 1.8$, find y when $x = 3.2$.
10. The surface area of a sphere varies directly as the square of the radius of the sphere. If the surface area is 16π square meters when the radius is 2 meters, what is the surface area when the radius is 5 meters? *(Section 8–9)*

Sketch the graphs of $y = 4x^2$ and $y = -4x^2$ on the same set of coordinate axes. Then sketch each of the following functions on the same set of axes. Give the coordinates of the vertex of each parabola. *(Section 8–10)*

11. $y = 4x^2 - 3$
12. $y = 4x^2 + 2$
13. $y = -4x^2 + 4$
14. $y = -4x^2 - 1$

In Exercises 15–17, sketch each parabola by graphing **(a)** the axis of symmetry, **(b)** the vertex, and **(c)** one other point on the parabola and the point symmetric to it with respect to the axis of the parabola (image point). *(Section 8–11)*

15. $y = x^2 + 3x$
16. $y = 3x^2 - 2x + 4$
17. $y = -5x^2 + 6x - 2$
18. The sum of the lengths of the height and base of a parallelogram is 25 centimeters. How long should the base be to maximize the area of the parallelogram? *(Section 8–12)*

BASIC: PARABOLAS

Problem: *Given the coefficients A, B, and C of a quadratic function $y = Ax^2 + Bx + C$, write a program which prints the vertex and axis of symmetry of the parabola, and determines whether it opens up or down.*

```
100 PRINT "ENTER A,B, AND C, THE COEFFICIENTS OF THE
110 PRINT "QUADRATIC FUNCTION A*X*X + B*X + C";
120 INPUT A, B, C
130 IF A <> 0 THEN 160
140 PRINT "NOT A QUADRATIC FUNCTION; PLEASE TRY AGAIN."
150 GOTO 120
160 LET X = -B/(2*A)
170 LET Y = A*X*X + B*X + C
180 PRINT "VERTEX IS (";X;",";Y;")"
190 PRINT "AXIS OF SYMMETRY IS X =";X
200 IF A < 0 THEN 230
210 PRINT "THE PARABOLA OPENS UPWARD."
220 GOTO 240
230 PRINT "THE PARABOLA OPENS DOWNWARD."
240 PRINT
250 PRINT "ANY MORE PARABOLAS (1 = YES, 0 = NO)";
260 INPUT Z
270 IF Z = 1 THEN 100
280 END
```

Output:

```
RUN
ENTER A, B, AND C, THE COEFFICIENTS OF THE
QUADRATIC FUNCTION A*X*X + B*X + C? 1,0,-16
VERTEX IS ( 0 , -16 )
AXIS OF SYMMETRY IS X = 0
THE PARABOLA OPENS UPWARD.

ANY MORE PARABOLAS (1 = YES, 0 = NO)? 1
ENTER A, B, AND C, THE COEFFICIENTS OF THE
QUADRATIC FUNCTION A*X*X + B*X + C? 0,5,-4
NOT A QUADRATIC FUNCTION; PLEASE TRY AGAIN.
? 10,-23,18
VERTEX IS ( 1.15 , 4.775 )
AXIS OF SYMMETRY IS X = 1.15
THE PARABOLA OPENS UPWARD.

ANY MORE PARABOLAS (1 = YES, 0 = NO)? 0
READY
```

Analysis

Statements 130–150: These statements function as a "bad data trap." If A = 0, the function is not quadratic. Also, if a zero value for A is not caught, division by zero will be attempted, resulting in an ERROR message.

Statements 160–190: The x value of the vertex is computed using the formula developed on page 301 of this chapter. Then, in statement 170, this x value is substituted into the equation of the function to calculate the y coordinate of the vertex. Statement 180 prints these coordinates as an ordered pair. Statement 190 prints the equation of the axis of symmetry, which is a vertical line through the vertex.

Statements 200–230: A positive value of A means that the parabola opens upward. A negative value of A means that it opens downward.

Statement 240: The purpose of this statement is to skip a line of output after the information about the parabola has been printed. (See the sample output.)

On computer keyboards there usually is no radical sign ($\sqrt{\ }$). Instead, in BASIC the SQR ("square root") function can be used to compute the square root of a nonnegative quantity. Here are examples of the SQR function in formulas.

```
25 LET Y = SQR(X)     30 LET C = SQR(A↑2 + B↑2)
```

EXERCISES

A In Exercises 1–8, write a BASIC program for each problem.

1. Given a real number x, print the square root of x or, if $x < 0$, print the message NO REAL SQUARE ROOT.

In Exercises 2 and 3, the input consists of the coordinates (x_1, y_1) and (x_2, y_2) of two points.

2. Compute and print the distance between the points.
3. Compute and print the coordinates of the midpoint of the segment joining the two points.
4. Given the lengths of any two sides of a right triangle, compute the length of the third side. The user will enter values for a, b, and c (c is the hypotenuse), with a 0 for the unknown side.
5. Given the area of a square, compute and print the length of its side.
6. Given the area of a circle, compute and print the length of its radius.
7. Given the length of the side of a square, compute and print the length of its diagonal.
8. Given the length of a side of an equilateral triangle, compute and print the area of the triangle.

Chapter Summary

IMPORTANT TERMS

Apparent solution (p. 278)
Axis of symmetry (p. 287)
Conjugates (p. 276)
Distance between points (p. 280)
Index (p. 270)
Like radicals (p. 275)
Maximum point (p. 287)
Minimum point (p. 287)
Parabola (p. 287)
Principal *n*th root (p. 270)
Quadratic function (p. 287)
Radical (p. 270)
Radical equation (p. 278)
Radicand (p. 270)
Variation as the square (p. 293)
Vertex (p. 287)

IMPORTANT IDEAS

1. Principal *n*th Root: For any real number a and any integer n, $n > 0$,
   a. $\sqrt[n]{a}$ is a positive number or 0 when a is positive or equal to 0:
   b. $\sqrt[n]{a}$ is a negative real number when a is negative and n is odd.

2. **Multiplication Theorem for Radicals:** For all real numbers a and b and for any integer n, $n > 1$,
$$\sqrt[n]{a} \cdot \sqrt[n]{b} = \sqrt[n]{ab}.$$

3. For real numbers a and b where $a \geq 0$ and $b \geq 0$, and for any integer n, $n > 1$,
$$\frac{\sqrt[n]{a}}{\sqrt[n]{b}} = \sqrt[n]{\frac{a}{b}}.$$

4. **Pythagorean Theorem:** In a right triangle, if the length of the hypotenuse is c and the lengths of the legs are a and b, then
$$c^2 = a^2 + b^2.$$

5. **Distance Formula for Two-Space:** The distance between the two points $P_1(x_1, y_1)$ and $P_2(x_2, y_2)$ is given by the formula
$$P_1P_2 = \sqrt{(x_1 - x_2)^2 + (y_1 - y_2)^2}$$

6. **Midpoint Formula:** The coordinates of the midpoint $M(x, y)$ of the line segment with endpoints $P_1(x_1, y_1)$ and $P_2(x_2, y_2)$ are
$$x = \frac{x_1 + x_2}{2} \quad \text{and} \quad y = \frac{y_1 + y_2}{2}$$

7. For the quadratic function $y = ax^2 + bx + c$, $a \neq 0$;
   a. Any two quadratic functions having the same value of a are congruent.
   When $a > 0$, the curve opens upward.
   When $a < 0$, the curve opens downward.
   When $|a| > 1$, the graph is steeper than the graph of $y = x^2$.
   When $|a| < 1$, the graph is less steep than the graph of $y = x^2$.
   b. Changes in the value of c move the curve up and down on its axis of symmetry. The y intercept of the graph is c.

c. Changes in the value of b move the curve to the right or left along the x axis.

d. The equation of the axis of symmetry is $x = -\frac{b}{2a}$.

e. The coordinates of the vertex are $\left(-\frac{b}{2a}, f\left(-\frac{b}{2a}\right)\right)$.

f. The maximum or minimum value of the function is $f\left(-\frac{b}{2a}\right)$.

Chapter Objectives and Review

Objective: *To simplify radicals (Sections 8–1 and 8–2)*

Simplify.

1. $\sqrt{81s^2}$ **2.** $\sqrt[3]{-8t^3}$ **3.** $\sqrt{a^{10}b^{14}}$ **4.** $\sqrt{\frac{49}{64w^4}}$

5. $\sqrt{18}$ **6.** $\sqrt{75b^3}$ **7.** $\sqrt[3]{54}$ **8.** $\sqrt[3]{-16w^6}$

Objective: *To multiply and divide radicals (Section 8–2)*

Multiply or divide as indicated. Simplify your answer.

9. $\sqrt{5a^3} \cdot \sqrt{5ab^2}$ **10.** $\sqrt[3]{16} \cdot \sqrt[3]{54x^3}$ **11.** $\frac{\sqrt{50}}{\sqrt{18}}$

12. $\frac{\sqrt{8w}}{\sqrt{2w}}$ **13.** $\sqrt{5} \div \sqrt{\frac{5}{16}}$ **14.** $\sqrt[3]{3cd^2} \cdot \sqrt[3]{9cd}$

Objective: *To add and subtract radicals (Section 8–3)*

Add or subtract as indicated.

15. $4\sqrt{5} + 3\sqrt{5}$ **16.** $3\sqrt{7} - \sqrt{14}$

17. $\sqrt{12} + \sqrt{18} - \sqrt{98}$ **18.** $\sqrt{125} - 3\sqrt{45} + 2\sqrt{20}$

Objective: *To rationalize the denominator of a fraction (Sections 8–2 and 8–3)*

Rationalize each denominator.

19. $\frac{2}{\sqrt{7}}$ **20.** $\frac{3}{\sqrt{6}}$ **21.** $\frac{2}{2 + \sqrt{5}}$ **22.** $\frac{\sqrt{5} + 3}{\sqrt{5} - 3}$

Objective: *To solve radical equations (Section 8–4)*

Solve and check.

23. $\sqrt{2b} = 8$ **24.** $\sqrt{3t + 1} = 4$ **25.** $\sqrt[4]{10w + 1} - 1 = 2$

26. $\sqrt{x - 2} = \sqrt{3x - 8}$ **27.** $\sqrt{4n} - 1 = \sqrt{n + 5}$ **28.** $\sqrt{11v + 1} = \sqrt{v} + 7$

Objective: *To use the distance and midpoint formulas (Section 8–5)*

Find the distance between the points with the given coordinates.

29. $A(4, 3)$; $B(7, 9)$ **30.** $C(3, -1)$; $D(-2, 8)$ **31.** $F(-6, -8)$; $G(1, 0)$

Find the coordinates of the midpoint of the line segment with the given endpoints.

32. $T(5, -1)$; $V(11, -19)$ **33.** $E(-12, -1)$; $H(-1, -3)$ **34.** $W(5, 0)$; $K(0, 5)$

35. One endpoint of a segment is $S(2, -2)$, and the midpoint is $T(-4, 6)$. Find the coordinates of the other endpoint.

36. Show that the quadrilateral with vertices $A(1, 1)$, $B(5, -2)$, $C(2, -6)$, and $D(-2, -3)$ has four sides of equal length.

Objective: *To prove theorems using coordinate geometry (Section 8–6)*

37. Prove that the median of a trapezoid is parallel to the bases. [*Hint:* Let the vertices be at $A(0, 0)$, $B(a, 0)$, $C(b, c)$, and $D(d, c)$.]

38. Prove that the diagonals of a square are perpendicular to each other and that they are equal in length.

Objective: *To sketch the graph of a quadratic function (Section 8–7)*

For each exercise, sketch the given quadratic functions on the same set of axes.

39. $y = 3x^2$; $y = 3x^2 + 2x$; $y = 3x^2 + 2x + 6$

40. $y = -\frac{1}{2}x^2$; $y = -\frac{1}{2}x^2 + 2x$; $y = -\frac{1}{2}x^2 + 2x - 3$

Objective: *To determine from a quadratic equation the coordinates of the vertex, the equation of the axis of symmetry, and whether the vertex is a maximum point or a minimum point (Sections 8–8, 8–10, 8–11)*

For each quadratic equation, write the equation of the axis of symmetry and the coordinates of the vertex. State whether the vertex is a maximum point or a minimum point.

41. $y = \frac{1}{4}x^2 - 3$ **42.** $y = -\frac{1}{2}x^2 - 2$ **43.** $y + 3x^2 = 7$

Objective: *To solve problems involving variation as the square (Section 8–9)*

44. If d varies directly as the square of t, and $d = 48$ when $t = 4$, find d when $t = 7$.

45. If V varies directly as the square of r and $V = 28\pi$ when $r = 2$, find V when $r = 3$.

Objective: *To solve maximum and minimum problems (Section 8–12)*

46. A rectangular pen is to be enclosed with 200 meters of fencing. What dimensions for the pen will give the largest area?

47. A projectile is shot vertically upward. Its distance d in meters above the starting point after t seconds is known to be $d = 210t - 4.9t^2$. How many seconds will it take the projectile to reach its highest point? How high will it go?

Chapter Test

Simplify.

1. $\sqrt[3]{81x^5y^{12}}$

2. $\pm\sqrt{(t-6)^6}$

Perform the indicated operations. Simplify your answer.

3. $\sqrt{3x^8} \cdot \sqrt{6x^2y^7}$

4. $\sqrt{18x^5} - 5\sqrt{8x^5}$

5. $3\sqrt{27} + 2\sqrt{24} - 6\sqrt{12}$

6. $\dfrac{2\sqrt{4x}}{\sqrt{10x^3y^5}}$

7. $(3 + \sqrt{5})^2$

8. Rationalize the denominator: $\dfrac{3}{\sqrt{2} - 6}$

Solve and check.

9. $\sqrt{2x - 5} = 3$

10. $\sqrt{6x - 8} + 2 = 0$

Find the midpoint of the segment whose endpoints have the given coordinates.

11. $A(-7, -3)$; $B(2, 4)$

12. $R(2, 2)$; $S(-3, 8)$

13. Show that the triangle with vertices $D(-1, 0)$, $E(5, 0)$, and $F(2, 3\sqrt{3})$ is equilateral.

14. If y varies directly as the square of x and $y = 18$ when $x = 6$, find y when $x = 4$.

For Exercises 15–17, use the function defined by $y = -2x^2 + 4x - 3$.

15. What is the equation of the axis of symmetry?

16. What are the coordinates of the vertex?

17. Is the vertex a maximum point or a minimum point?

For Exercises 18–19, use the functions defined by $y = \frac{1}{2}x^2$ and $y = -\frac{1}{2}x^2 + 3$.

18. How are the graphs of these functions alike?

19. How are the graphs different?

20. Josephine wishes to enclose three sides of a pen with fencing; the fourth side is a side of a barn. If she has 170 feet of fencing, what is the largest area that can be enclosed?

Preparing for College Entrance Tests

Some items on college entrance tests are designed so that one or more of several given statements are correct or no statement is correct. In such cases, you must examine **each** of the possibilities before deciding on the answer.

EXAMPLE If $n > 0$, which of the following statements are true?

I. $\sqrt{n^2} = n$ **II.** $\sqrt{n} > -\sqrt{n}$ **III.** $\sqrt{n} > \frac{1}{\sqrt{n}}$

(a) I only **(b)** I and II only

(c) I, II, and III **(d)** None of the statements are true.

Solution: I is true, since $\sqrt{n^2} = |n|$, and $|n| = n$ for $n > 0$.

II is true, since for $n > 0$, $\sqrt{n}$ is a positive real number, and $-\sqrt{n}$ is negative.

III is false for $0 < n \leq 1$. For example, $\sqrt{\frac{1}{4}} \not> \frac{1}{\sqrt{\frac{1}{4}}}$ **Answer: b**

Choose the best answer. Choose *a*, *b*, *c*, or *d*.

1. If $n < 0$, which of the following statements are true?

I. $\sqrt[3]{-n} < 0$ **II.** $\sqrt{1-n} < 1$ **III.** $\sqrt[5]{n^3} > 0$

(a) II only **(b)** I and II only

(c) I and III only **(d)** None of the statements are true.

2. If $\sqrt{k}$ is an integer, which of the following must be integers?

I. $\sqrt{\frac{k}{4}}$ **II.** $(\sqrt{5k})^2$ **III.** $\sqrt{9k}$

(a) I and II only **(b)** I and III only **(c)** II and III only **(d)** I, II, and III

3. If $x > 1$, which of the following decrease(s) as x increases?

I. $1 - \sqrt{x}$ **II.** $\frac{x}{x+1}$ **III.** $\frac{1}{1-x}$

(a) I only **(b)** II only **(c)** I and III only **(d)** None decrease.

4. $\frac{1}{1+\sqrt{2}} =$ __?__ (HINT: "None of these" means choices **a–c** are all incorrect.)

(a) $\frac{1}{3} - \frac{1}{3}\sqrt{2}$ **(b)** $\sqrt{2} - 1$ **(c)** $-1 - \sqrt{2}$ **(d)** None of these

5. $\sqrt{\frac{9}{16} + 1} =$ __?__ (HINT: Write $1\frac{9}{16}$ as a fraction.)

(a) $1\frac{3}{4}$ **(b)** $\frac{3}{4}$ **(c)** $\frac{5\sqrt{2}}{8}$ **(d)** None of these

CHAPTER **9**

Complex Numbers/ Quadratic Equations

9–1 Pure Imaginary Numbers

The linear equation

$$x + 1 = 0$$

has no solution over the set of positive real numbers. However, when the replacement set is extended to {real numbers}, the equation has one solution, -1.

Similarly, the equation

$$x^2 + 1 = 0$$

has no solution over the set of real numbers because there is no real number whose square is -1. To solve equations such as this, mathematicians invented the numbers, **i** and **−i.**

Definition

$$\mathbf{i^2 = -1} \quad \text{and} \quad \mathbf{(-i)^2 = -1}$$

By the definition of square root,

$$(\sqrt{-1})^2 = -1 \quad \text{and} \quad (-\sqrt{-1})^2 = -1.$$

So $\quad \mathbf{i = \sqrt{-1}} \quad$ and $\quad \mathbf{-i = -\sqrt{-1}}.$

Note that i is *not a variable;* it names a number, $\sqrt{-1}$. Numbers such as

$$\sqrt{-1} \qquad -\sqrt{-1} \qquad \sqrt{-5} \qquad -\sqrt{-\frac{1}{7}} \qquad -\sqrt{-49}$$

are *pure imaginary numbers*. A **pure imaginary number** is a square root of a negative number. Pure imaginary numbers are written as the product of a real number and i. This is called the **i-form** of the number.

Definition

For any real number a, where $-a < 0$,

$$\sqrt{-a} = \mathrm{i}\sqrt{a} = \sqrt{a} \cdot \mathrm{i}.$$

EXAMPLE 1 Write each pure imaginary number in i-form.

a. $\sqrt{-1}$ **b.** $\sqrt{-5}$ **c.** $\sqrt{-16}$ **d.** $-\sqrt{-25}$

Solutions:

a. $\sqrt{-1} = \mathbf{i}$ ← **By definition**

b. $\sqrt{-5} = \sqrt{-1} \cdot \sqrt{5} = \mathbf{i\sqrt{5}}$

c. $\sqrt{-16} = \sqrt{16} \cdot \mathrm{i} = \mathbf{4i}$

d. $-\sqrt{-25} = -\sqrt{25} \cdot \mathrm{i} = \mathbf{-5i}$

In Example 1b, $\sqrt{-5}$ is usually written as $\mathrm{i}\sqrt{5}$ to avoid the error of writing $\sqrt{5\mathrm{i}}$ for $\sqrt{5} \cdot \mathrm{i}$.

To add, subtract, multiply, or divide pure imaginary numbers, assume that they have all the properties of real numbers (except closure) listed on page 14. Before computing with these numbers, write them in i–form.

EXAMPLE 2 Add or subtract as indicated: **a.** $2i + 3i$ **b.** $\sqrt{-9} - 5i$

Solutions: **a.** $2i + 3i = (2 + 3)i$ ← **By the distributive postulate**

$= \mathbf{5i}$

b. $\sqrt{-9} - 5i = 3i - 5i$ ← **Write $\sqrt{-9}$ in i–form.**

$= (3 - 5)i = \mathbf{-2i}$

In multiplying pure imaginary numbers, you apply the definition of i.

EXAMPLE 3 Multiply. Express the product as a real number or in i–form.

a. $-5i \cdot \sqrt{-16}$ **b.** $\sqrt{-2} \cdot \sqrt{-18}$

Solutions: **a.** $-5i \cdot \sqrt{-16} = -5i \cdot 4i$ ← **Write in i–form.**

$= -20i^2$ ← **Replace i^2 with -1.**

$= -20(-1) = \mathbf{20}$

b. $\sqrt{-2} \cdot \sqrt{-18} = i\sqrt{2} \cdot 3i\sqrt{2}$

$= 3 \cdot (\sqrt{2})^2 \cdot i^2$

$= 3 \cdot 2 \cdot (-1) = \mathbf{-6}$

or $\sqrt{-2} \cdot \sqrt{-18} = i\sqrt{2} \cdot i\sqrt{18}$

$= i^2\sqrt{36}$

$= (-1)(6) = \mathbf{-6}$

Example 3b shows how important it is to write each number in i–form first. Thus, if you wrote

$$\sqrt{-2} \cdot \sqrt{-18} = \sqrt{(-2)(-18)} = \sqrt{36} = 6,$$

you would be applying properties that are true only for radicals denoting real numbers. That is why the answer, 6, is incorrect.

EXAMPLE 4 Divide. Rationalize denominators when necessary.

a. $-4i \div 5i$ **b.** $3 \div \sqrt{-12}$

Solutions: **a.** $-4i \div 5i = \frac{-4i}{5i}$ ← $\mathbf{\frac{i}{i} = 1}$

$= \mathbf{-\frac{4}{5}}$

b. $3 \div \sqrt{-12} = \frac{3}{2i\sqrt{3}}$ ← **Multiply both numerator and denominator by $i\sqrt{3}$ to obtain a rational number in the denominator.**

$= \frac{3}{2i\sqrt{3}} \cdot \frac{i\sqrt{3}}{i\sqrt{3}}$

$= \frac{3i\sqrt{3}}{2i^2(\sqrt{3})^2}$

$= \frac{3i\sqrt{3}}{2(-1)(3)} = \frac{3i\sqrt{3}}{-6}$, or $\mathbf{-\frac{i\sqrt{3}}{2}}$

CLASSROOM EXERCISES

Write in i–form.

1. $\sqrt{-49}$
2. $-\sqrt{-1}$
3. $-\sqrt{-49}$
4. $\sqrt{-\frac{1}{7}}$
5. $-\frac{1}{\sqrt{-3}}$

Add or subtract as indicated.

6. $3i + 5i$
7. $\sqrt{-1} - \sqrt{-4}$
8. $-\sqrt{-9} + \sqrt{-16}$
9. $-\sqrt{-144} - \sqrt{-49}$

Multiply. Express the product as a real number or in i–form.

10. $3i \cdot 5i$
11. $\sqrt{-9} \cdot \sqrt{-4}$
12. $(2i)(-3i)$
13. i^7

Divide. Express denominators as rational numbers.

14. $3i \div 5i$
15. $\sqrt{-16} \div \sqrt{-4}$
16. $\sqrt{-2} \div \sqrt{-1}$
17. $\sqrt{-3} \div \sqrt{-12}$

WRITTEN EXERCISES

A Write in i–form.

1. $\sqrt{-100}$
2. $\sqrt{-81}$
3. $\sqrt{-169}$
4. $\sqrt{-121}$
5. $-\sqrt{-100}$
6. $-\sqrt{-36}$
7. $-\sqrt{-64}$
8. $-3\sqrt{-4}$
9. $12\sqrt{-16}$
10. $\frac{1}{5}\sqrt{-25}$

Add or subtract as indicated.

11. $-2i + 3i$
12. $-2i + (-3i)$
13. $-5i + \sqrt{-4}$
14. $\sqrt{-9} + \sqrt{-16}$
15. $3i - 4i$
16. $-5i - 3i$
17. $\sqrt{-25} - 2i$
18. $-\sqrt{-100} - (-8i)$
19. $-\sqrt{-100} + \sqrt{-81}$
20. $2\sqrt{-4} + 3\sqrt{-9}$
21. $300i - 299i$
22. $18i - \sqrt{-81}$
23. $2\sqrt{-18} + 3\sqrt{-2}$
24. $\sqrt{-1} - 5\sqrt{-4}$
25. $3\sqrt{-5} - \sqrt{-125}$
26. $\sqrt{-75} + \sqrt{-147}$

Multiply. Express the product as a real number or in i–form.

27. $3i \cdot 4i$
28. $-3i \cdot 4i$
29. $(-3i)(-4i)$
30. $\sqrt{-9} \cdot 8i$
31. $-\sqrt{64} \cdot 3i$
32. $-\frac{2}{3} \cdot 9i$
33. $\sqrt{-2} \cdot \sqrt{-2}$
34. $\sqrt{-4} \cdot \sqrt{-4}$
35. $\sqrt{-2} \cdot \sqrt{-50}$
36. $\sqrt{-3} \cdot \sqrt{-12}$
37. $\sqrt{-4} \cdot \sqrt{-25}$
38. $\sqrt{-8} \cdot \sqrt{-18}$
39. $2\sqrt{-3} \cdot \sqrt{-10}$
40. $(3\sqrt{6})(6\sqrt{-24})$
41. $(2\sqrt{-12})(-3\sqrt{3})$
42. $(-2\sqrt{-8})(-5\sqrt{-2})$

Divide. Rationalize denominators when necessary.

43. $3i \div 2i$
44. $4i \div -3i$
45. $-\sqrt{-9} \div 3i$
46. $\sqrt{-100} \div -5i$
47. $-3\sqrt{-9} \div 3$
48. $-14i \div \frac{7}{2i}$
49. $\sqrt{-12} \div \sqrt{-3}$
50. $5 \div \sqrt{-20}$
51. $\sqrt{-24} \div \sqrt{-6}$
52. $-\sqrt{-75} \div \sqrt{-3}$
53. $2 \div -\sqrt{-50}$
54. $4 \div \sqrt{-4}$

B Perform the indicated operations.

55. $3i \cdot 5i \cdot 2i$ 56. $i^2 \cdot i^2$ 57. $i^3 \cdot i$ 58. $i^4 \cdot i$

59. $(\sqrt{-3})^2$ 60. $\sqrt{3}i + \sqrt{3}i$ 61. $(\sqrt{-5})^2 + 5$ 62. $-4 \div i^6$

Evaluate.

63. i^3 64. i^6 65. i^7 66. i^8 67. i^9 68. i^{10}

69. Find the repeating pattern in the powers of i from i to i^{12}.

C Complete.

70. $i^{4n+1} = \underline{\ ?\ }, n \in W$ 71. $i^{4n+2} = \underline{\ ?\ }, n \in W$

72. $i^{4n+3} = \underline{\ ?\ }, n \in W$ 73. $i^{4n+4} = \underline{\ ?\ }, n \in W$

Evaluate.

74. i^{35} 75. i^{78} 76. i^{121} 77. i^{64} 78. $\frac{1}{i^{64}}$ 79. $\frac{3}{i^{134}}$

80. Prove or disprove:

The set of pure imaginary numbers is closed with respect to addition.

81. Prove or disprove:

The set of pure imaginary numbers is closed with respect to multiplication.

82. Prove: $i \div i = 1$ (HINT: $a \div b = c$ if $a = bc$)

83. Is the set of pure imaginary numbers closed with respect to subtraction? Explain your answer.

84. Is the set of pure imaginary numbers closed with respect to division? Explain your answer.

85. Show that 1, −1, i, and −i are solutions of the equation $x^4 - 1 = 0$.

86. Show that if a and b are positive real numbers where $a \neq b$, then $\sqrt{-a^2} + \sqrt{-b^2}$ is a pure imaginary number.

REVIEW CAPSULE FOR SECTION 9-2

Add or subtract as indicated. *(Pages 275–277)*

1. $(7 + 2\sqrt{5}) + (3 + 9\sqrt{5})$ 2. $(-2 + 3\sqrt{3}) + (5 - 6\sqrt{3})$

3. $(-3 + 2\sqrt{7}) - (-3 + 2\sqrt{2})$ 4. $(4 + 2\sqrt{2}) - (5 - 6\sqrt{2})$

5. $\sqrt{32} - 5\sqrt{8} + 2\sqrt{50}$ 6. $3\sqrt{8} - \sqrt{32} + 2\sqrt{72}$

In Exercises 7–10, replace $\underline{\ ?\ }$ with $\subset$ or $\not\subset$ to make a true statement. *(Pages 6–9)*

7. The set of irrational numbers $\underline{\ ?\ }$ the set of real numbers
8. The set of irrational numbers $\underline{\ ?\ }$ the set of integers
9. The set of rational numbers $\underline{\ ?\ }$ the set of irrational numbers
10. The set of rational numbers $\underline{\ ?\ }$ the set of real numbers

9–2 Addition and Subtraction

A **complex number** is the sum of a real number and a pure imaginary number.

Definition

> A **complex number** is a number of the form $a + bi$, where a and b are real numbers and $i = \sqrt{-1}$.

The form $a + bi$ is the **standard form of a complex number.**

Complex Number	Standard Form	a	b
$2 - 3i$	$2 + (-3i)$	2	-3
$7i$	$0 + 7i$	0	7
0	$0 + 0i$	0	0
$-\frac{1}{2}$	$\left(-\frac{1}{2}\right) + 0i$	$-\frac{1}{2}$	0
$i\sqrt{5}$	$0 + \sqrt{5}i$	0	$\sqrt{5}$
$6 + \sqrt{-2}$	$6 + \sqrt{2}i$	6	$\sqrt{2}$

Any complex number for which $b \neq 0$ is an **imaginary number.** A complex number for which $a = 0$ is a **pure imaginary number.** When $b = 0$, the number is a real number. Thus, *every real number is also a complex number.* The diagram below illustrates the relationships between the set of complex numbers and some of its subsets.

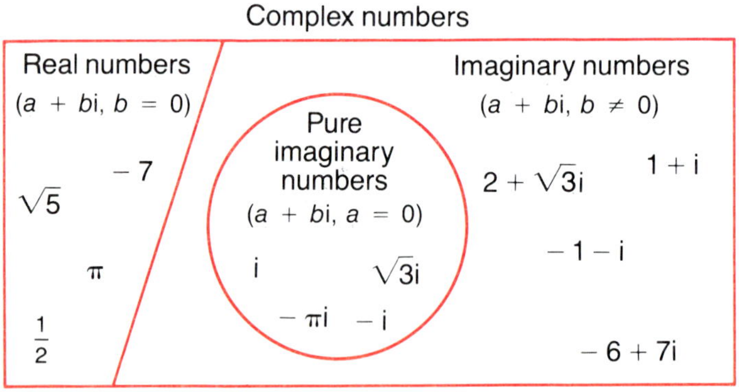

Since every real number is also a complex number, you should expect the definitions and properties of complex numbers to be true also for real numbers.

Definition

> **Equality of Two Complex Numbers**
>
> Two complex numbers, $a + bi$ and $c + di$, are equal if and only if $\boldsymbol{a = c}$ and $\boldsymbol{b = d}$.

EXAMPLE 1 Determine whether each pair of complex numbers is equal or unequal. Give a reason for each answer.

a. $4 + 5i$; $4 - 5i$ **b.** $4 + 5i$; $4 + \sqrt{-25}$

Solutions: **a.** $4 + 5i \neq 4 - 5i$ because $5 \neq -5$.

b. $4 + 5i = 4 + \sqrt{-25}$ because $4 + \sqrt{-25} = 4 + 5i$.

Addition of complex numbers is defined so that the familiar properties of addition over the set of real numbers are also true over the set of complex numbers.

Definition

Addition of Complex Numbers

For all real numbers a, b, c, and d,

$$(a + bi) + (c + di) = (a + c) + (b + d)i.$$

EXAMPLE 2 Add: $(-2 + 3i) + (5 - 6i)$

Solutions: $(-2 + 3i) + (5 - 6i) = (-2 + 5) + (3 - 6)i$ ← **By definition**

$= \mathbf{3 - 3i}$

Recall that the sum of a real number, a, and the identity element for addition, 0, is the real number, a. The identity element for addition over the set of complex numbers is also 0, which is written in standard form as $0 + 0i$.

Theorem 9–1

Additive Identity Theorem for Complex Numbers

The additive identity for the set of complex numbers is $\mathbf{0 + 0i}$.

Each complex number, $a + bi$ also has a unique additive inverse $-(a + bi)$ or $(-a - bi)$.

Theorem 9–2

Additive Inverse Theorem for Complex Numbers

For all real numbers a and b,

$$-(a + bi) = -a - bi.$$

You are asked to prove this theorem in the Written Exercises.

EXAMPLE 3 **a.** Find the additive inverse of $-4 + 3i$.

b. Find the sum of $-4 + 3i$ and its additive inverse.

Solutions: **a.** $-(-4 + 3i) = \mathbf{4 - 3i}$ ← **By Theorem 9–2**

b. $(-4 + 3i) + (4 - 3i) = (-4 + 4) + (3 - 3)i$ ← **By definition**

$= \mathbf{0 + 0i}$

As with real numbers, subtraction of complex numbers is defined in terms of addition.

Definition

Subtraction of Complex Numbers

To subtract a complex number, $a + bi$, where a and b are real numbers, add its additive inverse, $-a - bi$.

EXAMPLE 4 Subtract: $(4 + 2i) - (5 - 6i)$

Solution: $(4 + 2i) - (5 - 6i) = (4 + 2i) + (-5 + 6i)$ ← **By definition of subtraction**

$= (4 - 5) + (2 + 6)i$

$= \mathbf{-1 + 8i}$

CLASSROOM EXERCISES

Write each number in standard form. For each number, state the value of a and b.

1. $2 + \sqrt{-3}$ **2.** $\sqrt{-16}$ **3.** i^2 **4.** $1 - \sqrt{-1}$ **5.** $1 - \sqrt{-4}$

Determine whether each pair of complex numbers is equal or unequal. Give a reason for each answer.

6. $9 + 6i;\ 6 + 9i$ **7.** $1 - \sqrt{-1};\ 1 + \sqrt{-1}$ **8.** $5 - 2i;\ 5 - \sqrt{-4}$

Write the additive inverse of each complex number.

9. $4 - i$ **10.** $1 + \sqrt{-25}$ **11.** $-\sqrt{-16}$ **12.** $\sqrt{-9}$ **13.** $-7 + 6i$

Add or subtract as indicated.

14. $(-6 + 4i) + (-7 - 2i)$ **15.** $(3 + 5i) - (7 - 2i)$ **16.** $(4 + \sqrt{-4}) - (2 - 6i)$

17. $(8 - 13i) + (-18 + 13i)$ **18.** $-(-1 + 7i) - (1 - 7i)$ **19.** $(i^2 - i^3) + (i^3 - i^2)$

WRITTEN EXERCISES

A Write each number in standard form. For each number, state the value of a and b.

1. $4 + 0i$ **2.** $0 + 2i$ **3.** $4 + 2i$ **4.** $-3 + i$

5. $1 + 3i$ **6.** $3 - i$ **7.** $-2 - 2i$ **8.** 6

9. $-2i$ **10.** $\frac{4 + 2i}{2}$ **11.** $5(3 - 2i)$ **12.** $2(4 + \sqrt{-9})$

13. $8 + \sqrt{-5}$ **14.** $8 - \sqrt{-5}$ **15.** $i^3 + i^4$ **16.** $i^2 - i^6$

17. $-\frac{1}{2} + i^2$ **18.** $\sqrt{-7} - 5$ **19.** $i^3 + i^5$ **20.** $-i^5 + \sqrt{7}$

In Exercises 21–25, replace $\underline{\ ?\ }$ with $\subset$ or $\not\subset$ to make a true statement.

21. The set of real numbers $\underline{\ ?\ }$ the set of complex numbers.

22. The set of imaginary numbers $\underline{\ ?\ }$ the set of complex numbers.

23. The set of real numbers $\underline{\ ?\ }$ the set of imaginary numbers.

24. The set of pure imaginary numbers $\underline{\ ?\ }$ the set of imaginary numbers.

25. The set of pure imaginary numbers $\underline{\ ?\ }$ the set of real numbers.

Classify each pair of complex numbers as equal, *E*, or unequal, *U*. Give a reason for each answer.

26. $2 + i;\ 2 - i$ **27.** $2 + i;\ -2 - i$ **28.** $2 + i;\ \sqrt{4} + \sqrt{-4}$

29. $5i;\ i\sqrt{25}$ **30.** $-3i;\ 0 - \sqrt{-9}$ **31.** $\frac{1}{2};\ \frac{1}{2} + 0i$

32. When is the complex number, $a + bi$, a real number?

33. When is the complex number, $a + bi$, an imaginary number?

34. When is the complex number, $a + bi$, a pure imaginary number?

Add. Write each answer in standard form.

35. $9 + (6 - 2i)$ **36.** $(6 - 3i) + 4$ **37.** $(1 + 5i) + 3i$

38. $(3 + 2i) + (3 + 2i)$ **39.** $(4 + i) + (i - 4)$ **40.** $(2 - 3i) + (-2 + 3i)$

41. $(6 + 5i) + (-6 - 6i)$ **42.** $(-6 - 5i) + (-6 + i)$ **43.** $(4 + 0i) + (7 + 0i)$

Write the additive inverse of each complex number.

44. $3 - 2i$ **45.** $4 + 5i$ **46.** $-2 - 5i$ **47.** $-7 + 2i$

48. $3 + 0i$ **49.** $-5 + 0i$ **50.** $0 + 2i$ **51.** $-4i$

Subtract. Write each answer in standard form.

52. $(3 - 7i) - (6 + 2i)$ **53.** $(0 + 0i) - (2 + 5i)$ **54.** $(3 + 11i)\quad(\ 3\quad 11i)$

55. $(-2 - \frac{1}{2}i) - (-7 + 6i)$ **56.** $(1 - 2i) - 4i$ **57.** $7i - (6 - 3i)$

58. $9 - (6 - 2i)$ **59.** $(\sqrt{3} - i) - (\sqrt{3} + 2i)$ **60.** $(2\sqrt{3} - i) - i^2\sqrt{3}$

Add or subtract as indicated.

61. $(2 + \sqrt{-9}) + (5 - \sqrt{-9})$ **62.** $(3 + \sqrt{-7}) + (18 - \sqrt{-7})$ **63.** $(\sqrt{2} + 3i) - (\sqrt{2} - 4i)$

64. $(-2 + \sqrt{-25}) - (-9 + \sqrt{-36})$ **65.** $(a + bi) + (a - bi)$ **66.** $(a + bi) - (-a - bi)$

B

67. Use substitution to show that $\frac{3}{2}i$ satisfies the equation $4x^2 + 9 = 0$.

68. Show that the numbers $\frac{1}{3}i$ and $-\frac{1}{3}i$ satisfy the equation $9x^2 + 1 = 0$.

69. Show that the complex numbers $(5 + 2i)$ and $(5 - 2i)$ satisfy the equation $x^2 - 10x + 29 = 0$.

70. Show that i and $-i$ are reciprocals of each other; that is, show that

$$\mathbf{i \cdot (-i) = 1.}$$

In Exercises 71–78, find a number that can be added to each complex number such that the resulting sum is a real number.

71. $4 + 3i$ **72.** $8 - 4i$ **73.** $-2 + 6i$ **74.** $\sqrt{2} + 13i$

75. $4 + i\sqrt{3}$ **76.** $-4i$ **77.** $3 + 0i$ **78.** $\sqrt{7}$

C

79. Prove Theorem 9–1.

(HINT: You must prove two statements: $(a + bi) + (0 + 0i) = a + bi$, and if $(a + bi) + (c + di) = a + bi$, then $c = 0$ and $d = 0$.)

80. Prove Theorem 9–2.

81. Prove or disprove: *Addition of complex numbers is commutative.*

82. Prove or disprove: *Subtraction of complex numbers is commutative.*

83. Use examples to illustrate that addition of complex numbers is associative while subtraction is not.

84. Is the set of complex numbers closed under addition? Under subtraction? Explain.

CALCULATOR APPLICATIONS

Multiplying Radicals

To multiply radicals with a calculator, you use the square root and memory keys.

EXAMPLE Multiply: $(2\sqrt{13})(3\sqrt{3})$

SOLUTION Rewrite the problem so that each radical is the first factor in each product. (This may not be necessary on scientific calculators.)

$$(2\sqrt{13})(3\sqrt{3}) = (\sqrt{13} \cdot 2)(\sqrt{3} \cdot 3)$$

1 3 [√] [×] 2 [=] [M+] 3 [√] [×] 3 [×] [MR] [=] 37.469988

EXERCISES

Multiply.

1. $-\sqrt{8} \cdot \sqrt{15}$ **2.** $\sqrt{50} \cdot \sqrt{3}$ **3.** $(\sqrt{6})(-\sqrt{11})$ **4.** $(-\sqrt{7})(-\sqrt{8})$

5. $(3\sqrt{5})(7\sqrt{10})$ **6.** $(-5\sqrt{12})(6\sqrt{7})$ **7.** $(10\sqrt{6})(-8\sqrt{3})$ **8.** $(-2\sqrt{5})(-15\sqrt{30})$

REVIEW CAPSULE FOR SECTION 9–3

Rationalize each denominator. *(Pages 272–274)*

1. $\dfrac{1}{\sqrt{5}}$ **2.** $\dfrac{9}{3 - \sqrt{2}}$ **3.** $\dfrac{12}{\sqrt{3} - 3}$ **4.** $\dfrac{4 - 5\sqrt{2}}{7 + 3\sqrt{2}}$

9–3 Multiplication and Division

When you multiply two complex numbers, you use the distributive postulate twice.

EXAMPLE 1 Multiply: $(5 - 2i)(4 + 3i)$

Solution:

$$(5 - 2i)(4 + 3i) = 5(4 + 3i) - 2i(4 + 3i) \longleftarrow \text{By the distributive postulate}$$
$$= 5(4 + 3i) - 2i(4) - (2i)(3i)$$
$$= 20 + 15i - 8i - 6i^2 \longleftarrow i^2 = -1$$
$$= 20 + 7i + 6$$
$$= \mathbf{26 + 7i}$$

You can multiply two complex numbers as shown in Example 1 or you can apply the following definition.

Definition

Multiplication of Complex Numbers

For all real numbers a, b, c, and d,
$$(a + bi)(c + di) = (ac - bd) + (ad + bc)i.$$

The **multiplicative identity** over the set of complex numbers is $\mathbf{1 + 0i}$.

$$(a + bi)(1 + 0i) = a(1 + 0i) + bi(1 + 0i)$$
$$= a + bi$$

Also, by the definition of multiplication,

$$(a + bi)(a - bi) = (a^2 + b^2) + (-ab + ab)i$$
$$= a^2 + b^2. \longleftarrow \text{Real number}$$

The numbers $a + bi$ and $a - bi$ are called **complex conjugates.** Thus, the *product of a complex number and its conjugate is a real number.* This result can be expressed as a theorem.

Theorem 9–3

Complex Conjugate Theorem

For all real numbers a and b,
$$(a + bi)(a - bi) = a^2 + b^2.$$

Every nonzero complex number, $a + bi$, has a **multiplicative inverse,** or **reciprocal,** $\frac{1}{a + bi}$.

You can use Theorem 9–3 to express the reciprocal in standard form.

EXAMPLE 2 Express the reciprocal of 2 − 3i in standard form.

Solution: Multiply both the numerator and denominator of $\frac{1}{2-3i}$ by $2+3i$.

$$\frac{1}{2-3i}\cdot\frac{2+3i}{2+3i}=\frac{2+3i}{(2-3i)(2+3i)}$$ ← **2 + 3i is the conjugate of 2 − 3i.**

$$=\frac{2+3i}{2^2+3^2}$$ ← **By Theorem 9–3**

$$=\frac{2+3i}{13},\text{ or }\frac{2}{13}+\frac{3}{13}i$$ ← **Standard form**

The definition of division of complex numbers is similar to the definition for division of real numbers.

Definition

For all real numbers a, b, c, and d,

$$(a+bi)\div(c+di)=(a+bi)\cdot\frac{1}{c+di},\ (c+di)\neq 0.$$

From the definition and Theorem 9–3,

$$(a+bi)\div(c+di)=(a+bi)\left(\frac{1}{c+di}\right)\left(\frac{c-di}{c-di}\right)=\frac{a+bi}{c+di}\cdot\frac{c-di}{c-di}.$$

EXAMPLE 3 Express $\frac{4+2i}{3+5i}$ in standard form.

Solution:

$$\frac{4+2i}{3+5i}=\frac{4+2i}{3+5i}\cdot\frac{3-5i}{3-5i}$$ ← **(3 − 5i) is the conjugate of (3 + 5i).**

$$=\frac{4(3-5i)+2i(3-5i)}{3^2+5^2}$$

$$=\frac{12-20i+6i-10i^2}{34}$$

$$=\frac{22-14i}{34}$$

$$=\frac{2(11-7i)}{34}=\frac{11-7i}{17},\text{ or }\frac{11}{17}+\left(-\frac{7}{17}\right)i$$

CLASSROOM EXERCISES

Multiply.

1. $(2+3i)(4+5i)$
2. $(6-7i)(9+4i)$
3. $(4+9i)(1+0i)$

Express each quotient in standard form.

4. $\frac{5}{3i}$
5. $\frac{2}{2+i}$
6. $\frac{2-3i}{3+2i}$
7. $\frac{5+2i}{2-i}$

WRITTEN EXERCISES

A Multiply.

1. $(0 + i)(3 - 4i)$
2. $(2 + 0i)(5 + 8i)$
3. $(0 + 0i)(3 - 2i)$
4. $(4 + 8i)(2 - i)$
5. $(1 + i)(1 - i)$
6. $(7 + i)(7 + 7i)$
7. $(3 - i)(6 - i)$
8. $(2 + 2i)(1 + 2i)$
9. $(4 - 3i)(3 + 4i)$
10. $(\sqrt{2} + 2i)(1 + 0i)$
11. $(\frac{1}{2} + 2i)(\frac{1}{2} + 3i)$
12. $(\frac{1}{2} - i)(\frac{1}{2} + i)$
13. $(2 + 3i)(2 - 3i)$
14. $(a + bi)(a - bi)$
15. $(5 + i\sqrt{3})(5 + i\sqrt{3})$

Write the conjugate of each complex number.

16. $3 + 2i$
17. $4 - 5i$
18. $-3 + 6i$
19. $3 + i\sqrt{2}$
20. $4 - i\sqrt{7}$
21. $0 + 2i$
22. $-7i$
23. $8 + 0i$

Express the reciprocal of each complex number in standard form.

24. $2 + 3i$
25. $3 - 2i$
26. $-i$
27. $-5 + i$
28. $-3 - 2i$
29. $7 - i$
30. $-2 - 5i$
31. $-6 + 2i$

Express each quotient in standard form.

32. $\frac{4 + 2i}{1 + i}$
33. $\frac{6 - 5i}{2 + 3i}$
34. $\frac{3 + 5i}{2 + i}$
35. $\frac{-4 + 3i}{1 - 3i}$
36. $\frac{5 - \sqrt{-4}}{5 + \sqrt{-4}}$
37. $\frac{2 + 3\sqrt{-9}}{2 - 3\sqrt{-9}}$
38. $\frac{5 - \sqrt{-25}}{3 + \sqrt{-25}}$
39. $\frac{2 - \sqrt{-36}}{2 + \sqrt{-36}}$

B Find each product.

40. $(2 + 3i)(\frac{2}{13} - \frac{3}{13}i)$
41. $(2 + 3i)^2$
42. $(2 + 3i)^3$
43. $(1 - 3i)^3$

44. How are the two numbers in Exercise 40 related? What is the name for two numbers that have this special product?

In Exercises 45–52, use Theorem 9–3 to factor over the complex numbers.

45. $y^2 + 4$
46. $x^2 + y^2$
47. $7x^2 + 7y^2$
48. $9x^2 + 36$
49. $x^4 - 1$
50. $x^4 - y^4$
51. $1 - 256y^4$
52. $ax^2 + bx^4$

53. Show that $(a + bi)\left(\frac{a}{a^2 + b^2} + \frac{-b}{a^2 + b^2}i\right) = 1 + 0i$, or 1.

In Exercises 54–57, let $x = 2 - 3i$ and $y = -3 + 2i$. Express each quotient in standard form.

54. $\frac{xy}{5}$
55. $\frac{x^2 - y^2}{i}$
56. $\frac{2x}{y}$
57. $\frac{x}{y^2}$

58. Use the definition of multiplication of complex numbers to show that

$$\left[\frac{ac + bd}{c^2 + d^2} + \left(\frac{bc - ad}{c^2 + d^2}\right)i\right][c + di] = a + bi.$$

C

59. Explain why the reciprocal of i is $-i$.

60. Use the definition of multiplication of complex numbers to show that the product of zero and any complex number, $a + bi$, is zero.

61. Prove that multiplication of complex numbers is commutative.
For all $a, b, c, d \in R$, $(a + bi) \cdot (c + di) = (c + di) \cdot (a + bi)$.

62. Prove that multiplication of complex numbers is associative.
For all $a, b, c, d, e, f \in R$,
$[(a + bi) \cdot (c + di)] \cdot (e + fi) = (a + bi) \cdot [(c + di) \cdot (e + fi)]$.

63. Prove the distributive property for complex numbers.
For all $a, b, c, d, e, f \in R$,
$(a + bi) \cdot [(c + di) + (e + fi)] = (a + bi)(c + di) + (a + bi)(e + fi)$.

64. Prove: For all real numbers a and b, $(a + bi) + (-a - bi) = 0 + 0i$.

65. Prove: *The conjugate of the sum of two complex numbers equals the sum of the conjugates.*

66. Prove: *The conjugate of the product of two complex numbers equals the product of the conjugates.*

67. Prove: *The conjugate of the conjugate is the original complex number.*

Review

Perform the indicated operations. Rationalize denominators when necessary. *(Section 9–1)*

1. $2\sqrt{-9} + 5\sqrt{-36}$ **2.** $4\sqrt{-147} - 2\sqrt{-27}$ **3.** $16i - \sqrt{-25}$ **4.** $\sqrt{-100} + 3i$

5. $6i \cdot 5i$ **6.** $\sqrt{-4} \cdot 7i$ **7.** $(4\sqrt{-27})(2\sqrt{-12})$ **8.** $(7\sqrt{-6})(2\sqrt{150})$

9. $6i \div 3i$ **10.** $-20i \div \frac{5}{10i}$ **11.** $8 \div (-\sqrt{-32})$ **12.** $(-\sqrt{-5})^2 \div (2\sqrt{-10})$

Add or subtract as indicated. Express each answer in standard form. *(Section 9–2)*

13. $7 + (8 - 3i)$ **14.** $(9 + 4i) - (5 - 6i)$ **15.** $(10 - 6i) - (12 + 3i)$

16. $(-8 - 9i) + (6 + 10i)$ **17.** $(5 + 4i) + (-7 - 9i)$ **18.** $(-16 + 7i) - (-5 - 8i)$

Write the conjugate of each complex number. *(Section 9–3)*

19. $7 + i$ **20.** $21 - 3i$ **21.** $i - 5$ **22.** $4i + i^2$

Perform the indicated operations. Express each answer in standard form. *(Section 9–3)*

23. $(5 + 9i)(4 - 6i)$ **24.** $(7 + 0i)(3 - 5i)$ **25.** $(6 - i\sqrt{2})(8 + i\sqrt{2})$

26. $\frac{4 + 3i}{6 - 5i}$ **27.** $\frac{6 + \sqrt{-9}}{6 - \sqrt{-9}}$ **28.** $\frac{5 + \sqrt{-16}}{7 - \sqrt{-25}}$

9–4 Solving Incomplete Quadratic Equations

A quadratic equation of the form $ax^2 = c$ is **incomplete** because it does not have a term in x. Incomplete quadratic equations can be solved by applying the definition of square root.

EXAMPLE 1 Solve: $3x^2 + 5 = x^2 + 29$

Solution:

$3x^2 + 5 = x^2 + 29$ ← **Write in the form $ax^2 = c$.**

$2x^2 = 24$

$x^2 = 12$ ← **Apply the definition of square root.**

$x = \pm\sqrt{12} = \pm 2\sqrt{3}$ **Solution set:** $\{-2\sqrt{3}, 2\sqrt{3}\}$

When an incomplete quadratic equation is expressed in the form $x^2 = -c$ and $-c < 0$, the equation has no real number solutions. However, there are solutions over the set of complex numbers.

EXAMPLE 2 Solve: $5y^2 + 24 = 0$

Solution:

$5y^2 + 24 = 0$

$5y^2 = -24$

$y^2 = -\frac{24}{5}$

$y = \pm\sqrt{-\frac{24}{5}}$ ← $\sqrt{-\frac{24}{5}} = i\sqrt{\frac{4 \cdot 6}{5}} = 2i\sqrt{\frac{6}{5}}$

$y = \pm 2i\frac{\sqrt{6}}{\sqrt{5}}$ ← **Rationalize the denominator.**

$y = \pm 2i\frac{\sqrt{6}}{\sqrt{5}} \cdot \frac{\sqrt{5}}{\sqrt{5}} = \pm\frac{2i\sqrt{30}}{5}$ **Solution set:** $\left\{\frac{-2i\sqrt{30}}{5}, \frac{2i\sqrt{30}}{5}\right\}$

Follow a similar procedure to solve the equation in Example 3.

EXAMPLE 3 Solve: $(t + 2)^2 + 18 = 0$

Solution:

$(t + 2)^2 + 18 = 0$ ← **Solve for $(t + 2)^2$.**

$(t + 2)^2 = -18$

$(t + 2) = \pm\sqrt{-18}$ ← **If $x^2 = c$, then $x = \pm c$.**

$(t + 2) = \pm 3i\sqrt{2}$

$t + 2 = 3i\sqrt{2}$ or $t + 2 = -3i\sqrt{2}$

$t = -2 + 3i\sqrt{2}$ | $t = -2 - 3i\sqrt{2}$

Solution set: $\{-2 - 3i\sqrt{2}, -2 + 3i\sqrt{2}\}$

In Examples 1, 2, and 3, the check is left for you.

CLASSROOM EXERCISES

Solve each equation.

1. $x^2 = 9$
2. $x^2 = -9$
3. $2x^2 = 32$
4. $x^2 - 25 = 0$
5. $p^2 + 2 = 0$
6. $2x^2 - 7 = x^2 + 3$
7. $4x^2 + 3 = 3$
8. $(t - 3)^2 = -4$

WRITTEN EXERCISES

A Solve each equation. The replacement set is the set of complex numbers.

1. $3x^2 = 27$
2. $2x^2 = 72$
3. $2x^2 = 22$
4. $4x^2 - 36 = 0$
5. $7x^2 - 42 = 0$
6. $b^2 - 7 = 29$
7. $p^2 - 42 = 7$
8. $m^2 - 5 = 59$
9. $x^2 - 7 = 12$
10. $-x^2 + 41 = 0$
11. $2x^2 + 1 = 3$
12. $x^2 + 2 = 3$
13. $3x^2 - 5 = 2x^2 + 7$
14. $8x^2 + 9 = 3x^2 + 24$
15. $4x^2 - 5 = x^2 + 4$
16. $x^2 = -7$
17. $y^2 - 5 = -6$
18. $x^2 + 4 = 0$
19. $3x^2 = -21$
20. $(x - 5)^2 = 4$
21. $(x + 3)^2 = 16$
22. $(x + 1)^2 = 9$
23. $(2x + 3)^2 = 36$
24. $\frac{x}{5} = \frac{-8}{x}$
25. $\frac{x^2}{4} = -4$
26. $\left(x - \frac{3}{2}\right)^2 = \frac{25}{4}$
27. $\left(x + \frac{1}{2}\right)^2 = \frac{9}{4}$

B

28. $-2(x - 1)^2 = 8$
29. $3(2q + 3)^2 = -6$
30. $(x + \frac{3}{4})^2 = \frac{25}{16}$
31. $(x + \frac{1}{2})^2 = -\frac{1}{4}$
32. $(2x - 1)(3x - 2) = (x - 5)(x - 2)$
33. $(x + 2)^2 + (x - 2)^2 = 3(x + 2)(x - 2)$
34. $x(3x + 2) = (x + 1)^2$
35. $x^2 + 5x - 7 = (2x + 1)(x + 2)$

C

36. $p^2 + 2p + 1 = 4$
37. $t^2 + 6t + 9 = 16$
38. $4t^2 + 12t + 9 = 100$

Puzzle

Examine the following statements.

$$\sqrt{-\frac{9}{16}} = \sqrt{\frac{-9}{16}} = \frac{\sqrt{-9}}{\sqrt{16}} = \frac{3i}{4} = \frac{3}{4}i$$

$$\sqrt{-\frac{9}{16}} = \sqrt{\frac{9}{-16}} = \frac{\sqrt{9}}{\sqrt{-16}} = \frac{3}{4i} = -\frac{3}{4}i$$

Which is correct? Explain your answer.

REVIEW CAPSULE FOR SECTION 9–5

Square each binomial. *(Pages 193–196)*

1. $(x + 3)^2$
2. $(3x - 7)^2$
3. $(x + a)^2$
4. $(x - a)^2$

Factor. Express your answer as the square of a binomial. *(Pages 197–200)*

5. $t^2 + 2t + 1$
6. $9y^2 - 24y + 16$
7. $4a^2 + 4ab + b^2$
8. $p^4 - 2p^2 + 1$

9–5 Completing the Square

Recall that a **perfect square trinomial** is the square of a binomial. It has two equal binomial factors.

$$x^2 + 8x + 16 = (x + 4)(x + 4), \text{ or } (x + 4)^2$$

Now suppose that you are given $x^2 + 8x$ and asked to **"complete the square."** That is, you are asked to find a number that can be added to $x^2 + 8x$ to make it a perfect square trinomial. This is done by reversing the procedure above. That is, you take one–half of 8 and square the result.

$$x^2 + 8x + \left(\frac{1}{2} \cdot 8\right)^2 = x^2 + 8x + 4^2 = (x + 4)^2$$

Completing the Square

To obtain a perfect square trinomial from $x^2 + px$, add $\left(\frac{1}{2}p\right)^2$, or $\frac{1}{4}p^2$ to the binomial. That is,

$$x^2 + px + \left(\frac{1}{2}p\right)^2 = \left(x + \frac{p}{2}\right)^2.$$

EXAMPLE 1 Complete the square: $z^2 - 7z$

Solution: Find $\frac{1}{2}$ the coefficient of z. Square the result.

$\frac{1}{2}(-7) = -\frac{7}{2}$ and $\left(-\frac{7}{2}\right)^2 = \frac{49}{4}$ ← **Add $\frac{49}{4}$ to $z^2 - 7z$.**

$z^2 - 7z + \frac{49}{4} = \left(z - \frac{7}{2}\right)^2$ ← **Perfect square binomial**

The technique of completing the square can be used to solve quadratic equations. REMEMBER: Add $\left(\frac{1}{2}p\right)^2$ to *each* side of the equation.

EXAMPLE 2 Solve by completing the square: $x^2 - 6x - 20 = 0$

Solution: Write in the form $ax^2 + bx = c$.

$x^2 - 6x - 20 = 0$

$x^2 - 6x = 20$ ← **Add $\left[\frac{1}{2} \cdot (-6)\right]^2$ to each side.**

$x^2 - 6x + 9 = 20 + 9$ ← **$x^2 - 6x + 9 = (x - 3)^2$**

$(x - 3)^2 = 29$ ← **If $x^2 = c$, $x = \pm\sqrt{c}$.**

$x - 3 = \pm\sqrt{29}$

$x = 3 + \sqrt{29}$ <u>or</u> $x = 3 - \sqrt{29}$

The check is left for you. **Solution set: $\{3 - \sqrt{29}, 3 + \sqrt{29}\}$.**

When the coefficient of the second-degree term of a quadratic equation is not 1, each term of the equation must be divided by that coefficient before completing the square.

EXAMPLE 3 Solve: $3x^2 - 21 = 2x$

Solution:

$$3x^2 - 21 = 2x$$

$$3x^2 - 2x = 21 \quad \longleftarrow \quad \textbf{Divide each side by 3.}$$

$$x^2 - \frac{2}{3}x = \frac{21}{3} \quad \longleftarrow \quad \textbf{Add } \left[\frac{1}{2}\left(-\frac{2}{3}\right)\right]^2 \textbf{ to each side.}$$

$$x^2 - \frac{2}{3}x + \left[\frac{1}{2}\left(-\frac{2}{3}\right)\right]^2 = \frac{21}{3} + \left[\frac{1}{2}\left(-\frac{2}{3}\right)\right]^2$$

$$x^2 - \frac{2}{3}x + \frac{1}{9} = \frac{21}{3} + \frac{1}{9}$$

$$\left(x - \frac{1}{3}\right)^2 = \frac{64}{9} \quad \longleftarrow \quad \textbf{If } x^2 = c,\ x = \pm\sqrt{c}.$$

$$x - \frac{1}{3} = \pm\frac{8}{3}$$

$$x - \frac{1}{3} = \frac{8}{3} \quad \text{or} \quad x - \frac{1}{3} = -\frac{8}{3}$$

$$x = 3 \qquad\qquad x = -\frac{7}{3} \qquad \textbf{Solution set: } \left\{-\frac{7}{3}, 3\right\}$$

Solutions should always be checked in the original equation.

Steps for Completing the Square

1. Write an equivalent equation with only the second-degree term and the first-degree term on the left side of the equation. The coefficient of the second-degree term must be 1.
2. Add the square of one-half the coefficient of the first-degree term to each side of the equation.
3. Express the left side of the equation as a perfect square.
4. Solve for the variable.

CLASSROOM EXERCISES

Complete the square to form a perfect square trinomial. Express the trinomial as the square of a binomial.

1. $x^2 + 14x$ **2.** $a^2 - 16a$ **3.** $x^2 + 12x$ **4.** $x^2 + 5x$

Solve by completing the square.

5. $x^2 - 2x = 8$ **6.** $x^2 + 2x = -10$ **7.** $y^2 + 4y = 21$ **8.** $d^2 - 6d = 16$

WRITTEN EXERCISES

A Complete the square to form a perfect square trinomial. Express the trinomial as the square of a binomial.

1. $x^2 - 10x$ **2.** $n^2 - 16n$ **3.** $y^2 + y$ **4.** $x^2 + 3x$

5. $x^2 - 2bx$ **6.** $c^2 + \frac{3}{2}c$ **7.** $x^2 + \frac{b}{a}x$ **8.** $x^2 - \frac{b}{a}x$

Solve by completing the square. Check your solutions.

9. $x^2 + 10x = -24$ **10.** $r^2 + 4r = 21$ **11.** $x^2 - 2x = 15$

12. $x^2 - 8x = 48$ **13.** $x^2 = 8x - 15$ **14.** $x^2 - 9 = -8x$

15. $10a = 24 - a^2$ **16.** $x^2 - 6x + 5 = 0$ **17.** $b^2 - 2b = 48$

18. $y^2 = 6y + 7$ **19.** $x^2 - 9x = -20$ **20.** $x^2 + x = 20$

21. $c^2 - 9c = -18$ **22.** $x^2 + 3x = 18$ **23.** $x^2 - 5x = 50$

B

24. $x^2 - \frac{x}{3} = \frac{2}{3}$ **25.** $x^2 + \frac{x}{2} = \frac{3}{2}$ **26.** $4x^2 + x = 60$

27. $4x^2 + 12x = 7$ **28.** $6x^2 + x - 2 = 0$ **29.** $2x^2 + x - 6 = 0$

30. $3n^2 + 2n - 8 = 0$ **31.** $4n^2 + 19n - 5 = 0$ **32.** $6x^2 + 5x = 6$

33. $6x^2 - 7x = -1$ **34.** $x^2 - x - 1 = 0$ **35.** $x^2 + 3x - 5 = 0$

36. $2a^2 + 7a + 6 = 0$ **37.** $2a^2 - 7a + 6 = 0$ **38.** $x^2 + \frac{7x}{4} - \frac{1}{2} = 0$

C

39. $y^2 + y\sqrt{3} + 5 = 0$ **40.** $z^2 + 4z - 3 = 4\sqrt{3}$ **41.** $ax^2 + bx + c = 0$

42. $(x + 1)(2x + 3) = 4x^2 - 22$ **43.** $7(x + 2a)^2 + 3a^2 = 5a(7x + 23a)$

44. $\frac{3x - 1}{4x + 7} = 1 - \frac{6}{x + 7}$ **45.** $\frac{5x - 7}{7x - 5} = \frac{x - 5}{2x - 13}$

46. $x^2 - (4 + i)x + (5 - i) = 0$ **47.** $x^2 - (3 + 5)x + (6i - 2) = 0$

REVIEW CAPSULE FOR SECTION 9-6

Simplify. *(Pages 270–271, 318–322)*

1. $\sqrt{9 - 34}$ **2.** $-5 + \sqrt{24 - 25}$ **3.** $\frac{2 - \sqrt{64 - 48}}{3}$ **4.** $\frac{-4 - \sqrt{96 - 100}}{8}$

5. $\frac{-11 + \sqrt{121 - 240}}{10}$ **6.** $\frac{15 + \sqrt{225 - 220}}{10}$ **7.** $\frac{3 \pm \sqrt{9 - 20}}{2}$ **8.** $\frac{8 \pm \sqrt{64 + 512}}{-32}$

Evaluate $\frac{-b + \sqrt{b^2 - 4ac}}{2a}$ for the given values of a, b, and c. Express each answer in simplest radical form. *(Pages 2–5, 272–274)*

9. $a = 1,\ b = 3,\ c = -40$ **10.** $a = 1,\ b = 8,\ c = 16$ **11.** $a = 3,\ b = 5,\ c = 2$

12. $a = 2,\ b = 7,\ c = 1$ **13.** $a = 1,\ b = -3,\ c = -3$ **14.** $a = 2,\ b = -8,\ c = 1$

9–6 The Quadratic Formula

The solution set of any quadratic equation can be found by completing the square. When this method is applied to the general quadratic equation,

$$ax^2 + bx + c = 0,$$

the solutions are also general.

$$ax^2 + bx + c = 0 \quad \longleftarrow \text{ Divide each side by } a.$$

$$x^2 + \frac{b}{a}x + \frac{c}{a} = 0 \quad \longleftarrow \text{ Add } \left(-\frac{c}{a}\right) \text{ to each side.}$$

$$x^2 + \frac{b}{a}x = -\frac{c}{a} \quad \longleftarrow \text{ Add } \left(\frac{1}{2}\cdot\frac{b}{a}\right)^2 \text{ to each side.}$$

$$x^2 + \frac{b}{a}x + \left(\frac{1}{2}\cdot\frac{b}{a}\right)^2 = -\frac{c}{a} + \left(\frac{1}{2}\cdot\frac{b}{a}\right)^2$$

$$x^2 + \frac{b}{a}x + \frac{b^2}{4a^2} = -\frac{c}{a} + \frac{b^2}{4a^2}$$

$$\left(x + \frac{b}{2a}\right)^2 = \frac{b^2 - 4ac}{4a^2} \quad \longleftarrow \text{ If } x^2 = c,\ x = \pm\sqrt{c}.$$

$$x + \frac{b}{2a} = \pm\frac{\sqrt{b^2 - 4ac}}{2a}$$

$$x + \frac{b}{2a} = \frac{\sqrt{b^2 - 4ac}}{2a} \quad \text{or} \quad x + \frac{b}{2a} = -\frac{\sqrt{b^2 - 4ac}}{2a}$$

$$x = -\frac{b}{2a} + \frac{\sqrt{b^2 - 4ac}}{2a} \quad \Big| \quad x = -\frac{b}{2a} - \frac{\sqrt{b^2 - 4ac}}{2a}$$

$$x = \frac{-b + \sqrt{b^2 - 4ac}}{2a} \quad \Big| \quad x = \frac{-b - \sqrt{b^2 - 4ac}}{2a}$$

The general solutions are known as the *quadratic formula.*

> **Quadratic Formula**
>
> For any quadratic equation, $ax^2 + bx + c = 0$,
>
> $$x = \frac{-b + \sqrt{b^2 - 4ac}}{2a} \quad \text{or} \quad x = \frac{-b - \sqrt{b^2 - 4ac}}{2a}.$$

EXAMPLE 1 Solve and check: $x^2 + 5x = 6$

Solution: Express in the form $ax^2 + bx + c = 0$. Identify a, b, and c.

$$x^2 + 5x - 6 = 0 \quad \longleftarrow \quad a = 1,\ b = 5,\ c = -6$$

$$x = \frac{-5 \pm \sqrt{25 - 4(1)(-6)}}{2(1)} \quad \longleftarrow \text{ Substitute in the formula.}$$

$$x = \frac{-5 \pm \sqrt{49}}{2} \qquad \text{Thus, } x = \frac{-5 + \sqrt{49}}{2} \quad \text{or} \quad x = \frac{-5 - \sqrt{49}}{2}$$

$$x = 1 \quad \Big| \quad x = -6 \qquad \textbf{Solution set: } \{-6, 1\}$$

When the solutions of a quadratic equation are irrational numbers, it is customary to express any radicals in simplest form and to rationalize denominators.

EXAMPLE 2 Solve and check: $2x^2 - 3x = 7$

Solution: $2x^2 - 3x - 7 = 0$ ← **$a = 2$, $b = -3$, $c = -7$**

$$x = \frac{-(-3) \pm \sqrt{(-3)^2 - 4(2)(-7)}}{2(2)}$$ ← **Substitute in the formula.**

$$x = \frac{3 \pm \sqrt{9 + 56}}{4} = \frac{3 \pm \sqrt{65}}{4}$$

The check is left for you. **Solution set:** $\left\{\frac{3 - \sqrt{65}}{4}, \frac{3 + \sqrt{65}}{4}\right\}$

You can use a calculator or the Table of Square Roots on page 636 to find approximate values for the solutions in Example 2.

$$\frac{3 - \sqrt{65}}{4} \approx \frac{3 - 8.062}{4} = -1.27$$ ← **Nearest hundredth**

$$\frac{3 + \sqrt{65}}{4} \approx \frac{3 + 8.062}{4} = 2.77$$ ← **Nearest hundredth**

CLASSROOM EXERCISES

For each quadratic equation, state the values of *a*, *b*, and *c*.

1. $2x^2 + 5x + 6 = 0$ **2.** $x^2 - 6 = 0$ **3.** $x^2 + 2x + 3 = 0$ **4.** $5x^2 + 3x = 4$

5. $2y^2 + 5y + 12 = 0$ **6.** $3t - t^2 = 5$ **7.** $r^2 = 14 + 5r$ **8.** $y^2 = 1 - y$

WRITTEN EXERCISES

A In Exercises 1–18, solve by using the quadratic formula. Write irrational solutions in simplest radical form.

1. $x^2 - 5x + 6 = 0$ **2.** $x^2 - 8x + 15 = 0$ **3.** $x^2 - 3x - 10 = 0$

4. $x^2 - 4x - 12 = 0$ **5.** $x^2 + 4x - 12 = 0$ **6.** $p^2 + 8p + 16 = 0$

7. $y^2 - 3y = 2$ **8.** $y^2 + 3y + 1 = 0$ **9.** $y^2 = 7y - 11$

10. $y^2 - 5y + 5 = 0$ **11.** $x^2 + 6x = -9$ **12.** $2t^2 + 3t = 1$

13. $6x^2 - x - 12 = 0$ **14.** $9x^2 + 3x - 2 = 0$ **15.** $q^2 + q - 1 = 0$

16. $2x^2 + 5x + 2 = 0$ **17.** $3s^2 = 3s + 1$ **18.** $4x^2 + 12x = -9$

In Exercises 19–24, use a calculator or the Table of Square Roots on page 636 to approximate irrational solutions to the nearest hundredth.

19. $2x^2 - 3x + 1 = 0$ **20.** $4t^2 = 3 - 4t$ **21.** $x^2 + 4x - 6 = 0$

22. $x^2 - 2x - 4 = 0$ **23.** $x^2 + 3x + 1 = 0$ **24.** $x^2 + 8x + 5 = 0$

B Solve. Leave irrational solutions in simplest radical form.

25. $3x^2 - 3x = 1$

26. $2x^2 + 0.1x - 0.03 = 0$

27. $x^2 - 1 = (3x - 7)(x - 2)$

28. $2x + 1 = \frac{1}{x - 1}$

29. $\frac{2g + i}{g - i} = \frac{3g + 4i}{g + 3i}$

30. $\frac{x^2 - 4x}{6} - \frac{x - 3}{3} = 1$

C Solve. Be sure to check for apparent solutions.

31. $\frac{5x - 2}{2} - \frac{19x + 6}{2x} = \frac{3x - 2}{4}$

32. $\frac{1}{2 - x} - 1 = \frac{x - 6}{x^2 - 4} - \frac{1}{x + 2}$

33. $\frac{4}{c + 3} + \frac{1}{3} = \frac{4}{c - 3}$

34. $\frac{3y}{y + 4} - \frac{y - 12}{y^2 + y - 12} = \frac{2y}{y - 3}$

35. Find the three cube roots of 1. (HINT: Solve $x^3 - 1 = 0$.)

CALCULATOR APPLICATIONS

Evaluating Irrational Solutions

To evaluate an irrational solution on a calculator, you use the square root and memory keys.

EXAMPLE Evaluate $\frac{3 - \sqrt{65}}{4}$ to the nearest hundredth (Example 2 on page 333).

SOLUTION First evaluate the numerator. To do this, first find $\sqrt{65}$ and store.

6 5 [√] [M+] 3 [−] [MR] [=] [÷] 4 [=] -1.2655644

To the nearest hundredth, the answer is **−1.27.**

EXERCISES

Evaluate each of the following to the nearest hundredth.

1. $\frac{3 + \sqrt{65}}{4}$

2. $\frac{5 + \sqrt{42}}{2}$

3. $\frac{16 - \sqrt{54}}{2}$

4. $\frac{9 - \sqrt{7}}{8}$

5. $\frac{7 + \sqrt{5}}{2}$

6. $\frac{6 - \sqrt{3}}{6}$

7. $\frac{1 + \sqrt{73}}{36}$

8. $\frac{5 - \sqrt{8}}{2}$

REVIEW CAPSULE FOR SECTION 9–7

Sketch the triangle whose coordinates are given. Tell whether the triangle is a right triangle and, if so, write an equation relating the lengths of the legs to that of the hypotenuse. *(Pages 280–283)*

1. $A(3, -6)$; $B(-2, 3)$; $C(-2, -6)$

2. $A(8, -3)$; $B(6, -4)$; $C(4, 0)$

3. $A(-1, 0)$; $B(2, 1)$; $C(3, -2)$

4. $A(a, 0)$; $B(0, b)$; $C(0, 0)$

Problem Solving and Applications

9–7 Using Quadratic Equations

When you use a quadratic equation to solve a problem, one solution may not be physically possible. For example, the length of a side of a triangle cannot be negative. Any solution that does not "fit" with the conditions of the problem must be discarded.

EXAMPLE Jose, a long distance walker, and Ed, a jogger, plan to start their daily workout from the intersection of two straight paths that are perpendicular to each other. Jose walks at a rate of 2 kilometers per hour. Ed leaves 1 hour later than Jose, and jogs at a rate of 8 kilometers per hour. How long after Ed leaves will the two be 50 kilometers apart?

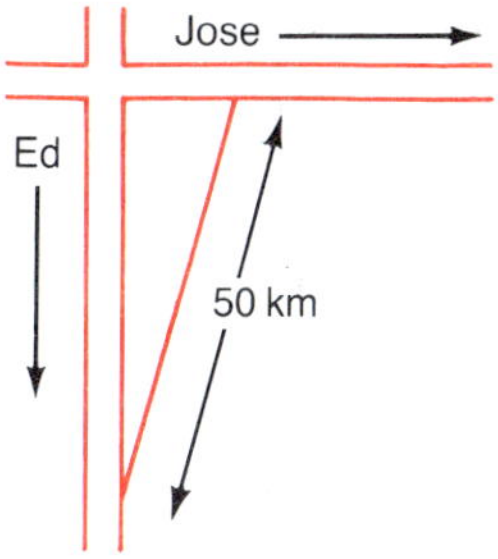

Solution: [1] Use Condition 1 (underscored once) to represent the variables.

Let t = the number of hours Ed jogged.
Then $t + 1$ = the number of hours Jose walked.

[2] Organize the information in a table.

	Rate (r)	Time (t)	Distance (d)
Ed	8 km/h	t	$8t$
Jose	2 km/h	$t+1$	$2(t+1)$

[3] Use the table and Condition 2 (underscored twice) to write an equation for the problem.

$$(8t)^2 + [2(t+1)]^2 = 50^2$$ ← **By the Pythagorean theorem (Theorem 8–3)**

$$64t^2 + 4(t^2 + 2t + 1) = 2500$$

$$64t^2 + 4t^2 + 8t + 4 = 2500$$

$$68t^2 + 8t - 2496 = 0$$ ← **Divide each side by 4.**

$$17t^2 + 2t - 624 = 0$$ ← **Use the quadratic formula.** $a = 17$, $b = 2$, $c = -624$

$$t = \frac{-2 \pm \sqrt{2^2 - 4(17)(-624)}}{2(17)}$$

$$t = \frac{-2 \pm \sqrt{42436}}{34}$$ ← **Use a calculator or use the Table of Square Roots.**

$$t = \frac{-2 \pm 206}{34}$$

$$t = \frac{-2 + 206}{34} \quad \text{or} \quad t = \frac{-2 - 206}{34}$$ ← **Time cannot be negative.**

$t = \frac{204}{34} = 6$ They will be 50 kilometers apart **6 hours** after Ed leaves.

CLASSROOM EXERCISES

In Exercises 1–4, use Condition 1 (underscored once) to represent the variables in each problem. Use Condition 2 (underscored twice) to write an equation for the problem.

1. The sum of two numbers is 13. The difference of the square of the larger and twice the square of the smaller is 14.
2. The difference of two numbers is 7. The sum of the numbers multiplied by the larger equals 400.
3. The perimeter of a rectangle is 18 meters. The area of the rectangle is 20 square meters.
4. One side of a right triangle is 3 meters less than the hypotenuse and 3 meters larger than the third side.

WORD PROBLEMS

A Solve each problem.

1. The sum of two numbers is 15. The sum of the squares of the numbers is 113. Find the numbers.
2. The difference of two numbers is 9. The sum of the squares of the numbers is 221. Find the numbers.
3. A rectangle has a perimeter of 6 centimeters and an area of 2 square centimeters. Find its dimensions.
4. The perimeter of a rectangle is 18 centimeters. When the width is increased by 12 centimeters and the length is decreased by 5 centimeters, the area of the resulting rectangle is twice that of the original rectangle. Find the dimensions of the original rectangle.
5. One leg of a right triangle is 2 centimeters longer than the other leg and 2 centimeters shorter than the hypotenuse. Find the length of each side.

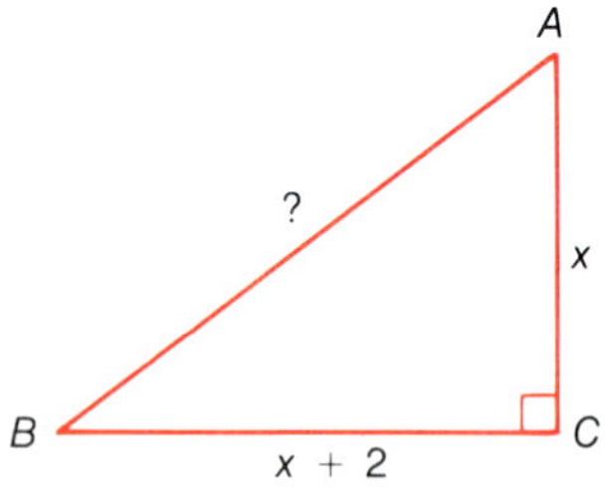

6. A diagonal of a square is 8 millimeters long. Find the width of the square.
7. The length of a rectangle is 5 centimeters more than its width. The diagonal of the rectangle is 25 centimeters. Find the dimensions.
8. Two hikers, Nancy and Jill, start from the point where two straight roads intersect at right angles. Jill starts two hours earlier than Nancy and hikes at a rate of 2 kilometers per hour. Nancy's rate is 3 kilometers per hour. How long after Nancy starts will they be 10 kilometers apart?
9. Bill and Wilma start from the same point and travel along straight roads that are at right angles to each other. Wilma travels 3 miles an hour faster than Bill. At the end of 2 hours, they are 30 miles apart. How fast is each traveling?

10. A ship sails due east from a certain point at a speed of 16 miles per hour. Two hours later, a second ship leaves from the same point and sails due north at a rate of 20 miles per hour. In how many hours after the first ship leaves will the two ships be 100 miles apart?

11. A playground, 40 meters wide by 60 meters long, is to be doubled in area by extending each side an equal amount. By how much should each side by extended?

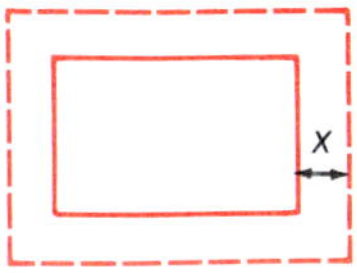

12. A rectangular lawn is 80 feet long and 60 feet wide. How wide a uniform strip must be cut around the edge when mowing the grass in order that one-half the area of the lawn be cut?

B

13. The sum of two numbers is 15 and the difference of their reciprocals is $\frac{1}{10}$. Find the numbers.

14. The difference of two numbers is 4 and the sum of their reciprocals is $\frac{3}{8}$. Find the numbers.

15. The area of the rectangular lot shown at the right is 2100 square meters. It takes 130 meters of fence to enclose three sides of the lot. Find the dimensions of the lot.

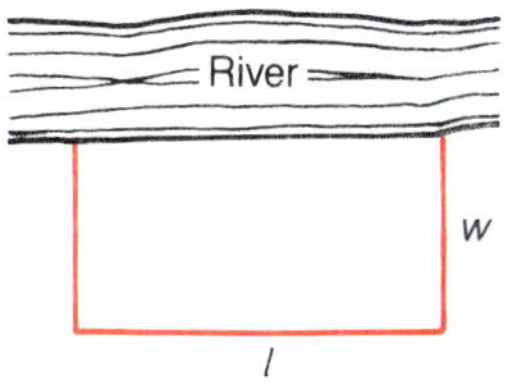

16. The diagonal of a rectangle is $2\sqrt{10}$ centimeters long. The area of the rectangle is 12 square meters. Find the dimensions of the rectangle.

17. The time required to travel from Smithton to York, a distance of 180 kilometers, was 1 hour more than the time required to make the return trip. The rate going was 15 kilometers per hour faster than the rate returning. Find the rate on the return trip.

18. Two trains traveled the same route of 300 miles. The speed of one train was 10 miles an hour faster than the other. Find the speed of each train if the faster train took $1\frac{1}{2}$ hours less time than the slower.

C

19. The square of one number is 14 more than twice the square of another number. If the second number is three less than the first, what are the two numbers? (HINT: There are two different solutions.)

20. The difference between a two-digit number and the number obtained by reversing its digits is 9. The product of the digits of the number is 6. Find the numbers.

21. Mr. Jones has two square pieces of tin of unequal size. The total area of the two pieces is 74 square centimeters. If the squares are placed side-by-side, the total perimeter is 38 centimeters. Find the length of a side of each square.

22. A group of small business owners pledged a total of $1000 a month to beautify a pedestrian mall. The owners agreed to split the cost equally. When five of them went out of business, the rest had to pay an additional $10 per month each. How many were in the original group?

Review

Solve. The replacement set is the set of complex numbers. *(Section 9–4)*

1. $2x^2 - 40 = 0$ **2.** $a^2 = -25$ **3.** $(b - \frac{1}{3})^2 = -\frac{1}{9}$ **4.** $-3(n - 1)^2 = 27$

Solve by completing the square. *(Section 9–5)*

5. $x^2 + 9x = -18$ **6.** $4y^2 + 5y - 7 = 0$ **7.** $3b^2 + 7b + 4 = 0$

Solve by using the quadratic formula. Write irrational solutions in simplest radical form. *(Section 9–6)*

8. $x^2 - 4x + 7 = 0$ **9.** $2x^2 - 5x - 9 = 0$ **10.** $x^2 - 2 = (3x + 4)(x - 1)$

11. The perimeter of a rectangle is 38 centimeters. When the width is decreased by 1 centimeter and the length is decreased by 5 centimeters, the area of the resulting rectangle is one half that of the original rectangle. Find the dimensions of the original rectangle. *(Section 9–7)*

Using Statistics:
Tables and Predicting

Insurance companies employ **actuaries** to gather and analyze statistics involving future losses due to accidents, fire, flood, death, etc.

The **Mortality Table** on page 339 was structured from statistics gathered by actuaries. The **Death Rate Per 1000** column can be used to predict the number of people who may be expected to die at a given age.

EXAMPLE 1 Of 20,000 females who are 40 years old how many can be expected to die before age 41?

Solution: Write a proportion, where n represents the number of females who are expected to die before age 41.

From the table, the rate is 2.80 per 1000. Thus,

$$\frac{2.80}{1000} = \frac{n}{20{,}000}$$

$$56{,}000 = 1000n$$

$$56 = n$$

Thus 56 out of 60,000 females who are 40 years old can be expected to die before age 41.

The actuaries compile this data in order to determine how much the company will have to pay out in claims.

From the information in Example 1, you can calculate the minimum amount the insurance company must collect to be able to pay out claims.

EXAMPLE 2 VNI Life and Casualty Co. has 20,000 female policyholders of age 40. Each policyholder has a $30,000 life insurance policy. Find the minimum amount in yearly premiums that must be collected from each policyholder.

Solution: [1] From Example, 1, you know that 56 of the women will die before age 41.

[2] Find the amount paid out.

$\$30,000 \times 56 = \$1,680,000$

[3] Divide by the number of policyholders.

$\$1,680,000 \div 20,000 = \mathbf{\$84}$

Thus, the minimum yearly premium is **$84**.

Age M	F	Death Rate Per 1,000	Expectation of Life
	0*	6.20	71.17
	1	1.67	70.61
	2	1.41	69.72
	3	1.35	68.82
	4	1.29	67.91
	5	1.24	67.00
	6	1.19	66.08
	7	1.15	65.16
	8	1.12	64.24
	9	1.11	63.31
	10	1.11	62.38
	11	1.12	61.45
	12	1.14	60.51
	13	1.17	59.58
	14	1.21	58.65
0*		7.08	68.30
1		1.76	67.78
2		1.52	66.90
3		1.46	66.00
3		1.40	65.10
5		1.35	64.19
6		1.30	63.27
7		1.26	62.35
8		1.23	61.43
9		1.21	60.51
10		1.21	59.58
11		1.23	58.65
12	15	1.26	57.72
13	16	1.32	56.80
14	17	1.39	55.87
15	18	1.46	54.95
16	19	1.54	54.03
17	20	1.62	53.11
18	21	1.69	52.19
19	22	1.74	51.28
20	23	1.79	50.37
21	24	1.83	49.46
22	25	1.86	48.55
23	26	1.89	47.64
24	27	1.91	46.73
25	28	1.93	45.82
26	29	1.96	44.90
27	30	1.99	43.99
28	31	2.03	43.08
29	32	2.08	42.16
30	33	2.13	41.25
31	34	2.19	40.34
32	35	2.25	39.43
33	36	2.32	38.51
34	37	2.40	37.60
35	38	2.51	36.69
36	39	2.64	35.78
37	40	2.80	34.88
38	41	3.01	33.97
39	42	3.25	33.07
40	43	3.53	32.18
41	44	3.84	31.29

* under 6 months

EXERCISES

1. Of 80,000 males who are 30 years old, how many can be expected to die before age 31?
2. Of 100,000 females who are 43 years old, how many can be expected to die before age 44?
3. Of 50,000 males who are 41 years old, how many can be expected to die before age 42?
4. Of 90,000 females who are 35 years old how many can be expected to die before age 36?
5. For every 500,000 females who are 39 years old, how many can be expected to die before age 40?
6. For every 500,000 males who are 39 years old, how many can be expected to die before age 40?

The Clover Insurance Company has 500 male policyholders and 8000 female policyholders who are 30 years old. Each has a $20,000 life insurance policy. Use this information for Exercises 7–10.

7. Find the minimum yearly premium for the men.
8. Find the minimum yearly premium for the women.
9. Find the minimum yearly premium if there were
   a. 8000 male policyholders.
   b. 5000 female policyholders.

REVIEW CAPSULE FOR SECTION 9–8

Give the x and y intercepts of the graph of each function. *(Pages 93–97)*

1. $y = x$ **2.** $y = 3x + 5$ **3.** $y = x^2 - 9$ **4.** $y = x^2 - 2x + 1$

Name all the sets, N, W, I, Q, Ir, R, and C to which each number belongs. *(Pages 318–322)*

5. $\sqrt[3]{8}$ **6.** $\sqrt[8]{-8}$ **7.** $\sqrt{8}$ **8.** $\sqrt{-8}$ **9.** $3i$ **10.** $5 - \sqrt{8 - 12}$

9–8 Roots of a Quadratic Equation

For a given function, f, any value of x in the domain of f which satisfies the equation $f(x) = 0$ is a **zero** of f. These values of x are also called the **roots of the equation $f(x) = 0$.**

Any real number zeros of the function are the x intercepts of the graph of the function. Recall that the **x intercepts** are the abscissas of the points where the graph crosses the x axis.

$g(x) = x^2 - 6x + 9$

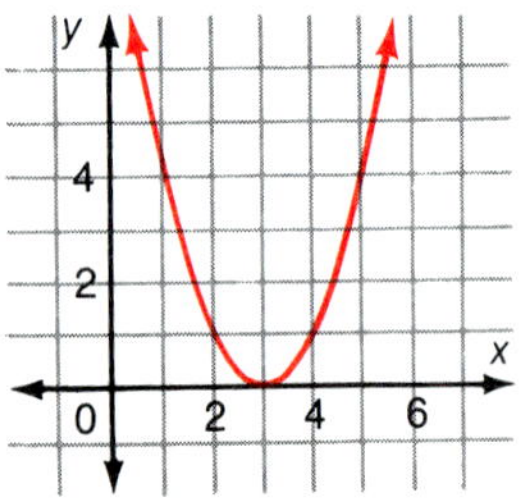

Figure 1

$h(x) = x^2 - x - 6$

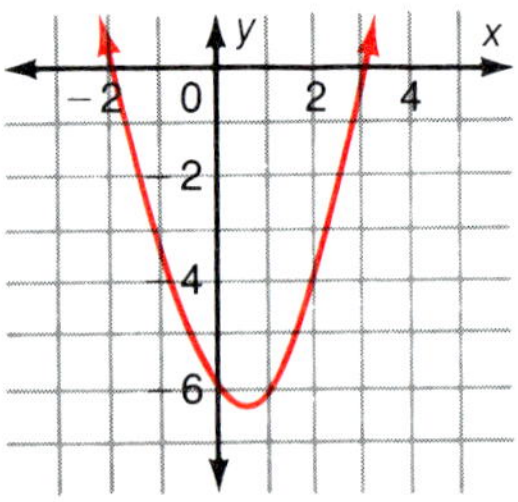

Figure 2

$f(x) = x^2 + 6$

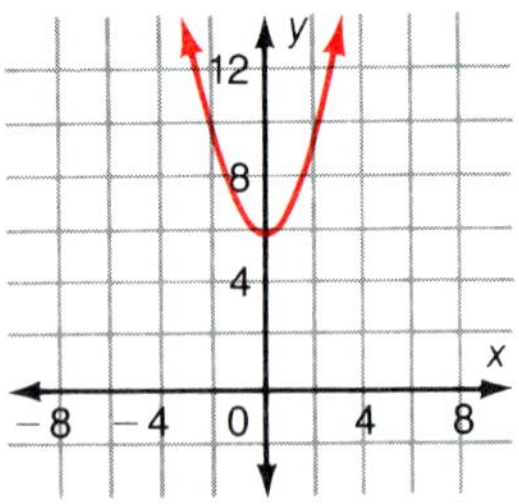

Figure 3

	$g(x) = x^2 - 6x + 9$	$h(x) = x^2 - x - 6$	$f(x) = x^2 + 6$
x intercepts	3	−2 and 3	None
Zero(s) of the function	3	−2 and 3	None
Solution(s) of the corresponding equation	3	−2 and 3	No real solutions

Since the solution set for every quadratic equation of the form $ax^2 + bx + c = 0$ is

$$\left\{\frac{-b + \sqrt{b^2 - 4ac}}{2a}, \frac{-b - \sqrt{b^2 - 4ac}}{2a}\right\},$$

the solutions or roots can be expressed as

$$r_1 = \frac{-b}{2a} + \frac{\sqrt{b^2 - 4ac}}{2a} \quad \text{or} \quad r_2 = \frac{-b}{2a} - \frac{\sqrt{b^2 - 4ac}}{2a}.$$

The expression $b^2 - 4ac$ is called the **discriminant** because it enables you to determine the nature of the solutions of a quadratic equation without having to solve the equation or graph the related quadratic function.

Nature of the Roots of a Quadratic Equation

For the quadratic equation $ax^2 + bx + c = 0$:

1. If $b^2 - 4ac = 0$ (see Figure 1) then $r_1 = r_2$, and the equation will have one real root (sometimes called a double root).
2. If $b^2 - 4ac > 0$ (see Figure 2), then r_1 and r_2 will be two distinct real numbers and the equation will have two, unequal, real roots. If $b^2 - 4ac$ is also the square of a rational number, these roots will be rational; otherwise they will be irrational.
3. If $b^2 - 4ac < 0$ (see Figure 3), then r_1 and r_2 will be two, distinct, imaginary numbers, and the equation will have no real roots.

EXAMPLE Use the discriminant to determine the nature of the roots of each quadratic equation.

a. $4x^4 + 4x + 1 = 0$ **b.** $x^2 + 2x + 5 = 0$

c. $2x^2 + x - 3 = 0$ **d.** $2x^2 + 2x - 3 = 0$

Solutions: Evaluate $b^2 - 4ac$ for each equation.

a. For $4x^4 + 4x + 1 = 0$, $a = 4$, $b = 4$, $c = 1$.

$b^2 - 4ac = 4^2 - 4(4)(1) = 0$

Since $b^2 - 4ac = 0$, there is **one real root.**

b. For $x^2 + 2x + 5 = 0$, $a = 1$, $b = 2$, $c = 5$.

$b^2 - 4ac = 2^2 - 4(1)(5) = -16$

Since $b^2 - 4ac < 0$, there are **no real roots.**

c. For $2x^2 + x - 3 = 0$, $a = 2$, $b = 1$, $c = -3$.

$b^2 - 4ac = 1^2 - 4(2)(-3) = 25$

Since $b^2 - 4ac > 0$, there are **two unequal, real roots.**

Since 25 is the square of a rational number, the roots are **rational numbers.**

d. For $2x^2 + 2x - 3 = 0$, $a = 2$, $b = 2$, $c = -3$.

$b^2 - 4ac = 2^2 - 4(2)(-3) = 28$

Since $b^2 - 4ac > 0$, there are **two, unequal, real roots.**

Since 28 is not the square of a rational number, the roots are **irrational numbers.**

In ***b*** of the Example, $b^2 - 4ac < 0$. Thus, $\sqrt{b^2 - 4ac}$ is the imaginary number, $\sqrt{-16}$. The roots of the equation are the **complex conjugates,**

$$\frac{-2 + \sqrt{-16}}{2} \quad \text{and} \quad \frac{-2 - \sqrt{-16}}{2}, \qquad \text{or} \qquad \frac{-2 \pm 4i}{2} = \mathbf{-1 \pm 2i.}$$

In fact, the imaginary roots of a quadratic equation with real coefficients **always occur as conjugate pairs.**

CLASSROOM EXERCISES

Use the discriminant to determine the nature of the roots of each quadratic equation. When an equation has real roots, tell whether they are rational or irrational numbers.

1. $x^2 + 3x - 4 = 0$
2. $4y^2 - 12y + 9 = 0$
3. $4m = 2 - 3m^2$
4. $6x^2 + 7x = 0$
5. $2y^2 - 7y = -3$
6. $2x^2 = 8 - 3x$

WRITTEN EXERCISES

A

Use the discriminant to determine the nature of the roots of each quadratic equation. When an equation has real roots, tell whether they are rational or irrational numbers.

1. $x^2 + 5x - 2 = 0$
2. $n^2 - 6n + 8 = 0$
3. $y^2 - 6y + 9 = 0$
4. $m^2 - 6m + 12 = 0$
5. $x^2 - 6x + 1 = 0$
6. $x^2 + 2x - 15 = 0$
7. $4a^2 + 1 = 4a$
8. $2 = 5y + y^2$
9. $2d^2 + 7d + 6 = 0$
10. $3m^2 - 7m = 2$
11. $2n^2 + 3 = 5n$
12. $4x^2 - 10x - 6 = 0$
13. $3d^2 - 6d = 8$
14. $6x^2 - 7x + 3 = 0$
15. $2a^2 - 3a = 5$
16. Describe the graph of $y = ax^2 + bx + c$ for $a, b, c \in R$ in terms of the number of points in which it intersects the x axis when $b^2 - 4ac = 0$, when $b^2 - 4ac > 0$, and when $b^2 - 4ac < 0$.
17. Determine, without graphing, whether the graph of $y = -4x^2 + 12x - 9$ intersects the x axis in exactly one point, in two points, or not at all.

Solve over the set of complex numbers.

18. $y^2 - 7y + 10 = 0$
19. $z^2 - z - 6 = 0$
20. $t^2 = t + 12$
21. $3a^2 - 8a + 4 = 0$
22. $2x^2 + x - 15 = 0$
23. $8x^2 + 2x - 1 = 0$

B

Determine the value(s) of k for which each equation will have the indicated roots.

24. $y^2 - 6y + k = 0$; one real number
25. $y^2 + ky + 16 = 0$; two distinct real numbers
26. $kn^2 - 10n + 5 = 0$; two imaginary numbers
27. $3n^2 - 12n + 2k = 0$; two imaginary numbers
28. $2x^2 - 20x + k = 0$; two distinct real numbers
29. $ka^2 + 8a + k = 0$; two distinct real numbers
30. $y^2 = -4ky - 1$; one real number
31. $k^2x^2 + 40x + 25 = 9$; two distinct real numbers
32. Find k in $12x^2 + 4kx + k = 0$ such that the graph of $y = 12x^2 + 4kx + k$ **(a)** is tangent to the x axis; **(b)** intersects the x axis in two points; and **(c)** has no intersection with the x axis.

9–9 Sum and Product of Roots

If r_1 and r_2 are the roots of the quadratic equation $ax^2 + bx + c = 0$, you can express the sum of the roots, $r_1 + r_2$, and the product of the roots, $r_1 \cdot r_2$, in terms of a, b, and c.

Theorem 9–4 If r_1 and r_2 are the roots of the quadratic equation, $ax^2 + bx + c = 0$, then

$$r_1 + r_2 = -\frac{b}{a} \quad \text{and} \quad r_1 \cdot r_2 = \frac{c}{a}.$$

You can use Theorem 9–4 to find the sum and product of the roots without finding the solution set.

EXAMPLE 1 Find the sum and product of the roots of $x^2 - 6x + 8 = 0$.

Solution: For $x^2 - 6x + 8 = 0$, $a = 1$, $b = -6$, $c = 8$.

$$r_1 + r_2 = -\frac{-6}{1} = 6 \longleftarrow r_1 + r_2 = -\frac{b}{a}$$

$$r_1 \cdot r_2 = \frac{8}{1} = 8 \longleftarrow r_1 \cdot r_2 = \frac{c}{a}$$

Theorem 9–4 can also be used to find a quadratic equation whose roots are given.

EXAMPLE 2 Find a quadratic equation whose solution set is $\{(2 + i\sqrt{3}), (2 - i\sqrt{3})\}$.

Solution: [1] Find the sum and product of the roots.

$$r_1 + r_2 = (2 + i\sqrt{3}) + (2 - i\sqrt{3}) = 4 = -\frac{b}{a}.$$

$$r_1 \cdot r_2 = (2 + i\sqrt{3})(2 - i\sqrt{3}) = 4 + 3 = 7 = \frac{c}{a}.$$

[2] Let $a = 1$, $b = -4$, and $c = 7$. **Equation:** $x^2 - 4x + 7 = 0$

Check: Show that $(2 + i\sqrt{3})$ and $(2 - i\sqrt{3})$ are roots of $x^2 - 4x + 7 = 0$ by using the quadratic formula. The check is left for you.

There are many other equations for which $r_1 + r_2 = 4$ and $r_1 \cdot r_2 = 7$. *Any equivalent equation* will have the *same roots.*

CLASSROOM EXERCISES

Find the sum and product of the roots of each equation.

1. $x^2 - 6x - 7 = 0$ **2.** $y^2 - 5y - 3 = 0$ **3.** $3a^2 - 2a - 6 = 0$

4. $5 + 2t^2 = 3t$ **5.** $2y^2 - 7y = 6$ **6.** $3r^2 - 17 = 0$

WRITTEN EXERCISES

A Find the sum and product of the roots of each equation.

1. $x^2 + 7x - 10 = 0$
2. $8x^2 - 6x + 1 = 0$
3. $y^2 - 2y + 9 = 0$
4. $m^2 - 4m + 1 = 0$
5. $3n^2 + 2n - 3 = 0$
6. $2t^2 - 3t - 5 = 0$
7. $a^2 + a + 1 = 0$
8. $5x^2 - 7x + 3 = 0$
9. $4r^2 - 3r + 9 = 0$

Write a quadratic equation having the given solution set.

10. $\{3, 5\}$
11. $\{-2, 7\}$
12. $\{2, -\frac{3}{4}\}$
13. $\{-4, 4\}$
14. $\{3i, -3i\}$
15. $\{\frac{1}{2}, \frac{1}{3}\}$
16. $\{2 + i, 2 - i\}$
17. $\{4\sqrt{3}, -4\sqrt{3}\}$
18. $\{2 + \sqrt{7}, 2 - \sqrt{7}\}$
19. $\{1 + \sqrt{5}, 1 - \sqrt{5}\}$
20. $\{1 + 2\sqrt{3}, 1 - 2\sqrt{3}\}$
21. $\{1 - 3i, 1 + 3i\}$
22. $\{2 + 6i, 2 - 6i\}$
23. $\left\{\frac{3 + i\sqrt{2}}{4}, \frac{3 - i\sqrt{2}}{4}\right\}$
24. $\left\{\frac{1 + i\sqrt{5}}{2}, \frac{1 - i\sqrt{5}}{2}\right\}$
25. $\left\{\frac{5 - 3i}{4}, \frac{5 + 3i}{4}\right\}$
26. $\{8 - \sqrt{5}, \sqrt{5}\}$

B Write a quadratic equation with integral coefficients whose solutions have the given sum and product.

27. Sum: 3; product: -40
28. Sum: 0; product: -9
29. Sum: -13; product: 40
30. Sum: -1; product: -30
31. Sum: $\frac{1}{2}$; product: $-\frac{3}{16}$
32. Sum: $-\frac{7}{10}$; product: -8

In Exercises 33–38, find k such that the given conditions are satisfied.

33. $x^2 + kx - 39 = 0$; $r_1 = 3$
34. $x^2 - 10x + k = 0$; $r_1 = 4$

C

35. $x^2 + kx + 40 = 0$; $r_1 - r_2 = 16$
36. $x^2 - 18x + k = 0$; $r_1 = 5r_2$
37. $5x^2 + kx = 4$; $r_1 = \frac{2}{5}$
38. $5x^2 + k^2x - 9x + 7 = 0$; $r_1 = r_2$
39. Find a quadratic function whose graph passes through the points $P(3, 0)$, $Q(6, 0)$, and $R(0, 6)$.
40. Prove Theorem 9–4.

REVIEW CAPSULE FOR SECTION 9–10

State the degree of each polynomial. *(Pages 188–189)*

1. $3x - 8$
2. $9x^3 - 7x^2 + 3x$
3. $6x^5 + 9x^4 - 8$
4. $xy - x$

Use synthetic division to find the quotient and the remainder. *(Pages 212–215)*

5. $x^4 - 2x^2 + 7x - 5 \div (x + 2)$
6. $3x^3 - 2x^2 - 7x - 5 \div (x - 2)$

Factor completely over the integers. *(Pages 206–208)*

7. $x^3 - 3x^2 - 18x + 40$
8. $x^3 + x^2 - 10x + 8$
9. $x^3 + x^2 - 11x + 10$

9–10 Roots of Polynomial Equations

The quadratic functions and equations studied in Chapters 8 and 9 are of degree 2. In general, a **polynomial function of degree n** is defined by

$$P(x) = a_0x^n + a_1x^{n-1} + a_2x^{n-2} + \cdots a_{n-1}x + a_n$$

where the coefficients, a_0, a_1, a_2, $\cdots$, are real numbers with $a_0 \neq 0$, and n is a positive integer. When $P(x) = 0$, the resulting equation is a **polynomial equation.**

The following important theorem concerns the roots of polynomial equations. The proof of this theorem is not given since it is beyond the scope of this book.

Theorem 9–5

Fundamental Theorem of Algebra

Every polynomial equation of degree greater than zero has **at least one real or imaginary zero.**

Theorem 9–6 follows from the Fundamental Theorem of Algebra.

Theorem 9–6

Number of Roots Theorem

Every polynomial $P(x)$ of degree n can be written as the product of n linear factors. That is, $P(x)$ of degree n can be written as

$$k(x - r_1)(x - r_2) \cdots (x - r_n).$$

Thus, **a polynomial equation of degree n has exactly n roots.**

When applying this theorem, you may have to count the same root more than once. For example, the equation

$$x^2 \quad 8x + 16 = 0$$ ← **$(x - 4)(x - 4) = 0$**

has a **multiple root,** 4. It is sometimes called a **double root.**

EXAMPLE 1 Find all the roots of $x^3 - 4x^2 + x + 6 = 0$, given that one root is 3.

Solution: Write the polynomial as a product of linear factors.
Use synthetic division to find one or more of the factors.

$$\begin{array}{r|rrr|r} 3 & 1 & -4 & 1 & 6 \\ & 0 & 3 & -3 & 6 \\ \hline & 1 & -1 & -2 & \mathbf{0} \end{array}$$ ← **The quotient is $(x^2 - x - 2)$.**

Thus, $x^3 - 4x^2 + x + 6 = (x - 3)(x^2 - x - 2)$ ← **Factor $(x^2 - x - 2)$.**

$$= (x - 3)(x - 2)(x + 1)$$

Since $(x - 3)(x - 2)(x + 1) = 0$, the roots are **3, 2,** and **−1.**

Recall from Section 9–8 that the imaginary roots of a quadratic equation with real coefficients occur in conjugate pairs. Theorem 9–7 tells you that this is also true for polynomial equations with real coefficients.

Theorem 9–7

Complex Conjugates Theorem

If a polynomial equation with real coefficients has $(a + bi)$ as a root, then $(a - bi)$ **is also a root** $(a$ and $b \in R, b \neq 0)$.

Given the roots of a polynomial equation, you can use them to write an equation.

EXAMPLE 2 Given that the roots of a polynomial equation are 5, 2, and $-3i$, write a polynomial equation of smallest degree and with integral coefficients having the given roots.

Solution: Since imaginary roots occur in pairs, $3i$ is also a root.

$(x - 5)(x - 2)(x - 3i)(x + 3i) = 0$ ← **By Theorems 9–6 and 9–7**

$(x^2 - 7x + 10)(x^2 + 9) = 0$

$\mathbf{x^4 - 7x^3 + 19x^2 - 63x + 90 = 0}$

The Rational Root Theorem can be used to identify possible roots of polynomial equations having integral coefficients.

Theorem 9–8

Rational Root Theorem

Let $a_0x^n + a_1x^{n-1} + \cdot \cdot \cdot \; a_{n-1}x + a_n = 0$ represent a polynomial equation of degree n with integral coefficients. If $\frac{p}{q}$ is a rational root (expressed in lowest terms) of the equation, then

p **is a factor of** a_n **and** q **is a factor of** a_0.

EXAMPLE 3 Find all rational roots of $3x^3 - 8x^2 + 3x + 2 = 0$.

Solution: By the Rational Root Theorem, if $\frac{p}{q}$ is a root of the equation, p is a factor of 2 and q is a factor of 3.

Thus, $p = \pm 1, \pm 2$ and $q = \pm 1, \pm 3$.

Possible rational roots: $\pm 1, \pm 2, \pm \frac{1}{3}, \pm \frac{2}{3}$ ← $\frac{p}{q}$

Test each possible rational root by synthetic division or by substitution. By synthetic division, you find that $-\frac{1}{3}$ is a root.

$$3x^3 - 8x^2 + 3x + 2 = (x + \tfrac{1}{3})(3x^2 - 9x + 6)$$

$$= (x + \tfrac{1}{3})(3)(x^2 - 3x + 2)$$

$$= 3(x + \tfrac{1}{3})(x - 1)(x - 2)$$

Rational roots: $\mathbf{-\frac{1}{3}, 1, 2}$

CLASSROOM EXERCISES

State the number of roots of each equation.

1. $y^2 + 6y + 5 = 0$
2. $x^3 - 2x^2 + 3 = 0$
3. $t^4 + 3t - 13 = 0$
4. $x^7 - 31x^2 = 0$
5. Two roots of $x^3 - 2x^2 + 9x - 18 = 0$ are 2 and $-3i$. What is the third root?
6. An equation of degree 3 with real coefficients has roots $2 - i$ and 3. What is the other root?

WRITTEN EXERCISES

A In Exercises 1–8, an equation and one or more of its roots is given. Find all the roots of the equation.

1. $x^3 - x^2 + 9x - 9 = 0$; 1, $3i$
2. $x^3 - 7x^2 - 3x + 81 = 0$; -3, $5 - i\sqrt{2}$
3. $x^3 + x^2 - 10x + 8 = 0$; 2
4. $2x^3 + x^2 - 13x + 6 = 0$; -3
5. $6x^3 + 3x^2 - 10x - 5 = 0$; $-\frac{1}{2}$
6. $2x^3 - 9x^2 + 2x + 1 = 0$; $\frac{1}{2}$
7. $x^4 + 8x^3 + 22x^2 + 23x + 6 = 0$; -2, -3
8. $x^4 + 5x^2 + 4 = 0$; i, $-2i$

In Exercises 9–20, write a polynomial equation of smallest degree and with integral coefficients having the given roots.

9. 1, -1, 2
10. -4, -4, 1
11. 4, -5, 2
12. $\frac{1}{2}$, $\frac{3}{2}$, -1
13. i, $-i$, 5
14. $3i$, $-3i$, -1
15. -2, $1 + i$, $1 - i$
16. -5, -2, $3i$
17. 2, 0, $1 + i$
18. $1 + i$, $1 + i$
19. 2, 2, $-i$
20. $1 + i\sqrt{3}$, $1 - i\sqrt{3}$

In Exercises 21–30, find all the rational roots of each equation.

21. $x^3 + 4x^2 + x - 2 = 0$
22. $x^3 - x^2 - 14x + 24 = 0$
23. $x^3 - 8x^2 + 5x + 14 = 0$
24. $x^3 - 2x^2 - 7x - 4 = 0$
25. $x^3 + 2x^2 - x - 2 = 0$
26. $x^3 - 4x + 10 = 0$
27. $x^4 + x^2 - 2 = 0$
28. $3x^4 - 2x^2 + 18 = 0$
29. $2x^3 + 5x^2 + 1 = 0$

B Find all the roots of each equation.

30. $x^3 + 3x^2 - 4x - 12 = 0$
31. $x^3 + 4x^2 - 17x - 60 = 0$
32. $3x^3 - x^2 - 6x + 2 = 0$
33. $2x^3 - 5x^2 + 1 = 0$
34. $x^4 + x^3 - x^2 - 7x - 6 = 0$
35. $3x^4 - 5x^3 - 5x^2 - 19x - 6 = 0$

C

36. Prove Theorem 9–6.
37. Prove: *If $P(x)$ is a polynomial of odd degree with real coefficients, then $P(x) = 0$ has at least one real root.*
38. Prove: *If $P(x)$ is a polynomial of even degree with real coefficients, then $P(x) = 0$ has an even number of real roots or no real roots.*
39. Prove: *If $P(x)$ is a polynomial with real coefficients and $P(a + bi) = 0$, then $P(a - bi) = 0$ for all real numbers a and b.*

9–11 Locating Real Zeros

Recall that any value of x in the domain of a function $f(x)$ for which $f(x) = 0$ is a **zero** of the function. You already know how to find the zeros of linear and quadratic functions. The Locater Theorem will help you to approximate the real zeros of a polynomial function of degree greater than 2.

Theorem 9–9

Locater Theorem

If f is a polynomial function and a and b are real numbers such that $f(a)$ and $f(b)$ have opposite signs, then there is at least one zero of f between a and b.

Geometrically, the Locater Theorem says that if the point $(a, f(a))$ is below the x axis and the point $(b, f(b))$ is above the x axis, then there must be at least one point between $x = a$ and $x = b$ where the graph intersects the x axis.

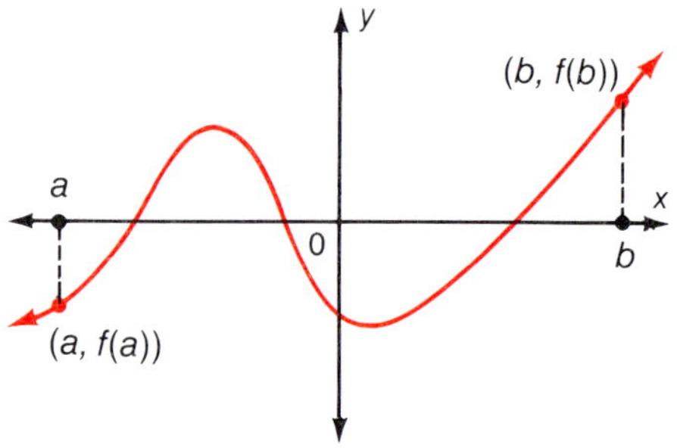

EXAMPLE Approximate to the nearest tenth a real zero of $f(x) = x^3 - 3x^2 + 3$.

Solution: [1] Use substitution to find two integers a and b for which $f(a)$ and $f(b)$ have opposite signs. By the Rational Root Theorem, possible rational roots are ± 1, ± 3.

c	-1	0	1	2	3
f**(c)**	-1	3	1	-1	3

Real zeros: Between $f(-1)$ and $f(0)$; Between $f(1)$ and $f(2)$; Between $f(2)$ and $f(3)$

[2] Choose one pair of integers, say -1 and 0, and "close in" on the zero by finding $f(c)$ for $-1 < c < 0$.

Since $f(-1) = -1$ is closer to 0 than $f(0) = 3$, choose values of c closer to -1.

c	-0.9	-0.8
f**(c)**	-0.159	0.568

← **Use a calculator.**

Since $f(-0.9) < 0$ and $f(-0.8) > 0$, there is a real zero between -0.9 and -0.8.

Since $f(-0.9)$ is closer to 0 than $f(-0.8)$, one real zero of the function is approximately $\mathbf{-0.9}$.

WRITTEN EXERCISES

A Locate the real zeros of each function between successive integers.

1. $f(x) = x^3 - 2x^2 + x - 3$

2. $f(x) = x^3 + x^2 - 5x - 5$

3. $f(x) = x^4 + x^3 - 5x^2 - 6x - 6$

4. $f(x) = x^4 + 3x^3 - 4x - 12$

5. $f(x) = x^4 - 3x^3 - 3x + 9$

6. $f(x) = x^4 + 4x^3 + 2x + 8$

Approximate the indicated real root to the nearest tenth.

7. $f(x) = x^3 - 12x + 3$; largest root

8. $f(x) = 3x^3 - 3x + 1$; smallest root

9. $f(x) = 2x^4 - 2x - 3$; smallest root

10. $f(x) = x^4 + 3x - 13$; largest root

B Approximate to the nearest tenth the real zeros of each function.

11. $f(x) = x^3 - 6x^2 + 11x - 5$

12. $f(x) = x^3 - 5$

13. $f(x) = x^4 - 4x^2 + 6$

14. $f(x) = 3x^4 - x^2 + x - 1$

15. $f(x) = x^5 - 6$

16. $f(x) = -x^3 + x^2 + 1$

C

17. Determine the range of values for k for which $f(x) = x^3 - 2x^2 + 4x - k$ has at least one zero between 0 and 1.

Review

Use the discriminant to determine the nature of the roots of each quadratic equation. When an equation has real roots, tell whether they are rational or irrational numbers. *(Section 9–8)*

1. $5c^2 - 3c = 2$

2. $2x^2 - 7x + 2 = 0$

3. $3b^2 + 2b + 5 = 0$

Determine the value(s) of k for which each equation will have the indicated roots. *(Section 9–8)*

4. $y^2 - 8y + k = 0$; one real number

5. $kx^2 + 10 = 0$; two imaginary numbers

Write a quadratic equation having the given solution set. *(Section 9–9)*

6. $\{3, -2\}$

7. $\{-4, 4\}$

8. $\{1 + \sqrt{2}, 1 - \sqrt{2}\}$

9. $\left\{\frac{2 - 5i}{3}, \frac{2 + 5i}{3}\right\}$

In Exercises 10–13, you are given the roots of a polynomial equation. Write a polynomial equation of smallest degree and with integral coefficients having the given roots. *(Section 9–10)*

10. 2, 3, −1

11. −3, −4, 2

12. $\frac{1}{3}, \frac{2}{3}, 4$

13. −4i, 3, 6

Approximate to the nearest tenth the real zeros of each function. *(Section 9–11)*

14. $f(x) = x^3 - 2x^2 + 3x - 1$

15. $f(x) = x^3 - 6$

16. $f(x) = x^4 - 4$

BASIC: SOLVING QUADRATIC EQUATIONS

Problem: *Given the values of a, b, and c in the equation $ax^2 + bx + c = 0$, write a program which computes and prints the roots, including complex roots.*

```
100 PRINT "FOR THE EQUATION A*X*X+B*X+C=0, WHAT ARE A,B,C";
110 INPUT A, B, C
120 LET D = B*B - 4*A*C
130 IF D = 0 THEN 190
140 IF D < 0 THEN 220
150 LET X1 = (-B + SQR(D))/(2*A)
160 LET X2 = (-B - SQR(D))/(2*A)
170 PRINT "THE TWO ROOTS ARE";X1;"AND ";X2
180 GOTO 250
190 LET X = -B/(2*A)
200 PRINT "DOUBLE ROOT IS";X
210 GOTO 250
220 LET R = -B/(2*A)
230 LET I = SQR(-D)/(2*A)
240 PRINT "COMPLEX ROOTS ARE ";R;" +";I;"I AND ";R;" -";I;"I"
250 PRINT
260 PRINT "ANY MORE EQUATIONS TO SOLVE (1 = YES, 0 = NO)";
270 INPUT Z
280 IF Z = 1 THEN 100
290 END
```

The following is the output from a sample run of the program above.

```
RUN
FOR THE EQUATION A*X*X*+B*X+C=0, WHAT ARE A,B,C? 1,0,-16
THE TWO ROOTS ARE 4 AND -4

ANY MORE EQUATIONS TO SOLVE (1 = YES, 0 = NO)? 1
FOR THE EQUATION A*X*X+B*X+C=0, WHAT ARE A,B,C? 5,4,2
COMPLEX ROOTS ARE -.4  + .489898 I AND -.4  - .489898 I

ANY MORE EQUATIONS TO SOLVE (1 = YES, 0 = NO)? 1
FOR THE EQUATION A*X*X+B*X+C=0, WHAT ARE A,B,C? 1,-4,4
DOUBLE ROOT IS 2

ANY MORE EQUATIONS TO SOLVE (1 = YES, 0 = NO)? 0
READY
```

Analysis

Statements 120–140: D is the discriminant of the equation. If D = 0, there is a double root. If D < 0, there are two complex (non-real) roots.

Statements 150–170: These statements compute and print the two real roots X1 and X2. The only difference between the roots is that X1 is computed using the positive square root of the discriminant, whereas X2 uses the negative square root.

Statements 190–200: These statements, which compute the double root when D = 0, could be combined into the following single statement.

```
190 PRINT "DOUBLE ROOT IS"; -B/(2*A)
```

Statements 220–240: BASIC cannot handle complex numbers as such. To the computer "I" is just another real variable. Likewise, a statement like `30 LET I = SQR(-1)` is meaningless since SQR handles only non-negative values. Instead the real part (R) and imaginary part (I) of the complex roots must be calculated separately (statements 220–230). Then statement 240 prints the real and imaginary parts with a "+" sign between them for the first root and a "–" between them for the second root. In both cases an "I" is printed after the imaginary part. (See the output for the second set of data in the sample run on page 350.)

EXERCISES

A

Write each equation in the form $ax^2 + bx + c = 0$. Then use the program on page 350 to find the roots.

1. $x^2 + 5x = 7$ **2.** $3x^2 = 8 + 2x$ **3.** $12 = 4x^2 + 9x$

4. $4x = 3x^2 + 6$ **5.** $4x^2 = 5x$ **6.** $24x^2 = 18$

Write a BASIC program for each problem.

7. For the same input as in the program on page 350, compute the sum and the product of the roots of the quadratic equation without actually calculating the roots.

8. Print the square roots of any real number, including imaginary roots.

In Exercises 9–12, the input is *a*, *b*, *c*, and *d* of the complex numbers $a + bi$ and $c + di$. Output should be printed as neatly as possible. For example, print `3 - 4 I` and not `3 + -4 I`; print `-5 I` and not `0 - 5 I`; print `-4` and not `-4 + 0 I`; `4 + I` and not `4 + 1 I`.

9. Print the sum of $a + bi$ and $c + di$.

10. Print the difference of $c + di$ subtracted from $a + bi$.

11. Print the product of the two complex numbers.

12. Print the quotient of $a + bi$ divided by $c + di$. If $c + di = 0$, print `QUOTIENT UNDEFINED.`

B

13. Given the two real roots of a quadratic equation, print the equation as neatly as possible. For example, print `X↑2 - 5 X + 3 = 0` and not `X↑2 + -5 X + 3 = 0`; print `X↑2 - 7 = 0` and not `X↑2 + 0 X + -7 = 0`.

Chapter Summary

IMPORTANT TERMS

Completing the square (p. 329)
Complex conjugates (p. 323)
Complex number (p. 318)
Discriminant (p. 341)
Double root (p. 345)
i–form (p. 314)
Incomplete quadratic equation (p. 327)
Multiple root (p. 345)
Polynomial equation (p. 345)
Polynomial function of degree n (p. 345)
Pure imaginary number (p. 314)
Root of an equation (p. 340)

IMPORTANT IDEAS

1. For any real number a, where $-a < 0$,
$$\sqrt{-a} = \mathrm{i}\sqrt{a} = \sqrt{a} \cdot \mathrm{i}.$$
2. Standard form of a complex number: $a + b\mathrm{i}$, where a and b are real numbers and $\mathrm{i} = \sqrt{-1}$.
3. Two complex numbers, $a + b\mathrm{i}$ and $c + d\mathrm{i}$, are equal if and only if $a = c$ and $b = d$.
4. For all real numbers a, b, c, and d,
$$(a + b\mathrm{i}) + (c + d\mathrm{i}) = (a + c) + (b + d)\mathrm{i}.$$
5. The additive identity for the set of complex numbers is $0 + 0\mathrm{i}$.
6. For all real numbers a and b,
$$-(a + b\mathrm{i}) = -a - b\mathrm{i}.$$
7. To subtract a complex number, $a + b\mathrm{i}$, where a and b are real numbers, add its additive inverse, $-a - b\mathrm{i}$.
8. For any real numbers a, b, c, and d,
$$(a + b\mathrm{i})(c + d\mathrm{i}) = (ac - bd) + (ad + bc)\mathrm{i}.$$
9. For all real numbers a and b,
$$(a + b\mathrm{i})(a - b\mathrm{i}) = a^2 + b^2.$$
10. For all real numbers a, b, c, and d,
$$(a + b\mathrm{i}) \div (c + d\mathrm{i}) = (a + b\mathrm{i}) \cdot \frac{1}{c + d\mathrm{i}}, \quad c + d\mathrm{i} \neq 0$$
$$= \frac{(a + b\mathrm{i})(c - d\mathrm{i})}{c^2 + d^2}$$
11. **Steps for Completing the Square**

    [1] Write an equivalent equation with only the second-degree term and the first-degree term on the left side of the equation. The coefficient of the second-degree term must be 1.

    [2] Add the square of one-half the coefficient of the first-degree term to each side of the equation.

    [3] Express the left side of the equation as a perfect square.

    [4] Solve for the variable.

12. **The Quadratic Formula:** For any quadratic equation, $ax^2 + bx + c = 0$,

$$x = \frac{-b + \sqrt{b^2 - 4ac}}{2a} \quad \text{or} \quad x = \frac{-b - \sqrt{b^2 - 4ac}}{2a}.$$

13. **Nature of the Roots of a Quadratic Equation:** For the quadratic equation $ax^2 + bx + c = 0$,

    a. If $b^2 - 4ac = 0$, then $r_1 = r_2$, and the equation will have one real root (sometimes called a double root).

    b. If $b^2 - 4ac > 0$, then r_1 and r_2 will be two distinct real numbers and the equation will have two unequal, real roots. If $b^2 - 4ac$ is also the square of a rational number, these roots will be rational; otherwise they will be irrational.

    c. If $b^2 - 4ac < 0$, then r_1 and r_2 will be two distinct, imaginary numbers, and the equation will have no real roots.

14. If r_1 and r_2 are the roots of the quadratic equation $ax^2 + bx + c = 0$, then

$$r_1 + r_2 = -\frac{b}{a} \quad \text{and} \quad r_1 \cdot r_2 = \frac{c}{a}.$$

15. **Fundamental Theorem of Algebra:** Every polynomial equation of degree greater than zero has at least one real or imaginary zero.

16. Every polynomial $P(x)$ of degree n can be written as the product of n linear factors. That is, $P(x)$ of degree n can be written as $k(x - r_1)(x - r_2) \cdots (x - r_n)$. Thus, a polynomial equation of degree n has exactly n roots.

17. **Complex Conjugates Theorem:** If a polynomial equation with real coefficients has $(a + b\text{i})$ as a root, then $(a - b\text{i})$ is also a root (a and $b \in \mathbf{R}$, $b \neq 0$).

18. **Rational Root Theorem:** Let $a_0x^n + a_1x^{n-1} + \cdots + a_{n-1}x + a_n = 0$ represent a polynomial equation of degree n with integral coefficients. If $\frac{p}{q}$ is a rational root (expressed in lowest terms) of the equation, then p is a factor of a_n and q is a factor of a_0.

19. **Locater Theorem:** If f is a polynomial function and a and b are real numbers such that $f(a)$ and $f(b)$ have opposite signs, then there is at least one zero of f between a and b.

Chapter Objectives and Review

Objective: *To add, subtract, multiply, and divide pure imaginary numbers (Section 9–1)*

Perform the indicated operations. Express each answer as a real number or in i–form.

1. $\text{i} + \sqrt{-3}$ **2.** $(3\text{i})(-4\text{i})$ **3.** $-20 \div \sqrt{-4}$ **4.** $(3\text{i})^2 + 3$ **5.** $-3 \div \text{i}^6$

Objective: *To add and subtract complex numbers (Section 9–2)*

Add or subtract as indicated. Express answers in standard form.

6. $3i + (-i + 3)$ **7.** $(3i + 1) - 2i$ **8.** $(2 - i) - (i + 2)$

9. $(4 + \sqrt{-5}) + (i - 2)$ **10.** $(\sqrt{2} + 7i) - (\sqrt{2} - 3i)$ **11.** $-2i + (3 - 4i)$

Objective: *To multiply and divide complex numbers (Section 9–3)*

Multiply or divide as indicated. Express answers in standard form.

12. $\frac{1}{2 + i}$ **13.** $\frac{29}{5 - 2i}$ **14.** $(2 + i)(i - 3)$ **15.** $\frac{2 + i}{3 - 4i}$

Objective: *To solve incomplete quadratic equations (Section 9–4)*

Solve each equation. The replacement set is the set of complex numbers.

16. $3x^2 = 75$ **17.** $2x^2 = -14$ **18.** $\frac{1}{7}y^2 = -7$

Objective: *To solve quadratic equations by completing the square (Section 9–5)*

Solve by completing the square.

19. $x^2 + 10x + 9 = 0$ **20.** $4y^2 + 12y = -9$ **21.** $2r^2 - 4r + 1 = 0$

Objective: *To use the quadratic formula to solve quadratic equations (Section 9–6)*

Use the quadratic formula to solve each equation.

22. $2x^2 - x - 3 = 0$ **23.** $3t - t^2 = 4$ **24.** $x^2 = 7x - 12$

Objective: *To use a quadratic equation to solve problems (Section 9–7)*

25. The sum of two numbers is 11. The sum of the squares of the numbers is 65. Find the numbers.

26. The perimeter of a rectangle is 22 yards. The area is 28 square yards. Find the length and width of the rectangle.

Objective: *To use the discriminant of a quadratic equation to determine the nature of the roots of a quadratic equation (Section 9–8)*

Use the discriminant to determine the nature of the roots of each quadratic equation. When an equation has real roots, tell whether they are rational or irrational numbers.

27. $3x^2 - 7x + 5 = 0$ **28.** $3r^2 - 5r + 2 = 0$ **29.** $7d^2 - 8d - 4 = 0$

Objective: *To find the sum and product of the roots of a quadratic equation without solving the equation (Section 9–9)*

Find the sum and product of the roots of each equation.

30. $8t^2 - 5t - 1 = 0$ **31.** $y^2 - 7y = 5$ **32.** $4r^2 - 5r - 1 = 0$

33. Write a quadratic equation having the solution set $\{3 + 7i, 3 - 7i\}$.

Objective: *To use the Number of Roots Theorem and the Rational Root Theorem to find the roots of a polynomial equation (Section 9–10)*

For Exercises 34–36, refer to the polynomial equation $9x^4 - 19x^2 + 2 = 0$.

34. How many roots does the equation have?

35. Identify all rational roots of the equation.

36. Given that $\sqrt{2}$ is a root, find all the remaining roots.

Objective: *To use the Locater Theorem to approximate the real zeros of a polynomial function (Section 9–11)*

37. Approximate the smallest root of $f(x) = x^3 - 7x^2 - 2x + 14$ to the nearest tenth.

Chapter Test

Perform the indicated operations. Express answers in standard form.

1. $3 - (5 + 2i)$ **2.** $(i + 2) + (3 - 4i)$ **3.** $i(4 - i)$

4. $2 \div 5i$ **5.** $2(10 - 12i)(5 + 6i)$ **6.** $(5i - 3) \div \sqrt{-2}$

7. Factor $25z^2 + 100$ completely over the complex numbers.

For Exercises 8–11, find all solutions of the given equations.

8. $z^2 - 4z = 10$ **9.** $3y^2 = 2y - 7$

10. $a^2 - 10a - 20 = 0$ **11.** $(r - 7)^2 = 36$

12. One leg of a right triangle is 3 inches longer than the other leg and 3 inches shorter than the hypotenuse. Find the length of each side.

13. The diagonal of a square field is 14 meters long. How long are the sides?

For each quadratic equation, determine whether there are two distinct real roots, one double root, or two complex roots. (Do not solve the equation.)

14. $3r^2 - 7r = 10$ **15.** $5t^2 - 2t + 3 = 0$

Preparing for College Entrance Tests

Instead of solving a given equation, it is sometimes easier to try each of the given "solutions" to see which one makes the equation true.

REMEMBER: Watch for clues that will enable you to eliminate one or more of the possibilities without extensive calculations.

EXAMPLE Which of the following numbers is in the solution set for $2x^2 - 23x + 30 = 0$?

(a) $-1\frac{1}{2}$ (b) $\frac{1}{2}$ (c) 2 (d) 10

Think: If $2x^2 - 23x + 30 = 0$, then $2x^2 = 23x - 30$.
But $2x^2$ is positive. Therefore, $23x - 30$ must be positive.
Thus, $23x - 30 > 0$, or $23x > 30$ and $x > 1$.
This eliminates choices (a) and (b). Test choices (c) and (d).

Solution: Test (c): $2(2)^2 - 23(2) + 30 = -8 \neq 0$ ← **2 is not a solution.**

Test (d): $2(10)^2 - 23(10) + 30 = 0$ ← **10 is a solution.** Answer: **d**

Choose the best answer. Choose *a*, *b*, *c*, or *d*.

1. Which of the following numbers is in the solution set for $2x^2 + 15x + 18 = 0$?
   (a) -6 (b) -1 (c) $\frac{1}{2}$ (d) 3
2. Which of the following numbers is in the solution set for $3x^2 - 11x + 10 = 0$?
   (a) -2 (b) 2 (c) $-\frac{5}{3}$ (d) 1
3. Which of the following numbers is in the solution set for $8x^2 + 10x - 3 = 0$?
   (a) 2 (b) 1 (c) $-\frac{1}{4}$ (d) $-\frac{3}{2}$
4. Which of the following numbers is in the solution set for $3x^2 - 8x - 3 = 0$?
   (a) -2 (b) -1 (c) $-\frac{1}{2}$ (d) 3
5. If $\dfrac{3}{4 + \dfrac{x+1}{x}} = \dfrac{3}{4}$, what is the value of x? (HINT: $\dfrac{x+1}{x}$ must equal 0.)
   (a) -1 (b) 0 (c) $\frac{1}{4}$ (d) 1
6. If $\dfrac{4}{6 + \dfrac{x-5}{x}} = \dfrac{2}{3}$, what is the value of x?
   (a) -5 (b) $-\frac{1}{5}$ (c) $\frac{1}{5}$ (d) 5
7. If $\dfrac{1}{3} + \dfrac{x+1}{3} = 0$, what is the value of x?
   (a) -3 (b) -2 (c) -1 (d) 0

CHAPTER **10**

Conic Sections

10–1 The Circle

Conic sections are formed by the intersection of a plane and a right circular cone. A **circle** is the conic section formed by the intersection of a right circular cone and a plane perpendicular to the axis of the cone.

A circle can also be defined as a set of points, or a **locus.**

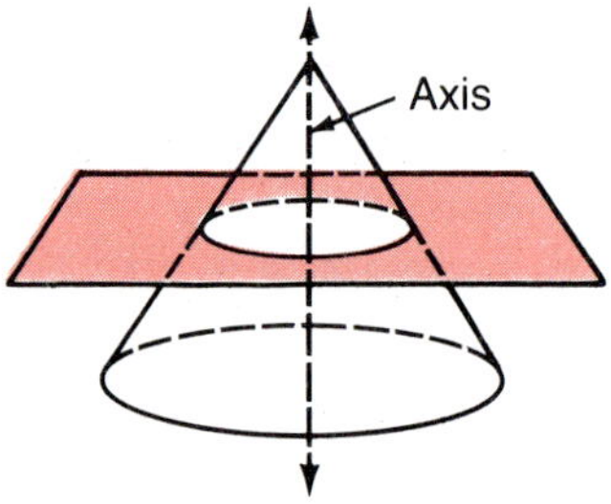

Definitions

> A **circle** is the locus of points in a plane at a given distance from a fixed point.
>
> The **radius,** r, of the circle is the given distance.
>
> The **center,** C, of the circle is the fixed point.

You can use these definitions and the distance formula to write an equation of the circle.

Theorem 10–1

> The equation of a circle with center at (0, 0) and radius, r, where $x, y, r \in R, r \geq 0$, is
>
> $$x^2 + y^2 = r.$$

Proof: Let $P(x, y)$ be any point on the circle at the right.

Then, by the distance formula,

$$OP = r = \sqrt{(x-0)^2 + (y-0)^2}$$

$$r = \sqrt{x^2 + y^2} \quad \leftarrow \text{Square both sides.}$$

$$r^2 = x^2 + y^2, \quad \text{or} \quad x^2 + y^2 = r^2$$

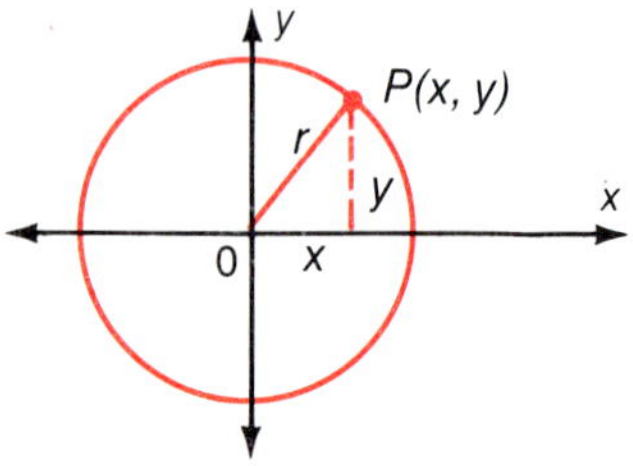

EXAMPLE 1 **a.** Write the equation of a circle with $r = 3$ and center, C, at (0, 0).

b. Graph the circle.

Solutions: **a.** By Theorem 10–2,

$$x^2 + y^2 = r^2 \quad \leftarrow \text{Replace } r \text{ with 3.}$$

$$x^2 + y^2 = 3^2$$

$$x^2 + y^2 = 9$$

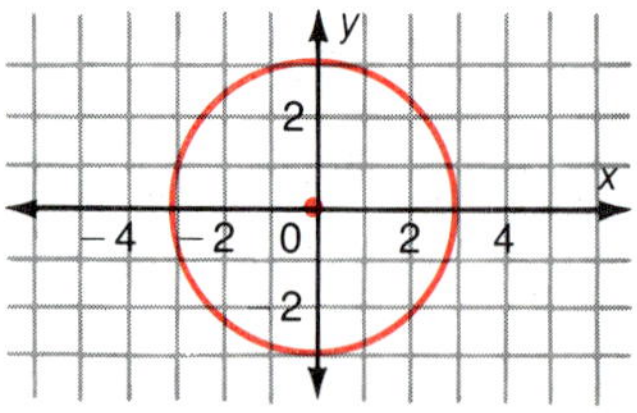

You can follow a similar procedure to find the equation of a circle whose center is *not* the origin.

Theorem 10–2

> The equation of a circle with center at (h, k) and radius, r, where $h, k, r \in R$, $r \geq 0$, is
> $$(x - h)^2 + (y - k)^2 = r^2.$$

You are asked to prove Theorem 10–2 in the Written Exercises.

EXAMPLE 2 **a.** Write an equation for a circle with $r = 5$ and center at (3, 6).
b. Graph the circle.

Solutions: Use Theorem 10–2 with $h = 3$, $k = 6$, and $r = 5$.

a. $(x - h)^2 + (y - k)^2 = r^2$

$(x - 3)^2 + (y - 6)^2 = 5^2$, or

$\mathbf{(x - 3)^2 + (y - 6)^2 = 25}$

b.

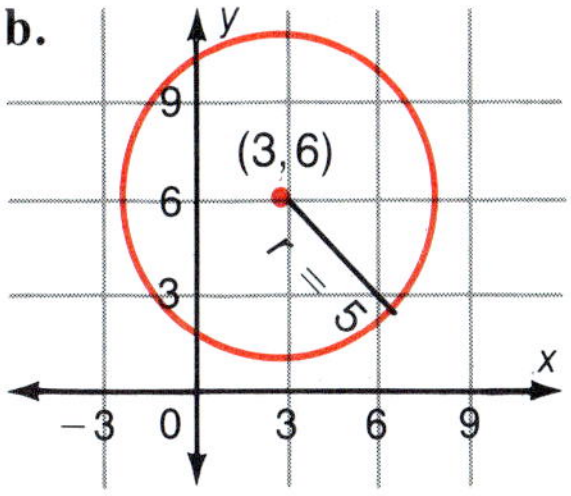

Simplifying $\quad (x - 3)^2 + (y - 6)^2 = 25 \quad$ 1
gives $\quad x^2 - 6x + y^2 - 12y + 20 = 0. \quad$ 2

To find the coordinates of the center and the length of the radius of a circle defined by an equation of the form of Equation 2, write it in the form of Equation 1. To do this, complete the square in x and y.

EXAMPLE 3 Find the center and radius of the circle defined by $x^2 + y^2 + 6x - 8y - 75 = 0$.

Solution: Add 75 to each side of the equation. Group the x terms and the y terms.

$(x^2 + 6x + \quad) + (y^2 - 8y + \quad) = 75$ ← **Now complete the squares on x and y.**

$(x^2 + 6x + 9) + (y^2 - 8y + 16) = 75 + 9 + 16$

$(x + 3)^2 + (y - 4)^2 = 100$ ← **Compare with $(x - h)^2 + (y - k)^2 = r^2$.**

Center: **(−3, 4)** Radius: $\sqrt{100}$, or **10**

CLASSROOM EXERCISES

Write the equation of each circle with the given center and radius.

1. $C(0, 0)$; $r = 8$ **2.** $C(2, -3)$; $r = 2$ **3.** $C(-1, -3)$; $r = \sqrt{2}$

Find the radius and the coordinates of the center of each circle.

4. $x^2 + y^2 = 4$ **5.** $x^2 + y^2 = 25$ **6.** $x^2 + y^2 - \sqrt{40} = 0$

7. $3x^2 + 3y^2 = 9$ **8.** $(x - 2)^2 + (y - 3)^2 = 16$

9. $(x + 4)^2 + (y - 5)^2 = 49$ **10.** $(x - 1)^2 + (y + 3)^2 = 36$

WRITTEN EXERCISES

A In Exercises 1–12:

a. Write an equation for a circle with the given radius and center.
b. Graph the circle.

1. $r = 4; C(0, 0)$
2. $r = 2; C(0, 0)$
3. $r = 1; C(3, 4)$
4. $r = 3; C(2, -3)$
5. $r = 5; C(-3, 2)$
6. $r = 1\frac{1}{2}; C(-5, -2)$
7. $r = 5; C(0, 6)$
8. $r = 6; C(\frac{1}{2}, -\frac{5}{2})$
9. $r = \frac{5}{2}; C(-7, 0)$
10. $r = 3; C(-8, 4)$
11. $r = 4; C(-\frac{3}{2}, \frac{1}{2})$
12. $r = \frac{3}{2}; C(8, 2)$

In Exercises 13–34, find the radius and the coordinates of the center of each circle.

13. $x^2 + y^2 = 36$
14. $x^2 + y^2 = 16$
15. $x^2 + y^2 = 7$
16. $x^2 + y^2 = 5$
17. $(x - 7)^2 + (y - 3)^2 = 4$
18. $(x - 4)^2 + (y + 5)^2 = 25$
19. $(x + 4)^2 + (y - 1)^2 = 8$
20. $(x + 6)^2 + (y + 2)^2 = 5$
21. $(x + \frac{1}{2})^2 + (y - \frac{1}{4})^2 = 9$
22. $(x + \frac{3}{2})^2 + (y + \frac{5}{2})^2 = 4$
23. $(x - \frac{1}{2})^2 + (y + \frac{1}{2})^2 = 6$
24. $(x + \frac{7}{2})^2 + (y - \frac{3}{2})^2 = 12$
25. $x^2 - 2x + y^2 = 0$
26. $x^2 + y^2 + 10y = 24$
27. $x^2 + y^2 + 6x + 8y = 15$
28. $x^2 + y^2 + 6x + 2y + 6 = 0$
29. $x^2 + 8x + y^2 - 4y = 16$
30. $x^2 + 4x + y^2 - 10y + 13 = 0$
31. $x^2 + y^2 - 6x + 8y = 24$
32. $4x^2 + 4y^2 + 16x - 8y = 24$
33. $x^2 + y^2 + 4y - 6 = 0$
34. $x^2 + y^2 + \frac{5}{2}x + 2y = 0$

B In Exercises 35–37, write the equation of a circle with the given center, *C*, and containing the given point, *P*.

35. $C(0, 0); P(3, 4)$
36. $C(2, 5); P(8, -3)$
37. $C(-3, -1); P(0, 4)$

In Exercises 38–40, write the equation of a circle whose diameter has the given endpoints. (HINT: The midpoint of a diameter is the center of a circle.)

38. $P(2, 3)$ and $Q(-6, 7)$
39. $P(-4, 5)$ and $Q(10, -7)$
40. $P(0, 4)$ and $Q(-2, -1)$
41. Prove Theorem 10–2.

C

42. Find an equation for the circle with center at $(1, 2)$ and tangent to the x axis.
43. Find an equation for the circle with center at $(-3, 2)$ and tangent to the y axis.
44. Find an equation for the circle with center at $(4, 4)$ and tangent to both coordinate axes.
45. Find an equation for the circle with a radius of 8 and tangent to both coordinate axes. Assume that the circle is in Quadrant I.
46. Find an equation for the circle with a radius of 5 and tangent to the x axis and to the line $y = x$. (HINT: There are four solutions.)

When the graph of a relation $f(x, y)$ is translated h units along the x axis and k units along the y axis, the equation of the relation in its new position is found by replacing x in the original relation with $x - h$ and by replacing y with $y - k$. Use this information in Exercises 47–48.

47. The circle with equation $x^2 + y^2 = 16$ is translated 4 units to the right and 2 units down. Find the equation of the circle in its new position.

48. The circle with equation $x^2 + y^2 = 9$ is translated 2 units to the right and 1 unit down. Find the equation of the circle in its new position.

49. The general form of a second degree equation in two variables is

$$Ax^2 + Bxy + Cy^2 + Dx + Ey + F = 0.$$

Suppose that $A = C \neq 0$, and $B = 0$. Show that the graph of this equation is a circle, or a point, or that it does not exist in the real number plane. (HINT: Complete the square on x and y.)

CALCULATOR APPLICATIONS

Evaluating Polynomials

Evaluating expressions of the form

$$a_n x^n + a_{n-1}x^{n-1} + a_{n-2}x^{n-2} + \cdots + a_1x^1 + a_0x^0$$

is greatly simplified by using the memory keys on a scientific calculator.

EXAMPLE Evaluate $x^4 - 5x^3 + 8x$ when $x = 2.3$.

SOLUTION The key step in the process is the storing of the x value.

2 . 3 [M+] [x^y] 4 [−] 5 [×] [MR] [x^y] 3 [+] 8 [×] [MR] [=] [-14.4509]

Storing the x value saves you from having to enter the x value each time it occurs in the expression.

EXERCISES

Evaluate each of the following when $x = 4.058$.

1. $6x^2 - 4x + 15$ **2.** $12x^3 + x^2 + 3x$ **3.** $6x^3 - 7x^2 - x + 2$ **4.** $x^4 - x^3 - x^2 - x$

REVIEW CAPSULE FOR SECTION 10–2

Simplify each radical. *(Pages 272–274)*

1. $\sqrt{18}$ **2.** $\sqrt{24}$ **3.** $7\sqrt{18}$ **4.** $2\sqrt{27}$ **5.** $9\sqrt{50}$ **6.** $5\sqrt{128}$

Solve and check. *(Pages 278–279)*

7. $\sqrt{x} = 9$ **8.** $\sqrt{2x} = 8$ **9.** $\sqrt{x - 1} = 4$ **10.** $x = \sqrt{x + 6}$

11. $x = \sqrt{2x - 1}$ **12.** $2x = \sqrt{7x - 3}$ **13.** $\sqrt{x + 4} + \sqrt{x - 3} = 7$ **14.** $\sqrt{x} - 4 = \sqrt{x + 8}$

10–2 The Ellipse

An **ellipse** is the conic section formed by the intersection of a right circular cone and a plane. If the plane that intersects the cone to form a circle is tilted (as shown in the figure), the *closed curve* formed by this intersection is an *ellipse*.

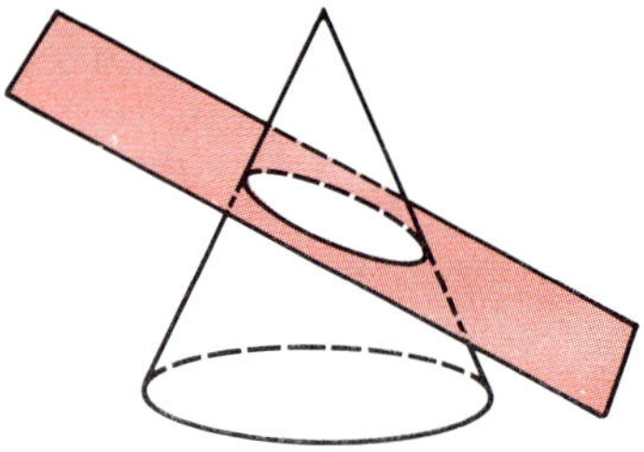

Figure 1

An ellipse can also be defined as a locus.

Definitions

> An **ellipse** is the locus of all points in a plane such that the sum of the distances from two fixed points to a point on the locus is constant.
> Each of the fixed points is a **focus** (plural: **foci**).

By using the definition of an ellipse and the distance formula, you can write the equation of an ellipse.

Consider an ellipse whose foci are $F(-4, 0)$ and $F'(4, 0)$ where

$$FP + F'P = 10.$$

Let $P(x, y)$ be any point on the ellipse. The figure at the right shows four different locations of P. In each case, $FP + F'P = 10$. From the distance formula,

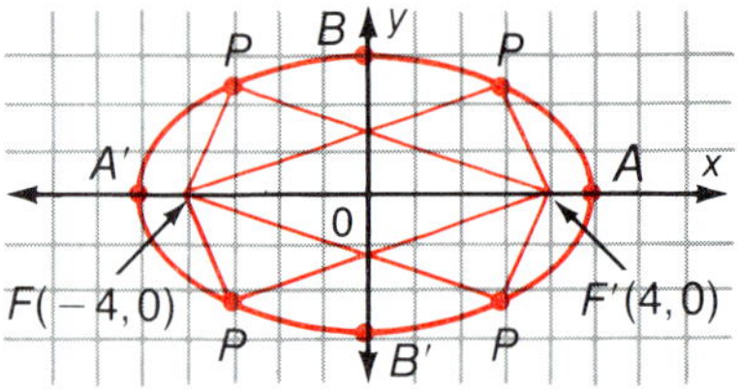

Figure 2

$$FP = \sqrt{(x+4)^2 + (y-0)^2} \text{ and } F'P = \sqrt{(x-4)^2 + (y-0)^2}.$$

$\sqrt{(x+4)^2 + y^2} + \sqrt{(x-4)^2 + y^2} = 10$ ← $FP + F'P = 10$

$\sqrt{(x-4)^2 + y^2} = 10 - \sqrt{(x+4)^2 + y^2}$ ← **Square both sides.**

$(x-4)^2 + y^2 = 100 - 20\sqrt{(x+4)^2 + y^2} + (x+4)^2 + y^2$

$x^2 - 8x + 16 + y^2 = 100 - 20\sqrt{(x+4)^2 + y^2} + x^2 + 8x + 16 + y^2$

$16x + 100 = 20\sqrt{(x+4)^2 + y^2}$ ← **Divide each side by 4.**

$4x + 25 = 5\sqrt{(x+4)^2 + y^2}$ ← **Square both sides.**

$16x^2 + 200x + 625 = 25x^2 + 200x + 400 + 25y^2$

$9x^2 + 25y^2 = 225$ ← **Divide each term by 225.**

$$\frac{x^2}{25} + \frac{y^2}{9} = 1$$

In Figure 2, $\overline{AA'}$ and $\overline{BB'}$ are **axes** of the ellipse. $\overline{AA'}$ is the **major axis** and is longer than $\overline{BB'}$, the **minor axis.** The point at which the major and minor axes intersect is the **center** of the ellipse.

Equation in Standard Form

Equation of an Ellipse

The standard form for the equation of an ellipse with center at (0, 0), major axis of length $2a$ (along the x axis), and minor axis of length $2b$, is

$$\frac{x^2}{a^2} + \frac{y^2}{b^2} = 1$$

where a is the length of the **semi-major axis** and b is the length of the **semi-minor axis.**

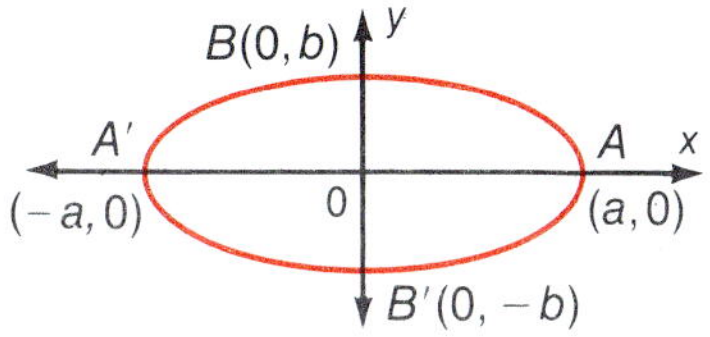

Notice that the x intercepts of the graph of an ellipse with center at (0, 0) and major axis along the x axis are a and $-a$; the y intercepts are b and $-b$.

EXAMPLE 1 Graph the ellipse $\frac{x^2}{16} + \frac{y^2}{9} = 1$.

Solution: From the equation, $a = 4$ and $b = 3$. Thus, the points $(\pm 4, 0)$ and $(0, \pm 3)$ are on the graph.

Since $a = 4$, the length of the semi-major axis is 4, and the set of possible values of x must be $\{-4 \le x \le 4\}$. Choose integral values of x and make a table.

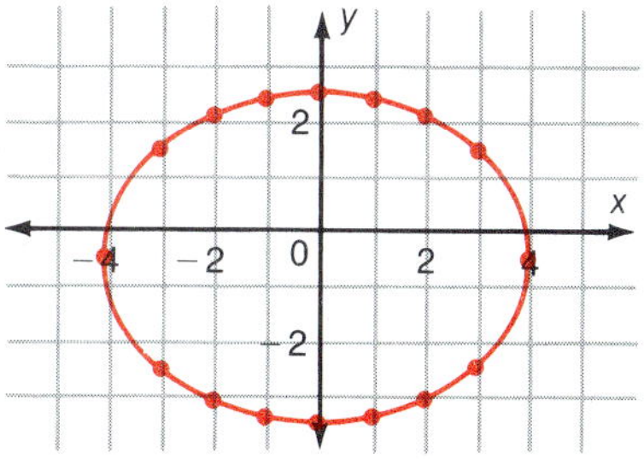

x	−4	−3	−2	−1	0	1	2	3	4
y	0	±2.0	±2.6	±2.9	±3.0	±2.9	±2.6	±2.0	0

The major axis of an ellipse can be on the y axis rather than on the x axis. REMEMBER: The major axis is *always the longer of the two axes.*

Equation in Standard Form

Equation of an Ellipse

The standard form for the equation of an ellipse with center at (0, 0), major axis of length $2a$ (along the y axis), and minor axis of length $2b$, is

$$\frac{x^2}{b^2} + \frac{y^2}{a^2} = 1$$

where a is the length of the semi-major axis and b is the length of the semi-minor axis.

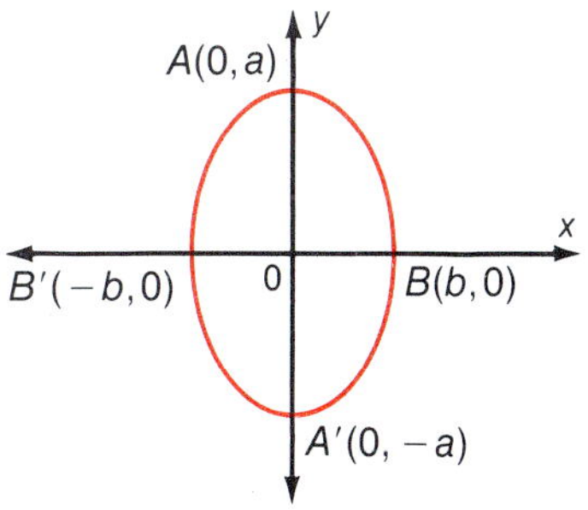

Notice that the x intercepts of the graph of an ellipse with center at (0, 0) and major axis along the y axis are b and $-b$; the y intercepts are a and $-a$.

EXAMPLE 2 Determine the x and y intercepts of the graph of each ellipse. Then tell whether the major axis is horizontal or vertical and find its length.

Solutions: Write the equations in standard form.

a.		b.
$16x^2 + 25y^2 = 400$		$49x^2 + 36y^2 = 1764$
$\frac{16x^2}{400} + \frac{25y^2}{400} = \frac{400}{400}$		$\frac{49x^2}{1764} + \frac{36y^2}{1764} = \frac{1764}{1764}$
$\frac{x^2}{25} + \frac{y^2}{16} = 1$	← **Standard form** →	$\frac{x^2}{36} + \frac{y^2}{49} = 1$
$a = 5,\ b = 4$		$a = 7,\ b = 6$
−5 and **5**	← ***x* intercepts** →	**−6** and **6**
−4 and **4**	← ***y* intercepts** →	**−7** and **7**
Major axis:	**Horizontal**	**Vertical**
Length of the major axis: **10**	← **2*a*** →	**14**

It can be shown that

$$c^2 = a^2 - b^2$$

where the coordinates of the foci are $(\pm c, 0)$ and the major axis is horizontal. When the major axis is vertical, the coordinates of the foci are $(0, \pm c)$. You are asked to prove this in the Written Exercises.

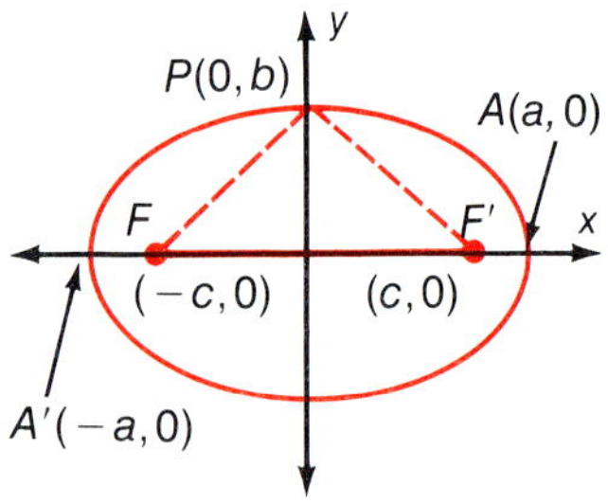

EXAMPLE 3 Determine the coordinates of the foci of the two ellipses in Example 2.

Solutions:

a. $\frac{x^2}{25} + \frac{y^2}{16} = 1$ ← $a = 5$; $b = 4$

$c^2 = a^2 - b^2$

$c^2 = 25 - 16$

$c^2 = 9$

$c = \pm 3$

Foci: **$(\pm 3, 0)$**

b. $\frac{x^2}{36} + \frac{y^2}{49} = 1$ ← $a = 7$; $b = 6$

$c^2 = a^2 - b^2$

$c^2 = 49 - 36$

$c^2 = 13$

$c = \pm\sqrt{13}$

Foci: **$(0, \pm\sqrt{13})$**

Summary

For either ellipse

$$\frac{x^2}{a^2} + \frac{y^2}{b^2} = 1 \quad \text{or} \quad \frac{x^2}{b^2} + \frac{y^2}{a^2} = 1,$$

(major axis along the x axis) (major axis along the y axis)

1. a is one-half the length of the major axis;
2. b is one-half the length of the minor axis;
3. c is the distance from each focus to the center of the ellipse and $c^2 = a^2 - b^2$.

CLASSROOM EXERCISES

Write the equation of each ellipse in standard form.

1. $x^2 + 4y^2 = 100$ **2.** $36x^2 + 4y^2 = 144$ **3.** $3x^2 + 2y^2 = 22$ **4.** $4x^2 + 5y^2 = 24$

Determine the x and y intercepts of each ellipse.

5. $\frac{x^2}{64} + \frac{y^2}{49} = 1$ **6.** $\frac{x^2}{144} + \frac{y^2}{169} = 1$ **7.** $9x^2 + 8y^2 = 72$ **8.** $5x^2 + 4y^2 = 20$

Find the length of the major and minor axes of each ellipse.

9. $x^2 + 4y^2 = 4$ **10.** $9x^2 + y^2 = 36$ **11.** $36x^2 + 4y^2 = 72$ **12.** $4x^2 + 8y^2 = 16$

WRITTEN EXERCISES

A In Exercises 1–13, sketch the graph of each ellipse.

1. $\frac{x^2}{9} + \frac{y^2}{4} = 1$ **2.** $\frac{x^2}{36} + \frac{y^2}{16} = 1$ **3.** $\frac{x^2}{4} + \frac{y^2}{16} = 1$ **4.** $\frac{x^2}{49} + \frac{y^2}{25} = 1$

5. $x^2 + 4y^2 = 100$ **6.** $9x^2 + y^2 = 81$ **7.** $4x^2 + 9y^2 = 36$

8. $4x^2 + 36y^2 = 144$ **9.** $16x^2 + 4y^2 = 64$ **10.** $25x^2 + 9y^2 = 225$

11. $x^2 + 9y^2 - 225 = 0$ **12.** $25x^2 + 16y^2 - 100 = 0$ **13.** $9x^2 + y^2 - 225 = 0$

For each ellipse in Exercises 14–23:

a. Determine the x and y intercepts of the graph.

b. Determine whether the major axis is horizontal or vertical.

c. Give the length of the major axis.

14. $\frac{x^2}{36} + \frac{y^2}{25} = 1$ **15.** $\frac{x^2}{64} + \frac{y^2}{81} = 1$ **16.** $\frac{x^2}{4} + \frac{y^2}{25} = 1$ **17.** $\frac{x^2}{100} + \frac{y^2}{81} = 1$

18. $x^2 + y^2 = 15$ **19.** $x^2 + 25y^2 = 100$ **20.** $4x^2 + y^2 = 16$

21. $2x^2 + y^2 = 8$ **22.** $6x^2 + 4y^2 = 36$ **23.** $16x^2 + 9y^2 = 144$

In Exercises 24–30, find the coordinates of the foci for each ellipse.

24. $x^2 + \frac{y^2}{4} = 1$ **25.** $\frac{x^2}{169} + \frac{y^2}{144} = 1$ **26.** $\frac{x^2}{121} + \frac{y^2}{64} = 1$ **27.** $\frac{x^2}{25} + y^2 = 1$

28. $4x^2 + y^2 = 16$ **29.** $4x^2 + 16y^2 = 16$ **30.** $x^2 + 9y^2 - 36 = 0$

B Given the following information, write an equation for each ellipse.

31. x intercepts: ± 5
y intercepts: ± 3

32. x intercepts: ± 2
y intercepts: ± 4

33. x intercepts: ± 10
y intercepts: $\pm \sqrt{5}$

34. x intercepts: $\pm \sqrt{7}$
y intercepts: $\pm \sqrt{3}$

35. Foci: $(\pm 4, 0)$
y intercepts: ± 3

36. Foci: $(0, \pm 6)$
x intercepts: ± 8

37. Foci at $(0, \pm 7)$; one endpoint of major axis at $(0, 9)$.

38. Major axis is horizontal and its length is 8; length of minor axis is 6; major and minor axes intersect at $(0, 0)$.

39. Major axis is vertical and its length is 12; length of minor axis is 4; major and minor axes intersect at $(0, 0)$.

The equations of an ellipse with center at (h, k) are

$$\frac{(x-h)^2}{a^2} + \frac{(y-k)^2}{b^2} = 1 \quad \text{and} \quad \frac{(x-h)^2}{b^2} + \frac{(y-k)^2}{a^2} = 1.$$

In Exercises 40–43, find the equation of the ellipse with the given center and given values for a and b.

40. $C(3, 6)$; $a = 4$, $b = 3$; major axis is horizontal

41. $C(-5, -6)$; $a = 9$, $b = 7$; major axis is vertical

42. $C(-4, 4)$; $a = 5$, $b = \sqrt{7}$; major axis is vertical

43. $C(-3, -1)$; $a = 8$, $b = \sqrt{5}$; major axis is horizontal

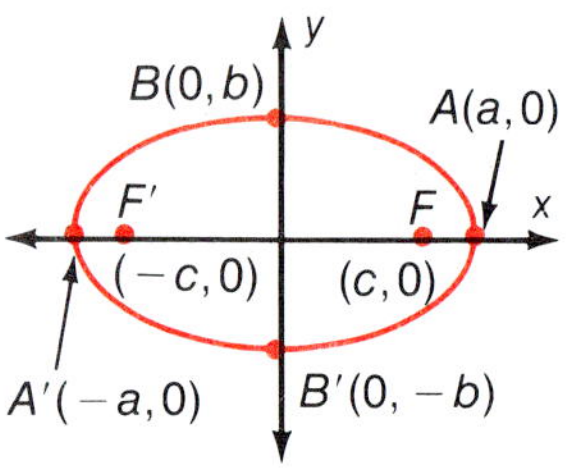

C

44. Use the figure at the right above to help you prove that

$$c^2 = a^2 - b^2$$

where the coordinates of the foci are $(\pm c, 0)$.

45. The ellipse with equation $25x^2 + 16y^2 - 400 = 0$ is translated eight units to the right and four units down. Write the equation for the ellipse in its new position.

46. The ellipse in the equation $25x^2 + 16y^2 - 400 = 0$ is translated 3 units to the left and 2 units down. Write the equation for the ellipse in its new position.

47. The ellipse in the equation $25x^2 + 16y^2 - 400 = 0$ is translated 2 units to the left and 3 units up. Write the equation for the ellipse in its new position.

REVIEW CAPSULE FOR SECTION 10–3

Find the distance between each pair of points. *(Pages 280–283)*

1. $P(2, 3)$ and $Q(2, 5)$
2. $P(2, 3)$ and $Q(-8, 3)$
3. $P(2, 3)$ and $Q(3, 2)$
4. $P(2, -3)$ and $Q(2, -8)$
5. $P(x, 8)$ and $Q(-4, 8)$
6. $P(x, 8)$ and $Q(4, 0)$

Complete the square to form a perfect square trinomial. Express the trinomial as the square of a binomial. *(Pages 329–331)*

7. $x^2 - 12x$
8. $y^2 + 10y$
9. $r^2 - 7r$
10. $t^2 + 11t$

Solve each equation by completing the square. Check your solutions. *(Pages 329–331)*

11. $x^2 - 2x - 3 = 0$
12. $r^2 - 6r = -2$
13. $z^2 + 4z = 16$
14. $10b = 24 - b^2$
15. $2y^2 = 3y - 2$
16. $3x^2 + 5x = 1$
17. $3t^2 - 2t - 6 = 0$
18. $2a^2 - 7a = -6$

10–3 The Parabola

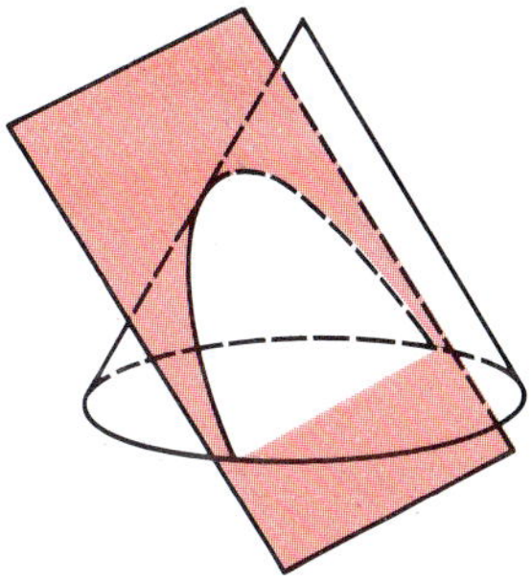

The **parabola** is the conic section formed by the intersection of a right circular cone and a plane. If the plane that intersects the cone to form an ellipse is tilted (as shown in the figure), the *open curve* formed is a parabola.

A parabola can also be defined as a locus.

Definitions

> A **parabola** is the locus of points in a plane that are the same distance from a given line and a fixed point not on the line.
>
> The **directrix** is the given line.
>
> The **focus** of the parabola is the fixed point.

You can use the definition of a parabola and the distance formula to find its equation.

EXAMPLE 1 The focus of the parabola at the right is at $F(3, 5)$. The equation of the directrix is $y = 1$. Find the equation of the parabola.

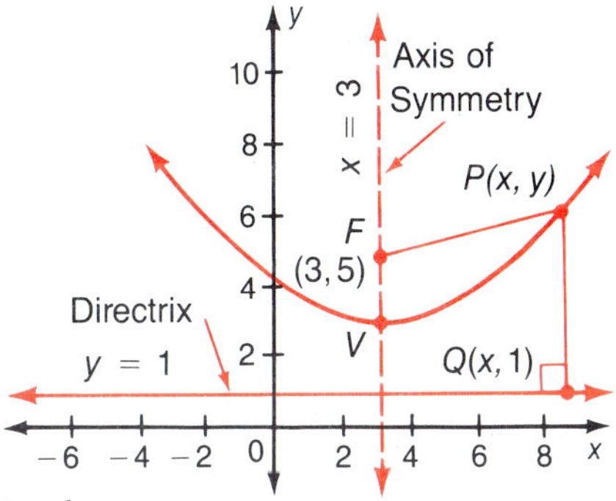

Solution: Let $P(x, y)$ be any point on the parabola. Use the distance formula to find PF and PQ.

$$PF = PQ$$

$$\sqrt{(x-3)^2 + (y-5)^2} = \sqrt{(x-x)^2 + (y-1)^2}$$

$$(x-3)^2 + (y-5)^2 = (y-1)^2 \quad \leftarrow \text{Solve for } y.$$

$$(x-3)^2 + y^2 - 10y + 25 = y^2 - 2y + 1$$

$$(x-3)^2 + 24 = 8y \quad \leftarrow \text{Divide each term by 8.}$$

$$\frac{1}{8}(x-3)^2 + 3 = y \quad \text{or} \quad y = \frac{1}{8}(x-3)^2 + 3$$

As the graph in Example 1 shows, the **axis of symmetry** of the parabola with equation $y = \frac{1}{8}(x - 3)^2 + 3$ is $x = 3$.

The **vertex** of the parabola is at **(3, 3)**.

The equation for a parabola can be written in one of these forms.

$$y = a(x - h)^2 + k \qquad \text{or} \qquad x = a(y - k)^2 + h$$

Summary

Parabola Facts		
Equation	$y = a(x-h)^2 + k$	$x = a(y-k)^2 + h$
Axis of Symmetry	$x = h$	$y = k$
Vertex	(h, k)	(h, k)
Focus	$\left(h, k + \frac{1}{4a}\right)$	$\left(h + \frac{1}{4a}, k\right)$
Directrix	$y = k - \frac{1}{4a}$	$x = h - \frac{1}{4a}$
$a > 0$	Opens upward.	Opens to the right.
$a < 0$	Opens downward.	Opens to the left.

Sometimes the equation of a parabola is written in the form $y = ax^2 + bx + c$ or $x = ay^2 + by + c$. To sketch the graph of the parabola defined by one of these equations, first complete the square on the second-degree term. Then compare the equation with the corresponding equation form in the table.

EXAMPLE 2 For the parabola defined by $x = \frac{1}{4}y^2 + \frac{1}{2}y + \frac{17}{4}$:

a. Find the coordinates of the vertex.

b. Find the equation of the axis of symmetry.

c. Find the coordinates of the focus.

d. Find the equation of the directrix.

e. Sketch the graph.

Solutions: Complete the square on y.

$$x = \frac{1}{4}(y^2 + 2y) + \frac{17}{4}$$

$$x = \frac{1}{4}(y^2 + 2y + 1) + \frac{17}{4} - \frac{1}{4}$$

$$x = \frac{1}{4}(y + 1)^2 + 4$$

a. Vertex: **(4, −1)** ← **(h, k)**

b. Axis of symmetry: $y = -1$ ← $y = k$

c. Focus: $\left(4 + \frac{1}{1}, -1\right)$, or **(5, −1)**

d. Directrix: $x = 4 - 1$, or $x = 3$

e. Since $a > 0$, the graph **opens to the right.** A sketch of the graph is shown.

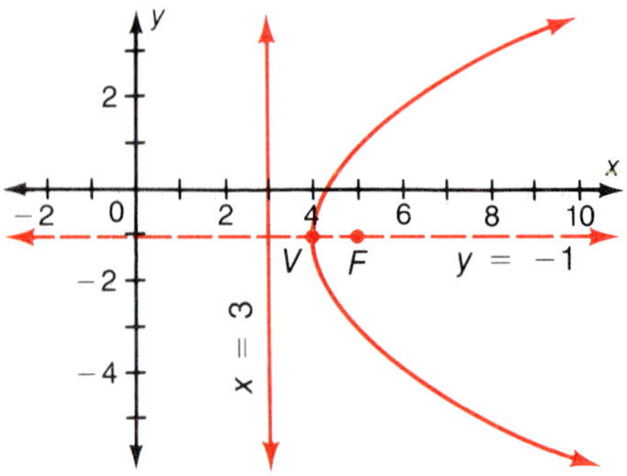

CLASSROOM EXERCISES

Identify each equation as that of a circle, an ellipse, or a parabola.

1. $x^2 + y^2 = 100$
2. $y = x^2 + 3$
3. $3x^2 + 4y^2 = 5$
4. $y^2 = 2x + 3$

Write each equation in the form $y = a(x - h)^2 + k$.

5. $y = x^2 + 2x + 6$
6. $y = x^2 - 6x + 7$
7. $x^2 = 5y$
8. $-3y = x^2$
9. $y = 3x^2 - 24x + 50$
10. $y = 7x^2 + 56x + 116$

Write each equation in the form $x = a(y - k)^2 + h$.

11. $x = y^2 - 6y$
12. $x = y^2 - 4y + 1$
13. $4x = y^2$
14. $x = y^2 + 2y + 11$
15. $x = 3y^2 + 6y + 2$
16. $x = 5y^2 - 20x + 22$

WRITTEN EXERCISES

A Find an equation for each parabola given the following information.

1. Focus: $(2, 0)$; directrix: $x = -2$
2. Focus: $(-3, 0)$; directrix: $x = 3$
3. Focus: $(0, -4)$; directrix: $y = 4$
4. Focus: $(0, 6)$; directrix: $y = -6$
5. Focus: $(5, 0)$; directrix: $x = -6$
6. Focus: $(-7, 0)$; directrix: $x = 2$
7. Focus: $(3, 0)$; directrix: $y = 5$
8. Focus: $(5, 4)$; directrix: $x = -6$

For each parabola in Exercises 9–28:

a. Find the coordinates of the vertex.
b. Find the equation of the axis of symmetry.
c. Find the coordinates of the focus.
d. Find the equation of the directrix.
e. Sketch the graph.

9. $x = \frac{1}{4}y^2$
10. $x = \frac{1}{8}y^2$
11. $y = \frac{1}{16}x^2$
12. $y = -\frac{1}{36}x^2$
13. $x = -\frac{1}{2}y^2$
14. $x^2 = 12y$
15. $y^2 = -8x$
16. $x^2 = 24y$
17. $y = (x - 2)^2 + 4$
18. $y = -(x - 1)^2 - 4$
19. $x = (y + 3)^2$
20. $x = -2(y - 1)^2$
21. $y = \frac{1}{6}(x + 2)^2$
22. $x = -8(y + 2)^2 + 3$
23. $x = y^2 + 6y + 9$
24. $y = x^2 + 8x - 9$
25. $y^2 + 4x = 8$
26. $x^2 - 8y = 4$
27. $y = 2x^2 - 12x + 14$
28. $x = y^2 - 6y$

B Find an equation for each parabola given the following information. Write the equation in the form $y = ax^2 + bx + c$ or $x = ay^2 + by + c$.

29. Vertex: $(-7, 3)$; focus: $(-3, 3)$
30. Vertex: $(-4, -6)$; focus: $(-4, -3)$
31. Focus: $(2, \frac{3}{2})$; directrix: $y = \frac{1}{2}$
32. Focus: $(\frac{3}{4}, 3)$; directrix: $x = \frac{5}{4}$
33. Vertex: $(\frac{5}{4}, -\frac{3}{2})$; focus: $(1, -\frac{3}{2})$
34. Vertex: $(-2, 3)$; focus: $(-2, 4)$
35. Vertex: $(-3, 1)$; focus: $(0, 1)$
36. Vertex: $(1, -3)$; focus: $(1, 0)$

The equation of a parabola can also be written in one of these forms.

$$y - k = a(x - h)^2 \qquad \text{or} \qquad x - h = a(y - k)^2$$

In Exercises 37–42, find the coordinates of the vertex and of the focus of each parabola. Give the equation of the axis of symmetry and of the directrix.

37. $(y - 2) = \frac{1}{8}(x - 3)^2$ **38.** $(x - 3) = -\frac{1}{8}(y + 2)^2$ **39.** $(y + 2) = -8(x - 3)^2$

40. $(x + 12) = \frac{1}{3}(y - 3)^2$ **41.** $(y - 9) = -4(x - \frac{5}{2})^2$ **42.** $(x + \frac{1}{2}) = \frac{1}{2}(y + 3)^2$

C The focus, F, of a parabola has coordinates $(p, 0)$. The equation of the directrix is $x = p$. Use this information in Exercises 43–44.

43. Show that, for $p > 0$, the equation of the parabola is $y^2 = 4px$ and that the parabola opens to the right.

44. Show that, for $p < 0$, the equation of the parabola is $y^2 = 4px$ and that the parabola opens to the left.

45. The parabola defined by $x = \frac{1}{2}y^2$ is translated 3 units to the left and 5 units up. Find the new equation.

46. The parabola defined by $y^2 = 4px$ is translated 5 units to the right and 2 units down. Find the new equation.

The equations of a parabola with vertex at (h, k) can also be written as follows.

$$y - k = \frac{1}{4p}(x - h)^2 \quad \longleftarrow \quad$$ **Focus at $(h, k + p)$**
Equation of directrix: $y = k - p$

$$x - h = \frac{1}{4p}(y - k)^2 \quad \longleftarrow \quad$$ **Focus at $(h + p, k)$**
Equation of directrix: $x = h - p$

In Exercises 47–50, use these general forms to find the vertex, focus, and directrix of each parabola. Then graph the parabola.

47. $x^2 - 4x + 4y + 12 = 0$ **48.** $y^2 + 4y - 2x + 2 = 0$

49. $2x^2 + 8x + y - 4 = 0$ **50.** $3y^2 + 6y + 2x - 5 = 0$

51. Show that the parabola with focus $(h, k + p)$ and directrix with equation $y = k - p$ passes through the points $P(h + 2p, k + p)$ and $Q(h - 2p, k + p)$.

52. Show that the line segment joining points P and Q in Exercise 51 has length $4|p|$.

REVIEW CAPSULE FOR SECTION 10–4

Write each equation in the form $\frac{x^2}{a^2} - \frac{y^2}{b^2} = 1$ by dividing each term by the constant on the right side of the equation. *(Pages 18–21)*

1. $4x^2 - 9y^2 = 36$ **2.** $4x^2 - 25y^2 = 100$ **3.** $x^2 - 4y^2 = 36$

4. $x^2 - y^2 = 25$ **5.** $2x^2 - y^2 = 20$ **6.** $x^2 - 25y^2 = 25$

7. $25x^2 - 36y^2 = 900$ **8.** $4x^2 - 20y^2 = 100$ **9.** $36x^2 - 4y^2 = 144$

10–4 The Hyperbola

The **hyperbola** is the conic section formed by the intersection of a right circular cone and a plane. If the plane that intersects the cone to form a parabola is tilted so that it is parallel to the axis of the cone, the open curve formed is a *branch of a hyperbola.*

A hyperbola can also be defined as a locus.

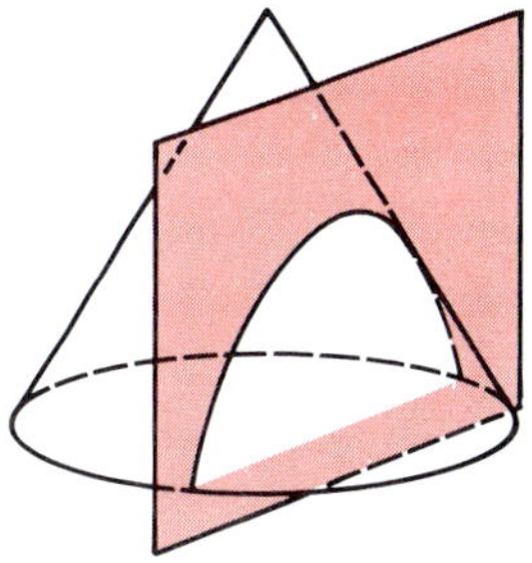

Figure 1

Definitions

> A **hyperbola** is the locus of all points in a plane such that the difference of the distances from two fixed points to a point of the locus is a constant.
>
> The **foci** are the fixed points.

You can use the definition of a hyperbola and the distance formula to find its equation. For example, the hyperbola at the right has foci at $F(0, 5)$ and $F'(0, -5)$.

Let $P(x, y)$ be a point on the parabola as shown. Then, since P is farther from F than from F', and

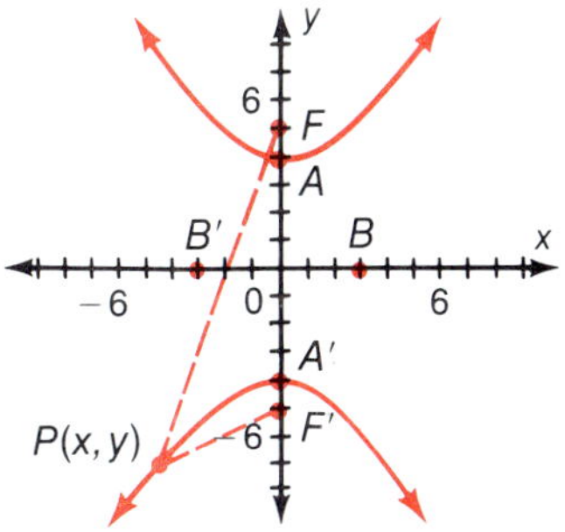

Figure 2

$$PF \quad - \quad PF' \quad = 8,$$

$$\sqrt{x^2 + (y-5)^2} - \sqrt{x^2 + (y+5)^2} = 8, \text{ or}$$

$$\sqrt{x^2 + (y-5)^2} = 8 + \sqrt{x^2 + (y+5)^2}.$$ ← **Square both sides.**

$$x^2 + (y-5)^2 = 64 + 16\sqrt{x^2 + (y+5)^2} + x^2 + (y+5)^2$$ ← **Simplify.**

$$-5y - 16 = 4\sqrt{x^2 + (y+5)^2}$$ ← **Square both sides.**

$$25y^2 + 160y + 256 = 16(x^2 + y^2 + 10y + 25)$$

$$\mathbf{9y^2 - 16x^2 = 144}, \quad \text{or} \quad \mathbf{\frac{y^2}{16} - \frac{x^2}{9} = 1}$$

In Figure 2, the **transverse axis,** $\overline{AA'}$, is 8 units long. The **conjugate axis,** $\overline{BB'}$, is 6 units long. The **center** of the hyperbola is at $(0, 0)$. The distance c from each focus to the center is $\sqrt{4^2 + 3^2} = 5$.

In general, the equation of a hyperbola can be written as

$\mathbf{\frac{y^2}{a^2} - \frac{x^2}{b^2} = 1}$**,** where the foci are at $F(0, c)$ and $F'(0, -c)$, or

$\mathbf{\frac{x^2}{a^2} - \frac{y^2}{b^2} = 1}$**,** where the foci are at $F(c, 0)$ and $F'(-c, 0)$.

Summary

Hyperbola Facts

For either hyperbola

$$\frac{y^2}{a^2} - \frac{x^2}{b^2} = 1 \qquad \text{or} \qquad \frac{x^2}{a^2} - \frac{y^2}{b^2} = 1$$

1. a is one-half the length of the transverse axis;
2. b is one-half the length of the conjugate axis;
3. c is the distance from each focus to the center and

$$c = \sqrt{a^2 + b^2}.$$

NOTE: For a hyperbola, the transverse axis is *not always longer* than the conjugate axis. Also, the foci of the hyperbola *always* lie on the line containing the transverse axis.

EXAMPLE 1 **a.** Sketch the graph of $\frac{x^2}{36} - \frac{y^2}{4} = 1$.

b. Give the coordinates of the foci.

Solutions: For this hyperbola, $a = 6$ and $b = 2$.

a. Draw the rectangle determined by the lines $x = 6$, $x = -6$, $y = 2$, and $y = -2$. Draw the diagonals of the rectangle and extend them.

Sketch the hyperbola. Since the transverse axis is along the x axis, the graph opens to the right and to the left. The branches of the graph come very close to the extended diagonals, but never intersect them.

The graph contains the points A and A'. These are **vertices** or **turning points.**

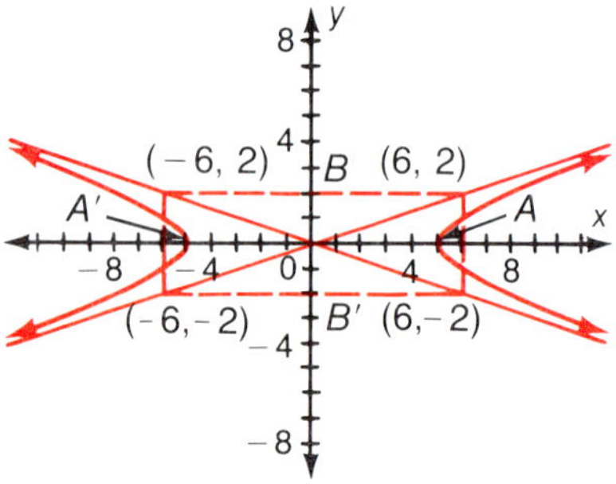

b. The coordinates of the foci are $F(2\sqrt{10}, 0)$ and $F'(-2\sqrt{10}, 0)$.

The extended diagonals are the **asymptotes** of the hyperbola. A hyperbola is asymptotic to (comes very close to, but never intersects) these two lines.

Hyperbola	Asymptotes
$\frac{x^2}{a^2} - \frac{y^2}{b^2} = 1$	$y = \frac{b}{a}x$ and $y = -\frac{b}{a}x$
$\frac{y^2}{a^2} - \frac{x^2}{a^2} = 1$	$y = \frac{a}{b}x$ and $y = -\frac{a}{b}x$

In Example 1, $a = 6$ and $b = 2$. Thus, the equations of the asymptotes are

$$y = \frac{2}{6}x \text{ or } y = \frac{1}{3}x \qquad \text{and} \qquad y = -\frac{2}{6}x \text{ or } y = -\frac{1}{3}x.$$

EXAMPLE 2 Find the center, vertices, foci, and asymptotes of the hyperbola defined by $16y^2 - 9x^2 = 144$.

Solution: Write the equation in the form $\frac{y^2}{a^2} - \frac{x^2}{b^2} = 1$.

$16y^2 - 9x^2 = 144$ ← **Divide each side by 144.**

$$\frac{y^2}{9} - \frac{x^2}{16} = 1$$

From the equation, $a = 3$, and $b = 4$.

$c = \sqrt{a^2 + v^2}$

$c = \sqrt{9 + 16}$

$c = 5$

Center: $C(0, 0)$

Vertices: $V(0, 3)$, $V'(0, -3)$

Foci: $F(0, 5)$, $F'(0, -5)$

Asymptotes: $y = \frac{3}{4}x$ and $y = -\frac{3}{4}x$.

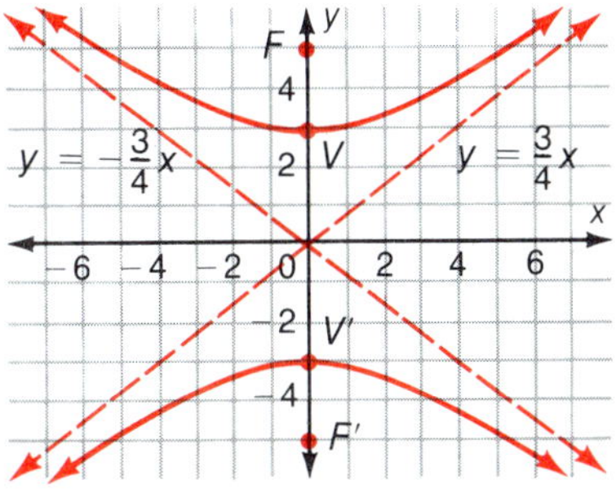

CLASSROOM EXERCISES

For each hyperbola in Exercises 1–12:

a. Give the coordinates of the foci.

b. Give the equations of the asymptotes.

1. $\frac{x^2}{9} - \frac{y^2}{16} = 1$ **2.** $\frac{x^2}{16} - \frac{y^2}{9} = 1$ **3.** $\frac{x^2}{4} - \frac{y^2}{4} = 1$ **4.** $\frac{y^2}{16} - \frac{x^2}{9} = 1$

5. $4y^2 - 25x^2 = 100$ **6.** $25x^2 - 4y^2 = 100$ **7.** $4x^2 - y^2 = 16$ **8.** $2y^2 - 3x^2 = 6$

9. $16x^2 - 9y^2 = 144$ **10.** $9x^2 - 16y^2 = 144$ **11.** $16y^2 - 9x^2 = 144$ **12.** $9x^2 - 4y^2 = 36$

WRITTEN EXERCISES

A

In Exercises 1–4:

a. Draw the rectangle whose diagonals are the asymptotes of the hyperbola and draw the asymptotes.

b. Sketch the hyperbola.

1. $\frac{x^2}{9} - \frac{y^2}{16} = 1$ **2.** $\frac{x^2}{16} - \frac{y^2}{9} = 1$ **3.** $\frac{x^2}{4} - \frac{y^2}{2} = 1$ **4.** $\frac{y^2}{16} - \frac{x^2}{9} = 1$

Graph each hyperbola. Write the equations of the asymptotes.

5. $\frac{x^2}{16} - \frac{y^2}{4} = 1$ **6.** $\frac{x^2}{25} - \frac{y^2}{9} = 1$ **7.** $\frac{y^2}{49} - \frac{x^2}{36} = 1$ **8.** $\frac{y^2}{64} - \frac{x^2}{25} = 1$

9. $9y^2 - x^2 = 81$

10. $x^2 - 4y^2 = 36$

11. $25x^2 - 4y^2 = 100$

12. $4y^2 - 9x^2 = 36$

13. $x^2 - y^2 = 9$

14. $x^2 - y^2 = 25$

15. $4x^2 - 16y^2 = 64$

16. $2y^2 - 8x^2 = 32$

17. $\frac{x^2}{9} - \frac{y^2}{16} = 1$

18. $\frac{y^2}{4} - \frac{x^2}{64} = 1$

19. $x^2 - 4y^2 = 16$

20. $9y^2 - x^2 = 9$

For each hyperbola in Exercises 21–32:

a. Give the coordinates of the vertices.

b. Give the coordinates of the foci.

c. Give the equations of the asymptotes.

21. $\frac{x^2}{16} - \frac{y^2}{16} = 1$

22. $\frac{x^2}{9} - \frac{y^2}{4} = 1$

23. $\frac{y^2}{4} - \frac{x^2}{9} = 1$

24. $\frac{y^2}{9} - \frac{x^2}{4} = 1$

25. $\frac{y^2}{16} - \frac{x^2}{16} = 1$

26. $\frac{y^2}{20} - \frac{x^2}{5} = 1$

27. $\frac{x^2}{1} - \frac{y^2}{16} = 1$

28. $\frac{y^2}{4} - x^2 = 1$

29. $y^2 - x^2 = 4$

30. $x^2 - 9y^2 = 36$

31. $4y^2 - x^2 = 16$

32. $25x^2 - y^2 = 25$

B Given the following information, write the equation of each hyperbola.

33. Center at $(0, 0)$; $a = 2$, $b = 5$; foci on the x axis

34. Center at $(0, 0)$; $a = 4$, $b = 4$; foci on the y axis

35. Center at $(0, 0)$; $a = 3$; $c = 5$; foci on the y axis

36. Center at $(0, 0)$; $b = 6$, $c = 10$; foci on the x axis

37. Foci: $F(10, 0)$; $F'(-10, 0)$; vertices: $V(6, 0)$, $V'(-6, 0)$

38. Foci: $F(0, 5)$, $F'(0, -5)$; vertices: $V(0, 4)$, $V'(0, -4)$

39. Vertices: $V(3, 0)$, $V'(-3, 0)$; asymptotes: $x = 3y$, $x = -3y$

40. Vertices: $V(6, 0)$, $V'(-6, 0)$; asymptotes: $6y + x = 0$, $6y - x = 0$

41. Foci: $F(0, 8)$, $F'(0, -8)$; $b = 4\sqrt{2}$

42. Foci: $F(7, 0)$, $F'(-7, 0)$; $a = 2$

C The equation of a hyperbola with center at (h, k) can be written as

$$\frac{x - h^2}{a^2} - \frac{(y - k)^2}{b^2} = 1 \quad \text{or} \quad \frac{(y - k)^2}{a^2} - \frac{(x - h)^2}{b^2} = 1.$$

Foci lie on a line parallel to the x axis. Foci lie on a line parallel to the y axis.

Use this information to sketch the graph of each equation.

43. $\frac{(x - 3)^2}{16} - \frac{(y - 2)^2}{9} = 1$

44. $\frac{(y - 3)^2}{49} - \frac{x^2}{25} = 1$

45. $\frac{(x - 1)^2}{9} - \frac{y^2}{4} = 1$

46. $(x + 4)^2 - (y + 6)^2 = 9$

47. $4(y - 5)^2 - 36(x - 1)^2 = 36$

48. $x^2 - (y + 6)^2 = 4$

49. The hyperbola with equation $\frac{x^2}{25} - \frac{y^2}{9} = 1$ is translated five units to the right and two units down. Find the equation of the hyperbola in its new position.

50. The hyperbola with equation $\frac{x^2}{25} - \frac{y^2}{9} = 1$ is translated four units to the left and two units up. Find the equation of the hyperbola in its new position.

10–5 The Rectangular Hyperbola

The hyperbola whose asymptotes are the x and y axes is defined by

$$xy = k.$$

If $k > 0$, the branches of the hyperbola lie in Quadrants I and III.
If $k < 0$, the branches of the hyperbola lie in Quadrants II and IV.

This special hyperbola is called a **rectangular hyperbola.**

EXAMPLE Graph the hyperbola $xy = 12$.

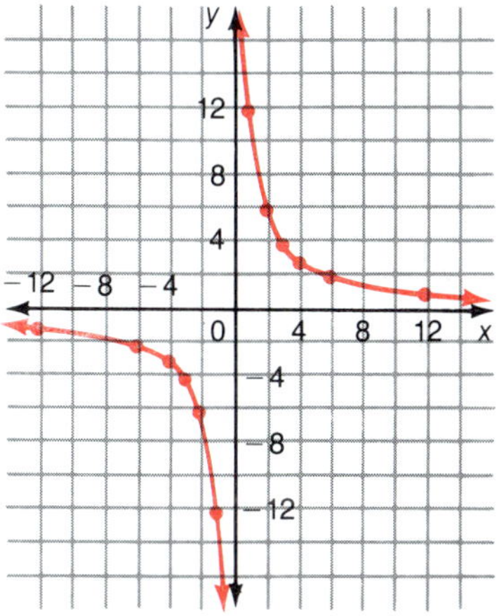

Solution: $y = \frac{12}{x}$ ← **Solve for *y*.**

From this equation, you can see that the set of possible values of x is

$$\{x: x \in \mathrm{R}, x \neq 0\}.$$

$y = \frac{12}{x}$ ← **Use integral values of x to find *y*, *x* ≠ 0.**

x	1	2	3	4	6	12	−1	−2	−4	−6	−12
y	12	6	4	3	2	1	−12	−6	−3	−2	−1

Join the points of the two branches.

The graph of $xy = -12$ is symmetric to that of $xy = 12$ with respect to either the x axis or the y axis. The foci and vertices of a rectangular hyperbola, $xy = k$, lie on the line $y = x$ when $k > 0$ and on the line $y = -x$ when $k < 0$.

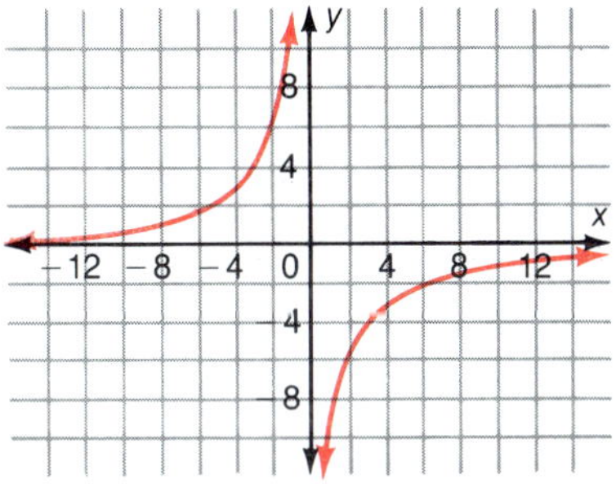

CLASSROOM EXERCISES

In which quadrants do the branches of each hyperbola lie?

1. $xy = -5$ **2.** $xy = 1$ **3.** $-xy = 4$ **4.** $xy = -3$ **5.** $-xy = -7$

Find the coordinates of the vertices of each hyperbola.

6. $xy = 16$ **7.** $xy = 25$ **8.** $xy = 10$ **9.** $xy = -15$ **10.** $xy = -16$

For Exercises 11–12, refer to the equation $xy = 4$.

11. Write the equations of the asymptotes.

12. Give the coordinates of the vertices.

WRITTEN EXERCISES

A Graph each hyperbola.

1. $xy = 18$
2. $xy = -18$
3. $xy = -7$
4. $xy = 15$
5. $y = \frac{4}{x}$
6. $x = \frac{8}{y}$
7. $x = \frac{-24}{y}$
8. $y = \frac{-20}{x}$
9. $xy - 9 = 0$
10. $xy - 6 = 0$
11. $xy + 9 = 0$
12. $xy + 6 = 0$

B Identify each equation as that of a line, a circle, an ellipse, a parabola, a hyperbola, or a rectangular hyperbola.

13. $x^2 + y^2 = 25$
14. $xy = 8$
15. $x + y = 7$
16. $x - y = 1$
17. $y = x^2 + 1$
18. $x^2 + 3y = 15$
19. $3x - y^2 = 5$
20. $4x^2 - 2y = 17$
21. $4x + 2y = 17$
22. $3x^2 = 27 - 7y$
23. $x^2 + 2y^2 = 2$
24. $x^2 + y^2 = 17$
25. $2x^2 + y^2 = 17$
26. $2y^2 + 3x^2 = 11$
27. $x^2 - y^2 = 4$
28. Use the locus definition of a hyperbola to find the equation of the hyperbola with foci at $(2, 0)$ and $(-2, 0)$ when the constant difference between any point of the hyperbola and the foci is $2\sqrt{2}$.

Review

Write the equation for the circle with the given radius and center. *(Section 10–1)*

1. $r = 4$; $C(1, 2)$
2. $r = 2$; $C(-4, 3)$
3. $r = 2\frac{1}{2}$; $C(-\frac{1}{2}, -\frac{3}{2})$

For each ellipse, determine whether the major axis is horizontal or vertical. Sketch each ellipse. *(Section 10–2)*

4. $\frac{x^2}{25} + \frac{y^2}{16} = 1$
5. $9x^2 + y^2 = 36$
6. $49x^2 + 9y^2 - 441 = 0$

For each parabola determine the coordinates of the vertex and of the focus. Then find the equations of the axis of symmetry and of the directrix. Sketch the graph. *(Section 10–3)*

7. $y = (x - 3)^2 + 5$
8. $y = x^2 + 9x - 8$
9. $x = y^2 - 5y$

For each hyperbola, find the coordinates of the center, of the vertices, and of the foci. Then write the equations of the asymptotes. *(Section 10–4)*

10. $\frac{x^2}{9} - \frac{y^2}{9} = 1$
11. $\frac{y^2}{16} - \frac{x^2}{9} = 1$
12. $\frac{x^2}{16} - y^2 = 1$

Graph each hyperbola. *(Section 10–5)*

13. $xy = 12$
14. $y = \frac{-3}{x}$
15. $x = \frac{-10}{y}$
16. $xy + 5 = 0$

Problem Solving and Applications

10–6 Inverse/Joint Variation

The graph of the rectangular hyperbola is used as a geometric model of the relationship between pressure, P, and volume, V, of a gas at constant temperature, k.

$$PV = k$$

As the value of P increases, the value of V decreases; as the value of V increases, the value of P decreases. This is an example of **inverse variation.**

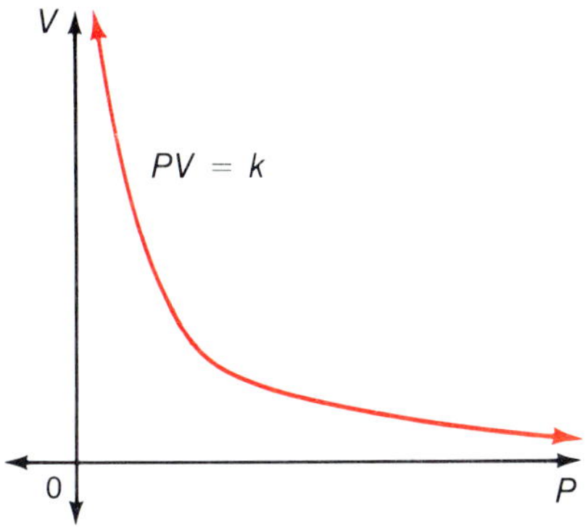

An inverse variation can be written as a proportion. For example, if $P_1(x_1, y_1)$ and $P_2(x_2, y_2)$ satisfy the equation,

$$xy = k,$$

then $\quad x_1y_1 = k \quad$ and $\quad x_2y_2 = k.$

Thus, $\quad x_1y_1 = x_2y_2 \quad$ or $\quad \frac{x_1}{x_2} = \frac{y_2}{y_1}.$

EXAMPLE 1 If x varies inversely as y and $y = 9$ when $x = 5$, find y when $x = 25$.

Solution: **Method 1:** Using the Definition

$xy = k$ ← **Replace x with 5 and y with 9.**

$\therefore 45 = k$

$xy = 45$ ← **Replace x with 25.**

$25y = 45$

$y = \frac{45}{25}$

$y = \frac{9}{5}$, or $1\frac{4}{5}$

Method 2: Using a Proportion

$\frac{x_1}{x_2} = \frac{y_2}{y_1}$ ← **$x_1 = 5$, $y_1 = 9$, $x_2 = 25$**

$\frac{5}{25} = \frac{y_2}{9}$

$45 = 25y_2$

$y_2 = \frac{45}{25}$

$y_2 = \frac{9}{5}$, or $1\frac{4}{5}$

When $y = \frac{k}{x^2}$, or $yx^2 = k$, then y is said to vary **inversely as the square of x.**

This inverse variation can also be written as a proportion.

$$\frac{y_1}{y_2} = \frac{x_2^2}{x_1^2}.$$

EXAMPLE 2 The intensity of light from a light source varies inversely as the square of the distance from the light source.

a. Write an equation for the variation. Use k as the constant, I for the intensity of light, and d for the distance.

b. If $I = 30$ units when $d = 6$ meters, find I when $d = 10$.

Solutions: **a.** $I = \frac{k}{d^2}$ ← **Inverse variation as the square**

b. $30 = \frac{k}{6^2}$ ← $I = 30$, $d = 6$

$k = 30 \times 36$, or 1080

$I = \frac{1080}{d^2}$ ← **Replace *d* with 10.**

$I = \frac{1080}{10^2} =$ **10.8 units**

The following equation, where x, y, and z are variables and k is a constant, expresses **joint variation.**

$$x = kyz$$

Example 3 combines inverse and joint variation **(combined variation).**

EXAMPLE 3 Heat loss, h, per hour through a glass window of a house on a cold day varies directly as the difference, d, between the inside and outside temperatures and the area, A, of the window, and inversely as the thickness, w, of the window.

a. Write an equation for the variation.

b. When the temperature difference is 40°C, there is a heat loss of 6000 calories in one hour through a window 100 × 150 centimeters and 0.25 centimeters thick. Find the heat loss in one hour through a window having the same area and 0.3 centimeters thick when the temperature difference is 20°C.

Solutions: **a.** $h = \frac{kdA}{w}$ ← **Joint and inverse variation.**

b. $h = \frac{kdA}{w}$ ← $h = 6000$; $d = 40$, $A = 100 \times 150$; $w = 0.25$

$6000 = \frac{40k(15{,}000)}{0.25}$

$6000 = 2{,}400{,}000k$

$0.0025 = k$

$h = \frac{0.0025dA}{w}$ ← $d = 20$; $A = 100 \times 150$; $w = 0.3$

$h = \frac{(0.0025)(20)(15{,}000)}{0.03} =$ **25,000** Heat loss: **25,000 calories**

CLASSROOM EXERCISES

Identify each equation as representing inverse variation, *I*, joint variation, *J*, or combined variation, *C*.

1. $xy = 16$
2. $x = \frac{8}{y}$
3. $\frac{y_1}{y_2} = \frac{x_2}{x_1}$
4. $R = \frac{1.35l}{d^2}$
5. $y = 17xz$
6. $y = \frac{3t}{r}$
7. $y = \frac{12}{x^2}$
8. $F = \frac{32m_1m_2}{r^2}$

WRITTEN EXERCISES

A In Exercises 1–6, *x* varies inversely as *y*.

1. $x = 4$ when $y = 12$. Find x when $y = 6$.
2. $x = 8$ when $y = 100$. Find y when $x = 2$.
3. $y = 20$ when $x = -4$. Find y when $x = 14$.
4. $x = 18$ when $y = -2\frac{1}{2}$. Find x when $y = -5$.
5. $y = 9$ when $x = 24$. Find y when $x = 16$.
6. $y = 10$ when $x = 45$. Find y when $x = 9$.

In Exercises 7–12, *y* varies inversely as the square of *x*.

7. $y = 2$ when $x = 4$. Find y when $x = 16$.
8. $x = 5$ when $y = 4$. Find x when $y = 6$.
9. $y = 8$ when $x = -1$. Find y when $x = 1$.
10. $y = 16$ when $x = \frac{1}{2}$. Find x when $y = \frac{1}{4}$.
11. $y = 15$ when $x = \frac{1}{3}$. Find y when $x = \frac{1}{6}$.
12. $x = -3$ when $y = 6$. Find x when $y = \frac{1}{2}$.

The formula for the area of a rectangular region with dimensions *l* and *w* is $A = lw$. Use this formula in Exercises 13–16.

13. Is $A = lw$ an example of inverse variation? Explain.
14. Let $A = 100$. Then $lw = 100$. When $l = 5$, $w =$ __?__. When $l = 10$, $w =$ __?__. When you choose a value of l and double it, the corresponding value of w is __?__.
15. For $lw = 100$, choose a value of l and triple it. The corresponding value for w is __?__.
16. For $lw = 100$, choose a value for l and multiply it by a. The corresponding value for w is __?__.

In Exercises 17–23, write an equation to represent each variation. Use *k* as the constant of variation.

17. y varies jointly as x and z.
18. y varies jointly as x and z and inversely as the square of m.
19. A varies jointly as m, n, and p.
20. y varies jointly as x and inversely as the square root of z.
21. y varies directly as x and z and inversely as the product of m and n.
22. y varies directly as x^2 and inversely as z^3.

23. y varies directly as the cube of x and inversely as the square of z.

24. If y varies as x and z, and $y = 12$ when $x = 1$ and $z = 4$, find y when $x = 4$ and $z = 2$.

25. If y varies jointly as z and the square of x, and $y = 15$ when $z = 4$ and $x = 2$, find y when $z = 4$ and $x = 6$.

26. If y varies jointly as x and z and inversely as w, and $y = 3$ when $x = 8$, $z = 3$ and $w = 4$, find y when $x = 6$, $z = 4$, and $w = 2$.

27. If y varies directly as x and inversely as the square of z, and $y = 2$ when $x = 5$ and $z = 3$, find y when $x = 25$ and $z = 5$.

APPLICATIONS: Using Inverse/Joint Variation

28. The Kellys took a 300-kilometer trip. Write an equation showing the relation between average velocity, v, and time traveled, t. Is this an example of inverse variation? Express this relation as (v_1, t_1) and (v_2, t_2). Make a table showing at least six pairs of numbers that satisfy the equation.

The law of the lever states that $WD = wd$, where w and W represent masses and d and D represent distances from the fulcrum. Use this law for Exercises 29–32.

29. Express this relation as a proportion.

30. Is this an example of direct or of inverse variation?

31. Find W when $w = 10$, $d = 5$, and $D = 20$.

32. A 70-kilogram person sits 2.5 meters from the center of a seesaw. At what distance from the center must a 90-kilogram person sit in order to balance the seesaw?

The weight, w, of a body varies inversely as the square of its distance, d, from the center of the earth. Use this relation in Exercises 33–35.

33. Express this relation as an equation using k as the constant of variation.

34. Express the relation as a proportion.

35. On the surface of the earth (4000 miles from the center), a certain body weighs 100 pounds. How much would it weigh 200 miles above the earth?

36. The intensity, I, of light varies inversely as the square of the distance, d, from the light source. How far should a book be placed from a light source if it is to receive four times as intense illumination as it received when it was 2 meters from the light?

37. Use the variation of Example 3 to find the heat loss through a window with dimensions 150×200 centimeters. The glass is 0.35 centimeters thick and the temperature difference is 30°C.

38. Use the variation of Example 3 to find the heat loss through a window with dimensions 150×200 centimeters. The glass is 0.25 centimeters thick and the temperature difference is 35°C.

B

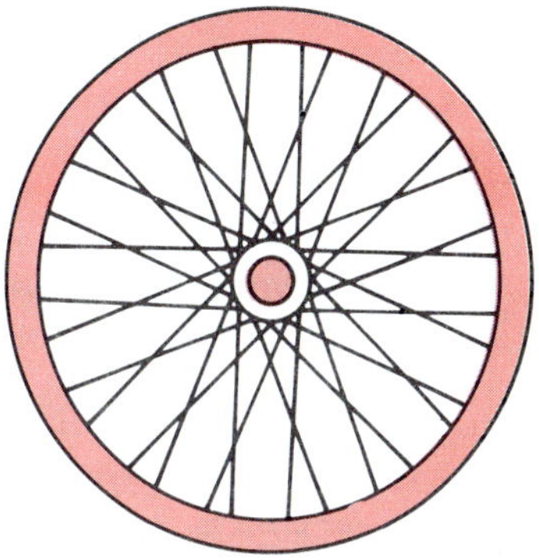

39. A bicycle wheel with a 26–inch diameter takes 10 revolutions to go a certain distance. How many revolutions will a wheel with a 20–inch diameter make in covering the same distance?

40. A pulley connects a 30–centimeter wheel with a 6–centimeter wheel. How many revolutions will the small wheel make for each revolution of the large wheel? How will the speeds of the two wheels compare?

41. The pressure of wind on a sail varies jointly as the area of the sail and the square of the velocity of the wind. When the velocity of the wind is 20 miles per hour, the pressure on 2 square feet of sail is 4 pounds. Find the velocity of the wind when the pressure on 9 square feet of sail is 32 pounds.

42. The number of rectangular tiles needed to cover a surface varies inversely as the length and width of a tile. It takes 270 tiles, each having dimensions of 2×5 centimeters, to cover a surface. How many tiles having dimensions of 3×6 centimeters are needed to cover the same surface?

43. The electrical resistance R of a cable varies directly as the length, l, of the cable and inversely as the square of its diameter, d. For a cable 1000 feet long and $\frac{1}{2}$ inch in diameter, the resistance is 0.08 ohms. Find the resistance for a cable with a diameter of 1 inch and a length of 500 feet.

C

44. The force of attraction, F, between two objects varies directly as the product of their masses (m and M), and inversely as the square of the distance, d, between the objects. The force $F = 6$ when $m = 64$, $M = 108$, and $d = 24$. Find m when $F = 96$, $M = 956$, and $d = 18$.

45. The safe load, s, on a horizontal beam supported at both ends varies jointly as the breadth, b, and the square of the depth, d. It also varies inversely as the length, l, between the supports. A certain beam, 8 feet long between supports, and whose breadth and depth are 2 inches and 6 inches respectively, can safely support 600 pounds. A $4'' \times 8''$ beam of the same material has a distance of 9 feet between supports. What is the safe load for the beam?

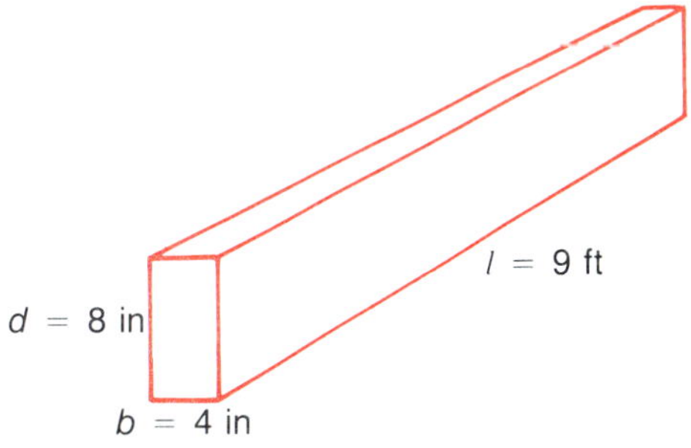

46. The least load, l, that will collapse a square oak pillar varies directly as the fourth power of its thickness, t, and inversely as the square of its height, h. A 4–inch pillar, 6 feet high, will be collapsed by a load of 98 tons. A square pillar of the same wood is 8 inches thick and 10 feet high. What is the least load that will collapse the pillar?

BASIC: ELLIPSES

Problem: *Given the equation of an ellipse centered at the origin, write a program which prints information about the ellipse.*

```
100 REM   GIVEN THE EQUATION OF AN ELLIPSE, THIS PROGRAM
110 REM   WILL PRINT INFORMATION ABOUT THE ELLIPSE.
120 PRINT "THE ELLIPSE IS CENTERED AT THE ORIGIN"
130 PRINT "WITH EQUATION X↑2/(A↑2) + Y↑2/(B↑2) = 1."
140 PRINT "WHAT ARE A AND B";
150 INPUT A, B
160 LET A = ABS(A)
170 LET B = ABS(B)
180 IF A < B THEN 260
190 LET C = SQR(A*A - B*B)
200 PRINT "THE LENGTH OF THE MAJOR AXIS =";2*A
210 PRINT "THE LENGTH OF THE MINOR AXIS =";2*B
220 PRINT "THE VERTICES ARE (";-A;", 0) AND (";A;", 0)"
230 IF C = 0 THEN 320
240 PRINT "THE FOCI ARE (";-C;", 0) AND (";C;", 0)"
250 GOTO 330
260 LET C = SQR(B*B - A*A)
270 PRINT "THE LENGTH OF THE MAJOR AXIS =";2*B
280 PRINT "THE LENGTH OF THE MINOR AXIS =";2*A
290 PRINT "THE VERTICES ARE (0,";B;") AND (0,";-B;")"
300 PRINT "THE FOCI ARE (0,";C;") AND (0,";-C;")"
310 GOTO 330
320 PRINT "THE CENTER OF THE CIRCLE IS (0, 0)."
330 PRINT
340 PRINT "ANY MORE ELLIPSES (1 = YES, 0 = NO)";
350 INPUT Z
360 IF Z = 1 THEN 120
370 END
```

The following is the output from a sample run of the program above.

Output:

```
RUN
THE ELLIPSE IS CENTERED AT THE ORIGIN
WITH EQUATION X↑2/(A↑2) + Y↑2/(B↑2) = 1.
WHAT ARE A AND B? 4,3
THE LENGTH OF THE MAJOR AXIS = 8
THE LENGTH OF THE MINOR AXIS = 6
THE VERTICES ARE (-4 , 0) AND ( 4 , 0)
THE FOCI ARE (-2.64575 , 0) AND ( 2.64575 , 0)
```

```
ANY MORE ELLIPSES (1 = YES, 0 = NO)? 1
THE ELLIPSE IS CENTERED AT THE ORIGIN
WITH EQUATION X↑2/(A↑2) + Y↑2/(B↑2) = 1.
WHAT ARE A AND B? 5, 5
THE LENGTH OF THE MAJOR AXIS = 10
THE LENGTH OF THE MINOR AXIS = 10
THE VERTICES ARE (-5 , 0) AND ( 5 , 0)
THE CENTER OF THE CIRCLE IS (0, 0).

ANY MORE ELLIPSES (1 = YES, 0 = NO)? 0
READY
```

Analysis

Statements 120–150: The user enters A and B, which are the square roots of the denominators of the X↑2 and Y↑2 terms respectively.

Statement 180: If A < B, then the major axis, vertices, and foci are on the y axis (statements 260–300).

Statements 190–240: When A ≥ B, the major axis, vertices, and foci are on the x axis. In statement 230, the program identifies the circle as the special case of the ellipse, where the foci fuse into the center (statement 320).

EXERCISES

A Put each equation into the form required by the program on page 382. Then enter the appropriate coefficients into the program to obtain information about each ellipse.

1. $5x^2 + 20y^2 = 80$ **2.** $25x^2 + 9y^2 = 225$ **3.** $x^2 + 25y^2 = 100$

4. $4x^2 + 4y^2 = 36$ **5.** $4x^2 = 16 - y^2$ **6.** $y^2 = 49 - x^2$

B Write a BASIC program for each problem.

7. Given the lengths of the major and minor axes of an ellipse centered at the origin, print its equation in the form X↑2/(A↑2)+Y↑2/(B↑2)=1. Also print the coordinates of the foci. To distinguish between the two types of ellipses, have the user enter a third number q. If $q = 1$, the major axis is on the x axis; if $q = 2$, the major axis is on the y axis.

8. For an ellipse centered at the origin, accept as input the coordinates of one focus and the length of the major axis. Print the equation of the ellipse. See Exercise 7 for the form of the output.

C

9. Given the equation of a parabola opening either vertically or horizontally, print the following information about the parabola: coordinates of the vertex and focus, whether the vertex is a minimum point or a maximum point, and the equations of the axis of symmetry and the directrix.

10–7 Systems of First- and Second-Degree Equations

The figures below show how the graphs of a system consisting of a linear equation and a second-degree equation can intersect.

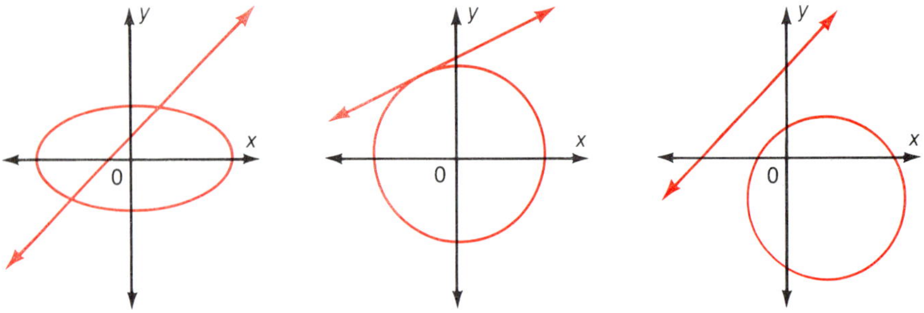

Two solutions **One solution** **No real solutions**

When the graphs of the system have two points in common, the system has two solutions. When the graphs have one point in common (the graphs are tangent), there is one distinct solution. When the graphs have no points in common, there are no real number solutions.

You can also use the substitution method to solve these systems.

EXAMPLE Solve and check: $\begin{cases} 4x^2 + 9y^2 = 36 & \textbf{1} \\ y - x = 3 & \textbf{2} \end{cases}$

Solution: [1] Solve the linear equation for y. ← $\mathbf{y = x + 3}$

[2] Replace y in Equation 1 with $(x + 3)$.

$$4x^2 + 9(x + 3)^2 = 36$$

$$4x^2 + 9(x^2 + 6x + 9) = 36$$

$$13x^2 + 54x + 45 = 0$$ ← **Use the quadratic formula to find x.**

$$x = -3 \quad \text{or} \quad x = -\frac{15}{13}$$ ← **Find y for each value of x.**

For $x = -3$, $y = 0$. For $x = -\frac{15}{13}$, $y = \frac{24}{13}$.

Check: Substitute both pairs in the original equation.

The solutions are $\mathbf{(-3, 0)}$ and $\mathbf{\left(-\frac{15}{13}, \frac{24}{13}\right)}$.

The substitution method identifies the ordered pairs (real or complex) which are solutions of the system. When using the graphing method, you sometimes have to approximate solutions because the exact solution is difficult to read from the graph.

Steps for Solving a Linear/Second-Degree System

1. Solve the linear equation for one of the variables.
2. Substitute the result in the second-degree equation.
3. Solve this second-degree equation for one variable.
4. Find the other solution by substituting both numbers found in Step 3 in the linear equation.
5. Check by substituting each number pair in both equations.

CLASSROOM EXERCISES

Without solving, tell whether each system has two solutions, one solution, or no real solution.

1. $\begin{cases} x^2 + y^2 = 1 \\ x = 1 \end{cases}$ **2.** $\begin{cases} x^2 + y^2 = 1 \\ y = 1 \end{cases}$ **3.** $\begin{cases} x^2 + y^2 = 4 \\ y = x \end{cases}$ **4.** $\begin{cases} x^2 + y^2 = 9 \\ y = -x \end{cases}$

5. $\begin{cases} y = x^2 \\ y = -3 \end{cases}$ **6.** $\begin{cases} y = -x^2 \\ y = -3 \end{cases}$ **7.** $\begin{cases} y = x^2 \\ y = 0 \end{cases}$ **8.** $\begin{cases} \frac{x^2}{9} + \frac{y^2}{4} = 1 \\ x = 3 \end{cases}$

WRITTEN EXERCISES

A Solve each system by graphing.

1. $\begin{cases} y = x^2 \\ y = x \end{cases}$ **2.** $\begin{cases} x^2 + y^2 = 9 \\ y = x + 3 \end{cases}$ **3.** $\begin{cases} x^2 - y^2 = 16 \\ x - y = 2 \end{cases}$ **4.** $\begin{cases} x^2 + 4y^2 = 36 \\ 2y = x - 6 \end{cases}$

Solve and check.

5. $\begin{cases} x^2 + y^2 = 25 \\ x - y = 1 \end{cases}$ **6.** $\begin{cases} y^2 - 9x = 0 \\ 2x - y = 2 \end{cases}$ **7.** $\begin{cases} x^2 + y^2 = 17 \\ x - 3y = 1 \end{cases}$ **8.** $\begin{cases} x^2 + y^2 = 25 \\ 2x - y = 2 \end{cases}$

9. $\begin{cases} x^2 + y^2 = 5 \\ 2x - y = 0 \end{cases}$ **10.** $\begin{cases} y^2 + 3x - 7 = 0 \\ y + 3x = 5 \end{cases}$ **11.** $\begin{cases} 2c + d = 7 \\ c^2 + cd - 12 = 0 \end{cases}$ **12.** $\begin{cases} x^2 + 3y = 16 \\ x + 3y = 10 \end{cases}$

13. $\begin{cases} x^2 + y^2 = 61 \\ 2x - y = 7 \end{cases}$ **14.** $\begin{cases} xy = 9 \\ 2x + 3y = 15 \end{cases}$ **15.** $\begin{cases} x^2 + y^2 = 116 \\ x + 2y = 18 \end{cases}$ **16.** $\begin{cases} 2x^2 + y^2 = 17 \\ 3x - y = 9 \end{cases}$

17. $\begin{cases} 2y^2 - 3x^2 = 11 \\ x - 7 = 3y \end{cases}$ **18.** $\begin{cases} x^2 - y^2 = 106 \\ x - y = 4 \end{cases}$ **19.** $\begin{cases} 2l^2 + 3k = 33 \\ 4l + k = -7 \end{cases}$ **20.** $\begin{cases} xy + x = 4 \\ 2y = 3x \end{cases}$

21. $\begin{cases} xy = 6 \\ 2y = 3x - 16 \end{cases}$ **22.** $\begin{cases} rs + r^2 = 24 \\ s + r = 34 \end{cases}$ **23.** $\begin{cases} 2a + b = 1 \\ 3a^2 + 12a + b = 8 \end{cases}$ **24.** $\begin{cases} xy = -10 \\ x - 2y = 12 \end{cases}$

25. $\begin{cases} y^2 + 2x = 17 \\ x + 4y = -8 \end{cases}$ **26.** $\begin{cases} 3x + 2y = -2 \\ xy + 8x = 4 \end{cases}$ **27.** $\begin{cases} 4x^2 + 9y^2 = 36 \\ 2x + 3y = 6 \end{cases}$ **28.** $\begin{cases} x^2 + y^2 = 50 \\ 7y + 9x = 70 \end{cases}$

B

29. $\begin{cases} x^2 - y^2 = p \\ x + y = q \end{cases}$ **30.** $\begin{cases} x^2 + xy + y^2 = 49 \\ x + y = 8 \end{cases}$ **31.** $\begin{cases} 6y^2 - 2x^2 = xy \\ 4x = 12 - 9y \end{cases}$ **32.** $\begin{cases} x^2 - y^2 = 3 \\ x - y = 1 \end{cases}$

REVIEW CAPSULE FOR SECTION 10-8

Solve for x^2. *(Pages 44–47)*

1. $x^2 + y^2 = 8$ **2.** $y - x^2 = 9$ **3.** $6x^2 + 12y^2 = 36$ **4.** $3x^2 - 4y^2 = 8$

5. $x^2 + 14 = 7y^2$ **6.** $4y - 2x^2 = 5$ **7.** $y^2 + 4x^2 = 8$ **8.** $6x^2 + 3y^2 - 2 = 0$

10-8 Systems of Two Second-Degree Equations

The figures below show some ways in which the graphs of a system consisting of two quadratic equations can intersect.

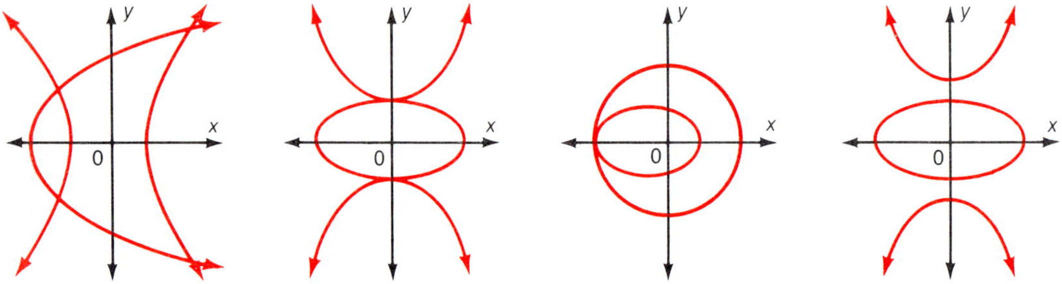

Four solutions **Two solutions** **One solution** **No real solutions**

Systems of equations involving one or more quadratic equations may have complex as well as real solutions. The graphs of such systems can be used to identify the real solutions only. Solving by the substitution method identifies both the real and complex solutions.

EXAMPLE Solve and check: $\begin{cases} 2x^2 - y^2 = 1 & \textbf{1} \\ x^2 + 2y^2 = 22 & \textbf{2} \end{cases}$

Solution: [1] Solve Equation **1** for y^2.

$$-y^2 = -1 - 2x^2$$
$$y^2 = 1 + 2x^2$$

[2] In Equation **2**, replace y^2 with $(1 + 2x^2)$.

$$x^2 + 2y^2 = 22$$
$$x^2 + 2(1 + 2x^2) = 22$$
$$x^2 + 4x^2 + 2 = 22$$
$$5x^2 = 20$$
$$x = \pm 2$$

Find y for each value of x.

When $x = 2$, $y = \pm 3$. When $x = -2$, $y = \pm 3$.

The check is left for you.

Solution set: $\{(2, 3), (2, -3), (-2, 3), (-2, -3)\}$.

CLASSROOM EXERCISES

Without solving, tell whether each system has at least one ordered pair of real numbers in its solution set. Answer *Yes* or *No.*

1. $\begin{cases} x^2 + y^2 = 25 \\ x^2 + y^2 = 9 \end{cases}$
2. $\begin{cases} x^2 - y^2 = 25 \\ x^2 + y^2 = 36 \end{cases}$
3. $\begin{cases} y = x^2 \\ x^2 + y^2 = 1 \end{cases}$
4. $\begin{cases} y = -x^2 \\ y = x^2 + 3 \end{cases}$
5. $\begin{cases} xy = 4 \\ x^2 + y^2 = 8 \end{cases}$
6. $\begin{cases} xy = 1 \\ 4x^2 + 9y^2 = 36 \end{cases}$
7. $\begin{cases} x^2 + y^2 = 16 \\ 4x^2 + 9y^2 = 36 \end{cases}$
8. $\begin{cases} y = x^2 \\ x^2 - y^2 = 16 \end{cases}$

WRITTEN EXERCISES

A Solve and check.

1. $\begin{cases} x^2 + y^2 = 9 \\ 2x^2 + 3y^2 = 18 \end{cases}$
2. $\begin{cases} 2x^2 - y^2 = -1 \\ x^2 + 2y^2 = 22 \end{cases}$
3. $\begin{cases} 4x^2 + y^2 = 17 \\ 3x^2 - 5y^2 = 7 \end{cases}$
4. $\begin{cases} a^2 - 3b^2 = 13 \\ 2a^2 - 8b^2 = 18 \end{cases}$
5. $\begin{cases} y = x^2 \\ y = -x^2 + 2 \end{cases}$
6. $\begin{cases} y = x^2 \\ x^2 + y^2 = 2 \end{cases}$
7. $\begin{cases} x = (y + 3)^2 \\ x^2 + y^2 = 9 \end{cases}$
8. $\begin{cases} \dfrac{x^2}{16} + \dfrac{y^2}{49} = 1 \\ x^2 + y^2 = 4 \end{cases}$
9. $\begin{cases} y = -x^2 \\ 9x^2 + 100y^2 = 900 \end{cases}$
10. $\begin{cases} y = x^2 \\ 16x^2 + 4y^2 = 64 \end{cases}$
11. $\begin{cases} x^2 + y^2 = 5 \\ y = x^2 - 3 \end{cases}$
12. $\begin{cases} 3x^2 + 5y^2 = 12 \\ 7x^2 - 3y^2 = -5 \end{cases}$

B

13. $\begin{cases} x^2 + y^2 = 2 \\ xy = 1 \end{cases}$
14. $\begin{cases} x^2 - y^2 = 12 \\ xy = 8 \end{cases}$
15. $\begin{cases} x^2 + y^2 = 25 \\ x + y^2 = 5 \end{cases}$
16. $\begin{cases} ax^2 + by^2 = 1 \\ bx^2 - ay^2 = 1 \end{cases}$
17. $\begin{cases} x^2 + y^2 = 29 \\ xy = 10 \end{cases}$
18. $\begin{cases} x = \dfrac{8}{y} \\ \dfrac{x^2}{9} - \dfrac{y^2}{4} = 1 \end{cases}$
19. $\begin{cases} y = \dfrac{10}{x} \\ \dfrac{y^2}{16} - \dfrac{x^2}{4} = 1 \end{cases}$
20. $\begin{cases} y^2 - 4 = 6x^2 \\ y + 2 = 3x^2 \end{cases}$

APPLICATIONS: Using Systems to Solve Word Problems

21. The sum of two numbers is 20 and the sum of their squares is 201. Find the numbers.
22. The fence around a rectangular plot of ground is 168 yards long. The area of the plot is 1728 square yards. How wide is the plot?
23. A dealer bought a shipment of shoes for \$480. He sold all but 5 pairs, making a profit of \$6 per pair. The total profit on the shipment was \$290. How many pairs of shoes did the dealer sell?
24. The product of a two-digit number and the tens digit is 384. The units digit is 2 less than the tens digit. What is the number?
25. A number of people shared the purchase price of a skiing chalet equally. If there had been two more persons, each would have paid \$12 less. If there had been three fewer persons, each would have paid \$24 more. How much did each person pay?
26. A motorist made a round-trip between two cities 270 miles apart. On the return trip, heavy rain reduced the motorist's average rate by 5 miles per hour and increased the time over the time going by $\frac{3}{4}$ hour. What was the rate of speed going?

10–9 Second-Degree Inequalities

To graph the solution set of an inequality of the second degree, you first graph its corresponding equation. Then you determine the set of points that satisfies the inequality.

EXAMPLE Graph $x^2 + y^2 < 25$.

Solution: [1] Draw the graph for $x^2 + y^2 = 25$. Use short dashes to show that points on the curve are not in the solution set.

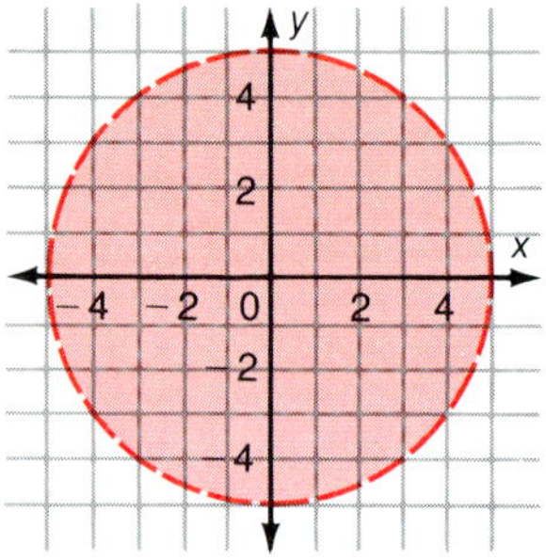

[2] Test points inside and outside the circle.

Inside: (0, 0)

$(0)^2 + (0)^2 < 25$? Yes ✓

Outside: (6, 2)

$(6)^2 + (2)^2 < 25$? No

The graph is the **set of points inside the circle.**

CLASSROOM EXERCISES

Test whether the given points satisfy the given equation.

1. $x^2 + y^2 > 36$
   $P(0, 0)$; $Q(4, 9)$
2. $xy \geq 15$
   $P(0, 0)$; $Q(-3, -5)$
3. $y > x^2$
   $H(0, 0)$; $R(-3, 12)$
4. $\frac{x^2}{25} + \frac{y^2}{16} \leq 1$
   $T(0, 0)$; $R(-5, -4)$
5. $\frac{x^2}{25} - \frac{y^2}{16} > 1$
   $S(-1, 1)$; $M(6, 9)$
6. $x^2 + 2y^2 < 3$
   $T(-\frac{1}{2}, -\frac{1}{4})$; $Q(5, 5)$

WRITTEN EXERCISES

A Graph each inequality.

1. $x^2 + y^2 > 36$
2. $x^2 + y^2 < 10$
3. $y > x^2$
4. $x < y^2$
5. $\frac{x^2}{16} + \frac{y^2}{9} > 1$
6. $\frac{x^2}{25} + \frac{y^2}{16} < 1$
7. $\frac{x^2}{4} - \frac{y^2}{9} > 1$
8. $\frac{y^2}{16} - \frac{x^2}{4} \leq 1$
9. $y \geq x^2 - 3x$
10. $xy < 16$
11. $x \geq -y^2$
12. $xy \geq -15$
13. $x^2 + 2y^2 < 3$
14. $x^2 + 4y^2 > 16$
15. $y - x^2 < -5$
16. $y + x^2 > 4$

B Graph the solution set of each system.

17. $\begin{cases} x^2 + y^2 \leq 25 \\ y \geq x^2 \end{cases}$
18. $\begin{cases} x^2 + y^2 \geq 25 \\ x^2 + y^2 \leq 36 \end{cases}$
19. $\begin{cases} y \geq -x^2 \\ x^2 + y^2 \leq 49 \end{cases}$
20. $\begin{cases} y \leq x^2 \\ x^2 + y^2 \leq 4 \end{cases}$

Review

1. If y varies jointly as x and z and inversely as d^2, and $y = 15$ when $x = 3$, $z = 5$, and $d = 2$, find y when $x = 9$, $z = 6$ and $d = 6$. *(Section 10–6)*
2. A bicycle wheel with a 24-inch diameter takes 20 revolutions to go a certain distance. How many revolutions will a wheel with an 18-inch diameter make in covering the same distance? *(Section 10–6)*

Solve and check. *(Sections 10–7, 10–8)*

3. $\begin{cases} x^2 + y^2 = 16 \\ x - y = 2 \end{cases}$
4. $\begin{cases} xy = 8 \\ 2x - y = 6 \end{cases}$
5. $\begin{cases} x - 2y = 8 \\ x^2 - y = 5 \end{cases}$
6. $\begin{cases} 4x^2 - 9y^2 = 36 \\ 2x - 3y = 6 \end{cases}$
7. $\begin{cases} x^2 + y^2 = 25 \\ y = -3x^2 + 9 \end{cases}$
8. $\begin{cases} y = x^2 \\ x^2 + y^2 = 4 \end{cases}$
9. $\begin{cases} xy = 10 \\ x^2 - y^2 = 16 \end{cases}$
10. $\begin{cases} x^2 + y^2 = 5 \\ xy = 4 \end{cases}$
11. The product of a two-digit number and the tens digit of the number is 172. The tens digit is 1 more than the units digit. What is the number? *(Section 10–8)*

Graph each inequality. *(Section 10–9)*

12. $x^2 + y^2 > 25$
13. $\frac{x^2}{16} + \frac{y^2}{9} < 1$
14. $x \geq y^2$
15. $y \leq -2x^2$

Chapter Summary

IMPORTANT TERMS

Asymptote (p. 372)
Center of a circle (p. 358)
Circle (p. 358)
Conic section (p. 358)
Conjugate axis (p. 371)
Directrix (p. 367)
Ellipse (p. 362)
Focus of an ellipse (p. 362)
Focus of a hyperbola (p. 371)
Focus of a parabola (p. 367)
Hyperbola (p. 371)
Inverse variation (p. 377)
Joint variation (p. 378)
Locus (p. 358)
Major axis (p. 362)
Minor axis (p. 362)
Parabola (p. 367)
Radius (p. 358)
Semi-major axis (p. 363)
Semi-minor axis (p. 363)
Transverse axis (p. 371)
Vertex of a hyperbola (p. 372)

IMPORTANT IDEAS

1. The equation of a circle with center at (h, k) and radius r, where $h, k, r \in \mathrm{R}$ and $r \geq 0$, is
$$(x - h)^2 + (y - k)^2 = r^2.$$
2. The standard form for the equation of an ellipse with center at (0, 0), major axis of length $2a$ (along the x axis), and minor axis of length $2b$ is
$$\frac{x^2}{a^2} + \frac{y^2}{b^2} = 1.$$

The standard form for the equation of an ellipse with center at $(0, 0)$, major axis of length $2a$ (along the y axis), and minor axis of length $2b$ is

$$\frac{x^2}{b^2} + \frac{y^2}{a^2} = 1.$$

For either ellipse, $c^2 = a^2 - b^2$ where c is the distance from each focus to the center of the ellipse.

3. The standard form for the equation of a parabola with vertex at (h, k) and axis of symmetry $x = h$ is

$$y = a(x - h)^2 + k$$

where the parabola opens upward when $a > 0$, and downward when $a < 0$.

The standard form for the equation of a parabola with vertex at (h, k) and axis of symmetry $y = k$ is

$$x = a(y - k)^2 + h$$

where the parabola opens to the right when $a > 0$ and to the left when $a < 0$.

4. **Parabola Facts:** See page 368.

5. The standard form for the equation of a hyperbola with foci at $F(c, 0)$ and $F'(-c, 0)$, transverse axis of length $2a$ (along the y axis), and conjugate axis of length $2b$ is

$$\frac{x^2}{a^2} - \frac{y^2}{b^2} = 1.$$

The standard form for the equation of a hyperbola with foci of $F(0, c)$ and $F'(0, -c)$, transverse axis of length $2a$ (along the x axis), and conjugate axis of length $2b$ is

$$\frac{y^2}{a^2} - \frac{x^2}{b^2} = 1.$$

For either hyperbola, $c = \sqrt{a^2 + b^2}$, where c is the distance from each focus to the center $(0, 0)$.

6. **Hyperbola Facts:** See page 372.

7. A rectangular hyperbola has an equation of the form $xy = k$. The asymptotes are the x and y axes.

8. Inverse variation is expressed by an equation of the form $xy = k$.

9. Joint variation is expressed by an equation of the form $x = kyz$.

10. **Steps for Solving a Linear/Second-Degree System:** See page 385.

11. To graph the solution set of an inequality of the second degree, you first graph its corresponding equation. Then you determine the set of points that satisfies the inequality.

Chapter Objectives and Review

Objective: *To write the equation of a circle and to graph it given the center and radius (Section 10–1)*

Write the equation of each circle with the given center and radius. Then sketch its graph.

1. $r = 2;\ (-1, 4)$ **2.** $r = 5;\ (3, 0)$ **3.** $r = 3;\ (-2, -3)$

Objective: *To write the length of the radius and the coordinates of the center of a circle given its equation (Section 10–1)*

For each equation of a circle, write the length of the radius and the coordinates of its center.

4. $x^2 + y^2 = 16$ **5.** $(x + 2)^2 + (y - 3)^2 = 36$

Objective: *To sketch the graph of an ellipse and to write the coordinates of the x intercepts, y intercepts, and foci from its equation (Section 10–2)*

Sketch the graph of each equation. Write the coordinates of the *x* intercepts, the *y* intercepts, and the foci.

6. $36x^2 + 9y^2 = 324$ **7.** $64x^2 + 4y^2 = 256$ **8.** $25x^2 + 16y^2 = 400$

Objective: *To write the equation of an ellipse given the x and y intercepts (Section 10–2)*

Write the equation of each ellipse given the following information.

9. x intercepts at $(3, 0)$ and $(-3, 0)$; y intercepts at $(0, 6)$ and $(0, -6)$

10. x intercepts at $(5, 0)$ and $(-5, 0)$; y intercepts at $(0, 4)$ and $(0, -4)$

Objective: *To sketch the graph of a parabola given its equation (Section 10–3)*

Sketch the graph of each equation. Label the focus, vertex, and directrix.

11. $x = \frac{1}{36}y^2$ **12.** $y = -\frac{1}{64}x^2$ **13.** $y^2 = -4x$

Objective: *To write the equation of a parabola given the focus and the directrix (Section 10–3)*

Write the equation of each parabola given the following information.

14. Focus: $(-2, 0)$; directrix: $x = 2$

15. Focus: $(0, 5)$; directrix: $y = -5$

Objective: *To sketch the graph of a hyperbola given its equation (Section 10–4)*

Sketch the graph of each equation.

16. $\frac{x^2}{36} - \frac{y^2}{4} = 1$ **17.** $\frac{y^2}{64} - \frac{x^2}{16} = 1$ **18.** $25x^2 - 9y^2 = 225$

Objective: *To write the equation of a hyperbola (Section 10–4)*

Write the equation of each hyperbola given the following information.

19. Center: $(0, 0)$; opens vertically; $a = 2$, $b = 3$

20. Foci: $(0, 4)$, $(0, -4)$; vertices: $(0, 3)$, $(0, -3)$

21. Vertices: $(2, 0)$, $(-2, 0)$; asymptotes: $x - 4y = 0$, $x + 4y = 0$

Objective: *To identify a given equation as that of a circle, ellipse, hyperbola, or rectangular hyperbola (Section 10–5)*

Identify each equation as that of a circle, an ellipse, a parabola, a hyperbola, or a rectangular hyperbola.

22. $xy = 15$

23. $\frac{x^2}{100} + \frac{y^2}{36} = 1$

24. $x = y^2 - 6y + 5$

25. $4x^2 + 9y^2 = 25$

26. $(x + 2)^2 + (y - 5)^2 = 10$

27. $y = -x^2 + 5x + 6$

Objective: *To sketch the graph of a rectangular hyperbola (Section 10–5)*

Sketch the graph of each equation.

28. $xy = 18$

29. $xy = -8$

30. $xy + 10 = 0$

31. $y = \frac{16}{x}$

Objective: *To solve problems involving joint, inverse, and combined variation (Section 10–6)*

For Exercises 32–34, use k as the constant of variation to write the equation that expresses each relation.

32. y varies inversely as x.

33. y varies jointly as a and b.

34. y varies inversely as the square of x.

35. If y varies inversely as x, and $x = 14$ when $y = 3$, find y when $x = 7$.

36. If y varies jointly as x and z, and $y = 20$ when $x = 4$ and $z = 2$, find y when $x = 3$ and $z = 1$.

37. The weight, w, of a body varies inversely as the square of its distance, d, from the center of the earth. A body weighs 300 pounds on the surface of the earth (4000 miles from the center). How much would it weigh 120 miles above the earth?

Objective: *To solve a system of a first- and a second-degree equation algebraically (Section 10–7)*

Solve each system algebraically.

38. $\begin{cases} 3x^2 + xy = 5 \\ 2x + y = 2 \end{cases}$

39. $\begin{cases} x^2 - y = 3 \\ x - y = -3 \end{cases}$

40. $\begin{cases} 9x^2 - 25y^2 = 225 \\ y - x = 5 \end{cases}$

41. $\begin{cases} x^2 + y^2 = 4 \\ x - 2y = 0 \end{cases}$

Objective: *To solve a system of two second-degree equations (Section 10–8)*

Solve each system.

42. $\begin{cases} x^2 + y^2 = 25 \\ 3x^2 - 2y^2 = 30 \end{cases}$

43. $\begin{cases} x^2 + 2y^2 = 18 \\ xy = 6 \end{cases}$

44. $\begin{cases} x^2 + y^2 = 25 \\ y = x^2 - 5 \end{cases}$

Objective: *To graph the solution set of an inequality of the second degree (Section 10–9)*

Graph each inequality.

45. $x^2 + y^2 < 16$

46. $y < x^2 + 3$

47. $4x^2 + 9y^2 \geq 36$

Chapter Test

Identify the conic section defined by each equation. Then graph the conic section.

1. $x^2 + y^2 = 36$

2. $xy = -5$

3. $4x^2 - 16y^2 = 144$

4. $\frac{x^2}{16} + \frac{y^2}{25} = 1$

5. $\frac{x^2}{25} - \frac{y^2}{49} = 1$

6. $y = (x - 1)^2 + 4$

7. Write an equation of the parabola with focus at $F(-1, 0)$ and directrix defined by the equation $x = 2$.

8. Write an equation for a circle with a radius 6 units long and center at $C(-2, 3)$.

9. Find the coordinates of the foci of the ellipse defined by the equation $25x^2 + 9y^2 = 225$.

Find the solution set of each system.

10. $\begin{cases} 3x^2 - 2y^2 = 15 \\ x - y = 1 \end{cases}$

11. $\begin{cases} 2x^2 + xy = 10 \\ x + 2y = 2 \end{cases}$

For Exercises 12–13, use k as the constant of variation to write an equation that expresses the given relationship.

12. a varies inversely as b.

13. x varies directly as y and inversely as t.

14. If a varies jointly as b and c, and $a = 16$ when $b = 10$ and $c = 5$, find a when $b = 9$ and $c = 4$.

15. The intensity, I, of light varies inversely as the square of the distance, d, from the light source. How far should a painting be placed from the light source if it is to receive 9 times as intense illumination as when it was 15 meters from the light?

More Challenging Problems

1. Under what conditions will one solution of $ax^2 + bx + c$ be the reciprocal of the other?
2. Solve: $|x|^2 + |3x| - 10 = 0$
3. Show that the points of intersection of $xy = 12$ and $x^2 + y^2 = 25$, if joined successively, will result in a rectangle.
4. If $\frac{a}{b} = \frac{4}{3}$ and $\frac{c}{d} = \frac{9}{14}$, what is the value of $\frac{3ac - bd}{4bd - 7ac}$?
5. Solve for x: $\frac{a + x}{a} - \frac{2x}{a + x} - \frac{1}{3} = \frac{x^2(a - x)}{a(a^2 - x^2)}$
6. Solve: $\frac{x + 3}{x + 6} - \frac{x + 6}{x + 9} = \frac{x + 2}{x + 5} - \frac{x + 5}{x + 8}$
7. The points $A(-2, 6)$, $B(-12, -2)$, and $C(0, 4)$ are on the graph of $y = ax^2 + bx + c$. Find a, b, and c.
8. Simplify: $\left(\sqrt[6]{\sqrt[3]{a^9}}\right)^4\left(\sqrt[6]{\sqrt{a^9}}\right)$
9. In a determinant, if two rows or two columns are identical, the determinant equals zero. Without expanding, find two solutions to this equation.

$$\begin{vmatrix} x^2 & 2x & 3 \\ 1 & 2 & 3 \\ 9 & -6 & 3 \end{vmatrix} = 0$$

10. Together Bob and Bill can plant a field of corn in 2 days. Bill and Bart can do the same job together in 4 days, and Bob and Bart can do it in 2.4 days. How long will Bob take to plant the field alone?

Review of Word Problems: Chapters 1–10

1. Five years from now, Mr. Morgan's age will be 3 times his son's present age. The sum of Mr. Morgan's present age and his son's age 5 years ago is 50. What is Mr. Morgan's age now? *(Section 1–7)*
2. A hockey team's yearly averages over the past four years were 0.430, 0.580, 0.375 and 0.605. What average must the team achieve this year in order to have a minimum five-year average of 0.500? *(Section 2–4)*
3. Out of a group of 320 people, 82 are skiers. At this rate, how many skiers could be expected in a group of 800 people? *(Section 3–7)*
4. Wingit airlines charges \$120 for a first-class ticket and \$104 for a coach ticket to fly from Hayer to Thayer. Total ticket receipts from a flight carrying 92 passengers were \$10,192. How many tickets of each kind were sold for the flight? *(Section 4–4)*
5. Yvonne Bailey has three times as much money invested at 16% as she does at 9%. Her total annual income from both investments is \$2280. How much did she invest at each rate? *(Section 4–4)*

6. One alloy of copper is 30% pure copper, while another is 18% pure copper. The two alloys are melted together to produce 50 kilograms of alloy containing 22% pure copper. How many kilograms were there of the 30% pure copper and of the 18% pure copper alloy? *(Section 4–4)*

7. A gardener mows a rectangular lawn by cutting a uniform strip around the edges until one third of the grass remains uncut. The lawn is 20 meters wide and 30 meters long. How wide is the strip? *(Section 6–10)*

8. A fishing club planned a trip for which the total costs were $288. When four additional people decided to go on the trip, the cost per person decreased by $6. How many went on the trip? *(Section 7–6)*

9. A speedboat travels 80 kilometers per hour in still water. The boat takes as long to travel 72 kilometers upstream as it does to travel 88 kilometers downstream. What is the speed of the current? *(Section 7–7)*

10. A large pipe can fill a pool in 4 hours; a smaller pipe can empty it in 6 hours. How long will it take to fill the pool if both pipes are open? *(Section 7–8)*

11. The distance, d, in feet an object will fall in 5 seconds is given approximately by $d = 16t^2$. A stone is dropped from a 800-foot cliff. How long will it take to hit the ground below? Round your answer to the nearest second. *(Section 8–9)*

12. Find two numbers whose sum is 20 and whose product is a maximum. *(Section 8–12)*

13. The vertical distance (height), d, of an object thrown upward at an initial velocity of 40 meters per second is given by $d = 40t - 4.9t^2$. How long will it take the object to reach its highest point? Round your answer to the nearest second. *(Section 8–12)*

14. A rectangular rug has a perimeter of 36 meters and an area of 72 square meters. Find the dimensions of the rug. *(Section 9–7)*

15. Two backpackers, Jane and Dan, hike the same distance of 15 miles. Dan hiked one mile an hour faster than Jane and Jane took $1\frac{1}{4}$ hours longer to hike the total distance. Find the rate of each person. *(Section 9–7)*

16. The frequency, f, of a radio wave varies inversely as the length, w, of the wave. A wave 200 meters long has a frequency of 1400 kilocycles per second. A second wave has a length of 800 meters. What is the frequency of the second wave? *(Section 10–6)*

17. The speed of two pulleys varies inversely as their diameters. A pulley with a diameter of 32 centimeters turns at 600 revolutions per minute. What is the speed of a pulley with a diameter of 24 centimeters? *(Section 10–6)*

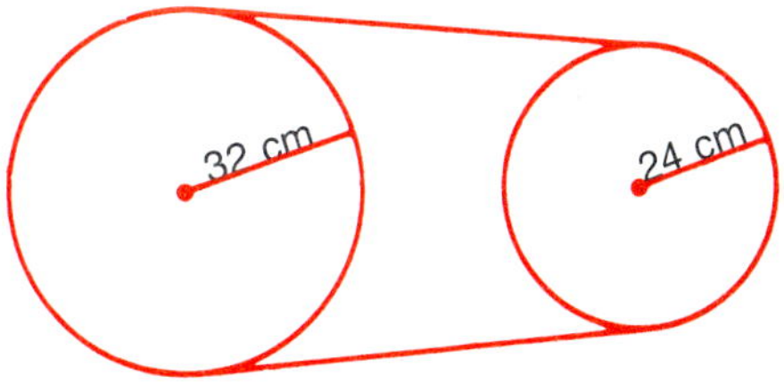

18. The sum of the squares of two numbers is 34. The difference between the squares of the numbers is 16. Find the numbers. *(Section 10–8)*

Cumulative Review: Chapters 8–10

Choose the best answer. Choose *a*, *b*, *c*, or *d*.

1. Write $\sqrt[3]{x^2} \cdot \sqrt[3]{x^5}$ in simplest radical form.
   a. $x\sqrt[6]{x}$ b. $x^2\sqrt[3]{x}$ c. $x\sqrt[3]{x}$ d. $x^3\sqrt[3]{x}$

2. Find the product and simplify: $(2\sqrt{3} - 5)(5\sqrt{3} + 7)$
   a. $5 - 11\sqrt{3}$ b. $-5 + 11\sqrt{3}$ c. $5 + 11\sqrt{3}$ d. $-(5 + 11\sqrt{3})$

3. Simplify: $-3\sqrt{54} - 2\sqrt{18} + 5\sqrt{96}$
   a. $\sqrt{3}(11\sqrt{2} - 6)$ b. $5\sqrt{6}$ c. $11\sqrt{6} - 6\sqrt{2}$ d. $\sqrt{6}(11 + 6\sqrt{2})$

4. Solve for x: $-\sqrt{x^2 + 3} = x + 1$
   a. $\{1\}$ b. $\{-1\}$ c. $\{-2\}$ d. $\{\ \}$ or ϕ

5. Find the coordinates of the midpoint of a line segment whose end points have coordinates $P(-5, 3)$ and $Q(-7, -1)$.
   a. $(-12, 2)$ b. $(-6, 1)$ c. $(1, 2)$ d. $(-1, -2)$

6. Find the distance between $A(-3, -7)$ and $B(1, -4)$.
   a. $5\sqrt{5}$ b. 25 c. 5 d. $\sqrt{137}$

7. Find the value of p such that $G(2, 2)$ is on the graph of $y = 2x^2 - 3x + p$.
   a. 12 b. 0 c. 4 d. -4

8. If y varies directly as the square of x and $y = 8$ when $x = 2$, find x when $y = 50$.
   a. 5 b. -5 c. 5 or -5 d. 5000

9. Which graph is congruent to the graph of $y = 2x^2$ with its vertex 3 units below that of $y = 2x^2$?
   a. $y = -2x^2 - 3$ b. $y = 2x^2 + 3$
   c. $y = -3x^2 + 2$ d. $y = 3x^2 - 2$

10. What are the coordinates of the vertex of $y = 3x^2 - 6x + 2$?
    a. $(-1, 11)$ b. $(-2, 26)$ c. $(2, 2)$ d. $(1, -1)$

11. The sum of two numbers is 10. Find the two numbers such that the sum of the square of the double of one and the square of the other is a minimum.
    a. 2, 8 b. 3, 7 c. 4, 6 d. 5, 5

12. Which number is a pure imaginary number?
    a. $-\sqrt{2}$ b. $-\sqrt{-x}$, $x < 0$ c. $\sqrt{-4}$ d. $\sqrt{6}$

13. Find the sum: $(2 - 3i) + (-5 + 2i)$
    a. -3 b. $7 - 5i$ c. $-4 + 12i$ d. $-(3 + i)$

14. Express in standard form: $\dfrac{2 + i}{3 - 2i}$
    a. $\dfrac{8}{13} + \dfrac{i}{13}$ b. $\dfrac{4}{13} + \dfrac{7i}{13}$ c. $4 + 7i$ d. $8 + i$

15. Which is the solution for $8x^2 + 40 = 0$?

a. $-\sqrt{5}$ or $\sqrt{5}$ b. $\sqrt{5}$ c. $i\sqrt{5}$ or $-i\sqrt{5}$ d. $i\sqrt{5}$

16. What should you do in order to complete the square for $x^2 - 6x = 12$?

a. Add 36 to both sides. b. Add 9 to both sides.
c. Subtract 36 from both sides. d. Subtract 9 from both sides.

17. Give the zeros of $f(x) = x^2 - 4x + 1$.

a. $2 - \sqrt{3}, 2 + \sqrt{3}$ b. $-2 + 2\sqrt{3}, 2 - 3\sqrt{3}$
c. $-2 + \sqrt{3}, -2 - \sqrt{3}$ d. $4 - 2\sqrt{3}, 4 + 2\sqrt{3}$

18. The difference of two numbers is 5 and the sum of their squares is 125. Find the numbers.

a. 5 and 10 b. $5\sqrt{5}$ and 0
c. -10 and -5 or 5 and 10 d. -10 and -5

19. For what value of c will $y^2 - 6y + c = 0$ have one real root?

a. 36 b. 9 c. 3 d. 6

20. Which of the following is the equation of a circle with center $A(-2, -5)$ and radius 4?

a. $(x - 2)^2 + (y - 5)^2 = 16$ b. $(x - 2)^2 + (y - 5)^2 = 4$
c. $(x + 2)^2 + (y + 5)^2 = 4$ d. $(x + 2)^2 + (y + 5)^2 = 16$

21. Which of the following is the equation of an ellipse with center at the origin, with x intercepts 3 and -3, and y intercepts 6 and -6?

a. $\frac{x^2}{9} + \frac{y^2}{36} = 1$ b. $\frac{x^2}{36} + \frac{y^2}{9} = 1$
c. $\frac{x^2}{9} - \frac{y^2}{36} = 1$ d. $\frac{x^2}{36} - \frac{y^2}{9} = 1$

22. What are the coordinates of the foci of $\frac{y^2}{49} - \frac{x^2}{36} = 1$?

a. $(7, 0)$ and $(-7, 0)$ b. $(0, 6)$ and $(0, -6)$
c. $(0, \sqrt{85})$ and $(0, -\sqrt{85})$ d. $(6, 0)$ and $(-6, 0)$

23. Which equation is *not* that of a hyperbola?

a. $\frac{x^2}{2} - \frac{y^2}{6} = 1$ b. $xy = -8$ c. $3y^2 - 9x^2 = 9$ d. $2x^2 + 3y^2 = 6$

24. If y varies directly as x and the square of z but varies inversely as t^3, write the equation expressing these relations.

a. $y = xz^2t^3$ b. $y = kxz^2t^3$ c. $y = \frac{kxz^2}{t^3}$ d. $y = \frac{xz^2}{t^3}$

25. Solve: $\begin{cases} x^2 - y^2 = 84 \\ x - y = 2 \end{cases}$

a. $\{(22, 20)\}$ b. $\{(20, 18)\}$
c. $\{(-22, 20); (22, 20)\}$ d. $\{(-20, -18); (20, 18)\}$

Preparing for College Entrance Tests

Choose the best answer. Choose *a*, *b*, *c*, or *d*.

1. If $\sqrt{n}$ is a rational number, which of the following must be rational numbers?

I. $\sqrt{100n}$ **II.** $\sqrt{\frac{n}{100}}$ **III.** $\sqrt{n + 100}$

(a) I only **(b)** I and II only **(c)** II and III only **(d)** I, II, and III

2. Which of the following numbers is in the solution set for $3x^2 - 77x + 120 = 0$?

(a) -5 **(b)** -1 **(c)** 1 **(d)** 24

3. If $6x + 9 = -\frac{3}{4}$, what is the value of $2x + 3$?

(a) $-\frac{13}{8}$ **(b)** $-\frac{1}{4}$ **(c)** $-\frac{3}{4}$ **(d)** $\frac{1}{4}$

4. If $y < -1$, which of the following increase(s) as y decreases?

I. $\frac{1}{1 - y}$ **II.** $\sqrt{1 - y}$ **III.** $\frac{1 - y}{y}$

(a) I only **(b)** II only **(c)** II and III only **(d)** None increase.

5. Which of the following numbers is in the solution set for $3x^2 + 44x + 65 = 0$?

(a) $-\frac{5}{3}$ **(b)** -1 **(c)** $\frac{1}{3}$ **(d)** $\frac{3}{11}$

6. If $a - b = 2$ and $a^2 + b^2 = 34$, find the value of ab.

(a) -17 **(b)** -15 **(c)** 15 **(d)** 30

7. $\sqrt{\frac{25}{16} - 1} = \underline{\ ?\ }$

(a) $\frac{1}{4}$ **(b)** $\frac{3}{4}$ **(c)** $\frac{\sqrt{6}}{2}$ **(d)** None of these

8. If $\dfrac{3}{2 + \dfrac{x + 3}{3}} = \dfrac{3}{4}$, what is the value of x?

(a) -3 **(b)** -2 **(c)** 0 **(d)** 3

9. If $\frac{x^2}{x^2 - 1} = \frac{x}{x^2 - 1}$, what is the value of x?

(a) 1 **(b)** 0 **(c)** -1 **(d)** None of these

10. If $\dfrac{2}{4 + \dfrac{x - 2}{4}} = \dfrac{1}{2}$, what is the value of x?

(a) -4 **(b)** 0 **(c)** 2 **(d)** 4

11. If $\sqrt{1} \cdot \sqrt{2} \cdot \sqrt{3} \cdot \sqrt{4} \cdot \sqrt{5} \cdot \sqrt{6} \cdot \sqrt{7} \cdot \sqrt{8} \cdot \sqrt{9} = 72\sqrt{70}$,
then $\sqrt{3} \cdot \sqrt{4} \cdot \sqrt{5} \cdot \sqrt{6} \cdot \sqrt{7} \cdot \sqrt{8} \cdot \sqrt{9} \cdot \sqrt{10} = \underline{\ ?\ }$

(a) $360\sqrt{14}$ **(b)** $720\sqrt{7}$ **(c)** $72\sqrt{70}$ **(d)** None of these

CHAPTER 11

Exponential and Logarithmic Functions

11-1 Negative Exponents

Recall that the following theorems (stated on pages 193 and 209) apply to exponents which are whole numbers.

$$a^m \cdot a^n = a^{m+n},\ a \neq 0$$

$$a^0 = 1,\ a \neq 0$$

These theorems can be extended so that any integer can be used as an exponent. To do this, we define exponents so that the same exponent theorems apply.

Assume that $a^m \cdot a^n = a^{m+n}$ also holds for negative integral exponents. Note how applying this property shows how to define a^{-m}, $a^{-m} \neq 0$, $a^m \neq 0$.

Using Number Exponents	**Using Variable Exponents**
$2^7 \cdot 2^{-7} = 2^{7+(-7)}$	$a^m \cdot a^{-m} = a^{m+(-m)}$
$= 2^0 = 1$	$a^m \cdot a^{-m} = a^0 = 1$
Thus, $2^7 \cdot 2^{-7} = 1$	Thus, $a^m \cdot a^{-m} = 1$
Then $2^{-7} = \frac{1}{2^7}$.	Then $a^{-m} = \frac{1}{a^m}$.

This leads to the definition of a^{-m}, $a \neq 0$.

Definition

For any real number a, $a \neq 0$ and any integer m,

$$a^{-m} = \frac{1}{a^m} \quad \text{and} \quad a^m = \frac{1}{a^{-m}}.$$

EXAMPLE 1 Write an equivalent expression with positive exponents and without parentheses for each of the following.

a. b^{-4} **b.** $\frac{1}{a^{-10}}$ **c.** mn^{-2} **d.** $(a+b)^{-1}$

Solutions: Apply the definitions of a^{-m} and a^m.

a. $b^{-4} = \frac{1}{b^4}$

b. $\frac{1}{a^{-10}} = \frac{1}{\frac{1}{a^{10}}} = 1 \cdot a^{10} = a^{10}$

c. $mn^{-2} = m \cdot \frac{1}{n^2} = \frac{m}{n^2}$

d. $(a+b)^{-1} = \frac{1}{(a+b)^1} = \frac{1}{a+b}$

Be careful when applying the definition to expressions such as $(3x)^{-2}$ and $3x^{-2}$.

$$(3x)^{-2} = \frac{1}{(3x)^2} = \frac{1}{9x^2} \quad \text{and} \quad 3x^{-2} = 3 \cdot \frac{1}{x^2} = \frac{3}{x^2}$$

Thus, $(3x)^{-2} \neq 3x^{-2}$.

The exponent property $(a^m)^n = a^{mn}$ (see Theorem 6–2, page 193) also holds for negative integral exponents.

EXAMPLE 2 Write an equivalent expression without negative exponents and without parentheses for $(2^{-3})^4$.

Solution:

Method 1

$$(2^{-3})^4 = \left(\frac{1}{2^3}\right)^4 \quad \leftarrow \textbf{By definition}$$
$$= \frac{1}{(2^3)^4}$$
$$= \frac{1}{2^{12}}, \text{ or } \frac{1}{4096}$$

Method 2

$$(2^{-3})^4 = (2)^{-3 \cdot 4} \quad \leftarrow (a^m)^n = a^{mn}$$
$$= (2)^{-12}$$
$$= \frac{1}{2^{12}}, \text{ or } \frac{1}{4096}$$

CLASSROOM EXERCISES

Write with positive exponents.

1. 3^{-2}
2. $\frac{1}{2^{-4}}$
3. $(2+3)^{-3}$
4. $2^{-3} + 4^{-2}$
5. $(2^{-3})^{-2}$
6. t^{-5}
7. $\frac{1}{s^{-2}}$
8. $(x+y)^{-4}$
9. $r^{-1} + q^{-4}$
10. $(a^{-4})^4$

WRITTEN EXERCISES

A Write an equivalent expression with positive exponents and without parentheses for each of the following. The replacement set for the variable is the set of complex numbers, excluding 0. (REMEMBER: i is not a variable.)

1. a^{-1}
2. a^{-2}
3. $\frac{1}{b^{-3}}$
4. $\frac{1}{m^{-8}}$
5. y^{-5}
6. $2x^{-3}$
7. $x^{-3}y^5$
8. $r^{-2}t^3$
9. a^2b^{-3}
10. $(2+3)^2$
11. $2^{-3} \cdot 6$
12. $5^3 \cdot 3^{-2}$
13. $(m+n)^{-1}$
14. $(r+s)^{-2}$
15. $(2+3)^{-2}$
16. $(1+2)^{-4}$
17. $\frac{1}{9^{-2}}$
18. $\frac{1}{a^{-3}}$
19. $\frac{1}{c^{-2}}$
20. $\frac{8}{y^{-1}}$
21. $\frac{2^{-3}}{5^{-1}}$
22. $\frac{a^{-1}}{b^{-4}}$
23. $\frac{m^{-4}}{t^{-3}}$
24. $\frac{s^{-3}}{q^{-5}}$
25. $\frac{n^2}{r^{-3}}$
26. $\frac{t^3}{s^{-5}}$
27. $\frac{x^{-4}y}{ab^{-3}}$
28. $\frac{3a^{-4}}{9c^{-3}}$
29. $\frac{5x^{-2}b^{-3}}{c^{-4}}$
30. $\frac{c^2}{d^{-1}}$
31. $\frac{y^{-2}}{x^3}$
32. $\frac{x^{-9}}{17}$
33. $\frac{4a^{-2}b^3}{9c^{-3}}$
34. $\frac{9r^{-3}t^5}{13q^{-3}}$
35. $\frac{an^{-5}t^{-5}}{5r^3z^{-1}}$
36. $(2^{-2})^3$
37. $(2^{-4})^{-1}$
38. $(3^{-2})^{-3}$
39. $(7^0)^{-2}$
40. $(2 \cdot 3)^2$
41. $(2x)^{-3}$
42. $(mn)^{-1}$
43. $(abc)^{-3}$
44. $(2\text{i})^0$
45. $(3+4\text{i})^0$
46. $(5\text{i})^{-2}$
47. $(5\text{i})^{-1}$
48. $(2-\text{i})^{-1}$
49. $(3+2\text{i})^{-2}$
50. $(1-3\text{i})^{-3}$

Write an equivalent expression with positive exponents and without parentheses for each of the following. The replacement set for the variables is the set of complex numbers, excluding 0. REMEMBER: i is not a variable.

51. $(\frac{1}{2})^{-1}$ **52.** $(\frac{3}{4})^{-1}$ **53.** $(\frac{2}{3})^{-1}$ **54.** $(\frac{1}{2}i)^{-3}$ **55.** $(\frac{2i}{3})^{-2}$

56. $\left(\frac{a}{b}\right)^{-1}$ **57.** $\left(\frac{r}{t}\right)^{-5}$ **58.** $\frac{(x+y)^{-1}}{9}$ **59.** $\frac{(xy^{-2})^{-1}}{(xy^{-2})^{-1}}$ **60.** $\frac{(1+2i)}{(1-2i)^{-2}}$

B In Exercises 61–70, evaluate each expression for $a = 2$ and $b = 3$.

61. a^0 **62.** $2a^0b^{-2}$ **63.** $3a^2b^{-1}$ **64.** $(ab)^{-2}$ **65.** $a^{-2}b^{-2}$

66. $(a+b)^0$ **67.** $a^0 + b^0$ **68.** $(\frac{2}{3} + 5b)^0$ **69.** $(a^0 + b^0)^2$ **70.** $(a^2 + b^2)^0$

In Exercises 71–80, evaluate each expression for $a = -1$ and $b = -2$.

71. $a^{-1} + b^{-1}$ **72.** $b^{-2} - a^{-2}$ **73.** $(a+b)^{-1}$ **74.** $(b-a)^{-2}$ **75.** $(a^0b^0)^3$

76. $[(a^2)^{-3}]^{-1}$ **77.** $[(b^{-1})^2]^{-3}$ **78.** $[(ab)^{-1}]^{-3}$ **79.** $[(a^{-2}b)^{-1}]^3$ **80.** $[(a^{-3})^0]^2$

C

81. Draw a graph for $y = 2^x$ where the replacement set for x is $\{\pm 5, \pm 4, \pm 3, \pm 2, \pm 1, 0\}$.

82. Draw a graph for $y = (-2)^x$ where the replacement set for x is $\{\pm 5, \pm 4, \pm 3, \pm 2, \pm 1, 0\}$. How does this graph differ from the graph of $y = 2^x$?

Puzzle

Complete each sequence.

1. 8, 5, 4, 9, 1, 7, 6, 3, 2, __?__

2. 11, 31, 71, 91, 32, 92, 13, 73, __?__

3. |1, Ƨ2, Ɛ3, ᖷ4, ꙅ5, ∂6, __?__

4. 1, 8, 11, 69, 88, 96, __?__

5. 61, 52, 63, 94, 46, 18, __?__

6. O, T, T, F, F, S, S, E, __?__

REVIEW CAPSULE FOR SECTION 11–2

Write each number using a power of 10.

Examples: **a.** $1{,}000{,}000 = 10^6$

b. $0.01 = \frac{1}{100} = \frac{1}{10^2}$

1. 100 **2.** 10,000 **3.** 1,000,000,000 **4.** 100,000 **5.** 0.1

6. 0.001 **7.** 0.000001 **8.** 0.0001 **9.** 1 **10.** 10

11–2 Scientific Notation

Scientific notation is a way of expressing very large and very small numbers as a product. For example, in scientific notation

253,000 is written as $\mathbf{2.53 \times 10^5}$.

Definition

> A number is written in **scientific notation** when it is written as the product of a number greater than or equal to 1 but less than 10 and an integral power of 10.

The following procedure will help you to express a number in scientific notation.

Procedure

> To write a number in scientific notation:
>
> [1] Using the digits of the number in the order in which they occur, write the first factor as a number a, where $1 \le a < 10$.
>
> [2] To find the second factor, divide the original number by the first factor.
>
> [3] Express the second factor as a power of 10.

EXAMPLE 1 Write each number in scientific notation.

a. 89,000,000 **b.** 0.00158

Solutions:

a. $89{,}000{,}000 = 8.9 \times \underline{\ ?\ }$ ← **Divide 89,000,000 by 8.9.**

$= 8.9 \times 10{,}000{,}000$ ← **Express 10,000,000 as a power of 10.**

$= \mathbf{8.9 \times 10^7}$

b. $0.00158 = 1.58 \times \underline{\ ?\ }$ ← **Divide 0.00158 by 1.58.**

$= 1.58 \times 0.001$ ← **Express 0.001 as a power of 10.**

$= \mathbf{1.58 \times 10^{-3}}$

Numbers written in scientific notation can be expressed in decimal form.

EXAMPLE 2 Write in decimal form.

a. 5.6×10^9 **b.** 1.9×10^{-4}

Solutions:

a. $5.6 \times 10^9 = 5.600{,}000{,}000$ ← **Move the decimal point 9 places to the right.**

$= \mathbf{5{,}600{,}000{,}000}$

b. $1.9 \times 10^{-4} = 0001.9$ ← **Move the decimal point 4 places to the left.**

$= \mathbf{0.00019}$

Scientific notation can be used to make some calculations easier.

EXAMPLE 3 Multiply or divide as indicated: **a.** $0.983 \times 517{,}000$ **b.** $\frac{275{,}000}{0.0025}$

Solutions: Write each factor in scientific notation.

a. $0.983 \times 517{,}000 = (9.83 \times 10^{-1})(5.17 \times 10^{5})$

$= (9.83 \times 5.17)(10^{-1} \times 10^{5})$

$= 50.8211 \times 10^{4}$

$= \mathbf{508{,}211}$

b. $\frac{275{,}000}{0.0025} = \frac{2.75 \times 10^{5}}{2.5 \times 10^{-3}}$

$= \frac{2.75}{2.5} \times 10^{5-(-3)}$

$= 1.1 \times 10^{8}$

$= \mathbf{110{,}000{,}000}$

In Example 3, you could use a calculator to find (9.83×5.17) and $\frac{2.75}{2.5}$.

CLASSROOM EXERCISES

Complete.

1. $255{,}000 = 2.55 \times$ _?_ **2.** $750 = 7.5 \times$ _?_ **3.** $0.075 = 7.5 \times$ _?_

4. $0.00486 =$ _?_ $\times 10^{-3}$ **5.** $80{,}000 =$ _?_ $\times 10^{4}$ **6.** $0.000015 =$ _?_ $\times 10^{-5}$

Write in scientific notation.

7. 526 **8.** 0.03 **9.** 2,250,000,000 **10.** 0.00000983

Write in decimal form.

11. 9.8×10^{4} **12.** 4.26×10^{-3} **13.** 2.18×10^{-8} **14.** 5.83×10^{6}

WRITTEN EXERCISES

A Write in scientific notation.

1. 190,000 **2.** 6981 **3.** 0.0059 **4.** 0.0000017

5. 0.0008 **6.** 0.1246 **7.** 98,472,000,000 **8.** 4,560,000,000,000

9. 2805 **10.** 28,050 **11.** 0.00283 **12.** 0.0703

13. 0.00000000007 **14.** 0.6666666 **15.** 10 **16.** 48,000,000,000

17. 0.000001 **18.** 0.0075 **19.** 790,000,000 **20.** 0.3

Write in decimal form.

21. 6×10^{-4} **22.** 1.5×10^{-8} **23.** 5.06×10^{10} **24.** 3.008×10^{3}

25. 9.018×10^{-6} **26.** 4.5×10^{7} **27.** 7.865×10^{8} **28.** 3.9×10^{-1}

29. 2.5×10^{-1} **30.** 1.44×10^{8} **31.** 5×10^{-4} **32.** 6.25×10^{-6}

Use scientific notation to perform the indicated operations. Write each answer in decimal form.

33. $0.63 \times 720{,}000$

34. $0.08 \times 380{,}000{,}000$

35. $84{,}000{,}000{,}000 \times 0.0047$

36. $5{,}610{,}000 \times 0.00000011$

37. $\frac{0.00000124}{0.0031}$

38. $\frac{465{,}000{,}000}{0.00015}$

39. $\frac{0.0004224}{0.24}$

40. $\frac{0.000056}{8{,}000{,}000{,}000{,}000}$

41. $\frac{0.00000072}{900{,}000{,}000{,}000}$

42. $\frac{126{,}000{,}000{,}000}{0.0000018}$

43. $\frac{2{,}070{,}000{,}000}{0.000023}$

44. $\frac{1{,}260{,}000{,}000}{0.0021}$

45. $\frac{0.00023}{5{,}520{,}000}$

APPLICATIONS: Using Scientific and Decimal Notation

In Exercises 46–55, write the numbers in scientific notation.

46. Light travels at the rate of approximately 300,000,000 meters per second.

47. The radius of the earth is about 6371 kilometers.

48. The sun is one of about 100 billion stars making up the Milky Way.

49. The distance light travels in a year is about 9,460,000,000,000 kilometers. (This is called a **light year.**)

50. The half-life of uranium is $4\frac{1}{2}$ billion years. (**Half-life** is a convenient term used by scientists for the time required for half of the atoms originally present in a radioactive element to disintegrate.)

51. Electromagnetic waves whose lengths are between 0.000036 centimeter and 0.000077 centimeter cause the sensation of light.

52. Special balances for weighing small quantities can weigh a quantity as small as 0.00000001 gram. Such a quantity of matter would contain more than 10,000,000,000,000 atoms.

53. A hydrogen atom weighs 0.000 000 000 000 000 000 000 001 67 gram.

54. The diameter of the planet Jupiter is about 143,000 kilometers.

55. One cubic centimeter of air at 0° Celsius and at a pressure of 76 centimeters of mercury has a mass of 0.00129 gram.

In Exercises 56–62, write the numbers in decimal form.

56. The mass of the sun is 2×10^{30} kilograms.

57. The volume of the earth is about 1.08×10^{27} cubic centimeters.

58. The temperature of the sun is 6×10^3 degrees Celsius.

59. One foot of copper wire expands 1.6×10^{-5} feet with an increase in temperature of 1° Celsius.

60. A helium atom has a diameter of 2.2×10^{-8} centimeters.

61. Light travels at the rate of 2.9979×10^8 meters per second.

62. The meter is defined as 1.65076373×10^6 wavelengths of the orange light in the spectrum of krypton 86.

11–3 Simplifying Expressions with Exponents

To simplify an algebraic expression involving integral exponents means to write the expression with whole number exponents.

EXAMPLE 1 Simplify: $\dfrac{a^{-m}}{b^{-m}}$

Solution: Method 1 $\dfrac{a^{-m}}{b^{-m}} = \dfrac{\dfrac{1}{a^m}}{\dfrac{1}{b^m}}$ ← **Express with positive exponents.**

$$= \frac{1}{a^m} \cdot \frac{b^m}{1} = \frac{b^m}{a^m}$$

Method 2 $\dfrac{a^{-m}}{b^{-m}} = \dfrac{a^{-m}}{b^{-m}} \cdot \dfrac{a^m b^m}{a^m b^m}$ ← $\dfrac{a^m b^m}{a^m b^m} = 1$

$$= \frac{a^0 b^m}{a^m b^0} = \frac{1 \cdot b^m}{a^m \cdot 1} = \frac{b^m}{a^m}$$

Recall that, by definition, $a^{-m} = \dfrac{1}{a^m}$ and $a^m = \dfrac{1}{a^{-m}}$. Applying the definition often makes it easier to simplify expressions.

EXAMPLE 2 Simplify: $3x^{-2}y - 2x^{-1}y^2 + 5 + \dfrac{7y^2}{x} - \dfrac{3y}{x^2}$, $x \neq 0$

Solution: Express with whole-number exponents. Then combine like terms.

$$3x^{-2}y - 2x^{-1}y^2 + 5 + \frac{7y^2}{x} - \frac{3y}{x^2} = \frac{3y}{x^2} - \frac{2y^2}{x} + 5 + \frac{7y^2}{x} - \frac{3y}{x^2}$$

$$= 5 + \frac{5y^2}{x}$$

Example 3 shows how to simplify a product involving negative exponents.

EXAMPLE 3 Simplify: $\dfrac{1}{(a^{-1} + b^{-1})}(a + b)$, where $a \neq 0$, $b \neq 0$.

Solution: $\left(\dfrac{1}{a^{-1} + b^{-1}}\right)(a + b) = \left(\dfrac{1}{\dfrac{1}{a} + \dfrac{1}{b}}\right)(a + b)$

$$= \left(\frac{1}{\dfrac{b + a}{ab}}\right)(a + b)$$ ← **Multiply by $\dfrac{ab}{ab}$.**

$$= \left(\frac{1}{\dfrac{a + b}{ab}}\right)(a + b)\left(\frac{ab}{ab}\right) = \frac{\overset{1}{\cancel{(a + b)}}ab}{\underset{1}{\cancel{(a + b)}}} = ab$$

CLASSROOM EXERCISES

Simplify. No denominator equals zero.

1. $\frac{a^{-2}}{b^{-3}}$ 2. $\frac{2ab^{-1}}{3b^{-3}d}$ 3. $\frac{(2x)^{-2}}{(3y)^{-3}}$ 4. $\frac{6x^{-2}}{2x^{-1}}$ 5. $\frac{x^2y^{-1}}{y^3}$

6. $2x^{-1}y + \frac{y}{x}$ 7. $5x^{-2}y + x^{-1} - \frac{6y}{x^2} + \frac{3}{x}$ 8. $\frac{1}{a^{-1}}(a - b)$

9. $(a^2 - b^2)(a - b)^{-1}$ 10. $(a^{-1} + b^{-1})\left(\frac{ab^2}{a^2 - b^2}\right)$ 11. $(a + b)^{-3}(a + b)^2$

WRITTEN EXERCISES

A Simplify. No denominator equals zero.

1. $\frac{x^{-1}y^{-1}}{z^{-3}}$ 2. $2^{-1}ab^{-3}$ 3. $a^{-2} + b^{-2}$ 4. $ab^{-3} + a^{-4}b^2$ 5. $3^{-5}ab^{-4}c^{-2}$

6. $\frac{(a + b)^{-3}}{4^{-1}c^3b^{-1}}$ 7. $\frac{x^{-2}y^3}{x^3y^{-2}}$ 8. $\frac{a^5b^{-3}}{a^{-1}b^4}$ 9. $\frac{2^{-1}a^{-1}}{3a^2b^{-1}}$ 10. $\frac{3^{-1}u^{-1}v^2}{2u^3v^{-4}}$

11. $\frac{p^{-4}g^5r^{-6}}{p^{-2}g^{-2}r^{-7}}$ 12. $\left(\frac{s^{-4}m^{-4}}{s^{-2}m}\right)^0$ 13. $\frac{4a}{a^{-1}} + \frac{3}{a^{-2}}$ 14. $\frac{3m}{n^{-1}} + \frac{2n}{m^{-1}}$ 15. $(r^{-1}s^2)^{-1}$

16. $(r^3s^{-4})^{-2}$ 17. $(rs^2)^2(rs)^{-1}$ 18. $\frac{(pq)^3}{(p^2q^{-1})^2}$ 19. $\frac{(2s)^{-2}t^4}{(2s)^{-3}t^{-1}}$ 20. $\left(\frac{9xy^{-2}}{3xy^3}\right)^{-3}$

21. $x^{-2} + 4x^{-1} + \frac{7}{x^3}$ 22. $3x^{-2} + \frac{5}{3}x^{-1}y^{-1} - \frac{4}{3}y^2$ 23. $ax^{-3} + bx^{-2} - cx^{-1}$

24. $\frac{1}{a^2} - 5(ab)^{-1} + 7b^{-2}$ 25. $\frac{9t^{-2}}{r} + 3rt^{-3} - \frac{3}{t^3}$ 26. $5x^{-2}y + x^{-1} - \frac{6y}{x^2} + \frac{3}{4}$

27. $\left(\frac{2}{x^{-1} - y^{-1}}\right)(y - x)$ 28. $(r + t)^2\left(\frac{1}{r^{-1} + t^{-1}}\right)$ 29. $\left(\frac{3}{a^{-1} - b^{-1}}\right)(b - a)^3$

30. $a^{-1}(b^{-1}a + c^{-2}a^2)$ 31. $(a^{-1} + b^{-1})(a + b)$ 32. $(a + b)^{-2}(a^2 - b^2)$

33. $(x + y)^{-2}(x^{-1} + y^{-1})$ 34. $(x^2 + y^2)(x^{-2} + y^{-2})$ 35. $x(x - x^{-1})$

B Write each of the following in such a way as to avoid the fraction form.

36. $\frac{1}{a^3}$ 37. $\frac{1}{x^2} + \frac{1}{y^2}$ 38. $\frac{2xy^3}{wz^2}$ 39. $\frac{2x^{-1}y^{-2}}{wz^{-2}}$

40. $\frac{x}{y^4}$ 41. $\frac{1}{(x + y)^2}$ 42. $\frac{2}{b^2} + \frac{3}{c^{-3}}$ 43. $\frac{3}{b^{-2}} + \frac{4}{c^4}$

Simplify.

44. $(x^{-2} + y^{-2})(x^{-2} + y^{-2})(x + y)$ 45. $\frac{x}{y^2} - 2x^{-1}y^{-1} + \frac{6y^{-2}}{x} - 3x^0$

46. $\frac{a^{-1} + b^{-1}}{a^{-2} + b^{-2}}$ 47. $\frac{x^{-1} + 3y^{-1}}{x^{-2} + 4x^{-1}y + 3y^{-2}}$ 48. $\left(\frac{1}{x} - \frac{1}{y}\right)(x - y) + \left(\frac{1}{x^{-1} - y^{-1}}\right)(x - y)$

49. Show that $(x^{-1} + y^{-1})^{-1} = x + y$ is a false statement.

50. Show that $(x + y)^{-1} = x^{-1} + y^{-1}$ is a false statement.

REVIEW CAPSULE FOR SECTION 11–4

Evaluate. *(Pages 270–271)*

1. $\pm\sqrt{36}$	**2.** $\sqrt[3]{125}$	**3.** $\sqrt[4]{16}$	**4.** $\sqrt{8}$	**5.** $-\sqrt[5]{-32}$
6. $\sqrt{9y^2}$	**7.** $\sqrt[3]{-64a^6b^{15}}$	**8.** $\sqrt{r^2t^2}$	**9.** $\sqrt[4]{(a+b)^8}$	**10.** $-\sqrt[3]{0.729s^9t^{15}v^{21}}$

11–4 Rational Exponents

In order to extend the exponent properties to include rational exponents, assume that this property holds for rational exponents.

$$(a^m)^n = a^{mn}$$

Using Number Exponents

$$(2^{\frac{1}{7}})^7 = 2^{\frac{1}{7}\cdot 7}$$
$$= 2^{\frac{7}{7}}$$
$$= 2$$

But, $(\sqrt[7]{2})^7 = 2.$

Thus, $\mathbf{2^{\frac{1}{7}} = \sqrt[7]{2}}.$

Using Variable Exponents

$$(a^{\frac{1}{n}})^n = a^{\frac{1}{n}\cdot n},\ a \neq 0,\ n \in \mathrm{N}$$
$$= a^{\frac{n}{n}}$$
$$= a$$

But, $(\sqrt[n]{a})^n = a.$

Thus, $\mathbf{a^{\frac{1}{n}} = \sqrt[n]{a}}.$

This leads to the following definition.

Definition

> For any real number a, $a \neq 0$ and $n \in \mathrm{N}$,
> $$\mathbf{a^{\frac{1}{n}} = \sqrt[n]{a}}.$$

Use this definition to evaluate expressions involving rational exponents.

EXAMPLE 1 Evaluate: **a.** $4^{\frac{1}{2}}$ **b.** $125^{\frac{1}{3}}$ **c.** $(-32)^{\frac{1}{5}}$

Solutions: Apply the definition.

a. $4^{\frac{1}{2}} = \sqrt{4}$
$= \mathbf{2}$

b. $125^{\frac{1}{3}} = \sqrt[3]{125}$
$= \mathbf{5}$

c. $(-32)^{\frac{1}{5}} = \sqrt[5]{-32}$
$= \mathbf{-2}$

In Example 1a, note that $4^{\frac{1}{2}} = 2$, not ± 2. According to the definition, $a^{\frac{1}{n}}$ is the **principal *n*th root of *a*** (see page 270).

EXAMPLE 2 Evaluate: $4^{\frac{3}{2}}$

Solution: First, express $4^{\frac{3}{2}}$ as $(4^{\frac{1}{2}})^3$ or as $(4^3)^{\frac{1}{2}}$.

Method 1
$4^{\frac{3}{2}} = (4^{\frac{1}{2}})^3$
$= (\sqrt{4})^3$ ← **Find $\sqrt{4}$.**
$= 2^3$, or **8**

Method 2
$4^{\frac{3}{2}} = (4^3)^{\frac{1}{2}}$ ← **Find 4^3.**
$= (64)^{\frac{1}{2}}$
$= \sqrt{64}$, or **8**

Example 2 illustrates two important ideas.

1. In the rational exponent $\frac{3}{2}$, the denominator, 2, tells you which root to take. The numerator, 3, tells how many factors you have.
2. You can raise the base to the exponent of the numerator either *before* or *after* you take the root shown in the denominator.

The properties of integral exponents are also true for rational exponents.

Theorem 11–1

Rational Exponent Theorem

For all $a \in R$, $a \neq 0$, $m \in I$, $n \in N$,

$$a^{\frac{m}{n}} = (a^{\frac{1}{n}})^m = (\sqrt[n]{a})^m, \quad \text{or} \quad a^{\frac{m}{n}} = (a^m)^{\frac{1}{n}} = \sqrt[n]{a^m}$$

except when n is even and $a^m < 0$.

In Example 3, note how the use of parentheses changes the value of an expression.

EXAMPLE 3 Evaluate: **a.** $(-64)^{-\frac{2}{3}}$ **b.** $-64^{-\frac{2}{3}}$

Solutions:

a. $(-64)^{-\frac{2}{3}} = \dfrac{1}{(-64)^{\frac{2}{3}}} = \dfrac{1}{(\sqrt[3]{-64})^2} = \dfrac{1}{(-4)^2} = \mathbf{\dfrac{1}{16}}$

b. $-64^{-\frac{2}{3}} = -\dfrac{1}{64^{\frac{2}{3}}} = -\dfrac{1}{(\sqrt[3]{64})^2} = -\dfrac{1}{(4)^2} = \mathbf{-\dfrac{1}{16}}$

Thus, $\mathbf{(-64)^{-\frac{2}{3}} \neq -64^{-\frac{2}{3}}}$.

CLASSROOM EXERCISES

Evaluate.

1. $25^{\frac{1}{2}}$ **2.** $(-27)^{\frac{1}{3}}$ **3.** $9^{\frac{3}{2}}$ **4.** $-64^{\frac{3}{2}}$ **5.** $4^{-\frac{3}{2}}$

6. $-125^{-\frac{1}{3}}$ **7.** $(-125)^{-\frac{1}{3}}$ **8.** $-8^{\frac{1}{3}}$ **9.** $81^{-\frac{1}{2}}$ **10.** $(-27)^{-\frac{4}{3}}$

WRITTEN EXERCISES

A Evaluate.

1. $9^{\frac{1}{2}}$ **2.** $125^{\frac{1}{3}}$ **3.** $16^{\frac{1}{4}}$ **4.** $-27^{\frac{1}{3}}$ **5.** $-36^{\frac{1}{2}}$

6. $(-27)^{\frac{1}{3}}$ **7.** $-(-27)^{\frac{1}{3}}$ **8.** 8^0 **9.** $8^{\frac{2}{3}}$ **10.** $8^{\frac{3}{5}}$

11. $8^{\frac{5}{3}}$ **12.** $81^{\frac{3}{4}}$ **13.** $-27^{\frac{3}{5}}$ **14.** $(-32)^{\frac{3}{5}}$ **15.** $4^{\frac{5}{2}}$

16. $25^{\frac{3}{2}}$ **17.** $16^{\frac{1}{2}}$ **18.** $64^{\frac{1}{3}}$ **19.** $64^{-\frac{2}{3}}$ **20.** $16^{-\frac{3}{4}}$

21. $-25^{-\frac{1}{2}}$ **22.** $27^{-\frac{4}{3}}$ **23.** $27^{-\frac{2}{3}}$ **24.** $27^{-\frac{1}{3}}$ **25.** $(-27)^{-\frac{4}{3}}$

26. $-(-27)^{-\frac{4}{3}}$ **27.** $(\frac{4}{9})^{-1}$ **28.** $(\frac{27}{8})^{-\frac{2}{3}}$ **29.** $(0.008)^{\frac{1}{3}}$ **30.** $(0.0025)^{\frac{1}{2}}$

B Write with radical signs.

Example: $3x^{\frac{2}{3}}y = 3y\sqrt[3]{x^2}$

31. $a^{\frac{1}{2}}$
32. $b^{\frac{1}{3}}$
33. $x^{\frac{1}{4}}$
34. $n^{\frac{2}{3}}$
35. $a^{\frac{3}{4}}$
36. $2x^{\frac{3}{4}}$
37. $3a^{\frac{1}{2}}$
38. $a^{\frac{1}{4}}b^{\frac{3}{4}}$
39. $2^{\frac{1}{2}}y^{\frac{1}{2}}$
40. $a^{\frac{1}{2}} + b^{\frac{1}{2}}$
41. $(x - y)^{\frac{1}{3}}$
42. $x^{\frac{1}{3}} - y^{\frac{1}{3}}$
43. $(-z)^{\frac{1}{3}}$
44. $-(a + b)^{\frac{3}{4}}$
45. $a^{\frac{3}{2}} + (-b)^{\frac{3}{2}}$

Write with fractional exponents.

46. $\sqrt[3]{x^2}$
47. $\sqrt[4]{y^3}$
48. $\sqrt{mn}$
49. $\sqrt[3]{mn^2}$
50. $\sqrt[4]{x^2y^3}$
51. $2\sqrt[4]{a^3}$
52. $a\sqrt[5]{b^2c^3}$
53. $2xy\sqrt{z}$
54. $a\sqrt[3]{b^2}$
55. $\sqrt{a + b}$
56. $\sqrt[4]{2x^5y^4}$
57. $\sqrt{4a - 9b}$
58. $\sqrt[3]{(-c)}$
59. $-\sqrt[3]{a^2b}$
60. $-\sqrt[5]{(-a)^3}$

Complete the table.

	Radical Form	Exponential Form	Numerical Value	Radical Equation	Exponential Equation
61.	$\sqrt{16}$	$16^{\frac{1}{2}}$	?	$\sqrt{16} = 4$	$16^{\frac{1}{2}} = 4$
62.	$(\sqrt{4})^3$	?	?	?	?
63.	?	$25^{\frac{1}{2}}$	?	?	?
64.	?	?	?	$\sqrt[3]{64} = 4$	?
65.	?	$8^{\frac{2}{3}}$	?	?	?
66.	$(\sqrt{9})^3$	?	?	?	?
67.	?	$16^{\frac{5}{4}}$	?	?	?
68.	?	?	?	?	$16^{\frac{3}{2}} = 64$
69.	?	$4^{-\frac{1}{2}}$	?	?	?
70.	$\frac{1}{(\sqrt[3]{8})}$	?	?	?	?
71.	$\frac{1}{(\sqrt[3]{8})^2}$	?	?	?	?

Evaluate when $a = 8$, $b = 1$, and $c = 4$.

72. $a^{\frac{2}{3}}b^{\frac{1}{2}}$
73. $(a + b)^{\frac{3}{2}}$
74. $a^{-\frac{1}{3}}c^{\frac{3}{2}}$
75. $(a - c)^{-\frac{1}{2}}$
76. $\left(\frac{a + b}{c}\right)^{-\frac{3}{2}}$
77. $(3a + b)^{\frac{1}{2}}$
78. $\left(\frac{a}{7c - b}\right)^{\frac{2}{3}}$
79. $(4a)^{-\frac{3}{5}}$
80. $(6c + b)^{\frac{3}{2}}$
81. $c^{\frac{1}{2}} + b^{\frac{1}{2}}$
82. $(c + b)^{\frac{1}{2}}$
83. $(2ac)^{-\frac{1}{2}}$

Write each of the following using rational and negative exponents. All denominators should equal 1.

84. $\sqrt{\frac{a}{b}}$
85. $\sqrt[3]{\frac{x^2}{y}}$
86. $\frac{x^3}{y}$
87. $\frac{m}{\sqrt{n}}$
88. $-\sqrt[5]{\frac{x^2}{y^3}}$

REVIEW CAPSULE FOR SECTION 11–5

Simplify. *(Pages 193–196, 209–211)*

1. $a^2 \cdot a^5$ **2.** $b \cdot b^2$ **3.** $(5x^2y)(-15xy^9)$ **4.** $(-17r^2s^3t^5)(-2r^4s^2t^8)$

5. $(3^2)^3$ **6.** $(a^0)^{18}$ **7.** $-(c^2d^5)^7$ **8.** $-t^3v^8 \div t^2v^6$

Multiply. Simplify the product when possible. *(Pages 272–274)*

9. $\sqrt{2} \cdot \sqrt{3}$ **10.** $\sqrt{8} \cdot \sqrt{12}$ **11.** $2\sqrt{3} \cdot 3\sqrt{5}$ **12.** $(\sqrt{7})(-\sqrt{7})$

Divide. Express denominators as rational numbers. *(Pages 272–274)*

13. $\frac{\sqrt{32}}{\sqrt{2}}$ **14.** $\frac{\sqrt{18x^3}}{\sqrt{6x}}$ **15.** $\frac{\sqrt{15}}{\sqrt{5}}$ **16.** $\frac{\sqrt[3]{25a^6}}{\sqrt[3]{250b^9}}$

11–5 Radicals and Exponents

The following exponent theorems are used to simplify radicals and to compute with radicals.

Exponent Theorems

For all real numbers a and b and all rational numbers m and n,

Theorem 11-2 $a^m \cdot a^n = a^{m+n}$ **Theorem 11-3** $(a^m)^n = a^{mn}$

Theorem 11-4 $\frac{a^m}{a^n} = a^{m-n}$ **Theorem 11-5** $(a^n)(b^n) = (a \cdot b)^n$

Theorem 11-6 $\frac{a^n}{b^n} = \left(\frac{a}{b}\right)^n,\ b \neq 0$

Theorems 11–2, 11–3, and 11–4 are familiar; Theorems 11–5 and 11–6 are new to you.

Theorem 11–5 says that $(2^3)(5^3) = (\mathbf{2} \cdot \mathbf{5})^3$.

Theorem 11–6 says that $\frac{4^3}{5^3} = \left(\frac{\mathbf{4}}{\mathbf{5}}\right)^3$.

You can use these theorems to simplify exponential expressions or to simplify radical expressions after they have been expressed in exponential form.

EXAMPLE 1 Simplify. Express each answer in simplified radical form.

a. $54^{\frac{1}{3}}$ **b.** $27^{\frac{1}{6}}$

Solutions:

a. $54^{\frac{1}{3}} = (27 \cdot 2)^{\frac{1}{3}}$ ← **27 is the largest factor of 54 that is a perfect cube.**

$= (27^{\frac{1}{3}})(2^{\frac{1}{3}})$

$= 3 \cdot 2^{\frac{1}{3}}$ ← **By Theorem 11–5**

$= \mathbf{3\sqrt[3]{2}}$

b. $27^{\frac{1}{6}} = (27)^{\frac{1}{6}}$

$= (3^3)^{\frac{1}{6}}$

$= 3^{\frac{3}{6}}$

$= 3^{\frac{1}{2}} = \mathbf{\sqrt{3}}$

It is sometimes helpful to find a common denominator of the exponents.

EXAMPLE 2 Simplify. Express each answer in simplified radical form.

a. $\sqrt{2} \cdot \sqrt[3]{3}$ **b.** $\sqrt[3]{9} \div \sqrt[4]{3}$

Solutions: Express with rational exponents. Then simplify.

a. $\sqrt{2} \cdot \sqrt[3]{3} = 2^{\frac{1}{2}} \cdot 3^{\frac{1}{3}}$ ← **Write each exponent with a common denominator.**

$= 2^{\frac{3}{6}} \cdot 3^{\frac{2}{6}}$

$= (2^3)^{\frac{1}{6}} \cdot (3^2)^{\frac{1}{6}}$ ← **Write as factors with equal exponents.**

$= 8^{\frac{1}{6}} \cdot 9^{\frac{1}{6}}$

$= (8 \cdot 9)^{\frac{1}{6}}$ ← **By Theorem 11–5**

$= 72^{\frac{1}{6}}$

$= \sqrt[6]{72}$

b. $\sqrt[3]{9} \div \sqrt[4]{3} = \dfrac{9^{\frac{1}{3}}}{3^{\frac{1}{4}}}$

$= \dfrac{9^{\frac{4}{12}}}{3^{\frac{3}{12}}}$

$= \left(\dfrac{9^4}{3^3}\right)^{\frac{1}{12}}$

$= \left[\dfrac{(3^2)^4}{3^3}\right]^{\frac{1}{12}}$

$= \left(\dfrac{3^8}{3^3}\right)^{\frac{1}{12}}$

$= \sqrt[12]{3^5}$

The properties of exponents are useful in solving some equations.

EXAMPLE 3 Solve: $x^{-\frac{2}{3}} = 4$

Solution:

$x^{-\frac{2}{3}} = 4$ ← **Raise each side to the same exponent, $-\frac{3}{2}$.**

$(x^{-\frac{2}{3}})^{-\frac{3}{2}} = (4)^{-\frac{3}{2}}$

$x = (4^3)^{-\frac{1}{2}}$ ← **By Theorem 11–3**

$x = \dfrac{1}{64^{\frac{1}{2}}}$ ← **To obtain all possible solutions, find both square roots.**

$= \pm\dfrac{1}{8}$ ← **Both solutions make the original equation true.**

The check is left for you. **Solution set:** $\{-\frac{1}{8}, \frac{1}{8}\}$

Some equations can be solved using either rational exponents or radicals.

EXAMPLE 4 Solve and check: $\sqrt{2x+7} - 2\sqrt{x} = -1$

Solution:

$\sqrt{2x+7} - 2\sqrt{x} = -1$ ← **Write with one radical on each side.**

$\sqrt{2x+7} = 2\sqrt{x} - 1$ ← **Square each side.**

$2x + 7 = 4x - 4\sqrt{x} + 1$ ← **Combine like terms.**

$4\sqrt{x} = 2x - 6$ ← **Divide each side by 2.**

$2\sqrt{x} = x - 3$ ← **Square each side.**

$4x = x^2 - 6x + 9$ ← **Combine like terms.**

$0 = x^2 - 10x + 9$

$0 = (x-1)(x-9)$ Thus, $x = 1$ or $x = 9$.

Check: $\sqrt{2x+7}-2\sqrt{x}=-1$ ← **Replace x with 1.**

$3-2\stackrel{?}{=}-1$

$1\stackrel{?}{=}-1$ No ← **Apparent solution**

$\sqrt{2x+7}-2\sqrt{x}=-1$ ← **Replace x with 9.**

$5-6\stackrel{?}{=}-1$

$-1\stackrel{?}{=}-1$ Yes ✓

Solution set: $\{9\}$

CLASSROOM EXERCISES

Simplify. Express each answer in simplified radical form.

1. $16^{\frac{1}{3}}$ **2.** $48^{\frac{1}{4}}$ **3.** $\sqrt[3]{24}\cdot\sqrt[3]{3}$ **4.** $\sqrt[3]{2}\cdot\sqrt{3}$ **5.** $\sqrt{2}\div\sqrt[3]{5}$

Solve and check.

6. $x^{\frac{1}{3}}=4$ **7.** $x^{-\frac{1}{2}}=4$ **8.** $\sqrt{3-x}=5$ **9.** $5\sqrt{x}=\sqrt{3-2x}$

WRITTEN EXERCISES

A Simplify. Express each answer in simplified radical form. Assume that variables represent positive real numbers.

1. $81^{\frac{1}{3}}$ **2.** $32^{\frac{1}{4}}$ **3.** $81^{\frac{1}{6}}$ **4.** $64^{\frac{1}{4}}$ **5.** $\sqrt[3]{128}$

6. $\sqrt[4]{9}$ **7.** $\sqrt[6]{8}$ **8.** $3\sqrt[4]{48}$ **9.** $2\sqrt[3]{250}$ **10.** $\sqrt[12]{125}$

11. $\sqrt{3}\cdot\sqrt[3]{2}$ **12.** $\sqrt{2}\cdot\sqrt[3]{2}$ **13.** $\sqrt[4]{3}\cdot\sqrt[3]{2}$ **14.** $\sqrt[5]{2}\cdot\sqrt{2}$

15. $\sqrt[3]{4}\cdot\sqrt[3]{2}$ **16.** $\sqrt[3]{2}\cdot\sqrt[3]{16}$ **17.** $\sqrt{2}\div\sqrt[3]{2}$ **18.** $\sqrt{8}\div\sqrt[3]{2}$

19. $\sqrt[4]{16}\div\sqrt[3]{2}$ **20.** $\sqrt[3]{25}\div\sqrt{5}$ **21.** $\sqrt[4]{125}\div\sqrt[3]{5}$ **22.** $\sqrt{32}\div\sqrt[4]{2}$

23. $(-4\sqrt{3})^2$ **24.** $(3\sqrt{2})^3$ **25.** $(6\sqrt[3]{9})(\frac{1}{2}\sqrt[3]{6})$ **26.** $\sqrt[5]{t^6}$

27. $\sqrt[6]{8y^9}$ **28.** $\sqrt[4]{4a^2}$ **29.** $\sqrt[6]{8s^{12}}$ **30.** $\sqrt{8x^2}\cdot\sqrt{9x^6y^2}$

Solve and check.

31. $x^{\frac{1}{2}}=7$ **32.** $n^{\frac{1}{3}}=2$ **33.** $x^{\frac{2}{3}}=4$ **34.** $y^{\frac{3}{2}}=8$

35. $2a^{\frac{2}{3}}=\frac{1}{8}$ **36.** $4x^{\frac{3}{2}}=-32$ **37.** $x^{-\frac{1}{2}}=2$ **38.** $y^{-\frac{1}{3}}=4$

39. $a^{-\frac{2}{3}}=25$ **40.** $8n^{-\frac{3}{2}}=1$ **41.** $\frac{1}{4}x^{-\frac{2}{3}}=16$ **42.** $y^{-3}-8=0$

43. $\sqrt{x+2}+4=x$ **44.** $\sqrt{x+2}+x=4$ **45.** $x-3=\sqrt{4x+9}$

46. $\sqrt{x+4}+2x=13$ **47.** $1-y=\sqrt{1-5y}$ **48.** $\sqrt{x^2-8}+x=4$

49. $2x-\sqrt{7x-3}=3$ **50.** $\sqrt{x+4}+\sqrt{x-3}=7$ **51.** $\sqrt{3x-5}-\sqrt{3x}=1$

B Simplify. Express each answer in exponential form. Assume that variables represent positive real numbers.

52. $\sqrt[2]{y}\cdot\sqrt[6]{y^5}$ **53.** $\sqrt[6]{t^5}\cdot\sqrt[2]{t}$ **54.** $\sqrt[3]{\sqrt{y}}$ **55.** $\sqrt{\sqrt[3]{r}}$

56. $\sqrt{\sqrt[3]{25}}$ **57.** $\sqrt{7\cdot\sqrt[3]{7}}$ **58.** $\sqrt[5]{\sqrt[3]{d^2}}$ **59.** $\sqrt[4]{\sqrt[3]{a^9}}$

Write without exponents. (HINT: $4^{0.5} = 4^{\frac{1}{2}}$)

60. $4^{0.5}$ **61.** $16^{0.75}$ **62.** $81^{1.25}$ **63.** $256^{0.125}$

Solve and check.

64. $4\sqrt{b-3} - \sqrt{b+2} = 5$ **65.** $\sqrt{3x+1} - \sqrt{6x+1} = -2$ **66.** $a - \sqrt{a+3} = 3$

67. $1 + \sqrt{x-1} = \sqrt{2x-1}$ **68.** $\frac{9}{\sqrt{2x-7}} = \sqrt{2x-7}$ **69.** $\sqrt{x} + \sqrt{1+x} = \frac{2}{\sqrt{1+x}}$

C Solve for the indicated variable.

70. $I = \frac{V}{\sqrt{R^2 + VL^2}}$, for R **71.** $f = \frac{1}{2\pi}\sqrt{\frac{1}{LC}}$, for C **72.** $T = \sqrt{\frac{\pi^2 R^3}{4mG}}$, for R

73. Prove: $\sqrt[n]{a} \cdot \sqrt[n]{b} = \sqrt[n]{ab}$, if $a \geq 0$, $b \geq 0$. (HINT: Show that $(\sqrt[n]{a} \cdot \sqrt[n]{b})^n = ab$.)

74. Prove: $\sqrt[n]{a} \div \sqrt[n]{b} = \sqrt[n]{\frac{a}{b}}$, if $a \geq 0$, $b \geq 0$.

Review

Write an equivalent expression with positive exponents and without parentheses for each of the following. The replacement set for the variables is the set of complex numbers, excluding 0. Remember that i is not a variable. *(Section 11–1)*

1. b^{-6} **2.** $\frac{9}{c^{-4}}$ **3.** $\frac{6x^{-2}y^3}{z^{-4}}$ **4.** $(1 + 2i)^{-2}$ **5.** $\frac{(ab^{-2})^{-4}}{(ab^{-2})^{-6}}$

Write in scientific notation. *(Section 11–2)*

6. 380,000 **7.** 0.00042 **8.** 0.0000006 **9.** 430,000,000,000

Write in decimal form. *(Section 11–2)*

10. 3×10^4 **11.** 2.2×10^{-5} **12.** 9.64×10^8 **13.** 4.25×10^{-8}

Simplify. Assume that no denominator equals zero. *(Section 11–3)*

14. $\frac{x^{-3}y^{-4}}{x^{-6}y^{-3}}$ **15.** $(s+t)^3\left(\frac{1}{s^{-1}+t^{-1}}\right)$ **16.** $4p^{-2}q + p^{-1} - \frac{2}{p^2} + \frac{p^{-2}q}{3}$

Evaluate. *(Section 11–4)*

17. $-81^{\frac{3}{4}}$ **18.** $81^{-\frac{1}{4}}$ **19.** $(-125)^{\frac{2}{3}}$ **20.** $16^{-\frac{3}{2}}$ **21.** $-8^{-\frac{4}{3}}$

Simplify. Express each answer in simplified radical form. Assume that variables represent positive real numbers. *(Section 11–5)*

22. $\sqrt{3} \cdot \sqrt[3]{4}$ **23.** $\sqrt{16} \div \sqrt[4]{2}$ **24.** $(8\sqrt[3]{9})(\frac{1}{4}\sqrt{6})$ **25.** $(9\sqrt[4]{a^2})(\frac{1}{3}\sqrt{ab})$

Solve and check. *(Section 11–5)*

26. $\sqrt{3t+1} + 3 = t$ **27.** $\sqrt{x-3} + 7 = x$ **28.** $\sqrt{2a+3} = \sqrt{a+1} + 1$

11–6 Exponential Functions and Equations

By defining the replacement set for x in the *exponential equation* $y = 2^x$ as the set of real numbers, you can graph the equation as a smooth curve in the coordinate plane. The graph at the right below shows that $y = 2^x$ is a function.

Definition

An **exponential function with base *a*** is a function defined by an equation of the form

$$y = a^x$$

where $a, x \in \text{R}$, $a \neq 1$, and $a > 0$.

Note that the domain of the exponential function is the set of real numbers, R; the range is $\{y: y > 0\}$.

Since the graph for $y = a^x$ when $a = 1$ is the same as the graph for $y = 1$, $y = 1^x$ is not usually considered to be an exponential function.

The exponential function $f(x) = 2^x$ has base 2. As x increases, the function increases; as x decreases, the function decreases. As x decreases, the curve gets closer and closer to the x axis, but never touches it. The x axis is an **asymptote** of the graph.

From the graph, note that a change of one unit in the value of x does *not always* result in a change of one unit in the value of y. When x is large, a small change in x implies a large change in y. Thus, the curve becomes steeper as x increases. The steepness of the graph of the exponential function $f(x) = a^x$ depends on the value of the base a.

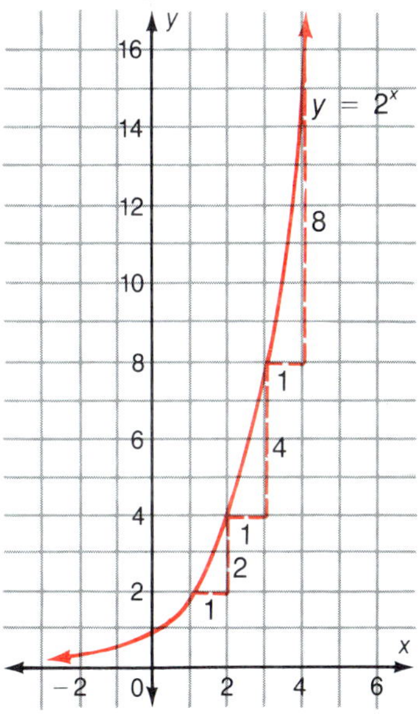

An **exponential equation** has a variable in an exponent. One method of solving an exponential equation such as $8 = 4^x$, is to express 8 and 4 with the same base. Then you can conclude that the exponents are equal.

For $a, m, k \in \text{R}$, $a > 0$, $a \neq 1$:

1. If $a^m = a^k$, then $m = k$.

2. If $m = k$, then $a^m = a^k$.

EXAMPLE Solve for x: **a.** $8 = 4^x$ **b.** $125^{x-1} = 25$

Solutions: Write each side of the equation as an exponential expression having the same base.

a. $8 = 4^x$ ← $8 = 2^3$; $4 = 2^2$

$2^3 = (2^2)^x$

$2^3 = 2^{2x}$

Thus, $3 = 2x$ ← If $a^m = a^k$, then $m = k$.

$\frac{3}{2} = x$

Solution set: $\left\{\frac{3}{2}\right\}$ ← The check is left for you. →

b. $125^{x-1} = 25$ ← $125 = 5^3$; $25 = 5^2$

$(5^3)^{x-1} = 5^2$

$5^{3x-3} = 5^2$

Thus, $3x - 3 = 2$

$3x = 5$ and $x = \frac{5}{3}$

Solution set: $\left\{\frac{5}{3}\right\}$

CLASSROOM EXERCISES

In Exercises 1–5, complete the tables.

1. $y = 10^x$

x	-3	-2	-1	0	$\frac{1}{2}$	1	$1\frac{1}{2}$
y	0.001	?	?	?	?	?	?

(NOTE: To find $y = 10^{1\frac{1}{2}}$, write $10^{1\frac{1}{2}}$ as $10^1 \cdot 10^{\frac{1}{2}}$. Then use the table on page 636 or a calculator to evaluate $10^{\frac{1}{2}}$.)

2. $y = 3^x$

x	-3	-2	-1	0	$\frac{1}{2}$	1	$1\frac{1}{2}$	2
y	$\frac{1}{27}$	?	?	?	?	?	?	?

3. $y = 4^x$

x	-3	-2	-1	0	$\frac{1}{2}$	1	$1\frac{1}{2}$	2	$2\frac{1}{2}$	3
y	$\frac{1}{64}$	?	?	?	?	?	?	?	?	?

4. $y = \left(\frac{1}{2}\right)^x$

x	-5	-4	-3	-2	-1	0	1	2	3	4	5
y	32	?	?	?	?	?	?	?	?	?	?

5. $y = \left(\frac{1}{3}\right)^x$

x	-3	-2	-1	0	1	2	3
y	27	?	?	?	?	?	?

Solve and check.

6. $2^x = 32$ **7.** $3^x = 81$ **8.** $9^x = 27$ **9.** $27^x = 81$

WRITTEN EXERCISES

A On the same coordinate axes, sketch the graphs of the following functions. Use the tables of values that you completed in the Classroom Exercises.

1. $y = 10^x$ **2.** $y = 3^x$ **3.** $y = 4^x$ **4.** $y = (\frac{1}{2})^x$ **5.** $y = (\frac{1}{3})^x$

Use the graphs in Exercises 1–5 to answer Exercises 6–12.

6. Do the graphs have any common points? If so, specify them.

7. Refer to the graphs of Exercises 1, 2, and 3 where $a > 1$. When the base, a, changes from 3 to 4 to 10, what happens to the shape of the curves?

8. Refer to the graphs of Exercises 4 and 5 where $a < 1$. As a changes from $\frac{1}{3}$ to $\frac{1}{2}$, what happens to the shape of the curves?

9. When $a < 1$, are the graphs asymptotic to the positive half of the x axis or to the negative half of the x axis?

10. When $a > 1$, are the graphs asymptotic to the positive half of the x axis or to the negative half of the x axis?

11. Will any graph in Exercises 1–5 intersect the x axis? Why or why not?

12. Is $y = a^x$ an exponential function when $a = 1$? Explain.

13. Sketch the graphs of $y = 2^x$ and $y = (\frac{1}{2})^x$ on the same coordinate axes.

14. Is there an axis of symmetry for the graphs in Exercise 13? If so, what is its equation?

15. Sketch the graphs of $y = 3^x$ and $y = (\frac{1}{3})^x$ on the same coordinate axes.

16. Do the pairs of graphs in Exercises 13 and 15 have the same axis of symmetry? If so, state its equation.

Solve and check.

17. $2^x = 8$ **18.** $4^x = 2$ **19.** $2^x = 32$ **20.** $8^x = 2$

21. $8^x = 4$ **22.** $25^x = 5$ **23.** $8^x = 32$ **24.** $27^x = 3$

25. $27^x = 9$ **26.** $2^x = \frac{1}{2}$ **27.** $3^x = \frac{1}{9}$ **28.** $9^{-x} = \frac{1}{27}$

29. $8^x = \frac{1}{8}$ **30.** $9^{x-2} = 27$ **31.** $2^{-x} = 16$ **32.** $2^{x+2} = 32$

B

33. $8^x = 0.5$ **34.** $4^x = 0.25$ **35.** $25^x = 0.04$ **36.** $25^x = 0.2$

37. An exponential function that is important in mathematical analysis is defined by $y = e^x$ where $e \approx 2.7182$. Sketch the graph of $y = e^x$ using the values given in the table below.

x	−4	−3	−2	−1	0	1	2	3
y	0.02	0.05	0.14	0.37	1	2.72	7.39	20.1

Use the graph in Exercise 37 to find approximations for the given powers of e.

38. $e^{\frac{1}{2}}$ **39.** $e^{\frac{3}{2}}$ **40.** $e^{-\frac{3}{2}}$ **41.** $e^{-\frac{1}{2}}$ **42.** $e^{\frac{5}{2}}$

C

43. Prove: The graphs of $y = a^x$ and $y = \left(\frac{1}{a}\right)^x$ are symmetric with respect to the y axis.

Solve and check.

44. $3^{7x-1} = 9^{-2}$ **45.** $16^{2x+3} = 8$ **46.** $\frac{1}{5^{x-3}} = \left(\frac{1}{25}\right)^{\frac{1}{4-x}}$ **47.** $5^{x^2+9} = 25^{-3x}$

REVIEW CAPSULE FOR SECTION 11–7

Sketch the graph of each function and its inverse on the same coordinate axes. *(Pages 178–181)*

1. $y = 3x - 2$ **2.** $y = 2x^2$ **3.** $y = x^2 - 4$ **4.** $y = |x|$

Solve for the variable. *(Pages 400–402, 411–414)*

5. $x^3 = 64$ **6.** $a^{\frac{3}{2}} = 8$ **7.** $10^{-3} = x$ **8.** $9^{-1} = n$

11–7 The Logarithmic Function

The *inverse* of the exponential function $y = 10^x$ is the function $\mathbf{x = 10^y}$. For convenience, this function is usually written as $y = \log_{10} x$ and is called the **logarithmic function.**

$y = \log_{10} x$ ← **Read:** **"y is the logarithm of x to the base 10."**

The table below gives corresponding coordinates for $y = 10^x$ and $y = \log_{10} x$. The numbers in the first column are exponents. These same numbers appear in the fourth column where they are logarithms. Thus, **a logarithm is an exponent.**

$y = 10^x$		$y = \log_{10} x$	
x	10^x	x	$\log_{10} x$
−3	0.001	0.001	−3
−2	0.01	0.01	−2
−1	0.1	0.1	−1
0	1	1	0
1	10	10	1
2	100	100	2
3	1000	1000	3

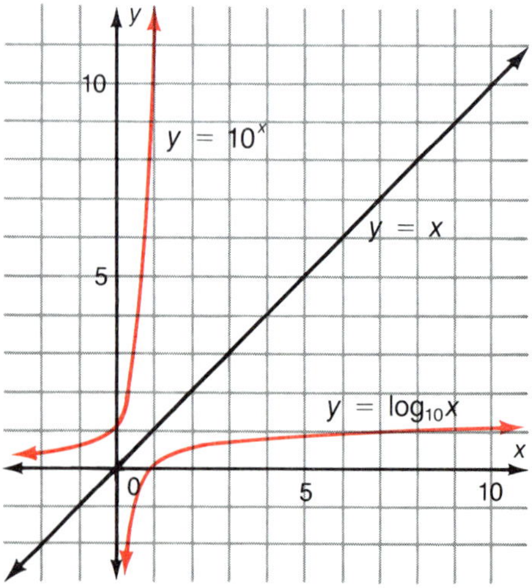

Compare the equivalent sentences in the table below.

Exponential Form	**Logarithmic Form**
$2^3 = 8$	$\log_2 8 = 3$
$10^3 = 1000$	$\log_{10} 1000 = 3$
$3^{-2} = \frac{1}{9}$	$\log_3 \frac{1}{9} = -2$
$3^5 = 243$	$\log_3 243 = 5$

This suggests the following definitions.

Definitions

> The **logarithmic function** $y = \log_a x$ is the inverse of the exponential function $y = a^x$, $a \neq 1$, $a > 0$. That is,
>
> $$y = \log_a x \text{ if and only if } x = a^y.$$
>
> A **logarithm** is a value of the logarithmic function.

REMEMBER: By definition, the base a can be *any positive number* a, $a \neq 1$.

EXAMPLE 1 Evaluate: **a.** $\log_{10} 1000$ **b.** $\log_{\frac{1}{2}} 32$

Solutions:

a. Let $y = \log_{10} 1000$.

Then $10^y = 1000$. ← **By definition**

Since $1000 = 10^3$,

$10^y = 10^3$ and $y = \mathbf{3}$.

b. Let $r = \log_{\frac{1}{2}} 32$.

Then $\left(\frac{1}{2}\right)^r = 32$.

Since $32 = 2^5 = \frac{1}{2^{-5}} = \left(\frac{1}{2}\right)^{-5}$,

$\left(\frac{1}{2}\right)^r = \left(\frac{1}{2}\right)^{-5}$ and $r = \mathbf{-5}$.

You can use the definition of the logarithmic function to solve logarithmic equations such as $\log_n \frac{1}{9} = -2$.

EXAMPLE 2 Solve for the variable: $\log_n \frac{1}{9} = -2$

Solution: $\log_n \frac{1}{9} = -2$ if and only if $n^{-2} = \frac{1}{9}$.

$\frac{1}{n^2} = \frac{1}{9}$ ← **By definition of a^{-m}**

$n^2 = 9$ ← **$1 \cdot 9 = n^2 \cdot 1$**

$n = 3$ Thus, $n = 3$ and $\log_3 \frac{1}{9} = -2$.

CLASSROOM EXERCISES

Express in logarithmic form.

1. $2^3 = 8$ **2.** $3^2 = 9$ **3.** $6^3 = 216$ **4.** $4^4 = 256$

5. $11^3 = 1331$ **6.** $10^{-2} = \frac{1}{100}$ **7.** $10^{-3} = \frac{1}{1000}$ **8.** $4^{0.5} = 2$

Express in exponential form.

9. $\log_2 16 = 4$ **10.** $\log_3 27 = 3$ **11.** $\log_8 64 = 2$ **12.** $\log_7 343 = 3$

13. $\log_{0.5} 0.25 = 2$ **14.** $\log_{10} 0.001 = -3$ **15.** $\log_2 \frac{1}{8} = -3$ **16.** $\log_{10} \frac{1}{10{,}000} = -4$

WRITTEN EXERCISES

A

1. What is the domain of $y = \log_{10} x$? What is the range?
2. Which axis is the asymptote for the graph of $y = \log_{10} x$?

Express in logarithmic form.

3. $2^6 = 64$
4. $8^{-2} = \frac{1}{64}$
5. $16^{\frac{3}{2}} = 64$
6. $18^{-2} = \frac{1}{324}$
7. $4^{-\frac{3}{2}} = \frac{1}{8}$
8. $10^{-1} = \frac{1}{10}$
9. $8^{\frac{1}{3}} = 2$
10. $10^{-3} = 0.001$

Express in exponential form.

11. $\log_{10} 1000 = 3$
12. $\log_9 81 = 2$
13. $\log_7 49 = 2$
14. $\log_2 1 = 0$
15. $\log_5 0.2 = -1$
16. $\log_{\frac{1}{3}} \frac{1}{27} = 3$
17. $\log_{10} 0.01 = -2$
18. $\log_{\frac{1}{2}} \frac{1}{4} = 2$

Evaluate.

19. $\log_{10} 10{,}000$
20. $\log_{10} 10$
21. $\log_2 8$
22. $\log_2 64$
23. $\log_2 32$
24. $\log_2 16$
25. $\log_3 1$
26. $\log_3 \frac{1}{9}$
27. $\log_{10} 0.01$
28. $\log_{10} 0.0001$
29. $\log_4 64$
30. $\log_6 36$
31. $\log_{10} 1$
32. $\log_{10} \sqrt{10}$
33. $\log_2 \frac{1}{4}$
34. $\log_3 \frac{1}{27}$
35. $\log_{\frac{1}{2}} 8$
36. $\log_{\frac{1}{2}} 16$
37. $\log_{\frac{1}{3}} 27$
38. $\log_{\frac{1}{3}} 81$

Solve for the variable.

39. $\log_a 64 = 3$
40. $\log_a 32 = 5$
41. $\log_a 2 = \frac{1}{2}$
42. $\log_a 8 = \frac{3}{2}$
43. $\log_3 x = 4$
44. $\log_2 x = 7$
45. $\log_{10} x = -3$
46. $\log_{10} 1 = y$
47. $\log_4 8 = y$
48. $\log_{27} 81 = y$
49. $\log_{\frac{1}{8}} \frac{1}{16} = y$
50. $\log_8 \frac{1}{32} = y$
51. $\log_n 4 = \frac{1}{3}$
52. $\log_n 81 = 4$
53. $\log_b 32 = 5$
54. $\log_3 a = \frac{1}{3}$
55. $\log_a 49 = -2$
56. $\log_{\frac{1}{2}} a = -6$
57. $\log_9 n = -1$
58. $\log_{10} 10 = x$

B

Use a large sheet of graph paper to graph the logarithmic functions in Exercises 59–61 on the same coordinate axes.

59. $y = \log_{10} x$

x	$\frac{1}{100}$	$\frac{1}{10}$	1	2	4	6	8	10	12	14	16	18	20
y	−2	−1	0	0.30	0.60	0.78	0.90	1	1.08	1.15	1.20	1.26	1.30

60. $y = \log_2 x$, for $x = \frac{1}{16}, \frac{1}{8}, \frac{1}{4}, \frac{1}{2}, 1, 2, 4, 8, 16$
61. $y = \log_3 x$, for $x = \frac{1}{27}, \frac{1}{9}, \frac{1}{3}, 1, 3, 9, 27$

Use the graphs for Exercises 59–61 to answer the following questions.

62. Do the graphs have a point in common? If so, which one(s)?
63. As the base for the logarithms changes from 2 to 3 to 10, describe the shape of the curves.

64. Is the logarithm of a negative number a real number?

65. For $x > 1$, is the logarithm of x a positive number or a negative number?

66. For $x = 1$, what is its logarithm with any base?

67. For $0 < x < 1$, is the logarithm of x a positive number or a negative number?

In Column A of Exercises 68–74 insert two consecutive powers of 10 to make a true statement. Then use this information to insert two consecutive integers in the blanks in Column B to make a true statement. The first one is done for you.

	Column A	Column B
68.	$\underline{10^2} < 562 < \underline{10^3}$;	$\underline{2} < \log_{10} 562 < \underline{3}$
69.	$\underline{?} < 50 < \underline{?}$;	$\underline{?} < \log_{10} 50 < \underline{?}$
70.	$\underline{?} < 30 < \underline{?}$;	$\underline{?} < \log_{10} 30 < \underline{?}$
71.	$\underline{?} < 15 < \underline{?}$;	$\underline{?} < \log_{10} 15 < \underline{?}$
72.	$\underline{?} < 4 < \underline{?}$;	$\underline{?} < \log_{10} 4 < \underline{?}$
73.	$\underline{?} < 697 < \underline{?}$;	$\underline{?} < \log_{10} 697 < \underline{?}$
74.	$\underline{?} < 1642 < \underline{?}$;	$\underline{?} < \log_{10} 1642 < \underline{?}$

CALCULATOR APPLICATIONS

Evaluating Radicals

The inverse, [INV], and power, [x^y], keys on the scientific calculator allow you to readily evaluate expressions that contain rational exponents.

EXAMPLE Evaluate: **a.** $(\sqrt[5]{27})^3$ **b.** $(3\sqrt[4]{9})(\frac{1}{2}\sqrt[3]{5})$

a. 2 7 [INV] [x^y] 5 [=] [x^y] 3 [=] → 7.2246741

b. 3 [×] 9 [INV] [x^y] 4 [×] . 5 [×] 5 [INV] [x^y] 3 [=] → 4.4426478

EXERCISES

Use a calculator to evaluate Written Exercises 6–25 on page 413.

REVIEW CAPSULE FOR SECTION 11–8

Complete. *(Pages 193–196, 400–402, 408–410)*

1. $a^n \cdot a^m = \underline{\ ?\ }$
2. $(a^x)^y = \underline{\ ?\ }$
3. $a^{-n} = \underline{\ ?\ }$
4. $(a \cdot b)^n = \underline{\ ?\ }$
5. $\left(\frac{a}{b}\right)^n = \underline{\ ?\ }$
6. $a^{\frac{1}{n}} = \underline{\ ?\ }$
7. $a^{\frac{p}{q}} = \underline{\ ?\ }$
8. $\frac{1}{a^{-n}} = \underline{\ ?\ }$

11–8 Properties of Logarithms

Since logarithms are exponents, their properties are similar to the properties of exponents.

Theorems 11–7, 11–8, 11–9

For positive real numbers M, N, and a, $a \neq 1$, and any real number p,

11–7	$\log_a (M \cdot N) = \log_a M + \log_a N$	**Product Theorem**
11–8	$\log_a \frac{M}{N} = \log_a M - \log_a N$	**Quotient Theorem**
11–9	$\log_a N^p = p \log_a N$	**Power Theorem**

Theorem 11–7 states that the logarithm of a product is the sum of the logarithms. To prove this theorem,

Proof: Let $\log_a M = x$ and $\log_a N = y$.

Then $a^x = M$ and $a^y = N$ ← **By definition**

Thus, $M \cdot N = a^x \cdot a^y$ ← **By substitution**

$= a^{x+y}$ ← **By Theorem 11–2**

So $\log_a (M \cdot N) = x + y$, and ← **By substitution and the definition of logarithm**

$\log_a (M \cdot N) = \log_a M + \log_a N$ ← **By substitution**

You are asked to prove Theorems 11–8 and 11–9 in the Written Exercises.

EXAMPLE 1 Express without using multiplication or division.

a. $\log_a 5x^2$ **b.** $\log_a \frac{10x}{y}$

Solutions: **a.** $\log_a 5x^2 = \log_a 5 + \log_a x^2$ ← **By Theorem 11–7**

$= \log_a 5 + 2 \log_a x$ ← **By Theorem 11–9**

b. $\log_a \frac{10x}{y} = \log_a (10x) - \log_a y$ ← **By Theorem 11–8**

$= \log_a 10 + \log_a x - \log_a y$ ← **By Theorem 11–7**

Logarithms with base 10 are called **common logarithms.** For convenience, $\log_{10} x$ is written $\log x$. Unless otherwise stated, the word "logarithm" means "common logarithm."

To find the common logarithm of a number between 1.00 and 9.99, you use the tables on pages 634 and 635. A portion of the table is shown below.

N	0	1	2	3	4	5	6	7	8	9
5.5	7404	7412	7419	7427	7435	7443	7451	7459	7466	7474
5.6	7482	7490	7497	7505	7513	7520	7528	7536	7543	7551
5.7	7559	7566	7574	7582	7589	7597	7604	7612	7619	7627
5.8	7634	7642	7649	7657	7664	7672	7679	7686	7694	7701
5.9	7709	7716	7723	7731	7738	7745	7752	7760	7767	7774

EXAMPLE 2 Find each logarithm. Use the portion of the table shown on page 422.

a. log 5.8 **b.** log 5.89

Solutions: **a.** Find 5.8 under "N."

Read the number under the "0" column.

log 5.8 = **0.7634**

N	0
5.8 →	7634

b. Find 5.8 under "N."

Read the number under the "9" column.

log 5.89 = **0.7701**

N	. . .	9
5.8 →	. . .	7701

Since the logarithms of the numbers in the table, except for 1, are irrational numbers, they *are approximated* by rational numbers. For convenience, however, we use the symbol "=" rather than "≈" when writing statements such as log 5.84 = 0.7664.

A table of logarithms can be used in two ways. Given a number N, you can find its logarithm. Also, given a logarithm (exponent), you can find the corresponding number N, called the **antilogarithm** (written: **antilog**).

Definition

> N = **antilog** L if and only if $L = \log_a$ N.

EXAMPLE 3 Find antilog 0.7767.

Solution: Find 7767 in the body of the table.

Read the corresponding "row" and "column."

N	. . .	8
5.9 ←	. . .	7767

Thus, antilog 0.7767 = **5.98.**

You can use logarithms to solve exponential equations.

EXAMPLE 4 Solve: $10^x = 3.26$

Solution: Write the equation in logarithmic form.

$10^x = 3.26$ if and only if $\log 3.26 = x$ ← **Find the antilog.**

$\mathbf{0.5132} = x$

Thus, log 3.26 = 0.5132, and $\mathbf{10^{0.5132} = 3.26}$.

CLASSROOM EXERCISES

Complete.

1. $\log M \cdot N =$ ___?___ **2.** $\log \dfrac{M}{N} =$ ___?___ **3.** $\log M^p =$ ___?___ **4.** $\log \sqrt[p]{M} =$ ___?___

Express without using multiplication or division.

5. $\log 27xy$ 6. $\log \frac{27x}{y}$ 7. $\log x^5$ 8. $\log \frac{9y^2}{z^2}$

Use the table of logarithms to find each of the following.

9. $\log 1.0$ 10. $\log 3.7$ 11. antilog 0.2553 12. antilog 0.8899

WRITTEN EXERCISES

A Express without using multiplication or division.

1. $\log (rst)$ 2. $\log x^5z$ 3. $\log \left(\frac{r}{s}\right)$ 4. $\log \left(\frac{\sqrt{x}}{z}\right)$

5. $\log rt^{\frac{1}{2}}$ 6. $\log (9\sqrt[3]{z})$ 7. $\log (ab)^3$ 8. $\log (rt)^{\frac{1}{2}}$

9. $\log \left(\frac{xy}{z^2}\right)$ 10. $\log \frac{\sqrt{a}}{3b}$ 11. $\log (\sqrt[3]{x} \cdot y^4)$ 12. $\log \sqrt[5]{\frac{a^4}{b^2}}$

Classify each statement as true, *T*, or false, *F*. When an equation is false, tell why it is false.

13. $\log \frac{10^5}{10^2} = \log 10^5 - \log 10^2$
14. $\frac{\log 10^5}{\log 10^2} = \log 10^5 - \log 10^2$
15. $\log 10^n = \log n \cdot 10$
16. $\log 10^n = n \log 10$
17. $\log \frac{10^n}{10^m} = \log 10^n - \log 10^m$
18. $\frac{\log 10^n}{\log 10^m} = \log 10^n - \log 10^m$
19. $\log M^{\frac{1}{2}} = \frac{1}{2} \log M$
20. $\log M^2 = 2 \log M$
21. $\frac{\log M}{\log N} = \log M - \log N$
22. $\log \frac{a}{b} = \frac{1}{b} \log a$
23. $\log \sqrt{45} = \sqrt{\log 45}$
24. $\log \sqrt[q]{m^p} = \frac{p}{q} \log m$
25. $\log a^2b = 2 \log a + \log b$
26. $\log \sqrt{ab} = \frac{1}{2} \log a + \log b$
27. If $5^4 = 625$, then $\log_4 625 = 5$.
28. $\log 1350 = \log 1.35 + \log 10^3$

Find each of the following. Use the table on pages 634–635.

29. $\log 6.75$ 30. $\log 4.32$ 31. $\log 1.76$ 32. $\log 7.84$

33. $\log 3.60$ 34. $\log 7.03$ 35. $\log 4.99$ 36. $\log 4.00$

37. $\log 1.06$ 38. $\log 8.63$ 39. $\log 6.01$ 40. $\log 9.65$

41. antilog 0.8993 42. antilog 0.9222 43. antilog 0.4639 44. antilog 0.0043

45. antilog 0.5729 46. antilog 0.7292 47. antilog 0.7396 48. antilog 0.9912

Use the table on pages 634–635 to find x.

49. antilog $0.8615 = x$ 50. antilog $0.0253 = x$ 51. antilog $0.6902 = x$ 52. $10^x = 4.85$

53. $10^x = 5.96$ 54. $10^x = 4.09$ 55. $10^x = 1.26$ 56. $10^x = 9.86$

57. $10^x = 0.4624$ 58. $\log x = 0.4624$ 59. $\log x = 0.7316$ 60. $\log x = 0.8639$

61. $10^{0.9294} = x$ 62. $10^{0.3464} = x$ 63. $10^{0.5378} = x$ 64. $10^{0.7846} = x$

B Given that $\log 2 = 0.3010$, $\log 3 = 0.4771$ and $\log 5 = 0.6990$, find each of the following without using tables.

Example: log 12 **Solution:** $\log 12 = \log (2^2 \cdot 3)$

$= 2 \log 2 + \log 3$

$= 2(0.3010) + 0.4771 =$ **1.0791** ⟵ **log 12**

65. log 30
66. log 24
67. log 45
68. log 60
69. log 0.75
70. $\log \sqrt{20}$
71. log 360
72. $\log \sqrt{500}$
73. Prove Theorem 11–8.
74. Prove Theorem 11–9.

C Solve each equation.

75. $\log (x + 3) + \log 2 = 0$
76. $\log 8 - \log (n - 4) = 1$
77. $3 \log_2 a = 6$
78. $\log_3 (2x + 3) = 4$

Using Statistics:
Population Sampling

In any sampling procedure, it is desirable to obtain a subset of the **universal set** U that represents the whole population as well as possible. This can only be achieved if every element in U has an equal chance of being chosen, that is, if the elements of the sample are chosen at random from U. Random sampling is a technique used for estimating the size of animal populations.

The size of an animal population is often of interest to the **biologist** who specializes in conservation of animal resources. Answers to questions such as: "Can the area sustain this size population?" or "Why is this population decreasing?" depend on a knowledge of the size of the population.

One method of sampling to estimate population size has been applied to "counting" the number of fish in a lake. The fish from a **random sample** are counted, tagged, and released. At a later time, another random sample is taken and the following formula is used.

$$\frac{\text{Tagged}}{\text{Second Sample}} \rightarrow \frac{s_1}{S} = \frac{F}{T} \leftarrow \frac{\text{First Sample}}{\text{Total Population}}$$

EXAMPLE A random sample of 500 fish was taken from a lake, tagged, and released. At a later time, another random sample was taken. This sample contained 800 fish, 38 of which were already tagged. Estimate the size of the population of fish in the lake.

Solution: Substitute $s_1 = 38$, $S = 800$, $F = 500$, in the formula to estimate T to the nearest hundred.

$$\frac{38}{800} = \frac{500}{T}$$

$$38T = 400{,}000$$

$$T = 10{,}526.319 \text{ or } 10{,}500$$

This is a rough estimate only. To be more accurate, several of these estimates should be made and their mean computed. This mean gets closer and closer to the actual number of fish in the lake as the number of sample estimates increases.

EXERCISES

In Exercises 1–4, compute an estimate for T to the nearest hundred.

1. $F = 900$, $S = 400$, $s_1 = 73$
2. $F = 850$, $S = 250$, $s_1 = 25$
3. $F = 2000$, $S = 800$, $s_1 = 40$
4. $F = 1000$, $S = 500$, $s_1 = 25$

Mr. Petri sampled and tagged 250 fish in early July. In early August he took three separate samples with the results shown at the right. Use this information for Exercises 5–9.

Sample	S	s_1
1	100	9
2	150	12
3	300	27

5. Find the estimate for T to the nearest hundred for each sample.
6. Find the mean of the three estimates.
7. Consider Samples 1, 2, and 3 as a single large sample. Compute T to the nearest hundred.
8. Compare the answers to Exercises 6 and 7.
9. What do you conclude?
10. To take a "straw" poll on an issue, a city editor ran a ballot in the newspaper, asking readers to mark the ballot and return it. Did his returns come from a random sample of the city population? Justify your answer.
11. A pollster thought he could conduct a poll in a certain city by calling every fifth name in the telephone book. He considered this a random sample. Do you agree? Justify your answer.
12. Explain how you would go about using a telephone book to get a random sample for use in a survey about political candidates.
13. Explain how you would go about getting a random sample of the persons in your school who eat cereal for breakfast.

11–9 Computing with Logarithms

In Section 11–8, you used the table on pages 634–635 to find the logarithms of numbers from 1.00 through 9.99. To find the logarithms of numbers greater than 10 or less than 1, you use scientific notation and the properties of logarithms. In particular, recall that for common logarithms,

$$\log 10^4 = 4 \qquad \log 10^{-2} = -2 \qquad \log 10 = 1.$$

EXAMPLE 1 Find log 13,600.

Solution: Express 13,600 in scientific notation.

$$13{,}600 = 1.36 \times 10^4$$

Then $\log 13{,}600 = \log (1.36 \times 10^4)$

$\log 13{,}600 = \log 1.36 + \log 10^4$ ← **By Theorem 11–7**

From the table, $\log 1.36 = 0.1335$.

From the definition of a logarithm, $\log 10^4 = 4$.

Thus, $\log 13{,}600 = 4 + 0.1335 =$ **4.1335.**

Every logarithm consists of two parts, the *characteristic* and the *mantissa*.

The **mantissa** is the logarithm of a number between 1 and 10.

The **characteristic** is the power of 10 by which that number is multiplied when it is expressed in scientific notation.

Characteristic

$$\log 13{,}600 = 4.\underline{1335}$$

Mantissa

Since a table of logarithms is actually a table of mantissas, the mantissa of a logarithm is *always kept nonnegative* so that it may be found in the table.

EXAMPLE 2 Find log 0.0342. Identify the characteristic and mantissa.

Solution: $0.0342 = 3.42 \times 10^{-2}$

$\log 0.0342 = \log 3.42 + \log 10^{-2}$

$= 0.5340 + (-2)$, or $-2 + 0.5340$

The characteristic is **−2;** the mantissa is **0.5340.**

In Example 2, the negative characteristic, −2, is *not added* to the positive mantissa, 0.5340.

The negative characteristic may be expressed in many ways.

$\log 0.0342 =$ **0.5340 − 2**

$\log 0.0342 =$ **8.5340 − 10** ← **8 − 10 = −2**

Usually, the difference with -10 is used. This is the **standard form.** However, it is sometimes helpful to use another difference, such as $18.5340 - 20$.

REMEMBER: Negative characteristics are associated with numbers between 0 and 1.

You can use the tables in reverse to find a number, given its logarithm.

EXAMPLE 3 Find antilog $9.3263 - 10$.

Solution: Characteristic: $9 - 10$, or -1 Mantissa: 0.3263

antilog $0.3263 = 2.12$ ← **From the table**

antilog $9.3263 - 10 = 2.12 \times 10^{-1} = 0.212$

Thus, antilog $9.3263 - 10 = \mathbf{0.212}$

Example 4 shows how to use logarithms to find a product.

EXAMPLE 4 Use logarithms to find 673×52.7.

Solution: $\log (673 \times 52.7) = \log 673 + \log 52.7$ ← **By Theorem 11–7**

$$\begin{aligned} \log 673 &= 2.8280 \\ \log 52.7 &= 1.7218 \\ \hline \log (673 \times 52.7) &= 4.5498 \end{aligned}$$

← **Find the antilogarithm.**

antilog $4.5498 = 35{,}500$

Thus, 673×52.7 is about **35,500.**

Check: Estimate the product and compare with 35,500.

Estimate: $673 \times 52.7 \approx 700 \times 50 = 35{,}000$ ✓

Example 5 shows that it is sometimes necessary to add 10 or a multiple of 10 and then to subtract this same number from a mantissa. This is done to keep the mantissa positive.

EXAMPLE 5 Use logarithms to find $8.23 \div 12.4$.

Solution: $\log (8.23 \div 12.4) = \log 8.23 - \log 12.4$

$$\begin{aligned} \log 8.23 &= 0.9154 \\ \log 12.4 &= 1.0934 \end{aligned}$$

← **Since 0.9154 < 1.0934, the mantissa will not be positive when you subtract.**

$$\log 8.23 = \mathbf{10.9154 - 10}$$ ← **Add 10 and subtract 10.**

$$\begin{aligned} \log 12.4 &= 1.0934 \\ \hline \log (8.23 \div 12.4) &= 9.8220 - 10 \end{aligned}$$

← **Find the antilog.**

antilog $(9.8220 - 10) = 0.664$ Thus, $8.23 \div 12.4$ is about **0.664.**

Check: $8.23 \div 12.4 \approx 8 \div 12 = \frac{2}{3}$, or about 0.6. ✓

The following example shows how to use logarithms to find a root.

EXAMPLE 6 Use logarithms to find $\sqrt[3]{0.734}$.

Solution: $\log \sqrt[3]{0.734} = \frac{1}{3}(\log 0.734)$

$\log 0.734 = 9.8657 - 10$ ← **Standard form**

$\frac{1}{3}(\log 0.734) = 3.2886 - 3\frac{1}{3}$

Since $\frac{1}{3}(10) = -3\frac{1}{3}$ and it is not possible to move the decimal point $3\frac{1}{3}$ places to the left, express $9.8657 - 10$ in a form that will produce an integer when it is multiplied by $\frac{1}{3}$.

$\log 0.734 = 29.8657 - 30$ ← **$29 - 30 = -1$, and $\frac{1}{3}(-30) = -10$**

$\frac{1}{3}(\log 0.734) = 9.9552 - 10$ ← **Find the antilogarithm.**

$\text{antilog } (9.9552 - 10) = 0.902$ Thus, $\sqrt[3]{0.734}$ is about **0.902.**

Check: $0.734 \approx 1$ and $\sqrt[3]{1} = 1$. The answer is reasonable.

CLASSROOM EXERCISES

State the characteristic of the logarithm of each number.

1. 7.86 **2.** 42 **3.** 325 **4.** 0.842 **5.** 0.00093

Find the logarithm of each number.

6. 1.23 **7.** 12.5 **8.** 0.124 **9.** 97,400 **10.** 0.000083

Find the antilogarithm of each number.

11. 2.7016 **12.** 1.9805 **13.** 9.8768 − 10 **14.** 7.3945 − 10

Subtract the lower logarithm from the upper one. Be sure to change the form of the upper logarithm so that the result for the mantissa is positive.

15. $\begin{array}{r} 2.3154 \\ \underline{4.2760} \end{array}$ **16.** $\begin{array}{r} 0.2391 \\ \underline{1.4562} \end{array}$ **17.** $\begin{array}{l} 2.2345 \\ \underline{9.6372 - 10} \end{array}$ **18.** $\begin{array}{r} 0.7292 \\ \underline{3.6730} \end{array}$

Perform the indicated operations on the given logarithms. Express each result in standard form.

19. 2.1389×4 **20.** $(9.3472 - 10) \times 2$ **21.** $6.3857 \div 2$ **22.** $(8.6134 - 10) \div 4$

23. $0.4728 \div 3$ **24.** $(8.5731 - 10) \times 3$ **25.** $2.2625 \div 5$ **26.** $(7.4825 - 10) \div 3$

WRITTEN EXERCISES

A Given that $\log 3.62 = 0.5587$, find the logarithm of each number.

1. 362 **2.** 36.2 **3.** 36,200 **4.** 0.0362 **5.** 0.362

Given that $\log 0.7230 = 9.8591 - 10$, find the logarithm of each number.

6. 0.07230 **7.** 7.230 **8.** 7230 **9.** 72.3 **10.** 723,000

Find the logarithm of each number. When the characteristic is negative, write it in standard form.

11. 9.83 **12.** 42.7 **13.** 83.40 **14.** 50 **15.** 432
16. 720.00 **17.** 5340 **18.** 9860 **19.** 0.00623 **20.** 0.00752
21. 0.00000456 **22.** 0.000095 **23.** 0.843 **24.** 0.777 **25.** 70,000

Find the antilog of each number.

26. 2.8351 **27.** 1.9818 **28.** 3.2672 **29.** $9.3118 - 10$
30. $6.7372 - 10$ **31.** $8.9533 - 10$ **32.** 1.7143 **33.** 2.9085
34. 4.4713 **35.** $7.8461 - 10$ **36.** $8.4841 - 10$ **37.** $9.7298 - 10$

Use logarithms to compute.

38. $(36.5)(24.9)$ **39.** $(53.7)(6.84)$ **40.** $(0.832)(0.0135)$ **41.** $(6.83)(9.67)(106)$

42. $\frac{738}{25.7}$ **43.** $\frac{69.3}{1.69}$ **44.** $\frac{0.342}{0.713}$ **45.** $\frac{29.6}{37.2}$

46. $(6.23)^4$ **47.** $(0.0041)^3$ **48.** $(8730)^{\frac{1}{2}}$ **49.** $(38.4)^{\frac{1}{3}}$

B

50. $\frac{(234)(48.3)}{0.816}$ **51.** $\frac{(2.39)^2}{\sqrt{0.00038}}$ **52.** $\frac{(0.831)\sqrt[3]{2.97}}{(0.638)^2}$ **53.** $\frac{(1.11)(\sqrt{1.36})}{(2.81)^3}$

CALCULATOR APPLICATIONS

Evaluating Radicals

You can use a scientific calculator to check your answers to Exercises 38–53 above, as well as to do more involved computations.

EXAMPLE Compute: $\sqrt[4]{\frac{46\sqrt{8.13}}{\sqrt{51.2}}}$

4 6 [×] 8 . 1 3 [√] [÷] 5 1 . 2 [√] [=] [INV] [x^y] 4 [=] 2.0691503

EXERCISES

Compute each of the following.

1. $\sqrt[3]{\frac{21.6}{\sqrt{3.42}}}$ **2.** $\frac{1}{(2)(76.5)}\sqrt{\frac{(5470)(980)}{0.0045}}$ **3.** $\sqrt{\frac{17.2\sqrt{28.1}}{\sqrt{3.47}}}$ **4.** $\frac{(2.96)^4\sqrt[3]{3.56}}{(2.05)^5}$

11–10 Interpolation

The table of logarithms in this book cannot be used directly when a number has more than three digits. But you can **interpolate** or "read between" the numbers. *Interpolation* is based on the properties of similar triangles as illustrated in the graph below.

The graph is an enlarged view of a portion of the graph of $y = \log x$. To find log 2.624 means to find the y coordinate of point C.

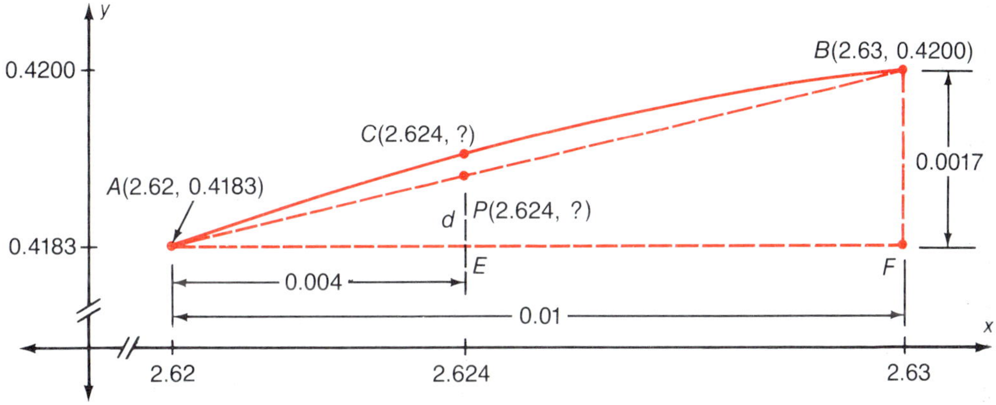

Since P is very close to C, the key to interpolation is the assumption that the length of arc AB is approximated by the length of segment AB. Thus, the y coordinate of P will provide a close approximation for the y coordinate of C. The y coordinate of P is $0.4183 + d$.

$$\triangle AEP \text{ is similar to } \triangle AFB.$$

Therefore, $\dfrac{AE}{AF} = \dfrac{EP}{FB}$, or $\dfrac{0.004}{0.01} = \dfrac{d}{0.0017}$.

Thus, $d \approx 0.0007$, and the y coordinate of P is approximately $0.4183 + 0.0007$, or 0.4190. The y coordinate of C is also approximately 0.4190. Thus,

$$\log 2.624 = 0.4190.$$

Example 1 shows how to find log 2.624 by interpolation. First, locate log 2.620 and log 2.630 in the table. Note that log 2.624 falls between these two values.

EXAMPLE 1 Find log 2.624.

Solution: Arrange your work in a table. Find the differences, set up a proportion, and solve for d.

		Number	Mantissa		
0.01	0.004	2.620	0.4183	d	0.0017
		2.624	$0.4183 + d$		
		2.630	0.4200		

$$\frac{0.004}{0.01} = \frac{d}{0.0017}$$ ← **Solve for *d*.**

$$d = 0.0007$$ ← **Round to the nearest ten-thousandth.**

$\log 2.624 = \log 2.620 + d$

$\log 2.624 = 0.4183 + 0.0007 = \mathbf{0.4190}$

To find log 2624, the procedure is the same as in Example 1. However, you would use 3 as the characteristic.

$$\log 2624 = 3.4190$$

Interpolation can also be used to find the antilogarithm of a number.

EXAMPLE 2 Find antilog 2.2611.

Solution: Ignore the characteristic for now and locate 0.2611 between entries in the table.

		Antilogarithm	Mantissa		
0.01	d	1.820	0.2601	0.0010	0.0024
		$1.82 + d$	0.2611		
		1.830	0.2625		

$$\frac{d}{0.01} = \frac{0.0010}{0.0024} \quad \text{and} \quad d = 0.004166 = 0.004$$

Thus, antilog $0.2611 = 1.82 + 0.004 = 1.824$.

The characteristic of the given logarithm, 2.2611, is 2.

Thus, antilog 2.2611 = **182.4.**

Logarithms can be used to find approximate real-number solutions for some exponential equations. The solution depends on this fact.

If two numbers are equal, their logarithms are equal.

EXAMPLE 3 Solve: $3^x = 150$

Solution:

$$3^x = 150$$

$$x \log 3 = \log 150$$

$$x = \frac{\log 150}{\log 3}$$

$$x = \frac{2.1761}{0.4771} = \mathbf{4.5611}$$ ← **Check whether the answer is reasonable.**

Check: $3^4 = 81$, $3^{4.5611} = 150$, $3^5 = 243$.

Thus, $4 < 4.5611 < 5$ and $81 < 150 < 243$. The answer is reasonable.

WRITTEN EXERCISES

A Use interpolation to find the logarithm of each number.

1. 6.233 **2.** 7.832 **3.** 56.25 **4.** 43.65 **5.** 237.6

6. 432.4 **7.** 0.02137 **8.** 0.6728 **9.** 0.1390 **10.** 0.005294

Use interpolation to find the antilogarithm of each number.

11. 1.2829 **12.** 2.7420 **13.** 0.1624 **14.** 0.8714 **15.** 3.1855

16. $9.5212 - 10$ **17.** $7.4015 - 10$ **18.** $8.9043 - 10$ **19.** $18.7478 - 20$

Use logarithms to compute.

20. 25.43×8.265 **21.** 9.382×27.36 **22.** 2.710×0.4931 **23.** 355.2×74.25

24. $0.9328 \div 0.02130$ **25.** $0.7239 \div 12.36$ **26.** $0.07215 \div 0.3272$ **27.** $8.763 \div 4.532$

28. $(25.34)^3$ **29.** $(4.361)^4$ **30.** $(5.506)^3$ **31.** $(6325)^2$

32. $\sqrt{294.6}$ **33.** $\sqrt[3]{92.83}$ **34.** $\sqrt[3]{0.02134}$ **35.** $\sqrt[3]{0.3468}$

When you use logarithms to compute with negative numbers as in Exercises 36–39, proceed as if the numbers were positive. Then determine the sign.

36. $(-6.321)(72.43) \div (36.45)$ **37.** $\sqrt[3]{-53.62}$ **38.** $(-2.344)^4$ **39.** $(-18.72)(-4.836)$

Solve each equation.

40. $4^x = 30$ **41.** $2^x = 25$ **42.** $5^y = 37.6$ **43.** $1.02^x = 3$

B

44. $5^{2x} = 72.5$ **45.** $3^{x+1} = 85.6$ **46.** $1.01^{24x} = 2$ **47.** $5^{3x-1} = 57.9$

Compute by logarithms. Give your answers to the nearest tenth. When this is not possible, given them to the nearest hundredth.

48. $\sqrt[3]{983.4} \times \sqrt[4]{4.216}$ **49.** $\sqrt[5]{8.926} \div \sqrt{1.635}$ **50.** $\dfrac{97.6 \times \sqrt{47.3}}{(315)^2}$ **51.** $\sqrt[3]{\dfrac{53.2 \times 7.87}{5.00}}$

C The **change of base formula** below relates the logarithms of a number N to two different bases, a and b.

$$\log_a N = \frac{\log_b N}{\log_b a}$$

In Exercises 52–55, express each logarithm as a common logarithm (with base 10).

52. $\log_2 12$ **53.** $\log_3 10$ **54.** $\log_5 8.5$ **55.** $\log_4 75$

56. Prove: *If N, a, and b are positive real numbers, $a \neq 1$, $b \neq 1$, then* $\log_a N = \dfrac{\log_a N}{\log_b a}$.

57. Prove: *If N, a, and b are positive real numbers, then* $\log_a b = \dfrac{1}{\log_b a}$.

Problem Solving and Applications

11–11 Compound Interest

The following formula can be used to find compound interest.

Formula

> If a sum of p dollars is invested at an interest rate r (expressed as a decimal), compounded k times a year for n years, the total amount A (principal and interest) at the end of n years is given by the formula
>
> $$A = p\left(1 + \frac{r}{k}\right)^{kn}.$$

EXAMPLE 1 Find what \$2000 will amount to in 10 years at 12% interest compounded semiannually.

Solution:

$A = p\left(1 + \frac{r}{k}\right)^{kn}$ ← **$p = \$2000$; $r = 0.12$; $k = 2$; $n = 10$**

$A = 2000(1.06)^{20}$ ← **Use logarithms to evaluate.**

$\log A = \log 2000 + 20 \log(1.06)$

$\log A = 3.3010 + 20(0.0253) = 3.807$

Use interpolation (or a calculator) to find antilog 3.807.

$A = 6412$ In 10 years, the amount will be about **\$6412.**

If a sum of p dollars is invested at an interest rate, r, and compounded continuously for n years, the total amount A at the end of n years is given by the formula,

$$A = pe^{rn}.$$

In this section, use $e = 2.71828$.

The compound interest formulas can also be solved for n or p.

EXAMPLE 2 What amount of money must be invested at 14% compounded continuously in order for the account to contain \$20,000 after 10 years?

Solution:

$A = pe^{rn}$ ← **$A = \$20{,}000$; $e = 2.71828$; $r = 0.14$; $n = 10$**

$20{,}000 = p(2.71828)^{1.4}$

$\log 20{,}000 = \log p + 1.4 \log 2.71828$

$\log p = \log 20{,}000 - 1.4 \log 2.71828$ ← **Interpolate or use a calculator.**

$\log p = 4.3010 - (1.4)(0.4343)$

$\log p = 4.3010 - 0.6080$

$\log p = 3.693$ ← **Find the antilogarithm.**

$p = 4932$ Thus, about **\$4932** must be invested.

WORD PROBLEMS

A Solve each problem.

1. Jim invested \$2000 at 12% interest compounded annually. What will the investment amount to in 8 years?
2. A deposit of \$150 today at 12% interest compounded annually will amount to what sum in 25 years?
3. Mr. Hill invested \$1000 at 15% interest compounded semiannually. How much will this investment earn for him in 10 years?
4. Diego can buy a used car for \$5000, or he can leave this sum in the bank at 6% interest compounded quarterly. Left in the bank, what would this amount to in 5 years?
5. The sum of \$2000 is deposited at 14% interest compounded quarterly. What will be the value of this investment in 6 years?
6. What principal will amount to \$5000 if invested at 14% interest compounded semiannually for 10 years?
7. What principal will yield \$10,000 in 12 years at 12% compounded semiannually?

B In Exercises 8–11, calculate the value of a \$1000 investment at a yearly interest rate of 10% compounded in the following ways.

8. annually
9. monthly
10. daily
11. continuously
12. Compare the results of Exercises 8–11. How does continuous compounding compare with monthly and daily compounding?
13. Compute the depreciation on a \$12,000-automobile at 25% deducted annually for 4 years. Use the formula for compound interest, but subtract the rate instead of adding it. The value of the automobile at the end of 4 years is $(0.75)^4$ times the original value.
14. Find the cost of an automobile bought 5 years ago and now worth \$1750, if the depreciation is figured at 30% per year.

C The "**effective yearly interest rate**" on \$1.00 invested at 8.25% compounded quarterly is the simple interest rate which would produce the same amount of interest in one year. For example, after one year, the \$1.00 mentioned above would amount to $1\left(1+\frac{0.0825}{4}\right)^4$, or **\$1.0851.**

Since \$1.00 earned \$0.0851 in 1 year, the effective yearly interest rate is **8.51%.**

15. Find the effective yearly interest rate on \$1.00 invested at 8.25% compounded monthly.
16. Find the effective yearly interest rate on \$1.00 invested at 13.8% compounded daily.
17. Find the effective yearly interest rate on \$1.00 invested at 11.5% compounded continuously.

BASIC: INTEGRAL EXPONENTS

Problem: *Given a real number b and an integer n, write a program which computes, if possible, b^n. Do not use the ↑ operator of* BASIC.

```
100 PRINT "THIS PROGRAM COMPUTES B↑N FOR ANY INTEGER N."
110 PRINT "WHAT ARE B AND N";
120 INPUT B, N
130 IF N = INT(N) THEN 170
140 PRINT "N MUST BE AN INTEGER. ENTER N AGAIN."
150 INPUT N
160 GOTO 130
170 IF N = 0 THEN 280
180 IF N > 0 THEN 220
190 IF B = 0 THEN 310
200 LET B = 1/B
210 LET N = -N
220 LET X = 1
230 FOR I = 1 TO N
240 LET X = X * B
250 NEXT I
260 PRINT "ANSWER IS";X
270 GOTO 320
280 IF B = 0 THEN 310
290 PRINT "ANSWER IS 1."
300 GOTO 320
310 PRINT "UNDEFINED EXPRESSION"
320 PRINT "ANY MORE POWERS (1 = YES, 0 = NO)";
330 INPUT Z
340 IF Z = 1 THEN 120
350 END
```

Output:

```
RUN
THIS PROGRAM COMPUTES B↑N FOR ANY INTEGER N.
WHAT ARE B AND N? 5,4
ANSWER IS 625

ANY MORE POWERS (1 = YES, 0 = NO)? 1
? 10,.5
N MUST BE AN INTEGER. ENTER N AGAIN.
? -2
ANSWER IS .01

ANY MORE POWERS (1 = YES, 0 = NO)? 0
READY
```

Analysis

Statements 130–160: The program tests whether N is an integer. If it is not, it makes the user enter the exponent again.

Statements 170 and 280–300: For N = 0 the only question is whether B = 0 also. If it is, the expression is undefined (statement 310). But if B ≠ 0, the answer is 1 (statement 290).

Statements 180 and 220–260: For a positive exponent the `FOR-NEXT` loop multiplies B times itself N times, accumulating the answer as X. Before the loop begins, X is set to 1 (statement 220). Otherwise it will be 0 by default and will remain 0 throughout the execution of the loop. Notice, however, that if B = 0, the loop will correctly compute any positive power of 0 as 0.

Statements 190–210: If the computer arrives at statement 190, then N < 0. If B = 0, the expression is undefined. But if B ≠ 0, the power is computed by multiplying the reciprocal of B times itself −N times. The same loop (statements 220–250) used for positive powers is executed, but first B is set equal to its own reciprocal (statement 200), and the sign of N is changed to make it positive (statement 210).

In the output for the program, one answer may be printed in a form such as `1.26765E+30`. This is `BASIC`'s version of scientific notation and means 1.26765×10^{30}. "E" stands for "exponent of ten." The computer will print answers in this form when they are very large or very small in absolute value. The number of significant figures (six in these examples) displayed will depend upon the computer being used.

EXERCISES

A Run the program on page 436 to evaluate the following expressions.

1. 10^0
2. 0^{56}
3. 0^0
4. 0^{-5}
5. 53^1
6. 2^{100}

Write a `BASIC` program for each problem.

7. Print a common logarithm table for the integers 1 through 20. Each version of `BASIC` has a common log function, although the abbreviation for the function varies with the computer. `LOG10(X)` and `CLG(X)` are widely used, but consult the manual for your machine.
8. Print a table of values for x and 2^x for $x = 1, 2, \cdots, 20$. For what value of x does your computer change to E–notation for 2^x?
9. Given a sum of p dollars invested at an interest rate, r (expressed as a decimal), compounded k times per year for n years, compute and print the total amount a (principal and interest) at the end of n years. Use the following formula. (See page 434.)

$$A = p\left(1 + \frac{r}{k}\right)^{kn}$$

Review

1. Sketch the graph of $y = 5^x$. *(Section 11–6)*

Solve and check. *(Section 11–6)*

2. $4^x = 32$ **3.** $125^x = 25$ **4.** $9^x = \frac{1}{27}$ **5.** $4^x = 0.125$

Evaluate. *(Section 11–7)*

6. $\log_4 16$ **7.** $\log_{81} \frac{1}{9}$ **8.** $\log_8 4$ **9.** $\log_5 1$

Solve for the variable. *(Section 11–7)*

10. $\log_5 t = -3$ **11.** $\log_b \frac{1}{16} = 4$ **12.** $\log_{27} 3 = s$ **13.** $\log_{27} r = -\frac{2}{3}$

Express without using multiplication or division. *(Section 11–8)*

14. $\log (3\sqrt[4]{8})$ **15.** $\log \left(\frac{4c}{d^2}\right)$ **16.** $\log \left(\frac{a^2}{b^5}\right)$ **17.** $\log \sqrt[4]{\frac{r^3}{s^5}}$

Find each of the following. Use the table on pages 634–635. *(Section 11–8)*

18. log 3.25 **19.** log 5.46 **20.** log 9.31 **21.** log 2.09

22. antilog 0.9047 **23.** antilog 0.4843 **24.** antilog 0.7007 **25.** antilog 0.3054

Use logarithms to compute. *(Sections 11–9, 11–10)*

26. $(32.8)(4.79)$ **27.** $\frac{(467)(2.1)}{568}$ **28.** $(24.6)^5$ **29.** $\sqrt[3]{0.0294}$

30. $(16.75)(286.4)$ **31.** $\frac{32.86}{2.43}$ **32.** $(6.237)^3$ **33.** $\sqrt{7.468}$

34. What principal will yield \$25,000 in 10 years at 14% compounded quarterly? *(Section 11–11)*

Chapter Summary

IMPORTANT TERMS

Antilogarithm (p. 423)
Characteristic (p. 427)
Common logarithm (p. 422)
Exponential equation (p. 415)
Exponential function (p. 415)
Logarithmic function (p. 419)
Mantissa (p. 427)
Scientific notation (p. 403)

IMPORTANT IDEAS

1. For any real number a, $a \neq 0$, and any integer m,

$$a^{-m} = \frac{1}{a^m} \quad \text{and} \quad a^m = \frac{1}{a^{-m}}.$$

2. To write a number in scientific notation:

   1 Using the digits of the number in the order in which they occur, write the first factor as a number a, where $1 \leq a < 10$.

   2 To find the second factor, divide the original number by the first factor.

   3 Express the second factor as a power of 10.

3. For any real number a, $a \neq 0$, and $n \in \mathbf{N}$,
$$a^{\frac{1}{n}} = \sqrt[n]{a}.$$

4. **Rational Exponent Theorem:** For all $a \in \mathbf{R}$, $a \neq 0$, $m \in \mathbf{I}$, $n \in \mathbf{N}$,
$$a^{\frac{m}{n}} = (a^{\frac{1}{n}})^m = (\sqrt[n]{a})^m, \qquad \text{or} \qquad a^{\frac{m}{n}} = (a^m)^{\frac{1}{n}} = \sqrt[n]{a^m},$$
except when n is even and $a^m < 0$.

5. For all real numbers a and b and all rational numbers m and n,
$$a^m \cdot a^n = a^{m+n} \qquad (a^m)^n = a^{mn}$$
$$\frac{a^m}{a^n} = a^{m-n} \qquad (a^n)(b^n) = (a \cdot b)^n$$
$$\frac{a^n}{b^n} = \left(\frac{a}{b}\right)^n, \quad b \neq 0$$

6. An exponential function with base a, where $a, x \in \mathbf{R}$, $a \neq 1$, and $a > 0$, is a function defined by an equation of the form
$$y = a^x.$$

7. $y = \log_a x$ if and only if $x = a^y$.

8. **Properties of Logarithms**
$$\log_a (M \cdot N) = \log_a M + \log_a N \qquad \log_a N^p = p \log_a N$$
$$\log_a \frac{M}{N} = \log_a M - \log_a N$$

9. $N = \text{antilog } L$ if and only if $L = \log_a N$.

10. Every logarithm consists of two parts, the characteristic and the mantissa. The **mantissa** is the logarithm of a number between 1 and 10. The **characteristic** is the power of 10 by which that number is multiplied when it is expressed in scientific notation.

Chapter Objectives and Review

Objective: *To rewrite an expression involving negative exponents in an equivalent form involving only positive exponents (Section 11–1)*

Write an equivalent expression with positive exponents and without parentheses.

1. s^{-4} **2.** $\frac{1}{z^{-5}}$ **3.** $\frac{t^3}{m^{-4}}$ **4.** $(t^2d)^{-2}$ **5.** $\left(\frac{3}{5}\right)^{-3}$

Objective: *To express a decimal number in scientific notation, and to express a number that is given in scientific notation in decimal form (Section 11–2)*

Write in scientific notation.

6. 7043 **7.** 79,283,100 **8.** 0.000502 **9.** 0.000000443

Write in decimal form.

10. 1.3×10^5 **11.** 3.7×10^{-6} **12.** 6.04×10^8 **13.** 2.04×10^{12}

Objective: *To simplify algebraic expressions involving integral exponents (Section 11–3)*

Simplify.

14. $\left(\frac{2}{5}\right)^{-2}$ **15.** $\frac{a^{-1}b^{-2}}{b^{-2}a^1}$ **16.** $(b^{-3})^{-2}$ **17.** $2^{-3} + 3^{-2}$ **18.** $(c^{-7})^0$

Objective: *To rewrite expressions involving radicals as expressions with rational number exponents, and to rewrite expressions with rational number exponents as expressions involving radicals (Section 11–4)*

Rewrite using rational exponents.

19. $\sqrt[3]{x^2}$ **20.** $\sqrt[4]{x^5}$ **21.** $\sqrt{3x - 1}$ **22.** $\frac{1}{\sqrt[3]{b}}$

Rewrite each as an expression involving radicals.

23. $b^{-\frac{3}{4}}$ **24.** $c^{\frac{1}{3}}$ **25.** $7^{\frac{2}{9}}$ **26.** $8^{\frac{5}{3}}$ **27.** $2^{\frac{1}{2}}y^{\frac{3}{2}}$

Objective: *To simplify radical expressions (Sections 11–4 and 11–5)*

Simplify.

28. $8^{\frac{1}{3}} + 81^{\frac{1}{4}}$ **29.** $108^{\frac{1}{3}}$ **30.** $3\sqrt{8} + 4\sqrt{18}$ **31.** $\sqrt{4} \cdot \sqrt[3]{5}$ **32.** $\sqrt{2} \cdot \sqrt[3]{2}$

Objective: *To solve radical and exponential equations (Sections 11–5 and 11–6)*

Solve and check.

33. $2\sqrt{x} = 7$ **34.** $\sqrt{x} = -3$ **35.** $\sqrt{x + 1} = 4$ **36.** $9^x = 27$ **37.** $4^x = \frac{1}{16}$

Objective: *To graph an exponential function and its inverse logarithmic function (Sections 11–6 and 11–7)*

38. Sketch the graphs of $y = 4^x$ and $y = \log_4 x$ on the same coordinate axes. State the domain and the range of each of these functions.

Objective: *To express an exponential equation in logarithmic form, and to express a logarithmic equation in exponential form (Section 11–7)*

Express in logarithmic form.

39. $64 = 4^3$ **40.** $y = 5^2$ **41.** $y = a^x$ **42.** $\frac{1}{16} = z^{-4}$

Express in exponential form.

43. $\log_3 81 = 4$ **44.** $y = \log_3 x$ **45.** $\log_2 \frac{1}{4} = -2$ **46.** $\log_{10} 0.001 = -3$

Objective: *To apply the properties of logarithms (Sections 11–8, 11–9, 11–10)*

Express without using multiplication or division.

47. $\log t^7p$ **48.** $\log \sqrt[4]{r^3}$ **49.** $\log\left(\frac{xy^2}{t^4}\right)$ **50.** $\log 2\sqrt{t^4}$

Use logarithms to compute.

51. 843×763 **52.** $1.77 \div 8.74$ **53.** $(38.42)^2$ **54.** $\sqrt[3]{2.334}$

Objective: *To use logarithms to solve problems involving compound interest (Section 11–11)*

55. If \$2000 is invested at 5% compounded annually, what will be its value after 10 years?

Chapter Test

Simplify.

1. $(3^3)^2$ **2.** $5^2 \div 5^6$ **3.** $5x^{-3}y - 4x^{-1}y$ **4.** $(a^3b^4)^{-2}$

5. Express 4,700,000,000 and 0.000005673 in scientific notation.

Evaluate.

6. $\left(\frac{2}{3}\right)^{-3}$ **7.** $(x + y)^0$ **8.** $49^{-\frac{1}{2}}$ **9.** $4^{-\frac{3}{2}}$

Solve and check.

10. $\sqrt{2x - 5} = 3$ **11.** $8^x = 32$ **12.** $5^x = \frac{1}{125}$

Complete.

13. If $\log_x 64 = 6$, then $x =$ __?__ **14.** $\log x^5y =$ __?__

Use logarithms to compute. Give answers to the nearest hundredth.

15. $75.6 \div 2.94$ **16.** $\frac{(43.2)^2 \times 124}{\sqrt{742}}$

Solve each equation.

17. $\log_{10} 10{,}000 = x$ **18.** $\log_a 64 = 2$ **19.** $\log_a 3 = \frac{1}{3}$

20. If \$2500 is invested at 4% interest for 10 years, compounded semiannually, what will be the amount at the end of the 10-year period?

Preparing for College Entrance Tests

For some questions on College Entrance tests, the given information may not be sufficient to enable you to determine the answer. One way to tell that a relationship "cannot be determined" is to show that the given information results in two statements that cannot *both* be true for *all* real numbers (contradiction).

EXAMPLE If m and p are integers with $m < 0$ and $p > 0$, compare m^p and p^m.

(a) $m^p > p^m$ **(b)** $m^p < p^m$ **(c)** $m^p = p^m$

(d) Cannot be determined from the given information

Think: From the Comparison Postulate (see page 62), either (a) is true, or (b) is true or (c) is true for any two real numbers.

Solution: (c) cannot be true since $m^p = p^m$ only when $m = p$, $m \neq 0$, $p \neq 0$.

Test (a): For $m = -1$ and $p = 2$, $(-1)^2 > 2^{-1}$. ← **Since** $1 > \frac{1}{2}$

Test (b): For $m = -1$ and $p = 3$, $(-1)^3 < 3^{-1}$. ← **Since** $-1 < \frac{1}{3}$

Thus, $m^p > p^m$ for $m = -1$ and $p = 2$ and $m^p < p^m$ for $m = -1$ and $p = 3$. Since both inequalities cannot be true for all real numbers, more information is needed. **Answer: d**

Choose the best answer. Choose *a*, *b*, *c*, or *d*.

1. If p and q are integers, with $p > 1$ and $q > 0$, compare p^q and $(-p)^{-q}$.
   (a) $p^q > (-p)^{-q}$ **(b)** $p^q < (-p)^{-q}$ **(c)** $p^q = (-p)^{-q}$
   (d) Cannot be determined from the given information

2. If a and b are integers, with $a < 0$ and $b > 0$, compare a^b and a^{-b}.
   (a) $a^b > a^{-b}$ **(b)** $a^b < a^{-b}$ **(c)** $a^b = a^{-b}$
   (d) Cannot be determined from the given information

3. If n is a real number, compare 1^{5n} and 1^{5+n}.
   (a) $1^{5n} > 1^{5+n}$ **(b)** $1^{5n} < 1^{5+n}$ **(c)** $1^{5n} = 1^{5+n}$
   (d) Cannot be determined from the given information.

4. If y is a positive integer, compare $(-1)^{2y}$ and $(-1)^{2y+1}$.
   (a) $(-1)^{2y} > (-1)^{2y+1}$ **(b)** $(-1)^{2y} < (-1)^{2y+1}$ **(c)** $(-1)^{2y} = (-1)^{2y+1}$
   (d) Cannot be determined from the given information

5. If x is a negative integer, compare x^{2x} and x^{2x+1}.
   (a) $x^{2x} > x^{2x+1}$ **(b)** $x^{2x} < x^{2x+1}$ **(c)** $x^{2x} = x^{2x+1}$
   (d) Cannot be determined from the given information

6. If a and b are positive integers and $a = b$, compare $(a-b)^{a+b}$ and $(a+b)^{a-b}$.
   (a) $(a-b)^{a+b} > (a+b)^{a-b}$ **(b)** $(a-b)^{a+b} < (a+b)^{a-b}$
   (c) $(a-b)^{a+b} = (a+b)^{a-b}$ **(d)** Cannot be determined from the given information

CHAPTER

12 Sequences and Series

12–1 Sequences

Each of these sets of numbers has a specific order and a specific pattern.

$$A = \{3, 5, 7, 9, 11, \cdots\}$$
$$B = \{2, 4, 8, 16, 32, \cdots\}$$
$$C = \{1, 4, 9, 16, 25, \cdots\}$$

By pairing each number of a set with a positive integer, you can find a **rule** to describe the pattern.

To find the rule means to find an expression for the nth number or ***term*** *of each set.*

The ***n*th term** is usually written as a_n (read: "a sub n").

EXAMPLE

a. Find the rule for the nth term in set A.

b. Use the rule to write the sixth, seventh, and eighth terms of set A.

Solutions: **a.** Pair each number with the corresponding positive integer.

$$\begin{array}{cccccccc} 1 & 2 & 3 & 4 & 5 & \cdots & n & \cdots \\ \updownarrow & \updownarrow & \updownarrow & \updownarrow & \updownarrow & & \updownarrow & \\ 3 & 5 & 7 & 9 & 11 & \cdots & a_n & \cdots \end{array}$$

← **Positive integer** (top row)

← **Set *A*** (bottom row)

Rule: Each number in set A is equal to one more than twice the positive integer with which it is paired. In symbols,

$$a_n = 2n + 1.$$

b. To find the sixth, seventh, and eighth terms means to find a_6, a_7, and a_8.

$a_6 = 2(6) + 1 = \mathbf{13}$ ← **Replace *n* with 6.**

$a_7 = 2(7) + 1 = \mathbf{15}$ ← **Replace *n* with 7.**

$a_8 = 2(8) + 1 = \mathbf{17}$ ← **Replace *n* with 8.**

The pairings can be written as a set of ordered pairs.

$$\{(1, 3), (2, 5), (3, 7), (4, 9), (5, 11), \cdots, (n, 2n + 1), \cdots\}$$

This set describes a function called a *sequence*.

Definitions

A **sequence** is a function whose domain is the set of positive integers.

The numbers contained in the range of the function are the **terms of the sequence.**

$$\{(1, a_1), (2, a_2), (3, a_3), (4, a_4), \cdots, (n, a_n), \cdots\}$$

Sets A–C are **infinite sequences,** because there is no last term. A sequence that has a last term is a **finite sequence.**

CLASSROOM EXERCISES

Verify that each rule gives the first three terms of the given sequence. Use the rule to write the next three terms.

1. $a_n = 3n$; 3, 6, 9, $\cdots$

2. $a_n = 2n - 3$; -1, 1, 3, $\cdots$

3. $a_n = n^2$; 1, 4, 9, $\cdots$

4. $a_n = 2^n$; 2, 4, 8, $\cdots$

Use the rule to write the first three terms of each sequence.

5. $a_n = 5n$

6. $a_n = n + \frac{1}{n}$

7. $a_n = n^2 + 1$

8. $a_n = n^3$

WRITTEN EXERCISES

A Find the rule for each sequence. Use the rule to write the next three terms.

1. 2, 4, 6, 8, $\cdots$

2. 4, 7, 10, 13, $\cdots$

3. 1, 2, 3, 4, $\cdots$

4. 0, -1, -2, -3, $\cdots$

5. 2, 5, 8, 11, $\cdots$

6. -2, -1, 0, 1, $\cdots$

7. -4, -2, 0, 2, $\cdots$

8. 1, 4, 16, 64, $\cdots$

9. 5, 9, 13, 17, $\cdots$

10. 3, 7, 11, 15, $\cdots$

11. 4, 5, 6, 7, $\cdots$

12. 1, 3, 5, 7, $\cdots$

13. 10, 15, 20, 25, $\cdots$

14. 2, 1, 0, -1, $\cdots$

15. -1, -4, -9, -16, $\cdots$

16. $1, \frac{1}{2}, \frac{1}{3}, \frac{1}{4}, \cdots$

17. $\frac{1}{2}, \frac{2}{3}, \frac{3}{4}, \frac{4}{5}, \cdots$

18. 2.5, 3, 3.5, 4, $\cdots$

19. 2, 5, 10, 17, $\cdots$

20. 1, 4, 9, 16, $\cdots$

21. $1, \frac{1}{4}, \frac{1}{9}, \frac{1}{16}, \cdots$

B

22. $2, 1, \frac{1}{2}, \frac{1}{4}, \cdots$

23. $1, \frac{1}{3}, \frac{1}{9}, \frac{1}{27}, \cdots$

24. 3, 6, 12, 24, $\cdots$

25. $1, a^2, a^4, a^6, \cdots$

26. $x, -x^2, x^3, -x^4, \cdots$

27. $x, x+2, x+4, x+6, \cdots$

Use the rule to write the first four terms of a sequence.

28. $a_n = n^2(n-1)$

29. $a_n = (-1)^{n+1}$

30. $a_n = \frac{n}{2}(n+1)$

31. $a_n = (-\frac{1}{2})^{n-1}$

32. $a_n = 5 + 3(n-1)$

33. $a_n = (-1)^{n-1}$

34. The first term of a sequence is -7 and every term after the first is 7 less than 3 times the preceding term. Find the nth term.

35. Write the first four terms of the sequence in Exercise 34.

36. The first term of a sequence is 6. Every term after the first is 3 more than the one that precedes it. Find the nth term.

REVIEW CAPSULE FOR SECTION 12-2

Solve and check. *(Pages 117–120)*

1. $\begin{cases} x - 3y = 7 \\ x - 5y = 13 \end{cases}$

2. $\begin{cases} 2x + y = 1 \\ 3x - 2y = 8 \end{cases}$

3. $\begin{cases} p + 5q = 11 \\ p - q = 5 \end{cases}$

4. $\begin{cases} a + 3b = 7 \\ a + b = 5 \end{cases}$

12–2 Arithmetic Sequences

If each term of a sequence is obtained by adding some fixed number to the preceding term, the sequence is called an *arithmetic sequence* or an *arithmetic progression.*

Definitions

An **arithmetic sequence** is a sequence in which the difference between successive terms is a constant.

The constant is called the **common difference, *d*. Thus,**

$$\boldsymbol{d = a_{n+1} - a_n.}$$

Consider an arithmetic sequence whose first term is a_1 and with common difference d. Then, by definition,

$$a_2 - a_1 = d \qquad \text{or} \qquad a_2 = a_1 + d.$$

Similarly,

$a_3 = a_2 + d$	$a_4 = a_3 + d$	$a_5 = a_4 + d$
$a_3 = (a_1 + d) + d$	$a_4 = (a_1 + 2d) + d$	$a_5 = (a_1 + 3d) + d$
$a_3 = a_1 + 2d$	$a_4 = a_1 + 3d$	$a_5 = a_1 + 4d$

Thus, the **general term, a_n,** of an arithmetic sequence whose first term is a_1 and with common difference d is

$$\boldsymbol{a_n = a_{n-1} + d} \qquad \text{or} \qquad \boldsymbol{a_n = a_1 + (n - 1)d.}$$

EXAMPLE 1 The first term of an arithmetic sequence is 3 and the common difference is 2. Find the twentieth term.

Solution: $a_n = a_1 + (n - 1)d$ ← **$a_1 = 3$; $d = 2$; $n = 20$**

$a_{20} = 3 + (20 - 1)2$

$a_{20} = \mathbf{41}$

Given any two terms of an arithmetic sequence you can find the first term and the common difference.

EXAMPLE 2 The sixth term of an arithmetic sequence is -2 and the twelfth term is -14. Find a_1 and d.

Solution: $\begin{cases} -2 = a_1 + 5d \\ -14 = a_1 + 11d \end{cases}$ ← **$a_6 = a_1 + (6 - 1)d$**; **$a_{12} = a_1 + (12 - 1)d$**

$\begin{cases} a_1 + 5d = -2 \\ a_1 + 11d = -14 \end{cases}$ ← **Solve this system to find a_1 and d.**

Thus, $a_1 = \mathbf{8}$ and $d = \mathbf{-2}$.

The terms between any two given terms of an arithmetic sequence are called **arithmetic means.** For example, in the sequence 2, 8, 14, 20, 26, · · · the terms 8, 14, and 20 are the three arithmetic means between 2 and 26.

EXAMPLE 3 Find three arithmetic means between 6 and 12.

Solution: Consider 6 and 12 as the first and fifth terms of the sequence.

$$6, \underline{\ ?\ }, \underline{\ ?\ }, \underline{\ ?\ }, 12, \cdots$$

Thus, $a_5 = a_1 + (5-1)d$. ← **$a_5 = 12$; $a_1 = 6$**

$$12 = 6 + 4d$$

$$\frac{3}{2} = d$$

$a_2 = a_1 + d$	$a_3 = a_1 + 2d$	$a_4 = a_1 + 3d$
$a_2 = 6 + \frac{3}{2}$	$a_3 = 6 + 2\left(\frac{3}{2}\right)$	$a_4 = 6 + 3\left(\frac{3}{2}\right)$
$a_2 = 7\frac{1}{2}$	$a_3 = 9$	$a_4 = 10\frac{1}{2}$

Arithmetic means: $7\frac{1}{2}$, 9, $10\frac{1}{2}$

CLASSROOM EXERCISES

Classify each sequence as *arithmetic* or *non-arithmetic.* For each arithmetic sequence, find the difference between successive terms.

1. $-6, -2, 2, 6, 10, \cdots$

2. $\frac{1}{3}, \frac{1}{9}, \frac{1}{27}, \frac{1}{81}, \frac{1}{243}, \cdots$

3. $6, 3, 0, -3, -6, \cdots$

4. $1, \frac{3}{2}, 2, \frac{5}{2}, 3, \cdots$

5. $2x, x, 0, -x, -2x, \cdots$

6. $x + 1, 2x, 3x - 1, 4x - 2, \cdots$

WRITTEN EXERCISES

A Write the first five terms of each arithmetic sequence.

1. $a_1 = 1$, $d = 5$

2. $a_1 = -5$, $d = 6$

3. $a_1 = -30$, $d = -2$

4. $a_1 = \frac{3}{4}$, $d = -\frac{1}{4}$

5. $a_1 = 2.3$, $d = 1.6$

6. $a_1 = 25b$, $d = -5b$

7. $a_1 = c$, $d = x - c$

8. $a_1 = x$, $d = x + m$

9. The first term of an arithmetic sequence is 2. The common difference is 5. Find a_6.

10. The first term of an arithmetic sequence is 16. The common difference is -2. Find a_{17}.

11. In an arithmetic sequence $a_1 = 20$, and $d = \frac{5}{2}$. Find a_{11}.

12. In an arithmetic sequence $a_1 = -6$, and $d = \frac{2}{3}$. Find a_{10}.

13. Find the 10th term of 4, 5, 6, $\cdots$

14. Find the 20th term of 5, 7, 9, $\cdots$

15. Find the 18th term of 18, 14, 10, $\cdots$

16. Find the 24th term of -3, 0, 3, $\cdots$

17. Find the 9th term of a, $a+2$, $a+4$, $\cdots$

18. Find the 11th term of $2\sqrt{3}$, $\sqrt{3}$, 0, $\cdots$

19. Find the 25th term of 5, $4\frac{1}{3}$, $3\frac{2}{3}$, $\cdots$

20. Find the 15th term of 0.5, 2, 3.5, $\cdots$

21. The fourth term of an arithmetic sequence is 13 and the sixth term is 7. Find the first term and the common difference.

22. The third term of an arithmetic sequence is 5 and the eighth term is 20. Find the first term, the common difference, and the first three terms of the sequence.

23. The seventh term of an arithmetic sequence is 6 and the thirteenth term is -18. Find the first five terms of the sequence.

24. The fourth term of an arithmetic sequence is 9 and the seventh term is 10. Find the twentieth term.

25. Find two arithmetic means between 3 and 15.

26. Find two arithmetic means between 4 and 80.

27. Find three arithmetic means between 6 and 11.

28. Find four arithmetic means between 4 and 179.

29. Find five arithmetic means between 2 and 11.

30. Find the arithmetic mean between x and y.

APPLICATIONS: Using Arithmetic Sequences

31. The top row of a pile of logs contains 6 logs, the row below the top one contains 7 logs, the third row from the top contains 8 logs, and so on. If there are 45 rows, how many logs are there in the bottom row?

32. Tuition costs for a certain university amount to \$8000 per year. Suppose that these costs increase by \$500 per year. How much will tuition costs be in 13 years?

33. In Turner City, the fine for a first parking offense is \$15. The city adds \$25 to the fine for each further offense. What is the fine for the sixth parking offense?

34. The Acme Theater has 54 seats in the last row at the back. Each of the other rows has 2 fewer seats than the row before it. If there are 17 rows of seats, how many seats are there in the first row?

35. A person wearing contact lenses is told to wear them 3 hours the first day. The lenses are to be worn $\frac{1}{2}$ hour longer each succeeding day, up to a maximum of 12 hours. How many days will it take to reach the 12–hour maximum?

36. The cost of renting a power tool is \$25 for the first day. The cost then decreases \$1 per day up to a maximum of 7 days. What is the rental cost on the seventh day?

B

37. To drill a water well, the Deland Corporation charges \$6.00 per foot for each of the first 10 feet drilled. The cost then increases to \$7.00 per foot for the next 10 feet, \$8.00 per foot for the next 10 feet, and so on. How much will it cost to drill the last 10 feet of a 160-foot well?

38. A consumer buys a foreign car for \$8500. Rising import taxes increase the price of the car by \$225 each year. How many years will it take for the price of the car to exceed \$10,000?

39. The cost for printing 1000 copies of a 128-page book is \$2500. The cost for printing each additional eight pages is \$72. How much will it cost to print 1000 copies of a 208-page book?

40. A traveling salesperson receives a 5% commission on each of the first 5 vacuum cleaners sold, a 6% commission on each of the next 3 sold, a 7% commission on each of the next 3, and so on. If the vacuum cleaners are selling for \$400 each and Mary Ings sells 23, what is her commission on the last three sold?

41. Felipe drove his car 9400 kilometers during the first year of operation. Operating costs for that year were 14¢ per kilometer. In each of the following years, Felipe drove the car 850 kilometers more than the preceeding year while operating costs per kilometer increased by 3¢ each year. How much did it cost Felipe to run the car in its fifth year?

42. The first term in an arithmetic sequence is 38 and the last term is 1. The sequence has 75 terms. Find the common difference.

43. How many terms are there in an arithmetic sequence if the first term is 100, the last term is -14, and the common difference is -2?

44. Which term in the sequence $5, 2, -1, -4, \cdots$ is -85?

45. How many even integers are there between 12 and 238?
(HINT: $a_1 = 14$, $a_n = 236$, and $d = 2$.)

46. How many odd integers are there between 133 and 351?

47. How many integers divisible by 3 are there between 1 and 100?

48. Find the value of x so that $x + 1$, $3x - 2$, and $4x + 5$ form an arithmetic sequence.

C

49. Prove: *Finding one arithmetic mean between two numbers is the same as finding the "average" of the numbers.*

50. Prove: *Adding corresponding terms of two arithmetic sequences forms another arithmetic sequence. Illustrate with an example.*

51. Prove: *Multiplying corresponding terms of two arithmetic sequences does not, in general, form another arithmetic sequence.*

52. Prove: *If each term of an arithmetic sequence is multiplied by a constant, the resulting sequence is an arithmetic sequence.*

REVIEW CAPSULE FOR SECTION 12–3

Simplify. *(Pages 2–5, 270–271, 408–410)*

1. $\sqrt[3]{27}$ **2.** $\sqrt[4]{256}$ **3.** $(81)^{\frac{1}{3}}$ **4.** $16^{\frac{3}{2}}$ **5.** 5^4 **6.** $(\frac{1}{2})^5$

7. $\left(-\frac{2}{3}\right)^5$ **8.** $(-3)^6$ **9.** $\frac{1}{\sqrt[4]{16}}$ **10.** $(0.03)^2$ **11.** $(-1)^7$ **12.** $3 \cdot 2^5$

12–3 Geometric Sequences

In arithmetic sequences, there is a common difference, d, between successive terms. A sequence in which the ratio of successive terms is the same constant, r (called a **common ratio**), is a *geometric sequence* or a *geometric progression.*

Definition

A **geometric sequence** is one in which the ratio, r, of successive terms is always the same number.

$$\frac{a_{n+1}}{a_n} = r$$

EXAMPLE 1 The first term of a geometric sequence is a_1. Find the next four terms.

Solution: $\frac{a_2}{a_1} = r$ or $a_2 = a_1 r$ ← **By definition**

Similarly: $a_3 = a_2 r$ $a_4 = a_3 r$ $a_5 = a_4 r$

$a_3 = (a_1 r)r$ $a_4 = (a_1 r^2)r$ $a_5 = (a_1 r^3)r$

$\mathbf{a_3 = a_1 r^2}$ $\mathbf{a_4 = a_1 r^3}$ $\mathbf{a_5 = a_1 r^4}$

Generalizing from Example 1, you have the expression for the nth term of a geometric sequence.

$$a_n = a_{n-1} r, \quad \text{or} \quad a_n = a_1 r^{n-1}$$

The symbol for the first term is often given without the subscript. That is, $a_1 = a$. Thus,

$$a_n = ar^{n-1}.$$

EXAMPLE 2 Find the eighth term of 4, 8, 16, 32, $\cdots$.

Solution: $a_n = a \cdot r^{n-1}$ ← **$a = 4$; $r = 2$; $n = 8$**

$a_8 = 4 \cdot (2^{8-1})$

$a_8 = 4 \cdot (2^7)$

$a_8 = 4(128) = \mathbf{512}$

The terms between any two given terms of a geometric sequence are called the **geometric means.**

EXAMPLE 3 Find three geometric means between 3 and $\frac{3}{16}$.

Solution: First find r.

$a_n = ar^{n-1}$ ← $a_n = \frac{3}{16}$; $a = 3$; $n = 5$

$\frac{3}{16} = 3r^4$

$r = \pm\frac{1}{\sqrt[4]{16}} = \pm\frac{1}{2}$ ← **To obtain all possible real number solutions, use both the positive and negative roots.**

$a_2 = 3(\frac{1}{2}) = \frac{3}{2}$ or $a_2 = 3(-\frac{1}{2}) = -\frac{3}{2}$

$a_3 = 3(\frac{1}{2})^2 = \frac{3}{4}$ or $a_3 = 3(-\frac{1}{2})^2 = \frac{3}{4}$

$a_4 = 3(\frac{1}{2})^3 = \frac{3}{8}$ or $a_4 = 3(-\frac{1}{2})^3 = -\frac{3}{8}$

Thus, either $\mathbf{\frac{3}{2}, \frac{3}{4}, \frac{3}{8}}$ or $\mathbf{-\frac{3}{2}, \frac{3}{4}, -\frac{3}{8}}$ may be inserted between 3 and $\frac{3}{16}$ to form a geometric sequence.

CLASSROOM EXERCISES

Find the common ratio for each geometric sequence.

1. $3, 9, 27, \cdots$
2. $8, 2, \frac{1}{2}, \frac{1}{8}, \cdots$
3. $ar, ar^2, ar^3, \cdots$

Find the missing term in each geometric sequence.

4. $2, 8, 32,$ __?__ $\cdots$
5. __?__, $3, 1, \frac{1}{3} \cdots$
6. $\sqrt{3}$, __?__, $3\sqrt{3}, 9 \cdots$

WRITTEN EXERCISES

A

In Exercises 1–9, the first term, *a*, and the common ratio, *r*, are given for each sequence. Write the first four terms.

1. $a = 2, r = 3$
2. $a = 2, r = -2$
3. $a = \frac{1}{2}, r = 4$
4. $a = -\frac{1}{2}, r = -2$
5. $a = \frac{1}{4}, r = 2$
6. $a = -27, r = \frac{2}{3}$
7. $a = 0.7, r = -4$
8. $a = \sqrt{2}, r = \sqrt{2}$
9. $a = i, r = i$

Find the specified term of each geometric sequence.

10. 8th term: $1, 2, 4, \cdots$
11. 8th term: $-1, -2, -4, \cdots$
12. 6th term: $-6, -3, -\frac{3}{2}, \cdots$
13. 14th term: $x, -x, x, \cdots$
14. 5th term: $-8, 16, -32, \cdots$
15. 7th term: $5\frac{1}{3}, -8, 12, \cdots$
16. 9th term: $1, 1.03, (1.03)^2, \cdots$
17. 6th term: $1, 0.01, 0.0001, \cdots$
18. 10th term: $\sqrt{2}, 2, 2\sqrt{2}, \cdots$
19. 6th term: $96, 24, 6, \cdots$
20. 9th term: $0.3, 0.03, 0.003, \cdots$
21. 12th term: $i, -1, -i, \cdots$

Find the indicated geometric means.

22. One positive geometric mean between 3 and 48
23. One negative geometric mean between -3 and -27
24. Three geometric means between 3 and 48
25. Five geometric means between -2 and -128
26. Four geometric means between $\sqrt{2}$ and 8

A single geometric mean between two numbers is called the **geometric mean**, or the **mean proportional**, of the two numbers.

27. Find the mean proportional between $\sqrt{6}$ and $\sqrt{24}$.
28. The geometric mean between two numbers is 15. One number is 9. Find the the other number.

APPLICATIONS: Using Geometric Sequences

29. Each year the value of a car is 70% of its value of the previous year. At the end of the first year, the value of the car was \$6000. What was its value at the end of 3 years?
30. A tank contains 8000 liters of water. Each day, one half of the remaining water is removed. How much water will be in the tank on the eighth day?
31. A culture contains 200 bacteria. The number of bacteria in the culture doubles every two hours. Find the number of bacteria at the end of 6 hours.
32. The holder of the first ticket drawn in a lottery will win \$5000. The holders of each of the three succeeding tickets will win one-half as much as the preceeding winner. What amount is paid to the holder of the fourth ticket?
33. In payment for a favor done, a mathematician asked that the squares on a checkerboard (64 squares) be filled as follows:

    1¢ on the first square, 2¢ on the second square, 4¢ on the third square, and so on.

    What sum of money is placed on the tenth square?
34. A car which originally cost \$8200 depreciates each year $\frac{1}{4}$ of its value of the preceding year. What is the value of the car at the end of 5 years?

B

35. A tennis ball, dropped from a height of 3.06 meters, rebounds on each bounce one-third of the distance of its previous bounce. How high does it bounce on the sixth bounce?
36. A vacuum pump is used to remove air from a bell jar. The pump removes $\frac{1}{12}$ of the air at each stroke. How much air remains after 20 strokes? Express your answer as a per cent, rounded to the nearest whole number.
37. Which term of the geometric sequence $5, -10, 20, \cdots$ is 320?
38. Which term of the geometric sequence $\frac{1}{4}, \frac{1}{2}, 1, \cdots$ is 128?
39. The first term of a geometric sequence is 3 and the sixth term is 96. Find the twelfth term.

40. The figure at the right contains a large square with a side of length 12 and a succession of smaller squares formed by connecting midpoints of consecutive sides. The areas of the squares form a geometric sequence.

Find the common ratio and the area of the shaded square.

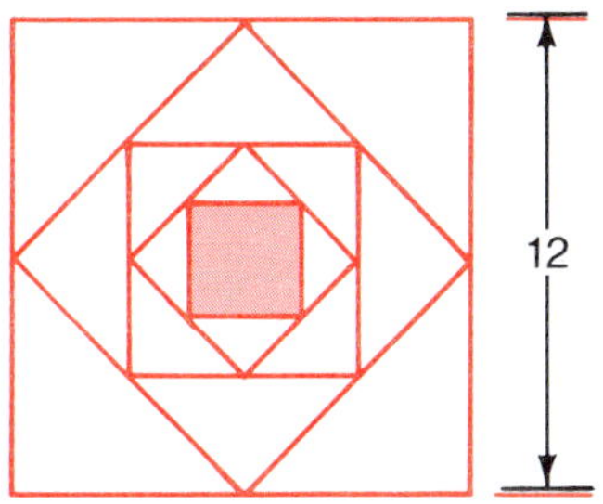

41. Find the value of x so that $x + 1$, $x - 1$, and $x - 2$ form a geometric sequence.

C

42. There are 3 amoebas in a jar. They double every minute. If the jar is completely filled in 20 minutes, when is the jar half full?

43. Prove: *If* a_1, a_2, a_3 *form a geometric sequence, then* a_1^2, a_2^2, a_3^2, *form a geometric sequence.*

Review

Find the rule for each sequence. Use the rule to write the next three terms. *(Section 12–1)*

1. 5, 8, 13, 20, · · ·
2. $-2, 4, -8, 16, \cdots$
3. $2, 1\frac{1}{4}, 1\frac{1}{9}, 1\frac{1}{16}, \cdots$
4. The first term of an arithmetic sequence is 5 and the common difference is 2. Find the tenth term. *(Section 12–2)*
5. The fourth term of an arithmetic sequence is 3 and the eighth term is 5. Find the fiftieth term. *(Section 12–2)*
6. Mr. Miller begins work for Company Z with a starting salary of $15,000 per year. He gets an automatic raise of $300 every 3 months. What is Mr. Miller's salary at the end of 5 years? *(Section 12–2)*

Find the specified term of each geometric sequence. *(Section 12–3)*

7. 10th term: 5, 10, 20, · · ·
8. 5th term: $\frac{2}{3}, \frac{1}{2}, \frac{3}{8}, \cdots$
9. Find 4 geometric means between 2 and 6250. *(Section 12–3)*
10. Jeanine owns 20 shares of Contex, Inc. If the stock splits (she receives 2 shares for every share she owns) every December for 5 years, how many shares will she own at the end of the fifth year? *(Section 12–3)*

REVIEW CAPSULE FOR SECTION 12–4

Evaluate each expression for $k = 1, 2, 3,$ and 4. *(Pages 2–5)*

1. $6k$
2. $3k + 10$
3. $\frac{1}{2}k - 3$
4. k^2
5. 3^k
6. 5^{k+1}
7. $ki^2 + k$
8. $\frac{1}{k}$
9. $\frac{1}{k+1}$
10. $k^2 - \frac{1}{k^2+1}$

12–4 Arithmetic Series

You can use the terms of a sequence to write a *series*.

Sequence	Series
$1, 2, 3, \cdots n, \cdots$	$1 + 2 + 3 + \cdots n + \cdots$
$1, 4, 7, \cdots (3n - 2), \cdots$	$1 + 4 + 7 + \cdots + (3n - 2) + \cdots$
$1, 4, 9, \cdots n^2, \cdots$	$1 + 4 + 9 + \cdots + n^2 + \cdots$
$2, 4, 8, \cdots 2^n, \cdots$	$2 + 4 + 8 + \cdots + 2^n + \cdots$

Definition

A **series** is the indicated sum of the terms of a sequence.

The sum of the first n terms of a series is called the **partial sum, S_n.** For example, for the series

$$2 + 4 + 6 + \cdots + 2n + \cdots,$$

$S_1 = 2$ $\quad$ $S_2 = 2 + 4 = 6$ $\quad$ $S_3 = 2 + 4 + 6 = 12$ ← **Partial sum**

The Greek letter Σ (sigma) can be used to represent a sum. For example, the third partial sum, S_3, of the series above can also be written as

$$\sum_{k=1}^{3} 2k.$$

This is read "the sum of the series with general term $2k$ from $k = 1$ to $k = 3$." The letter k is called the **index of summation.**

EXAMPLE 1 **a.** Write $\sum_{k=1}^{4} (4k - 3)$ in expanded form.

b. Find S_4.

Solutions: **a.** $\sum_{k=1}^{4} (4k - 3) = [4(1) - 3] + [4(2) - 3] + [4(3) - 3] + [4(4) - 3]$
$= \mathbf{1} + \mathbf{5} + \mathbf{9} + \mathbf{13}$

b. From a, $S_4 = \sum_{h=1}^{4} (4k - 3) = 1 + 5 + 9 + 13 = \mathbf{28}$

An **arithmetic series** is the indicated sum of an arithmetic sequence.

$$1 + 2 + 3 + \cdots + n, \quad \text{or} \quad \sum_{k=1}^{n} k$$ ← **Arithmetic series**

To find a general formula for the sum of an arithmetic series, you first write the general form of the series in expanded form. Then you write the sum in reverse order and add.

$$S_n = a_1 + (a_1 + d) + (a_1 + 2d) + \cdots + [a_1 + (n-1)]d \quad \longleftarrow \text{General form}$$
$$S_n = [a_1 + (n-1)d] + [a_1 + (n-2)d] + \cdots + a_1 \quad \longleftarrow \text{Reverse order}$$

$$2S_n = [2a_1 + (n-1)d] + [2a_1 + (n-1)d] + \cdots + [2a_1 + (n-1)d]$$
$$2S_n = n[2a_1 + (n-1)d]$$
$$S_n = \frac{n}{2}[2a_1 + (n-1)d]$$

Theorem 12–1

Sum of an Arithmetic Series

If S_n represents the sum of the arithmetic series

$a_1 + (a_1 + d) + (a_1 + 2d) + \cdots + a_1 + (n-1)d$, then

$$S_n = \frac{n}{2}[2a_1 + (n-1)d].$$

Example 2 shows how to use this formula.

EXAMPLE 2 Find the sum of the first fifteen terms of $2 + 5 + 8 + \cdots$.

Solution: $S_n = \frac{n}{2}[2a_1 + (n-1)d] \quad \longleftarrow \quad a_1 = 2;\ n = 15;\ d = 3$

$$S_{15} = \frac{15}{2}[2(2) + 14(3)] = \frac{15(4 + 42)}{2} = \mathbf{345}$$

The formula, $$S_n = \frac{n}{2}(a_1 + a_n),$$

where a_1 is the first term and a_n is the last term, can be derived from Theorem 12–1. It can be used to find the sum of n terms of an arithmetic series when you know, or can find, the first and last terms of the series.

EXAMPLE 3 Find the indicated sum: $\sum_{k=1}^{10} (7 - 2k)$

Solution: $n = 10,\ a_1 = 7 - 2(1) = 5;\ a_{10} = 7 - 2(10) = -13$

$$S_n = \frac{n}{2}(a_1 + a_n)$$

$$S_{10} = \frac{10}{2}[5 + (-13)] = \mathbf{-40} \quad \longleftarrow \text{Sum of 10 terms}$$

CLASSROOM EXERCISES

In Exercises 1–3, write the Σ notation for each statement.

1. The sum of the series with general term $k + 4$ from $k = 1$ to $k = 11$
2. The sum of the series with general term k^3 from $k = 1$ to $k = 8$
3. The sum of the series with general term $2k^2 - 1$ from $k = 1$ to $k = 15$

Write each partial sum in expanded form.

4. $\sum_{k=1}^{3} (k+6)$ 5. $\sum_{k=1}^{4} 5k$ 6. $\sum_{k=1}^{3} k^2$ 7. $\sum_{k=1}^{5} \frac{1}{k}$

Find the sum of the terms as indicated.

8. $3+5+7\cdots$; to eight terms
9. $17+15+13+\cdots$; to five terms
10. $5+8+11+\cdots$; to ten terms
11. $-5-3-1-\cdots$; to twelve terms

WRITTEN EXERCISES

A In Exercises 1–8, write each partial sum in expanded form.

1. $\sum_{k=1}^{4} (3k+2)$ 2. $\sum_{k=1}^{5} (8-2k)$ 3. $\sum_{k=1}^{10} \frac{1}{3}k$ 4. $\sum_{k=1}^{6} \left(\frac{2}{5}k-\frac{1}{5}\right)$

5. $\sum_{k=1}^{4} 3^{k-1}$ 6. $\sum_{k=1}^{5} (-2)^{k-1}$ 7. $\sum_{k=1}^{3} (3^k-2^k)$ 8. $\sum_{k=1}^{3} (2^k+2^{-k})$

Use Exercises 1–8 to evaluate the indicated partial sum.

9. S_4 in Exercise 1
10. S_5 in Exercise 2
11. S_4 in Exercise 3
12. S_3 in Exercise 4
13. S_4 in Exercise 5
14. S_5 in Exercise 6

Find the sum of the terms as indicated.

15. $-7-5-3-\cdots$ to sixteen terms
16. $7+12+17+\cdots$ to ten terms
17. $-13-11-9-\cdots$ to twenty terms
18. $-10-5-0+\cdots$ to ten terms
19. $7+11+15+\cdots$ to fifteen terms
20. $14+10+6+\cdots$ to eight terms
21. $5+4\frac{1}{2}+4+\cdots$ to twenty terms
22. $-7-9-11-\cdots$ to seven terms
23. $2\sqrt{2}+\sqrt{2}+0-\cdots$ to eight terms
24. $-6-4-2\cdots$ to six terms
25. $-6-2+2+\cdots$ to twelve terms
26. $\frac{1}{2}+2+3\frac{1}{2}+\cdots$ to twenty terms
27. $\frac{1}{4}+\frac{1}{2}+\frac{3}{4}+\cdots$ to ten terms
28. $3+\frac{7}{2}+4+\cdots$ to fifteen terms
29. $x+0-x-\cdots$ to eighteen terms
30. $2x+5x+8x+\cdots$ to fifteen terms
31. $2b+5b+8b+\cdots$ to twelve terms
32. $(a+7)+(a+4)+(a+1)+\cdots$ to nine terms
33. $(a+b)+(a+2b)+(a+3b)+\cdots$ to ten terms
34. $(2x-5y)+(3x-4y)+(4x-3y)+\cdots$ to ten terms

35. $\sum_{k=1}^{6} (3k+5)$ 36. $\sum_{k=1}^{8} (7k-27)$ 37. $\sum_{k=1}^{10} (2k-1)$ 38. $\sum_{k=1}^{5} (8-2k)$

39. $\sum_{k=1}^{4} (1-2k)$ 40. $\sum_{k=1}^{5} (90-k)$ 41. $\sum_{k=1}^{20} (17-k)$ 42. $\sum_{k=1}^{15} \left(\frac{2}{5}k-\frac{1}{5}\right)$

43. $\sum_{k=2}^{5} (5k-22)$ 44. $\sum_{k=5}^{10} -k$ 45. $\sum_{k=5}^{8} (3k-4)$ 46. $\sum_{k=40}^{45} (8-k)$

APPLICATIONS: Using Arithmetic Series

47. In a lecture hall, the front row has 18 seats. Each succeeding row has 4 more seats than the row ahead of it. How many seats are there in the first 12 rows?

48. Marge decided to save for a ski trip. She saved one quarter on January 1, two quarters on January 2, three quarters on January 3, and so on. How many quarters had she saved after she added those for January 31?

49. To pay for a motorcycle, Ray makes a down payment of $450 and 18 monthly payments which equal $100 the first month and $5 less for each succeeding payment. Find the total amount paid.

50. The sequence 16, 48, 80, · · · represents the distance in feet an object will fall during the first second, the second second, the third second, and so on. How far will the object fall in the tenth second? How far will if fall in the first ten seconds?

51. The sequence 2, 5, 8, · · · shows how fast a certain object rolls down an inclined plane. That is, it rolls 2 meters the first second, 5 meters the next second, and so on. How far will it roll in the twelfth second? How far will it roll in twelve seconds?

52. A contest winner will receive the prize money over 24 months. The first month's payment is $5000. Each succeeding payment will be $100 less than the preceding one. How much will the contest winner receive in all?

53. A bean picker on a truck farm picks 60 pounds of beans the first day. On each of 4 succeeding days, he picks 15 more pounds than on the preceding day. How much does he receive at the end of the 5 days if he is paid at the rate of 15¢ per pound?

54. A ball, rolling down an inclined plane, travels 3.16 meters the first second, and, during each succeeding second, 6.32 meters more than the preceding second. How far will the ball roll in 10 seconds?

55. During a free fall, a parachutist falls 9.8 meters farther each second than the distance traveled the previous second. A parachutist falls 8 meters the first second. How far will she fall in 15 seconds?

B

56. How long will it take the parachutist in Exercise 55 to fall 5000 meters? Round your answer to the nearest whole second.

57. Ladders are often trapezoidal in shape with regular changes in the length of consecutive rungs. The bottom rung of a 20–rung, trapezoidal ladder is 80 centimeters long and the top rung is 50 centimeters long. Find the total length of the rungs.

Write the arithmetic series that results from each procedure.

58. Eight arithmetic means are inserted between 6 and 60.

59. Five arithmetic means are inserted between 24 and 240.

60. Four arithmetic means are inserted between x and y.

Find the missing data for each arithmetic series.

	a_1	a_n	n	d	S_n
61.	2	31	?	?	165
62.	−1	?	14	?	441
63.	5	−33	20	?	?
64.	?	?	40	$\frac{1}{2}$	630
65.	?	−113	25	?	−1325

C

66. Use Theorem 12–1 to prove that $S_n = \frac{n}{2}(a_1 + a_n)$.

67. Find the sum of the integers between 1 and 100 that are not divisible by 3.

68. If $a_1 + a_2 + a_3 + a_4 + a_5 + \cdots$ is an arithmetic series, prove that the terms $a_1 + a_3 + a_5 + \cdots$ also form an arithmetic series.

CALCULATOR APPLICATIONS

Sum of an Arithmetic Series

The parenthesis keys on a scientific calculator simplify the process for finding the sum of an arithmetic series.

EXAMPLE Find the sum of the first eighteen terms of $2 + 5 + 8 + \cdots$.

SOLUTION $S_n = \frac{18}{2}[2(2) + 17(3)]$ ← **$n = 18$; $a_1 = 2$; $n - 1 = 17$; $d = 3$**

1 8 [÷] 2 [×] [(] 2 [×] 2 [+] 1 7 [×] 3 [)] [=] 495.

EXERCISES

Check your answers to Exercises 9, 11, 12, and 15–46 on page 456.

REVIEW CAPSULE FOR SECTION 12–5

Perform the indicated operations. *(Pages 2–5)*

1. $3(2)^6$ **2.** $(-4)^7$ **3.** $6\left(\frac{1}{2}\right)^5$ **4.** $128\left(\frac{1}{2}\right)^{11}$

5. $27\left(\frac{1}{3}\right)^7$ **6.** $4(2)^5 - 4$ **7.** $\frac{1}{8}(2)^{16} - \frac{1}{8}$ **8.** $\frac{2(-2)^6 - 2}{-3}$

12–5 Geometric Series

A *geometric series* is related to a geometric sequence in the same way that an arithmetic series is related to an arithmetic sequence.

Definition

A **geometric series** is the indicated sum of the terms of a geometric sequence.

The general term of a geometric series is the same as the general term of the corresponding geometric sequence.

$$a_n = a_1 \cdot r^{n-1}$$

A formula for the sum of a geometric series can be found.

$$S_n = a_1 + a_1r + a_1r^2 + \cdots + a_1r^{n-1} \quad \longleftarrow \text{Sum of } n \text{ terms}$$

$$rS_n = \quad a_1r + a_1r^2 + a_1r^3 + \cdots + a_1r^n \quad \longleftarrow \text{Multiply by } r.$$

$$rS_n = \quad a_1r + a_1r^2 + a_1r^3 + \cdots + a_1r^{n-1} + a_1r^n \quad \longleftarrow \text{Find } rS_n - S_n.$$

$$-S_n = -a_1 - a_1r - a_1r^2 - \cdots - a_1r^{n-2} - a_1r^{n-1}$$

$$rS_n - S_n = a_1r^n - a_1$$

$$S_n(r-1) = a_1r^n - a_1$$

$$S_n = \frac{a_1(r^n - 1)}{r - 1} \quad \longleftarrow r \neq 1$$

Theorem 12–2

Sum of a Geometric Series

If S_n represents the sum of the geometric series, then

$$S_n = a_1 + a_1r + a_1r^2 + \cdots + a_1r^{n-1}, \quad \text{or} \quad S_n = \frac{a_1(r^n - 1)}{r - 1}, \; r \neq 1.$$

EXAMPLE 1 Find the sum of the series $2 + 6 + 18 + 54 + 162$.

Solution: $S_n = \dfrac{a_1(r^n - 1)}{r - 1} \quad \longleftarrow \quad n = 5;\ a_1 = 2;\ r = 3$

$$S_5 = \frac{2(3^5 - 1)}{3 - 1} = \frac{2(242)}{2} = \mathbf{242}$$

A geometric series can also be written in summation (Σ) notation.

EXAMPLE 2 Find the indicated sum: $\displaystyle\sum_{k=1}^{6} \left(\frac{1}{2}\right)^{k+1}$

Solution: $S_n = \dfrac{a_1(r^n - 1)}{r - 1} \quad \longleftarrow \quad a_1 = \frac{1}{4},\ r = \frac{1}{2};\ n = 6$

$$S_6 = \frac{\frac{1}{4}\left[\left(\frac{1}{2}\right)^6 - 1\right]}{\frac{1}{2} - 1} = \frac{\frac{1}{4}\left(-\frac{63}{64}\right)}{-\frac{1}{2}} = \mathbf{\frac{63}{128}}$$

CLASSROOM EXERCISES

For each geometric series, give the value of a_1, r, and n.

1. $1 + 4 + 16 + 64 + 256$
2. $4 + 20 + 100 + 500$
3. $6 + 3 + \frac{3}{2} + \frac{3}{4} + \frac{3}{8}$
4. $\frac{27}{2} + 9 + 6 + 4 + \frac{8}{3} + \frac{16}{9}$
5. $\sum_{k=1}^{4} 2^k$
6. $\sum_{k=1}^{6} 10^k$
7. $\sum_{k=1}^{6} \left(\frac{1}{2}\right)^{k-1}$
8. $\sum_{k=1}^{n} 3(-2)^k$

WRITTEN EXERCISES

A Find the sum of each series.

1. $2 + 1 + \frac{1}{2} + \frac{1}{4} + \frac{1}{8} + \frac{1}{16}$
2. $12 - 4 + \frac{4}{3} - \frac{4}{9} + \frac{4}{27}$
3. $1 - 1 + 1 - 1 + 1 - 1 + 1 - 1$
4. $1.34 + 0.67 + 0.335 + 0.1675$

Find the sum to the required number of terms for each geometric series.

5. $4 + 8 + 16 + \cdots$ to five terms
6. $2 - 8 + 32 - \cdots$ to five terms
7. $12 - 6 + 3 - \cdots$ to eight terms
8. $3 + \frac{9}{2} + \frac{27}{4} + \cdots$ to six terms
9. $2 - 10a + 50a^2 - \cdots$ to five terms
10. $\frac{1}{2} - \frac{1}{4} + \frac{1}{8} - \cdots$ to six terms
11. $\sum_{k=1}^{3} \left(\frac{1}{3}\right)^k$
12. $\sum_{k=1}^{11} 2^k$
13. $\sum_{k=1}^{4} (3^{k-1})$
14. $\sum_{k=1}^{4} 24\left(-\frac{1}{2}\right)^k$
15. $\sum_{k=1}^{5} (-2)^k$
16. $\sum_{k=1}^{6} 2^{-k}$
17. $\sum_{k=1}^{8} (2^{k-4})$
18. $\sum_{k=1}^{5} \left(\frac{2}{3}\right)^k$

Find the indicated partial sum.

19. $25 + 45 + 81 + \cdots ; S_6$
20. $6.4 + 9.6 + 14.4 + \cdots ; S_{10}$

Find the number of terms in each series.

21. $6 + 12 + 24 + \cdots + 384$
22. $64 + 96 + 144 + \cdots + 729$

APPLICATIONS: Using Geometric Series

23. How much would you have saved at the end of seven days if you had set aside 64¢ on the first day, 96¢ on the second day, \$1.44 on the third day, and so on through the period?
24. A ball is dropped to the ground from a height of 30 meters. If it rebounds to a height of 10 meters on the first bounce, $3\frac{1}{3}$ meters on the second bounce and so on, what is the total distance traveled by the ball at the time it hits the ground the sixth time?
25. The population of a city of 100,000 increases 10% for each of ten years. What will the population be after ten years?
26. During each year, the value of a computer decreases 20%. If the computer cost \$40,000, what is its value after four years?
27. A certain city lost $\frac{1}{2}$% of its population each year for 20 years. Its population was 1,000,000 ten years ago. What is its population now?

28. On a merit salary plan, yearly raises are determined as a per cent of present salary. Suppose that a worker starts at a yearly salary of \$15,000 and receives merit increases of 8.5% for 10 years. What is the worker's salary for the tenth year? What is the worker's total salary over the ten years?

29. A piece of paper is 0.0025 centimeter thick. Assuming that the result is twice as thick after folding, how thick will the paper be if it is folded 15 times?

B

30. In place of a yearly salary of \$100,000, a baseball pitcher agreed to accept \$1 for his first win, \$2 for his second win, \$4 for his third win, and so on. How many games must he win for his earnings to exceed \$100,000?

C

31. Find the sum of the first n terms of $1 + x^3 + x^6 + x^9 + \cdots$.

32. Find the sum of all seven terms of a geometric progression when the sum of the first two terms is 20, and the sum of the last two terms is $-\frac{5}{8}$.

33. Find the first five terms of the geometric series with $r = \frac{1}{2}$, $n = 5$, and $S_n = \frac{31}{2}$.

CALCULATOR APPLICATIONS

Sum of a Geometric Series

The parenthesis and power keys on a scientific calculator simplify the process for finding the sum of a geometric series.

EXAMPLE Find the sum of the series 3 + 9 + 27 + 81 + 243 + 729.

SOLUTION $S_n = \frac{3(3^6 - 1)}{2}$ ← $a_1 = 3$; $r = 3$; $n = 6$

3 [×] [(] 3 [x^y] 6 [−] 1 [)] [÷] 2 [=] **1092.**

EXERCISES

Check your answers to Exercises 5–7, 12–17, and 19–20 on page 460.

REVIEW CAPSULE FOR SECTION 12-6

Express each repeating decimal as a fraction.

Example: $0.\overline{15}$ **Solution:** $N = 0.151515\cdots$

$$100N = 15.151515\cdots$$ ← **Multiply by 10^n where n is the number of digits that repeat.**

$$-N = -0.151515\cdots$$ ← **Subtract.**

$$99N = 15, \text{ and } N = \frac{15}{99} \text{ or } \frac{5}{33}$$

1. $0.\overline{8}$ **2.** $0.\overline{4}$ **3.** $0.\overline{5}$ **4.** $0.\overline{12}$ **5.** $0.\overline{03}$ **6.** $0.\overline{06}$

12–6 The Sum of an Infinite Geometric Series

The formulas for the sum of a finite geometric series,

$$S_n = \frac{a_1(r^n - 1)}{r - 1}, r \neq 1 \qquad \text{or} \qquad S_n = \frac{a_1(1 - r^n)}{1 - r}, r \neq 1,$$

can be used to find the partial sums of an infinite geometric series, such as $\frac{1}{2} + \frac{1}{4} + \frac{1}{8} + \frac{1}{16} + \cdots$.

EXAMPLE 1 Find the partial sums S_4 and S_5 for $\frac{1}{2} + \frac{1}{4} + \frac{1}{8} + \frac{1}{16} + \cdots$.

Solutions: $S_n = \dfrac{a_1(1 - r^n)}{1 - r}$ ← $|r| < 1$

← $a = \frac{1}{2}, r = \frac{1}{2}$ →

$$S_4 = \frac{\frac{1}{2}\left(1 - \left(\frac{1}{2}\right)^4\right)}{1 - \frac{1}{2}} \qquad S_5 = \frac{\frac{1}{2}\left(1 - \left(\frac{1}{2}\right)^5\right)}{1 - \frac{1}{2}}$$

$$= \frac{\frac{1}{2}\left(1 - \frac{1}{16}\right)}{\frac{1}{2}} \qquad = \frac{\frac{1}{2}\left(1 - \frac{1}{32}\right)}{\frac{1}{2}}$$

$$= \frac{\frac{1}{2}\left(\frac{15}{16}\right)}{\frac{1}{2}} = \mathbf{\frac{15}{16}} \qquad = \frac{\frac{1}{2}\left(\frac{31}{32}\right)}{\frac{1}{2}} = \mathbf{\frac{31}{32}}$$

In Example 1, r^n becomes small as you use larger and larger values of n. (NOTE: This will not occur unless $|r| < 1$.) In fact, r^n is near zero when n is very large. Therefore, the value of $(1 - r^n)$ nears 1 as r^n nears zero.)

This suggests that the formula

$$S = \frac{a_1}{1 - r}$$

can be used to find the sum of an infinite geometric series when $|r| < 1$. Thus, you only need to know a_1 and r to find S. For example, the sum of the infinite geometric series in Example 1 is 1.

$$S = \frac{\frac{1}{2}}{1 - \frac{1}{2}} = \frac{\frac{1}{2}}{\frac{1}{2}} = 1$$

This sum, 1, is called the **limit of the infinite geometric series,**

$$\frac{1}{2} + \frac{1}{4} + \frac{1}{8} + \frac{1}{16} + \cdots.$$

You can think of the **limit of a series** as a certain number that the sum of the series will approach, but never actually reach, as you add more and more terms.

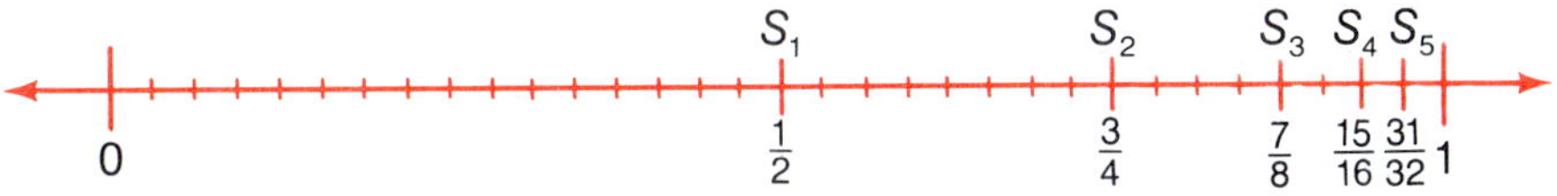

EXAMPLE 2 Find the sum of $5 - \frac{5}{3} + \frac{5}{9} - \frac{5}{27} + \cdots$.

Solution: $S = \dfrac{5}{1 + \frac{1}{3}}$ ⟵ $a_1 = 5,\ r = -\frac{1}{3}$

$= \dfrac{5}{\frac{4}{3}} = \dfrac{15}{4} = 3\frac{3}{4}$

CLASSROOM EXERCISES

For each infinite geometric series, find a_1 and r. Then find S_1, S_3, and S_5.

1. $36 + 18 + 9 + \cdots$ **2.** $18 + 6 + 2 \cdots$ **3.** $1 + \frac{1}{4} + \frac{1}{16} + \cdots$

4. $2 - \frac{1}{2} + \frac{1}{8} - \cdots$ **5.** $\frac{1}{10} + \frac{1}{100} + \frac{1}{1000} + \cdots$ **6.** $0.5 + 0.05 + 0.005 + \cdots$

WRITTEN EXERCISES

A In Exercises 1–6, find the indicated partial sums for the geometric series $6 + 3 + \frac{3}{2} + \cdots$.

1. S_3 **2.** S_4 **3.** S_5 **4.** S_6 **5.** S_7 **6.** S_n

Find the sum of each infinite geometric series.

7. $4 + 2 + 1 + \cdots$ **8.** $12 + 4 + \frac{4}{3} + \cdots$ **9.** $5 + \frac{5}{3} + \frac{5}{9} + \cdots$

10. $6 + 4 + \frac{8}{3} + \cdots$ **11.** $100 + 60 + 46 + \cdots$ **12.** $1 - \frac{1}{3} + \frac{1}{9} - \cdots$

13. $0.3 + 0.03 + 0.003 + \cdots$ $(a = 0.3,\ r = 0.1)$

14. $0.6 + 0.06 + 0.006 + \cdots$ $(a = 0.6,\ r = 0.1)$

15. Find the value of the repeating decimal $0.151515\cdots$. (HINT: This is the same as finding the sum of $0.15 + 0.0015 + 0.000015 + \cdots$. Any repeating decimal can be expressed as a fraction by this method.)

Express each repeating decimal as a fraction (see Exercise 15). The bar indicates the digits that repeat.

16. $0.\overline{8}$ **17.** $0.\overline{4}$ **18.** $0.\overline{5}$ **19.** $0.\overline{12}$ **20.** $0.0\overline{3}$

21. $0.\overline{06}$ **22.** $0.\overline{18}$ **23.** $0.\overline{270}$ **24.** $0.\overline{135}$ **25.** $0.0\overline{75}$

APPLICATIONS: Using Infinite Geometric Series

26. A ball is dropped from a height of 12 meters. It falls straight down and each rebound is $\frac{1}{3}$ of the distance it fell (the distance includes falling and rebounding). How far will the ball travel before coming to rest?

27. A ball is thrown vertically upward a distance of 54 meters. After hitting the ground, it rebounds $\frac{2}{3}$ the distance fallen and continues to rebound in the same manner. What distance does the ball cover before coming to rest?

28. A pendulum swings through an arc of 20 centimeters on the first swing, $\frac{7}{8}$ this distance on the second swing, and continues to swing $\frac{7}{8}$ of the previous distance in each succeeding swing. If this process continues until the pendulum comes to rest, what distance is covered by the pendulum?

29. The output of an oil well decreases by 15% each year. In 1980, the well produced 100,000 barrels of oil. What is the expected maximum output of this well?

30. The productive capability of a field that receives no fertilizer decreases by 12% each year that corn is grown on it. One field produced 38 bushels of corn the first year. What is the maximum yield in bushels that can be expected if no fertilizer is added to the field and corn is grown each year?

31. A drill bit loses its edge at the rate of 5% per meter of drilling. Can the bit be used to drill through a rock 30 meters thick?

B

32. The midpoints of the sides of a 12–centimeter equilateral triangle are connected to form a second inscribed triangle, whose sides are in turn connected to form a third equilateral triangle, and so on. If this process is continued indefinitely, find the sum of the perimeters of the triangles. (See the figure at the left below.)

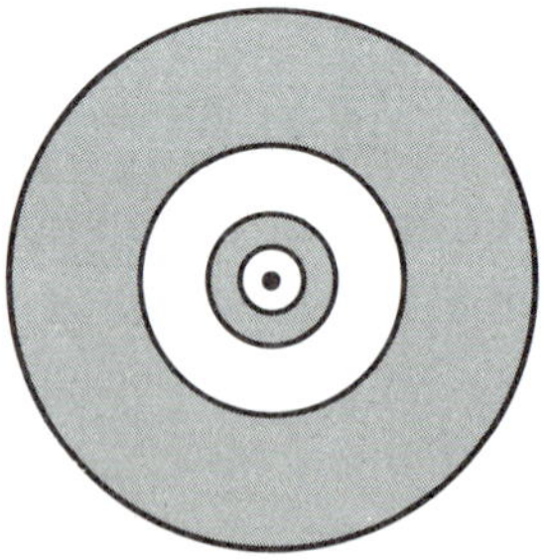

33. An abstract piece of art depicts an infinite series of concentric circles (circles having the same center). The largest of these circles has a circumference of 20 decimeters and each successive circle has a circumference of $\frac{1}{2}$ the circumference of each preceding circle. Find the sum of the circumferences of the infinite series. (See the figure at the right above.)

34. The sides of a square are 25 meters long. Successive squares are inscribed within squares by joining the midpoints of the sides. Find the sum of the areas of the squares.

12–7 Binomial Theorem

A product or power rewritten as a sum is called an **expansion.** Here are some expansions of $(a+b)^n$.

$$n=0: \quad (a+b)^0 = 1$$
$$n=1: \quad (a+b)^1 = a+b$$
$$n=2: \quad (a+b)^2 = a^2 + 2ab + b^2$$
$$n=3: \quad (a+b)^3 = a^3 + 3a^2b + 3ab^2 + b^3$$
$$n=4: \quad (a+b)^4 = a^4 + 4a^3b + 6a^2b^2 + 4ab^3 + b^4$$

The expansion of a binomial has patterns such as the following that simplify finding $(a+b)^n$.

$$a+b = \quad a + b$$
$$(a+b)^2 = \quad a^2 + 2ab + b^2$$
$$(a+b)^3 = \quad a^3 + 3a^2b + 3ab^2 + b^3$$
$$(a+b)^4 = \quad a^4 + 4a^3b + 6a^2b^2 + 4ab^3 + b^4$$

The numerical coefficients form a triangular array of numbers that is called **Pascal's Triangle.**

n		
0	1	← $(a+b)^0 = 1$
1	1 1	
2	1 2 1	
3	1 3 3 1	
4	1 4 6 4 1	

Note that the first and last number of each row is 1.

Each of the other numbers is found by adding the numbers that are above it and just to its left and to its right. For example, the first 4 in the fifth row is the sum of the 3 and 1 above, and on either side of, the 4.

EXAMPLE 1 Use Pascal's triangle to write the coefficients for $(a+b)^5$.

Solution:

1 4 6 4 1	←	**Write the coefficients for the fifth row.**
1 5 10 10 5 1	←	**Write 1 for the first and last numbers. Add to find the other numbers.**

There is another way to write Pascal's triangle.

0	1					
1	1	$\frac{1}{1}$				
2	1	$\frac{2}{1}$	$\frac{2 \cdot 1}{1 \cdot 2}$			
3	1	$\frac{3}{1}$	$\frac{3 \cdot 2}{1 \cdot 2}$	$\frac{3 \cdot 2 \cdot 1}{1 \cdot 2 \cdot 3}$		
4	1	$\frac{4}{1}$	$\frac{4 \cdot 3}{1 \cdot 2}$	$\frac{4 \cdot 3 \cdot 2}{1 \cdot 2 \cdot 3}$	$\frac{4 \cdot 3 \cdot 2 \cdot 1}{1 \cdot 2 \cdot 3 \cdot 4}$	
5	1	$\frac{5}{1}$	$\frac{5 \cdot 4}{1 \cdot 2}$	$\frac{5 \cdot 4 \cdot 3}{1 \cdot 2 \cdot 3}$	$\frac{5 \cdot 4 \cdot 3 \cdot 2}{1 \cdot 2 \cdot 3 \cdot 4}$	$\frac{5 \cdot 4 \cdot 3 \cdot 2 \cdot 1}{1 \cdot 2 \cdot 3 \cdot 4 \cdot 5}$

Using this method, you can write the coefficients for any row without writing the previous rows.

Summary

Coefficients of $(a + b)^n$

1. The coefficient of the first and last term is 1.
2. The number of terms in the expansion of $(a+b)^n$ is one more than n. Thus, there are $(n + 1)$ *terms.*
3. The coefficient of the second term is a fraction with $(2 - 1)$, or 1 factor, in the numerator and denominator; the third term has $(3 - 1)$, or 2 factors, in the numerator and denominator. That is, the coefficient of the rth term has $(r - 1)$ factors in the numerator and denominator.

EXAMPLE 2 Write the coefficients for the expansion of $(a + b)^5$.

Solution:

Term	1	2	3	4	5	6
	1	$\frac{5}{1}$	$\frac{5 \cdot 4}{1 \cdot 2}$	$\frac{5 \cdot 4 \cdot 3}{1 \cdot 2 \cdot 3}$	$\frac{5 \cdot 4 \cdot 3 \cdot 2}{1 \cdot 2 \cdot 3 \cdot 4}$	$\frac{5 \cdot 4 \cdot 3 \cdot 2 \cdot 1}{1 \cdot 2 \cdot 3 \cdot 4 \cdot 5}$
	↓	↓	↓	↓	↓	↓
	1	**5**	**10**	**10**	**5**	**1**

The pattern for the variable factors of each term is easy to see.

n	**Variable factors for the expansion of $(a + b)^n$**					
1	a	b				
2	a^2	ab	b^2			
3	a^3	a^2b	ab^2	b^3		
4	a^4	a^3b	a^2b^2	ab^3	b^4	
5	a^5	a^4b	a^3b^2	a^2b^3	ab^4	b^5

Summary

Variable Factors for $(a + b)^n$

1. For the first term, the exponent of a is n.
2. The exponent of a decreases by 1 for each term. In the last, or $(n + 1)$st term the exponent for a is 0. Since $a^0 = 1$, the a does not usually appear in the last term.
3. The exponent of b starts with b^0 for the first term, and increases by 1 for each term. The exponent of b for the rth term is $r - 1$.
4. In any term, the sum of the exponents is n.

These patterns for coefficients and variable factors are summarized below.

Theorem 12–3: Binomial Theorem

For any positive integer n,

$$(a + b)^n = a^n + \frac{n}{1}a^{n-1}b + \frac{n(n-1)}{1 \cdot 2}a^{n-2}b^2 + \frac{n(n-1)(n-2)}{1 \cdot 2 \cdot 3}a^{n-3}b^3 + \cdots + b^n.$$

EXAMPLE 3 Write the expansion of $(2x + y)^5$. Simplify.

Solution:

$a^5 + 5a^4b + 10a^3b^2 + 10a^2b^3 + 5ab^4 + b^5$ ← **Write the expansion of $(a + b)^5$.**

$(2x)^5 + 5(2x)^4y + 10(2x)^3y^2 + 10(2x)^2y^3 + 5(2x)y^4 + y^5$ ← **Replace a with $2x$ and b with y.**

$32x^5 + 80x^4y + 80x^3y^2 + 40x^2y^3 + 10xy^4 + y^5$ ← **Simplify.**

Thus, $(2x + y)^5 = \mathbf{32x^5 + 80x^4y + 80x^3y^2 + 40x^2y^3 + 10xy^4 + y^5}$.

Consider the third term of each expression in the table.

$(a + b)^n$	Third Term	Exponent b	Exponent a	Number of Factors in the Denominator	Number of Factors in the Numerator
$(a + b)^3$	$\frac{3 \cdot 2}{1 \cdot 2}ab^2$	2	$(3 - 2)$	2	2
$(a + b)^4$	$\frac{4 \cdot 3}{1 \cdot 2}a^2b^2$	2	$(4 - 2)$	2	2
$(a + b)^5$	$\frac{5 \cdot 4}{1 \cdot 2}a^3b^2$	2	$(5 - 2)$	2	2

Note that the exponent of b is one less than the number of the term. The exponent of a is then (n − the exponent of b).

EXAMPLE 4 Write the expansion of $(m - 2n)^4$. Simplify.

Solution:

$a^4 + 4a^3b + 6a^2b^2 + 4ab^3 + b^4$ ← **Write the expansion of $(a + b)^4$.**

$m^4 + 4m^3(-2n) + 6m^2(-2n)^2 + 4m(-2n)^3 + (-2n)^4$ ← **Replace a with m and b with $(-2n)$.**

$\mathbf{m^4 - 8m^3n + 24m^2n^2 - 32mn^3 + 16n^4}$ ← **Simplify.**

CLASSROOM EXERCISES

State the number of terms in each expression.

1. $(x+y)^4$
2. $(x-y)^6$
3. $(2x+y)^{11}$
4. $(x-\frac{1}{2}y)^{17}$

Write the fourth term of each binomial.

5. $(x-y)^6$
6. $(x+y)^6$
7. $(2x+y)^6$
8. $(x-\frac{1}{2}y)^6$

WRITTEN EXERCISES

A Write the coefficients for each expansion. Then write the variable factors.

1. $(a+b)^6$
2. $(x+y)^4$
3. $(r+s)^7$
4. $(c-d)^3$

Write each expansion. Simplify.

5. $(x+y)^5$
6. $(c+d)^7$
7. $(m-n)^6$
8. $(x+3)^5$
9. $(2r+1)^4$
10. $(w-4)^3$
11. $(3m+2)^5$
12. $(2x-3)^4$
13. $(3x+2y)^4$
14. $(a-2b)^4$
15. $(2a-3b)^5$
16. $(x+0.2)^3$

Find the required term.

17. 5th term of $(a+b)^7$
18. 4th term of $(m+n)^6$
19. 3rd term of $(r-s)^7$
20. 6th term of $(2x+y)^{10}$
21. 2nd term of $(3-\frac{1}{2}x)^4$
22. 7th term of $(y-b)^6$

B Write each expansion. Simplify.

23. $(3s^2+t)^5$
24. $(1+i)^7$
25. $(1-i)^7$
26. $(4x^2-3y^2)^5$
27. $\left(x+\frac{1}{2}\right)^4$
28. $\left(2y-\frac{1}{2}\right)^5$
29. $\left(\frac{a}{b}+\frac{b}{a}\right)^6$
30. $\left(1-\frac{m}{n}\right)^4$
31. Find the 8th term of $(2x^2-y)^9$.
32. Find the 4th term of $\left(\frac{x}{2}+4y\right)^8$.
33. Find the middle term of $(2a-b)^6$.
34. Find the term with n^{10} in $(m+n^2)^5$.
35. Find the third term of $(y+\sqrt{2})^6$.
36. Find the third term of $\left(\frac{1}{x}+\frac{1}{y}\right)^5$.
37. Find the middle term of $(2t^2+\frac{1}{2})^4$.
38. Find the two middle terms of $(2a-\sqrt{b})^5$.

C Use the Binomial Theorem to find the first four terms. Simplify.

39. $(1.04)^5$
40. $(2.03)^6$
41. $(3+n)^n$
42. $(2-n)^n$

Isaac Newton showed that the Binomial Theorem holds when *n* is a rational number. The resulting expansions are infinite series.

Give the first three terms of each expansion.

43. $(1+y)^{-3}$
44. $(y+4)^{\frac{1}{2}}$
45. $(b-2)^{-\frac{1}{2}}$
46. $(a+b)^{-\frac{1}{3}}$

BASIC: SEQUENCES AND SUBSCRIPTED VARIABLES

In BASIC the subscripted variables a_1, a_2, a_3, etc., are written as A(1), A(2), A(3), or, in general, as A(I). Any letter may represent a subscripted variable and any other letter may represent the subscript. Furthermore, the subscript may be an expression, such as Z(N+1).

Problem: *Given the first term a and common difference d of an arithmetic sequence, write a program which prints the first n terms, where the user enters n. Also print the sum of the first n terms.*

```
100 DIM T(200)
110 PRINT "GIVE THE FIRST TERM AND COMMON DIFFERENCE"
120 PRINT "OF AN ARITHMETIC SEQUENCE."
130 INPUT A, D
140 PRINT "HOW MANY TERMS (<= 200) DO YOU WANT COMPUTED";
150 INPUT N
160 IF N <= 0 THEN 180
165 IF N > 200 THEN 180
170 IF N = INT(N) THEN 200
180 PRINT "NUMBER OF TERMS MUST BE A POSITIVE INTEGER
<= 200"
190 GOTO 150
200 LET T(1) = A
210 LET S = T(1)
220 FOR I = 2 TO N
230 LET T(I) = T(I - 1) + D
240 LET S = S + T(I)
250 NEXT I
260 PRINT "THE FIRST";N;"TERMS ARE:"
270 FOR I = 1 TO N
280 PRINT T(I);
290 NEXT I
300 PRINT
310 PRINT "THE SUM OF THE FIRST";N;"TERMS IS";S
320 PRINT
330 PRINT "ANY MORE SEQUENCES (1 = YES, 0 = NO)"
340 INPUT Z
350 IF Z = 1 THEN 110
360 END
```

Analysis

Statement 100: DIM is short for **DIM**ension. Any time a subscripted variable appears in a program, a DIM statement is needed to reserve memory space for the list of values (called an **array**) that will make up the sequence. In this program the array T may contain as many as 200 values.

Statements 160–170: These statements test whether N is a positive integer ≤ 200. If it is not, the user must enter N again.

Statements 200–250: The first term of the sequence, T(1), is set equal to A. The sum, S, is also started at T(1). Then a loop computes and adds the remaining terms. Each term after the first is calculated by adding D to the previous term (statement 220). The sum is continually updated (statement 240).

Statements 260–290: This loop prints the terms of the progression. The semicolon at the end of statement 280 causes the terms to be printed across a line of output rather than one per line down the screen or page.

Statements 300–310: Statement 300 causes the computer to get off the line on which it was printing the terms before printing the sum in statement 310.

EXERCISES

A Run the program on page 469 for the following values of a, d, and n.

1. $a = 5;\ d = 4;\ n = 10$ **2.** $a = 4.5;\ d = -0.5;\ n = 12$ **3.** $a = 45;\ d = -6;\ n = 5$

In Exercises 4–6, revise the program on page 469 in the manner indicated.

4. Combine into one loop the loops for computing the terms of the sequence and printing the terms.

5. Rewrite the program so that no subscripted variable is used.

6. Print, in three labeled columns, the number of the term, the term itself, and the sum of the terms up to and including that term.

B Write a BASIC program for each problem.

7. Given the first term and common ratio of a geometric sequence, print the first n terms, where the user also enters n.

8. Expand the program in Exercise 7 so that it also prints the sum of the first n terms.

9. Given the first term and common ratio of a geometric series, print the sum of the infinite series, if there is a sum.

The **Fibonacci sequence** begins 1, 1, 2, 3, 5, · · · . Each term after the first two equals the sum of the previous two terms.

10. Print the first n terms of the Fibonacci sequence, where the user enters n.

11. Print the sum of the first n terms of the Fibonacci sequence, where the user enters n.

C

12. Print the first n rows of Pascal's Triangle, where the user enters n. (See page 465.)

Review

Write each partial sum in expanded form. *(Section 12–4)*

1. $\sum_{k=1}^{5} (2 + 4k)$ **2.** $\sum_{k=1}^{3} (k^2 - 2)$ **3.** $\sum_{k=1}^{7} \frac{1}{2}k^2$ **4.** $\sum_{k=1}^{4} (\frac{1}{3}k - \frac{2}{3})$

Find the sum of the terms as indicated. *(Section 12–4)*

5. $-3 + 1 + 5 + \cdots$ to twenty terms **6.** $2\sqrt{2} + 5\sqrt{2} + 8\sqrt{2} + \cdots$ to fifteen terms

7. A pile of bricks has 100 in the bottom layer, 96 in the next layer, 92 in the next, and so on. How many bricks are in the pile if it has 10 layers? *(Section 12–4)*

Find the sum to the required number of terms for each geometric series. *(Section 12–5)*

8. $\sum_{k=1}^{5} (\frac{1}{2})^{k-1}$ **9.** $\sum_{k=1}^{6} (\frac{3}{4})k$ **10.** $\sum_{k=1}^{4} (-1)^k$ **11.** $\sum_{k=1}^{8} (3)^{k-2}$

12. The population of a city of 50,000 increases 5% for each of 20 years. What will the population be after 20 years? *(Section 12–5)*

Find the sum of each infinite geometric series. *(Section 12–6)*

13. $3 + 1 + \frac{1}{3} + \cdots$ **14.** $4 - 2 + 1 - \cdots$ **15.** $25 + 5 + 1 + \cdots$

Write each expansion. Simplify. *(Section 12–7)*

16. $(x + y)^6$ **17.** $(c - d)^4$ **18.** $(2s + t)^5$ **19.** $(3w - 0.5z)^3$

Find the required term. *(Section 12–7)*

20. 4th term of $(2a + 3b)^7$ **21.** 6th term of $(2x - \frac{1}{4})^8$

22. The middle term of $(x - \sqrt{2})^8$ **23.** The middle term of $(2d^2 \quad 3b^3)^6$

Chaper Summary

IMPORTANT TERMS

Arithmetic means (p. 447)
Arithmetic sequence (p. 446)
Arithmetic series (p. 454)
Common difference (p. 446)
Common ratio (p. 450)
Finite sequence (p. 444)
Geometric means (p. 451)
Geometric sequence (p. 450)
Geometric series (p. 459)
Index of summation (p. 454)
Infinite sequence (p. 444)
Limit of an infinite geometric series (p. 462)
*n*th term (p. 444)
Partial sum (p. 454)
Pascal's triangle (p. 465)
Sequence (p. 444)
Series (p. 454)

IMPORTANT IDEAS

1. **Sum of an Arithmetic Series:** If S_n represents the sum of the arithmetic series $a_1 + (a_1 + d) + (a_1 + 2d) + \cdots + a_1 + (n-1)d$, then

$$S_n = \frac{n}{2}[2a_1 + (n-1)d].$$

2. **Sum of a Geometric Series:** If S_n represents the sum of the geometric series $a_1 + a_1r + a_1r^2 + \cdots + a_1r^{n-1}$, then

$$S_n = \frac{a_1(r^n - 1)}{r - 1}, \ r \neq 1.$$

3. For an infinite geometric series with first term a_1 and common ratio r, $|r| < 1$, the sum is given by

$$S = \frac{a_1}{1 - r}.$$

4. **Coefficients in the expansion of $(a + b)^n$**

   a. The coefficient of the first term and of the last term is 1.

   b. The number of terms in the expansion of $(a + b)^n$ is $(n + 1)$.

   c. The coefficient of the rth term has $(r - 1)$ factors in the numerator and in the denominator.

5. **Variable factors in the expansion of $(a + b)^n$**

   a. For the first term, the exponent of a is n.

   b. The exponent of a decreases by 1 for each term. In the last, or $(n+1)$st, term the exponent of a is 0.

   c. The exponent of b starts with b^0 for the first term and increases by 1 for each term. The exponent of b for the rth term is $r - 1$.

   d. In any term, the sum of the exponents is n.

6. **Binomial Theorem:** For any positive integer n,

$$(a+b)^n = a^n + \frac{n}{1}a^{n-1}b + \frac{n(n-1)}{1 \cdot 2}a^{n-2}b^2 + \frac{n(n-1)(n-2)}{1 \cdot 2 \cdot 3}a^{n-3}b^3 + \cdots + b^n.$$

Chapter Objectives and Review

Objective: *To use the rule for a sequence to find specified terms of the sequence (Section 12–1)*

Find the rule for each sequence. Use the rule to write the next three terms.

1. 3, 6, 9, 12, $\cdots$

2. $\frac{1}{2}$, 1, $\frac{3}{2}$, 2, $\cdots$

3. 1, -2, 3, -4, $\cdots$

Use the rule to write the first four terms of each sequence.

4. $a_n = 4(n - 1)$

5. $a_n = 3 + \frac{n}{2}$

6. $a_n = 7 - 3n$

Objective: *To write the terms of an arithmetic sequence given the first term and the common difference (Section 12–2)*

Write the first five terms of the arithmetic sequence having first term a_1 and common difference d.

7. $a_1 = 7,\ d = \frac{3}{4}$ 8. $a_1 = -2,\ d = 3$ 9. $a_1 = \frac{5}{2},\ d = \frac{3}{2}$ 10. $a_1 = 3,\ d = -4$

11. The cost of renting a color television is \$32 for the first month. The cost then decreases by \$2 per month up to a maximum of five months. What is the rental cost for the fifth month?

Objective: *To write the terms of a geometric sequence given the first term and the common ratio (Section 12–3)*

Write the first five terms of the geometric sequence having first term a and common ratio r.

12. $a = 1,\ r = 3$ 13. $a = 2,\ r = \frac{1}{3}$ 14. $a = \frac{1}{4},\ r = 5$ 15. $a = \frac{1}{5},\ r = -\frac{1}{3}$

16. Rick signs a five-year lease for an apartment. The lease calls for a rent of \$300 per month the first year, and an increase of 8% in each of the remaining years of the lease. What will Rick's monthly rent be during the last year of the lease?

17. Find one positive geometric mean between 5 and 45.

18. Find two geometric means between 2 and −51.

Objective: *To find partial sums of arithmetic series (Section 12–4)*

Write each partial sum in expanded form. Then evaluate the partial sum.

19. $\sum_{k=1}^{7} (3 - k)$ 20. $\sum_{k=1}^{6} \frac{1}{2}k$ 21. $\sum_{k=1}^{4} (-2)^{k+1}$ 22. $\sum_{k=1}^{5} (\frac{2}{3}k - \frac{1}{3})$

23. A concert hall has 25 rows. The first row has 16 seats, and each succeeding row has two more seats than the row in front of it. How many seats are there in the hall?

Objective: *To find the sum of a finite geometric series (Section 12–5)*

Find the sum of each series.

24. $8 - 4 + 2 - 1 + \frac{1}{2}$ 25. $3 + 1 + \frac{1}{3} + \frac{1}{9} + \frac{1}{27} + \frac{1}{81}$

Find the sum to the required number of terms for each geometric series.

26. $3 + 6 + 12 + \cdots$ to six terms 27. $2 - \frac{2}{3} + \frac{2}{9} - \cdots$ to eight terms

28. $\sum_{k=1}^{4} (\frac{1}{4})k$ 29. $\sum_{k=1}^{13} (-1)^k$ 30. $\sum_{k=1}^{3} (\frac{2}{5})^k$ 31. $\sum_{k=1}^{4} 3(-\frac{1}{2})^k$

32. When a certain antibiotic capsule is swallowed, it is gradually absorbed into the bloodstream. After each minute, the amount not yet absorbed is $\frac{9}{10}$ of the amount that remained at the end of the preceding minute. How much remains to be absorbed after six minutes?

Objective: *To find the sum of an infinite geometric series (Section 12–6)*

Find the sum of each infinite geometric series.

33. $5 + \frac{5}{3} + \frac{5}{9} + \cdots$

34. $9 + 6 + 4 + \cdots$

35. $2 - \frac{2}{3} + \frac{2}{9} - \cdots$

36. A pendulum swings through an arc of 28 meters on the first swing, $\frac{13}{15}$ of this distance on the second swing, and continues to swing $\frac{13}{15}$ of the previous distance on each succeeding swing. If this process continues until the pendulum comes to rest, what distance is covered by the pendulum?

Objective: *To use the Binomial Theorem (Section 12–7)*

Write each expansion. Simplify.

37. $(r + 3)^7$

38. $(z - 5)^6$

39. $(2y + 1)^5$

40. $(w - 2t)^4$

Find the required term.

41. 5th term of $(x - y)^6$

42. 4th term of $(r - 3)^5$

43. 6th term of $(d + 2t)^8$

Chapter Test

Identify each sentence as *arithmetic, geometric,* or *neither*. Then list the next three terms for each arithmetic and geometric sequence.

1. $\frac{1}{5}, -\frac{1}{10}, \frac{1}{20}, -\frac{1}{40}, \cdots$

2. $\frac{1}{3}, 1, \frac{5}{3}, \frac{7}{3}, \cdots$

3. $\frac{1}{2}, \frac{3}{4}, \frac{7}{8}, \frac{15}{16}, \cdots$

4. $-\frac{1}{2}, \frac{1}{4}, -\frac{1}{8}, \frac{1}{16}, \cdots$

5. Find the 20th term in the sequence $-3, -1, 1, 3, 5, \cdots$

6. Find the 10th term in the sequence $-2, 4, -8, 16, \cdots$

7. Find the sum of the odd integers between 30 and 70.

8. Insert three arithmetic means between 4 and 12.

9. Insert two geometric means between 10 and $-\frac{2}{25}$.

10. Find the first five terms in the sequence of partial sums of the following series. That is, find S_1, S_2, S_3, S_4, S_5.

$$\frac{1}{2} + \frac{2}{3} + \frac{3}{4} + \frac{4}{5} + \cdots + \frac{n}{n+1} + \cdots$$

11. Find the sum of the infinite geometric series $-25 - 5 - 1 - \frac{1}{5} \cdots$

12. State the number of terms in the expansion of $(w - 7)^{11}$.

13. Write the expansion of $(2 - q)^6$.

14. Gloria decides to save to buy a \$250 suit. She plans to save \$1.00 today, \$1.50 tomorrow, \$2.00 the next day, and so on. How much will she have saved after thirty days?

15. A population of 300 amoebas increases by 50% every hour. How many amoebas will there be after five hours?

Review of Word Problems: Chapters 1–12

1. In a certain area, $3\frac{3}{4}$ inches of snow fell in 2 hours. At this rate, how much snow would fall in 5 hours? *(Section 3–7)*
2. The units digit of a two-digit number is two more than three times the tens digit. When the digits are reversed, the new number is thirty less than four times the original number. Find the original number. *(Section 4–4)*
3. A nurse mixes a 2% boric acid solution with an 8% boric acid solution to produce 3 liters of a 6% boric acid solution. How many liters were there of the 2% boric acid solution and of the 8% boric acid solution? *(Section 4–4)*
4. A farmer wants to build a chicken yard against the side of a barn as shown. The farmer has 20 meters of fencing with which to enclose a total area of 50 square meters. Find the dimensions of the yard. *(Section 6–10)*

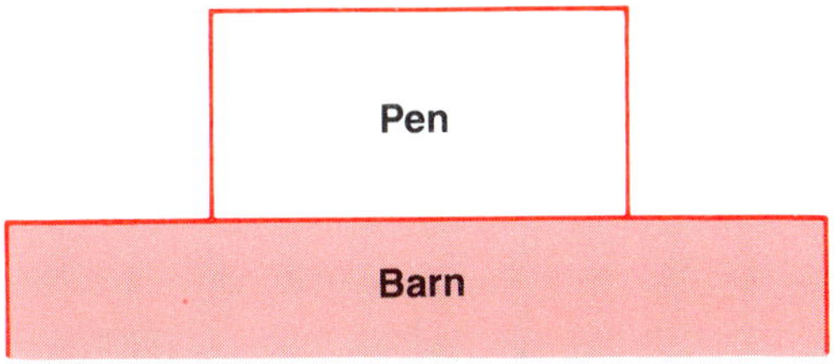

5. Alice, Joan, and Greg can do a piece of work in 4 days. Alice can do the same piece of work in 10 days and Greg in 12 days. How many days will it take Joan to do the work alone? *(Section 7–8)*
6. What principal will amount to \$8000 if invested at 12% interest compounded quarterly for 5 years? *(Section 11–11)*
7. The diameter of a tree is 32 centimeters. The diameter increases by 6 centimeters each year. How many years will it take the diameter to reach 88 centimeters? *(Section 12–2)*
8. A builder's profit from the sale of the first house in a development is \$750. For each additional house sold in the development, the profit is \$750 more than the profit received from the sale of the previous house. How many houses were sold if the profit on the last house amounted to \$13,500? *(Section 12–2)*
9. A sailboat which originally cost \$15,000 depreciates each year 20% of its value in the previous year. What is the value of the sailboat after 4 years? *(Section 12–3)*
10. The bottom row of a pile of logs contains 53 logs, the row above the bottom one contains 51 logs, the third row from the bottom contains 49 logs, and so on. The top row contains one log. How many logs are there? *(Section 12–4)*
11. A cell divides into two cells every 50 seconds. There are 8 cells at the start of an experiment. How many cells will there be at the end of 5 minutes? *(Section 12–5)*
12. A rotating flywheel rotates 800 revolutions in the first minute, $\frac{9}{10}$ as many times in the second minute, and continues to rotate $\frac{9}{10}$ as many times in each succeeding minute. If this process continues until the flywheel comes to rest, how many revolutions will the flywheel make? *(Section 12–6)*

Cumulative Review: Chapters 1–12

Choose the best answer. Choose *a*, *b*, *c*, or *d*.

1. Which statement is false?
   a. The sentence $5x - 3 = 2$ is neither true nor false.
   b. $-x$ is always a negative number.
   c. $a + b = b + a$ for all real numbers a and b.
   d. $|-x| \geq 0$

2. Which set is a subset of every set of real numbers?
   a. N b. W c. Q d. ϕ

3. What is the solution of $3x + \frac{5}{2} = -9x - \frac{7}{2}$ over the set of rational numbers?
   a. $\{-\frac{1}{2}\}$ b. $\{2\}$ c. $\{-2\}$ d. ϕ

4. Today Mary is one-fifth the age of her mother. In two years Mary will be one-fourth the age of her mother. How old is Mary's mother now?
   a. 20 b. 25 c. 30 d. 35

5. What is the solution for $3x - 5 < -7x + 10$?
   a. $x > \frac{5}{4}$ b. $x < \frac{5}{4}$ c. $x < \frac{3}{2}$ d. $x > 1$

6. Pedro had math scores of 82, 89, 73 and 90 on four tests. His final test counts for two test scores. What must Pedro get on the final to have an average of 80?
   a. 66 b. 73 c. 80 d. 90

7. Solve: $|-3x + 2| > 5$
   a. $x < -1$ <u>and</u> $x > \frac{7}{3}$ b. $x < -1$ <u>or</u> $x > \frac{7}{3}$
   c. $x < \frac{7}{3}$ <u>and</u> $x > -1$ d. $x < \frac{7}{3}$ <u>or</u> $x > -1$

8. How are the graphs of the lines $5x - 3y = 7$ and $6y - 10x = 14$ related?
   a. They are parallel. b. They are perpendicular.
   c. They are coincident. d. They intersect in only one point.

9. Which line contains the point of intersection of $3x - 2y = 5$ and $x + 2y = -3$?
   a. $2x - 4y = 5$ b. $4x = -3$
   c. $2x - 4y = 8$ d. $-8y = 10$

10. y varies directly as x and $y = \frac{2}{3}$ when $x = -\frac{1}{2}$. Write the equation for the variation.
    a. $3y = x$ b. $3y = -4x$ c. $2y = -3x$ d. $y = -\frac{2}{3}x$

11. The domain of the function defined by $y = \sqrt{x^2 + 25}$ is the set of real numbers. What is its range?
    a. {Real numbers} b. {Real numbers greater than 0}
    c. {Real numbers between 5 and -5} d. {Real numbers greater than or equal to 5}

12. Let $g(x) = mx$, where $m > 0$.
Which statement is false?

a. $g(a) + g(b) = g(a + b)$
b. $g(a) - g(b) = g(a - b)$
c. $g(a) \cdot g(b) = g(a \cdot b)$
d. $g(5a) = 5g(a)$

13. If $f(x) = 2x + 1$ and $g(x) = \frac{1}{2}x - \frac{1}{2}$, which statement is true?

a. $f[g(1)] = 2$
b. $g[f(-3)] = 0$
c. $g[f(x)] = x^2$
d. $f[g(x)] = x$

14. Simplify: $(3x^2 - 2x) - (x^4 - 2x - 1) + (5x^2 - 2)$

a. $-(x^4 - 8x^2 + 1)$
b. $-x^4 + 8x^2 - 4x - 3$
c. $7x^2 - 1$
d. $x^4 + 8x^2 + 1$

15. Which of the following is a factor of $6x^2 - x - 2$?

a. $x - 2$ b. $2x - 1$ c. $3x - 2$ d. $x + 1$

16. Expand: $(3x - 2y)^2$

a. $9x^2 - 4y^2$
b. $9x^2 - 6xy - 4y^2$
c. $9x^2 - 12xy + 4y^2$
d. $9x^2 - 12xy + 4y^2$

17. Suppose that $x - 2$ is *not* a factor of $P(x)$.
Which statement is true?

a. $P(-2) = 0$
b. $P(2) = 0$
c. $P(x) \div (x - 2) \neq Q(x)$
d. $x + 2$ is not a factor of $P(x)$.

18. The numerator of a fraction is one more than triple the denominator. Adding 4 to each makes the numerator double the denominator. Find the numerator.

a. 10 b. 3 c. 14 d. 7

19. Solve: $\frac{3}{x^2 - 12} = \frac{3}{3x - 2}$

a. $\{-2\}$ b. $\{2\sqrt{3}, -2\sqrt{2}\}$ c. $\{5, -2\}$ d. $\{\frac{2}{3}\}$

20. John can build a birdhouse in 2 hours; his son John Peter can build it in 5 hours. If they work together for an hour, how long will it take John Peter to complete the task?

a. $\frac{1}{2}$ hr b. 1 hr c. $1\frac{1}{2}$ hr d. 2 hr

21. Write $\sqrt{2x^4} \cdot \sqrt{4x^2}$ in simplest radical form.

a. $2x^3\sqrt{2}$ b. $2x^2\sqrt{2}$ c. $2|x|^2\sqrt{2}$ d. $2|x|^3\sqrt{2}$

22. Simplify: $5\sqrt{45} + 3\sqrt{20} - 13\sqrt{5}$

a. $2\sqrt{5} + 3\sqrt{20}$ b. $8\sqrt{5}$ c. $16\sqrt{5}$ d. $5\sqrt{45} - \sqrt{5}$

23. What is the distance between $A(-3, 5)$ and $B(4, -3)$?

a. $AB = \sqrt{113}$ b. $AB = \sqrt{15}$ c. $AB = 1$ d. $AB = 1.1$

24. The weight of sheet aluminum varies as the square of its width. When the width is 5, the weight is 10 grams. If the width is doubled, what is the weight?

a. 20 grams b. 30 grams c. 40 grams d. 50 grams

25. Express in standard form: $\dfrac{-3 + 2i}{1 - i}$

a. $-3 - 2i$ b. $3 - 2i$ c. $-\dfrac{5}{2} - \dfrac{i}{2}$ d. $\frac{3}{2} + i$

26. What are the zeros of $f(x) = 2x^2 + 3x - 4$?

a. $\dfrac{-3 \pm \sqrt{41}}{2}$ b. $\dfrac{3 \pm \sqrt{41}}{2}$ c. $\dfrac{-3 \pm \sqrt{41}}{4}$ d. $\dfrac{3 \pm \sqrt{41}}{4}$

27. The difference of two numbers is 3; the difference of their squares is 39. Find the larger number.

a. 8 b. 7 c. 6 d. 5

28. Which statement is false?

a. $3x^2 + 3y^2 = 27$ is the equation of a circle.

b. $3x^2 - 3y^2 = 27$ is the equation of a circle.

c. $3x^2 + 9(y + 2)^2 = 27$ is the equation of an ellipse.

d. $3x^2 - 9(y + 2)^2 = 27$ is the equation of a hyperbola.

29. For the graph of $y = x^2 + 2x + 6$, which statement is false?

a. The graph is a parabola. b. The graph has a minimum point.

c. The graph opens upward. d. The vertex is at $(1, -5)$.

30. Which of the following is the graph of $\dfrac{x^2}{4} + \dfrac{y^2}{16} \leq 1$?

a.

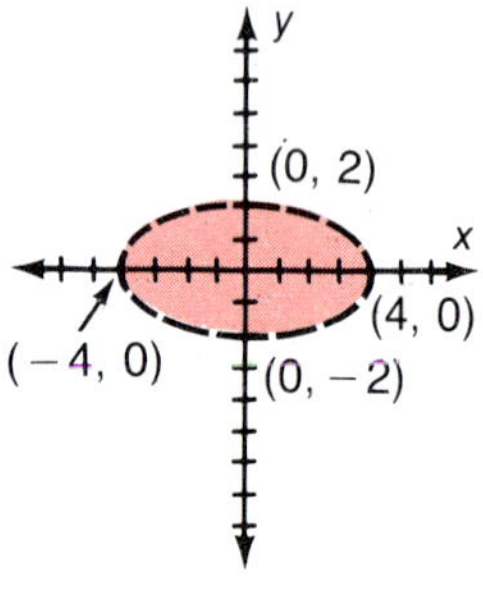

b.

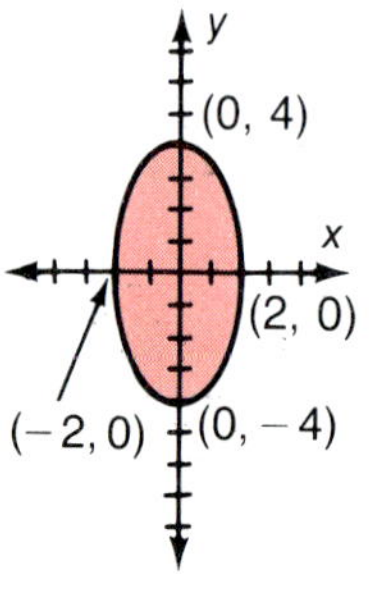

c.

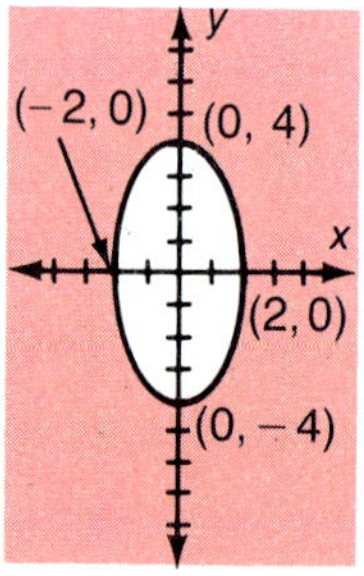

d.

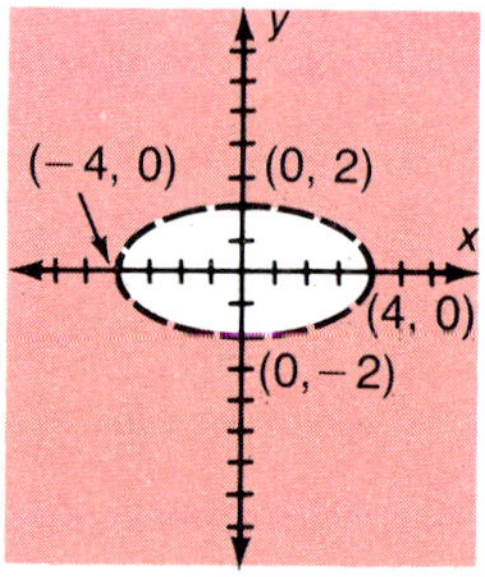

31. Express using positive exponents only: $(x^{-2} + y^{-2})\dfrac{(x^{-2}y^{-1})^{-2}}{x^2 + y^2}$

a. $\dfrac{x^6y^4}{(x^2 + y^2)^2}$ b. x^2 c. $\dfrac{(x^2 + y^2)^2}{x^6y^4}$ d. $\dfrac{1}{x^2}$

32. Write 0.003521 in scientific notation.

a. 3.521×10^3 b. 3.521×10^4 c. 3.521×10^{-3} d. 3.521×10^{-4}

33. Write in simplest exponential form: $\dfrac{x^{-3} \cdot x^{3y-2} \cdot x^{\frac{1}{2}}}{x^{2y-1} \cdot x^{-\frac{1}{2}}}$

a. x^{y-3} b. x^{3y-1} c. $\dfrac{x^{2y-1}}{x^3}$ d. $\dfrac{1}{x^{y+2}}$

34. Solve: $\sqrt{x-5} + \sqrt{2x-2} = 6$

a. $x = 201$ b. $x = 9$ c. $x = 9$ or 201 d. No solution

35. Solve: $(2x-3)^{\frac{1}{3}} = 16^{\frac{1}{6}}$

a. $\frac{13}{2}$ b. $\frac{7}{2}$ c. $\frac{1}{2}$ d. No solution

36. Which expression is equivalent to $\log 5 - \log 7 + 3 \log 2$?

a. $\log 5 \cdot 7 \cdot 6$ b. $\log 35 \cdot 2^{\frac{1}{3}}$ c. $\log(\frac{5}{7} \cdot 2^{\frac{1}{3}})$ d. $\log(\frac{5}{7} \cdot 2^3)$

37. Which expression is equivalent to $\log(15\sqrt[3]{p})$?

a. $\log 15 + \log 3p$ b. $\log 45 + \log p$

c. $\log 15 + \frac{1}{3} \log p$ d. $\log 15 + 3 \log p$

38. Solve $\log(x-5) + \log 2 = 1$ for x.

a. $\{5\}$ b. $\{10\}$ c. $\{15\}$ d. No solution

39. What are the next three terms of 3, 4, 6, 9, 13, · · · ?

a. 17, 22, 28 b. 18, 24, 31 c. 17, 23, 30 d. 18, 25, 33

40. Which of the following is an arithmetic sequence?

a. 7, 14, 28, 56, 112, · · · b. 1, 3, 6, 10, 15, · · ·

c. 5, 8, 11, 14, 17, · · · d. 2, 3, 5, 7, 11, · · ·

41. Which of the following is a geometric sequence?

a. $-24, -12, -6, -3, \cdots$ b. 1, 1, 2, 3, 5, 8, · · ·

c. 2, 7, 12, 17, · · · d. $\frac{1}{2}, 1, \frac{3}{2}, 2, \frac{5}{2}$

42. Find three arithmetic means between 6 and 14.

a. $8\frac{2}{3}, 11\frac{1}{3}, 14$ b. 8, 10, 12 c. $7\frac{2}{3}, 10\frac{1}{3}, 13$ d. $8\frac{1}{2}, 10, 12\frac{1}{2}$

43. Find 2 geometric means between 1 and 5.

a. $2\frac{1}{3}, 3\frac{2}{3}$ b. $2, 3\frac{1}{2}$ c. $5\frac{1}{3}, 5\frac{2}{3}$ d. $\sqrt[3]{5}, \sqrt[3]{25}$

44. What is the fifth term of $(2x+y)^9$?

a. $\frac{9 \cdot 8 \cdot 7 \cdot 6}{1 \cdot 2 \cdot 3 \cdot 4}(2x)^5y^4$ b. $9 \cdot 2 \cdot 7 \cdot 32x^4y^5$

c. $\frac{9 \cdot 8 \cdot 7}{1 \cdot 2 \cdot 3}(2x)^4y^5$ d. $\frac{9 \cdot 8 \cdot 7 \cdot 6 \cdot 5}{1 \cdot 2 \cdot 3 \cdot 4 \cdot 5}(2x)^6y^3$

45. Find $\sum_{n=1}^{10}(4n-3)$.

a. 230 b. 180 c. 210 d. 190

46. Find $\sum_{n=1}^{10}(\frac{1}{2})^{n+1}$

a. $\frac{512}{1024}$ b. $\frac{1023}{2048}$ c. $\frac{2047}{4096}$ d. $\frac{1}{2}$

47. Find $\sum_{n=1}^{\infty}(\frac{4}{3})^n$.

a. 4 b. -4 c. 2 d. No sum

Preparing for College Entrance Tests

Choose the best answer. Choose *a*, *b*, *c*, or *d*.

1. If x is a real number, compare 1^{3x+1} and 1^{3x}.

(a) $1^{3x+1} > 1^{3x}$ **(b)** $1^{3x+1} < 1^{3x}$ **(c)** $1^{3x+1} = 1^{3x}$

(d) Cannot be determined from the given information

2. If $\frac{3}{4}x - \frac{1}{2} = 1$, what is the value of $3x - 2$?

(a) $\frac{1}{4}$ **(b)** 1 **(c)** 2 **(d)** 4

3. If a and b are positive integers, and $a > b$, compare $(a-b)^{a+b}$ and $(a+b)^{a-b}$.

(a) $(a-b)^{a+b} > (a+b)^{a-b}$ **(b)** $(a-b)^{a+b} < (a+b)^{a-b}$

(c) $(a-b)^{a+b} = (a+b)^{a-b}$ **(d)** Cannot be determined from the given information

4. If $3a - 4b = 4$ and $3a + 4b = \frac{1}{2}$, what is the value of $9a^2 - 16b^2$?

(a) 2 **(b)** 4 **(c)** $6\frac{1}{4}$ **(d)** 32

5. If n is an integer, which of the following must be integers?

I. 2^{-n} **II.** 2^{2n} **III.** 2^{n^2}

(a) II only **(b)** III only **(c)** I and III only **(d)** None of these

6. If n is a negative integer, compare $(-2)^{2n}$ and $(-2)^{2n+1}$.

(a) $(-2)^{2n} > (-2)^{2n+1}$ **(b)** $(-2)^{2n} < (-2)^{2n+1}$ **(c)** $(-2)^{2n} = (-2)^{2n+1}$

(d) Cannot be determined from the given information

7. Which of the following numbers is in the solution set for $16x^2 - 64x + 15 = 0$?

(a) -2 **(b)** $-\frac{3}{4}$ **(c)** 1 **(d)** $3\frac{3}{4}$

8. $\dfrac{3^3 + 3^5}{3^3} = $ ___?___

(a) 243 **(b)** 27 **(c)** 10 **(d)** None of these

9. For real numbers a, b, c, and d, $4a = 5b$ and $20b = 12c$. Then $\dfrac{a}{c} = $ ___?___

(a) $\frac{3}{4}$ **(b)** $1\frac{1}{3}$ **(c)** $2\frac{1}{12}$ **(d)** 12

10. If p is a real number, compare $\dfrac{3^p}{3^{p+1} + 3^p}$ and $\dfrac{1}{3}$.

(a) $\dfrac{3^p}{3^{p+1} + 3^p} > \dfrac{1}{3}$ **(b)** $\dfrac{3^p}{3^{p+1} + 3^p} < \dfrac{1}{3}$ **(c)** $\dfrac{3^p}{3^{p+1} + 3^p} = \dfrac{1}{3}$

(d) Cannot be determined from the given information

11. $\dfrac{2^{-2} + 2^{-2}}{2^{-2} \cdot 2^{-2}} = $ ___?___

(a) $\frac{1}{8}$ **(b)** $\frac{1}{2}$ **(c)** 1 **(d)** 8

CHAPTER **13**

Systems of Sentences: Three Variables

13–1 Graphing in Three-Space

To construct a coordinate system in three-space, begin with the xy plane in a *horizontal* position. Through the origin, 0, construct a vertical z axis that is perpendicular to both the x and y axes. Thus, the three axes are perpendicular to each other.

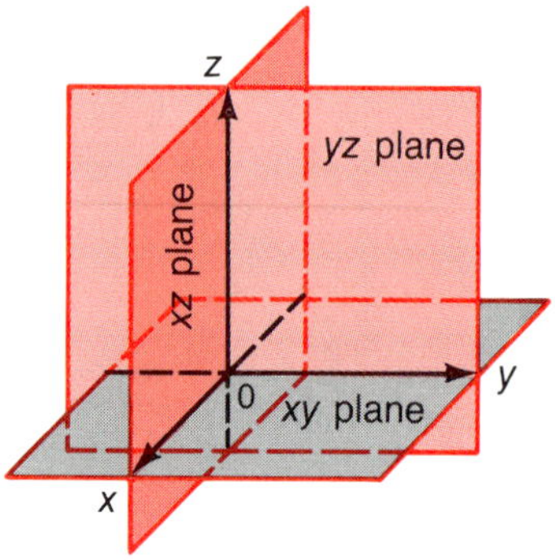

Taken in pairs, the three axes form three coordinate planes: the xy plane, the xz plane and the yz plane. You use an ordered triple, (x, y, z) to locate a point in three-space.

EXAMPLE 1 Graph each point in three-space.

a. $A(4, 5, 6)$ **b.** $B(-3, -4, 2)$

Solutions: **a.** $A(4, 5, 6)$

[1] Move 4 units from 0 along the positive x axis.

[2] Move 5 units in the direction of the positive y axis.

[3] Move 6 units in the direction of the positive z axis.

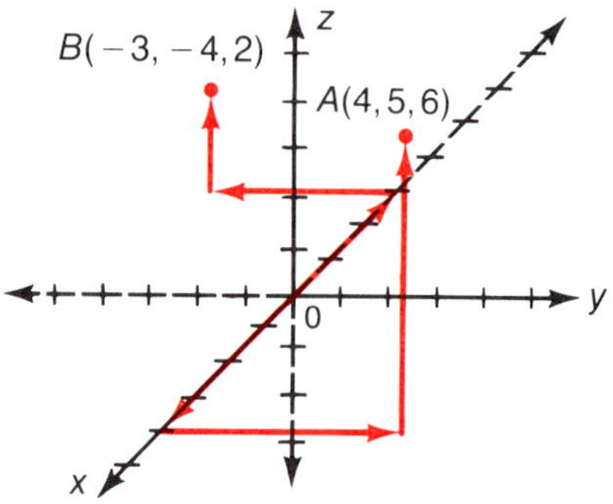

b. $B(-3, -4, 2)$

[1] Move 3 units from 0 along the negative x axis.

[2] Move 4 units in the direction of the negative y axis.

[3] Move 2 units in the direction of the positive z axis.

NOTE: When graphing, the angle between the x and y axes is drawn as 135° instead of 90°. This gives spatial perspective to the drawing. Also, the units of length on the y and z axes are drawn the same, while the units of the x axis are $\frac{2}{3}$ the unit on either of the other two. This helps to give "depth" to the drawing. The negative portion of each axis is often shown as a dashed line.

It can be proved that the graph of a linear equation in three variables is a plane, and conversely, that every plane is the graph of a linear equation in three variables. To graph an equation such as

$$x + y + z = 8$$

it is convenient to find the points in which the plane intersects each axis. That is, find the x, y, and z intercepts.

EXAMPLE 2 Graph: $x + y + z = 8$

Solution:

x intercept: $y = 0$ and $z = 0$; $\therefore x = 8$

y intercept: $x = 0$ and $z = 0$; $\therefore y = 8$

z intercept: $x = 0$ and $y = 0$; $\therefore z = 8$

Graph the intercepts:

(8, 0, 0) (0, 8, 0) (0, 0, 8)

Draw the line segments connecting the points and shade the triangle as shown.

The triangle is part of the graph of the plane with equation $x + y + z = 8$.

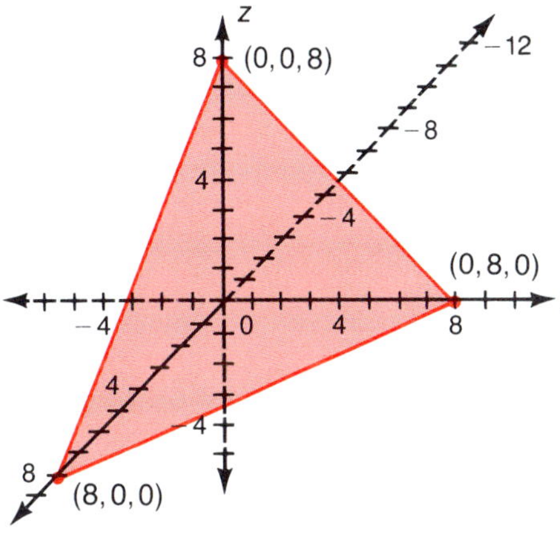

CLASSROOM EXERCISES

Name the ordered triple that represents each point.

1. F
2. E
3. C
4. D
5. G
6. J
7. B
8. A
9. H

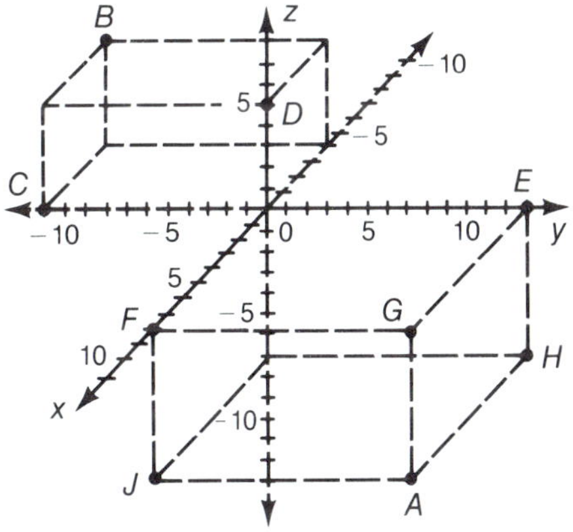

Name the coordinate axis on which the given point lies.

10. (0, 0, 0)
11. (5, 0, 0)
12. $(0, \frac{1}{2}, 0)$
13. (0, 0, −6)

Give the x, y, and z intercepts for the graph of each equation.

14. $x - y + z = 10$
15. $x - y - z = 6$
16. $3x - 2y + z = 5$
17. $2x - 3y - 6z = 12$

WRITTEN EXERCISES

A Graph each point in three-space.

1. $A(3, 4, 5)$
2. $B(-2, 4, -3)$
3. $C(-1, -6, 6)$
4. $P(1, 4, 3)$
5. $Q(4, 8, -2)$
6. $R(2, -2, -1)$
7. $T(0, 1, 2)$
8. $V(4, 8, -2)$
9. $S(1, 1, 2)$
10. $E(1, -1, 3)$
11. $F(-5, 7, 0)$
12. $G(-1, -1, 0)$

Graph the plane represented by each equation.

13. $x + y + z = 4$
14. $x + y - z = 2$
15. $2x + 6y - 2z = 4$
16. $3x + 2y + z = 6$
17. $4x + 2y - z = 4$
18. $x + 3y - z = -3$

19. Write the equation of the xy plane.

20. Write the equation of the xz plane.

21. Write the equation of the yz plane.

B Graph the plane represented by each equation.

22. $2y - 5z + 10 = 0$ 23. $4x - 2y = 0$ 24. $2y - 5z = -10$

25. $z = 4$ 26. $x = -2$ 27. $y = 3$

C Find the equation of the plane satisfying the given conditions.

28. Parallel to the plane $5x - 2y + 7z + 1 = 0$ and containing the point $B(2, 1, 0)$.

29. Parallel to the plane $-2x + 3y - 4z - 5 = 0$ and containing the point $P(1, -3, -2)$.

30. Parallel to the plane $7x - 3y - 2z + 1 = 0$ and containing the point $T(1, 0, 2)$.

REVIEW CAPSULE FOR SECTION 13–2

Determine whether the given ordered pair is a solution of the given system. *(Pages 114–116)*

1. $\begin{cases} 3x + 4y = 18 \\ 5x - y = 7 \end{cases}$; $(2, 3)$

2. $\begin{cases} 3a + 5b = 9 \\ 4b = 22 - 5a \end{cases}$; $(2, -1)$

3. $\begin{cases} 3q = 5 - t \\ t = q - 7 \end{cases}$; $(3, -4)$

13–2 Solving Linear Systems: Three Variables

Consider a linear system of three equations in three variables.

$$\begin{cases} x + y + z = 10 & \longleftarrow \textbf{Figure 1} \\ x - y = 0 & \longleftarrow \textbf{Figure 2} \\ z = 5 & \longleftarrow \textbf{Figure 3} \end{cases}$$

Each equation has a plane in three-space as its graph.

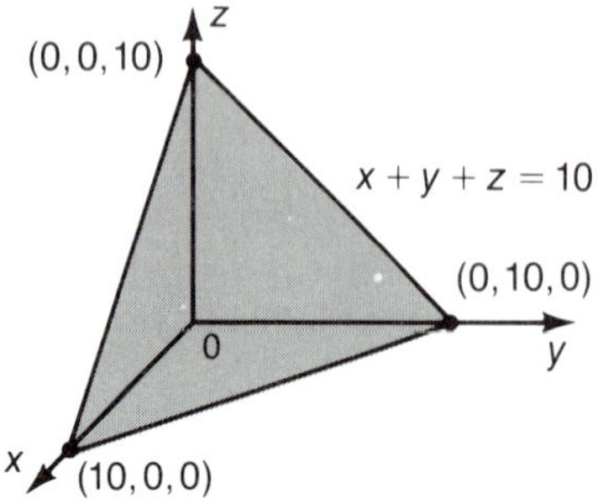

Figure 1

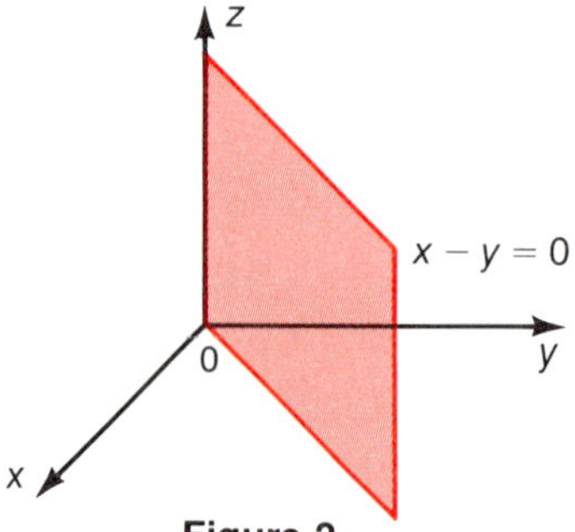

Figure 2

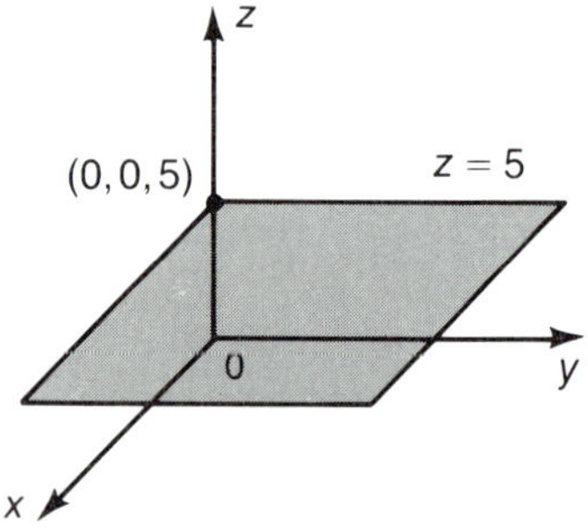

Figure 3

The solution to the system is represented geometrically by the intersection of the three planes.

If that intersection is a single point, then the coordinates of that point, an **ordered triple**, satisfy each equation of the system.

Thus, the solution set of such a system consists of a single ordered triple.

Two methods are shown in Example 1 for finding the coordinate of the point of intersection of this system.

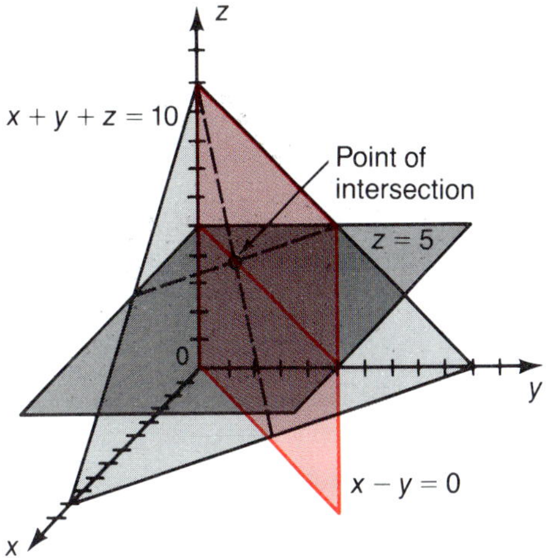

EXAMPLE 1 Solve and check: $\begin{cases} x+y+z=10 & \mathbf{1} \\ x-y=0 & \mathbf{2} \\ z=5 & \mathbf{3} \end{cases}$

Solution:

Method 1

$\begin{cases} 2x+z=10 & \mathbf{4a} \\ z=5 & \mathbf{3} \end{cases}$ ← **Equation 1 + Equation 2**

Equations 4a and 3 form a system equivalent to the given system. Replace z in Equation 4a with 5.

$$2x+5=10$$
$$x=2\tfrac{1}{2}$$
$$y=2\tfrac{1}{2}$$
$$z=5$$

Method 2

$\begin{cases} x+y+5=10 & \mathbf{4b} \\ x-y=0 & \mathbf{2} \end{cases}$ ← **Replace z in Equation 1 with 5.**

Equations 4b and 2 form a system equivalent to the given system.

$$2x=5$$ ← **Equation 2 + Equation 4b**
$$x=2\tfrac{1}{2}$$
$$y=2\tfrac{1}{2}$$
$$z=5$$

The check is left for you. **Solution set:** $\{(2\tfrac{1}{2}, 2\tfrac{1}{2}, 5)\}$

In general, the **solution set** of a system of three linear equations in three variables is the set of ordered triples that satisfy all three equations.

The solution set contains:

(i) One ordered triple when the three planes intersect in one point (see Figure 1 on page 486).

(ii) Infinitely many ordered triples when the three planes intersect in one line or when they coincide (see Figures 2 and 3 on page 486).

Consistent Systems

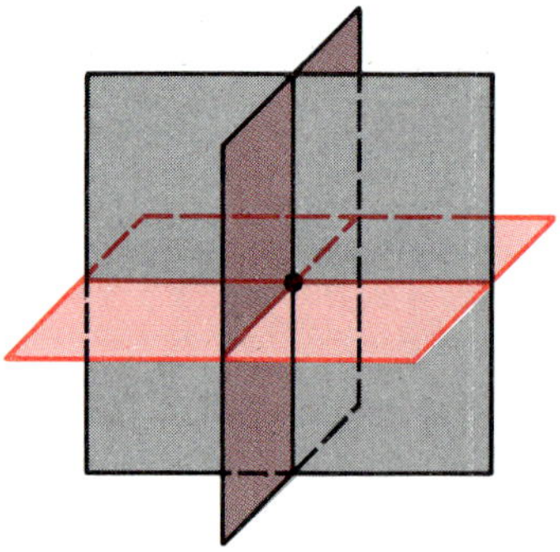

Figure 1

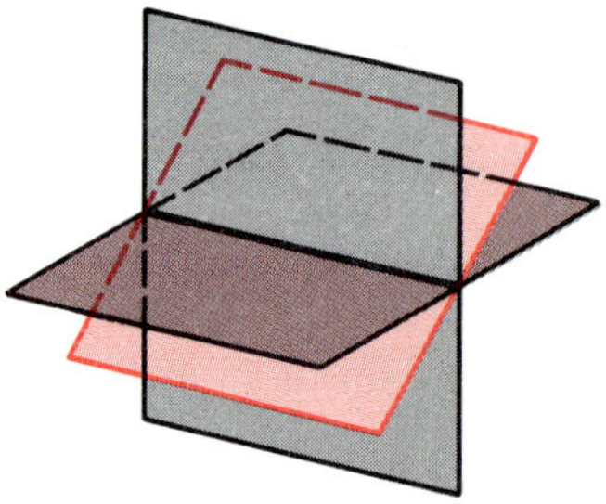

Figure 2

Figure 3

(iii) No ordered triples when

a. the three planes are parallel (see Figure 4 below).

b. two planes are coincident and parallel to the third plane (see Figure 5 below).

c. the three planes intersect in three parallel lines (see Figure 6 below).

d. two planes are parallel and intersect the third plane in two parallel lines (see Figure 7 below).

Inconsistent Systems

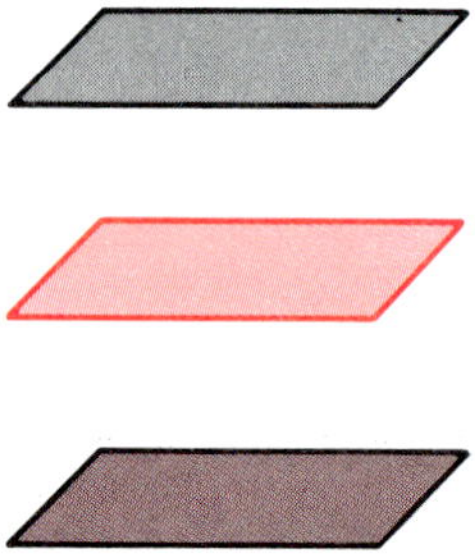

Figure 4

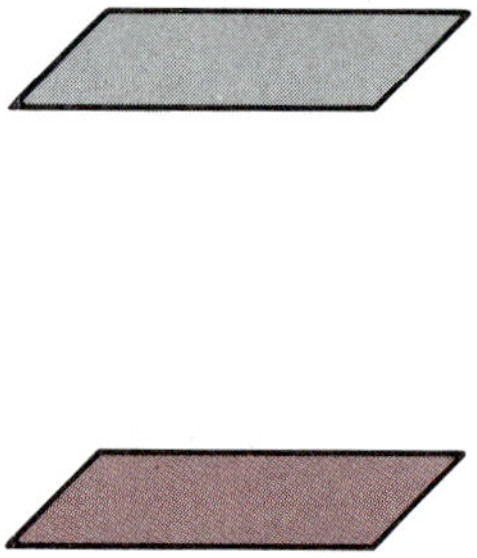

Figure 5

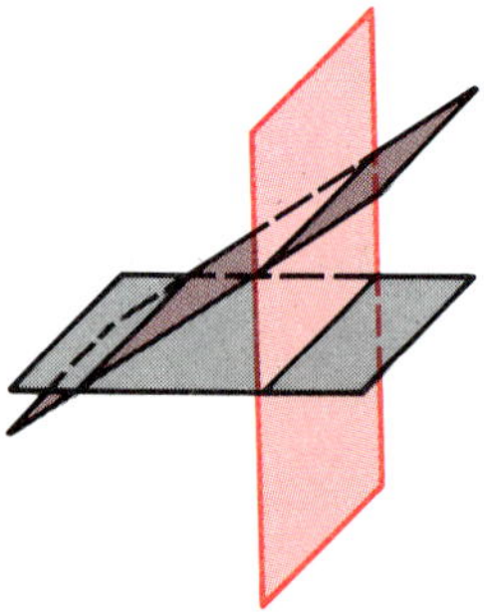

Figure 6

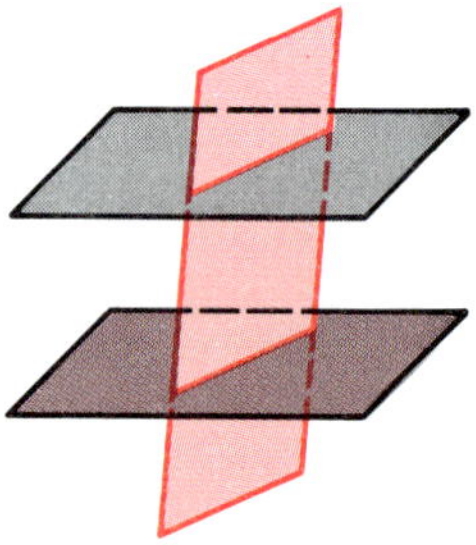

Figure 7

EXAMPLE 2 Solve and check: $\begin{cases} x - 2y + 3z = -3 & \textbf{1} \\ 2x - 3y - z = 7 & \textbf{2} \\ 3x + y - 2z = 6 & \textbf{3} \end{cases}$

Solution: [1] Eliminate any one variable to obtain an equivalent system in two variables.

Eliminate y from **1** and **3**.

$$\begin{array}{rcl} x - 2y + 3z = -3 & & \textbf{1} \\ 6x - 2y - 4z = 12 & 2 \times & \textbf{3} \\ \hline 7x \qquad - z = 9 & & \textbf{4} \end{array}$$

Eliminate y from **2** and **3**.

$$\begin{array}{rcl} 2x - 3y - z = 7 & & \textbf{2} \\ 9x + 3y - 6z = 18 & 3 \times & \textbf{3} \\ \hline 11x \qquad - 7z = 25 & & \textbf{5} \end{array}$$

$$\begin{cases} 7x - z = 9 & \textbf{4} \\ 11x - 7z = 25 & \textbf{5} \end{cases}$$

2 Eliminate z from the new system and solve for x. Then use equation **4** to determine z.

$$\begin{array}{rcll} -49x + 7z &=& -63 & -7 \times \mathbf{4} \\ 11x - 7z &=& 25 & \mathbf{5} \\ \hline -38x &=& -38 & \\ x &=& 1 & \end{array} \qquad \begin{array}{rcll} 7x - z &=& 9 & \mathbf{4} \\ 7(1) - z &=& 9 & \\ 7 - z &=& 9 & \\ z &=& -2 & \end{array}$$

3 Substitute $x = 1$ and $z = -2$ in **1**, **2**, or **3** to determine y.

$$\begin{aligned} 1 - 2y + 3(-2) &= -3 \\ -2y &= 2 \\ y &= -1 \end{aligned}$$

Check:

$$\begin{array}{ccc} x - 2y + 3z = -3 & 2x - 3y - z = 7 & 3x + y - 2z = 6 \\ 1 - 2(-1) + 3(-2) \stackrel{?}{=} -3 & 2(1) - 3(-1) - (-2) \stackrel{?}{=} 7 & 3(1) + (-1) - 2(-2) \stackrel{?}{=} 6 \\ -3 \stackrel{?}{=} -3 \checkmark & 7 \stackrel{?}{=} 7 \checkmark & 6 \stackrel{?}{=} 6 \checkmark \end{array}$$

Solution set: $\{(1, -1, -2)\}$

Summary

A system of three linear equations in three variables is:

1. **Consistent** and **independent** when the solution set contains one ordered triple.
2. **Consistent** and **dependent** when the solution set contains many ordered triples.
3. **Inconsistent** when the solution set is ϕ, the empty set.

CLASSROOM EXERCISES

Solve each system.

1. $\begin{cases} x + y + z = 8 \\ x - y - z = 2 \\ z = 3 \end{cases}$

2. $\begin{cases} x + y + z = 12 \\ y = 1 \\ x - y - z = 0 \end{cases}$

3. $\begin{cases} x - y + z = -8 \\ y = 4 \\ x + y - z = 6 \end{cases}$

WRITTEN EXERCISES

A Solve and check.

1. $\begin{cases} 2x - 3y - 4z = -21 \\ -4x + 2y - 3z = -14 \\ -3x - 4y + 2z = -10 \end{cases}$

2. $\begin{cases} 6x - y - 3z = 2 \\ -3x + y - 3z = 1 \\ -2x + 3y + z = -6 \end{cases}$

3. $\begin{cases} 3x + 2y = z - 7 \\ 5x + 3y = -12 + 2z \\ 2x + 3y = -5 + z \end{cases}$

4. $\begin{cases} 2x + 5y = -5 + 3z \\ 3x + 7z = 15 - 2y \\ 5x + 6z = 34 + 4y \end{cases}$

5. $\begin{cases} 2x + 3y + 4z = 8 \\ 4x + 9y + 8z = 17 \\ 6x + 12y + 16z = 31 \end{cases}$

6. $\begin{cases} 2x + 3y = -2 \\ 4y + 2z = -10 \\ 3x + 5z = 1 \end{cases}$

Solve and check.

7. $\begin{cases} 3x + 4z = 22 \\ y - \frac{z}{2} = 1 \\ 5x + 3y = 19 \end{cases}$

8. $\begin{cases} x + y + z = 180 \\ \frac{x}{4} + \frac{y}{2} + \frac{z}{3} = 60 \\ 2y + 3z = 330 \end{cases}$

9. $\begin{cases} 2x - y + 3z = -9 \\ x + 3y - z = 10 \\ 3x + y - z = 8 \end{cases}$

B

10. $\begin{cases} 0.5x + 0.8y + 0.9z = 32 \\ x + 0.6y + 0.4z = 26 \\ 2x + 0.3y + 0.2z = 31 \end{cases}$

11. $\begin{cases} 0.5a + 0.3b = 2.2 \\ 1.2c - 8.5b = -24.4 \\ 3.3c + 1.3a = 29 \end{cases}$

12. $\begin{cases} 3x + 2y + z = 7.7 \\ 2x - y - z = 3.3 \\ 5x - 4y - 2z = 5.5 \end{cases}$

13. $\begin{cases} 3x - 6y + 3z = 4 \\ x - 2y + z = 1 \\ 2x - 4y + 2z = 5 \end{cases}$

14. $\begin{cases} x - \frac{4}{3}y - \frac{1}{3}z = 1 \\ y + z = 6 \\ -2x - \frac{5}{3}y = 5 \end{cases}$

15. $\begin{cases} 0.25x + 0.5y + 3z = 2 \\ 0.75x - 1.5y - z = 0 \\ x + 2y - 4z = 8 \end{cases}$

Problem Solving and Applications

13–3 Solving Problems with Three Variables

Systems of linear equations in three variables can be used to solve problems.

EXAMPLE A store offers a special on cotton, orlon, and wool ski sweaters. On the first day, 5 cotton, 4 orlon, and 8 wool sweaters were sold. On the second day, 4 cotton, 3 orlon, and 6 wool sweaters were sold. On the third day, 3 cotton, 5 orlon, and 4 wool sweaters were sold. Total sales for the three days were \$210, \$160, and \$146 respectively. What was the sale price for each sweater?

Solution:

[1] Let x = the sale price of a cotton sweater.
Let y = the sale price of an orlon sweater.
Let z = the sale price of a wool sweater. ← **Represent the variables.**

[2] Organize the information in a table.

Days	Value of Sweaters Sold		
	Cotton	**Orlon**	**Wool**
First	$5x$	$4y$	$8z$
Second	$4x$	$3y$	$6z$
Third	$3x$	$5y$	$4z$

[3] Write a system of equations. $\begin{cases} 5x + 4y + 8z = 210 \\ 4x + 3y + 6z = 160 \\ 3x + 5y + 4z = 146 \end{cases}$

[4] Solve for x, y, and z. ← **The check is left for you.**

$x = 10$, $y = 12$, $z = 14$

Cotton sweaters were on sale for **\$10,** orlon sweaters for **\$12,** and wool sweaters for **\$14.**

CLASSROOM EXERCISES

Use three variables to represent the unknowns. Then write an equation to represent the statement.

1. The sum of the digits of a three-digit number is 19.
2. The total capacity of three trucks is 108 cubic meters.
3. Three typists working together can type a certain manuscript in 3 days.
4. The sum of the angles of a triangle is 180°.
5. A farmer has 100 acres of land on which to plant corn, oats, and beans.

WORD PROBLEMS

A Use a system of three equations in three variables to solve each problem.

1. A woman invested a total of $40,000 in three business ventures: a bowling alley, a diner, and a laundromat. In a recent year, the bowling alley returned a profit of 3%, the diner returned a profit of 8%, and the laundromat a profit of 12%. Total income from the three investments was $2850. The income from the diner was the same as the income from the laundromat. Find the amount invested in each venture.
2. A ceramics artisan makes cups, pitchers, and vases. The cost is $0.60 per cup, $0.25 per pitcher, and $2.00 per vase. His total cost for the 15 items made per day is $9.00. It takes the artisan 20 minutes to make a cup, 10 minutes to make a pitcher, and 30 minutes to make a vase. The total time worked per day is 4 hours. How many items of each type are produced each day?
3. The sum of the digits of a certain three-digit number is 14. The sum of the hundreds digit and tens digit is 11. When the digits are reversed, the new number is 198 less than the original number. Find the original number.
4. The sum of the digits of a certain three-digit number is 20. If 300 is subtracted from the number, the hundreds digit and the units digit of the new number will be the same. The units digit is one-third the sum of the tens digit and the hundreds digit. Find the original number.
5. The sum of the three digits of a number is 13. If the number, decreased by 8, is divided by the sum of the units and tens digits, the quotient is 25. When the digits are reversed, the new number exceeds the original number by 99. Find the original number.
6. Find the three angles of a triangle if the sum of the first angle and twice the second equals the third angle, and if four times the second angle is 15° more than the third.
7. The perimeter of a triangle is 38 centimeters. The longest side is four centimeters less than twice the middle side. It is also two centimeters less than the sum of the other two sides. Find the three sides.

8. Three trucks together haul 78 cubic meters, 81 cubic meters, and 69 cubic meters in three days. Find the capacity of each truck if they haul the following number of loads each day. First day: 4, 3, and 5 loads; second day: 5, 4, and 4 loads; third day: 3, 5, and 3 loads.

B

9. Mary and Tiffany can do a job in 10 days, Mary and Cecile in 12 days, and Tiffany and Cecile in 20 days. How many days would it take each person working alone to do the job?

10. John, Joe, and Art working together can complete a job in 3 days. John and Joe working together take $3\frac{3}{4}$ days to complete the job. Joe and Art working together take 6 days to complete the job. How many days would it take each person working alone to complete the job?

13–4 Solving Linear Systems: Augmented Matrix Method

A system of linear equations in two or more variables can also be solved by writing a matrix for the system. A **matrix** is a rectangular array of numbers. The elements of a matrix are arranged in rows and columns and are usually enclosed in brackets. A matrix with m rows and n columns is an $\boldsymbol{m \times n}$ matrix. The **dimensions** of the matrix are m and n.

Two-by-Two Matrix (2×2)

Column 1 Column 2

Row 1 →
Row 2 →

$$\begin{bmatrix} 1 & 2 \\ -1 & 4 \end{bmatrix}$$

Three-by-Three Matrix (3×3)

$$\begin{bmatrix} 1 & 0 & 0 \\ 0 & 1 & 0 \\ 0 & 0 & 1 \end{bmatrix}$$

Column Matrix (3×1)

$$\begin{bmatrix} 1 \\ -8 \\ 6\frac{1}{2} \end{bmatrix}$$

Row Matrix (1×4)

$$\begin{bmatrix} -7 & 3 & -2\frac{2}{3} & 5 \end{bmatrix}$$

The matrix corresponding to a system of three linear equations in three variables is a 3×4 **augmented matrix.**

$$\begin{cases} 12x - y + 12z = 6 \\ 2x + y - 2z = -4 \\ 9x + 2y + 3z = 3 \end{cases} \longrightarrow \begin{bmatrix} 12 & -1 & 12 & 6 \\ 2 & 1 & -2 & -4 \\ 9 & 2 & 3 & 3 \end{bmatrix}$$

The entries in each row of the matrix correspond to the coefficients of the three variables of an equation in the system together with the constant term. Each column of the matrix lists the coefficients of a given variable or the constant terms. REMEMBER: The equation of the system must be written in the form $Ax + By + Cz = D$ before writing the corresponding augmented matrix.

Three operations can be performed on the rows of a matrix. These "row operations" produce matrices which represent equivalent linear systems. Thus, the matrices are also equivalent.

Row Operations for a Matrix

1. Multiply or divide any row by a nonzero real number.
2. Add the corresponding numbers in any two rows. This sum can replace either row, but not both. (This corresponds to adding equations).
3. Interchange any two rows.

In performing these operations, the goal is to obtain an equivalent matrix in triangular form. The solution to the system can be determined from the equivalent system represented by the matrix.

$$\begin{bmatrix} a_1 & 0 & 0 & C_1 \\ b_1 & b_2 & 0 & C_2 \\ c_1 & c_2 & c_3 & C_3 \end{bmatrix} \quad \text{or} \quad \begin{bmatrix} a_1 & a_2 & a_3 & C_1 \\ b_1 & b_2 & 0 & C_2 \\ c_1 & 0 & 0 & C_3 \end{bmatrix}$$

EXAMPLE Solve by using the augmented matrix method.

$$\begin{cases} 12x - y + 12z = 6 \\ 2x + y - 2z = -4 \\ 9x + 2y + 3z = 3 \end{cases}$$

Solution: Write a matrix for the system.

$$\begin{bmatrix} 12 & -1 & 12 & 6 \\ 2 & 1 & -2 & -4 \\ 9 & 2 & 3 & 3 \end{bmatrix} \begin{matrix} 1 \\ 2 \\ 3 \end{matrix}$$

← **To eliminate the second 12 in Row 1, multiply Row 2 by 6 and add it to Row 1. Replace Row 1 with the result.**

$$\begin{bmatrix} 24 & 5 & 0 & -18 \\ 2 & 1 & -2 & -4 \\ 9 & 2 & 3 & 3 \end{bmatrix} \begin{matrix} 1 \\ 2 \\ 3 \end{matrix}$$

← **To eliminate the −2 in Row 2, multiply Row 2 by 3 and Row 3 by 2. Add these products. Replace Row 2 with the result.**

$$\begin{bmatrix} 24 & 5 & 0 & -18 \\ 24 & 7 & 0 & -6 \\ 9 & 2 & 3 & 3 \end{bmatrix} \begin{matrix} 1 \\ 2 \\ 3 \end{matrix}$$

← **To eliminate the 5 in Row 1, multiply Row 1 by 7 and Row 2 by −5. Add these products. Replace Row 1 with the result.**

$$\begin{bmatrix} 48 & 0 & 0 & -96 \\ 24 & 7 & 0 & -6 \\ 9 & 2 & 3 & 3 \end{bmatrix} \begin{matrix} 1 \\ 2 \\ 3 \end{matrix}$$

← **The matrix is now in triangular form.**

Write the system represented by the augmented matrix.

$$\begin{cases} 48x = -96 & 1 \\ 24x + 7y = -6 & 2 \\ 9x + 2y + 3z = 3 & 3 \end{cases}$$

From Equation 1, $48x = -96$. ⟶ $x = -2$

In Equation 2, replace x with -2. ⟶ $y = 6$

In Equation 3, replace x with -2 and y with 6. ⟶ $z = 3$

The check is left for you. **Solution set: {(−2, 6, 3)}**

CLASSROOM EXERCISES

Write the matrix associated with each linear system.

1. $\begin{cases} 2x - y + 3z = -9 \\ x + 3y - z = 10 \\ 3x + y - z = 8 \end{cases}$

2. $\begin{cases} x + y + z = 1 \\ 2x - 3y - 2z = -4 \\ 3x - 2y - z = -2 \end{cases}$

3. $\begin{cases} x + 2y - z = -3 \\ -2x - 4y + 2z = 6 \\ 3x + 6y - 3z = -9 \end{cases}$

WRITTEN EXERCISES

A Use the augmented matrix method to solve each system.

1. $\begin{cases} 2x - y + 3z = 16 \\ 3x - 5y + 4z = -6 \\ 4x + 2y - 3z = 28 \end{cases}$

2. $\begin{cases} 3x - 2y + 7z = 80 \\ 5x + 3y - 4z = 2 \\ 2x + 5y + z = 42 \end{cases}$

3. $\begin{cases} 2x + y + z = 6 \\ 3x + 3y - 2z = -5 \\ -x + y - z = -5 \end{cases}$

4. $\begin{cases} 3x + 2y - z = 5 \\ 2x + y - 2z = -7 \\ 3x - 2y + 3z = 21 \end{cases}$

5. $\begin{cases} 6x + y - z = -2 \\ 2x + 5y - z = 2 \\ x + 2y + z = 5 \end{cases}$

6. $\begin{cases} 2x + 2y + z = 3 \\ 3x + 2y - 2z = -1 \\ 2x - 3y - z = 22 \end{cases}$

7. $\begin{cases} x + y + z = 6 \\ 2x + y - z = 2 \\ 3x - 2y + 2z = 10 \end{cases}$

8. $\begin{cases} 2x + 3y - 2z = 18 \\ 5x - 6y + z = 21 \\ 4y - 2z = 6 \end{cases}$

9. $\begin{cases} 3x + 2y = 12 \\ 4z + 3y = 25 \\ 5x + 2y = 16 \end{cases}$

10. $\begin{cases} 4x - 3y = 1 \\ 6x - 8z = 1 \\ 2y - 4z = 0 \end{cases}$

11. $\begin{cases} 5x - y - z = -1 \\ 2x + 3y - z = -4 \\ x - 7y + z = 7 \end{cases}$

12. $\begin{cases} 2x + 5y - 3z = -1 \\ -x + y + 3z = 6 \\ x - 2y + 8z = 1 \end{cases}$

B

13. The row operations can be used to obtain an equivalent matrix in row–echelon form (shown at the right). In this form, the solution to the linear system can be read directly from the matrix. Thus, $x = D_1$, $y = D_2$, and $z = D_3$. Write the matrices in Exercises 1, 2, and 3 in row-echelon form.

$$\begin{bmatrix} 1 & 0 & 0 & D_1 \\ 0 & 1 & 0 & D_2 \\ 0 & 0 & 1 & D_3 \end{bmatrix}$$

Review

Graph each point in three-space. *(Section 13–1)*

1. $A(2, 1, -3)$
2. $H(4, -2, 5)$
3. $D(0, -1, 3)$
4. $F(-4, 0, -2)$

Solve and check. *(Section 13–2)*

5. $\begin{cases} 2x + y - z = 11 \\ x - y + 3z = -4 \\ -x + 4y + z = 12 \end{cases}$

6. $\begin{cases} x - 5y + z = 16 \\ -3x + 2y - 4z = -28 \\ 2x + 6y + 3z = 6 \end{cases}$

7. $\begin{cases} x + \frac{3}{2}y - 4z = 3 \\ \frac{7}{2}x - 4y + 5z = 20 \\ 3x - 2y + 6z = 18 \end{cases}$

8. The sum of the digits of a three-digit number is 12. The sum of the units digit and the hundreds digit equals the tens digit. When the digits are reversed, the new number is 198 more than the original number. Find the original number. *(Section 13–3)*

Use the augmented matrix method to solve each system. *(Section 13–4)*

9. $\begin{cases} 4x + 3y + z = 2 \\ 3x - 2y + 6z = -6 \\ -2x + 4y - 7z = 9 \end{cases}$

10. $\begin{cases} 5x - 2y + 4z = -1 \\ 2x + 7y - 2z = 14 \\ -4x - 3y + z = -16 \end{cases}$

11. $\begin{cases} x - y + z = 5 \\ 3x + 2y - z = 9 \\ 2x - y + 4z = 6 \end{cases}$

13–5 Addition and Multiplication of Matrices

Recall that you add complex numbers by adding corresponding parts.

$$(a+bi)+(c+di)=(a+c)+(b+d)i$$

Matrices of the same dimension are added in a similar way. Two matrices have the **same dimension** when they have the same number of rows and columns.

$$\begin{bmatrix} a_1 & b_1 \\ a_2 & b_2 \end{bmatrix} + \begin{bmatrix} c_1 & d_1 \\ c_2 & d_2 \end{bmatrix} = \begin{bmatrix} a_1+c_1 & b_1+d_1 \\ a_2+c_2 & b_2+d_2 \end{bmatrix}$$

Definition

> When A and B are matrices of the **same dimension,** $A+B$ can be found by adding corresponding elements of A and B.

EXAMPLE 1 Find $A+B$ when $A=\begin{bmatrix} 3 & 2 \\ -5 & 4 \end{bmatrix}$ and $B=\begin{bmatrix} -2 & 1 \\ 3 & -3 \end{bmatrix}$.

Solution:

$$\begin{bmatrix} 3 & 2 \\ -5 & 4 \end{bmatrix} + \begin{bmatrix} -2 & 1 \\ 3 & -3 \end{bmatrix} = \begin{bmatrix} 3-2 & 2+1 \\ -5+3 & 4-3 \end{bmatrix}$$

$$= \begin{bmatrix} \mathbf{1} & \mathbf{3} \\ \mathbf{-2} & \mathbf{1} \end{bmatrix}$$

Matrices of different dimensions do not have corresponding elements. Thus, addition is *not defined* for matrices with different dimensions.

A matrix can be multiplied by a constant called a **scalar.**

Definition

> The product of a scalar k and an $m \times n$ matrix, A, is an $m \times n$ matrix, kA. Each element in kA is equal to k multiplied by the corresponding element in A.

EXAMPLE 2 Find kA for $k=-3$ and $A=\begin{bmatrix} -3 & 6 & -9 \\ 0 & 4 & -3 \end{bmatrix}$.

Solution:

$$-3\begin{bmatrix} -3 & 6 & -9 \\ 0 & 4 & -3 \end{bmatrix} = \begin{bmatrix} (-3)(-3) & (-3)(6) & (-3)(-9) \\ (-3)(0) & (-3)(4) & (-3)(-3) \end{bmatrix}$$

$$= \begin{bmatrix} \mathbf{9} & \mathbf{-18} & \mathbf{27} \\ \mathbf{0} & \mathbf{-12} & \mathbf{9} \end{bmatrix}$$

In order to multiply two matrices, the *number of elements in the rows* of the first matrix **must equal** the *number of elements in the columns* of the second matrix. Matrix multiplication is "row by column" multiplication.

$$A_{1\times 2} = \begin{bmatrix} -1 & 4 \end{bmatrix} \qquad B_{2\times 4} = \begin{bmatrix} 4 & 1 & 0 & 1 \\ 2 & 0 & 2 & 0 \end{bmatrix}$$

These must be the same.

EXAMPLE 3 Find $AB (A \times B)$ for $A = [2 \quad -1 \quad -3]$ and $B = \begin{bmatrix} 3 \\ 1 \\ 5 \end{bmatrix}$.

Solution: Multiply the first element in the row matrix by the first element in the column matrix.

Continue this procedure with the second elements, and so on.

$$\begin{aligned} [2 \quad -1 \quad -3] \times \begin{bmatrix} 3 \\ 1 \\ 5 \end{bmatrix} &= [(2 \times 3) + (-1 \times 1) + (-3 \times 5)] \\ &= [6 + (-1) + (-15)] \\ &= [\mathbf{-10}] \end{aligned}$$

Notice that the product of a 1×3 matrix and a 3×1 matrix is a 1×1 matrix.

$$A_{1\times 3} \times B_{3\times 1} = C_{\underline{1\times 1}}$$

Definition

For $A = \begin{bmatrix} a_1 & b_1 \\ a_2 & b_2 \end{bmatrix}$ and $B = \begin{bmatrix} c_1 & d_1 \\ c_2 & d_2 \end{bmatrix}$,

$$\begin{aligned} A \times B = AB &= \begin{bmatrix} a_1 & b_1 \\ a_2 & b_2 \end{bmatrix} \times \begin{bmatrix} c_1 & d_1 \\ c_2 & d_2 \end{bmatrix} \\ &= \begin{bmatrix} a_1c_1 + b_1c_2 & a_1d_1 + b_1d_2 \\ a_2c_1 + b_2c_2 & a_2d_1 + b_2d_2 \end{bmatrix} \end{aligned}$$

EXAMPLE 3 Find AB for $A = \begin{bmatrix} 2 & -1 & -3 \\ 1 & 0 & 5 \end{bmatrix}$ and $B = \begin{bmatrix} 3 & 2 \\ 1 & -3 \\ 5 & 1 \end{bmatrix}$

Solution: Multiply each row of matrix A by each column in matrix B.

$$\begin{aligned} \begin{bmatrix} 2 & -1 & -3 \\ 1 & 0 & 5 \end{bmatrix} \times \begin{bmatrix} 3 & 2 \\ 1 & -3 \\ 5 & 1 \end{bmatrix} &= \begin{bmatrix} (2\times 3) + (-1\times 1) + (-3\times 5) & (2\times 2) + (-1\times -3) + (-3\times 1) \\ (1\times 3) + (0\times 1) + (5\times 5) & (1\times 2) + (0\times -3) + (5\times 1) \end{bmatrix} \\ &= \begin{bmatrix} \mathbf{-10} & \mathbf{4} \\ \mathbf{28} & \mathbf{7} \end{bmatrix} \end{aligned}$$

Notice that the product of a 2×3 matrix and a 3×2 matrix is a 2×2 matrix.

$$A_{2\times 3} \times B_{3\times 2} = C_{\underline{2\times 2}}$$

CLASSROOM EXERCISES

Perform the indicated operations.

1. $[1 \quad -3] + [3 \quad 2]$

2. $\begin{bmatrix} 3 & -2 \\ 4 & 1 \end{bmatrix} + \begin{bmatrix} 2 & 0 \\ -1 & 4 \end{bmatrix}$

3. $\begin{bmatrix} 1 & 0 \\ 0 & 1 \end{bmatrix} + \begin{bmatrix} -2 & 1 \\ 3 & -5 \end{bmatrix}$

4. $3[6 \quad 10]$

5. $-3\begin{bmatrix} 9 & -1 \\ 6 & 5 \end{bmatrix}$

6. $\frac{3}{4}\begin{bmatrix} 8 & -16 & 4 \\ 12 & -12 & -8 \end{bmatrix}$

7. $[4 \quad 1 \quad 3 \quad 5] \times \begin{bmatrix} 1 \\ 2 \\ 3 \\ 4 \end{bmatrix}$

8. $\begin{bmatrix} 3 & 5 \\ 0 & 1 \end{bmatrix} \times \begin{bmatrix} 0 & 1 \\ 2 & 3 \end{bmatrix}$

9. $\begin{bmatrix} 2 & -3 & 0 & 1 \\ 4 & 8 & 1 & 3 \end{bmatrix} \times \begin{bmatrix} 1 & 2 & 3 \\ 3 & 2 & 1 \\ 1 & 3 & 2 \\ 3 & 1 & 2 \end{bmatrix}$

In Exercises 10–15, give the dimensions of matrix *C*.

10. $A_{1\times4} \times B_{4\times2} = C$

11. $A_{5\times4} \times C = B_{5\times2}$

12. $C \times B_{1\times3} = A_{3\times3}$

13. $A_{2\times4} \times C = B_{2\times1}$

14. $A_{3\times3} \times B_{3\times1} = C$

15. $C \times B_{2\times1} = A_{2\times1}$

WRITTEN EXERCISES

A Add when possible. When the matrices cannot be added, write *NP*.

1. $\begin{bmatrix} 1 \\ 0 \\ 1 \end{bmatrix} + \begin{bmatrix} 3 \\ 1 \\ 2 \end{bmatrix}$

2. $\begin{bmatrix} 1 \\ -1 \\ 2 \end{bmatrix} + \begin{bmatrix} 2 \\ 1 \\ 5 \end{bmatrix}$

3. $\begin{bmatrix} 1 \\ 2 \end{bmatrix} + \begin{bmatrix} 2 \\ 1 \\ 5 \end{bmatrix}$

4. $\begin{bmatrix} 1 & 2 & 3 \\ 4 & 5 & 6 \end{bmatrix} + \begin{bmatrix} 3 & 2 & 1 \\ 5 & 4 & 6 \end{bmatrix}$

5. $\begin{bmatrix} 1 & 2 & -2 \\ 2 & 1 & -3 \end{bmatrix} + \begin{bmatrix} 1 & -5 \\ 3 & -7 \end{bmatrix}$

6. $[2] + [-1]$

7. $[1 \quad 2 \quad 3 \quad 4] + \begin{bmatrix} 4 \\ 3 \\ 2 \\ 1 \end{bmatrix}$

8. $\begin{bmatrix} 2 & 3 \\ 1 & 5 \end{bmatrix} + \begin{bmatrix} 1 & 3 \\ -2 & 1 \end{bmatrix}$

9. $\begin{bmatrix} 1 & 3 \\ 4 & 6 \end{bmatrix} + \begin{bmatrix} -1 & -3 \\ -4 & -6 \end{bmatrix}$

10. $\begin{bmatrix} 1 & 2 & 5 \\ 7 & 9 & 1 \\ -3 & 2 & 5 \end{bmatrix} + \begin{bmatrix} 2 & -1 & 3 \\ -6 & -10 & 5 \\ 2 & -3 & 1 \end{bmatrix}$

11. $\begin{bmatrix} -3 & -6 & 9 \\ 4 & -3 & 0 \\ 8 & -2 & 3 \end{bmatrix} + \begin{bmatrix} 6 & 9 & -4 \\ -11 & 13 & -8 \\ 20 & 4 & -2 \end{bmatrix}$

Use matrices A, B, and C to find each product in Exercises 12–17.

$$A = \begin{bmatrix} 0 & 3 \\ 4 & -1 \end{bmatrix} \qquad B = \begin{bmatrix} 1 & 2 \\ 6 & 0 \\ -1 & 4 \end{bmatrix} \qquad C = \begin{bmatrix} 2 & -2 & 1 \\ 4 & 6 & -3 \end{bmatrix}$$

12. $4A$ **13.** $3B$ **14.** $6C$ **15.** $-5A$ **16.** $-\frac{1}{2}B$ **17.** $-C$

Multiply when possible. When the matrices cannot be multiplied, write *NP*.

18. $\begin{bmatrix} 1 \\ 0 \\ 1 \end{bmatrix} \times \begin{bmatrix} 5 & -1 & 2 \end{bmatrix}$

19. $\begin{bmatrix} 2 & 1 & 5 \end{bmatrix} \times \begin{bmatrix} -2 \\ -3 \\ 7 \end{bmatrix}$

20. $\begin{bmatrix} 2 & 3 \\ 1 & 5 \end{bmatrix} \times \begin{bmatrix} 3 & 5 & 1 \\ 2 & 1 & 3 \end{bmatrix}$

21. $\begin{bmatrix} 3 \\ -5 \end{bmatrix} \div \begin{bmatrix} 1 & 7 \end{bmatrix}$

22. $\begin{bmatrix} 1 \\ 2 \\ 3 \end{bmatrix} \times \begin{bmatrix} 1 & 2 & 3 \end{bmatrix}$

23. $\begin{bmatrix} 3 & 4 \\ 1 & 5 \end{bmatrix} \times \begin{bmatrix} 5 & -4 \\ -1 & 3 \end{bmatrix}$

24. $\begin{bmatrix} 1 & 2 \\ 3 & 4 \end{bmatrix} \times \begin{bmatrix} 1 & 2 \\ -3 & 4 \end{bmatrix}$

25. $\begin{bmatrix} 3 & -2 \\ 1 & 4 \end{bmatrix} \times \begin{bmatrix} 4 & 5 \\ 2 & -6 \end{bmatrix}$

26. $\begin{bmatrix} -1 \\ 3 \\ 4 \end{bmatrix} \times \begin{bmatrix} 2 & 3 & 1 \\ 1 & 5 & 4 \end{bmatrix}$

B

27. $8\begin{bmatrix} 1 & 0 & 0 \\ 0 & 1 & 0 \\ 0 & 0 & 1 \end{bmatrix}\begin{bmatrix} 1 & 3 & 2 \\ 1 & 4 & 1 \\ 5 & 1 & 2 \end{bmatrix}$

28. $\begin{bmatrix} 1 & -2 \\ 3 & 4 \\ 2 & 5 \\ -1 & 6 \end{bmatrix}\begin{bmatrix} 3 & 1 & 2 \\ -1 & 2 & 4 \end{bmatrix}$

29. $r\begin{bmatrix} a_1 & a_2 \\ b_1 & b_2 \end{bmatrix}\begin{bmatrix} 0 & 1 \\ 1 & 0 \end{bmatrix}$

30. $\begin{bmatrix} 2 & 3 & -1 \\ 1 & 1 & 1 \\ 0 & 2 & -1 \end{bmatrix}\begin{bmatrix} 3 & -1 & -4 \\ -1 & 2 & 3 \\ -2 & 4 & 1 \end{bmatrix}$

31. $\begin{bmatrix} 1 & 0 & 0 \\ 0 & 0 & 1 \\ 0 & 1 & 0 \end{bmatrix}\begin{bmatrix} a & b & c \\ d & e & f \\ g & h & i \end{bmatrix}$

32. $\begin{bmatrix} 1 & 0 & 0 \\ 0 & 2 & 0 \\ 0 & 0 & 3 \end{bmatrix}\begin{bmatrix} a & b & c \\ d & e & f \\ g & h & i \end{bmatrix}$

Matrix subtraction can be defined by using the scalar −1.

If A and B are $m \times n$ matrices, then

$$\mathbf{A - B = A + (-1)B} \quad \text{or} \quad \mathbf{A - B = A + (-B)},$$

where the entries in $-B$ are the negatives of the entries in B.

Use this definition in Exercises 33–35.

33. $\begin{bmatrix} 2 \\ 3 \end{bmatrix} - \begin{bmatrix} -1 \\ 8 \end{bmatrix}$

34. $\begin{bmatrix} 6 & 7 \\ 5 & -1 \end{bmatrix} - \begin{bmatrix} 3 & 4 \\ -5 & 1 \end{bmatrix}$

35. $\begin{bmatrix} 0 & 4 & -2 \\ -5 & 1 & 3 \end{bmatrix} - \begin{bmatrix} 6 & -7 & -1 \\ 0 & -4 & 8 \end{bmatrix}$

Two matrices are **equal** if and only if they have the same dimensions and all their corresponding entries are equal.

Use this definition to solve for x and y in Exercises 36–41.

36. $\begin{bmatrix} x & 4y \end{bmatrix} = \begin{bmatrix} y+5 & 2x+10 \end{bmatrix}$

37. $\begin{bmatrix} 2x \\ 5 \\ 0 \end{bmatrix} = \begin{bmatrix} -10 \\ y-x \\ y \end{bmatrix}$

38. $\begin{bmatrix} x+y \\ x-y \end{bmatrix} = \begin{bmatrix} 9 \\ 4 \end{bmatrix}$

39. $\begin{bmatrix} 4 & -1 \\ 3 & 2 \end{bmatrix}\begin{bmatrix} x \\ y \end{bmatrix} = \begin{bmatrix} 5 \\ 12 \end{bmatrix}$

40. $\begin{bmatrix} 2 & -1 \\ 5 & 2 \end{bmatrix}\begin{bmatrix} x \\ y \end{bmatrix} = \begin{bmatrix} 16 \\ 7 \end{bmatrix}$

41. $\begin{bmatrix} 5 & -2 \\ 7 & 3 \end{bmatrix}\begin{bmatrix} x \\ y \end{bmatrix} = \begin{bmatrix} 19 \\ 15 \end{bmatrix}$

42. For $A = \begin{bmatrix} 1 & -1 \\ 0 & 3 \end{bmatrix}$, $B = \begin{bmatrix} 3 & 2 \\ 1 & 5 \end{bmatrix}$, and $C = \begin{bmatrix} 0 & 1 \\ -3 & 2 \end{bmatrix}$, show that $A(BC) = (AB)C$.

43. For the matrices defined in Exercise 42, show that $(A+B)+C = A+(B+C)$.

44. For the matrices defined in Exercise 42, determine whether $(A+B)-C = A+(B-C)$.

In Exercises 45–46, let $A = \begin{bmatrix} a_1 & b_1 \\ a_2 & b_2 \end{bmatrix}$, and $O = \begin{bmatrix} 0 & 0 \\ 0 & 0 \end{bmatrix}$.

Matrix O is the 2×2 zero matrix, or additive identity matrix. For any $m \times n$ matrix, the **zero matrix** is the $m \times n$ matrix each of whose entries is 0.

45. Show that $A + O = O + A = A$.

46. The matrix $-A$ is the **additive inverse** of A since the sum of A and $-A$ is the zero matrix. Show that $A + (-A) = O$.

C

47. Is matrix addition for 2×2 matrices commutative? Use three examples to illustrate your answer.

48. Is matrix multiplication for 2×2 matrices commutative? Use three examples to illustrate your answer.

49. Find A^3 when $A = \begin{bmatrix} 1 & 0 & 0 \\ 0 & 1 & 0 \\ 0 & 0 & 1 \end{bmatrix}$.

Problem Solving and Applications

13–6 Using Matrices

Matrices are useful in dealing with data that can be arranged in tables. For example, in Mrs. Flotsam's factory, boxes for four different assortments of nuts, seeds, and raisins are prepared: Sampler, Assorted, Super Mix, and Nature's Delight. The ingredients for these assortments include cashews, walnuts, pecans, sunflower seeds, sesame seeds, and raisins.

The number of grams of each ingredient for each assortment can be written as a matrix.

	Cashews	Walnuts	Pecans	Sunflower Seeds	Sesame Seeds	Raisins
Sampler	[2	3	3	3	8	2]
Assorted	[5	3	5	0	18	8]
Super Mix	[0	0	0	8	32	0]
Nature's Delight	[10	5	5	2	3	10]

EXAMPLE 1 Mrs. Flotsam produces an equal number of boxes of each assortment. To produce one box of each assortment, how much of each ingredient does she need on hand?

Solution: Solve by adding the four matrices above. The result is:

$$[\mathbf{17} \quad \mathbf{11} \quad \mathbf{13} \quad \mathbf{13} \quad \mathbf{61} \quad \mathbf{20}]$$

The matrix at the right gives the cost in cents per gram for each ingredient used.

7	cashews
5	walnuts
8	pecans
3	sunflower seeds
3	sesame seeds
5	raisins

EXAMPLE 2 Find the total cost of the ingredients in each assortment.

Solution: Solve by multiplying matrices.

$$\begin{bmatrix} 2 & 3 & 3 & 3 & 8 & 2 \\ 5 & 3 & 5 & 0 & 18 & 8 \\ 0 & 0 & 0 & 8 & 32 & 0 \\ 10 & 5 & 5 & 2 & 3 & 10 \end{bmatrix} \times \begin{bmatrix} 7 \\ 5 \\ 8 \\ 3 \\ 3 \\ 5 \end{bmatrix} = \begin{bmatrix} 14 + 15 + 24 + 9 + 24 + 10 \\ 35 + 15 + 40 + 0 + 54 + 40 \\ 0 + 0 + 0 + 24 + 96 + 0 \\ 70 + 25 + 40 + 6 + 9 + 50 \end{bmatrix}$$

$$= \begin{bmatrix} \mathbf{96} \\ \mathbf{184} \\ \mathbf{120} \\ \mathbf{200} \end{bmatrix} \begin{matrix} \text{Sampler} \\ \text{Assorted} \\ \text{Super Mix} \\ \text{Nature's Delight} \end{matrix}$$

A **network** is a set of points connected by arcs. Examples of networks include railway systems, airline routes, highways, telephone systems, etc. A **directed network** is one which shows the directions of travel possible along each arc.

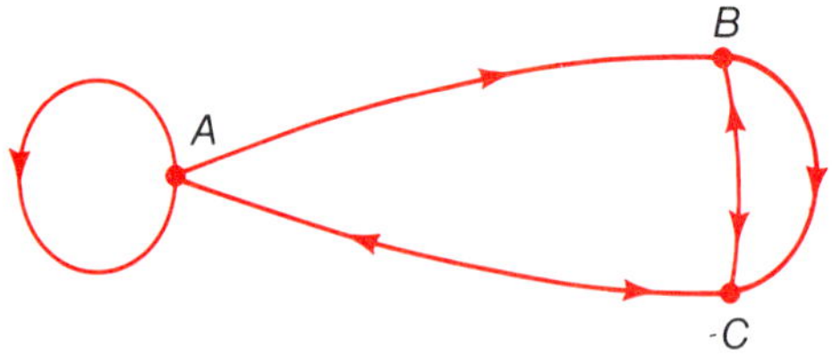

A matrix is a convenient way to represent a network. The 1 in the first row and first column represents "one path from A to A." Similarly, the 2 in row 2 represents "two paths from B to C."

From \ To	A	B	C
A	1	1	1
B	0	0	2
C	1	1	0

EXAMPLE 3 Construct a matrix that represents the directed network below.

Solution: Since there are 4 elements for "From" and 4 for "To," it will be a 4×4 matrix. For the first row, there are no paths from A to A, one path from A to B, 1 path from A to C, and 1 path from A to D. The remaining rows are constructed in a similar manner.

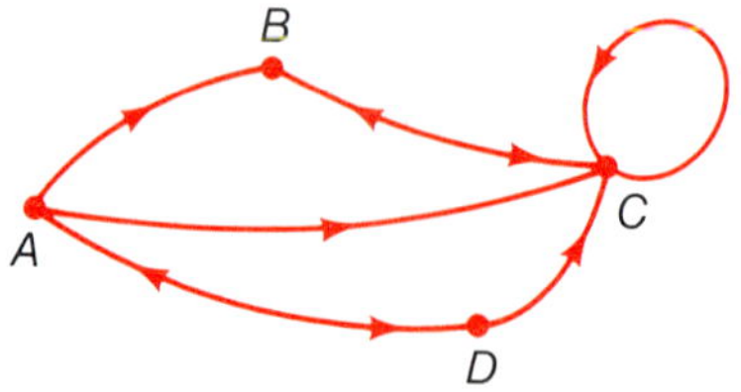

From \ To	A	B	C	D
A	**0**	**1**	**1**	**1**
B	**0**	**0**	**1**	**0**
C	**0**	**1**	**1**	**0**
D	**1**	**0**	**1**	**0**

CLASSROOM EXERCISES

A book-of-the-month club offers 3 packages of books, 1, 2, and 3. Each package consists of novels, poems, and plays. Package 1 contains 3 novels, 1 book of poems, and 1 book of plays. Package 2 contains 2 novels, 2 books of poems, and 1 book of plays. Package 3 contains 2 novels, 1 book of poems, and 2 books of plays. Each book is available in hardcover and paperback editions. A novel costs $12 in hardback and $4 in paperback; a book of poems costs $8 in hardback and $5 in paperback; a book of plays costs $10 in hardback and $6 in paperback. Use this information in Exercises 1–4.

1. Complete matrix A, which shows the contents of each package.

Content

	Novels	Poems	Plays
$A =$ 1			
2			
3			

2. Complete matrix B, which gives the price of each kind of book in the package.

Cost

	Hardcover	Paperback	
$B =$			Novels
			Poems
			Plays

3. Use matrix multiplication to find the cost of each hardcover package.
4. Use matrix multiplication to find the cost of each paperback package.

WORD PROBLEMS

A

1. Three "do it yourself" radio kits are available from Benton Harbor, Michigan. The beginner's kit include 3 valves, 2 coils, 1 speaker, 9 resistors, and 6 capacitors. The intermediate kit includes 8 valves, 6 coils, 2 speakers, 25 resistors, and 24 capacitors. The advanced kit has 8 coils, 2 speakers, 6 transistors, 23 resistors, and 16 capacitors. Represent this information in three 1×6 matrices. Add to determine the number of each item needed.

2. Suppose the materials in Exercise 1 have the following costs: valves – $0.80; coils – $1.20; speakers – $4.20; resistors – $0.15; capacitors – $0.25; transistors – $1.20. Find the value of the materials in each kit.
3. Four classes in a middle school are named Mars, Pluto, Saturn and Venus. The way the 7th grade students from 3 feeder schools are placed in each class is described by the following matrices.

$$\begin{array}{c} \\ \text{Boys} \\ \text{Girls} \end{array} \begin{array}{c} \begin{array}{cccc} M & P & S & V \end{array} \\ \begin{bmatrix} 3 & 2 & 5 & 5 \\ 4 & 6 & 2 & 4 \end{bmatrix} \end{array} \quad \begin{array}{c} \begin{array}{cccc} M & P & S & V \end{array} \\ \begin{bmatrix} 5 & 4 & 2 & 1 \\ 4 & 3 & 5 & 5 \end{bmatrix} \end{array} \quad \begin{array}{c} \begin{array}{cccc} M & P & S & V \end{array} \\ \begin{bmatrix} 3 & 6 & 3 & 4 \\ 3 & 2 & 6 & 5 \end{bmatrix} \end{array}$$

Feeder School 1 **Feeder School 2** **Feeder School 3**

Find the matrix showing the total distribution of boys and girls by class.

4. In a track meet, 5 points is awarded for a first place, 3 for second place, and 1 for third place. Mudville and Tarrytown competed. The matrices show their numbers of first, second, and third place finishes.

	1st	2nd	3rd
Mudville	6	3	7

	1st	2nd	3rd
Tarrytown	4	7	3

Which team gained the most points?

5. Six boys competed in a diving contest. The table shows the scores received on each of the four dives. The second table shows the degree of difficulty for each dive performed. The score for a dive is the points awarded times the degree of difficulty. Determine the order of finish in this competition.

	I	II	III	IV
Alan	8	9	9	7
Bob	7	8	6	5
Claude	7	7	5	9
Dan	6	9	7	6
Edgar	8	8	7	8
Fred	9	6	7	7

	A	B	C	D	E	F
I	1.4	1.5	1.3	1.6	1.5	1.2
II	1.6	1.4	1.7	1.4	1.6	1.5
III	1.5	1.9	1.9	2.0	1.9	1.7
IV	1.7	2.1	1.4	1.8	2.0	1.9

Degree of Difficulty

Write a matrix for each network.

6.

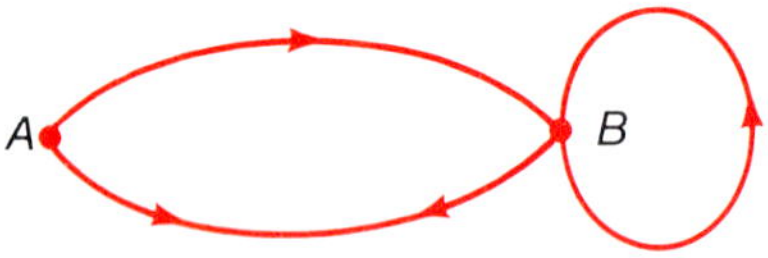

7.

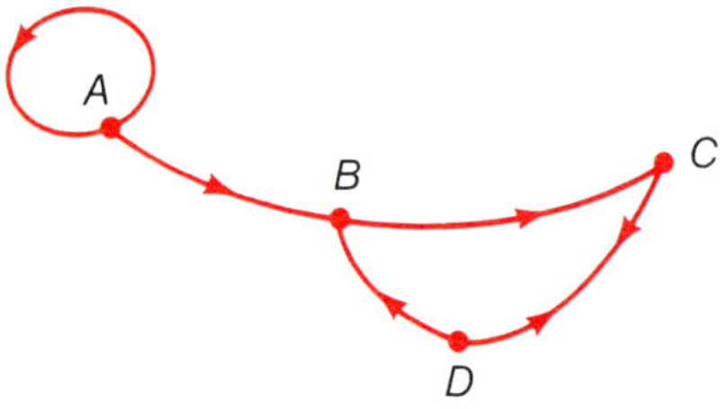

8.

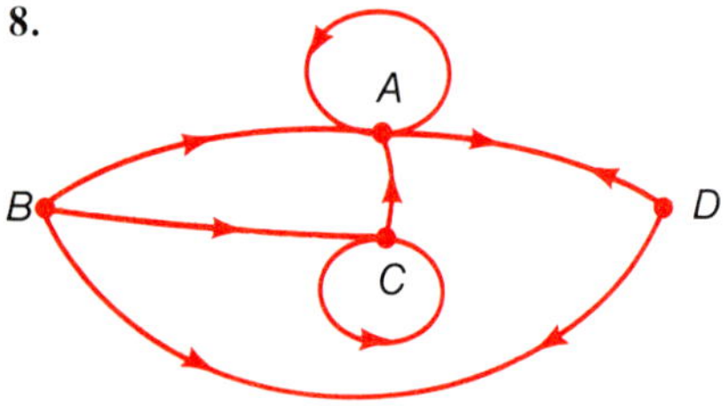

9.

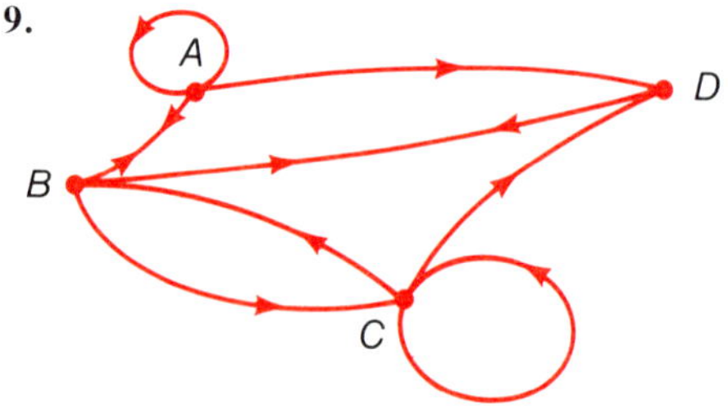

B

10. Draw a directed network that is represented by the matrix at the right.

$$\begin{bmatrix} 2 & 2 & 1 \\ 1 & 0 & 1 \\ 0 & 1 & 1 \end{bmatrix}$$

11. For the directed network at the right, the matrix $\begin{bmatrix} 0 & 1 \\ 1 & 1 \end{bmatrix}$ gives the "one-stage journeys." A "two-stage journey" is one such as A to B to A or A to B to B. List all the two-stage journeys for the network.

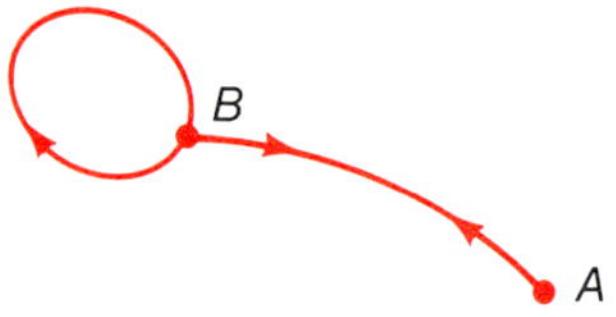

12. Write the two-stage journey matrix for Exercise 11.

13. Find $\begin{bmatrix} 0 & 1 \\ 1 & 1 \end{bmatrix} \times \begin{bmatrix} 0 & 1 \\ 1 & 1 \end{bmatrix}$ and compare it with the result in Exercise 12.

14. Find the two-stage journeys for each directed network in Exercises 6, 7, 8 and 9.

C

15. Explain with illustrations why multiplying single-stage journey matrices produces two-stage journey matrices.

REVIEW CAPSULE FOR SECTION 13–7

Evaluate each determinant. *(Pages 134–137)*

1. $\begin{vmatrix} 1 & -1 \\ 1 & 1 \end{vmatrix}$
2. $\begin{vmatrix} 3 & 0 \\ -5 & -2 \end{vmatrix}$
3. $\begin{vmatrix} -2 & 4 \\ 5 & -6 \end{vmatrix}$
4. $\begin{vmatrix} d & e \\ f & g \end{vmatrix}$
5. $\begin{vmatrix} a_1 & b_1 \\ a_2 & b_2 \end{vmatrix}$

13–7 Inverse of a Matrix

Recall that the identity element for multiplication of real numbers is 1, because the product of any real number and 1 is the real number.

$$-\sqrt{3} \cdot 1 = -\sqrt{3} \qquad 2^{10} \cdot 1 = 2^{10} \qquad 0.3\overline{3} \cdot 1 = 0.3\overline{3}$$

Similarly, the identity matrix for multiplication of matrices is the matrix I, such that $AI = IA = A$. For example, since

$$\begin{bmatrix} a_1 & b_1 \\ a_2 & b_2 \end{bmatrix}\begin{bmatrix} 1 & 0 \\ 0 & 1 \end{bmatrix} = \begin{bmatrix} a_1 & b_1 \\ a_2 & b_2 \end{bmatrix},$$

the matrix $\begin{bmatrix} \mathbf{1} & \mathbf{0} \\ \mathbf{0} & \mathbf{1} \end{bmatrix}$ is the *identity matrix* under multiplication for a 2×2 matrix.

Definition

> The **identity matrix,** I_n, of order n, is the square matrix whose elements in the main diagonal, from upper left to lower right, are 1's. All other elements are 0's.

A matrix having the same number of rows and columns is a **square matrix.**

EXAMPLE 1 Write the identity matrix under multiplication for a fourth order matrix.

Solution:

$$\begin{bmatrix} \mathbf{1} & \mathbf{0} & \mathbf{0} & \mathbf{0} \\ \mathbf{0} & \mathbf{1} & \mathbf{0} & \mathbf{0} \\ \mathbf{0} & \mathbf{0} & \mathbf{1} & \mathbf{0} \\ \mathbf{0} & \mathbf{0} & \mathbf{0} & \mathbf{1} \end{bmatrix}$$

← **Square matrix**

← **Main diagonal**

Some square matrices have multiplicative inverses. For example, for a nonzero matrix of order 2,

let $A = \begin{bmatrix} a_1 & b_1 \\ a_2 & b_2 \end{bmatrix}$ and $A^{-1} = \begin{bmatrix} x_1 & y_1 \\ x_2 & y_2 \end{bmatrix}$.

Then $AA^{-1} = I$, or $\begin{bmatrix} a_1 & b_1 \\ a_2 & b_2 \end{bmatrix}\begin{bmatrix} x_1 & y_1 \\ x_2 & y_2 \end{bmatrix} = \begin{bmatrix} 1 & 0 \\ 0 & 1 \end{bmatrix}$. **1**

Thus, $\begin{bmatrix} a_1x_1 + b_1x_2 & a_1y_1 + b_1y_2 \\ a_2x_1 + b_2x_2 & a_2y_1 + b_2y_2 \end{bmatrix} = \begin{bmatrix} 1 & 0 \\ 0 & 1 \end{bmatrix}$. **2**

From matrix equation **2**, you can write two systems.

$$\begin{cases} a_1x_1 + b_1x_2 = 1 \\ a_2x_1 + b_2x_2 = 0 \end{cases} \qquad \begin{cases} a_1y_1 + b_1y_2 = 0 \\ a_2y_1 + b_2y_2 = 1 \end{cases}$$

Now solve the systems for x_1, y_1, x_2, and y_2.

$$x_1 = \frac{b_2}{a_1b_2 - a_2b_1} \qquad y_1 = \frac{-b_1}{a_1b_2 - a_2b_1}$$

$$x_2 = \frac{-a_2}{a_1b_2 - a_2b_1} \qquad y_2 = \frac{a_1}{a_1b_2 - a_2b_1}$$

In all cases, the denominators are identical. Further, the expression,

$$a_1b_2 - a_2b_1 = \begin{vmatrix} a_1 & b_1 \\ a_2 & b_2 \end{vmatrix}.$$ ← **Determinant of A**

Thus, $A^{-1} = \begin{bmatrix} x_1 & y_1 \\ x_2 & y_2 \end{bmatrix} = \frac{1}{\begin{vmatrix} a_1 & b_1 \\ a_2 & b_2 \end{vmatrix}} \times \begin{bmatrix} b_2 & -b_1 \\ -a_2 & a_1 \end{bmatrix}$

Theorem 13-1

> For any 2×2 matrix A, where $A = \begin{bmatrix} a_1 & b_1 \\ a_2 & b_2 \end{bmatrix}$,
>
> $$A^{-1} = \frac{1}{\begin{vmatrix} a_1 & b_1 \\ a_2 & b_2 \end{vmatrix}}\begin{bmatrix} b_2 & -b_1 \\ -a_2 & a_1 \end{bmatrix}, \text{ where } \begin{vmatrix} a_1 & b_1 \\ a_2 & b_2 \end{vmatrix} \neq 0.$$

NOTE: The determinant of A cannot equal 0. If it does, the inverse does not exist.

EXAMPLE 2 **a.** For $A = \begin{bmatrix} 5 & 2 \\ -2 & 1 \end{bmatrix}$, find A^{-1}.

b. Show that $AA^{-1} = I$.

Solutions: **a.** Find the determinant of A. Then find A^{-1}.

$$\begin{vmatrix} 5 & 2 \\ -2 & 1 \end{vmatrix} = 5 - (-4) = 9$$

$$A^{-1} = \frac{1}{9}\begin{bmatrix} 1 & -2 \\ 2 & 5 \end{bmatrix} = \begin{bmatrix} \frac{1}{9} & -\frac{2}{9} \\ \frac{2}{9} & \frac{5}{9} \end{bmatrix}$$

b. $AA^{-1} = \begin{bmatrix} 5 & 2 \\ -2 & 1 \end{bmatrix}\begin{bmatrix} \frac{1}{9} & -\frac{2}{9} \\ \frac{2}{9} & \frac{5}{9} \end{bmatrix} = \begin{bmatrix} \frac{5}{9}+\frac{4}{9} & -\frac{10}{9}+\frac{10}{9} \\ -\frac{2}{9}+\frac{2}{9} & \frac{4}{9}+\frac{5}{9} \end{bmatrix} = \begin{bmatrix} \mathbf{1} & \mathbf{0} \\ \mathbf{0} & \mathbf{1} \end{bmatrix}$

CLASSROOM EXERCISES

Write the determinant for each matrix.

1. $\begin{bmatrix} 2 & 1 \\ -3 & 2 \end{bmatrix}$ **2.** $\begin{bmatrix} -1 & 4 \\ 3 & -6 \end{bmatrix}$ **3.** $\begin{bmatrix} -9 & 3 \\ 14 & -3 \end{bmatrix}$ **4.** $\begin{bmatrix} -2 & -1 \\ 9 & 4 \end{bmatrix}$

For each matrix A, find A^{-1}.

5. $A = \begin{bmatrix} 0 & 1 \\ -6 & 2 \end{bmatrix}$ **6.** $A = \begin{bmatrix} 3 & -1 \\ 4 & 2 \end{bmatrix}$ **7.** $A = \begin{bmatrix} 2 & 1 \\ -1 & 3 \end{bmatrix}$ **8.** $A = \begin{bmatrix} 3 & 0 \\ 4 & -5 \end{bmatrix}$

WRITTEN EXERCISES

A Write the indicated identity matrix under multiplication.

1. I_2 **2.** I_4 **3.** I_5 **4.** I_3 **5.** I_6

For each matrix A, find A^{-1}. Then show that $AA^{-1} = I$. If A^{-1} does not exist, write N.

6. $A = \begin{bmatrix} 0 & 1 \\ 1 & 1 \end{bmatrix}$ **7.** $A = \begin{bmatrix} -2 & 1 \\ 0 & 6 \end{bmatrix}$ **8.** $A = \begin{bmatrix} 1 & 5 \\ 1 & -2 \end{bmatrix}$ **9.** $A = \begin{bmatrix} 2 & 2 \\ 1 & 1 \end{bmatrix}$

10. $A = \begin{bmatrix} 5 & -3 \\ 2 & -4 \end{bmatrix}$ **11.** $A = \begin{bmatrix} 1 & 2 \\ 2 & 1 \end{bmatrix}$ **12.** $A = \begin{bmatrix} \frac{1}{2} & -1 \\ \frac{2}{3} & \frac{3}{4} \end{bmatrix}$ **13.** $A = \begin{bmatrix} 1 & -\frac{1}{4} \\ \frac{2}{5} & \frac{3}{10} \end{bmatrix}$

B Solve for the matrix X.

Example: $\begin{bmatrix} 2 & 0 \\ 1 & 3 \end{bmatrix}X = \begin{bmatrix} -2 & 6 \\ -1 & 9 \end{bmatrix}$

Solution: To get X alone on the left side of the equation, **"left"**-multiply each side by the inverse of $\begin{bmatrix} 2 & 0 \\ 1 & 3 \end{bmatrix}$.

[1] Find the inverse of A where $A = \begin{bmatrix} 2 & 0 \\ 1 & 3 \end{bmatrix}$. $A^{-1} = \begin{bmatrix} \frac{1}{2} & 0 \\ -\frac{1}{6} & \frac{1}{3} \end{bmatrix}$

[2] $\begin{bmatrix} \frac{1}{2} & 0 \\ -\frac{1}{6} & \frac{1}{3} \end{bmatrix}\begin{bmatrix} 2 & 0 \\ 1 & 3 \end{bmatrix}X = \begin{bmatrix} \frac{1}{2} & 0 \\ -\frac{1}{6} & \frac{1}{3} \end{bmatrix}\begin{bmatrix} -2 & 6 \\ -1 & 9 \end{bmatrix}$

$\begin{bmatrix} 1 & 0 \\ 0 & 1 \end{bmatrix}X = \begin{bmatrix} -1 & 3 \\ 0 & 2 \end{bmatrix}$ and $X = \begin{bmatrix} \mathbf{-1} & \mathbf{3} \\ \mathbf{0} & \mathbf{2} \end{bmatrix}$

14. $\begin{bmatrix} 2 & -1 \\ 3 & 5 \end{bmatrix} X = \begin{bmatrix} 4 & 3 \\ 1 & -2 \end{bmatrix}$

15. $\begin{bmatrix} 2 & 0 \\ 1 & 3 \end{bmatrix} X = \begin{bmatrix} -2 & 6 \\ -1 & 9 \end{bmatrix}$

16. $\begin{bmatrix} 2 & 3 \\ 1 & 2 \end{bmatrix} X = \begin{bmatrix} 1 & 2 \\ 3 & 7 \end{bmatrix}$

17. $\begin{bmatrix} 4 & 3 \\ -2 & -2 \end{bmatrix} X = \begin{bmatrix} 2 & 3 \\ 1 & 2 \end{bmatrix}$

Solve for x and y,

18. $\begin{bmatrix} 3 & 1 \\ 2 & 1 \end{bmatrix}\begin{bmatrix} x \\ y \end{bmatrix} = \begin{bmatrix} 8 \\ 5 \end{bmatrix}$

19. $\begin{bmatrix} 3 & 2 \\ 2 & -3 \end{bmatrix}\begin{bmatrix} x \\ y \end{bmatrix} = \begin{bmatrix} 1 \\ 18 \end{bmatrix}$

20. $\begin{bmatrix} 5 & -1 \\ 2 & 3 \end{bmatrix}\begin{bmatrix} x \\ y \end{bmatrix} = \begin{bmatrix} -29 \\ 2 \end{bmatrix}$

21. $\begin{bmatrix} 3 & 4 \\ -1 & 2 \end{bmatrix}\begin{bmatrix} x \\ y \end{bmatrix} = \begin{bmatrix} 10 \\ 0 \end{bmatrix}$

22. $\begin{bmatrix} 3 & 6 \\ -1 & -2 \end{bmatrix}\begin{bmatrix} x \\ y \end{bmatrix} = \begin{bmatrix} 5 \\ -4 \end{bmatrix}$

23. $\begin{bmatrix} 2 & 5 \\ 1 & -2 \end{bmatrix}\begin{bmatrix} x \\ y \end{bmatrix} = \begin{bmatrix} 6 \\ -5 \end{bmatrix}$

C Solve each system. Use the method of inverse matrices.

24. $\begin{cases} 5x - 2y = 19 \\ 7x + 3y = 15 \end{cases}$

25. $\begin{cases} x - y = 5 \\ 2x + 3y = 10 \end{cases}$

26. $\begin{cases} -x + y = 1 \\ -2x + 3y = 0 \end{cases}$

27. $\begin{cases} 3x - 7y = 5 \\ 2x + y = 9 \end{cases}$

Review

In Exercises 1–4, use the matrices A, B, C, and D to perform the indicated operations. If the operation is not possible, write *NP*. *(Section 13–5)*

$A = \begin{bmatrix} 3 & -2 \\ 0 & 4 \end{bmatrix}$ $B = \begin{bmatrix} 6 \\ 2 \\ -7 \end{bmatrix}$ $C = \begin{bmatrix} 1 & 2 & 4 \\ -2 & 3 & 6 \\ 0 & 4 & -5 \end{bmatrix}$ $D = \begin{bmatrix} 4 & 3 & 6 \\ -1 & -2 & 2 \\ 0 & 5 & -1 \end{bmatrix}$

1. $C + D$ 2. $3A$ 3. CB 4. DB 5. CD 6. $D - C$

7. In June, an appliance company shipped 30 dishwashers, 20 refrigerators, and 15 freezers to Albany. In July, they shipped 45 dishwashers, 10 refrigerators, and 25 freezers. In August, the company shipped 15 dishwashers, 30 refrigerators and 10 freezers. Represent this information in three 1×6 matrixes. Find the total number of each item shipped. *(Section 13–6).*

For each matrix A, find A^{-1}. Then show that $AA^{-1} = I$. If A^{-1} does not exist, write *N*. *(Section 13–7)*

8. $A = \begin{bmatrix} 1 & 2 \\ -3 & 1 \end{bmatrix}$

9. $A = \begin{bmatrix} 3 & 2 \\ 9 & 6 \end{bmatrix}$

10. $A = \begin{bmatrix} -3 & 4 \\ 0 & 2 \end{bmatrix}$

11. $A = \begin{bmatrix} \frac{1}{2} & 3 \\ \frac{1}{3} & -4 \end{bmatrix}$

COMPUTER APPLICATIONS

BASIC: MATRICES

In BASIC, as in algebra, a matrix is named by a capital letter. To refer to a particular element of the matrix, two subscripts are needed. As was done for elements of sequences, the subscripts are placed in parentheses. Thus, $A_{i,j}$ becomes A(I,J) in BASIC. Likewise, $M_{3,4}$ is M(3,4).

Problem: *Write a program to multiply, if possible, two matrices.*

```
100 REM   THIS PROGRAM MULTIPLIES MATRIX A
110 REM   (DIMENSIONS M X N) TIMES MATRIX
120 REM   B (P X Q) TO GIVE MATRIX C (M X Q).
130 REM
140 DIM A(10,10), B(10,10), C(10,10)
150 READ M, N
160 PRINT "MATRIX A ="
170 FOR I = 1 TO M
180 FOR J = 1 TO N
190 READ A(I,J)
200 PRINT A(I,J);TAB(W*J/N);
210 NEXT J
220 PRINT
230 NEXT I
240 PRINT
250 READ P, Q
260 PRINT "MATRIX B ="
270 FOR I = 1 TO P
280 FOR J = 1 TO Q
290 READ B(I,J)
300 PRINT B(I,J);TAB(W*J/Q);
310 NEXT J
320 PRINT
330 NEXT I
340 PRINT
350 IF N = P THEN 390
360 PRINT "CANNOT MULTIPLY"
370 PRINT
380 GOTO 150
390 PRINT "A X B ="
400 FOR I = 1 TO M
410 FOR J = 1 TO Q
420 LET C(I,J) = 0
430 FOR K = 1 TO N
440 LET C(I,J) = C(I,J)+A(I,K)*B(K,J)
450 NEXT K
460 PRINT C(I,J);TAB(W*J/Q);
470 NEXT J
480 PRINT
490 NEXT I
500 PRINT
510 GOTO 150
520 DATA 2,2
530 DATA 3,-4,2,1
540 DATA 2,2
550 DATA 1,0,-1,3
560 DATA 2,4
570 DATA 2,0,-1,1,5,-2,0,3
580 DATA 4,2
590 DATA 0,1,3,0,-1,-1,4,2
600 DATA 2,3
610 DATA 10,-.5,1,0,4,8
620 DATA 2,3
630 DATA 0,1,0,4,-5,1.5
640 END
```

Replace each W in this program with the number of spaces in a line of your screen or printer.

Analysis

Statement 150: All previous programs in this text have used the INPUT statement. But for longer lists of data, such as the elements of matrices, the READ command is better. In this approach the values of variables are listed in DATA statements in the program itself. When the program on page 505 is executed, the computer reaches statement 150. It selects the first two numbers in the DATA (2 and 2) as the values of M and N. Then each time a READ is executed, the next item(s) in the DATA are used. The program continues until there is no more data, at which point an OUT OF DATA message is printed, followed by READY.

Statements 160–240: This section reads and prints matrix A. Statements 180–210 form a "nested loop" inside the I-loop (statements 170–230). I represents, in turn, each row of matrix A. J represents the columns. So in the DATA (statement 530, for example) the elements of A are listed in row order. The purpose of the TAB formula in statements 200, 300 and 460 is to space the elements of each matrix in columns across the width of the screen or printer.

Statements 250–340: These statements read and print matrix B.

Statements 350–380: If N ≠ P, then the matrices cannot be multiplied and a message is printed to this effect.

Statements 390–510: Here matrices A and B are multiplied and the product C printed. A K-loop (statements 430–450) is nested inside the J-loop (statements 410–470), which in turn is nested in the I-loop (statements 400–490). The function of the K-loop is to cause the appropriate elements to be multiplied: the first element in a row of A times the first element in a column of B, and so on.

EXERCISES

A For the program on page 505, write the DATA statements (statements 520 and following) needed to input the matrices below.

1. $\begin{bmatrix} 2 & 0 \\ -1 & 3 \end{bmatrix}\begin{bmatrix} 1 & 5 \\ 0 & -6 \end{bmatrix}$ **2.** $\begin{bmatrix} 5 & 1 \\ 0 & -3 \end{bmatrix}\begin{bmatrix} -7 & 2 & 1 \\ 6 & 0 & -6 \end{bmatrix}$ **3.** $\begin{bmatrix} 16 & -4 & 0 \end{bmatrix}\begin{bmatrix} 0 & 2 \\ -3 & 5 \end{bmatrix}$

Write a BASIC program for each problem.

4. Given matrices A and B, print A and B. Then print their sum or, if they do not have the same dimensions, print CANNOT BE ADDED.

5. Given a matrix and a scalar, print the matrix and the product of the matrix and the scalar.

6. Given n, print the $n \times n$ multiplicative identity matrix I.

7. Given a 2×2 matrix, print the matrix and its determinant.

8. Given a 2×2 matrix, print the matrix and its multiplicative inverse, if it has one.

Chapter Summary

IMPORTANT TERMS

Augmented matrix (p. 490)
Dimensions of a matrix (p. 490)
Directed network (p. 498)
Identity matrix (p. 501)
Matrix (p. 490)
Network (p. 498)
Ordered triple (p. 485)
Scalar (p. 493)
Square matrix (p. 501)
xy plane (p. 482)
xz plane (p. 482)
yz plane (p. 482)

IMPORTANT IDEAS

1. The graph of a linear equation in three variables is a plane in three–space, and every plane in three–space is the graph of a linear equation in three variables.

2. The solution set of a system of three linear equations in three variables contains:
   a. One ordered triple when the three planes intersect in one point.
   b. Infinitely many ordered triples when the three planes intersect in one line or when they coincide.
   c. No ordered triples when
      (i) the three planes are parallel.
      (ii) two planes are coincident and parallel to the third plane.
      (iii) the three planes intersect in three parallel lines.
      (iv) two planes are parallel and intersect the third plane in two parallel lines.

3. A system of three linear equations in three variables is:
   a. Consistent and independent when the solution set contains one ordered triple.
   b. Consistent and dependent when the solution set contains many ordered triples.
   c. Inconsistent when the solution set is ϕ, the empty set.

4. **Row Operations for a Matrix**

   [1] Multiply or divide any row by a nonzero real number.

   [2] Add the corresponding numbers in any two rows. This sum can replace either row, but not both. (This corresponds to adding equations.)

   [3] Interchange any two rows.

5. **Matrix method for solving a system of linear equations in two or more variables:**

   [1] Write the augmented matrix of the system.

   [2] Use row operations to obtain an equivalent matrix in triangular form.

   [3] Determine the solution to the original system from the equivalent system represented by the new matrix.

6. To add matrices of the *same dimension,* add their corresponding elements.

7. To multiply a scalar k by a matrix A, multiply each element of A by k.

8. In order to multiply two matrices, the number of elements in the rows of the first matrix must equal the number of elements in the columns of the second matrix. Matrix multiplication is "row by column" multiplication (see pages 493–497).

9. For any 2×2 matrix A, where $A = \begin{bmatrix} a_1 & b_1 \\ a_2 & b_2 \end{bmatrix}$,

   $$A^{-1} \text{ (inverse of } A) = \frac{1}{\begin{vmatrix} a_1 & b_1 \\ a_2 & b_2 \end{vmatrix}} \begin{bmatrix} b_2 & -b_1 \\ -a_2 & a_1 \end{bmatrix}, \text{ where } \begin{vmatrix} a_1 & b_1 \\ a_2 & b_2 \end{vmatrix} \neq 0.$$

Chapter Objectives and Review

Objective: *To graph points and planes in three-space (Section 13–1)*

Graph each point in three-space.

1. $A(4, 1, -7)$ **2.** $B(2, 0, 3)$ **3.** $C(0, -2, 6)$ **4.** $D(-1, 3, 2)$

5. $P(4, 4, -4)$ **6.** $Q(-4, 4, 4)$ **7.** $R(1, \frac{1}{2}, 3)$ **8.** $S(2, 1, 0)$

Find the x, y, and z intercepts of the plane represented by each equation. Then graph the equation.

9. $x + y - z = 3$ **10.** $2x - y + z = 7$ **11.** $3x - 4y + 2z = 2$

Objective: *To solve systems of linear equations in three variables by the elimination method (Section 13–2)*

Solve and check.

12. $\begin{cases} 2x - y + 3z = 6 \\ -x + 4y - 2z = -1 \\ 3x - 7y + 3z = -1 \end{cases}$ **13.** $\begin{cases} x - 5y + 6z = 9 \\ 2x + 5y - 5z = 1 \\ -x + 3y + 3z = 0 \end{cases}$ **14.** $\begin{cases} x + y - z = 9 \\ 2x - y + z = -3 \\ y + z = 1 \end{cases}$

Objective: *To use a system of equations in three variables to solve problems (Section 13–3)*

15. The sum of the digits of a three-digit number is 13. The sum of the first two digits is 1 more than half the third digit. If 100 is subtracted from the number, the first two digits of the resulting number will be the same. Find the original number.

Objective: *To use the augmented matrix method to solve systems of equations in three variables (Section 13–4)*

Solve by using the augmented matrix method.

16. $\begin{cases} -x + 2y - z = 5 \\ 3x - 3y + z = 2 \\ -2x + y - 2z = 7 \end{cases}$

17. $\begin{cases} 2x - 4y + 3z = 3 \\ x - y - 3z = 0 \\ -x + 3y + z = 4 \end{cases}$

18. $\begin{cases} 2x + 10y - 3z = 3 \\ -x - 4y + z = 0 \\ 3x + 10y + 4z = 3 \end{cases}$

Objective: *To perform addition, multiplication, and scalar multiplication on matrices (Section 13–5)*

Use the matrices *A*, *B*, *C*, and *D* to perform the indicated operations. If the operation is not possible, write *NP*.

$A = \begin{bmatrix} 1 & 3 \\ -2 & 0 \end{bmatrix}$ $B = \begin{bmatrix} 3 & 0 \\ 4 & 1 \\ -1 & 6 \end{bmatrix}$ $C = \begin{bmatrix} 3 & 2 & 0 \\ 1 & 4 & 5 \end{bmatrix}$ $D = \begin{bmatrix} 0 & 6 \\ 3 & 1 \\ 5 & -2 \end{bmatrix}$

19. $5A$ **20.** $A + B$ **21.** CB **22.** $B + (2D)$

Objective: *To use matrices to solve problems (Section 13–6)*

23. The Stitch-In-Time Company manufactures three sewing kits. The small kit contains 6 spools of thread, 1 package of needles, 2 thimbles, and 1 pair of scissors. The large kit contains 12 spools of thread, 3 packages of needles, 2 thimbles, 2 pairs of scissors, and 2 needle-threaders. The traveler's kit contains 4 spools of thread, 1 package of needles, 1 pair of scissors, 1 needle-threader, and 4 safety pins.

   a. Represent this information in three 1×6 matrices. Add to determine the number of each item needed.

   b. The materials in the kits have the following costs: spool of thread – \$0.12; package of needles – \$0.18; thimble – \$0.06; pair of scissors – \$1.02; needle-threader – \$0.08; safety pin – \$0.03. Find the value of the materials in each kit.

Objective: *To find the inverse of a 2×2 matrix (Section 13–7)*

For each *A*, find A^{-1}. If A^{-1} does not exist, write *N*.

24. $A = \begin{bmatrix} 1 & 3 \\ -2 & 5 \end{bmatrix}$ **25.** $A = \begin{bmatrix} 4 & 1 \\ 0 & 8 \end{bmatrix}$ **26.** $A = \begin{bmatrix} 6 & 3 \\ -2 & -1 \end{bmatrix}$ **27.** $A = \begin{bmatrix} 7 & 1 \\ 3 & 0 \end{bmatrix}$

Chapter Test

Name the coordinate axis on which each point lies.

1. $T(0, 3, 0)$
2. $M(0, 0, 5)$
3. $N(\frac{1}{3}, 0, 0)$

Graph each point in three-space.

4. $R(3, -2, 1)$
5. $S(-3, -1, 7)$
6. $W(4, 1, 2)$

In Exercises 7 and 8, use the equation $x - 3y + z = 5$.

7. Write the x, y, and z intercepts for the graph of the equation.

8. Graph the plane represented by the equation.

9. Solve and check: $\begin{cases} 6x + y - z = 0 \\ 3x - 2y - 2z = -5 \\ -x - y + 2z = 10 \end{cases}$

10. The sum of the digits of a three-digit number is 10. The third digit is one more than the second. If the third digit is multiplied by 3 and the result added to the second digit, the first digit is obtained. Find the number.

Use the matrices A, B, and C to perform the operations in Exercises 11–14. If an operation is not possible, write *NP*.

$$A = \begin{bmatrix} 1 & 0 \\ 0 & -2 \end{bmatrix} \qquad B = \begin{bmatrix} 1 & 3 & 1 \\ 0 & 2 & 5 \end{bmatrix} \qquad C = \begin{bmatrix} 2 & 3 \\ -3 & 5 \end{bmatrix}$$

11. $A + C$
12. $B + C$
13. $3B$
14. CB

15. To renovate a house, the owner is employing 4 carpenters, 5 painters, and 2 plumbers. Carpenters are paid \$12 per hour, painters \$9 per hour, and plumbers \$15 per hour.

    a. Represent this information in two matrices.

    b. Use matrix multiplication to determine the total cost of each hour of renovation work.

CHAPTER 14 Probability

14–1 Permutations

You can arrange the four letters **a, b, c,** and **d** in the following ways.

abcd,	**abdc,**	**acbd,**	**acdb,**	**adbc,**	**adcb,**
bacd,	**badc,**	**bdac,**	**bdca,**	**bcad,**	**bcda,**
cabd,	**cadb,**	**cbad,**	**cbda,**	**cdab,**	**cdba,**
dabc,	**dacb,**	**dbac,**	**dbca,**	**dcab,**	**dcba**

Each is an ordered arrangement. That is, **abcd** is not the same as **acbd** or **abdc** or **adbc.** An ordered arrangement is a *permutation.*

Definition

> A **permutation** of a set of elements is an arrangement of a specified number of those elements in a definite order.

In ordering the four letters, there are four choices for the first position. For each choice for the first position, there are three choices for the second position. For each choice for the second position, there are two choices for the third position. For each choice for the third position, there is one choice for the fourth position. Thus, there are $4 \cdot 3 \cdot 2 \cdot 1$ or 24 permutations for the four letters **a, b, c,** and **d.**

Position	1	2	3	4	
Number of Choices	4	3	2	1	→ $\mathbf{4 \cdot 3 \cdot 2 \cdot 1 = 24}$

The number of permutations for the five letters **a, b, c, d,** and **e,** is

$$\mathbf{5 \cdot 4 \cdot 3 \cdot 2 \cdot 1 = 120.}$$

The number of permutations for n distinct objects is

$$n(n-1)(n-2)(n-3)(n-4) \cdot \cdots \cdot (3)\,(2)\,(1).$$

Symbolically, the number of permutations of n elements taken n at a time (this means you have n elements and n positions) is ${}_nP_n$. Thus,

$${}_nP_n = n(n-1)(n-2)(n-3)(n-4) \cdot \cdots \cdot (3)\,(2)\,(1).$$

Note that ${}_nP_n$ is the product of the integer n and *all* positive integers less than n. The symbol $n!$ (read "*n* **factorial**") represents this product.

$${}_nP_n = n!$$

The Fundamental Principle of Counting can help you to find the number of permutations of n objects taken r at a time.

Postulate 14–1

> **Fundamental Principle of Counting**
>
> If one event can occur in m different ways, and if, after it occurs or at the same time, a second event can occur in n ways, then together the two events can occur in $m \cdot n$ different ways.

Now consider n elements and r positions, where $r < n$.

EXAMPLE 1 In how many ways can 5 of 8 people be seated in a row of 5 vacant chairs?

Solution:

Position	1	2	3	4	5
Number of Choices	8	7	6	5	4

In all, there are $8 \cdot 7 \cdot 6 \cdot 5 \cdot 4$ permutations when 8 elements are taken 5 at a time. The symbol ${}_8P_5$ represents this.

${}_8P_5 = 8 \cdot 7 \cdot 6 \cdot 5 \cdot 4 = \mathbf{6720}$

Symbolically, the number of permutations of n elements taken r at a time is ${}_nP_r$. To generalize, match each position with the number of choices for that position.

Position	1	2	3	4	$\cdots$	r
Number of Choices	n	$n-1$	$n-2$	$n-3$	$\cdots$	$n-(r-1)$

Thus, there are $n-(r-1)$ or $n-r+1$ choices for the rth position.

$${}_nP_r = n(n-1)(n-2)\cdots(n-r+1)$$ ← **Fundamental Principle of Counting**

This formula can be simplified by multiplying by $\frac{(n-r)!}{(n-r)!}$, which is 1.

$${}_nP_r = \frac{n(n-1)(n-2)\cdots(n-r+1)(n-r)!}{(n-r)!}$$

$${}_nP_r = \frac{n(n-1)(n-2)\cdots(n-r+1)(n-r)(n-r-1)\cdots(3)(2)(1)}{(n-r)!}$$

$${}_nP_r = \frac{n!}{(n-r)!}$$

EXAMPLE 2 You have 10 different books to choose from. How many permutations can be formed on a shelf that holds 6 books?

Solution:

$${}_{10}P_6 = \frac{10!}{(10-6)!} \quad \leftarrow \quad n = 10,\ r = 6$$

$$= \frac{10 \cdot 9 \cdot 8 \cdot 7 \cdot 6 \cdot 5 \cdot 4 \cdot 3 \cdot 2 \cdot 1}{4 \cdot 3 \cdot 2 \cdot 1}$$

$$= 10 \cdot 9 \cdot 8 \cdot 7 \cdot 6 \cdot 5 = \mathbf{151{,}200}$$

If n is replaced by r in ${}_nP_r$, $\quad {}_nP_r = \frac{n!}{(n-r)!}$

$${}_rP_r = \frac{r!}{(r-r)!} = \frac{r!}{0!}$$

Since ${}_rP_r = r!$, then $\frac{r!}{0!} = r!$ Thus, $0!$ must equal 1 for this to be true.

Definition

The number **0! is equal to 1.**

CLASSROOM EXERCISES

1. Write the 6 permutations of the letters **x, y,** and **z.**
2. Use the letters **a, b, c, d** to write the 12 permutations of two letters each.

In Exercises 3–7, evaluate each expression.

3. $7!$ 4. $8! - 4!$ 5. $\frac{5!}{3!}$ 6. $\frac{10!}{(10-2)!}$ 7. $\frac{n!}{(n-1)!}$ when $n = 9$

WRITTEN EXERCISES

A In Exercises 1–6, evaluate each expression.

1. ${}_3P_3$ 2. ${}_7P_2$ 3. $\frac{{}_6P_2}{{}_3P_2}$ 4. $\frac{{}_5P_5}{{}_5P_2}$ 5. $\frac{{}_6P_6}{{}_3P_3}$ 6. $\frac{{}_4P_2 \cdot {}_5P_3}{{}_3P_2}$

7. In how many ways can 6 out of 8 people be seated in 6 empty seats?
8. In how many ways can 6 girls occupy a row of 6 seats?
9. In how many ways can 9 different books be arranged on a shelf?
10. In how many ways can three cars be parked in 10 parking spaces?
11. Each of 10 basketball players can play any of the 5 positions on the team. In how many ways may the coach arrange the team?
12. How many permutations of the letters in the word FACTOR can be formed by using all of the letters?
13. How many permutations of 4 letters each can be formed from the letters of the word NUMBERS?
14. How many positive four-digit integers can be formed from the digits **1, 2, 3,** and **4,** when no digit is repeated in a numeral?
15. In how many ways can 6 out of 10 candles be arranged in a row on a mantle?
16. In how many ways can you choose 2 ice cream flavors out of 26 possible flavors?
17. How many positive four-digit integers can be formed by using the digits **1, 2, 3, 4, 5, 6,** and **7** when no digit is repeated in any numeral?
18. In Exercise 17, how many are possible when repetitions are allowed?
19. How many license-plate numbers consisting of two letters followed by four digits are possible when repetition of numbers and letters is allowed?
20. In Exercise 19, how many license-plate numbers are possible when repetition of numbers and letters is not allowed?

B

21. How many positive, five-digit, even integers can be formed with the digits **2, 3, 5, 7,** and **9** when no repetitions are allowed?
22. How many different positive integers greater than 2000 can be formed by using **1, 2, 3,** and **4** when each is used just once in every numeral?

23. How many permutations of the letters in the word REPLICA end in a vowel?

24. How many positive integers less than 100,000 contain the digits **3, 4, 5,** and **6** in the order given?

25. Find the number of batting orders for nine baseball players when the best two batters must be in the second and third positions.

26. A committee of seven persons line up for a photograph with the chairman in the middle. In how many ways can the committee line up?

27. In a family picture, a mother and father stand at the ends of a row with their three children between them. In how many ways can the family be arranged for the photograph?

28. In how many ways can four boys and four girls be lined up for a photograph if the boys must alternate with the girls?

29. In how many ways can six people be lined up for a photograph if Ralph must stand immediately to the left of Bernice?

30. In how many ways can six people be lined up for a photograph if Edgar and Elaine must stand next to each other?

31. In how many ways can nine people be lined up for a photograph if Clark and Billy must not stand next to each other?

C

32. Show that ${}_5P_3 = 5({}_4P_2)$.

33. Show that ${}_7P_4 = 7({}_6P_3)$.

34. Show that ${}_nP_r = n({}_{n-1}P_{r-1})$

35. If ${}_nP_4 = 4({}_nP_3)$, find n.

36. If ${}_nP_5 = 42({}_nP_3)$, find n.

37. If ${}_nP_5 = 6({}_{n-1}P_4)$, find n.

38. If ${}_nP_6 = 12{}_nP_4$, find n.

39. If ${}_nP_8 = 8{}_{n-1}P_6$, find n.

14–2 Permutations of Elements Not All Different

If the three E's in the word EERIE are distinguished from one another as E_1, E_2, and E_3, then the number of permutations of the 5 letters is ${}_5P_5 = 5!$, or 120. However, E_1, E_2, and E_3 cannot really be distinguished. That is,

$$E_1E_2E_3RI, \quad E_2E_1E_3RI, \quad E_2E_3E_1RI, \text{ and so on,}$$

are the same.

Using the Fundamental Principle of Counting there are $5 \cdot 4$ or 20 different arrangements if you just consider R and I in the 5 positions.

R I _ _ _	I R _ _ _	_ R I _ _	_ I R _ _
R _ I _ _	I _ R _ _	_ R _ I _	_ I _ R _
R _ _ I _	I _ _ R _	_ R _ _ I	_ I _ _ R
R _ _ _ I	I _ _ _ R		
_ _ R I _	_ _ I R _	_ _ _ R I	_ _ _ I R
_ _ R _ I	_ _ I _ R		

For *each* one of the distinct arrangements on page 515, you can rearrange the three E's in 3! ways. Thus, if T represents the number of distinct arrangements of the 5 letters, then

$$3!T = {}_5P_5$$

$$T = \frac{{}_5P_5}{3!}$$

$$T = \frac{5!}{3!} = \frac{120}{6} = \mathbf{20}$$

In general, the number of distinct permutations of n things, where p of these are alike, is

$$T = \frac{n!}{p!},\ p \leq n.$$

Further, if there is more than one set of *like* elements, say p things alike, q things alike, r things alike, and so on, then

$$T = \frac{n!}{p!q!r! \cdots},\ p \leq n,\ q \leq n,\ r \leq n.$$

EXAMPLE How many different permutations can you make with the letters in CONNECTICUT?

Solution: $T = \frac{11!}{3!2!2!}$ ← **$n = 11$ letters**; ← **$p = 3$ C's, $q = 2$ N's, $r = 2$ T's**

$$= \frac{11 \cdot 10 \cdot 9 \cdot 8 \cdot 7 \cdot 6 \cdot 5 \cdot 4 \cdot 3 \cdot 2 \cdot 1}{3 \cdot 2 \cdot 1 \cdot 2 \cdot 1 \cdot 2 \cdot 1}$$

$$= 11 \cdot 10 \cdot 9 \cdot 8 \cdot 7 \cdot 6 \cdot 5 = \mathbf{1,663,200}$$

CLASSROOM EXERCISES

1. Write the 24 different arrangements of the letters of the word A_1REA_2, counting A_1 and A_2 as different letters.
2. How many different permutations can you make with the letters in AREA?

In Exercises 3–8, evaluate each expression.

3. $\frac{8!}{3!}$
4. $\frac{6!}{2!}$
5. $\frac{12!}{3!3!}$
6. $\frac{{}_{10}P_{10}}{4!}$
7. $\frac{{}_{11}P_{11}}{3!2!2!}$
8. $\frac{21!}{19!3!}$

WRITTEN EXERCISES

A In Exercises 1–8, find the number of permutations that can be made from the letters of the given word.

1. COLORADO
2. MICHIGAN
3. SATELLITE
4. MASSACHUSETTS
5. OHIO
6. LONDON
7. CANADA
8. NEPTUNE
9. How many permutations can be made from the digits in the numeral 121221?

10. How many distinct ways can 10 balloons be arranged in a row when 3 are red, 2 are blue, 2 are green, and the rest are yellow?

11. At a safety program, a traffic instructor wishes to display 9 traffic signs. In how many different ways can she display the signs in a row when there are four identical stop signs, two identical direction signs, and three identical speed limit signs?

12. Each of 7 cheerleaders stands in a line holding a pompon. There are 3 white pompons, 2 blue ones, and the rest are yellow. In how many distinguishable ways can the pompons be arranged?

13. A father loses a bet to his 6 children and agrees to distribute the coins in his pockets among them. He has 3 dimes, 2 quarters, and one half dollar. How many ways can he distribute the coins equally?

14. In how many ways can 3 blue pennants, 2 red pennants, and a yellow pennant be arranged on a staff?

In Exercises 15–20, use the numeral **1213.**

15. How many permutations can be made from these digits?

16. How many permutations can be made when the first digit must be 2?

17. How many permutations can be made when the number formed must be greater than 2000?

18. How many permutations can be made when the number formed must be an even number?

19. How many permutations can be made when the number formed must be an odd number?

20. How many permutations can be made when the number formed must be a multiple of 3?

B

21. In how many different ways can 12 books be arranged on a shelf when 5 of of them are identical books with blue covers, 4 are identical books with green covers, and 3 are identical books with black covers?

22. In Exercise 21, how many arrangements are there when books of the same color must be grouped together?

23. How many different line-ups, each consisting of 2 forwards, 2 guards, and one center, can be made from a basketball squad of 5 forwards, 4 guards, and 2 centers?

At the head table at a banquet are seated two senators, two governors, and three mayors. In Exercises 24–26, find the number of ways in which these seven people can be seated under the conditions described.

24. A governor is at each end.

25. A mayor is at each end and the senators are in consecutive seats.

26. The mayors and governors are in alternating seats.

C An arrangement of elements in a circle is called a **circular permutation.** If n elements are arranged in a circle, any one of the n elements may be selected as the first position. Then the other $n - 1$ elements can be arranged in the remaining positions. Hence, the total number of distinguishable circular permutations is

$$\frac{n!}{n} = \frac{n(n-1)!}{n} = (n-1)!$$

27. Find the number of ways 6 people can be seated at a round table.

28. In how many ways can a family of 4 be seated at a round table?

29. Ten members of a football team huddle in a circle around the quarterback to hear the next play. In how many ways can they arrange themselves?

30. In how many ways can 5 different keys be arranged on a key ring? (HINT: Objects such as keys can be flipped over to produce arrangements identical to those originally counted by $(n-1)!$. For such objects, there are only half the usual number of circular permutations, $\frac{(n-1)!}{2}$.)

31. In how many ways can 8 objects be fastened to a circular piece of wire?

32. In how many ways can 4 boys and 4 girls be seated alternately in a circle?

33. In how many ways could King Arthur and six of his court sit at the Round Table when one of the chairs was the throne chair for King Arthur, and there were six other chairs?

34. How many v ays are there to seat ten persons at a round table when two of these people must sit next to each other?

A five-card hand is dealt from a bridge deck of 52 cards. Find the number of ways in Exercises 35–38 in which each event can occur.

35. The hand has exactly one pair of sevens.

36. The hand has exactly one pair of sevens or a pair of aces.

37. The hand has at least one pair of sevens.

38. The hand has at least one pair.

14–3 Combinations

The number of permutations of the letters **a, b,** and **c** is 6.

abc **acb** **bac** **bca** **cab** **cba**

However, the permutations listed above consist of only one combination. That is, **abc, acb, bac,** and so on are the same *combination,* because each contains the elements **a, b,** and **c.**

Definition

> An r-element subset of a set having n elements is a **combination** of the n elements taken r at a time.

For the three letters **a**, **b**, and **c**, there is only one combination but 3! permutations. For four different letters there is one combination but 4! permutations. In general, there are $r!$ permutations for each combination of r objects.

Choose a set of r objects from n objects. The number of ways in which this can be done is represented by ${}_nC_r$ or $\binom{n}{r}$ (read: "the combination of n objects taken r at a time"). The r objects can be arranged in $r!$ ways. From the Fundamental Principal of Counting, the number of arrangements (permutations) of all n objects taken r at a time is

$${}_nC_r \cdot r! = {}_nP_r \quad \longleftarrow \textbf{Solve for } {}_nC_r.$$

Thus,
$${}_nC_r = \frac{{}_nP_r}{r!} = \frac{n!}{r!(n-r)!}$$

EXAMPLE In how many ways can a committee of 3 be chosen from 30 students?

Solution:

$${}_{30}C_3 \text{ or } \binom{30}{3} = \frac{30!}{3!(30-3)!} \quad \longleftarrow n = 30,\ r = 3$$

$$= \frac{30 \cdot 29 \cdot 28 \cdot \cdots \cdot 2 \cdot 1}{3 \cdot 2 \cdot 1 \cdot 27 \cdot 26 \cdot \cdots \cdot 1}$$

$$= \frac{30 \cdot 29 \cdot 28}{3 \cdot 2 \cdot 1} = \mathbf{4060}$$

CLASSROOM EXERCISES

1. List all the two-person committees that can be formed from three students named Fran, José, and Elmer.

In Exercises 2–11, evaluate each expression.

2. ${}_3C_2$
3. ${}_3C_1$
4. ${}_5C_3$
5. ${}_8C_2$
6. ${}_9C_9$
7. $\binom{8}{5}$
8. $\binom{5}{2}$
9. ${}_4C_2 \cdot {}_6C_3$
10. $\binom{7}{2}\binom{7}{5}$
11. $\dfrac{{}_6C_1 \cdot {}_7C_2}{{}_8C_3}$

WRITTEN EXERCISES

A Determine the number of ways that each selection can be made.

1. Two student officers from a group of 100
2. A group of 4 from a group of 7
3. Five boys for a basketball team from a physical education class of 30 boys
4. Three cheerleaders from a squad of 9
5. One winner out of 45 contestants
6. Six ice cream flavors out of 11 flavors
7. Five students out of 30 students
8. Two sports cars out of 14 models of sports cars

9. You may choose five classmates out of a group of 15 to accompany you on a camping trip. In how many ways could you choose the group of five?

10. In how many ways can you draw 56 marbles from a bag that contains 59 marbles?

11. A sample of ten light bulbs is to be chosen from a set of 100 light bulbs. In how many ways can the sample be selected?

12. In how many ways can you draw two cards from a deck of 52 cards?

13. In how many ways can you choose a committee of five out of a group of nine people?

14. How many lines are determined by ten points, no three of which are on the same line?

15. In how many ways can you choose a committee of five girls and seven boys from a group of ten girls and eleven boys? (HINT: Find the number of combinations for girls, then for boys. Then find the product.)

B

16. In how many ways can you choose a group of four or more from a group of seven?

17. How many committees of four can be chosen from twelve students? How many of these will include a given student? How many will exclude a given student?

18. Show that ${}_nC_n = 1$ and that ${}_nC_1 = n$.

19. Show that ${}_nC_r = {}_nC_{n-r}$.

The committees in Exercises 20–23 are being formed from a group of twelve Republicans and ten Democrats.

20. How many different committees of six can be formed if at least four are Democrats?

21. How many different committees of six can be formed if no more than four are Democrats?

22. How many different committees of six can be formed if at least three are Republicans?

23. How many different committees of six can be formed if no more than three are Republicans?

24. In how many ways can a person invite one or more of four friends to dinner?

25. From a group of six persons, how many committees of three can be formed if two of the six people cannot be on the same committee?

26. How many diagonals can be drawn in an octagon?

On page 467, the Binomial Theorem shows how you can expand $(a + b)^n$ to $n + 1$ terms. The theorem can also be written using combinations.

$$(a + b)^n = {}_nC_0\, a^n + {}_nC_1\, a^{n-1} b + {}_nC_2\, a^{n-2} b^2 + {}_nC_3\, a^{n-3} b^3 + \cdots + {}_nC_n\, b^n$$
$$= \binom{n}{0}a^n + \binom{n}{1}a^{n-1} b + \binom{n}{2}a^{n-2} b^2 + \binom{n}{3}a^{n-3} b^3 + \cdots + \binom{n}{n}b^n$$

The **rth term** of $(a+b)^n$ can be written as

$${}_nC_{r-1}\, a^{n-r+1}\, b^{r-1} \quad \text{or as} \quad \binom{n}{r-1} a^{n-r+1}\, b^{r-1}.$$

In Exercises 27–30, use combinations to expand the given binomial.

27. $(a+b)^4$
28. $(a+b)^6$
29. $(a-b)^5$
30. $(a+2b)^3$

In Exercises 31–42, use combinations to write the indicated term of the given binomial.

31. $(a+b)^7$; 4th term
32. $(2a+b)^6$; 5th term
33. $(x-y)^8$; 7th term
34. $(2x-3y)^9$; 6th term
35. $(3-\frac{1}{9}x)^4$; 2nd term
36. $(x^2-y^2)^5$; 3rd term
37. $(x+\frac{4}{3})^7$; last term
38. $(2x-4z)^{12}$; 4th term
39. $(1+0.01)^8$; 6th term
40. $(2a-b)^6$; middle term
41. $\left(\frac{y}{4}+\frac{4}{y}\right)^{10}$; last term
42. $\left(\frac{y}{4}+\frac{4}{y}\right)^{10}$; 4th term

Review

1. Evaluate ${}_7P_3$ and ${}_5P_5$. *(Section 14–1)*
2. Seven men and women are to line up for flu shots. In how many ways can the line form? *(Section 14–1)*
3. In how many ways can 4 out of 7 people be seated in 4 empty seats? *(Section 14–1)*
4. How many different permutations can you make with the letters in OOSTERBAAN? *(Section 14–2)*
5. In a group of seven balloons, 2 are red, 3 are blue, and the rest are white. In how many distinct ways can they be arranged in a row? *(Section 14–2)*
6. Evaluate ${}_7C_3$ and ${}_5C_5$. *(Section 14–3)*
7. How many committees of four members can be formed from a group of 9 possible members? *(Section 14–3)*

REVIEW CAPSULE FOR SECTION 14–4

List all of the numbers 1 through 6 that correspond to the following possible outcomes of a single roll of a die.

1. An even number
2. An odd number
3. A prime number
4. A composite (non–prime) number
5. A perfect square
6. A multiple of 3

14–4 Probability

When you toss a coin, there are 2 possible outcomes: Heads or Tails.

The chances of tossing a head are one out of two, or $\frac{1}{2}$. Similarly, the chances of tossing a tail are one out of two, or $\frac{1}{2}$.

The ratio of the number of successful outcomes to the number of possible outcomes is the **probability** that an outcome will occur.

The probability that you will get neither a head nor a tail when you toss a coin is zero.

Postulate 14–2

> **The probability of an outcome that cannot occur is 0.**

The probability that you will get either a head or a tail is one.

Postulate 14–3

> **The probability of an outcome that must occur is 1.**

A probability is a number p, where $0 \le p \le 1$.

EXAMPLE 1 In rolling a die, what is the probability of getting a 5?

Solution: 1, 2, 3, 4, 5, 6 ← **Possible outcomes**

5 ← **Successful outcomes**

$p = \frac{1}{6}$ ← $\frac{\textbf{Number of successful outcomes}}{\textbf{Number of possible outcomes}}$

In probability, the set of possible outcomes is called a **sample space.** Each outcome or element in a sample space is a **sample point.**

EXAMPLE 2 In rolling a die, what is the probability of getting an even number?

Solution: 1, 2, 3, 4, 5, 6 ← **Possible outcomes**

2, 4, 6 ← **Successful outcomes**

$p = \frac{3}{6} = \frac{1}{2}$ ← $\frac{\textbf{Number of successful outcomes}}{\textbf{Number of possible outcomes}}$

In Example 2, each of the successful outcomes, 2, 4, or 6, is equally likely, and they cannot occur together. Because of this, you can find the probability of each successful outcome and then add these 3 probabilities to find the probability of rolling either a 2, a 4, or a 6.

Probability of Rolling a 2	4	6	Probability of Rolling a 2, 4, or 6.
$\frac{1}{6}$	$\frac{1}{6}$	$\frac{1}{6}$	$\frac{1}{6}+\frac{1}{6}+\frac{1}{6}=\frac{3}{6}=\frac{1}{2}$

Postulate 14–4

> If two or more outcomes cannot occur simultaneously, the probability of any one of the outcomes occurring is the sum of the probabilities that each will occur.

In Examples 1 and 2, the set of successful outcomes is a subset of the sample space. The subset is called an **event.**

In Example 1, the event $\{5\}$ is a **simple event** because it is only one element of the sample space. In Example 2, the event $\{2, 4, 6\}$ is the union of three simple events, $\{2\}$, $\{4\}$, and $\{6\}$.

CLASSROOM EXERCISES

Exercises 1–6 refer to one toss of a die. Find each probability.

1. Getting an odd number
2. Getting a composite number
3. Getting a number greater than 1
4. Getting a number less than 4
5. Getting a prime number
6. Getting a multiple of 3

WRITTEN EXERCISES

A Exercises 1–7 refer to one toss of a die.

1. What is the probability of getting a 4?
2. What is the probability of getting a 1 or a 6?
3. What is the probability of getting a number less than 5?
4. What is the probability of getting a prime number?
5. What is the probability of getting a composite number?
6. What is the probability of getting 1, 2, 3, 4, 5, or 6?
7. What is the probability of getting an 8?
8. A bridge deck consists of two black suits (spades and clubs) and two red suits (hearts and diamonds). Each suit contains an ace, a king, a queen, a jack and one each of cards numbered 2, 3, 4, 5, 6, 7, 8, 9, 10. What is the probability of drawing an ace in a single draw from a bridge deck of cards?

In Exercises 9–12, one card is drawn from a bridge deck.

9. What is the probability of drawing a spade?
10. What is the probability of drawing a black card?

11. What is the probability of drawing a red 7?

12. What is the probability of drawing a 2 or a 3?

13. Ten slips of paper numbered from 6 to 15 are placed in a hat. One slip of paper is drawn. What is the probability that the number on the slip drawn is a prime?

14. A bag contains four red cubes and five black cubes. What is the probability that a cube drawn at random is blue?

15. In Exercise 14, what is the probability that the cube drawn is red or black?

In Exercises 16–19, a bag contains two red tokens, three white tokens, and four black tokens. What is the probability of each event?

16. Choosing a red token

17. Choosing a white token

18. Choosing a black token

19. Choosing a red, white, or black token

In Exercises 20–25, a bag contains three red cubes and six white cubes. What is the probability of each event?

20. Choosing a red cube

21. Choosing a white cube

22. Choosing neither a red nor a white cube

23. Choosing either a white cube or a red cube

24. Choosing a cube that is *not* white

25. Choosing a cube that is *not* red

B

26. C and D are distinct points on $\overline{AB}$ and C is the midpoint of $\overline{AB}$. What is the probability that D is the midpoint of $\overline{AB}$?

27. P is a point on the bisector of angle ABC. What is the probability that P is equidistant from the sides of the angle?

28. Five books labeled **a, b, c, d, e** are placed on a shelf at random. What is the probability that they will be placed in the order **b, d, e, c, a?**

In Exercises 29–31, a pair of dice is tossed once.

29. What is the total number of possible outcomes in the sample space?

30. What is the probability of getting a 6?

31. What is the probability of getting a 2 or a 6?

For large sample spaces, it is often convenient to use combinations to find the number of elements.

Example: A box contains four red cubes and seven white cubes. Three cubes are drawn. Find the probability that the three cubes are red.

Solution: The three cubes can be chosen in ${}_{11}C_3$ or 165 ways.

Three red cubes can be chosen in ${}_4C_3$ or 3 ways.

$$P = \frac{{}_4C_3}{{}_{11}C_3} = \frac{3}{165} = \frac{1}{55}$$

32. For the box and cubes of the Example above, find the probability that the three cubes are white.

33. For the box and cubes of the Example above, find the probability that the three cubes are of the same color.

34. Find the probability of dealing a five-card hand that contains the ace, king, queen, jack, and ten of spades from a bridge deck of 52 cards.

35. Find the probability of dealing a five-card hand that contains the ace, king, queen, jack, and ten of the same suit (spades, clubs, hearts, or diamonds) from a bridge deck of 52 cards.

In Exercises 36–40, a committee of six members is chosen from a group of twelve Republicans and ten Democrats. Find the indicated probability.

36. That there will be no Republicans

37. That there will be no Democrats

38. That there will be at least three Republicans

39. That there will be no more than two Democrats

40. That there will be exactly three Republicans and three Democrats

REVIEW CAPSULE FOR SECTION 14–5

A bridge deck contains 52 cards. Given that $A =$ the set of 13 hearts, $B =$ the set of 4 kings, $C =$ the set of 4 queens, and $D =$ the set of 12 face cards, list the elements of each set below. *(Pages 6–9)*

1. $B \cup C$ **2.** $A \cup B$ **3.** $B \cup D$ **4.** $A \cap B$ **5.** $B \cap C$

14–5 Mutually Exclusive Events

For a sample space with n equally likely outcomes, the probability of a simple event is $\frac{1}{n}$. Suppose that in the same sample space, an event E has m outcomes. Then, $p(E) = \frac{m}{n}$.

EXAMPLE 1 Six disks of the same size are labeled as shown in the figure and placed in a box. What is the probability that a disk drawn at random is labeled A?

Solution:

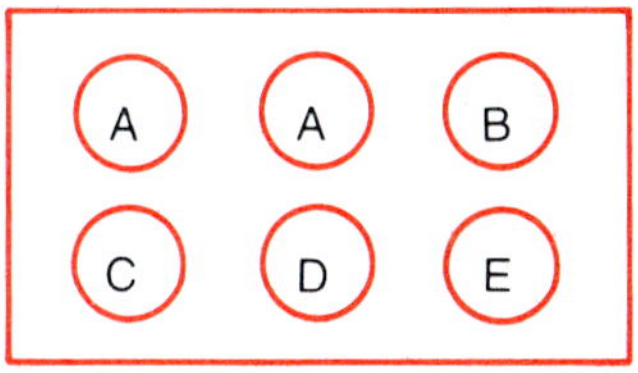

Since two disks are labeled **A**, refer to them as $\mathbf{A}_1$ and $\mathbf{A}_2$. The sample space has 6 equally likely outcomes.

Sample space: {$\mathbf{A}_1$, $\mathbf{A}_2$, **B**, **C**, **D**, **E**}

Event E (drawing an **A**): {$\mathbf{A}_1$, $\mathbf{A}_2$} ← **Successful outcomes**

$p(E) = \frac{2}{6} = \frac{1}{3}$ ← $\frac{\textbf{Number of successful outcomes}}{\textbf{Number of possible outcomes}}$

Events are **mutually exclusive** if they have no elements in common. Thus, in Example 2, events E_1 and E_2 are mutually exclusive.

EXAMPLE 2 Five red tokens labeled **1, 2, 3, 4, 5** and five white tokens labeled **3, 4, 5, 6, and 7** are placed in a box. What is the probability of drawing an odd-numbered red token or an even-numbered white token?

Solution

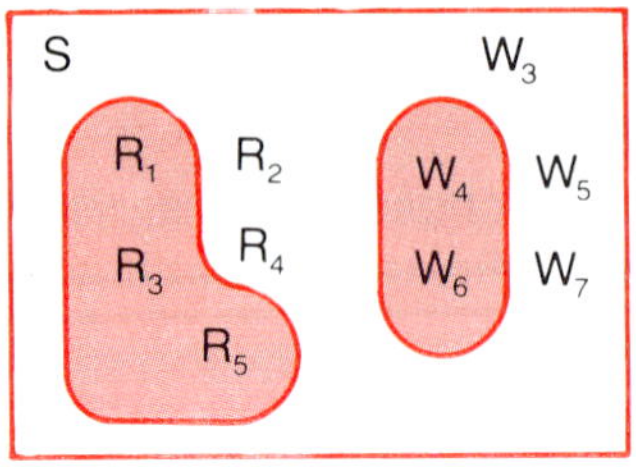

Sample space:
$\{\mathbf{R_1, R_2, R_3, R_4, R_5, W_3, W_4, W_5, W_6, W_7}\}$

Event E_1 (drawing an odd-numbered red token):
$\{\mathbf{R_1, R_3, R_5}\}$

Event E_2 (drawing an even-numbered white token):
$\{\mathbf{W_4, W_6}\}$

$E_1 \cup E_2$: $\{\mathbf{R_1, R_3, R_5, W_4, W_6}\}$

$p(E_1 \cup E_2) = \frac{5}{10} = \mathbf{\frac{1}{2}}$

In Example 3, the events E_1 and E_2 have common elements. Thus, they are not mutually exclusive events.

EXAMPLE 3 What is the probability of drawing a red token or an even-numbered token from the sample space in Example 2?

Solution

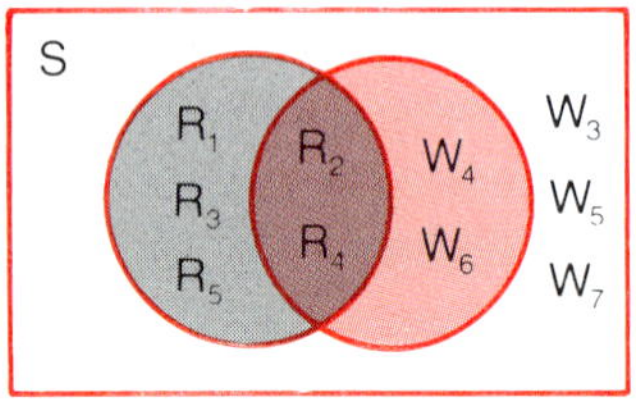

Sample space:
$\{\mathbf{R_1, R_2, R_3, R_4, R_5, W_3, W_4, W_5, W_6, W_7}\}$

E_1 (drawing a red token): $\{\mathbf{R_1, R_2, R_3, R_4, R_5}\}$

E_2 (drawing an even-numbered token): $\{\mathbf{R_2, R_4, W_4, W_6}\}$

$E_1 \cup E_2$:
$\{\mathbf{R_1, R_2, R_3, R_4, R_5, W_4, W_6}\}$ ← **Common elements are listed only once.**

$p(E_1 \cup E_2) = \mathbf{\frac{7}{10}}$

In Example 3, note that

$$p(E_1) = \frac{5}{10},\ p(E_2) = \frac{4}{10}, \text{ and } p(E_1 \cap E_2) = \frac{2}{10}.$$

Thus, $p(E_1 \cup E_2) = \left(\frac{5}{10} + \frac{4}{10}\right) - \frac{2}{10} = \mathbf{\frac{7}{10}}$.

This suggests the following theorem.

Theorem 14–1

General Addition Rule

For any two events E_1 and E_2 for a given sample space,

$$\mathbf{p(E_1 \cup E_2) = p(E_1) + p(E_2) - p(E_1 \cap E_2).}$$

For mutually exclusive events, $p(E_1 \cap E_2) = 0$. Thus, for these events,

$$\mathbf{p(E_1 \cup E_2) = p(E_1) + p(E_2).}$$

EXAMPLE 4 There are 52 cards in a bridge deck. One card is drawn. What is the probability of drawing a club or a queen?

Solution: Sample space: {52 cards}

E_1 (drawing a club): {Thirteen clubs}

E_2 (drawing a queen): {Four queens}

$E_1 \cap E_2$: {Queen of clubs}

$p(E_1 \cup E_2) = p(E_1) + p(E_2) - p(E_1 \cap E_2)$ ← **Theorem 14–1**

$$p(E_1 \cup E_2) = \frac{13}{52} + \frac{4}{52} - \frac{1}{52}$$

$$= \frac{16}{52}, \text{ or } \frac{4}{13}$$

CLASSROOM EXERCISES

One card is drawn from a bridge deck of 52 cards. Identity each of the following pairs of events as *mutually exclusive* or *not mutually exclusive.*

1. E_1: a spade; E_2: a diamond
2. E_1: a queen; E_2: a heart
3. E_1: a face card; E_2: a card number that is at least **2**

Six slips of red paper are numbered **2, 3, 4, 5, 8, 9** and four slips of blue paper are numbered **3, 6, 8, 10** and placed in a hat. The slips are shuffled and one slip is drawn at random. Identity each of the following pairs of events as *mutually exclusive* or *not mutually exclusive.*

4. E_1: a blue slip; E_2: a red slip
5. E_1: a red slip; E_2: a slip number that is **8** or **10**
6. E_1: a blue slip; E_2: a slip number that is a prime number
7. Indicate the sample space for Exercises 1–3.
8. Indicate the sample space for Exercises 4–6.

WRITTEN EXERCISES

A

1. Jim has five nickels, four dimes and six pennies in his pocket. What is the probability that a coin he draws at random is a penny?
2. A committee consists of three women and two men. A committee representative is chosen at random. What is the probability of choosing a woman?

In Exercises 3–11, one card is drawn from a bridge deck of 52 cards. Find each probability.

3. Drawing a spade <u>or</u> a diamond
4. Drawing a seven <u>or</u> a ten
5. Drawing a club <u>or</u> a spade
6. Drawing an ace <u>or</u> a two

7. Drawing a numbered or a face card

8. Drawing a ten or a fifteen

9. Drawing a queen or a heart

10. Drawing a six or a club

11. Drawing a face or a spade

In Exercises 12–14, refer to one spin of a spinner (shown at the right). Find each probability.

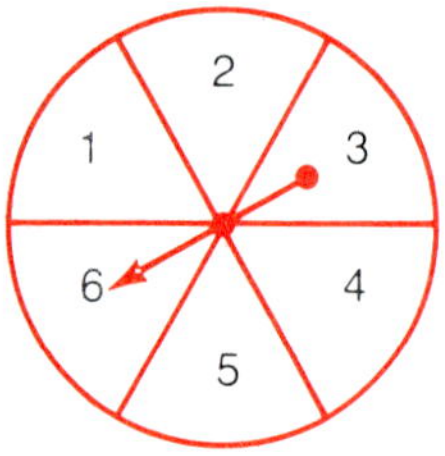

12. Getting a multiple of 3

13. Getting an even number

14. Getting an even number or a multiple of 3

15. A single die is thrown. What is the probability of getting an even number or a number less than 4?

16. A collection of books contains 5 French books, 6 physics books, 4 mathematics books, and 7 English books. You choose one book at random. What is the probability that it is a French or a mathematics book?

In Exercises 17–20, one license plate is randomly selected from fifty plates numbered from 1 to 50. Find the probabilities.

17. The number begins with a 2 or a 3

18. The number is divisible by 3 or 6

19. The number ends in a 2 or a 3

20. The number ends in a 0 or a 5

B

21. A pair of dice is thrown and the outcomes are recorded. Determine the probability of throwing a double or getting a sum of 7.

22. In Exercise 16, what is the probability that the book chosen is either a physics, a mathematics, or an English book?

23. Mr. Montorez was born October n. What is the probability that n is a multiple of 3 or that n satisfies the equation

$$n^3 - 10n^2 + 27n - 18 = 0?$$

Each of the possible three-digit numbers less than 300 that can be written using only the digits 1, 2, 3 is written on a blank card. In Exercises 24–29, one card is selected at random.

24. List the sample space.

25. Determine the probability that the number ends in 1 or that it is divisible by 3.

26. Determine the probability that the tens digit of the number is 2 or that the number ends in 3.

27. Determine the probability that the units digit is a multiple of the hundreds digit or that the number is odd.

28. Determine the probability that the number is an even number or that the number is divisible by 6.

29. Determine the probability that two digits are alike or that three digits are alike.

14–6 Conditional Probability

Suppose that one die is tossed. There are 6 equally likely outcomes in the sample space. Let E_2 = the top face shows a 3. Then

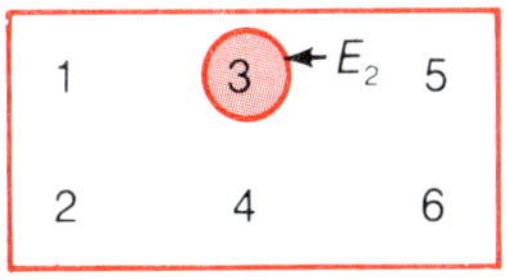

$$p(E_2) = \frac{1}{6}.$$

Now suppose that one die is tossed and you are told that the top face shows an odd number. This subset of the original sample space contains 3 equally likely outcomes. Let E_1 = top face shows an odd number. Then,

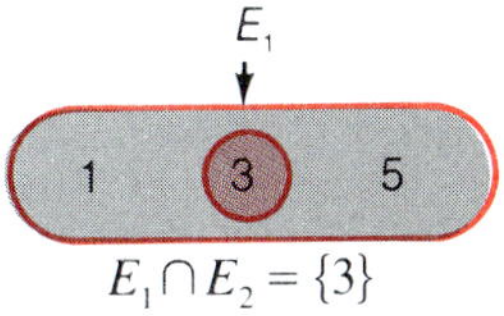

$$p(E_2 \text{ occurs given that } E_1 \text{ occurs}) = \frac{1}{3}.$$

In terms of the original sample space,

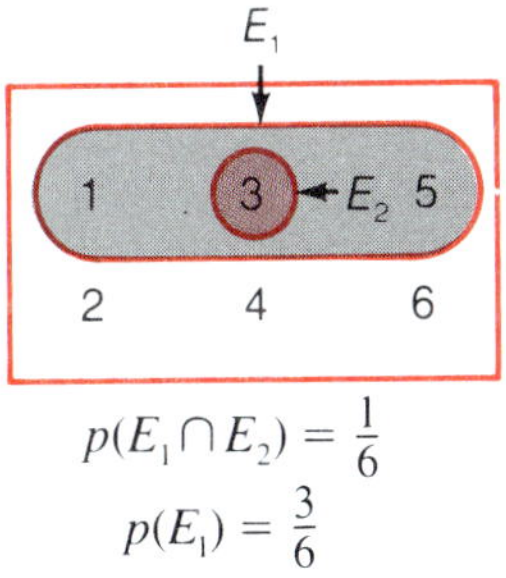

$$p\begin{pmatrix} E_2 \text{ occurs given} \\ \text{that } E_1 \text{ occurs} \end{pmatrix} = \frac{p(E_1 \cap E_2)}{p(E_1)} = \frac{\frac{1}{6}}{\frac{3}{6}} = \frac{1}{3}$$

In probability, it is usual to write $p(E_2$ occurs given that E_1 occurs) as $p(E_2|E_1)$. The probability, $p(E_2|E_1)$, is called the **conditional probability of E_2 given E_1.** Note that $p(E_2) \neq p(E_2|E_1)$.

Definition

> The **conditional probability** of an event E_2 when E_1 has occurred is the ratio of the probability of $E_1 \cap E_2$ to the probability of E_1.
>
> $$p(E_2 \mid E_1) = \frac{p(E_1 \cap E_2)}{p(E_1)},\ p(E_1) \neq 0.$$

EXAMPLE 1 A pair of dice are tossed. What is the probability that their sum is greater than 8 given that the numbers on the top faces match?

Solution: The sample space contains the 36 ordered pairs shown at the right.

Let E_2 = sum is greater than 8.
Let E_1 = numbers on the top faces match.

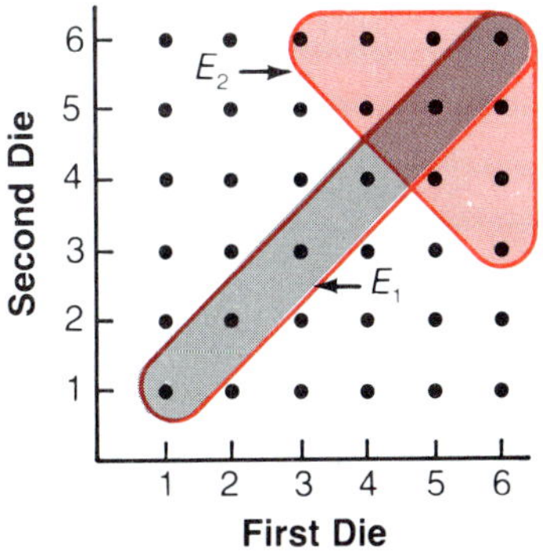

Then, $p(E_1) = \frac{6}{36}$ and $p(E_1 \cap E_2) = \frac{2}{36}$.

$$p(E_2 \mid E_1) = \frac{p(E_1 \cap E_2)}{p(E_1)} = \frac{\frac{2}{36}}{\frac{6}{36}} = \frac{2}{6}, \text{ or } \frac{1}{3}.$$

Since $p(E_1) \neq 0$ in the definition of conditional probability, you can multiply both sides of the equation to obtain this theorem.

Theorem 14–2

$$p(E_1 \cap E_2) = p(E_1) \cdot p(E_2 | E_1)$$

EXAMPLE 2 A box contains 6 disks as shown. One disk is drawn but not replaced. Then a second is drawn. What is the probability that the first disk is labeled **A** and the second disk is labeled **B**?

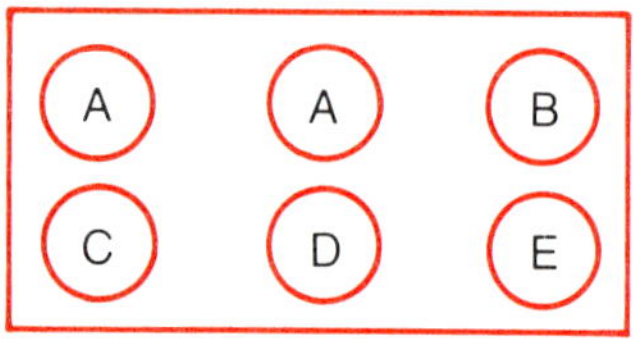

Solution: On the first draw, one of 6 disks can be chosen. On the second draw, one of 5 disks can be chosen. Thus, there are 6×5, or 30 elements in the sample space.

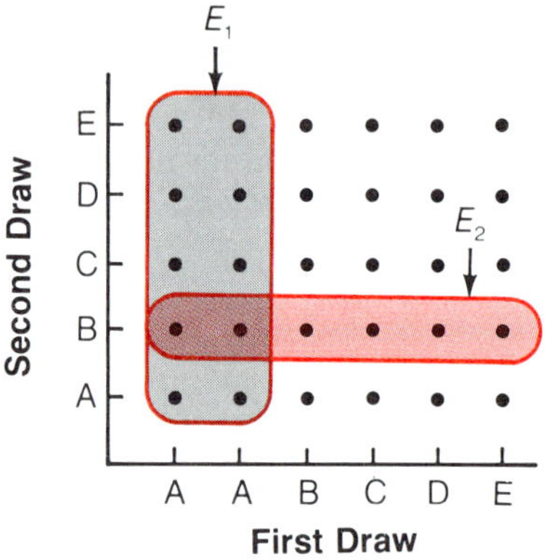

Let E_1 = first disk drawn is **A**.

Let E_2 = second disk drawn is **B**.

Then $p(E_1) = \frac{10}{30} = \frac{1}{3}$.

$$p(E_2 | E_1) = \frac{2}{10} = \frac{1}{5}$$

$$p(E_1 \cap E_2) = p(E_1) \cdot p(E_2 | E_1)$$

$$= \frac{1}{3} \cdot \frac{1}{5} = \mathbf{\frac{1}{15}}$$

In Example 2, suppose that the first disk is replaced. Then the second draw does *not depend* on the first.

EXAMPLE 3 In Example 2, the first disk is replaced after it is drawn. Then a second disk is drawn. Find the probability that the first disk is labeled **A** and the second disk is labeled **B**.

Solution: On each draw, one of 6 disks can be chosen. Thus, there are 6×6, or 36 elements in the sample space.
Let E_1 and E_2 be the events defined in Example 2.

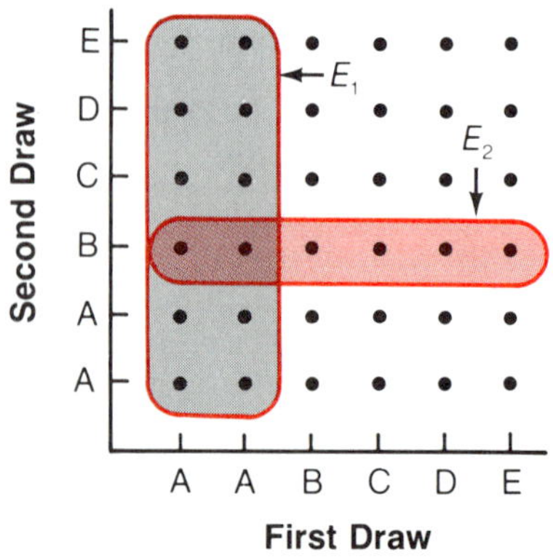

$$p(E_1) = \frac{12}{36} = \frac{1}{3}$$

$$p(E_2 | E_1) = \frac{2}{12} = \frac{1}{6}$$

$$p(E_1 \cap E_2) = p(E_1) \cdot p(E_2 | E_1)$$

$$= \frac{1}{3} \cdot \frac{1}{6} = \mathbf{\frac{1}{18}}$$

In Example 3,

$$p(E_2) = \frac{6}{36} = \frac{1}{6} = p(E_2 \mid E_1).$$

This suggests the following definition of *independent* and *dependent* events.

Definition

> Two events (or trials) E_1 and E_2 are **independent** *if and only if*
> $$\mathbf{p(E_1 \cap E_2) = p(E_1) \cdot p(E_2).}$$
> Otherwise, E_1 and E_2 are **dependent.**

CLASSROOM EXERCISES

In Exercises 1–4, find $P(E_1 \cap E_2)$.

1. $P(E_1) = \frac{3}{4}$; $P(E_2 \mid E_1) = \frac{1}{6}$
2. $P(E_1) = \frac{1}{6}$; $P(E_2 \mid E_1) = \frac{1}{8}$
3. $P(E_1) = 0.1$; $P(E_2 \mid E_1) = 0.4$
4. $P(E_2) = 0.3$; $P(E_1 \mid E_2) = 0.2$

In Exercises 5–8, indicate whether events E_1 and E_2 are *dependent* or *independent.*

5. $p(E_1) = \frac{1}{2}$
   $p(E_2) = \frac{1}{3}$
   $p(E_1 \cap E_2) = \frac{1}{6}$
6. $p(E_1) = 0.2$
   $p(E_2) = 0.4$
   $p(E_1 \cap E_2) = 0.12$
7. $p(E_1) = \frac{2}{7}$
   $p(E_2 \mid E_1) = \frac{1}{2}$
   $p(E_1 \cap E_2) = \frac{1}{7}$
8. $p(E_1) = 0.15$
   $p(E_2) = 0.3$
   $p(E_2 \cap E_1) = 0.045$

WRITTEN EXERCISES

A

In Exercises 1–2, one card is drawn from a bridge deck. Find the probability that the card is a black queen, given that the indicated information is known.

1. The card is a face card
2. The card is a spade

In Exercises 3–4, a red die and a green die are tossed. Find the probability that the sum of the dice is 9 or greater, given that the indicated information is known.

3. The red die shows a 4
4. The green die shows a 5
5. One card is drawn from a bridge deck. Without replacing the first card, a second card is drawn. Find the probability that the first card is red and the second card is black.
6. A box contains 3 red dishes and 4 blue dishes. One dish is drawn and without replacing the first dish, a second dish is drawn. Find the probability that both dishes are red.
7. Two cards are drawn simultaneously from a bridge deck. Find the probability that the first card is a king and the second card an ace.

8. A box contains 3 red, 4 blue, and 5 green disks. Three disks are drawn at the same time. Find the probability that all three disks are green.

In Exercises 9–10, refer to the spinner at the right. Find the probability of obtaining the indicated event.

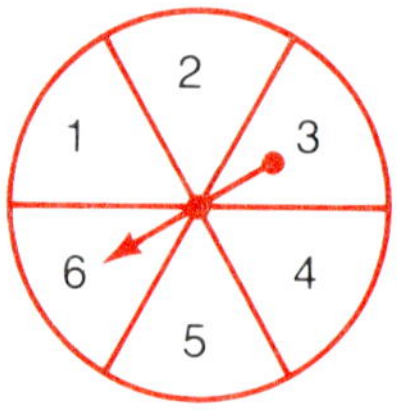

9. Two successive 5's

10. A 5 followed by an odd number

In Exercises 11–12, a green die and a red die are tossed. Find the probability of the indicated event.

11. The red die shows a 6 and the sum of the dice is greater than 10

12. The red die is an odd number and the green die shows a 4

13. A coin and a die are tossed. Find the probability that the coin shows a head and that the die shows a 2.

14. One coin is tossed. Then a die is tossed twice. Find the probability that the coin shows a tail and that the die shows a 6 on both tosses.

In Exercises 15–16, a box contains 12 disks numbered from 1 to 12. One disk is drawn and replaced. Then a disk is drawn. Find each probability.

15. The same disk is drawn each time

16. Both disks show an odd number

In Exercises 17–19, a box contains 10 disks numbered from 1 to 10. A disk is drawn and is not replaced. Then a second disk is drawn. Find the probability of the indicated event.

17. The first disk shows an odd number and the second shows an even number.

18. The first disk shows an even number and the second shows an odd number.

19. One disk is numbered with an odd number, the other with an even number.

B

20. A pair of dice is rolled three times. Find the probability that a doublet (two ones, two twos, two threes, etc.) is rolled exactly twice. (HINT: Consider three mutually exclusive events.)

21. Two cards are drawn from a bridge deck. Find the probability that the two cards are of the same color.

22. A three-person committee is to be chosen from a group of 6 boys and 4 girls. Find the probability that the committee consists of two boys and one girl.

C

23. A die is tossed three times. Find the probability that it shows a four at least once.

24. A coin is tossed six times. Find the probability of getting exactly three heads.

Using Statistics:
The Binomial Distribution

Mechanical engineers test manufactured products to insure that standards are maintained. This testing process gives rise to a statistical model called the **binomial distribution.** This model can be used to predict the success or failure of a given manufactured item. Certain conditions are necessary in order to have a binomial distribution.

Conditions

1. There must be exactly n independent, identical trials.
2. Each trial must result in either success or failure.
3. The probability, p, of success on each trial must stay the same.
4. There must be a definite number of successes, x, desired.

EXAMPLE A company that manufactures tires tests 10 out of every 1000 tires it produces. The tires are driven for 40,000 miles. If the tire does not blow out over this distance, the trial is considered a success. If it does blow out, the trial is a failure. After 20 trials, the average number of successes is 8.

a. Find the probability that exactly 8 tires will not blow out in the next trial.

b. Find the probability that at most 3 tires will blow out.

Solution: The average number of successes in 20 trials of 10 tires each is 8, so $p = \frac{8}{10} = 0.8$. Thus q, the probability of failure is $1 - p = 1 - 0.8 = 0.2$. Since this problem satisfies the conditions for a binomial distribution, you can use the binomial theorem on page 467 to find the solutions. The table below shows the probabilities for r successes where $0 \leq r \leq 10$, and the probability of r successes is ${}_{10}C_r p^r q^{10-r}$.

r	${}_{10}C_r$	p^r	q^{10-r}	*Probability*
0	1	$(0.8)^0$	$(0.2)^{10}$	0.0000001
1	10	$(0.8)^1$	$(0.2)^9$	0.000004
2	45	$(0.8)^2$	$(0.2)^8$	0.000074
3	120	$(0.8)^3$	$(0.2)^7$	0.000786
4	210	$(0.8)^4$	$(0.2)^6$	0.005505
5	252	$(0.8)^5$	$(0.2)^5$	0.026423
6	210	$(0.8)^6$	$(0.2)^4$	0.088077
7	120	$(0.8)^7$	$(0.2)^3$	0.201320
8	45	$(0.8)^8$	$(0.2)^2$	0.301980
9	10	$(0.8)^9$	$(0.2)^1$	0.268420
10	1	$(0.8)^{10}$	$(0.2)^0$	0.107370

From the table, you can see that there is only 1 chance in ten million that all 10 tires in a sample will blow out at or before 40,000 miles of use. However, there is a 10.7% chance that all 10 tires will be free of blowouts over the same distance.

a. The probability that exactly 8 tires will not blow out is the probability of 8 successes. From the table $P = \mathbf{0.301980}$. Thus, there is a **30.2%** chance that 8 tires will be blow-out free.

b. The probability that at most 3 tires will blow out is the probability that at least 7 tires will not blow out, or at least 7 successes. This is the probability of exactly 7 + exactly 8 + exactly 9 + exactly 10, successes.

From the table: $0.201320 + 0.301980 + 0.268420 + 0.107370 = P = \mathbf{0.879090}$
Thus there is an **87.9%** chance that at most 3 tires will blow out in the next trial.

EXERCISES

In Exercises 1–4, use the information in the table on page 533.

1. What is the probability that exactly 6 tires will not blow out?
2. What is the probability that exactly 3 tires will blow out?
3. What is the probability that at least 3 but at most 6 tires will not blow out?
4. What is the probability that at least 3 but at most 6 tires will blow out?
5. A student says that he can identify a 6-cylinder car by its sound. What is the probability of identifying 8 out of 10 cars correctly? If he does this, will you agree that he has a special talent? Why?
6. A magician says that by mental telepathy she can select the aces from the high cards (ace, king, queen, jack) in a deck. She draws a sample of 5 cards out of the 16 face cards, replacing and reshuffling the cards after each draw, and picks 3 aces. What is the probability that she could do this with no special ability?
7. A baseball player's batting average is .300. What is the probability of his getting exactly 3 hits in 4 times at bat?
8. If you answer 10 true-false questions by guessing, what is the probability that you will be correct on 7 of them?
9. There are 10 multiple choice questions on an exam. Each question has 4 alternative answers and you do them by guessing. What is the probability that you will be correct on 7 out of 10 questions?
10. It is known from past experience that 3% of all valves produced by a certain company are defective. If 10 valves produced by the company are selected at random, find the probability that 6 of them are defective.
11. A corporation claims that 95% of its trains arrive on time. If 10 trains are picked at random, what is the probability that at least 9 of them are on time?

Review

In Exercises 1–6, the four faces of a regular tetrahedron are numbered from 1 to 4. The tetrahedron is tossed and the number on the bottom face is noted.

Find the probability of obtaining the indicated number on the bottom face. *(Section 14–4)*

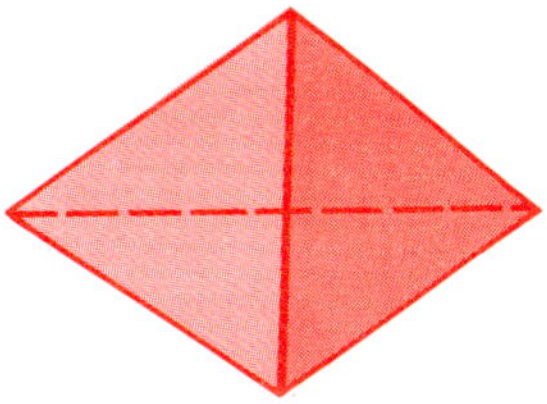

1. 4 **2.** A prime number **3.** A counting number

4. 1 or 3 **5.** 2 or 5 **6.** 6

In Exercises 7–12, one card is drawn from a bridge deck of 52 cards. Find the probability of each indicated drawing. *(Section 14–5)*

7. A queen or a king **8.** A heart or a club **9.** A nine or a two

10. A heart or a two **11.** A club or a face card **12.** A jack or a diamond

In Exercises 13–17, two cards are drawn in succession from a bridge deck of 52 cards. The first card is replaced before the second card is drawn. Find each probability. *(Section 14–6)*

13. Drawing two kings **14.** Drawing two spades **15.** Drawing two red cards

16. Drawing a 10, then a heart **17.** Drawing the same card twice

Chapter Summary

IMPORTANT TERMS

Combination (p. 518)
Conditional probability (p. 529)
Dependent events (p. 531)
Event (p. 523)
Factorial (p. 512)
Independent events (p. 531)
Mutually exclusive events (p. 526)
Permutation (p. 512)
Probability (p. 522)
Sample point (p. 522)
Sample space (p. 522)
Simple event (p. 523)

IMPORTANT IDEAS

1. **The Fundamental Principle of Counting:** If one event can occur in m different ways, and if, after it occurs or at the same time, a second event can occur in n ways, then the two events can occur in $m \cdot n$ different ways.
2. The number of permutations of n elements taken r at a time is given by

$$_nP_r = \frac{n!}{(n-r)!}.$$

3. The number of distinct permutations of n things, where p of them are alike, q things are alike, r things are alike, and so on is given by

$$T = \frac{n!}{p!q!r! \cdot \cdot \cdot}, p \leq n, q \leq n, r \leq n, \cdot \cdot \cdot .$$

4. The number of combinations of n objects taken r at a time is given by

$$_nC_r = \frac{n!}{r!(n-r)!}.$$

5. The probability of an outcome that cannot occur is 0.

6. The probability of an outcome that must occur is 1.

7. **Mutually Exclusive Events:** If two or more outcomes cannot occur simultaneously, the probability of any one of the outcomes occurring is the sum of the probabilities that each will occur.

$$p(E_1 \cup E_2) = p(E_1) + p(E_2).$$

8. **General Addition Rule:** For any two events E_1 and E_2 for a given sample space,

$$p(E_1 \cup E_2) = p(E_1) + p(E_2) - p(E_1 \cap E_2).$$

9. Events E_1 and E_2 are **independent** if and only if

$$p(E_1 \cap E_2) = p(E_1) \cdot p(E_2).$$

Chapter Objectives and Review

Objective: *To calculate the number of permutations of a set of elements (Sections 14–1 and 14–2)*

1. Write all the permutations of three elements that can be formed from **a, b, c.**
2. How many four-digit integers can be formed from the digits **0, 1, 2, 3, 4, 5** and **6** when no digit is repeated in an integer?
3. In Exercise 2, how many four-digit integers are possible when repetitions are allowed?
4. In how many ways can seven different books be arranged on a shelf?
5. How many permutations can be made from the letters of the word STATISTICS?
6. In how many ways can three identical mathematics books, four identical physics books, and seven identical English books be placed on a shelf?

Objective: *To calculate the number of combinations of a set of elements (Section 14–3)*

7. How many combinations of two elements can be formed from **a, b, c, d?**
8. In how many ways can you choose a committee of three boys and five girls from a group of seven boys and six girls?
9. Five boys are to be chosen from a group of twelve to make one basketball team. In how many ways can the team be chosen?
10. In Exercise 9, how many of the ways will include Phil, a member of the group?

Objective: *To compute the probability that an event will occur (Section 14–4)*

In Exercises 11–14, the sample space is {1, 2, 3, 4, 5, 6, 7, 8}. Compute the probability of each event.

11. What is the probability of drawing a 3 in one random selection?

12. What is the probability of drawing an even number?

13. What is the probability of drawing one of 1, 2, 3, 4, 5, 6, 7, 8?

14. What is the probability of drawing 0?

Objective: *To compute the probability that at least one of two or more events in a given sample space will occur (Section 14–5)*

In Exercises 15–16, a single card is drawn from a deck of 52 cards.

15. What is the probability that it will be an ace or a face card?

16. What is the probability that it will be a ten or a diamond?

In Exercises 17–18, a single die is thrown.

17. What is the probability of getting a 3 or a 4?

18. What is the probability of getting an odd number or a multiple of three?

In Exercises 19–20, a pair of dice is thrown.

19. What is the probability of getting a 3 or a 4?

20. What is the probability of getting a double or a 4?

Objective: *To compute the probability that two dependent events will both occur or that two independent events will both occur (Section 14–6)*

21. A card is drawn from a bridge deck of 52 cards. Then, without replacing the first card, a second card is drawn. Find the probability that the first card is a queen and the second card a spade.

22. Twenty slips of paper numbered 1 to 20 are placed in a hat. A slip of paper is drawn from the hat and immediately replaced. A second draw from the hat is then made. Find the probability that the number on the first slip is a multiple of 3 and the number on the second slip is odd.

Chapter Test

1. In how many ways can you arrange five people in a straight line?

2. How many 5-digit numbers can be formed from the integers **1, 2, 3, 5, 7, 9** when no integer may be used more than once?

3. How many permutations can be made of the letters of the word HAWAII?

Evaluate each expression.

4. $\frac{8!}{7!}$ **5.** $\frac{6!}{4!2!}$ **6.** $0!$ **7.** ${}_5C_3$ **8.** $\frac{1}{{}_7C_4}$

9. How many committees of three people can you form from a set of eight people?
10. A sample space for tossing three coins consists of 8 points representing the possible outcomes. What is the probability of each point?

The numbers from 1 through 15 are printed on 15 tokens, one number per token. In Exercises 11–14, one token is drawn at random. Find each probability.

11. The number on the token is divisible by 5
12. The number on the token is an even number
13. The number on the token is an odd number
14. The number on the token is a perfect square

In Exercises 15–18, two dice are thrown. The number of dots on the top face of each die is then noted. Find each probability.

15. One die shows either an odd number <u>or</u> a prime number
16. The sum of the numbers is less than <u>or</u> greater than 3
17. The sum of the numbers is greater than or equal to 4 <u>or</u> the sum is 12
18. The sum of the numbers is an even number <u>or</u> a multiple of 3

In Exercises 19–20, two successive draws of a card are made from a bridge deck of 52 cards.

19. The first card is replaced in the deck before the second draw is made. Find the probability that the same card is drawn both times.
20. The first card is not replaced in the deck before the second draw is made. Find the probability that a heart is drawn both times.

More Challenging Problems

1. At a party, 21 handshakes were exchanged. Each person shook the hand of every other person present. How many people were at the party?
2. In how many ways can 11 people be divided into two groups of 7 and 4?
3. There are nine algebra books, six English books, and three history books on a shelf. Find the number of ways that you could choose three algebra books, three English books, and three history books.
4. At a party, four men and four women are to sit at a round table. The host takes his seat and the guests are seated at random. What is the probability that the men and women will be seated alternately?
5. Six coins are tossed simultaneously. What is the probability of getting exactly three heads?
6. Show that: ${}_nC_r + {}_nC_{r-1} = {}_{n+1}C_r$. 7. Determine n such that $5{}_nC_5 = {}_nC_4$.
8. How many triangles can be formed with nine points in a plane as vertices, when no three points are on a straight line?

CHAPTER **15**

Trigonometric Functions

15-1 Angles

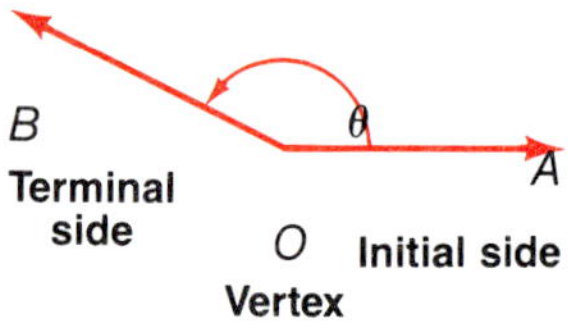

Figure 1

Angle AOB at the right is formed by rotating a ray OA about its endpoint, O, in a fixed plane. The initial position of the ray is the **initial side** of the angle ray OA. The terminal position of the ray is the **terminal side** of the angle ray OB.

The amount and direction of rotation is the measure of the angle represented by θ (theta) in Figure 1. Angles with measures of 20°, −20°, 370°, and −370° are shown below.

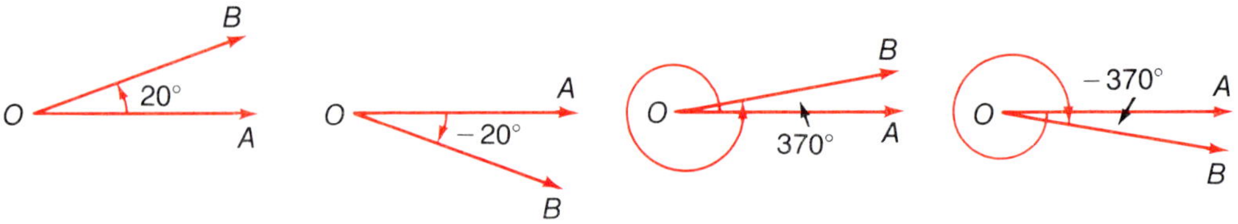

Counterclockwise	Clockwise	Counterclockwise	Clockwise
20°	−20°	370°	−370°

NOTE: When the measure of the angle is positive, the rotation is in a *counterclockwise* direction. When the measure of the angle is negative, the rotation is in a *clockwise* direction.

An angle of 360° is one full counterclockwise rotation. The initial and terminal sides coincide (see the figure at the left below).

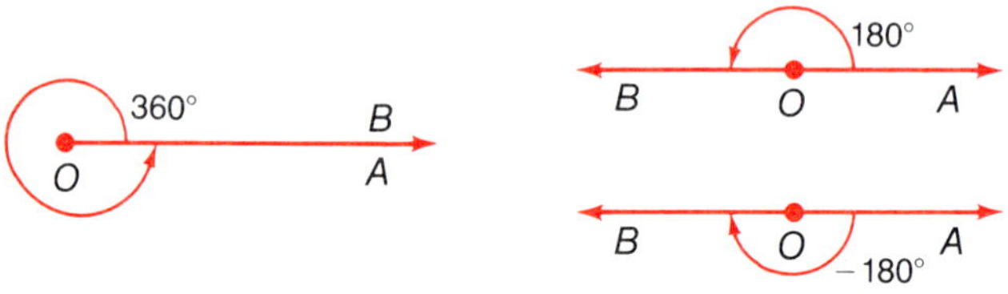

An angle of 180° is one-half a counterclockwise rotation. An angle of −180° is one-half a clockwise rotation (see the figure at the right above).

An angle is in **standard position** in the coordinate plane when its vertex is at the origin O and its initial side is on the positive x axis. Angle AOB is a *first-quadrant angle* because its terminal side is in Quadrant I. An angle is a **quadrantal angle** when the terminal side is on an axis. In Figure 2, $\angle AOC$ is a quadrantal angle.

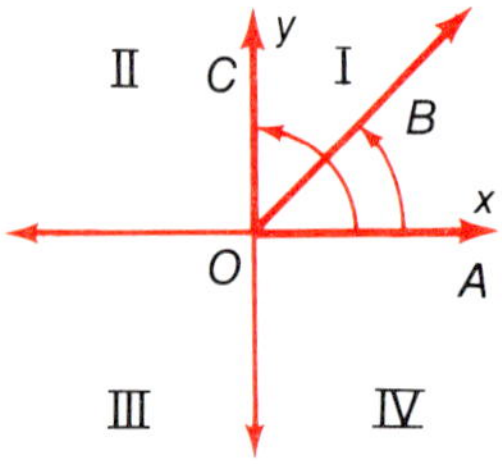

Figure 2

Angles in standard position whose initial and terminal sides coincide are **coterminal angles.** The measures of coterminal angles differ by an integral multiple of 360°.

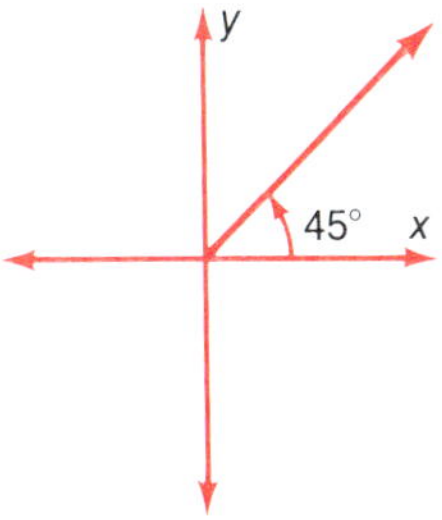

45°

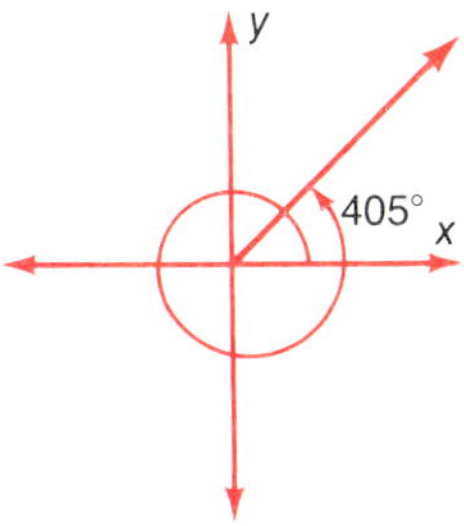

$1 \cdot 360° + 45° = \mathbf{405°}$

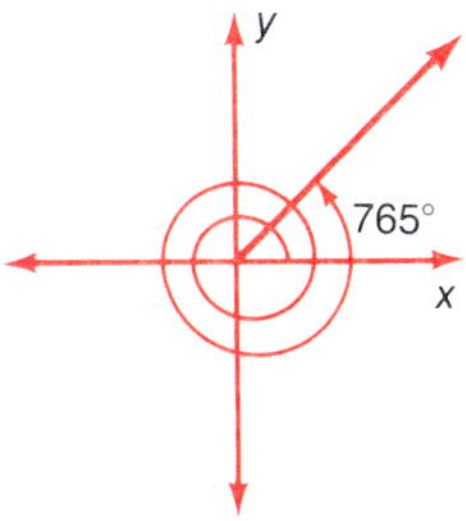

$2 \cdot 360° + 45° = \mathbf{765°}$

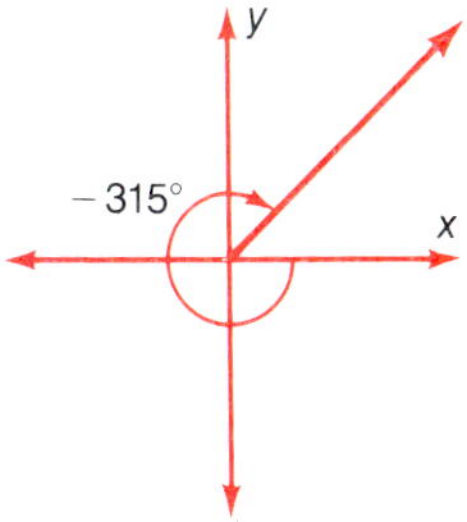

$(-1) \cdot 360° + 45° = \mathbf{-315°}$

Thus, for any angle with measure θ, the measures of its coterminal angles are

$$\mathbf{n \cdot 360° + \theta}, \text{ where } n \text{ is an integer.}$$

EXAMPLE Find three coterminal angles for $\theta = 30°$. Use the given values for n.

a. $n = 3$ **b.** $n = -2$ **c.** $n = -3$

Solutions:
a. $3 \cdot 360° + 30° = \mathbf{1110°}$
b. $(-2) \cdot 360° + 30° = \mathbf{-690°}$
c. $(-3) \cdot 360° + 30° = \mathbf{-1050°}$

CLASSROOM EXERCISES

In Exercises 1–6, sketch each angle in standard position.

1. 30° **2.** 60° **3.** −45° **4.** −15° **5.** −270° **6.** 120°

Sketch a coterminal angle for each angle. Let $n = -1$.

7. 60° **8.** 100° **9.** −15° **10.** −80° **11.** 185° **12.** −145°

WRITTEN EXERCISES

A In Exercises 1–12, sketch each angle in standard position.

1. 90° **2.** −90° **3.** 15° **4.** 45° **5.** −30° **6.** −60°
7. 135° **8.** 180° **9.** 270° **10.** −180° **11.** 315° **12.** 450°

In Exercises 13–20, the rotation of an angle is given. Sketch the angle in standard position, and give its measure in degrees.

13. $\frac{1}{4}$ rotation, counterclockwise
14. $\frac{1}{2}$ rotation, clockwise
15. $\frac{1}{3}$ rotation, counterclockwise
16. 1 rotation, counterclockwise
17. $2\frac{1}{4}$ rotations, counterclockwise
18. $1\frac{1}{3}$ rotations, clockwise
19. $\frac{1}{4}$ rotation, clockwise
20. 2 rotations, clockwise

Name the quadrant in which the terminal side of each angle lies.

21. 98° **22.** 156° **23.** 350° **24.** 735° **25.** −240° **26.** −315°

For each angle, find three coterminal angles. Let $n = -1$, 2, and 5.

27. 30° **28.** −60° **29.** 120° **30.** −199° **31.** 270° **32.** 320°

B

APPLICATIONS: Using Angles in Standard Position

33. Through how many meters does a wheel of radius 6 centimeters roll in 3 revolutions? (HINT: Find the circumference. Use $\pi = 3.14$)

34. How many revolutions does a wheel 8 meters in diameter make in going 1 kilometer?

35. A merry-go-round 20 meters in diameter makes 3 revolutions in 1 minute. What is the speed in meters per second of a boy standing at the edge?

15–2 Trigonometric Functions Defined

Think of a point $P(x, y)$ on the terminal side of an angle with measure θ. The distance from the origin to point P is the length of the **radius vector, r.**

The triangle formed by drawing a perpendicular from point P on the radius vector to the x axis is called the **reference triangle.**

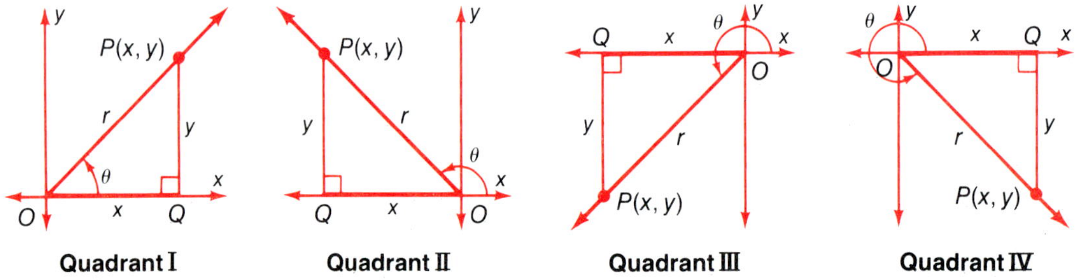

The ratios $\frac{y}{r}$, $\frac{x}{r}$, and $\frac{y}{x}$ have special names.

Definitions

Let $P(x, y)$ be any point (not the origin) on the terminal side of an angle with measure θ in standard position and let r be the radius vector. Then

sine θ (abbreviated: **sin θ**) $= \frac{y}{r}$,

cosine θ (abbreviated: **cos θ**) $= \frac{x}{r}$,

tangent θ (abbreviated: **tan θ**) $= \frac{y}{x}$, $x \neq 0$.

Recall that in similar triangles corresponding sides have equal ratios. Thus, as long as the measure of an angle in standard position remains the same, the ratios $\frac{y}{r}$, $\frac{x}{r}$, and $\frac{y}{x}$ will not be affected by a change in the position of point P on the terminal side of the angle.

Theorem 15–1

> For an angle with measure θ in standard position, each of the ratios $\sin\theta$, $\cos\theta$, and $\tan\theta$ is the same for every point P on the terminal of the angle.

Proof: Triangles OP_1Q_1 and OP_2Q_2 are similar. Therefore,

$$\frac{y_1}{r_1} = \frac{y_2}{r_2} = \sin\theta,$$

$$\frac{x_2}{r_1} = \frac{x_2}{r_2} = \cos\theta, \text{ and}$$

$$\frac{y_1}{x_1} = \frac{y_2}{x_2} = \tan\theta.$$

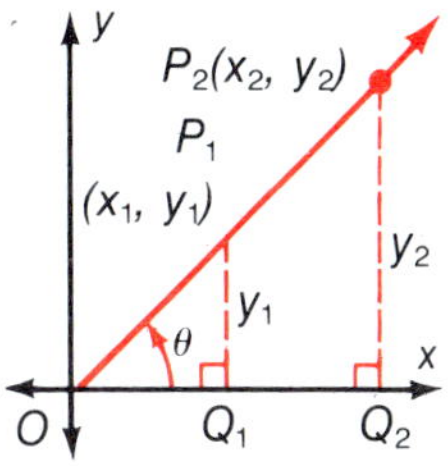

By Theorem 15–1, the sine, cosine, and tangent ratios depend *only* on the measure of the angle. Further, for each given θ, each ordered pair $(\theta, \sin\theta)$, $(\theta, \cos\theta)$, and $(\theta, \tan\theta)$ is **unique.** Thus, $\sin\theta$, $\cos\theta$, and $\tan\theta$ are functions.

The table below shows how these functions increase or decrease for values of θ where $0° \le \theta \le 180°$.

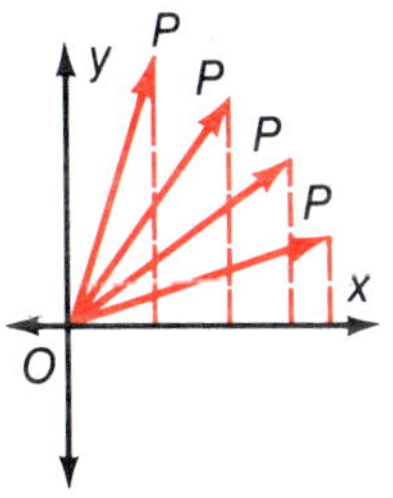

Angle Measure	Function	Changes in the Function
$0° \le \theta < 90°$	sine	Increasing from 0 to nearly 1
	cosine	Decreasing from 1 to nearly 0
	tangent	Increasing from 0 to very large values
$\theta = 90°$	sine	$\sin 90° = 1$
	cosine	$\cos 90° = 0$
	tangent	$\tan 90°$ is not defined.
$90° < \theta \le 180°$	sine	Decreasing from nearly 1 to 0
	cosine	Decreasing from nearly 0 to -1
	tangent	Increasing

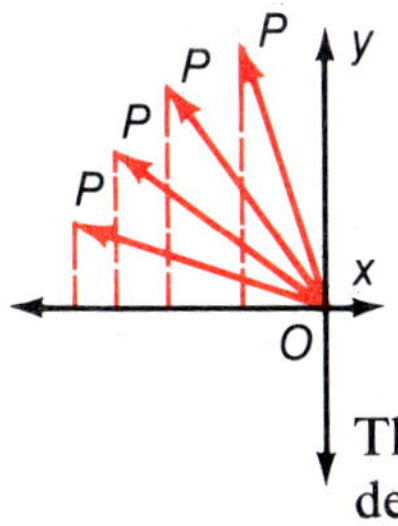

The table at the top of page 544 shows how these functions increase or decrease for values of θ where $180° < \theta \le 360°$.

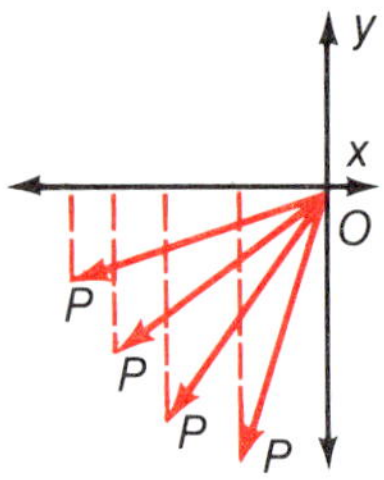

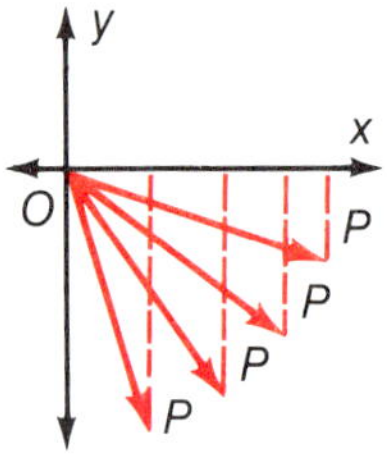

Angle Measure	Function	Changes in the Function
$180° < \theta < 270°$	sine	Decreasing from 0 to nearly -1
	cosine	Increasing from -1 to nearly 0
	tangent	Increasing from 0 to very large values
$270°$	sine	$\sin 270° = -1$
	cosine	$\cos 270° = 0$
	tangent	$\tan 270°$ is not defined.
$270° < \theta \leq 360°$	sine	Increasing from -1 to 0
	cosine	Increasing from 0 to 1
	tangent	Increasing

The domain of the sine and cosine functions is the set of real numbers. However, certain real numbers are excluded from the domain of the tangent function. For example, when $\theta = 90°$, $\tan \theta = \frac{1}{0}$. Since division by 0 is undefined, $\tan \theta$ is not defined for $\theta = 90°$.

EXAMPLE 1 Evaluate $\sin \theta$, $\cos \theta$, and $\tan \theta$ for angle QOP.

Solution: $r = \overline{OP} = \sqrt{5^2 + 6^2} = \sqrt{61}$

$$\sin \theta = \frac{y}{r} = \frac{\mathbf{6}}{\mathbf{\sqrt{61}}} \qquad \cos \theta = \frac{x}{r} = \frac{\mathbf{5}}{\mathbf{\sqrt{61}}} \qquad \tan \theta = \frac{y}{x} = \frac{\mathbf{6}}{\mathbf{5}}$$

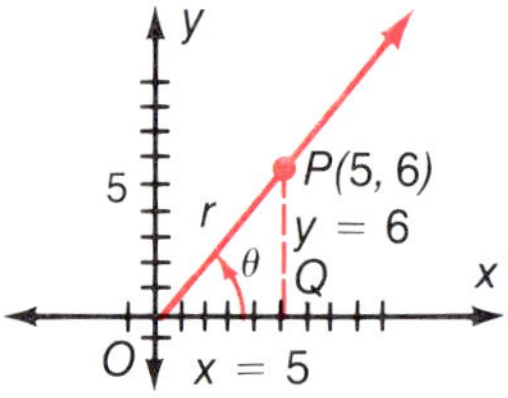

NOTE: In Example 2, you find the length of the radius vector r by applying the Pythagorean Theorem. That is, if $P(x, y)$ is any point on the terminal side of an angle θ, and r is the distance from the origin to point P, then

$$r^2 = x^2 + y^2, \quad \text{or} \quad r = \sqrt{x^2 + y^2}$$ ← **Distance is positive.**

EXAMPLE 2 Sketch the angle whose terminal side contains $P(-5, 12)$. Find the quadrant of the angle and use the reference triangle to find $\sin \theta$, $\cos \theta$, and $\tan \theta$.

Solution: The angle is in Quadrant II.

$r = \sqrt{(-5)^2 + 12^2} = 13$

$$\sin \theta = \frac{y}{r} = \frac{\mathbf{12}}{\mathbf{13}} \qquad \cos \theta = \frac{x}{r} = \frac{-5}{13} = -\frac{\mathbf{5}}{\mathbf{13}} \qquad \tan \theta = \frac{y}{x} = \frac{12}{-5} = -\frac{\mathbf{12}}{\mathbf{5}}$$

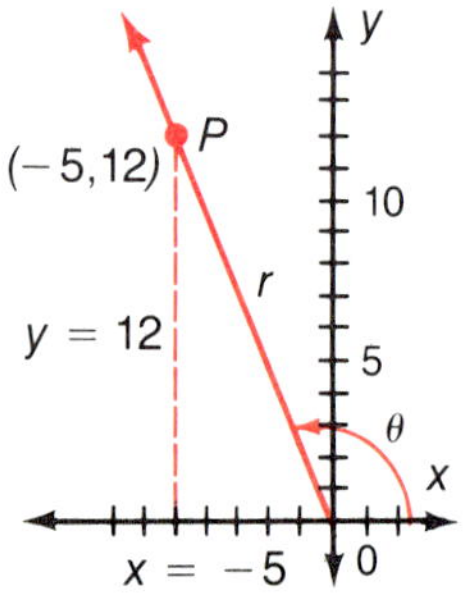

CLASSROOM EXERCISES

Evaluate sin θ, cos θ, and tan θ where θ is an angle in standard position whose terminal side contains the given point.

1. $P(3, 4)$ **2.** $Q(5, 12)$ **3.** $R(-5, -12)$ **4.** $T(5, -12)$ **5.** $V(-1, -1)$

Sketch the angle whose terminal side contains the given point. Give the quadrant of the angle and use the reference triangle to find sin θ, cos θ, and tan θ.

6. $Q(6, 8)$ **7.** $P(-6, 8)$ **8.** $N(6, -8)$ **9.** $R(-6, -8)$ **10.** $U(-12, -10)$

WRITTEN EXERCISES

A In Exercises 1–20, the terminal side of an angle with measure θ passes through the given point. Sketch the reference triangle. Evaluate sin θ, cos θ, and tan θ. Simplify all radicals.

1. $P(-3, 4)$ **2.** $Q(10, -24)$ **3.** $T(-6, 8)$ **4.** $U(-6, -8)$ **5.** $R(6, -8)$
6. $S(8, -6)$ **7.** $N(12, -5)$ **8.** $T(-12, -5)$ **9.** $S(5, 10)$ **10.** $V(-6, 4)$
11. $T(-12, -10)$ **12.** $P(3, 4)$ **13.** $N(-1, -1)$ **14.** $T(-12, 5)$ **15.** $P(3, -4)$
16. $R(0, 7)$ **17.** $T(-2, 3)$ **18.** $Q(6, -3)$ **19.** $R(-6, 3)$ **20.** $V(-23, 7)$

In Exercises 21–24, θ is in Quadrant I. Find sin θ and cos θ.

21. $\tan\theta = \frac{2}{3}$ **22.** $\tan\theta = \frac{5}{12}$ **23.** $\tan\theta = \frac{3}{4}$ **24.** $\tan\theta = \frac{15}{8}$

Replace each __?__ with + or − to indicate whether the values of the given functions are positive or negative in the indicated quadrants.

	I	II	III	IV
25. sin θ	__?__	__?__	__?__	__?__
26. cos θ	__?__	__?__	__?__	__?__
27. tan θ	__?__	__?__	__?__	__?__

B State the quadrants in which θ lies under the given conditions.

28. $\sin\theta > 0$

29. $\cos\theta < 0$

30. $\sin\theta > 0$ and $\cos\theta < 0$

31. $\tan\theta > 0$

32. $\tan\theta > 0$ and $\cos\theta < 0$

33. $\sin\theta < 0$ and $\tan\theta > 0$

Complete.

Example: $\sin\theta = \frac{1}{2}$; $\cos\theta < 0$

$\tan\theta =$ __?__

Solution: $\cos\theta = \frac{x}{r} = \frac{-\sqrt{3}}{2} = -\frac{\sqrt{3}}{2}$

$\tan\theta = \frac{y}{x} = \frac{1}{-\sqrt{3}} = -\frac{\sqrt{3}}{3}$

34. $\sin\theta = -\frac{1}{2}$; $\cos\theta < 0$; $\tan\theta =$ __?__

35. $\sin\theta = -\frac{1}{2}$; $\cos\theta > 0$; $\tan\theta =$ __?__

36. $\tan\theta = 1$; $\cos\theta < 0$; $\sin\theta =$ __?__

37. $\cos\theta = \frac{3}{5}$; $\tan\theta < 0$; $\sin\theta =$ __?__

Classify each statement as true, *T*, or false, *F*.

38. As θ increases from 0° to 90°, tan θ increases from 1 without bound.

39. As θ varies from 0° to 360°, sin θ takes all values from -1 to 1.

40. As θ varies from 0° to 360°, the range of cos θ is the set of all numbers from 0 to -1.

41. As θ decreases from 180° to 90°, sin θ decreases from 1 to 0.

42. As θ increases from 180° to 270°, tan θ decreases from 0 without bound.

15-3 Functions of Special Angles

Recall from your study of geometry that 30°-60°-90° and 45°-45°-90° right triangles have sides whose lengths are related. These relationships are shown in Figure 1. In the figure, s represents any positive real number.

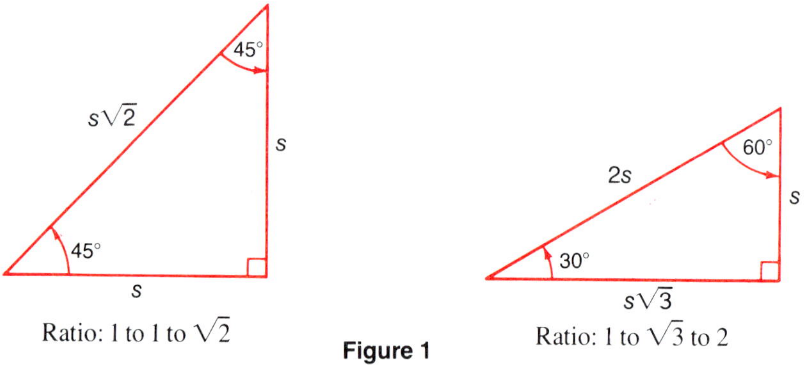

Figure 1

Finding the unknown sides and angles in a triangle is called **solving the triangle.**

EXAMPLE 1 The length of the hypotenuse of a 45°-45°-90° triangle is 16 meters. Solve the triangle.

Solution: **Method 1** Let x = the length of each leg.

$x^2 + x^2 = 16^2$ ← **Pythagorean Theorem**

$2x^2 = 256$

$x^2 = 128$

$x = \mathbf{8\sqrt{2}}$ ← **The length of each leg is $8\sqrt{2}$ m.**

Method 2 From Figure 1, s is the length of a leg and $s\sqrt{2}$ is the length of the hypotenuse.

$s\sqrt{2} = 16$ ← **Solve for s.**

$s = \frac{16}{\sqrt{2}}$ ← **Rationalize the denominator.**

$s = \mathbf{8\sqrt{2}}$

It is usually more convenient to use the special right triangle relationships when solving 30°-60°-90° or 45°-45°-90° triangles.

EXAMPLE 2 The length of the hypotenuse of a 30°-60°-90° triangle is 10 centimeters. Solve the triangle.

Solution: From Figure 1, the leg opposite the 30° angle is one–half the hypotenuse.

$\frac{1}{2}(10) = 5$ ← **The length of one leg is 5 cm.**

$s\sqrt{3} = \mathbf{5\sqrt{3}}$ ← **The length of the other leg is $5\sqrt{3}$ cm.**

The relationships of the length of the sides of 30°-60°-90° and 45°-45°-90° triangles are used to find the trigonometric functions of angles whose reference triangles have a 30°–, a 60°–, or a 45°– angle. It is helpful to sketch the angle and the reference triangle.

EXAMPLE 3 Find sin 135°, cos 135°, and tan 135°.

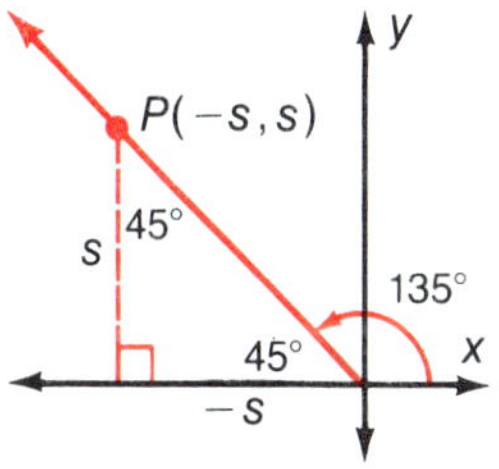

Solution: The reference triangle is 45°-45°-90°. By Figure 1, the sides of the reference triangle have the same measures, s, and $r = s\sqrt{2}$.

Thus, the coordinates of P are $(-s, s)$, because P is in Quadrant II.

$$\sin 135° = \frac{s}{s\sqrt{2}} = \frac{1}{\sqrt{2}} = \mathbf{\frac{\sqrt{2}}{2}}$$

$$\cos 135° = \frac{-s}{s\sqrt{2}} = \mathbf{-\frac{\sqrt{2}}{2}}$$

$$\tan 135° = \frac{s}{-s} = \mathbf{-1}$$

Example 4 shows that it is sometimes useful to choose a *convenient value* for the radius vector, r.

EXAMPLE 4 Find sin 330°, cos 330°, and tan 330°.

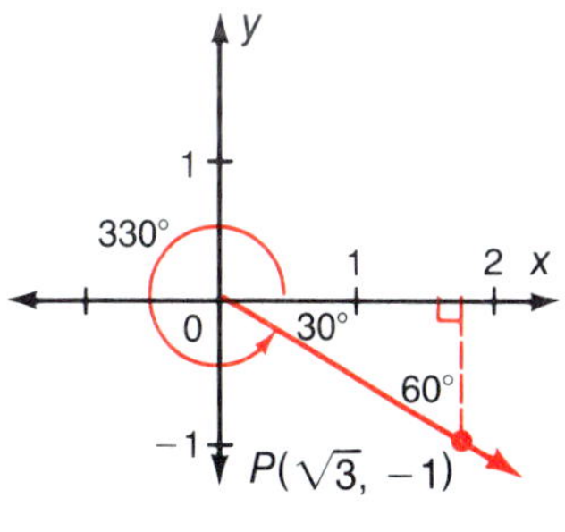

Solution: As the figure shows, the reference triangle is a 30°-60°-90°- triangle. By Figure 1, the lengths of its sides are $2s$, s, and $s\sqrt{3}$. Let s = 1. Then the length of the radius vector of point P on the terminal side of the 330° angle is 2. Thus, the coordinates of P are $(\sqrt{3}, -1)$. So,

$$\sin 330° = \frac{-1}{2} = \mathbf{-\frac{1}{2}}$$

$$\cos 330° = \mathbf{\frac{\sqrt{3}}{2}}$$

$$\tan 330° \frac{-1}{\sqrt{3}} = \mathbf{-\frac{\sqrt{3}}{3}}$$

CLASSROOM EXERCISES

1. The length of the hypotenuse of a 30°-60°-90° triangle is 14 centimeters. Solve the triangle.
2. The length of the hypotenuse of a 45°-45°-90° triangle is 20 centimeters. Solve the triangle.
3. The length of the hypotenuse of a 45°-45°-90° triangle is $10\sqrt{2}$. Solve the triangle.
4. A point P lies on the terminal side of a 30° angle in standard position and the length of the radius vector of P is 2. Find the coordinates of P.
5. Use the results of Exercise 4 to find sin 30°, cos 30°, and tan 30°. Leave your answers in radical form. Rationalize any irrational denominators.

WRITTEN EXERCISES

A In Exercises 1–6, the length of the hypotenuse, c, of each triangle is given. Solve the triangle.

1. 30°-60°-90° triangle; $c = 8$
2. 45°-45°-90° triangle; $c = 4$
3. 45°-45°-90° triangle; $c = 11\sqrt{2}$
4. 30°-60°-90° triangle; $c = 8\sqrt{3}$
5. 45°-45°-90° triangle; $c = 19$
6. 30°-60°-90° triangle; $c = 15$
7. A point P lies on the terminal side of a 60° angle that is in standard position. The length of the radius vector of P is 2.
   a. Find the coordinates of P.
   b. Use the results of Exercise 7a to find sin 60°, cos 60°, and tan 60°.
8. A point P lies on the terminal side of a 45° angle that is in standard position. The x coordinate of point P is 1.
   a. Find the y coordinate and the length of the radius vector of P.
   b. Use the results of 8**a** to find sin 45°, cos 45°, and tan 45°.
9. A point P lies on the terminal side of a 315° angle that is in standard position. The x coordinate of point P is 1.
   a. Find the y coordinate and the length of the radius vector of P.
   b. Use the results of 9**a** to find sin 315°, cos 315°, and tan 315°.
10. A point P lies on the terminal side of a 225°-angle in standard position. The length of the radius vector of P is 2.
    a. Find the coordinates of P.
    b. Use the results of 10**a** to find sin 225°, cos 225°, and tan 225°.
11. A point P lies on the terminal side of a 120°-angle in standard position. The radius vector of P is 2.
    a. Find the coordinates of P.
    b. Use the results of 11**a** to find sin 120°, cos 120°, and tan 120°.

In Exercises 12–15, θ is an angle in standard position and P is a point on the terminal side of θ. The length of the radius vector of P is 2. Complete the table.

	θ	$P(x, y)$	$\sin\theta$	$\cos\theta$	$\tan\theta$
12.	150°	?	?	?	?
13.	210°	?	?	?	?
14.	240°	?	?	?	?
15.	300°	?	?	?	?

B In Exercises 16–25, use a reference triangle to evaluate each function.

16. cos 405° **17.** sin 570° **18.** tan 420° **19.** sin 330° **20.** tan 405°

21. cos 660° **22.** tan 585° **23.** tan 870° **24.** cos 495° **25.** sin 780°

26. Show that $\sin^2\theta + \cos^2\theta = 1$. (HINT: $\sin^2\theta$ means $(\sin\theta)^2$.)

27. Show that $\tan 60° = \dfrac{\sin 60°}{\cos 60°}$.

15–4 Trigonometric Functions in Right Triangles

The reference triangle of every angle in the coordinate plane that is not a quadrantal angle is a right triangle. The length of the hypotenuse is r, and the lengths of the sides opposite, and adjacent to, the acute angle with measure θ are $|y|$ and $|x|$, where x and y are the coordinates of P. By substituting for x, y and r in the coordinate system definitions, the right-triangle definitions can be stated as follows.

$$\sin\theta = \frac{\textbf{length of side opposite } \theta}{\textbf{length of hypotenuse}}$$

$$\cos\theta = \frac{\textbf{length of side adjacent } \theta}{\textbf{length of hypotenuse}}$$

$$\tan\theta = \frac{\textbf{length of side opposite } \theta}{\textbf{length of side adjacent } \theta}$$

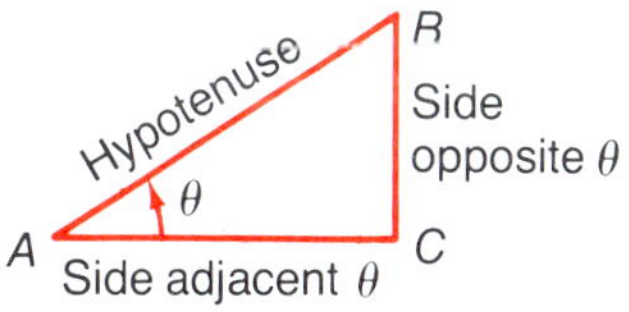

The tables on pages 629–633 give the sine, cosine, and tangent function values for angles of 0° to 90° to the nearest ten-thousandth. Angles that measure 45° or less are read in the left column using the headings at the top of the table. Angles of 45° or greater are read in the right column using the headings at the bottom of the table. (NOTE: 60 minutes (60′) = 1°.)

Although the values in the table are approximations, we shall use the symbol "=" when writing statements involving these values because it is more convenient.

EXAMPLE 1 In right triangle ABC, the length of the hypotenuse c is 220 meters and the measure of $\angle A$ is $28°40'$ ($40'$ is read "forty minutes"). Find the length of b, the leg opposite $\angle B$. Round your answer to the nearest meter.

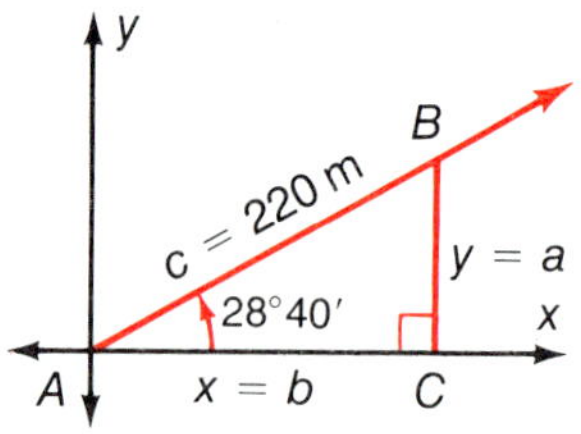

Solution:

$$\cos A = \frac{b}{c}$$ ← **By definition**

$$\cos 28°40' = \frac{b}{220}$$

$$b = 220 \cos 28°40'$$ ← **To find cos 28°40′, use the tables or use a calculator.**

$$b = 220\ (0.8774) = 193.028$$ Thus, b is about **193 meters** long.

In Example 2, the right–triangle definitions are applied to $\angle B$.

EXAMPLE 2 In right triangle ABC, the measure of $\angle B$ is $51°30'$ and the length of b is 15 feet. How long is side a? Round your answer to the nearest foot.

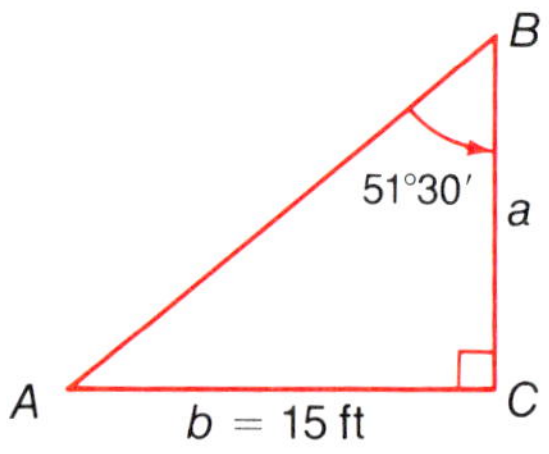

Solution:

$$\tan B = \frac{b}{a}$$ ← **By definition**

$$\tan 51°30' = \frac{15}{a}$$

$$a = \frac{15}{\tan 51°30'}$$ ← **To find tan 51°30′, use the tables or use a calculator.**

$$a = \frac{15}{1.2572} = 11.931$$ Thus, a is about **12 feet** long.

CLASSROOM EXERCISES

Find the value of each of the following.

1. $\sin 14°50'$ **2.** $\tan 27°20'$ **3.** $\cos 67°30'$ **4.** $\tan 45°10'$

In Exercises 4–8, θ is an angle in Quadrant I. Use the tables on pages 629–633 to find θ.

5. $\sin \theta = 0.4014$ **6.** $\tan \theta = 1.5798$ **7.** $\cos \theta = 0.6539$ **8.** $\sin \theta = 0.1016$

WRITTEN EXERCISES

A Refer to $\triangle ABC$ for Exercises 1–7.

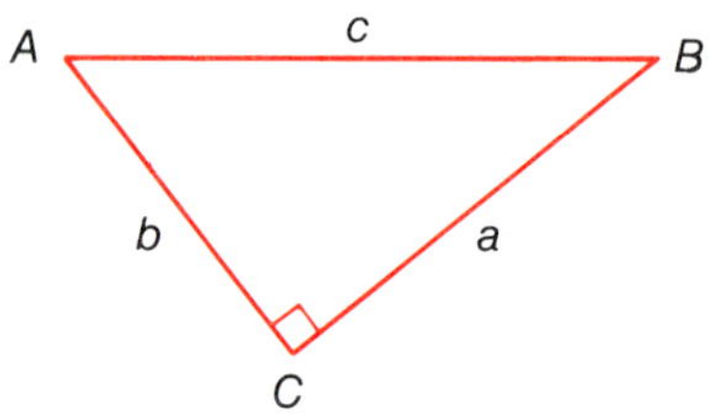

1. Given b and the measure of $\angle B$, which trigonometric function would you use to find a?

2. In Exercise 1, which trigonometric function would you use to find c?

3. Given the measure of $\angle A$ and the value of b, which trigonometric function would you use to find c?

4. In Exercise 3, which trigonometric function would you use to find a?
5. Given the measure of $\angle B$ and the value of a, which trigonometric function would you use to find c?
6. In Exercise 5, which trigonometric function would you use to find b?
7. Given the values of a and b, which trigonometric function would you use to find B (the measure of $\angle B$)?

In Exercises 8–13, use the tables on pages 629–633 to find the length of the unknown side. Round your answer to the nearest whole number. Refer to triangle ABC at the right.

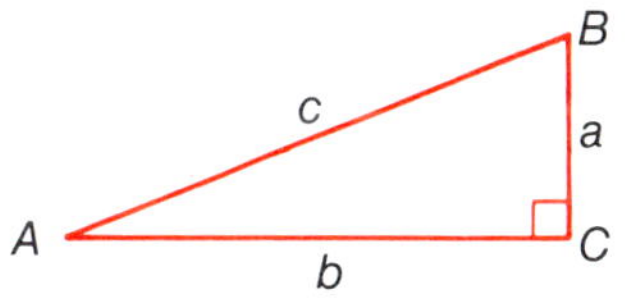

8. $A = 75°$; $c = 21$ m; $b =$ __?__
9. $B = 52°$; $a = 32$ m; $b =$ __?__
10. $A = 35°40'$; $c = 20$ m; $a =$ __?__
11. $B = 81°30'$; $c = 40$ m; $a =$ __?__
12. $B = 15°10'$; $c = 16$ m; $b =$ __?__
13. $A = 41°40'$; $b = 3$ m; $a =$ __?__

B In Exercises 14–25, refer to right $\triangle ABC$ to find the indicated values. Leave your answers in radical form.

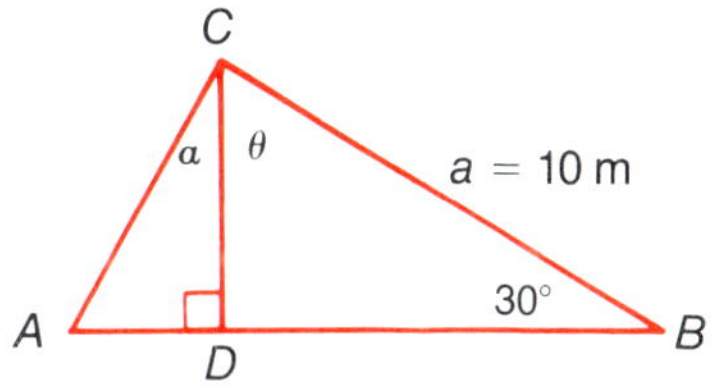

14. $\theta =$ __?__
15. $\alpha =$ __?__
16. $A =$ __?__
17. $\sin B$
18. $\cos \theta$
19. $\cos B$
20. $\cos A$
21. $\sin \alpha$
22. $\sin \theta$
23. Complete: $\sin B =$ __?__ $= \cos(90° - B) = \cos$ __?__
24. Complete: $\cos A =$ __?__ $= \sin(90° - A) = \sin$ __?__
25. Complete: $\sin \theta = \cos(90° -$ __?__ $) = \cos$ __?__

APPLICATIONS: Solving Right Triangles

26. A bridge joins two points, B and C, on the opposite banks of Endless River. A highway engineer stands at point A, 300 feet from C, and measures $\angle CAB$ to be $36°30'$. Find the length of the line of sight from the engineer to point B on the opposite bank of the river (see the figure at the left below).

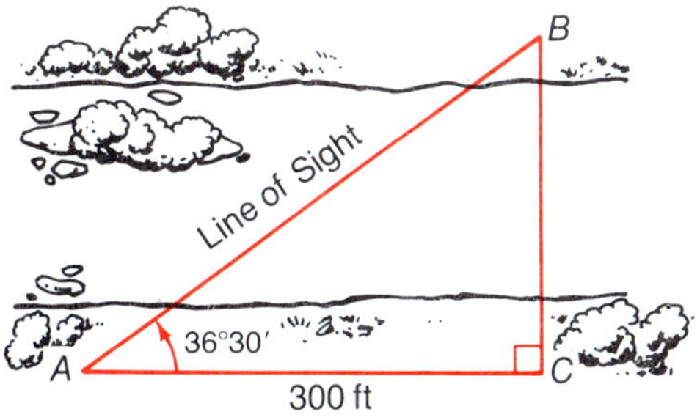

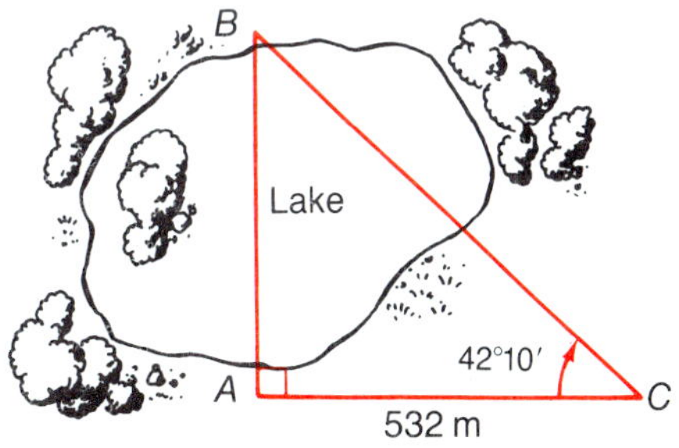

27. The distance AB shown in the figure at the right above can be found by placing stakes at A, B, and C in such a way as to make $\angle A$ a right angle, and by measuring $\overline{AC}$ and $\angle C$. $\overline{AC}$ is 532 meters and $\angle C$ is $42°10'$. How long is $\overline{AB}$?

28. A guy wire 40 meters long runs from the ground to the top of a pole. It makes a 64°–angle with the line drawn to the foot of the pole. Find the height of the pole (see the figure at the left below).

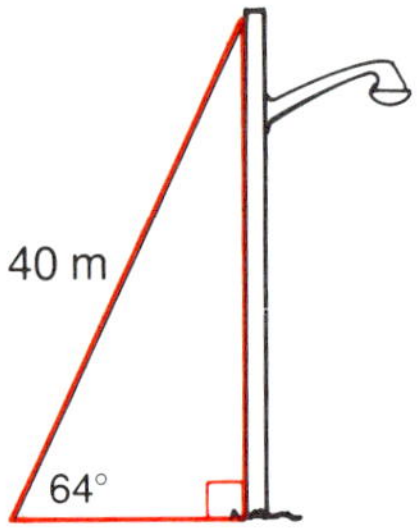

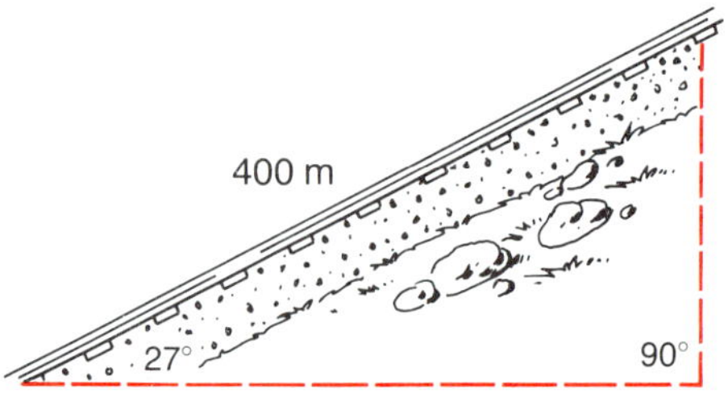

29. The railroad that runs to the summit of Pikes Peak makes, at the steepest place, a 27°–angle with the horizontal. How many meters would you rise in going 400 meters up this track (see the figure at the right above)?

30. A ladder 6 meters long leans against a building and makes an angle of 68° with the ground (see the figure at the left below). How far from the building is the base of the ladder?

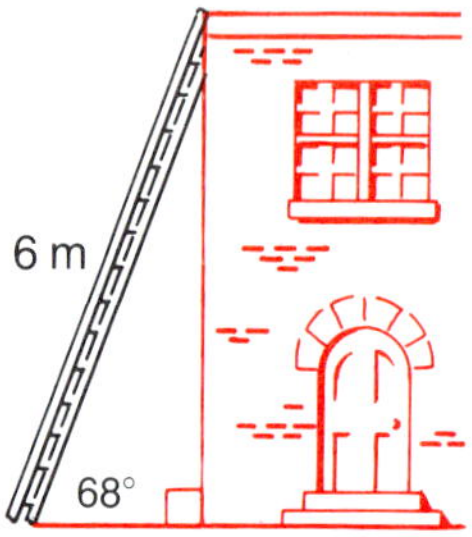

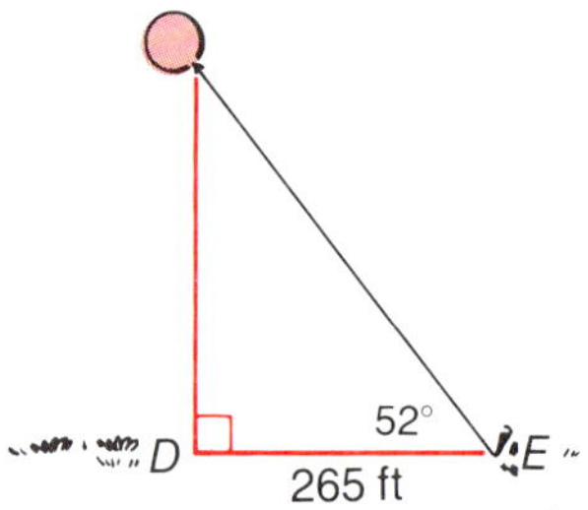

31. A balloon is anchored at point E by a cable (see the figure at the right above). The cable makes an angle of 52° with the ground. Point D, on the ground directly under the balloon, is 265 feet from E. Find the length of the cable.

32. An 8–meter pole is leaning against a tree (see the figure at the left below). The pole makes an angle of 34° with the tree. Find the distance d from the foot of the pole to the foot of the tree.

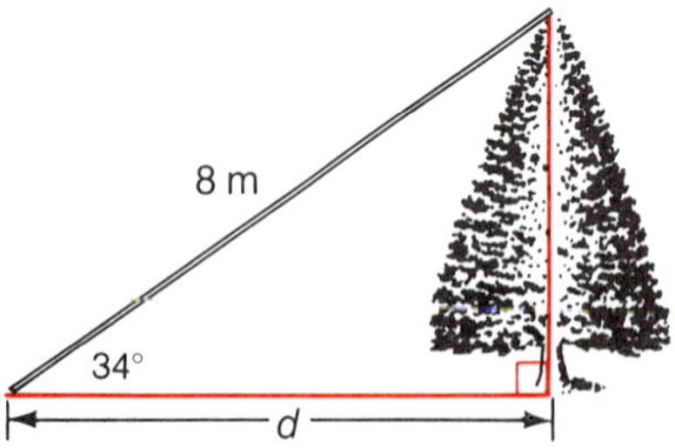

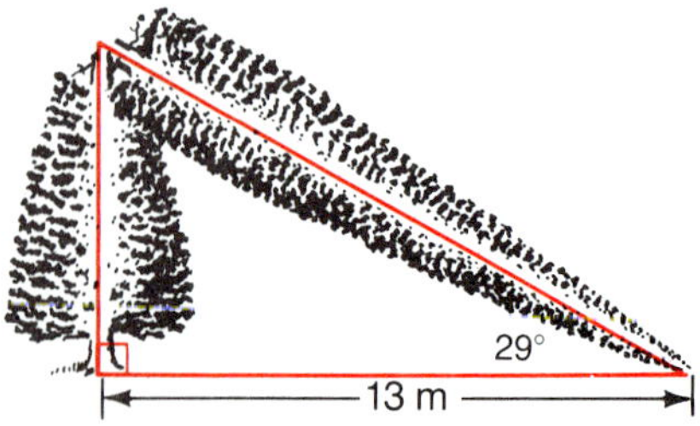

33. A tree is broken by the wind. The top of the tree touches the ground 13 meters from the base and makes an angle of 29° with the ground (see the figure at the right above). Find the original height of the tree.

CALCULATOR APPLICATIONS

Trigonometric Functions

To evaluate a trigonometric function, first enter the measure of the angle. If minutes are involved, change the minutes to degree measure.

EXAMPLE 1 Evaluate: $b = 220 \cos 28°40'$ (Example 1 on page 550)

SOLUTION $220 \cos 28°40' = 220 \cos 28.6667°$ ← 4 0 [÷] 6 0 **(4 places)**

2 2 0 [×] 2 8 . 6 6 6 7 [cos] [=] 193.03353

EXAMPLE 2 Evaluate: $a = \frac{15}{\tan 51°30'}$ (Example 2 on page 550)

SOLUTION **1 5 [÷] 5 1 . 5 [tan] [=]** 11.931539

EXERCISES

Use a calculator to check your answers to Exercises 8–13 on page 551.

Puzzle

Simplify each fraction. Then generalize and explain.

1. $\frac{1+3}{5+7}$ **2.** $\frac{1+3+5}{7+9+11}$ **3.** $\frac{1+3+5+7+9+11}{13+15+17+19+21+23}$

4. $\frac{1+3+5+\cdots+95+97+99}{101+103+105+\cdots+195+197+199}$

REVIEW CAPSULE FOR SECTION 15–5

Complete each statement. *(Pages 542–546)*

1. As θ increases from 0° to 90°, sin θ __?__ from 0 to 1.

2. As θ increases from 90° to 180°, cos θ decreases from __?__ to __?__.

3. Sin θ decreases from 1 to 0 as θ increases from __?__ to __?__.

4. As θ increases from 180° to 270°, __?__ increases from 0 to very large numbers.

5. For θ in Quadrant III, sin θ is __?__. (increasing/decreasing)

6. For θ in Quadrant IV, tan θ is __?__. (increasing/decreasing)

7. For θ in Quadrant II, tan θ is __?__. (increasing/decreasing)

8. For θ in Quadrant I, cos θ is __?__. (increasing/decreasing)

15–5 Interpolation

Interpolation for trigonometric functions involves the same principle as interpolation for logarithms.

In Chapter 16, you will see that the graphs of trigonometric functions are not straight lines. For small portions of the graphs, however, a straight line approximates the graphs quite well.

EXAMPLE 1 Evaluate tan 76°33′.

Solution: Use the table on page 629 to interpolate.

10 [3 [	76°30′	From the table →	4.1653] d] 0.0540
	76°33′		?
	76°40′	From the table →	4.2193

$\frac{3}{10} = \frac{d}{0.0540}$ ← **Write the ratios. Solve for *d*.**

$d = \frac{3 \times 0.0540}{10}$, or $d = 0.0162$

$\tan 76°33' = \tan 76°30' + d$

$\tan 76°33' = 4.1653 + 0.0162 = \mathbf{4.1815}$

Recall that in the first quadrant the cosine function decreases as the measure of the angle increases. This fact is used in Example 2.

EXAMPLE 2 Evaluate cos 32°17′.

Solution:

10 [7 [	32°10′	From the table →	0.8465] d] 0.0015
	32°17′		?
	32°20′	From the table →	0.8450

$\frac{7}{10} = \frac{d}{0.0015}$ ← **Write the ratios. Solve for *d*.**

$d = \frac{7 \times 0.0015}{10}$

$d = 0.0011$

The cosine of 32°17′ is smaller than the cosine of 32°10′.

Thus, $\cos 32°17' = \cos 32°10' - d$

$\cos 32°17' = 0.8465 - 0.0011 = \mathbf{0.8454}$

Given the value of a trigonometric function, you can also use interpolation to find the corresponding angle measure.

EXAMPLE 3 $\tan B = 1.3456$. Find B to the nearest minute.

Solution:

10 (d (	$53°20'$		1.3432	**0.0024**	**0.0082**
	B		1.3456		
	$53°30'$		1.3514		

$$\frac{d}{10} = \frac{0.0024}{0.0082}$$

$$d = \frac{0.0024 \times 10}{0.0082}$$

$d = 2.92$, or about $3'$

Thus, $B = 53°20' + d$

$B = 53°20' + 3'$

$B = \mathbf{53°23'}$

CLASSROOM EXERCISES

In Exercises 1–6, tell whether, when interpolating, you would add or subtract d from the smaller of the two given angles.

1. $\sin \theta$, where $3°10' < \theta < 3°20'$
2. $\cos \theta$, where $39° < \theta < 39°10'$
3. $\tan \theta$, where $24°30' < \theta < 24°40'$
4. $\tan \theta$, where $16°20' < \theta < 16°30'$
5. $\sin \theta$, where $75°40' < \theta < 75°50'$
6. $\cos \theta$, where $83°40' < \theta < 83°50'$

WRITTEN EXERCISES

A Evaluate. Use linear interpolation and the tables on pages 629–633.

1. $\tan 34°23'$
2. $\tan 51°36'$
3. $\tan 75°29'$
4. $\tan 77°14'$
5. $\tan 42°42'$
6. $\tan 88°52'$
7. $\sin 42°18'$
8. $\sin 64°23'$
9. $\sin 79°42'$
10. $\sin 68°15'$
11. $\sin 18°46'$
12. $\cos 17°08'$
13. $\cos 22°12'$
14. $\cos 40°36'$
15. $\cos 64°45'$
16. $\cos 2°43'$

In Exercises 17–28, find θ to the nearest minute. Use linear interpolation and the tables on pages 629–633.

17. $\tan \theta = 0.7325$
18. $\tan \theta = 1.9371$
19. $\tan \theta = 0.3147$
20. $\tan \theta = 1.705$
21. $\sin \theta = 0.8537$
22. $\sin \theta = 0.8670$
23. $\sin \theta = 0.1111$
24. $\sin \theta = 0.3291$
25. $\cos \theta = 0.7561$
26. $\cos \theta = 0.3890$
27. $\cos \theta = 0.5132$
28. $\cos \theta = 0.3333$

APPLICATIONS: Solving Right Triangles

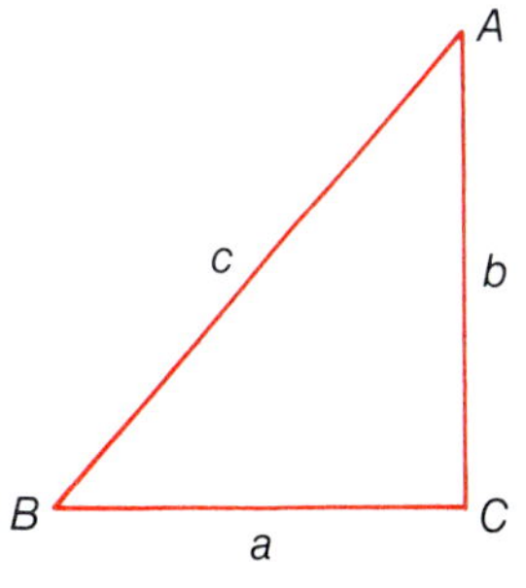

Refer to triangle ABC at the right in Exercises 29–32.

29. Find a when $B = 48°13'$ and $c = 735$ centimeters.
30. Find b when $B = 36°45'$ and $c = 124$ centimeters.
31. Find b when $A = 52°28'$ and $a = 452$ centimeters.
32. Find c when $A = 24°22'$ and $a = 505$ centimeters.

B

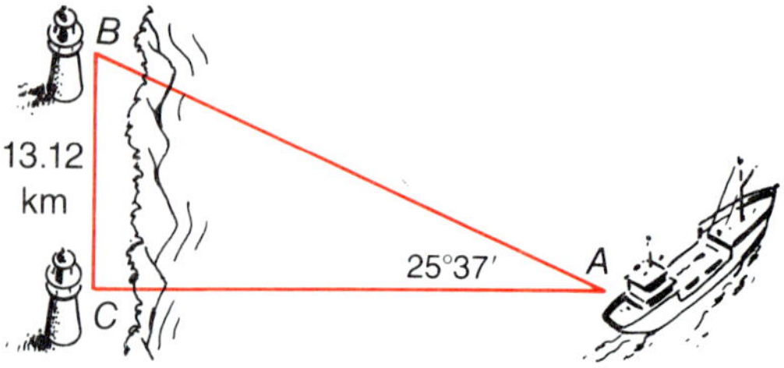

33. Lighthouse C is due west from ship A. Lighthouse B, which is due north of C, is so situated that angle CAB measures $25°37'$. The distance from C to B is 13.12 kilometers. How far is the ship from C?

34. In the triangle at the left below, sides AC and BC are each 462 centimeters long and side AB is 302 centimeters. Find the measures of angles A and B to the nearest minute.

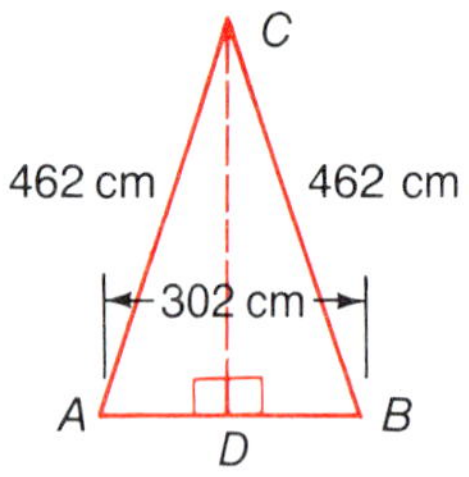

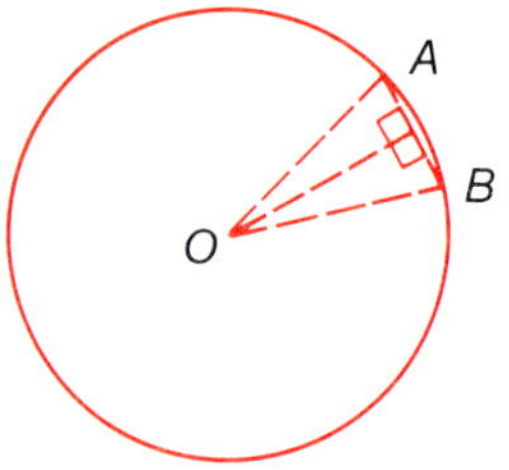

35. In the figure at the right above, the diameter of circle O is 7.000 decimeters and the length of chord AB is 2.000 decimeters. Find the measure of angle AOB to the nearest minute.

C

36. A cylindrical wire with a radius of 0.5000 centimeters is placed in a V-block. Find the distance from B to O. Round your answer to the nearest ten-thousandth.

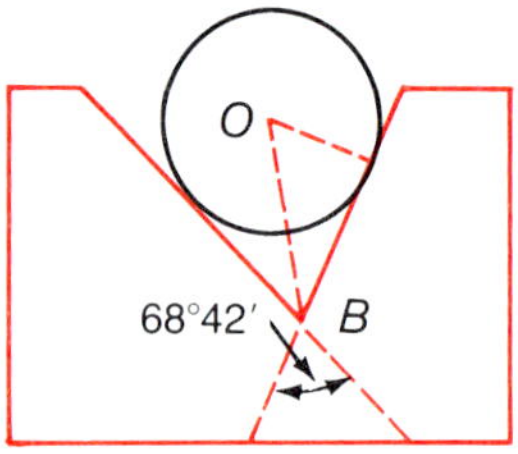

Review

Sketch each angle in standard position. *(Section 15–1)*

1. 70° **2.** 120° **3.** 315° **4.** 540°

5. In which quadrant does the terminal side of a 343° angle lie? *(Section 15–1)*

6. Find three coterminal angles for $\theta = 40°$. Let $n = 1, 2$, and -1. *(Section 15–1)*

7. Define $\sin\theta$, $\cos\theta$ and $\tan\theta$ for $P(x, y)$. *(Section 15–2)*

8. Find $\sin\theta$, $\cos\theta$, and $\tan\theta$ when θ is an angle in standard position whose terminal side contains the point $P(7, 24)$. *(Section 15–2)*

In Exercises 9 and 10, the length of the hypotenuse, c, of each triangle is given. Solve the triangle. *(Section 15–3)*

9. 30°-60°-90° triangle; $c = 12$ **10.** 45°-45°-90° triangle; $c = 6$

In Exercises 11–14, use a reference triangle. Find each value. *(Section 15–3)*

11. tan 30° **12.** cos 135° **13.** sin 330° **14.** tan 240°

15. In $\triangle ABC$, $A = 35°20'$ and $c = 40$ m. Find the length of side a to the nearest whole number. *(Section 15–4)*

16. Find θ to the nearest minute when $\cos \theta = 0.7561$. *(Section 15–5)*

17. Find a when $B = 24°22'$ and $c = 312$ centimeters. *(Section 15–5)*

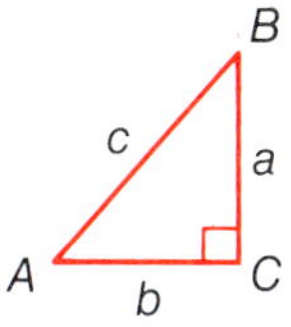

15–6 Angles with Negative Measures

You can also use the definition of the trigonometric functions to find function values for angles with negative measures.

EXAMPLE 1 Evaluate: **a.** sin (−45°) **b.** cos (−45°) **c.** tan (−45°) **d.** sin (−210°) **e.** cos (−210°) **f.** tan (−210°)

Solutions:

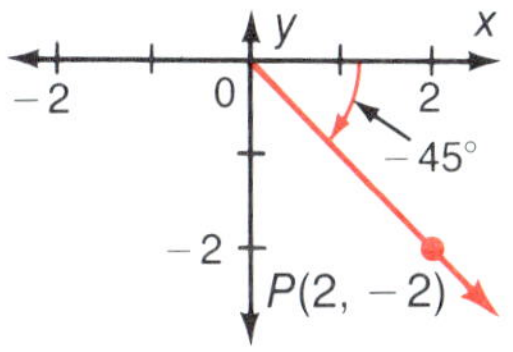

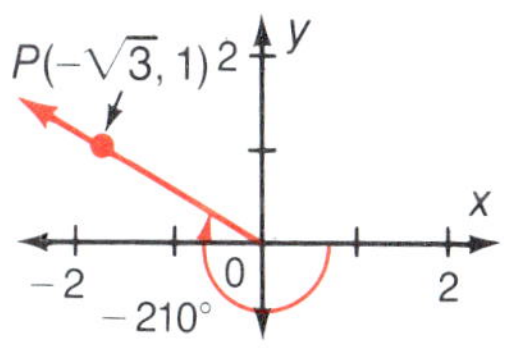

a. $\sin(-45°) = \frac{y}{r} = \frac{-2}{2\sqrt{2}} = -\frac{\sqrt{2}}{2}$

b. $\cos(-45°) = \frac{x}{r} = \frac{2}{2\sqrt{2}} = \frac{\sqrt{2}}{2}$

c. $\tan(-45°) = \frac{y}{x} = \frac{-2}{2} = -1$

d. $\sin(-210°) = \frac{1}{2}$

e. $\cos(-210°) = \frac{-\sqrt{3}}{2} = -\frac{\sqrt{3}}{2}$

f. $\tan(-210°) = \frac{1}{-\sqrt{3}} = -\frac{\sqrt{3}}{3}$

For any angle QOP whose measure is θ, there is exactly one angle whose measure is $-\theta$. If P' is chosen so that $\overline{OP'}$ is congruent to $\overline{OP}$, the reference triangles for angles QOP and QOP' are congruent. In each of four possible cases shown, $x' = x$, $y' = -y$, and $r' = r$.

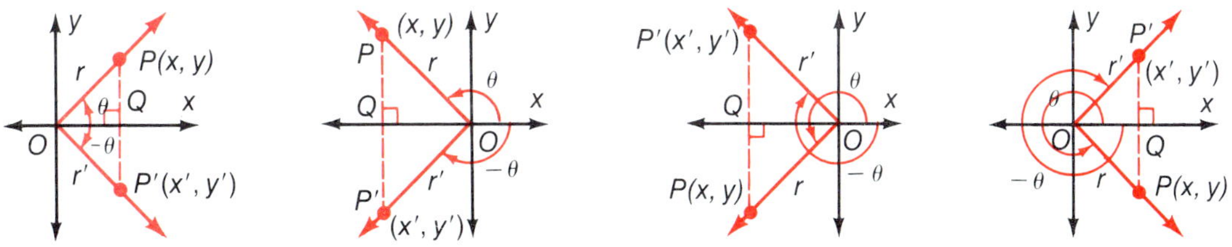

Therefore, $\sin(-\theta) = \frac{y'}{r'} = \frac{-y}{r} = -\sin\theta$

$\cos(-\theta) = \frac{x'}{r'} = \frac{x}{r} = \cos\theta$ and $\tan(-\theta) = \frac{y'}{x'} = \frac{-y}{x} = -\tan\theta.$

Theorem 15–2

> For any angles with measure θ and $-\theta$,
> $$\sin(-\theta) = -\sin\theta$$
> $$\cos(-\theta) = \cos\theta$$
> $$\tan(-\theta) = -\tan\theta$$

Example 2 shows how you can use this theorem to find function values of angles with negative measures.

EXAMPLE 2 Evaluate each of the following. Use the tables on pages 629–633.

a. $\sin(-60)$ **b.** $\cos(-60^\circ)$ **c.** $\tan(-60^\circ)$

Solutions: Sketch an angle of (-60°).

Since the measure of the angle is negative, rotate the terminal side of $\overrightarrow{OB}$ clockwise.

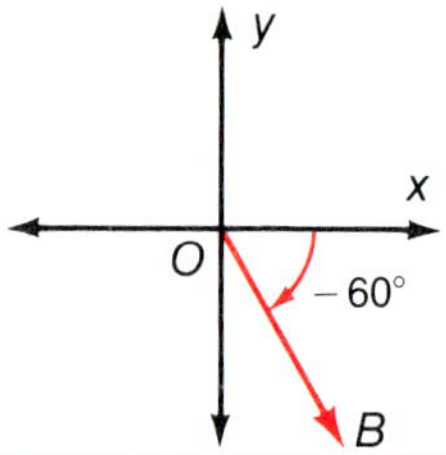

$\sin(-60^\circ) = -\sin 60^\circ = \mathbf{-0.8660}$

$\cos(-60^\circ) = \cos 60^\circ = \mathbf{0.5000}$

$\tan(-60^\circ) = -\tan 60^\circ = \mathbf{-1.7321}$

CLASSROOM EXERCISES

Sketch each angle.

1. (-30°) **2.** (-90°) **3.** (-200°) **4.** (-185°) **5.** (-300°)
6. (-480°) **7.** (-560°) **8.** (-405°) **9.** (-270°) **10.** (-660°)

WRITTEN EXERCISES

A In Exercises 1–18, first sketch the indicated angle. Then use the tables on pages 629–633 to find the values of the sine, cosine, and tangent functions for the angle.

1. -45° **2.** -33° **3.** -165° **4.** -300° **5.** -261° **6.** -85°
7. -200° **8.** -110° **9.** -75° **10.** -50° **11.** -15° **12.** -90°
13. -180° **14.** -270° **15.** -520° **16.** -495° **17.** -315° **18.** -630°

B In Exercises 19–21, show that each statement is true.

19. $\sin(-135^\circ) = \sin(-45^\circ)\cos(-90^\circ) + \cos(-45^\circ)\sin(-90^\circ)$

20. $\cos(-150^\circ) = \cos(-30^\circ)\cos(-120^\circ) - \sin(-30^\circ)\sin(-120^\circ)$

21. $\sin(-240^\circ) = -\sin 240^\circ$

22. Given that $A = \theta$ and $B = 180^\circ - \theta$, how are the reference triangles of $\angle A$ and $\angle B$ related when $\theta < 180^\circ$?

23. In Exercises 22, how are the triangles related when $\theta > 180^\circ$?

15–7 Reduction Formulas

In Figure 1, $A = \theta$ and $B = 180° - \theta$. Thus, the reference triangles of angles A and B are similar for any points P and P' chosen on the terminal sides of $\angle A$ and $\angle B$, respectively.

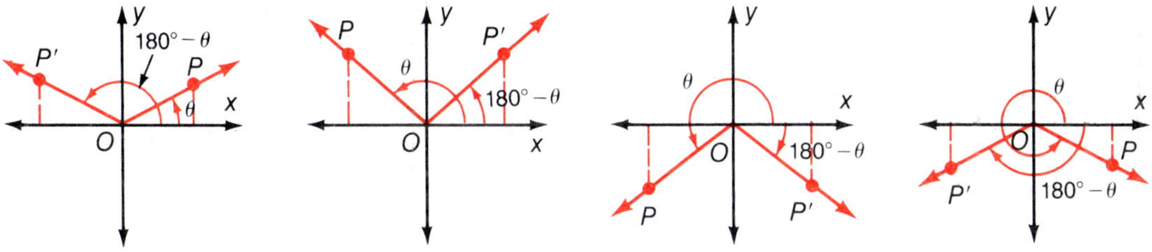

Figure 1

Figure 1 suggests the following theorem.

Theorem 15–3

For any angle with measure θ,

$$\sin (180° - \theta) = \sin \theta$$
$$\cos (180° - \theta) = -\cos \theta$$
$$\tan (180° - \theta) = -\tan \theta$$

In Figure 2, suppose $A = \theta$ and $B = 180° + \theta$. Again, the reference triangles for angles A and B are similar regardless of how points P and P' are chosen.

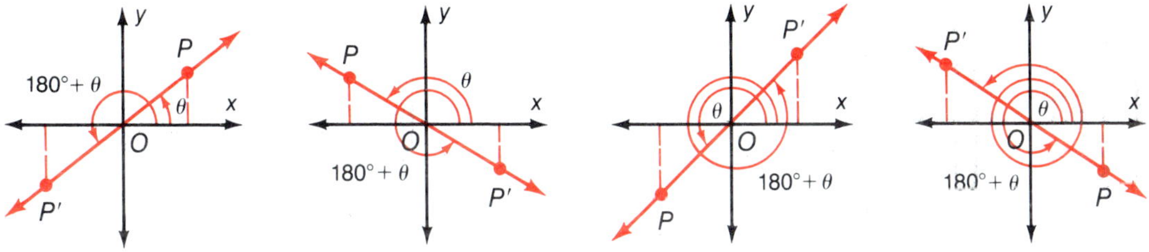

Figure 2

Figure 2 suggests this theorem.

Theorem 15–4

For any angle with measure θ,

$$\sin (180° + \theta) = -\sin \theta$$
$$\cos (180° + \theta) = -\cos \theta$$
$$\tan (180° + \theta) = \tan \theta$$

Theorem 15–5 states a similar relationship between angles with measures θ and $(360° - \theta)$, as well as between angles with measures θ and $(360° + \theta)$.

Theorem 15–5

> For any angle with measure θ,
>
> $\sin(360^\circ - \theta) = -\sin\theta \qquad \sin(360^\circ + \theta) = \sin\theta$
>
> $\cos(360^\circ - \theta) = \cos\theta \qquad \cos(360^\circ + \theta) = \cos\theta$
>
> $\tan(360^\circ - \theta) = -\tan\theta \qquad \tan(360^\circ + \theta) = \tan\theta$

A different relationship exists between the values of the sine, cosine, and tangent functions for the angles whose measures are θ and $(90^\circ - \theta)$.

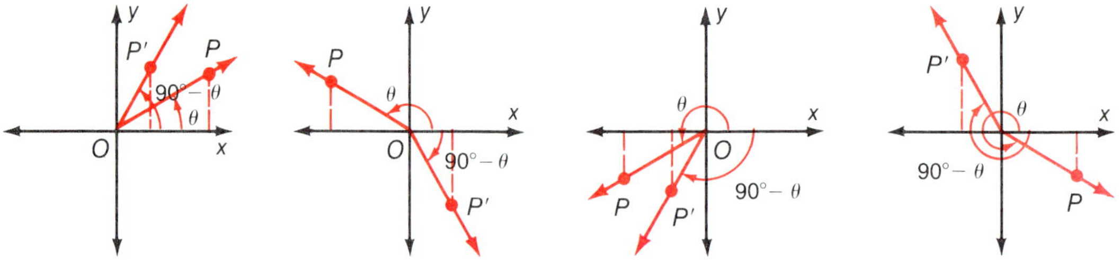

Figure 3

When points P and P' are chosen on the terminal sides of the angles so that $OP' = OP$, the reference triangles are congruent right triangles with $r' = r$, $x' = y$, and $y' = x$.

When P and P' are chosen so that $OP \neq OP'$, the reference triangles of the angles with measures θ and $(90^\circ - \theta)$ are similar and $\frac{r'}{r} = \frac{x'}{y} = \frac{y'}{x}$. Thus, the following relationships hold in both instances.

Theorem 15–6

> For any angle with measure θ,
>
> $\sin(90^\circ - \theta) = \cos\theta$
>
> $\cos(90^\circ - \theta) = \sin\theta$
>
> $\tan(90^\circ - \theta) = \frac{1}{\tan\theta}$

Consider the angles whose measures are θ and $(90^\circ + \theta)$.

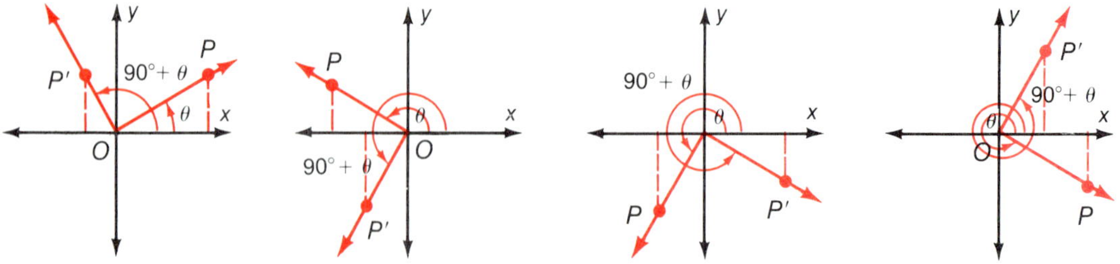

Figure 4

When P and P' are chosen so that $r' = r$, and when the coordinates of P are (x, y), the coordinates of P' are $(-y, x)$.

Theorem 15–7

> For any angle with measure θ,
>
> $$\sin(90° + \theta) = \cos\theta,$$
> $$\cos(90° + \theta) = -\sin\theta,$$
> $$\tan(90° + \theta) = -\frac{1}{\tan\theta}.$$

You can use the theorems in this section to evaluate trigonometric functions.

EXAMPLE Evaluate cos 159°30′. Use the tables on pages 629–633.

Solution:

$\cos 159°30' = \cos(180° - 20°30')$ ← **By Theorem 15–3**

$\cos 159°30' = \cos 20°30'$ ← $\cos(180° - \theta) = -\cos\theta$

$= \mathbf{-0.9367}$

You can use the "ACTS" method to help you remember whether the values of the trigonometric functions are positive or negative in all four quadrants. The drawing at the right indicates which functions are positive in each quadrant. For example, $\sin\theta$ is positive in Quadrants I and II only.

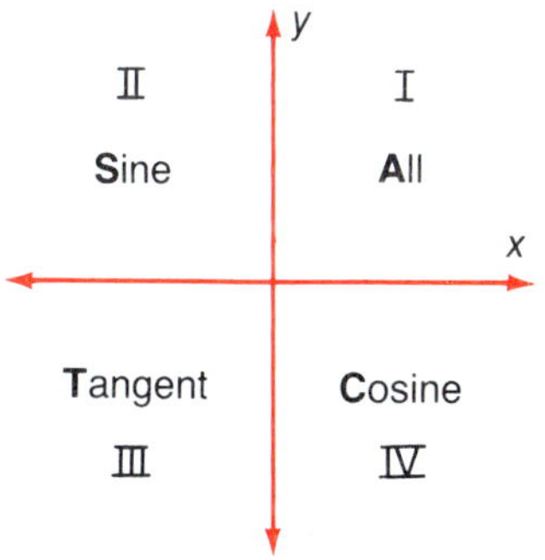

CLASSROOM EXERCISES

Use the Reduction Formulas to complete each statement.

1. $\cos(180° - \theta) =$ _?_
2. $\sin(180° + \theta) =$ _?_
3. $\tan(360° - \theta) =$ _?_
4. $\sin(90° - \theta) =$ _?_
5. $\cos(90° + \theta) =$ _?_
6. $\tan(90° + \theta) =$ _?_
7. $\cos(360° - 25°) = \cos$ _?_
8. $\sin(90° + 15°) = \cos$ _?_
9. $\tan(180° - 75°) = -\tan$ _?_
10. $\sin(180° + 60°) = -\sin$ _?_

In Exercises 11–30, indicate whether the value of each trigonometric function is positive, *P*, or negative, *N*.

11. cos 112°	**12.** sin 91°	**13.** tan 215°	**14.** cos 305°
15. tan 150°	**16.** cos 99°	**17.** sin 194°	**18.** sin 286°
19. tan 312°	**20.** cos 345°	**21.** sin 252°	**22.** cos 98°
23. cos 172°	**24.** sin 303°	**25.** tan 104°	**26.** sin 182°
27. tan 327°	**28.** sin 216°	**29.** tan 244°	**30.** cos 359°

WRITTEN EXERCISES

A Express in the form $(180° - \theta)$.

1. 125°30′	**2.** 145°40′	**3.** 136°10′	**4.** 160°40′	**5.** 116°32′	**6.** 134°26′
7. 99°58′	**8.** 156°45′	**9.** 172°23′	**10.** 127°14′	**11.** 144°06′	**12.** 98°11′

Express in the form $(180° + \theta)$.

13. 217°	**14.** 200°	**15.** 236°	**16.** 245°	**17.** 225°20′	**18.** 237°40′
19. 211°50′	**20.** 253°10′	**21.** 250°17′	**22.** 267°54′	**23.** 269°04′	**24.** 222°25′

Express in the form $(360° - \theta)$.

25. 275°	**26.** 283°	**27.** 294°	**28.** 307°	**29.** 302°10′	**30.** 271°50′
31. 326°40′	**32.** 296°30′	**33.** 345°36′	**34.** 351°15′	**35.** 357°24′	**36.** 289°39′

Express in the form $(90° + \theta)$.

37. 119°	**38.** 172°	**39.** 135°	**40.** 158°	**41.** 135°10′	**42.** 121°20′
43. 148°50′	**44.** 104°40′	**45.** 123°14′	**46.** 162°36′	**47.** 149°57′	**48.** 121°03′

Evaluate. Use the tables on pages 629–633.

49. sin 214°	**50.** tan 129°	**51.** cos 98°	**52.** tan 303°
53. cos 143°30′	**54.** sin 152°40′	**55.** cos 227°10′	**56.** sin 316°50′
57. tan 304°20′	**58.** sin 200°15′	**59.** tan 94°36′	**60.** cos 176°23′

Use a sketch to show that Theorem 15–5 is true for each value of θ.

61. 30°	**62.** 45°	**63.** 135°	**64.** 210°	**65.** 270°20′	**66.** 300°50′

B

67. 168°30′	**68.** 247°10′	**69.** 352°11′	**70.** 118°26′	**71.** 216°08′	**72.** 136°59′

Show that Theorem 15–6 is true for each value of θ.

73. 10°	**74.** 25°	**75.** 52°	**76.** 86°	**77.** 73°	**78.** 61°

REVIEW CAPSULE FOR SECTION 15–8

Simplify each expression. *(Pages 235–241)*

1. $\dfrac{\frac{1}{x}}{y}$ **2.** $x \div \dfrac{y}{x}$ **3.** $\dfrac{\frac{r}{\sqrt{r}}}{y}$ **4.** $\dfrac{x^2 + y}{\frac{1}{x} + \frac{1}{y}}$ **5.** $\dfrac{2}{x + \frac{1}{y}}$

Simplify. Rationalize the denominator. *(Pages 272–274)*

6. $\dfrac{6\sqrt{8}}{\sqrt{6}}$ **7.** $\dfrac{5\sqrt{17}}{\sqrt{12}}$ **8.** $\dfrac{\sqrt{11}}{2\sqrt{6}}$ **9.** $\dfrac{5\sqrt{44}}{\sqrt{18}}$ **10.** $\dfrac{7}{6 - \sqrt{30}}$

15–8 Other Trigonometric Functions

The reciprocals of the sine, cosine, and tangent ratios define the following functions.

Definitions

$$\textbf{cotangent } \theta \textbf{ (cot } \theta\textbf{)} = \frac{1}{\tan\theta} = \frac{1}{\frac{y}{x}} = \frac{x}{y}, \ \tan\theta \neq 0$$

$$\textbf{secant } \theta \textbf{ (sec } \theta\textbf{)} = \frac{1}{\cos\theta} = \frac{1}{\frac{x}{r}} = \frac{r}{x}, \ \cos\theta \neq 0$$

$$\textbf{cosecant } \theta \textbf{ (csc } \theta\textbf{)} = \frac{1}{\sin\theta} = \frac{1}{\frac{y}{r}} = \frac{r}{y}, \ \sin\theta \neq 0$$

EXAMPLE 1 Evaluate: **a.** cot 30° **b.** sec 30° **c.** csc 30°

Solutions:

a. $\cot 30° = \frac{1}{\tan 30°} = \frac{1}{\frac{1}{\sqrt{3}}} = \sqrt{3}$

b. $\sec 30° = \frac{1}{\cos 30°} = \frac{1}{\frac{\sqrt{3}}{2}} = \frac{2}{\sqrt{3}} = \frac{2\sqrt{3}}{3}$

c. $\csc 30° = \frac{1}{\sin 30°} = \frac{1}{\frac{1}{2}} = 2$

EXAMPLE 2 Evaluate sec 57°.

Solution: $\sec 57° = \frac{1}{\cos 57°} = \frac{1}{0.5446}$ ← **From the tables**

$= \mathbf{1.8362}$

EXAMPLE 3 Point $A(12, 5)$ is on the terminal side of an angle θ in standard position. Evaluate the six trigonometric functions of θ.

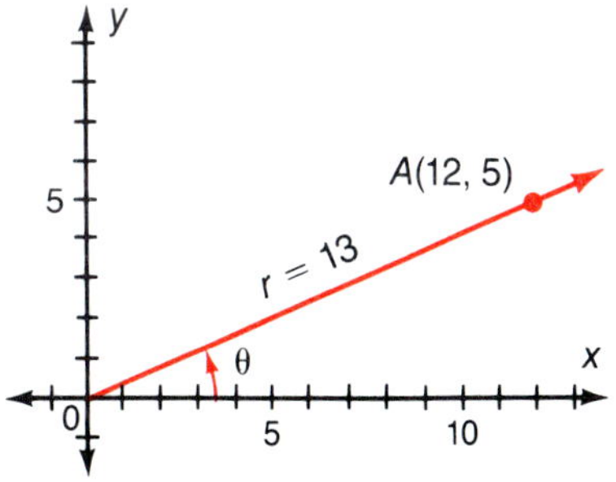

Solution: $r = \sqrt{144 + 25} = 13 \qquad x = 12, y = 5$

$\sin\theta = \frac{\mathbf{5}}{\mathbf{13}} \qquad \csc\theta = \frac{\mathbf{13}}{\mathbf{5}}$

$\cos\theta = \frac{\mathbf{12}}{\mathbf{13}} \qquad \sec\theta = \frac{\mathbf{13}}{\mathbf{12}}$

$\tan\theta = \frac{\mathbf{5}}{\mathbf{12}} \qquad \cot\theta = \frac{\mathbf{12}}{\mathbf{5}}$

WRITTEN EXERCISES

A Use the tables on pages 629–633 to evaluate sec θ, cot θ, and csc θ for each given angle θ in Exercises 1–6. When a function value is not defined, write *ND*.

1. 75° **2.** 68° **3.** 55° **4.** 14° **5.** 27° **6.** 33°

Evaluate. Use the tables on pages 629–633.

7. csc 52° **8.** cot 80° **9.** sec 40° **10.** cot 10°

11. sec 73° **12.** csc 15° **13.** cot 110° **14.** sec 245°

15. csc 340° **16.** sec 83°10′ **17.** csc 210°20′ **18.** cot 326°14′

19. cot 240°18′ **20.** sec 108°09′ **21.** cot 315°24′ **22.** csc 292°53′

In Exercises 23–27, the given point P lies on the terminal side of an angle θ in standard position. Evaluate the six trigonometric functions of θ.

23. $P(-24, 7)$ **24.** $P(4, -5)$ **25.** $P(-5, -12)$ **26.** $P(4, 0)$ **27.** $P(\sqrt{3}, -1)$

In Exercises 28–33, use the given information to find θ where $0° \le \theta \le 90°$.

28. sec $\theta = 1.236$ **29.** cot $\theta = 1.8165$ **30.** cot $\theta = 3.9617$

31. csc $\theta = 5.575$ **32.** sec $\theta = 1.187$ **33.** csc $\theta = 1.122$

B In which quadrant must θ lie under the given conditions?

34. sec $\theta < 0$ and cot $\theta > 0$ **35.** csc $\theta > 0$ and cot $\theta < 0$

36. sec $\theta > 0$ and cot $\theta > 0$ **37.** sec $\theta > 0$ and csc $\theta < 0$

Problem Solving and Applications

15–9 Angles of Elevation and Depression

The trigonometric functions are useful in calculating distances. In many of these instances, an angle is determined by a horizontal line and a *line of sight.*

In the figure at the right, angle BEP is the **angle of elevation** of E from P. To measure $\angle BEP$, a surveyor places his transit at E and first points the telescope horizontally toward the flagpole. Then the telescope is pointed to the top of the pole, P.

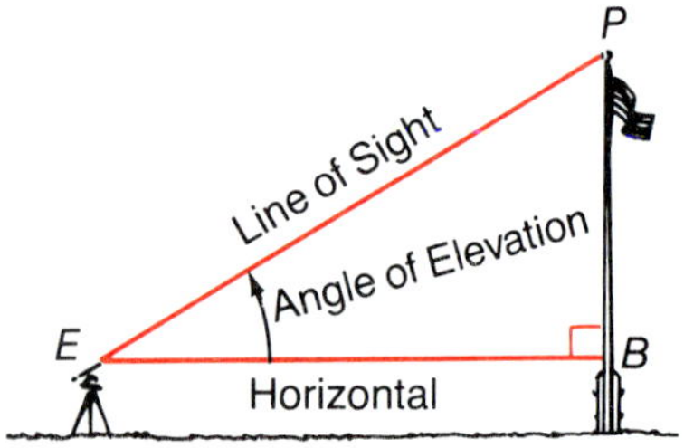

The figure at the right shows an **angle of depression.** When the telescope is turned from the horizontal position to DM (looking at the persons in the boat), it turns through the angle of depression, HDM.

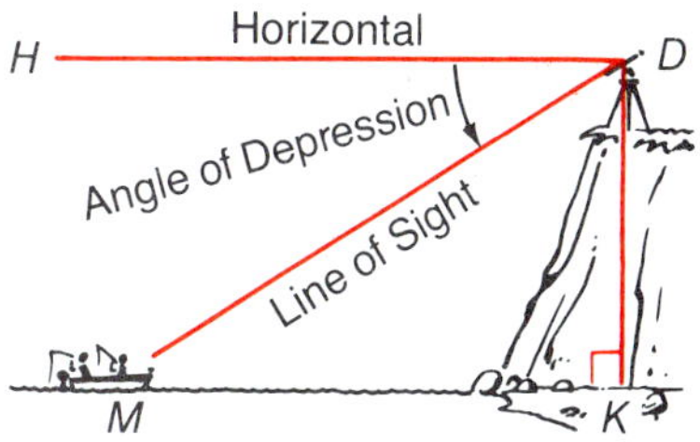

EXAMPLE 1 At a point 5 meters from the base of a building, the angle of elevation of the top of the building is 80°. Find the height of the building. Round your answer to the nearest meter.

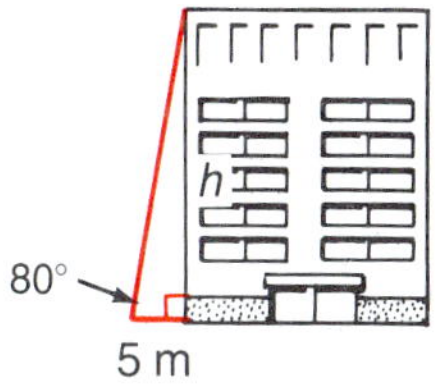

Solution: $\tan 80° = \frac{h}{5}$

$h = 5 \tan 80°$

$h = 5(5.6713) = 28.3565$ The building is about **28 meters** high.

Example 2 shows how an angle of depression is used in solving a problem involving right triangles.

EXAMPLE 2 From the top of a lighthouse, the angle of depression of a boat is 38°. The height of the lighthouse is 126 feet. Find, to the nearest foot, the line-of-sight distance from the lighthouse to the boat.

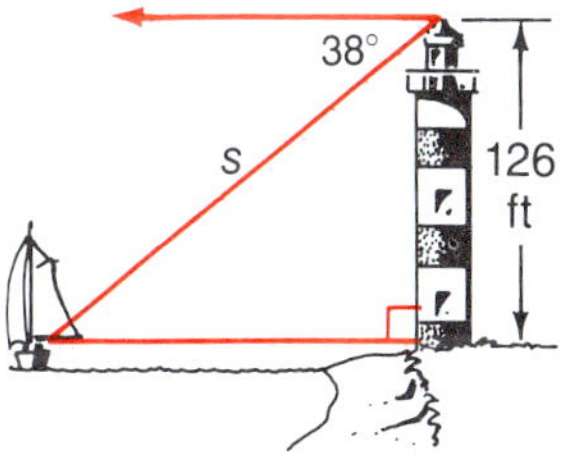

Solution: $\sin 38° = \frac{126}{s}$

$s \cdot \sin 38° = 126$

$s = \frac{126}{\sin 38°}$

$s = \frac{126}{0.6157} = 204.64$ The distance is about **205 feet.**

CLASSROOM EXERCISES

Refer to the rectangle at the right in Exercises 1–6.

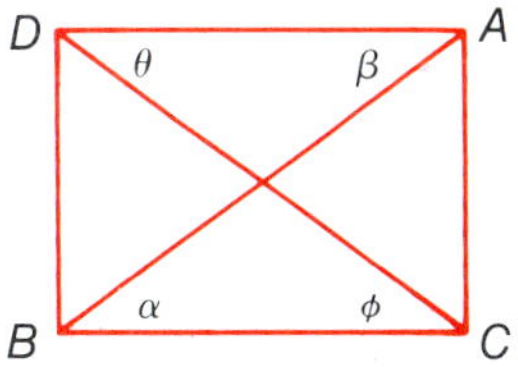

1. Suppose that you are looking from B to A. Name the angle of elevation.
2. Suppose that you are looking from C to D. Name the angle of elevation.
3. Suppose that you are looking from D to C. Name the angle of depression.
4. Suppose that you are looking from A to B. Name the angle of depression.
5. Given that $\alpha = 40°$, find β.
6. Given that $\theta = 50°$, find ϕ.

WORD PROBLEMS

A

1. From the top of a cliff 800 meters high, the angle of depression to the base of a log cabin is 37°20′. Find the distance from the cabin to the foot of the cliff. Round your answer to the nearest meter.

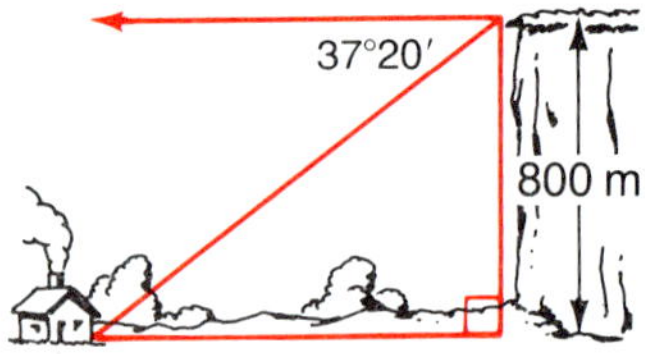

2. An airplane over the Atlantic Ocean is 150,000 feet from shore. The angle of depression of the shore is 13°20′. Find the altitude, h, of the plane to the nearest foot. (See the figure at the left below.)

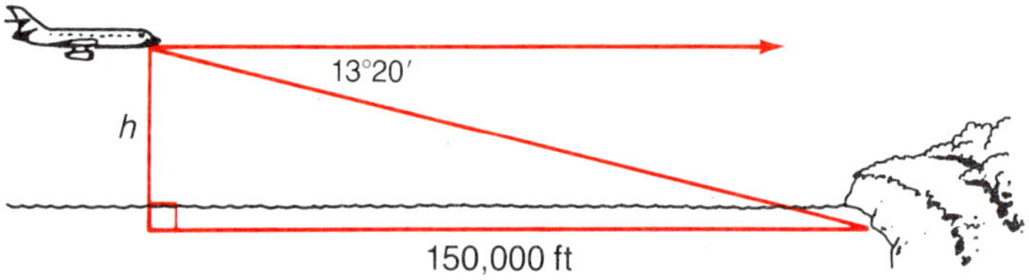

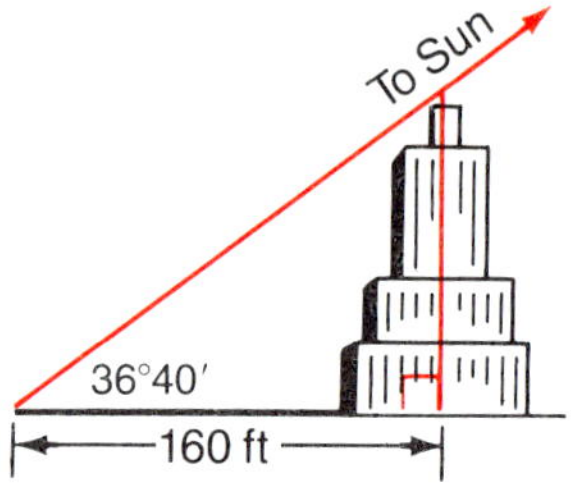

3. A building casts a shadow of 160 feet. From the end of the shadow the angle of elevation of the sun is 36°40′. Find the height of the building to the nearest foot. (See the figure at the right above.)

4. Two mountain stations are connected by a cable car. The angle of depression of the cable is 50°. The vertical distance between the stations is 1417 meters. Find the length of the cable. Round your answer to the nearest meter. (See the figure at the left below.)

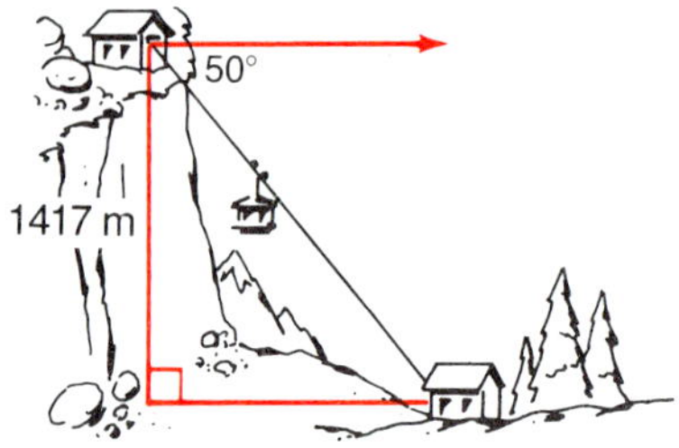

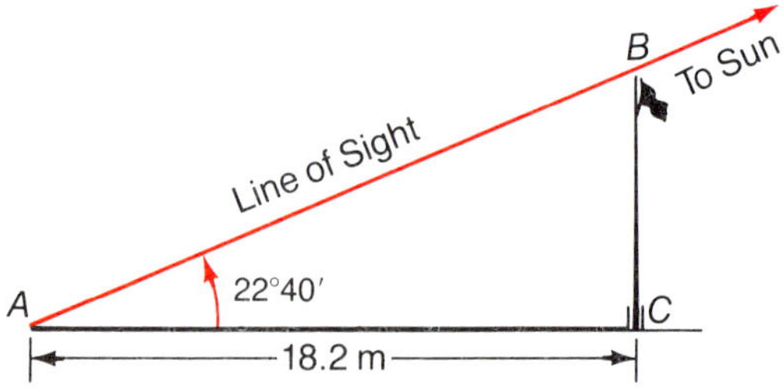

5. A flagpole casts a shadow 18.2 meters long. From the end of the shadow, the angle of elevation of the sun is 22°40′. Find the height of the flagpole to the nearest meter. (See the figure at the right above.)

B

6. The Hirsch Building and the County Hospital are 38 meters apart. From a window in the Hirsch Building, the angle of elevation of the top of the hospital is 73°. From the same window the angle of depression of the ground at the base of the hospital is 64°. Find the height of the hospital to the nearest meter. (See the figure at the right.)

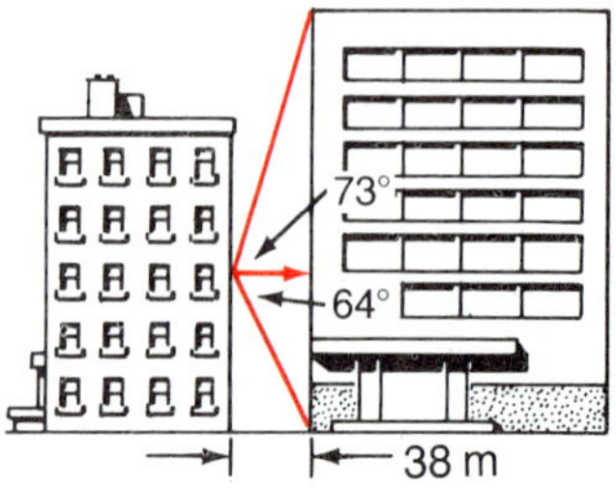

7. Two boats are observed from a tower 75 meters above a lake. The angles of depression are 12°30′ and 7°10′. How far apart, to the nearest meter, are the boats? (Refer to the figure at the left below.)

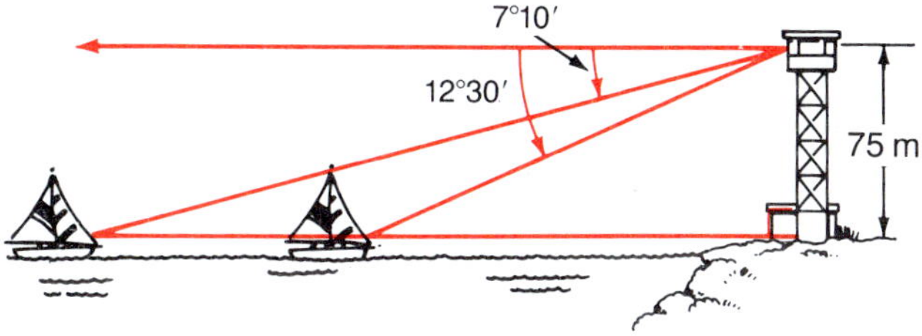

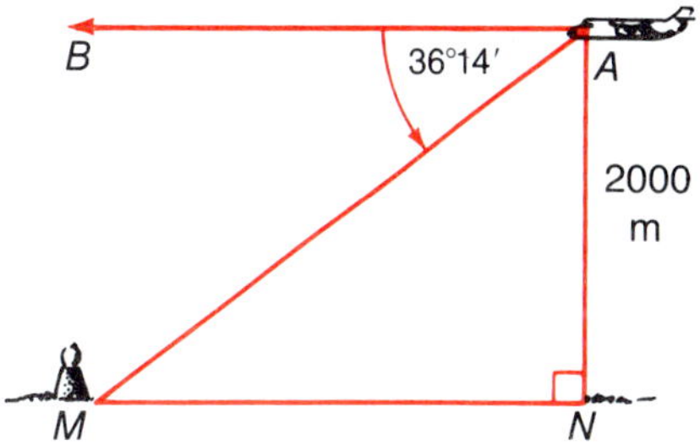

C

8. An aviator observes the measure of the angle of depression of a marker to be 36°14′. The plane is 2000 meters above the ground. How far from the marker is the point on the ground directly under the plane? (Refer to the figure at the right above.)

9. A camp director wishes to buy a new rope for a flagpole at the camp. At a point 4 meters from the foot of the pole, she measured the angle of elevation of the top of the pole and found it to be 82°36′. What length of rope should she buy if she wishes it to be double the distance from the top of the pole to the spot from which she measured the angle?

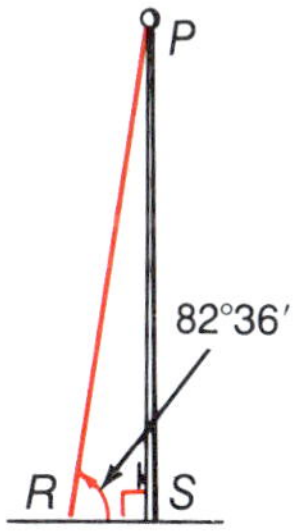

15–10 Law of Sines

Recall that **solving a triangle** means finding the measures of the unknown sides and angles of the triangle. You have already solved right triangles by using the Pythagorean Theorem and the definitions of trigonometric functions. These techniques can be extended to the solution of any triangle.

EXAMPLE 1 Find, to the nearest meter, the width of Sun Tan Lake from point A to point B. The length of $\overline{AC}$, or b, equals 95 meters, and the measures of angle B and angle C are 39° and 65°, respectively.

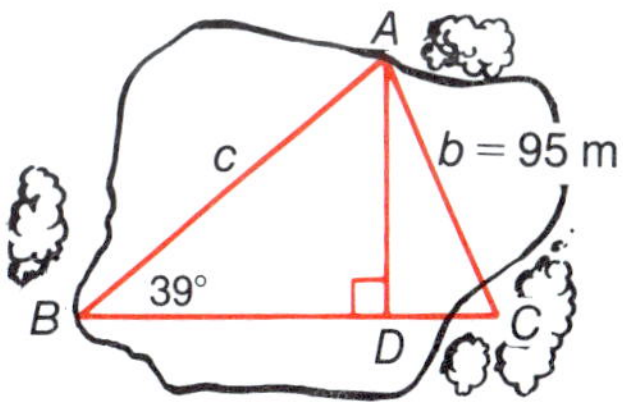

Solution: From point A, construct $\overline{AD}$ perpendicular to $\overline{BC}$.

In right triangle ADB,	In right triangle ADC,
$\sin B = \frac{h}{c}$	$\sin C = \frac{h}{b}$
$h = c \sin B$	$h = b \sin C$

Thus, $b \sin C = c \sin B$ or $\dfrac{b}{\sin B} = \dfrac{c}{\sin C}$.

Since the measures of $\angle B$, $\angle C$, and b are known, this equation can be used to calculate c, the distance from A to B.

$$\frac{95}{\sin 39^\circ} = \frac{c}{\sin 65^\circ}, \quad \text{or} \quad c = \frac{95 \cdot \sin 65^\circ}{\sin 39^\circ}$$

Thus, $c = \dfrac{95(0.9063)}{0.6293}$. ← **From the tables on pages 629–633.**

$c = 136.81$

The width is about **137 meters.**

Example 1 suggests the following theorem.

Theorem 15–8

Law of Sines

In any triangle, the ratio of the length of a side to the sine of the angle opposite that side is the same for each side–angle pair. That is, in any $\triangle ABC$,

$$\frac{a}{\sin A} = \frac{b}{\sin B} = \frac{c}{\sin C}.$$

To prove the Law of Sines, two cases must be considered. Case I, where $\triangle ABC$ is an acute triangle, is proved below. Case II, where $\triangle ABC$ is an obtuse triangle, is left for the Written Exercises.

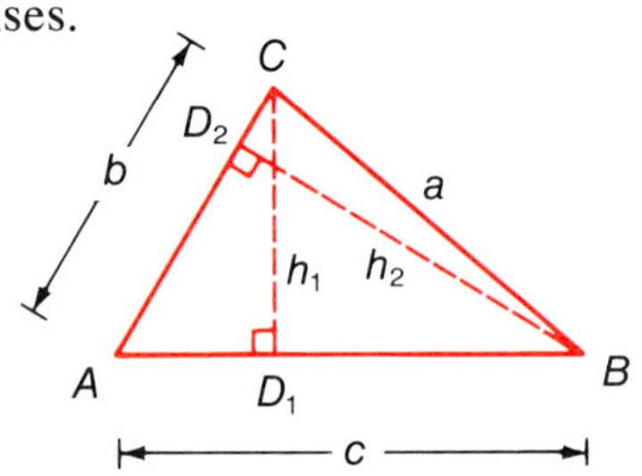

Case I: Acute Triangles

Proof: Each angle of $\triangle ABC$ is an acute angle.

The measures of the altitudes from C and B are h_1 and h_2, respectively.

In right triangles ACD_1 and BCD_1,

$\dfrac{h_1}{b} = \sin A$ and $\dfrac{h_1}{a} = \sin B$.

Hence, $h_1 = b \sin A$ and $h_1 = a \sin B$, so $b \sin A = a \sin B$.

Thus, $\dfrac{b}{\sin B} = \dfrac{a}{\sin A}$.

In right triangles ABD_2 and CBD_2,

$\dfrac{h_2}{c} = \sin A$ and $\dfrac{h_2}{a} = \sin C$.

Thus, $\dfrac{a}{\sin A} = \dfrac{c}{\sin C}$.

Therefore, $\dfrac{a}{\sin A} = \dfrac{b}{\sin B} = \dfrac{c}{\sin C}$.

EXAMPLE 2 In $\triangle ABC$, $A = 35^\circ$, $C = 115^\circ$, and $b = 250$ units. Find sides AB and BC. Round answers to the nearest unit.

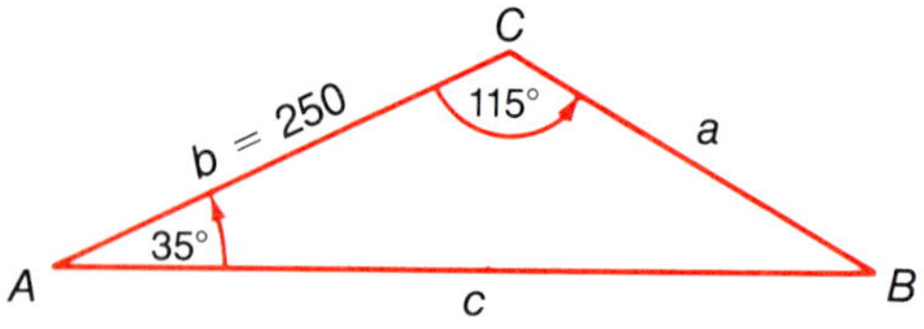

Solution: First find the angle opposite the known side, b.

$B = 180° - (115° + 35°) = 30°$

Now use the Law of Sines.

$$\frac{c}{\sin C} = \frac{b}{\sin B}, \text{ or } \frac{c}{\sin 115°} = \frac{250}{\sin 30°}$$ ← **Solve for c.**

$$c = \frac{250 \sin 115°}{\sin 30°}$$

$$c = \frac{250 \cdot \sin 65°}{\sin 30°}$$ ← **sin 115° = sin(180° − 115°)**

$$c = \frac{250\ (0.9063)}{0.5000}$$ ← **From the tables on pages 629–633.**

$c = 453.15$ Thus, side AB is about **453** units long.

Use the Law of Sines again to find a.

$$\frac{a}{\sin A} = \frac{b}{\sin B}, \text{ or } \frac{a}{\sin 35°} = \frac{250}{\sin 30°}$$

$$a = \frac{250 \sin 35°}{\sin 30°}$$

$$a = \frac{250\ (0.5736)}{0.5000}$$ ← **From the tables on pages 629–633.**

$a = 286.8$ Thus, side BC is about **287** units long.

CLASSROOM EXERCISES

Express each of the following as a function of a first-quadrant angle.

1. $\sin 105°$ **2.** $\cos 135°$ **3.** $\cos 118°$ **4.** $\sin 170°$ **5.** $\cos 150°$

Evaluate. Use the tables on pages 629–633.

6. $\sin 125°$ **7.** $\cos 120°$ **8.** $\cos 125°$ **9.** $\sin 101°$ **10.** $\sin 130°$

WRITTEN EXERCISES

A Solve each triangle.

1. $a = 13, A = 41°, B = 75°$

2. $A = 71°, a = 20, C = 62°$

3. $A = 71°, B = 42°, c = 15$

4. $a = 12, B = 110°, c = 35°$

5. $b = 503, A = 15°, B = 105°$

6. $B = 125°, A = 28°, b = 14$

7. $c = 16.5, A = 38°, C = 54°$

8. $b = 14.4, A = 72°, C = 19°$

9. $b = 224, A = 21°10', B = 84°40'$

10. $c = 916, A = 15°40', B = 60°30'$

11. $A = 101°, c = 37°10', a = 23$

12. $B = 152°, b = 95, C = 12°10'$

13. $a = 1.50, B = 32°30', C = 54°50'$

14. $a = 75.36, A = 18°25', C = 32°05'$

APPLICATIONS: Using the Law of Sines

15. To find the distance from a point A to a point B across a river, a base line $\overline{AC}$ is established (see the figure at the left below). $\overline{AC}$ is 495 meters long. Angles BAC and BCA are 89° and 55° respectively. Find the distance from A to B.

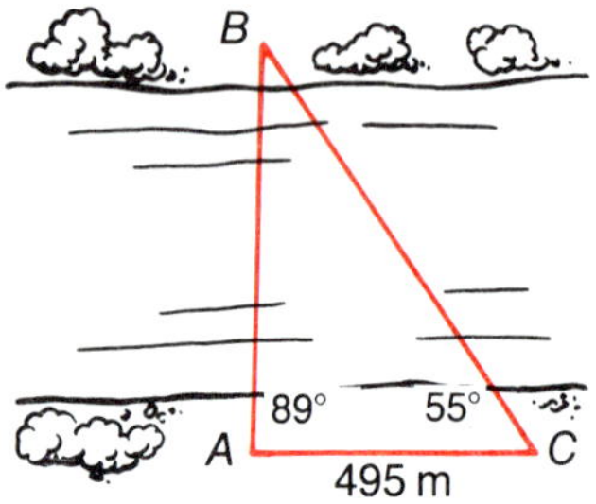

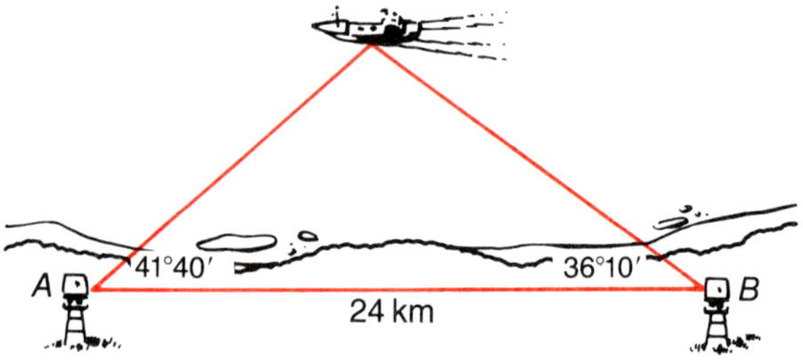

16. A ship at sea is sighted from two observation posts, A and B, on shore (see the the figure at the right above). Points A and B are 24 kilometers apart. The measure of the angle at A between $\overline{AB}$ and the ship is 41°40′. The angle at B is 36°10′. Find the distance, to the nearest tenth of a kilometer, from observation post A to the ship.

B

17. Two points, M and N, are separated by a swamp. A base line $\overline{MK}$ is established on one side of the swamp. $\overline{MK}$ is 180 meters in length. The angles NMK and MKN are measured and found to be 44° and 62°, respectively. Find the distance between M and N.

18. Two angles of a triangle are 30° and 55° and the longest side is 34 meters. Find the length of the shortest side. (HINT: The shortest side is opposite the smallest angle.)

19. Two ranger stations located 10 kilometers apart receive a distress call from a camper. Electronic equipment allows them to determine that the camper is at an angle of 71° from the first station and 100° from the second. Each of these angles has as one side the line segment connecting the stations. Which of the stations is closer to the camper? How far away is it from the camper?

20. A tree stands at point C across a river from point A. A base line $\overline{AB}$ is established on one side of the river. The measure of $\overline{AB}$ is 80 meters. The measure of $\angle BAC$ is 54°20′ and that of $\angle CBA$ is 74°10′. The angle of elevation of the top of the tree from A measures 10°20′. Find the height of the tree.

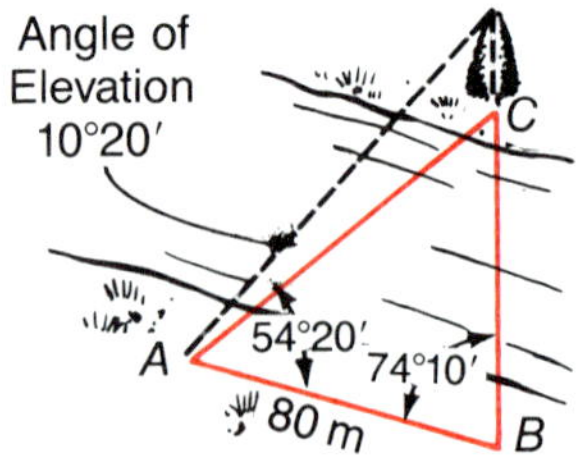

21. From the top and bottom of a tower 28 meters high, the angles of depression of a ship are 18°40′ and 14°20′, respectively. What is the distance of the ship from the foot of the tower? (HINT: The distance is not measured along a horizontal segment in this case.)

22. In $\triangle ABC$, $\angle B$ is an obtuse angle and the measures of the altitudes from C and B are h_1 and h_2, respectively. Prove Case II of the Law of Sines. (HINT: $\angle CBD_1 = 180° - B$, and $\sin(180° - B) = \sin B$.)

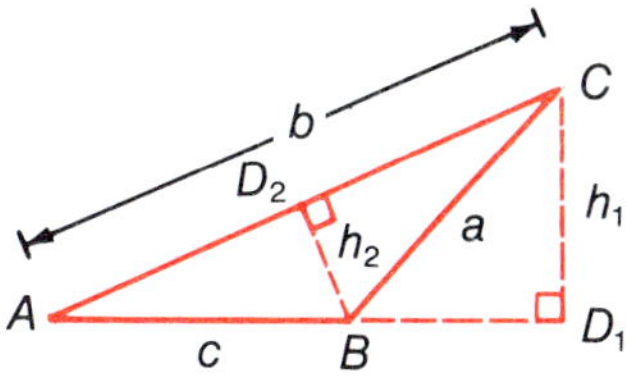

C You can use the Law of Sines to solve a triangle when you know the lengths of two of its sides and the measure of an angle opposite one of them. In this case there may be no, one, or two solutions.

The diagrams below illustrate the various possibilities, given *A, b,* and *a*. They are divided into two cases:

$$A < 90° \quad \text{and} \quad A \geq 90°.$$

Case I: $A < 90°$

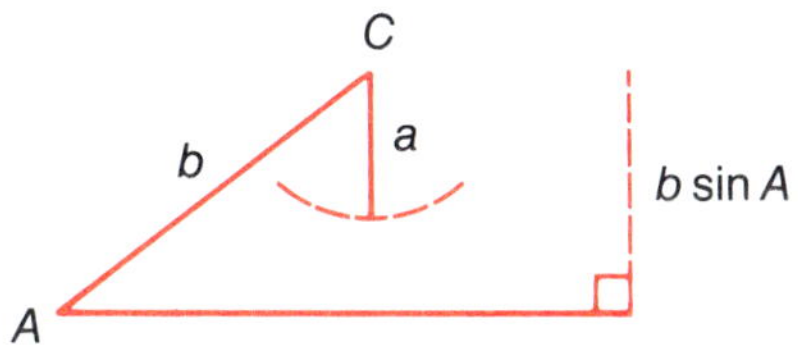

No solution: $a < b \sin A$

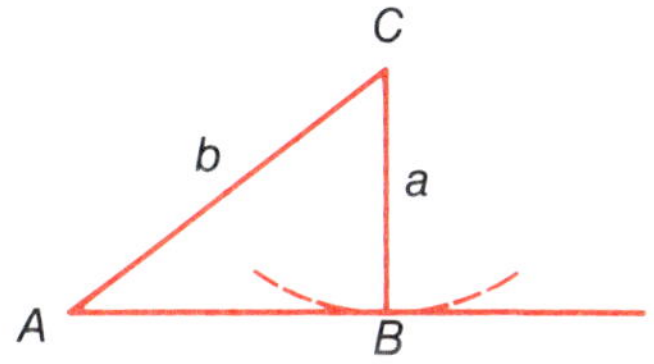

One solution: $a = b \sin A$

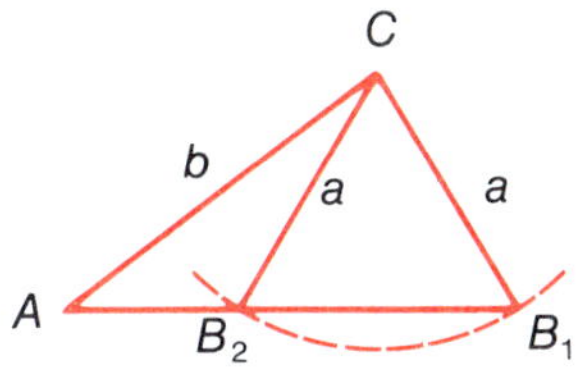

Two solutions: $a < b$ and $a > b \sin A$

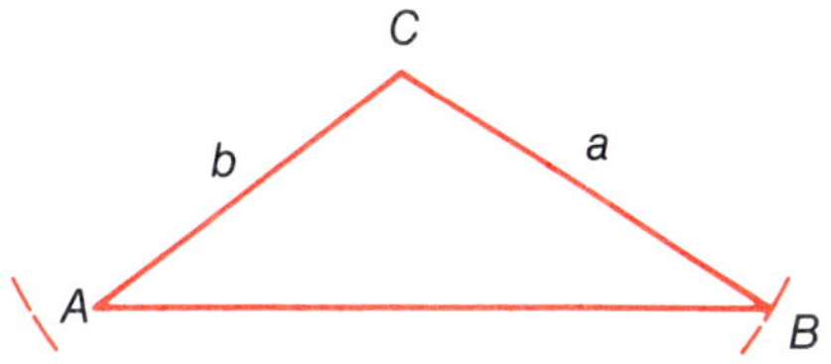

One solution: $a \geq b$

Case II: $A \geqq 90°$

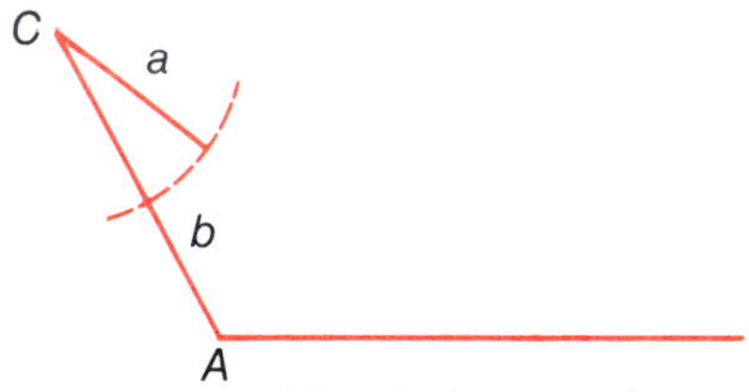

No solution: $a \leq b$

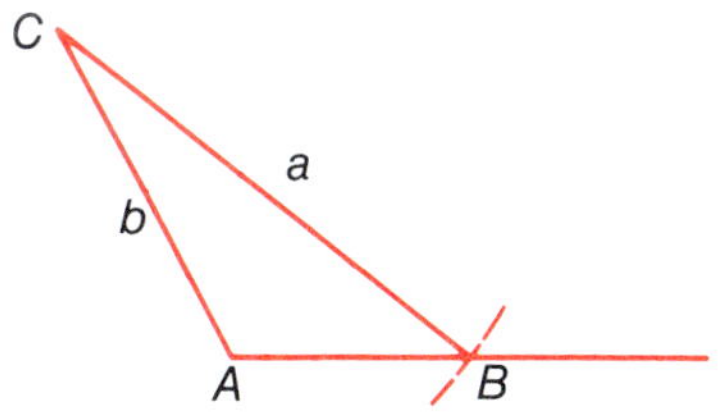

One solution: $a \geq b$

In Exercises 23–28, sketch and label $\triangle ABC$. Then determine the number of possible solutions.

23. $A = 30°,\ a = 6,\ b = 12$

24. $A = 30°,\ a = 8,\ b = 12$

25. $A = 30°,\ a = 4,\ b = 12$

26. $A = 73°,\ a = 8,\ b = 8$

27. $A = 42°,\ a = 5,\ b = 7$

28. $A = 93°,\ a = 4,\ b = 8$

29. The distance from the earth E to the sun S is approximately 1.5×10^8 kilometers. The distance from Venus to the sun is about 1.1×10^8 kilometers. The angle observed on the earth between Venus and the sun is 20°. Find the distance from the earth to Venus. (HINT: There are two possible positions for Venus, indicated by V_1 and V_2.)

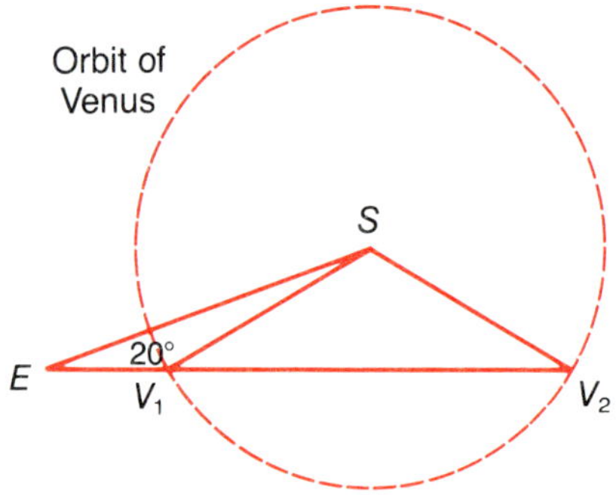

CALCULATOR APPLICATIONS

Law of Sines

The Example below shows how to use a scientific calculator to solve for *a*, *b*, or *c* in the Law of Sines.

EXAMPLE $c = \dfrac{95 \cdot \sin 65°}{\sin 39°}$; $c =$ ___?___ (Example 1, pages 567–568)

SOLUTION 9 5 [×] 6 5 [sin] [÷]   86.09924

3 9 [sin] [=]   136.81305

EXERCISES

Use a calculator to do Exercises 1–14 on page 569.

15–11 Law of Cosines

When only two sides and the included angle of a triangle are known or when only three sides are known, you cannot use the Law of Sines. In these cases, you use the *Law of Cosines*.

Theorem 15–9

> **Law of Cosines**
>
> In any triangle, the square of a side is equal to the sum of the squares of the other two sides minus twice the product of these two sides and the cosine of their included angle. In $\triangle ABC$,
>
> $$a^2 = b^2 + c^2 - 2bc \cos A$$
> $$b^2 = a^2 + c^2 - 2ac \cos B$$
> $$c^2 = a^2 + b^2 - 2ab \cos C.$$

Case I of the Law of Cosines, where $\triangle ABC$ is an acute triangle, is proved below. Case II, where $\triangle ABC$ is an obtuse triangle, is left for the Exercises.

Case I: Acute Triangles

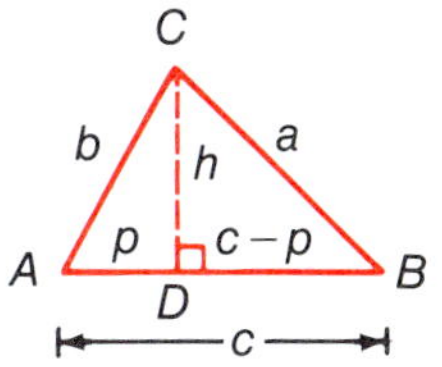

Proof: Suppose $\triangle ABC$ is acute and $\overline{CD} \perp \overline{AB}$.

In $\triangle CDB$, $a^2 = h^2 + (c - p)^2$ **1**

In $\triangle CAD$, $h^2 = b^2 - p^2$ **2**

$$a^2 = b^2 - p^2 + c^2 - 2cp + p^2$$

$$a^2 = b^2 + c^2 - 2cp \quad \mathbf{3}$$

← **Substitute for h^2 in Equation 1.**

But $\cos A = \frac{p}{b}$ or $p = b \cos A$ **4**

← **Substitute for p in Equation 3.**

Thus, $a^2 = b^2 + c^2 - 2bc \cos A$

Since a represents any side of a triangle, the formulas for b^2 and c^2 may be derived by interchanging the letters.

EXAMPLE 1 In $\triangle ABC$, $A = 40°$, $b = 10$, and $c = 20$. Use the tables on pages 629–633 to find a to the nearest tenth.

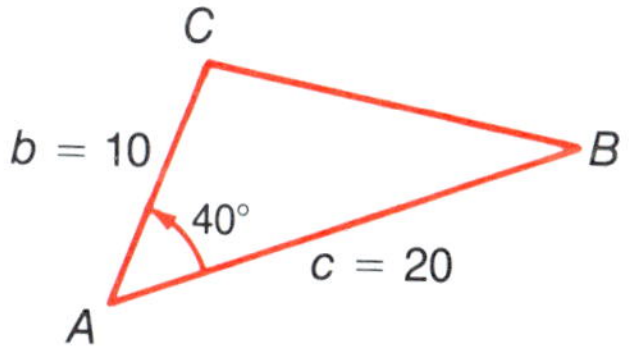

Solution: Since two sides and the included angle are given, use the Law of Cosines to find a.

$$a^2 = b^2 + c^2 - 2bc \cos A$$
$$= (10)^2 + (20)^2 - 2 \cdot 10 \cdot 20 \cdot \cos 40°$$
$$= 100 + 400 - 400(0.7660)$$
$$= 500 - 306.4$$
$$= 193.6$$

Thus, $a = \sqrt{193.6}$, or about **13.9.**

EXAMPLE 2 The distance at the ground between the sides of the roof on an A-frame cabin is 20 meters. The length of each side is 12 meters. Use the tables on pages 629–633 to find the measure of the angle between these sides.

Solution: Since the three sides of the triangle are known, use the Law of Cosines to find $\angle A$.

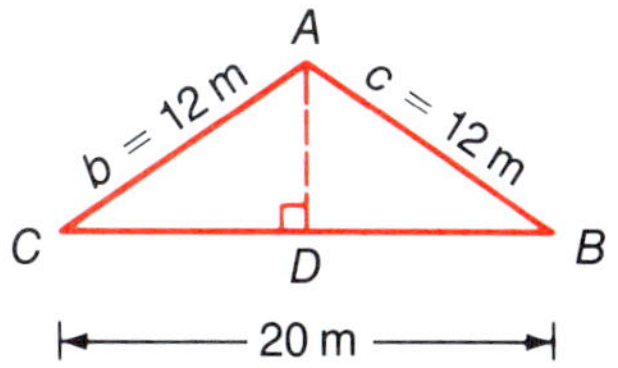

$$a^2 = b^2 + c^2 - 2bc \cos A$$

$$\cos A = \frac{b^2 + c^2 - a^2}{2bc}$$

$$= \frac{(12)^2 + (12)^2 - (20)^2}{2(12)(12)} = -0.3889$$

For $\cos A < 0$, $\angle A$ is either in Quadrant II or Quadrant III. From the conditions of the problem, $\angle A$ must be in Quadrant II.

Thus, $\angle A$ is about **112″50′.**

Sometimes a triangle is solved by using the Law of Cosines and the Law of Sines.

EXAMPLE 3 In $\triangle ABC$, $a = 14$, $b = 11$, and $c = 21$. Use the tables on pages 629–633 to solve the triangle.

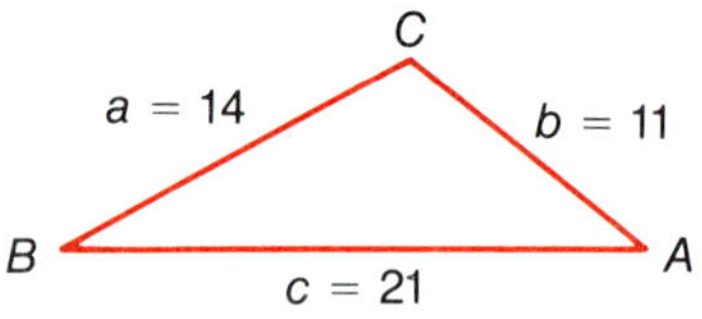

Solution:

$$a^2 = b^2 + c^2 - 2bc \cos A$$

$$\cos A = \frac{b^2 + c^2 - a^2}{2bc}$$

$$\cos A = \frac{11^2 + 21^2 - 14^2}{2 \cdot 11 \cdot 21} = 0.7922$$ ← **Since cos *A* is positive, *A* is in Quadrant I.**

Therefore, $A = \mathbf{37°40'}$. ← **To the nearest ten minutes**

Next, find B. Use the Law of Sines.

$$\frac{b}{\sin B} = \frac{a}{\sin A}$$ ← **Solve for sin *B*.**

$$\sin B = b\,\frac{\sin A}{a}$$

$$= 11\left(\frac{\sin 37°40'}{14}\right) = 11\left(\frac{0.6111}{14}\right) = 0.4802$$

Therefore, $B = \mathbf{28°40'}$, to the nearest ten minutes.

Finally, $C = 180° - (A + B) = 180° - (37°40' + 28°40')$.

Therefore, $C = \mathbf{113°40'}$, to the nearest ten minutes.

CLASSROOM EXERCISES

In Exercises 1–4, use the tables on pages 629–633.

1. In triangle ABC, $b = 5$, $c = 4$, and $\cos A = \frac{1}{8}$. Find a.
2. In triangle PQR, $p = 3$, $q = 4$, and $R = 120°$. Find r.
3. In triangle DEF, $d = 5$, $e = 7$, and $f = 6$. Find $\cos E$.
4. In triangle GHW, $g = 2$, $h = 4$, and $w = 5$. Find $\cos G$.

WRITTEN EXERCISES

A

In Exercises 1–21 and 24–28, use the tables on pages 629–633 to give the lengths of sides to the nearest tenth, and the measures of angles to the nearest ten minutes.

In Exercises 1–8, find the indicated side of $\triangle ABC$.

1. $a = 5$, $b = 8$, $C = 35°$; $c =$ __?__
2. $b = 7$, $c = 10$, $A = 51°$; $a =$ __?__
3. $b = 9$, $c = 11$, $A = 123°$; $a =$ __?__
4. $a = 7$, $c = 5$, $B = 152°$; $b =$ __?__
5. $a = 8$, $b = 6$, $C = 60°$; $c =$ __?__
6. $a = 9$, $c = 5$, $B = 120°$; $b =$ __?__
7. $b = 6$, $c = 9$, $A = 49°20'$; $a =$ __?__
8. $b = 5$, $a = 6$, $C = 81°40'$; $c =$ __?__

In Exercises 9–14, find the measure of the indicated angle of $\triangle ABC$.

9. $a = 5$, $b = 7$, $c = 4$; $A =$ _?_

10. $a = 3$, $b = 7$, $c = 5$; $B =$ _?_

11. $a = 1$, $b = 2$, $c = 2$; $C =$ _?_

12. $a = 5$, $b = 4$, $c = 3$; $A =$ _?_

13. $a = 10$, $b = 12$, $c = 14$; $B =$ _?_

14. $a = 7$, $b = 8$, $c = 3$; $C =$ _?_

In Exercises 15–20, solve $\triangle ABC$.

15. $b = 40$, $c = 45$, $A = 51°$

16. $a = 20$, $c = 24$, $B = 47°$

17. $a = 5$, $b = 6$, $c = 7$

18. $a = 5$, $b = 12$, $c = 13$

19. $b = 16$, $c = 19$, $A = 35°25'$

20. $b = 13$, $a = 11$, $C = 76°53'$

APPLICATIONS: Using the Law of Cosines

21. The lengths of two sides and one diagonal of a parallelogram are 8 meters, 13 meters, and 20 meters, respectively. What is the measure of each angle of the parallelogram?

22. The lengths of the sides of a triangle are 8, 9, and 13 centimeters. Without using tables, determine whether the largest angle is acute or obtuse.

23. Why is the formula $c^2 = a^2 + b^2 - 2ab \cos C$ equivalent to the Pythagorean Theorem when $C = 90°$?

B

24. A triangular lot has sides of 215,185, and 125 meters. Find the measures of the angles at its corners.

25. From point C, both ends A and B of a railroad tunnel are visible. If $AC = 165$ meters, $BC = 115$ meters, and $C = 74°$, find AB, the length of the tunnel.

26. The distances from a boat B to two points A and C on the shore are known to be 100 meters and 80 meters respectively, and $\angle ABC = 55°$. Find AC.

27. The radius of a circle is 20 centimeters. Two radii, OX and OY, form an angle of 115°. How long is the chord XY?

28. Two sides and a diagonal of a parallelogram are 7, 9, and 15 feet respectively. Find the measures of the angles of the parallelogram.

C

29. Show that $b^2 = a^2 + c^2 - 2ac \cos B$, and $c^2 = a^2 + b^2 - 2ab \cos C$. (HINT: Refer to the proof for $a^2 = b^2 + c^2 - 2bc \cos A$, on page 573.)

30. The measure of the angle of elevation to the top of a radar tower from point A is θ. From point B, which is c feet closer to the tower than A, the measure of the angle of elevation is ϕ (Greek letter "phi"). Show that the height of the tower above $\overline{AB}$ is $\dfrac{c \tan \theta \tan \phi}{\tan \phi - \tan \theta}$.

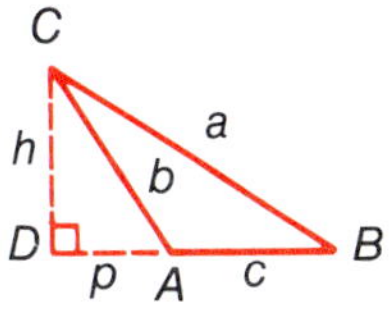

31. Given $\triangle ABC$ with the obtuse angle A and altitude h, prove Case II of the Law of Cosines. (HINT: Let $\alpha = m\angle CAD$. Then $\angle A = (180° - \alpha)$.)

Review

In Exercises 1–6, first sketch the indicated angle. Then use the tables on pages 629–633 to find the value of the sine, cosine, and tangent functions for each angle. *(Section 15–6)*

1. $-135°$ **2.** $-75°$ **3.** $-210°$ **4.** $-320°$ **5.** $-375°$

Evaluate. Use the tables on pages 629–633. *(Section 15–7)*

6. $\tan 149°$ **7.** $\sin 315°20'$ **8.** $\cos 179°50'$ **9.** $\tan 200°15'$

In Exercises 10–13, the given point P lies in the terminal side of an angle θ in standard position. Evaluate the six trigonometric functions of θ. *(Section 15–8)*

10. $P(24, -7)$ **11.** $P(-4, -5)$ **12.** $P(5, -12)$ **13.** $P(-\sqrt{3}, 1)$

14. From point A at the base of a mountain, the angle of elevation of the mountain's peak is 20°. A cable car traveling from point A at the base to point B at the peak covers a distance of 1350 meters. How far above ground level is the peak of the mountain? *(Section 15–9)*

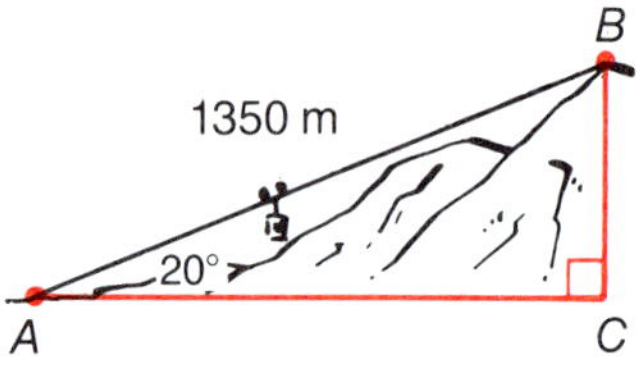

15. In $\triangle ABC$, $A = 25°$, $B = 125°$, and $c = 150$ centimeters. Find side AC. *(Section 15–10)*

16. Dodgeville is 10 kilometers west of Moon City. Cornetta is 12 kilometers from Dodgeville and 13 kilometers from Moon City. Find the measure of the angle formed by an imaginary line drawn from Cornetta to Dodgeville to Moon City. *(Section 15–11)*

CALCULATOR APPLICATIONS

Law of Cosines

The Example below shows how to use a scientific calculator to solve for *a*, *b*, or *c* in the Law of Cosines.

EXAMPLE $a = \sqrt{10^2 + 20^2 - 2 \cdot 10 \cdot 20 \cos 40°}$; $a = \underline{\ ?\ }$ (Example 1, page 573)

SOLUTION 1 0 [x^2] [+] 2 0 [x^2] [−] 2 [×] 1 0 [×] 2 0 [×] → 400.

4 0 [cos] [=] [$\sqrt{\ }$] → 13.913383

EXERCISES

Use a calculator to do Exercises 15–20 on page 575.

Chapter Summary

IMPORTANT TERMS

Angle (p. 540)
Angle of depression (p. 565)
Angle of elevation (p. 564)
Cosecant ratio (p. 563)
Cosine ratio (p. 549)
Cotangent ratio (p. 563)
Coterminal angles (p. 540)
Initial side of an angle (p. 540)
Interpolation (p. 554)
Quadrantal angle (p. 540)
Radius vector (p. 542)
Reference triangle (p. 542)
Secant ratio (p. 563)
Sine ratio (p. 549)
Solving a triangle (p. 546)
Tangent ratio (p. 549)
Terminal side of an angle (p. 540)

IMPORTANT IDEAS

1. The measures of two or more coterminal angles are given by $n \cdot 360° + \theta$, where n is an integer.

2. For an angle with measure θ in standard position, each of the ratios $\sin \theta$. $\cos \theta$, and $\tan \theta$ is the same for every point P on the terminal side of the angle.

3. For any angles with measures θ and $-\theta$,

   a. $\sin(-\theta) = -\sin \theta$ **b.** $\cos(-\theta) = \cos \theta$ **c.** $\tan(-\theta) = -\tan \theta$

4. **Reduction Formulas:** The following formulas are true for any angle with measure θ.

   a. $\sin(180° - \theta) = \sin \theta$
   $\cos(180° - \theta) = -\cos \theta$
   $\tan(180° - \theta) = -\tan \theta$

   b. $\sin(180° + \theta) = \sin \theta$
   $\cos(180° + \theta) = -\sin \theta$
   $\tan(180° + \theta) = \tan \theta$

   c. $\sin(360° - \theta) = -\sin \theta$
   $\cos(360° - \theta) = \cos \theta$
   $\tan(360° - \theta) = -\tan \theta$

   d. $\sin(360° + \theta) = \sin \theta$
   $\cos(360° + \theta) = \cos \theta$
   $\tan(360° + \theta) = \tan \theta$

   e. $\sin(90° - \theta) = \cos \theta$
   $\cos(90° - \theta) = \sin \theta$
   $\tan(90° - \theta) = \dfrac{1}{\tan \theta}$

   f. $\sin(90° + \theta) = \cos \theta$
   $\cos(90° + \theta) = -\sin \theta$
   $\tan(90° + \theta) = -\dfrac{1}{\tan \theta}$

5. **Law of Sines:** In any triangle, the ratio of the length of a side to the sine of the angle opposite that side is the same for each side-angle pair. That is, in any $\triangle ABC$,

$$\frac{a}{\sin A} = \frac{b}{\sin B} = \frac{c}{\sin C}.$$

6. **Law of Cosines:** In any triangle, the square of a side is equal to the sum of the square of the other two sides minus twice the product of these sides and the cosine of their included angle. In $\triangle ABC$,

$$a^2 = b^2 + c^2 - 2bc \cos A$$
$$b^2 = a^2 + c^2 - 2ac \cos B$$
$$c^2 = a^2 + b^2 - 2ab \cos C.$$

Chapter Objectives and Review

Objective: *To sketch an angle in standard position (Section 15–1)*

Sketch each angle in standard position.

1. $60°$ **2.** $-45°$ **3.** $30°$ **4.** $-15°$ **5.** $-135°$ **6.** $225°$

Objective: *To find angles that are coterminal with a given angle (Section 15–1)*

In Exercises 7–12, find three coterminal angles for each angle. Let $n = 2, -1$, and -2.

7. $50°$ **8.** $15°$ **9.** $-30°$ **10.** $-90°$ **11.** $310°$ **12.** $-225°$

Objective: *To evaluate sin θ, cos θ, and tan θ, given a point on the terminal side of an angle (Section 15–2)*

In Exercises 13–17, the terminal side of an angle with measure θ passes through the given point. Sketch the reference triangle. Evaluate $\sin \theta$, $\cos \theta$, and $\tan \theta$. Simplify all radicals.

13. $P(4, 3)$ **14.** $Q(-5, 12)$ **15.** $R(-2, -2)$ **16.** $S(4, 8)$ **17.** $T(2, -7)$

Objective: *To use the relationship between the sides of 30°–60°–90° triangles and the relationship between the sides of 45°–45°–90° triangles to evaluate the sine, cosine, and tangent of certain angles (Section 15–3)*

In Exercises 18–19, the length, c, of the hypotenuse of each triangle is given. Solve the triangle.

18. $30°$–$60°$–$90°$ triangle: $c = 4$ **19.** $45°$–$45°$–$90°$ triangle: $c = \sqrt{2}$

In Exercises 20–22, point P lies on the terminal side of a 210°-angle in standard position. The y coordinate of point P is -1. Find each of these.

20. $\sin 210°$ **21.** $\cos 210°$ **22.** $\tan 210°$

Objective: *To use trigonometric functions in triangles (Section 15–4)*

In Exercises 23–24, find the length of the unknown side. Use the table on pages 629–633. Round your answer to the nearest whole number. Refer to triangle ABC.

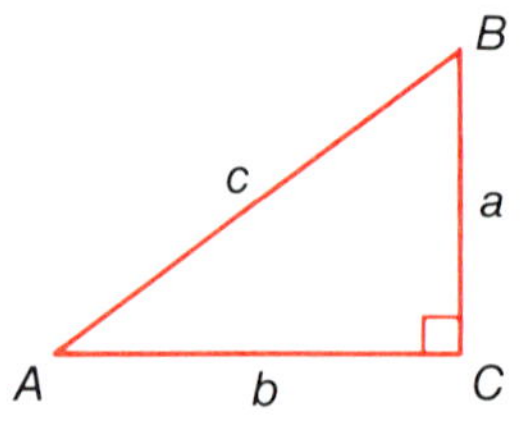

23. $A = 40°$; $c = 15$ m; $b =$ __?__

24. $B = 72°50'$; $b = 52$ m; $a =$ __?__

Objective: *To use tables and interpolation to evaluate sin θ, cos θ, and tan θ (Section 15–5)*

Evaluate.

25. $\cos 41°47'$ **26.** $\tan 24°16'$ **27.** $\sin 34°16'$ **28.** $\sin 78°24'$

In Exercises 29–32, find θ to the nearest minute.

29. $\sin \theta = 0.4500$ **30.** $\cos \theta = 0.7710$ **31.** $\tan \theta = 1.2500$ **32.** $\sin \theta = 0.9500$

Objective: *To evaluate the sine, cosine, and tangent of negative angles (Section 15–6)*

In Exercises 33–38, find the values of the sine, cosine, and tangent functions for each angle.

33. $-30°$ **34.** $-135°$ **35.** $-240°$ **36.** $-235°$ **37.** $-20°$ **38.** $-100°$

Objective: *To apply the Reduction Formulas (Section 15–7)*

Evaluate. Use the tables on pages 629–633.

39. $\sin 240°$ **40.** $\cos 240°$ **41.** $\tan 167°$ **42.** $\cos 195°$ **43.** $\sin 145°$

Objective: *To evaluate cot θ, sec θ, and csc θ for a given value of θ (Section 15–8)*

Evaluate. Use the tables on pages 629–633.

44. $\cot 42°$ **45.** $\sec 75°$ **46.** $\csc 10°$ **47.** $\cot 71°20'$ **48.** $\sec 80°40'$

Objective: *To find the measure of an angle, given its cotangent, secant, or cosecant (Section 15–8)*

In Exercises 49–52, find θ, where $0° \leq \theta \leq 90°$.

49. $\cot \theta = 1.8807$ **50.** $\sec \theta = 1.048$ **51.** $\csc \theta = 1.382$ **52.** $\cot \theta = 0.9380$

Objective: *To use the angles of elevation and depression to solve problems (Section 15–9)*

53. A salvage ship using sonar finds the angle of depression of wreckage on the ocean floor to be 13°10′. The charts show that in this region the ocean floor is 35 meters below the surface. How far must a diver lowered from the salvage ship travel along the ocean floor to reach the wreckage?

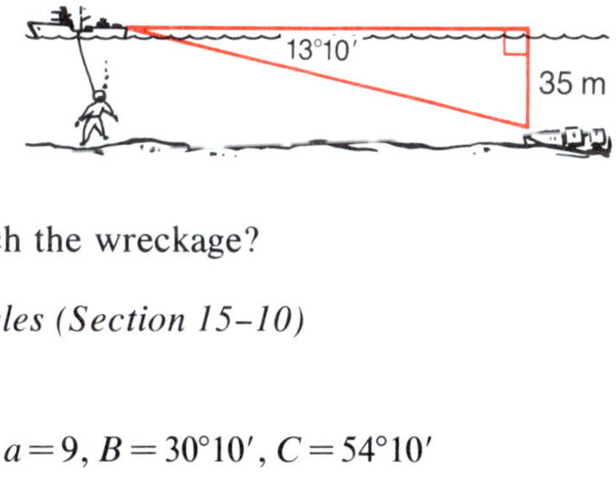

Objective: *To use the Law of Sines to solve triangles (Section 15–10)*

Solve each triangle.

54. $A = 40°$; $C = 53°$; $c = 20$ **55.** $a = 9$, $B = 30°10'$, $C = 54°10'$

56. Two travelers, driving along a highway, spot their destination in the distance at point C. They stop at the side of the road to measure angle A. They then travel 5 more miles along the road, stop again, and measure angle B. How far are the travelers now from their destination?

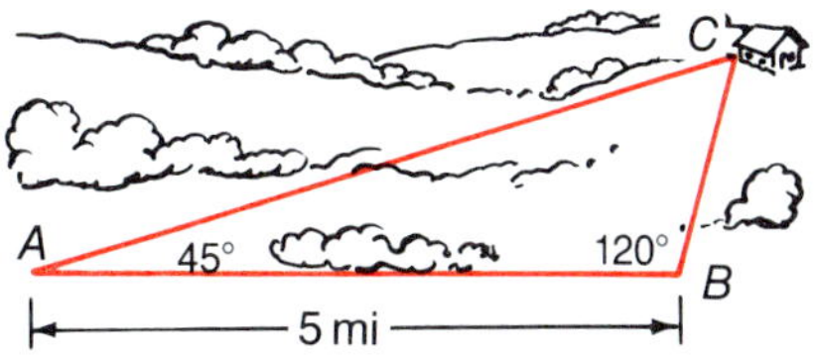

Objective: *To use the Law of Cosines to solve triangles (Section 15–11)*

Find the indicated side or angle.

57. $A = 53°$, $b = 20$, $c = 35$, $a = \underline{\ ?\ }$ **58.** $a = 7$, $b = 10$, $c = 5$; $B = \underline{\ ?\ }$

Chapter Test

In Exercises 1–6, find two angles coterminal with the given angle.

1. 35° **2.** 97° **3.** 180° **4.** −100° **5.** 290° **6.** −390°

In Exercises 7–11, the terminal side of an angle with measure θ passes through the given point. Sketch the reference triangle. Evaluate sin θ, cos θ and tan θ. Simplify all radicals.

7. $P(12, 5)$ **8.** $P(4, -4)$ **9.** $P(-3, -4)$ **10.** $P(-1, 2)$ **11.** $P(1, -24)$

In Exercises 12–14, find the length of the indicated side. Refer to triangle *ABC*.

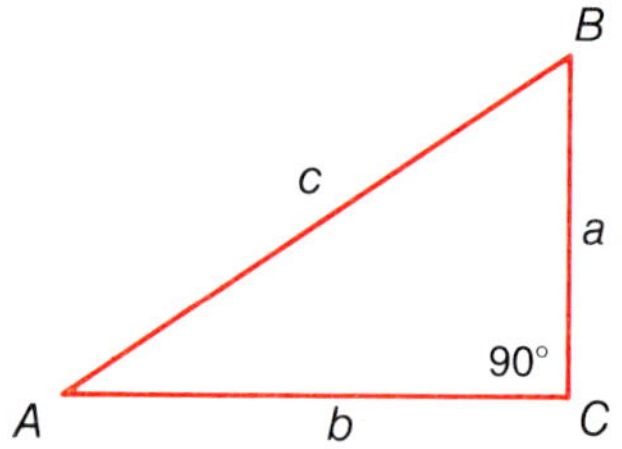

12. $A = 45°$; $a = 4$; $c = \underline{\ ?\ }$

13. $B = 60°$; $B = \sqrt{75}$; $a = \underline{\ ?\ }$

14. $A = 20°$; $c = 16$ m; $b = \underline{\ ?\ }$ Use the tables on pages 629–633. Round your answer to the nearest whole number.

15. Evaluate cos 63°20′. Use the tables on pages 629–633.

16. Evaluate sin 37°43′. Use interpolation and the tables on pages 629–633.

17. Evaluate tan (−315°).

Evaluate. Use the tables on pages 629–633.

18. tan 225° **19.** cos 185° **20.** sin 130° **21.** sec 28° **22.** cot 17°

23. In the figure at the right, a kite string 90 meters long makes a 48° angle with the horizontal. Find, to the nearest meter, the distance from the kite to the ground. Use the tables on pages 629–633.

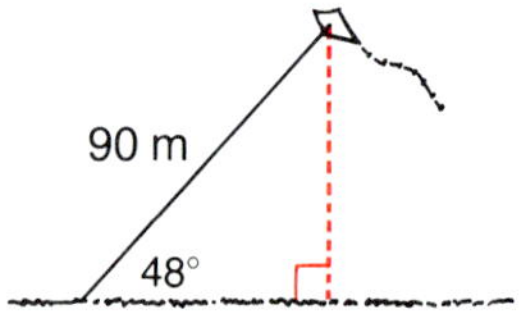

24. Two lighthouses at points *A* and *B* are 40 kilometers apart. Each has visual contact with a freighter at point *C*. (See the figure at the right.). How far is the freighter from point *A*?

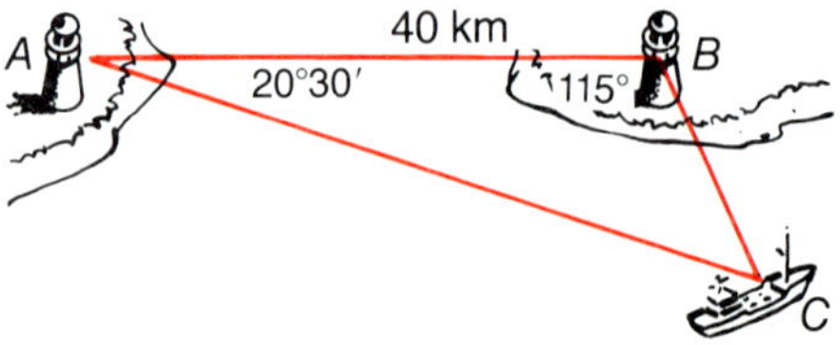

25. Two space-monitoring stations are located at two different points *A* and *B* on the earth's surface. Angle *ACB*, at the earth's center, is 110°. The radius of the earth is about 3960 miles. Find the straight-line distance between the two stations.

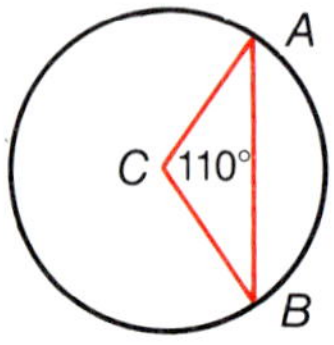

CHAPTER **16**

More Topics in Trigonometry

16–1 Vectors

The complete description of certain quantities, such as velocity, requires both a direction and a magnitude. Such quantities are called **vectors.**

Mathematically, a vector can be considered as a **directed line segment.** The direction of a vector is given by its position in the plane.

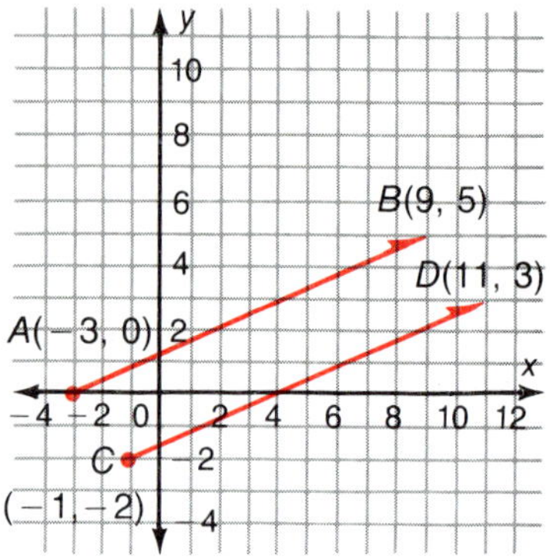

In the figure at the right, vectors AB and CD have the same direction. In the figure, vector AB, written $\overrightarrow{AB}$, has **initial point** $A(-3, 0)$ and **terminal point** $B(9, 5)$. The length or **magnitude** of $\overrightarrow{AB}$ is symbolized as $|\overrightarrow{AB}|$. The distance formula is used to find the magnitude of a vector.

EXAMPLE 1 Refer to the figure above to find the magnitude of each vector.

a. $\overrightarrow{AB}$ **b.** $\overrightarrow{CD}$

Solutions: Use the Distance Formula.

a. $|\overrightarrow{AB}| = \sqrt{[9-(-3)]^2 + [5-0]^2}$
$= \sqrt{144 + 25}$
$= \sqrt{169}$
$= \mathbf{13}$

b. $|\overrightarrow{CD}| = \sqrt{[11-(-1)]^2 + [3-(-2)]^2}$
$= \sqrt{144 + 25}$
$= \sqrt{169}$
$= \mathbf{13}$

In Example 1, $|\overrightarrow{AB}| = |\overrightarrow{CD}|$. It can also be shown that $\overrightarrow{AB}$ and $\overrightarrow{CD}$ have the same direction. (You are asked to show this in the Written Exercises.)

Vectors with the same magnitude and direction, such as $\overrightarrow{AB}$ and $\overrightarrow{CD}$, are called **equivalent vectors.**

In the figure at the right, $OCAB$ is a rectangle. $\overrightarrow{OB}$ is the ***x* component** and $\overrightarrow{OC}$ is the ***y* component** of $\overrightarrow{OA}$. The angle θ is the positive angle that $\overrightarrow{OA}$ makes with the positive x axis.

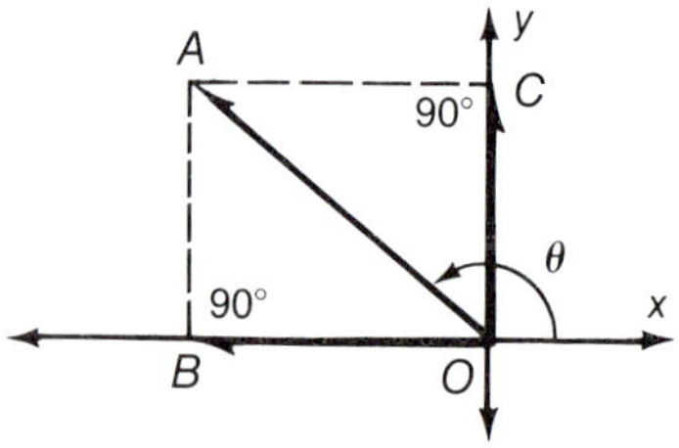

When you know the magnitude of a vector and the value of θ, you can find the magnitude of the x and y components.

EXAMPLE 2 In the figure at the right, $\overrightarrow{OB}$ and $\overrightarrow{OC}$ are the x and y components of $\overrightarrow{OA}$. $|\overrightarrow{OA}| = 100$ and $\theta = 160°$. Find each of the following to the nearest whole number.

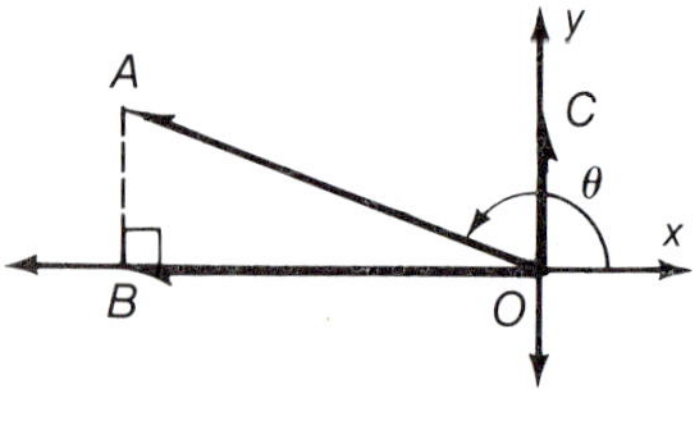

a. $|\overrightarrow{OB}|$ **b.** $|\overrightarrow{OC}|$

Solutions: **a.** Since $\overrightarrow{OB}$ is the x component of $\overrightarrow{OA}$, $m\angle OBA = 90°$.
Also, $m\angle AOB = 180° - 160° = 20°$. Thus, in right triangle OBA,

$$\frac{|\overrightarrow{OB}|}{|\overrightarrow{OA}|} = \cos 20°$$

$$|\overrightarrow{OB}| = |\overrightarrow{OA}| \cos 20°$$

$$|\overrightarrow{OB}| = 100 \cos 20°$$

$$|\overrightarrow{OB}| = 100(0.9397)$$

$$|\overrightarrow{OB}| = 93.97, \text{ or about } \mathbf{94}$$

b. Similarly, in right triangle OBA,

$$\frac{|\overrightarrow{BA}|}{|\overrightarrow{OA}|} = \sin 20°$$

$$|\overrightarrow{BA}| = |\overrightarrow{OA}| \sin 20°$$

$$|\overrightarrow{BA}| = 100(0.3420)$$

$$|\overrightarrow{BA}| = 34.20, \text{ or about } \mathbf{34}$$

Since $\overrightarrow{BA}$ and $\overrightarrow{OC}$ have the same magnitude and direction, they are equivalent. Thus, $|\overrightarrow{OC}| = |\overrightarrow{BA}| = \mathbf{34}$.

In Example 2, $\overrightarrow{OA}$ is called the **resultant** or **vector sum** of $\overrightarrow{OB}$ and $\overrightarrow{OC}$. In symbols, this is written

$$\overrightarrow{OA} = \overrightarrow{OB} + \overrightarrow{OC}.$$

Since $\overrightarrow{OC} = \overrightarrow{BA}$, the vector sum can also be written as $\overrightarrow{OA} = \overrightarrow{OB} + \overrightarrow{BA}$. The diagram for these two vector sums are shown below.

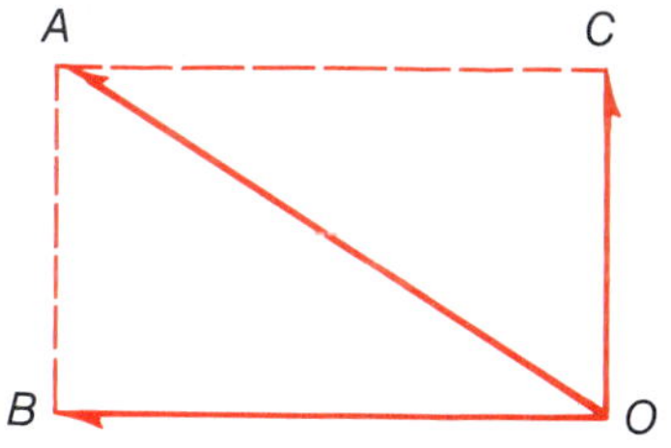

$\overrightarrow{OA} = \overrightarrow{OB} + \overrightarrow{OC}$

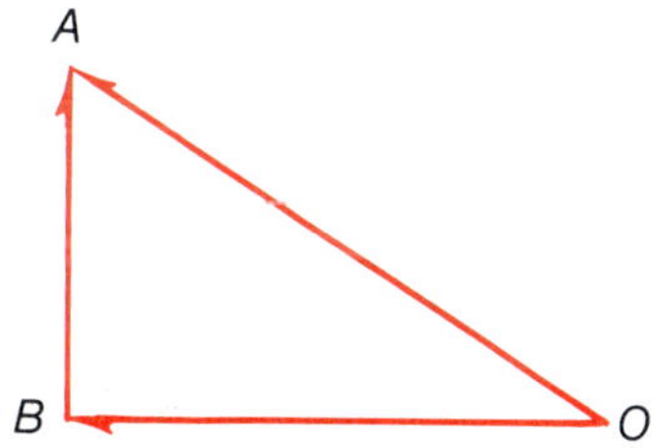

$\overrightarrow{OA} = \overrightarrow{OB} + \overrightarrow{BA}$

Thus, the x and y components of $\overrightarrow{OA}$ can be represented either as two adjacent sides of a rectangle or as the two legs of a right triangle.

A vector can be used to represent a *change of position,* or **displacement.**

EXAMPLE 3 A hiker walks east for 3 kilometers and then walks south for 4 kilometers.

a. Find the magnitude of the hiker's displacement.

b. Find the positive angle, θ, that the displacement makes with the x axis.

Solutions: **a.** The displacement is represented by $\overrightarrow{OA}$. The magnitude of the displacement is $|\overrightarrow{OA}|$. From the diagram,

$|\overrightarrow{OB}| = 3$, $|\overrightarrow{BA}| = 4$.

Thus, $|\overrightarrow{OA}| = \sqrt{3^2 + 4^2}$

$= \sqrt{25}$, or **5**.

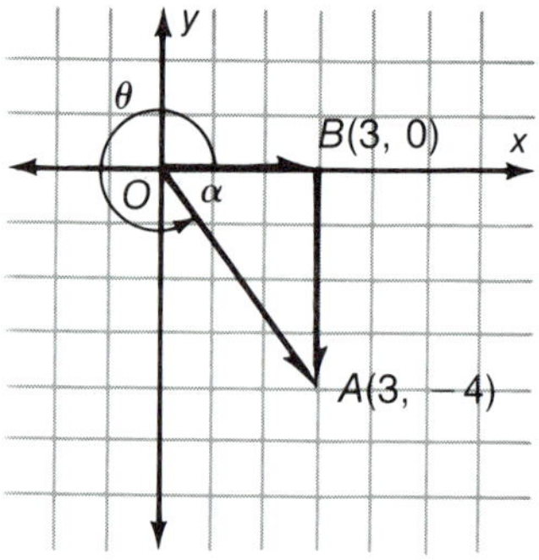

b. To find θ, let $\alpha = m\angle BOA$. Then

$$\cos \alpha = \frac{|\overrightarrow{OB}|}{|\overrightarrow{OA}|}$$

$$\cos \alpha = \frac{3}{5} = 0.6000$$

$\alpha = 53°$, to the nearest degree.

Therefore, $\theta = 360 - 53°$, or **307°**, to the nearest degree.

CLASSROOM EXERCISES

In the figure at the right, *OBAC* is a rectangle. Name each vector.

1. The x component of $\overrightarrow{OA}$.
2. The y component of $\overrightarrow{OA}$.
3. The vector sum of $\overrightarrow{OB} + \overrightarrow{OC}$.
4. The vector equivalent to $\overrightarrow{OC}$.
5. The vector sum $\overrightarrow{OB} + \overrightarrow{BA}$.

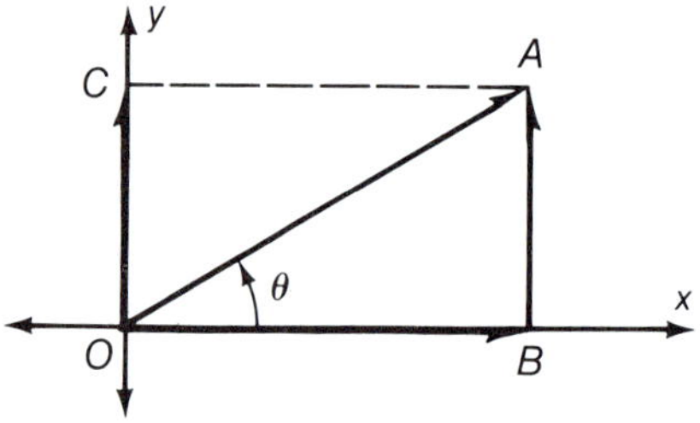

In the figure for Exercises 1–5, suppose that $\theta = 30°$ and $|\overrightarrow{OA}| = 10$. Find these.

6. m$\angle OBA$
7. $|\overrightarrow{OB}|$
8. $|\overrightarrow{BA}|$
9. $|\overrightarrow{OC}|$

WRITTEN EXERCISES

A In Exercises 1–8, find $|\overrightarrow{AB}|$ to the nearest tenth.

1. $A(4, 2)$; $B(3, 1)$
2. $A(-4, -2)$; $B(-3, 1)$
3. $A(5, -2)$; $B(1, -5)$
4. $A(1, -3)$; $B(2, -3)$
5. $A(-3, -7)$; $B(2, 1)$
6. $A(-3, 5)$; $B(-2, -5)$
7. $A(2, 5)$; $B(-7, -3)$
8. $A(1, 0)$; $B(-3, 5)$

In Exercises 9–11, $\overrightarrow{OA}$ has initial point $O(0, 0)$ and forms an angle θ with the positive x axis. Find $|\overrightarrow{OB}|$ and $|\overrightarrow{OC}|$, the x and y components of $\overrightarrow{OA}$, to the nearest whole number.

9. $|\overrightarrow{OA}| = 5$; $\theta = 120°$
10. $|\overrightarrow{OA}| = 3$; $\theta = 17°$
11. $|\overrightarrow{OA}| = 8$; $\theta = 290°$

In Exercises 12–15, a bicyclist travels on the course that is described. Find the magnitude of the bicyclist's displacement and, to the nearest 10 minutes, the positive angle θ that the bicyclist's displacement makes with the x axis.

12. Three miles east; four miles north

13. Six miles west; 8 miles north

14. 20 km west; 15 km south

15. 24 km east; 18 km south

16. Show that $\overrightarrow{AB}$ and $\overrightarrow{CD}$ in Example 1 have the same direction. (HINT: Show that the two vectors have the same slope.)

B

In Exercises 17–19, refer to the figure at the right. Then find each of the following to the nearest whole number.

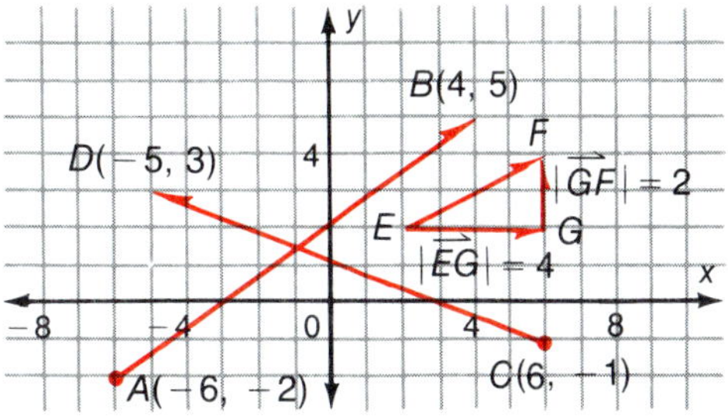

17. The magnitude of $\overrightarrow{AB}$.

18. The magnitude of $\overrightarrow{CD}$.

19. The magnitude of $\overrightarrow{EF}$.

C

20. The components of a vector need not lie along the x and y axes. In the figure at the right, $\overrightarrow{OC}$ and $\overrightarrow{OB}$ are component vectors of $\overrightarrow{OA}$. $|\overrightarrow{OC}| = 10$, $|\overrightarrow{OB}| = 15$, $\theta_1 = 120°$, and $\theta_2 = 30°$. Find $|\overrightarrow{OA}|$ to the nearest tenth.

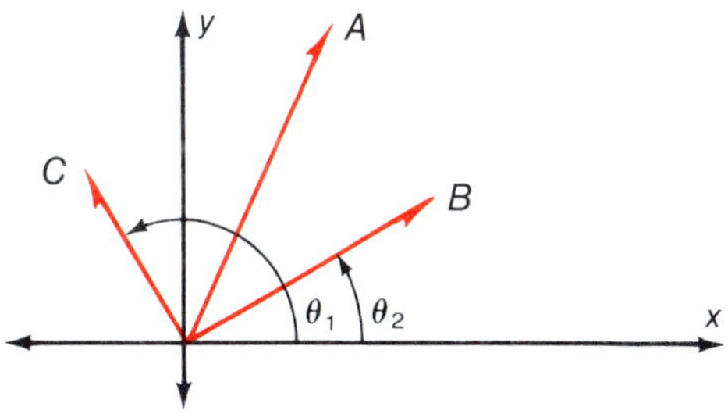

Problem Solving and Applications

16–2 Velocity Vectors

In the previous section, a vector, $\overrightarrow{OA}$, was represented as the vector sum of two component vectors along the x axis and y axis, respectively. As illustrated below, *any* two vectors can be used to form a vector sum. In Figure 2, note that $\overrightarrow{OA}$ is a diagonal of parallelogram $OBAC$.

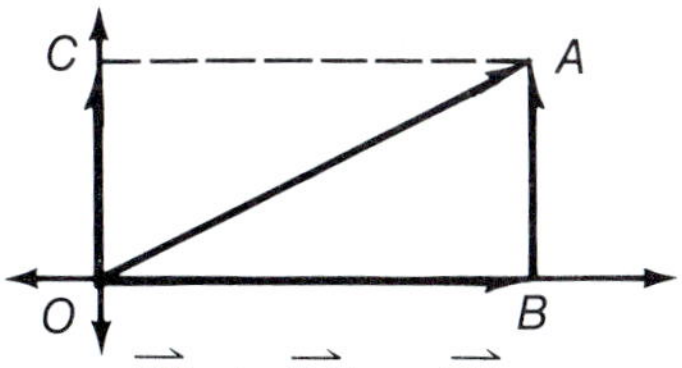

$\overrightarrow{OB} + \overrightarrow{OC} = \overrightarrow{OA}$

(m $\angle$ OBA = 90°)

Figure 1

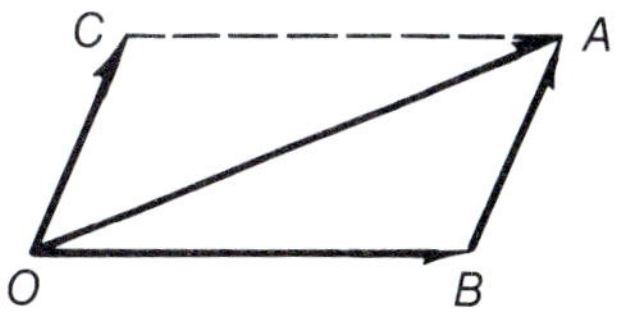

$\overrightarrow{OB} + \overrightarrow{OC} = \overrightarrow{OA}$

(m $\angle$ OBA ≠ 90°)

Figure 2

Another example of a vector quantity is *velocity*. The **velocity** of an object is the time rate of change of its *displacement* (see previous section). As with all vectors, velocity may be the resultant of two component velocities. The velocity of an airplane relative to the ground is the resultant of the wind velocity and of the velocity produced by the plane's engines (air velocity). The magnitude of the velocity relative to the ground is called the **ground speed.** The magnitude and direction of the velocity produced by the engines are called the **air speed** and **heading,** respectively. The heading is measured as a clockwise angle formed by the air velocity vector and the North–South axis.

EXAMPLE 1 Figure 1 at the left below shows the component velocities for an airplane with air speed 481.6 kilometers per hour on heading 85°12′. The wind velocity is 40 kilometers per hour in the direction 180°. Find, to the nearest unit, the ground speed of the airplane.

Solution: Draw $\overrightarrow{BC}$ equivalent to $\overrightarrow{AH}$. (See Figure 2.)

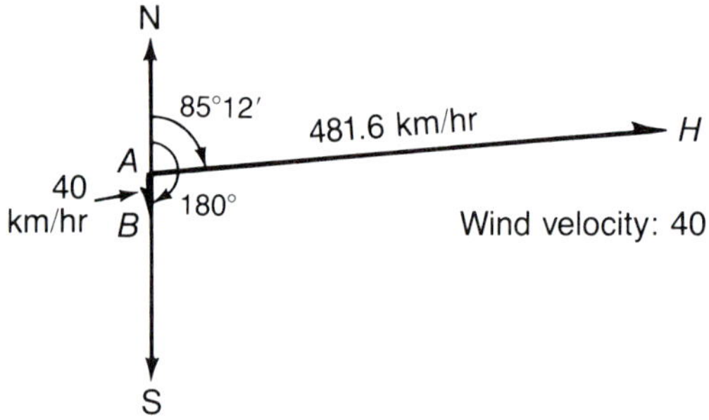

Figure 1

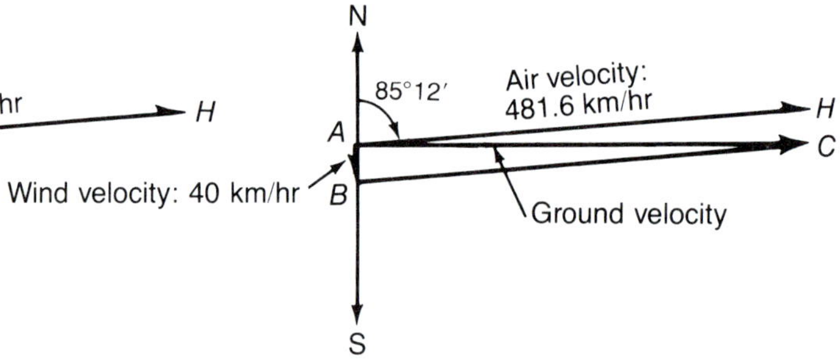

Figure 2

Then, $\overrightarrow{AC} = \overrightarrow{AB} + \overrightarrow{AH}$ represents the resultant ground velocity. Therefore, $|\overrightarrow{AC}|$ is the ground speed. Since $\overrightarrow{AH}$ and $\overrightarrow{BC}$ are equivalent,

$$|\overrightarrow{BC}| = |\overrightarrow{AH}| = 481.6 \text{ and m } \angle ABC = \text{m } \angle NAH = 85°12'.$$

From the Law of Cosines,

$$|\overrightarrow{AC}|^2 = |\overrightarrow{AB}|^2 + |\overrightarrow{BC}|^2 - 2|\overrightarrow{AB}||\overrightarrow{BC}| \cos 85°12',$$

or

$$|\overrightarrow{AC}| = \sqrt{|\overrightarrow{AB}|^2 + |\overrightarrow{BC}|^2 - 2|\overrightarrow{AB}||\overrightarrow{BC}| \cos 85°12'}$$

$$|\overrightarrow{AC}| = \sqrt{40^2 + (481.6)^2 - 2(40)(481.6)(0.0837)} = 479.9.$$

Thus, the ground speed is about**480 km/hr.**

The direction of motion of a ship or airplane is its **course.** The course of an airplane gives the direction of its actual (ground) velocity. In Example 2, m $\angle NAC$ is the course of the airplane.

EXAMPLE 2 Find, to the nearest degree, the course of the airplane in Example 1.

Solution: The course is m∠*NAC*. Since m∠*NAC* + m∠*CAB* = m∠*NAB* = 180°,

$$m\angle NAC = 180° - m\angle CAB.$$

To find m∠*CAB*, use the Law of Sines. Let m∠*CAB* = θ.

Then, $\dfrac{|\overrightarrow{BC}|}{\sin\theta} = \dfrac{|\overrightarrow{AC}|}{\sin 85°12'}$.

Thus, $\sin\theta = \dfrac{|\overrightarrow{BC}|\sin 85°12'}{|\overrightarrow{AC}|}$ ← **Use the tables or use a calculator.**

$= \dfrac{481.6(0.9965)}{480} = 0.9998.$

Thus, θ = 89°. ← **To the nearest degree**

Therefore the course is about 180° − 89°, or **91°.**

CLASSROOM EXERCISES

In Exercises 1–5, refer to Example 1 and to the figures at the right. Write *True* or *False* for each given vector sum.

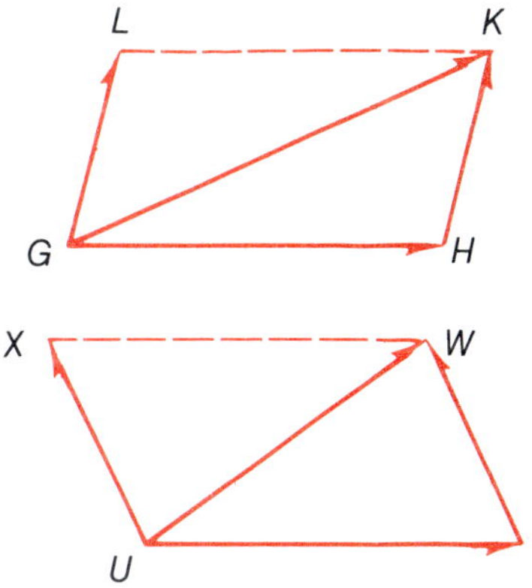

1. $\overrightarrow{GH} + \overrightarrow{HK} = \overrightarrow{GK}$
2. $\overrightarrow{UW} + \overrightarrow{UX} = \overrightarrow{VW}$
3. $\overrightarrow{GL} + \overrightarrow{GH} = \overrightarrow{GK}$
4. $\overrightarrow{UW} = \overrightarrow{UV} + \overrightarrow{VW}$
5. Ground velocity + wind velocity = air velocity

WORD PROBLEMS

A

In Exercises 1–3, the air velocity of an airplane is shown together with the velocity of the wind. Find the ground speed of the airplane.

1. **2.** **3.**

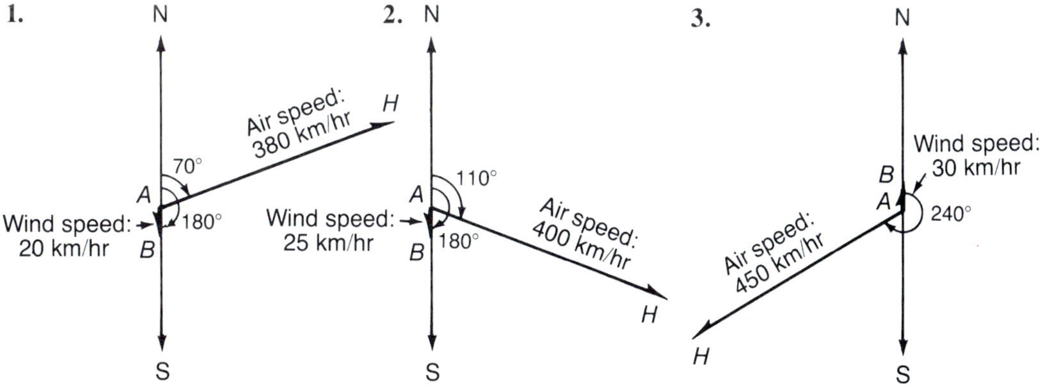

4–6. Find, to the nearest unit, the course of each airplane in Exercises 1–3.

B

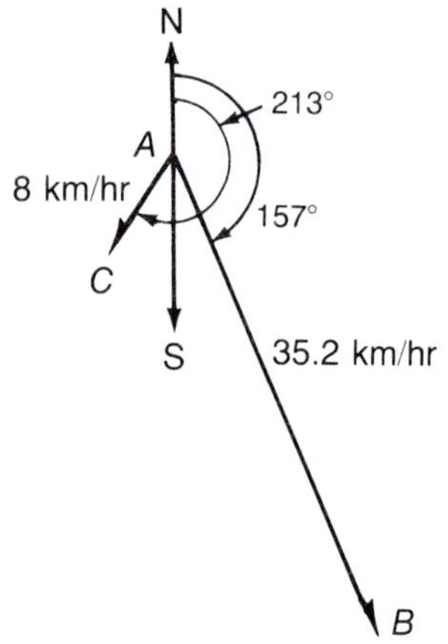

7. Find, to the nearest unit, the ground speed of an airplane with airspeed 480 km/hr and heading 90° if the wind velocity is 40 km/hr in the direction 180°.

8. Find, to the nearest degree, the course of the airplane in Exercise 7.

9. The velocity produced by the engines of a ship on a heading 157° has magnitude 35.2 km/hr. The water current has a velocity of 8 km/hr in the direction 213°. Find, to the nearest unit, the magnitude of the actual (observed) velocity of the ship.

10. Find the course of the ship in Exercise 9. Give your answer to the nearest degree.

C

11. The angle between the heading and course of a ship or airplane is called the **angle of drift.** Find, to the nearest degree, the angle of drift of an airplane with air speed 240 km/hr on heading 90° in a 65 km/hr wind in the direction 0° (refer to the figure at the left below).

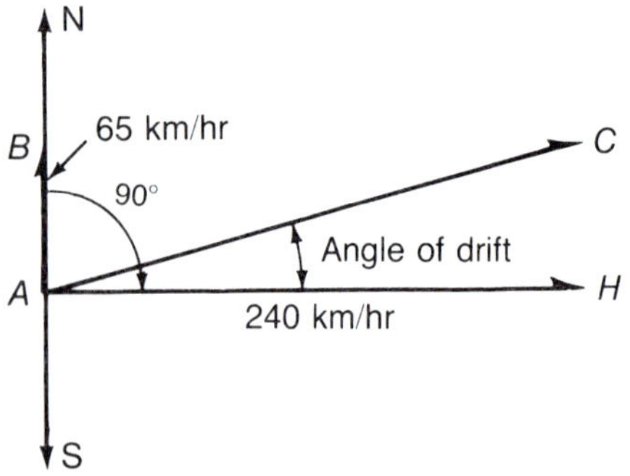

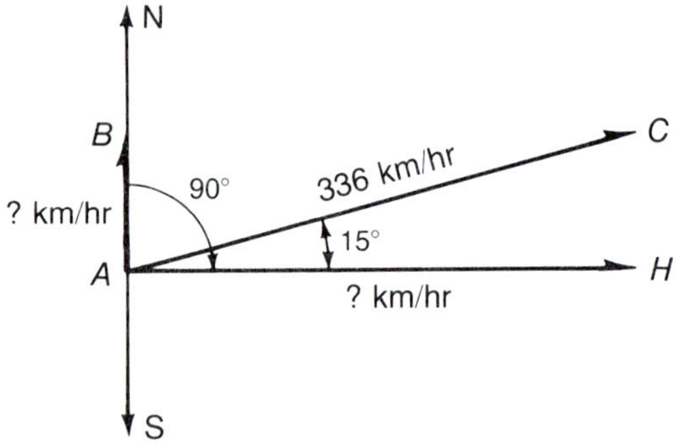

12. Find, to the nearest unit, the ground speed of the airplane in Exercise 11.

13. A wind in the direction 0° causes a 15°–angle of drift for an airplane on heading 90° (refer to the figure at the right above). The ground speed of the plane is 336 km/hr. Find the air speed of the plane and the velocity of the wind. Give your answers to the nearest unit.

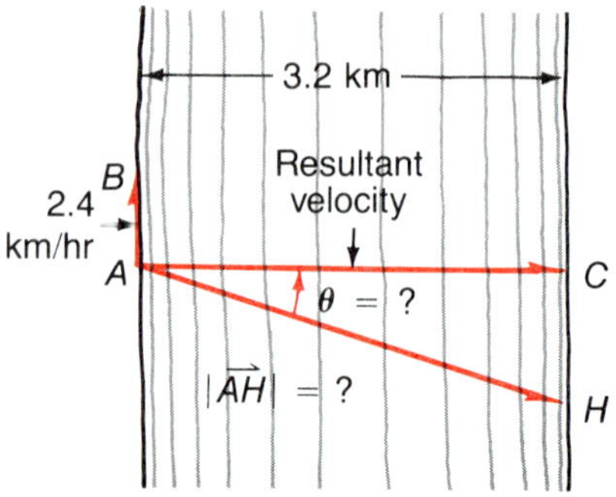

14. A river 3.2 kilometers wide flows due north at the rate of 2.4 km/hr. A small boat must travel directly across the river in 15 minutes. Refer to the figure at the right to find, to the nearest degree, the angle of drift, θ.

15. Find, to the nearest unit, the value of $|\overrightarrow{AH}|$, the magnitude of the velocity that must be maintained by the boat in Exercise 14. (HINT: Write the magnitude of the desired resultant velocity in km/hr.)

16. An airplane must fly at a ground speed of 450 km/hr on course 170° to be on schedule. The wind velocity is 25 km/hr in the direction 40°. Find the necessary heading to the nearest degree and the air speed to the nearest unit.

17. Point A is located on the western bank of a river, and point C is located downstream on the eastern bank. The river is of uniform width and flows due south at a rate of 2.4 km/hr. The line–of–sight distance from A to C is 762 meters. A boat crosses from A to C in 10 minutes. The boat's heading is 90°. Find, to the nearest unit, $|\overrightarrow{AH}|$, the magnitude of the velocity to be maintained by the boat.

18. Find, to the nearest degree, the angle of drift, θ, of the boat in Exercise 17.

16–3 Radian Measure

In mathematics, it is often convenient to use a unit of angular measure called a **radian.** In the figure at the right, if the measure of arc AB intercepted by angle AOB is equal to radius r, then the measure of the central angle AOB is one radian.

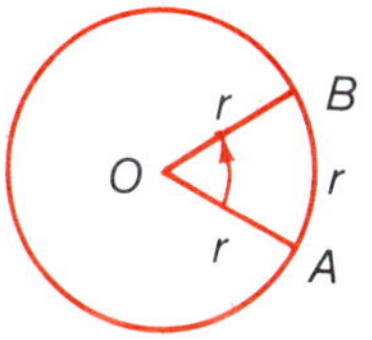

Definition

An angle has a measure of **one radian** if, when its vertex is placed at the center of a circle, it intercepts an arc equal in length to the radius of the circle.

In a circle with radius 3 units, an arc of 6 units subtends a central angle of $\frac{6}{3}$, or 2 radians. Thus, in a circle with radius r units, a central angle intercepting an arc of s units has a measure of $\frac{s}{r}$ **radians.**

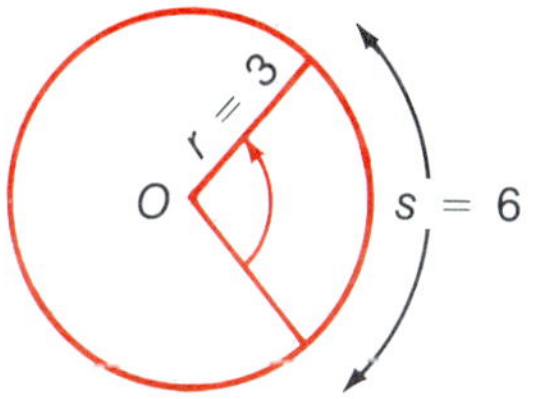

Consider a circle with radius r, and an arc of the circle of length C, the circumference. Then $C = 2\pi r$, and

$$\frac{s}{r} = \frac{2\pi r}{r} \quad \leftarrow \text{Length of arc} / \text{Radius}$$

$$= 2\pi \text{ radians} \quad \leftarrow \text{Number of radians in a circle}$$

Recall that there are 360° in a circle. Thus,

$$\mathbf{2\pi \text{ radians} = 360°} \quad \text{or} \quad \mathbf{\pi \text{ radians} = 180°}.$$

Since π radians = 180°,

$$\mathbf{1 \text{ radian} = \frac{180°}{\pi}} \quad \text{and} \quad \mathbf{1° = \frac{\pi}{180} \text{ radians.}}$$

Therefore, 1 radian is about **57°17′44.8″** and 1° is about **0.0174533 radian.**

EXAMPLE a. $315° = \underline{\quad ? \quad}$ radians b. $\frac{5\pi}{18}$ radians $= \underline{\quad ? \quad}°$.

Solutions:

a. $180° = \pi$ radians

$1° = \frac{\pi}{180}$ radians

$315° = 315\left(\frac{\pi}{180}\right)$

$315° = \frac{315\pi}{180}$, or $\mathbf{\frac{7\pi}{4}}$ **radians**

b. π radians $= 180°$

1 radian $= \left(\frac{180}{\pi}\right)°$

$\frac{5\pi}{18}$ radians $= \frac{5\pi}{18}\left(\frac{180}{\pi}\right)°$

$\frac{5\pi}{18}$ radians $= \mathbf{50°}$

CLASSROOM EXERCISES

Complete each table.

	Degrees	Radians
1.	45°	?
2.	?	$\frac{\pi}{12}$
3.	−60°	?
4.	?	$-\frac{\pi}{2}$

	Degrees	Radians
5.	?	$\frac{\pi}{6}$
6.	150°	?
7.	?	$-\frac{4\pi}{3}$
8.	−270°	?

	Degrees	Radians
9.	720°	?
10.	?	-5π
11.	−540°	?
12.	?	$\frac{7\pi}{2}$

WRITTEN EXERCISES

A In Exercises 1–8, circle *O* has a radius, *r*, and a central angle *POQ* that intercepts $\overset{\frown}{QP}$. Use the given values of *r* and lengths of $\overset{\frown}{QP}$ to find the radian measure of $\angle POQ$. Round each answer to the nearest tenth.

1. $r = 8$ cm; length of $\overset{\frown}{QP} = 12$ cm
2. $r = 1$ m; length of $\overset{\frown}{QP} = 2$ m
3. $r = 10$ m; length of $\overset{\frown}{QP} = 5$ m
4. $r = 10$ cm; length of $\overset{\frown}{QP} = 40$ cm
5. $r = 20$ cm; length of $\overset{\frown}{QP} = 8$ cm
6. $r = 4$ m; length of $\overset{\frown}{QP} = 20$ m
7. $r = 15$ m; length of $\overset{\frown}{QP} = 1200$ cm
8. $r = 6$ km; length of $\overset{\frown}{QP} = 40$ m

Change each degree measure to radian measure in terms of π.

9. 360° **10.** 270° **11.** 90° **12.** 330° **13.** −225° **14.** −30°
15. −45° **16.** −180° **17.** 135° **18.** −210° **19.** −195° **20.** 105°
21. 420° **22.** −390° **23.** 600° **24.** −630° **25.** −13° **26.** 41°

Change each radian measure to degree measure.

27. $\frac{\pi}{3}$ **28.** π **29.** $\frac{5\pi}{6}$ **30.** $\frac{2\pi}{3}$ **31.** $\frac{\pi}{12}$ **32.** $-\frac{\pi}{4}$
33. $\frac{5\pi}{3}$ **34.** 3π **35.** $-\frac{7\pi}{4}$ **36.** -4π **37.** $-\frac{11\pi}{15}$ **38.** $\frac{7\pi}{6}$
39. $\frac{\pi}{10}$ **40.** $-\frac{\pi}{15}$ **41.** 7π **42.** $\frac{9\pi}{2}$ **43.** $\frac{5\pi}{2}$ **44.** -11π

B Change to radian measure in terms of π.

45. 13°45′ **46.** −10°30′ **47.** −48°15′ **48.** 116.85°

APPLICATIONS: Measuring Angles

49. How many complete rotations does the second hand of a clock make in 3 minutes?

50. Through how many degrees does the second hand of a watch rotate in 3 minutes?

51. Through how many radians does the second hand of a watch rotate in 3 minutes?

52. What part of a complete rotation does the minute hand of a clock make in 3 minutes?

53. Through how many degrees does the minute hand of a clock rotate in 3 minutes?

54. Through how many radians does the minute hand of a clock rotate in 3 minutes?

55. What part of a complete rotation does the hour hand of a clock make in 3 minutes?

56. Through how many degrees does the hour hand of a clock rotate in 3 minutes?

Given a circle with radius r and central angle θ, you can use the formula $s = r\theta$ to find the length of the arc intercepted by θ, where θ is measured in radians.

EXAMPLE: Find the length, s, of the arc intercepted by a central angle of 30° in a circle with a 12-centimeter radius.

SOLUTION: $\theta = 30(\frac{\pi}{180}) = \frac{\pi}{6}$ radians

$s = r\theta = 12(\frac{\pi}{6}) = 2\pi \approx 2(3.14) = \mathbf{6.28\ cm}$

57. Find the length of the arc intercepted by a central angle of 120° in a wheel of radius 10 centimeters.

58. Find, in terms of π, the radius of a circle in which a central angle of 20° intercepts an arc 9 centimeters long.

59. Find the length of the arc traversed in 20 minutes by the end of a 10-centimeter minute hand on a clock.

C

60. For any angle with degree measure θ, the measures of its coterminal angles are **$n \cdot 360° + \theta$, where n is an integer.** Write a similar expression for the radian measures of angles coterminal with an angle θ in radians.

Use the formula found in Exercise 60 to find the radian measure of an angle coterminal with angle θ under the given number of rotations.

61. $\theta = \frac{\pi}{3}$; 2 counterclockwise

62. $\theta = \frac{\pi}{4}$; 1 counterclockwise

63. $\theta = \frac{\pi}{3}$; 2 clockwise

64. $\theta = -\frac{\pi}{6}$; 3 clockwise

65. $\theta = \frac{3\pi}{4}$; 1 clockwise

66. $\theta = -\frac{3\pi}{2}$; counterclockwise

16–4 Radian Measure and Trigonometric Functions

In the figure at the right, the measure of angle AOB is 50° or 0.8727 radian.

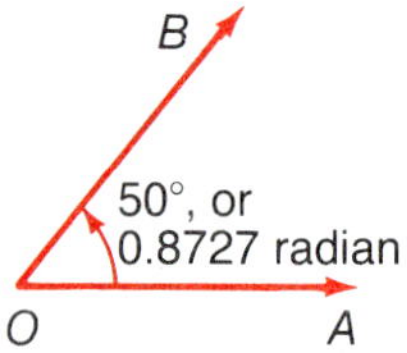

Recall that for every angle of measure θ, there is one and only one value for $\sin\theta$, $\cos\theta$, and $\tan\theta$. Thus, for $\angle AOB$,

$\sin 0.8727 = \sin 50° = 0.7660$; $\cos 0.8727 = \cos 50° = 0.6428$;

$\tan 0.8727 = \tan 50° = 1.1918$.

When you are asked to find the trigonometric functions of an angle expressed in terms of π, it is sometimes easier to change from radian measure to degree measure first. Then you evaluate the trigonometric functions.

EXAMPLE Evaluate: **a.** $\cos\frac{3\pi}{2}$ **b.** $\tan\left(-\frac{3\pi}{4}\right)$

Solutions: **a.** $\cos\frac{3\pi}{2}$ means $\cos\left(\frac{3\pi}{2}\text{ radians}\right)$.

$\cos\frac{3\pi}{2} = \cos 270° = \mathbf{0}$

b. $\tan\left(-\frac{3\pi}{4}\right) = \tan(-135°)$

$= \mathbf{-1}$

CLASSROOM EXERCISES

Use the tables on pages 629–633 to find the indicated values.

	θ Deg.	θ Rad.	Sin θ	Cos θ	Tan θ	Cot θ	Sec θ	Csc θ
1.	15°	0.2618	?	?	0.2679	?	?	3.864
2.	26°30′	?	?	0.8949	?	?	1.117	?
3.	?	0.5614	0.5324	?	?	?	?	?
4.	?	0.0145	?	?	?	68.750	?	?

WRITTEN EXERCISES

A In Exercises 1 and 2, write *True* or *False* for each statement. (The degree measure is rounded to the nearest ten minutes.)

1. $\sin 0.5 = \sin 28°40' = 0.4797$

2. $\csc(-0.6) = \csc(-34°20') = -1.773$

Use the tables on pages 629–633 to evaluate each function.

3. $\sin 0.0902$ **4.** $\cos 1.2072$ **5.** $\cot 0.8348$ **6.** $\tan 1.4835$

7. $\csc 1.1519$ **8.** $\sin 0.5061$ **9.** $\sec 1.2857$ **10.** $\tan 0.9977$

Evaluate.

11. $\sin \frac{\pi}{2}$
12. $\cos \frac{\pi}{3}$
13. $\sec \pi$
14. $\csc \frac{\pi}{4}$
15. $\cot \frac{\pi}{4}$
16. $\tan \frac{3\pi}{4}$
17. $\sec \frac{2\pi}{3}$
18. $\csc \frac{7\pi}{6}$
19. $\sec(-\frac{11\pi}{6})$
20. $\cos(-\frac{11\pi}{6})$
21. $\cos(-\frac{\pi}{6})$
22. $\csc(-\frac{3\pi}{2})$

B

23. $\sin \frac{\pi}{4} \cos \frac{\pi}{6} + \cos \frac{\pi}{4} \sin \frac{\pi}{6}$
24. $\cos \frac{\pi}{3} \cos \frac{3\pi}{4} + \sin \frac{\pi}{3} \sin \frac{3\pi}{4}$
25. $\cos(-\frac{\pi}{6}) \cos \frac{2\pi}{3} - \sin(-\frac{\pi}{6}) \sin \frac{2\pi}{3}$
26. $\sin(-\frac{\pi}{4}) \cos(-\frac{\pi}{3}) - \cos(-\frac{\pi}{4}) \sin(-\frac{\pi}{3})$

C In Exercises 27–34, find all radian measures ($0 \leq \theta \leq 2\pi$) that make each equation true. (HINT: Each equation has *at least two* such radian measures.)

27. $\sin \theta = 0$
28. $\cos \theta = \frac{1}{2}$
29. $\cos \theta = -\frac{\sqrt{2}}{2}$
30. $\sin \theta = \frac{\sqrt{3}}{2}$
31. $\tan \theta = -1$
32. $\cot \theta = -\sqrt{3}$
33. $\sec \theta = 2$
34. $\sec \theta = -\sqrt{2}$
35. Show that $\sin \frac{\pi}{2} = 2 \sin \frac{\pi}{4} \cos \frac{\pi}{4}$.

APPLICATIONS: Using Radian Measure

At time t(in seconds), the displacement d(in cm), velocity v(in cm/sec), and acceleration a(in cm/sec^2) of a weight hanging on an oscillating spring are given by

$$d = \frac{1}{3\pi} \cos 3\pi t, \quad v = -\sin 3\pi t, \quad \text{and} \quad a = -3\pi \cos 3\pi t.$$

Find d, v, and a for the following values of t.

36. 0
37. $\frac{1}{3}$
38. $\frac{1}{6}$
39. $\frac{1}{12}$
40. 1
41. $2\frac{1}{2}$

CALCULATOR APPLICATIONS

Radian Measure

The Examples below show how to use a scientific calculator having a radian mode key to obtain the trigonometric functions of radian measures directly. To place the calculator in the radian mode, you press the [RAD] key or, on some calculators, a special switch.

EXAMPLE 1 Find tan 1.3526.

SOLUTION 1 . 3 5 2 6 [tan]

4.5100646 RAD

EXAMPLE 2 Find sec 0.1367. ← **Use the reciprocal key [1/x] and the cosine function.**

SOLUTION . 1 3 6 7 [cos] [1/x]

1.0094168 RAD

EXERCISES

Use a calculator to do Exercises 1–10 on page 592.

16–5 Graphs of sin x and cos x

To graph the function $f(\theta) = \sin \theta$ in the coordinate plane, you begin by making a table of values for the ordered pairs $(\theta, \sin \theta)$. However, in order to graph on the customary xy axes, x will be used to represent θ and y will be used to represent $f(\theta)$, or $\sin \theta$. The values of x are given in radian measure.

To graph $y = \sin x$, use the tables on pages 629–633 to make a table of values for $0 \le x \le 2\pi$. Round each value to the nearest hundredth. Then plot the points and join them with a smooth curve.

EXAMPLE 1 Sketch the graph $y = \sin x$, where $0 \le x \le 2\pi$.

Solution: Make a table.

x	0	$\frac{\pi}{6}$	$\frac{\pi}{4}$	1	$\frac{\pi}{3}$		2	$\frac{2\pi}{3}$	$\frac{3\pi}{4}$	$\frac{5\pi}{6}$	3	π
y = sin x	0	0.50	0.71	0.84	0.87	1	0.91	0.87	0.71	0.50	0.14	0

x	$\frac{7\pi}{6}$	$\frac{5\pi}{4}$	4	$\frac{4\pi}{3}$	$\frac{3\pi}{2}$	5	$\frac{5\pi}{3}$	$\frac{7\pi}{4}$	$\frac{11\pi}{6}$	6	2π
y = sin x	−0.50	−0.71	−0.76	−0.87	−1	−0.96	−0.87	−0.71	−0.50	−0.28	0

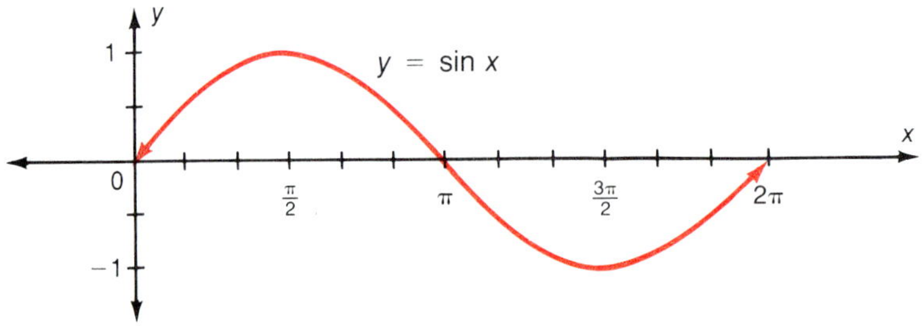

You can use a similar procedure to graph the cosine function.

EXAMPLE 2 Sketch the graph of $y = \cos x$, where $0 \le x \le 2\pi$.

Solution: Make a table.

x	0	$\frac{\pi}{6}$	$\frac{\pi}{4}$	1	$\frac{\pi}{3}$	$\frac{\pi}{2}$	2	$\frac{2\pi}{3}$	$\frac{3\pi}{4}$	$\frac{5\pi}{6}$	3	π
y = cos x	1	0.87	0.71	0.54	0.50	0	−0.42	−0.50	−0.71	−0.87	−0.99	−1

x	$\frac{7\pi}{6}$	$\frac{5\pi}{4}$	4	$\frac{4\pi}{3}$	$\frac{3\pi}{2}$	5	$\frac{5\pi}{3}$	$\frac{7\pi}{4}$	$\frac{11\pi}{6}$	6	2π
y = cos x	−0.87	−0.71	−0.65	−0.50	0	0.23	0.50	0.71	0.87	0.96	1

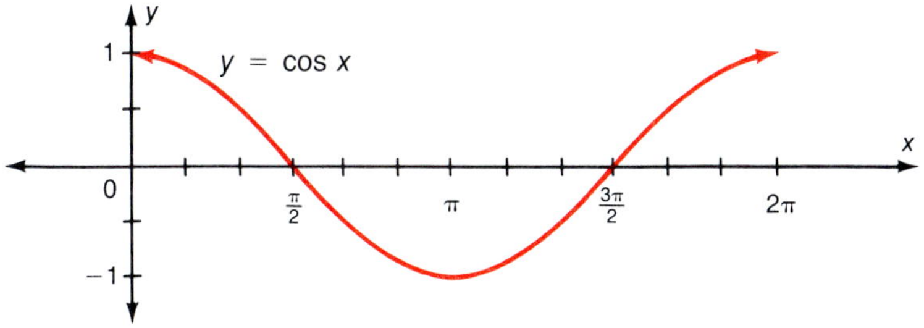

Although the domain of x in the graphs for the sine and cosine functions is $0 \le x \le 2\pi$, the graphs can be extended for $x < 0$ and $x > 2\pi$. Thus, the domain of both $y = \sin x$ and $y = \cos x$ is the set of real numbers. The range of each function is the set of real numbers such that $-1 \le y \le 1$.

EXAMPLE 3 Sketch the graph of $y = 2 \sin x$, where $0 \le x \le 2\pi$.

Solution: For each value of x between 0 and 2π, the corresponding value of y is *twice* the value of y in $y = \sin x$. Therefore, multiply each y value in the table of $y = \sin x$ by 2, and plot the resulting pairs.

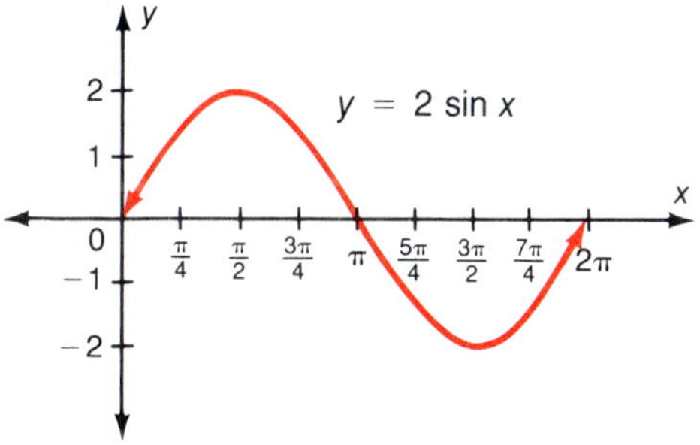

To graph $y = A \sin x$ or $y = A \cos x$, the y values are A *times those in* $y = \sin x$ or $y = \cos x$. The number $|A|$ is the **amplitude.**

When you know the amplitude and the general shape of the sine and cosine curves, you can sketch the graphs of functions of the form $y = A \sin x$ and $y = A \cos x$.

EXAMPLE 4 Sketch the graph of $y = 2 \cos x$ for $-2\pi \le x \le 2\pi$.

Solution: Amplitude: 2 General shape: $y = \cos x$

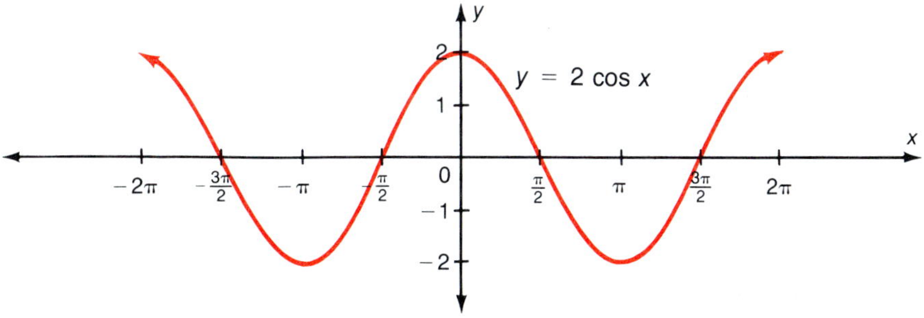

In Example 4, the graph $y = 2 \cos x$ in the interval $-2\pi \le x \le 0$ is the same as the graph in the interval $0 \le x \le 2\pi$. After each interval of 2π, the graphs of the sine and cosine functions repeat. The distance you need to travel along the x axis before the graph repeats itself for the first time is the **period** of the function. Thus, 2π is the period of the sine and cosine functions. A function that repeats in this way is called a **periodic function.**

Definition

> If there is a smallest positive number p such that $f(p + x) = f(x)$ for every x in the domain of f, then p is the **period** of the function f and f is a **periodic function.**

CLASSROOM EXERCISES

Write the period and amplitude of each function.

1. $y = 3 \sin x$ **2.** $y = 7 \cos x$ **3.** $y = \frac{1}{2} \cos x$

4. $y = \frac{1}{8} \sin x$ **5.** $y = \frac{1}{4} \cos x$ **6.** $y = \frac{2}{3} \sin x$

7. $y = -2 \sin x$ **8.** $y = -1000 \sin x$ **9.** $y = -\cos x$

10. $y = \pi \sin x$ **11.** $y = -\frac{1}{6} \sin x$ **12.** $y = -\frac{1}{2} \cos x$

In Exercises 13–21, complete each table.

	x	cos x	2 cos x
	0	1	2
13.	$\frac{\pi}{2}$	?	?
14.	π	?	?
15.	$\frac{3\pi}{2}$	?	?

	x	sin x	$\frac{1}{3}$ sin x
	0	0	0
16.	$\frac{\pi}{2}$	?	?
17.	π	?	?
18.	$\frac{3\pi}{2}$	?	?

	x	cos x	−cos x
	0	1	−1
19.	$\frac{\pi}{2}$	?	?
20.	π	?	?
21.	$\frac{3\pi}{2}$	?	?

WRITTEN EXERCISES

A

For each function in Exercises 1–8, make a table of the pairs of real numbers (x, y), where $0 \le x \le 2\pi$. Then sketch the graph of each function.

1. $y = 4 \cos x$ **2.** $y = -2 \sin x$ **3.** $y = \frac{1}{2} \sin x$ **4.** $y = \frac{1}{3} \cos x$

5. $y = \frac{2}{3} \cos x$ **6.** $y = -\frac{1}{2} \sin x$ **7.** $y = -\sin x$ **8.** $y = -3 \cos x$

In Exercises 9–12, sketch the graphs of the functions on the same coordinate axes where $0 \le x \le 2\pi$.

9. $y = \cos x$ **10.** $y = 4 \cos x$ **11.** $y = 2 \cos x$ **12.** $y = \frac{1}{2} \cos x$

In Exercises 13–16, sketch the graphs of the functions on the same coordinate axes where $0 \le x \le 2\pi$.

13. $y = \sin x$ **14.** $y = 2 \sin x$ **15.** $y = -2 \sin x$ **16.** $y = -4 \sin x$

B

17. Make a table of values and then sketch the graph of $y = \sin 2x$, where $-2\pi \le x \le 2\pi$. Note that $y = \sin 2x$ is *not* the same as $y = 2 \sin x$.
18. How does the amplitude of $y = \sin 2x$ compare with that of $y = \sin x$?
19. How does the period of $y = \sin 2x$ compare with that of $y = \sin x$?
20. Make a table of values and draw the graph of $y = \cos 2x$, where $-2\pi \le x \le 2\pi$.
21. How does the amplitude of $y = \cos 2x$ compare with that of $y = \cos x$?
22. How do the periods of $y = \cos 2x$ and $y = \cos x$ compare?

C

The period of $y = \sin (Bx)$ or $y = \cos (Bx)$ is $\left|\frac{1}{B}\right|$ times the period of $y = \sin x$ or $y = \cos x$.

Find the period of each of the following.

23. $y = \cos 3x$
24. $y = \sin 5x$
25. $y = \sin \frac{1}{2} x$
26. $y = \cos \frac{1}{4} x$
27. $y = \cos (-2x)$
28. $y = \sin (-3x)$
29. $y = \sin (-\frac{1}{3} x)$
30. $y = \sin 0.5x$
31. $y = \cos (-0.25x)$

In Exercises 32–34, graph each function for $-2\pi \le x \le 2\pi$.

32. $y = \cos x + 2$
33. $y = \cos x - 2$
34. $y = 2 \sin x + \frac{1}{2}$
35. Compare the amplitude and period of the functions in Exercises **32** and **33**.
36. Write the amplitude and period of the function in Exercise **34**.
37. Graph the function $y = |\sin x|$ for $0 \le x \le 2\pi$.

In the figure below, the functions $y = \sin x$ and $y = \cos x$ are graphed on the same coordinate axes over the interval $0 \le x \le 2\pi$.

Refer to this figure in Exercises 38–43 to estimate the value(s) of x in the given interval that satisfies each statement. Some statements may be satisfied by no value of x.

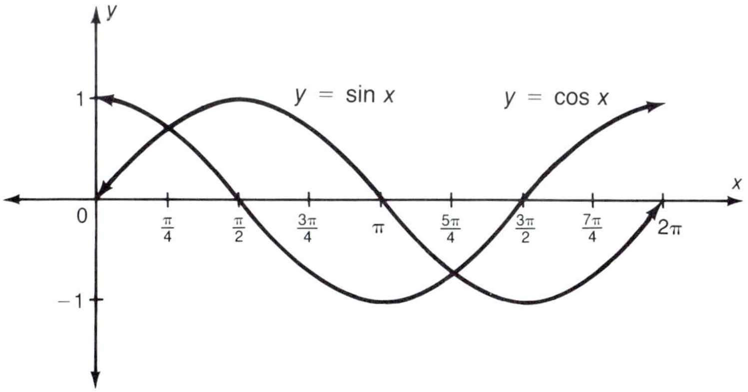

38. $\cos x = \sin x$
39. $\cos x = -\sin x$
40. $\sin x > \cos x$
41. $\sin x < \cos x$
42. $\sin x + \cos x = 1$
43. $\sin x + \cos x = 2$

16–6 Graphs of tan x, cot x, sec x, csc x

The sine and cosine functions are the basic trigonometric functions. The remaining trigonometric functions can be expressed in terms of these two. For example, by definition,

$$\sin\theta = \frac{y}{r} \quad \text{and} \quad \cos\theta = \frac{x}{r}.$$

Thus,
$$\frac{\sin\theta}{\cos\theta} = \frac{\frac{y}{r}}{\frac{x}{r}} = \frac{y}{x} = \tan\theta.$$

Therefore,
$$\boldsymbol{\tan\theta = \frac{\sin\theta}{\cos\theta}, \ \cos\theta \neq 0.}$$

The equation above shows that $\tan\theta$ is not defined for $\cos\theta = 0$, that is, for

$$\theta = \pm\frac{\pi}{2}, \pm\frac{3\pi}{2}, \ldots, \frac{(2n+1)\pi}{2}, n \in \text{I}.$$

Near these values of θ, $\tan\theta$ increases or decreases without bound. Thus, amplitude is not defined for $f(\theta) = \tan\theta$. Recall that in order to graph on the customary xy axes, x is used to represent θ and y is used to represent $f(\theta)$.

EXAMPLE 1 Graph $y = \tan x$ for $-\pi \leq x \leq 2\pi$.

Solution: Make a table.

x	tan x
$-\pi$	0
$-\frac{3\pi}{4}$	1.0
$-\frac{\pi}{2}$	Not defined
-1.0	-1.6
$-\frac{\pi}{4}$	-1.0
0	0
$\frac{\pi}{6}$	0.58
$\frac{\pi}{4}$	1.0
1.0	1.6
$\frac{\pi}{3}$	1.7
$\frac{\pi}{2}$	Not defined

x	tan x
2.0	-2.2
3.0	-0.14
π	0
$\frac{5\pi}{4}$	1.0
4.0	1.2
$\frac{3\pi}{2}$	Not defined
5.0	-3.4
$\frac{7\pi}{4}$	-1.0
6.0	-0.29
2π	0

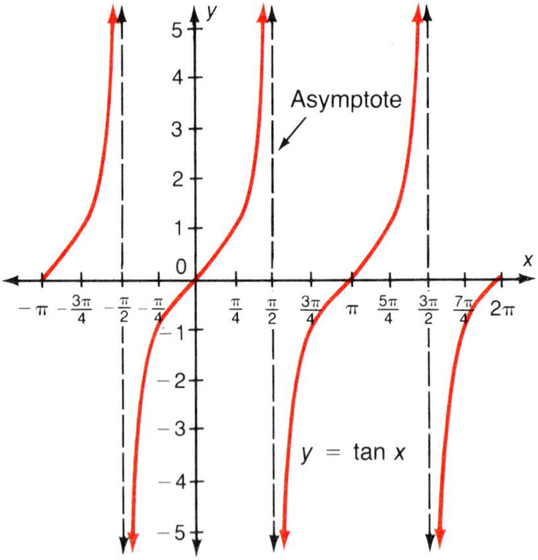

The graph in Example 1 shows that the range of $y = \tan x$ is the set of real numbers. The period of $y = \tan x$ is π, because the graph repeats at intervals of π units. That is, $\mathbf{\tan (\pi + \theta) = \tan \theta}$. The dashed lines in the graphs are **asymptotes.** The graph of $y = \tan x$ nears but does not intersect the asymptotes.

Unlike the sine and cosine functions, the tangent function has "breaks" in its graph. The sine and cosine are **continuous** functions. The function $y = \tan x$ is **discontinuous** at

$$x = \pm\frac{\pi}{2}, \pm\frac{3\pi}{2}, \ldots, \frac{(2n+1)\pi}{2}, n \in \text{I}.$$

Recall that the cotangent, secant, and cosecant functions are reciprocals of the tangent, cosine, and sine functions respectively.

$$\cot x = \frac{1}{\tan x}, \tan x \neq 0$$

$$\sec x = \frac{1}{\cos x}, \cos x \neq 0$$

$$\csc x = \frac{1}{\sin x}, \sin x \neq 0$$

Each function has the same period as its reciprocal. Each of these functions is discontinuous at values of x for which its reciprocal is zero.

Function	Values for which the function is not defined	Asymptotes
$y = \cot x$	$x = n\pi, n \in \text{I}$	$x = n\pi, n \in \text{I}$
$y = \sec x$	$x = \left(\frac{2n+1}{2}\right)\pi, n \in \text{I}$	$x = \left(\frac{2n+1}{2}\right)\pi, n \in \text{I}$
$y = \csc x$	$x = n\pi, n \in \text{I}$	$x = n\pi, n \in \text{I}$

Near those values of x for which each function is not defined, the values of the functions are unbounded. Hence, amplitude is *not* defined for these functions.

The graph of the cotangent function can be sketched by first sketching the graph of $y = \tan x$ and then estimating the reciprocals of $\tan x$. You can sketch the graphs of $y = \sec x$ and $y = \csc x$ in a similar way.

EXAMPLE 2 Graph each function.

a. $y = \cot x$ **b.** $y = \sec x$ **c.** $y = \csc x$

Solutions: **a.**

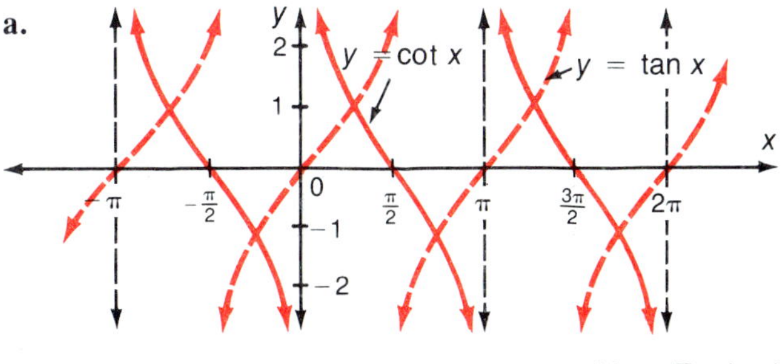

b.

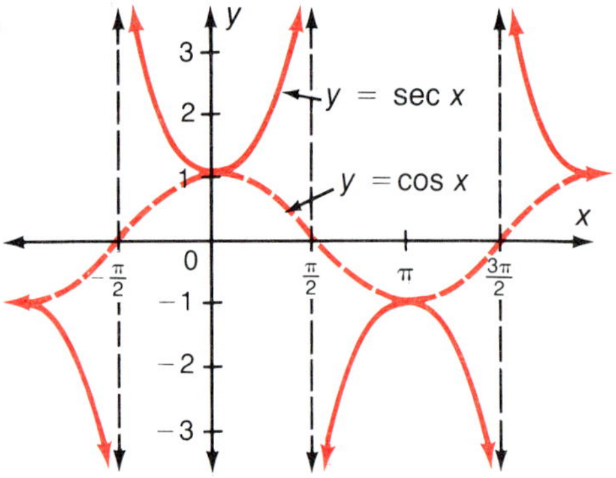

c.

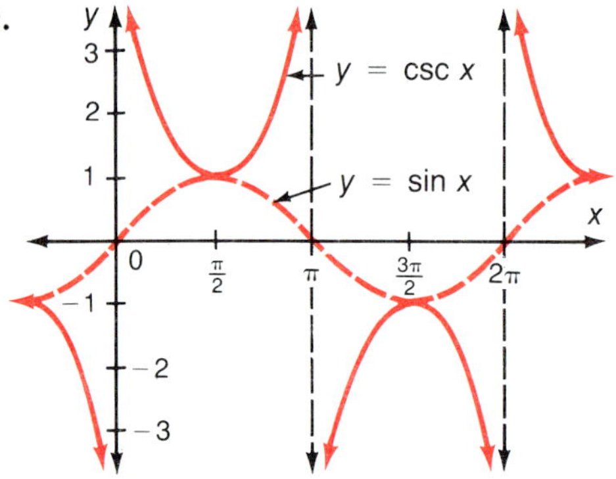

The following table summarizes some important properties of the six trigonometric functions.

Summary of Important Properties

Function	Domain	Range	Period	Continuity
$y = \sin x$	All real numbers x	$\{y: -1 \le y \le 1, y \in R\}$	2π	Continuous
$y = \cos x$	All real numbers x	$\{y: -1 \le y \le 1, y \in R\}$	2π	Continuous
$y = \tan x$	All real numbers x, except $x = \left(\frac{2n+1}{2}\right)\pi,\ n \in I$	All real numbers y	π	Discontinuous at $x = \left(\frac{2n+1}{2}\right)\pi,\ n \in I$
$y = \cot x$	All real numbers x, except $x = n\pi,\ n \in I$	All real numbers y	π	Discontinuous at $x = n\pi,\ n \in I$
$y = \sec x$	All real numbers x, except $x = \left(\frac{2n+1}{2}\right)\pi \in I$	All real numbers y, except $-1 < y < 1$	2π	Discontinuous at $x = \left(\frac{2n+1}{2}\right)\pi,\ n \in I$
$y = \csc x$	All real numbers x, except $x = n\pi,\ n \in I$	All real numbers y, except $-1 < y < 1$	2π	Discontinuous at $x = n\pi,\ n \in I$

CLASSROOM EXERCISES

Find the indicated values. Round each value to the nearest hundredth.

$y = 2 \tan x$

	x	2 tan x
1.	$-\pi$	?
2.	$-\frac{\pi}{2}$	?
3.	$\frac{\pi}{6}$	?
4.	$\frac{3\pi}{2}$	?
5.	$\frac{7\pi}{4}$	?

$y = \cot x$

	x	cot x
6.	$-\pi$	?
7.	$-\frac{\pi}{2}$	?
8.	$\frac{\pi}{6}$	?
9.	$\frac{3\pi}{2}$	?
10.	$\frac{7\pi}{4}$	?

$y = -\sec x$

	x	−sec x
11.	0	?
12.	$\frac{\pi}{3}$	?
13.	$\frac{\pi}{2}$	?
14.	π	?
15.	$\frac{5\pi}{4}$	?

WRITTEN EXERCISES

A Sketch the graph of each function for $-2\pi \le x \le 2\pi$.

1. $y = \tan x$ **2.** $y = \cot x$ **3.** $y = \sec x$ **4.** $y = \csc x$

5. $y = 2 \tan x$ **6.** $y = \frac{1}{2} \tan x$ **7.** $y = -\tan x$ **8.** $y = 2 \csc x$

9. $y = \frac{1}{3} \sec x$ **10.** $y = -\frac{1}{2} \sec x$ **11.** $y = \frac{1}{2} \cot x$ **12.** $y = 3 \cot x$

Write the equations of the asymptotes for each function where $-2\pi \le x \le 2\pi$.

13. $y = \tan x$ **14.** $y = \cot x$ **15.** $y = \sec x$ **16.** $y = \csc x$

Complete each statement.

17. Since the period of $y = \cot x$ is π, $\cot(\pi + \theta) =$ __?__

18. Since the period of $y = \sec x$ is 2π, $\sec(2\pi + \theta) =$ __?__

19. Since the period of $y = \csc x$ is 2π, $\csc(2\pi + \theta) =$ __?__

20. How do the periods of $y = \sin x$ and $y = \csc x$ compare?

21. How do the periods of $y = \csc x$ and $y = \sec x$ compare?

B

22. Make a table of values for $y = \tan(\frac{1}{4}x)$ and graph the function.

23. What is the period of $y = \tan(\frac{1}{4}x)$?

24. How does the period of $y = \tan(\frac{1}{4}x)$ compare with the period of $y = \tan x$?

25. Write a general rule for the period of $y = \tan(Bx)$.

In Exercises 26–29, find the period of each function and sketch the graph over the interval indicated.

26. $y = \tan 3x;\ -\pi \le x \le \pi$ **27.** $y = \tan \frac{1}{3}x;\ -2\pi \le x \le 2\pi$

28. $y = \frac{1}{2} \tan 3x;\ -\frac{\pi}{2} < x < \frac{\pi}{2}$ **29.** $y = 2 \tan \frac{1}{3}x;\ -\frac{3}{2}\pi < x < \frac{3}{2}\pi$

Sketch the graph of each function for $0 \le x \le 2\pi$.

30. $y = \cot 2x$ **31.** $y = \csc 2x$ **32.** $y = \cot \frac{1}{2}x$ **33.** $y = \sec \frac{1}{2}x$

C Graph each pair of functions on the same coordinate axes over the given interval.

34. $y = 2 \sin x;\ y = 1 + 2 \sin x;\ 0 \le x \le 2\pi$

35. $y = \sin 2x;\ y = 1 + \sin 2x;\ 0 \le x \le 2\pi$

36. $y = \cos 3x;\ y = -1 + \cos 3x;\ 0 < x < 2\pi$

37. $y = \sin \frac{1}{2}x;\ y = \sin(\frac{\pi}{2} + \frac{1}{2}x);\ -2\pi \le x \le 2\pi$

38. $y = \cos 4x;\ y = 1 + \cos 4x;\ -2\pi \le x \le 2\pi$

39. $y = 2 \cos \frac{1}{3}x;\ y = -\frac{1}{2} \cos \frac{1}{3}x;\ 0 \le x \le 2\pi$

40. $y = 2 \sin x;\ y = 2 \sin(x + \frac{\pi}{4});\ 0 \le x \le 2\pi$

BASIC: SOLVING TRIANGLES

An obvious use of the computer is the solution of triangles.

Problem: *Given the lengths of two sides of a triangle and the degree measure of the included angle, solve the triangle.*

The program below shows the steps for solving the triangle. The figure on the right is used as a reference in writing this program. A, B, and C represent the lengths of the sides. A1, B1, and C1 are the measures of the angles in degrees.

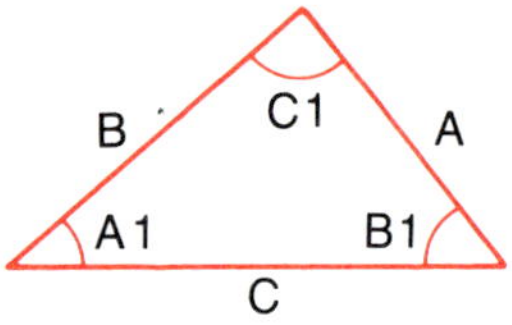


```
100 READ B,C,A1
110 LET R1 = A1*3.14159/180
120 LET A = SQR(B↑2 + C↑2 - 2*B*C*COS(R1))
130 IF C < B THEN 230
140 LET S = B*SIN(R1)/A          ← S is the sine of B1.
150 LET D = SQR(1 - S↑2)         ← D is the cosine of B1.
160 LET R2 = ATN(S/D)
170 LET B1 = R2*180/3.14159
180 LET C1 = 180 - (A1 + B1)
190 PRINT "SIDES ARE:";A;B;C
200 PRINT "ANGLES ARE:";A1;B1;C1
210 PRINT
220 GOTO 100
230 LET S = C*SIN(R1)/A          ← S is the sine of C1.
240 LET D = SQR(1 - S↑2)         ← D is the cosine of C1.
250 LET R2 = ATN(S/D)
260 LET C1 = R2*180/3.14159
270 LET B1 = 180 - (A1 + C1)
280 GOTO 190
290 DATA 5,8,35,6,9,49.33,7,5,152   ← B = 5, 6 and 7; C = 8, 9, and 5;
                                      A1 = 35, 49.33, and 152
300 END
```

Output:

```
SIDES ARE:     4.84436    5    8
ANGLES ARE:     35    36.2994     108.701

SIDES ARE:     6.82761    6    9
ANGLES ARE:     49.33    41.8004     88.8696

SIDES ARE:     11.6536    7    5
ANGLES ARE:     152    16.3795     11.6205

OUT OF DATA AT LINE 100
```

Analysis

Statement 110: This statement changes degree measure to radian measure.

Statement 120: This statement applies the Law of Cosines.

Statement 130: This statement involves a decision. If C < B, then C1 is an acute angle, and the computer jumps to statements 230–260 to find the measure of C1. If C > B, then B1 is an acute angle, and the computer uses statements 140–170 to find the measure of B1.

Statement 140: This statement applies the Law of Sines. (See also statement 230.)

Statements 150 and 240: These statements apply Pythagorean Identity 6, $\sin^2\theta + \cos^2\theta = 1$. (See page 606.)

Statements 160 and 250: These statements find the radian measures of B1 and C1, respectively. ATN is short for "arctangent."

EXERCISES

A Run the program on page 602 for the following values of *B, C,* and *a*.

1. $B = 9$; $C = 12$; $a = 27°$
2. $B = 15.5$; $C = 7$; $a = 100°$
3. $B = 4.25$; $C = 4.25$; $a = 117°$
4. $B = 11$; $C = 8$; $a = 54.5°$

In each of Exercises 5–7, a formula for the area of a triangle is given. Write a program for the area based on the given formula.

5. $K = \frac{1}{2}ab \sin C$
6. $K = \frac{1}{2} \cdot \frac{c^2 \sin A \sin B}{\sin C}$
7. $K = \sqrt{s(s-a)(s-b)(s-c)}$, where $s = \frac{a+b+c}{2}$
8. Given three positive numbers, write a program to decide whether the numbers could be the lengths of three sides of a triangle.

In each of Exercises 9–12, write a program for solving a right triangle. In each exercise, you are given the measure of two parts of a right triangle, where angle measures are in degrees.

9. Hypotenuse and an acute angle
10. A leg and an acute angle
11. Two legs
12. Hypotenuse and a leg

B In each of Exercises 13–16, write a program for solving a triangle. In each exercise you are given the measures of three parts, where angle measures are in degrees.

13. One side and two angles
14. Two angles and the included side
15. Three sides
16. Two sides and an angle

Review

1. A hiker walks north for 5 kilometers and then walks west for 8 kilometers. Find the magnitude of the hiker's displacement. *(Section 16–1)*

2. In the figure at the right, the air velocity of a plane is shown together with the wind velocity. Find the ground speed of the plane. *(Section 16–2)*

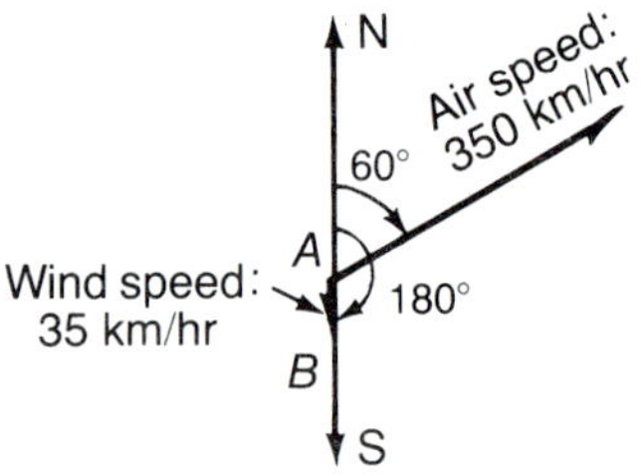

Change each degree measure to radian measure in terms of π. *(Section 16–3)*

3. $720°$
4. $120°$
5. $-270°$
6. $-135°$
7. $700°$
8. $-108°$

Change each radian measure to degree measure. *(Section 16–3)*

9. $\frac{\pi}{6}$
10. $\frac{3\pi}{8}$
11. -6π
12. $-\frac{5\pi}{12}$
13. $-\frac{11\pi}{2}$
14. $\frac{2\pi}{15}$

Use the tables on pages 629–633 to evaluate each function. *(Section 16–4)*

15. $\cos 0.1949$
16. $\sin 1.2770$
17. $\sec 0.3985$
18. $\tan 0.8378$

Evaluate. *(Section 16–4)*

19. $\cos \frac{\pi}{2}$
20. $\sin \frac{3\pi}{4}$
21. $\tan \frac{\pi}{4}$
22. $\cot \frac{\pi}{6}$
23. $\sec \frac{\pi}{3}$
24. $\csc \left(-\frac{5\pi}{6}\right)$
25. $\sin (-\pi)$
26. $\cos \left(-\frac{11\pi}{6}\right)$

Sketch the graphs of the following functions, where $-2\pi \le x \le 2\pi$. *(Sections 16–5 and 16–6)*

27. $y = 3 \cos x$
28. $y = -\frac{1}{4} \sin x$
29. $y = \cot x$
30. $y = -2 \tan x$

Write the equation of the asymptotes for each function, where $-2\pi \le x \le 2\pi$. *(Section 16–6)*

31. $y = -\tan x$
32. $y = -\csc x$
33. $y = -\sec x$
34. $y = 2 \cot x$

16–7 Fundamental Identities

In trigonometry as in algebra, conditional equations are true only for *particular* values of the variable. For example, for $0° < \theta < 360°$, if

$$\sin \theta = \frac{1}{2}, \quad \text{then} \quad \theta = 30° \quad \underline{\text{or}} \quad \theta = 150°.$$

Trigonometric identities are equations which are true for all values of the variable for which the trigonometric functions in the identity are defined. For example, you can prove that

$$\boldsymbol{\cot \theta = \frac{1}{\tan \theta}, \ \tan \theta \neq 0,}$$

is an identity by applying the definitions of the trigonometric functions.

The eight basic trigonometric identities are listed in this section. They are called the **Fundamental Identities.**

Theorem 16–1

Reciprocal Identities

1. $\csc\theta = \dfrac{1}{\sin\theta},\ \sin\theta \neq 0$
2. $\sec\theta = \dfrac{1}{\cos\theta},\ \cos\theta \neq 0$
3. $\cot\theta = \dfrac{1}{\tan\theta},\ \tan\theta \neq 0$

EXAMPLE 1 Prove Identity 1: $\csc\theta = \dfrac{1}{\sin\theta}$

Proof:

$\csc\theta = \dfrac{r}{y}$ ← **By definition**

$= \dfrac{\frac{r}{r}}{\frac{y}{r}}$ ← **Divide numerator and denominator by r, $r \neq 0$.**

$= \dfrac{1}{\frac{y}{r}}$

$= \dfrac{1}{\sin\theta},\ \sin\theta \neq 0$ ← **Since $\sin\theta = \dfrac{y}{r}$**

You are asked to prove Identities 2 and 3 in the Exercises.

Theorem 16–2

Ratio Identities

4. $\tan\theta = \dfrac{\sin\theta}{\cos\theta},\ \cos\theta \neq 0$
5. $\cot\theta = \dfrac{\cos\theta}{\sin\theta},\ \sin\theta \neq 0$

EXAMPLE 2 Prove Identity 4: $\tan\theta = \dfrac{\sin\theta}{\cos\theta}$

Proof:

$\tan\theta = \dfrac{y}{x}$ ← **By definition**

$= \dfrac{\frac{y}{r}}{\frac{x}{r}} = \dfrac{\sin\theta}{\cos\theta}$ ← **By definition**

You are asked to prove Identity 5 in the Exercises.

The symbol "$\sin^2\theta$" found in the following theorem, means $(\sin\theta)^2$. For example, if $\sin\theta = \frac{1}{2}$, then $\sin^2\theta = (\frac{1}{2})^2$, or $\frac{1}{4}$.

Theorem 16–3

Pythagorean Identities

For all replacements of θ for which the functions are defined,

6. $\sin^2\theta + \cos^2\theta = 1$

7. $1 + \cot^2\theta = \csc^2\theta$

8. $\tan^2\theta + 1 = \sec^2\theta$

EXAMPLE 3 Prove Identity 6: $\sin^2\theta + \cos^2\theta = 1$

Proof:

$y^2 + x^2 = r^2$ ← **True for any point (x, y) on the radius vector of θ**

$\frac{y^2}{r^2} + \frac{x^2}{r^2} = \frac{r^2}{r^2}$

$\left(\frac{y}{r}\right)^2 + \left(\frac{x}{r}\right)^2 = 1$ ← **$\sin\theta = \frac{y}{r}$; $\cos\theta = \frac{x}{r}$**

$\sin^2\theta + \cos^2\theta = 1$

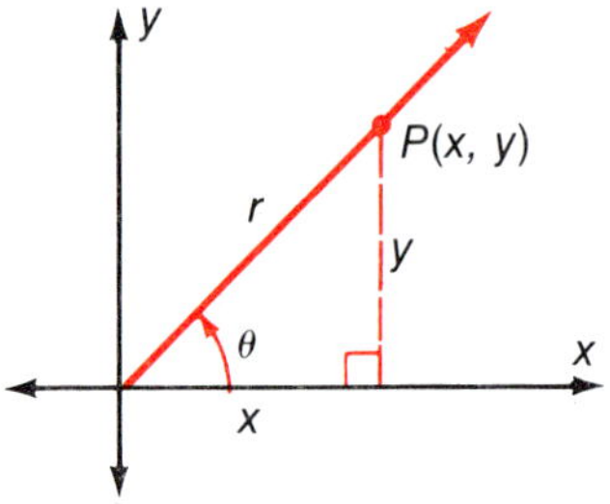

Identities 7 and 8 may be proved by following a similar procedure or by using the **Reciprocal Identities (1–3)** and the **Ratio Identities (4–5).**

EXAMPLE 4 Write $\csc\theta - \cos\theta\cot\theta$ in terms of $\sin\theta$.

Solution:

$\csc\theta - \cos\theta\cot\theta = \frac{1}{\sin\theta} - \cos\theta\left(\frac{\cos\theta}{\sin\theta}\right)$ ← **By Identities 1 and 5**

$= \frac{1 - \cos^2\theta}{\sin\theta}$

$= \frac{\sin^2\theta}{\sin\theta}$ ← **By Identity 6**

$= \sin\theta$

CLASSROOM EXERCISES

Write an equivalent expression in terms of $\sin\theta$, $\cos\theta$, or both.

1. $\csc\theta$ **2.** $\sec\theta$ **3.** $\cot\theta$ **4.** $\frac{\tan\theta}{\csc\theta}$

Write an equivalent expression in terms of $\sec\theta$, $\csc\theta$, or both.

5. $\frac{1}{\cos\theta}$ **6.** $\frac{1}{\sin\theta}$ **7.** $1 + \tan^2\theta$ **8.** $\frac{1 + \cot^2\theta}{\cos^2\theta}$

Write an equivalent expression that is a single term.

9. $1 - \cos^2\theta$ **10.** $1 - \sin^2\theta$ **11.** $\sec^2\theta - 1$ **12.** $1 - \csc^2\theta$

Write $\tan^2\theta - 2\sec\theta\tan\theta$ as described.

13. In terms of $\sin\theta$ and $\cos\theta$ **14.** In terms of $\tan\theta$ **15.** In terms of $\cot\theta$

WRITTEN EXERCISES

A

1. Prove: $\sec \theta = \dfrac{1}{\cos \theta}$, $\cos \theta \neq 0$.

2. Prove: $\cot \theta = \dfrac{1}{\tan \theta}$, $\tan \theta \neq 0$.

3. Prove: $\cot \theta = \dfrac{\cos \theta}{\sin \theta}$, $\sin \theta \neq 0$.

4. Prove: $\tan^2\theta + 1 = \sec^2\theta$.

In Exercises 5–20, write as an equivalent single trigonometric function such as $\cos \theta$, $-\cos \theta$, $\sin \theta$, $-\sin \theta$, $\tan^2\theta$, $-\sec \theta$, etc., or as 1 or -1.

5. $\cot \theta \cdot \sin \theta$
6. $\tan \theta \cdot \cot \theta$
7. $\csc \theta \cdot \cos \theta$
8. $\csc \theta \cdot \cos \theta \cdot \tan \theta$
9. $\dfrac{\sin^2\theta}{\cos^2\theta}$
10. $\dfrac{\cos^2\theta - 1}{\sin^2\theta}$
11. $\dfrac{1}{\sec^2\theta}$
12. $\dfrac{\tan \theta}{\sec \theta}$
13. $\dfrac{\cos \theta}{\cot \theta}$
14. $\dfrac{\cot \theta}{\csc \theta}$
15. $\dfrac{\cos^2\theta}{1 - \cos^2\theta}$
16. $\dfrac{1 + \tan^2\theta}{1 + \cot^2\theta}$
17. $\dfrac{\csc \theta}{\cot \theta}$
18. $\csc^2\theta - \cot^2\theta$
19. $\dfrac{\tan^2\theta - \sin^2\theta}{\tan^2\theta \sin^2\theta}$
20. $\dfrac{\sin^4\theta - \cos^4\theta + 1}{2}$

Express in terms of $\sin \theta$, $\cos \theta$, or both. Simplify.

21. $\cot \theta + \tan \theta$
22. $\sec \theta - \tan \theta$
23. $\dfrac{1 + \tan \theta}{\sec \theta}$
24. $\dfrac{\csc \theta}{\cot \theta + \tan \theta}$

In Exercises 25–28 write each expression in terms of $\cos \theta$.

25. $\sin \theta$
26. $\csc \theta$
27. $\tan \theta$
28. $\cot \theta$

B

Express in terms of the given function.

29. $\dfrac{\tan \theta}{\sin \theta}$; $\sec \theta$

30. $\tan \theta \cot^2\theta$; $\tan \theta$

31. $\sin^2 x + \cos^2 x + \tan^2 x$; $\cos x$

32. $\sec^2\theta - \tan^2\theta + \cot^2\theta$; $\sin \theta$

33. $\cos \theta \sec \theta - \cos^2\theta$; $\cos \theta$

34. $\csc \theta(\csc^2\theta - \cot^2\theta)$; $\sin \theta$

35. Write $\dfrac{\sin^2\theta + \cos^2\theta}{\sin \theta + \cos \theta}$ in terms of $\sec \theta$ and $\csc \theta$. Simplify.

36. Write $\dfrac{\cot \theta \tan \theta - \cos^2\theta}{\cos \theta \tan \theta}$ in terms of $\sin \theta$.

37. Write $(\cos \theta \cot^2\theta \sec \theta)(\tan^2\theta \csc^2\theta - 1)$ in terms of $\sin \theta$ and $\cos \theta$ only.

38. Write $\tan^2\theta + \csc^2\theta + 1$ in terms of $\sin \theta$ and $\cos \theta$ only.

C

39. Write $1 + \tan^2\theta - \dfrac{\sin^2\theta}{\csc^2\theta - 1}$ in terms of $\cos \theta$.

In Exercises 40–42, determine whether the given equation is an identity. Prove your answer.

40. $\sin \theta + \cos \theta = 1$
41. $\sin 2\theta = 2 \sin \theta$
42. $\tan \theta + 1 = \sec \theta$

16–8 Sum and Difference Identities

It can be readily shown that the cosine of a difference *does not* equal the difference of cosines. For example,

$$\cos\left(\frac{\pi}{3}-\frac{\pi}{6}\right) \neq \cos\frac{\pi}{3}-\cos\frac{\pi}{6}.$$

Proof: $\cos\left(\frac{\pi}{3}-\frac{\pi}{6}\right)=\cos\frac{\pi}{6}=\frac{\sqrt{3}}{2}$ $\qquad \cos\frac{\pi}{3}-\cos\frac{\pi}{6}=\frac{1}{2}-\frac{\sqrt{3}}{2}$

Since $\frac{\sqrt{3}}{2}\neq\frac{1}{2}-\frac{\sqrt{3}}{2}$, $\cos\left(\frac{\pi}{3}-\frac{\pi}{6}\right)\neq\cos\frac{\pi}{3}-\cos\frac{\pi}{6}$.

The correct relationship is the *Difference Identity for Cosine.*

Theorem 16–4

Difference Identity for Cosine

$$\cos(\alpha-\beta)=\cos\alpha\cos\beta+\sin\alpha\sin\beta$$

Proof: Recall from Section 15–2 that if $P(x, y)$ is on the terminal side of an angle θ in standard position, then

$$x = r\cos\theta \qquad \text{and} \qquad y = r\sin\theta,$$

where r is the distance from the origin to point P.

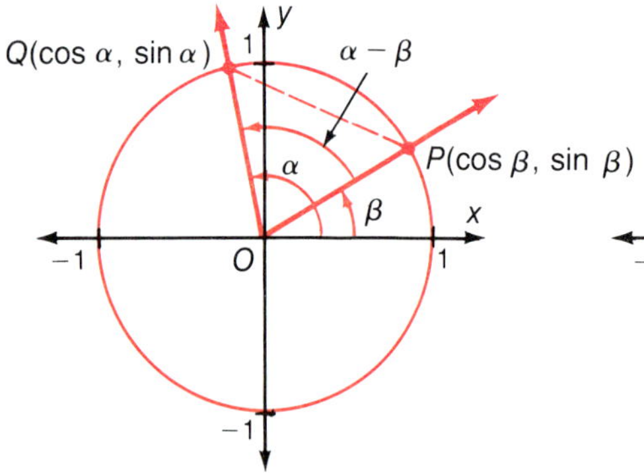

Figure 1

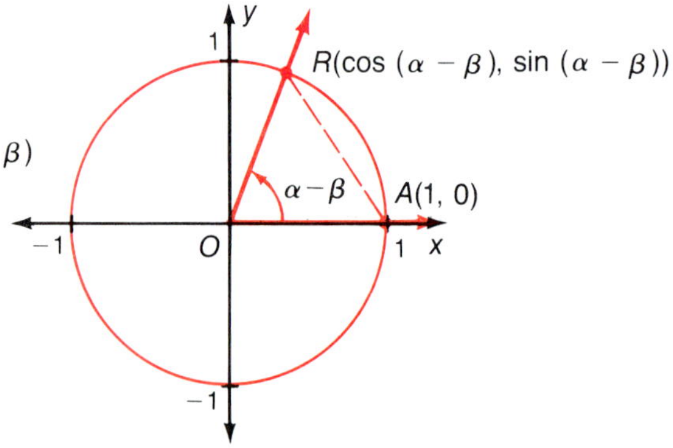

Figure 2

In Figure **1**, α (alpha) and β (beta) are angles in standard position. Points Q and P are on the terminal sides of α and β, respectively, such that $OQ = OP = 1$. Thus, Q and P are points on the circle with center at O and radius $r = 1$ (called the **unit circle**). Therefore, Q and P have coordinates as shown. The difference $\alpha - \beta$ is $\angle POQ$.

In Figure **2**, $\alpha - \beta$ is placed in standard position and R is a point on the terminal side such that $OR = 1$. Therefore, R is on the unit circle and has coordinates as shown.

Use the distance formula to find $(PQ)^2$ and $(AR)^2$.

$(PQ)^2 = (\cos \alpha - \cos \beta)^2 + (\sin \alpha - \sin \beta)^2$

$(AR)^2 = [\cos(\alpha - \beta) - 1]^2 + [\sin(\alpha - \beta) - 0]^2$

Angles POQ and AOR are congruent central angles of the unit circle. Thus, their chords, $\overline{PQ}$ and $\overline{AR}$, are equal in length.
Since $PQ = AR$, $(PQ)^2 = (AR)^2$. Thus,

$$(\cos \alpha - \cos \beta)^2 + (\sin \alpha - \sin \beta)^2 = [\cos(\alpha - \beta) - 1]^2 + [\sin(\alpha - \beta) - 0]^2.$$

Expand the left side.

$(\cos^2 \alpha - 2 \cos \alpha \cos \beta + \cos^2 \beta) + (\sin^2 \alpha - 2 \sin \alpha \sin \beta + \sin^2 \beta) =$

$(\cos^2 \alpha + \sin^2 \alpha) + (\cos^2 \beta + \sin^2 \beta) - 2(\cos \alpha \cos \beta + \sin \alpha \sin \beta) =$

$1 + 1 - 2(\cos \alpha \cos \beta + \sin \alpha \sin \beta) = 2 - 2(\cos \alpha \cos \beta + \sin \alpha \sin \beta)$

Now expand the right side.

$\cos^2(\alpha - \beta) - 2 \cos(\alpha - \beta) + 1 + \sin^2(\alpha - \beta) =$

$[\cos^2(\alpha - \beta) + \sin^2(\alpha - \beta)] + 1 - 2 \cos(\alpha - \beta) =$

$1 + 1 - 2 \cos(\alpha - \beta) = 2 - 2 \cos(\alpha - \beta)$

Thus,

$2 - 2(\cos \alpha \cos \beta + \sin \alpha \sin \beta) = 2 - 2 \cos(\alpha - \beta)$,

or $\quad 2 \cos(\alpha - \beta) = 2(\cos \alpha \cos \beta + \sin \alpha \sin \beta)$.

Therefore, $\boldsymbol{\cos(\alpha - \beta) = \cos \alpha \cos \beta + \sin \alpha \sin \beta}$.

Theorem 16–4 can be used to prove the *Sum Identity for Cosine.*

Theorem 16–5

Sum Identity for Cosine

$$\boldsymbol{\cos(\alpha + \beta) = \cos \alpha \cos \beta - \sin \alpha \sin \beta}$$

Proof: $\cos(\alpha + \beta) = \cos[\alpha - (-\beta)]$

$= \cos \alpha \cos(-\beta) + \sin \alpha \sin(-\beta)$

$= \cos \alpha \cos \beta + \sin \alpha(-\sin \beta)$ ← **$\cos(-\beta) = \cos \beta$; $\sin(-\beta) = -\sin \beta$**

$= \boldsymbol{\cos \alpha \cos \beta - \sin \alpha \sin \beta}$

Theorem 16–5 can also be used to prove the *Sum and Difference Identities for Sine.*

Theorem 16–6
Theorem 16–7

Sum and Difference Identities for Sine

$$\boldsymbol{\sin(\alpha + \beta) = \sin \alpha \cos \beta + \cos \alpha \sin \beta}$$

$$\boldsymbol{\sin(\alpha - \beta) = \sin \alpha \cos \beta - \cos \alpha \sin \beta}$$

You are asked to prove these theorems in the Written Exercises.

When an angle can be expressed as the sum or difference of two "special" angles, such as 30°, 45°, 60°, 90°, 180°, and so on, then its sine and cosine can be evaluated without using tables.

EXAMPLE 1 Evaluate cos 150° by using Theorem 16–4.

Solution: $\cos 150° = \cos(180° - 30°)$ ← $\alpha = 180°;\ \beta = 30°$

$$= \cos 180° \cos 30° + \sin 180° \sin 30°$$

$$= -1 \cdot \frac{\sqrt{3}}{2} + 0 \cdot \frac{1}{2} = -\frac{\sqrt{3}}{2}$$

Note that the given angle must be renamed before the sum or difference identity can be used.

EXAMPLE 2 Evaluate without tables. Write radicals in simplest form.

a. $\cos 75°$ **b.** $\sin \frac{\pi}{12}$

Solutions:

a. $\cos 75° = \cos(45° + 30°)$

$$= \cos 45° \cos 30° - \sin 45° \sin 30°$$

$$= \frac{\sqrt{2}}{2} \cdot \frac{\sqrt{3}}{2} - \frac{\sqrt{2}}{2} \cdot \frac{1}{2}$$

$$= \frac{\sqrt{6}}{4} - \frac{\sqrt{2}}{4}$$

$$= \frac{1}{4}(\sqrt{6} - \sqrt{2})$$

b. $\sin \frac{\pi}{12} = \sin\left(\frac{\pi}{4} - \frac{\pi}{6}\right)$

$$= \sin \frac{\pi}{4} \cos \frac{\pi}{6} - \cos \frac{\pi}{4} \sin \frac{\pi}{6}$$

$$= \frac{\sqrt{2}}{2} \cdot \frac{\sqrt{3}}{2} - \frac{\sqrt{2}}{2} \cdot \frac{1}{2}$$

$$= \frac{\sqrt{6}}{4} - \frac{\sqrt{2}}{4}$$

$$= \frac{1}{4}(\sqrt{6} - \sqrt{2})$$

CLASSROOM EXERCISES

Write each angle measure as the sum or difference of the measures of angles whose sines and cosines are known.

1. 105° **2.** 75° **3.** 15° **4.** 60° **5.** 135° **6.** 120°

7. $\frac{\pi}{3}$ **8.** $\frac{3\pi}{4}$ **9.** $\frac{5\pi}{12}$ **10.** $\frac{7\pi}{3}$ **11.** 240° **12.** 315°

WRITTEN EXERCISES

A Use the Sum and Difference Identities to evaluate the cosine of each angle measure.

1. 120° **2.** 105° **3.** 210° **4.** 300° **5.** 315° **6.** 15°

7. $\frac{\pi}{12}$ **8.** $\frac{5\pi}{6}$ **9.** $\frac{4\pi}{3}$ **10.** $\frac{5\pi}{12}$ **11.** $\frac{8\pi}{3}$ **12.** $\frac{3\pi}{4}$

Use the Sum and Difference Identities to evaluate the sine of each angle measure.

13. $120°$ **14.** $135°$ **15.** $150°$ **16.** $225°$ **17.** $330°$ **18.** $240°$

19. $\frac{5\pi}{2}$ **20.** $\frac{13\pi}{12}$ **21.** $\frac{7\pi}{4}$ **22.** $\frac{11\pi}{6}$ **23.** $\frac{5\pi}{3}$ **24.** $\frac{5\pi}{4}$

Use the tables to evaluate $\cos(\alpha + \beta)$ and $\cos(\alpha - \beta)$.

25. $\alpha = 30°, \beta = 30°$ **26.** $\alpha = 90°, \beta = 20°$ **27.** $\alpha = 63°, \beta = 50°$ **28.** $\alpha = 42°, \beta = 13°$

B In Exercises 29–32, express in terms of θ.

29. $\sin(45° + \theta)$ **30.** $\cos(45° + \theta)$ **31.** $\sin(270° - \theta)$ **32.** $\cos(270° - \theta)$

33. Without using tables, show that $\cos(\alpha + \beta) = -\frac{56}{65}$, where $0 < \alpha < 90°$, $90° < \beta < 180°$, $\cos\alpha = \frac{3}{5}$, and $\sin\beta = \frac{5}{13}$. (HINT: First find $\sin\alpha$ and $\cos\beta$.)

34. Given: $\sin\alpha = \frac{3}{5}$, $\cos\beta = \frac{5}{13}$, $0 < \alpha < 90°$, $0 < \beta < 90°$, Evaluate $\cos(\alpha + \beta)$ without using tables.

35. Given: $\sin\alpha = \frac{5}{13}$, $\cos\beta = -\frac{4}{5}$, $\frac{\pi}{2} < \alpha < \pi$, $\pi < \beta < \frac{3\pi}{2}$. Evaluate $\cos(\alpha - \beta)$ without using tables.

36. Given: $\sin\alpha = \frac{12}{13}$, $\tan\beta = \frac{4}{3}$, $0 < \alpha < \frac{\pi}{2}$, $0 < \beta < \frac{\pi}{2}$. Evaluate $\sin(\alpha + \beta)$ without using tables.

37. Given: $\cos\alpha = -\frac{3}{5}$, $\tan\beta = -\frac{5}{12}$, neither α nor β is in Quadrant II. Evaluate $\sin(\alpha - \beta)$ without using tables.

Use Theorems 16–4 through 16–6 to verify these Reduction Formulas.

38. $\cos(90° - \theta) = \sin\theta$ **39.** $\sin(90° - \theta) = \cos\theta$ **40.** $\sin(90° + \theta) = \cos\theta$

41. $\sin(\pi - \theta) = \sin\theta$ **42.** $\cos(\pi - \theta) = -\cos\theta$ **43.** $\cos(\pi + \theta) = -\cos\theta$

44. $\cos(270° + \theta) = \sin\theta$ **45.** $\cos(\frac{3\pi}{2} - \theta) = -\sin\theta$ **46.** $\cos(360° - \theta) = \cos\theta$

C

47. Prove the **Double Angle Identity for Sine:** $\sin 2\alpha = 2\sin\alpha\cos\alpha$ (HINT: Let $\beta = \alpha$.)

48. Prove the **Double Angle Identities for Cosine:**
i. $\cos 2\alpha = \cos^2\alpha - \sin^2\alpha$
ii. $\cos 2\alpha = 2\cos^2\alpha - 1$
iii. $\cos 2\alpha = 1 - 2\sin^2\alpha$

In Exercises 49–51, use the given information and the Double Angle Identities for Sine and Cosine to evaluate $\sin 2\alpha$ and $\cos 2\alpha$ without using tables.

49. $\sin\alpha = \frac{4}{5}$; $0 < \alpha < \frac{\pi}{2}$ **50.** $\sin\alpha = \frac{2}{\sqrt{5}}$; $0 < 2 < \frac{\pi}{2}$ **51.** $\tan\alpha = -\frac{5}{12}$; $\frac{\pi}{2} < \alpha < \pi$

52. Prove Theorem 16–6. (HINT: Refer to the Reduction Formulas of Exercises 38–46. Begin by letting $\theta = (\alpha + \beta)$ in one of the formulas. Then apply Theorem 16–4.

53. Prove Theorem 16–7.

16–9 Proving Trigonometric Identities

You can use the Fundamental Identities to prove that other trigonometric equations are also identities. Examples 1 and 2 exhibit a basic strategy for proving identities.

Strategy 1: Use known identities to replace the expressions on one side of the equation with equivalent expressions. Continue this procedure until the expressions on both sides of the equation are the same.

EXAMPLE 1 Prove: $\cos\theta = \sin\theta\cot\theta$

Proof:

$\cos\theta$	$\sin\theta\cot\theta$	← $\cot\theta = \dfrac{\cos\theta}{\sin\theta}$
	$\sin\theta\cdot\dfrac{\cos\theta}{\sin\theta}$	
$\cos\theta$ =	$\cos\theta$	

Since the last equation is an identity, $\cos\theta = \sin\theta\cot\theta$ is an identity.

In general, change trigonometric functions to sines and cosines as was done in Example 1. In Example 2, note that $\cos^2\theta - 1 = -\sin^2\theta$ is simply another way of writing $\sin^2\theta + \cos^2\theta = 1$.

EXAMPLE 2 Prove: $\tan^2\theta - \sin^2\theta = (\cos\theta - \sec\theta)^2$

Proof:

$\tan^2\theta - \sin^2\theta$	$(\cos\theta - \sec\theta)^2$	← Square the right side.
	$\cos^2\theta - 2\cos\theta\sec\theta + \sec^2\theta$	← $\sec\theta = \dfrac{1}{\cos\theta}$
	$\cos^2\theta - 2\cos\theta\left(\dfrac{1}{\cos\theta}\right) + \sec^2\theta$	
	$\cos^2\theta - 2 + \sec^2\theta$	
	$\cos^2\theta - 1 - 1 + \sec^2\theta$	
	$(\cos^2\theta - 1) + (-1 + \sec^2\theta)$	← $\cos^2\theta - 1 = -\sin^2\theta$; $-1 + \sec^2\theta = \tan^2\theta$
	$-\sin^2\theta + \tan^2\theta$	
$\tan^2\theta - \sin^2\theta$ =	$\tan^2\theta - \sin^2\theta$	

Example 3 shows that you can sometimes rewrite either the expression on the left side or the expression on the right side.

EXAMPLE 3 Prove: $1 - 2\sin^2\theta = 2\cos^2\theta - 1$

Proof:

Working with the Left Side

$1 - 2\sin^2\theta$	$2\cos^2\theta - 1$
$1 - 2(1 - \cos^2\theta)$	
$1 - 2 + 2\cos^2\theta$	
$2\cos^2\theta - 1$ =	$2\cos^2\theta - 1$

Working with the Right Side

$1 - 2\sin^2\theta$	$2\cos^2\theta - 1$
	$2(1 - \sin^2\theta) - 1$
	$2 - 2\sin^2\theta - 1$
$1 - 2\sin^2\theta$ =	$1 - 2\sin^2\theta$

A second strategy for solving identities is shown in Example 4. Use this strategy only when Strategy 1 does not work.

Strategy 2: Use known identities to replace the expressions on both sides of the equation. Continue this procedure until the expressions on both sides of the equation are the same.

EXAMPLE 4 Prove: $\tan\theta \cdot \sin\theta = \sec\theta - \cos\theta$

Proof:

$\tan\theta \cdot \sin\theta$	$\sec\theta - \cos\theta$
$\frac{\sin\theta}{\cos\theta} \cdot \sin\theta$	$\frac{1}{\cos\theta} - \cos\theta$

$$\frac{\sin^2\theta}{\cos\theta} = \frac{1-\cos^2\theta}{\cos\theta}$$

$$\frac{\sin^2\theta}{\cos\theta} = \frac{\sin^2\theta}{\cos\theta}$$

Note that the vertical line is used to emphasize that each side of the equation is done independently.

CLASSROOM EXERCISES

Prove each identity.

1. $\csc^2\theta = \frac{1}{1-\cos^2}$
2. $1 + \tan^2\theta = \frac{1}{1-\sin^2\theta}$
3. $\sin^2\theta = (1-\sin^2\theta)\tan^2\theta$
4. $\sin\theta = \tan\theta \cdot \cos\theta$
5. $\sin^4\theta - \cos^4\theta = \sin^2\theta - \cos^2\theta$
6. $\tan\theta \cdot \sin\theta = \sec\theta - \cos\theta$

WRITTEN EXERCISES

A Prove each identity.

1. $\tan\theta = \sin\theta\sec\theta$
2. $\cot\theta = \cos\theta\csc\theta$
3. $\tan^2\theta = \frac{1-\cos^2\theta}{\cos^2\theta}$
4. $\sec^2\theta = \frac{\sin^2\theta + \cos^2\theta}{\cos^2\theta}$
5. $\tan^2\theta = \sec^2\theta - 1$
6. $\cot^2\theta = \csc^2\theta - 1$
7. $\csc\theta = \frac{\cot\theta}{\cos\theta}$
8. $\frac{1}{\sec^2\theta} + \frac{1}{\csc^2\theta} = 1$
9. $\csc^2\theta\tan^2\theta - 1 = \tan^2\theta$
10. $\frac{\sec\theta}{\cos\theta} - \frac{\tan\theta}{\cot\theta} = 1$
11. $(1-\tan\theta)^2 = \sec^2\theta - 2\tan\theta$
12. $(1-\sin^2\theta)(1+\tan^2\theta) = 1$
13. $\frac{\cos^2\theta}{\sin\theta} + \sin\theta = \csc\theta$
14. $\tan\theta + \cot\theta = \sec\theta\csc\theta$

Prove each identity.

15. $\dfrac{\tan \theta}{1 - \cos^2 \theta} = \dfrac{\sec \theta}{\sin \theta}$

16. $\dfrac{\cot \theta}{\cos \theta} + \dfrac{\sec \theta}{\cot \theta} = \sec^2 \theta \csc \theta$

17. $\dfrac{\cos x - \sin x}{\cos x} = 1 - \tan x$

18. $\dfrac{\cot \theta + 1}{\cot \theta} = 1 + \tan \theta$

19. $\tan x\ (\tan x + \cot x) = \sec^2 x$

20. $(\sec \theta - \tan \theta)(\sec \theta + \tan \theta) = 1$

21. $\sec^4 x - \tan^4 x = \sec^2 x + \tan^2 x$

22. $\sin^4 x + 2 \sin^2 x \cos^2 x + \cos^4 x = 1$

23. $\dfrac{\sin^4 \theta - \cos^4 \theta}{1 - \cot^4 \theta} = \sin^4 \theta$

24. $\dfrac{1 - 2 \sin x - 3 \sin^2 x}{\cos^2 x} = \dfrac{1 - 3 \sin x}{1 - \sin x}$

25. $\csc \theta + \dfrac{\tan \theta}{\sin \theta} - \sec \theta = \dfrac{\cot \theta}{\cos \theta}$

26. $\dfrac{\sin^3 x + \cos^3 x}{1 - 2 \cos^2 x} = \dfrac{\sec x - \sin x}{\tan x - 1}$

B Use the Sum Identities for the sine and cosine functions (Theorems 16–4 through 16–7 on pages 608–609) along with the eight Fundamental Identities (pages 605–606) to prove that each statement is an identity.

27. $\tan(\alpha + \beta) = \dfrac{\tan \alpha + \tan \beta}{1 - \tan \alpha \tan \beta}$ (HINT: Use $\tan \alpha = \dfrac{\sin \alpha}{\cos \alpha}$ and $\tan \beta = \dfrac{\sin \beta}{\cos \beta}$.)

28. $\tan(\alpha - \beta) = \dfrac{\tan \alpha - \tan \beta}{1 + \tan \alpha \tan \beta}$ (HINT: Use $\tan[\alpha - \beta] = \tan[\alpha + (-\beta)]$ and Exercise 27).

29. $\tan 2\alpha = \dfrac{2 \tan \alpha}{1 - \tan^2 \alpha}$

C The formulas in Exercises 30–34 are called **Half-Angle Identities.** Prove each identity.

30. $\sin \dfrac{\alpha}{2} = \pm\sqrt{\dfrac{1 - \cos \alpha}{2}}$ (HINT: Use one of the Double Angle Identities on page 611.)

31. $\cos \dfrac{\alpha}{2} = \pm\sqrt{\dfrac{1 + \cos \alpha}{2}}$

32. $\tan \dfrac{\alpha}{2} = \pm\sqrt{\dfrac{1 - \cos \alpha}{1 + \cos \alpha}}$

33. $\tan \dfrac{\alpha}{2} = 1\sqrt{\dfrac{1 - \cos^2 \alpha}{(1 + \cos \alpha)^2}} = \pm\dfrac{\sin \alpha}{1 + \cos \alpha}$

34. $\tan \dfrac{\alpha}{2} = \dfrac{1 - \cos \alpha}{\sin \alpha}$

16–10 Inverse Trigonometric Functions

The graph of the inverse of a relation can be obtained by drawing the graph of $y = x$ and finding the "mirror image" of the original curve with respect to the graph of $y = x$. In Figure 1 on the next page, this is done for the function $y = \sin x$. The resulting mirror image is the **inverse** of $y = \sin x$ and is denoted by ***y = arc sin x.*** Check several points such as A, B, and C of $y = \sin x$ to see that they have image points A', B', and C', respectively, on the curve $y = \text{arc} \sin x$.

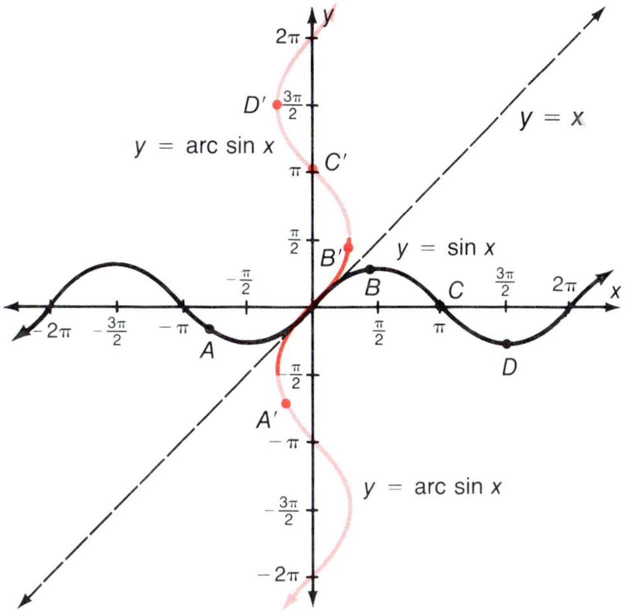

Figure 1

Definition

> **Inverse Sine Relation**
>
> $y = \textbf{arc sin } x$ means $x = \textbf{sin } y$.

A good way to think about the inverse of the sine function is that $y = \text{arc sin } x$ means "y is the measure of an angle whose sine is x."

EXAMPLE 1 Find arc sin $\frac{1}{2}$.

Solution: $y = \text{arc sin } \frac{1}{2}$ means $\frac{1}{2} = \sin y$.

$\sin y = \frac{1}{2}$ ← **Solve for y.**

$y = \frac{\pi}{6}, \frac{5\pi}{6}, -\frac{7\pi}{6}, -\frac{11\pi}{6}$, and so on.

Since solving $\sin y = \frac{1}{2}$ gives an infinite number of solutions, $y = \text{arc sin } x$ is not a function. To get an inverse that *is* a function, you must restrict the range of $y = \text{arc sin } x$. The numbers in the restricted range, $-\frac{\pi}{2} \le y \le \frac{\pi}{2}$, are called the **principal values** of the inverse sine relation. When the word "arc" in $y = \text{Arc sin } x$ is written with a capital A, the principal value is desired. The graph of $y = \text{Arc sin } x$ is shown as the heavy portion of the curve $y = \text{arc sin } x$ in Figure 1 above.

Definition

Principal Value of The Inverse Sine Relation

$y = \text{arc sin } x$ for $-1 \le x \le 1$ and $-\frac{\pi}{2} \le y \le \frac{\pi}{2}$ is the **principal value of the inverse sine relation** and is written $y = \text{Arc sin } x$.

EXAMPLE 2 Find $\text{Arc sin } \frac{1}{2}$.

Solution: $\text{Arc sin } \frac{1}{2}$ is the principal value of $y = \text{arc sin } \frac{1}{2}$.

Thus, $\text{Arc sin } \frac{1}{2} = \frac{\pi}{6}$ ← $-\frac{\pi}{2} \le \frac{\pi}{6} \le \frac{\pi}{2}$

The functions $y = \cos x$ and $y = \tan x$ also have inverses.

Definitions

Inverse Cosine Relation

$y = \text{arc cos } x$ means $x = \cos y$.

$y = \text{arc cos } x$ for $-1 \le x \le 1$ and $0 \le y \le \pi$ is the **principal value of the inverse cosine relation** and is written $y = \text{Arc cos } x$.

Inverse Tangent Relation

$y = \text{arc tan } x$ means $x = \tan y$.

$y = \text{arc tan } x$ where x is any real number and $-\frac{\pi}{2} < y < \frac{\pi}{2}$ is the **principal value of the inverse tangent relation**, written $y = \text{Arc tan } x$.

EXAMPLE 3 Evaluate: sin (Arc tan 8).

Solution: Begin by evaluating the function in parentheses. Let $\alpha = \text{Arc tan } 8$.

Then $\tan \alpha = \frac{8}{1}$

and $y = 8,\ x = 1$ ← $\tan \alpha = \frac{y}{x}$

Thus, $r = \sqrt{8^2 + 1^2} = \sqrt{65}$. ← **Pythagorean Theorem**

Therefore, $\sin \alpha = \frac{y}{r} = \frac{8}{\sqrt{65}}$ and $\sin(\text{Arc tan } 8) = \frac{8}{\sqrt{65}}$.

The superscript -1 is sometimes used to denote the inverse. Thus, $\text{Cos}^{-1} x$ means **Arc cos** x, and $\cos^{-1} x$ means **arc cos** x.

CLASSROOM EXERCISES

Evaluate.

1. $\text{Arc sin } (-\frac{\sqrt{2}}{2})$
2. $\text{Arc tan } (-1)$
3. $\text{Arc cos } (-1)$
4. $\text{Arc sin } 0$
5. $\text{Arc cos } 0$
6. $\text{Arc tan } 1.3764$
7. $\text{Arc cos } \frac{\sqrt{3}}{2}$
8. $\text{Arc sin } 0.7009$

WRITTEN EXERCISES

A Evaluate.

1. Arc cos 0.4848
2. Arc sin 0.6691
3. Arc sin $(-\frac{\sqrt{3}}{2})$
4. Arc tan 1
5. Arc sin $\frac{\sqrt{2}}{2}$
6. Arc cos 0.5422
7. Arc cos $(-\frac{\sqrt{2}}{2})$
8. Arc tan (-0.1763)
9. Arc sin $(-\frac{1}{2})$
10. Arc tan $\sqrt{3}$
11. Arc sin (-1)
12. Arc sin (0.7330)
13. sin (Arc sin $\frac{1}{2}$)
14. cos (Arc sin $\frac{1}{2}$)
15. cos (Arc cos $\frac{\sqrt{3}}{2}$)
16. cos (Arc sin $\frac{\sqrt{3}}{2}$)
17. sin (Arc cos $\frac{\sqrt{3}}{2}$)
18. sin (Arc cos $\frac{1}{2}$)
19. sin (Arc cos 0)
20. cos (Arc sin 1)
21. sin (Arc cos $\frac{\sqrt{2}}{2}$)
22. cos (Arc cos 1)
23. cos (Arc sin 0)
24. sin (Arc cos 0)
25. sin [Arc cos (-1)]
26. cos [Arc sin $(-\frac{\sqrt{2}}{2})$]
27. sin [Arc cos $(-\frac{1}{2})$]
28. sin [Arc sin $(-\frac{1}{2})$]
29. cos [Arc cos $(-\frac{\sqrt{3}}{2})$]
30. cos [Arc sin (-1)]
31. Arc cos 0.9272
32. sin (Arc cos $\frac{1}{3}$)
33. cos (Arc tan 3)
34. tan [Arc sin $(-\frac{1}{6})$]
35. cos (Arc sin $\frac{2}{3}$)
36. tan (Arc tan 5)
37. sin (Arc sin $\frac{4}{5}$)
38. sin [Arc tan (-2)]
39. tan [Arc cos $(-\frac{12}{13})$]
40. sec (Arc sin $\frac{1}{2}$)
41. cot (Arc cos $\frac{3}{5}$)
42. csc [Arc tan (-4)]
43. csc (Arc sin $\frac{3}{7}$)
44. cot (Arc tan $\frac{2}{5}$)
45. sec (Arc cos $\frac{1}{7}$)
46. cot (Arc sin $\frac{2}{3}$)
47. sec [Arc tan (-7)]
48. csc [Arc cos $(-\frac{3}{4})$]

B Sketch the graph of each inverse over the interval stated. Indicate the portion of each curve that represents the principal values.

49. $y = \text{arc cos } x; -\pi \le y \le 2\pi$
50. $y = \text{arc tan } x; -2\pi \le y \le 2\pi$
51. $y = \text{arc sin } \frac{1}{2}x; -\pi \le y \le \pi$
52. $y = \text{arc sin } 2x; -\pi \le y \le \pi$

16–11 Trigonometric Equations

To solve a conditional trigonometric equation such as $\sin x = \frac{1}{2}$, you find all real values of x such that $\sin x = \frac{1}{2}$. Since $y = \sin x$ is a periodic function with period 2π, this graph shows that there are two solutions for each period.

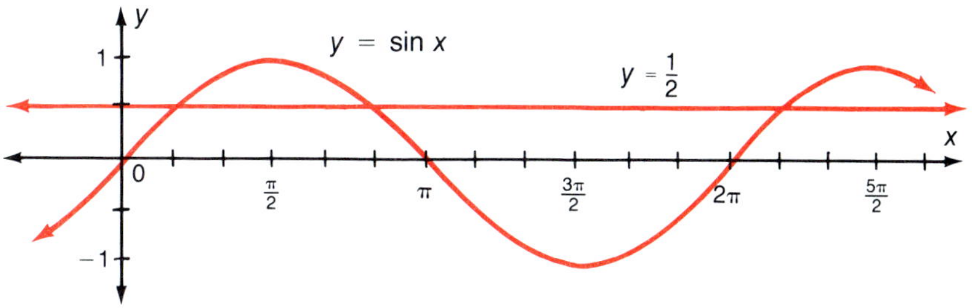

EXAMPLE 1 Solve: $\sin x = \frac{1}{2}$, where x is any real number.

Solution: $\sin x = \frac{1}{2}$

$$x = \frac{\pi}{6}, \frac{5\pi}{6}, 2\pi + \frac{\pi}{6}, 2\pi + \frac{5\pi}{6}, 4\pi + \frac{\pi}{6}, 4\pi + \frac{5\pi}{6}, \ldots$$

Any integral multiple of 2π added to $\frac{\pi}{6}$ and $\frac{5\pi}{6}$ will give solutions of $\sin x = \frac{1}{2}$.

Therefore, the solutions can be written as $\boldsymbol{2k\pi + \frac{\pi}{6}}$ and $\boldsymbol{2k\pi + \frac{5\pi}{6}}$, where k is an integer.

Two basic techniques that are useful in solving trigonometric equations are *factoring* and *applying known identities*. Example 2 shows how both of these techniques are used. Example 2 also illustrates finding the solution set over a restricted domain, such as $0 \le x < 2\pi$.

EXAMPLE 2 Solve $\frac{1 - \cos x}{\sin x} = 1$, where $0 \le x \le 2\pi$.

Solution:

$$\frac{1 - \cos x}{\sin x} = 1$$

$$1 - \cos x = \sin x$$ ← **Square each side.**

$$(1 - \cos x)^2 = (\sin x)^2$$

$$1 - 2\cos x + \cos^2 x = \sin^2 x$$

$$1 - 2\cos x + \cos^2 x = 1 - \cos^2 x$$

$$2\cos^2 x - 2\cos x = 0$$

$$2\cos x(\cos x - 1) = 0$$

$$2\cos x = 0 \quad \underline{\text{or}} \quad \cos x - 1 = 0$$

$$\cos x = 0 \quad \underline{\text{or}} \quad \cos x = 1$$

For $\cos x = 0$ and $0 \le x < 2\pi$, $x = \frac{\pi}{2}$ or $x = \frac{3\pi}{2}$.

For $\cos x = 1$ and $0 \le x < 2\pi$, $x = 0$. Thus, the three *possible* values for x in the interval $0 \le x < 2\pi$ are 0, $\frac{\pi}{2}$, and $\frac{3\pi}{2}$.

Check:

$x = 0$: $\frac{1 - \cos 0}{\sin 0} = \frac{1 - 1}{0} = \frac{0}{0}$ ← **$\frac{0}{0}$ is undefined, so $x = 0$ is not a solution.**

$x = \frac{\pi}{2}$: $\frac{1 - \cos \frac{\pi}{2}}{\sin \frac{\pi}{2}} = \frac{1 - 0}{1} = 1$ ← **The original equation is true for $x = \frac{\pi}{2}$.**

$x = \frac{3\pi}{2}$: $\frac{1 - \cos \frac{3\pi}{2}}{\sin \frac{3\pi}{2}} = \frac{1 - 0}{-1} = -1$ ← **The original equation is false for $x = \frac{3\pi}{2}$.**

Thus, the solution is $\boldsymbol{\frac{\pi}{2}}$.

CLASSROOM EXERCISES

Solve for x.

1. $\cos x = 1$
2. $\sin x = 0$
3. $\tan x = \sqrt{3}$
4. $2 \sin x - \sqrt{3} = 0$

Solve for x over the given interval.

5. $\sqrt{3} \sec x + 2 = 0 \ (-\frac{\pi}{2} \leq x \leq \frac{\pi}{2})$
6. $2 \cos^2 x = \cos x \ (-\frac{\pi}{2} \leq x \leq \frac{\pi}{2})$
7. $\tan x - 1 = 0 \ (0 \leq x < 2\pi)$
8. $4 \sin^2 x = 1 \ (0 \leq x < 2\pi)$

WRITTEN EXERCISES

A Solve for x, where $0 \leq x \leq 2\pi$.

1. $\cos x = 0$
2. $3 \cot x + \sqrt{3} = 0$
3. $(\sin x - 1)(2 \sin x + 1) = 0$
4. $(2 \cos x + 1)(\cos x - 1) = 0$
5. $3 \tan^2 x - \sqrt{3} \tan x = 0$
6. $\sec^2 x + 2 \sec x = 0$
7. $\cos x = \frac{\sqrt{2}}{2}$
8. $\cot x = 0$
9. $\sqrt{3} \cot x + 1 = 0$
10. $2 \sin x + \sqrt{3} = 0$
11. $2 \sin^2 x - \sin x - 1 = 0$
12. $2 \tan^2 x - 3 \sec x + 3 = 0$
13. $\sin^2 x = \frac{1}{2}$
14. $\tan^2 x = 1$
15. $3 \sin^2 x - \cos^2 x = 0$
16. $\sqrt{3} \csc^2 x + 2 \csc x = 0$
17. $\sin x = \cos x$
18. $\cos x = 3 \cos x - 2$
19. $4 \sin^2 x - 4 \sin x + 1 = 0$
20. $\dfrac{\sin x}{1 + \cos x} = 1$

B

21. $\cos 2x + \sin x = 1$
22. $\sin 2x + \cos x = 0$
23. $4 \tan x + \sin 2x = 0$
24. $\sin 2x = 2 \sin x$
25. $\tan 2x \cot x - 3 = 0$
26. $\cos^2 x + \cos 2x = \frac{5}{4}$
27. $\sin (\frac{\pi}{4} + x) - \sin (\frac{\pi}{4} - x) = \frac{\sqrt{2}}{2}$
28. $\cos (\frac{\pi}{4} + x) + \cos (\frac{\pi}{4} - x) = 1$

Use the tables on pages 629–633 to find all solutions of each equation to the nearest minute, where $0 \leq \theta < 2\pi$.

29. $4 \sin \theta - 2 = 1$
30. $\tan \theta = -2.1290$
31. $\cos \theta = 0.1263$
32. $2 \sec 0 = -3$

C Solve for x where $0 \leq x < 2\pi$.

33. $\cos 2x + 3 \cos x - 1 = 0$
34. $|\sin x| = \frac{1}{2}$
35. $\sin 3x + \sin x = 0$
36. $6 \cos^2 x + 5 \cos x + 1 = 0$
37. $2 \tan x - 2 \cot x = -3$
38. $2 \sin^3 x - \sin x = 0$
39. $\sin 2x + \cot 3x = 0$
40. $4 \sin^4 x + \sin^2 x = 3$

16–12 Complex Numbers in Polar Form

Every complex number $a + bi$ can also be written as an ordered pair (a, b), where a is the **real part** and b is the **imaginary part** (the coefficient of i).

Standard Form	$2 + 4i$	$2 - 4i$	$0 + 0i$	$0 + \sqrt{3}i$	$\pi + 0i$
Ordered Pair Form	$(2, 4)$	$(2, -4)$	$(0, 0)$	$(0, \sqrt{3})$	$(\pi, 0)$

Thus, each complex number can be associated with a point on the coordinate plane called the **complex number plane.** The **real axis** represents complex numbers of the form $(a, 0)$. The **imaginary axis** represents complex numbers of the form $(0, b)$. In this plane, a complex number can be represented by a **point** or by a **vector** drawn from the origin to the point.

A vector representing $a + bi$ or (a, b) has both length *and* direction. The symbol $|a + bi|$, read "the *absolute value* of $a + bi$," represents the length. You find the length or **magnitude** by applying the distance formula.

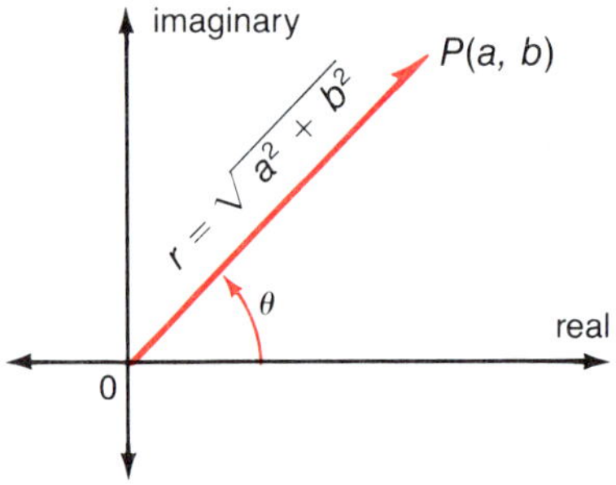

length: $(a, b) = |a + bi| = \sqrt{a^2 + b^2}$

The direction of a vector is indicated by the angle determined by the positive real axis and the vector. The measure, θ, of this angle is called the **argument** of the complex number.

$$\textbf{argument: } (a + bi) = \theta$$

Any of the trigonometric functions can be used to find θ. However, since a and b are known and $\tan\theta = \frac{b}{a}$, θ can be found by applying one of the following.

$$\text{For } a > 0,\ \theta = \text{Arc tan}\,\frac{b}{a}. \qquad \textbf{Equation 1}$$

$$\text{For } a < 0,\ \theta = \pi + \text{Arc tan}\,\frac{b}{a}. \qquad \textbf{Equation 2}$$

$$\text{For } a = 0 \text{ and } b > 0,\ \theta = \frac{\pi}{2}. \qquad \text{For } a = 0 \text{ and } b < 0,\ \theta = \frac{3\pi}{2}.$$

$$\text{For } a = b = 0,\ \theta \text{ can have any real value.}$$

Note that two equations are used to define θ because the range of Arc tangent is $-\frac{\pi}{2} < \theta < \frac{\pi}{2}$ (Quadrants I and IV). The second equation is used for angles that terminate in Quadrants II and III.

From the graph at the right, note that

$$\cos \theta = \frac{a}{r} \text{ and } \sin \theta = \frac{b}{r}.$$

Then $a = r \cos \theta$ and $b = r \sin \theta$.

Thus, $a + bi = r \cos \theta + ir \sin \theta$,

or $\boldsymbol{a + bi = r(\cos \theta + i \sin \theta)}$.

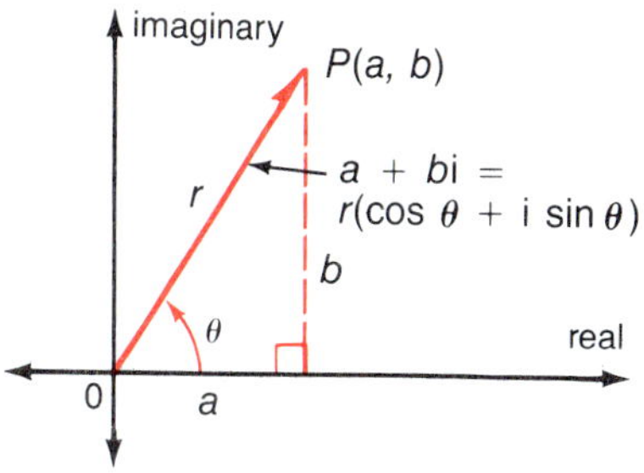

From the equation above, the expression $\boldsymbol{r(\cos \theta + i \sin \theta)}$ is called the **polar form** of the complex number, $\boldsymbol{a + bi}$. The numbers r and θ, written (r, θ), are the **polar coordinates** of the point P.

EXAMPLE 1 Write $-2\sqrt{3} + 2i$ in polar form.

Solution: Find r.

$r = \sqrt{(-2\sqrt{3})^2 + 2^2}$ ← $\boldsymbol{r = \sqrt{a^2 + b^2}}$

$= \sqrt{12 + 4}$

$= \sqrt{16}$, or 4

$\theta = \pi + \text{Arc tan}\left(\frac{2}{-2\sqrt{3}}\right)$ ← **Since $a = -2\sqrt{3}$, $a < 0$.**

$= \pi + \text{Arc tan}\left(-\frac{1}{\sqrt{3}}\right)$

$= \pi - \frac{\pi}{6} = \frac{5\pi}{6}$ Thus, $-2\sqrt{3} + 2i = \boldsymbol{4\left(\cos \frac{5\pi}{6} + i \sin \frac{5\pi}{6}\right)}$.

Example 2 shows that, given the polar form of a complex number, you can write the complex number in standard form.

EXAMPLE 2 Write each complex number in standard form.

a. $\cos 6° + i \sin 6°$ **b.** $5(\cos 30° + i \sin 30°)$

Solutions: **a.** $a + bi = r(\cos \theta + i \sin \theta)$

$= 1(\cos 6° + i \sin 6°)$

$= \boldsymbol{0.9945 + 0.1045i}$

b. $a + bi = r(\cos \theta + i \sin \theta)$

$= 5(\cos 30° + i \sin 30°)$

$= 5\left(\frac{\sqrt{3}}{2} + i\left(\frac{1}{2}\right)\right)$

$= \boldsymbol{\frac{5\sqrt{3}}{2} + \frac{5}{2} i}$

CLASSROOM EXERCISES

In Exercises 1–4, write the magnitude of each vector.

1. $1 + i$ **2.** $\frac{1}{2} - \frac{\sqrt{3}}{2} i$ **3.** $-1 - i$ **4.** $-\sqrt{3} + i$

5–8. Find the argument of each complex number in Exercises 1–4.

WRITTEN EXERCISES

A Write the polar form of each complex number in radians.

1. $4 + 4i$
2. $-1 + \sqrt{3}i$
3. $\frac{\sqrt{3}}{2} - \frac{1}{2}i$
4. $-1 + 0i$
5. $0 + i$
6. $-3 - 3\sqrt{3}i$
7. $\sqrt{3} - i$
8. $\sqrt{2} + \sqrt{2}i$
9. $0 + 3i$
10. $-2 + 0i$
11. $-\sqrt{3} - i$
12. $1 + i$
13. $5 + 5i$
14. $-1 + \sqrt{3}i$
15. $-2 - 2\sqrt{3}i$
16. Graph each complex number in Exercises 1, 3, 4, 5, 9, 10, and 11 as a vector in the complex plane.

In Exercises 17–24, write each complex number in standard form.

17. $\frac{1}{2}(\cos 60° + i \sin 60°)$
18. $2(\cos 0° + i \sin 0°)$
19. $3(\cos 330° + i \sin 330°)$
20. $4(\cos 225° + i \sin 225°)$
21. $\sqrt{2}[\cos(-45°) + i \sin(-45°)]$
22. $\cos(-\frac{\pi}{6}) + i \sin(-\frac{\pi}{6})$
23. $2(\cos \frac{3\pi}{4} + i \sin \frac{3\pi}{4})$
24. $4(\cos \frac{\pi}{2} + i \sin \frac{\pi}{2})$

B

25. $2[\cos(\text{Arc tan } 6.3138) + i \sin(\text{Arc tan } 6.3138)]$
26. $3[\cos(\text{Arc tan } 0.3249) + i \sin(\text{Arc tan } 0.3249)]$
27. $1[\cos(\text{Arc tan } 0) + i \sin(\text{Arc tan } 0)]$

Review

In Exercises 1–6, write as an equivalent trigonometric function such as $\cos\theta$, $-\cos\theta$, $\sin\theta$, $\tan^2\theta$, $-\sec\theta$, etc., or as 1 or -1. *(Section 16–7)*

1. $\frac{1}{\csc^2\theta}$
2. $\frac{\cos^2\theta}{\sin^2\theta}$
3. $\sin^2\theta + \sec^2\theta + \cos^2\theta - 2$
4. $\frac{\sin\theta}{\tan\theta}$
5. $\cos^4\theta - \sin^4\theta + \sin^2\theta$
6. $\sec\theta - \sin\theta\tan\theta$

Use the Sum and Difference Identities to evaluate the following functions. *(Section 16–8)*

7. $\sin 105°$
8. $\cos 135°$
9. $\sin 15°$
10. $\sin 300°$
11. $\cos 150°$
12. $\sin \frac{3\pi}{4}$
13. $\cos \frac{7\pi}{12}$
14. $\cos \frac{11\pi}{6}$
15. $\sin \frac{7\pi}{4}$
16. $\cos \frac{8\pi}{3}$

Prove each identity. *(Section 16–9)*

17. $\cot\theta(\cot\theta + \tan\theta) = \csc^2\theta$
18. $\frac{1 + \cot^2\theta}{1 + \tan^2\theta} = \cot^2\theta$
19. $\sec\theta(\sec\theta - \cos\theta) = \tan^2\theta$
20. $\frac{\sin\theta}{1 + \cos\theta} + \cot\theta = \csc\theta$
21. $\frac{\cos\theta}{1 - \sin\theta} - \frac{\cos\theta}{1 + \sin\theta} = 2\tan\theta$
22. $\frac{1}{1 + \sin\theta} + \frac{1}{1 - \sin\theta} = 2\sec^2\theta$

Evaluate. *(Section 16–10)*

23. Arc sin 0.5736 **24.** Arc cos $\frac{\sqrt{3}}{2}$ **25.** Arc tan 0

26. $\sin(\text{Arc tan}\frac{1}{\sqrt{3}})$ **27.** $\tan(\text{Arc sin}\frac{3}{5})$ **28.** $\cos[\text{Arc sin}(-\frac{4}{7})]$

Solve for x, where $0 \le x \le 2\pi$. *(Section 16–11)*

29. $\tan x = -1$ **30.** $2 \sin^2 x = 1$ **31.** $2 \cos x - \sin^2 x = -2$

Where the polar form of each complex number. *(Section 16–12)*

32. $0 - 3i$ **33.** $1 - i$ **34.** $\sqrt{3} + i$ **35.** $\sqrt{2} - \sqrt{2}i$

Write each complex number in standard form. *(Section 16–12)*

36. $6(\cos 315° + i \sin 315°)$ **37.** $\frac{1}{3}[\cos(-\frac{2\pi}{3}) + i \sin(-\frac{2\pi}{3})]$

Chapter Summary

IMPORTANT TERMS

Air speed (p. 586)
Amplitude (p. 595)
Asymptote (p. 599)
Complex number
  Argument of (p. 620)
  Imaginary part of (p. 620)
  Magnitude of (p. 620)
  Polar form of (p. 621)
  Real part of (p. 620)
Complex number plane (p. 620)
Continuous function (p. 599)
Course (p. 586)
Discontinuous function (p. 599)
Displacement (p. 583)
Equivalent vectors (p. 582)
Fundamental Identities (p. 605)
Ground speed (p. 586)
Heading (p. 586)
Imaginary axis (p. 620)
Inverse relations
  arc cosine (p. 616)
  arc sine (p. 615)
  arc tangent (p. 616)
Inverse functions
  Arc cosine (p. 616)
  Arc sine (p. 616)
  Arc tangent (p. 616)
Magnitude of a vector (p. 582)
Period (p. 596)
Periodic function (p. 596)
Polar coordinates (p. 621)
Principal value of the
  Inverse cosine (p. 616)
  Inverse sine (p. 616)
  Inverse tangent (p. 616)
Radian (p. 589)
Real axis (p. 620)
Resultant (p. 583)
Trigonometric Identities (p. 604)
Unit circle (p. 608)
Vector (p. 582)
Vector sum (p. 583)
Velocity (p. 586)
x component of a vector (p. 582)
y component of a vector (p. 582)

IMPORTANT IDEAS

1. **Vectors** can be used to represent physical quantities such as displacement (change of position) and velocity.
2. The **radian measure** of one complete revolution is 2π.

3. The **graph of a trigonometric function** is periodic, that is, the basic shape of a trigonometric curve repeats itself indefinitely.

4. **Reciprocal Identities:**

   a. $\csc \theta = \dfrac{1}{\sin \theta}$ **b.** $\sec \theta = \dfrac{1}{\cos \theta}$ **c.** $\cot \theta = \dfrac{1}{\tan \theta}$

5. **Ratio Identities:**

   a. $\tan \theta = \dfrac{\sin \theta}{\cos \theta}$ **b.** $\cot \theta = \dfrac{\cos \theta}{\sin \theta}$

6. **Pythagorean Identities:**

   a. $\sin^2 \theta + \cos^2 \theta = 1$ **b.** $1 + \cot^2 \theta = \csc^2 \theta$ **c.** $\tan^2 \theta + 1 = \sec^2 \theta$

7. **Sum Identities for Sine and Cosine:**

   a. $\sin(\alpha + \beta) = \sin \alpha \cos \beta + \cos \alpha \sin \beta$

   b. $\cos(\alpha + \beta) = \cos \alpha \cos \beta - \sin \alpha \sin \beta$

8. **Difference Identities for Sine and Cosine:**

   a. $\sin(\alpha - \beta) = \sin \alpha \cos \beta - \cos \alpha \sin \beta$

   b. $\cos(\alpha - \beta) = \cos \alpha \cos \beta + \sin \alpha \sin \beta$

9. An **inverse trigonometric relation** is obtained from an inverse trigonometric relation by restricting the range of the relation.

10. A **complex number** is a vector. It can be associated with a point (a, b) in the complex number plane.

Chapter Objectives and Review

Objective: *To use vectors to solve problems (Sections 16–1 and 16–2)*

In Exercises 1–3, $\overrightarrow{OA}$ has initial point (0, 0) and forms an angle θ with the positive x axis. Find $|\overrightarrow{OB}|$ and $|\overrightarrow{OC}|$, the x and y components of $\overrightarrow{OA}$. Round your answer to the nearest whole number.

1. $|\overrightarrow{OA}| = 6;\ \theta = 150°$
2. $|\overrightarrow{OA}| = 2;\ \theta = 23°$
3. $|\overrightarrow{OA}| = 12;\ \theta = 245°$

4. The air velocity of an airplane is pictured at the right together with the velocity of the wind. Find the ground speed of the airplane.

5. Find, to the nearest unit, the course of the airplane in Exercise 4.

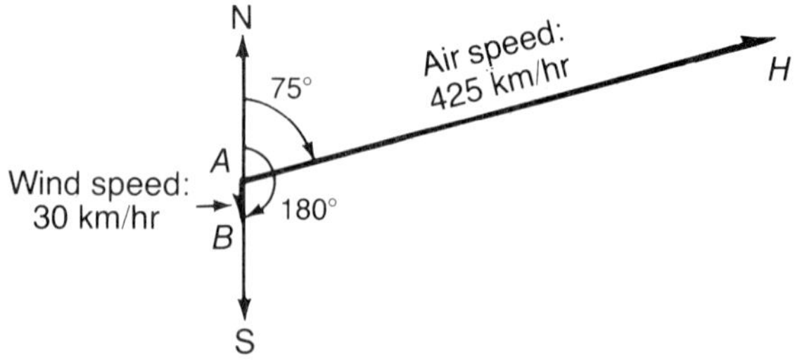

Objective: *To change degree measure to radian measure and radian measure to degree measure (Section 16–3)*

Change each degree measure to radian measure in terms of π.

6. 180°
7. 60°
8. $-30°$
9. $-225°$
10. 70°
11. $-100°$

Change each radian measure to degree measure.

12. $\frac{\pi}{2}$ **13.** 2π **14.** $-\frac{3\pi}{4}$ **15.** $-\frac{\pi}{6}$ **16.** $-\frac{7\pi}{6}$ **17.** $\frac{13\pi}{15}$

Objective: *To evaluate trigonometric functions when the measure of the angle is expressed in radians (Section 16–4)*

Evaluate.

18. $\sin \frac{\pi}{4}$ **19.** $\cos(-\frac{2\pi}{3})$ **20.** $\tan(-\frac{5\pi}{4})$ **21.** $\cot \frac{\pi}{2}$ **22.** $\sec \frac{5\pi}{6}$

Objective: *To graph trigonometric functions (Sections 16–5 and 16–6)*

Sketch the graph of the following functions, where $-2\pi \leq x \leq 2\pi$.

23. $y = \cos x$ **24.** $y = -\frac{1}{2} \sin x$ **25.** $y = 2 \sec x$ **26.** $y = -\tan x$

Objective: *To prove trigonometric identities (Sections 16–7 and 16–9)*

Prove each identity.

27. $(1 - \cos^2 \theta)(1 + \cot^2 \theta) = 1$

28. $\sec \theta \csc \theta - \tan \theta = \cot \theta$

29. $\sin^4 \theta - 2 \sin^2 \theta + 1 = \cos^4 \theta$

30. $\frac{1 + \sin \theta}{\cot^2 \theta} = \frac{\sin \theta}{\csc \theta - 1}$

Objective: *To use the Sum and Difference Identities to evaluate trigonometric functions (Section 16–8)*

Use the Sum and Difference Identities to evaluate the following functions.

31. $\sin 75°$ **32.** $\cos 225°$ **33.** $\sin \frac{11\pi}{12}$ **34.** $\sin \frac{8\pi}{3}$ **35.** $\cos \frac{13\pi}{12}$

Objective: *To evaluate expressions that involve inverse trigonometric functions (Section 16–10)*

Evaluate.

36. $\text{Arc cos } \frac{\sqrt{2}}{2}$ **37.** $\text{Arc sin } 1$ **38.** $\text{Arc tan}(-\sqrt{3})$

39. $\cos(\text{Arc sin } 0)$ **40.** $\sin(\text{Arc tan } \frac{\sqrt{3}}{3})$ **41.** $\tan[\text{Arc cos}(-\frac{2}{3})]$

Objective: *To solve trigonometric equations (Section 16–11)*

Solve for x, where $0 \leq x < 2\pi$.

42. $2 \sin x + 1 = 0$ **43.** $\tan^2 x = \tan x$ **44.** $\sec x = \csc x$

Objective: *To change a complex number from standard form to polar form and from polar form to standard form (Section 16–12)*

Write the polar form of each complex number.

45. $1 + i$ **46.** $0 - i$ **47.** $-2 + 2i$ **48.** $\frac{1}{2} + \frac{\sqrt{3}}{2} i$ **49.** $4 - 4\sqrt{3}i$

Write each complex number in standard form.

50. $3(\cos 0° - i \sin 0°)$ **51.** $\frac{1}{2}(\cos 120° + i \sin 120°)$ **52.** $\cos \frac{3\pi}{4} - i \sin \frac{3\pi}{4}$

Chapter Test

A bicyclist travels 20 kilometers west and 15 kilometers south.

1. Find the magnitude of the bicyclist's displacement.
2. Find, to the nearest ten degrees, the positive angle θ that the bicyclist's displacement makes with the x axis.

Change each degree measure to radian measure in terms of π.

3. $30°$
4. $-18°$
5. $-150°$
6. $-90°$

Change each radian measure to degree measure.

7. $\frac{\pi}{4}$
8. 5π
9. $-\frac{5\pi}{4}$
10. $-\frac{11\pi}{12}$

Evaluate.

11. $\cos \frac{3\pi}{4}$
12. $\sin \frac{7\pi}{6}$
13. $\tan(-\frac{5\pi}{6})$
14. $\csc(-\pi)$
15. Sketch the graph of $y = -\frac{1}{2} \cos x$, where $0 \leq x \leq 2\pi$.

Prove each identity.

16. $1 + \cos \theta = \frac{\sin^2 \theta}{1 - \cos \theta}$
17. $\sec^2 \theta + \csc^2 \theta = (\tan \theta + \cot \theta)^2$
18. Use one or more of the Sum and Difference Identities to evaluate $\sin \frac{5\pi}{12}$.
19. Evaluate $\cos(\text{Arc} \sin \frac{\sqrt{2}}{2})$.
20. Write the polar form of $\frac{\sqrt{3}}{2} - \frac{1}{2}i$.

More Challenging Problems

1. Graph $y = \sin (2x + \frac{\pi}{2})$ and $y = \sin 2x$ on the same coordinate axes. How is the graph of $y = \sin (2x + \frac{\pi}{2})$ related to the graph of $y = \sin 2x$?
2. Graph $y = \sin (3x + \frac{\pi}{2})$ and $y = \sin 3x$ on the same coordinate system. How are the two graphs related?

In Exercises 3–6, prove each identity.

3. $\sin \alpha \cos \beta = \frac{1}{2}[\sin (\alpha + \beta) + \sin (\alpha - \beta)]$
4. $\cos \alpha \sin \beta = \frac{1}{2}[\sin (\alpha + \beta) - \sin (\alpha - \beta)]$
5. $\cos \alpha \cos \beta = \frac{1}{2}[\cos (\alpha + \beta) + \cos (\alpha - \beta)]$
6. $\sin \alpha \sin \beta = \frac{1}{2}[\cos (\alpha - \beta) - \cos (\alpha + \beta)]$
7. Evaluate $\cos (\text{Arc} \tan 2 + \text{Arc} \sec 3)$.
8. Prove $\text{Arc} \tan x + \text{Arc} \cot x = \frac{\pi}{2}$.
9. Prove that the area, K, of a triangle is given by the formula $K = \frac{1}{2}bc \sin A$.

Cumulative Review: Chapters 13–16

Choose the best answer. Choose *a*, *b*, *c*, or *d*.

1. Write the equation of the plane that intersects the x axis at 2, the y axis at 3, and the z axis at $\frac{1}{2}$.
   a. $x + y + z = 12$ **b.** $2x + 3y + 12z = 6$
   c. $3x + 2y + 12z = 6$ **d.** $3x + 2y + 12z = 12$

2. Solve: $\begin{cases} 2x + y + z = 1 \\ x + 2y + z = -2 \\ x + y + 2z = -3 \end{cases}$
   a. $(2, -1, -2)$ **b.** $(1, 1, -2)$ **c.** $(-1, 1, 1)$ **d.** $(-4, 1, 2)$

3. In a three-digit number, the sum of the digits is 19. The hundreds digit is twice the tens digit, and the ones digit is 3 more than the tens digit. What is the tens digit?
   a. 1 **b.** 2 **c.** 3 **d.** 4

4. Find the sum: $\begin{bmatrix} 2 & -8 & 7 \\ 1 & 3 & -4 \end{bmatrix} + \begin{bmatrix} -3 & 9 & -2 \\ 3 & -3 & 2 \end{bmatrix}$
   a. $\begin{bmatrix} 5 & -17 & 9 \\ -2 & 6 & -6 \end{bmatrix}$ **b.** $\begin{bmatrix} -1 & 1 & 5 \\ 4 & 0 & -2 \end{bmatrix}$ **c.** $\begin{bmatrix} -92 \\ -14 \end{bmatrix}$ **d.** $\begin{bmatrix} 1 & 0 & 5 \\ 0 & 1 & 8 \end{bmatrix}$

5. Find the product: $\begin{bmatrix} 2 & -1 \\ 3 & 4 \end{bmatrix} \times \begin{bmatrix} 2 & -1 & 5 \\ 3 & 4 & 1 \end{bmatrix}$
   a. $\begin{bmatrix} 1 & 0 & 5 \\ 0 & 1 & 1 \end{bmatrix}$ **b.** $\begin{bmatrix} 1 & -16 \\ 18 & 13 \end{bmatrix}$ **c.** $\begin{bmatrix} 1 & -6 & 9 \\ 18 & 13 & 19 \end{bmatrix}$ **d.** $\begin{bmatrix} 0 & 0 & 5 \\ 0 & 0 & 1 \end{bmatrix}$

6. In how many ways can 3 of 8 bowling trophies be placed on a shelf?
   a. $8!$ **b.** $3!$ **c.** $\frac{8!}{5!}$ **d.** $\frac{8!}{3!\,5!}$

7. In how many ways can 8 flags be hung on a pole if there are 3 red flags, 2 green flags, and 3 others of different colors?
   a. $\frac{8!}{3!\,2!}$ **b.** $\frac{8!}{3!}$ **c.** $\frac{8!}{2!}$ **d.** $\frac{8!}{3!\,2!\,3!}$

8. A committee of 8 is to be selected from 20 eligible persons. How many such committees are possible?
   a. $\frac{20!}{12!}$ **b.** $\frac{20!}{8!}$ **c.** $\frac{12!}{8!}$ **d.** $\frac{20!}{12!\,8!}$

9. Which is not equal to ${}_{20}C_{13}$?
   a. ${}_{20}C_7$ **b.** $\binom{20}{13}$ **c.** $\frac{20!}{13!}$ **d.** $\frac{20!}{(20-7)!\,13!}$

10. One card is drawn from a standard 52-card deck. What is the probability that the card is red or a spade?
    a. $\frac{1}{2}$ **b.** $\frac{1}{4}$ **c.** 1 **d.** $\frac{3}{4}$

11. $P(2, -5)$ is a point on the radius vector of θ, an angle in standard position. Which statement is false?

a. $\sin\theta = \frac{-5\sqrt{29}}{29}$ b. $\tan\theta = -2.5$

c. $\cos\theta = \frac{2}{\sqrt{29}}$ d. $\sin\theta = \frac{-5\sqrt{21}}{21}$

12. Which statement is true?

a. $\sin 45° = \frac{\sqrt{2}}{2}$ b. $\cos 30° = -\frac{\sqrt{3}}{2}$

c. $\tan 135° = 1$ d. $\sin 120° = -\frac{\sqrt{3}}{2}$

13. In right triangle ABC, $a = 5$ and $m\angle A = 41°$. Which statement is true?

a. $5\cos 41° = c$ b. $b\tan 49° = 5$

c. $5\cos 49° = c$ d. $c^2 = 5^2 + (\sin 41°)^2$

14. Which statement is *not true* for some angle θ?

a. $\sin(-\theta) = -\sin\theta$ b. $\cos(180° - \theta) = -\cos\theta$

c. $\tan(360° - \theta) = -\tan\theta$ d. $\cos(90° - \theta) = -\sin\theta$

15. Find the component in the x direction for vector OB.

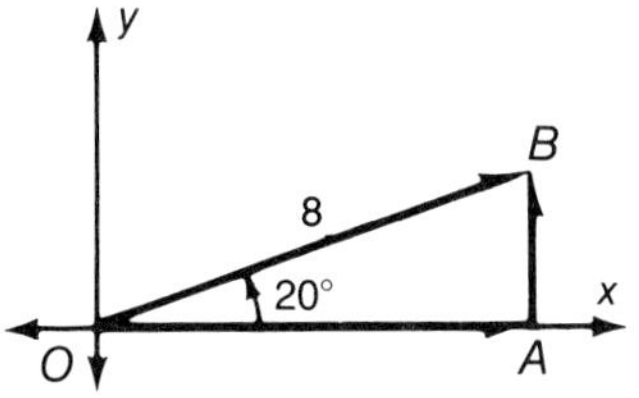

a. 8 b. $8\tan 20°$

c. $8\cos 20°$ d. $8\sin 20°$

16. A speed boat heads due north at 15 km/hr across a river traveling 2 km/hr due east. What is the actual course of the boat to the nearest degree?

a. 7° b. 7.5° c. 8° d. 8.5°

17. Evaluate $\cos\frac{2\pi}{3}$.

a. -0.5 b. 0 c. 0.5 d. 0.866

18. Which statement is true for the function $y = -3\sin\theta$?

a. The period is π. b. The amplitude is 3.

c. The amplitude is -3. d. The period is $\frac{\pi}{2}$.

19. Which expression is equivalent to $\frac{\tan^2\theta - \sin^2\theta}{\tan^2\theta\sin^2\theta}$?

a. $\tan^2\theta$ b. $\cos^2\theta$ c. $\sin^2\theta$ d. 1

20. Given that $\sin 20° = 0.3420$ and $\sin 16° = 0.2756$, find $\sin 4°$.

a. $\sin 20° - \sin 16°$ b. $0.3420\cos 20° - 0.2756\cos 16°$

b. $\cos 20° - \sin 20°$ d. $0.3420\cos 16° - 0.2756\cos 20°$

Table of Values of the Trigonometric Functions

θ Deg.	θ Rad.	Sin θ	Cos θ	Tan θ	Cot θ	Sec θ	Csc θ		
0° 00′	.0000	.0000	1.0000	.0000		1.000		1.5708	90° 00′
10′	.0029	.0029	1.0000	.0029	343.77	1.000	343.8	1.5679	50′
20′	.0058	.0058	1.0000	.0058	171.89	1.000	171.9	1.5650	40′
30′	.0087	.0087	1.0000	.0087	114.59	1.000	114.6	1.5621	30′
40′	.0116	.0116	.9999	.0116	85.940	1.000	85.95	1.5592	20′
50′	.0145	.0145	.9999	.0145	68.750	1.000	68.76	1.5563	10′
1° 00′	.0175	.0175	.9998	.0175	57.290	1.000	57.30	1.5533	89° 00′
10′	.0204	.0204	.9998	.0204	49.104	1.000	49.11	1.5504	50′
20′	.0233	.0233	.9997	.0233	42.964	1.000	42.98	1.5475	40′
30′	.0262	.0262	.9997	.0262	38.188	1.000	38.20	1.5446	30′
40′	.0291	.0291	.9996	.0291	34.368	1.000	34.38	1.5417	20′
50′	.0320	.0320	.9995	.0320	31.242	1.001	31.26	1.5388	10′
2° 00′	.0349	.0349	.9994	.0349	28.636	1.001	28.65	1.5359	88° 00′
10′	.0378	.0378	.9993	.0378	26.432	1.001	26.45	1.5330	50′
20′	.0407	.0407	.9992	.0407	24.542	1.001	24.56	1.5301	40′
30′	.0436	.0436	.9990	.0437	22.904	1.001	22.93	1.5272	30′
40′	.0465	.0465	.9989	.0466	21.470	1.001	21.49	1.5243	20′
50′	.0495	.0494	.9988	.0495	20.206	1.001	20.23	1.5213	10′
3° 00′	.0524	.0523	.9986	.0524	19.081	1.001	19.11	1.5184	87° 00′
10′	.0553	.0552	.9985	.0553	18.075	1.002	18.10	1.5155	50′
20′	.0582	.0581	.9983	.0582	17.169	1.002	17.20	1.5126	40′
30′	.0611	.0610	.9981	.0612	16.350	1.002	16.38	1.5097	30′
40′	.0640	.0640	.9980	.0641	15.605	1.002	15.64	1.5068	20′
50′	.0669	.0669	.9978	.0670	14.924	1.002	14.96	1.5039	10′
4° 00′	.0698	.0698	.9976	.0699	14.301	1.002	14.34	1.5010	86° 00′
10′	.0727	.0727	.9974	.0729	13.727	1.003	13.76	1.4981	50′
20′	.0756	.0756	.9971	.0758	13.197	1.003	13.23	1.4952	40′
30′	.0785	.0785	.9969	.0787	12.706	1.003	12.75	1.4923	30′
40′	.0814	.0814	.9967	.0816	12.251	1.003	12.29	1.4893	20′
50′	.0844	.0843	.9964	.0846	11.826	1.004	11.87	1.4864	10′
5° 00′	.0873	.0872	.9962	.0875	11.430	1.004	11.47	1.4835	85° 00′
10′	.0902	.0901	.9959	.0904	11.059	1.004	11.10	1.4806	50′
20′	.0931	.0929	.9957	.0934	10.712	1.004	10.76	1.4777	40′
30′	.0960	.0958	.9954	.0963	10.385	1.005	10.43	1.4748	30′
40′	.0989	.0987	.9951	.0992	10.078	1.005	10.13	1.4719	20′
50′	.1018	.1016	.9948	.1022	9.7882	1.005	9.839	1.4690	10′
6° 00′	.1047	.1045	.9945	.1051	9.5144	1.006	9.567	1.4661	84° 00′
10′	.1076	.1074	.9942	.1080	9.2553	1.006	9.309	1.4632	50′
20′	.1105	.1103	.9939	.1110	9.0098	1.006	9.065	1.4603	40′
30′	.1134	.1132	.9936	.1139	8.7769	1.006	8.834	1.4573	30′
40′	.1164	.1161	.9932	.1169	8.5555	1.007	8.614	1.4544	20′
50′	.1193	.1190	.9929	.1198	8.3450	1.007	8.405	1.4515	10′
7° 00′	.1222	.1219	.9925	.1228	8.1443	1.008	8.206	1.4486	83° 00′
10′	.1251	.1248	.9922	.1257	7.9530	1.008	8.016	1.4457	50′
20′	.1280	.1276	.9918	.1287	7.7704	1.008	7.834	1.4428	40′
30′	.1309	.1305	.9914	.1317	7.5958	1.009	7.661	1.4399	30′
40′	.1338	.1334	.9911	.1346	7.4287	1.009	7.496	1.4370	20′
50′	.1367	.1363	.9907	.1376	7.2687	1.009	7.337	1.4341	10′
8° 00′	.1396	.1392	.9903	.1405	7.1154	1.010	7.185	1.4312	82° 00′
10′	.1425	.1421	.9899	.1435	6.9682	1.010	7.040	1.4283	50′
20′	.1454	.1449	.9894	.1465	6.8269	1.011	6.900	1.4254	40′
30′	.1484	.1478	.9890	.1495	6.6912	1.011	6.765	1.4224	30′
40′	.1513	.1507	.9886	.1524	6.5606	1.012	6.636	1.4195	20′
50′	.1542	.1536	.9881	.1554	6.4348	1.012	6.512	1.4166	10′
9° 00′	.1571	.1564	.9877	.1584	6.3138	1.012	6.392	1.4137	81° 00′
		Cos θ	Sin θ	Cot θ	Tan θ	Csc θ	Sec θ	θ Rad.	θ Deg.

Table of Values of the Trigonometric Functions

θ Deg.	θ Rad.	Sin θ	Cos θ	Tan θ	Cot θ	Sec θ	Csc θ		
9° 00′	.1571	.1564	.9877	.1584	6.3138	1.012	6.392	1.4137	81° 00′
10′	.1600	.1593	.9872	.1614	6.1970	1.013	6.277	1.4108	50′
20′	.1629	.1622	.9868	.1644	6.0844	1.013	6.166	1.4079	40′
30′	.1658	.1650	.9863	.1673	5.9758	1.014	6.059	1.4050	30′
40′	.1687	.1679	.9858	.1703	5.8708	1.014	5.955	1.4021	20′
50′	.1716	.1708	.9853	.1733	5.7694	1.015	5.855	1.3992	10′
10° 00′	.1745	.1736	.9848	.1763	5.6713	1.015	5.759	1.3963	80° 00′
10′	.1774	.1765	.9843	.1793	5.5764	1.016	5.665	1.3934	50′
20′	.1804	.1794	.9838	.1823	5.4845	1.016	5.575	1.3904	40′
30′	.1833	.1822	.9833	.1853	5.3955	1.017	5.487	1.3875	30′
40′	.1862	.1851	.9827	.1883	5.3093	1.018	5.403	1.3846	20′
50′	.1891	.1880	.9822	.1914	5.2257	1.018	5.320	1.3817	10′
11° 00′	.1920	.1908	.9816	.1944	5.1446	1.019	5.241	1.3788	79° 00′
10′	.1949	.1937	.9811	.1974	5.0658	1.019	5.164	1.3759	50′
20′	.1978	.1965	.9805	.2004	4.9894	1.020	5.089	1.3730	40′
30′	.2007	.1994	.9799	.2035	4.9152	1.020	5.016	1.3701	30′
40′	.2036	.2022	.9793	.2065	4.8430	1.021	4.945	1.3672	20′
50′	.2065	.2051	.9787	.2095	4.7729	1.022	4.876	1.3643	10′
12° 00′	.2094	.2079	.9781	.2126	4.7046	1.022	4.810	1.3614	78° 00′
10′	.2123	.2108	.9775	.2156	4.6382	1.023	4.745	1.3584	50′
20′	.2153	.2136	.9769	.2186	4.5736	1.024	4.682	1.3555	40′
30′	.2182	.2164	.9763	.2217	4.5107	1.024	4.620	1.3526	30′
40′	.2211	.2193	.9757	.2247	4.4494	1.025	4.560	1.3497	20′
50′	.2240	.2221	.9750	.2278	4.3897	1.026	4.502	1.3468	10′
13° 00′	.2269	.2250	.9744	.2309	4.3315	1.026	4.445	1.3439	77° 00′
10′	.2298	.2278	.9737	.2339	4.2747	1.027	4.390	1.3410	50′
20′	.2327	.2306	.9730	.2370	4.2193	1.028	4.336	1.3381	40′
30′	.2356	.2334	.9724	.2401	4.1653	1.028	4.284	1.3352	30′
40′	.2385	.2363	.9717	.2432	4.1126	1.029	4.232	1.3323	20′
50′	.2414	.2391	.9710	.2462	4.0611	1.030	4.182	1.3294	10′
14° 00′	.2443	.2419	.9703	.2493	4.0108	1.031	4.134	1.3265	76° 00′
10′	.2473	.2447	.9696	.2524	3.9617	1.031	4.086	1.3235	50′
20′	.2502	.2476	.9689	.2555	3.9136	1.032	4.039	1.3206	40′
30′	.2531	.2504	.9681	.2586	3.8667	1.033	3.994	1.3177	30′
40′	.2560	.2532	.9674	.2617	3.8208	1.034	3.950	1.3148	20′
50′	.2589	.2560	.9667	.2648	3.7760	1.034	3.906	1.3119	10′
15° 00′	.2618	.2588	.9659	.2679	3.7321	1.035	3.864	1.3090	75° 00′
10′	.2647	.2616	.9652	.2711	3.6891	1.036	3.822	1.3061	50′
20′	.2676	.2644	.9644	.2742	3.6470	1.037	3.782	1.3032	40′
30′	.2705	.2672	.9636	.2773	3.6059	1.038	3.742	1.3003	30′
40′	.2734	.2700	.9628	.2805	3.5656	1.039	3.703	1.2974	20′
50′	.2763	.2728	.9621	.2836	3.5261	1.039	3.665	1.2945	10′
16° 00′	.2793	.2756	.9613	.2867	3.4874	1.040	3.628	1.2915	74° 00′
10′	.2822	.2784	.9605	.2899	3.4495	1.041	3.592	1.2886	50′
20′	.2851	.2812	.9596	.2931	3.4124	1.042	3.556	1.2857	40′
30′	.2880	.2840	.9588	.2962	3.3759	1.043	3.521	1.2828	30′
40′	.2909	.2868	.9580	.2994	3.3402	1.044	3.487	1.2799	20′
50′	.2938	.2896	.9572	.3026	3.3052	1.045	3.453	1.2770	10′
17° 00′	.2967	.2924	.9563	.3057	3.2709	1.046	3.420	1.2741	73° 00′
10′	.2996	.2952	.9555	.3089	3.2371	1.047	3.388	1.2712	50′
20′	.3025	.2979	.9546	.3121	3.2041	1.048	3.356	1.2683	40′
30′	.3054	.3007	.9537	.3153	3.1716	1.049	3.326	1.2654	30′
40′	.3083	.3035	.9528	.3185	3.1397	1.049	3.295	1.2625	20′
50′	.3113	.3062	.9520	.3217	3.1084	1.050	3.265	1.2595	10′
18° 00′	.3142	.3090	.9511	.3249	3.0777	1.051	3.236	1.2566	72° 00′
		Cos θ	Sin θ	Cot θ	Tan θ	Csc θ	Sec θ	θ Rad.	θ Deg.

Table of Values of the Trigonometric Functions

θ Deg.	θ Rad.	Sin θ	Cos θ	Tan θ	Cot θ	Sec θ	Csc θ		
18° 00′	.3142	.3090	.9511	.3249	3.0777	1.051	3.236	1.2566	72° 00′
10′	.3171	.3118	.9502	.3281	3.0475	1.052	3.207	1.2537	50′
20′	.3200	.3145	.9492	.3314	3.0178	1.053	3.179	1.2508	40′
30′	.3229	.3173	.9483	.3346	2.9887	1.054	3.152	1.2479	30′
40′	.3258	.3201	.9474	.3378	2.9600	1.056	3.124	1.2450	20′
50′	.3287	.3228	.9465	.3411	2.9319	1.057	3.098	1.2421	10′
19° 00′	.3316	.3256	.9455	.3443	2.9042	1.058	3.072	1.2392	71° 00′
10′	.3345	.3283	.9446	.3476	2.8770	1.059	3.046	1.2363	50′
20′	.3374	.3311	.9436	.3508	2.8502	1.060	3.021	1.2334	40′
30′	.3403	.3338	.9426	.3541	2.8239	1.061	2.996	1.2305	30′
40′	.3432	.3365	.9417	.3574	2.7980	1.062	2.971	1.2275	20′
50′	.3462	.3393	.9407	.3607	2.7725	1.063	2.947	1.2246	10′
20° 00′	.3491	.3420	.9397	.3640	2.7475	1.064	2.924	1.2217	70° 00′
10′	.3520	.3448	.9387	.3673	2.7228	1.065	2.901	1.2188	50′
20′	.3549	.3475	.9377	.3706	2.6985	1.066	2.878	1.2159	40′
30′	.3578	.3502	.9367	.3739	2.6746	1.068	2.855	1.2130	30′
40′	.3607	.3529	.9356	.3772	2.6511	1.069	2.833	1.2101	20′
50′	.3636	.3557	.9346	.3805	2.6279	1.070	2.812	1.2072	10′
21° 00′	.3665	.3584	.9336	.3839	2.6051	1.071	2.790	1.2043	69° 00′
10′	.3694	.3611	.9325	.3872	2.5826	1.072	2.769	1.2014	50′
20′	.3723	.3638	.9315	.3906	2.5605	1.074	2.749	1.1985	40′
30′	.3752	.3665	.9304	.3939	2.5386	1.075	2.729	1.1956	30′
40′	.3782	.3692	.9293	.3973	2.5172	1.076	2.709	1.1926	20′
50′	.3811	.3719	.9283	.4006	2.4960	1.077	2.689	1.1897	10′
22° 00′	.3840	.3746	.9272	.4040	2.4751	1.079	2.669	1.1868	68° 00′
10′	.3869	.3773	.9261	.4074	2.4545	1.080	2.650	1.1839	50′
20′	.3898	.3800	.9250	.4108	2.4342	1.081	2.632	1.1810	40′
30′	.3927	.3827	.9239	.4142	2.4142	1.082	2.613	1.1781	30′
40′	.3956	.3854	.9228	.4176	2.3945	1.084	2.595	1.1752	20′
50′	.3985	.3881	.9216	.4210	2.3750	1.085	2.577	1.1723	10′
23° 00′	.4014	.3907	.9205	.4245	2.3559	1.086	2.559	1.1694	67° 00′
10′	.4043	.3934	.9194	.4279	2.3369	1.088	2.542	1.1665	50′
20′	.4072	.3961	.9182	.4314	2.3183	1.089	2.525	1.1636	40′
30′	.4102	.3987	.9171	.4348	2.2998	1.090	2.508	1.1606	30′
40′	.4131	.4014	.9159	.4383	2.2817	1.092	2.491	1.1577	20′
50′	.4160	.4041	.9147	.4417	2.2637	1.093	2.475	1.1548	10′
24° 00′	.4189	.4067	.9135	.4452	2.2460	1.095	2.459	1.1519	66° 00′
10′	.4218	.4094	.9124	.4487	2.2286	1.096	2.443	1.1490	50′
20′	.4247	.4120	.9112	.4522	2.2113	1.097	2.427	1.1461	40′
30′	.4276	.4147	.9100	.4557	2.1943	1.099	2.411	1.1432	30′
40′	.4305	.4173	.9088	.4592	2.1775	1.100	2.396	1.1403	20′
50′	.4334	.4200	.9075	.4628	2.1609	1.102	2.381	1.1374	10′
25° 00′	.4363	.4226	.9063	.4663	2.1445	1.103	2.366	1.1345	65° 00′
10′	.4392	.4253	.9051	.4699	2.1283	1.105	2.352	1.1316	50′
20′	.4422	.4279	.9038	.4734	2.1123	1.106	2.337	1.1286	40′
30′	.4451	.4305	.9026	.4770	2.0965	1.108	2.323	1.1257	30′
40′	.4480	.4331	.9013	.4806	2.0809	1.109	2.309	1.1228	20′
50′	.4509	.4358	.9001	.4841	2.0655	1.111	2.295	1.1199	10′
26° 00′	.4538	.4384	.8988	.4877	2.0503	1.113	2.281	1.1170	64° 00′
10′	.4567	.4410	.8975	.4913	2.0353	1.114	2.268	1.1141	50′
20′	.4596	.4436	.8962	.4950	2.0204	1.116	2.254	1.1112	40′
30′	.4625	.4462	.8949	.4986	2.0057	1.117	2.241	1.1083	30′
40′	.4654	.4488	.8936	.5022	1.9912	1.119	2.228	1.1054	20′
50′	.4683	.4514	.8923	.5059	1.9768	1.121	2.215	1.1025	10′
27° 00′	.4712	.4540	.8910	.5095	1.9626	1.122	2.203	1.0996	63° 00′
		Cos θ	Sin θ	Cot θ	Tan θ	Csc θ	Sec θ	θ Rad.	θ Deg.

Table of Values of the Trigonometric Functions

θ Deg.	θ Rad.	Sin θ	Cos θ	Tan θ	Cot θ	Sec θ	Csc θ		
27° 00′	.4712	.4540	.8910	.5095	1.9626	1.122	2.203	1.0996	63° 00′
10′	.4741	.4566	.8897	.5132	1.9486	1.124	2.190	1.0966	50′
20′	.4771	.4592	.8884	.5169	1.9347	1.126	2.178	1.0937	40′
30′	.4800	.4617	.8870	.5206	1.9210	1.127	2.166	1.0908	30′
40′	.4829	.4643	.8857	.5243	1.9074	1.129	2.154	1.0879	20′
50′	.4858	.4669	.8843	.5280	1.8940	1.131	2.142	1.0850	10′
28° 00′	.4887	.4695	.8829	.5317	1.8807	1.133	2.130	1.0821	62° 00′
10′	.4916	.4720	.8816	.5354	1.8676	1.134	2.118	1.0792	50′
20′	.4945	.4746	.8802	.5392	1.8546	1.136	2.107	1.0763	40′
30′	.4974	.4772	.8788	.5430	1.8418	1.138	2.096	1.0734	30′
40′	.5003	.4797	.8774	.5467	1.8291	1.140	2.085	1.0705	20′
50′	.5032	.4823	.8760	.5505	1.8165	1.142	2.074	1.0676	10′
29° 00′	.5061	.4848	.8746	.5543	1.8040	1.143	2.063	1.0647	61° 00′
10′	.5091	.4874	.8732	.5581	1.7917	1.145	2.052	1.0617	50′
20′	.5120	.4899	.8718	.5619	1.7796	1.147	2.041	1.0588	40′
30′	.5149	.4924	.8704	.5658	1.7675	1.149	2.031	1.0559	30′
40′	.5178	.4950	.8689	.5696	1.7556	1.151	2.020	1.0530	20′
50′	.5207	.4975	.8675	.5735	1.7437	1.153	2.010	1.0501	10′
30° 00′	.5236	.5000	.8660	.5774	1.7321	1.155	2.000	1.0472	60° 00′
10′	.5265	.5025	.8646	.5812	1.7205	1.157	1.990	1.0443	50′
20′	.5294	.5050	.8631	.5851	1.7090	1.159	1.980	1.0414	40′
30′	.5323	.5075	.8616	.5890	1.6977	1.161	1.970	1.0385	30′
40′	.5352	.5100	.8601	.5930	1.6864	1.163	1.961	1.0356	20′
50′	.5381	.5125	.8587	.5969	1.6753	1.165	1.951	1.0327	10′
31° 00′	.5411	.5150	.8572	.6009	1.6643	1.167	1.942	1.0297	59° 00′
10′	.5440	.5175	.8557	.6048	1.6534	1.169	1.932	1.0268	50′
20′	.5469	.5200	.8542	.6088	1.6426	1.171	1.923	1.0239	40′
30′	.5498	.5225	.8526	.6128	1.6319	1.173	1.914	1.0210	30′
40′	.5527	.5250	.8511	.6168	1.6212	1.175	1.905	1.0181	20′
50′	.5556	.5275	.8496	.6208	1.6107	1.177	1.896	1.0152	10′
32° 00′	.5585	.5299	.8480	.6249	1.6003	1.179	1.887	1.0123	58° 00′
10′	.5614	.5324	.8465	.6289	1.5900	1.181	1.878	1.0094	50′
20′	.5643	.5348	.8450	.6330	1.5798	1.184	1.870	1.0065	40′
30′	.5672	.5373	.8434	.6371	1.5697	1.186	1.861	1.0036	30′
40′	.5701	.5398	.8418	.6412	1.5597	1.188	1.853	1.0007	20′
50′	.5730	.5422	.8403	.6453	1.5497	1.190	1.844	.9977	10′
33° 00′	.5760	.5446	.8387	.6494	1.5399	1.192	1.836	.9948	57° 00′
10′	.5789	.5471	.8371	.6536	1.5301	1.195	1.828	.9919	50′
20′	.5818	.5495	.8355	.6577	1.5204	1.197	1.820	.9890	40′
30′	.5847	.5519	.8339	.6619	1.5108	1.199	1.812	.9861	30′
40′	.5876	.5544	.8323	.6661	1.5013	1.202	1.804	.9832	20′
50′	.5905	.5568	.8307	.6703	1.4919	1.204	1.796	.9803	10′
34° 00′	.5934	.5592	.8290	.6745	1.4826	1.206	1.788	.9774	56° 00′
10′	.5963	.5616	.8274	.6787	1.4733	1.209	1.781	.9745	50′
20′	.5992	.5640	.8258	.6830	1.4641	1.211	1.773	.9716	40′
30′	.6021	.5664	.8241	.6873	1.4550	1.213	1.766	.9687	30′
40′	.6050	.5688	.8225	.6916	1.4460	1.216	1.758	.9657	20′
50′	.6080	.5712	.8208	.6959	1.4370	1.218	1.751	.9628	10′
35° 00′	.6109	.5736	.8192	.7002	1.4281	1.221	1.743	.9599	55° 00′
10′	.6138	.5760	.8175	.7046	1.4193	1.223	1.736	.9570	50′
20′	.6167	.5783	.8158	.7089	1.4106	1.226	1.729	.9541	40′
30′	.6196	.5807	.8141	.7133	1.4019	1.228	1.722	.9512	30′
40′	.6225	.5831	.8124	.7177	1.3934	1.231	1.715	.9483	20′
50′	.6254	.5854	.8107	.7221	1.3848	1.233	1.708	.9454	10′
36° 00′	.6283	.5878	.8090	.7265	1.3764	1.236	1.701	.9425	54° 00′
		Cos θ	Sin θ	Cot θ	Tan θ	Csc θ	Sec θ	θ Rad.	θ Deg.

Table of Values of the Trigonometric Functions

θ Deg.	θ Rad.	Sin θ	Cos θ	Tan θ	Cot θ	Sec θ	Csc θ		
36° 00′	.6283	.5878	.8090	.7265	1.3764	1.236	1.701	.9425	**54° 00′**
10′	.6312	.5901	.8073	.7310	1.3680	1.239	1.695	.9396	**50′**
20′	.6341	.5925	.8056	.7355	1.3597	1.241	1.688	.9367	**40′**
30′	.6370	.5948	.8039	.7400	1.3514	1.244	1.681	.9338	**30′**
40′	.6400	.5972	.8021	.7445	1.3432	1.247	1.675	.9308	**20′**
50′	.6429	.5995	.8004	.7490	1.3351	1.249	1.668	.9279	**10′**
37° 00′	.6458	.6018	.7986	.7536	1.3270	1.252	1.662	.9250	**53° 00′**
10′	.6487	.6041	.7969	.7581	1.3190	1.255	1.655	.9221	**50′**
20′	.6516	.6065	.7951	.7627	1.3111	1.258	1.649	.9192	**40′**
30′	.6545	.6088	.7934	.7673	1.3032	1.260	1.643	.9163	**30′**
40′	.6574	.6111	.7916	.7720	1.2954	1.263	1.636	.9134	**20′**
50′	.6603	.6134	.7898	.7766	1.2876	1.266	1.630	.9105	**10′**
38° 00′	.6632	.6157	.7880	.7813	1.2799	1.269	1.624	.9076	**52° 00′**
10′	.6661	.6180	.7862	.7860	1.2723	1.272	1.618	.9047	**50′**
20′	.6690	.6202	.7844	.7907	1.2647	1.275	1.612	.9018	**40′**
30′	.6720	.6225	.7826	.7954	1.2572	1.278	1.606	.8988	**30′**
40′	.6749	.6248	.7808	.8002	1.2497	1.281	1.601	.8959	**20′**
50′	.6778	.6271	.7790	.8050	1.2423	1.284	1.595	.8930	**10′**
39° 00′	.6807	.6293	.7771	.8098	1.2349	1.287	1.589	.8901	**51° 00′**
10′	.6836	.6316	.7753	.8146	1.2276	1.290	1.583	.8872	**50′**
20′	.6865	.6338	.7735	.8195	1.2203	1.293	1.578	.8843	**40′**
30′	.6894	.6361	.7716	.8243	1.2131	1.296	1.572	.8814	**30′**
40′	.6923	.6383	.7698	.8292	1.2059	1.299	1.567	.8785	**20′**
50′	.6952	.6406	.7679	.8342	1.1988	1.302	1.561	.8756	**10′**
40° 00′	.6981	.6428	.7660	.8391	1.1918	1.305	1.556	.8727	**50° 00′**
10′	.7010	.6450	.7642	.8441	1.1847	1.309	1.550	.8698	**50′**
20′	.7039	.6472	.7623	.8491	1.1778	1.312	1.545	.8668	**40′**
30′	.7069	.6494	.7604	.8541	1.1708	1.315	1.540	.8639	**30′**
40′	.7098	.6517	.7585	.8591	1.1640	1.318	1.535	.8610	**20′**
50′	.7127	.6539	.7566	.8642	1.1571	1.322	1.529	.8581	**10′**
41° 00′	.7156	.6561	.7547	.8693	1.1504	1.325	1.524	.8552	**49° 00′**
10′	.7185	.6583	.7528	.8744	1.1436	1.328	1.519	.8523	**50′**
20′	.7214	.6604	.7509	.8796	1.1369	1.332	1.514	.8494	**40′**
30′	.7243	.6626	.7490	.8847	1.1303	1.335	1.509	.8465	**30′**
40′	.7272	.6648	.7470	.8899	1.1237	1.339	1.504	.8436	**20′**
50′	.7301	.6670	.7451	.8952	1.1171	1.342	1.499	.8407	**10′**
42° 00′	.7330	.6691	.7431	.9004	1.1106	1.346	1.494	.8378	**48° 00′**
10′	.7359	.6713	.7412	.9057	1.1041	1.349	1.490	.8348	**50′**
20′	.7389	.6734	.7392	.9110	1.0977	1.353	1.485	.8319	**40′**
30′	.7418	.6756	.7373	.9163	1.0913	1.356	1.480	.8290	**30′**
40′	.7447	.6777	.7353	.9217	1.0850	1.360	1.476	.8261	**20′**
50′	.7476	.6799	.7333	.9271	1.0786	1.364	1.471	.8232	**10′**
43° 00′	.7505	.6820	.7314	.9325	1.0724	1.367	1.466	.8203	**47° 00′**
10′	.7534	.6841	.7294	.9380	1.0661	1.371	1.462	.8174	**50′**
20′	.7563	.6862	.7274	.9435	1.0599	1.375	1.457	.8145	**40′**
30′	.7592	.6884	.7254	.9490	1.0538	1.379	1.453	.8116	**30′**
40′	.7621	.6905	.7234	.9545	1.0477	1.382	1.448	.8087	**20′**
50′	.7650	.6926	.7214	.9601	1.0416	1.386	1.444	.8058	**10′**
44° 00′	.7679	.6947	.7193	.9657	1.0355	1.390	1.440	.8029	**46° 00′**
10′	.7709	.6967	.7173	.9713	1.0295	1.394	1.435	.7999	**50′**
20′	.7738	.6988	.7153	.9770	1.0235	1.398	1.431	.7970	**40′**
30′	.7767	.7009	.7133	.9827	1.0176	1.402	1.427	.7941	**30′**
40′	.7796	.7030	.7112	.9884	1.0117	1.406	1.423	.7912	**20′**
50′	.7825	.7050	.7092	.9942	1.0058	1.410	1.418	.7883	**10′**
45° 00′	.7854	.7071	.7071	1.0000	1.0000	1.414	1.414	.7854	**45° 00′**
		Cos θ	**Sin θ**	**Cot θ**	**Tan θ**	**Csc θ**	**Sec θ**	**θ Rad.**	**θ Deg.**

Table of Common Logarithms

N	0	1	2	3	4	5	6	7	8	9
1.0	0000	0043	0086	0128	0170	0212	0253	0294	0334	0374
1.1	0414	0453	0492	0531	0569	0607	0645	0682	0719	0755
1.2	0792	0828	0864	0899	0934	0969	1004	1038	1072	1106
1.3	1139	1173	1206	1239	1271	1303	1335	1367	1399	1430
1.4	1461	1492	1523	1553	1584	1614	1644	1673	1703	1732
1.5	1761	1790	1818	1847	1875	1903	1931	1959	1987	2014
1.6	2041	2068	2095	2122	2148	2175	2201	2227	2253	2279
1.7	2304	2330	2355	2380	2405	2430	2455	2480	2504	2529
1.8	2553	2577	2601	2625	2648	2672	2695	2718	2742	2765
1.9	2788	2810	2833	2856	2878	2900	2923	2945	2967	2989
2.0	3010	3032	3054	3075	3096	3118	3139	3160	3181	3201
2.1	3222	3243	3263	3284	3304	3324	3345	3365	3385	3404
2.2	3424	3444	3464	3483	3502	3522	3541	3560	3579	3598
2.3	3617	3636	3655	3674	3692	3711	3729	3747	3766	3784
2.4	3802	3820	3838	3856	3874	3892	3909	3927	3945	3962
2.5	3979	3997	4014	4031	4048	4065	4082	4099	4116	4133
2.6	4150	4166	4183	4200	4216	4232	4249	4265	4281	4298
2.7	4314	4330	4346	4362	4378	4393	4409	4425	4440	4456
2.8	4472	4487	4502	4518	4533	4548	4564	4579	4594	4609
2.9	4624	4639	4654	4669	4683	4698	4713	4728	4742	4757
3.0	4771	4786	4800	4814	4829	4843	4857	4871	4886	4900
3.1	4914	4928	4942	4955	4969	4983	4997	5011	5024	5038
3.2	5051	5065	5079	5092	5105	5119	5132	5145	5159	5172
3.3	5185	5198	5211	5224	5237	5250	5263	5276	5289	5302
3.4	5315	5328	5340	5353	5366	5378	5391	5403	5416	5428
3.5	5441	5453	5465	5478	5490	5502	5514	5527	5539	5551
3.6	5563	5575	5587	5599	5611	5623	5635	5647	5658	5670
3.7	5682	5694	5705	5717	5729	5740	5752	5763	5775	5786
3.8	5798	5809	5821	5832	5843	5855	5866	5877	5888	5899
3.9	5911	5922	5933	5944	5955	5966	5977	5988	5999	6010
4.0	6021	6031	6042	6053	6064	6075	6085	6096	6107	6117
4.1	6128	6138	6149	6160	6170	6180	6191	6201	6212	6222
4.2	6232	6243	6253	6263	6274	6284	6294	6304	6314	6325
4.3	6335	6345	6355	6365	6375	6385	6395	6405	6415	6425
4.4	6435	6444	6454	6464	6474	6484	6493	6503	6513	6522
4.5	6532	6542	6551	6561	6571	6580	6590	6599	6609	6618
4.6	6628	6637	6646	6656	6665	6675	6684	6693	6702	6712
4.7	6721	6730	6739	6749	6758	6767	6776	6785	6794	6803
4.8	6812	6821	6830	6839	6848	6857	6866	6875	6884	6893
4.9	6902	6911	6920	6928	6937	6946	6955	6964	6972	6981
5.0	6990	6998	7007	7016	7024	7033	7042	7050	7059	7067
5.1	7076	7084	7093	7101	7110	7118	7126	7135	7143	7152
5.2	7160	7168	7177	7185	7193	7202	7210	7218	7226	7235
5.3	7243	7251	7259	7267	7275	7284	7292	7300	7308	7316
5.4	7324	7332	7340	7348	7356	7364	7372	7380	7388	7396

Table of Common Logarithms

N	0	1	2	3	4	5	6	7	8	9
5.5	7404	7412	7419	7427	7435	7443	7451	7459	7466	7474
5.6	7482	7490	7497	7505	7513	7520	7528	7536	7543	7551
5.7	7559	7566	7574	7582	7589	7597	7604	7612	7619	7627
5.8	7634	7642	7649	7657	7664	7672	7679	7686	7694	7701
5.9	7709	7716	7723	7731	7738	7745	7752	7760	7767	7774
6.0	7782	7789	7796	7803	7810	7818	7825	7832	7839	7846
6.1	7853	7860	7868	7875	7882	7889	7896	7903	7910	7917
6.2	7924	7931	7938	7945	7952	7959	7966	7973	7980	7987
6.3	7993	8000	8007	8014	8021	8028	8035	8041	8048	8055
6.4	8062	8069	8075	8082	8089	8096	8102	8109	8116	8122
6.5	8129	8136	8142	8149	8156	8162	8169	8176	8182	8189
6.6	8195	8202	8209	8215	8222	8228	8235	8241	8248	8254
6.7	8261	8267	8274	8280	8287	8293	8299	8306	8312	8319
6.8	8325	8331	8338	8344	8351	8357	8363	8370	8376	8382
6.9	8388	8395	8401	8407	8414	8420	8426	8432	8439	8445
7.0	8451	8457	8463	8470	8476	8482	8488	8494	8500	8506
7.1	8513	8519	8525	8531	8537	8543	8549	8555	8561	8567
7.2	8573	8579	8585	8591	8597	8603	8609	8615	8621	8627
7.3	8633	8639	8645	8651	8657	8663	8669	8675	8681	8686
7.4	8692	8698	8704	8710	8716	8722	8727	8733	8739	8745
7.5	8751	8756	8762	8768	8774	8779	8785	8791	8797	8802
7.6	8808	8814	8820	8825	8831	8837	8842	8848	8854	8859
7.7	8865	8871	8876	8882	8887	8893	8899	8904	8910	8915
7.8	8921	8927	8932	8938	8943	8949	8954	8960	8965	8971
7.9	8976	8982	8987	8993	8998	9004	9009	9015	9020	9025
8.0	9031	9036	9042	9047	9053	9058	9063	9069	9074	9079
8.1	9085	9090	9096	9101	9106	9112	9117	9122	9128	9133
8.2	9138	9143	9149	9154	9159	9165	9170	9175	9180	9186
8.3	9191	9196	9201	9206	9212	9217	9222	9227	9232	9238
8.4	9243	9248	9253	9258	9263	9269	9274	9279	9284	9289
8.5	9294	9299	9304	9309	9315	9320	9325	9330	9335	9340
8.6	9345	9350	9355	9360	9365	9370	9375	9380	9385	9390
8.7	9395	9400	9405	9410	9415	9420	9425	9430	9435	9440
8.8	9445	9450	9455	9460	9465	9469	9474	9479	9484	9489
8.9	9494	9499	9504	9509	9513	9518	9523	9528	9533	9538
9.0	9542	9547	9552	9557	9562	9566	9571	9576	9581	9586
9.1	9590	9595	9600	9605	9609	9614	9619	9624	9628	9633
9.2	9638	9643	9647	9652	9657	9661	9666	9671	9675	9680
9.3	9685	9689	9694	9699	9703	9708	9713	9717	9722	9727
9.4	9731	9736	9741	9745	9750	9754	9759	9763	9768	9773
9.5	9777	9782	9786	9791	9795	9800	9805	9809	9814	9818
9.6	9823	9827	9832	9836	9841	9845	9850	9854	9859	9863
9.7	9868	9872	9877	9881	9886	9890	9894	9899	9903	9908
9.8	9912	9917	9921	9926	9930	9934	9939	9943	9948	9952
9.9	9956	9961	9965	9969	9974	9978	9983	9987	9991	9996

Table of Squares, Cubes, Square and Cube Roots

No.	Squares	Cubes	Square Roots	Cube Roots	No.	Squares	Cubes	Square Roots	Cube Roots
1	1	1	1.000	1.000	51	2,601	132,651	7.141	3.708
2	4	8	1.414	1.260	52	2,704	140,608	7.211	3.733
3	9	27	1.732	1.442	53	2,809	148,877	7.280	3.756
4	16	64	2.000	1.587	54	2,916	157,464	7.348	3.780
5	25	125	2.236	1.710	55	3,025	166,375	7.416	3.803
6	36	216	2.449	1.817	56	3,136	175,616	7.483	3.826
7	49	343	2.646	1.913	57	3,249	185,193	7.550	3.849
8	64	512	2.828	2.000	58	3,364	195,112	7.616	3.871
9	81	729	3.000	2.080	59	3,481	205,379	7.681	3.893
10	100	1,000	3.162	2.154	60	3,600	216,000	7.746	3.915
11	121	1,331	3.317	2.224	61	3,721	226,981	7.810	3.936
12	144	1,728	3.464	2.289	62	3,844	238,328	7.874	3.958
13	169	2,197	3.606	2.351	63	3,969	250,047	7.937	3.979
14	196	2,744	3.742	2.410	64	4,096	262,144	8.000	4.000
15	225	3,375	3.873	2.466	65	4,225	274,625	8.062	4.021
16	256	4,096	4.000	2.520	66	4,356	287,496	8.124	4.041
17	289	4,913	4.123	2.571	67	4,489	300,763	8.185	4.062
18	324	5,832	4.243	2.621	68	4,624	314,432	8.246	4.082
19	361	6,859	4.359	2.668	69	4,761	328,509	8.307	4.102
20	400	8,000	4.472	2.714	70	4,900	343,000	8.367	4.121
21	441	9,261	4.583	2.759	71	5,041	357,911	8.426	4.141
22	484	10,648	4.690	2.802	72	5,184	373,248	8.485	4.160
23	529	12,167	4.796	2.844	73	5,329	389,017	8.544	4.179
24	576	13,824	4.899	2.884	74	5,476	405,224	8.602	4.198
25	625	15,625	5.000	2.924	75	5,625	421,875	8.660	4.217
26	676	17,576	5.099	2.962	76	5,776	438,976	8.718	4.236
27	729	19,683	5.196	3.000	77	5,929	456,533	8.775	4.254
28	784	21,952	5.292	3.037	78	6,084	474,552	8.832	4.273
29	841	24,389	5.385	3.072	79	6,241	493,039	8.888	4.291
30	900	27,000	5.477	3.107	80	6,400	512,000	8.944	4.309
31	961	29,791	5.568	3.141	81	6,561	531,441	9.000	4.327
32	1,024	32,768	5.657	3.175	82	6,724	551,368	9.055	4.344
33	1,089	35,937	5.745	3.208	83	6,889	571,787	9.110	4.362
34	1,156	39,304	5.831	3.240	84	7,056	592,704	9.165	4.380
35	1,225	42,875	5.916	3.271	85	7,225	614,125	9.220	4.397
36	1,296	46,656	6.000	3.302	86	7,396	636,056	9.274	4.414
37	1,369	50,653	6.083	3.332	87	7,569	658,503	9.327	4.431
38	1,444	54,872	6.164	3.362	88	7,744	681,472	9.381	4.448
39	1,521	59,319	6.245	3.391	89	7,921	704,969	9.434	4.465
40	1,600	64,000	6.325	3.420	90	8,100	729,000	9.487	4.481
41	1,681	68,921	6.403	3.448	91	8,281	753,571	9.539	4.498
42	1,764	74,088	6.481	3.476	92	8,464	778,688	9.592	4.514
43	1,849	79,507	6.557	3.503	93	8,649	804,357	9.644	4.531
44	1,936	85,184	6.633	3.530	94	8,836	830,584	9.695	4.547
45	2,025	91,125	6.708	3.557	95	9,025	857,375	9.747	4.563
46	2,116	97,336	6.782	3.583	96	9,216	884,736	9.798	4.579
47	2,209	103,823	6.856	3.609	97	9,409	912,673	9.849	4.595
48	2,304	110,592	6.928	3.634	98	9,604	941,192	9.899	4.610
49	2,401	117,649	7.000	3.659	99	9,801	970,299	9.950	4.626
50	2,500	125,000	7.071	3.684	100	10,000	1,000,000	10.000	4.642

GLOSSARY

Abscissa The first number, x, in an ordered pair (x, y) is the *abscissa*. (Page 74)

Absolute value The *absolute value* of x, written $|x|$, is x for $x \geq 0$ and $-x$ for $x < 0$. (Page 10)

Angle An *angle* is a rotation of a ray about its endpoint from an initial position to a terminal position. (Page 540)

Angle of depression When a surveying instrument is turned from the horizontal position down to the object being sighted, it turns through an *angle of depression.* (Page 565)

Angle of elevation When a surveying instrument is turned from the horizontal position up to the object being sighted, it turns through an *angle of elevation*. (Page 564)

Antilogarithm If $L = \log_a N$, then N is the *antilogarithm* of L. (Page 423)

Arithmetic means The terms between any two given terms of an arithmetic sequence are called *arithmetic means*. (Page 447)

Arithmetic sequence An *arithmetic sequence* is a sequence in which the difference between successive terms is a constant. (Page 446)

Arithmetic series An *arithmetic series* is the indicated sum of an arithmetic sequence. (Page 454)

Asymptotes Lines that a curve nears but does not intersect are called *asymptotes*. (Page 599)

Characteristic In a logarithm, the *characteristic* is the power of 10 by which that number is multiplied when it is expressed in scientific notation. (Page 427)

Circle A *circle* is the locus of points in a plane at a given distance from a fixed point. (Page 358)

Combination An r-element subset of a set having n elements is a *combination* of the n elements taken r at a time. (Page 518)

Common difference The difference obtained by subtracting any term in an arithmetic sequence from the following term is a constant called the *common difference*. (Page 446)

Complex conjugate The numbers $a + bi$ and $a - bi$ are called *complex conjugates.* (Page 323)

Complex number A *complex number* is a number of the form $a + bi$, where a and b are real numbers and $i = \sqrt{-1}$. (Page 318)

Composite functions If g and f are functions such that the range of f is in the domain of g, then the *composite function* $g \circ f$ can be described as $[g \circ f](x) = g(f(x))$. (Page 174)

Conic section *Conic sections* are formed by the intersection of a plane and a right circular cone. (Page 358)

Consistent system A *consistent system* of equations or inequalities is one whose solution set contains at least one ordered pair. (Page 128)

Coordinates The *coordinate* of a point on a line is the number associated with the point. The *coordinates* of a point in a plane are the numbers in the ordered pair associated with the point. (Page 74)

Cosecant ratio Let a point $P(x, y)$ be on the terminal side of an angle with measure θ in standard position and let r be the radius vector. Then *cosecant* θ (*csc* θ) $= \frac{r}{y}$, $y \neq 0$. (Page 563)

Cosine ratio Let a point $P(x, y)$ be on the terminal side of an angle with measure θ in standard position and let r be the radius vector. Then *cosine* θ (*cos* θ) $= \frac{x}{r}$. (Page 542)

Cotangent ratio Let a point $P(x, y)$ be on the terminal side of an angle with measure θ in standard position and let r be the radius vector. Then *cotangent* θ (*cot* θ) $= \frac{x}{y}$, $y \neq 0$. (Page 563)

Coterminal angles Angles in standard position whose initial and terminal sides coincide are *coterminal angles*. (Page 540)

Degree of a monomial The *degree of a monomial* is the sum of the exponents of its variables. (Page 188)

Degree of a polynomial The *degree of a polynomial* is the greatest degree of its terms. (Page 188)

Dependent system A system of linear equations whose solution set is infinite is called a *dependent system*. (Page 128)

Determinant For all real numbers a_1, a_2, b_1, b_2, $\begin{vmatrix} a_1 & b_1 \\ a_2 & b_2 \end{vmatrix} = a_1b_2 - a_2b_1$. The array of numbers set off with bars is called a *determinant*. (Page 134)

Direct variation A *direct variation*, or a *linear variation*, is defined by an equation of the form $y = mx$ or $y = kx$, where m (or k) is a nonzero constant. (Page 98)

Discriminant The *discriminant* of the quadratic equation $ax^2 + bx + c = 0$ is the number $b^2 - 4ac$. (Page 341)

Domain For a relation, the set of first coordinates is the *domain* of the relation. (Page 160)

Ellipse An *ellipse* is the locus of all points in a plane such that the sum of the distances from two fixed points to a point on the locus is constant. (Page 362)

Equivalent equations *Equivalent equations* are equations that have the same solution. (Page 18)

Equivalent systems Systems of equations that have the same solution set are *equivalent systems*. (Page 117)

Equivalent vectors Vectors with the same magnitude and direction are called *equivalent vectors*. (Page 582)

Event The set of successful outcomes is a subset of the sample space. The subset is called an *event*. (Page 523)

Exponential function An *exponential function* with base a is a function defined by an equation of the form $y = a^x$, where a and x are real numbers and $a \neq 1$, and $a > 0$. (Page 415)

Factorial The symbol $n!$ is read "*n factorial*." $n! = {}_nP_n = n(n-1)(n-2)(n-3) \cdots 3 \cdot 2 \cdot 1$. (Page 512)

Function A *function* is a relation such that for every element in the domain there is exactly one element of the range. (Page 161)

General form of the linear equation Any line in the coordinate plane can be written in the form $Ax + By = C$, where A, B, and C are real numbers and A and B are not both zero. Also, $Ax + By = C$, where A, B, and C are real numbers and A and B are not both zero, is the equation of a line in the coordinate plane. (Page 95)

Geometric means The terms between any two given terms of a geometric sequence are called the *geometric means*. (Page 451)

Geometric sequence A *geometric sequence* is one in which the ratio r of successive terms is always the same number. (Page 450)

Geometric series A *geometric series* is the indicated sum of the terms of a geometric sequence. (Page 459)

Greatest common factor The *greatest common factor*, GCF, of two expressions is the greatest expression that is a factor of both expressions. (Page 197)

Hyperbola A *hyperbola* is the locus of all points in a plane such that the difference of the distances from two fixed points to a point of the locus is a constant. (Page 371)

Identity matrix The *identity matrix*, I_n, of order n is the square matrix whose elements in the main diagonal, from upper left to lower right, are ones. All other elements are zeros. (Page 501)

Imaginary number Any complex number for which $b \neq 0$ is an *imaginary number*. (Page 318)

Inconsistent system An *inconsistent system* of equations or inequalities is one whose solution set is the empty set. (Page 128)

Integers The set of *integers* is the union of the set of whole numbers with the additive inverses, or opposites, of the natural numbers. (Page 6)

Inverse relation Relations Q and S are *inverse relations* if and only if for every ordered pair (x, y) in Q, there is an ordered pair (y, x) in S. (Page 178)

Inverse variation If the product of two variables is a nonzero constant, the variables are an *inverse variation*. The equation for the inverse variation may be written as $xy = k$ or $y = \frac{k}{x}$, where k is the constant of variation. (Page 377)

Irrational number A number that cannot be written as the ratio of an integer and a natural number is an *irrational number*. Irrational numbers cannot be represented as terminating or nonterminating repeating decimals. (Page 7)

Joint variation *Joint variation* is expressed by the equation $x = kyz$, where x, y, and z are variables and k is a constant. This equation means that x varies jointly as y and z. (Page 378)

Law of Cosines In any triangle, the square of a side is equal to the sum of the squares of the other two sides minus twice the product of these two sides and the cosine of their included angle. (Page 572)

Law of Sines In any triangle, the ratio of the length of a side to the sine of the angle opposite that side is the same for each side-angle pair. (Page 568)

Least common denominator The *least common denominator*, LCD, of two or more rational expressions is the product of their greatest common factor and the other factors. (Page 238)

Linear function A function that can be expressed in the form $y = mx + b$ or $f(x) = mx + b$ is a *linear function*. (Page 166)

Logarithmic function The inverse of the exponential function $y = 10^x$ is the function $y = \log_{10} x$, and is called the *logarithmic function*. (Page 418)

Mantissa The *mantissa* is the logarithm of a number between 1 and 10. (Page 427)

Mapping A *mapping* is a correspondence that associates the elements of two sets. (Page 165)

Matrix A *matrix* is a rectangular array of numbers. (Page 490)

Monomial A *monomial* in the variable x is an expression of the form ax^n, where a is a real number and n is a whole number. (Page 188)

Mutually exclusive events Events are *mutually exclusive* if they have no elements in common. (Page 526)

Natural number The set of *natural numbers* is the set of numbers whose members are 1 and every number found by adding 1 to a member of the set. (Page 6)

Open sentence Sentences with variables are *open sentences*. They cannot be classified as true or false until you know which numbers can replace the variables. (Page 18)

Ordinate The second number, y, in an ordered pair (x, y) is the *ordinate*. (Page 74)

Parabola A *parabola* is the locus of points in a plane that are the same distance from a given line and a fixed point not on the line. (Page 367)

Periodic function If there is a smallest positive number p such that $f(p + x) = f(x)$ for every x in the domain of f, then p is the *period* of the function f and f is a *periodic function*. (Page 596)

Permutation A *permutation* of a set of elements is an arrangement of a specified number of those elements in a definite order. (Page 512)

Polar Form of a Complex Number The expression $r(\cos\theta + i \sin\theta)$ is the *polar form* of the complex number $a + bi$. (Page 621)

Polynomial A *polynomial* is a monomial or the sum of monomials. (Page 188)

Prime polynomial A polynomial that has no polynomial factors other than itself, its opposite, and 1, is a *prime polynomial*. (Page 197)

Principal nth root For any real number a and any integer n, $n > 1$,

a. $\sqrt[n]{a}$ is a positive real number or 0 when a is positive or equal to 0.

b. $\sqrt[n]{a}$ is a negative number when a is negative and n is odd. (Page 270)

Probability The ratio of the number of successful outcomes to the number of possible outcomes is the *probability* that an outcome will occur. (Page 522)

Pure imaginary number A *pure imaginary number*

is a square root of a negative number. (Page 314)

Quadrantal angle An angle whose terminal side is on the x axis or the y axis is a *quadrantal angle.* (Page 540)

Quadratic equation An equation that can be written in the form $ax^2 + bx + c = 0$, where a, b, and c are complex numbers and $a \neq 0$, is a *quadratic equation*. (Page 332)

Quadratic formula For any quadratic equation $ax^2 + bx + c = 0$, the solutions are $x = \frac{-b \pm \sqrt{b^2 - 4ac}}{2a}$. This is called the *quadratic formula*. (Page 332)

Quadratic function A *quadratic function* is a function that may be defined by $y = ax^2 + bx + c$, where a, b, and c are real number constants and $a \neq 0$. (Page 287)

Radian measure An angle has a measure of *one radian* if, when its vertex is placed at the center of a circle, it intercepts an arc equal in length to the radius of the circle. (Page 589)

Radical The expression $\sqrt{25}$ is called a *radical.* (Page 270)

Radicand In the expression $\sqrt{25}$, 25 is the *radicand*. (Page 270)

Radius vector In trigonometry, the distance from the origin to a point on the terminal side of an angle in standard position is the *radius vector.* (Page 542)

Range The set of second elements of a relation is the *range*. (Page 160)

Rational expression A *rational expression* is an expression of the form $\frac{P}{Q}$, where P and Q are polynomials and $Q \neq 0$. (Page 230)

Rational numbers A *rational number* is a number that can be expressed in the form $\frac{a}{b}$, where a is an integer and b is a natural number. (Page 6)

Real numbers The union of the set of rational numbers with the set of irrational numbers is the set of *real numbers*. (Page 7)

Reference triangle The triangle formed by drawing a perpendicular from a point on the radius vector to the x axis is a *reference triangle*. (Page 542)

Relation A *relation* is a set of ordered pairs. (Page 160)

Replacement set The *replacement set* for the variables in an algebraic expression or in a formula may contain both positive and negative real numbers. (Page 10)

Sample space In probability, the set of possible outcomes is the *sample space*. (Page 522)

Scientific notation A number is written in *scientific notation* when it is written as the product of a number greater than or equal to 1 but less than 10 and an integral power of 10. (Page 403)

Secant ratio Let a point $P(x, y)$ be on the terminal side of an angle with measure θ in standard position and let r be the radius vector. Then *secant* θ ($\sec \theta$) $= \frac{r}{x}$, $x \neq 0$. (Page 563)

Sequence A *sequence* is a function whose domain is the set of positive integers. (Page 444)

Series A *series* is the indicated sum of the terms of a sequence. (Page 454)

Sine ratio Let a point $P(x, y)$ be on the terminal side of an angle with measure θ in standard position and let r be the radius vector. Then *sine* θ ($\sin \theta$) $= \frac{y}{r}$. (Page 542)

Slope The *slope*, m, of a nonvertical line that contains the points $P_1(x_1, y_1)$ and $P_2(x_2, y_2)$ is $m = \frac{y_2 - y_1}{x_2 - x_2}$ or $m = \frac{y_1 - y_2}{x_1 - x_2}$. (Page 80)

Solution set The *solution set* of an open sentence is the set of numbers from the replacement set (or domain) of the variable that makes the sentence true. (Page 18)

Solution set of a system The *solution set of a system of sentences* is the set of all ordered pairs that makes all the sentences true. (Page 114)

Standard position An angle is in *standard position*

in the coordinate plane when its vertex is at the origin and its initial side is on the positive x axis. (Page 540)

System of equations A set of two or more equations is called a *system of equations*. (Page 114)

Tangent ratio Let a point $P(x, y)$ be on the terminal side of an angle with measure θ in standard position and let r be the radius vector. Then *tangent* θ (*tan* θ) $= \frac{y}{x}$, $x \neq 0$. (Page 542)

Variable A *variable* is a symbol that represents any element of a specified replacement set. (Page 2)

Vector A line segment with both direction and magnitude is called a *vector*. (Page 582)

Whole numbers The set of *whole numbers* is the set that is the union of the set of natural numbers and zero. (Page 6)

Zero of a function For a given function f, any value of x in the domain of f which satisfies the equation $f(x) = 0$ is a *zero* of f. (Page 340)

INDEX

Boldfaced *numerals indicate the pages that contain formal or informal definitions.*

Answers to Selected Exercises

CHAPTER 1 REAL NUMBERS

Page 3 Classroom Exercises 1. 23 3. 51 5. 9 7. $1\frac{11}{16}$ 9. $-\frac{255}{256}$ 11. 1 13. $\frac{1}{2}$ 15. 1 17. $\frac{7}{72}$ 19. 2.24

Page 3 Written Exercises 1. 15 3. 63 5. $11\frac{3}{4}$ 7. $24\frac{1}{2}$ 9. 10 11. 729 13. 5 15. 11 17. 32 19. 152 21. −480 23. 32 25. 969 27. 188 29. −51 31. $87\frac{1}{2}$ 33. $\frac{1}{512}$ 35. \$7885 37. \$6619 39. \$5796 41. 189 43. 161 45. 191 47. 80 ft 49. 144 ft 51. 80 ft 53. 144 ft 55. 0.18 57. 0.18 59. \$1210 61. 819,200

Page 5 Puzzle See Solution Key.

Page 8 Classroom Exercises 1. T 3. T 5. NT 7. $-2\frac{2}{3} \in Q, R$ 9. $17.08 \in Q, R$ 11. $\pi \in Ir, R$ 13. $\{-1, 0, 1, 2$ 15. $\{x : x \geq 1\}, x \in R$

Page 8 Written Exercises 1. $0.4\overline{5}$; T 3. $0.\overline{5}$; NT 5. −0.875; T 7. $0.\overline{142857}$; NT 9. 1.5; T 11. $-5.\overline{3}$; NT 13. $0.\overline{27}$; NT 15. $0 \in W, I, Q, R$ 17. $\frac{1}{2} \in Q, R$ 19. $\sqrt{5} \in Ir, R$ 21. $1.8 \in Q, R$ 23. $\sqrt{3} \in Ir, R$ 25. $0.\overline{5} \in Q, R$ 27. $25\frac{1}{4} \in Q, R$

29. 

31. 

33. 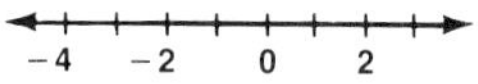

35. 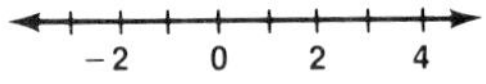

37. 

39. R 41. I 43. N 45. W 47. $\subset$ 49. $\not\subset$ 51. $\not\subset$ 53. $\not\subset$ 55. $\{-6, -4, 4\}$ 57. $\{2, 16\}$ 59. $\{-6, -4, -1\frac{3}{4}, 4, 3.\overline{14}\}$ 61. $\{0, 2, \pi, 16\}$ 63. $-6, -2.1, -1.75\}$ 65. $\{4, 3.\overline{14}, \sqrt{11}\}$ 67. $\{-6, -4\}$ 69. $\{\pi, \sqrt{11}\}$ 71. $\{\sqrt{11}\}$

Page 11 Classroom Exercises 1. 12 3. 12 5. 0 7. −6 9. −82 11. −1 13. 32 15. −32 17. −8 19. a + b 21. 360 23. 10 25. −55 27. 6 29. 12

Page 12 Written Exercises 1. 10 3. −21 5. 18 7. 3 9. 3 11. 13 13. −16 15. 20 17. 32 19. $\frac{2}{3}$ 21. $-1\frac{1}{8}$ 23. −2.3 25. $-11\frac{1}{8}$ 27. 9 29. 5.7 31. 68.93 33. 10 35. −48 37. 7.6 39. −16.5 41. −15 43. −27 45. −0.05 47. −45 49. 18 51. 9 53. 256 55. −1 57. 152 59. −28 61. 36 63. 33 km/hr 65. 42° 67. 4534 years 69. 5568 Btu/hr 71. −2552 Btu/hr 73. 6728 Btu/hr 75. A positive Btu indicates that heat is flowing from the inside to the outside. (Heat loss) A negative Btu indicates that heat is flowing from the outside to the inside. (Heat gain) 77. $-1\frac{3}{4}$ 79. −33 81. $-16\frac{5}{8}$ 83. $a = d = s^2; b = c = -s^2$ 85. $a = b = c = d = -\frac{rs}{t}$ 87. ∅ 89. $\{-1, -3\}$ 91. ∅ 93. ∅ 95. $\{-5, -4, -3, -2, -1, 0\}$ 97. $\{-1, 0, 1\}$

Page 15 Classroom Exercises 1. d 3. a 5. c 7. 21t 9. 16s + 5 11. −2q + 42 13. −13c − 9 15. −5y − 5 17. −4r + 10

Page 16 Written Exercises 1. Identity Post. for Add. 3. Mult. Inv. Post. 5. Identity Post. for Mult. 7. Assoc. Post. for Mult. 9. Identity Post. for Add. 11. Mult. Inv. Post. 13. 19s 15. $10x^2$ 17. −14x 19. −2b 21. 31t + 2 23. −7x − 8 25. 10a + 10b 27. −3x − 3y 29. 15p + 18q 31. −8a − b 33. 2y + 12 35. b − 2 37. m − 1 39. −4p + 11 41. f + 17 43. −64x + 41 45. a + 26 47. 0 49. No; $(24 \div 6) \div 2 \neq 24 \div (6 \div 2)$; $2 \neq 8$ 51. Yes; 0 53. Yes (Examples will vary.) 55. No (Examples will vary.) 57. No; 1 + 1 = 2

Page 17 Review 1. 299 2. T 3. NT 4. T 5. NT 6. NT 7. NT 8. 9. 6 10. −15 11. 52 12. 77 13. Add. Inv. Post. 14. Identity Post. for Mult. 15. Comm. Post. for Mult. 16. Mult. Inv. Post. 17. Identity Post. for Add. 18. Comm. Post. for Add.

Page 17 Calculator Exercises 1. −37 3. 344 5. 28.44 7. 240 9. 4108 11. −9274.625

Page 18 Review Capsule for Section 1-5 1. 12 2. −4 3. 2.7 4. $\frac{1}{4}$ 5. 5 6. 72 7. −20 8. −21

Page 20 Classroom Exercises 1. ∅ 3. 3; $\frac{1}{7}$ 5. −16; $\frac{1}{7}$ 7. 3; 4 9. 3; 4 11. $\{16\}$ 13. $\{3\}$

Page 20 Written Exercises 1. $\{5\}$ 3. $\{1\}$ 5. $\{0\}$ 7. ∅ 9. ∅ 11. $\{-2, 2\}$ 13. $\{-1, 1\}$ 15. $\{-2\}$ 17. $\{0.65\}$ 19. $\{3\}$ 21. $\{2\frac{1}{6}\}$ 23. $\{-12\}$ 25. $\{-12\}$ 27. $\{1.5\}$ 29. $\{18\}$ 31. $\{1.6\}$ 33. $\{-7\}$ 35. $\{4\frac{1}{2}\}$ 37. $\{-4\}$ 39. $\{-5\}$ 41. $\{11\}$ 43. $\{-5\frac{4}{5}\}$ 45. $\{1\frac{10}{11}\}$ 47. $\{4\}$ 49. $\{\frac{5}{6}\}$ 51. $\{0.2\}$ 53. $\{-\frac{1}{2}\}$ 55. $\{1\}$ 57. $\{-3, 0\}$ 59. ∅ 61. ∅ 63. ∅ 65. ∅ 67. $\{-\frac{1}{10}\}$ 69. $\{0\}$

Page 21 Calculator Exercises 1. $\{-5\}$ 3. $\{14\}$ 5. $\{-6\}$

Page 23 Classroom Exercises 1. j − 12 3. n + 20 5. $\frac{2}{3}v + 1$ 7. $\frac{100}{n}$ 9. 3a − 500

Page 24 Word Problems 1. 2s + 6 3. r − 3 5. 150c 7. 3v + 20 9. $\frac{c}{21}$ 11. Coach: \$375; 1st class: \$625 13. 9 oz 15. \$59.84 17. Population: 31,137 19. 37 cm × 37 cm × 22 cm 21. 66°; 33°; 81° 23. 20 years old 25. 4 cm; 8 cm; 5 cm 27. \$20 29. Each side of each parcel is 700 feet.

Page 27 Classroom Exercises 1. 18 − t 3. 18 − 5 5. 21 + y 7. 98 9. t + 1; t + 2; t + 3 11. t − 2; t − 4

Page 28 Word Problems 1. Jerry is 40 and Laura is 10. 3. Eva is 24; son is 6. 5. −22; −23 7. 73; 74; 75 9. Length: 10 m; width: 9 m 11. 671 km; 673 km; 675 km 13. t = 11 years 15. 23; 25; 27 17. 5; 7; 9 19. No 21. 751 tickets; 753 tickets

Page 29 Puzzle See Solution Key.

Page 29 Review Capsule for Section 1-8 1. Comm. Post. for Add. 2. Identity Post. for Mult. 3. Add. Inv. Post. 4. Assoc. for Mult. 5. Identity Post. for Add. 6. Add. Inv. Post. 7. Distributive 8. Assoc. Post. for Mult.

Page 31 Classroom Exercises **1.** Reasons: 1. Given; 2. Add. Prop. for Equality; 3. Assoc. Post. for Add.; 4. Add. Inv. Post.; 5. Identity for Add.

Page 32 Written Exercises **1.** Reasons: 1. Given; 2. Def. of Add. Inv.; 3. Def. of Sub.; 4. Comm. Post. for Add.; 5. Ident. Post. for Add.; 6. Transitive Post. **3.** Reasons: 1. Given; 2. Def. of Mult. Inv.; 3. Closure Post. for Mult., Assoc. Post. for Mult.; 4. Inv. Post. for Mult.; 5. Identity Post. for Mult.; 6. Transitive Post. **5.** Reasons: 1. Given; 2. Closure Post. for Add.; 3. Def. of Add. Inv.; 4. Closure Post. for Add.; 5. Assoc. Post. for Add.; 6. Add. Inv. Post.; 7. Identity Post. for Add.; 8. Transitive Post. **7.** Reasons: 1. Given; 2. Def. of Add. Inv.; 3. Add. Inv. Post.; 4. Identity Post. for Add.; 5. Substitution; 6. Assoc. Post. for Add.; 7. Add. Inv. Post.; 8. Substitution; 9. Identity Post. for Add.; 10. Transitive Post. **9.** Reasons: 1. Given; 2. Def. of Add. Inv.; 3. Distributive; 4. Ex 7; 5. Subst. In Exercises 11-20, statements are followed by the reasons. **11.** 1. a is an even number, b is odd. (Given) 2. Let a = 2k (Def. of even number) 3. Let b = 2n + 1 (Def. of odd number); 4. a + b = 2k + 2n + 1 (Add. Prop. for Equations); 5. a + b = 2(k + n) + 1 (Distributive); 6. 2(k + n) + 1 is an odd number. (Def. of an odd number); 7. a + b is an odd number. (Transitive Post.)
13. 1. a + c = b + d, c = d (Given); 2. a + d = b + d (Subst.); 3. (a + d) + (−d) = (b + d) + (−d) (Add. Prop. for Eqs.); 4. a + (d + (−d)) = b + (d + (−d)) (Assoc. Post. for Add.); 5. a + 0 = b + 0 (Add. Inv. Post.); 6. a = b (Ident. Post. for Add.)
15. 1. x + 2 = 9 (Given); 2. (x + 2) + (−2) = 9 + (−2) (Add. Prop. for Eqs.); 3. x + (2 + (−2)) = 9 + (−2) (Assoc. Post. for Add.); 4. x + 0 = 9 + (−2) (Add. Inv. Post.); 5. x = 9 + (−2) (Identity Post. for Add.); 6. x = 7 (Add. of Integers)
17. 1. x = a + (−b) (Given); 2. x + b = [a + (−b)] + b (Assoc. Prop. for Add.); 3. x + b = a + [(−b) + b] (Assoc. Prop. for Add.); 4. x + b = a + 0 (Add. Inv. Post.); 5. x + b = a (Ident. Post. for Add.) **19.** 1. a = c • b (Given); 2. $\frac{1}{b} \in R$ (Def. of Mult. Inv.); 3. $\frac{1}{b} \cdot a = (c \cdot b) \cdot \frac{1}{b}$ (Mult. Prop. for Eqs.); 4. $\frac{1 \cdot a}{b} = c \cdot (b \cdot \frac{1}{b})$ (Assoc. Post. for Mult.); 5. $\frac{1 \cdot a}{b} = c \cdot 1$ (Mult. Inv. Post.); 6. $\frac{a}{b} = c$ (Ident. Post. for Mult.) **21.** 1. a = b, c = d (Given); 2. a + c = b + c (Add. Prop. for Eqs.); 3. a + c = b + d (Substitution)

Page 36 Computer Exercises See Solution Key.

Page 37 Review 1. 12 2. 6 3. −2 4. −3 5. $2\frac{1}{2}$ 6. 2 7. \$0.05 8. Jake: 15; Jeff: 17 9. Reasons: 2. Def.; Add. Post. of Eqs.; 3. Assoc., for Add.; 4. Distributive; 5. Closure, for Add.; 6. Def. of odd integer; 7. Transitive Post.

Page 38 Chapter Objectives and Review 1. $14\frac{1}{120}$ 3. 25 5. 80 ft/sec 7. 48 ft/sec 9. 0 ft/sec 11. $0.8\overline{3}$; NT 13. $0.\overline{428571}$; NT 15. $-0.\overline{45}$; NT 17. −17 ∈ I, Q, R 19. $\sqrt{3} \in$ Ir, R 21. $0.\overline{51} \in$ Q, R 23. $\frac{0}{\pi} \in$ W, I, Q, R 25. −2 27. 6 29. −4 31. 16 33. −1.12 35. $2\frac{1}{4}$ 37. Mult. Inverse Post. 39. Comm. Post. for Add. 41. −5t 43. 2x 45. 4 47. −0.9 49. $-\frac{19}{25}$ 51. 4 yards 53. −48; −46; 44

Page 40 Chapter Test 1. 219 3. 144 5. 0.292929 · · · ∈ Q, R 7. $\pi(\sqrt{2} + 3) \in$ Ir, R 9. −132 11. Distributive 13. Identity Post. for Add. 15. {8} 17. {8} 19. 79; 81; 83

Page 41 Preparing for College Entrance Tests 1. a 3. b 5. b 7. c 9. a 11. c 13. b

CHAPTER 2 EQUATIONS AND INEQUALITIES

Page 44 Classroom Exercises 1. $y = \frac{g - f}{e}$, e ≠ 0 3. $y = \frac{c - d - 1}{a + b}$, a ≠ −b 5. E = IR 7. $n = \frac{s + 360}{180}$ 9. $h = \frac{2A}{a + b}$, a ≠ −b

Page 45 Written Exercises 1. $x = \frac{2c}{a - b}$, a ≠ b 3. $x = \frac{c}{rt}$, r ≠ 0, t ≠ 0 5. $x = \frac{t - s}{r}$, r ≠ 0 7. $x = \frac{m^2 - 9}{m}$, m ≠ 0 9. $x = \frac{c}{a - b}$, a ≠ b 11. x = b(d − f − c) 13. $x = \frac{ae}{5}$ 15. $x = \frac{9h}{f}$, f ≠ 0 17. $x = \frac{r + s}{b}$, b ≠ 0 19. $g = \frac{2s}{t^2}$, t ≠ 0

21. $p=\frac{A}{1+rt}$, $rt \neq -1$ 23. $a=S(1-r)$ 25. $h=\frac{3V}{\pi r^2}$, $r \neq 0$ 27. $v=\frac{2S+gt^2}{2t}$, $t \neq 0$ 29. $\ell=\frac{2S-an}{n}$, $n \neq 0$ 31. $r=\frac{E^2t}{JWh}$, $J \neq 0, W \neq 0, h \neq 0$ 33. $p=\frac{ra}{v^2h}$, $v \neq 0, h \neq 0$ 35. $E=\frac{RM}{I}$, $I \neq 0$ 37. $R=\frac{W}{I^2}$, $I \neq 0$ 39. $a=\frac{c-2000}{12.5}$

41. $v=\frac{s+16t^2}{t}$, $t \neq 0$ 43. $A=\frac{rS}{1780d}$, $d \neq 0$ 45. $F=\frac{r^2}{30b}$, $b \neq 0$ 47. $F=\frac{9}{5}C+32$ 49. $b_2=\frac{2A-hb_1}{h}$, $h \neq 0$ 51. $g=\frac{2s}{t^2}$, $t \neq 0$ 53. $n=\frac{S+360}{180}$ 55. $t=\frac{A-p}{pr}$, $p \neq 0, r \neq 0$ 57. $i=\frac{H+1.13Ao}{1.13A}$, $A \neq 0$ 59. $I=\frac{TB(n+1)}{24}$ 61. $s=\frac{4D}{\pi b^2 n}$, $b \neq 0, n \neq 0$ 63. $S=\frac{F+P-RP}{R-1}$ 65. 1. $a, b, c \in R, a \neq 0, ax+b=c$ (Given); 2. $-b \in R$ (Def. of Add. Inv.); 3. $ax+b+(-b)=c+(-b)$ (Add. Prop. of Equality); 4. $ax+[b+(-b)]=c+(-b)$ (Assoc. Post. for Add.); 5. $ax+0=c+(-b)$ (Add. Inverse Post.); 6. $ax=c+(-b)$ (Ident. Post. for Add.); 7. $ax=c-b$ (Def. of Sub.); 8. $\frac{1}{a} \in R$ (Def. of Mult. Inv.); 9. $\frac{1}{a}(ax)=\frac{1}{a}(c-b)$ (Mult. Prop. of Equality); 10. $(\frac{1}{a} \cdot a)x=\frac{1}{a}(c-b)$ (Assoc. Post. for Mult.); 11. $1 \cdot x = \frac{1}{a}(c-b)$ (Mult. Inv. Post.); 12. $x=\frac{c-b}{a}$ (Ident. Post. for Mult. and Def. of Division) 67. 1. $a, b \in R; a \neq 0; ax+b=0$ (Given); 2. $-b \in R$ (Def. of Add. Inv.); 3. $(ax+b)+(-b)=0+(-b)$ (Add. Prop. for Equations); 4. $ax+[b+(-b)]=0+(-b)$ (Assoc. Post. for Add.); 5. $ax+0=0+(-b)$ (Add. Inv. Post.); 6. $ax=-b$ (Ident. Post. for Add.)

Page 47 Review Capsule for Section 2-2 1. 3.8 2. 4 3. −3 4. −7 5. $\{2, -2\}$ 6. $\{12, -12\}$ 7. $\{\frac{1}{2}, -\frac{1}{2}\}$ 8. $\{0\}$ 9. $\{3, -3\}$ 10. $\{24, -24\}$ 11. $\{3, -3\}$ 12. $\{3, -3\}$

Page 49 Classroom Exercises 1. $x=9$ or $x=-9$ 3. $n+7=3$ or $n+7=-3$ 5. $7t=35$ or $7t=-35$ 7. $6d=\frac{3}{2}$ or $6d=-\frac{3}{2}$ 9. $\{6, -4\}$ 11. $\{4, -12\}$ 13. $\{14, -18\}$ 15. $\{\frac{1}{2}, -\frac{1}{2}\}$ 17. $\{\frac{1}{5}\}$ 19. $\{-\frac{7}{10}\}$

Page 49 Written Exercises 1. $\{-2, 2\}$ 3. $\emptyset$ 5. $\{3, \frac{1}{3}\}$ 7. $\{\frac{10}{13}, -\frac{8}{3}\}$ 9. $\emptyset$ 11. $\{1, -1\}$ 13. $\{22, -26\}$ 15. $\{\frac{28}{3}, -\frac{32}{3}\}$ 17. $\{7\}$ 19. $\{2\}$ 21. $\emptyset$ 23. $\{3, 15\}$ The graphs can be found in the Solution Key. 25. $\{12, -36\}$ 27. $\emptyset$ 29. $\{\frac{1}{8}, -\frac{1}{8}\}$ 31. $\{-\frac{3}{5}, -\frac{6}{5}\}$ 33. $\{3, -\frac{3}{11}\}$ 35. $\{12, \frac{12}{13}\}$ 37. 1 39. If $x \geq 0$, $|x|=x$ and $|-x|=x$; Both ≥ 0. If $x<0$, $|x|=-x$ and $|-x|=-x$; Both >0. Thus, $|x|=|-x|$. 41. $\{2, \frac{4}{3}\}$ 43. $\emptyset$

Page 50 Review 1. $x=\frac{t}{3}$ 2. $x=-\frac{b}{c}$, $c \neq 0$ 3. $x=\frac{4p}{m+n}$, $m \neq -n$ 4. $x=\frac{4y}{3}$ 5. $x=\frac{t}{7}$ 6. $x=\frac{(p+c)m}{4}$ 7. $R=\frac{pV}{nT}$, $n \neq 0, T \neq 0$ 8. $r^2=\frac{A}{4\pi}$ 9. $R=\frac{R_1R_2}{R_1+R_2}$, $R_1=-R_2$ 10. $\{-\frac{20}{3}, 10\}$ 11. $\{1, -1\}$ 12. $\{2, \frac{2}{9}\}$

Page 50 Review Capsule for Section 2-3 The graphs can be found in the Solution Key. 1. All points to the right of 4. 2. All points to the left of 0. 3. All points to the right of $\frac{1}{2}$. 4. All points to the left of −2.5. 5. All points to the left of and including −1. 6. All points to the right of and including $6\frac{1}{3}$. 7. All points to the left of and including −0.5. 8. All points to the right of and including $-7\frac{2}{3}$.

Page 52 Classroom Exercises 1. $n+4+(-4)>16+(-4)$ 3. $r-\frac{1}{2}+(\frac{1}{2})<-2\frac{1}{2}+(\frac{1}{2})$ 5. $3t+5+(-5)>11+(-5)$ 7. $7a+\frac{2}{3}+(-\frac{2}{3})<10+(-\frac{2}{3})$ 9. Same 11. Reverse 13. Add (10) 15. Mult. $(-\frac{1}{5})$ 17. Mult. (−1)

Page 53 Written Exercises The graphs for Ex. 1-32 can be found in the Solution Key. 1. $s>12$ 3. $m<-3$ 5. $t>6\frac{2}{3}$ 7. $m<1$ 9. $p<-4$ 11. $q>0$ 13. $s>6$ 15. $w>-1\frac{1}{2}$ 17. $q \geq 2$ 19. $p \geq -6$ 21. $c \leq -1.7$ 23. $c \leq \frac{1}{2}$ 25. $t>1$ 27. $w \leq -6\frac{1}{2}$ 29. all real nos. 31. $n>\frac{1}{26}$ 33. $\{1, 2\}$ 35. 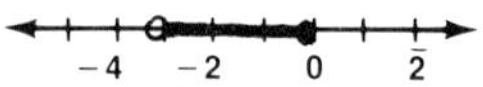

37. $\emptyset$ 39. 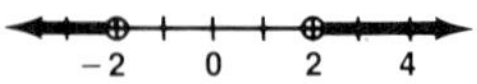

41. 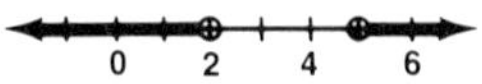

43. 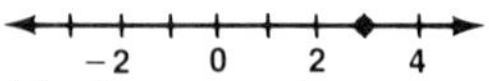

Page 55 Classroom Exercises 1. $w \geq 29$ 3. $2{,}500{,}000<s<3{,}125{,}000$ 5. $13<\ell<15$ 7. $s \geq 30$ 9. $500 \leq m \leq 10{,}000$

Page 56 Written Exercises 1. At least 20,000 3. A minimum of 86 5. At least .234 7. Maximum length: 24 meters 9. Charity: \$75,000; each grandchild: \$25,000 11. 24

Page 57 Review Capsule for Section 2-5 1. $\{5, -5\}$ 2. $\{7, -7\}$ 3. $\{12, -7\}$ 4. $\{5, -7\frac{1}{2}\}$ 5. $\emptyset$ 6. $\{4, -4\}$ 7. $\{7, -3\}$ 8. $\emptyset$

Page 58 Classroom Exercises 1. $y > 6$ or $y < -6$ 3. $2x - 1 > 5$ or $2x - 1 < -5$ 5. $\frac{z}{3} < 9$ and $\frac{z}{3} > -9$ 7. $t - 3 \leq 5$ and $t - 3z \geq 5$ 9. c 11. b

Page 58 Written Exercises 1. $x < 4$ and $x > -4$; C 3. $c - 3 \leq 5$ and $c - 3 \geq -5$; C 5. $2t - 3 \geq 7$ or $2t - 3 \leq -7$; D 7. $2a + 3 \leq 1$ and $2a + 3 \geq -1$; C 9. $\{n : n > 6 \text{ or } n < -6\}$ 11. $\{x : 2 < x < 4\}$ 13. $\{m : 2 < m < 10\}$ 15. $\{a : -2 \leq a \leq -1\}$ 17. $\{t : 1 \leq t \leq 4\}$ 19. $\{x : x > 1 \text{ or } x < -9\}$ 21. $\{d : d > 9 \text{ or } d < -6\}$ 23. $\{m : -2 \leq m \leq 12\}$ 25. $\{n : -3 < n < -\frac{1}{3}\}$ 27. $\{c : -2 \leq c \leq 1\frac{1}{3}\}$ 29. $\{y : -27 \leq y \leq 12\frac{3}{5}\}$ 31. $\{a : -1 \leq a \leq 7\}$ 33. $\{x : -1\frac{1}{3} < x < \frac{2}{3}\}$ 35. $\{x : -3 \leq x \leq 9\}$ 37. $|t| < 3$ 39. $|a| > 4$ 41. $\{x : x \leq -1 \text{ or } x \geq 1 \text{ and } -4 < x < 4\}$ 43. $\{x : x < -1 \text{ or } x > 5 \text{ and } -3 < x < 7\}$ 45. $\{y : y < -1 \text{ or } y > 4 \text{ and } -2 < y < 5\}$

Page 59 Puzzle See Solution Key.

Page 59 Review Capsule for Section 2-6 1. Add. Inv. Post. 2. Add. Prop. for Inequalities 3. Mult. Prop. for Ineq., $(c < 0)$ 4. Mult. Prop. for Ineq., $(c > 0)$ 5. Transitive Post. 6. Assoc. Post. for Mult. 7. Def. of Add. Inverse. 8. Add. Prop. of Equality

Page 60 Using Statistics 1. $-22°$ F, $-30°$ C 3. $-13°$ F, $-25°$ C 5. $-11°$ F, $-24°$ C 7. $-59°$ F, $-51°$ C 9. a. $-8.9°$ C b. $7.8°$ C 11. a. $-8.1°$ C b. $13.1°$ C 13. a. $-46°$ C b. $25.5°$ C

Page 64 Classroom Exercises 1. $= -1$; Def. of "less than" 3. $=$; Def. of "less than" 5. $= q$; Comparison Post. 7. $-19 + c = 0$; Def. of "less than" 9. $< t$, > 0; Def. of "less than"

Page 64 Written Exercises **1.** For real numbers a and b, if $a < b$ then $(b - a)$ is positive. For real numbers a and b, if $(b - a)$ is positive then $a < b$. **3.** For real numbers a and b, if $b < a$ then $-(b - a)$ is positive. For real numbers a and b, if $-(b - a)$ is positive then $b < a$. **5.** For real numbers a and b, if $a \geq b$ then $a > b$ or $a = b$. For real numbers a and b, if $a > b$ or $a = b$ then $a \geq b$. **7.** Given: $a, b \in R$; $a - b < 0$; Prove: $a < b$; Reasons: 1. Given; 2. Def. of subtr.; 3. Add. Prop. for Ineqs.; 4. Assoc. Post. for Add.; 5. Add. Inv. Post.; 6. Ident. Post. for Add. **9.** Given: $a \in R$; $a < 0$; Prove: $-a > 0$; Reasons: 1. Given; 2. Add. Prop. for Ineqs.; 3. Add. Inv. Post.; 4. Ident. Post. for Add. and Equivalent Statements **11.** Given: $a \in R$; $a \neq 0$; Prove: $a^2 > 0$; Reasons: 1. Given; 2. Comparison Post.; 3. Positive Product Theorem; 4. Positive Product Theorem; 5. Substitution **13.** Given: $a, b, c, d \in R$; $a > b$; $c > d$; Prove: $ac > bd$; Reasons: 1. Given; 2. Mult. Prop. for Ineq., $(c > 0)$; 3. Mult. Prop. for Ineq., $(c > 0)$; 4. Trans. Theorem for Order **15.** 1. $a, b \in R$; $a - b > 0$ (Given); 2. $(a - b) + b > 0 + b$ (Add. Prop. for Inequalities); 3. $a + [(-b) + b] > 0 + b$ (Assoc. Post. for Add.); 4. $a + 0 > 0 + b$ (Add. Inv. Post.); 5. $a > b$ (Ident. Post. for Add.) **17.** Not true. Examples will vary. **19.** Not true. Examples will vary. **21.** Given: $a \in R$; $a < 0$; Prove: $|a| = -a$; Proof: 1. $a \in R$; $a < 0$ (Given); 2. $a(-1) > 0(-1)$ (Mult. Prop. for Ineqs., $c < 0$); 3. $-a > 0$ (Ident. Post. for Mult. and Zero Mult. Prop.); 4. $|a| = -a$ (Def. of Absolute Value) **23.** Given: $a, b \in R$; $a > 0$; $b > 0$; Prove: $ab > 0$; Proof: 1. $a, b \in R$; $a > 0$; $b > 0$ (Given); 2. $a \cdot b > 0 \cdot b$ (Mult. Prop. for Ineqs., $c > 0$); 3. $ab > 0$ (Zero Mult. Prop.) **25.** Given: $a, b \in R$; $a < 0$; $b < 0$; $a < b$; Prove: $a^2 > b^2$; Proof: 1. $a, b \in R$; $a < 0$; $b < 0$; $a < b$ (Given); 2. $a \cdot a > b \cdot a$ (Mult. Prop. for Ineqs., $c < 0$); 3. $a \cdot a > a \cdot b$ (Comm. Post. for Mult.); 4. $a \cdot b > b \cdot b$ (Mult. Prop. for Ineqs., $c < 0$); 5. $a \cdot a > b \cdot b$ (Trans. Th. of Order); 6. $a^2 > b^2$ (Def. of Exponents) **27.** Given: $a, b, c \in R$; $a < b$; $c < 0$; Prove: $ac > bc$; Proof: 1. $a, b, c \in R$; $a < b$; $c < 0$ (Given); 2. $a + q = b$ (Def. of "Less Than"); 3. $(a + q)c = bc$ (Mult, Prop. for Eqs.); 4. $ac + qc = bc$ (Dist. Prop.); 5. qc is negative. (Negative Product Theorem); 6. $-qc$ is positive. (Def. of Opposites); 7. $(ac + qc) + (-qc) = bc + (-qc)$ (Add. Prop. for Eqs.); 8. $ac + (qc + (-qc)) = bc + (-qc)$ (Assoc. Post. for Add.); 9. $ac + 0 = bc + (-qc)$ (Add. Inv. Post.); 10. $ac = bc + (-qc)$ (Ident. Post. for Add.); 11. $bc + (-qc) = ac$ (Symmetric Prop. for Eqs.); 12. $bc < ac$, $(ac > bc)$ (Def. of "Less Than" and Equivalent Expressions)

Page 66 Review 1. $x < 3$ 2. $w \geq 1$ 3. $y \geq -9\frac{2}{3}$ 4. 4; 8; 24 5. Width: 6 m; Length: 10 m 6. $\{x : x < -1 \text{ or } x > 1\}$ 7. $\{t : -2 \leq t \leq 4\}$ 8. $\{y : -1\frac{3}{4} < y < 2\frac{1}{4}\}$ 9. $\{p : p \leq -4\frac{1}{2} \text{ or } p \geq 8\frac{1}{2}\}$ 10. 1. $a, b, c \in R$; $a + c < b + c$ (Given); 2. $(a + c) + (-c) < (b + c) + (-c)$ (Add. Prop. for Ineq.); 3. $a + [c + (-c)] < b + [c + (-c)]$ (Assoc. Post. for Add.); 4. $a + 0 < b + 0$ (Add. Inv. Post.); 5. $a < b$ (Identity Post. for Add.)

Page 66 Puzzle See Solution Key.

Page 69 Chapter Objectives and Review 1. $x = p - q + r$ 3. $x = \frac{u - v}{a}$; $a \neq 0$ 5. $h = \frac{v^2}{2g}$; $g \neq 0$ 7. $\{-27, 27\}$ 9. $\{\frac{52}{15}, -\frac{52}{15}\}$ 11. $\{-\frac{23}{3}, 13\}$ 13. $\{1, -\frac{27}{11}\}$ 15. $\{x : x > 7\}$ 17. $\{p : p \leq -\frac{1}{2}\}$ 19. Passenger: 85 cubic feet

Luggage: 10 cubic feet 21. Max. width: 12.1 cm, Max. length: 22.9 cm 23. $\{t : 3 < t < 11\}$ 25. $\{x : x \leq 2$ or $x \geq 4\frac{2}{3}\}$ 27. 1. $a, b, c \in R; \frac{1}{c} < 0; ac > bc$ (Given); 2. $(ac)\frac{1}{c} < (bc)\frac{1}{c}$ (Mult. Prop. for Ineq., $(c < 0)$); 3. $a(c \cdot \frac{1}{c}) < b(c \cdot \frac{1}{c})$ (Assoc. Post. for Mult.); 4. $a \cdot 1 < b \cdot 1$ (Mult. Inverse Post.); 5. $a < b$ (Ident. Post. for Mult.)

Page 71 Chapter Test 1. $x = \frac{\pi q - \pi t}{p}, p \neq 0$ 3. $v = 0.4$ 5. $\{2, -2\}$ 7. $\{x : x \geq 2\}$ 9. $\{t : -2\frac{1}{2} \leq t \leq 3\frac{1}{2}\}$

Page 72 Preparing for College Entrance Tests 1. b 3. d 5. b 7. d 9. b 11. c 13. e 15. b

CHAPTER 3 GRAPHING

Page 74 Classroom Exercises 1. Abscissa: −1; Ordinate: 5 3. Abscissa: 0; Ordinate: −8 5. Abscissa: 9; Ordinate: −2 7. III 9. II 11. y axis 13. x axis 15. x and y axes 17. (4, 1) 19. (0, −6) 21. (5, −6) 23. (−6, 0) 25. (4, 5) 27. (−6, 5)

Page 75 Written Exercises 1. 3 right, 5 up; I 3. 3 right, 2 down; IV 5. 6 right; No quadrant, x axis 7. Located at the origin. 9. 4 up; No quadrant, y axis 11. 4 left, $1\frac{1}{2}$ up; II 13. (3, 5) 15. (−7, 3) 17. $(-4, 5\frac{1}{2})$ 19. $(2\frac{1}{2}, -5\frac{1}{2})$ 21. (0, −4) 23. (0, 3) 25. (0, 0) 27. A straight line parallel to and 8 units above the x axis; A straight line parallel to and 2 units below the x axis; x axis 29. A straight line through (0, 0) and (1, 1) 31. D(2, 4) 33. D: (−4, 4), (0, 4), (−8, 0) 35. No point; Are parallel. 37. Two rays starting at the origin, one passing through (1, 1) and the other through (−1, 1).

Page 76 Puzzle See Solution Key.

Page 76 Review Capsule for Section 3-2 1. $a = p - b - c$ 2. $h = \frac{V}{\ell w}; \ell \neq 0, w \neq 0$ 3. $m = \frac{E}{c^2}; c \neq 0$ 4. $r = \frac{A - P}{Pt}$; $P \neq 0, t \neq 0$ 5. $T = \frac{E}{am} + t; a \neq 0, m \neq 0$ 6. $h = \frac{V}{\pi(R^2 - r^2)}; R^2 \neq r^2$

Page 78 Classroom Exercises 1. $y = 3x - 9$ 3. $y = -2x$ 5. No 7. Yes 9. Yes 11. Vertical 13. Horizontal 15. Horizontal

Page 78 Written Exercises 1. Line passing through (2, 1), (5, 3) 3. Line passing through (1, 9), (−1, −6) 5. Line passing through (2, 1), (5, −1) 7. Line passing through (2, 2), (−2, 0) 9. Line passing through (3, 1), (0, −1) 11. Line passing through (6, 1), (2, 2) 13. Graph would be V-shaped with the vertex of the V at $(-\frac{1}{2}, 0)$. The ray on the right would have the equation $y = 2x + 1$ and pass through another point (1, 3). The other ray would have the equation $y = -2x - 1$ and pass through (−2, 3). 15. Graph would be V-shaped with the vertex at (0, 0). The ray on the right would have the equation $y = 5x$ and pass through another point (1, 5). The other ray would have the equation $y = -5x$ and pass through the point (1, −5). 17. $k = 6$ 19. $k = 3$ 21. $k = 2$ 23. $3x - y = 3$, A line passing through (1, 0), (2, 3). $-x + y = 1$, A line passing through (1, 2), (−1, 0), intersection (2, 3) 25. $2y - x = 3$, A line passing through (1, 2), (−1, 1). $y - 5 = 0$, A line passing through (0, 5), (1, 5), intersection at (7, 5)

Page 79 Calculator Exercises 1. 1026173.1 3. 1.0108849

Page 79 Review Capsule for Section 3-3 1. 3 2. 7 3. 7 4. 0 5. −2 6. 3 7. $-x - 2$ 8. $-\frac{5}{2}$ 9. $\frac{-3}{2x + 1}$ 10. $\frac{5}{2x + 1}$

Page 81 Classroom Exercises 1. $\frac{5}{3}$ 3. 0 5. $-\frac{1}{2}$ 7. −3 9. Not defined. 11. Positive slope

Page 82 Written Exercises 1. $m = 1$ 3. $m = -3$ 5. $m = \frac{1}{2}$ 7. $m = \frac{2}{7}$ 9. $m_1 = \frac{1}{6}; m_2 = \frac{5}{7}; m_3 = \frac{5}{3}; m_4 = 3; m_5$ is undefined; $m_6 = -3; m_7 = -\frac{6}{5}; m_8 = -\frac{2}{3}; m_9 = -\frac{1}{3}; m_{10} = -\frac{1}{7}$ 11. Negative; increasing 13. Position which coincides with y axis. 15. Line passing through (2, 1) and (6, 4). 17. Line passing through (5, −2) and (8, −4). 19. Line passing through (0, 0) and (7, −3). 21. Line passing through (−2, 4) and (1, 8). 23. Line passing through (5, 2) and parallel to y axis. 25. $x = \frac{11}{4}$ 27. $m_{AB} = m_{BC} = m_{AC} = \frac{3}{5}$ 29. m is undefined. 31. $m = \frac{4b}{a}$ 33. $x = 11$

Page 83 Puzzle See Solution Key.

Page 83 Review Capsule for Section 3-4 1. Line passing through (0, 0) and (2, 1). 2. Line passing through (0, −3) and (2, 1). 3. Line passing through (0, 5) and (2, 6). 4. Line passing through (0, 6) and (−9, 0).

Page 85 Classroom Exercises 1. $P = 16{,}000 - 1280t$ 3. $J = 500 + 75t$

Page 85 Word Problems 1. a. $V = 12{,}000 - 1800t$; b. Line passing through (0, 12,000) and (5, 3000); c. 5 years; d. 7 yrs. 3. a. $A = 20{,}000 + 3000t$; b. Line passing through (0, 23,000), (2, 26,000) and (5, 35,000); c. \$50,000; d. \$65,000; e. \$74,000

5. a. $N = 1{,}800{,}000 + 180{,}000t$; b. Line passing through (0, 1,800,000) and (3, 2,340,000); c. 2,000,000 messages; d. 3,000,000 messages 7. a. $c = 1 + 1.25m$; b. Line passing through (0, 1) and (4, 6); c. 3 miles; d. $12 9. a. $P = 30{,}000 + 1200t$; b. Line passing through (0, 30,000) and (2, 32,400); c. $1200; d. $4\frac{1}{6}$ years

Page 87 Review 1. 6 right, 2 up; I 2. 2 left, 4 up; II 3. 3 right, 1 up; I 4. 4 left, 3 down; III 5. Line passing through (0, 2) and (2, 5). 6. Line passing through (−1, 1) and (−5, 2). 7. Graph would be V-shaped with the vertex at $(-\frac{2}{3}, 0)$. The ray on the right would have the equation $y = 3x + 2$ and pass through another point (2, 8). The left ray would have the equation $y = -3x - 2$ and point (−2, 4). 8. $m = -\frac{4}{3}$ 9. $m = -\frac{9}{8}$ 10. $m = \frac{9}{2}$ 11. Line passing through (2, 3), (3, 7). 12. Line passing through (−1, 2), (0, −1). 13. Line passing through (2, −4), (5, −6). 14. $x = -44$ 15. $A = 40{,}000 + 7200t$; Line passing through (0, 40,000) and (2, 54,400); $76,000

Page 87 Review Capsule for Section 3-5 1. −1 2. −1 3. −1 4. −1 5. $-\frac{1}{8}$ 6. $\frac{9}{5}$ 7. $-\frac{5}{4h}$ 8. $\frac{q}{p}$

Page 88 Using Statistics 1. mean: 53; median: 54; Since they are so close, either measure can be used. 3. mean: 166; median: 170; The data is skewed, so the median is better.

Page 91 Classroom Exercises 1. $-\frac{1}{6}$ 2. $-\frac{5}{3}$ 3. $\frac{4}{5}$ 4. $\frac{1}{2}$ 5. $\frac{6}{5}$ 6. −4 7. $-\frac{4}{3}$ 8. 0

Page 91 Written Exercises 1. Parallel 3. Perpendicular 5. The slope of $\overline{PQ}$ and $\overline{AB}$ is $\frac{3}{4}$. Thus, the lines are parallel. 7. The slope of $\overline{DE}$ is $\frac{2}{9}$; and for $\overline{QT}$ it is $-\frac{9}{2}$. Since the slopes are negative reciprocals, $\overline{DE}$ and $\overline{QT}$ are perpendicular. 9. $x = 4\frac{1}{2}$ 11. $k = -3\frac{5}{7}$ 13. The slope of $\overline{AC}$ is $\frac{5}{4}$ and for $\overline{BD}$ it is $-\frac{4}{5}$. Since the slopes are negative reciprocals, $\overline{AC}$ and $\overline{BD}$ are perpendicular. 15. The slope of $\overline{PQ}$ and $\overline{RS}$ is $-\frac{2}{11}$. The slope of $\overline{QR}$ and $\overline{PS}$ is $\frac{4}{3}$. Thus, the opposite sides are parallel. 17. $m = \frac{1}{2}$ 19. $a = 3$ 21. Reasons: 1. Given; 2. One and only one line may be drawn from a point perpendicular to a line.; 3. If two parallel lines are cut by a transversal, corresponding angles are congruent.; 4. Perpendicular lines form right angles and definition of right triangle.; 5. If two right triangles have an acute angle of one equal to an acute angle of the other, the triangles are similar.; 6. Corresponding sides of similar triangles are proportional.; 7. Definition of slope and substitution.

Page 95 Classroom Exercises 1. $y - 11 = 3(x - 2)$ 3. $y - 3 = -\frac{1}{2}(x + 2)$ 5. −3; 1 7. −1; 0 9. $y = -5x + 1$ 11. $y = -\frac{1}{2}x$ 13. $2x + y = 7$ 15. $-3x + 2y = -1$

Page 96 Written Exercises 1. $y - 2 = 3(x - 5)$ 3. $y - 7 = -3(x + 2)$ 5. $y - 3 = -\frac{3}{2}(x + 2)$ 7. $y + 4 = 0$ 9. $y - 4 = -\frac{4}{3}(x + 2)$ 11. $y - 3 = x + 2$ 13. $y = -2x + 1$ 15. $y = -\frac{1}{2}x - 1$ 17. $y = 3x + \frac{1}{2}$ 19. $y = -\frac{1}{3}x + 1$ 21. $y = -2x + 4$, $m = -2$; y intercept (0, 4); Line passing through (0, 4) and (1, 2). 23. $y = \frac{1}{2}x$; $m = \frac{1}{2}$; (0, 0); Line passing through (0, 0) and (2, 1). 25. $y = -\frac{3}{4}x + 3$; $m = -\frac{3}{4}$; (0, 3); Line passing through (0, 3) and (4, 0). 27. $y = -\frac{1}{3}x$; $m = -\frac{1}{3}$; (0, 0); Line passing through (0, 0) and (3, −1). 29. $-5x + 2y = -13$ 31. $-7x + 12y = 11$ 33. $x + 5y = 0$ 35. $-3x + 5y = 0$ 37. P 39. D 41. $y = 2x + 8$, $y = 2x + 5$, same slope = 2, therefore the lines are parallel. 43. $y - 2x = 8$, $y = -\frac{1}{2}x + \frac{1}{2}$, $m_1 \cdot m_2 = (2)(-\frac{1}{2}) = -1$, therefore the lines are perpendicular. 45. $y = -\frac{2}{3}x$ 47. $y = -\frac{3}{2}x - 13$ 49. $x = 3$ 51. $x = -7$ 53. $y = \frac{1}{7}x - \frac{10}{7}$ 55. $y = \frac{1}{7}x + \frac{40}{7}$ 57. $y = 3x - 26$; $y = x - 4$; $y = \frac{1}{2}x + \frac{3}{2}$ 59. $mx - y = mx_1 - y_1$; $A = m$; $B = -1$; $C = mx_1 - y_1$ 61. $bx + ay = ab$; $A = b$; $B = a$; $C = ab$

Page 97 Review Capsule for Section 3-7 1. $y = 3x$ 2. Increase 3. Decrease 4. $y = -2x$ 5. Decrease 6. Increase

Page 100 Classroom Exercises 1. Yes 3. Yes 5. Yes 7. $-\frac{3}{2}$ 9. 3 11. $T = kn$ 13. $B = kA$ 15. $v = kT$

Page 100 Written Exercises 1. −16 3. $\frac{27}{5}$ 5. 2.1 7. $C = kL$ 9. $A = kh$ 11. $P = ks$ 13. a. $-\frac{3}{2}, -\frac{3}{2}, -\frac{3}{2}, -\frac{2}{3}$ b. Not a direct variation. 15. a. $5, \frac{1}{5}, 5, 5$ b. Not a direct variation. 17. $p = 1689.6$ 19. $\frac{5}{6}$ liter 21. 12.5 kg 23. 2.1 cm of snow 25. 565 people 27. $x = 15$ 29. $x = 4$ 31. $x = 3$ 33. $x = \frac{1}{8}$ 35. $x = \frac{8}{9}$ 37. $x = -14\frac{1}{2}$ 39. $x = 6$ 41. Decrease 43. Increase 45. If $\frac{x}{y} = \frac{a}{b}$, $xb = ay$. Then adding $(-ab)$ to both sides $xb + (-ab) = ay + (-ab)$, $xb - ab = ay - ab$. $b(x - a) = a(y - b)$; Let a and $(y - b)$ be the means and b and $(x - a)$ be the extremes. $\frac{x - a}{y - b} = \frac{a}{b}$ 47. If $\frac{x}{y} = \frac{a}{b}$, $xb = ay$. Let a and y be the extremes and x and b the means. $\frac{y}{x} = \frac{b}{a}$.

Page 102 **Calculator Exercises** 1. \$812.75 3. \$1797.07 5. \$3684.44

Page 102 **Review Capsule for Section 3-8** 1. $x < 3$; open circle at 3 and an arrow to the left 2. $x \le -9$; closed circle at -9 and an arrow to the left 3. $x > 9$; open circle at 9 and an arrow to the right 4. $x \le 3$; closed circle at 3 and an arrow to the left 5. $x \ge -7\frac{1}{2}$; closed circle at $-7\frac{1}{2}$ and an arrow to the right 6. $x \ge 9$; closed circle at 9 and an arrow to the right

Page 104 **Classroom Exercises** 1. (3, 8), (4, 2) 3. (−3, −6), (0, 0), $(\frac{1}{5}, -5)$ 5. O 7. C 9. $x \ge 2$ 11. $y \ge x$

Page 104 **Written Exercises** 1. Right 3. Above 5. Left 7. Above 9. Below 11. Above 13. Above 15. Below 17. Above 19. Below In Exercises 21-26, the graph is the intersection of the graphs of each pair of inequalities. 21. The set of points above the boundary $y = -2$ <u>and</u> to the left of the boundary $x = 3$. The boundaries are not included. 23. The set of points below and including the boundary $y = 4$ <u>and</u> above but not including the boundary $x + 3 = 6$. 25. The set of points in Quadrant III including the boundaries. Note that (0, 0) is not included. 27. A circle with center at (0, 0) and intersecting the axes at (4, 0), (0, 4), (−4, 0), and (0, −4). 29. A circle with center at (3, −4) and containing (3, 2), (−3, −4), (3, −10), and (9, −4). 31. The open half-plane below $y = -x$.

Page 105 **Review** 1. The slope of $\overline{ST}$ and $\overline{AB}$ is −3. Thus, $\overline{ST}$ and $\overline{AB}$ are parallel. 2. $x = 4$ 3. $y = 3x + 1$; $3x - y = -1$ 4. $y = \frac{1}{2}x + 4$; $x - 2y = -8$ 5. $y = -6x - 2$; $6x + y = -2$ 6. $y = -\frac{1}{3}x + 3$; $x + 3y = 9$ 7. $y = -10$ 8. $y = 2$ 9. \$1305 10. The closed half-plane to the right of $x = -3$. 11. The closed half-plane above $y = 4$. 12. The closed half-plane to the left of $y = -2x + 3$. 13. The closed half-plane below $x = -3y - 6$.

Page 107 **Computer Exercises** See Solution Key.

Page 109 **Chapter Objectives and Review** 1. Right 2, up 6; I 3. Left $\frac{1}{2}$, –; none, x axis 5. Left 6, down 4; III 7. Line passing through (0, −5) and $(\frac{5}{3}, 0)$. 9. Line passing through (1, 3) and (−1, −4). 11. Line passing through (0, 0) and (2, 1). 13. $\frac{1}{4}$ 15. $\frac{1}{13}$ 17. Graph would be a ray starting at (0, 12,000) and passing through (3, 14,160). 19. Parallel, since both lines have a slope of 2. 21. $y = -6x + 8$ 23. $y = -2x + 9$ 25. $d = 15$ 27. The open half-plane below $x = 3y + 4$. 29. The closed half-plane to the right of $y = 2x + 2$. 31. The open half-plane above $y = |x| + 3$.

Page 110 **Chapter Test** 1. $\frac{3}{4}$ 3. Line passing through (1, 1) and $(0, 3\frac{1}{2})$. 5. Graph is V-shaped with the vertex at $(0, -\frac{1}{2})$ and point on right ray (1, 0) and on left ray (−1, 0). 7. $y = \frac{3}{2}x - 4$ 9. The slope of $\overleftrightarrow{AB}$ is $\frac{2}{3}$, the slope of $\overleftrightarrow{CD}$ is $-\frac{3}{2}$, $\frac{2}{3} \cdot (-\frac{3}{2}) = -1$, therefore $\overleftrightarrow{AB}$ is perpendicular to $\overleftrightarrow{CD}$. 11. $y + 4 = -\frac{3}{2}(x - 4)$ 13. $y = \frac{1}{7}x - 2$ 15. The open half-plane below $x + 2y = -6$.

Page 111 **More Challenging Problems** 1. $x = -\frac{12}{5}$ or $x = \frac{-1 \pm \sqrt{33}}{4}$ 3. $F = -2$ 5. 3456 sheep

Page 111 **Preparing for College Entrance Tests** 1. b 3. b 5. d 7. b 9. d 11. d

CHAPTER 4 SYSTEMS OF SENTENCES

Page 115 **Classroom Exercises** 1. No 3. No 5. $xy = 6$, $2y - x = 1$

Page 115 **Written Exercises** 1. $\{(\frac{3}{2}, -\frac{3}{2})\}$ 3. $\{(-4, 6)\}$ 5. $\{(-3, 7)\}$ 7. $\{(\frac{1}{2}, -\frac{2}{3})\}$ 9. $\emptyset$ 11. $\begin{cases} n_1 + n_2 = 12 \\ n_1 - n_2 = 3 \end{cases}$; $\{(7\frac{1}{2}, 4\frac{1}{2})\}$ 13. $\begin{cases} p + 2q = 34 \\ p - 4q = 4 \end{cases}$; $\{(24, 5)\}$ 15. $\begin{cases} r^2 - s^2 = 0 \\ r + s = 0 \end{cases}$ 17. $\begin{cases} (\ell + 2)w = \ell w + 12 \\ \ell(w - 1) = \ell w - 8 \end{cases}$; $\ell = 8$; $w = 6$ 19. $\begin{cases} x + y = 22 \\ 0.10x + 0.25y = 3.25 \end{cases}$; 15 dimes, 7 quarters 21. $\begin{cases} c + d = 17 \\ 2c - d = -2 \end{cases}$; 5 and 12

Page 116 **Review Capsule for Section 4-2** 1. $-7x + 2y = -3$ 2. $3x + 5y = 4$ 3. $2x + y = 1$ 4. $4x - 7y = -3$ 5. $2, -\frac{1}{2}$ 6. $-\frac{2}{3}, \frac{3}{2}$ 7. $\frac{3}{8}, -\frac{8}{3}$ 8. $-1\frac{3}{4}, \frac{4}{7}$ 9. $-\frac{1}{7}, 7$ 10. $\frac{1}{5}, -5$

Page 119 **Classroom Exercises** 1. 2, 1 3. −1, 7 5. −1, 2 7. 1, 2 9. $\{(4, 2)\}$ 11. $\{(-3, 2)\}$

Page 119 **Written Exercises** 1. $\{(2, 3)\}$ 3. $\{(-2, -3)\}$ 5. $\{(2, 7)\}$ 7. $\{(4, 1)\}$ 9. $\{(4, 5)\}$ 11. $\{(-5, 2)\}$ 13. $\{(\frac{10}{7}, -\frac{13}{7})\}$ 15. $\{(3, 0)\}$ 17. $\emptyset$ 19. $\{(\frac{83}{7}, \frac{38}{7})\}$ 21. $\{(-\frac{52}{5}, -\frac{66}{5})\}$ 23. $\{(-\frac{5}{3}, \frac{3}{2})\}$ 25. $\{(-\frac{2}{3}, \frac{10}{3})\}$ 27. $\{(-2, 3)\}$ 29. $\{(\frac{a + b}{2}, \frac{a - b}{2})\}$ 31. $\{(2a, 3a)\}$ 33. $\{(-\frac{3c}{14}, -\frac{8c}{7})\}$ 35. $\{(-1, 2)\}$; $\{(2, 5)\}$; $\{(3, -1)\}$; $\{(-1, 2), (2, 5), (3, -1)\}$ 37. $y = -2x$ 39. a. $x - 2y - 7 = 0$ and would contain (3, −2) and (5, −1); b. $x + 3y = -3$ and would contain (3, −2) and (0, −1); c. $3x - y - 11 = 0$ and would contain (3, −2) and (0, −11)

41. Solving the system yields the solution $\{(3, -2)\}$. By substituting this solution into the equation $r(4x + 2y - 8) + s(-3x - 4y + 1) = 0$, this results in the equation $0 + 0 = 0$. Therefore, the resulting line contains the point.

Page 121 Classroom Exercises 1. $y = 6 - x$ 3. $y = -6x + 42$ 5. $x = y + 8$ 7. $x = \frac{2y + 6}{3}$

Page 122 Written Exercises 1. $\{(2, 3)\}$ 3. $\{(6, -1)\}$ 5. $\{(4, -1)\}$ 7. $\{(4, 2)\}$ 9. $\{(\frac{-34}{7}, \frac{5}{7})\}$ 11. $\{(5, -2)\}$ 13. $\{(-6, -8)\}$ 15. $\{(14, 15)\}$ 17. $\{(0.2, 0.3)\}$ 19. $x - 5y = -7$ 21. $2x + 9y = 0$

Page 122 Review Capsule for Section 4-4 1. $t + u = 16$ 2. $10.50x + 12.90y = 775$ 3. $x + y = 42$ 4. $10x + 25y = 1890$ 5. $10.50r + 11.00t = 1615.00$

Page 124 Classroom Exercises 1. Apples: $0.20x, Oranges: $0.25y; $x + y = 24$; $0.20x + 0.25y + 1.95 = 7.50$ 3. Dimes: $0.10x; Nickels: $0.05y; $\begin{cases} x + y = 30 \\ 0.10x + 0.05y = 2.10 \end{cases}$

Page 125 Word Problems 1. $\frac{1}{3}$ ounce of 50¢ seed; $\frac{2}{3}$ ounce of 35¢ seed 3. 8 uprights; 8 spinets 5. 4 oz of 18% solution; 8 oz of 45% solution 7. 23 and 29 9. 75 11. 16 13. 128° and 52° 15. Quarters: 17; Dimes: 11 17. 80 quarters; 170 dimes 19. $12,000 at 12%; $8000 at 14% 21. $8000 at 12% and $16,000 at 16% 23. Boat: 6 km/hr; Current: 3 km/hr 25. Ferry: 4.2 mph; Current: 0.6 mph

Page 127 Calculator Exercises $3 \times (6 + 3) = 27$; $6 \times (6 - 3) = 18$ $2 \times (4.2 + .6) = 9.6$; $2 \times (4.2 - .6) = 7.2$

Page 127 Review Capsule for Section 4-5 1. -3, $(0, 6)$ 2. $\frac{1}{5}$, $(0, 0)$ 3. $-\frac{4}{3}$, $(0, 4)$ 4. $\frac{3}{2}$, $(0, -2)$ 5. Same line 6. NP 7. NP 8. P

Page 129 Classroom Exercises 1. I 3. C, D

Page 130 Written Exercises 1. I 3. C 5. C 7. C 9. $m_1 = -1$, $b_1 = 6$; $m_2 = 1$, $b_2 = -2$; one solution 11. $m_1 = -3$, $b_1 = 5$; $m_2 = -3$, $b_2 = 5$; infinite number of solutions 13. $m_1 = 1$, $b_1 = 0$; $m_2 = 1$, $b_2 = \frac{9}{4}$; no solution 15. $m_1 = -1$, $b_1 = 12$; $m_2 = 1$, $b_2 = -2$; one solution 17. $\{(x, y) : 9x - 2y = 54\}$ 19. $\{(-3, -4)\}$ 21. $\{(12, -24)\}$ 23. inconsistent

25. Solve system for x to get $x = \frac{b_2 - b_1}{m_2 - m_1}$. Since $m_1 \neq m_2$, division yields unique real number. 27. $a = -12$

Page 131 Review 1. $\{(4, 1)\}$ 2. $\{(-2, 2)\}$ 3. $\{(3, -3)\}$ 4. $\{(-1, -4)\}$ 5. $\{(-2, 7)\}$ 6. $\{(\frac{7}{2}, -\frac{1}{2})\}$ 7. $\{(-10, 8)\}$ 8. $\{(\frac{53}{41}, \frac{81}{41})\}$ 9. $\{(4, -1)\}$ 10. $\{(-3, 5)\}$ 11. $\{(-6, -2)\}$ 12. $\{(\frac{24}{5}, -\frac{32}{5})\}$ 13. 25% salt solution: 40 g; 10% salt solution: 60 g 14. 38 15. 15%: $6000; 12%: $14,000 16. $m_1 = -1$, $b_1 = 0$; $m_2 = -1$, $b_2 = 0$; an infinite number of solutions 17. $m_1 = 1$, $b_1 = 0$; $m_2 = 1$, $b_2 = 7$; no solution 18. $m_1 = \frac{3}{2}$, $b_1 = 2$; $m_2 = \frac{2}{3}$, $b_2 = 2$; one solution 19. $m_1 = -5$, $b_1 = 6$; $m_2 = -5$, $b_2 = -6$; no solution

Page 131 Review Capsule for Section 4-6 1. $6x + 2y = 9$ 2. $5x + y = 12$ 3. $4x + y = 3$ 4. $-8x + 3y = -12$ 5. $gbf - gec$ 6. $-afh + cdh$ 7. $-bed + bgf + cde - cgf$

Page 133 Using Statistics

1.

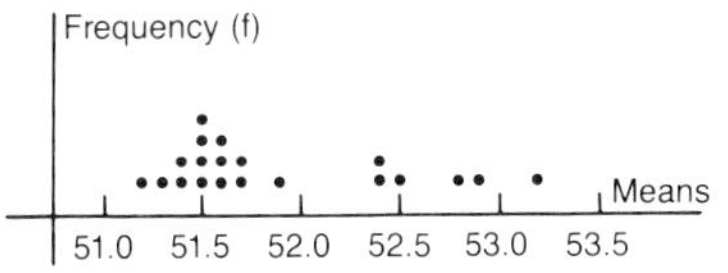

5.

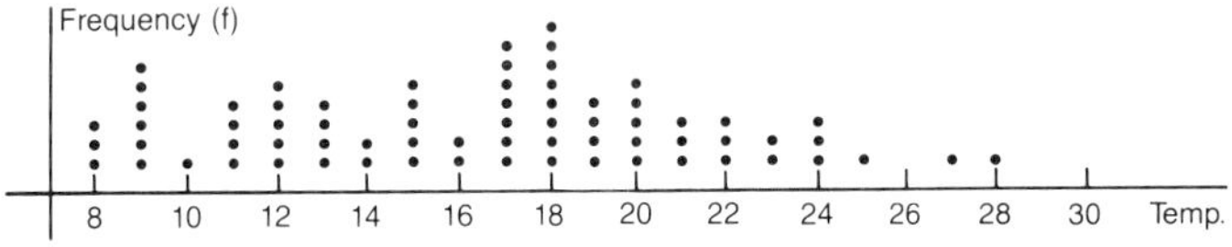

3. No 7. Average

Page 136 Classroom Exercises 1. 41 3. 7 5. $x = \frac{\begin{vmatrix} 8 & -1 \\ 9 & 2 \end{vmatrix}}{\begin{vmatrix} 2 & -1 \\ 1 & 2 \end{vmatrix}}$, $y = \frac{\begin{vmatrix} 2 & 8 \\ 1 & 9 \end{vmatrix}}{\begin{vmatrix} 2 & -1 \\ 1 & 2 \end{vmatrix}}$ 7. $x = \frac{\begin{vmatrix} 10 & 4 \\ 0 & 2 \end{vmatrix}}{\begin{vmatrix} 3 & 4 \\ -1 & 2 \end{vmatrix}}$, $y = \frac{\begin{vmatrix} 3 & 10 \\ -1 & 0 \end{vmatrix}}{\begin{vmatrix} 3 & 4 \\ -1 & 2 \end{vmatrix}}$

Page 136 Written Exercises 1. 3 3. -28 5. 0 7. $p^2 + 3$ 9. $t = 2$ 11. $t = \frac{-13}{3}$ 13. $\{(\frac{-3}{13}, \frac{-11}{13})\}$ 15. $\{(\frac{10}{3}, -\frac{2}{3})\}$ 17. $\{(\frac{11}{7}, \frac{2}{7})\}$ 19. $\{(\frac{5}{7}, \frac{80}{21})\}$ 21. 5. $\{(5, 2)\}$ 6. $\{(-\frac{29}{7}, -\frac{46}{7})\}$ 7. $\{(2, 1)\}$ 8. $\{(1, 3)\}$ 23. Dependent 25. Dependent 27. 20 dimes, 18 quarters 29. $\frac{18}{24}$ 31. $\{(\frac{1}{4}, \frac{1}{3})\}$ 33. $\{(-\frac{1}{2}, \frac{2}{3})\}$

35. Proof: 1. Lines defined by $a_1x + b_1y = c_1$ and $a_2x + b_2y = c_2$ are parallel. (Given) 2. $m_1 = -\frac{a_1}{b_1}$, $m_2 = -\frac{a_2}{b_2}$; (Def. of slope. 3. $-\frac{a_1}{b_1} = -\frac{a_2}{b_2}$; (Slopes of parallel lines are equal.) 4. $-a_1b_2 = -a_2b_1$; (Mult. Prop. for Eqs.) 5. $\begin{vmatrix} a_1 & b_1 \\ a_2 & b_2 \end{vmatrix} = a_1b_2 - a_2b_1$; (Def. of Determinant.) 6. $a_1b_2 + (-a_2b_1) = a_1b_2 - a_2b_1$; (Def. of Subtraction) 7. $a_1b_2 + (-a_1b_2) = a_1b_2 - a_2b_1$; (Substitution) 8. $0 = a_1b_2 - a_2b_1$; (Add. Inv. Post.) 9. $0 = \begin{vmatrix} a_1 & b_1 \\ a_2 & b_2 \end{vmatrix}$; (Substitution)
10. $\begin{vmatrix} a_1 & b_1 \\ a_2 & b_2 \end{vmatrix} = 0$; (Symmetry Post.)

Page 137 Calculator Exercises 13. {−0.2307692, −0.8461538} 15. {3.3333333, −0.6666666}
17. {1.5714286, 0.2857142} 19. {0.7142857, 3.8095238}

Page 140 Classroom Exercises 1. $x = \frac{\begin{vmatrix} 3 & 1 & -1 \\ 0 & -1 & 4 \\ 6 & -3 & 2 \end{vmatrix}}{\begin{vmatrix} 2 & 1 & -1 \\ 4 & -1 & 4 \\ 0 & -3 & 2 \end{vmatrix}}$; $y = \frac{\begin{vmatrix} 2 & 3 & -1 \\ 4 & 0 & 4 \\ 0 & 6 & 2 \end{vmatrix}}{\begin{vmatrix} 2 & 1 & -1 \\ 4 & -1 & 4 \\ 0 & -3 & 2 \end{vmatrix}}$; $z = \frac{\begin{vmatrix} 2 & 1 & 3 \\ 4 & -1 & 0 \\ 0 & 3 & 6 \end{vmatrix}}{\begin{vmatrix} 2 & 1 & -1 \\ 4 & -1 & 4 \\ 0 & -3 & 2 \end{vmatrix}}$ 3. $x = \frac{\begin{vmatrix} 6 & -1 & 2 \\ 3 & 1 & 1 \\ -9 & -1 & 3 \end{vmatrix}}{\begin{vmatrix} 1 & -1 & 2 \\ 2 & 1 & 1 \\ 1 & -1 & 3 \end{vmatrix}}$

$y = \frac{\begin{vmatrix} 1 & 6 & 2 \\ 2 & 3 & 1 \\ 1 & -9 & 3 \end{vmatrix}}{\begin{vmatrix} 1 & -1 & 2 \\ 2 & 1 & 1 \\ 1 & -1 & 3 \end{vmatrix}}$; $z = \frac{\begin{vmatrix} 1 & -1 & 6 \\ 2 & 1 & 3 \\ 1 & -1 & -9 \end{vmatrix}}{\begin{vmatrix} 1 & -1 & 2 \\ 2 & 1 & 1 \\ 1 & -1 & 3 \end{vmatrix}}$

Page 140 Written Exercises 1. −41 3. 61 5. z = 3 7. q = 1 9. {(5, −3, 5)} 11. $\{(-\frac{4}{3}, 0, \frac{20}{3})\}$ 13. {(8, 27, 13)}
15. {(5, −8, −10)} 17. Length: 8 cm; Width: 7 cm; Height: 3 cm 19. 20¢ - stamps: 13; 5¢ - stamps: 40; 3¢ - stamps: 40
21. 5 full-page ads, 10 half-page ads; 25 quarter-page ads 23. Inconsistent

Page 141 Puzzle See Solution Key.

Page 141 Review Capsule for Section 4-8 1. $\{(x, y) : x = 3\}$ 2. $\{(x, y) : x = -2\}$ 3. $\{(x, y) : x > 0 \text{ and } y > 0\}$
4. $\{(x, y) : x > 0 \text{ and } y < 0\}$

Page 144 Classroom Exercises 1. 1, 4 3. 5, 7 5. 1, 7 7. 7

Page 144 Written Exercises

1.

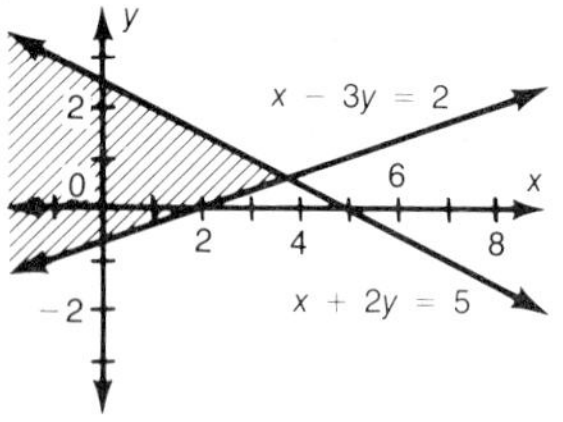

3.

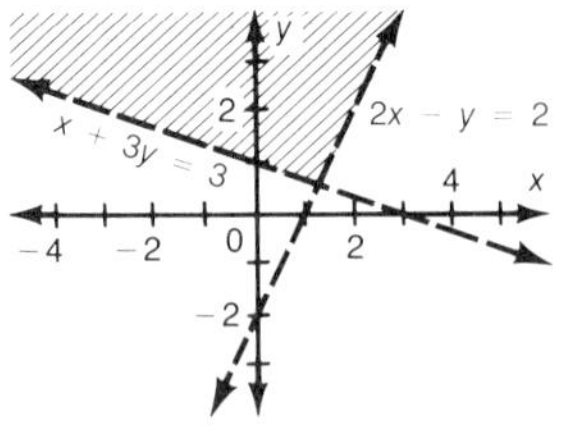

5.

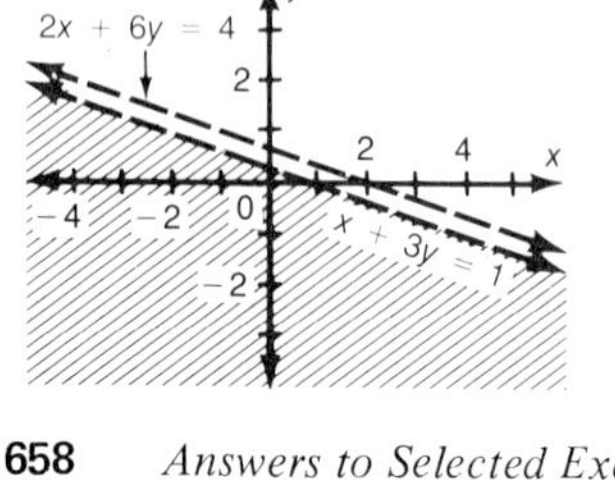

7.

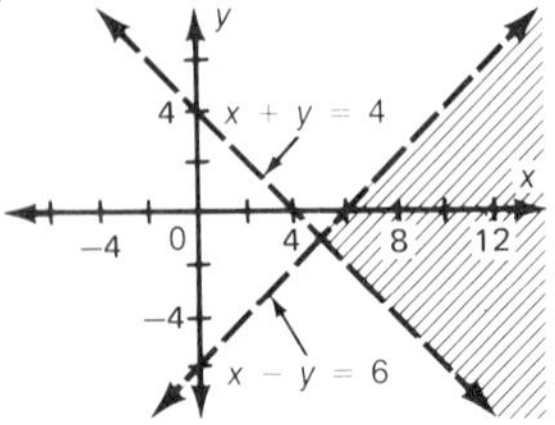

9.

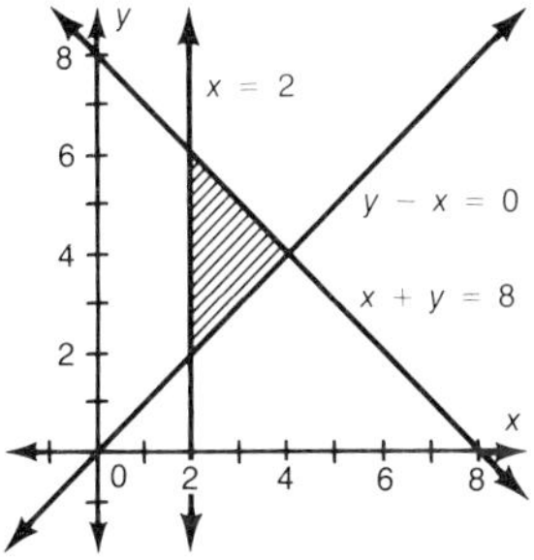

11.

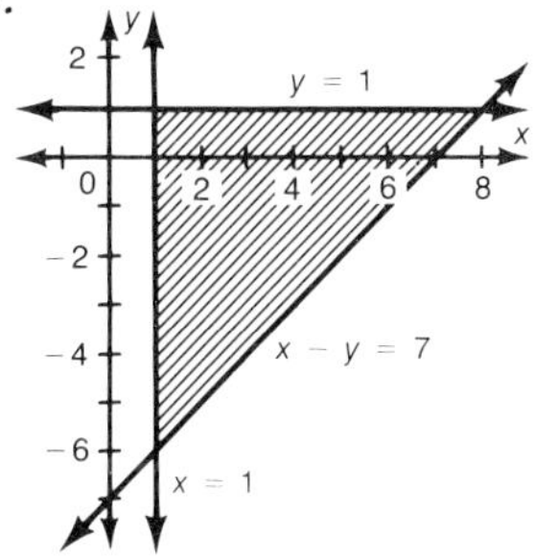

13.

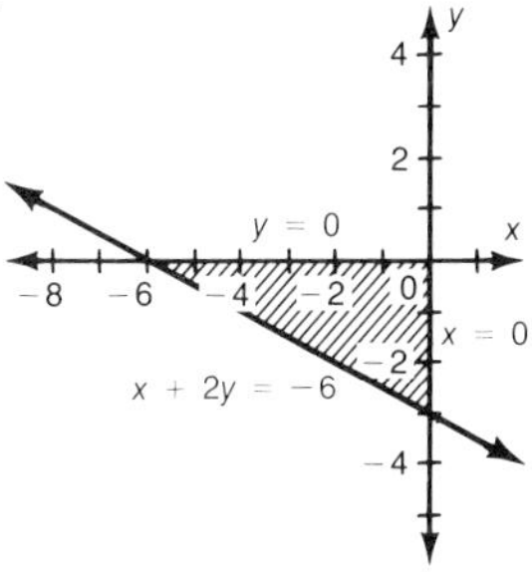

15.

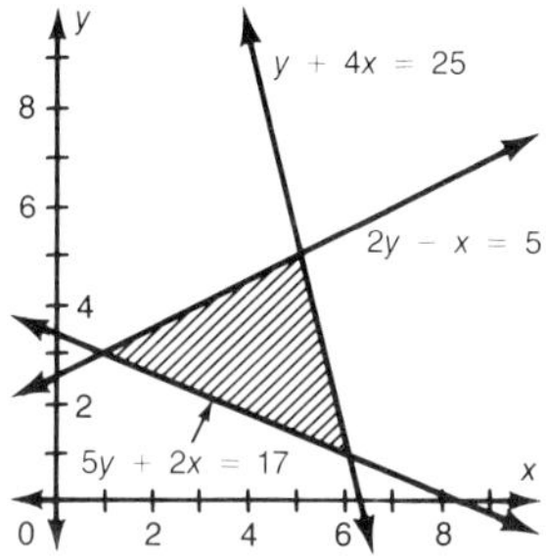

17. ∅

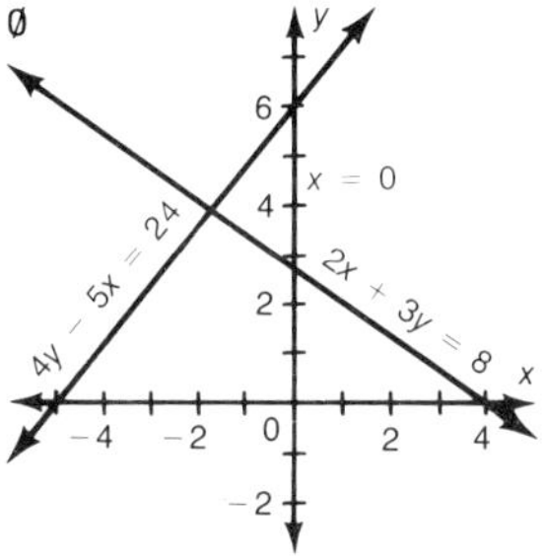

19.

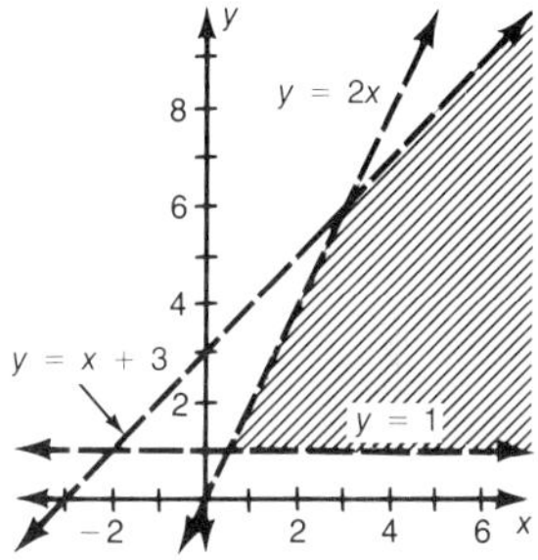

21.

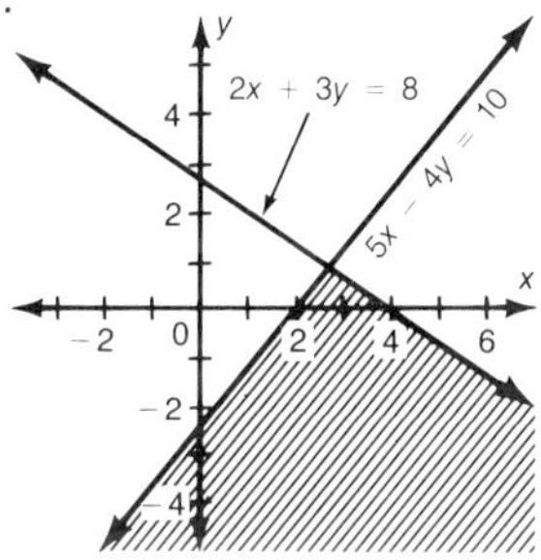

23.

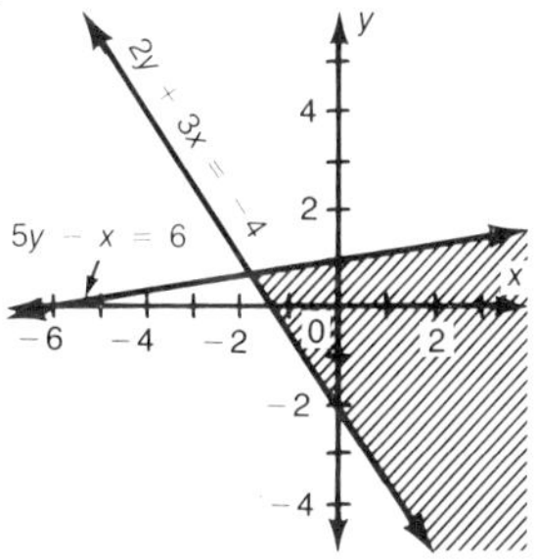

25.

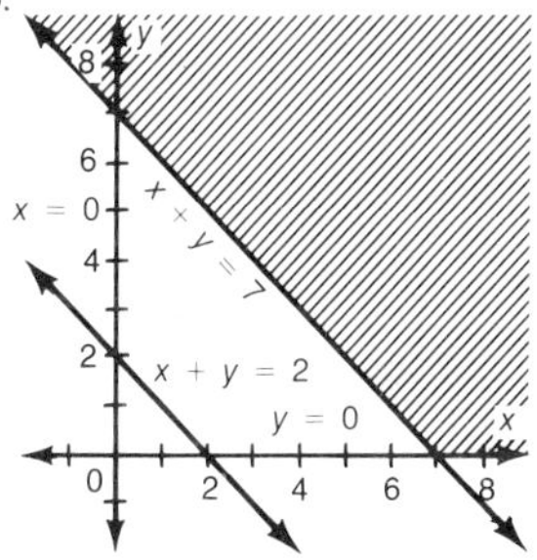

27.

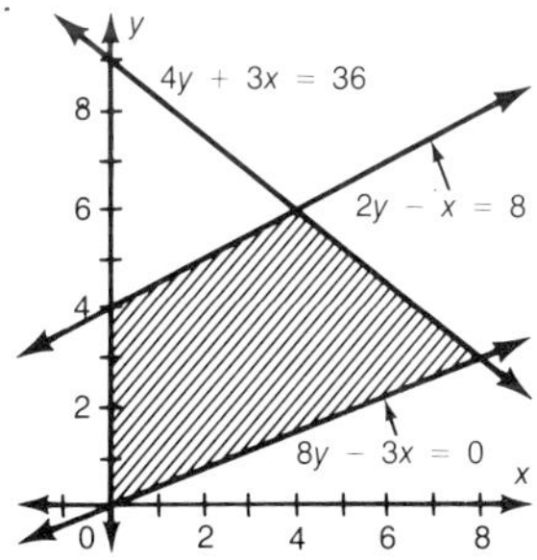

29.

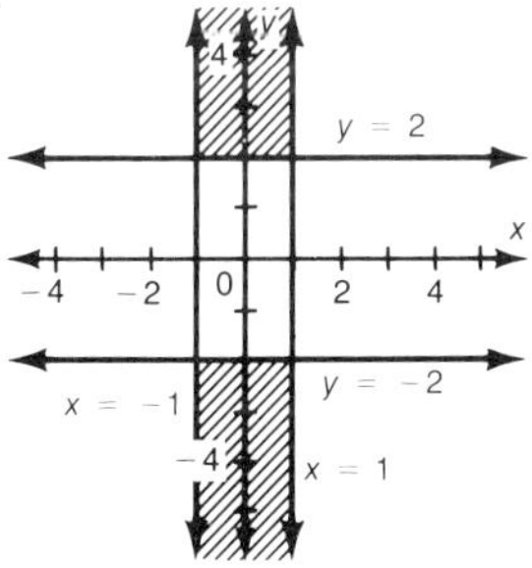

31. ∅ $|x + y| \geq -1$ will be true for all values of x and y. $|x + y| < -1$ will not be true for any values of x and y.

Page 146 Classroom Exercises 1. Maximum at P(5, 1) 3. Maximum at $P(1\frac{1}{2}, 0)$

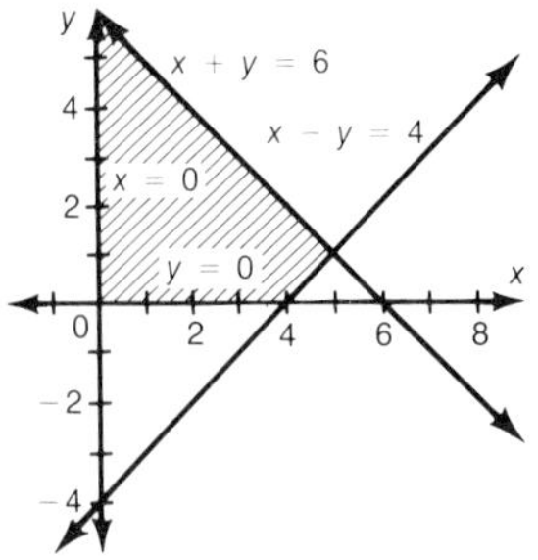

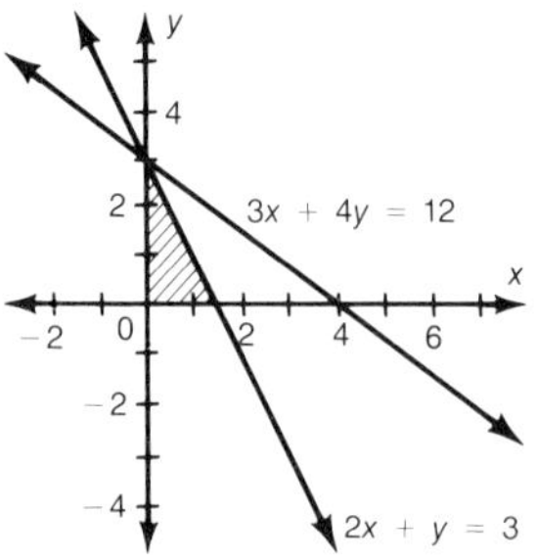

Page 146 Word Problems

1. x = 4; y = 1

3. 818 square feet

5. Fancy legs: 13; Plain legs: 17

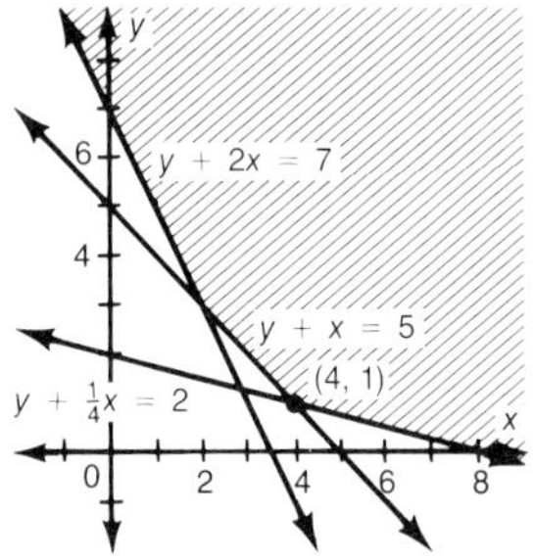

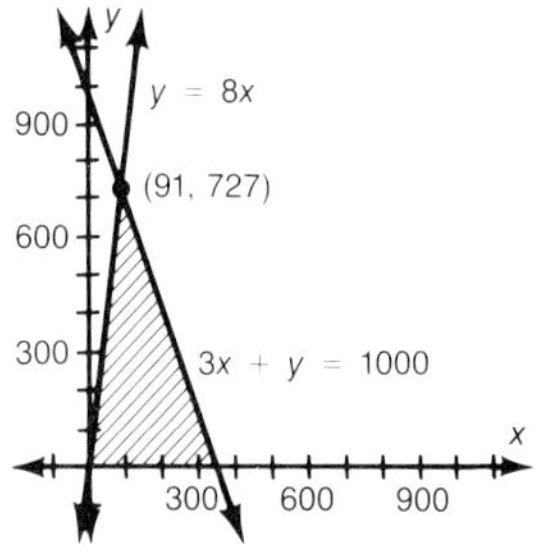

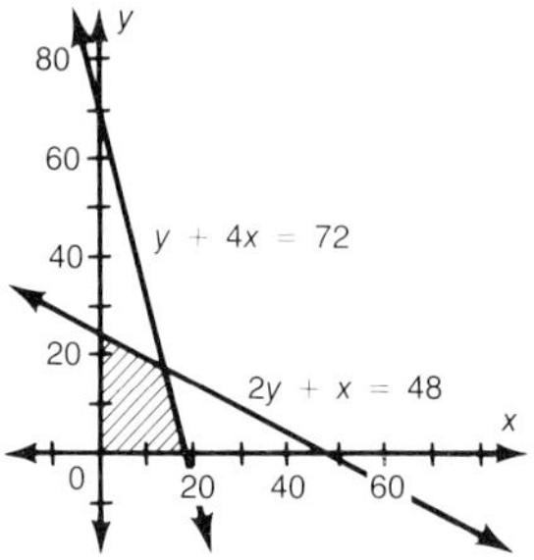

Page 147 Review 1. 2 2. −51 3. 0 4. 0.11 5. $\{(7, 9)\}$ 6. 2 7. 120 8. 1584 9. −485 10. Ø 11. Points in Quadrants I, II, and III, that are on and below the graph of $x - y \leq 6$ and also are on or above the line $x - 2y \leq 3$. 12. The points in Quadrants I and II that lie above the graph of the line $x - 3y = -3$ and also above the graph of the line $2x + 5y = 1$. 13. Points on the sides of, and within, the small triangle formed by the graphs of $x \geq -1$, $3x + 7y = -4$, $2x + 5y = -3$, and the y axis. 14. Points on the sides of, and within, the small triangle formed by the graphs of $x + 2y = 2$, $x - y = -1$, and $2x - y = 1$. 15. 1 cup of oatmeal, 1 cup of protein

Page 151 Chapter Objectives and Review 1. $\{(-1, -1)\}$ 3. $\{(2, -5)\}$ 5. $\{(0, 6)\}$ 7. $\{(\frac{124}{81}, \frac{59}{81})\}$ 9. $\{(-\frac{23}{2}, -4)\}$ 11. $\{(\frac{95}{9}, \frac{67}{9})\}$ 13. 65 large size cars were sold and 35 compacts were sold. 15. L = 32 meters, w = 8 meters 17. 12% investment: \$2000; 9% investment: \$1500 19. C 21. C 23. $\{(5, -5)\}$ 25. $\{(2, -2)\}$ 27. $\{(5, -1, -2)\}$ 29. $\{(10, 5, -15)\}$

31.

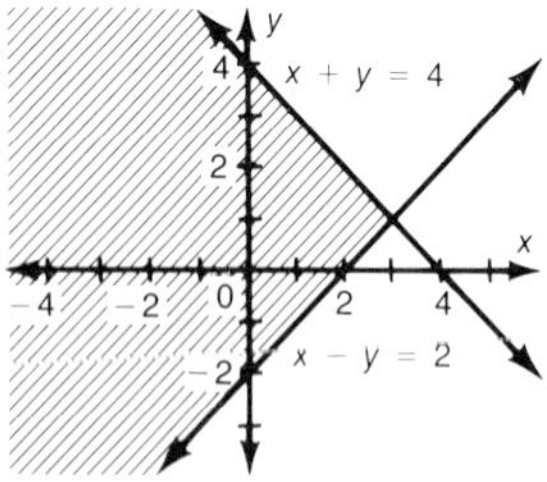

33.

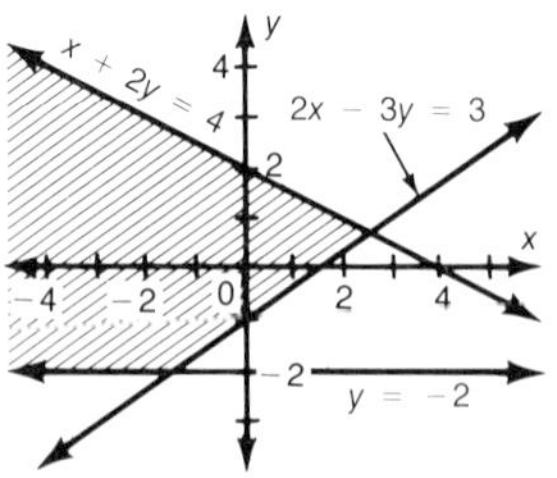

35. Belts: 0; Handbags: 5

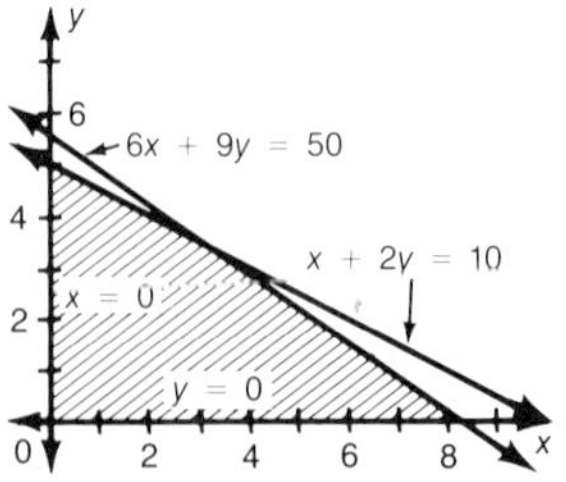

Page 153 Chapter Test 1. $\{(-2, \frac{[illegible]}{2})\}$ 3. $\{(-\frac{5}{2}, -15)\}$ 5. Certificate of Deposit: \$2700; Money Market Fund: \$1800 7. $\{(-3, -7)\}$

9.

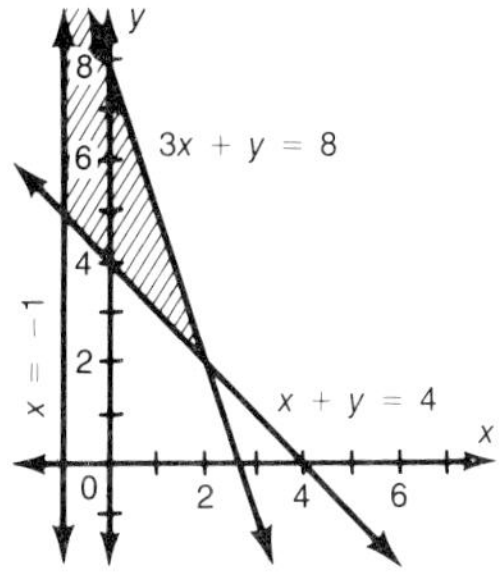

Page 154 Review of Word Problems: Chapters 1-4 1. width: 5 m 3. −18; −19; −20 5. a. V = 85,000 + 10,200t b. 6 years 7. Gofar: 3200; Sure-Par: 1050 9. 27°; 153° 11. at 12%: \$8000; at 16%: \$9500 13. Washing machines: 12; dryers: 30

Page 155 Cumulative Review: Chapters 1-4 1. d 3. a 5. b 7. c 9. c 11. c 13. a 15. a 17. b 19. c 21. b 23. d 25. c

Page 157 Preparing for College Entrance Tests 1. b 3. c 5. a 7. a 9. d 11. c 13. b 15. d 17. b 19. d 21. c 23. c

CHAPTER 5 RELATIONS AND FUNCTIONS

Page 162 Classroom Exercises 1. D: {1, 2, 5, 8}; R: {4, 3, 6, 13} 3. D: {−1, 2}; R: {1, 4, 7, 9} 5. Function: Each element of the domain is paired with one element in the range. 7. Not a function: A vertical line would intersect more than one point on the graph.

Page 162 Written Exercises 1. D: {1, 3, 5, 8}; R: {4, 8, 12, 16} 3. D: {−2, −1, 0, 1, 2}; R: {2, 1, 0} 5. D: {−2, −1, 0, 1}; R: {2, 1, 0, −1} 7. R: {7, 14, 21} 9. R: {8, 5, 2, −1} 11. R: {3, 6, 9} 13. R: {−9} 15. R: {−4, −2, −1, 0} 17. Function. Each element in the domain has exactly one element in the range. 19. Function. Each element in the domain has exactly one element in the range. 21. Not a function, an element in the domain has more than one element in the range. 23. Not a function. 25. Not a function. 27. Function 29. Not a function.

31. Function

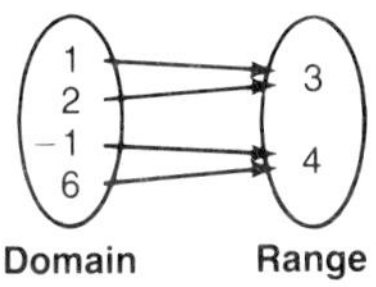

33. Function

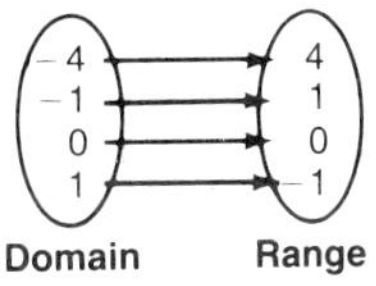

35. Function

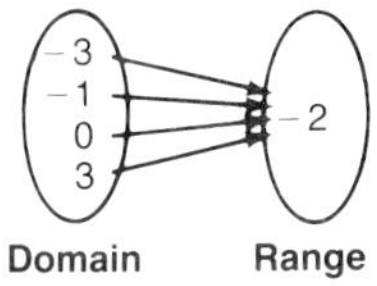

In Exercises 37-44, each relation is a function. Each graph consists of the following points.
37. (−3, −3); (−2, −2); (−1, −1); (0, 0); (1, 1); (3, 3)
39. (−3, −2); (−2, −1); (−1, 0); (0, 1); (1, 2); (2, 3); (3, 4)
41. (−3, −9); (−2, −7); (−1, −5); (0, −3); (1, −1); (2, 1); (3, 3)
43. (−3, 9); (−2, 4); (−1, 1); (0, 0); (1, 1); (2, 4); (3, 9)

45. D: {−3, −2, −1, 1, 2}; R: {3, 2, 1, −2, −3}; Not a function 47. D: $\{x : 0 \le x \le 3\}$; R: $\{y : 0 \le y \le 3\}$; Function 49. D: $\{x : x \in R\}$; R; $\{y : y \in R\}$; Function 51. D: $\{x : x \in R\}$; R: $\{y : y \in R\}$; Function 53. D: $\{x : x \in R\}$; R: {−4}; Function 55. D: $\{x : x \in R\}$; R: $\{y : y \ge 0\}$; Function 57. a. The graph is V-shaped and opens upward. The vertex is at (0, 3) and contains (−1, 4) and (1, 4). b. $\{y : y \ge 3\}$ 59. a. Draw a dashed line for y = x. The graph consists of all the points below the graph of y = x. b. $\{y : y \in R\}$ 61. a. Draw the graph of y = 2x − 3. The graph consists of all the points on and below the graph of y = 2x − 3. b. $\{y : y \in R\}$ 63. a. Draw dashed lines for y = 3 and y = −3. The graph consists of all the points between the graphs of y = 3 and y = −3. b. $\{y : -3 < y < 3\}$ 65. a. The graph is diamond-shaped. The vertices are (1, 0), (0, 1), (−1, 0), and (0, −1). b. $\{y : -1 \le y \le 1\}$

Page 164 Calculator Exercises 1. {12.2, −0.8, −3.8, −28.05} 3. {−40, −30, −6.875, 5} 5. {−6.3975, −6.0515625, −5.7, −5.62125}

Page 164 Review Capsule for Section 5-2 1. 9 2. −236 3. 4 4. $9 - 5c^2$ 5. $9 - 5d^2$ 6. 41 7. 34 8. 1 9. $3a^2 - 2a + 1$ 10. $3b^2 + 2b + 1$

Page 167 Classroom Exercises 1. f(x) = x + 2 3. f(x) = −x 5. 2 7. 6 9. 7 11. 8 13. −19 15. $8 - 3t^2$

Page 167 Written Exercises 1. $f : x \to 2x + 1, f(x) = 2x + 1$ 3. $f : x \to x^2 - 1, f(x) = x^2 - 1$ 5. $f : x \to \frac{1}{x}, f(x) = \frac{1}{x}$ 7. -20 9. -2 11. 1 13. 0 15. 5 17. $-\frac{7}{9}$ 19. $-\frac{1}{3}$ 21. $-\frac{2}{3}$ 23. $-\frac{2t-1}{3}$ 25. 2 27. 3.5 29. π 31. 0 33. 3 35. -4 37. C 39. D 41. A 43. Shape is a V with vertex at (0, 1); Point (1, 2) is on the right ray and $(-1, 2)$ is on the left ray.

45.

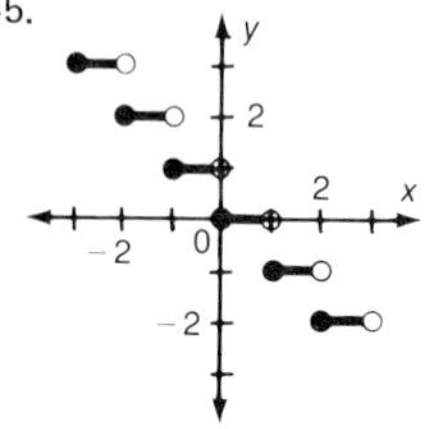

47.

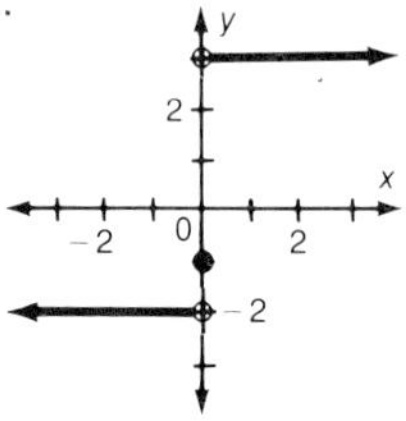

49.

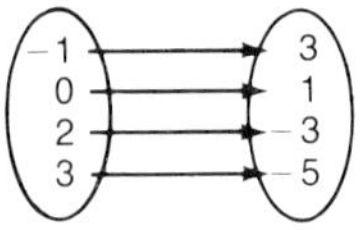

51.

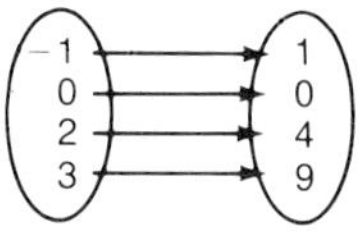

53.

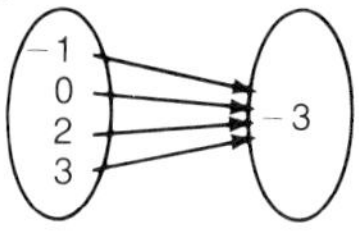

55. $\{1, 2, 3, 4, 5, 6\}$ 57. Yes 59. $\{-4, -3, -2, -1, 0, 1, 2, 3, 4, 5, 6\}$ 61. Yes 63. $f(x) = -x$ 65. $f(x) = 0$ 67. $f(x) = 2x - 1$ 69. $f(x) = x^2$ 71. 0 73. 1 75. 1.6 77. $3x^2 - 1$ 79. $-3x - 1$ 81. $f(x) = x, f(x) = |x|$ 83. $f(x + h) = f(x) + f(h)$; Let $x = -h$, $f(-h + h) = f(-h) + f(h)$, $f(0) = f(-h) + f(h)$. Adding $-f(h)$ to both sides and simplifying, we have $f(-h) = -f(h)$.

Page 169 Puzzle See Solution Key.

Page 169 Review Capsule for Section 5-3 1. 1 2. $\frac{9}{2}$ 3. $\frac{a-b}{2}$ 4. $y - 5 = \frac{1}{2}(x - 3)$ or $y - 4 = \frac{1}{2}(x - 1)$ 5. $y + 1 = 1(x - 0)$ or $y + 2 = 1(x + 1)$ 6. $y - 9 = 1(x - 9)$ or $y + 4 = 1(x + 4)$

Page 171 Written Exercises 1. A line containing (0.7, 5), (2, 13). 3. $t = 3.7875$ seconds 5. A line containing (21, 3000), (15, 3180). 7. 2130 calories 9. $-12\frac{1}{3}°C > T > -45\frac{2}{3}°C$ 11. \$320 13. \$120 15. \$2.07 17. \$1.56 19. 190 ft/sec 21. 8 seconds 23. 11 seconds 25. 3 seconds

Page 172 Review Capsule for Section 5-4 1. $D = \{x : x \in R\}$; $R = \{y : y \in R\}$ 2. $D = \{x : x \in R\}$; $R = \{y : y \leq 0\}$ 3. $D = \{x : x \in R\}$; $R = \{y : y \geq 0\}$ 4. $D = \{x : x \in R\}$; $R = \{y : y = 1\}$ 5. $2t - 1$ 6. $2t + 1$ 7. $2t - 7$ 8. $2t^2 - 1$ 9. $-2t - 1$ 10. $-2t^2 - 1$

Page 173 Review 1. $\{1, 0, -4\}$ 2. $\{-7, -3, -1, 3\}$ 3. $\{-4\}$ 4. $\{-1, 0, 1, -2\}$ 5. Function, each element of domain has exactly one element in range. 6. Not a function, element in domain has more than one element in range. 7. Function, each element in domain has exactly one element in range. 8. Not a function, vertical line would intersect it in two points. 9. Function, vertical line intersects it in one point. 10. Not a function, vertical line intersects it in two points. 11. 0 12. 9 13. -1 14. -8 15. 12.6 16. $-2x + |-x|$ 17. A straight line containing (0, 0.95) and (−100, 0.6) 18. $100.6T - 0.95p - 95 = 0$ 19. 0.947° 20. 0.944°

Page 175 Classroom Exercises

1. a.

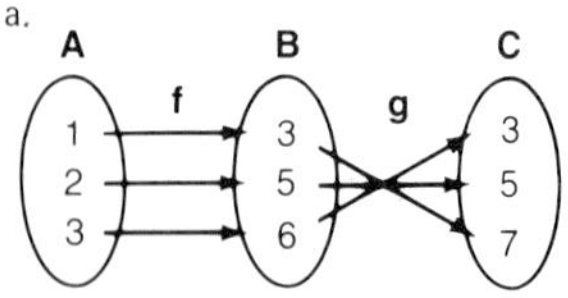

b. $\{(1, 7), (2, 5), (3, 3)\}$

3. a.

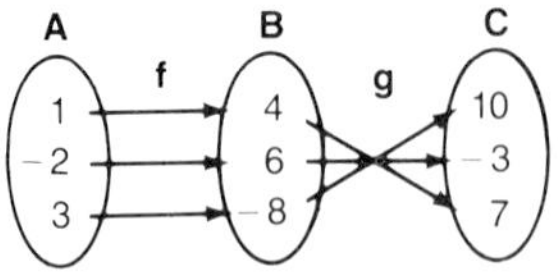

b. $\{(1, 7), (-2, -3), (3, 10)\}$

5. 1. $D = \{1, 2, 3\}$; $R = \{3, 5, 7\}$
2. $D = \{1, 2, 3\}$; $R = \{6, 7\}$
3. $D = \{1, -2, 3\}$; $R = \{10, -3, 7\}$
4. $D = \{-5, 1, 2\}$; $R = \{1, -5, 2\}$

Page 176 Written Exercises 1. $x - 1$; 2; -3; $a - 1$ 3. $-3x - 2$; -11; 4; $-3a - 2$ 5. $-x$; -3; 2; $-a$ 7. $-\frac{1}{5x+2}$; $-\frac{1}{17}$; $\frac{1}{8}$; $-\frac{1}{5a+2}$ 9. $x + 1$; 0; 6; $a + b + 1$ 11. $2 - x$; 3; -3; $2 - a - b$ 13. $\frac{1}{x-2}$; $-\frac{1}{3}$; $\frac{1}{3}$; $\frac{1}{a+b-2}$ 15. $-2|x| - 3$; -5; -13; $-2|a + b| - 3$ 17. $-3x + 1$; $-3x + 3$ 19. $\frac{2}{x} - 3$; $\frac{1}{2x-3}$ 21. $x + 7$; $x + 7$ 23. $\frac{1}{2-|x|}$; $2 - \left|\frac{1}{x}\right|$ 25. $-x$, $-x$;

$f(g(x)) = g(f(x))$ 27. $-\frac{1}{x}, -\frac{1}{x}$; $f(g(x)) = g(f(x))$ 29. $-\frac{x}{2}, -\frac{x}{2}$; $f(g(x)) = g(f(x))$ 31. $10 - x, -x$; $f(g(x)) \neq g(f(x))$ 33. $d(p) = 75p$ 35. $F = \frac{n}{4} + 38$ 37. Since $n = 4F - 38$, $g(h(c)) = \frac{4(9c + 160)}{5} - 38$ 39. $h(g(x)) = x^2$; A parabola with vertex at (0, 0) and passing through (1, 1) and (−1, 1). 41. $h(f(x)) = -(2x - 1) = -2x + 1$; The graph is a line through (0, 1), (1, −1). 43. $\frac{1}{x}$ 45. $1 - x$ 47. $1 - x$ 49. Answers will vary. 51. $g(x) = -\frac{x}{2}$

Page 177 Review Capsule for Section 5-5 1. $y = -3x - 1$ 2. $y = \frac{3}{2}x$ 3. $y = 5x - 6$ 4. $y = -3x + 5$ 5. $y = 5x$ 6. $y = -2x + 1$ 7. $y = \frac{3}{2}x$ 8. $y = -x + 3$

Page 179 Classroom Exercises 1. {(1, 1), (3, 2), (4, 3), (5, 4)}; Yes 3. {(2, 1), (1, 2), (2, 3), (3, 4)}; No 5. $y = -\frac{1}{3}x - \frac{1}{3}$ 7. $x = 9$

Page 180 Written Exercises 1. {(3, −1), (3, −2), (3, −4), (0, −5)}, No 3. $\left\{\left(\frac{2}{3}, \frac{1}{2}\right), \left(\frac{3}{4}, \frac{2}{3}\right), \left(\frac{4}{5}, \frac{3}{4}\right), \left(\frac{5}{6}, \frac{4}{5}\right)\right\}$, Yes 5. $y = x - 2$ 7. $y = \frac{1}{2}x$ 9. $y = -\frac{1}{3}x - \frac{4}{3}$ 11. $y = 2x$ 13. $x = 3$ 15. Inverses 17. Not inverses 19. Not inverses 21. Not inverses 23. Not inverses 25. Graph the given points for f; f^{-1}, graph {(1, 2), (3, 1), (−2, 5), (−4, −3)} 27. The graph of f contains (0, −3) and (1, −1). For f^{-1} graph (−3, 0) and (−1, 1). 29. The graph of f contains (1, 1) and (2, 2). For f^{-1} graph (1, 1) and (2, 2). 31. The graph of f is $\wedge$ shaped with the right ray containing (0, 0), (2, −2) and the left ray (0, 0), (−2, −2); the graph of f^{-1} has an upper ray with points (0, 0), (−2, 2) and lower ray with points (0, 0), (−2, −2). 33. If the inverse is a function, then no horizontal line intersects the graph of the relation in more than one point. 35. Not a function 37. Function 39. Not a function 41. Function 43. No, not symmetric to line $y = x$. 45. Answers will vary.

Page 182 Review 1. $g(f(x)) = 2x + 3$; $g(f(3)) = 9$; $g(f(-2)) = -1$; $g(f(a)) = 2a + 3$; $g(f(-b)) = -2b + 3$ 2. $g(f(x)) = \frac{1}{3}x$; $g(f(3)) = 1$; $g(f(-2)) = -\frac{2}{3}$; $g(f(a)) = \frac{1}{3}a$; $g(f(-b)) = -\frac{1}{3}b$ 3. $g(f(x)) = -\frac{1}{x}$; $g(f(-3)) = -\frac{1}{3}$; $g(f(-2)) = \frac{1}{2}$; $g(f(a)) = -\frac{1}{a}$; $g(f(-b)) = \frac{1}{b}$ 4. $g(f(x)) = \frac{5}{2}x - 1$; $g(f(3)) = 6\frac{1}{2}$; $g(f(-2)) = -6$; $g(f(a)) = \frac{5}{2}a - 1$; $g(f(-b)) = -\frac{5}{2}b - 1$ 5. {(−2, −2), (5, −1), (9, 0), (5, 3)}; Not a function 6. {(−1, −3), (−3, 1), (7, 4), (2, 9)}; Function 7. The graph of f consists of three points: (−4, 4), (−2, 2), and (0, 0). The graph of f^{-1} consists of the points (0, 0), (2, −2), and (4, −4). 8. The graph of f would contain (0, 0), (1, −2) and be a line. Points on the line representing f^{-1} are (0, 0), (−2, 1). 9. The graph of f is a line that contains (0, 1) and (3, 2). Points on the line prepresenting f^{-1} are (1, 0) and (2, 3). 10. Points on the line f are (0, −2), (−2, 0) and points on f^{-1} are the same points. 11. The graph of f would be V-shaped and opening upward with the vertex at (2, 0). The right ray contains (3, 1) and the left ray contains (1, 1). f^{-1} is $<$ shaped with the vertex at (0, 2). The bottom ray contains (1, 1). The top ray contains (1, 3) 12. f is a line containing (1, 2), (5, 2) while the line representing f^{-1} contains (2, 1), (2, 5).

Page 183 Chapter Objectives and Review 1. Yes. Each element in the domain has exactly one element in the range. 3. Yes. Each element in the domain has exactly one element in the range. 5. No. Vertical has two points of intersection. 7. Yes. One point intersection. 9. 0 11. $-\frac{7}{16}$ 13. $\frac{-t - 2}{4}$ 15. A line containing (0, 2) and (2, 0) 17. A line containing (0, 7) and (5, 7). 19. A line containing (6, 5000), (3, 7500) 21. $1666.67 23. $g(f(x)) = -x - 2$; $g(f(-4)) = 2$; $g(f(8)) = -10$; $g(f(t)) = -t - 2$; $g(f(-t)) = t - 2$ 25. $g(f(x)) = x$; $g(f(-4)) = -4$; $g(f(8)) = 8$; $g(f(t)) = t$; $g(f(-t)) = -t$ 27. {(4, −6), (4, 3), (4, 0), (4, 5)}; Not a function 29. $y = -x + 2$ 31. $x = -8$

Page 184 Chapter Test 1. Yes 3. 0 5. $\frac{5}{6}$ 7. $-\frac{1}{3}x + \frac{2}{3}$ 9. The graph is V-shaped with the vertex at (3, 0) and the right ray containing (4, 1) and the left ray containing (0, 3). 11. $r = 9.8t + 3$ 13. $g(f(x)) = 4x$; $g(f(2)) = 8$; $g(f(-4)) = -16$; $g(f(-t)) = -4t$; $g(f(s + t)) = 4(s + t)$ 15. $g(f(x)) = 17$; $g(f(2)) = 17$; $g(f(-4)) = 17$; $g(f(-t)) = 17$; $g(f(s + t)) = 17$ 17. $y = 4x$ 19. $x = -5$

Page 185 More Challenging Problems 1. $3 \leq x \leq 6$ or $-4 \leq x \leq -1$ 3. $f(f(x)) = x$ 5. −5 7. Quadrants I and II 9. $n = 2$; $c = 6$, $b = -1$ or $c = -6$, $b = 1$ 11. 100 yards

Page 186 Preparing for College Entrance Tests 1. b 3. b 5. d

CHAPTER 6 POLYNOMIALS

Page 189 Classroom Exercises 1. P; def. of polynomial 3. P; def. of polynomial 5. P; def. of polynomial 7. P; def. of polynomial 9. P; def. of polynomial 11. NP; The exponent is not a whole number.

Page 189 Written Exercises 1. P; def. of polynomial 3. P; def. of polynomial 5. P; def. of polynomial 7. P; def. of polynomial 9. P; def. of polynomial 11. NP; The exponent is not a whole number. 13. NP; The exponent is not a whole number. 15. P; def. of polynomial 17. NP; The exponent is not a whole number. 19. P; def. of polynomial 21. monomial; 3 23. trinomial; 2 25. binomial; 2 27. trinomial; 3 29. 3, 2, 7; over the set of integers. 31. $\sqrt{2}, -3$; over the set of real numbers. 33. $\frac{1}{2}, -\frac{1}{3}, -\frac{1}{4}$; over the set of rational numbers. 35. $y^3 + 3xy^2 + 3x^2y + x^3$ 37. $2xy + x + z$

Page 191 Classroom Exercises 1. $4r^2s, 2r^2s$ 3. $-2u, -6u$ 5. $7x^2yz, x^2yz$ 7. $15zt^2$ 9. $12n - 5$ 11. $10y + 1$ 13. $-2x + 20$ 15. $-2t^2 + 15q^2 - 1$ 17. $3n^2 - 9nw + 8w^2$

Page 191 Written Exercises 1. $6y^2 + 6y + 11$ 3. $a^3 + 4a^2 - 6$ 5. $7x^2$ 7. $5.5rt^2$ 9. $16m^2n$ 11. $2q^2r$ 13. $3x + 8$ 15. $3x + 2$ 17. $9x^2 - 5x + 3$ 19. $12y^3 - 11y^2 + 5y$ 21. $25a^3 - 5a^2 - 14a + 11$ 23. $a^2 - 1$ 25. $-5x^2 + 8x - 18$ 27. $4r + 6$ 29. $-6p - 6$ 31. $-7x - 11y + 12$ 33. $6y^4 - 13y^3 + 9y^2 - 30y + 1$ 35. 0 37. $x^2 - 3xy + y^2$ 39. $2x - 3y + z$ 41. $2x^2y^5$ 43. $-2x^2 + 10xy - 7y^2$ 45. $3k^2 - 2k - 1$ 47. d 49. a 51. Ex. 46. $P + Q = 2x^2 - x + 9$; $Q + P = 2x^2 - x + 9$ Ex. 47. $P + 0 = 3x + 5$; $P = 3x + 5$ Ex. 48. $P + (-P) = 0$ Ex. 49. $P + Q = 2x^2 - x + 9$, which is a polynomial by definition. Ex. 50. $(P + Q) + R = 3x^2 + 4x + 15$; $P + (Q + R) = 3x^2 + 4x + 15$ 53. Answers will vary.

Page 195 Classroom Exercises 1. $-6x^3y^4$ 3. $50r^5t^4$ 5. $x^2 + 12x + 35$ 7. $2d^2 - 19d + 24$ 9. $4t^2 + 4t + 1$ 11. $d^2 - 36$

Page 195 Written Exercises 1. $-6x^{10}$ 3. $-56s^3t^7$ 5. $-6m^3n^3$ 7. $-18c^4d^2$ 9. $-200s^6t^6$ 11. $5m^2n - 3mn^2$ 13. $-r^5t^3 + r^3t^7$ 15. $6x^2 - 7xy^2$ 17. $12x^2y^3 - 4x^3y^3 + 20xy^2$ 19. $24e^3f^2$ 21. $-6a^5b^5$ 23. $6x^2 + 7x - 20$ 25. $10x^2 - 19x + 6$ 27. $a^4 - a^2b^2 - 2b^4$ 29. $4m^2 - 28m + 49$ 31. $x^2 - 2xy + y^2$ 33. $49t^2 - 42rt + 9r^2$ 35. $x^2 - y^2$ 37. $6c^2 - 11cd - 7d^2$ 39. $2x^3 + 10x^2 - 30x + 18$ 41. $r^3 - 3r^2 + 3r - 1$ 43. $16a^4 - 81b^4$ 45. The postulate e. does not hold.

Page 196 Review 1. P 2. P 3. P 4. P 5. $2x^2$; 2 6. $4x^3y^2$; 5 7. $3x^2y^6$; 8 8. $3y^3$; 3 9. $4x^3 + 3x^2 + 5x + 2$ 10. $-8t^3 - 2t^2 - t + 16$ 11. $2m^2 + 7m + 1$ 12. $10a^4 + 6a^2 - 2a - 17$ 13. $-8x^{10}$ 14. $944{,}784t^{12}v^{12}$ 15. $24abc^2 + 20abcd - 84abd^2$

Page 196 Review Capsule for Section 6-4 1. a 2. b 3. 4b 4. x 5. x 6. 7rs 7. 3uw 8. $3s^2$ 9. 5b 10. 10a 11. $5m^3$ 12. 2xy 13. $\frac{1}{6}t^2$ 14. $10q^2$ 15. $1.1cd^2$

Page 198 Classroom Exercises 1. 7a 3. $6rt^3$ 5. $3x^2y^2$ 7. $x^2y^2(x^2 - y)$ 9. $a(a + 3)^2$ 11. $(s + 8)(s - 8)$ 13. $3a^2(a + 1) - 4(a + 1) = (a + 1)(3a^2 - 4)$

Page 199 Written Exercises 1. 6 3. 2c 5. yz 7. 5y 9. 5r 11. $2c^2(c - 1)$ 13. $x(3x - 1)$ 15. $5x(x^2 - 2x + 7)$ 17. $8(y + 2b)$ 19. $9r(r + 2)$ 21. $2(x^2 + 4)$ 23. $5ax(1 - a)$ 25. $3a(a^2 - ab - b^3)$ 27. $x^2(2 + 3x + 4x^2y)$ 29. $(x + y)(a - b)$ 31. $(2x - 5)(3 - a)$ 33. $(x + y)(c + b)$ 35. $(1 - y)(1 + y^2)$ 37. $(x - 9)^2$ 39. $(r - 5)^2$ 41. $(5x - 6)^2$ 43. $(s - 4t)^2$ 45. $(2n + 3)^2$ 47. $(6m + 1)^2$ 49. $(x + 2)(x - 2)$ 51. $(3x + b)(3x - b)$ 53. $(3x + 2)(3x - 2)$ 55. $(x^2 + y)(x^2 - y)$ 57. $(5a^2 + 2)(5a^2 - 2)$ 59. $(13 + 7x)(13 - 7x)$ 61. $(3 + 4b^2)(3 - 4b^2)$ 63. $(4x + 1)(4x - 1)$ 65. $(xy + a)(xy - a)$ 67. $(12 + t)(12 - t)$ 69. $(3c + 1)(2c - 1)$ 71. $(7 + 6z)(7 - 6z)$ 73. $(3s + r)(1 - 2s)$ 75. $6rt(5r - 4t + 6r^2)$ 77. $(5c - 1)^2$ 79. $(8e + 3)^2$ 81. $(S + s)(S - s)$; 1.89 square meters 83. 251.2 square centimeters 85. $(b + 1)(b^2 + 1)$ 87. $(x + y)^2(x - y)$ 89. $(x - y)(a + b)$ 91. $(x + y)(a - b)$ 93. $(a + b + c)(a + b - c)$ 95. $(2x + 3y + 3z)(2x + 3y - 3z)$ 97. $(m + n + a)(m + n - a)$ 99. $(1 + x + y)(1 - x - y)$

Page 201 Review Capsule for Section 6-5 1. a = 7 and b = 5 2. a = −7 and b = −5 3. a = −8 and b = 5 4. a = −8 and b = −10 5. a = 3 and b = 5 6. a = −6 and b = −9 7. a = 6 and b = −7 8. a = −3 and b = 7 9. a = −9 and b = 7 10. a = 2 and b = 27

Page 203 Classroom Exercises 1. 5 3. $(5c + 4)(2c - 3)$ 5. $(3a + 1)(a + 2)$

Page 203 Written Exercises 1. $(x - 7)(x + 6)$ 3. $(r + 6)(r - 2)$ 5. $(e + 10)(e - 9)$ 7. $(m - 8)(m - 1)$ 9. $(s - 9)(s - 2)$ 11. $(x + 7)(x - 5)$ 13. $(2y - 5)(y + 1)$ 15. $(3a - 2)(a + 4)$ 17. $(2c - 5)(c + 6)$ 19. $(3b + 2)(b + 1)$ 21. $(n - 6)(n - 1)$ 23. $(2y + 3)(y - 5)$ 25. $(x^2 - 3y)(x^2 + 2y)$ 27. $(a^2 - 3y^2)(a^2 + y^2)$ 29. $(x^4 + 7)(x^4 - 3)$ 31. $(x + \frac{1}{4})(x + \frac{1}{2})$ 33. $(x + 0.3)(x + 0.2)$ 35. $(a + 0.8)(a - 0.3)$ 37. $(x + a)(x + b)$ 39. $(z + k)(z + m)$ 41. $(c + d - 9)(c + d + 2)$ 43. $(2y^{2x} - 5)(y^{2x} + 1)$ 45. $(3 - c^{3n})(19 - c^{3n})$

Page 203 Puzzle See Solution Key.

Page 204 Using Statistics 1. b. The line connects (20, 8) and (110, 58). c. $y = \frac{5}{9}x - 3\frac{1}{9}$ d. 36 3. b. The line connects (150, 65) and (400, 200). c. $y = 0.54x - 16$ d. $92,000

Page 207 Classroom Exercises 1. $(x-2)(x^2+2x+4)$ 3. $(a+1)(a^2-a+1)$ 5. $11(a^2+4b^2)$ 7. $10(3y+10)(3y-10)$ 9. $12z(2z+1)(2z-1)$ 11. $-7(a+3)^2$ 13. $b(2b-3)(3b+5)$ 15. $-3x^3(x+3)(x-3)$

Page 207 Written Exercises 1. $(m-n)(m^2+mn+n^2)$ 3. $(a-3b)(a^2+3ab+9b^2)$ 5. $(ab+3c)(a^2b^2-3abc+9c^2)$ 7. $(1-4y)(1+4y+16y^2)$ 9. $(z^2+5)(z^4+5z^2+25)$ 11. $(y+4z)(y^2-4yz+16z^2)$ 13. $3(g-8)(g+1)$ 15. $2k(3k-5)^2$ 17. $2(r+10)(r-10)$ 19. $2(x+4y)(x-4y)$ 21. $(3+s)(2-s)$ 23. $(x^2+9)(x+3)(x-3)$ 25. $a(2x+1)^2$ 27. $4(a^2+4b^2)$ 29. $a(b+9a)(b+5a)$ 31. $-3c(d+7)(d+2)$ 33. $(3y^2+7)(y^2-5)$ 35. $(m+n+1)(m-n+1)$ 37. $(a+b)(a-b-3)$ 39. $(2a+1)(a-2)(a-1)$ 41. $6b(a-b)$ 43. $(a^2+b^2)(a^4-a^2b^2+b^4)$ 45. $2c(2c+d)(4c^2-2cd+d^2)$ 47. $2x^2(2x+3y)(4x^2-6xy+9y^2)$ 49. $2a(2a-3)(4a^2+6a+9)$ 51. $(7x^n+1)^2$ 53. $(3z^{2n}-1)(z^{2n}-3)$ 55. $(x+\sqrt{3})(x-\sqrt{3})$ 57. $(\sqrt{5}b+\sqrt{7})(\sqrt{5}b-\sqrt{7})$ 59. $(x+2\sqrt{2})(x-2\sqrt{2})(x^2+5)$ 61. $(\sqrt{7}y+\sqrt{11}b)(\sqrt{7}y-\sqrt{11}b)$

Page 208 Review 1. $(4c+3d)(4c-3d)$ 2. $(5a+4)^2$ 3. $(d+e+f)(d-e-f)$ 4. $(a+c)(b^2+2)$ 5. $(4c-1)(3b-1)$ 6. $b(b-2)(c+1)$ 7. $(x-7)(x+4)$ 8. $(2d^2+3f)(d^2-4f)$ 9. $3(2r-7s)(r-s)$ 10. $(x^2+4)(x^2-2)$ 11. $(2a+3b)(a-5b)$ 12. $(3x^3-2y)(2x^3+7y)$ 13. $(2t-3v)(4t^2+6tv+9v^2)$ 14. $6(c^2+4d^2)(c+2d)(c-2d)$ 15. $(a+b+d)(a+b-d)$

Page 208 Review Capsule for Section 6-7 1. $\frac{7}{5}$ 2. $-\frac{23}{3}$ 3. $\frac{26}{27}$ 4. $-\frac{219}{4}$

Page 211 Classroom Exercises 1. x^3 3. $\frac{1}{m^5}$ 5. $x+3$ 7. $2ab+3b-3a$

Page 211 Written Exercises 1. a^2 3. $\frac{3y^3}{4}$ 5. $11-\frac{5}{g}$ 7. $-m+2$ 9. $-9a^3-4b^3+3b^2$ 11. $-5r^2+7r-\frac{35}{4}$ 13. $b-3$ 15. $x+11$ 17. $3x+4$ 19. $2m-1, R=3$ 21. $4a+3, R=5$ 23. x^2-2x+4 25. $5y^2+3y-4$ 27. $a^2-a+3, R=7$ 29. $6x^2-19x+75, R=-296$ 31. $4y^2-10y+25$ 33. $-2n^3+3n^2-2n+8, R=-16$ 35. $x^n-1+\frac{1}{x}$ 37. $a^2-(b+c)a+(b+c)^2+\frac{-3b^2c-3bc^2}{a+b+c}$ 39. $2x^3+3x^2+4x+5+\frac{6x-5}{x^2-2x+1}$

41. Proof: 1. $\frac{a^m}{a^n}, m > n$ (Given) 2. $n < m$ (Def. of Ineq.) 3. $n+r=m, r \in R$ (Def. of Less Than); 4. $\frac{a^m}{a^n}=\frac{a^{n+r}}{a^n}$ (Subst. Prop.) 5. $\frac{a^m}{a^n}=\frac{a^n \cdot a^r}{a^n}$ (Exponent Th. for Mult.) 6. $\frac{a^m}{a^n}=\frac{1}{a^n}(a^n \cdot a^r)$ (Def. of Div.) 7. $\frac{a^m}{a^n}=(\frac{1}{a^n}\cdot a^n)\cdot a^r$ (Assoc. Post. for Mult.) 8. $\frac{a^m}{a^n}=1\cdot a^r$ (Mult. Inv. Post.) 9. $\frac{a^m}{a^n}=a^r$ (Ident. Post. for Mult.) 10. $r=m-n$ (Add. Prop. for Eqs.) 11. $a^r=a^{m-n}$ (Subst.) 12. $\frac{a^m}{a^n}=a^{m-n}$ (Transitive Post.)

43. Proof: 1. $a^m \in R$ (Given) 2. $\frac{a^m}{a^m}=a^m\cdot\frac{1}{a^m}$ (Def. of Div.) 3. $\frac{a^m}{a^m}=1$ (Mult. Inv. Post.) 4. $a^0=1$ (Def. of a^0) 5. $\frac{a^m}{a^m}=a^0$ (Trans. Post. or Subst.) 45. No 47. $k=-10$ 49. $k=1$

Page 214 Classroom Exercises 1. -16 3. 99 5. No 7. No 9. No

Page 214 Written Exercises 1. 1 3. 25 5. 35 7. $y^2-y+2, R=-24$ 9. $2b^2-b-7, R=-16$ 11. $2x^3+3x^2+3x+1, R=0$ 13. $6y^2+14y+6, R=0$ 15 Yes 17. No 19. No 21. $(x-6)(x-3)$ 23. $(z-3)(z-2)$ 25 $(y-1)(y-4)(y-3)$ 27. $(y-1)(y-7)(y-3)$ 29. $(x+2)^2(x-2)^2$ 31. $(x+2)(x-3)(x+5)(x-4)$ 33 $(a+1)(a^2-a+1)$ 35. $(c-1)(c^2+c+1)$ 37. $(a-3b)(a^2+3ab+9b^2)$ 39. $(ab+3c)(a^2b^2-3abc+9c^2)$ 41. $k=-18$ 43. $k=5$ 45. $k=-5$ <u>or</u> $k=1$

Page 215 Review Capsule for Section 6-9 1. $12-x$ 2. $y; y-3$ 3. $2t-5; t$ 4. $g; g+2; g+4$ 5. $z; z+2$ 6. $x^2-10x+25$ 7. $64-16x+x^2$ 8. $2b^2+2b+1$ 9. $2y^2+6y+9$ 10. $12r+36$

Page 217 Classroom Exercises 1. $\{9, -7\}$ 3. $\{\frac{1}{5}, -\frac{1}{5}\}$ 5. $\{\frac{5}{3}, -\frac{1}{2}\}$ 7. $\{-\frac{10}{13}, \frac{10}{13}\}$ 9. $\{-\frac{1}{2}, 3\}$

Page 217 Written Exercises 1. $\{3, -2\}$ 3. $\{5, 4\}$ 5. $\{-\frac{1}{2}, \frac{1}{2}\}$ 7. $\{\frac{1}{2}, -5\}$ 9. $\{11, -11\}$ 11. $\{-\frac{2}{3}, -\frac{1}{2}\}$ 13. $\{-1, -2, 3\}$ 15. $\{2, 1, -1\}$ 17. $\{1, 2, -2\}$ 19. $\{2, \frac{1}{2}, -3\}$ 21. $\{-3, 3, -1, 1\}$ 23. 13 and 10 25. Smaller number: -24 <u>or</u> 20; larger number: -20 <u>or</u> 24 27. 15, 17, 19 <u>or</u> $-3, -1, 1$ 29. Proof: 1. a and b are real numbers, $a\cdot b=0$ (Given) 2. If $a \neq 0, \frac{1}{a} \in R$ (Def. of Mult. Inv.) 3. $\frac{1}{a}(a\cdot b)=\frac{1}{a}\cdot 0$ (Mult. Prop. for Eqs.) 4. $(\frac{1}{a}\cdot a)b=\frac{1}{a}\cdot 0$ (Assoc. Post. for Mult.) 5. $1\cdot b=\frac{1}{a}\cdot 0$ (Mult. Inv. Post.) 6. $b=\frac{1}{a}\cdot 0$ (Ident. Post. for Mult.) 7. $b=0$ (Zero Mult. Post.) 8. If $b \neq 0, \frac{1}{b} \in R$ (Def. of Mult. Inv.) 9. $(a\cdot b)\cdot\frac{1}{b}=0\cdot\frac{1}{b}$ (Mult. Prop. for Eqs.) 10. $a(b\cdot\frac{1}{b})=0\cdot\frac{1}{b}$

(Assoc. Post. for Mult.) 11. $a \cdot 1 = 0 \cdot \frac{1}{b}$ (Mult. Inv. Post.) 12. $a = 0 \cdot \frac{1}{b}$ (Ident. Post. for Mult.) 13. $a = 0$ (Zero Mult. Post.)

Page 218 Calculator Exercises 1. $(x-2)$ 3. $(x-1)(x+5)(x-5)$ 5. $(x-4)(x+3)(x+5)$ 7. $(x-2)(x-4)(x+3)$

Page 220 Word Problems 1. 15 cm 3. 4 cm 5. 10 m 7. 4 km

Page 223 Review 1. b^3 2. $\frac{3}{a^2}$ 3. $14-6x^2$ 4. $-3y^5+2y^3-4y$ 5. $x-10$ 6. $2b^2-4b-9$ 7. Yes 8. Yes 9. $(a-6)(a^2-2a+3)$ 10. $(x-2)(x-6)(x+2)$ 11. $x=-4$ or $x=4$ 12. $x=-5$ or $x=2$ 13. $b=3$ or $b=5$ or $b=-2$ 14. 6 and 4 15. 2 m

Page 224 Chapter Objectives and Review 1. $-5x^3$; 3 3. $-y^3z^6$; 9 5. $-4t^4+t^3-2t^2+7t+5$ 7. $6w^5+6u^2w-2u$ 9. $-3a^4b^6$ 11. $2x^2+3x-20$ 13. $y^3-9y^2+27y-27$ 15. $3x(3-xy)$ 17. $s^2(r^3+3s-6r)$ 19. $(z-2)(z^2+5)$ 21. $(9+4t)(9-4t)$ 23. $(b+7)(b-1)$ 25. $(7x-5)(2x+1)$ 27. $(4+y^2)(16-4y^2+y^4)$ 29. $2r(r-3)(r^2+3r+9)$ 31. $\frac{3}{a^4}$ 33. $z+7z^2$ 35. $2x+3$ 37. 27 39. $b^3-3b^2+6b-13$, $R=38$ 41. Not a factor. 43. $x=-2$ or $x=3$ or $x=1$ 45. 3 m

Page 226 Chapter Test 1. $3x^5$; 5 3. $-3x^2-3x-7$ 5. $2t^4v^2-6tv^3+2t^3v^3-t^2v^2$ 7. $3b^2(a+2)$ 9. $(6x-1)(x+7)$ 11. $(x+2)(x^2+3)$ 13. $3x-2$, $R=4$ 15. -49 17. Yes 19. $t=\frac{2}{3}$ or $t=-2$

Page 227 More Challenging Problems 1. If $a \neq b$, $a-b \neq 0$; $(a-b)^2 = a^2-2ab+b^2$; $a^2-2ab+b^2 > 0$; $a^2+b^2 > 2ab$ 3. 6 5. 64 7. 5 9. $x=7$ 11. 1806 13. 5.808 or $5\frac{101}{125}$ hr

Page 228 Preparing for College Entrance Tests 1. c 3. b 5. a 7. d 9. b 11. c

CHAPTER 7 RATIONAL EXPRESSIONS

Page 231 Classroom Exercises 1. $\frac{7}{4}$ 3. $\frac{b^4}{a^2}$ 5. $\frac{2-b}{b}$ 7. $\frac{1}{3}$

Page 231 Written Exercises 1. $\frac{6}{y}$ 3. $\frac{y}{z}$ 5. $\frac{r(x-y)}{t}$ 7. $\frac{4(a+b)}{2a+b}$ 9. $\frac{ax^2}{3}$ 11. $\frac{5y}{2}$ 13. $\frac{x-y}{x+y}$ 15. $\frac{x}{x-2}$ 17. -2 19. $-a$ 21. $-a-b$ 23. $-\frac{2}{3}$ 25. $-\frac{y+3}{3}$ 27. $\frac{5x-y}{4x-y}$ 29. $\frac{3t(t-4)}{t-5}$ 31. $\frac{a+b}{a-b}$ 33. $\frac{t+5r}{4t+3r}$ 35. $\frac{a-b}{a^2-ab+b^2}$ 37. $\frac{5x-6}{9x-10}$ 39. $\frac{3(x+y)}{2x}$ 41. $\frac{x^4-x^3y+x^2y^2-xy^3+y^4}{x^2-xy+y^2}$ 43. $\frac{2a^2-1}{a+1}$ 45. $\frac{a}{a-2}$ 47. $(x+y)$ is not a common factor of x^2+y^2. It cannot be simplified. 49. $(a+b)$ is not a factor of $(a+b+c)$. It cannot be simplifed. 51. The numerator and denominator do not have a common factor. It cannot be simplified. 53. Since $(a+b) \div (a+b) = 1$ and $(x+y) \div (x+y) = 1$, the expression simplifies to 1, which is the correct answer. 55. $x=-2$ or $x=2$ 57. $y=-\frac{4}{3}$ or $y=\frac{5}{2}$ 59. $\frac{P}{Q}=\frac{R}{S}$ if R is a multiple of P and S is the same multiple of Q. Examples will vary.

Page 232 Puzzle See Solution Key.

Page 234 Classroom Exercises 1. $\frac{x-3}{6}$ 3. $\frac{3(3x-5)}{2}$ 5. $\frac{1}{3(a+b)}$ 7. r^2

Page 234 Written Exercises 1. $\frac{1}{2a}$ 3. $\frac{3(x+3)}{5}$ 5. $\frac{x(3x-5)}{2}$ 7. $\frac{2a+3}{4}$ 9. $\frac{2x(x-10)}{5}$ 11. $2a^3y^2$ 13. $\frac{8b}{3a}$ 15. $\frac{y}{ab}$ 17. $-d$ 19. $-\frac{x^2+y^2}{x-y}$ 21. $\frac{x^2-xy+y^2}{(x+y)(x^2+xy+y^2)}$ 23. -1 25. $-\frac{a}{2}$ 27. $\frac{2r+1}{r(2r-1)}$ 29. 1 31. $\frac{1}{x-y}$ 33. $\frac{(t-6)(t-5)(t+5)}{t(t-3)^2(t+3)}$ 35. $\frac{x^4-x^2y^2+y^4}{x^2+xy+y}$ 37. $\frac{Q}{P}$, $P \neq 0$

Page 236 Classroom Exercises 1. $\frac{4}{3}$ 3. $\frac{5}{2(x+3)}$ 5. $\frac{5a}{a-b}$

Page 236 Written Exercises 1. $\frac{7x}{81}$ 3. -1 5. 1 7. $\frac{2x}{3}$ 9. $\frac{7}{x}$ 11. $\frac{x^2y^2(a+b)}{3}$ 13. $\frac{(a+4)(a-7)}{3}$ 15. $\frac{2a-3b}{2}$ 17. $-\frac{2a}{a+4}$ 19. $\frac{(x+3)^2}{x+1}$ 21. $\frac{3(a+5)^3}{a}$ 23. $\frac{d}{(d-2)(d-6)}$ 25. $\frac{-x-1}{x-3}$ 27. $\frac{-x-5}{x+1}$ 29. $\frac{x+5y}{x+2y}$ 31. $(2x-3y)^2$ 33. For rational expressions $\frac{P}{Q}$, $\frac{R}{S}$, and $\frac{T}{U}$ where $\frac{R}{S} \neq 0$, $\frac{P}{Q} \div \frac{R}{S} = \frac{T}{U}$ means $\frac{T}{U} \cdot \frac{R}{S} = \frac{P}{Q}$

Page 237 Review Capsule for Section 7-4 1. $-17m$ 2. $-9n+9$ 3. $-y-1$ 4. $-x+9y$ 5. $2r-2s$ 6. 6 7. $-a^2$ 8. $4y^2+4y+1$

Page 239 Classroom Exercises 1. $\frac{30}{x+y}$ 3. $3x$ 5. $\frac{4x-2}{x-4}$ 7. a. $15a^2$; b. $\frac{10a}{15a^2}, \frac{9}{15a^2}$ 9. a. 20 b. $\frac{5(x+2)}{20}, \frac{4(x-5)}{20}$ 11. a. $(2x-5)(2x+5)$ b. $\frac{(2x-3)(2x+5)}{(2x-5)(2x+5)}, \frac{25x^2}{(2x-5)(2x+5)}$

Page 240 Written Exercises 1. $\frac{25m}{m^2-n^2}$ 3. $\frac{x-10}{y}$ 5. $\frac{-6n+2}{5}$ 7. $\frac{-6x+3}{a+b}$ 9. $\frac{2}{13}$ 11. $6x^3y^2$ 13. $a^3(a-b)$ 15. $3(x+2)^2$ 17. $(2x-5)(2x+5)$ 19. $\frac{ay+bx}{x^2y^2}$ 21. $\frac{9+16n}{24mn^2}$ 23. $\frac{4mr-2ms}{r^2-s^2}$ 25. $\frac{4b-3a}{12}$ 27. $\frac{m+19}{m^2-3m-10}$ 29. $\frac{26}{21c-84}$ 31. $\frac{9z-26}{3z^2-16z+20}$ 33. $\frac{37x+3y}{x^2-y^2}$ 35. $\frac{3c+8}{c^2-4}$ 37. $\frac{9n^2}{3n-1}$ 39. $\frac{38}{15x^2-2x-24}$ 41. $\frac{x-2y^2+y}{x+y}$ 43. $\frac{2x-4}{x+2}$ 45. $\frac{-t-11}{(t+2)(t-2)(t-1)}$ 47. $\frac{15n-6}{(n-4)^2(n+2)}$ 49. $\frac{10n^2+28n+20}{(2n-3)(3n+4)(n+2)}$ 51. $\frac{1}{t^2-3t+2}$ 53. $\frac{22}{39}$ 55. $\frac{48-8m}{120+3m}$ 57. $\frac{ab}{b-1}$ 59. $\frac{c}{c+d}$ 61. $\frac{abc}{bc+ac+ab}$ 63. $\frac{t-1}{t+1}$ 65. $\frac{m^2-5m+13}{(m+7)(m+2)(m-5)}$ 67. 1. $\frac{P}{Q}-\frac{R}{Q}=\frac{P-R}{Q}$ (Def. of Subt. for Rational Expressions) 2. $\frac{P-R}{Q}=\frac{P+-R}{Q}$ (Def. of Subt. of Polynomials) 3. $\frac{P+-R}{Q}=\frac{P}{Q}+\frac{-R}{Q}$ (Dist. Post.) 4. $\frac{P}{Q}-\frac{R}{Q}=\frac{P}{Q}+\frac{-R}{Q}$ (Trans. Post.) 69. $\frac{Ax+B}{(x-1)(x-2)}=\frac{rx-2r+tx-t}{(x-1)(x-2)}$; Since the denominators are equal, $Ax+B=rx-2r+tx-t=x(r+t)-2r-t$. Therefore, adding $r+t=A$ and $-2r-t=B$ yields $r=-A-B$. Adding $2r+2t=2A$ and $-2r-t=B$ yields $t=2A+B$.

Page 242 Using Statistics 1. Class A: $\sigma=15.5$; Class B: $\sigma=2.6$. They differ because the scores in Class A are much more dispersed than those in Class B. 3. 16% 5. 500 7. 16,300 9. 19,500

Page 244 Review 1. $\frac{8}{x^2}$ 2. $\frac{b}{2d^3}$ 3. $\frac{x-3}{2(x-4)}$ 4. $\frac{x^3-y^3}{2(x+y)^2}$ 5. $\frac{8s^4t}{3rv}$ 6. $\frac{b(b+4)^2}{b-4}$ 7. $\frac{x(x+7)}{(x-2)}$ 8. 12 9. $\frac{a^2+2ab-a-2b}{2a+6b}$ 10. $-3b+5$ 11. $\frac{15+z}{2z-5}$ 12. $\frac{-7x+30}{6x^2+11x-10}$ 13. $\frac{-x+12}{(x+6)(x-1)(x+4)}$

Page 244 Calculator Exercises 37. 0.5675675 40. 2 43. −8.5

Page 244 Review Capsule for Section 7-5 1. $10x$ 2. $2n^2+20n+50$ 3. $-2x^2+16x-6$ 4. $x^2-4x-21$

Page 246 Classroom Exercises 1. 6 3. 7a 5. 3y

Page 246 Written Exercises 1. $n=36$ 3. $x=18$ 5. $n=8$ 7. $n=18$ 9. $n=-5$ 11. $x=-40$ 13. $x=\frac{19}{3}$ 15. $x=5$ 17. $z=6$ 19. $a=\frac{19}{2}$ 21. $x=4$ 23. $x=8$ 25. $x=-1$ 27. $y=-1$ 29. Empty set 31. $x=\frac{2}{5}$ 33. $n=-\frac{35}{2}$ 35. $x=17a$ 37. When both sides of an equation are multiplied by a nonzero polynomial, the equation formed is equivalent to the original equation. Therefore, if r is in the solution set of the original equation, then r is in the solution set of any equation that is equivalent to the original equation.

Page 247 Review Capsule for Section 7-6 1. x = the numerator. Then $x-2$ = the denominator. $\frac{x}{x-2}$ 2. x = the numerator. Then $71-x$ = the denominator. $\frac{x}{71-x}$ 3. x = the denominator. Then $3x-1$ = the numerator. $\frac{3x-1}{x}$ 4. x = the numerator. Then $2x+5$ = the denominator. $\frac{x}{2x+5}$ 5. $2x$ = the numerator. Then $5x$ = the denominator. $\frac{2x}{5x}$

Page 249 Classroom Exercises For each exercise, the answers are given in this order: original fraction; new fraction; equation. 1. $\frac{x}{x+5}; \frac{x+3}{x+9}; \frac{x+3}{x+9}=\frac{1}{2}$ 3. $\frac{x}{x+1}; \frac{x+y}{x+1+y}; \frac{x+y}{x+1+y}=\frac{8}{9}$

Page 249 Word Problems 1. $\frac{15}{7}$ 3. $\frac{25}{70}$ 5. 3.5 cm 7. \$60,000 at 12%; \$40,000 at 11% 9. 1 11. $-\frac{2}{3}$, 1 13. 8 15. 6000

Page 251 Review Capsule for Section 7-7 1. 80t km 2. 3r miles 3. $\frac{1000}{t}$ km/hr 4. $\frac{500}{r}$ hr

Page 252 Classroom Exercises 1. Rate: 6; Time: $\frac{y}{6}$; Equation: $\frac{y}{6}=\frac{y}{8}+\frac{1}{2}$

Page 253 Word Problems 1. 960 kilometers 3. 60 km/hr 5. 400 km/hr 7. 50 km 9. 20 mi/hr 11. 50 mi/hr 13. 10 km/hr 15. $27\frac{3}{11}$ minutes 17. $10\frac{10}{11}$ minutes

Page 254 Calculator Exercises 1. 1.3529412 3. 2.7333333 5. 4.2982456

Page 256 Classroom Exercises 1. $A=\frac{1}{4}t$ 3. $A=\frac{1}{5}t+\frac{1}{4}t$

Page 257 Word Problems 1. $1\frac{5}{7}$ minutes 3. $4\frac{2}{5}$ days 5. 6 minutes 7. 12 minutes 9. 6 days 11. $1\frac{13}{15}$ hours 13. 15 hours 15. $2\frac{2}{17}$ hours

Page 258 Review 1. $b = \frac{43}{6}$ 2. $c = \frac{73}{11}$ 3. $a = -6$ or $a = 2$ 4. $\frac{7}{11}$ 5. 20 6. 40 mi/hr 7. $2\frac{2}{5}$ hours 8. $10\frac{1}{2}$ days

Page 261 Chapter Objectives and Review 1. $\frac{x}{3}$ 3. $\frac{-a^2+2a-4}{a-2}$ 5. $\frac{6}{b^2}$ 7. $\frac{t+1}{t^2(t-1)}$ 9. $\frac{b}{3a-5b}$ 11. $\frac{x}{z^2}$ 13. $\frac{-3r^3+11r^2s-10rs^2+4s^3}{(3r^2-5rs+2s^2)(r^2-rs-2s^2)}$ 15. $x = 14$ 17. $t = 2$ 19. $x = \frac{14}{5}$ or $x = 7$ 21. $\frac{8}{13}$ 23. $7\frac{1}{3}$ miles 25. $4\frac{19}{20}$ hours

Page 263 Chapter Test 1. $\frac{tz^2}{2}$ 3. $\frac{3}{4}$ 5. $\frac{16}{6t+3}$ 7. $\frac{10}{t+2q}$ 9. $\frac{-3y^3-10y^2-25y-41}{y^4-16}$ 11. $x = \frac{13}{7}$ or $x = -4$ 13. $\frac{17}{35}$ 15. 3 hours

Page 264 Review of Word Problems 1. Gwen: 10 years old 3. Speed of plane going: 200 mi/hr; speed of plane returning: 400 mi/hr 5. $a = 4d$ 7. 1896 entrants 9. 8 and 14 or −8 and −14 11. 2; 5 13. 3 days

Page 265 Cumulative Review 1. d 3. b 5. c 7. a 9. d 11. d 13. c 15. d 17. a 19. d 21. b

Page 267 Preparing for College Entrance Tests 1. a 3. a 5. c 7. b 9. a 11. d 13. c 15. c 17. b 19. d 21. b 23. d 25. b

CHAPTER 8 RADICALS/QUADRATIC EQUATIONS

Page 271 Classroom Exercises 1. 11 3. $-10|b|$ 5. −2 7. $-|z|$ 9. −2 11. $12|x|y^2$ 13. −54 15. $|z+5|$

Page 271 Written Exercises 1. 4 3. 6 5. −10 7. $\frac{3}{5}$ 9. 8 11. −6 13. 4 15. 7 17. −5 19. 2 21. $|x|$ 23. x^2 25. $2|a|$ 27. $\pm 5a^2$ 29. $2x$ 31. $-5b^5$ 33. $|c^5d^6|$ 35. $-0.3b^2$ 37. $-|y-9|$ 39. $|a+1|$ 41. $|t-1|$ 43. $6b^{18}$ 45. $\left|\frac{6ab^2}{7m^3}\right|$ 47. $\frac{(a+b)^2}{(a-b)^2}$ 49. $2(x-y)^2$ 51. $x-y$ 53. $-(r^3+s^3)$ 55. $2|x^a|$ 57. x^{4b-8} 59. $|x+y|$

Page 271 Review Capsule for Section 8-2 1. $4 \cdot 5$ 2. $25 \cdot 6$ 3. $9s^2t^4 \cdot t$ 4. $81x^2 \cdot 2xy$ 5. $4a^4b^6c^8 \cdot 3ac$ 6. $25a^2b^6 \cdot 10a$ 7. $81a^4b^2 \cdot b$ 8. $25x^6y^6 \cdot 15y$ 9. $36p^4q^4 \cdot 6q$ 10. $9c^2d^4 \cdot 6d$ 11. $64x^{10} \cdot 5$ 12. $16x^2y^8 \cdot 2xy$ 13. $25r^8t^8 \cdot 35r$ 14. $a^2b^4 \cdot 5ab$ 15. $49a^{2n} \cdot 1$

Page 273 Classroom Exercises 1. $\sqrt{15}$ 3. 3 5. $|b|\sqrt{ab}$ 7. $\sqrt{6ab}$ 9. $\sqrt{2}$ 11. $|b|\sqrt{2}$ 13. $\frac{\sqrt{14}}{4}$ 15. $\frac{\sqrt{3}}{3}$ 17. $\frac{\sqrt{30}}{5}$

Page 273 Written Exercises 1. $\sqrt{10}$ 3. 5 5. 30 7. $24\sqrt{3}$ 9. $60\sqrt{6}$ 11. 36 13. $y^2\sqrt{y}$ 15. $2z^2\sqrt{5}$ 17. $2|a|\sqrt{b}$ 19. $4|x|\sqrt{3x}$ 21. $15|x|\sqrt{2xy}$ 23. $3|xy|\sqrt{6x}$ 25. 2 27. 210 29. $|ab|$ 31. $5a^2b^3\sqrt[3]{5}$ 33. $\sqrt{17}$ 35. $2\sqrt{5}$ 37. $\sqrt{2}$ 39. 3 41. 3 43. $\sqrt{3}$ 45. $2\sqrt{2}$ 47. 10 49. $\sqrt{7}$ 51. $\frac{\sqrt{5}}{5}$ 53. $\frac{\sqrt{10}}{10}$ 55. $\frac{\sqrt{33}}{11}$ 57. $\frac{\sqrt{a}}{|a|}$ 59. $\frac{2\sqrt{q}}{|q|}$ 61. $\frac{\sqrt{3xy}}{|xy|}$ 63. $\frac{\sqrt{39}}{3}$ 65. $\frac{16}{5}$ 67. $\frac{11\sqrt{2}}{12}$ 69. $\sqrt{3y}$ 71. $|a|\sqrt{5a}$ 73. $2|x|$ 75. 3 77. 32 79. $3\sqrt{3}-3\sqrt{2}$ 81. $2\sqrt{7}-7$ 83. 3 85. $\frac{2}{3}$ 87. 1. $a, b \in R$ (Given) 2. $(\sqrt{a}\sqrt{a})(\sqrt{b}\sqrt{b}) = (\sqrt{a})^2(\sqrt{b})^2$ (Def. of Exponent) 3. $(\sqrt{a}\sqrt{b})(\sqrt{a}\sqrt{b}) = (\sqrt{a})^2(\sqrt{b})^2$ (Comm. and Assoc. Post. for Mult.) 4. $(\sqrt{a}\sqrt{b})^2 = (\sqrt{a})^2(\sqrt{b})^2$ (Def. of Exponent) 5. $(\sqrt{a}\sqrt{b})^2 = a \cdot b$ (Power Theorem for Exponents) 6. $\sqrt{a}\sqrt{b} = \sqrt{a \cdot b}$ (Def. of Square Root) 89. Suppose $\sqrt{a^2+b^2} > a+b$. Then squaring, we get $a^2+b^2 > a^2+2ab+b^2$. Since a and b are nonnegative, 2ab is either positive or zero. If $2ab > 0$, $a^2+b^2 < a^2+2ab+b^2$. If $2ab = 0$, $a^2+b^2 = a^2+2ab+b^2$. Both cases contradict the supposition. Thus, the supposition is false and it follows that $\sqrt{a^2+b^2} \le a+b$.

Page 275 Review Capsule for Section 8-3 1. $-2a^3+7a^2+2ab+4b^2$ 2. $-6r^3+6r+10$ 3. $2x^3-2x^2$ 4. $-4q^2+3q-9$ 5. x^2-y^2 6. $81q^2-49r^2$ 7. $2x^2-15x-8$

Page 276 Classroom Exercises 1. $7\sqrt{7}$ 3. $6\sqrt{2}$ 5. 23 7. $-2+2\sqrt{5}$ 9. $\frac{5-5\sqrt{7}}{-6}$ 11. $a\sqrt{3}+a\sqrt{2}$

Page 276 Written Exercises 1. $2\sqrt{3}$ 3. $\sqrt{a}$ 5. $4\sqrt{3}$ 7. $3\sqrt{2}$ 9. $-8\sqrt{3}$ 11. $19\sqrt{3}$ 13. $21\sqrt{2}$ 15. $(32|b|-5b^2)\sqrt{b}$ 17. $11\sqrt{5}$ 19. $11\sqrt{2}+7\sqrt{3}$ 21. $3-21\sqrt{2}$ 23. $\frac{25\sqrt{3}-6\sqrt{5}}{30}$ 25. −3 27. $\frac{3}{14}\sqrt{7y}$

29. $2\sqrt[3]{3} + \sqrt[3]{4}$ 31. $17\sqrt[3]{2}$ 33. $24 + 7\sqrt{6}$ 35. $62 - 20\sqrt{6}$ 37. −11 39. −6 41. −2 43. $-\frac{11}{4}$ 45. $\frac{3 + \sqrt{5}}{2}$
47. $2(\sqrt{7} - \sqrt{5})$ 49. $\frac{5 + 3\sqrt{3}}{-2}$ 51. $\frac{43 + 19\sqrt{2}}{23}$ 53. $\frac{15 + 14\sqrt{7} - 21\sqrt{5} - 2\sqrt{35}}{-17}$ 55. $2 - \sqrt{3}$ 57. $\frac{4\sqrt{ab}}{|a| - |b|}$
59. $\frac{-\sqrt{r + 1}}{|r + 1|}$ 61. $(y + 5\sqrt{2})(y + 3\sqrt{2})$ 63. $(t + \sqrt{5})^2$ 65. $x = \frac{-1 + \sqrt{5}}{2}$

Page 279 Classroom Exercises 1. x = 9 3. y = 46 5. x = 9 7. q = 3

Page 279 Written Exercises 1. x = 36 3. $y = \frac{1}{3}$ 5. x = 25 7. n = 7 9. x = 6 11. empy set 13. $y = \frac{49}{2}$ 15. $x = \frac{17}{4}$
17. ∅ 19. y = 23 21. x = 15 23. a = −64 25. x = −5 or x = 5 27. x = 4 29. ∅ 31. x = 1 33. x = 0 35. z = 5
37. ∅ 39. $s = \frac{1}{2}gt^2$ 41. z = 5

Page 279 Review Capsule for Section 8-5 1. 7 2. 7 3. 8 4. 25 5. $5\sqrt{5}$ 6. −4

Page 281 Classroom Exercises 1. 5 3. 4 5. $7\sqrt{2}$ 7. (3, 3) 9. $(\frac{7}{2}, \frac{1}{2})$

Page 281 Written Exercises 1. $\sqrt{10}$ 3. $\sqrt{26}$ 5. 10 7. $\sqrt{2}$ 9. 9 11. $\sqrt{(a - c)^2 + (b - d)^2}$ 13. 48 15. PQ $= 3\sqrt{10}$, QR $= 3\sqrt{5}$, and RP $= 3\sqrt{5}$. Since QR = RP, it is isosceles. 17. CD = 5, DE $= 2\sqrt{5}$, and EC = 5. Since CD = EC, the triangle is isosceles. 19. $\sqrt{97}$ 21. PR $= \sqrt{130}$; QS $= \sqrt{130}$ 23. $(-\frac{7}{2}, 2)$ 25. (−3, −5) 27. $(-3, -\frac{5}{2})$

29. $(\frac{9}{4}, -\frac{25}{4})$ 31. $(0, \frac{a + b}{2})$ 33. (0, 8) 35. (0, −2) 37. Let M be the midpoint of $\overline{AC}$ and N be the midpoint of $\overline{AB}$. Then M(−2, 4) and N(1, 1) are the endpoints of $\overline{MN}$. $MN = \sqrt{(-2 - 1)^2 + (4 - 1)^2} = 3\sqrt{2}$. $\frac{1}{2}BC$ $\frac{1}{2}\sqrt{(-2 - (-8))^2 + (-4 - 2)^2} = 3\sqrt{2}$. Thus, $MN = \frac{1}{2}BC$. 39. $PQ = \sqrt{(-2 - 2)^2 + (1 - 2)^2} = \sqrt{17}$, $QR = \sqrt{(2 + 1)^2 + (2 + 1)^2}$ $= 3\sqrt{2}$, $RP = \sqrt{(-1 + 2)^2 + (-1 - 1)^2} = \sqrt{5}$. Thus, PQ + RP ≠ QR. 41. $r = \sqrt{x^2 + y^2}$ 43. $\overline{BC}$ and $\overline{AD}$ are the nonparallel sides of the trapezoid. If M is the midpoint of $\overline{BC}$ and N is the midpoint of $\overline{AD}$, then $M = (-\frac{5}{2}, 1)$ and $N = (\frac{7}{2}, 1)$. Thus, MN $= \sqrt{(-\frac{5}{2} - \frac{7}{2})^2 + (1 - 1)^2} = 6$. The parallel sides are $\overline{AB}$ and $\overline{DC}$. AB = 3 and DC = 9. Therefore, $\frac{1}{2}(AB + DC) = \frac{1}{2}(3 + 9)$ = 6. Since MN = 6 and $\frac{1}{2}(AB + DC) = 6$, $MN = \frac{1}{2}(AB + DC)$. 45. $\sqrt{37}$ 47. (8, 8) 49. $(\frac{mx_2 + nx_1}{m + n}, \frac{my_2 + ny_1}{m + n})$

51. Draw any segment with endpoints $P_1(x_1, y_1)$ and $P_2(x_2, y_2)$. Let $M(\frac{x_1 + x_2}{2}, \frac{y_1 + y_2}{2})$ be a point on P_1P_2. Use the distance formula to show that $P_1M = \frac{1}{2}P_1P_2$ and $MP_2 = \frac{1}{2}P_1P_2$.

Page 284 Review Capsule for Section 8-6 1. $\frac{3}{2}$ 2. $\frac{3}{5}$ 3. $\frac{7}{5}$ 4. $-\frac{1}{8}$ 5. $\frac{13}{3}$ 6. 3 7. −1 8. $-\frac{1}{4}$ 9. $\frac{1}{6}$

Page 285 Classroom Exercises 1. (0, 0), (a, 0), (a, a), (0, a) 3. (0, 0), (a, 0), (a + b, c), (b, c) 5. (0, 0), (2a, 0), (a, b)

Page 285 Written Exercises 1. Slope of $\overline{AB}$ = 0. Slope of $\overline{DC}$ = 0. Therefore, $\overline{AD}$ is parallel to $\overline{DC}$. Slope of $\overline{AD} = \frac{b}{c}$. Slope of $\overline{BC} = \frac{-b}{c}$. Since the slopes are not equal, $\overline{AD}$ is not parallel to $\overline{BC}$. 3. $AD = \sqrt{c^2 + b^2}$ and $BC = \sqrt{c^2 + b^2}$
5. AB = | a | and $MN = \frac{1}{2}|a|$. Therefore $MN = \frac{1}{2}AB$. 7. Let A(0, 0), B(0, b), C(a, b), and D(a, 0) be the vertices of a rectangle. Then the diagonals of the rectangle are $\overline{BD}$ and $\overline{AC}$. Use the distance formula to show that BD = AC. 9. Let A(0, 0), B(b, c), C(a + b, c) and D(a, 0) be the vertices of a parallelogram. $\overline{AD}$ and $\overline{BC}$ are opposite sides and $\overline{AB}$ and $\overline{DC}$ are opposite sides. Use the distance formula to show that AD = BC and that AB = DC. 11. The slope of y = x is 1. The slope of PQ is −1. Thus, they are perpendicular. The midpoint of PQ is $(\frac{a + b}{2}, \frac{a + b}{2})$, which is a solution to y = x. Thus, y = x bisects $\overline{PQ}$. 13. Let the vertices of the isosceles triangle be A(0, 0), B(a, 0), and $C(\frac{a}{2}, b)$. The median to the base intersects $\overline{AB}$ at at $M(\frac{a}{2}, 0)$, the midpoint of the base. The slope of $\overline{CM}$, the median, is $\frac{b}{0}$, which is undefined. Thus, $\overline{CM}$ is parallel to the y axis. The slope of $\overline{AB}$, the base, is 0. Thus, $\overline{AB}$ and $\overline{CM}$ are perpendicular. 15. Let A(0, 0), C(2a, 0), and B(2b, 2c) be the vertices of a triangle. Then R(b, c) is the midpoint of $\overline{AB}$, S(a + b, c) is the midpoint of $\overline{BC}$, and T(a, 0) is the midpoint of AC. Let P(x, y) be the point of intersection of the medians $\overline{AS}$, $\overline{BT}$, and $\overline{CR}$. To find the values of x and y: 1. Find the slope of each median. 2. Use the point-slope form to write an equation for each median. 3. Solve the system of equations for x and y. 4. Show that $PA = \frac{2}{3}AS$, $PB = \frac{2}{3}BT$, and $PC = \frac{2}{3}CR$.

Page 286 Review 1. −12 2. $6x^2$ 3. $2c^6d^2$ 4. $\frac{9a^4|b^5|}{5c^2|d|}$ 5. $-(a + b)^3$ 6. $90\sqrt{2}$ 7. $10c^4$ 8. $2y^2|z|$

9. $3\sqrt{5} - 10$ 10. $\frac{3\sqrt{3}}{2}$ 11. $\frac{3\sqrt{15}}{5}$ 12. $2\sqrt{y}$ 13. $x\sqrt{5}$ 14. 5 15. $3|b|$ 16. $22\sqrt{5}$ 17. $8\sqrt[3]{3}$ 18. $\frac{3\sqrt{5a}}{10}$ 19. $\frac{2\sqrt[3]{9}}{5}$ 20. 40 21. $36|a| - 25|b|$ 22. $6\sqrt{2} - 6$ 23. $\frac{8\sqrt{3} + 40 + \sqrt{6} + 5\sqrt{2}}{-22}$ 24. $\frac{(2\sqrt{2} - 12 + 3\sqrt{10} - 18\sqrt{5})}{-34}$ 25. $4 - \sqrt{15}$ 26. b = 24 27. x = 5 28. ∅ 29. 10 30. Let A(0, 0), B(2c, 0), C(2d, 2e), and D(2b, 2c) be the vertices of a quadrilateral. Then M(b + d, e) is the midpoint of $\overline{DC}$; N(c + d, e) is the midpoint of $\overline{BC}$; P(c, 0) is the midpoint of $\overline{AB}$; and Q(b, c) is the midpoint of $\overline{AD}$. Let S be the midpoint of $\overline{MP}$. Then S has coordinates $(\frac{b+d+c}{2}, \frac{c+e}{2})$. Let T be the midpoint of $\overline{QN}$. Then T has coordinates $(\frac{b+c+d}{2}, \frac{c+e}{2})$. Since the coordinates of S and T are the same, $\overline{MP}$ and $\overline{QN}$ bisect each other.

Page 286 Review Capsule for Section 8-7 1. 1 2. −25 3. $\frac{7}{4}$ 4. 10 5. 10 6. $\frac{105}{16}$ or $6\frac{9}{16}$

Page 287 Classroom Exercises 1. L 3. L 5. Q

Page 288 Written Exercises 1. (−5, 26), (−4, 17), (−3, 10), (−2, 5), (−1, 2), (0, 1), (1, 2), (2, 5), (3, 10), (4, 17), (5, 26) 3. (−5, 56), (−4, 42), (−3, 30), (−2, 20), (−1, 12), (0, 6), (1, 2), (2, 0), (3, 0), (4, 2), (5, 6) 5. c = −5 7. b = −4 9. A 11. B 13. The three parabolas have the same size and shape, but they are positioned differently with respect to the x and y axes. All parabolas open upward. For $y = 2x^2$, the vertex is (0, 0) and the axis of symmetry is the y axis. For $y = 2x^2 - 8x + 8$, the vertex is (2, 0), and x = 2 is the axis of symmetry. For $y = 2x^2 - 8x + 10$, the vertex is (2, 2), and x = 2 is the axis of symmetry. 15. Both parabolas have the same vertex (0, c), and both have the y axis as the axis of symmetry. If $a_1 > 0$ or $a_2 > 0$, the parabola will open upward. If $a_1 < 0$ or $a_2 < 0$, the parabola will open downward. The absolute value of a determines the steepness of the parabola. The greater the absolute value of a, the steeper the curve. 17. one second; 2.04 seconds 19. about 16.33 seconds; 326.5 m

Page 289 Calculator Exercises 1. 6.7476077 3. 7.7796351 5. 38.601048 7. 2.2967226 9. 2.9974033 11. 5.2165245

Page 289 Review Capsule for Section 8-8
1. Yes 2. Yes 3. Yes 4. No

Page 291 Classroom Exercises 1. U 3. D 5. U 7. (−4, 80), (−3, 45), (−2, 20), (−1, 5), (0, 0), (1, 5), (2, 20), (3, 45), (4, 80) 9. (−4, 48), (−3, 27), (−2, 12), (−1, 3), (0, 0), (1, 3), (2, 12), (3, 27), (4, 48)

Page 291 Written Exercises 1. $y = -3x^2$ 3. $y = 2x^2$ 5. e, c, f, d, a, b 7. (0, 0); (0, 6) 9. They have the same size and shape, the same vertex, (0, 0), and the same axis of symmetry, the y axis. However, the graphs open in opposite directions. 11. Axis of symmetry: x = 0; vertex: (0, 0); minimum 13. Axis of symmetry: x = 0; vertex: (0, 0); maximum 15. Axis of symmetry: x = 0; vertex: (0, 0); minimum 17. Axis of symmetry: x = 0; vertex: (0, 0); maximum 19. The axis of symmetry is the y axis, and the points (−2, 6) and $(3, 13\frac{1}{2})$ are also on the graph. 21. The y axis is the axis of symmetry, and the points (−1, 5) and (−2, 8) are also on the graph. 23. $x = ay^2$ 25. If a < 0, then x < 0 and it will open to the left. 27. The x axis is the axis of symmetry and the parabola opens to the left. The vertex is (0, 0). The points (−8, 2) and (−8, −2) are on the graph.

Page 292 Review Capsule for Section 8-9 1. y = 100 2 y = 36 3. $t = \frac{1}{3}$ 4. $\frac{P_1}{s_1} = \frac{P_2}{s_2}$

Page 293 Classroom Exercises 1. 2 3. −1 5. $\frac{45}{8}$ 7. $-\frac{320}{9}$

Page 294 Written Exercises 1. 48 3. 75 5. 9 7. ±4 9. $\frac{E_1}{V_1^2} = \frac{E_2}{V_2^2}$ 11. A increases 13. The area is multiplied by 9. 15. 9k; 7k; 5k; 3k; k; −k; −3k; −5k; −7k; −9k; No 17. As x increases by 1 in $y = kx^2$, the second differences do not change for y. Each difference is 2k. 19. $E = kv^2$ 21. d is multiplied by 4; d is multiplied by 9; d is multiplied by $\frac{1}{4}$ 23. 45 square inches 25. 4 times as great; The stopping time is multiplied by 4. 27. approximately 6.4 seconds

Page 295 Puzzle See Solution Key.

Page 295 Review Capsule for Section 8-10 1. 8; 21; 1 2. −21; −15; $-2t^2 + 13t - 21$ 3. 0 4. 0 5. 3 6. −7

Page 297 Classroom Exercises 1. f_1, f_4, and f_5; f_2 and f_6 3. (0, 0) for f_1, (0, 1) for f_2, (0, 0) for f_3, (0, −1) for f_4, (0, 6) for f_5, and (0, 4) for f_6

Page 297 Written Exercises 1. $f_1(x)$: (0, −3); $f_2(x)$: (0, 0); $f_3(x)$: (0, 6) 3. (0, −4) 5. (0, 8) 7. Yes, because |3| = |−3| = 3. 9. Opens upward; vertex: (0, −3) 11. Opens upward; vertex: (0, 6) 13. Opens downward; vertex: $(0, -\frac{1}{2})$ 15. The vertex will be (3, −13); $g(x) = x^2 - 6x - 4$ 17. They are congruent and have the same axis of symmetry. They have different vertices and open in opposite directions. 19. No change in shape; vertex moves up 4; opens in opposite direction. 21. $c = -13$ 23. The graphs are congruent. They are not functions, as they fail the vertical line test. 25. Slide the graph of $x = ay^2 + by$ c units to the right to obtain $x = ay^2 + by + c$. To obtain $x = ay^2 + by - c$, slide $x = ay^2 + by$ c units to the left.

Page 299 Review Capsule for Section 8-11 1. -9 2. 12 3. $-5t^2 + 13t - 7$ 4. $(ax - 2ah + b)x + (ah - b)h + c$ 5. $35 + 8\sqrt{5}$ 6. $2\sqrt{3}$

Page 301 Classroom Exercises 1. $x = 0$ 3. $x = 2$ 5. (0, 1); minimum 7. (0, −4); minimum 9. The three graphs have the y axis as the axis of symmetry. The vertices for $y = -2x^2$, $y = -2x^2 + x - 3$, and $y = 2x^2 + x - 3$ are (0, 0), (0, −3), and (0, −3), respectively. The vertex is a maximum for $y = -2x^2$ and $y = -2x^2 + x - 3$. The vertex is a minimum for $y = 2x^2 + x - 3$. $y = -2x^2 + x - 3$ and $y = 2x^2 + x - 3$ are symmetric with respect to the line $y = -3$.

Page 301 Written Exercises 1. $x = \frac{1}{2}$; $V(\frac{1}{2}, -\frac{9}{4})$; minimum 3. $x = 0$; V(0, 4); minimum 5. $x = -\frac{1}{2}$; $V(-\frac{1}{2}, -\frac{9}{4})$; minimum 7. $x = -\frac{1}{6}$; $V(-\frac{1}{6}, \frac{25}{12})$; maximum 9. $x = -1$; V(−1, −2); minimum 11. $x = \frac{3}{10}$; $V(\frac{3}{10}, \frac{29}{20})$; maximum 13. a. $x = 2$ b. (2, −4) c. (1, −3), (3, −3) 15. a. $x = \frac{1}{2}$ b. $(\frac{1}{2}, -\frac{25}{2})$ c. (0, −12), (1, −12) 17. a. $x = \frac{1}{2}$ b. $(\frac{1}{2}, -4)$ c. (0, −3), (1, −3) 19. The axis of symmetry is shifted 4 units to the left. The vertex changes from (2, −7) to (−2, −7). 21. $y = -3x^2 + 24x + 4$ 23. $y = a(x - h)^2 + k$ is equivalent to $y = ax^2 + (-2ah)x + (ah^2 + k)$. The abscissa of the vertex of this equation is $-\frac{(-2ah)}{2a} = h$. The ordinate is $-\frac{(-2ah)^2}{4a} + (ah^2 + k)$ which simplifes to k.

Page 302 Calculator Exercises 12. −6.25 13. −4.0 14. −6.25 15. −12.5 16. −9.375 17. −4

Page 304 Computer Exercises See Solution Key.

Page 305 Classroom Exercises 1. minimum 3. maximum 5. $f(n) = 16n - n^2$ 7. $f(n) = 2n^2 - 40n + 400$

Page 306 Word Problems 1. $112\frac{1}{2}\,m^2$ 3. 4592 m; 30.6 seconds 5. 10 and 10 7. 15 and 15 9. height: 3 cm; width: 6 cm 11. Each side is $\frac{P}{4}$ units

Page 307 Review 1. (−6, 38), (−5, 27), (−4, 18), (−3, 11), (−2, 6), (−1, 3), (0, 2), (1, 3), (2, 6), (3, 11), (4, 18), (5, 27), (6, 38) 2. (−6, 16), (−5, 9), (−4, 4), (−3, 1), (−2, 0), (−1, 1), (0, 4), (1, 9), (2, 16), (3, 25), (4, 36), (5, 49), (6, 64) 3. (−6, 14), (−5, 6), (−4, 0), (−3, −4), (−2, −6), (−1, −6), (0, −4), (1, 0), (2, 6), (3, 14), (4, 24), (5, 36), (6, 50) 4. $b = 4$ 5. $c = 12$ 6. $y = 3x^2$ 7. It opens upward and has the y axis as its axis of symmetry. (−1, 7) and (−2, 12) are on the graph. 8. 105 9. 4.6 to the nearest tenth 10. $100\pi\,m^2$ In Exercises 11-14, the graphs are congruent to each other and congruent to $y = 4x^2$ and $y = -4x^2$. All graphs have the y axis as the axis of symmetry. $y = 4x^2$, $y = 4x^2 - 3$, and $y = 4x^2 + 2$ open upward. $y = -4x^2$, $y = -4x^2 + 4$, and $y = -4x^2 - 1$ open downward. The vertex of $y = 4x^2$ and $y = -4x^2$ is (0, 0). 11. (0, −3) 12. (0, 2) 13. (0, 4) 14. (0, −1) 15. $x = -\frac{3}{2}$; V(0, 0); (1, 4); (−4, 4) 16. $x = \frac{1}{3}$; $V(\frac{1}{3}, 3\frac{2}{3})$; (0, 4), $(\frac{2}{3}, 4)$ 17. $x = \frac{3}{5}$; $V(\frac{3}{5}, -\frac{1}{5})$; (0, −2), $(\frac{6}{5}, -2)$ 18. $12\frac{1}{2}$ cm

Page 309 Chapter Objectives and Review 1. $9|s|$ 3. $|a^5b^7|$ 5. $3\sqrt{2}$ 7. $3\sqrt[3]{2}$ 9. $5a^2|b|$ 11. $\frac{5}{3}$ 13. 4 15. $7\sqrt{5}$ 17. $2\sqrt{3} - 4\sqrt{2}$ 19. $\frac{2\sqrt{7}}{7}$ 21. $-4 + 2\sqrt{5}$ 23. $b = 32$ 25. $w = 8$ 27. $n = 4$ 29. $3\sqrt{5}$ 31. $\sqrt{113}$ 33. $(-6\frac{1}{2}, -2)$ 35 (−10, 14) 37. Let M be the midpoint of side AD and N be the midpoint of side BC. M has coordinates $(\frac{d}{2}, \frac{c}{2})$ and N has coordinates $(\frac{a+b}{2}, \frac{c}{2})$. The slope of median MN is 0. The slope of each base is also 0. Thus, the median is parallel to the bases. 39. The graphs are congruent and all open upward. The y axis is the axis of symmetry for $y = 3x^2$. Its vertex is (0, 0). $x = -\frac{1}{3}$ is the axis of symmetry for $y = 3x^2 + 2x$ and $y = 3x^2 + 2x + 6$. The vertices for these graphs are $(-\frac{1}{3}, -\frac{1}{3})$ and $(-\frac{1}{3}, 5\frac{2}{3})$, respectively. 41. $x = 0$; V(0, −3); minimum 43. $x = 0$; V(0, 7); maximum 45. 63π 47. about 21.4 seconds; 2250 m

Page 311 Chapter Test 1. $3xy^4\sqrt[3]{3x^2}$ 3. $3|x^5y^3|\sqrt{2y}$ 5. $-3\sqrt{3} + 4\sqrt{6}$ 7. $14 + 6\sqrt{5}$ 9. $x = 7$ 11. $(-2\frac{1}{2}, \frac{1}{2})$ 13. Use the distance formula to show that DE = EF = FD = 6. 15. $x = 1$ 17. maximum 19. They have different vertices and open in opposite directions. $y = \frac{1}{2}x^2$ opens upward; $y = -\frac{1}{2}x^2 + 3$ opens downward

Page 312 Preparing for College Entrance Tests 1. d 3. a 5. d

CHAPTER 9 COMPLEX NUMBERS/QUADRATIC EQUATIONS

Page 316 Classroom Exercises 1. $7i$ 3. $-7i$ 5. $\frac{i\sqrt{3}}{3}$ 7. $-i$ 9. $-19i$ 11. -6 13. $-i$ 15. 2 17. $\frac{1}{2}$

Page 316 Written Exercises 1. $10i$ 3. $13i$ 5. $-10i$ 7. $-8i$ 9. $48i$ 11. $-i$ 13. $-3i$ 15. $-i$ 17. $3i$ 19. $-i$ 21. i 23. $9i\sqrt{2}$ 25. $-2i\sqrt{5}$ 27. -12 29. -12 31. 24 33. -2 35. -10 37. -10 39. $-2\sqrt{30}$ 41. $-36i$ 43. $\frac{3}{2}$ 45. -1 47. $-3i$ 49. 2 51. 2 53. $\frac{i\sqrt{2}}{5}$ 55. $-30i$ 57. 1 59. -3 61. 0 63. $-i$ 65. $-i$ 67. i 69. $i, -1, -i$, and 1 71. -1 73. 1 75. -1 77. 1 79. -3 81. For all pure imaginary numbers ai and bi, $ai \cdot bi = abi^2$. But $abi^2 = ab(-1) = -ab$, which is a real number. Therefore, the set is not closed. 83. No, $ai - bi$ can be written as $ai + (-bi) = [a + (-b)]i$. $[a + (-b)]i$ is a pure imaginary number when $b \neq a$. However, when $b = a$, it is a real number. 85. To show that $1, -1, i$, and $-i$ are solutions, replace x with each of the values.

Page 317 Review Capsule for Section 9-2 1. $10 + 11\sqrt{5}$ 2. $3 - 3\sqrt{3}$ 3. $2\sqrt{7} - 2\sqrt{2}$ 4. $-1 + 8\sqrt{2}$ 5. $4\sqrt{2}$ 6. $14\sqrt{2}$ 7. $\subset$ 8. $\not\subset$ 9. $\not\subset$ 10. $\subset$

Page 320 Classroom Exercises 1. $2 + \sqrt{3}i$; $a = 2$; $b = \sqrt{3}$ 3. $-1 + 0i$; $a = -1$; $b = 0$ 5. $1 + (-2i)$; $a = 1$; $b = -2$ 7. $\neq$; $1 \neq -1$ 9. $-4 + i$ 11. $4i$ 13. $7 - 6i$ 15. $-4 + 7i$ 17. -10 19. 0

Page 320 Written Exercises Exercises 1-5 are already in standard form. 1. $a = 4$; $b = 0$ 3. $a = 4$, $b = 2$ 5. $a = 1$; $b = 3$ 7. $-2 + (-2i)$; $a = -2$; $b = -2$ 9. $0 + (-2i)$; $a = 0$; $b = -2$ 11. $15 + (-10i)$; $a = 15$; $b = -10$ 13. $8 + \sqrt{5}i$; $a = 8$; $b = \sqrt{5}$ 15. $1 + (-i)$; $a = 1$; $b = -1$ 17. $-\frac{3}{2} + 0i$; $a = -\frac{3}{2}$; $b = 0$ 19. $-i + i = 0 + 0i$; $a = 0$; $b = 0$ 21. $\subset$ 23. $\not\subset$ 25. $\not\subset$ 27. U; $2 \neq -2$ and $i \neq -i$ 29. E; $i\sqrt{25} = 5i$ 31. E; $\frac{1}{2} + 0i = \frac{1}{2}$ 33. $b \neq 0$ 35. $15 + (-2i)$ 37. $1 + 8i$ 39. $0 + 2i$

41. $0 + (-i)$ 43. $11 + 0i$ 45. $-4 - 5i$ 47. $7 - 2i$ 49. 5 51. $4i$ 53. $-2 + (-5i)$ 55. $5 + (-6\frac{1}{2}i)$ 57. $-6 + 10i$ 59. $2\sqrt{3} + (-3i)$ 61. 7 63. $7i$ 65. $2a$ 67. $4(\frac{3}{2}i)^2 + 9 = 4(\frac{9}{4})i^2 + 9 = -9 + 9 = 0$ 69. Substitute the given values into the equation. 71. $x - 3i$ 73. $x - 6i$ 75. $x - i\sqrt{3}$ 77. Add any real number. 79. $(a + bi) + (0 + 0i) = (a + 0) + (b + 0)i$ (Def. of Add.) $(a + 0) + (b + 0)i = a + bi$ (Zero Post.) Thus, $(a + bi) + (0 + 0i) = a + bi$. $(a + bi) + (c + di) = a + bi$ (Given) $(a + bi) + (c + di) = (a + c) + (b + d)i$ by (Def. of Add.) $(a + c) + (b + d)i = a + bi$ (Substitution Post.) $(a + c) = a$ and $(b + d) = bi$ (Def. of complex no.) If $a + c = a$, then $c = 0$ (Add. Ident. Post.) Similarly, if $(b + d)i = b$, then $d = 0$. 81. $(a + bi) + (c + di) = (a + c) + (b + d)i$ (Def. of Add.) $(a + c)(b + d)i = (c + a) + (d + b)i$ (Comm. Post.) $(c + a) + (d + b)i = (c + di) + (a + bi)$ (Def. of complex no.) Thus, addition of complex numbers is commutative. 83 Answers will vary.

Page 322 Calculator Exercises 1. 10.954451 3. −8.1240384 5. 148.49242 7. −339.41125

Page 322 Review Capsule for Section 9-3 1. $\frac{\sqrt{5}}{5}$ 2. $\frac{27 + 9\sqrt{2}}{7}$ 3. $-2\sqrt{3} - 6$ 4. $\frac{58 - 47\sqrt{2}}{31}$

Page 324 Classroom Exercises 1. $-7 + 22i$ 3. $4 + 9i$ 5. $\frac{4 - 2i^2}{5}$ 7. $\frac{8 + 9i}{5}$

Page 325 Written Exercises 1. $4 + 3i$ 3. 0 5. 2 7. $17 - 9i$ 9. $24 + 7i$ 11. $-\frac{23}{4} + \frac{5}{2}i$ 13. 13 15. $22 + 10\sqrt{3}i$ 17. $4 + 5i$ 19. $3 - \sqrt{2}i$ 21. $-2i$ 23. 8 25. $\frac{3}{13} + \frac{2}{3}i$ 27. $-\frac{5}{26} - \frac{1}{26}i$ 29. $\frac{7}{50} + \frac{1}{50}i$ 31. $-\frac{3}{20} - \frac{1}{20}i$ 33. $-\frac{3}{13} - \frac{28}{13}i$ 35. $-\frac{13}{10} - \frac{9}{10}i$ 37. $-\frac{77}{85} + \frac{36}{85}i$ 39. $-\frac{4}{5} - \frac{3}{5}i$ 41. $-5 + 12i$ 43. $-26 + 18i$ 45. $(y + 2i)(y - 2i)$ 47. $7(x + yi)(x - yi)$ 49. $(x + 1)(x - 1)(x + i)(x - i)$ 51. $(1 + 4y)(1 - 4y)(1 + 4yi)(1 - 4yi)$ 53. Perform the indicated computation. 55. $0 - 10i$ 57. $\frac{46}{169} + \frac{9}{169}i$ 59. Because the product of i and $-i$ is 1. 61. 1. $(a + bi)(c + di) = a(c + di) + bi(c + di)$ (Def. of Mult. of complex no.) 2. $= (ac + adi) + [bci + (bi)(di)]$ (Dist. Prop.) 3. $= (ac + bci) + [adi + (bi)(di)]$ (Comm. Prop. of Add.) 4. $= c(a + bi) + di(a + bi)$ (Dist. Prop.) 5. $= (c + di)(a + bi)$ (Dist. Prop.) 63. 1. $(a + bi)[(c + di) + (e + fi)] = (a + bi)[(c + e) + (d + f)i]$ (Def. of Add. of complex no.) 2. $= a[(c + e) + (d + f)i] + bi[(c + e) + (d + f)i]$ (Def. of Mult. of complex no.) 3. $= a(c + e) + a(d + f)i + bi(c + e) + bi^2(d + f)$ (Dist. Prop.) 4. $= ac + ae + adi + afi + bci + bei - bd - bf$ (Dist. Prop.) 5. $= (ac - bd) + (bc + ad)i + (ae - bf) + (be + af)i$ (Comm. and Assoc. Prop.) 6. $= (a + bi)(c + di) + (a + bi)(e + fi)$ (Mult. of complex no.) 65. $(a + bi) + (c + di) = (a + c) + (b + d)i$ (Def. of Add.) The conjugate is $(a + c) - (b + d)i$ The conjugate of $(a + bi)$ is $(a - bi)$ and the conjugate of $(c + di)$ is $(c - di)$. Thus, $(a - bi) + (c - di) = (a + c) + (-b - d)i = (a + c) - (b + d)i$. 67. The conjugate of $(c + di)$ is $(c - di)$. The conjugate of $(c - di)$ is $(c + di)$, which equals the original number.

Page 326 Review 1. $36i$ 2. $22i\sqrt{3}$ 3. $11i$ 4. $13i$ 5. -30 6. -14 7. -144 8. $420i$ 9. 2 10. 40 11. $i\sqrt{2}$

12. $-\frac{\sqrt{2}}{4}$ 13. $15-3i$ 14. $4+2i$ 15. $-2-9i$ 16. $-2+i$ 17. $-2-5i$ 18. $-11+15i$ 19. $(7+i)(7-i)$ 20. $21+3i$ 21. $-5-i$ 22. $-1-4i$ 23. $74+6i$ 24. $21-35i$ 25. $50-2i\sqrt{2}$ 26. $\frac{9+38i}{61}$ 27. $\frac{3+4i}{5}$ 28. $\frac{15+53i}{74}$

Page 328 Classroom Exercises 1. $x=\pm 3$ 3. $x=\pm 4$ 5. $p=\pm i\sqrt{2}$ 7. $x=0$

Page 328 Written Exercises 1. $x=\pm 3$ 3. $x=\pm\sqrt{11}$ 5. $x=\pm\sqrt{6}$ 7. $p=\pm 7$ 9. $x=\pm\sqrt{19}$ 11. $x=\pm 1$ 13. $x=\pm 2\sqrt{3}$ 15. $x=\pm\sqrt{3}$ 17. $y=\pm i$ 19. $x=\pm i\sqrt{7}$ 21. $x=1$ or $x=-7$ 23. $x=\frac{3}{2}$ or $x=-\frac{9}{2}$ 25. $x=\pm 2i$ 27. $x=1$ or $x=-2$ 29. $q=\frac{-3\pm i\sqrt{2}}{2}$ 31. $x=-\frac{1}{2}\pm\frac{1}{2}i$ 33. $x=\pm 2\sqrt{5}$ 35. $x=\pm 3i$ 37. $t=1$ or $t=-7$

Page 328 Puzzle See Solution Key.

Page 328 Review Capsule for Section 9-5 1. x^2+6x+9 2. $9x^2-42x+49$ 3. $x^2+2ax+a^2$ 4. $x^2-2ax+a^2$ 5. $(t+1)^2$ 6. $(3y-4)^2$ 7. $(2a+b)^2$ 8. $(p^2-1)^2$

Page 330 Classroom Exercises 1. $(x+7)^2$ 3. $(x+6)^2$ 5. $x=4$ or $x=-2$ 7. $y=7$ or $y=3$

Page 331 Written Exercises 1. $x^2-10x+25$; $(x-5)^2$ 3. $y^2+y+\frac{1}{4}$; $(y+\frac{1}{2})^2$ 5. $x^2-2bx+b^2$; $(x-b)^2$ 7. $x^2+\frac{b}{a}x+\frac{b^2}{4a}$; $(x+\frac{b}{2a})^2$ 9. $x=-4$ or $x=-6$ 11. $x=-3$ or $x=5$ 13. $x=5$ or $x=3$ 15. $a=-12$ or $a=2$ 17. $b=-6$ or $b=8$ 19. $x=5$ or $x=4$ 21. $c=3$ or $c=6$ 23. $x=-5$ or $x=10$ 25. $x=1$ or $x=-\frac{3}{2}$ 27. $x=-\frac{7}{2}$ or $x=\frac{1}{2}$ 29. $x=\frac{3}{2}$ or $x=-2$ 31. $n=\frac{1}{4}$ or $n=-5$ 33. $x=1$ or $x=\frac{1}{6}$ 35. $x=\frac{-3\pm\sqrt{29}}{2}$ 37. $a=\frac{3}{2}$ or $a=2$ 39. $y=\frac{-\sqrt{3}\pm i\sqrt{17}}{2}$ 41. $x=\frac{-b\pm\sqrt{b^2-4ac}}{2a}$ 43. $x=\frac{3\pm a\sqrt{57}}{2}$ 45. $x=2$ or $x=11$ 47. $x=4\pm\sqrt{18-6i}$

Page 331 Review Capsule for Section 9-6 1. $5i$ 2. $-5+i$ 3. $\frac{2-4i}{3}$ 4. $\frac{-2-i}{4}$ 5. $\frac{-11+i\sqrt{119}}{10}$ 6. $\frac{15+\sqrt{5}}{10}$ 7. $\frac{3\pm i\sqrt{11}}{2}$ 8. $\frac{-1\pm 3}{4}$ 9. 5 10. -4 11. $-\frac{2}{3}$ 12. $\frac{-7+\sqrt{41}}{4}$ 13. $\frac{3+\sqrt{21}}{2}$ 14. $\frac{4+\sqrt{14}}{2}$

Page 333 Classroom Exercises 1. $a=2$, $b=5$, and $c=6$ 3. $a=1$, $b=2$, $c=3$ 5. $a=2$, $b=5$, $c=12$ 7. $a=1$, $b=-5$, and $c=-14$

Page 333 Written Exercises 1. $x=3$ or $x=2$ 3. $x=5$ or $x=-2$ 5. $x=2$ or $x=-6$ 7. $y=\frac{3\pm\sqrt{17}}{2}$ 9. $y=\frac{7\pm\sqrt{5}}{2}$ 11. $x=-3$ 13. $x=\frac{3}{2}$ or $x=-\frac{4}{3}$ 15. $q=\frac{-1\pm\sqrt{5}}{2}$ 17. $s=\frac{3\pm\sqrt{21}}{6}$ 19. $x=1$ or $x=\frac{1}{2}$ 21. $x\approx 1.16$ or $x\approx -5.16$ 23. $x\approx -0.38$ or $x\approx -2.62$ 25. $x=\frac{3\pm\sqrt{21}}{6}$ 27. $x=\frac{3}{2}$ or $x=5$ 29. $x=-i$ or $x=7i$ 31. $x=6$ or $x=-\frac{2}{7}$ 33. $c=\pm 9$ 35. $x=\frac{-1\pm i\sqrt{3}}{2}$

Page 334 Calculator Exercises 1. 2.77 3. 4.33 5. 4.62 7. 0.27

Page 334 Review Capsule for Section 9-7 Each of Exercises 1-4 is a right triangle. 1. $5^2+9^2=14^2$ 2. $(\sqrt{20})^2+(\sqrt{15})^2=5^2$ 3. $(\sqrt{10})^2+(\sqrt{10})^2=(\sqrt{20})^2$ 4. $(\sqrt{a^2+b^2})^2=a^2+b^2$

Page 336 Classroom Exercises 1. x; $13-x$; $x^2-2(13-x)^2=14$ 3. ℓ; $9-\ell$; $\ell(9-\ell)=20$

Page 336 Word Problems 1. 7 and 8 3. 2 cm by 1 cm 5. AC = 6 cm, BC = 8 cm, AB = 10 cm 7. 15 cm by 20 cm 9. Bill: 9mi/hr; Wilma: 12 mi/hr 11. 10 m 13. 30 and −15 or 5 and 10 15. 35 m by 60 m or 30 m by 70 m 17. 45 km/hr 19. 8 and 5 or 4 and 1 21. $\frac{38\pm\sqrt{231}}{8}$ cm and $\frac{19\pm\sqrt{231}}{4}$ cm

Page 338 Review 1. $x=\pm 2\sqrt{5}$ 2. $a=\pm 5i$ 3. $b=\frac{1\pm i}{3}$ 4. $n=1\pm 3i$ 5. $x=3$ or $x=-6$ 6. $y=\frac{-5\pm\sqrt{37}}{8}$ 7. $b=-1$ or $b=-\frac{4}{3}$ 8. $x=2-i\sqrt{3}$ 9. $x=\frac{5\pm\sqrt{97}}{4}$ 10. $x=\frac{-1\pm\sqrt{17}}{4}$ 11. 7 cm by 12 cm or 4 cm by 15 cm

Page 339 Using Statistics 1. 170 3. 192 5. 1320 7. \$42.60 9. \$42.60

Page 340 Review Capsule for Section 9-8 1. (0, 0) 2. x intercept: $(-\frac{5}{3}, 0)$; y intercept: (0, 5) 3. 2 x intercepts: (3, 0) and (−3, 0); y intercept: (0, −9) 4. x intercept: (1, 0); y intercept: (0, 1) 5. N; W; I; Q; R; C 6. Ir and C 7. Ir, R, and C 8. C 9. C 10. C

Page 342 Classroom Exercises 1. 2 real rational roots 3. 2 real irrational roots 5. 2 real rational roots

Page 342 Written Exercises 1. 2 real irrational roots 3. 1 real rational root 5. 2 real irrational roots 7. 1 real rational root 9. 2 real rational roots 11. 2 real rational roots 13. 2 real irrational roots 15. 2 real rational roots 17. exactly one point 19. $z = 3$ or $z = -2$ 21. $a = 2$ or $a = \frac{2}{3}$ 23. $x = \frac{1}{4}$ or $x = -\frac{1}{2}$ 25. $k > 8$ or $k < -8$ 27. $k > 6$ 29. $-4 < k < 4$ 31. $-5 < k < 5$

Page 343 Classroom Exercises 1. 6; −7 3. $\frac{2}{3}$; −2 5. $\frac{7}{2}$; −3

Page 344 Written Exercises 1. −7; −10 3. 2; 9 5. $-\frac{2}{3}$; −1 7. −1; 1 9. $\frac{3}{4}; \frac{9}{4}$ 11. $x^2 - 5x - 14 = 0$ 13. $x^2 - 16 = 0$ 15. $x^2 - \frac{5}{6}x + \frac{1}{6} = 0$ 17. $x^2 - 48 = 0$ 19. $x^2 - 2x - 4 = 0$ 21. $x^2 - 2x + 10 = 0$ 23. $x^2 - \frac{3}{2}x + \frac{11}{16} = 0$ 25. $x^2 - \frac{5}{2}x + \frac{17}{8} = 0$ 27. $x^2 - 3x - 40 = 0$ 29. $x^2 + 13x + 40 = 0$ 31. $16x^2 - 8x - 3 = 0$ 33. $k = 10$ 35. $k = \pm 4\sqrt{26}$ 37. $k = 8$ 39. $f(x) = \frac{1}{3}x^2 - 3x + 6$

Page 344 Review Capsule for Section 9-10 1. 1 2. 3 3. 5 4. 2 5. $x^3 - 2x^2 + 2x + 3$, $R = -11$ 6. $3x^2 + 4x + 1$, $R = -3$ 7. $(x - 2)(x - 5)(x + 4)$ 8. $(x - 1)(x + 4)(x - 2)$ 9. $(x - 2)(x^2 + 3x - 5)$

Page 347 Classroom Exercises 1. 2 3. 4 5. 3i

Page 347 Written Exercises 1. 1, 3i, −3i 3. −4, 1, 2 5. $-\frac{1}{2}, \pm\frac{\sqrt{15}}{3}$ 7. −2, −3, $\frac{-3 \pm \sqrt{5}}{2}$ 9. $x^3 - 2x^2 - x + 2 = 0$ 11. $x^3 - x^2 - 22x + 40 = 0$ 13. $x^3 - 5x^2 + x - 5 = 0$ 15. $x^3 - 2x + 4 = 0$ 17. $x^4 - 4x^3 + 6x^2 - 4x = 0$ 19. $x^4 - 4x^3 + 5x^2 - 4x + 4 = 0$ 21. 1 23. −1, −2, −7 25. 2, −1, 1 27. 1, −1 29. None 31. −5, −3, 4 33. $1 \pm \sqrt{2}, \frac{1}{2}$ 35. $\frac{-1 \pm i\sqrt{7}}{2}, -\frac{1}{3}, 3$ 37. Every polynomial, P(x), of degree n has exactly n roots. If P(x) = 0 has any imaginary roots, the number of such roots is divisible by two. If n is odd, then the greatest number of imaginary roots that P(x) = 0 can have is n − 1. Thus, if P(x) is a polynomial of odd degree with real coefficients, then P(x) = 0 has at least one real root. 39. If P(a + bi) = 0, a + bi is a root of P(x) = 0. (Definition) Then a + bi is a complex number whose conjugate is equal to zero. But, the conjugate of zero is zero, since the conjugate of a real number is the number itself, and a − bi is the conjugate of a + bi by definition. Thus, P(a − bi) = 0.

Page 349 Computer Exercises See Solution Key.

Page 351 Written Exercises 1. Between f(2) and f(3) 3. Between f(−3) and f(−2) and between f(2) and f(3). 5. Between f(1) and f(2) and at f(3). 7. 0.3 9. −0.1712 11. 0.7 13. no real zeros 15. 1.4 17. $-k < k < 3 - k$

Page 351 Review 1. 2 real rational roots 2. 2 real irrational roots 3. 2 imaginary roots 4. $k = 16$ 5. $k > -\frac{1}{4}$ 6. $x^2 - x - 6 = 0$ 7. $x^2 - 16 = 0$ 8. $x^2 - 2x - 1 = 0$ 9. $9x^2 - 12x + 29 = 0$ 10. $x^3 - 4x^2 + x + 6 = 0$ 11. $x^3 + 5x^2 - 2x - 24 = 0$ 12. $9x^3 - 45x^2 - 34x - 8 = 0$ 13. $x^4 - 9x^3 + 18x^2 + 16x^2 - 144x - 288 = 0$ 14. 0.4 15. 1.8 16. −1.6 and 1.6

Page 353 Chapter Objectives and Review 1. $i + i\sqrt{3}$ 3. i 5. 3 7. 1 + i 9. $2 + (\sqrt{5} + 1)i$ 11. 3 − 6i 13. 5 + 2i 15. $\frac{2}{25} + \frac{11}{25}i$ 17. $x = \pm i\sqrt{7}$ 19. $x = -1$ or $x = -9$ 21. $r = \frac{2 \pm \sqrt{2}}{2}$ 23. $t = \frac{3 \pm i\sqrt{7}}{2}$ 25. 4 and 7 27. two imaginary roots 29. 2 imaginary roots 31. 7; −5 33. $x^2 - 6x + 58 = 0$ 35. $-\frac{1}{3}$ and $\frac{1}{3}$ 37. −1.4

Page 355 Chapter Test 1. −2 − 2i 3. 1 + 4i 5. 244 + 0i 7. $25(z + 2i)(z - 2i)$ 9. $y = \frac{1 \pm \sqrt{22}}{3}$ 11. $r = 13$ or $r = 1$ 13. $7\sqrt{2}m$ 15. 2 distinct complex roots

Page 356 Preparing for College Entrance Tests 1. a 3. d 5. c 7. d

CHAPTER 10 CONIC SECTIONS

Page 359 Classroom Exercises 1. $x^2 + y^2 = 64$ 3. $(x + 1)^2 + (y + 3)^2 = 2$ 5. (0, 0); r = 5 7. (0, 0); $r = \sqrt{3}$ 9. (−4, 5); r = 7

Page 360 Written Exercises 1. $x^2 + y^2 = 16$ 3. $(x - 3)^2 + (y - 4)^2 = 1$ 5. $(x + 3)^2 + (y - 2)^2 = 25$ 7. $x^2 + (y - 6)^2 = 25$ 9. $(x + 7)^2 + y^2 = \frac{25}{4}$ 11. $(x + \frac{3}{2})^2 + (y - \frac{1}{2})^2 = 16$ 13. (0, 0); r = 6 15. (0, 0); $r = \sqrt{7}$ 17. (7, 3); r = 2 19. (−4, 1); $r = 2\sqrt{2}$ 21. $(-\frac{1}{2}, \frac{1}{4})$; r = 3 23. $(\frac{1}{2}, -\frac{1}{2})$; $r = \sqrt{6}$ 25. (1, 0); r = 1 27. (−3, −4); $r = 2\sqrt{10}$ 29. (−4, 2); r = 6 31. (3, −4); r = 7 33. (0, −2); $r = \sqrt{10}$ 35. $x^2 + y^2 = 25$ 37. $(x + 3)^2 + (y + 1)^2 = 34$ 39. $(x - 3)^2 + (y + 1)^2 = 85$ 41. Let P(x, y) be any point on the circle whose center is C(h, k). Then by the distance formula, $CP = r = \sqrt{(x - h)^2 + (y - k)^2}$. Squaring both sides of the equation gives $r^2 = (x - h)^2 + (y - k)^2$.

43. $(x + 3)^2 + (y - 2)^2 = 9$ 45. $(x - 8)^2 + (y - 8)^2 = 64$ 47. $(x - 4)^2 + (y + 2)^2 = 16$ 49. If A = C and B = 0, then $Ax^2 + Ay^2 + Dx + Ey + F = 0$ or $x^2 + y^2 + \frac{D}{A}x + \frac{E}{A}y + \frac{F}{A} = 0$; $(x^2 + \frac{D}{A}x) + (y^2 + \frac{E}{A}y) = -\frac{F}{A}$; Completing the square yields $(x + \frac{D}{2A})^2 + (y + \frac{E}{2A})^2 = \frac{-4AF + D^2 + E^2}{4A^2}$. Let R equal the right side of the equation. If $r > 0$, then the graph of the equation is a circle. If R = 0, then the radius is 0 and the graph of the equation is a point. If $R < 0$, then the graph does not exist because the radius is an imaginary number.

Page 361 Calculator Exercises 1. 97.572184 3. 283.61783

Page 361 Review Capsule for Section 10-2 1. $3\sqrt{2}$ 2. $2\sqrt{6}$ 3. $21\sqrt{2}$ 4. $6\sqrt{3}$ 5. $45\sqrt{2}$ 6. $40\sqrt{2}$ 7. $x = 81$ 8. $x = 32$ 9. $x = 17$ 10. $x = 3$ 11. $x = 1$ 12. $x = \frac{3}{4}$ or $x = 1$ 13. $x = 12$ 14. $\emptyset$

Page 365 Classroom Exercises 1. $\frac{x^2}{100} + \frac{y^2}{25} = 1$ 3. $\frac{x^2}{\frac{22}{3}} + \frac{y^2}{11} = 1$ 5. x intercepts: (−8, 0) and (8, 0); y intercepts: (0, −7) and (0, 7) 7. x intercepts: $(-2\sqrt{2}, 0)$ and $(2\sqrt{2}, 0)$; y intercepts: (0, −3) and (0, 3) 9. Major axis: 4; minor axis: 2 11. Major axis: $6\sqrt{2}$; minor axis: $2\sqrt{2}$

Page 365 Written Exercises 1. Center: (0, 0); x intercepts: (−3, 0) and (3, 0); y intercepts: (0, −2) and (0, 2) 3. Center: (0, 0); x intercepts: (−2, 0) and (2, 0); y intercepts: (0, −4) and (0, 4) 5. Center: (0, 0); x intercepts: (−10, 0) and (10, 0); y intercepts: (0, −5) and (0, 5) 7. Center: (0, 0); x intercepts: (−3, 0) and (3, 0); y intercepts: (0, −2) and (0, 2) 9. Center: (0, 0); x intercepts: (2, 0) and (−2, 0); y intercepts: (0, 4) and (0, −4) 11. Center: (0, 0); x intercepts: (−15, 0) and (15, 0); y intercepts: (−5, 0) and (5, 0) 13. Center: (0, 0); x intercepts: (−5, 0) and (5, 0); y intercepts: (0, −15) and (0, 15) 15. a. x intercepts: (−8, 0) and (8, 0); y intercepts: (0, −9) and (0, 9) b. vertical c. 18 17. a. x intercepts: (−10, 0) and (10, 0); y intercepts: (0, −9) and (0, 9) b. horizontal c. 20 19. a. x intercepts: (−10, 0) and (10, 0); y intercepts: (0, −2) and (0, 2) b. horizontal c. 20 21. a. x intercepts: (−2, 0) and (2, 0); y intercepts: $(0, -2\sqrt{2})$ and $(0, 2\sqrt{2})$ b. vertical c. $4\sqrt{2}$ 23. a. x intercepts: (−3, 0) and (3, 0); y intercepts: (0, −4) and (0, 4) b. vertical c. 8 25. (±5, 0) 27. $(\pm 2\sqrt{6}, 0)$ 29. $(\pm\sqrt{3}, 0)$ 31. $\frac{x^2}{25} + \frac{y^2}{9} = 1$ 33. $\frac{x^2}{10} + \frac{y^2}{5} = 1$ 35. $\frac{x^2}{25} + \frac{y^2}{9} = 1$ 37. $\frac{x^2}{32} + \frac{y^2}{81} = 1$ 39. $\frac{x^2}{4} + \frac{y^2}{36} = 1$ 41. $\frac{(x + 5)^2}{49} + \frac{(y + 6)^2}{81} = 1$ 43. $\frac{(x + 3)^2}{64} + \frac{(y + 1)^2}{5} = 1$ 45. $\frac{(x - 8)^2}{16} + \frac{(y + 4)^2}{25} = 1$ 47. $\frac{(x + 2)^2}{16} + \frac{(y - 3)^2}{25} = 1$

Page 366 Review Capsule for Section 10-3 1. 2 2. 10 3. $\sqrt{2}$ 4. 5 5. $x + 4$ 6. $\sqrt{x^2 - 8x + 80}$ 7. $x^2 + 12x + 36$; $(x + 6)^2$ 8. $y^2 + 10y + 25$; $(y + 5)^2$ 9. $r^2 - 7r + \frac{49}{4}$; $(r - \frac{7}{2})^2$ 10. $t^2 + 11t + \frac{121}{4}$; $(t + \frac{11}{2})^2$ 11. $x = 3$ or $x = -1$ 12. $r = 3 \pm \sqrt{7}$ 13. $z = -2 \pm 2\sqrt{5}$ 14. $b = 2$ or $b = -12$ 15. $y = 2$ or $y = -\frac{1}{2}$ 16. $x = \frac{5 \pm \sqrt{37}}{6}$

Page 369 Classroom Exercises 1. circle 3. ellipse 5. $y = (x + 1)^2 + 5$ 7. $y = \frac{1}{5}(x - 0)^2 + 0$ 9. $y = 3(x - 4)^2 + 2$ 11. $x = (y - 3)^2 - 9$ 13. $x = \frac{1}{4}(y - 0)^2 + 0$ 15. $x = 3(y + 1)^2 - 1$

Page 369 Written Exercises 1. $x = \frac{1}{8}y^2$ 3. $y = -\frac{1}{16}x^2$ 5. $x = \frac{1}{22}y^2 - \frac{1}{2}$ 7. $y = -\frac{1}{10}(x - 3)^2 + \frac{5}{2}$ 9. (0, 0); y = 0; (1, 0); x = −1; Opens to the right 11. (0, 0); x = 0: (0, 4); y = −4; Opens upward 13. (0, 0); y = 0; $(-\frac{1}{2}, 0)$; $x = \frac{1}{2}$; Opens to the left 15. (0, 0); y = 0; (−2, 0); x = 2; Opens to the left 17. (2, 4); x = 2; $(2, \frac{17}{4})$; $y = \frac{15}{14}$; Opens upward 19. (0, −3); y = −3; $(\frac{1}{4}, -3)$; $x = -\frac{1}{4}$; Opens to the right 21. (−2, 0); x = −2; $(-2, \frac{3}{2})$; $y = -\frac{3}{2}$; Opens upward 23. (0, −3); y = −3; $(\frac{1}{4}, -3)$; $x = -\frac{1}{4}$; Opens to the right 25. (2, 0); y = 0; (1, 0); x = 3; Opens to the left 27. (3, −11); x = 3; $(3, -\frac{87}{8})$; $y = -\frac{89}{8}$; Opens upward 29. $x = \frac{1}{16}y^2 - \frac{3}{8}y - \frac{103}{16}$ 31. $y = \frac{1}{2}x^2 - 2x + 3$ 33. $x = \frac{1}{2}y^2 + \frac{3}{2}y + \frac{23}{8}$ 35. $x = \frac{1}{6}y^2 - \frac{1}{3}y - \frac{13}{3}$ 37. Vertex: (3, 2); Focus: (3, 4); Directrix: y = 0 39. Vertex: (−3, 2); Focus: $(-3, \frac{63}{32})$; Axis of symmetry: x = −3; Directrix: $y = \frac{65}{32}$ 41. Vertex: $(\frac{5}{2}, 9)$; Focus: $(\frac{5}{2}, \frac{143}{16})$; Axis of symmetry: $x = \frac{5}{2}$; Directrix: $y = \frac{145}{16}$ 43. Let P(x, y) be any point on the parabola. The distance from P to the directrix is $x + p$. Therefore, $\sqrt{(x - p)^2 + y^2} = \sqrt{(x + p)^2}$; $(x - p)^2 + y^2 = (x + p)^2$; $x^2 - 2px + p^2 + y^2 = x^2 + 2px + p^2$; and $y^2 = 4px$. If $p > 0$, then $4p > 0$ and the parabola opens to the right, since $a > 0$. 45. $x = \frac{1}{2}(y - 5)^2 - 3$ 47. Vertex: (2, −2); Focus: (2, −3); Directrix: y = −1; Graph: opens downward; y intercept: $(0, -3)$ 49. Vertex: $(-2, 12)$; Focus: $(-2, \frac{95}{8})$; Directrix: $y = \frac{97}{8}$; Graph: opens downward; y intercept: $(0, 4)$; x intercepts: $(-2 + \sqrt{6}, 0)$ and $(-2 - \sqrt{6}, 0)$. 51. Use the distance formula to write an equation for the parabola with focus $(h, k + p)$ and directrix with equation $y = k - p$. Then show that $P(h + 2p, k + p)$ and $Q(h - 2p, k + p)$ are solutions of the equation.

Page 370 Review Capsule for Section 10-4 1. $\frac{x^2}{9} - \frac{y^2}{4} = 1$ 2. $\frac{x^2}{25} - \frac{y^2}{4} = 1$ 3. $\frac{x^2}{36} - \frac{y^2}{9} = 1$ 4. $\frac{x^2}{25} - \frac{y^2}{25} = 1$ 5. $\frac{x^2}{10} - \frac{y^2}{20} = 1$ 6. $\frac{x^2}{25} - \frac{y^2}{1} = 1$ 7. $\frac{x^2}{36} - \frac{y^2}{25} = 1$ 8. $\frac{x^2}{25} - \frac{y^2}{5} = 1$ 9. $\frac{x^2}{4} - \frac{y^2}{36} = 1$

Page 373 Classroom Exercises 1. Foci: $(5, 0)$, $(-5, 0)$; asymptotes: $y = \frac{4}{3}x$, $y = -\frac{4}{3}x$ 3. Foci: $(2\sqrt{2}, 0)$, $(-2\sqrt{2}, 0)$; asymptotes: $y = x$, $y = -x$ 5. Foci: $(0, \sqrt{29})$, $(0, -\sqrt{29})$; asymptotes: $y = \frac{5}{2}x$, $y = -\frac{5}{2}x$ 7. Foci: $(2\sqrt{5}, 0)$, $(-2\sqrt{5}, 0)$; asymptotes: $y = 2x$, $y = -2x$ 9. Foci: $(5, 0)$, $(-5, 0)$; asymptotes: $y = \frac{4}{3}x$, $y = -\frac{4}{3}x$ 11. Foci: $(0, 5)$, $(0, -5)$; asymptotes: $y = \frac{3}{4}x$, $y = -\frac{3}{4}x$

Page 373 Written Exercises 1. Rectangle: determined by $x = 3$, $x = -3$, $y = 4$, and $y = -4$; vertices: $(3, 4)$, $(3, -4)$, $(-3, 4)$, $(-3, -4)$; diagonals: $y = \frac{4}{3}x$, $y = -\frac{4}{3}x$; Hyperbola: opens to the left and to the right; vertices: $(3, 0)$ and $(-3, 0)$

3. Rectangle: determined by $x = 2$, $x = -2$, $y = \sqrt{2}$, and $y = -\sqrt{2}$; vertices: $(-2, \sqrt{2})$, $(-2, -\sqrt{2})$, $(2, \sqrt{2})$, $(2, -\sqrt{2})$; diagonals: $y = \frac{\sqrt{2}}{2}$, $y = -\frac{\sqrt{2}}{2}$. Hyperbola with vertices at $(2, 0)$ and $(-2, 0)$; opens to the left and to the right. 5. Vertices: $(4, 0)$, $(-4, 0)$, opens to the left and to the right; asymptotes: $y = \frac{1}{2}x$, $y = -\frac{1}{2}x$ 7. Opens upward and downward. Vertices: $(0, 7)$, $(0, -7)$; asymptotes: $y = \frac{7}{6}x$, $y = -\frac{7}{6}x$ 9. Opens upward and downward; vertices: $(0, 3)$, $(0, -3)$; asymptotes: $y = \frac{1}{3}x$, $y = -\frac{1}{3}x$ 11. Opens to the left and to the right; vertices: $(2, 0)$, $(-2, 0)$; asymptotes: $y = \frac{5}{2}x$, $y = -\frac{5}{2}x$ 13. Opens to the right and to the left; vertices: $(3, 0)$, $(-3, 0)$; asymptotes: $y = x$, $y = -x$ 15. Opens to the left and to the right; vertices: $(4, 0)$, $(-4, 0)$; asymptotes: $y = \frac{1}{2}x$, $y = -\frac{1}{2}x$ 17. Opens to the left and to the right; vertices: $(3, 0)$, $(-3, 0)$; asymptotes: $y = \frac{4}{3}x$, $y = -\frac{4}{3}x$ 19. Opens to the left and to the right; vertices: $(4, 0)$, $(-4, 0)$; asymptotes: $y = \frac{1}{2}x$, $y = -\frac{1}{2}x$

21. Vertices: $(4, 0)$, $(-4, 0)$; Foci: $(4\sqrt{2}, 0)$, $(-4\sqrt{2}, 0)$; asymptotes: $y = x$ and $y = -x$ 23. Vertices: $(0, 2)$ and $(0, -2)$; Foci: $(0, \sqrt{13})$, and $(0, -\sqrt{13})$; asymptotes: $y = \frac{2}{3}x$, $y = -\frac{2}{3}x$ 25. Vertices: $(0, 4)$, $(0, -4)$; Foci: $(0, 4\sqrt{2})$, $(0, -4\sqrt{2})$; asymptotes: $y = x$, $y = -x$ 27. Vertices: $(1, 0)$, $(-1, 0)$; Foci: $(\sqrt{17}, 0)$, $(-\sqrt{17}, 0)$; asymptotes: $y = 4x$, $y = -4x$ 29. Vertices: $(0, 2)$, $(0, -2)$; Foci: $(0, 2\sqrt{2})$, $(0, -2\sqrt{2})$; asymptotes: $y = x$, $y = -x$ 31. Vertices: $(0, 2)$, $(0, -2)$; Foci: $(0, 2\sqrt{5})$, $(0, -2\sqrt{5})$; asymptotes: $y = \frac{1}{2}x$, $y = -\frac{1}{2}x$ 33. $\frac{x^2}{4} - \frac{y^2}{25} = 1$ 35. $\frac{y^2}{9} - \frac{x^2}{25} = 1$ 37. $\frac{x^2}{36} - \frac{b^2}{64} = 1$ 39. $\frac{x^2}{9} - \frac{y^2}{1} = 1$ 41. $\frac{y^2}{32} - \frac{x^2}{32} = 1$ 43. Center: $(3, 2)$; vertices: $(7, 2)$ and $(-1, 2)$. The hyperbola opens to the left and to the right. x intercepts: $(7, 0)$, $(-7, 0)$; y intercepts: $(0, 5)$, $(0, -5)$; asymptotes: the line that passes through $(7, 5)$ and $(-1, -1)$ and the line that passes through $(7, -1)$ and $(-1, 5)$. 45. Center: $(1, 0)$; vertices: $(4, 0)$, $(-2, 0)$; opens to the left and to the right; asymptotes: the line that passes through $(4, -2)$ and $(-2, 2)$ and the line that passes through $(-2, -2)$ and $(4, 2)$. 47. $\frac{(y - 5)^2}{9} - \frac{(x - 1)^2}{1} = 1$; Center: $(1, 5)$; vertices: $(1, 8)$, $(1, 2)$; opens upward and downward; the asymptotes: the line that passes through $(2, -2)$ and $(0, 8)$ and the line that passes through $(0, 2)$ and $(2, 8)$. 49. $\frac{(x - 5)^2}{25} - \frac{(y + 2)^2}{9} = 1$; Center: $(5, -2)$; vertices: $(10, -2)$, $(0, -2)$; asymptotes: the line that passes through $(0, 1)$ and $(10, -5)$ and the line that passes through $(10, 1)$ and $(0, -5)$; opens to the left and to the right.

Page 375 Classroom Exercises 1. II and IV 3. II and IV 5. I and III 7. $(5, 5)$, $(-5, -5)$ 9. $(\sqrt{15}, -\sqrt{15})$, $(\sqrt{15}, \sqrt{15})$ 11. $x = 0$, $y = 0$

Page 376 Written Exercises 1. $xy = 18$ is a rectangular hyperbola in Quadrants I and III. Points: $(-1, -18)$, $(-2, -9)$, $(-3, -6)$, $(-9, -2)$, $(-18, -1)$, $(1, 18)$, $(2, 9)$, $(3, 6)$, $(9, 2)$, $(18, 1)$ 3. Rectangular hyperbola in Quadrants II and IV. Points: $(-1, 7)$, $(-2, \frac{7}{2})$, $(-3, \frac{7}{3})$, $(-5, \frac{7}{5})$, $(-7, 1)$, $(1, 7)$, $(2, -\frac{7}{2})$, $(3, -\frac{7}{3})$, $(5, -\frac{7}{5})$, $(7, -1)$ 5. Rectangular hyperbola in Quadrants I and III. Points: $(-1, -4)$, $(-2, -2)$, $(-4, -1)$, $(1, 4)$, $(2, 2)$, $(4, 1)$ 7. Rectangular hyperbola in Quadrants II and IV. Points: $(-24, 1)$, $(-12, 2)$, $(-8, 3)$, $(-6, 4)$, $(-3, 8)$, $(-2, 12)$, $(-1, 24)$, $(1, 24)$, $(2, 12)$, $(3, 8)$, $(6, 4)$, $(8, 3)$, $(12, 2)$,

(24, 1) 9. Rectangular hyperbola in Quadrants I and III. Points: (−1, −9), (−3, −3), (−9, −1), (1, 9), (3, 3), (9, 1) 11. Rectangular hyperbola in Quadrants II and IV. Points: (−1, 9), (−3, 3), (−9, 1), (1, −9), (3, −3), (9, −1) 13. circle 15. line 17. parabola 19. parabola 21. line 23. ellipse 25. ellipse 27. hyperbola

Page 376 Review 1. $(x-1)^2 + (y-2)^2 = 16$ 2. $(x+4)^2 + (y-3)^2 = 4$ 3. $(x+\frac{1}{2})^2 + (y+\frac{3}{2})^2 = \frac{25}{4}$ 4. Horizontal; Center: (0, 0); x intercepts: (−5, 0), (5, 0); y intercepts: (0, −4), (0, 4) 5. Vertical; Center: (0, 0); x intercepts: (2, 0), (−2, 0); y intercepts: (0, 6), (0, −6) 6. Vertical; Center: (0, 0); x intercepts: (−3, 0), (3, 0); y intercepts: (0, −7), (0, 7) 7. Vertex: (3, 5); Focus: $(3, \frac{21}{4})$; axis of symmetry: $x = 3$; directrix: $y = \frac{19}{4}$; The graph opens upward. 8. Vertex: $(-\frac{9}{2}, -\frac{113}{4})$; Focus: $(\frac{9}{2}, -28)$; axis of symmetry: $x = -\frac{9}{2}$; directrix: $y = -\frac{57}{2}$; Graph opens upward. 9. Vertex: $(-\frac{25}{4}, \frac{5}{2})$; axis of symmetry: $y = \frac{5}{2}$; Focus: $(-6, \frac{5}{2})$; directrix: $x = -\frac{13}{2}$; Graph opens to the right. 10. Center: (0, 0); vertices: (3, 0), (−3, 0); Foci: $(3\sqrt{2}, 0)$, $(-3\sqrt{2}, 0)$; asymptotes: $y = x$, $y = -x$ 11. Center: (0, 0); vertices: (0, 4), (0, −4); Foci: (0, 5), (0, −5); asymptotes: $y = \frac{3}{4}x$, $y = -\frac{3}{4}x$ 12. Center: (0, 0); vertices: (4, 0), (−4, 0); Foci: $(\sqrt{17}, 0)$, $(-\sqrt{17}, 0)$; asymptotes: $y = \frac{1}{4}x$, $y = -\frac{1}{4}x$ 13. A rectangular hyperbola with branches in Quadrants I and III. Points: (−12, −1), (−6, −2), (−4, −3), (−3, −4), (−2, −6), (−1, −12), (1, 12), (2, 6), (3, 4), (4, 3), (6, 2), (12, 1) 14. A rectangular hyperbola with branches in Quadrants II and IV. Points: (−3, 1), $(-2, \frac{3}{2})$, (−1, 3), (1, −3), $(2, -\frac{3}{2})$, (3, −1) 15. A rectangular hyperbola with branches in Quadrants II and IV. Points: (−10, 1), $(-7, \frac{10}{7})$, (−5, 2), (−2, 5), (−1, 10), (1, −10), (2, −5), (5, −2), $(7, -\frac{10}{7})$, (10, −1) 16. A rectangular hyperbola with branches in Quadrants II and IV. Points: (−5, 1), $(-3, \frac{5}{3})$, $(-2, \frac{5}{2})$, (−1, 5), (1, −5), $(2, -\frac{5}{2})$, $(3, -\frac{5}{3})$, (5, −1)

Page 379 Classroom Exercises 1. I 3. I 5. J 7. i

Page 379 Written Exercises 1. 2 3. $5\frac{5}{7}$ 5. $13\frac{1}{2}$ 7. $\frac{1}{8}$ 9. 8 11. 60 13. Yes. If A is constant and if ℓ increases in value, then w must decrease in value. If ℓ decreases in value, then w must increase in value. 15. divided by 3 17. $y = kxz$ 19. $A = kmnp$ 21. $y = \frac{kxz}{mn}$ 23. $y = \frac{kx^3}{z^2}$ 25. 225 27. $\frac{18}{5}$ 29. $\frac{W}{w} = \frac{d}{D}$, or $\frac{w}{W} = \frac{D}{d}$ 31. $\frac{5}{2}$ 33. $w = \frac{k}{d^2}$ 35. about 90.7 lb 37. about 6429 calories 39. 13 revolutions 41. about 27 miles 43. 0.01 ohms 45. 1896.3 lb

Page 383 Computer Exercises See Solution Key.

Page 385 Classroom Exercises 1. one 3. two 5. no 7. one

Page 385 Written Exercises 1. (0, 0), (1, 1) 3. (5, 3) 5. (4, 3), (−3, −4) 7. $(-\frac{19}{5}, -\frac{8}{5})$, (4, 1) 9. (1, 2), (−1, −2) 11. (4, −1), (3, 1) 13. $(-\frac{2}{5}, -\frac{39}{5})$, (6, 5) 15. $(-\frac{14}{5}, \frac{52}{5})$, (10, 4) 17. $(\frac{-14+3\sqrt{19}}{25}, \frac{-63+\sqrt{19}}{25})$, $(\frac{-14-3\sqrt{19}}{25}, \frac{-63-\sqrt{19}}{25})$ 19. (−43, 9), (5, −3) 21. $(-\frac{2}{3}, -9)$, (6, 1) 23. $(\frac{-5+\sqrt{46}}{3}, \frac{13-2\sqrt{46}}{3})$, $(\frac{-5-\sqrt{46}}{3}, \frac{13+2\sqrt{46}}{3})$ 25. (−52, 11), (4, −3) 27. (0, 2), (3, 0) 29. $(\frac{p+q^2}{2q}, \frac{q^2-p}{2q})$ 31. $(\frac{6}{5}, \frac{4}{5})$, (−24, 12)

Page 386 Review Capsule for Section 10-8 1. $x^2 = 8 - y^2$ 2. $x^2 = y - 9$ 3. $x^2 = 6 - 2y^2$ 4. $x^2 = \frac{4}{3}y^2 + \frac{8}{3}$ 5. $x^2 = 7y^2 - 14$ 6. $x^2 = 2y - \frac{5}{2}$ 7. $x^2 = 2 - \frac{1}{4}y^2$ 8. $x^2 = \frac{1}{3} - \frac{1}{2}y^2$

Page 387 Classroom Exercises 1. No 3. Yes 5. Yes 7. No

Page 387 Written Exercises 1. (3, 0), (−3, 0) 3. (2, 1), (2, −1), (−2, 1), (−2, −1) 5. (1, 1), (−1, 1) 7. (−3, 0) 9. $\left(\sqrt{-2+2\sqrt{5}}, -2 \pm \sqrt{5}\right)$, $\left(-\sqrt{-2+2\sqrt{5}}, -2 \pm \sqrt{5}\right)$ 11. (1, −2), (−1, −2), (2, 1), (−2, 1) 13. (1, 1), (−1, −1) 15. (5, 0), (−4, 3), (−4, −3) 17. (5, 2), (−5, −2), (2, 5), (−2, −5) 19. $\left(\sqrt{-2 \pm \sqrt{29}}, 2\sqrt{2 \pm \sqrt{29}}\right)$, $\left(-\sqrt{-2 \pm \sqrt{29}}, -2\sqrt{2 \pm \sqrt{29}}\right)$ 21. $10 + \frac{1}{2}\sqrt{2}$ and $10 - \frac{1}{2}\sqrt{2}$ or the numbers are $10 - \frac{1}{2}\sqrt{2}$ and $10 + \frac{1}{2}\sqrt{2}$ 23. 55 pairs 25. $120

Page 388 Classroom Exercises 1. No; Yes 3. No; Yes 5. No; No

Page 388 Written Exercises 1. The graph is the set of points outside the circle with center at (0, 0) and radius 6.

3. The graph is the set of points inside the parabola with vertex at (0, 0) and opening upward. 5. The graph is the set of points outside the ellipse with center at (0, 0) and intercepts at (4, 0), (−4, 0), (0, 3), and (0, −3). 7. The graph is the set of points inside both branches of the hyperbola with vertices at (2, 0) and (−2, 0) and center at (0, 0), and opening left and right. 9. The graph is the set of points on and inside the parabola with vertex at $(\frac{3}{2}, -\frac{9}{4})$, opening upward, and x intercepts at (0, 0) and (3, 0). 11. The graph is the set of points on and outside the parabola with vertex at (0, 0) and opening to the left. 13. The graph is the set of points inside the ellipse, $\frac{x^2}{3} + \frac{y^2}{\frac{3}{2}} = 1$, with center at (0, 0) and intercepts at $(\sqrt{3}, 0)$, $(-\sqrt{3}, 0)$, $(0, \frac{1}{2}\sqrt{6})$, and $(0, -\frac{1}{2}\sqrt{6})$. 15. The graph is the set of points outside the parabola, $y = x^2 - 5$, with vertex at (0, −5), opening upward, and with x intercepts at $(\sqrt{5}, 0)$ and $(-\sqrt{5}, 0)$. 17. Draw the circle with center at (0, 0) and radius of 5. Draw the parabola with vertex (0, 0) passing through (2, 4) and (−2, 4). Shade the region inside the parabola that is also inside the circle. 19. Draw the circle with center at (0, 0) and radius of 7. Draw the parabola with vertex at (0, 0) passing through (2, −4) and (−2, −4). Shade the region inside the circle that is not inside the parabola.

Page 389 Review 1. 6 3. $(1 + \sqrt{7}, -1 + \sqrt{7})$, $(1 - \sqrt{7}, -1 - \sqrt{7})$ 5. $(\frac{1}{4} + \frac{1}{4}\sqrt{17}, -\frac{31}{8} + \frac{1}{8}\sqrt{17})$, $(\frac{1}{4} - \frac{1}{4}\sqrt{17}, -\frac{31}{8} - \frac{1}{8}\sqrt{17})$ 7. $\left(\sqrt{\frac{53 \pm \sqrt{793}}{18}}, \frac{1 - \sqrt{793}}{6}\right)$, $\left(-\sqrt{\frac{53 \pm \sqrt{793}}{18}}, \frac{1 + \sqrt{793}}{6}\right)$ 9. $\left(\sqrt{8 \pm 2\sqrt{41}}, \sqrt{-8 \pm 2\sqrt{41}}\right)$, $\left(-\sqrt{8 \pm 2\sqrt{41}}, -\sqrt{-8 \pm 2\sqrt{41}}\right)$ 11. 43 13. The graph is the set of points inside the ellipse with center at (0, 0) and intercepts at (4, 0), (−4, 0), (0, 3), and (0, −3). 15. The graph is the set of points on and inside the parabola with vertex at (0, 0) and opening downward.

Page 391 Chapter Objectives and Review 1. The circle has y intercepts at (0, 2) and (0, 6) and center at (−1, 4). 3. The circle is tangent to the x axis at (−2, 0) and has y intercepts at (0, −0.8) and (0, −5.2) approximately and center at (−2, −3). 5. r = 6; center: (−2, 3) 7. Ellipse: x intercepts: (2, 0) and (−2, 0); y intercepts: (0, 8), (0, −8); Foci: $(0, 2\sqrt{15})$, $(0, -2\sqrt{15})$. 9. $\frac{x^2}{9} + \frac{y^2}{36} = 1$ 11. Parabola: Opens to the right; Focus: (9, 0); vertex: (0, 0); directrix: x = −9 13. Parabola: Opens to the left; Focus: (−1, 0); vertex: (0, 0); directrix: x = 1 15. $y = \frac{1}{20}x^2$ 17. Opens upward and downward; vertices: (0, 8), (0, −8); asymptotes: y = 2x, y = −2x 19. $\frac{y^2}{4} - \frac{x^2}{9} = 1$ 21. $\frac{x^2}{4} - 4y^2 = 1$ 23. ellipse 25. ellipse 27. parabola 29. Rectangular hyperbola with branches in Quadrants II and IV; vertices: $(2\sqrt{2}, -2\sqrt{2})$, $(-2\sqrt{2}, 2\sqrt{2})$ 31. Rectangular hyperbola with branches in Quadrants I and III; vertices: (4, 4), (−4, −4) 33. y = kab, or $k = \frac{y}{ab}$ 35. $y_2 = 6$ 37. about 282.8 pounds 39. (3, 6), (−2, 1) 41. $(\frac{4}{5}\sqrt{5}, \frac{2}{5}\sqrt{5})$, $(-\frac{4}{5}\sqrt{5}, -\frac{2}{5}\sqrt{5})$ 43. $(2\sqrt{3}, \sqrt{3})$, $(-2\sqrt{3}, -\sqrt{3})$, $(\sqrt{6}, \sqrt{6})$, $(-\sqrt{6}, -\sqrt{6})$ 45. The graph is the set of points inside the circle with center at (0, 0) and radius 4. 47. The graph is the set of points on and outside the ellipse with center at (0, 0) and intercepts at (3, 0), (−3, 0), (0, 2), and (0, −2).

Page 393 Chapter Test 1. Circle: center: (0, 0); radius 6 3. Hyperbola; opening to the left and the right; vertices: (6, 0), (−6, 0); asymptotes: $y = \frac{1}{2}x$, $y = -\frac{1}{2}x$. 5. Hyperbola: opening to the left and to the right; vertices: (5, 0), (−5, 0); asymptotes: $y = \frac{7}{5}x$, $y = -\frac{7}{5}x$ 7. $x = -\frac{1}{6}y^2 + \frac{1}{2}$ 9. Foci: (0, 4), (0, −4) 11. $(\frac{-1 - \sqrt{61}}{3}, \frac{7 + \sqrt{61}}{3})$, $(\frac{-1 + \sqrt{61}}{3}, \frac{7 - \sqrt{61}}{3})$ 13. $\frac{xt}{y} = k$ 15. 5 m

Page 394 More Challenging Problems 1. When a = c. 3. Solving simultaneously or graphing, the points of intersection are: A(3, 4), B(4, 3), C(−3, −4), D(−4, −3). Slope of $\overline{AB}$ is −1. Slope of $\overline{BC}$ is 1. Slope of $\overline{CD}$ is −1. Slope of $\overline{DA}$ is 1. Opposite sides are parallel and adjacent sides are perpendicular. Therefore, figure is a rectangle. 5. x = 2a 7. $a = -\frac{3}{20}$; $b = -\frac{13}{10}$; c = 4 9. {1, 3}

Page 394 Review of Word Problems: Chapters 1-10 1. 40 years old 3. 205 skiers 5. at 9%: $4000; at 16%: $12,000 7. Width: 5 m 9. 8 km per hour 11. 7 seconds 13. 4 seconds 15. Dan: 4 miles per hour; Jane: 3 miles per hour 17. 450 revolutions per minute

Page 396 Cumulative Review: Chapters 8-10 1. b 3. c 5. b 7. b 9. a 11. a 13. d 15. c 17. a 19. b 21. a 23. d 25. a

Page 398 Preparing for College Entrance Tests 1. b 3. b 5. a 7. b 9. b 11. a

CHAPTER 11 EXPONENTIAL AND LOGARITHMIC FUNCTIONS

Page 401 Classroom Exercises 1. $\frac{1}{9}$ 3. $\frac{1}{125}$ 5. 64 7. s^2 9. $\frac{q^4+r^4}{q^4r^4}$

Page 401 Written Exercises 1. $\frac{1}{a}$ 3. b^3 5. $\frac{1}{y^5}$ 7. $\frac{y^5}{x^3}$ 9. $\frac{a^2}{b^3}$ 11. $\frac{6}{2^3}$, or $\frac{6}{8}=\frac{3}{4}$ 13. $\frac{1}{m+n}$ 15. $\frac{1}{5^2}$, or $\frac{1}{25}$ 17. 9^2, or 81 19. e^2 21. $\frac{5}{2^3}$, or $\frac{5}{8}$ 23. $\frac{t^3}{m^4}$ 25. n^2r^3 27. $\frac{b^3y}{ax^4}$ 29. $\frac{5c^4}{b^3x^2}$ 31. $\frac{1}{x^3y^2}$ 33. $\frac{4b^3c^3}{9a^2}$ 35. $\frac{az^4}{5n^5t^5}$ 37. 2^4, or 16 39. $7^0=1$ 41. $\frac{1}{8x^3}$ 43. $\frac{1}{a^3b^3c^3}$ 45. 1 47. $\frac{1}{5i}$, or $-\frac{i}{5}$ 49. $\frac{1}{5+12i}$, or $\frac{5-12i}{169}$ 51. 2 53. $\frac{3}{2}$ 55. $-\frac{9}{4}$ 57. $\frac{t^5}{r^5}$ 59. 1 61. 1 63. 4 65. $\frac{1}{36}$ 67. 2 69. 4 71. $-1\frac{1}{2}$ 73. $-\frac{1}{3}$ 75. -1 77. 64 79. $\frac{1}{8}$

81. The graph contains $(-5, \frac{1}{32})$, $(-4, \frac{1}{16})$, $(-3, \frac{1}{8})$, $(-2, \frac{1}{4})$, $(-1, \frac{1}{2})$, (0, 1), (1, 2), (2, 4), (3, 8), (4, 16), (5, 32).

Page 402 Puzzle See Solution Key.

Page 402 Review Capsule for Section 11-2 1. 10^2 2. 10^4 3. 10^9 4. 10^5 5. $\frac{1}{10}$, or $\frac{1}{10^1}$ 6. $\frac{1}{10^3}$ 7. $\frac{1}{10^6}$ 8. $\frac{1}{10^4}$ 9. 10^0 10. 10^1

Page 404 Classroom Exercises 1. 10^5 3. 10^{-2} 5. 8 7. 5.26×10^2 9. 2.25×10^9 11. 98,000 13. 0.0000000218

Page 404 Written Exercises 1. 1.9×10^5 3. 5.9×10^{-3} 5. 8×10^{-4} 7. 9.8472×10^{10} 9. 2.805×10^3 11. 2.83×10^{-3} 13. 7×10^{-11} 15. 1×10^1 17. 1×10^{-6} 19. 7.9×10^8 21. 0.0006 23. 50,600,000,000 25. 0.000009018 27. 786,500,000 29. 0.25 31. 0.0005 33. 453,600 35. 394,800,000 37. 0.0004 39. 176,000 41. 0.000000000000000000008 43. 90,000,000,000,000 45. 0.00000000004166666, or $0.0000000004166\overline{6}$ 47. 6.371×10^3 49. 9.46×10^{12} 51. 3.6×10^{-5}; 7.7×10^{-5} 53. 1.67×10^{-24} 55. 7.6×10^1; 1.29×10^{-3} 57. 1,080,000,000,000,000,000,000,000,000 59. 0.000016 61. 299,790,000

Page 407 Classroom Exercises 1. $\frac{b^3}{a^2}$ 3. $\frac{27y^3}{8x^2}$ 5. $\frac{x^2}{y^4}$ 7. $-\frac{y}{x^2}+\frac{4}{x}$ 9. $a+b$ 11. $\frac{1}{a+b}$

Page 407 Written Exercises 1. $\frac{z^3}{xy}$ 3. $\frac{b^2+a^2}{a^2b^2}$ 5. $\frac{a}{3^5b^4c^2}$, or $\frac{a}{243b^4c^2}$ 7. $\frac{y^5}{x^5}$ 9. $\frac{b}{6a^3}$ 11. $\frac{g^7r}{p^2}$ 13. $7a^2$ 15. $\frac{r}{s^2}$ 17. rs^3 19. $2st^5$ 21. $\frac{4x^2+x+7}{x^3}$ 23. $\frac{a+bx-cx^2}{x^3}$ 25. $\frac{9t+3r^2-3r}{rt^3}$ 27. $2xy$ 29. $3ab(b-a)^2$ 31. $\frac{(a+b)^2}{ab}$ 33. $\frac{1}{xy(x+y)}$ 35. x^2-1 37. $x^{-2}+y^{-2}$ 39. $2w^{-1}x^{-1}y^{-2}z^2$ 41. $(x+y)^{-2}$ 43. $3b^2+4c^{-4}$ 45. $\frac{x^2-3xy^2-2y+6}{xy^2}$ 47. $\frac{3x^2y+xy^2}{3x^2+4xy^3+y^2}$ 49. $(x^{-1}+y^{-1})^{-1}=\frac{1}{\frac{1}{x}+\frac{1}{y}}=\frac{1}{\frac{y+x}{xy}}=\frac{xy}{y+x}$. Since $xy \neq y+x$ for all values of x and y, $(x^{-1}+y^{-1})^{-1} \neq y+x$.

Page 408 Review Capsule for Section 11-4 1. ± 6 2. 5 3. 2 4. $2\sqrt{2}$ 5. 2 6. $3|y|$ 7. $-4a^2b^5$ 8. $|rt|$ 9. $(a+b)^2$ 10. $-0.9s^3t^5v^7$

Page 409 Classroom Exercises 1. 5 3. 27 5. $\frac{1}{8}$ 7. $-\frac{1}{5}$ 9. $\frac{1}{9}$

Page 409 Written Exercises 1. 3 3. 2 5. -6 7. 3 9. 4 11. 32 13. $-3\sqrt[5]{81}$ 15. 32 17. 4 19. 16 21. $-\frac{1}{5}$ 23. 9 25. $\frac{1}{81}$ 27. $\frac{9}{4}$ 29. 0.2 31. $\sqrt{a}$ 33. $\sqrt[4]{x}$ 35. $\sqrt[4]{a^3}$ 37. $3\sqrt{a}$ 39. $\sqrt{2y}$ 41. $\sqrt[3]{x-y}$ 43. $\sqrt[3]{-z}$ 45. $a\sqrt{a}+bi\sqrt{b}$ 47. $y^{\frac{3}{4}}$ 49. $m^{\frac{1}{3}}n^{\frac{2}{3}}$ 51. $2a^{\frac{3}{4}}$ 53. $2xyz^{\frac{1}{2}}$ 55. $(a+b)^{\frac{1}{2}}$ 57. $(4a-9b)^{\frac{1}{2}}$ 59. $-a^{\frac{2}{3}}b^{\frac{1}{3}}$ 61. 4 63. $\sqrt{25}$; 5; $\sqrt{25}=5$; $25^{\frac{1}{2}}=5$ 65. $(\sqrt[3]{8})^2$; 4; $(\sqrt[3]{8})^2=4$; $8^{\frac{2}{3}}=4$ 67. $(\sqrt[4]{16})^5$; 32; $(\sqrt[4]{16})^5=32$; $16^{\frac{5}{4}}=32$ 69. $\sqrt{4}$; 2; $\sqrt{4}=2$; $4^{\frac{1}{2}}=2$ 71. $8^{-\frac{2}{3}}$; $\frac{1}{4}$; $\frac{1}{(\sqrt[3]{8})^2}=\frac{1}{4}$; $8^{-\frac{2}{3}}=\frac{1}{4}$ 73. 27 75. $\frac{1}{2}$ 77. 5 79. $\frac{1}{8}$ 81. 3 83. $\frac{1}{8}$ 85. $x^{\frac{2}{3}}y^{-\frac{1}{3}}$ 87. $mn^{-\frac{1}{2}}$

Page 411 Review Capsule for Section 11-5 1. a^7 2. b^3 3. $-75x^3y^{10}$ 4. $34r^6s^5t^{13}$ 5. 3^6, or 729 6. 1 7. $-c^{14}d^{35}$ 8. $-tv^2$ 9. $\sqrt{6}$ 10. $4\sqrt{6}$ 11. $6\sqrt{15}$ 12. -7 13. 4 14. $\sqrt{3}\,|x|$ 15. $\sqrt{3}$ 16. $\frac{a^2}{10b^3}\sqrt[3]{100}$

Page 413 Classroom Exercises 1. $2\sqrt[3]{2}$ 3. $2\sqrt[3]{9}$ 5. $\sqrt[6]{\frac{8}{25}}$ 7. $\{\frac{1}{16}\}$ 9. $\{\frac{1}{9}\}$

Page 413 Written Exercises 1. $3\sqrt[3]{3}$ 3. $(\sqrt[3]{3})^2$ 5. $4\sqrt[3]{2}$ 7. $\sqrt{2}$ 9. $10\sqrt[3]{2}$ 11. $\sqrt[6]{108}$ 13. $\sqrt[12]{432}$ 15. 2

17. $\sqrt[6]{2}$ 19. $\sqrt[3]{4}$ 21. $\sqrt[12]{5^5}$, or $\sqrt[12]{3125}$ 23. 48 25. $9\sqrt[3]{2}$ 27. $y\sqrt[6]{8y^3}$ 29. $s^2\sqrt{2}$ 31. $x = 49$ 33. $x = 2^3$ 35. $a = \pm\frac{1}{64}$ 37. $x = \frac{1}{4}$ 39. $a = \frac{1}{125}$ 41. $x = \frac{1}{512}$ 43. $x = 7$ 45. $x = 10$ 47. $y = 0$ or $y = -3$ 49. $x = 4$ 51. ∅ 53. $t^{\frac{4}{3}}$ 55. $r^{\frac{1}{6}}$ 57. $7^{\frac{2}{3}}$ 59. $a^{\frac{3}{4}}$ 61. 8 63. 2 65. $x = 8$ 67. $x = 5$ or $x = 1$ 69. $x = \frac{1}{3}$ 71. $C = \frac{1}{4\pi^2 f^2 L}$ 73. By definition, $\sqrt[n]{a} = a^{\frac{1}{n}}$ and $\sqrt[n]{b} = b^{\frac{1}{n}}$. Then $(a^{\frac{1}{n}} \cdot b^{\frac{1}{n}})^n = ab$. Therefore, $[(\sqrt[n]{a} \cdot \sqrt[n]{b})^n]^{\frac{1}{n}} = (ab)^{\frac{1}{n}}$; $\sqrt[n]{a} \cdot \sqrt[n]{b} = (ab)^{\frac{1}{n}}$, or $\sqrt[n]{a} \cdot \sqrt[n]{b} = \sqrt[n]{ab}$.

Page 414 Review 1. $\frac{1}{b^6}$ 2. $9c^4$ 3. $\frac{6y^3}{x^2z^4}$ 4. $\frac{-3-4i}{25}$ 5. $\frac{a^2}{b^4}$ 6. 3.8×10^5 7. 4.2×10^{-4} 8. 6×10^{-7} 9. 4.3×10^{11} 10. 30,000 11. 0.000022 12. 964,000,000 13. 0.0000000425 14. $\frac{x^3}{y}$ 15. $st(s+t)^2$ 16. $\frac{13q + 3p - 6}{3p^2}$ 17. -27 18. $\frac{1}{3}$ 19. 25 20. $\frac{1}{64}$ 21. $\frac{1}{16}$ 22. $\sqrt[6]{432}$ 23. $\sqrt[4]{8}$ 24. $6\sqrt[6]{24}$ 25. $3a\sqrt{b}$ 26. $t = 8$ 27. $x = \frac{15 + \sqrt{17}}{2}$ 28. $a = 3$ or $a = -1$

Page 416 Classroom Exercises 1. 0.01; 0.1; 1; 3.162; 10; 31.62 3. $\frac{1}{16}$; $\frac{1}{4}$; 1; 2; 4; 8; 16; 32; 64 5. 9; 3; 1; $\frac{1}{3}$; $\frac{1}{9}$; $\frac{1}{27}$

Page 416 Written Exercises In Exercises 1-5, the graphs are smooth curves that pass through the points listed in Classroom Exercises 1-5 on page 416. The graph of $y = (\frac{1}{3})^x$ is the least steep and the graph of $y = 10^x$ is the steepest. 7. Becomes steeper. 9. Positive half 11. No 13. The graph for $y = 2^x$ contains: $(-5, \frac{1}{32})$, $(-4, \frac{1}{16})$, $(-3, \frac{1}{8})$, $(-2, \frac{1}{4})$, $(-1, \frac{1}{2})$, (0, 1), (1, 2), (2, 4), (3, 8), (4, 16), (5, 32). The graph for $y = (\frac{1}{2})^x$ contains the points listed in Classroom Exercise 4 on page 416. 15. For a listing of the points through which the graphs of $y = 3^x$ and $y = (\frac{1}{3})^x$ pass, see the tables for Classroom Exercises 2 and 5, respectively, on page 416. 17. $x = 3$ 19. $x = 5$ 21. $x = \frac{2}{3}$ 23. $x = \frac{5}{3}$ 25. $x = \frac{2}{3}$ 27. $x = -2$ 29. $x = -1$ 31. $x = -4$ 33. $x = -\frac{1}{3}$ 35. $x = -1$ 37. The graph contains (−4, 0.02), (−3, 0.05), (−2, 0.14), (−1, 0.37), (0, 1), (1, 2.72), (2, 7.39), (3, 20.1). 39. 4.5 41. 0.6 43. Show that the y axis is the perpendicular bisector of the segment joining any pair of corresponding points P on $y = a^x$ and Q on $y = (\frac{1}{a})^x$. For m, any real value of x, P has coordinates (m, a^m). For $-m$, any real value of x, Q has coordinates $(-m, a^m)$. Since the y coordinates of P and Q are the same, P and Q lie on a line parallel to the x axis. Thus, the y axis is perpendicular to $\overline{PQ}$. The midpoint of $\overline{PQ}$ has coordinates $(0, a^m)$. Therefore, the y axis is the perpendicular bisector of $\overline{PQ}$ and the graphs are symmetric with respect to the y axis. 45. $x = \frac{15}{8}$ 47. $x = -3$

Page 418 Review Capsule for Section 11-7 1. The graph of $y = 3x - 2$ contains (0, −2) and (−2, 0). The graph of $x = 3y - 2$ contains $(0, \frac{2}{3})$ and $(-\frac{2}{3}, 0)$. 2. The graph of $y = 2x^2$ is a parabola that opens upward with vertex at (0, 0) and containing (1, 2), (−1, 2), (2, 8), (−2, 8). The graph of $x = 2y^2$ is a parabola that opens to the right with vertex at (0, 0) and containing (2, 1), (2, −1), (8, 2), (8, −2). 3. The graph of $y = x^2 - 4$ is a parabola, that opens upward, with vertex at (0, −4), and with x intercepts at (2, 0) and (−2, 0). The graph of $x = y^2 - 4$ is a parabola that opens to the right with vertex at (−4, 0) and with y intercepts at (0, 2) and (0, −2). 4. The graph of $y = |x|$ is V-shaped. The vertex is (0, 0) and the graph contains (−1, 1) and (1, 1). The graph of $x = |y|$ is V-shaped and opens to the right. The vertex is (0, 0) and the graph contains (1, 1) and (1, −1). Thus, the two graphs have a common side in Quadrant I. 5. $x = 4$ 6. $a = 4$ 7. $x = 0.001$ 8. $n = \frac{1}{9}$

Page 419 Classroom Exercises 1. $\log_2 8 = 3$ 3. $\log_6 216 = 3$ 5. $\log_{11} 1331 = 3$ 7. $\log_{10} \frac{1}{1000} = -3$ 9. $2^4 = 16$ 11. $8^2 = 64$ 13. $0.5^2 = 0.25$ 15. $2^{-3} = \frac{1}{8}$

Page 420 Written Exercises 1. positive real numbers; real numbers 3. $\log_2 64 = 6$ 5. $\log_{16} 64 = \frac{3}{2}$ 7. $\log_4 \frac{1}{8} = -\frac{3}{2}$ 9. $\log_8 2 = \frac{1}{3}$ 11. $10^3 = 1000$ 13. $7^2 = 49$ 15. $5^{-1} = 0.2$ 17. $10^{-2} = 0.01$ 19. 4 21. 3 23. 5 25. 0 27. −1 29. 3 31. 0 33. −2 35. −3 37. −3 39. $a = 4$ 41. $a = 4$ 43. $x = 81$ 45. $x = 0.001$ 47. $y = \frac{3}{2}$ 49. $y = \frac{4}{3}$ 51. $n = 64$ 53. $b = 2$ 55. $a = \frac{1}{7}$ 57. $n = \frac{1}{9}$ 59. The graph of $y = \log_{10} x$ contains the points listed in the table on page 420. 61. The graph contains: $(\frac{1}{27}, -3)$, $(\frac{1}{9}, -2)$, $(\frac{1}{3}, -1)$, (1, 0), (3, 1), (9, 2), (27, 3). 63. The curves rise less steeply to the right. 65. positive 67. negative 69. $10^1 < 50 < 10^2$; $1 < \log_{10} 50 < 2$ 71. 10^1; 10^2; 1; 2 73. 10^2; 10^3; 2; 3

Page 421 Calculator Exercises 7. 1.4142136 9. 12.59921 11. 2.1822472 13. 1.6581493 15. 1.9999999

17. 1.122462 19. 1.5874011 21. 1.9554085 23. 48 25. 11.339289

Page 421 Review Capsule for Section 11-8 1. a^{m+n} 2. a^{xy} 3. $\frac{1}{a^n}$ 4. $a^n b^n$ 5. $\frac{a^n}{b^n}$ 6. $\sqrt[n]{a}$ 7. $\sqrt[q]{a^p}$ 8. a^n

Page 423 Classroom Exercises 1. log M + log N 3. p log M 5. log 27 + log x + log y 7. 5 log x 9. 0.0000, or 0 11. 0.9248 13. 5.2 15. 1.8

Page 424 Written Exercises 1. log r + log s + log t 3. log r − log s 5. log r + $\frac{1}{2}$ log t 7. 3 log a + 3 log b 9. log x + log y − 2 log z 11. $\frac{1}{3}$ log x + 4 log y 13. True 15. False; by the power theorem, log 10^n = n log 10 which is not the same as log n · 10. 17. True 19. True 21. False; by the quotient theorem log $\frac{M}{N}$ = log M − log N. 23. False; log $\sqrt{45}$ = log $45^{\frac{1}{2}}$ = $\frac{1}{2}$ log 45 by the power theorem. 25. True 27. False; if 5^4 = 625, then $\log_5 625 = 4$. 29. 0.8293 31. 0.2455 33. 0.5563 35. 0.6981 37. 0.0253 39. 0.7789 41. 7.93 43. 2.91 45. 3.74 47. 5.49 49. x = 7.27 51. x = 4.90 53. x = 0.7752 55. 0.1004 57. x = 2.90 59. 5.39 61. x = 8.50 63. x = 3.45 65. 1.4771 67. 1.6532 69. −0.1249 71. 2.5562 73. Let $\log_a M = x$ and $\log_a N = y$, then $a^x = M$ and $a^y = N$, by def. of logarithm. Thus, $\frac{M}{N} = \frac{a^x}{a^y} = a^{x-y}$, by substitution and the quotient of powers theorem. Therefore, $\log_a \frac{M}{N} = x - y$, by definition of logarithm, or $\log_a \frac{M}{N}$ = log M − log N, by substitution. 75. $x = -2\frac{1}{2}$ 77. a = 4

Page 426 Using Statistics 1. 4900 3. 40,000 5. Sample 1: 2800; Sample 2: 3100; Sample 3: 2800 7. 2900 9. You can estimate the population by taking several samples or one large sample after the first sample. 11. No. Each name will not have the same chance of being chosen.

Page 429 Classroom Exercises 1. 0 3. 2 5. −4 7. 1.0969 9. 4.9886 11. 503 13. 0.753 15. 8.0394 − 10 17. 2.5973 19. 8.5556 21. 3.1929 23. 0.1576 25. 0.4525

Page 430 Written Exercises 1. 2.5587 3. 4.5587 5. 9.05587 − 10 7. 0.8591 9. 1.8591 11. 0.9926 13. 1.9212 15. 2.6355 17. 3.7275 19. 7.7945 − 10 21. 4.6590 − 10 23. 9.9258 − 10 25. 4.8451 27. 95.9 29. 0.205 31. 0.0898 33. 810 35. 0.00702 37. 0.537 39. 367 41. 7000 43. 41.0 45. 0.796 47. 6.89×10^{-8} 49. 3.37 51. 293 53. 0.0583

Page 430 Calculator Exercises 1. 2.2688907 3. 6.9961391

Page 433 Written Exercises 1. 0.7947 3. 1.7501 5. 2.3758 7. 8.3298 − 10 9. 9.1430 − 10 11. 19.18 13. 1.453 15. 1532.9 17. 0.002521 19. 0.05595 21. 256.8 23. 26,375 25. 0.05856 27. 1.934 29. 361.8 31. 40,010,000 33. 4.529 35. 0.7024 37. −3.771 39. 90.52 41. 4.6442 43. 55.48 45. 3.051 47. 1.1739 49. 1.2 51. 4.4 53. $\frac{\log 10}{\log 3}$ 55. $\frac{\log 75}{\log 4}$ 57. Let $x = \log_a b$, then $a^x = b$. $x \log_b a = \log_b b$ by the quotient theorem, and if two numbers are equal, their logarithms are equal. Substituting for x yields $\log_a b \cdot \log_b a = \log_b b$, or $\log_a b \log_b a = 1$, since $b^1 = b$. Thus, $\log_a b = \frac{1}{\log_b a}$.

Page 435 Word Problems 1. \$4950 3. \$3246 5. \$4556 7. \$2471 9. \$1105 11. \$1105 13. \$8200 15. 8.57% 17. 12.19%

Page 437 Computer Exercises See Solution Key.

Page 438 Review 1. The graph of $y = 5^x$ contains: $(-3, \frac{1}{125})$, $(-2, \frac{1}{25})$, $(-1, \frac{1}{5})$, (0, 1), (1, 5), (2, 25), (3, 125) 2. $x = \frac{5}{2}$ 3. $x = \frac{2}{3}$ 4. $x = -\frac{3}{2}$ 5. $x = -\frac{3}{2}$ 6. 2 7. $-\frac{1}{2}$ 8. $\frac{2}{3}$ 9. 0 10. $t = \frac{1}{125}$ 11. $b = \frac{1}{2}$ 12. $s = \frac{1}{3}$ 13. $r = 2^{-\frac{2}{3}}$ 14. log 3 + 4 log 8 15. log 4 + log c − 2 log d 16. 2 log a − 5 log b 17. $\frac{3}{4}$ log x − $\frac{5}{4}$ log a 18. 0.5119 19. 0.7372 20. 0.9689 21. 0.3201 22. 8.03 23. 3.05 24. 5.02 25. 2.02 26. 157.1 27. 1.727 28. 9,005,000 29. 0.3086 30. 4797 31. 13.52 32. 242.7 33. 2.733 34. \$6337

Page 439 Chapter Objectives and Review 1. $\frac{1}{s^4}$ 3. t^3m^4 5. $\frac{125}{27}$ 7. 7.9283×10^7 9. 4.43×10^{-7} 11. 0.0000037 13. 2,040,000,000,000 15. $\frac{1}{a^2b^4}$ 17. $\frac{17}{72}$ 19. $x^{\frac{2}{3}}$ 21. $(3x-1)^{\frac{1}{2}}$ 23. $\frac{1}{\sqrt[4]{b^3}}$ 25. $\sqrt[9]{7^2}$ 27. $\sqrt{2y^3}$ 29. $3\sqrt[3]{4}$ 31. $2\sqrt[3]{5}$ 33. $x = \frac{49}{4}$ 35. x = 15 37. x = −2 39. $\log_4 64 = 3$ 41. $\log_a y = x$ 43. $3^4 = 81$ 45. $2^{-2} = \frac{1}{4}$

47. log p + 7 log t 49. log x + 2log y − 4 log t 51. 643,100 53. 1476 55. $3258

Page 441 Chapter Test 1. 729 3. $\frac{5y - 4x^2y}{x^3}$ 5. 4.7×10^9; 5.673×10^{-6} 7. 1 9. $\frac{1}{8}$ 11. $x = \frac{5}{3}$ 13. 2 15. 25.72 17. x = 4 19. a = 27

Page 442 Preparing for College Entrance Tests 1. a 3. c 5. a

CHAPTER 12 SEQUENCES AND SERIES

Page 445 Classroom Exercises 1. 12, 15, 18 3. 16, 25, 36 5. 5, 10, 15 7. 2, 5, 10

Page 445 Written Exercises 1. $a_n = 2n$; 10, 12, 14 3. $a_n = n$; 5, 6, 7 5. $a_n = 3n - 1$; 14, 17, 20 7. $a_n = 2n - 6$; 4, 6, 8 9. $a_n = 4n + 1$; 21, 25, 29 11. $a_n = n + 3$; 8, 9, 10 13. $a_n = 5 + 5n$; 30, 35, 40 15. $a_n = -n^2$; −25, −36, −49 17. $a_n = \frac{n}{n+1}$; $\frac{5}{6}, \frac{6}{7}, \frac{7}{8}$ 19. $a_n = 1 + n^2$; 26, 37, 50 21. $a_n = \frac{1^n}{n^2}$; $\frac{1}{25}, \frac{1}{36}, \frac{1}{49}$ 23. $a_n = 3^{1-n}$; $\frac{1}{81}, \frac{1}{243}, \frac{1}{729}$ 25. $a_n = a^{2n-2}$; a^8, a^{10}, a^{12} 27. $a_n = x + (2n - 2)$; x + 8, x + 10, x + 12 29. 1; −1;1; −1 31. 1; $-\frac{1}{2}$; $\frac{1}{4}$; $-\frac{1}{8}$ 33. 1; −1; 1; −1 35. −7, −4, −1, 2

Page 445 Review Capsule for Section 12-2 1. x = −2, y = −3 2. $x = \frac{10}{7}, y = -\frac{13}{7}$ 3. x = 6, y = 1 4. x = 4, y = 1

Page 447 Classroom Exercises 1. Arithmetic; d = 4 3. Arithmetic; d = 3 5. Arithmetic; d = x

Page 447 Written Exercises 1. 1, 6, 11, 16, 21 3. −30, −32, −34, −36, −38 5. 2.3, 3.9, 5.5, 7.1, 8.7 7. c; x; 2x − c; 3x − 2c; 4x − 3c 9. 27 11. 45 13. 13 15. −50 17. a + 16 19. −11 21. $a_1 = 22, d = -3$ 23. 30, 26, 22, 18, 14 25. 7, 11 27. $7\frac{1}{4}, 8\frac{1}{2}, 9\frac{3}{4}$ 29. $3\frac{1}{2}, 5, 6\frac{1}{2}, 8, 9\frac{1}{2}$ 31. 50 33. $140 35. 19 37. $210 39. $3220 41. $2912 43. n = 58 45. 112 47. 33 49. Let $x = a_1$ and $y = a_3$. Then the arithmetic mean between x and y is $\frac{x+y}{2}$, which is the average of the numbers x and y. Therefore, finding one arithmetic mean between two numbers is the same as finding the average of the numbers. 51. Let a = 5, 7, 9, 11, . . ., b = 6, 9, 12, 15, . . ., and x = a • b. Therefore, x = 30, 63, 108, 165, ... The difference between x_2 and x_1 is 33; between x_3 and x_2 is 45; between x_4 and x_3 is 57; and so on. Since the difference between consecutive terms of x is not constant, x is not an arithmetic sequence.

Page 450 Review Capsule for Section 12-3 1. 3 2. 4 3. $3\sqrt[3]{3}$ 4. 64 5. 625 6. $\frac{1}{32}$ 7. $-\frac{32}{243}$ 8. 729 9. $\frac{1}{2}$ 10. 0.0009 11. −1 12. 96

Page 451 Classroom Exercises 1. 3 3. r 5. 9

Page 451 Written Exercises 1. 2, 6, 18, 54 3. $\frac{1}{2}$, 2, 8, 32 5. $\frac{1}{4}, \frac{1}{2}$, 1, 2 7. 0.7, −2.8, 11.2, −44.8 9. i, −1, −i, 1 11. −128 13. −x 15. $60\frac{3}{4}$ 17. 0.0000000001 19. $\frac{3}{32}$ 21. 1 23. −9 25. −4, −8, −16, −32, −64, or 4, −8, 16, −32, 64 27. $\sqrt{12} = 2\sqrt{3}$, or $-\sqrt{12} = -2\sqrt{3}$ 29. $2940 31. 1600 33. $5.12 35. 0.004 meter 37. 7th 39. 6144 41. x = 3 43. If a_1, a_2, a_3 form a geometric sequence, then the ratio between successive terms is a constant, r. Hence, $\frac{a_2}{a_1} = r$ and $\frac{a_3}{a_2} = r$. If $\frac{a_2^2}{a_1^2} = \frac{a_3^2}{a_2^2}$, then a_1^2, a_2^2, a_3^2 form a geometric sequence. Since $\frac{a_2^2}{a_1^2} = r^2$ and $\frac{a_3^2}{a_2^2} = r^2$, there is a constant ratio, r^2, between successive terms. Therefore, it is a geometric sequence.

Page 453 Review 1. $4 + n^2$; 29, 40, 53 2. $(-2)^n$; −32, 64, −128 3. $\frac{n^2+1}{n^2}$; $1\frac{1}{25}, 1\frac{1}{36}, 1\frac{1}{49}$ 4. 23 5. $21\frac{4}{5}$ 6. $18,600 7. 2560 8. $\frac{27}{128}$ 9. 10, 50, 250, and 1250 10. 640

Page 453 Review Capsule for Section 12-4 1. 6; 12; 18; 24 2. 13; 16; 19; 22 3. $-2\frac{1}{2}$; −2; $-1\frac{1}{2}$; −1 4. 1; 4; 9; 16 5. 3; 9; 27; 81 6. 25; 125; 625; 3125 7. 0 8. 1; $\frac{1}{2}$; $\frac{1}{3}$; $\frac{1}{4}$ 9. $\frac{1}{2}$; $\frac{1}{3}$; $\frac{1}{4}$; $\frac{1}{5}$ 10. $\frac{1}{2}$; $3\frac{4}{5}$; $8\frac{9}{10}$; $15\frac{16}{17}$

Page 455 Classroom Exercises 1. $\sum_{k=1}^{11} (k + 4)$ 3. $\sum_{k=1}^{15} (2k^2 - 1)$ 5. 5 + 10 + 15 + 20 7. $1 + \frac{1}{2} + \frac{1}{3} + \frac{1}{4} + \frac{1}{5}$ 9. 65 11. 72

Page 456 Written Exercises 1. 5 + 8 + 11 + 14 3. $\frac{1}{3} + \frac{2}{3} + 1 + \frac{4}{3} + \frac{5}{3} + 2 + \frac{7}{3} + \frac{8}{3} + 3 + \frac{10}{3}$ 5. 1 + 3 + 9 + 27 7. 1, 5, 19 9. 38 11. $\frac{10}{3}$, or $3\frac{1}{3}$ 13. 40 15. 128 17. 120 19. 525 21. 5 23. $-12\sqrt{2}$ 25. 192 27. $13\frac{3}{4}$ 29. −135x 31. 222b 33. 10a + 55b 35. 93 37. 100 39. −16 41. 130 43. $-22\frac{1}{2}$ 45. 62 47. 480 49. $1485 51. 35 m; 222 m 53. $67.50 55. 1149 m 57. 1300 cm 59. 24 + 60 + 96 + 132 + 168 + 204 + 240 61. n = 10; $d = \frac{29}{9}$ 63. d = −2; −280 65. $a_1 = 7$; d = −5 67. 3266

Page 458 Calculator Exercises See the answers to Exercises 9-12 and 15-28, and 35-46 on page 456.

Page 458 Review Capsule for Section 12-5 1. 192 2. $-16{,}384$ 3. $\frac{3}{16}$ 4. $\frac{1}{16}$ 5. $\frac{1}{81}$ 6. 124 7. $8191\frac{7}{8}$ 8. -42

Page 460 Classroom Exercises 1. $a_1 = 1; r = 4; n = 5$ 3. $a_1 = 6; r = \frac{1}{2}; n = 5$ 5. $a_1 = 2; r = 2; n = 4$ 7. $a_1 = 1; r = \frac{1}{2}; n = 6$

Page 460 Written Exercises 1. $\frac{63}{16}$, or $3\frac{15}{16}$ 3. 0 5. 124 7. $\frac{255}{32}$, or $7\frac{31}{32}$ 9. $\frac{2(3125a^5 + 1)}{5a + 1}$, or $\frac{6250a^5 + 2}{5a + 1}$ 11. $\frac{13}{27}$ 13. 40 15. -22 17. $\frac{255}{8}$, or $31\frac{7}{8}$ 19. 1031.632 21. 7 23. \$20.59 25. 260,000 27. 951,110 29. 81.92 cm 31. $\frac{x^{3n} - 1}{x^3 - 1}$ 33. $8 + 4 + 2 + 1 + \frac{1}{2}$

Page 461 Calculator Exercises See the answers for Exercises 5-7, 12-17, and 19-20 on page 460.

Page 461 Review Capsule for Section 12-6 1. $\frac{8}{9}$ 2. $\frac{4}{9}$ 3. $\frac{5}{9}$ 4. $\frac{4}{33}$ 5. $\frac{1}{33}$ 6. $\frac{2}{33}$

Page 463 Classroom Exercises The answers to Exercises 1-6 are given in the order: a_1, r, S_1, S_3 S_5. 1. 36; $\frac{1}{2}$; 36; 63; $\frac{279}{4}$, or $69\frac{3}{4}$ 3. 1; $\frac{1}{4}$; 1; $\frac{21}{16}$ or $1\frac{5}{16}$; $\frac{341}{256}$, or $1\frac{85}{256}$ 5. $\frac{1}{10}$; $\frac{1}{10}$; $\frac{1}{10}$; $\frac{111}{1000}$; $\frac{11{,}111}{100{,}000}$

Page 463 Written Exercises 1. $\frac{21}{2}$, or $10\frac{1}{2}$ 3. $\frac{93}{8}$, or $11\frac{5}{8}$ 5. $\frac{381}{32}$, or $11\frac{29}{32}$ 7. 8 9. $\frac{15}{2}$, or $7\frac{1}{2}$ 11. 250 13. $\frac{1}{3}$ 15. $\frac{5}{3}$ 17. $\frac{4}{9}$ 19. $\frac{4}{33}$ 21. $\frac{2}{33}$ 23. $\frac{10}{37}$ 25. $\frac{25}{333}$ 27. 324 m 29. $666{,}666\frac{2}{3}$ barrels 31. Cannot 33. 40 dm

Page 468 Classroom Exercises 1. 5 3. 12 5. $-20x^3y^3$ 7. $160x^3y^3$

Page 468 Written Exercises 1. 1, 6, 15, 20, 15, 6, 1; $a^6, a^5b, a^4b^2, a^3b^3, a^2b^4, ab^5, b^6$ 3. 1, 7, 21, 35, 35, 21, 7, 1; $r^7, r^6s, r^5s^2, r^4s^3, r^3s^4, r^2s^5, rs^6, s^7$ 5. $x^5 + 5x^4y + 10x^3y^2 + 10x^2y^3 + 5xy^4 + y^5$ 7. $m^6 - 6m^5n + 15m^4n^2 - 20m^3n^3 + 15m^2n^4 - 6mn^5 + n^6$ 9. $16r^4 + 32r^3 + 24r^2 + 8r + 1$ 11. $243m^5 + 810m^4 + 1080m^3 + 720m^2 + 240m + 32$ 13. $81x^4 + 216x^3y + 216x^2y^2 + 96xy^3 + 16y^4$ 15. $32a^5 - 240a^4b + 720a^3b^2 - 1080a^2b^3 + 810ab^4 - 243b^5$ 17. $35a^3b^4$ 19. $21r^5s^2$ 21. $-54x$ 23. $243s^{10} + 405s^8t + 270s^6t^2 + 90s^4t^3 + 15s^2t^4 + t^5$ 25. $8 + 8i$ 27. $x^4 + 2x^3 + \frac{3}{2}x^2 + \frac{1}{2}x + \frac{1}{16}$ 29. $(\frac{a}{b})^6 + 6(\frac{a}{b})^4 + 15(\frac{a}{b})^2 + 20 + 15(\frac{b}{a})^2 + 6(\frac{b}{a})^4 + (\frac{b}{a})^6$ 31. $-144x^4y^7$ 33. $-160a^3b^3$ 35. $30y^4$ 37. $\frac{3}{2}t^2$ 39. ≈ 1.21664 41. $\approx 3^n + 3^{n-1}n^2 + \frac{3^{n-2}}{2}n^3(n-1) + \frac{3^{n-3}}{6}n^4(n-1)(n-2)$ 43. $1 - 3y + 6y^2$ 45. $b^{-\frac{1}{2}} - b^{-\frac{3}{2}} - 2b^{-\frac{5}{2}}$

Page 470 Computer Exercises See Solution Key.

Page 471 Review 1. $6 + 10 + 14 + 18 + 22$ 2. $-1 + 2 + 7$ 3. $\frac{1}{2} + 2 + \frac{9}{2} + 8 + \frac{25}{2} + 18 + \frac{49}{2}$ 4. $-\frac{1}{3} + 0 + \frac{1}{3} + \frac{2}{3}$ 5. 700 6. $345\sqrt{2}$ 7. 820 8. $\frac{31}{32}$ 9. $\frac{10{,}101}{4096}$, or $2\frac{1909}{4096}$ 10. 0 11. 0 12. 132,700 13. $\frac{9}{2}$, or $4\frac{1}{2}$ 14. $\frac{8}{3}$, or $2\frac{2}{3}$ 15. $\frac{125}{4}$, or $31\frac{1}{4}$ 16. $x^6 + 6x^5y + 15x^4y^2 + 20x^3y^3 + 15x^2y^4 + 6xy^5 + y^6$ 17. $c^4 - 4c^3d + 6c^2d^2 - 4cd^3 + d^4$ 18. $32s^5 + 80s^4t + 80s^3t^2 + 40s^2t^3 + 10st^4 + t^5$ 19. $27w^3 - 4.5w^2z + 2.25wz^2 - 0.125z^3$ 20. $15{,}120a^4b^3$ 21. $-\frac{7}{16}x^3$ 22. $-140\sqrt{2}x^6$ 23. $-3240d^6b^6$

Page 472 Chapter Objectives and Review 1. $a_n = 3n$ 3. $a_n = n$ if n is odd and $a_n = -n$ if n is even. 5. $3\frac{1}{2}$; 4; $4\frac{1}{2}$; 5 7. 7, $7\frac{3}{4}$, $8\frac{1}{2}$, $9\frac{1}{4}$, 10 9. $\frac{5}{2}$, 4, $\frac{13}{2}$, 8, $\frac{19}{2}$ 11. \$24 13. 2; $\frac{2}{3}$; $\frac{2}{9}$; $\frac{2}{27}$; $\frac{2}{81}$ 15. $\frac{1}{5}$; $-\frac{1}{15}$; $\frac{1}{45}$; $-\frac{1}{135}$; $\frac{1}{405}$ 17. 15 19. $2 + 1 + 0 - 1 - 2 - 3 - 4$; -7 21. $4 - 8 + 16 - 32$; -20 23. 1000 25. $\frac{364}{81}$, or $4\frac{40}{81}$ 27. $\frac{3280}{2187}$, or $1\frac{1093}{2187}$ 29. -1 31. $-\frac{15}{16}$ 33. $\frac{15}{2}$, or $7\frac{1}{2}$ 35. $\frac{6}{5}$, or $1\frac{1}{5}$ 37. $r^7 + 21r^6 + 189r^5 + 945r^4 + 2835r^3 + 5103r^2 + 5103r + 2187$ 39. $32y^5 + 80y^4 + 80y^3 + 40y^2 + 10y + 1$ 41. $-15x^2y^4$ 43. $1792d^3t^5$

Page 474 Chapter Test 1. Geometric; $\frac{1}{80}$, $\frac{1}{160}$, $\frac{1}{320}$ 3. Neither 5. 35 7. 1000 9. -2 and $\frac{2}{5}$ 11. $-\frac{125}{4}$, or $-31\frac{1}{4}$ 13. $64 - 192q + 240q^2 - 160q^3 + 60q^4 - 12q^5 + q^6$ 15. 2278

Page 475 Review of Word Problems: Chapters 1-12 1. $18\frac{3}{4}$ inches 3. at 2%: 1 liter; at 8%: 2 liters 5. 15 days 7. $9\frac{1}{3}$ years 9. \$6144 11. 712 cells

Page 476 Cumulative Review: Chapters 1-12 1. b 3. a 5. c 7. b 9. c 11. d 13. d 15. c 17. c 19. c 21. d 23. a 25. c 27. a 29. d 31. b 33. a 35. b 37. c 39. b 41. a 43. d 45. d 47. d

Page 480 Preparing for College Entrance Tests 1. c 3. d 5. d 7. d 9. a 11. d

CHAPTER 13 SYSTEMS OF SENTENCES: THREE VARIABLES

Page 483 Classroom Exercises 1. (8, 0, 0) 3. (0, −11, 0) 5. (8, 13, 0) 7. (−4, −11, 5) 9. (0, 13, −7) 11. x axis 13. z axis 15. (6, 0, 0), (0, −6, 0), (0, 0, −6) 17. (6, 0, 0), (0, −4, 0), (0, 0, −2)

Page 483 Written Exercises For Exercises 1-12, the moves for plotting each point are described. The number of units and the direction are given. 1. 3 units from 0, direction: pos. x axis; 4 units, direction: pos. y axis; 5 units, direction: pos. z axis 3. 1 unit from 0, direction: neg. x axis; 6 units, direction: neg. y axis; 6 units, direction: pos. z axis 5. 4 units from 0, direction: pos. x axis; 8 units, direction: pos. y axis; 2 units, direction: neg. z axis 7. 1 unit from 0, direction: pos. y axis; 2 units, direction: pos. z axis 9. 1 unit from 0, direction: pos. x axis; 1 unit, direction: pos. y axis; 2 units, direction: pos. z axis 11. 5 units from 0, direction: neg. x axis; 7 units, pos. y axis For Exs. 13, 15, 17, the intercepts are given. Draw the line segments connecting these points and shade the interior. 13. (4, 0, 0), (0, 4, 0), (0, 0, 4) 15. (2, 0, 0), $(0, \frac{2}{3}, 0)$, (0, 0, −2) 17. (1, 0, 0), (0, 2, 0), (0, 0, −4) 19. z = 0 21. x = 0 23. The graph is similar to Fig. 2 on pg. 484. The plane contains (0, 0, 0) and (1, 2, 0). 25. A plane through (0, 0, 4) and parallel to the xy-plane. 27. A plane through (0, 3, 0) and parallel to the xz plane. 29. $-2x + 3y - 4z + 3 = 0$

Page 484 Review Capsule for Section 13-2 1. Yes 2. No 3. Yes

Page 487 Classroom Exercises 1. x = 5, y = 0, z = 3 3. x = −1, y = 4, z = −3

Page 487 Written Exercises 1. x = 2, y = 3, z = 4 3. x = −2, y = 0, z = 1 5. $x = \frac{1}{2}, y = \frac{1}{3}, z = \frac{3}{2}$ 7. x = 2, y = 3, z = 4 9. x = 1, y = 2, z = −3 11. x = 2, y = 4, z = 8 13. ∅ 15. x = 4, y = 2, z = 0

Page 489 Classroom Exercises 1. h + t + 0 = 19 3. x + y + z = 3 5. x + y + z = 100

Page 489 Written Exercises 1. x = $15,000 invested in the bowling alley. y = $15,000 invested in the diner. z = $10,000 invested in the laundromat. 3. 563 5. 283 7. x = 9 cm, y = 11 cm and z = 18 cm 9. Mary: 15 days; Tiffany: 30 days; Cecile: 60 days

Page 492 Classroom Exercises 1. $\begin{bmatrix} 2 & -1 & 3 & -9 \\ 1 & 3 & -1 & 10 \\ 3 & 1 & -1 & 8 \end{bmatrix}$ 3. $\begin{bmatrix} 1 & 2 & -1 & -3 \\ -2 & -4 & 2 & 6 \\ 3 & 6 & -3 & -9 \end{bmatrix}$

Page 492 Written Exercises 1. x = 6, y = 8, z = 4 3. x = 1, y = 0, z = 4 5. x = 0, y = 1, z = 3 7. x = 2, y = 1, z = 3 9. x = 2, y = 3, z = 4 11. $x = \frac{4}{17}z - \frac{7}{17}$; $y = \frac{3}{17}z - \frac{18}{17}$; z is any complex number 13. 1. $\begin{bmatrix} 1 & 0 & 0 & 6 \\ 0 & 1 & 0 & 8 \\ 0 & 0 & 1 & 4 \end{bmatrix}$ 2. $\begin{bmatrix} 1 & 0 & 0 & 6 \\ 0 & 1 & 0 & 4 \\ 0 & 0 & 1 & 10 \end{bmatrix}$ 3. $\begin{bmatrix} 1 & 0 & 0 & 1 \\ 0 & 1 & 0 & 0 \\ 0 & 0 & 1 & 4 \end{bmatrix}$

Page 492 Review For Exercises 1-4, the moves for plotting each point are described. The number of units and the direction are given. 1. 2 units from 0, direction: pos. x axis; 1 unit, direction: pos. y axis; 3 units, direction: neg. z axis 2. 4 units from 0, direction: pos. x axis; 2 units, direction: neg. y axis; 5 units, direction: pos. z axis 3. 1 unit from 0, direction: neg. y axis; 3 units, direction: pos. z axis 4. 4 units from 0, direction: neg. x axis; 2 units, direction: neg. z axis 5. x = 3, y = 4, z = −1 6. x = 0, y = −2, z = 6 7. x = 5, y = 0, $z = \frac{1}{2}$ 8. 264 9. x = −2, y = 3, z = 1 10. x = 3, y = 0, z = −4 11. x = 4, y = −2, z = −1

Page 495 Classroom Exercises 1. [4 −1] 3. $\begin{bmatrix} 3 & -1 \\ -3 & 6 \end{bmatrix}$ 5. $\begin{bmatrix} -27 & 3 \\ -18 & -15 \end{bmatrix}$ 7. [35] 9. $\begin{bmatrix} -4 & -1 & 5 \\ 38 & 30 & 28 \end{bmatrix}$ 11. 4 × 2 13. 4 × 1 15. 2 × 2

Page 495 Written Exercises 1. $\begin{bmatrix} 4 \\ 1 \\ 3 \end{bmatrix}$ 3. NP 5. NP 7. NP 9. $\begin{bmatrix} 0 & 0 \\ 0 & 0 \end{bmatrix}$ 11. $\begin{bmatrix} 3 & 3 & 5 \\ -7 & 10 & -8 \\ 28 & 2 & 1 \end{bmatrix}$ 13. $\begin{bmatrix} 3 & 6 \\ 18 & 0 \\ -3 & 12 \end{bmatrix}$ 15. $\begin{bmatrix} 0 & -15 \\ -20 & 5 \end{bmatrix}$ 17. $\begin{bmatrix} -2 & 2 & -1 \\ -4 & -6 & 3 \end{bmatrix}$ 19. [28] 21. NP 23. $\begin{bmatrix} 11 & 0 \\ 0 & 11 \end{bmatrix}$ 25. $\begin{bmatrix} 8 & 27 \\ 12 & -19 \end{bmatrix}$ 27. $\begin{bmatrix} 8 & 24 & 16 \\ 8 & 32 & 8 \\ 40 & 8 & 16 \end{bmatrix}$ 29. $\begin{bmatrix} ra_2 & ra_1 \\ rb_2 & rb_1 \end{bmatrix}$ 31. $\begin{bmatrix} a & b & c \\ g & h & i \\ d & e & f \end{bmatrix}$ 33. $\begin{bmatrix} 3 \\ -5 \end{bmatrix}$ 35. $\begin{bmatrix} -6 & 11 & -1 \\ -5 & 5 & -5 \end{bmatrix}$ 37. x = −5, y = 0 39. x = 2; y = 3 41. x = 3, y = −2

43. $(A + B) + C = A + (B + C) = \begin{bmatrix} 4 & 2 \\ 4 & 10 \end{bmatrix}$ 45. $A + 0 = \begin{bmatrix} a_1 & b_1 \\ a_2 & b_2 \end{bmatrix} + \begin{bmatrix} 0 & 0 \\ 0 & 0 \end{bmatrix} = \begin{bmatrix} a_1 & b_1 \\ a_2 & b_2 \end{bmatrix}$; $0 + A = \begin{bmatrix} 0 & 0 \\ 0 & 0 \end{bmatrix} + \begin{bmatrix} a_1 & b_1 \\ a_2 & b_2 \end{bmatrix} = \begin{bmatrix} a_1 & b_1 \\ a_2 & b_2 \end{bmatrix}$

47. Yes 49. $\begin{matrix} 1 & 0 & 0 \\ 0 & 1 & 0 \\ 0 & 0 & 1 \end{matrix}$

Page 499 Classroom Exercises 1. $A = \begin{bmatrix} 3 & 1 & 1 \\ 2 & 2 & 1 \\ 2 & 1 & 2 \end{bmatrix}$ 3. 1 costs \$54; 2 costs \$50; 3 costs \$42

Page 499 Written Exercises 1. Beginner's Kit: [3 2 1 9 6 0]; Intermediate Kit: [8 6 2 25 24 0]; Advanced Kit: [0 8 2 23 16 6]; Total Needed: [11 16 5 57 46 6] 3. $\begin{bmatrix} 11 & 12 & 10 & 10 \\ 11 & 11 & 13 & 14 \end{bmatrix}$ 5. Edgar, Alan, Dan, Fred, Bob, Claude

7. From (rows) / To (columns): $\begin{array}{c|cccc} & A & B & C & D \\ \hline A & 1 & 1 & 0 & 0 \\ B & 0 & 0 & 1 & 0 \\ C & 0 & 0 & 0 & 1 \\ D & 0 & 1 & 1 & 0 \end{array}$ 9. From (rows) / To (columns): $\begin{array}{c|cccc} & A & B & C & D \\ \hline A & 1 & 1 & 0 & 1 \\ B & 1 & 0 & 1 & 1 \\ C & 0 & 1 & 1 & 0 \\ D & 0 & 1 & 0 & 0 \end{array}$

11. A to B to A, A to B to B, B to A to B, B to B to A, B to B to B

13. Same

15. $\begin{array}{c} \\ A \\ B \\ C \end{array}\begin{array}{ccc} A & B & C \\ \hline a_1 & b_1 & c_1 \\ a_2 & b_2 & c_2 \\ a_3 & b_3 & c_3 \end{array} \quad \begin{array}{c} \\ A \\ B \\ C \end{array}\begin{array}{ccc} A & B & C \\ \hline a_1 & b_1 & c_1 \\ a_2 & b_2 & c_2 \\ a_3 & b_3 & c_3 \end{array} = [a_1a_1 + b_1a_2 + c_1a_3]$

a_1a_1 is the two-stage journey A to A to A
b_1a_2 is the two-stage journey A to B to A
c_1a_3 is the two-stage journey A to C to A
Adding these three together gives all the two-stage journeys from A back to A.

Page 501 Review Capsule for Section 13-7 1. 2 2. −6 3. −4 4. $dg - ef$ 5. $a_1b_2 - a_2b_1$

Page 503 Classroom Exercises 1. 7 3. −15 5. $\begin{bmatrix} \frac{1}{3} & -\frac{1}{6} \\ 1 & 0 \end{bmatrix}$ 7. $\begin{bmatrix} \frac{3}{7} & -\frac{1}{7} \\ \frac{1}{7} & \frac{2}{7} \end{bmatrix}$

Page 503 Written Exercises 1. $\begin{bmatrix} 1 & 0 \\ 0 & 1 \end{bmatrix}$ 3. $\begin{bmatrix} 1 & 0 & 0 & 0 & 0 \\ 0 & 1 & 0 & 0 & 0 \\ 0 & 0 & 1 & 0 & 0 \\ 0 & 0 & 0 & 1 & 0 \\ 0 & 0 & 0 & 0 & 1 \end{bmatrix}$ 5. $\begin{bmatrix} 1 & 0 & 0 & 0 & 0 & 0 \\ 0 & 1 & 0 & 0 & 0 & 0 \\ 0 & 0 & 1 & 0 & 0 & 0 \\ 0 & 0 & 0 & 1 & 0 & 0 \\ 0 & 0 & 0 & 0 & 1 & 0 \\ 0 & 0 & 0 & 0 & 0 & 1 \end{bmatrix}$ 7. $\begin{bmatrix} -\frac{1}{2} & \frac{1}{12} \\ 0 & \frac{1}{6} \end{bmatrix}$ 9. N

11. $\begin{bmatrix} -\frac{1}{3} & \frac{2}{3} \\ \frac{2}{3} & -\frac{1}{3} \end{bmatrix}$ 13. $\begin{bmatrix} \frac{3}{4} & \frac{5}{8} \\ -1 & \frac{5}{2} \end{bmatrix}$ 15. $\begin{bmatrix} -2 & 0 \\ 0 & -2 \end{bmatrix}$ 17. $\begin{bmatrix} \frac{7}{2} & 6 \\ -4 & -7 \end{bmatrix}$ 19. $x = 3$; $y = -4$ 21. $x = 2$; $y = 1$ 23. $x = -\frac{13}{9}$; $y = \frac{16}{9}$ 25. $x = 5$; $y = 0$ 27. $x = 4$; $y = 1$

Page 504 Review 1. $\begin{bmatrix} 5 & 5 & 10 \\ -3 & 1 & 8 \\ 0 & 9 & -6 \end{bmatrix}$ 2. $\begin{bmatrix} 9 & -6 \\ 0 & 12 \end{bmatrix}$ 3. $\begin{bmatrix} -18 \\ -48 \\ 48 \end{bmatrix}$ 4. $\begin{bmatrix} -12 \\ 24 \\ 17 \end{bmatrix}$ 5. $\begin{bmatrix} 2 & 19 & 6 \\ -11 & 18 & -12 \\ -4 & -33 & 13 \end{bmatrix}$ 6. $\begin{bmatrix} 3 & 1 & 2 \\ 1 & -5 & -4 \\ 0 & 1 & 4 \end{bmatrix}$

7. June: [30 20 15]; July: [45 10 25]; Aug.: [15 30 10]; Total: [90 60 50] 8. $\begin{bmatrix} \frac{1}{7} & -\frac{2}{7} \\ \frac{3}{7} & \frac{1}{7} \end{bmatrix}$ 9. N

10. $\begin{bmatrix} -\frac{1}{3} & \frac{2}{3} \\ 0 & \frac{1}{2} \end{bmatrix}$ 11. $\begin{bmatrix} \frac{4}{3} & 1 \\ \frac{1}{3} & -\frac{1}{6} \end{bmatrix}$

Page 506 Computer Exercises See Solution Key.

Page 508 Chapter Objectives and Review For Exercises 1-8, the moves for plotting each point are described. The number of units and the direction are given. 1. 4 units from 0, direction: pos. x axis; 1 unit, direction: pos. y axis; 7 units, direction: neg. z axis 3. 2 units from 0, direction: neg. y axis; 6 units, direction: pos. z axis 5. 4 units from 0, direction: pos. x axis; 4 units, direction: pos. y axis; 4 units, direction: neg. z axis 7. 1 unit from 0, direction: pos. x axis; $\frac{1}{2}$ unit, direction: pos. y axis; 3 units, direction: pos. z axis 9. (3, 0, 0); (0, 3, 0); (0, 0, −3) 11. $(\frac{2}{3}, 0, 0)$; $(0, -\frac{1}{2}, 0)$; (0, 0, 1) 13. $x = 3$, $y = 0$, $z = 1$ 15. 328 17. $x = 6$, $y = 3$, $z = 1$ 19. $\begin{bmatrix} 5 & 15 \\ -10 & 0 \end{bmatrix}$ 21. $\begin{bmatrix} 17 & 2 \\ 14 & 34 \end{bmatrix}$ 23. a. Small: [6 1 2 1 0 0]; Large: [12 3 2 2 2 0]; Travel: [4 1 0 1 1 4]; Total: [22 5 4 4 3 4] b. Value of small kit: \$2.04; large kit: \$4.30; Travel kit: \$1.88

25. $\begin{bmatrix} \frac{1}{4} & -\frac{1}{32} \\ 0 & \frac{1}{8} \end{bmatrix}$ 27. $\begin{bmatrix} 0 & \frac{1}{3} \\ 1 & -\frac{7}{3} \end{bmatrix}$

Page 510 Chapter Test 1. y axis 3. x axis 5. 3 units from 0, direction: neg. x axis; 1 unit, direction: pos. y axis; 7 units, direction: pos. z axis 7. x intercept, (5, 0, 0); y intercept, $(0, -\frac{5}{3}, 0)$; z intercept, (0, 0, 5) 9. $x = 1$, $y = -1$, $z = 5$
11. $\begin{bmatrix} 3 & 3 \\ -3 & 3 \end{bmatrix}$ 13. $\begin{bmatrix} 3 & 9 & 3 \\ 0 & 6 & 15 \end{bmatrix}$ 15. a.

Carpenters	Painters	plumbers
[4	5	2]

	Cost
carpenters	12
painters	9
plumbers	15

b. $123

CHAPTER 14 PROBABILITY

Page 514 Classroom Exercises 1. xyz, yzx, zxy, xzy, yxz, zyx 3. 5040 5. 20 7. 9

Page 514 Written Exercises 1. 6 3. 5 5. 120 7. 20,160 9. 362,880 11. 30,240 13. 840 15. 151,200 17. 840 19. 6,760,000 21. 24 23. 1440 25. 10,080 27. 12 29. 120 31. 10,080 33. ${}_7P_4 = 840; 7({}_6P_3) = 7(120) = 840; {}_7P_4 = 7({}_6P_3)$ 35. $n = 7$ 37. $n = 6$ 39. $n = 8$

Page 516 Classroom Exercises 1. A_1REA_2, A_1RA_2E, A_1ERA_2, A_1EA_2R, A_1A_2RE, REA_1A_2, A_1A_2ER, RA_1EA_2, RA_1A_2E, RA_2A_1E, RA_2EA_1, REA_2A_1, EA_1RA_2, EA_1A_2R, ERA_1A_2, ERA_2A_1, EA_2A_1R, EA_2RA_1, A_2A_1RE, A_2A_1ER, A_2RA_1E, A_2REA_1, A_2EA_1R, A_2ERA_1 3. 6720 5. 13,305,600 7. 1,663,200

Page 516 Written Exercises 1. 6720 3. 45,360 5. 12 7. 120 9. 20 11. 1260 13. 60 15. 12 17. 6 19. 6 21. 27,720 23. 480 25. 12 27. 120 29. 362,880 31. 2520 33. 720 35. One way 37. 2 ways

Page 519 Classroom Exercises 1. Fran, José; Fran, Elmer; José, Elmer 3. 3 5. 28 7. 56 9. 120 11. $\frac{9}{4}$

Page 519 Written Exercises 1. 4950 3. 142,506 5. 45 7. 142,506 9. 3003 11. $\approx 1.731 \times 10^{13}$ 13. 126 15. 83,160 17. 495 19. ${}_nC_{n-r} = \frac{n!}{(n-r)!(n-(n-r))!} = \frac{n!}{(n-r)!r!} = \frac{n!}{r!(n-r)!} = {}_nC_r$ 21. 71,379 23. 43,494 25. 10 27. $a^4 + 4a^3b + 6a^2b^2 + 4ab^3 + b^4$ 29. $a^5 - 5a^4b + 10a^3b^2 - 10a^2b^3 + 5ab^4 - b^5$ 31. $35a^4b^3$ 33. $28x^2y^6$ 35. $-12x$ 37. $\frac{16384}{2187}$ 39. 56.0000000056 41. $\frac{4^{10}}{y^{10}}$

Page 521 Review 1. 210; 120 2. 5040 3. 840 4. 907,200 5. 210 6. 35; 1 7. 126

Page 521 Review Capsule for Section 14-4 1. 2, 4, 6, 2. 1, 3, 5 3. 2, 3, 5 4. 4, 6 5. 4 6. 3, 6

Page 523 Using Statistics 1. P = 0.088077 or 8.8% 3. P = 0.120791 or 12.1% 5. 0.044; yes; The probability of doing this is so small, he must have a special talent. 7. 0.0756 9. 0.003 11. 0.914

Page 525 Classroom Exercises 1. $\frac{1}{2}$ 3. $\frac{5}{6}$ 5. $\frac{1}{2}$

Page 525 Written Exercises 1. $\frac{1}{6}$ 3. $\frac{2}{3}$ 5. $\frac{1}{3}$ 7. 0 9. $\frac{1}{4}$ 11. $\frac{1}{26}$ 13. $\frac{3}{10}$ 15. 1 17. $\frac{1}{3}$ 19. 1 21. $\frac{2}{3}$ 23. 1 25. $\frac{2}{3}$ 27. 1 29. 36 31. $\frac{1}{6}$ 33. $\frac{59}{220}$ 35. $\frac{1}{649740}$ 37. $\frac{44}{3553}$ 39. $\frac{10373}{24871}$

Page 527 Review Capsule for Section 14-5 1. The set of 4 kings and 4 queens. 2. The set of 13 hearts and 4 kings, one of which is also a heart. 3. The set of 12 face cards. 4. The king that is a heart. 5. ∅

Page 529 Classroom Exercises 1. mutually exclusive 3. mutually exclusive 5. not mutually exclusive 7. 1. $E_1 = \{S_2, S_3, S_4, S_5, S_6, S_7, S_8, S_9, S_{10}, S_J, S_Q, S_K, S_A\}$ $E_2 = \{D_2, D_3, D_4, D_5, D_6, D_7, D_8, D_9, D_{10}, D_J, D_Q, D_K, D_A\}$
2. $E_1 = \{S_Q, C_Q, H_Q, D_Q\}$ $E_2 = \{H_2, H_3, H_4, H_5, H_6, H_7, H_8, H_9, H_{10}, H_J, H_Q, H_K, H_A\}$ 3. $E_1 = \{H_J, H_Q, H_K, D_J, D_Q, D_K, S_J, S_Q, S_K, C_J, C_Q, C_K\}$ $E_2 = \{H_2, H_3, H_4, H_5, H_6, H_7, H_8, H_9, H_{10}, D_2, D_3, D_4, D_5, D_6, D_7, D_8, D_9, D_{10}, S_2, S_3, S_4, S_5, S_6, S_7, S_8, S_9, S_{10}, C_2, C_3, C_4, C_5, C_6, C_7, C_8, C_9, C_{10}\}$

Page 529 Written Exercises 1. $\frac{2}{5}$ 3. $\frac{1}{2}$ 5. $\frac{1}{2}$ 7. 1 9. $\frac{4}{13}$ 11. $\frac{11}{26}$ 13. $\frac{1}{2}$ 15. $\frac{5}{6}$ 17. $\frac{11}{25}$ (Remember to include the numbers 2 and 3.) 19. $\frac{1}{5}$ 21. $\frac{1}{3}$ 23. $\frac{11}{31}$ 25. $\frac{5}{9}$ 27. $\frac{5}{6}$ 29. $\frac{13}{18}$

Page 533 Classroom Exercises 1. $\frac{1}{8}$ 3. 0.04 5. independent 7. dependent

Page 533 Written Exercises 1. $\frac{1}{6}$ 3. $\frac{1}{3}$ 5. $\frac{13}{51}$ 7. $\frac{4}{663}$ 9. $\frac{1}{12}$ 11. $\frac{1}{3}$ 13. $\frac{1}{12}$ 15. $\frac{1}{144}$ 17. $\frac{5}{18}$ 19. $\frac{5}{9}$ 21. $\frac{25}{102}$ 23. $\frac{1}{6}$

Page 535 Review 1. $\frac{1}{4}$ 2. $\frac{1}{2}$ 3. 1 4. $\frac{1}{2}$ 5. $\frac{1}{4}$ 6. 0 7. $\frac{2}{13}$ 8. $\frac{1}{2}$ 9. $\frac{2}{13}$ 10. $\frac{4}{13}$ 11. $\frac{11}{26}$ 12. $\frac{4}{13}$ 13. $\frac{1}{169}$ 14. $\frac{1}{16}$ 15. $\frac{1}{4}$ 16. $\frac{1}{52}$ 17. $\frac{1}{52}$

Page 536 Chapter Objectives and Review 1. abc, acb, bca, bac, cab, cba 3. 2058 5. 50,400 7. 6 9. 792 11. $\frac{1}{8}$ 13. 1 15. $\frac{4}{13}$ 17. $\frac{1}{3}$ 19. $\frac{5}{36}$ 21. $\frac{1}{51}$

Page 537 Chapter Test 1. 120 3. 180 5. 15 7. 10 9. 56 11. $\frac{1}{5}$ 13. $\frac{8}{15}$ 15. $\frac{5}{6}$ 17. $\frac{33}{36}$ 19. $\frac{1}{52}$

Page 538 More Challenging Problems 1. 7 3. 1680 5. $\frac{5}{16}$ 7. $n = 5$

CHAPTER 15 TRIGONOMETRIC FUNCTIONS

Page 541 Classroom Exercises 1. one-twelfth rotation, counterclockwise 3. one-eighth rotation, clockwise 5. three-fourths rotation, clockwise 7. $-300°$ 9. $-345°$ 11. $-175°$

Page 541 Written Exercises 1. one-fourth rotation, counterclockwise 3. one-twenty-fourth rotation, counterclockwise 5. one-twelfth rotation, clockwise 7. three-eighths rotation, counterclockwise 9. three-fourths rotation, counterclockwise 11. seven-eighths rotation, counterclockwise 13. 90° 15. 120° 17. 810° 19. $-90°$ 21. second 23. fourth 25. second 27. -330; 750; 1830 29. -240; 840; 1920 31. -90; 990; 2070 33. 1.13 35. 3.14 m/sec

Page 545 Classroom Exercises The answers to Exercises 1-10 are given in the order: sin, cos, tan. 1. $\frac{4}{5}; \frac{3}{5}; \frac{4}{3}$ 3. $-\frac{12}{13}; -\frac{5}{13}; \frac{12}{5}$ 5. $\frac{-\sqrt{2}}{2}; \frac{-\sqrt{2}}{2}; 1$ 7. second; $\frac{4}{5}; -\frac{3}{5}; -\frac{4}{3}$ 9. third; $-\frac{4}{5}; -\frac{3}{5}; \frac{4}{3}$

Page 545 Written Exercises The answers to Exercises 1-20 are given in the order: sin, cos, tan. 1. $\frac{4}{5}; -\frac{3}{5}; -\frac{4}{3}$ 3. $\frac{4}{5}; -\frac{3}{5}; -\frac{4}{3}$ 5. $-\frac{4}{5}; \frac{3}{5}; -\frac{4}{3}$ 7. $-\frac{5}{13}; \frac{12}{13}; -\frac{5}{12}$ 9. $\frac{2\sqrt{5}}{5}; \frac{\sqrt{5}}{5}; 2$ 11. $-\frac{5\sqrt{61}}{61}; -\frac{6\sqrt{61}}{61}; \frac{5}{6}$ 13. $-\frac{\sqrt{2}}{2}; -\frac{\sqrt{2}}{2}; 1$ 15. $-\frac{4}{5}; \frac{3}{5}; -\frac{4}{3}$ 17. $\frac{3\sqrt{13}}{13}; -\frac{2\sqrt{13}}{13}; -\frac{3}{2}$ 19. $\frac{\sqrt{5}}{5}; -\frac{2\sqrt{5}}{5}; -\frac{1}{2}$ 21. $\frac{2\sqrt{13}}{13}; \frac{3\sqrt{13}}{13}$ 23. $\frac{3}{5}; \frac{4}{5}$ 25. +; +; −; − 27. +; −; +; − 29. second and third 31. first and second 33. third quadrant 35. $-\frac{\sqrt{3}}{2}$ 37. $-\frac{4}{5}$ 39. T 41. F

Page 548 Classroom Exercises 1. Leg opposite the 30° angle: 7; other leg: $7\sqrt{3}$ 3. $s = 10$ 5. $\frac{1}{2}; \frac{\sqrt{3}}{2}; \frac{\sqrt{3}}{3}$

Page 548 Written Exercises 1. $4; 4\sqrt{3}$ 3. 11 5. $\frac{19\sqrt{2}}{2}$ 7. a. $P(1, \sqrt{3})$ b. $\frac{\sqrt{3}}{2}; \frac{1}{2}; \sqrt{3}$ 9. a. $-1; \sqrt{2}$ b. $-\frac{\sqrt{2}}{2}; \frac{\sqrt{2}}{2}; -1$ 11. a. $P(-1, \sqrt{3})$ b. $\frac{\sqrt{3}}{2}; -\frac{1}{2}; -\sqrt{3}$ 13. $(-\sqrt{3}, -1); (-\frac{1}{2}; -\frac{\sqrt{3}}{2}); \frac{\sqrt{3}}{3}$ 15. $(1, -\sqrt{3}); -\frac{\sqrt{3}}{2}; \frac{1}{2}; -\sqrt{3}$ 17. $-\frac{1}{2}$ 19. $-\frac{1}{2}$ 21. $\frac{1}{2}$ 23. $-\frac{\sqrt{3}}{3}$ 25. $\frac{\sqrt{3}}{2}$ 27. $\tan 60° = \sqrt{3}$, $\sin 60° = \frac{\sqrt{3}}{2}$, $\cos 60° = \frac{1}{2}$; thus, $\frac{\sin 60°}{\cos 60°}$

$$= \frac{\frac{\sqrt{3}}{2}}{\frac{1}{2}} = \frac{\sqrt{3}}{2} \cdot \frac{2}{1} = \sqrt{3} = \tan 60°; \text{ Therfore, } \tan 60° = \frac{\sin 60°}{\cos 60°}.$$

Page 550 Classroom Exercises 1. 0.2560 3. 0.3827 5. 23° 40′ 7. 49° 10′

Page 550 Written Exercises 1. tan B 3. cos A 5. cos B 7. tan B 9. 41 11. 6 13. 3 15. 30° 17. $\frac{1}{2}$ 19. $\frac{\sqrt{3}}{2}$ 21. $\frac{1}{2}$ 23. $\frac{1}{2}; \theta$ 25. $\theta; \beta$ or α 27. 481.83 m 29. 181.6 m 31. 339.18 ft 33. 22.07 m

Page 553 Calculator Exercises See the answers to Exercises 9-14 on page 551.

Page 553 Puzzle See Solution Key.

Page 553 Review Capsule for Section 15-5 1. increases 2. 0 to −1 3. 90° to 180° 4. tangent 5. decreasing 6. increasing 7. increasing 8. decreasing

Page 555 Classroom Exercises 1. add 3. add 5. add

Page 555 Written Exercises 1. 0.6843 3. 3.8621 5. 0.9228 7. 0.6730 9. 0.9839 11. 0.3217 13. 0.9259 15. 0.4266 17. 36° 13′ 19. 17° 28′ 21. 58° 37′ 23. 6° 23′ 25. 40° 53′ 27. 59° 7′ 29. 490 cm 31. 347 cm 33. 27.36 km 35. 33° 12′

Page 556 Review 1. seven thirty-sixths rotation, counterclockwise 2. one-third rotation, counterclockwise 3. seven-eighths rotation, counterclockwise 4. one and one-half rotation, counterclockwise 5. fourth 6. 400°; 760°; −320° 7. $\sin\theta = \frac{y}{r}$; $\cos\theta = \frac{k}{r}$; $\tan\theta = \frac{y}{x}$ 8. $\frac{24}{25}; \frac{7}{25}; \frac{24}{7}$ 9. $6; 6\sqrt{3}$ 10. $3\sqrt{2}; 3\sqrt{2}$ 11. $\frac{\sqrt{3}}{3}$ 12. $-\frac{\sqrt{2}}{2}$ 13. $-\frac{1}{2}$ 14. $-\frac{\sqrt{3}}{3}$ 15. 23 m 16. 40° 53′ 17. 284 cm

Page 558 Classroom Exercises Each rotation in Exercises 1-10 is clockwise. 1. one-twelfth rotation 3. five-ninths rotation 5. five-sixths rotation 7. one and five-ninths rotation 9. three-fourths rotation

Page 558 Written Exercises The answers for Exs. 1-18 are given in the order: sin, cos, tan. 1. −0.7071; 0.7071; −1.000 3. −0.2588; −0.9659; 0.2679 5. 0.9877; −0.1564; −6.3138 7. 0.3420; −0.9397; −0.3640 9. −0.9659; 0.2588; −3.7321 11. −0.2588; 0.9659; −0.2679 13. 0; −1.000; 0 15. −0.3420; −0.9397; 0.3640 17. 0.7071; 0.7071; 1.000 19. $\sin(-135°) = -0.7071$; $\sin(-45°)\cos(-90°) + \cos(-45°)\sin(-90°) = (-0.7071)(0) + (0.7071)(-1) = -0.7071$ 21. $\sin(-240°) = 0.8660$; $-\sin 240° = -(-0.8660) = 0.8660$ 23. The triangles are similar in quadrants III and IV.

Page 561 Classroom Exercises 1. $-\cos\theta$ 3. $-\tan\theta$ 5. $-\sin\theta$ 7. cos 25° 9. −tan 75° 11. N 13. P 15. N 17. N 19. N 21. N 23. N 25. N 27. N 29. P

Page 562 Written Exercises 1. (180° − 54° 30′) 3. (180° − 43° 50′) 5. (180° − 63° 28′) 7. (180° − 80° 02′) 9. (180° − 52° 46′) 11. (180° − 35° 54′) 13. (180° + 37°) 15. (180° + 56°) 17. (180° + 45° 20′) 19. (180° + 31° 50′) 21. (180° + 70° 17′) 23. (180° + 89° 04′) 25. (360° − 85°) 27. (360° − 66°) 29. (360° − 57° 50′) 31. (360° − 33° 20′) 33. (360° − 14° 24′) 35. (360° − 2° 36′) 37. (90° + 29°) 39. (90° + 45°) 41. (90° + 45° 10′) 43. (90° + 58° 50′) 45. (90° + 33° 14′) 47. (90° + 59° 57′) 49. −0.5592 51. −0.1392 53. −0.8039 55. −0.6799 57. −1.4641 59. −12.4288 In Exercises 61-66, draw the given angles. 61. 30°; 330° or −30° 63. 135°; 225° or −135° 65. 270° 20′; 89° 40′ or −270° 20′ 67. 168° 30′; 191° 30′ or −168° 30′ 69. 352° 11′; 7° 49′ or −352° 11′ 71. 216° 08′; 143° 52′ or −216° 08′ 73. $\sin(90° - 10°) = \sin 80° = 0.9848 = \cos 10°$; $\cos(90° - 10°) = \cos 80° = 0.1736 = \sin 10°$; $\tan(90° - 10°) = \tan 80° = 5.6713, \frac{1}{\tan 10°} = \frac{1}{0.1763} = 5.6713$ 75. $\sin(90° - 52°) = \sin 38° = 0.6157 = \cos 52°$; $\cos(90° - 52°) = \cos 38° = 0.7880 = \sin 52°$; $\tan(90° - 52°) = \tan 38° = 0.7813, \frac{1}{\tan 52°} = \frac{1}{1.2799} = 0.7813$ 77. $\sin(90° - 73°) = \sin 17° = 0.2924 = \cos 73°$; $\cos(90° - 73°) = \cos 17° = 0.9563 = \sin 73°$; $\tan(90° - 73°) = \tan 17° = 0.3057, \frac{1}{\tan 73°} = \frac{1}{3.2709} = 0.3057$

Page 562 Review Capsule for Section 15-8 1. $\frac{y}{x}$ 2. $\frac{x^2}{y}$ 3. $y\sqrt{r}$ 4. $\frac{xy(x^2+y)}{x+y}$ 5. $\frac{2y}{xy+1}$ 6. $4\sqrt{3}$ 7. $\frac{5\sqrt{51}}{6}$ 8. $\frac{\sqrt{66}}{12}$ 9. $\frac{5\sqrt{22}}{3}$ 10. $7 + \frac{7}{6}\sqrt{3}$

Page 564 Written Exercises The answers to Exercises 1-6 are given in the order: sec, cot, csc. 1. 3.864; 0.2679; 1.035 3. 1.743; 0.7002; 1.221 5. 1.122; 1.9626; 2.203 7. 1.269 9. 1.305 11. 3.420 13. −0.3640 15. −2.924 17. −1.980 19. 0.5704 21. −1.014 The answers to Exs. 23-27 are given in the order: sin, csc, cos, sec, tan cot. 23. $\frac{7}{25}, \frac{25}{7}, -\frac{24}{25}, -\frac{25}{24}, -\frac{7}{24}, -\frac{24}{7}$ 25. $-\frac{12}{13}, -\frac{13}{12}, -\frac{5}{13}, -\frac{13}{5}, \frac{12}{5}, \frac{5}{12}$ 27. $\frac{1}{2}, 2; \frac{\sqrt{3}}{2}, \frac{2\sqrt{3}}{3}, \frac{-\sqrt{3}}{3}, -\sqrt{3}$ 29. 28° 50′ 31. 10° 20′ 33. 63° 35. second 37. fourth

Page 565 Classroom Exercises 1. α 3. θ 5. 40°

Page 566 Word Problems 1. 1049 m 3. 119 feet 5. 8 m 7. 258 m 9. 62 m

Page 569 Classroom Exercises 1. sin 75° 3. −cos 62° 5. −cos 30° 7. −0.5000 9. 0.9816

Page 569 Written Exercises 1. b = 19; c = 18, C = 64° 3. C = 67°; a = 15; b = 11 5. a = 135; C = 60°; 451 7. a = 12.6; B = 88°; b = 20.4 9. a = 81; C = 74° 10′; c = 216 11. B = 41° 50′; b = 16; c = 14 13. A = 92° 40′; b = 0.8; c = 1.2 15. 690 m 17. 165 m 19. The station forming the 100° angle; 60 km 21. 351 m 23. One 25. None 27. Two 29. $SV_1 = 4.4 \times 10^7$ km; $SV_2 = 2.4 \times 10^8$ km

Page 572 Calculator Exercises See the answers to Exercises 1-14 on page 569.

Page 574 Classroom Exercises 1. 6 3. 0.2

Page 574 Written Exercises 1. 4.8 3. 17.6 5. 7.2 7. 6.8 9. A = 44° 20′ 11. 75° 30′ 13. 57° 10′ 15. a = 36.9;

B = 57° 30′; C = 71° 30′ 17. A = 44° 20′; B = 57° 10′; C = 78° 30′ 19. a = 11; B = 57° 20′; C = 87° 20′ 21. 36° 40′; B = 22° 50′; 143° 20′ 23. cos 90° = 0, thus $c^2 = a^2 + b^2 - 2ab \cos 90°$ becomes $c^2 = a^2 + b^2$. 25. 173.2 m 27. xy = 33.7 cm 29. In acute triangle ABC, draw $\overline{AD}$ perpendicular to side BC. Let h = length of $\overline{AD}$ and p = length of $\overline{BD}$; then a − p = length of $\overline{CD}$. In triangle ADC, $b^2 = h^2 + (a - p)^2$. In triangle ADB, $h^2 = c^2 - p^2$; $b^2 = c^2 - p^2 + a^2 - 2ap + p^2 = c^2 + a^2 - 2ap$; $\cos B = \frac{p}{c}$ or $p = c \cos B$. Thus, $b^2 = a^2 + c^2 - 2ac \cos B$. In acute triangle ABC, draw $\overline{BD}$ perpendicular to side AC. Let h = length of $\overline{BD}$ and p = length of $\overline{CD}$; then b − p = length of $\overline{AD}$. In triangle ADB, $c^2 = h^2 + (b - p)^2$. In triangle CDB, $h^2 = a^2 - p^2$; $c^2 = a^2 - p^2 + b^2 - 2bp + p^2 = a^2 + b^2 - 2bp$; $\cos C = \frac{p}{a}$ or $p = a \cos C$. Thus, $c^2 = a^2 + b^2 - 2ab \cos C$. 31. $a^2 = h^2 + (c + p)^2$; $h^2 = b^2 - p^2$; $a^2 = b^2 + c^2 + 2cp$; $\cos \alpha = \frac{p}{b}$ so $p = b \cos \alpha$. $\angle A = 180° - \alpha$; $\cos A = \cos (180° - \alpha) = -\cos \alpha$, so $\cos \alpha = -\cos A$ then $p = -b \cos A$. Thus, $a^2 = b^2 + c^2 - 2bc \cos A$.

Page 576 Review The answers to Exercises 1-5 are given in the order: sin, cos, tan. 1. −0.7071; −0.7071; 1 2. −0.9659; −0.2588; −3.7321 3. 0.5000; −0.8660; −0.5774 4. 0.6428; 0.7660; 0.8391 5. −0.2588; 0.9659; −0.2679 6. −0.6009 7. −0.7030 8. −1.0000 9. 0.3689 The answers to Exercises 10-13 are given in the order: sin, csc, cos, sec, tan, cot. 10. $-\frac{7}{25}; -\frac{25}{7}; \frac{24}{25}; \frac{25}{24}; -\frac{7}{24}; -\frac{24}{7}$ 11. $-\frac{5\sqrt{41}}{41}; -\frac{\sqrt{41}}{5}; -\frac{4\sqrt{41}}{41}; -\frac{\sqrt{41}}{4}; \frac{5}{4}; \frac{4}{5}$ 12. $-\frac{12}{13}; -\frac{13}{12}; \frac{5}{13}; \frac{13}{5}; -\frac{12}{5}; -\frac{5}{12}$ 13. $\frac{1}{2}; 2; -\frac{\sqrt{3}}{2}; -\frac{2\sqrt{3}}{3}; -\frac{\sqrt{3}}{3}; -\sqrt{3}$ 14. 461.7 meters 15. 245.7 cm 16. 71° 47′

Page 576 Calculator Exercises See the answers to Exercises 15-20 on page 575.

Page 578 Chapter Objectives and Review 1. One-sixth rotation, counterclockwise 3. One-twelfth rotation, counterclockwise 5. Three-eighths rotation, clockwise 7. 770°; −310°; −670° 9. 690°; −390°; −750° 11. 1030°; −50°; −410° 13. $\frac{3}{5}; \frac{4}{5}; \frac{3}{4}$ 15. $-\frac{\sqrt{2}}{2}; -\frac{\sqrt{2}}{2}; 1$ 17. $-\frac{7\sqrt{53}}{53}; \frac{2\sqrt{53}}{53}; -\frac{7}{2}$ 19. 1; 1 21. $-\frac{\sqrt{3}}{2}$ 23. 11.5 m 25. 0.7457 27. 0.5630 29. 26° 44′ 31. 51° 20′ 33. −0.5000; 0.8660; −0.5774 35. 0.8660; −0.5000; −1.7321 37. −0.3420; 0.9397; −0.3640 39. −0.8660 41. −0.3609 43. 0.5736 45. 3.864 47. 0.3378 49. 28° 51. 46° 20′ 53. 149.6 meters 55. A = 95° 40′; b = 4.5; c = 7.3 57. 28

Page 580 Chapter Test 1. 395°, 325° 3. 540°, 180° 5. 650°, 70° 7. $\frac{5}{13}, \frac{12}{13}, \frac{5}{12}$ 9. $-\frac{4}{5}, -\frac{3}{5}, \frac{4}{3}$ 11. $-\frac{24\sqrt{577}}{577}, \frac{\sqrt{577}}{577}$ −24 13. 5 15. 0.4488 17. 1 19. −0.9962 21. 1.133 23. 67 m 25. 6487.7 mi

CHAPTER 16 MORE TOPICS IN TRIGONOMETRY

Page 584 Classroom Exercises 1. $\overrightarrow{OB}$ 3. $\overrightarrow{OA}$ 5. $\overrightarrow{OA}$ 7. 9 9. 5

Page 584 Written Exercises 1. $\sqrt{2}$ 3. 5 5. 10.9 7. 12.0 9. −3; 4 11. 3; −8 13. 10; 53° 10′ 15. 30; 323° 10′ 17. x: 10; y: 7 19. 4

Page 587 Classroom Exercises 1. True 3. True 5. False

Page 587 Word Problems 1. 374 km/hr 3. 436 km/hr 5. 113° 7. 482 km/hr 9. 40 km/hr 11. 15° 13. Wind: 87 km/hr; air speed: 325 km/hr 15. 4 km /hr 17. 5 km/hr

Page 590 Classroom Exercises 1. $\frac{\pi}{4}$ 3. -3π 5. 30° 7. −240° 9. 4π 11. -3π

Page 590 Written Exercises 1. 1.5 3. 0.5 5. 0.4 7. 80 9. 2π 11. $\frac{\pi}{2}$ 13. $-\frac{5\pi}{4}$ 15. $-\frac{\pi}{4}$ 17. $\frac{3\pi}{4}$ 19. $-\frac{13\pi}{12}$ 21. $\frac{7\pi}{3}$ 23. $\frac{10\pi}{3}$ 25. $-\frac{13\pi}{180}$ 27. 60° 29. 150° 31. −15° 33. 300° 35. −315° 37. −132° 39. 18° 41. 1260° 43. 450° 45. $\frac{11\pi}{144}$ 47. $-\frac{193\pi}{720}$ 49. 3 51. 6π 53. 18° 55. $\frac{1}{240}$ 57. 20.93 cm 59. 20.93 cm 61. $\frac{13\pi}{3}$ 63. $-\frac{11\pi}{3}$ 65. $-\frac{5\pi}{4}$

Page 592 Classroom Exercises 1. 0.2588; 0.9659; 3.7321; 1.035 3. 32° 10′; 0.8465; 0.6289; 1.5900; 1.181; 1.878

Page 592 Written Exercises 1. True 3. 0.0901 5. 0.9057 7. 1.095 9. 3.556 11. 1.000 13. −1.000 15. 1.000 17. −2.000 19. 1.155 21. 0.8660 23. 0.9659 25. 0 27. 0, π 29. $\frac{3\pi}{4}, \frac{5\pi}{4}$ 31. $\frac{3\pi}{4}, \frac{7\pi}{4}$ 33. $\frac{\pi}{3}, \frac{5\pi}{3}$ 35. $\sin \frac{\pi}{2} = 1$; $2 \sin \frac{\pi}{4} \cos \frac{\pi}{4} = 2(\frac{\sqrt{2}}{2})(\frac{\sqrt{2}}{2}) = 1$ 37. $-\frac{1}{3\pi}$; 0; 3π 39. $\frac{\sqrt{2}}{6\pi}; -\frac{\sqrt{2}}{2}; -\frac{3\pi\sqrt{2}}{2}$ 41. 0; 1; 0

Page 593 Calculator Exercises See the answers for Exercises 1-10 on page 592.

Page 596 Classroom Exercises 1. $2\pi; 3$ 3. $2\pi; \frac{1}{2}$ 5. $2\pi; \frac{1}{4}$ 7. $2\pi; 2$ 9. $2\pi; 1$ 11. $2\pi; \frac{1}{6}$ 13. 0; 0 15. 0; 0 17. 0; 0 19. 0; 0 21. 0; 0

Page 596 Written Exercises For Exercises 1-16, multiply each y value in the table for y = cos x or y = sin x by the coefficient of the given function and plot the resulting pairs. 17. $(-2\pi, 0)$, $(-\frac{7\pi}{4}, 1)$, $(-\frac{3\pi}{2}, 0)$, $(-\frac{5\pi}{4}, -1)$, $(-\pi, 0)$, $(-\frac{3\pi}{4}, 1)$, $(-\frac{\pi}{2}, 0)$, $(-\frac{\pi}{4}, -1)$, $(0, 0)$, $(\frac{\pi}{4}, 1)$, $(\frac{\pi}{2}, 0)$, $(\frac{3\pi}{4}, -1)$, $(\pi, 0)$ $(\frac{5\pi}{4}, 1)$, $(\frac{3\pi}{2}, 0)$ $(\frac{7\pi}{4}, -1)$, $(2\pi, 0)$ 19. The period of y = sin x is 2π and the period of y = sin 2x is π. 21. They are the same. 23. $\frac{2\pi}{3}$ 25. 4π 27. π 29. 6π 31. 8π 33. Subtract 2 from each y value in the table for y = cos x. Thus, for x = 0, y = −1; for x = π, y = −3 and so on. 35. The amplitude and period are the same. 37. For $0 \le x \le \pi$, use the y values in the table in Example 1 on page 594. For $\pi < x \le 2\pi$, multiply the y values for this range by −1. 39. $\frac{3\pi}{4}, \frac{7\pi}{4}$ 41. $0 < x < \frac{\pi}{4}$ and $\frac{5\pi}{4} < x < 2\pi$ 43. No values.

Page 600 Classroom Exercises 1. 0 3. 1.16 5. −2 7. 0 9. 0 11. −1 13. Not defined 15. 1.41

Page 601 Written Exercises 1. See page 598. 3. See Example 2 on page 599. 5. Multiply each y value in the table on page 598 by 2. 7. Multiply each y value in the table on page 599 by −1. 9. Multiply the reciprocals of the y values in the table for y = cos x by $\frac{1}{3}$. Thus, for x = 0, y = $\frac{1}{3}$. 11. Each y value in the graph of y = cot x (see Example 2 on page 599) is multiplied by $\frac{1}{2}$. Thus, for $x = \frac{\pi}{4}$, $y = \frac{1}{2}$ 13. $x = \frac{\pi}{2}$, $x = \frac{3\pi}{2}$, $x = -\frac{\pi}{2}$, $x = -\frac{3\pi}{2}$ 15. $x = \frac{\pi}{2}$, $x = \frac{3\pi}{2}$, $x = -\frac{\pi}{2}$, $x = -\frac{3\pi}{2}$ 17. $\cot \theta$ 19. $\csc \theta$ 21. They are the same. 23. 4π 25. $\frac{\pi}{|B|}$ 27. Period: 3π; asymptotes are at the following values of x: $\pm\frac{3\pi}{2}$; x intercept: 0 29. Period: 3π; asymptotes are at the following values of x: $\pm\frac{3\pi}{2}$; x intercept: 0 31. Period: π; asymptotes are at the following values of x: $0, \frac{\pi}{2}, \pi, \frac{3\pi}{2}, 2\pi$ 33. Period: 4π; asymptote at $x = \pi$ 35. y = sin 2x: (0, 0), $(\frac{\pi}{4}, 1)$, $(\frac{\pi}{2}, 0)$, $(\frac{3\pi}{4}, -1)$, $(\pi, 0)$, $(\frac{5\pi}{4}, 1)$, $(\frac{3\pi}{2}, 0)$, $(\frac{7\pi}{4}, -1)$, $(2\pi, 0)$ For y = 1 + sin 2x, add 1 to each y value of y = sin 2x. 37. $y = \sin \frac{1}{2}x$: $(-2\pi, -1)$, $(-\frac{3\pi}{2}, -0.71)$, $(-\pi, -1)$, $(-\frac{\pi}{2}, -0.71)$, $(0, 0)$, $(\frac{\pi}{2}, 0.71)$, $(\pi, 1)$, $(\frac{3\pi}{4}, 0.71)$, $(2\pi, 0)$ For $y = \sin(\frac{\pi}{2} + \frac{1}{2}x)$, shift the graph of $y = \sin \frac{1}{2}x$ to the left by π units. 39. $y = 2 \cos \frac{1}{3}x$: (0, 2), $(\frac{\pi}{4}, 1.8)$, $(\frac{\pi}{2}, 1.7)$, $(\frac{3\pi}{4}, 1.4)$, $(\pi, 1)$, $(\frac{5\pi}{4}, 0.52)$, $(\frac{3\pi}{2}, 0)$, $(\frac{7\pi}{4}, -0.52)$, $(2\pi, 1)$ $y = -\frac{1}{2} \cos \frac{1}{3}x$: $(0, -\frac{1}{2})$, $(\frac{\pi}{4}, -0.49)$, $(\frac{\pi}{2}, -0.44)$, $(\frac{3\pi}{4}, -0.35)$, $(\pi, -0.25)$, $(\frac{5\pi}{4}, -0.13)$, $(\frac{7\pi}{4}, 0.25)$, $(2\pi, 0.35)$

Page 603 Computer Exercises See the Solution Key.

Page 604 Review 1. 9.4 km 2. 333.9 km/hr 3. 4π 4. $\frac{2\pi}{3}$ 5. $-\frac{3\pi}{2}$ 6. $\frac{3\pi}{4}$ 7. $\frac{35\pi}{9}$ 8. $\frac{3\pi}{5}$ 9. 30° 10. 67° 30′ 11. −1080° 12. −75° 13. −990° 14. 24° 15. 0.9811 16. 0.9572 17. 1.085 18. 1.1106 19. 0 20. 0.7071 21. 1.0000 22. 1.7321 23. 2.0000 24. −2.0000 25. 0 26. 0.8660 27. Multiply each y value in the table for y = cos x by 3. 28. Multiply each y value in the table for y = sin x by $-\frac{1}{4}$. 29. See Example 2 on page 599. 30. Multiply each y value in the table for y = tan x by −2. 31. $x = -\frac{3\pi}{2}$, $x = -\frac{\pi}{2}$, $x = \frac{\pi}{2}$, $x = \frac{3\pi}{2}$ 32. $x = -2\pi$, $x = -\pi$, $x = 0$, $x = \pi$, $x = 2\pi$ 33. $x = -\frac{3\pi}{2}$, $x = -\frac{\pi}{2}$, $x = \frac{\pi}{2}$, $x = \frac{3\pi}{2}$ 34. $x = -2\pi$, $x = -\pi$, $x = 0$, $x = \pi$, $x = 2\pi$

Page 606 Classroom Exercises 1. $\frac{1}{\sin \theta}$ 3. $\frac{\cos \theta}{\sin \theta}$ 5. $\sec \theta$ 7. $\sec^2 \theta$ 9. $\sin^2 \theta$ 11. $\tan^2 \theta$ 13. $\frac{\sin^2 \theta - 2 \sin \theta}{\cos^2 \theta}$

15. $\frac{1}{\cot^2 \theta} - \frac{2\sqrt{\frac{1}{\cot^2 \theta} + 1}}{\cot \theta}$

Page 607 Written Exercises 1. $\sec \theta = \frac{r}{x} = \frac{\frac{r}{r}}{\frac{x}{r}} = \frac{1}{\frac{x}{r}} = \frac{1}{\cos \theta}$ 3. $\cot \theta = \frac{x}{y} = \frac{\frac{x}{r}}{\frac{y}{r}} = \frac{\cos \theta}{\sin \theta}$ 5. $\cos \theta$ 7. $\cot \theta$ 9. $\tan^2 \theta$ 11. $\cos^2 \theta$ 13. $\sin \theta$ 15. $\cot^2 \theta$ 17. $\sec \theta$ 19. 1 21. $\frac{1}{\sin \theta \cos \theta}$ 23. $\cos \theta + \sin \theta$ 25. $\sqrt{1 - \cos^2 \theta}$

27. $\frac{\sqrt{1-\cos^2\theta}}{\cos\theta}$ 29. $\sec\theta$ 31. $\frac{1}{\cos^2 x}$ 33. $1-\cos^2\theta$ 35. $\frac{\csc\theta\sec\theta}{\sec\theta+\csc\theta}$ 37. $\frac{\cos^2\theta}{\sin^2\theta}$ 39. $2+\cos^2\theta$ 41. Not an identity. $\sin 2\theta$ will always have values between -1 and 1 but $2\sin\theta$ can have values between -2 and 2.

Page 610 Classroom Exercises 1. $60°+45°$ 3. $45°-30°$ 5. $90°+45°$ 7. $\frac{\pi}{6}+\frac{\pi}{6}$ 9. $\frac{\pi}{4}+\frac{\pi}{6}$ 11. $180°+60°$

Page 610 Written Exercises 1. $-\frac{1}{2}$ 3. $-\frac{\sqrt{3}}{2}$ 5. $\frac{\sqrt{2}}{2}$ 7. $\frac{1}{4}(\sqrt{6}+\sqrt{2})$ 9. $-\frac{1}{2}$ 11. $-\frac{1}{2}$ 13. $\frac{\sqrt{3}}{2}$ 15. $\frac{1}{2}$ 17. $-\frac{1}{2}$ 19. 1 21. $-\frac{\sqrt{2}}{2}$ 23. $-\frac{\sqrt{3}}{2}$ 25. 0.5; 1.0 27. -0.3907; 0.9743 29. $\frac{\sqrt{2}}{2}(\cos\theta+\sin\theta)$ 31. $-\cos\theta$ 33. $(\frac{3}{5})(-\frac{12}{13})-(\frac{4}{5})(\frac{5}{13})=-\frac{56}{65}$ 35. $\frac{33}{65}$ 37. $-\frac{63}{65}$ 39. $\sin 90°\cos\theta-\cos 90°\sin\theta = 1\cdot\cos\theta-0\cdot\sin\theta=\cos\theta$ 41. $\sin\pi\cos\theta-\cos\pi\sin\theta=\sin\theta$ 43. $\cos\pi\cos\theta+\sin\pi\sin\theta=-\cos\theta$ 45. $\cos\frac{3\pi}{2}\cos\theta+\sin\frac{3\pi}{2}\sin\theta=-\sin\theta$ 47. $\sin 2\alpha=\sin(\alpha+\alpha)=\sin\alpha\cos\alpha+\cos\alpha\sin\alpha=2\sin\alpha\cos\alpha$ 49. $\sin 2\alpha=\frac{24}{25}$; $\cos 2\alpha=-\frac{7}{25}$ 51. $\sin 2\alpha=-\frac{120}{169}$; $\cos 2\alpha=\frac{119}{169}$ 53. $\sin(\alpha-\beta)=\sin[\alpha+(-\beta)]=\sin\alpha\cos(-\beta)+\cos\alpha\sin(-\beta)=\sin\alpha\cos\beta-\cos\alpha\sin\beta$

Page 613 Classroom Exercises 1. $\frac{1}{1-\cos^2\theta}=\frac{1}{\sin^2\theta}=\csc^2\theta$ 3. $(1-\sin^2\theta)\tan^2\theta=\cos^2\theta\,(\frac{\sin^2\theta}{\cos^2\theta})=\sin^2\theta$
5. $\sin^4\theta-\cos^4\theta=(\sin^2\theta+\cos^2\theta)(\sin^2\theta-\cos^2\theta)=1(\sin^2\theta-\cos^2\theta)=\sin^2\theta-\cos^2\theta$

Page 613 Written Exercises 1. $\sin\theta\sec\theta=\sin\theta\cdot\frac{1}{\cos\theta}=\frac{\sin\theta}{\cos\theta}=\tan\theta$ 3. $\frac{1-\cos^2\theta}{\cos^2\theta}=\frac{\sin^2\theta}{\cos^2\theta}=\tan^2\theta$ 5. $\sec^2\theta-1=(\tan^2\theta+1)-1=\tan^2\theta$ 7. $\frac{\cot\theta}{\cos\theta}=\frac{\cos\theta}{\sin\theta}\cdot\frac{1}{\cos\theta}=\frac{1}{\sin\theta}=\csc\theta$ 9. $\csc^2\theta\tan^2\theta-1=(\frac{1}{\sin^2\theta}\cdot\frac{\sin^2\theta}{\cos^2\theta})-1=\frac{1}{\cos^2\theta}-1=\sec^2\theta-1=\tan^2\theta$ 11. $(1-\tan\theta)^2=1-2\tan\theta+\tan^2\theta=(1+\tan^2\theta)-2\tan\theta=\sec^2\theta-2\tan\theta$ 13. $\frac{\cos^2\theta}{\sin\theta}+\sin\theta=\frac{\cos^2\theta+\sin^2\theta}{\sin\theta}=\frac{1}{\sin\theta}=\csc\theta$ 15. $\frac{\tan\theta}{1-\cos^2\theta}=\frac{\tan\theta}{\sin^2\theta}=\frac{\sin\theta}{\cos\theta}\cdot\frac{1}{\sin^2\theta}=\frac{1}{\cos\theta\sin\theta}=\frac{\sec\theta}{\sin\theta}$ 17. $\frac{\cos x-\sin x}{\cos x}=\frac{\cos x}{\cos x}-\frac{\sin x}{\cos x}=1-\tan x$ 19. $\tan x(\tan x+\cot x)=\tan^2 x+1=\sec^2 x$ 21. $\sec^4 x-\tan^4 x=(\sec^2 x-\tan^2 x)(\sec^2 x+\tan^2 x)=1(\sec^2 x+\tan^2 x)=\sec^2 x+\tan^2 x$ 23. $\frac{\sin^4\theta-\cos^4\theta}{1-\cot^4\theta}=\frac{\sin^4\theta-\cos^4\theta}{1-\frac{\cos^4\theta}{\sin^4\theta}}=\frac{\sin^4\theta-\cos^4\theta}{\frac{\sin^4\theta-\cos^4\theta}{\sin^4\theta}}=\sin^4\theta$

25. $\csc\theta+\frac{\tan\theta}{\sin\theta}-\sec\theta=\csc\theta+\frac{\frac{\sin\theta}{\cos\theta}}{\sin\theta}-\sec\theta=\frac{1}{\sin\theta}+\frac{1}{\cos\theta}-\frac{1}{\cos\theta}=\frac{1}{\sin\theta}\cdot\frac{\cos\theta}{\cos\theta}=\frac{\cos\theta}{\sin\theta}\cdot\frac{1}{\cos\theta}=\frac{\cot\theta}{\cos\theta}$

27. $$\frac{\tan\alpha+\tan\beta}{1-\tan\alpha\tan\beta}=\frac{\frac{\sin\alpha}{\cos\alpha}+\frac{\sin\beta}{\cos\beta}}{1-\frac{\sin\alpha}{\cos\alpha}\cdot\frac{\sin\beta}{\cos\beta}}=\frac{\frac{\sin\alpha\cos\beta+\sin\beta\cos\alpha}{\cos\alpha\cos\beta}}{\frac{\cos\alpha\cos\beta-\sin\alpha\sin\beta}{\cos\alpha\cos\beta}}=\frac{\sin\alpha\cos\beta+\sin\beta\cos\alpha}{\cos\alpha\cos\beta-\sin\alpha\sin\beta}=\frac{\sin(\alpha+\beta)}{\cos(\alpha+\beta)}=\tan(\alpha+\beta)$$

29. $\tan 2\alpha=\tan(\alpha+\alpha)=\frac{\tan\alpha+\tan\alpha}{1-\tan\alpha\tan\alpha}=\frac{2\tan\alpha}{1-\tan^2\alpha}$ 31. $\cos(\frac{\theta}{2}+\frac{\theta}{2})=2\cos^2\frac{\theta}{2}-1$; $\cos\theta=2\cos^2\frac{\theta}{2}-1$; $\cos^2\frac{\theta}{2}=\frac{\cos\theta+1}{2}$; Thus, $\cos\frac{\theta}{2}=\pm\sqrt{\frac{1+\cos\theta}{2}}$. 33. $\tan\frac{\alpha}{2}=\pm\sqrt{\frac{1-\cos\alpha}{1+\cos\alpha}}\cdot\sqrt{\frac{1+\cos\alpha}{1+\cos\alpha}}=\pm\sqrt{\frac{1-\cos^2\alpha}{(1+\cos\alpha)^2}}=\pm\sqrt{\frac{\sin^2\alpha}{(1+\cos\alpha)^2}}=\pm\frac{\sin\alpha}{1+\cos\alpha}$

Page 616 Classroom Exercises 1. $-\frac{\pi}{4}$ 3. π 5. $\frac{\pi}{2}$ 7. $\frac{\pi}{6}$

Page 617 Written Exercises 1. 61° or 1.0647 radians 3. $-\frac{\pi}{3}$ or $-60°$ 5. 45° or $\frac{\pi}{4}$ 7. $\frac{3\pi}{4}$ or 135° 9. $-30°$ or $-\frac{\pi}{6}$ 11. $-\frac{\pi}{2}$ or $-90°$ 13. $\frac{1}{2}$ 15. $\frac{\sqrt{3}}{2}$ 17. $\frac{1}{2}$ 19. 1 21. $\frac{\sqrt{2}}{2}$ 23. 1 25. 0 27. $\frac{\sqrt{3}}{2}$ 29. $-\frac{\sqrt{3}}{2}$ 31. 22° or 0.3840 radians 33. $\frac{\sqrt{10}}{10}$ 35. $\frac{\sqrt{5}}{3}$ 37. $\frac{4}{5}$ 39. $-\frac{5}{12}$ 41. $\frac{3}{4}$ 43. $\frac{7}{3}$ 45. 7 47. $5\sqrt{2}$ 49. To graph $y = \text{arc}\cos x$, interchange the coordinates of the points on the graph of $y=\cos x$. For example, the point $(2\pi, 1)$ on the graph of $y=\cos x$ becomes the point $(1, 2\pi)$ on the graph of $y=\text{arc}\cos x$. The range of the principal values is $0 \le y \le \pi$ and the domain is $-1\le x\le 1$. 51. To graph $y=\text{arc}\sin\frac{1}{2}x$, interchange the coordinates of the points on the graph of $y=\sin\frac{1}{2}x$. For example, the point $(-\pi, -1)$ on the graph of $y = \sin \frac{1}{2}x$ becomes the point $(-1, -\pi)$ on the graph of $y = \text{arc sin } \frac{1}{2}x$. The range of the principal values is $-\pi \leq y \leq \pi$ and the domain is $-1 \leq x \leq 1$.

Page 619 Classroom Exercises 1. $x = 2k\pi$, where k is any integer. 3. $2k\pi + \frac{\pi}{3}$ and $2k\pi + \frac{4\pi}{3}$, where k is any integer. 5. ϕ 7. $x = \frac{\pi}{4}, \frac{5\pi}{4}$

Page 619 Written Exercises 1. $x = \frac{\pi}{2}, \frac{3\pi}{2}$ 3. $x = \frac{\pi}{2}, \frac{7\pi}{6}, \frac{11\pi}{6}$ 5. $x = 0, \frac{\pi}{6}, \pi, \frac{7\pi}{6}, 2\pi$ 7. $x = \frac{\pi}{4}, \frac{7\pi}{4}$ 9. $x = \frac{2\pi}{3}, \frac{5\pi}{3}$ 11. $x = \frac{\pi}{2}, \frac{7\pi}{6}, \frac{11\pi}{6}$ 13. $x = \frac{\pi}{4}, \frac{3\pi}{4}, \frac{5\pi}{4}, \frac{7\pi}{4}$ 15. $x = \frac{\pi}{6}, \frac{5\pi}{6}, \frac{7\pi}{6}, \frac{11\pi}{6}$ 17. $x = \frac{\pi}{4}, \frac{5\pi}{4}$ 19. $x = \frac{\pi}{6}, \frac{5\pi}{6}$ 21. $x = 0, \frac{\pi}{6}, \pi, \frac{5\pi}{6}, 2\pi$ 23. $x = 0, \pi, 2\pi$ 25. $x = \frac{\pi}{6}, \frac{5\pi}{6}, \frac{7\pi}{6}, \frac{11\pi}{6}$ 27. $x = \frac{\pi}{6}, \frac{5\pi}{6}$ 29. 48° 35′ or 131° 25′ 31. 82° 45′ or 277° 15′

33. $x = \frac{\pi}{3}, \frac{5\pi}{3}$ 35. $x = 0, \frac{\pi}{2}, \pi, \frac{3\pi}{2}, 2\pi$ 37. x = 26° 34′; 116° 34′; 206° 34′; 296° 34′ 39. $x = \frac{\pi}{8}$

Page 621 Classroom Exercises 1. $\sqrt{2}$ 3. $\sqrt{2}$ 5. $\frac{\pi}{4}$ 7. $\frac{5\pi}{4}$

Page 622 Written Exercises 1. $4\sqrt{2}(\cos \frac{\pi}{4} + i \sin \frac{\pi}{4})$ 3. $\frac{1}{2}(\cos \frac{3\pi}{2} + i \sin \frac{3\pi}{2})$ 5. $\cos \frac{\pi}{2} + i \sin \frac{\pi}{2}$ 7. $2(\cos \frac{\pi}{6} + i \sin \frac{\pi}{6})$ 9. $3(\cos \frac{\pi}{2} + i \sin \frac{\pi}{2})$ 11. $2(\cos \frac{7\pi}{6} + i \sin \frac{7\pi}{6})$ 13. $5\sqrt{2}(\cos \frac{\pi}{4} + i \sin \frac{\pi}{4})$ 15. $4(\cos \frac{4\pi}{3} + i \sin \frac{4\pi}{3})$ 17. $1 + \sqrt{3}i$ 19. $\frac{3\sqrt{3}}{2} - \frac{1}{2}i$ 21. $1 - i$ 23. $-\sqrt{2} + \sqrt{2}i$ 25. $0.3128 + 1.9754i$ 27. $1 + 0i$

Page 622 Review 1. $\sin^2 \theta$ 2. $\cot^2 \theta$ 3. $\tan^2 \theta$ 4. $\cos \theta$ 5. $\cos^2 \theta$ 6. $\cos \theta$ 7. $\frac{1}{4}(\sqrt{6} + \sqrt{2})$ 8. $-\frac{\sqrt{2}}{2}$ 9. $\frac{1}{4}(\sqrt{6} - \sqrt{2})$ 10. $-\frac{\sqrt{3}}{2}$ 11. $-\frac{\sqrt{3}}{2}$ 12. $-\frac{\sqrt{2}}{2}$ 13. $\frac{1}{4}(\sqrt{2} - \sqrt{6})$ 14. $-\frac{\sqrt{3}}{2}$ 15. $-\frac{\sqrt{2}}{2}$ 16. $-\frac{1}{2}$ 17. $\cot \theta(\cot \theta + \tan \theta) = \cot^2 \theta + 1 = \csc^2 \theta$ 18. $\frac{1 + \cot^2 \theta}{1 + \tan^2 \theta} = \frac{\csc^2 \theta}{\sec^2 \theta} = \cot^2 \theta$ 19. $\sec \theta(\sec \theta - \cos \theta) = \sec^2 \theta - 1 = \tan^2 \theta$ 20. $\frac{\sin \theta}{1 + \cos \theta} + \cot \theta = \frac{\sin \theta}{1 + \cos \theta} + \frac{\cos \theta}{\sin \theta} = \frac{\sin^2 \theta + \cos^2 \theta + \cos \theta}{\sin \theta(1 + \cos \theta)} = \frac{1 + \cos \theta}{\sin \theta(1 + \cos \theta)} = \frac{1}{\sin \theta} = \csc \theta$ 21. $\frac{\cos \theta}{1 - \sin \theta} - \frac{\cos \theta}{1 + \sin \theta} = \frac{(\cos \theta + \sin \theta \cos \theta) - (\cos \theta - \sin \theta \cos \theta)}{1 - \sin^2 \theta} = \frac{2 \sin \theta \cos \theta}{\cos^2 \theta} = \frac{2 \sin \theta}{\cos \theta} = 2 \tan \theta$ 22. $\frac{1}{1 + \sin \theta} + \frac{1}{1 - \sin \theta} = \frac{(1 - \sin \theta) + (1 + \sin \theta)}{1 - \sin^2 \theta} = \frac{2}{\cos^2 \theta} = 2 \sec^2 \theta$ 23. 35° 24. 0° 25. 0° 26. $\frac{2\sqrt{5}}{5}$ 27. $\frac{3}{4}$ 28. $\frac{\sqrt{33}}{7}$ 29. $x = \frac{3\pi}{4}, \frac{7\pi}{4}$ 30. $x = \frac{\pi}{4}, \frac{3\pi}{4}, \frac{5\pi}{4}, \frac{7\pi}{4}$ 31. $x = \pi$ 32. $3(\cos \frac{3\pi}{2} + i \sin \frac{3\pi}{2})$ 33. $\sqrt{2}(\cos \frac{7\pi}{4} + i \sin \frac{7\pi}{4})$ 34. $2(\cos \frac{\pi}{6} + i \sin \frac{\pi}{6})$ 35. $2(\cos \frac{7\pi}{4} + i \sin \frac{7\pi}{4})$ 36. $3\sqrt{2} - 3\sqrt{2}i$ 37. $-\frac{1}{6} - \frac{\sqrt{3}}{6}i$

Page 624 Chapter Objectives and Review 1. 5.2; 3 3. 5.1; 10.9 5. 79° 7. $\frac{\pi}{3}$ 9. $-\frac{5\pi}{4}$ 11. $-\frac{5\pi}{9}$ 13. 360° 15. −30° 17. 156° 19. $-\frac{1}{2}$ 21. 0 23. See page 595. 25. Multiply the reciprocals of the y values in the table for y = cos x by 2. 27. $(1 - \cos^2 \theta)(1 + \cot^2 \theta) = (\sin^2 \theta)(\csc^2 \theta) = 1$ 29. $\sin^4 \theta - 2 \sin^2 \theta + 1 = (\sin^2 \theta - 1)^2 = (\cos^2 \theta)^2 = \cos^4 \theta$ 31. $\frac{1}{4}(\sqrt{2} + \sqrt{6})$ 33. $\frac{1}{4}(\sqrt{6} - \sqrt{2})$ 35. $-\frac{1}{4}(\sqrt{2} + \sqrt{6})$ 37. $\frac{\pi}{2}$ 39. 1 41. $-\frac{\sqrt{5}}{2}$ 43. $x = 0, \frac{\pi}{4}, \pi, \frac{5\pi}{4}, 2\pi$ 45. $\sqrt{2}(\cos \frac{\pi}{4} + i \sin \frac{\pi}{4})$ 47. $2\sqrt{2}(\cos \frac{3\pi}{4} + i \sin \frac{3\pi}{4})$ 49. $8(\cos \frac{5\pi}{3} + i \sin \frac{5\pi}{3})$ 51. $-\frac{1}{4} + \frac{\sqrt{3}}{4}i$

Page 626 Chapter Test 1. 25 km 3. $\frac{\pi}{6}$ 5. $-\frac{5\pi}{6}$ 7. 45° 9. −225° 11. $-\frac{\sqrt{2}}{2}$ 13. $\frac{\sqrt{3}}{3}$ 15. The graph is the graph of y = cos x shifted 180° along the x axis with the amplitude of $\frac{1}{2}$. 17. $(\tan \theta + \cot \theta)^2 = \tan^2 \theta + 2 \tan \theta \cot \theta + \cot^2 \theta = \tan^2 \theta + 2 + \cot^2 \theta = (\tan^2 \theta + 1) + (\cot^2 \theta + 1) = \sec^2 \theta + \csc^2 \theta$ 19. $\frac{\sqrt{2}}{2}$

Page 626 More Challenging Problems 1. Both graphs have a period of π and an amplitude of 1. Their shape is the same but the graph of $y = \sin (2x + \frac{\pi}{2})$ "leads" the graph of $y = \sin 2x$ by $\frac{\pi}{4}$. 3. $\frac{1}{2}[\sin (\alpha + \beta) + \sin (\alpha - \beta)] = \frac{1}{2}[(\sin \alpha \cos \beta + \cos \alpha \sin \beta) + (\sin \alpha \cos \beta - \cos \alpha \sin \beta)] = \frac{1}{2}(2 \sin \alpha \cos \beta) = \sin \alpha \cos \beta$ 5. $\frac{1}{2}[\cos (\alpha + \beta) + \cos (\alpha - \beta)] = \frac{1}{2}[(\cos \alpha \cos \beta - \sin \alpha \sin \beta) + (\cos \alpha \cos \beta + \sin \alpha \sin \beta)] = \frac{1}{2}(2 \cos \alpha \cos \beta) = \cos \alpha \cos \beta$ 7. cos (Arc tan 2 + Arc sec 3) = cos (Arc tan 2) cos (Arc sec 3)

$-\sin(\text{Arc tan } 2)\sin(\text{Arc sec } 3) = (\frac{\sqrt{5}}{5} \cdot \frac{1}{3}) - (\frac{2\sqrt{5}}{5} \cdot \frac{2\sqrt{2}}{3}) = \frac{\sqrt{5} - 4\sqrt{10}}{15}$ 9. Draw triangle ABC with altitude CD, where D is on side AB. Area of triangle BDC = $\frac{1}{2}ph$; area of triangle ADC = $\frac{1}{2}(c - p)h$; area of triangle ABC = K = $\frac{1}{2}ph + \frac{1}{2}(c - p)h = \frac{1}{2}h[p + (c - p)] = \frac{1}{2}hc$; since $\sin A = \frac{h}{b}$, $h = b \sin A$; Thus, $K = \frac{1}{2}bc \sin A$

Page 627 Cumulative Review: Chapters 13-16 1. c 3. d 5. c 7. a 9. c 11. d 13. b 15. c 17. a 19. d

A 2
B 3
C 4
D 5
E 6
F 7
G 8
H 9
I 0
J 1